	Object	Mass Center Coordinates and Volume V	Moments of Inertia About Indicated Axes
Four Special Cases **1.** If $r = 0$: solid cylinder	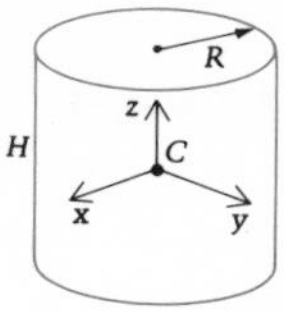	$(0, 0, 0)$ $V = \pi R^2 H$	$I_x = I_y = \frac{mR^2}{4} + \frac{mH^2}{12}$ $I_z = \frac{mR^2}{2}$
2. If $r \approx R$: cylindrical shell	R = average radius 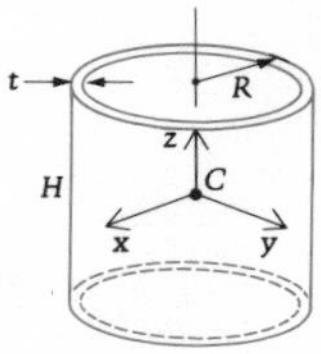	$(0, 0, 0)$ $V = 2\pi RtH$	
3. If $H \ll r$: annular disk	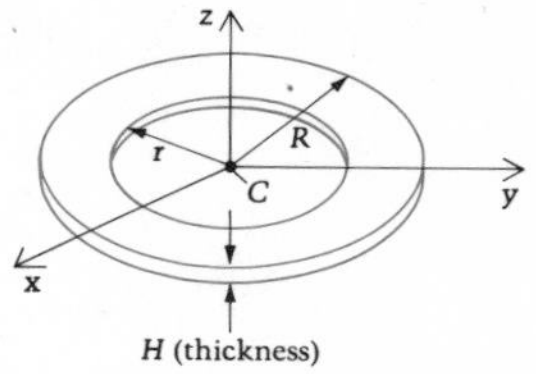	$(0, 0, 0)$ $V = \pi(R^2 - r^2)H$	$I_x = I_y \approx \frac{m}{4}(R^2 + r^2)$ $I_z = \frac{m(R^2 + r^2)}{2}$
4. If $H \ll r$ and $r \approx R$: thin ring with rectangular cross section (area)	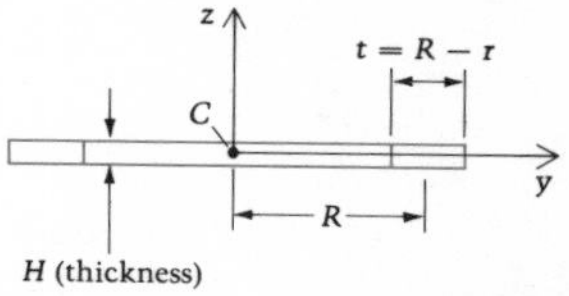	$(0, 0, 0)$ $V = 2\pi RtH$ $= 2\pi R(A)$ $(A = tH$ = area of cross section)	$I_x = I_y \approx \frac{mR^2}{2}$ $I_z \approx mR^2$ (Note: These apply for *any* cross-sectional shape as long as the ring is thin.)
triangular plate	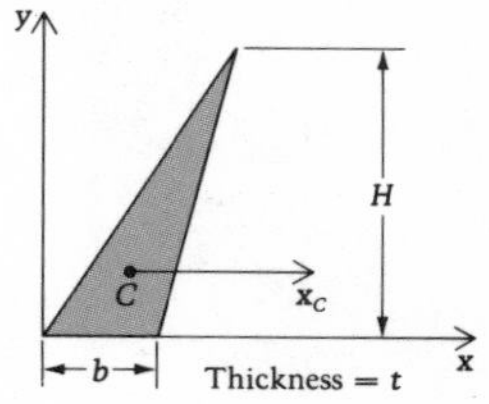	(We are interested here in properties for any base axis, such as x in the figure.) $y_C = \frac{H}{3}$ $V = \frac{1}{2}bHt$	$I_x = \frac{mH^2}{6}$ $I_{x_C} = \frac{mH^2}{18}$
elliptical plate	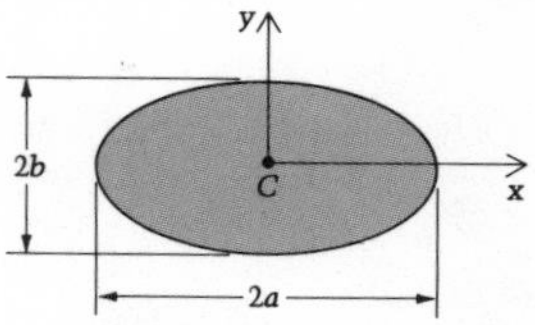 Thickness = t in z direction	$(0, 0, 0)$ $V = \pi abt$	$I_x = \frac{mb^2}{4}$ $I_y = \frac{ma^2}{4}$ $I_z = \frac{m(a^2 + b^2)}{4}$
paraboloidal plate	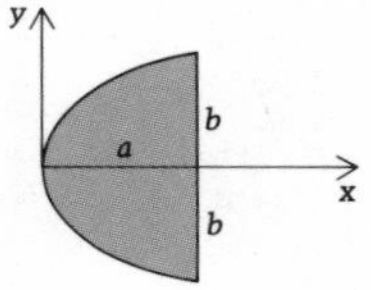 Thickness = t in z direction	$\left(\frac{3a}{5}, 0, 0\right)$ $V = \frac{4}{3}abt$	$I_x = \frac{mb^2}{5}$ $I_y = \frac{3ma^2}{7}$ $I_z = \frac{m(15a^2 + 7b^2)}{35}$

continued inside back cover

An Introduction to Dynamics

AN INTRODUCTION TO

Dynamics

David J. McGill and Wilton W. King

Georgia Institute of Technology

BROOKS/COLE ENGINEERING DIVISION

Monterey, California

[redacted]

Brooks/Cole Engineering Division
A Division of Wadsworth, Inc.

Printed in the United States of America

10 9 8 7 6 5 4 3 2 1

Library of Congress Cataloging in Publication Data:

McGill, David J.,
Engineering Mechanics.

Includes index.
1. Dynamics. I. King, Wilton W.,
II. Title.
TA352.M385 1984 620.1'04 83-25283

ISBN 0-534-02933-7

Sponsoring Editor: Ray Kingman
Production Services Coordinator: Bill Murdock
Production: Mary Forkner, Publication Alternatives
Manuscript Editor: Don Yoder
Interior Design: Al Burkhardt
Cover Design: Katherine Minerva
Cover Photo: Index/Stone International, Paul Brierley
Illustrations: Georg Klatt, Design Graphic
Typesetting: Allservice Phototypesetting

To our wives, Carolyn and Kay

Preface

This book is one of a two-volume set covering two of the basic topics of mechanics: statics and dynamics. In this second volume we have written what we believe to be a concise yet accurate introduction to engineering dynamics. The work is the outgrowth of our lecture notes, developed over nearly twenty years together at Georgia Tech and prior to that in teaching and study at VPI, LSU, the University of Virginia, and the University of Kansas.

We have tried to follow one basic guideline—to write the book in the same way we teach the course. To this end, we have written many explanatory footnotes. Moreover, you will find a number of text questions interspersed throughout the text. The answers to these questions are given at the ends of the chapters for your convenience. These are the kinds of questions we ask of students in class, and we believe they will serve a good purpose on the printed page as well. To make the most of these questions, treat them as serious homework as you read, and look up the answers only after you have an answer of your own in mind. The goals of these questions are to encourage thinking about tricky points and to emphasize the basic principles of the subject.

We have also written a set of approximately a dozen classroom review questions and answers at the close of Chapters 1 to 7. These true-false questions are designed for classroom discussion, should the teacher wish to use them. There are also more than 1100 homework problems, of varying degrees of difficulty, in the text. The answers to the odd-numbered problems constitute Appendix E in the back of the book.

Although the vast majority of examples and problems in kinetics deal with rigid bodies, several examples and exercises treat deformable bodies to make the student aware of the applicability of most of the principles to this broader class of problems.

The reader is encouraged to begin by reading Appendix A on units. Some of the examples and problems in the book are presented in SI (Système International) metric units and some in the traditional U.S. engineering system. While America is slowly and painfully converting to SI units, our consulting activities make it clear to us that a great amount of engineering work is still being performed in the U.S. system of units. Most engineers still tend to think in pounds instead of newtons and in feet instead of meters. The truth is that students will become much better engineers, scientists, and scholars if they are thoroughly familiar with both systems, especially during the next ten to fifteen years.

Dynamics is a subject rich in its varied applications; therefore, it is important that the student develop a feel for realistically modeling an engineering situation. Consequently, we have included a large number of actual engineering problems among the examples and exercises. Being aware of the assumptions and accompanying limitations of the model and the solution method is a valuable skill that can only be developed by sweating over many problems outside the classroom atmosphere. Only in this way can a student develop the insight and creativity that must be brought to bear on engineering problems.

Kinematics of the particle, or of a material point of a body, is covered in Chapter 1. The associated kinetics of particles and mass centers of bodies follows logically in Chapter 2. Here it will be seen that we have not dwelt at all upon the "point-mass" model of a body. Since the engineering student will be dealing with bodies of finite dimensions, we believe that it is important to present equations of motion valid for such bodies as quickly as possible. Thus Euler's laws have been used as the basis for kinetics; this provides for a compact presentation of general principles without, in our opinion and experience, any loss of understanding on the student's part. This is not meant to deprecate the point-mass model, which surely plays an important role in classical physics and can be utilized in a number of engineering problems. As we shall see in Chapter 2, however, these problems may be attacked directly through the equation of motion of the mass center of a body without detracting from the view that the body has finite dimensions. Trajectory problems, sometimes placed with particle kinematics, will be found in this kinetics chapter also, since a law of motion is essential to their formulation.

The rigid body moving in a plane is treated in detail in the center of the book—the kinematics in Chapter 3 and the kinetics in Chapters 4 and 5. In Chapter 3 the topic of rolling is discussed after both the velocity and acceleration equations relating two points of a rigid body have been covered. This presentation gives the student time to digest the meanings of these important equations and to develop a feel for angular velocity and angular acceleration before being hit with rolling (or "no slip") conditions. Rolling is an important concept in the kinematics of rigid bodies, and we have found that students learn the concept much more easily if it is the only topic being covered at the time.

Chapter 4 approaches plane kinetics from the equations of motion, written with the aid of a free-body diagram of the body being studied—that is, a sketch of the body depicting all the external forces and couples but excluding any vectors expressing acceleration. Thus the free-body diagram means the same thing in dynamics as it does in statics, facilitating the student's transition to the more difficult subject. We are encouraged in this approach by a recent national survey of dynamics faculties undertaken by Dr. L. Glenn Kraige of VPI. He reports that the overwhelming majority of respondents prefer this "pure free-body-diagram method."

Mass centers (Chapter 2) and moments and products of inertia (Chapter 4) are covered right where they appear in the development of kinetics. This presentation gives students an appreciation of these concepts, as well as a sense of history, as they encounter them along the same paths that were traveled by the old masters such as Newton and Euler. Further, we treat the equations of velocity and acceleration of a point moving relative to two frames of reference ("moving frames") in Chapters 3 (plane kinematics) and 6 (three-dimensional, or general kinematics), after the student has been properly introduced to the angular velocity vector in these chapters.

Chapter 5 is dedicated to solving plane kinetics problems of rigid bodies with certain special yet general solutions (or integrals) of the equations of Chapters 2 and 4. These are known as the principles of work and kinetic energy, impulse and momentum, and angular impulse and angular momentum.

Chapters 6 and 7 deal comprehensively with the kinematics and kinetics, respectively, of rigid bodies in three dimensions—also known as the general motion of rigid bodies. There is no natural linear extension from plane to general motion, and the culprit is the angular velocity vector $\boldsymbol{\omega}$, which depends in a much more complicated way than "$\dot{\theta}\hat{\mathbf{k}}$" on the angles used to orient the body in three dimensions. We have found that if students understand the angular velocity vector $\boldsymbol{\omega}$, they will have little trouble with the general motion of rigid bodies. Thus we begin Chapter 6 with a study of $\boldsymbol{\omega}$ and its properties. In three dimensions, the definition of angular velocity is motivationally developed through the relationship between derivatives of a vector in two different frames of reference. While this point of view is normally associated with more advanced texts (e.g., Kane, Milne), we have found that college students at the junior level are quite capable of appreciating and exploiting the power of this approach. In particular, it allows the student to attack, in an orderly way, intimidating problems such as motions of gear systems and those of universal joints connecting noncollinear shafts.

Chapter 8 is an introduction to three special topics in the area of dynamics: vibrations, mass redistribution problems, and central force motion. For the student who is further interested in one or more of these topics, more complete treatments of them are readily available in the literature of dynamics.

Our experience has been that students should be discouraged from falling into the trap of regarding dynamics as a system of recipes and gimmicks appropriate to various classes of problems. In our view, as teachers and writers, we are shortchanging students if we fail to teach them how to understand and apply what is in fact a very small number of principles to all the problems they may be required to solve, instead of memorizing a large number of different solution techniques depending upon "what the picture looks like." Consequently, while we have included a wide variety of examples and exercises (which should be considered integral parts of the text), no attempt has been made to be encyclopedic.

The support of Dean William M. Sangster and of the Georgia Tech Foundation, which provided financial assistance at critical times in this project, are most gratefully acknowledged.

We thank in a special way our longtime department head, Dr. Milton E. Raville, for his encouragement in so many ways over our years together at Georgia Tech. We also thank Peggy Varalla, Betty Hunnicutt, and Vivian Tucker of Dr. Raville's staff for assisting us with this project in many ways.

We are greatly indebted to our friend and colleague, Professor Andrew Marris, for many conversations over nearly 20 years that have enlightened our thinking about the subjects of kinematics and kinetics. We also gratefully acknowledge the University of London and the University of British Columbia, who have given us permission to use their Examination Problems, mostly from the 1950s, as exercises in the book. About 100 of them are to be found among the 1116 problems in the text. Again, we thank Dr. Marris (who once studied at the University of London and taught at British Columbia) for providing us with these problems.

Other specific help with the book was provided by our friend and colleague Bill Johnston, who suggested the form of presentation of Section 8.2. Three friends from industry each suggested a practical example: Leroy Fuss, Sal Calabrese, and C. J. Mayeux. To these we also owe a debt of gratitude.

Several of our former students provided invaluable help in proofreading the manuscript and checking solutions. Their names are Jay Humphrey, John Donniacuo, Connie Watson, Keith Eubanks, David Kim, and Steve Skinner, and we are grateful to each of them. We also thank our reviewers: Nicholas J. Altiero, Michigan State University; Dwight Bushnell, Oregon State University; K. L. DeVries, University of Utah; Oscar W. Dillon, Jr., University of Kentucky; Kerry Havner and Patrick McDonald, North Carolina State University; L. Glenn Kraige, Virginia Polytechnic Institute; William Lee, U.S. Naval Academy; Lawrence Nelson, California Polytechnic State University; George C.

Richardson, Rutgers University, Norman C. Small, University of South Florida, and especially Larry Mack, University of Texas, Austin.

We also wish to acknowledge in a general way the following friends and colleagues, from discussions with whom we have learned much about mechanics through the years: Jerry Anderson and Jim Aberson (Bell Laboratories); George Webb (Christopher Newport College); Steve Passman (Sandia Laboratories); and Bob Shreeves, Don Berghaus, Mike Bernard, Hyland Chen, Jack Clark, Art Koblasz, Dick Kunz, Bill Lnenicka, John Papastavridis, George Rentzepis, George Simitses, Charles Ueng, Don Vawter, Ray Vito, James Wang, Gerry Wempner, and Wan-Lee Yin, all of the School of Engineering Science and Mechanics at Georgia Tech.

We are very grateful to our sponsoring editor, Ray Kingman, and to Brooks/Cole Engineering Division, for constant support throughout the project. In addition we thank the following people and their staffs for their help and excellent work: Mary Forkner (production), Georg Klatt (illustrations), and Don Yoder (editing).

And finally we express our sincerest thanks to Betty Mitchell, who has cheerfully, neatly, and accurately typed the manuscript through so many rewrites that she now knows the subject by osmosis.

David J. McGill

Wilton W. King

Contents

CHAPTER

1

Kinematics of Material Points or Particles

chapter outline ►

1.1 Introduction

Dynamics is the general name given to the study of the motions of bodies and the forces that accompany or cause those motions. The branch of the subject that deals only with considerations of space and time is called **kinematics.** The branch that deals with the relationships between forces and motions is called **kinetics,** but since the force-motion relationships involve kinematic considerations, it is necessary to study kinematics first.

In this chapter we present some fundamentals of the kinematics of a material point or, equivalently, an infinitesimal element of material. We shall often use the term **particle** for such an element. As will be seen in Chapter 2, however, the word *particle* is often used in a broader sense to denote a piece of material sufficiently small that the locations of its material points need not be distinguished. Of course this definition is vague enough so that, for *some* purposes, a truck or a space vehicle or even a planet might be adequately modeled as a particle.

The derivative of a vector function of a scalar (usually time) plays a fundamental role in kinematics. Section 1.2 is devoted to that topic, with particular emphasis given to the way in which the derivative is tied to a frame of reference.

Position, velocity, and acceleration vectors are defined in Section 1.3. The forms that these vectors take in the simplest case, rectilinear motion, are considered in Section 1.4. Three-dimensional motion described by rectangular and cylindrical coordinates is presented in Sections 1.5 and 1.6. In Section 1.7 we investigate the relationship of velocity and acceleration to the geometry of the path being traversed by the point (or particle).

1.2 Reference Frames and Vector Derivatives

In the next section and throughout the book, we are going to be differentiating vectors; the derivative of the position vector of a point will be its velocity, for example. Thus in this preliminary section it seems wise to examine the concept of the derivative of a vector $\mathbf{A}$, which is a function of time t. The definition of $d\mathbf{A}/dt$, which is also commonly written as $\dot{\mathbf{A}}$, is deceptively simple:

$$\frac{d\mathbf{A}}{dt} \equiv \lim_{\Delta t \to 0} \left[\frac{\mathbf{A}(t + \Delta t) - \mathbf{A}(t)}{\Delta t} \right] \tag{1.1}$$

This definition closely parallels the definition of the derivative of a scalar, such as $dy(x)/dx$, as found in any calculus text. But what we must realize about a vector is that it can change with time in *two* ways—in direction as well as in magnitude. This means that $\dot{\mathbf{A}}$ is intrinsically tied to the frame of reference in which the derivative is taken.

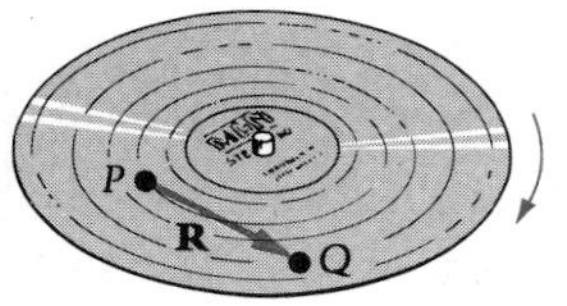

Figure 1.1

To illustrate this idea, consider the two points P and Q on the surface of the phonograph record in Figure 1.1. The record rests on a turntable that revolves in the indicated direction at, say, $33\frac{1}{3}$ rpm.

Suppose we call **R** the vector that is the directed line segment from P to Q, and inquire about the rate at which **R** changes with time as the turntable rotates. Even though we perceive the record and turntable to behave as a rigid body so that the distance between P and Q (that is, the magnitude of **R**) is constant, most of us would judge $d\mathbf{R}/dt$ to be nonzero, owing to the varying direction of **R**. This conclusion follows from our automatically having adopted the building (or earth) as our frame of reference. If we were able to ride on the turntable and blind ourselves to the surroundings, however, our perception would be that **R** is a constant vector and consequently has a vanishing derivative. Thus **R**, relative to the turntable, is a constant vector and, relative to the building, is a constant-magnitude but varying-direction vector. It is therefore seen that $d\mathbf{R}/dt$ cannot be evaluated except by specific association with a **frame of reference,** which is nothing more nor less than a **rigid body.** We shall discuss the frame of reference concept further in the next section and again in Chapter 3.

We shall be needing several vector derivative relationships that have analogs in the calculus of scalars; these relationships follow directly from the definition (Equation 1.1). If α is a scalar and **A** and **B** are vector functions of t, then

$$\frac{d}{dt}(\alpha\mathbf{A}) = \left(\frac{d\alpha}{dt}\right)\mathbf{A} + \alpha\left(\frac{d\mathbf{A}}{dt}\right) \tag{1.2}$$

$$\frac{d}{dt}(\mathbf{A} + \mathbf{B}) = \frac{d\mathbf{A}}{dt} + \frac{d\mathbf{B}}{dt} \tag{1.3}$$

$$\frac{d}{dt}(\mathbf{A} \cdot \mathbf{B}) = \left(\frac{d\mathbf{A}}{dt}\right) \cdot \mathbf{B} + \mathbf{A} \cdot \left(\frac{d\mathbf{B}}{dt}\right) \tag{1.4}$$

$$\frac{d}{dt}(\mathbf{A} \times \mathbf{B}) = \left(\frac{d\mathbf{A}}{dt}\right) \times \mathbf{B} + \mathbf{A} \times \left(\frac{d\mathbf{B}}{dt}\right) \tag{1.5}$$

The first and second of these equations allow us to be more specific about the manner in which differentiation is linked to a frame of reference. Suppose that $\hat{\mathbf{e}}_1$, $\hat{\mathbf{e}}_2$, $\hat{\mathbf{e}}_3$ are mutually perpendicular unit vectors* and A_1, A_2, A_3 are the corresponding components of a vector **A** so that

$$\mathbf{A} = A_1\hat{\mathbf{e}}_1 + A_2\hat{\mathbf{e}}_2 + A_3\hat{\mathbf{e}}_3 \tag{1.6}$$

*Note that we could use any set of base vectors (that is, linearly independent reference vectors) here, in which case A_1, A_2, A_3 are not necessarily orthogonal components of **A**. Equation (1.6) simply illustrates the most common choice of scalars and base vectors. When this is the case, the magnitude of **A**, written $|\mathbf{A}|$ or sometimes A, is $\sqrt{A_1^2 + A_2^2 + A_3^2}$.

If $\mathcal{F}$ is the frame of reference* and we denote the derivative of $\mathbf{A}$ relative to $\mathcal{F}$ by ${}^{\mathcal{F}}d\mathbf{A}/dt$, then

$$\frac{{}^{\mathcal{F}}d\mathbf{A}}{dt} = \left(\frac{dA_1}{dt}\right)^{\dagger}\hat{\mathbf{e}}_1 + A_1\,{}^{\mathcal{F}}\!\left(\frac{d\hat{\mathbf{e}}_1}{dt}\right) + \left(\frac{dA_2}{dt}\right)\hat{\mathbf{e}}_2 + A_2\,{}^{\mathcal{F}}\!\left(\frac{d\hat{\mathbf{e}}_2}{dt}\right) + \left(\frac{dA_3}{dt}\right)\hat{\mathbf{e}}_3 + A_3\,{}^{\mathcal{F}}\!\left(\frac{d\hat{\mathbf{e}}_3}{dt}\right) \tag{1.7}$$

Now if we choose $\hat{\mathbf{e}}_1$, $\hat{\mathbf{e}}_2$, $\hat{\mathbf{e}}_3$ to have fixed directions in $\mathcal{F}$, they are each constant there and

$$\frac{{}^{\mathcal{F}}d\mathbf{A}}{dt} = \frac{dA_1}{dt}\hat{\mathbf{e}}_1 + \frac{dA_2}{dt}\hat{\mathbf{e}}_2 + \frac{dA_3}{dt}\hat{\mathbf{e}}_3 \tag{1.8}$$

which is the most straightforward way to express the derivative of a vector and its intrinsic association with a frame of reference. We now give one example of the use of Equation (1.8) and, assuming the reader to be familiar with Equations (1.1) to (1.5), then move on to Section 1.3 and the task of describing the motion of a point (particle) P.

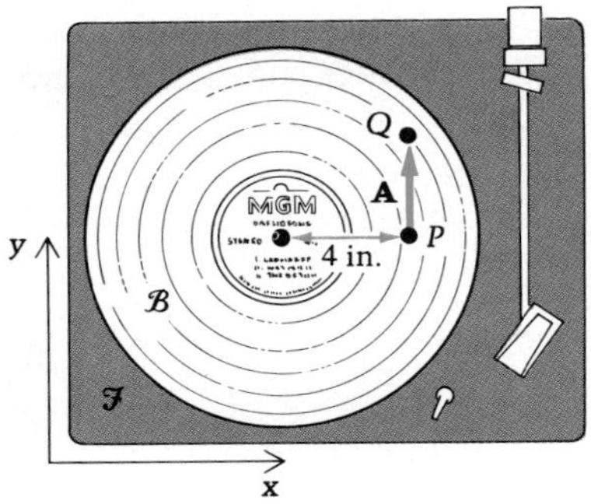

EXAMPLE 1.1

If the distance from P to Q on the $33\frac{1}{3}$ rpm record in the figure is 3 in. and if $\mathbf{A}$ is the vector from P to Q, find ${}^{\mathcal{F}}\dot{\mathbf{A}}$, where the frame $\mathcal{F}$ is the cabinet of the stereo in which the axes (x, y, z) are embedded. It is also given that the line PQ is in the indicated position (parallel to y) when $t = 0$.

SOLUTION

At a later time t (in seconds), the vector $\mathbf{A}$ is seen to make an angle $\theta(t)$ with y of

$$\theta = (33\tfrac{1}{3})\left(\frac{2\pi}{60}\right)t \text{ rad}$$

$$= 3.49t \text{ rad}$$

(Continued)

*Throughout the book, frames (rigid bodies) are denoted by capital script letters.

†Note that the derivatives of the scalar components of $\mathbf{A}$, such as dA_1/dt, need not be "tagged" since they are the same in any frame.

The vector $\mathbf{A}$, expressed in terms of the unit vectors $\hat{\mathbf{i}}$ and $\hat{\mathbf{j}}$ in the respective directions of $\mathbf{x}$ and y, has the following form:

$$\mathbf{A} = 3(\sin\theta\hat{\mathbf{i}} + \cos\theta\hat{\mathbf{j}}) \text{ in.}$$

Noting that the unit vectors do not change in direction in $\mathcal{F}$, we obtain, using Equation (1.8),

$$\frac{{}^{\mathcal{F}}d\mathbf{A}}{dt} = {}^{\mathcal{F}}\dot{\mathbf{A}} = 3\cos\theta\frac{d\theta}{dt}\hat{\mathbf{i}} - 3\sin\theta\frac{d\theta}{dt}\hat{\mathbf{j}}$$

$$= 3\cos(3.49t)(3.49)\hat{\mathbf{i}} - 3\sin(3.49t)(3.49)\hat{\mathbf{j}}$$

$$= 10.5(\cos\theta\hat{\mathbf{i}} - \sin\theta\hat{\mathbf{j}}) \text{ in./sec}$$

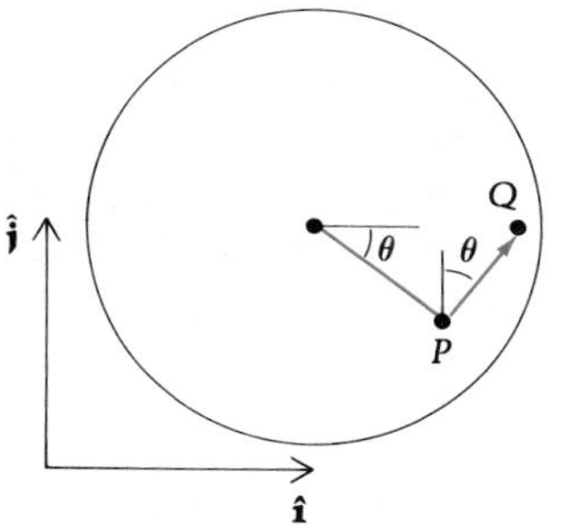

We see from this result, for example, that:

1. At $\theta = 0$, ${}^{\mathcal{F}}\dot{\mathbf{A}}$ is in the $\mathbf{x}$ direction.
2. At $\theta = \pi/2$, ${}^{\mathcal{F}}\dot{\mathbf{A}}$ is in the $-y$ direction.
3. At $\theta = \pi$, ${}^{\mathcal{F}}\dot{\mathbf{A}}$ is in the $-\mathbf{x}$ direction.
4. At $\theta = 3\pi/2$, ${}^{\mathcal{F}}\dot{\mathbf{A}}$ is in the y direction.

In all four cases, and at all intermediate angles as well, the derivative of $\mathbf{A}$ in $\mathcal{F}$ is seen to be that of the cross product:

$$[\dot{\theta}(-\hat{\mathbf{k}})] \times \mathbf{A}$$

The bracketed vector represents what will come to be called the ***angular velocity*** of the record ($\mathcal{B}$) in the reference frame (stereo cabinet) $\mathcal{F}$. In later chapters we shall see that it is precisely this cross product that must be added to ${}^{\mathcal{B}}\dot{\mathbf{A}}$ to obtain ${}^{\mathcal{F}}\dot{\mathbf{A}}$. Here, of course, $\mathbf{A}$ is constant relative to the turntable $\mathcal{B}$ so that its derivative in $\mathcal{B}$ (that is, ${}^{\mathcal{B}}\dot{\mathbf{A}}$) vanishes.

P R O B L E M S / Section 1.2

In Problems 1.1–1.8, $\hat{\mathbf{i}}$, $\hat{\mathbf{j}}$, $\hat{\mathbf{k}}$ are mutually perpendicular unit vectors having directions fixed in the frame of reference. In each case t is time measured in seconds ("s" in SI units). Determine at $t = 3$ s the rate of change (with respect to time) of vector $\mathbf{L}$.

1.1 $\mathbf{L} = 2\hat{\mathbf{i}} + 3t^2\hat{\mathbf{j}} - 8t\hat{\mathbf{k}}$ kg-m/s

1.2 $\mathbf{L} = 20\sin(\pi t/4)\hat{\mathbf{i}} - 20\cos(\pi t/4)\hat{\mathbf{k}}$ slug-ft/sec

1.3 $\mathbf{L} = 50\hat{\mathbf{i}} + 60\ln t\hat{\mathbf{j}}$ kg-m/s

1.4 $\mathbf{L} = -100e^{-t/2}\hat{\mathbf{i}} + 20t\hat{\mathbf{j}} - 5t^2\hat{\mathbf{k}}$ slug-ft/sec

1.5 $\mathbf{L} = 20\cosh(\pi t/4)\hat{\mathbf{i}} + 20\sinh(\pi t/4)\hat{\mathbf{k}}$ kg-m/s

1.6 $\mathbf{L} = 5t\hat{\mathbf{i}} - \dfrac{12}{t^2}\hat{\mathbf{j}} + \dfrac{6}{t^3}\hat{\mathbf{k}}$ slug-ft/sec

1.7 $\mathbf{L} = -2te^{-t^2}\hat{\mathbf{i}} + 3^t\hat{\mathbf{j}}$ kg-m/s

1.8 $\mathbf{L} = e^{-6t}(5\hat{\mathbf{i}} + 8t\hat{\mathbf{j}})$ slug-ft/sec

1.9–1.16 If the vectors enumerated in Problems 1.1–1.8 represent various forces $\mathbf{F}$, find the integral of each force over the time interval from $t = 2$ through 5 sec. Let the metric units become newtons and the U.S. units become pounds.

1.3 Position, Velocity, and Acceleration

In this short but important section, we present the definitions of the position, velocity, and acceleration vectors of a material point P as it moves relative to a frame of reference $\mathcal{I}$. It is important to mention that while a frame of reference is usually identified by the material constituting the reference body (for example, the earth, the moon, or the body of a truck), the frame is actually composed of all those material points *plus* the points generated by a rigid extension of the body to all of space. Thus, for example, we refer to a point on the centerline of a straight pipe as a point in (or of) the pipe.

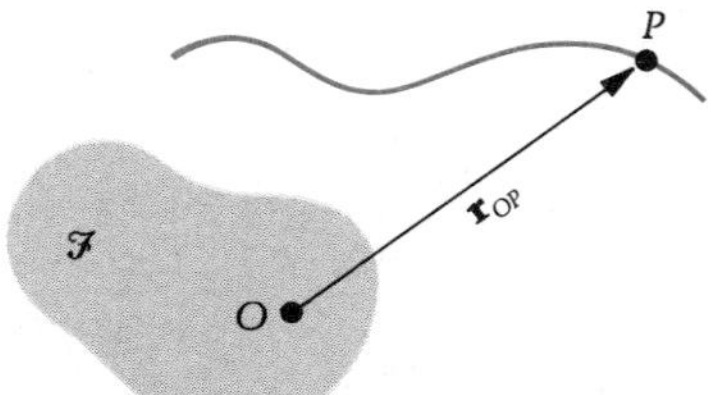

Figure 1.2 Position vector for P in $\mathcal{I}$.

We now consider a point P as it moves along a path as shown in Figure 1.2. The **path** is the locus of points of $\mathcal{I}$ that P occupies as time passes. If we select a point O of $\mathcal{I}$ to be our reference point (or origin), then the depicted vector from O to P is called a **position vector** for P in $\mathcal{I}$ and is written $\mathbf{r}_{OP}$.

The first and second derivatives (with respect to time) of the position vector are respectively called the **velocity** ($\mathbf{v}_P$) and **acceleration** ($\mathbf{a}_P$) of point P in $\mathcal{I}$:

$$\mathbf{v}_P = \frac{d\mathbf{r}_{OP}}{dt} = \dot{\mathbf{r}}_{OP} \quad \text{(The magnitude of } \mathbf{v}_P \text{ is called the } \textbf{speed} \text{ of } P.) \tag{1.9}$$

$$\mathbf{a}_P = \frac{d^2\mathbf{r}_{OP}}{dt^2} = \ddot{\mathbf{r}}_{OP} = \frac{d\mathbf{v}_P}{dt} = \dot{\mathbf{v}}_P \tag{1.10}$$

The derivatives in Equations (1.9) and (1.10) are calculated in frame $\mathcal{I}$, the only frame under consideration here. Later, however, we shall sometimes find it necessary to specify the frame in which derivatives, velocities, and accelerations are to be computed; we shall then tag the derivatives as in Equation (1.8) and write

$$\mathbf{v}_{P/\mathcal{I}} = {}^{\mathcal{I}}\dot{\mathbf{r}}_{OP} \tag{1.11}$$

Whenever there is just one frame involved, we shall omit the $\mathcal{I}$ on both sides and write an equation such as (1.11) in the form of (1.9).

Throughout the text we have inserted questions for the reader to think about. (The answers are at the ends of the chapters.) Here is the first question:

Question 1.1 Do the velocity and acceleration of a point P depend upon: (a) the choice of reference frame? (b) the position vector origin selected in the frame?

At this point it is reasonable to wonder why we have not chosen to introduce time derivatives of the position vector higher than the second. The reason is that the relationships between forces and motions do not involve those higher derivatives. As we shall see later when we study kinetics, if we know the accelerations of the particles making up a body, the force-motion laws will yield the external forces; conversely, if we know the external forces we can calculate the accelerations and then, by integrating twice, the position vectors. The force-motion laws turn out to be valid only in certain frames of reference; for that reason writers sometimes refer to motion relative to such a frame as ***absolute motion***. We have not used the word ***absolute*** here because we wish to emphasize that kinematics inherently expresses relationships of geometry and time, independent of any laws linking forces and motions. *Thus in kinematics all frames of reference are of the same importance.*

Finally, we note that positions (or locations) of points are normally established through the use of a coordinate system. The ways in which positions, velocities, and accelerations are expressed in two of the most common systems, rectangular and cylindrical, are presented in the next three sections.

PROBLEMS / Section 1.3

1.17 Show that the velocity (and therefore the acceleration also) of a point P in a frame $\mathcal{F}$ does not depend on the choice of the origin. *Hint:* Differentiate the following relationship in $\mathcal{F}$ (see Figure P1.17):

$$\mathbf{r}_{OP} = \mathbf{r}_{OO'} + \mathbf{r}_{O'P}$$

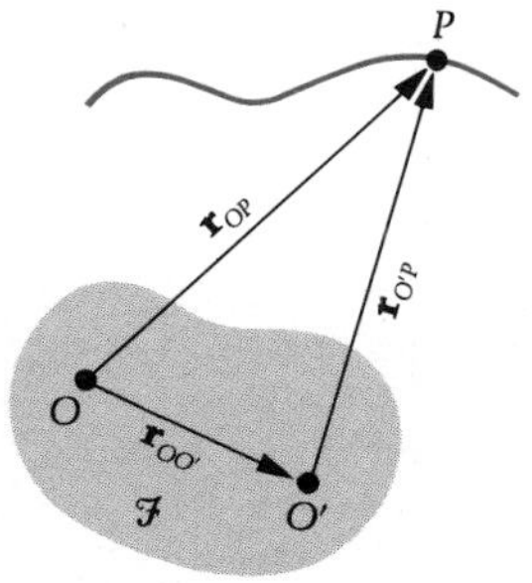

Figure P1.17

In Problems 1.18–1.22, $\hat{\mathbf{i}}$, $\hat{\mathbf{j}}$, $\hat{\mathbf{k}}$ are mutually perpendicular unit vectors having directions fixed in the frame of reference; $\mathbf{v}_P$ is the velocity of a point P moving in the frame; t is time measured in seconds. Determine at $t = 2$ s the acceleration of the point for the velocity given.

1.18 $\mathbf{v}_P = 2t\hat{\mathbf{i}} - 3t^2\hat{\mathbf{k}}$ m/s

1.19 $\mathbf{v}_P = 20e^{-0.1t}\left(\sin\dfrac{\pi t}{4}\hat{\mathbf{i}} - \cos\dfrac{\pi t}{4}\hat{\mathbf{j}}\right)$ ft/sec

1.20 $\mathbf{v}_P = \dfrac{\hat{\mathbf{i}}}{t} - \dfrac{1}{2t^2}\hat{\mathbf{j}} + \dfrac{1}{3t^3}\hat{\mathbf{k}}$ m/s

1.21 $\mathbf{v}_P = t\left(\sin\dfrac{\pi t}{4}\hat{\mathbf{i}} + \cos\dfrac{\pi t}{4}\hat{\mathbf{j}}\right)$ ft/sec

1.22 $\mathbf{v}_P = 5e^{-0.1t}\hat{\mathbf{i}} - 4e^{-0.4t}\hat{\mathbf{k}}$ m/s

1.23–1.27 The **displacement** of a point over a time interval t_1 to t_2 is defined to be the difference of the position vectors—that is, $\mathbf{r}(t_2) - \mathbf{r}(t_1)$. For the cases enumerated in Problems 1.18–1.22, find the displacement and the magnitude of the displacement over the interval $t = 4$ s to $t = 6$ s.

1.4 Kinematics of a Point in Rectilinear Motion

In this section we study problems in which point P moves along a straight line in the reference frame $\mathcal{F}$. This situation is called **rectilinear motion,** and the position of P may be expressed with a single coordinate x measured along the fixed line ℓ on which P moves (see Figure 1.3).

A position vector for P is simply

$$\mathbf{r}_{OP} = x\hat{\mathbf{i}} \tag{1.12}$$

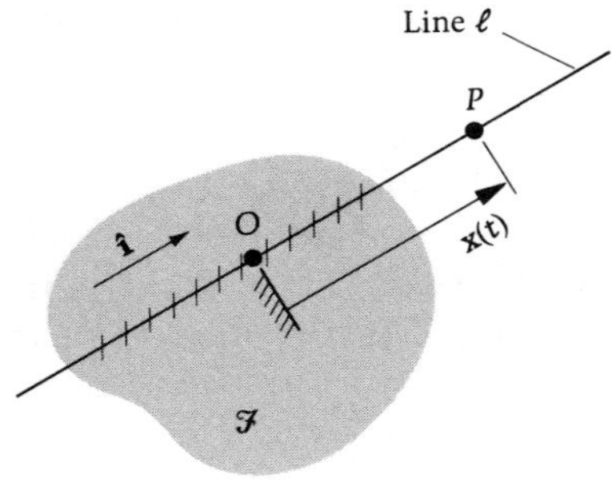

Figure 1.3

in which the unit vector $\hat{\mathbf{i}}$ is parallel to the line as shown in Figure 1.3 and hence does not change in either magnitude or direction in $\mathcal{F}$. Therefore P has the following very simple velocity and acceleration expressions:

$$\mathbf{v}_P = \dot{\mathbf{r}}_{OP} = \dot{x}\hat{\mathbf{i}} \tag{1.13}$$

$$\mathbf{a}_P = \ddot{\mathbf{r}}_{OP} = \ddot{x}\hat{\mathbf{i}} \tag{1.14}$$

In rectilinear motion, there are three interesting cases worthy of special note:

1. Acceleration is a function of time, $f(t)$.
2. Acceleration is a function of velocity, $g(v)$, where $v = \dot{x}$.*
3. Acceleration is a function of position, $h(x)$.

In each case, we can go far with general integrations. We shall consider each case in turn and give an example.

First, if $\ddot{x} = f(t)$, then

$$\frac{d\dot{x}}{dt} = f(t) \Rightarrow \dot{x} = \int f(t)\,dt + C_1 \Rightarrow x = \int\int f(t)\,dt + C_1 t + C_2 \tag{1.15}$$

in which C_1 and C_2 are to be determined by the initial conditions on velocity and position, respectively, once the problem (and thus $f(t)$) is stated and the indefinite integrals are performed. Alternatively, we might know the values of x at two times, instead of one position and one velocity. In any case, we need two constants.

EXAMPLE 1.2

The acceleration of a point P in rectilinear motion is given by the equation $\ddot{x} = 5t^2$ m/s^2, with initial conditions $\dot{x}(0) = 2$ m/s and $x(0) = -7$ m. Find $x(t)$.

(Continued)

*In this section on one-dimensional motion, we make no distinction between v and $\dot{x}$; that is, v can be negative and thus need not be the velocity magnitude. In plane and general (three-dimensional) motion, however, v and a often denote the magnitudes of the velocity and acceleration vectors.

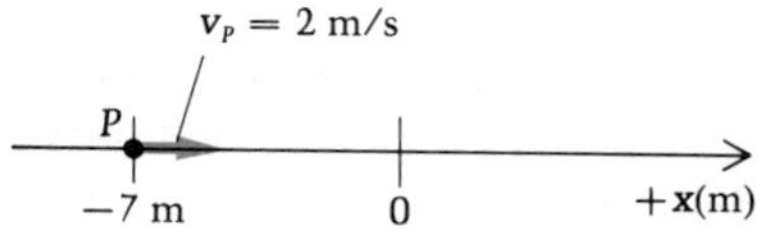

SOLUTION

We note that this is the problem of a point moving with a quadratically varying acceleration magnitude and with the following initial conditions, which are the position and velocity of P at $t = 0$:
Integrating as above,

$$\dot{x} = \int 5t^2\,dt + C_1 = \tfrac{5}{3}t^3 + C_1 \text{ m/s}$$

And integrating once more,

$$x = \tfrac{5}{12}t^4 + C_1 t + C_2 \text{ m}$$

The constants are found from the initial conditions to be $C_1 = 2$ m/s and $C_2 = -7$ m, as follows:

$$\dot{x}(0) = 2 = (\tfrac{5}{3})(0)^3 + C_1 \Rightarrow C_1 = 2 \text{ m/s}$$

$$x(0) = -7 = (\tfrac{5}{12})(0)^4 + 2(0) + C_2 \Rightarrow C_2 = -7 \text{ m}$$

Thus the motion of the point P is given by the integrated function of time:

$$x = \tfrac{5}{12}t^4 + 2t - 7 \text{ m}$$

Second, if $\ddot{x} = g(v)$, then

$$\ddot{x} = \frac{dv}{dt} = g(v) \Rightarrow \frac{dv}{g(v)} = dt \Rightarrow t + C_3 = \int \frac{dv}{g(v)}$$

If the integral can be found as a function $p(v)$, then we may be able to solve the equation $p(v) = t + C_3$ for the velocity:

$$v = q(t)$$

If so, then

$$v = \frac{dx}{dt} = q(t) \Rightarrow dx = q(t)\,dt$$

so that

$$x = \int q(t)\,dt + C_4 \tag{1.16}$$

This procedure should become clearer with the following example.

EXAMPLE 1.3

Suppose that the acceleration of a point P in one-dimensional motion is proportional to velocity according to $\ddot{x} = -2v$ m/s^2 with the same initial conditions as in the previous example. Solve for the motion $x(t)$.

(Continued)

SOLUTION

$$\ddot{x} = \frac{dv}{dt} = -2v \Rightarrow \int \frac{dv}{-2v} = \int dt + C_3$$

so that, integrating,* we get

$$t + C_3 = \frac{-\ln v}{2} \Rightarrow v = e^{-2t - 2C_3}$$

Since $v = 2$ when $t = 0$, then $C_3 = (-\ln 2)/2$ and

$$\frac{dx}{dt} = v = e^{-2t + \ln 2} = 2e^{-2t} \text{ m/s}$$

Therefore

$$x = \int 2e^{-2t}\, dt + C_4 = -e^{-2t} + C_4 \text{ m}$$

But $x = -7$ m when $t = 0$ s gives $C_4 = -6$ m, and so we obtain our solution:

$$x = -6 - e^{-2t} \text{ m}$$

The third case occurs when acceleration is a function of position, $a = \ddot{x} = h(x)$. Then we may combine $a = \dot{v}$ and $v = \dot{x}$ to obtain the useful relation

$$a\frac{dx}{dt} = v\frac{dv}{dt}$$

so that

$$\int a\, dx = \int v\, dv + C_5$$

And so, for this case,

$$\int h(x)\, dx = \int v\, dv + C_5 = \frac{v^2}{2} + C_5 \qquad (1.17)$$

If the integral $\int h(x)\, dx$ may be written in closed form as $r(x)$, then

$$\frac{v^2}{2} + C_5 = r(x) \Rightarrow v^2 = 2r(x) - 2C_5$$

and thus

$$\frac{dx}{dt} = v = \sqrt{2r(x) - 2C_5}$$

from which another integration yields

$$\int \frac{dx}{\sqrt{2r(x) - 2C_5}} = t + C_6 \qquad (1.18)$$

*This problem could also be solved by first integrating the linear differential equation $\dot{v} + 2v = 0$, observing that Ae^{-2t} is the general solution.

If the left side of this equation can be written as $s(\mathbf{x})$, then the equation

$$s(\mathbf{x}) = t + C_6 \tag{1.19}$$

may be solvable for $\mathbf{x}(t)$, as in the following example.

EXAMPLE 1.4

Let $\ddot{\mathbf{x}} = h(\mathbf{x}) = -4\mathbf{x}\ \mathrm{m/s^2}$. Find $\mathbf{x}(t)$ if the initial conditions are the same as in Examples 1.2 and 1.3.

SOLUTION

We are to solve the equation

$$\ddot{\mathbf{x}} + 4\mathbf{x} = 0$$

Actually we know that the solution to this equation, by the theory of differential equations, is $\mathbf{x} = A \sin 2t + B \cos 2t$—which, with $\mathbf{x}(0) = -7$ m and $\dot{\mathbf{x}}(0) = 2$ m/s, becomes $\mathbf{x} = \sin 2t - 7 \cos 2t$ meters. But let us obtain this result by using the procedure described above, which applies even when $h(\mathbf{x})$ is *not* linear. Carrying out the first integration, following the suggested steps, we get

$$\int -4\mathbf{x}\, d\mathbf{x} = \frac{\mathbf{v}^2}{2} + C_5$$

$$\mathbf{v}^2 = -4\mathbf{x}^2 - 2C_5 = 200 - 4\mathbf{x}^2$$

$$\mathbf{v} = \frac{d\mathbf{x}}{dt} = \sqrt{200 - 4\mathbf{x}^2} = 2\sqrt{50 - \mathbf{x}^2}\ \mathrm{m/s}$$

where C_5 has been found by using $\mathbf{v} = 2$ m/s and $\mathbf{x} = -7$ m at $t = 0$. Then

$$\frac{d\mathbf{x}}{\sqrt{50 - \mathbf{x}^2}} = 2\, dt$$

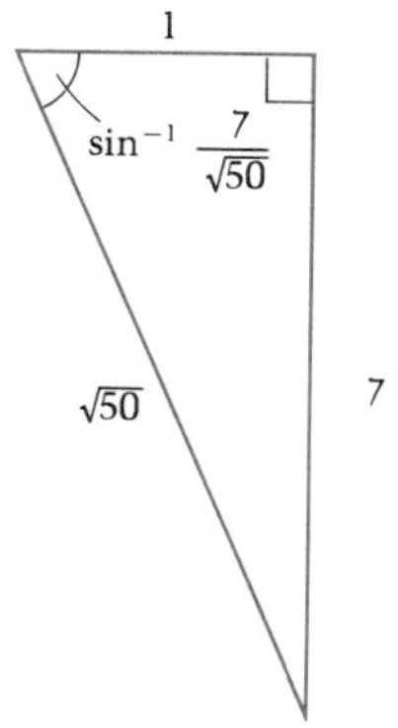

Integrating a second time, we get

$$\sin^{-1} \frac{\mathbf{x}}{\sqrt{50}} = 2t + C_6 = 2t + \sin^{-1}\left(\frac{-7}{\sqrt{50}}\right)$$

in which $\mathbf{x} = -7$ when $t = 0$ yielded the value of C_6. Therefore

$$\frac{\mathbf{x}}{\sqrt{50}} = \sin\left[2t - \sin^{-1}\left(\frac{7}{\sqrt{50}}\right)\right] \mathrm{m}$$

which is the same result as $\mathbf{x} = \sin 2t - 7 \cos 2t$, as the reader may easily show with the help of the triangle and trigonometry identity shown in the diagram:

$$\sin(A - B) = \sin A \cos B - \cos A \sin B$$

Sometimes there are special conditions in a problem that require ingenuity in expressing the kinematics. If there is an inextensible rope, string, cable, or cord present, for example, we may have to express the constancy of length mathematically. This is the case in the following example.

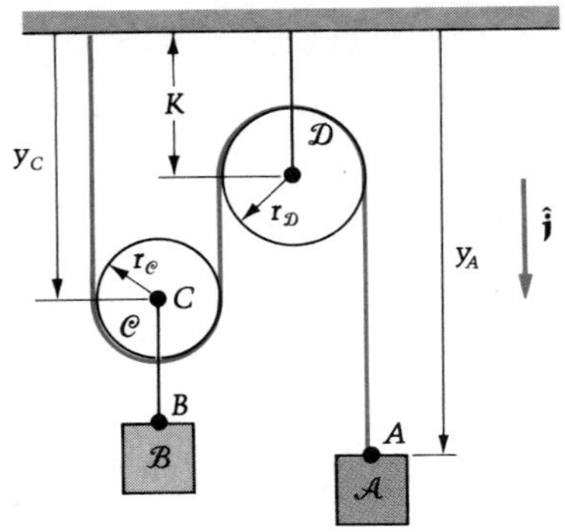

EXAMPLE 1.5

Point A of block $\mathcal{A}$ travels downward with $\mathbf{v}_A = 3t^2 \downarrow$ m/s. Find the velocity of point B of block $\mathcal{B}$ when $t = 4$ s.

SOLUTION

The length L of the rope that passes around both small pulleys is a constant. This is a constraint equation that must be used in the solution. The procedure is as follows:

$$L = y_C + \pi r_\mathcal{C} + (y_C - K) + \pi r_\mathcal{D} + (y_A - K)\ \text{m}$$

Differentiating and noting that L, π, $r_\mathcal{C}$, $r_\mathcal{D}$, and K are constants, we get

$$0 = 2\dot{y}_C + \dot{y}_A \Rightarrow \dot{y}_C = -\frac{\dot{y}_A}{2} = -\frac{3}{2}t^2$$

The velocities of C and B are equal since both points move on the same path with a constant length separating them. Hence

$$\mathbf{v}_B = -\frac{3}{2}t^2\hat{\mathbf{j}}\ \text{m/s} \qquad \text{(Note that } C \text{ moves upward since } \hat{\mathbf{j}} \text{ is downward!)}$$

Therefore

$$\mathbf{v}_B\big|_{t=4} = -24\hat{\mathbf{j}}\ \text{m/s} \qquad \text{or} \qquad 24\uparrow \text{m/s}$$

In problems of rectilinear kinematics in which the acceleration is a known function of time (Case 1), we sometimes use what is called the v-t diagram. We shall give just one example of its use because it is a method somewhat limited in application. (We discuss this shortcoming at the end of the example.)

EXAMPLE 1.6

A point P moves on a line, starting from rest at the origin with constant acceleration of $0.8\ \mathrm{m/s^2}$ to the right. After 10 s, the acceleration of P is suddenly reversed to $0.2\ \mathrm{m/s^2}$ to the left. Determine the total time elapsed when P is again passing through the origin.

SOLUTION

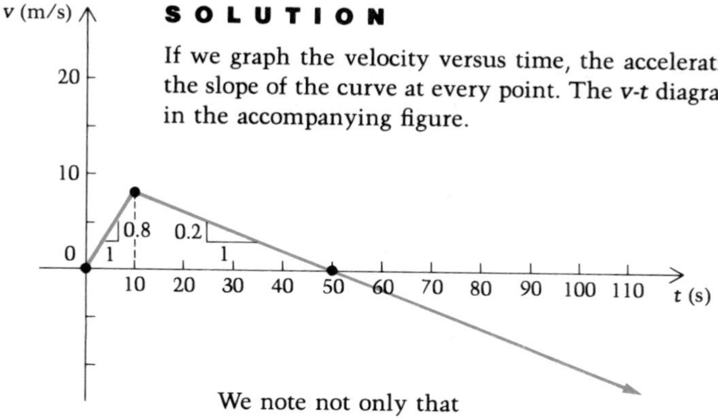

If we graph the velocity versus time, the acceleration (dv/dt) will of course be the slope of the curve at every point. The v-t diagram for this problem is shown in the accompanying figure.

We note not only that

$$a = \frac{dv}{dt} = \text{slope of diagram}$$

but also that

$$x = \int v\, dt + \text{constant}$$

Hence the change in the displacement x between any two times is nothing more than the area beneath the v-t diagram between those points. Thus four points, or times, are important in the diagram for this problem:

t_1 = starting time (in this case zero)

t_2 = time when acceleration changes (given to be 10 s)

t_3 = time when velocity has been reduced to zero (deceleration causes P to stop before moving in opposite direction)

t_4 = required total time elapsed before point P is again at origin

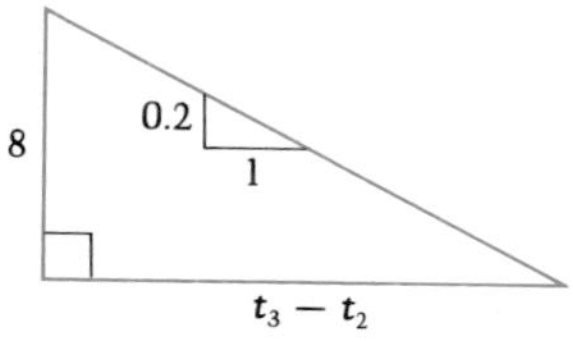

The velocity at time t_2 is seen to be $0.8\ \mathrm{m/s} \times 10\ \mathrm{s} = 8\ \mathrm{m/s}$. To find the time interval $t_3 - t_2$, we use the similar triangles shown at the left.

$$\frac{0.2}{1} = \frac{8}{t_3 - t_2} \Rightarrow t_3 - t_2 = 40$$

$$t_3 = 50\ \mathrm{s}$$

The total distance traveled before the point (momentarily) stops is thus

S_1 = area of 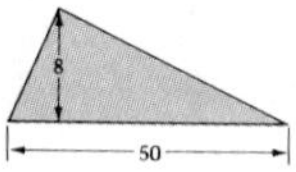 $= \frac{1}{2}(50\ \mathrm{s})(8\ \mathrm{m/s}) = 200\ \mathrm{m}$

(Continued)

This is the distance traveled by the point in the positive direction (to the right).

The point will be back at $\mathbf{x} = 0$ when the absolute value of the negative area ***beneath*** the t axis (the distance traveled back to the left) equals the 200 m traveled to the right (represented by the area ***above*** the axis):

$$\underbrace{\tfrac{1}{2}(t_4 - 50)}_{\text{base of triangle}}\,\underbrace{[0.2(t_4 - 50)]}_{\text{height of triangle}} = 200$$

which can be rewritten as

$$t_4^2 - 100t_4 + 500 = 0$$

The only root of this equation larger than 50 s is $t_4 = 94.7$ s, and this is the answer to the problem.

An alternative approach to the preceding *v-t* diagram solution is as follows. Integrating the acceleration during the time interval $0 \leq t < 10$ s, with $\mathbf{x}$ during this interval called $\mathbf{x}_1$,

$$\ddot{\mathbf{x}}_1 = 0.8 \Longrightarrow \dot{\mathbf{x}}_1 = 0.8t + C_1 = 0.8t \text{ m/s} \qquad (\text{since } \dot{\mathbf{x}}_1 = 0 \text{ at } t = 0)$$

Integrating again (over the same interval), we get

$$\mathbf{x}_1 = 0.4t^2 + C_2 = 0.4t^2 \text{ m} \qquad (\text{since } \mathbf{x}_1 = 0 \text{ at } t = 0)$$

Thus at $t = 10$ s, by substitution,

$$\mathbf{x}_1 = 40 \text{ m} \qquad \text{and} \qquad \dot{\mathbf{x}}_1 = 8 \text{ m/s}$$

Next, after the deceleration starts, using $\mathbf{x}_2$ in this interval,

$$\ddot{\mathbf{x}}_2 = -0.2 \Longrightarrow \dot{\mathbf{x}}_2 = -0.2t + C_3 \qquad (\text{for } t \geq 10 \text{ s})$$

and since $\dot{\mathbf{x}}_2 = 8$ m/s when $t = 10$ s, we obtain $C_3 = 10$. Therefore

$$\dot{\mathbf{x}}_2 = -0.2t + 10 \text{ m/s}$$

Integrating again, we get

$$\mathbf{x}_2 = -0.1t^2 + 10t + C_4 \text{ m}$$

And with $\mathbf{x}_2 = 40$ m when $t = 10$ s, then $C_4 = -50$ m:

$$\mathbf{x}_2 = -0.1t^2 + 10t - 50 \text{ m}$$

When $\mathbf{x}_2 = 0$, we can solve for the time; the equation is the same as in the *v-t* diagram solution:

$$t_4^2 - 100t_4 + 500 = 0$$

Of the roots, $t_4 = 5.28$ and 94.7 s, only the latter is valid since 5.28 s occurs prior to the change of acceleration expressions.

Even though both approaches yield the correct answer of 94.7 s in the preceding example, we must recommend the latter approach of integrating the accelerations and matching velocities and positions between intervals. The reason is that when we are faced with ***nonconstant*** accelerations, the **v-t** diagram approach requires us to find areas under curves, the formulas for which are not ordinarily memorized.

It is interesting, in using the equations, to start a new time measurement t_2 at the beginning of the second interval:

$$\ddot{x}_2 = -0.2 \text{ m/s}^2 \Rightarrow \dot{x}_2 = -0.2t_2 + C_3 = -0.2t_2 + 8 \text{ m/s}$$

Integrating again, we get

$$x_2 = -0.1t_2^2 + 8t_2 + C_4 = -0.1t_2^2 + 8t_2 + 40 \text{ m}$$

Then $x_2 = 0$ yields the equation

$$t_2^2 - 80t_2 - 400 = 0$$

which has the positive root $t_2 = 84.7$—which, added to the 10-s duration of the first interval, gives again 94.7 s of total time elapsed. It is slightly easier to calculate the integration constants with this approach of "starting time over" than to use the same t throughout. The only price we pay for this convenience is that we must add the times at the end.

EXAMPLE 1.7

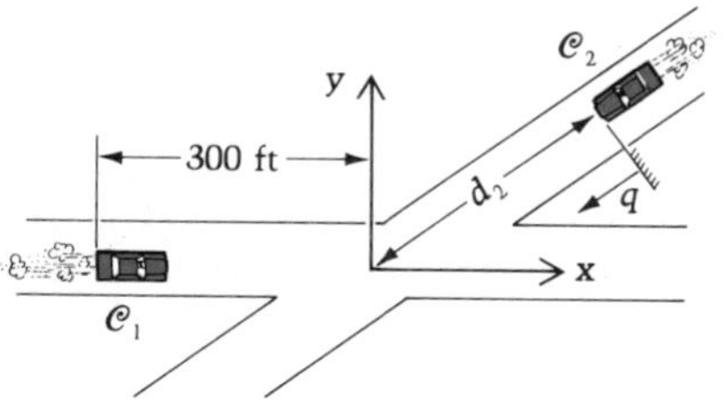

Two cars in a demolition derby are approaching a common point (the origin in the figure), each at 55 mph in a straight line as indicated. Car $\mathscr{C}_1$ does not speed up or slow down; the driver of $\mathscr{C}_2$ applies the brakes. Find the smallest rate of deceleration of $\mathscr{C}_2$ that will barely avoid a collision if

a. $d_2 = 200$ ft
b. $d_2 = 100$ ft

SOLUTION

Placing the origin at the point of intersection of the two cars' paths, we have the following for $\mathscr{C}_1$:

$$\dot{x}_1 = 55 \text{ mph}\left(\frac{88 \text{ ft/sec}}{60 \text{ mph}}\right) = 80.7 \text{ ft/sec}$$

$$x_1 = 80.7t + C_1 \text{ ft}$$

(Continued)

Using the initial condition that $\mathbf{x}_1 = -300$ ft when $t = 0$, we get

$$\mathbf{x}_1 = 80.7t - 300 \text{ ft}$$

The back of car $\mathcal{C}_1$ will be at the origin (point of possible collision) when $\mathbf{x} = 0$:

$$0 = 80.7t_0 - 300$$

$$t_0 = 3.72 \text{ sec}$$

Now let us study the motion of car $\mathcal{C}_2$. We use the coordinate q as shown for this car. Calling the unknown deceleration K, we obtain

$$\ddot{q} = -K \text{ ft/sec}^2$$

so that

$$\dot{q} = -Kt + C_2 = -Kt + 80.7 \text{ ft/sec}$$

and

$$q = \frac{-Kt^2}{2} + 80.7t + C_3 \text{ ft}$$

But $C_3 = 0$, since $q = 0$ at $t = 0$.

Next we see that at 3.72 sec the position of $\mathcal{C}_2$ is

$$q = \frac{-K(3.72^2)}{2} + 80.7(3.72)$$

$$= -6.92K + 300 \text{ ft}$$

Finally, car $\mathcal{C}_2$ just passes the rear of $\mathcal{C}_1$ if q is d_2 at this time:

$$d_2 = -6.92K + 300 \text{ ft}$$

Hence:

a. If $d_2 = 200$ ft, then $K = 14.5$ ft/sec^2.
b. If $d_2 = 100$ ft, then $K = 28.9$ ft/sec^2 (close to 1 g).

Note also that if $d_2 = 300$ ft, then $K = 0$; this is because *no* deceleration is needed for the same distances at the same speeds. Further, if $d_2 > 300$ then K is negative, meaning that car $\mathcal{C}_2$ would have to *accelerate* to arrive at the intersection at the same time as car $\mathcal{C}_1$.

EXAMPLE 1.8

A man walks along a sidewalk at 4 ft/sec. (See the figure.) Determine the speed of the top of his shadow relative to (a) the ground and (b) a frame moving along with the man.

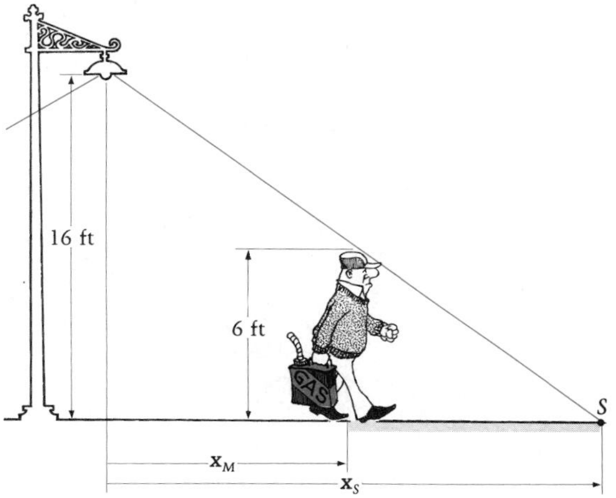

SOLUTION

The $\mathbf{x}$ coordinate of the man* is obtained from

$$\dot{\mathbf{x}}_M = 4 \text{ ft/sec}$$

$$\mathbf{x}_M = 4t + c_1 = 4t \text{ ft}$$

where $t = 0$ is chosen to be the time when the man passes beneath the streetlight so that $c_1 = 0$. Next, by similar triangles, we may relate $\mathbf{x}_S$ to $\mathbf{x}_M$:

$$\frac{\mathbf{x}_S}{16} = \frac{\mathbf{x}_S - \mathbf{x}_M}{6} = \frac{\mathbf{x}_S - 4t}{6}$$

or

$$6\mathbf{x}_S = 16\mathbf{x}_S - 64t$$

or

$$\mathbf{x}_S = 6.4t \text{ ft}$$

Thus the shadow's speed relative to the ground is $\dot{\mathbf{x}}_S = 6.4$ ft/sec; relative to a frame traveling with the man it moves forward at $\dot{\mathbf{x}}_S - \dot{\mathbf{x}}_M = 2.4$ ft/sec.

*The man is treated as a particle here, with representative point 'M.'

EXAMPLE 1.9

The acceleration of each point of the body $\mathcal{A}$ shown in the figure is 2 m/s^2 up the inclined plane. In particular, point A has this acceleration. Its velocity at $t = 0$ is 2 m/s (also up the plane). The cord is assumed to be inextensible, and body $\mathcal{B}$ has negligibly small dimensions. Point D of body $\mathcal{D}$ has $\dot{x}_D = 1$ m/s = constant (up the plane). Find the time at which $\mathcal{B}$ leaves $\mathcal{D}$.

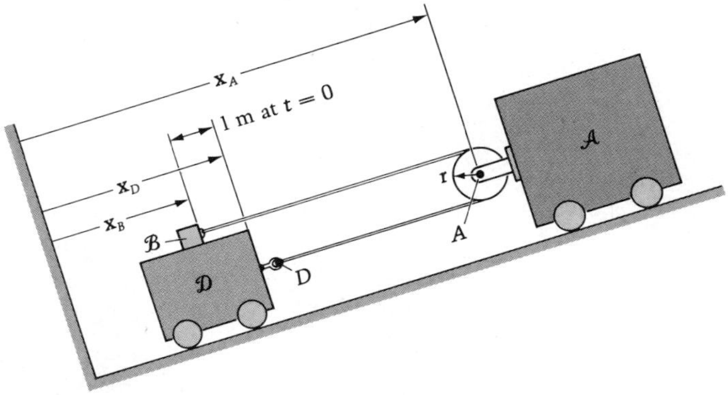

SOLUTION

Let x_B, x_D, and x_A be measured up the plane, locating points of the three translating bodies as shown. We emphasize again that the center of the pulley has the same **v** and **a** as all points of $\mathcal{A}$.

We begin the solution by integrating the known acceleration of A:

$$\ddot{x}_A = 2 \Rightarrow \dot{x}_A = 2t + C_1 \text{ m/s}$$

Since our initial condition is that $\dot{x}_A = 2$ m/s when $t = 0$, we see that $C_1 =$ 2 m/s. Hence the velocity of A is known for all time t:

$$\dot{x}_A = 2t + 2 \text{ m/s}$$

Next we must express the kinematic constraint that the cord has constant length ℓ:

$$\ell = (x_A - x_D) + \pi r + (x_A - x_B) = 2x_A - x_B - x_D + \pi r \text{ m}$$

Differentiating, and noting that ℓ and r are constants, we obtain the relationship between the three velocities:

$$0 = 2\dot{x}_A - \dot{x}_B - \dot{x}_D$$

Thus if we substitute the expressions for $\dot{x}_A$ and $\dot{x}_D$, we obtain the speed of B:

$$0 = 2(2t + 2) - \dot{x}_B - 1 \Rightarrow \dot{x}_B = 4t + 3 \text{ m/s}$$

Integrating, we get

$$x_B = 2t^2 + 3t + C_2 \text{ m}$$

(Continued)

and integrating $\dot{x}_D = 1$ gives

$$x_D = t + C_3 \text{ m}$$

But from the figure, $x_D - x_B = 1$ m at $t = 0$, so that

$$C_3 - C_2 = 1 \Rightarrow C_3 = 1 + C_2 \text{ m}$$

Thus

$$x_D = t + C_2 + 1 \text{ m}$$

Finally, B leaves D when $x_D = x_B$:

$$t + C_2 + 1 = 2t^2 + 3t + C_2$$

$$2t^2 + 2t - 1 = 0$$

Thus by the quadratic formula,

$$t = 0.366 \text{ s}$$

Note that the angle of the plane, gravity, and friction are all unimportant in this kinematics problem. Whatever forces on $\mathcal{A}$ and $\mathcal{D}$ are needed to give them $\ddot{x}_A = 2$ m/s^2 and $\dot{x}_D = 1$ m/s are presumed to be present; finding forces from given motions and vice versa is what we shall be studying in kinetics in Chapters 2, 4, and 7.

EXAMPLE 1.10

A point B starts from rest at the origin at $t = 0$ and accelerates at a constant rate k m/s^2 in rectilinear motion. After 6 s, the acceleration changes to the time-dependent function $0.006t_2^2$ m/s^2 in the *opposite* direction, where $t_2 = 0$ when $t = 6$ s. If the point stops at $t = 26$ s (from the starting time) and reverses direction, find the acceleration k during the first interval and the distance traveled by B before it reverses direction. Then find the total time elapsed before B passes back through the origin.

SOLUTION

We begin the solution by determining the motion ($x_1(t)$) during the first time interval; we integrate the acceleration to obtain the velocity and then again to get the position:

$$\ddot{x}_1 = k \text{ m/s}^2$$

$$\dot{x}_1 = kt_1 + c_1 = kt_1 \text{ m/s} \qquad (\text{since } \dot{x}_1 = 0 \text{ when } t_1 = 0)$$

$$x_1 = \frac{kt_1^2}{2} + c_2 = \frac{kt_1^2}{2} \text{ m} \qquad (\text{since } x_1 = 0 \text{ when } t_1 = 0)$$

(Continued)

At $t_1 = 6$ s, the acceleration changes to a negative value and point B "decelerates." At the beginning of this second interval, the speed and position of B are given by the "ending" values during the first interval. These values are $\dot{\mathbf{x}}_1$ and $\mathbf{x}_1$ at $t_1 = 6$:

$$\mathbf{x}_2\big|_{t_2=0} = \mathbf{x}_1\big|_{t_1=6} = \frac{k6^2}{2} = 18k \text{ m}$$

$$\dot{\mathbf{x}}_2\big|_{t_2=0} = \dot{\mathbf{x}}_1\big|_{t_1=6} = 6k \text{ m/s}$$

Note that we start time t_2 at the beginning of the second interval. (This strategy simplifies the calculation of constants during the second interval, as we shall see.) During the second interval,

$$\ddot{\mathbf{x}}_2 = -0.006t^2 \text{ m/s}^2$$

where we note that the minus sign is needed to express the ***deceleration***. Integrating, we get

$$\begin{aligned}\dot{\mathbf{x}}_2 &= -0.002t_2^3 + c_3 \\ &= -0.002t_2^3 + 6k \text{ m/s}\end{aligned}$$

since $\dot{\mathbf{x}}_2 = 6k$ m/s when $t_2 = 0$. Integrating a second time, we get

$$\begin{aligned}\mathbf{x}_2 &= -0.0005t_2^4 + 6kt_2 + c_4 \\ &= -0.0005t_2^4 + 6kt_2 + 18k \text{ m}\end{aligned}$$

where c_4 was computed by using the initial condition that $\mathbf{x}_2 = 18k$ meters when $t_2 = 0$.

Now we use the fact that $\dot{\mathbf{x}}_2$ is zero at time $t_2 = 26 - 6 = 20$ s; this strategy will allow us to determine k:

$$0 = -0.002(20^3) + 6k$$

$$k = 2.67 \text{ m/s}^2$$

Substituting k into the $\mathbf{x}_2$ expression at $t_2 = 20$ s gives us the position of B at the "turnaround":

$$\begin{aligned}\mathbf{x}_{2\,\text{STOP}} &= -0.0005(20^4) + 6(2.67)(20) + 18(2.67) \\ &= 288 \text{ m}\end{aligned}$$

Finally, to obtain the time $t_{2\,\text{END}}$ when B is passing back through the origin we set

$$\mathbf{x}_2 = 0 = -0.0005t_{2\,\text{END}}^4 + 6(2.67)t_{2\,\text{END}} + 18(2.67)$$

Rewriting, we get

$$t_{2\,\text{END}}^4 - 32{,}000t_{2\,\text{END}} - 96{,}100 = 0 \tag{1}$$

(Continued)

On a calculator, the only positive root to this equation* is found in a matter of minutes to be (to three significant figures):

$$t_{2\,\mathrm{END}} = 32.7 \text{ s}$$

The total time is $t_{2\,\mathrm{END}}$ plus the duration of the first interval, or 38.7 s.

Before we leave this problem, we wish to note that during the first time interval, ***while the acceleration is constant,***

$$\mathbf{x} = \frac{kt^2}{2} + v_0 t + \mathbf{x}_0 \text{ m} \tag{1.20}$$

where

$$\mathbf{x}_0 = \mathbf{x}(0) \text{ m}$$

$$v_0 = \dot{\mathbf{x}}(0) \text{ m/s}$$

Letting $v = \dot{\mathbf{x}}$, we note further that

$$v = kt + v_0 \text{ m/s} \tag{1.21}$$

and eliminating t we obtain†

$$v^2 = v_0^2 + 2k(\mathbf{x} - \mathbf{x}_0) \text{ m}^2/\text{s}^2 \tag{1.22}$$

This expression gives us the magnitude of the velocity in terms of displacement. Most students have used this relationship in high school or perhaps elementary college physics. There is sometimes a tendency, however, to forget the conditions under which it is valid; it holds only for ***rectilinear*** motion with ***constant acceleration***. Thus it could not be used during the second interval of the preceding example, nor could the equations for $\mathbf{x}$ and v from which it was derived.

PROBLEMS / Section 1.4

1.28 A point P starts from rest and accelerates uniformly (meaning $\ddot{\mathbf{x}}$ = constant) to a speed of 88 ft/sec after traveling 120 ft. Find the acceleration of P.

1.29 If in the preceding problem a braking deceleration of 3 ft/sec^2 is experienced beginning when P is at 120 ft, determine the time and distance required for stopping.

*Descartes' rule of signs tells us that the maximum number of positive real roots to Equation (1) is one (the number of changes in sign on the left-hand side). And there will be exactly one because the left side is negative at $t_{2\,\mathrm{END}} = 0$ and positive for large values of $t_{2\,\mathrm{END}}$.

†We can also obtain Equation (1.22) by integrating the handy relation $a\,dx = v\,dv$ for the case when a is constant and then evaluating the integration constant using $v = v_0$ when $\mathbf{x} = \mathbf{x}_0$.

1.30 A speeder zooms past a parked police car at a constant speed of 70 mph (Figure P1.30). Then, 3 sec later, the policewoman starts accelerating from rest at 10 ft/sec^2 until her velocity is 85 mph. How long does it take her to overtake the speeding car if it neither slows down nor speeds up?

Figure P1.30

1.31 In the preceding problem, suppose the speeder sees the policewoman 10 sec after she begins to move, and decelerates at 3 ft/sec^2. How long does it take the policewoman to *pass* the car if she is actually chasing a faster speeder ahead of it?

1.32 A train travels from one city to another which is 134 miles away. It accelerates from rest to a maximum speed of 100 mph in 4 min, averaging 65 mph during this time interval. It maintains maximum velocity until just before arrival, when it decelerates to rest at an average speed during the deceleration of 40 mph. If the total travel time was 85 min, find the deceleration interval.

1.33 A car $\mathcal{C}$ is 40 ft behind a truck $\mathcal{T}$; both are moving at 55 mph. (See Figure P1.33.) Suddenly the truck driver slams on his brakes after seeing an obstruction in the road ahead, and he decelerates at 10 ft/sec^2. Then, 2 sec later, the driver of the car reacts by slamming on *his* brakes, giving his car a deceleration a_C. Find the minimum value of a_C for which the car will not collide with the truck. *Hint:* Enforce $x_T > x_C$ for *all* time t before the vehicles are stopped.

Figure P1.33

1.34 A point P moves rectilinearly with velocity vector $\mathbf{v}_P = (4t^3 - 5 \sinh 0.09t)\hat{\mathbf{i}}$ m/s. At $t = 0$, the point is 2 m to the left of the origin. Find the position and acceleration vectors of P when $t = 10$ s.

1.35 Two people moving at 2.5 ft/sec to the right are using a rope to drag the box $\mathcal{B}$ along the ground at the lower level (Figure P1.35). Determine the speed of $\mathcal{B}$ as a function of the angle θ between the rope and the vertical.

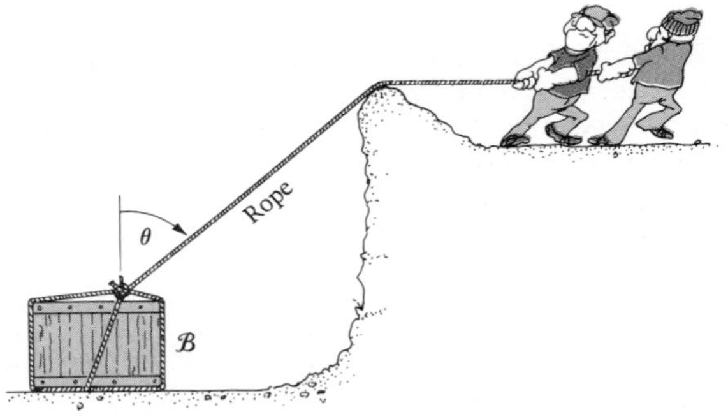

Figure P1.35

1.36 The airplane in Figure P1.36 travels at constant speed at a constant altitude. The radar tracks the plane and computes the distance D, the angle θ, and the rate of change of θ ($\dot{\theta}$) at all times. In terms of θ, $\dot{\theta}$, and D, find the speed of the airplane.

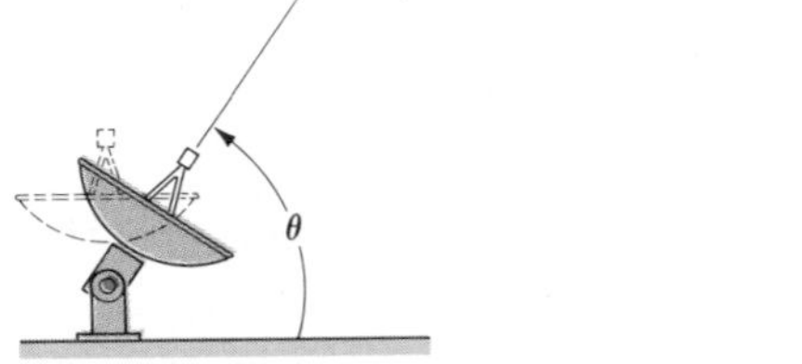

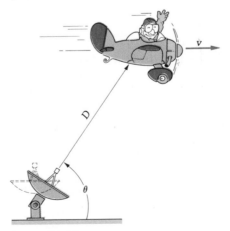

Figure P1.36

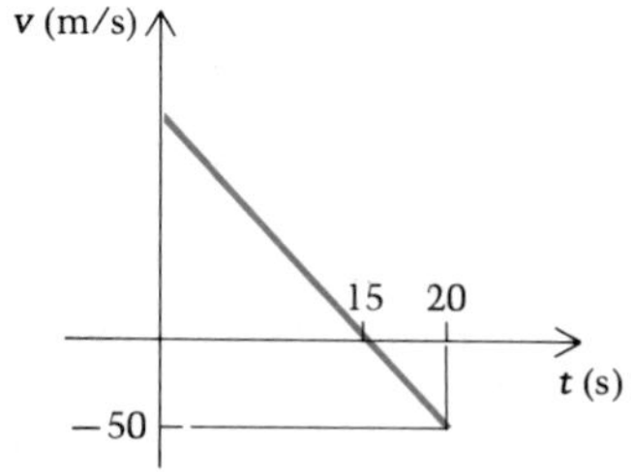

Figure P1.38

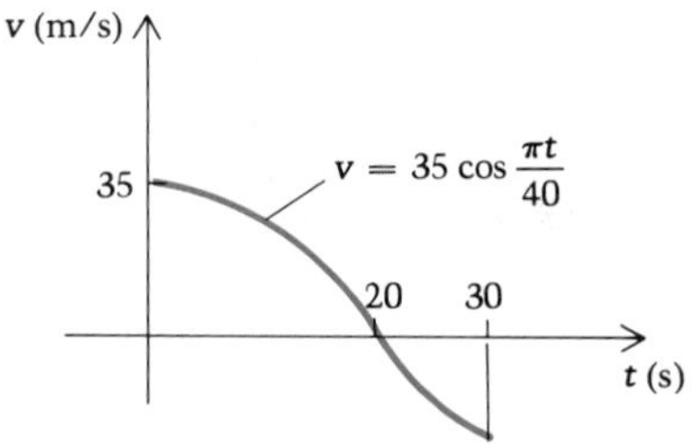

Figure P1.39

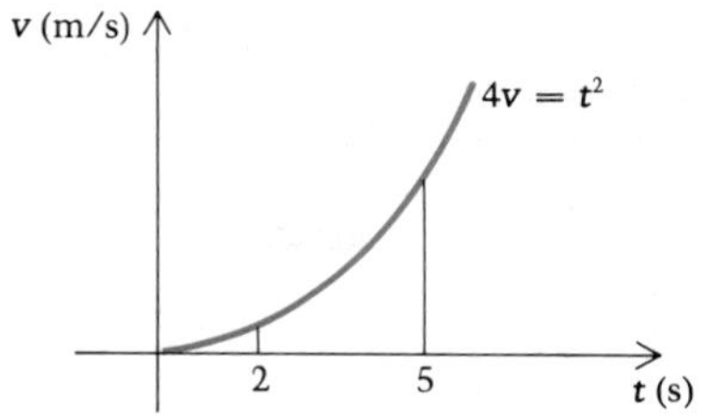

Figure P1.40

1.37 A point begins at rest at $\mathbf{x} = 0$ and experiences constant acceleration to the right for 10 s. It then continues at constant velocity for 8 more seconds. In the third phase of its motion, it decelerates at 5 m/s^2 and is observed to be passing again through the origin when the total time of travel equals 28 s. Determine the acceleration in the first 10 s.

In Problems 1.38–1.40, the graph depicts the velocity of a point P in rectilinear motion. Draw curves showing the position ($\mathbf{x}(t)$) and acceleration ($a(t)$) of P if the point is at the indicated position $\mathbf{x}_0$ at $t = 0$.

1.38 $\mathbf{x}_0 = -1125$ m
Time interval: $0 \le t \le 20$ s (See Figure P1.38.)

1.39 $\mathbf{x}_0 = 10$ m
Time interval: $0 \le t \le 30$ s (See Figure P1.39.)

1.40 $\mathbf{x}_0 = 10$ m
Time interval: $2 \le t \le 5$ s (See Figure P1.40.)

1.41 A point P moves on a line. The acceleration of P is given by $\mathbf{a}_P = \ddot{x}_P\hat{\mathbf{i}} = (3t^2 - 30t + 56)\hat{\mathbf{i}}$ m/s^2. The velocity of P at $t = 0$ is $-60\hat{\mathbf{i}}$ m/s, with the point at $\mathbf{x}_P = 7$ m at that time. Find the distance traveled by P in the time interval $t = 0$ to $t = 13$ s.

1.42 The velocity of a particle moving along a horizontal path is proportional to its distance from a fixed point on the path. When $t = 0$, the particle is 1 ft to the right of the fixed point. When $v = 20$ ft/sec to the right, $a = 5$ ft/sec^2 to the right. Determine the position of the particle when $t = 4$ sec. (See Figure P1.42.)

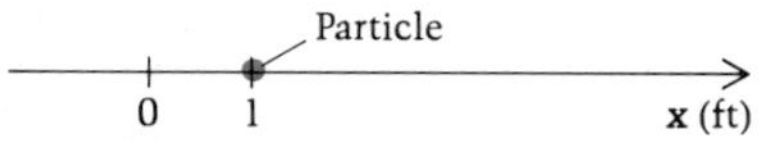

Figure P1.42

1.43 Point B of block $\mathcal{B}$ has a constant acceleration of 10 m/s^2 upward. At the instant shown in Figure P1.43, it is 30 m below the level of point A of $\mathcal{A}$. At this time, $\mathbf{v}_A$ and $\mathbf{v}_B$ are zero. Determine the velocities of A and B as they pass each other.

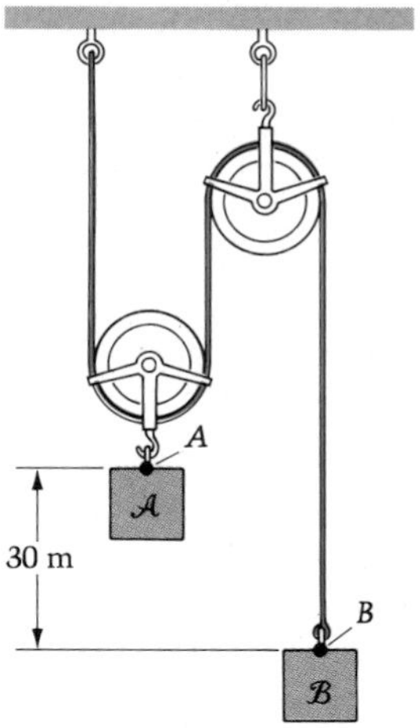

Figure P1.43

1.44 The position of a point P on a line is given by the equation $\mathbf{x} = t\sin(\pi t/2)$. The point starts moving at $t = 0$. Find the total distance traveled by P when it passes through the origin (counting the start as the first pass) for the third time.

1.45 Block $\mathcal{A}$ has $\mathbf{v}_A = 10$ m/s to the right at $t = 0$ and a constant acceleration of 2 m/s^2 to the left. Find the distance traveled by block $\mathcal{B}$ during the interval $t = 0$ to 8 s. (See Figure P1.45.)

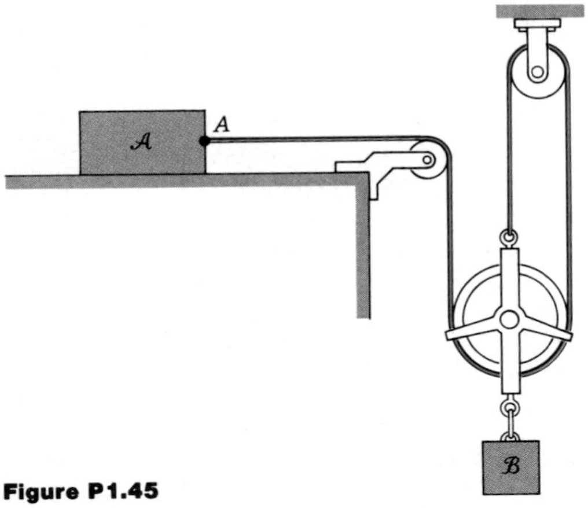

Figure P1.45

1.46 The accelerations of the translating blocks $\mathcal{A}$ and $\mathcal{B}$ are 2 m/s$^2\downarrow$ and 3 m/s$^2\uparrow$, respectively. (See Figure P1.46.) The entire system is at rest at the given instant. Find how long it will take for block $\mathcal{C}$ to hit the ground. (Do not assume that pulleys $\mathcal{P}_1$ and $\mathcal{P}_2$ remain at the same level!)

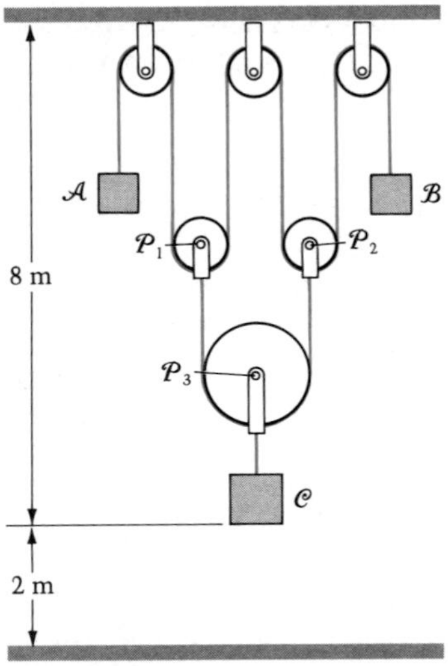

Figure P1.46

1.47 The cord shown in Figure P1.47, attached to the wall at D, passes around a small pulley fixed to $\mathcal{B}$ at B; it then passes around another small pulley $\mathcal{P}$ and ends at point A of body $\mathcal{A}$. The cord is 44 m long, and the system is being held at rest in the given position. Suddenly point B is forced to move to the right with constant acceleration $a_B = 2$ m/s^2. Determine the velocity of A just before it reaches the pulley.

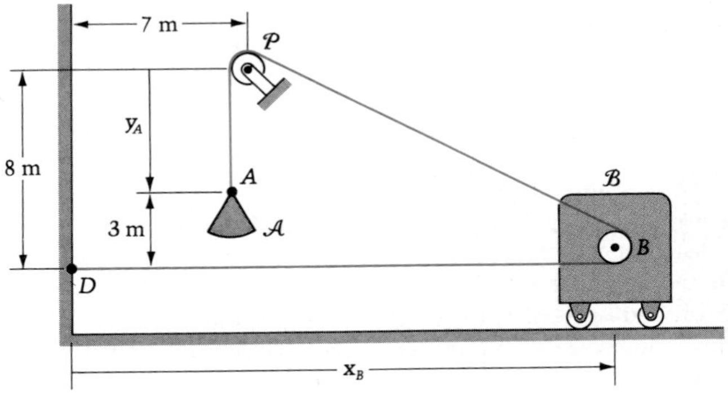

Figure P1.47

1.48 A car is traveling at 55 mph on a straight road. The driver applies her brakes for 6 sec, producing a constant deceleration of 5 ft/sec^2, and then immediately accelerates at 2 ft/sec^2. How long does it take for the car to return to its original velocity?

1.49 In the preceding problem, suppose the acceleration following the braking is not constant but is instead given by $\ddot{x} = 0.6t$ ft/sec^2. *Now* how long does it take to return to 55 mph?

1.50 A particle has a linearly varying rectilinear acceleration of $\mathbf{a} = \ddot{x}\hat{\mathbf{i}} = 12t\hat{\mathbf{i}}$ m/s^2. Two observations of the particle's motion are made: Its velocity at $t = 1$ s is $\dot{x}\hat{\mathbf{i}} = 2\hat{\mathbf{i}}$ m/s, and its position at $t = 2$ s is given by $\mathbf{x} = 3\hat{\mathbf{i}}$ meters.

a. Find the displacement of the particle at $t = 5$ s relative to where it was at $t = 0$.

b. Determine the distance traveled by the particle over the same time interval.

1.51 Suppose an airplane touches down smoothly on a runway at 60 mph. If it then decelerates to a stop at the constant deceleration rate of 10 ft/sec^2, find the required length of runway.

1.52 A particle moving on a straight line is subject to an acceleration directly proportional to its distance from a fixed point P on the line and directed toward P. Initially the particle is 5 ft to the left of P and moving to the right with a velocity of 24 ft/sec. If the particle momentarily comes to rest 13 ft to the right of P, find its velocity as it passes through P.

1.53 A particle moving on the $\mathbf{x}$ axis has an acceleration always directed to the origin. The magnitude of the acceleration is nine times the distance from the origin. When the particle is 6 m to the left of the origin, it has a velocity of 3 m/s to the right. Find the time for the particle to get from this position to the origin.

1.54 An automobile passes a point P at a speed of 80 mph. At P it begins to decelerate at a rate that varies linearly with time. If after 5 sec the car has slowed to 50 mph, what distance has it traveled?

1.55 Work the preceding problem, but suppose the deceleration varies as the square of time. The other information is the same.

1.56 A slider block $\mathcal{S}$ moves rectilinearly in a slot (see Figure P1.56) with an acceleration given by

$$\mathbf{a}_s = \ddot{\mathbf{x}}\hat{\mathbf{i}} = -\pi^2 \sin \pi t \hat{\mathbf{i}} \text{ m/s}^2$$

Find the motion $\mathbf{x}(t)$ of the slider block if at $t = 0$:

a. It is passing through the origin, and
b. It has velocity $\dot{\mathbf{x}}\hat{\mathbf{i}} = 2\pi\hat{\mathbf{i}}$ m/s.

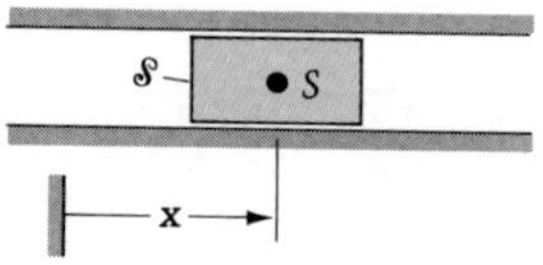

Figure P1.56

1.57 A professional tennis player's serve has been clocked at 147 ft/sec. If the distance between the end lines is 78 ft, estimate the time interval during which his opponent must react in order to return the serve. (Assume these heights: serve, 7 ft; net, 3 ft; receive, 3.5 ft.)

1.58 A train is traveling at 70 km/hr. If its brakes give the train a constant deceleration of 0.5 m/s^2, find the distance from the station where the brakes should be applied so that the train will come to a stop at the station. How long will it take the train to stop?

1.59 A point P has an acceleration that is position-dependent according to the equation $\ddot{\mathbf{x}} = 5\mathbf{x}$ m/s^2. Determine the velocity of P as a function of its displacement $\mathbf{x}$ if P is at 0.3 m with $\dot{\mathbf{x}} = 0.6$ m/s when $t = 0$.

1.60 In the preceding problem, find the time required for P to reach the point $\mathbf{x} = 4$ m.

1.61 A hot rod enthusiast accelerates his dragster along a straight drag strip at a constant rate of acceleration from zero to 120 mph. Then he immediately decelerates at a constant rate to a stop. He finds that he has traveled a total distance of $\frac{1}{4}$ mi from start to stop. How much time passes from the instant he starts to the time he stops?

1.62 Two cars start from rest at the same location and at the same instant and race along a straight track. Car $\mathcal{A}$ accelerates at 6.6 ft/sec^2 to a speed of 90 mph and then runs at a constant speed. Car $\mathcal{B}$ accelerates at 4.4 ft/sec^2 to a speed of 96 mph and then runs at a constant speed.

a. Which car will win the 3-mi race, and by what distance?
b. What will be the maximum lead of $\mathcal{A}$ over $\mathcal{B}$?
c. How far will the cars have traveled when $\mathcal{B}$ passes $\mathcal{A}$?

1.63 If the acceleration in one-dimensional motion is $\ddot{\mathbf{x}}\hat{\mathbf{i}} = K\hat{\mathbf{i}} =$ constant, then for the three cases considered in this section:

a. $\ddot{\mathbf{x}} = f(t) = K$
b. $\ddot{\mathbf{x}} = g(v) = K$
c. $\ddot{\mathbf{x}} = h(\mathbf{x}) = K$

Demonstrate that the general solutions all give the same answer for $\mathbf{x}(t)$ if the various integration constants are matched and renamed.

1.5 Rectangular Cartesian Coordinates

In this section we merely add the y and z components of position to the rectilinear ($\mathbf{x}$) component studied in the preceding section. This step allows the point P to move on a curve in three-dimensional space instead

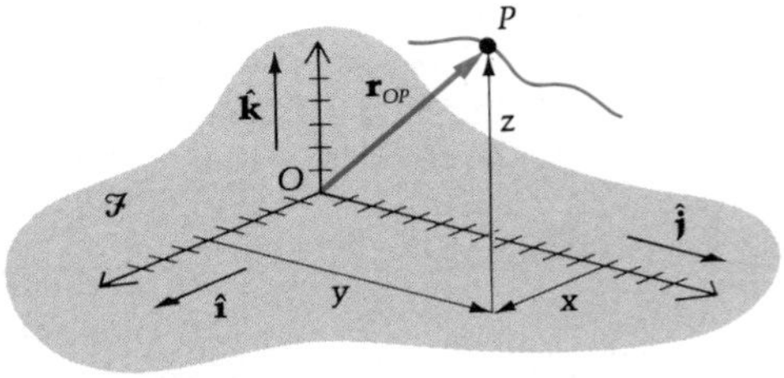

Figure 1.4 Rectangular Cartesian coordinates of P.

of being constrained to movement along a straight line in the reference frame $\mathcal{F}$.

Suppose that P is in a state of general (three-dimensional) motion in frame $\mathcal{F}$. We may study this motion by embedding a set of orthogonal axes in $\mathcal{F}$ as shown in Figure 1.4. The position vector of point P may then be expressed as

$$\mathbf{r}_{OP} = x\hat{\mathbf{i}} + y\hat{\mathbf{j}} + z\hat{\mathbf{k}} \tag{1.23}$$

in which (x, y, z) are **rectangular Cartesian coordinates** of P measured along the embedded axes and $(\hat{\mathbf{i}}, \hat{\mathbf{j}}, \hat{\mathbf{k}})$ are unit vectors respectively parallel to these axes (Figure 1.4). Using the basic definitions (Equations 1.9 and 1.10), we may differentiate $\mathbf{r}_{OP}$ and obtain expressions for velocity and acceleration in rectangular Cartesian coordinates:

$$\mathbf{v}_P = \dot{x}\hat{\mathbf{i}} + \dot{y}\hat{\mathbf{j}} + \dot{z}\hat{\mathbf{k}} \tag{1.24}$$

$$\mathbf{a}_P = \ddot{x}\hat{\mathbf{i}} + \ddot{y}\hat{\mathbf{j}} + \ddot{z}\hat{\mathbf{k}} \tag{1.25}$$

We shall now consider examples in which points move in two and three dimensions.

EXAMPLE 1.11

The position vector of a point P is given as

$$\mathbf{r}_{OP} = 2t\hat{\mathbf{i}} + t^3\hat{\mathbf{j}} + 3t^2\hat{\mathbf{k}} \text{ ft}$$

Find the velocity and acceleration of P at $t = 1$ sec.

SOLUTION

Differentiating the position vector, we obtain the velocity vector of P:

$$\mathbf{v}_P = 2\hat{\mathbf{i}} + 3t^2\hat{\mathbf{j}} + 6t\hat{\mathbf{k}} \text{ ft/sec}$$

Another derivative yields the acceleration of P:

$$\mathbf{a}_P = 6t\hat{\mathbf{j}} + 6\hat{\mathbf{k}} \text{ ft/sec}^2$$

(Continued)

Therefore, at $t = 1$ sec, the velocity and acceleration of P are

$$\mathbf{v}_P|_{t=1} = 2\hat{\mathbf{i}} + 3\hat{\mathbf{j}} + 6\hat{\mathbf{k}} = 7\left(\frac{2\hat{\mathbf{i}} + 3\hat{\mathbf{j}} + 6\hat{\mathbf{k}}}{\sqrt{2^2 + 3^2 + 6^2}}\right) \text{ ft/sec}$$

$$\mathbf{a}_P|_{t=1} = 6\hat{\mathbf{j}} + 6\hat{\mathbf{k}} = 6\sqrt{2}\left(\frac{\hat{\mathbf{j}} + \hat{\mathbf{k}}}{\sqrt{2}}\right) \text{ ft/sec}^2$$

Note that the speed (magnitude of velocity) of P at $t = 1$ is $|\mathbf{v}_P| = 7$ ft/sec and the magnitude of the acceleration at $t = 1$ is $6\sqrt{2}$ ft/sec^2. We shall return to this example in Section 1.7.

We see from the previous example that if the position vector of P is known as a function of time, it is a very simple matter to obtain the velocity and acceleration of the point. In the following example we are given the acceleration of P and asked for its *position*. Since this problem requires integration instead of differentiation, initial conditions enter the picture. These conditions allow us to compute the constants of integration.

EXAMPLE 1.12

A point Q has the acceleration vector

$$\mathbf{a}_Q = 4\hat{\mathbf{i}} - 6t\hat{\mathbf{j}} + \sin 0.2t\hat{\mathbf{k}} \text{ m/s}^2$$

At $t = 0$, the point Q is located at $(x, y, z) = (1, 3, -5)$ m and has a velocity vector of $2\hat{\mathbf{i}} - 7\hat{\mathbf{j}} + 3.4\hat{\mathbf{k}}$ m/s. When $t = 3$ s, find the speed of Q and its distance from the starting point.

SOLUTION

Integrating, we get

$$\mathbf{v}_Q = 4t\hat{\mathbf{i}} - 3t^2\hat{\mathbf{j}} - 5\cos 0.2t\hat{\mathbf{k}} + \mathbf{c} \text{ m/s}$$

in which $\mathbf{c}$ is a vector constant. Using the initial condition for velocity at $t = 0$, we obtain

$$\mathbf{v}_Q|_{t=0} = 0\hat{\mathbf{i}} - 0\hat{\mathbf{j}} - 5\hat{\mathbf{k}} + \mathbf{c} = 2\hat{\mathbf{i}} - 7\hat{\mathbf{j}} + 3.4\hat{\mathbf{k}} \text{ m/s}$$

so that

$$\mathbf{c} = 2\hat{\mathbf{i}} - 7\hat{\mathbf{j}} + 8.4\hat{\mathbf{k}} \text{ m/s}$$

Therefore

$$\mathbf{v}_Q = (4t + 2)\hat{\mathbf{i}} - (3t^2 + 7)\hat{\mathbf{j}} + (8.4 - 5\cos 0.2t)\hat{\mathbf{k}} \text{ m/s}$$

(Continued)

Integrating again, we get

$$\mathbf{r}_{OQ} = (2t^2 + 2t)\hat{\mathbf{i}} - (t^3 + 7t)\hat{\mathbf{j}} + (8.4t - 25 \sin 0.2t)\hat{\mathbf{k}} + \mathbf{c}' \text{ m}$$

where $\mathbf{c}'$ is another vector constant, evaluated below from the initial condition for the *position* of Q at $t = 0$:

$$\mathbf{r}_{OQ}|_{t=0} = 0\hat{\mathbf{i}} - 0\hat{\mathbf{j}} + 0\hat{\mathbf{k}} + \mathbf{c}' = \hat{\mathbf{i}} + 3\hat{\mathbf{j}} - 5\hat{\mathbf{k}} \text{ m}$$

so that

$$\mathbf{c}' = \hat{\mathbf{i}} + 3\hat{\mathbf{j}} - 5\hat{\mathbf{k}} \text{ m}$$

and thus

$$\mathbf{r}_{OQ} = (2t^2 + 2t + 1)\hat{\mathbf{i}} - (t^3 + 7t - 3)\hat{\mathbf{j}} + (8.4t - 25 \sin 0.2t - 5)\hat{\mathbf{k}} \text{ m}$$

Substituting the required time, $t = 3$ s, into the expressions for $\mathbf{v}_Q$ and $\mathbf{r}_{OQ}$ gives the answers:

$$\mathbf{v}_Q|_{t=3} = 14\hat{\mathbf{i}} - 34\hat{\mathbf{j}} + (8.4 - 5 \cos 0.6)\hat{\mathbf{k}}$$

$$= 14\hat{\mathbf{i}} - 34\hat{\mathbf{j}} + 4.27\hat{\mathbf{k}} \text{ m/s}$$

Thus the speed of Q is given by

$$v_Q|_{t=3} = \sqrt{14^2 + (-34)^2 + 4.27^2} = 37.0 \text{ m/s}$$

Continuing, we have

$$\mathbf{r}_{OQ}|_{t=3} = 25\hat{\mathbf{i}} - 45\hat{\mathbf{j}} + (20.2 - 25 \sin 0.6)\hat{\mathbf{k}}$$

$$= 25\hat{\mathbf{i}} - 45\hat{\mathbf{j}} + 6.08\hat{\mathbf{k}} \text{ m}$$

The distance d between Q and its starting point is therefore given by

$$d = |\mathbf{r}_{OQ}(3) - \mathbf{r}_{OQ}(0)|$$

$$= \sqrt{(25 - 1)^2 + (-45 - 3)^2 + [6.08 - (-5)]^2}$$

$$= 54.8 \text{ m}$$

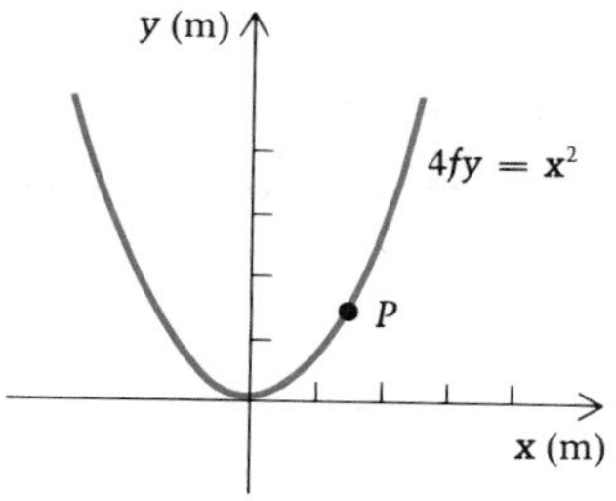

EXAMPLE 1.13

The point P travels on the parabola (with focal distance $f = \frac{1}{2}$ m) at the constant speed of 0.2 m/s. Determine the acceleration of P: (a) as a function of $\mathbf{x}$ and (b) at $\mathbf{x} = 2$ m.

SOLUTION

We may obtain the velocity components by differentiating:

$$2y = \mathbf{x}^2$$

$$2\dot{y} = 2\mathbf{x}\dot{\mathbf{x}} \Rightarrow \dot{y} = \mathbf{x}\dot{\mathbf{x}}$$

Thus

$$\mathbf{v}_P = \dot{\mathbf{x}}\hat{\mathbf{i}} + \dot{y}\hat{\mathbf{j}} = \dot{\mathbf{x}}\hat{\mathbf{i}} + \mathbf{x}\dot{\mathbf{x}}\hat{\mathbf{j}} \text{ m/s} \tag{1}$$

Differentiating Equation (1) gives us the acceleration of P:

$$\mathbf{a}_P = \ddot{\mathbf{x}}\hat{\mathbf{i}} + (\dot{\mathbf{x}}^2 + \mathbf{x}\ddot{\mathbf{x}})\hat{\mathbf{j}} \text{ m/s}^2 \tag{2}$$

Since $|\mathbf{v}_P|$, or v_P, is constant, we have

$$v_P = 0.2 = \sqrt{\dot{\mathbf{x}}^2 + (\mathbf{x}\dot{\mathbf{x}})^2} \text{ m/s}$$

$$\dot{\mathbf{x}} = \frac{0.2}{\sqrt{1 + \mathbf{x}^2}} \text{ m/s} \tag{3}$$

We also see from (2) that we need $\ddot{\mathbf{x}}$; differentiating (3), we get

$$\ddot{\mathbf{x}} = \frac{-0.2\mathbf{x}\dot{\mathbf{x}}}{(1 + \mathbf{x}^2)^{3/2}} = \frac{-0.2\left(\dfrac{0.2}{\sqrt{1 + \mathbf{x}^2}}\right)\mathbf{x}}{(1 + \mathbf{x}^2)^{3/2}} \text{ m/s}^2$$

or

$$\ddot{\mathbf{x}} = \frac{-0.04\mathbf{x}}{(1 + \mathbf{x}^2)^2} \text{ m/s}^2 \tag{4}$$

Substituting (3) and (4) into (2), we get

$$\mathbf{a}_P = \frac{-0.04\mathbf{x}}{(1 + \mathbf{x}^2)^2}\hat{\mathbf{i}} + \left[\frac{0.04}{1 + \mathbf{x}^2} - \frac{0.04\mathbf{x}^2}{(1 + \mathbf{x}^2)^2}\right]\hat{\mathbf{j}} \text{ m/s}^2$$

When $\mathbf{x} = 2$ m,

$$\mathbf{a}_P = \frac{-0.04(2)}{5^2}\hat{\mathbf{i}} + \left[\frac{0.04}{5} - \frac{0.04(2^2)}{5^2}\right]\hat{\mathbf{j}}$$

$$= -0.0032\hat{\mathbf{i}} + 0.0016\hat{\mathbf{j}} \text{ m/s}^2$$

EXAMPLE 1.14

The curve AB on block $\mathcal{B}$ is a parabola whose vertex is at A. Its equation is $x^2 = (64/3)y$. The block $\mathcal{B}$ is pushed to the left with a constant velocity of 10 ft/sec. The rod $\mathcal{R}$ slides on the parabola so that the plate $\mathcal{P}$ is forced upward. Find the acceleration of plate $\mathcal{P}$.

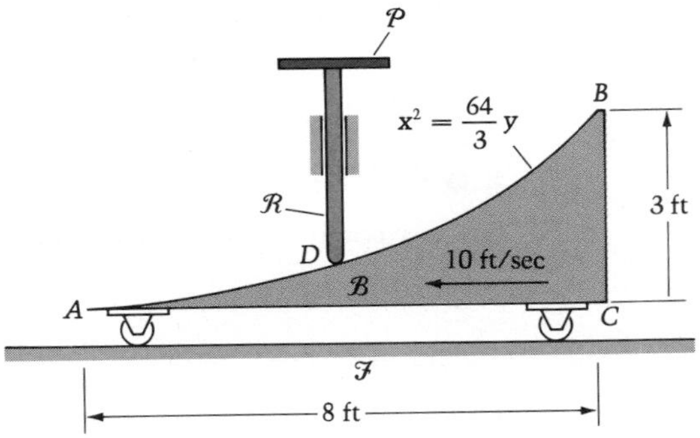

SOLUTION

We first note that plate $\mathcal{P}$ and rod $\mathcal{R}$ together constitute a single body, each of whose points has one-dimensional (y) motion. The velocities and accelerations of all these points are therefore the same. We shall then focus on point D, the lowest point of $\mathcal{R}$, which is in contact with $\mathcal{B}$.

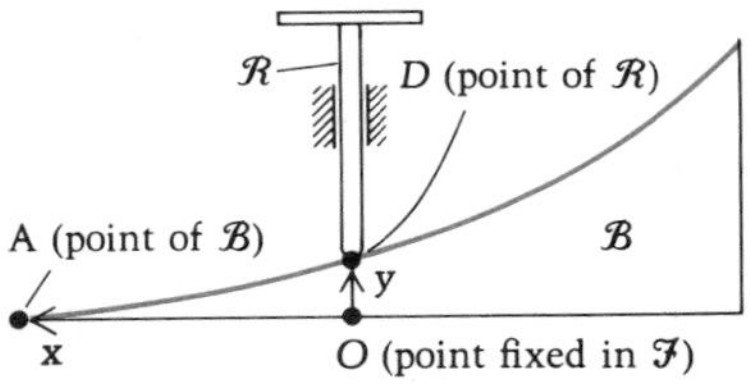

Defining the ground to be the reference frame $\mathcal{I}$, we establish its origin at O as shown in the diagram. Furthermore, we define x and y in the directions shown, and we let them respectively measure the positions of the left end A of $\mathcal{B}$ and the point D of $\mathcal{R}$. Therefore, noting that these coordinates match those in the equation of the curve,

$$\mathbf{r}_{OA} = x\hat{\mathbf{i}} \qquad \text{and} \qquad \mathbf{r}_{OD} = y\hat{\mathbf{j}} \text{ ft} \tag{1}$$

But because D always rests on the parabolic surface of $\mathcal{B}$, $y = (3/64)x^2$ so that

$$\mathbf{r}_{OD} = \tfrac{3}{64}x^2\hat{\mathbf{j}} \text{ ft} \tag{2}$$

To get the acceleration of D, we first find its velocity:

$$\mathbf{v}_D = \tfrac{3}{32}x\dot{x}\hat{\mathbf{j}} \text{ ft/sec} \tag{3}$$

To obtain $\dot{x}$, we differentiate $\mathbf{r}_{OA}$ from Equation (1):

$$\dot{\mathbf{r}}_{OA} = \mathbf{v}_A = \dot{x}\hat{\mathbf{i}} = 10\hat{\mathbf{i}} \text{ ft} \tag{4}$$

since it is given that all points of the body $\mathcal{B}$ have the constant velocity 10 ft/sec to the left.

Substitution of $\dot{x} = 10$ into Equation (3) then gives

$$\mathbf{v}_D = \tfrac{3}{32}x(10)\hat{\mathbf{j}} = \tfrac{15}{16}x\hat{\mathbf{j}} \text{ ft/sec} \tag{5}$$

(Continued)

and we see that the velocity of D depends upon x. Differentiating $\mathbf{v}_D$ will give us the acceleration of D:

$$\mathbf{a}_D = \frac{15}{16}\dot{x}\hat{\mathbf{j}} = \frac{150}{16}\hat{\mathbf{j}} = 9.38\hat{\mathbf{j}} \text{ ft/sec}^2 \tag{6}$$

Equation (6) gives the acceleration of all the points of the plate. Note that the acceleration of D is a constant.

Question 1.2 Would $\mathbf{a}_D$ be a constant if instead of being quadratically shaped, the inclined surface were (a) flat or (b) cubic?

In closing this section, we remark that the simple forms of Equations (1.24) and (1.25) are due to the fact that the unit vectors $\hat{\mathbf{i}}$, $\hat{\mathbf{j}}$, $\hat{\mathbf{k}}$ remain constant in both magnitude and direction when the axes are fixed in the frame of reference. For *planar* applications (Chapter 3), we shall set the z component of velocity identically to zero, obtaining the following for a point in plane motion (moving only in a plane parallel to the xy plane):

$$\mathbf{r}_{OP} = x\hat{\mathbf{i}} + y\hat{\mathbf{j}} + z\hat{\mathbf{k}} \qquad \text{(where } z \text{ is constant)} \tag{1.26}$$

$$\mathbf{v}_P = \dot{x}\hat{\mathbf{i}} + \dot{y}\hat{\mathbf{j}} \tag{1.27}$$

$$\mathbf{a}_P = \ddot{x}\hat{\mathbf{i}} + \ddot{y}\hat{\mathbf{j}} \tag{1.28}$$

PROBLEMS / Section 1.5

1.64 A point moves on a path, with a position vector as a function of time given by $\mathbf{r}_{OP} = \sin 2t\hat{\mathbf{i}} + 3t\hat{\mathbf{j}} + e^{6t}\hat{\mathbf{k}}$, in units of meters when t is in seconds. Find:

a. The speed of the point at $t = 0$.

b. Its acceleration at $t = \pi/2$ s.

c. The component of the velocity vector, at $t = 0$, which is parallel to the line ℓ in the xy plane given by $y = \frac{5}{12}x - 6$ and having the directed sense shown in Figure P1.64.

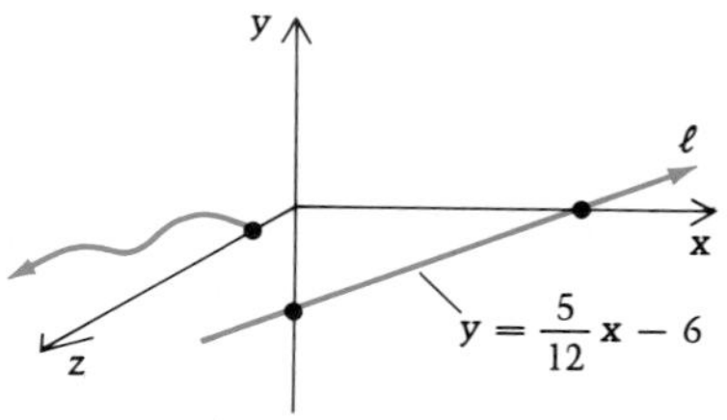

Figure P1.64

1.65 The acceleration of a point is given by

$$\mathbf{a}_P = 6t\hat{\mathbf{i}} + 12t^2\hat{\mathbf{j}} - 4\hat{\mathbf{k}} \text{ m/s}^2$$

At $t = 0$, the initial conditions are that $\mathbf{v}_P = 2\hat{\mathbf{i}}$ m/s and $\mathbf{r}_{OP} = \hat{\mathbf{i}} + 3\hat{\mathbf{j}} + 9\hat{\mathbf{k}}$ meters. Find the position vector of P at $t = 5$ s, and determine how far P then is from its position at $t = 0$.

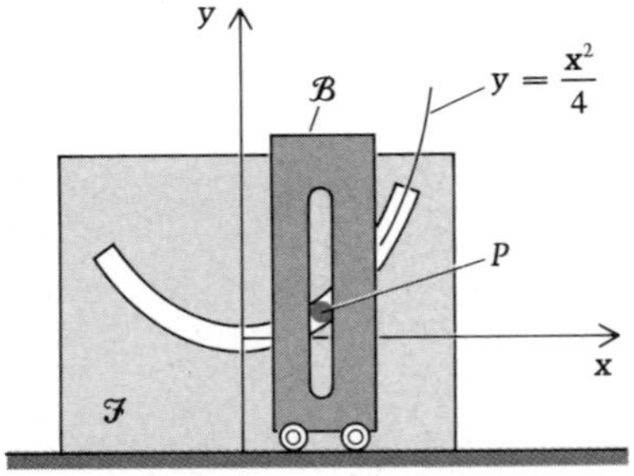

Figure P1.66

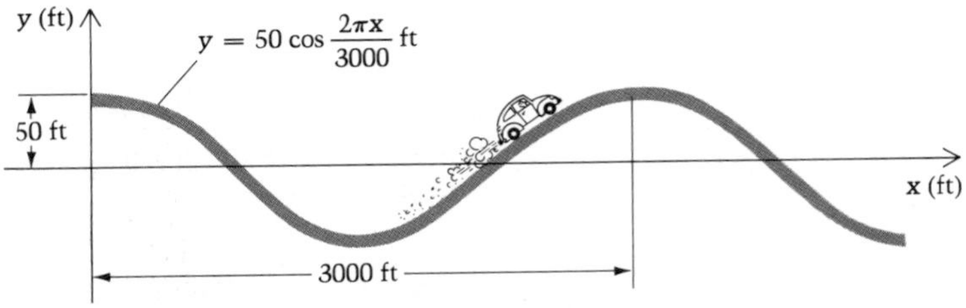

Figure P1.67

1.66 The pin P shown in Figure P1.66 moves in a parabolic slot cut in the reference frame $\mathcal{F}$ and is guided by the vertical slot in body $\mathcal{B}$. For body $\mathcal{B}$, $x = 0.05t^3$ m locates the centerline of its slot.

a. Find the acceleration of P at $t = 5$ s.
b. Find the time(s) when the x and y components of $\mathbf{a}_P$ are equal.

1.67 A car travels on a section of highway that approximates the cosine curve in Figure P1.67. If the driver maintains a constant speed of 55 mph, determine his x and y components of velocity when $x = 2500$ ft.

1.68 A car travels along the highway of the preceding problem with a constant x-component of velocity of 54.9 mph. Over what sections of the highway does the driver exceed the speed limit of 55 mph?

1.69 Determine the minimum acceleration of the car in Problem 1.67. Where on the curve is this acceleration experienced?

1.70 Find the maximum acceleration of the car in Problem 1.68. Where does it occur on the curve?

1.71 A point P travels on a path and has the following coordinates as functions of time t (in seconds):

$$x = 12 \cos \frac{\pi t}{2} \text{ m} \qquad y = 8 \sin \frac{\pi t}{2} \text{ m}$$

$$z = 0$$

a. Find the velocity $\mathbf{v}_P(t)$ and acceleration $\mathbf{a}_P(t)$ of P.
b. Find the position, velocity, and acceleration of P when $t = 4$ s.
c. Eliminate the time t from the x and y expressions and obtain the equation of the path of P.

1.72 Describe the path of a point P that has the following rectangular Cartesian coordinates as functions of time: $x = a \cos \omega t$, $y = a \sin \omega t$, and $z = bt$, where a, b, and ω are constants. Identify the meanings of the three constants.

1.73 For the following values of the constants, find the velocity of P at $t = 5$ s in the preceding problem: $a = 2$ m, $b = 0.5$ m/s, and $\omega = 1.2$ rad/s.

In Problems 1.74–1.78 a point P travels on the curve with a constant x component of velocity, $\dot{x} = 3$ in./sec. Each starts on the curve at $x = 1$ when $t = 0$. Find the velocity vector of P when $t = 10$ sec in each case.

1.74 Logarithmic curve

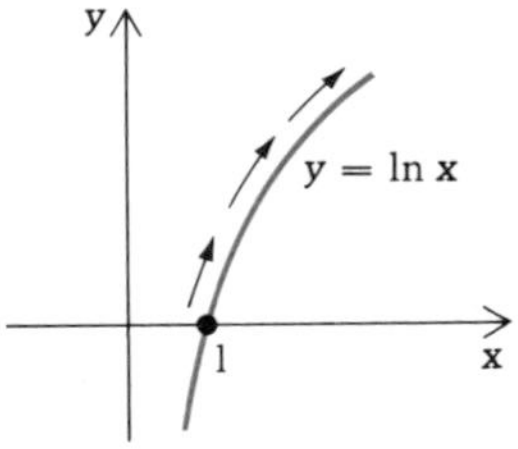

1.75 Exponential curve

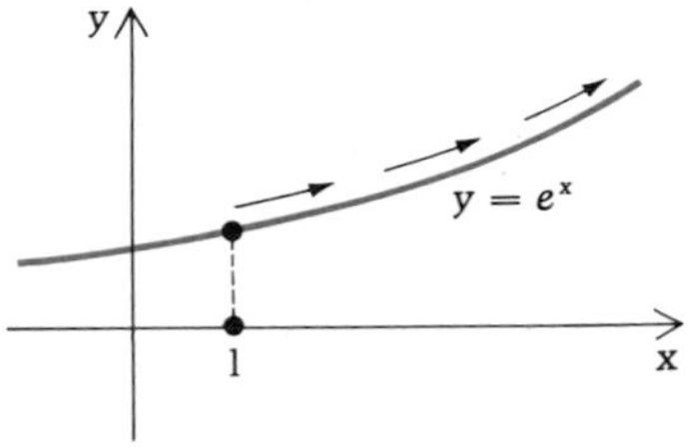

1.76 Parabola tangent to axes ($a = 1$)

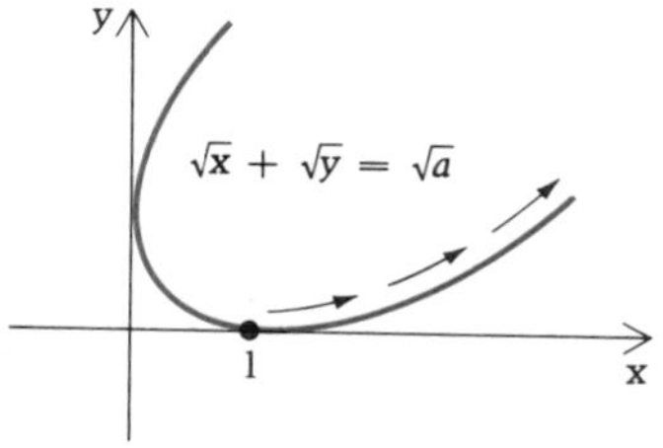

1.77 First-quadrant branch of rectangular hyperbola

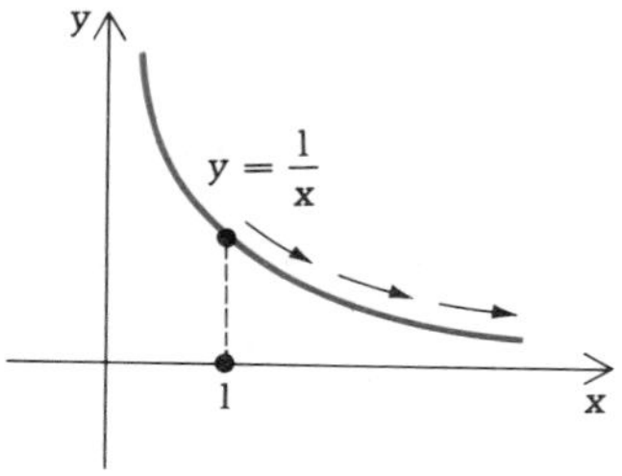

1.78 First-quadrant branch of semicubical parabola

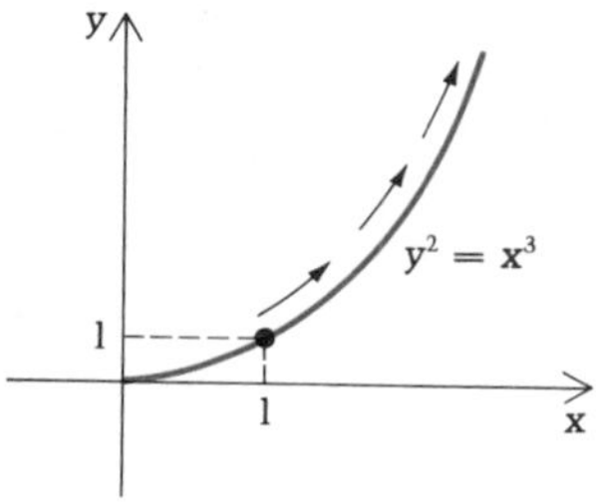

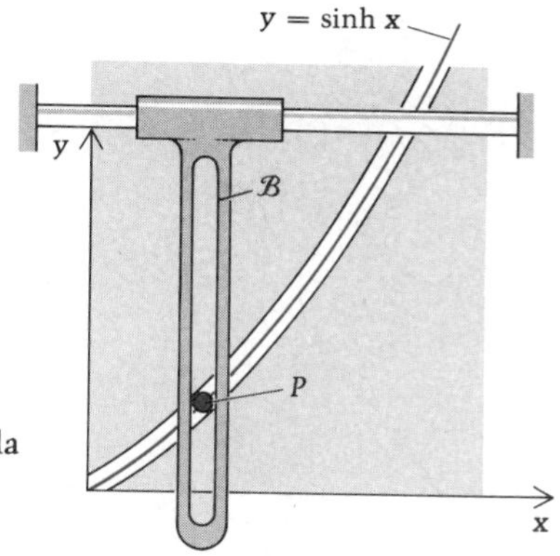

Figure P1.80

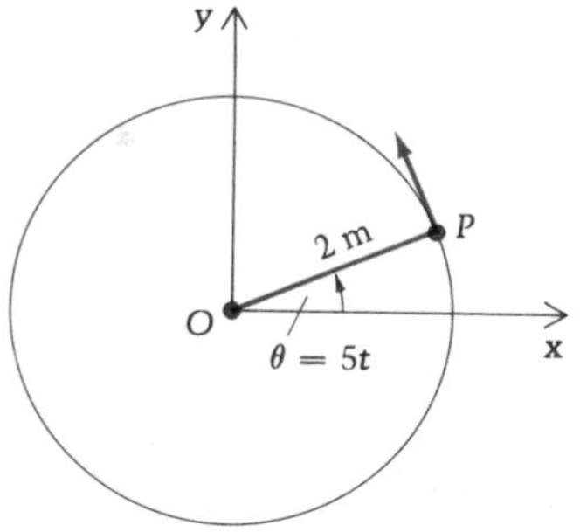

Figure P1.81

1.79 The center C of the roller in the horizontal slot of Figure P1.79 has a motion given by $x = 0.2t^2$ m. Find the velocity of point B. *Hint:* Write the Pythagorean theorem and differentiate the equation, noting that the length L of the rod connecting the rollers is constant.

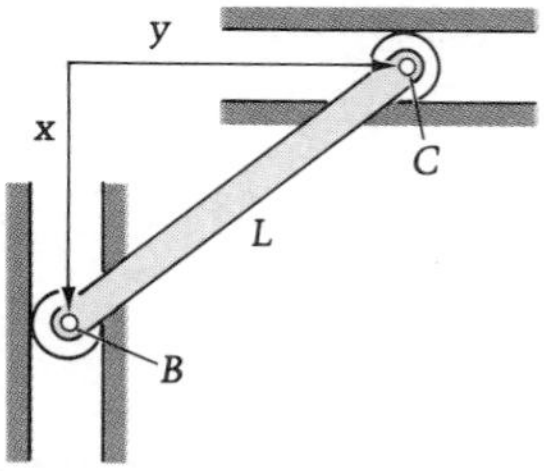

Figure P1.79

1.80 A pin P moves in a slot that is cut in the shape of a hyperbolic sine as shown in Figure P1.80. It is guided along by the vertical slotted body $\mathcal{B}$, all the points of which have velocity 0.08 m/s to the right. Find $\mathbf{v}_P$ and $\mathbf{a}_P$ when $x = 0.2$ m.

1.81 A point P moves on a circle in the direction shown in Figure P1.81. Express $\mathbf{r}_{OP}$ in (x, y, z) coordinates and differentiate to obtain $\mathbf{v}_P$ and $\mathbf{a}_P$. (Angle θ in radians.)

1.82 Repeat the preceding problem. In this case, however, the angle θ increases quadratically, instead of linearly, with time according to $\theta = 3t^2$ rad.

1.83 Describe precisely the path of a particle's motion if its xy coordinates are given by $(2.5t^2 + 7, 6t^2 + 9)$ meters when t is in seconds.

1.84 The velocity of point A in Figure P1.84 is a constant 2 m/s to the right. Find the velocity of B when $x = 10$ m. *Hint:* Use the law of cosines to relate the sides of triangle ABC; then differentiate this relationship.

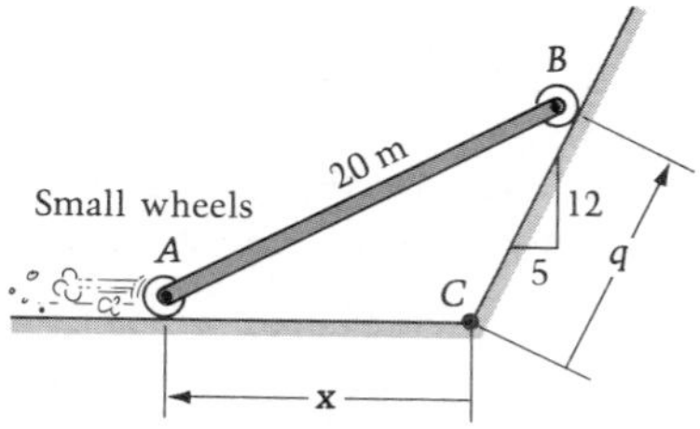

Figure P1.84

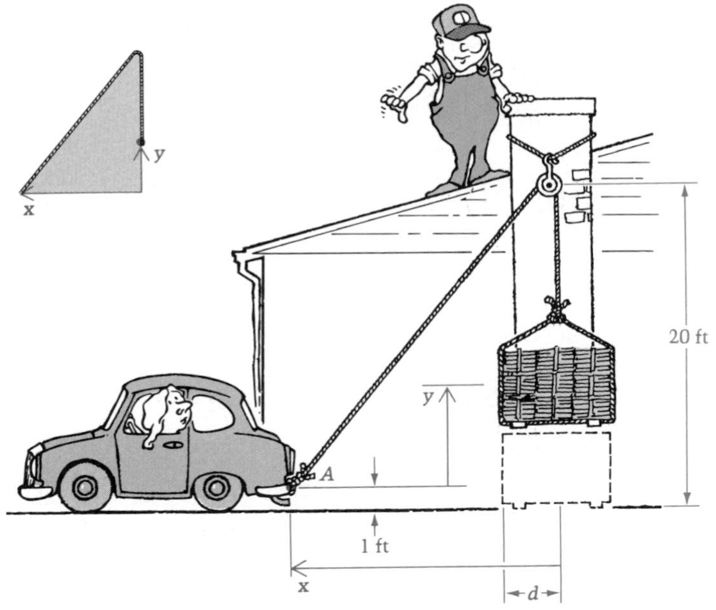

Figure P1.86

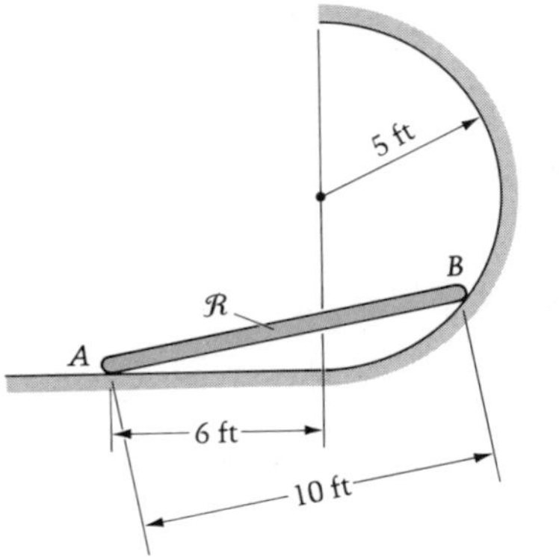

Figure P1.87

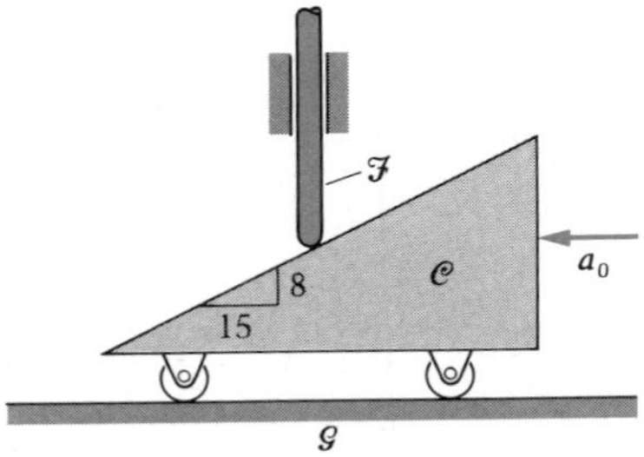

Figure P1.91

1.85 In Example 1.14, let the equation of the incline be given by $\mathbf{x}^3 = (512/3)y$. If the motion starts when $\mathbf{x} = y = 0$, find the acceleration of the plate $\mathcal{P}$ when $y = 2$ ft.

1.86 A man and his daughter have figured out an ingenious way to hoist 8000 lb of shingles onto their roof, several bundles at a time. They have rigged a pulley onto a frame around the chimney (Figure P1.86) and will use the car to raise the weights. When the bumper of the car is at $\mathbf{x} = 0$ (neglect d), the pallet of shingles is on the ground with no slack in the rope. While the car is traveling to the left at a constant speed of $v_A = 2$ mph, find the velocity and acceleration of the shingles as a function of $\mathbf{x}$. Do this by using the triangle to the left of the figure to express y as a function of $\mathbf{x}$; then differentiate the result.

1.87 The rigid rod $\mathcal{R}$ in Figure P1.87 moves so that its ends, A and B, remain in contact with the surfaces. If, at the instant shown, the velocity of A is 0.5 ft/sec to the right, find the velocity of B.

1.88 In Problem 1.87, find the acceleration of B at the instant in question if the acceleration of A is 1.0 ft/sec^2 to the left at that time.

1.89 The motion of a particle P is given by $\mathbf{x} = C \cosh kt$ ft and $y = C \sinh kt$ ft, where C and k are constants. Find the equation of the path of P by eliminating time t.

1.90 In the preceding problem, find the speed of P as a function of the distance r $(= \sqrt{\mathbf{x}^2 + y^2})$ from the origin to P.

1.91 The wedge-shaped cam $\mathcal{C}$ in Figure P1.91 is moving to the left with constant acceleration a_0. Find the acceleration of the follower $\mathcal{F}$.

1.92 The moving pin P of a rotating crank has a location defined by

$$\mathbf{x} = 20 \cos \pi t \text{ m}$$

$$y = 20 \sin \pi t \text{ m}$$

Find the velocity of P when $t = 0, \frac{1}{2}, \frac{3}{2}$, and 2 s.

1.93 In the preceding problem, find the acceleration of P at the same instants of time.

1.94 A particle P moves in the xy plane. The motion of P is given by

$$x = 30t + 6 \text{ ft}$$
$$y = 20t - 7 \text{ ft}$$

Find the equation of the path of P.

1.95 Repeat Problem 1.94 if:

$$x = 5t \text{ m}$$
$$y = 25t^2 \text{ m}$$

1.96 Repeat Problem 1.94 if:

$$x = 2 + 3 \sin t \text{ ft}$$
$$y = 4 \cos t \text{ ft}$$

1.97 The acceleration of a point is given by $\mathbf{a}_P = 6t\hat{\mathbf{i}} + 12t^2\hat{\mathbf{j}} - 4\hat{\mathbf{k}}$ m/s². The initial conditions are that $\mathbf{v}_P = 2\hat{\mathbf{i}}$ m/s and $\mathbf{r}_{OP} = \hat{\mathbf{i}} + 3\hat{\mathbf{j}} + 9\hat{\mathbf{k}}$ meters at $t = 0$. Find the position vector of P at $t = 10$ s, and determine how far P is from its original position at that time.

1.98–1.102 Find the acceleration vectors at $t = 10$ s of the points whose motions are described in Problems 1.74–1.78.

1.103 Two points P and Q have position vectors in a reference frame that are given by $\mathbf{r}_{OP} = 50t\hat{\mathbf{i}}$ meters and $\mathbf{r}_{OQ} = 40\hat{\mathbf{i}} - 20t\hat{\mathbf{j}}$ meters. Find the minimum distance between P and Q and the time at which this occurs.

1.6 Cylindrical Coordinates

If a point P is moving in such a way that its projection into the xy plane is more easily described with polar (r and θ) coordinates than with x and y, then we may use **cylindrical coordinates** to advantage. These coordinates are nothing more than the **polar coordinates** r and θ together with an "axial" coordinate z. Thus r and θ locate the projection point of P in a plane, while z gives the distance of P *from* the plane.

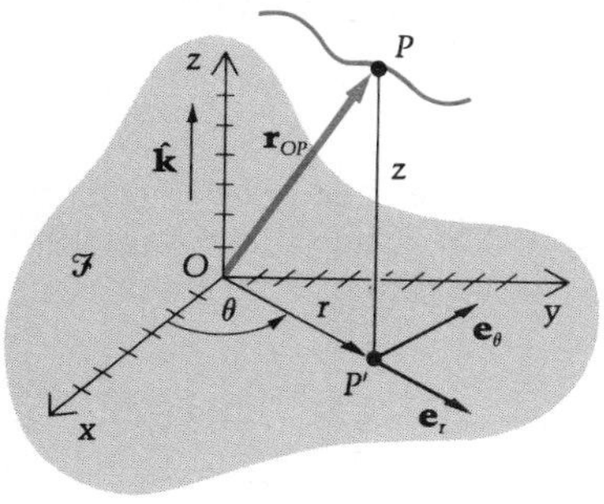

Figure 1.5 Cylindrical coordinates of P.

Embedding the same set of rectangular axes (x, y, z) in the reference frame $\mathcal{F}$ as we did in the preceding section, we now show the coordinates r and θ as well (Figure 1.5). Note that P' is the projection of P into the plane xy. From Figure 1.5 we see that the unit vectors $\hat{\mathbf{e}}_r$ and $\hat{\mathbf{e}}_\theta$ are drawn in the xy plane and that:

1. The direction of $\hat{\mathbf{e}}_r$ is that of OP'.
2. $\hat{\mathbf{e}}_\theta$ is normal to $\hat{\mathbf{e}}_r$ in the direction of increasing θ.

It will be helpful later in the section to note carefully at this point that $\hat{\mathbf{e}}_r$ and $\hat{\mathbf{e}}_\theta$ change (in direction) with changes in θ, but not with r or z. Thus if the point P moves along either a radial line or a vertical line, the two unit vectors remain the same. But if P moves in such a way as to alter θ, then the directions of $\hat{\mathbf{e}}_r$ and $\hat{\mathbf{e}}_\theta$ will vary.

The rectangular and cylindrical coordinates (both having z in common) are related through

$$x = r \cos \theta$$
$$y = r \sin \theta \tag{1.29}$$

which can be differentiated to produce, by Equations (1.24) and (1.25), formulas for velocity and acceleration in terms of the cylindrical coordinates (and their derivatives) and the unit vectors $\hat{\mathbf{i}}$, $\hat{\mathbf{j}}$, and $\hat{\mathbf{k}}$. It is usually more desirable, however, to express the velocity and acceleration in terms of the set of unit vectors ($\hat{\mathbf{e}}_r$, $\hat{\mathbf{e}}_\theta$, $\hat{\mathbf{k}}$), which are naturally associated with cylindrical coordinates. Thus it is useful to express a position vector $\mathbf{r}_{OP}$ as

$$\mathbf{r}_{OP} = r\hat{\mathbf{e}}_r + z\hat{\mathbf{k}} \tag{1.30}$$

Question 1.3 Why is there no $\hat{\mathbf{e}}_\theta$ term in Equation (1.30)?

Differentiating Equation (1.30), we obtain the velocity of P:

$$\mathbf{v}_P = \dot{r}\hat{\mathbf{e}}_r + r\dot{\hat{\mathbf{e}}}_r + \dot{z}\hat{\mathbf{k}} \tag{1.31}$$

To evaluate $\dot{\hat{\mathbf{e}}}_r$ we note from Figure 1.6 that

$$\hat{\mathbf{e}}_r = \cos\theta\hat{\mathbf{i}} + \sin\theta\hat{\mathbf{j}} \tag{1.32a}$$

$$\hat{\mathbf{e}}_\theta = -\sin\theta\hat{\mathbf{i}} + \cos\theta\hat{\mathbf{j}} \tag{1.32b}$$

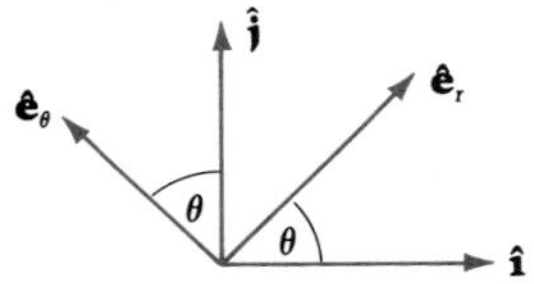

Figure 1.6

Hence

$$\dot{\hat{\mathbf{e}}}_r = \dot{\theta}(-\sin\theta\hat{\mathbf{i}} + \cos\theta\hat{\mathbf{j}}) = \dot{\theta}\hat{\mathbf{e}}_\theta \tag{1.33}$$

and thus the velocity in cylindrical coordinates takes the form

$$\mathbf{v}_P = \dot{r}\hat{\mathbf{e}}_r + r\dot{\theta}\hat{\mathbf{e}}_\theta + \dot{z}\hat{\mathbf{k}} \tag{1.34}$$

Differentiating again, we get

$$\mathbf{a}_P = \ddot{r}\hat{\mathbf{e}}_r + \dot{r}\dot{\hat{\mathbf{e}}}_r + \dot{r}\dot{\theta}\hat{\mathbf{e}}_\theta + r\ddot{\theta}\hat{\mathbf{e}}_\theta + r\dot{\theta}\dot{\hat{\mathbf{e}}}_\theta + \ddot{z}\hat{\mathbf{k}} \tag{1.35}$$

Using Equations (1.32), we find that

$$\dot{\hat{\mathbf{e}}}_\theta = \dot{\theta}(-\cos\theta\hat{\mathbf{i}} - \sin\theta\hat{\mathbf{j}}) = -\dot{\theta}\hat{\mathbf{e}}_r \tag{1.36}$$

Thus the acceleration expression in cylindrical coordinates is

$$\mathbf{a}_P = (\ddot{r} - r\dot{\theta}^2)\hat{\mathbf{e}}_r + (r\ddot{\theta} + 2\dot{r}\dot{\theta})\hat{\mathbf{e}}_\theta + \ddot{z}\hat{\mathbf{k}} \tag{1.37}$$

In the special case for which the motion is in a plane defined by $z =$ constant, we have $\dot{z} = \ddot{z} = 0$. In this case we need only the polar coordinates r and θ, and the directions of $\hat{\mathbf{e}}_r$ and $\hat{\mathbf{e}}_\theta$ are said to be **radial** and **transverse,** respectively.

Question 1.4 Why is $|\mathbf{v}_P| \neq \dot{r}$? Why is $|\mathbf{a}_P| \neq \ddot{r}$?

Before turning to the examples of this section, we return briefly to calculation of the derivatives of the unit vectors $\hat{\mathbf{e}}_r$ and $\hat{\mathbf{e}}_\theta$. We note that each derivative turns out to be perpendicular to the vector being differentiated. To understand why this is the case, we consider an alternative derivation of the formula for $\dot{\hat{\mathbf{e}}}_r$. The time dependence of $\hat{\mathbf{e}}_r$ is due to the time dependence of the coordinate θ on which $\hat{\mathbf{e}}_r$ depends explicitly; thus we can write

$$\dot{\hat{\mathbf{e}}}_r = \frac{d\hat{\mathbf{e}}_r}{d\theta}\dot{\theta}$$

Let us study the derivative $d\hat{\mathbf{e}}_r/d\theta$. By definition,

$$\frac{d\hat{\mathbf{e}}_r}{d\theta} = \lim_{\Delta\theta \to 0}\left[\frac{\hat{\mathbf{e}}_r(\theta + \Delta\theta) - \hat{\mathbf{e}}_r(\theta)}{\Delta\theta}\right] = \lim_{\Delta\theta \to 0}\left(\frac{\Delta\hat{\mathbf{e}}_r}{\Delta\theta}\right)$$

With the aid of Figure 1.7 we can see that:

1. The direction of $\Delta\hat{\mathbf{e}}_r$ (and hence $\Delta\hat{\mathbf{e}}_r/\Delta\theta$) approaches that of $\hat{\mathbf{e}}_\theta$ as $\Delta\theta$ approaches zero.
2. The magnitude of $\Delta\hat{\mathbf{e}}_r/\Delta\theta$ is $[2(1)\sin(\Delta\theta/2)]/\Delta\theta$, which approaches unity as $\Delta\theta$ approaches zero.

Thus $d\hat{\mathbf{e}}_r/d\theta = \hat{\mathbf{e}}_\theta$, and we obtain $\dot{\hat{\mathbf{e}}}_r = \dot{\theta}\hat{\mathbf{e}}_\theta$ in agreement with Equation (1.33). The reader may wish to sketch a similar geometric proof of Equation (1.36).

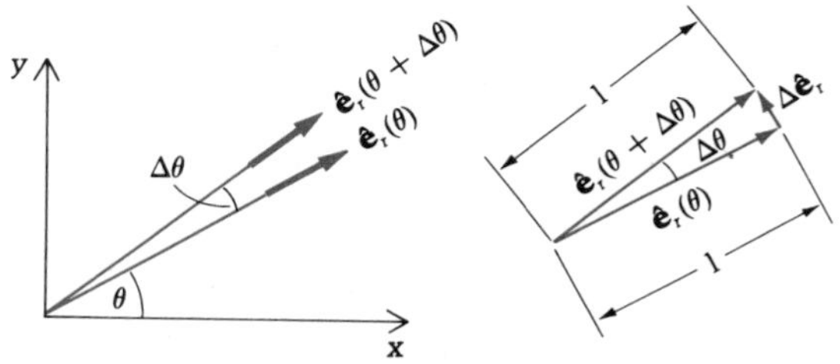

Figure 1.7 Change in $\hat{\mathbf{e}}_r$ as θ changes.

This mutual orthogonality of a vector and its derivative, incidentally, is not just restricted to unit vectors. It is in fact a property of all vectors of constant magnitude. We can show that this is the case by noting that if $\mathbf{A}$ is such a vector, then

$$\mathbf{A} \cdot \mathbf{A} = |\mathbf{A}|^2 = \text{constant}$$

and thus

$$\frac{d}{dt}(\mathbf{A} \cdot \mathbf{A}) = 0$$

or

$$\frac{d\mathbf{A}}{dt} \cdot \mathbf{A} + \mathbf{A} \cdot \frac{d\mathbf{A}}{dt} = 0$$

or

$$2\mathbf{A} \cdot \frac{d\mathbf{A}}{dt} = 0 \tag{1.38}$$

Hence, provided that neither the vector nor its derivative vanishes, the vector and the derivative are mutually perpendicular. This is a result we shall make use of frequently throughout the book. We now proceed to some examples of velocity and acceleration in cylindrical coordinates.

EXAMPLE 1.15

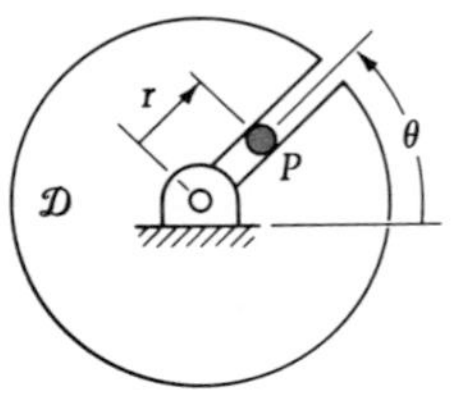

A pin P moves outward with respect to a horizontal circular disk, and its radial coordinate r is given as a function of time by $r = 3t^2/2$ meters. The disk $\mathcal{D}$ turns with the time-dependent angle $\theta = 4t^2/3$ rad. Find the velocity and acceleration of P at $t = 1$ s.

SOLUTION

From Equation (1.34) we have

$$\mathbf{v}_P = \dot{r}\hat{\mathbf{e}}_r + r\dot{\theta}\hat{\mathbf{e}}_\theta + \dot{z}\hat{\mathbf{k}}$$

$$= 3t\hat{\mathbf{e}}_r + \left(\frac{3}{2}t^2\right)\left(\frac{8}{3}t\right)\hat{\mathbf{e}}_\theta + \mathbf{0}$$

$$= 3t\hat{\mathbf{e}}_r + 4t^3\hat{\mathbf{e}}_\theta \text{ m/s}$$

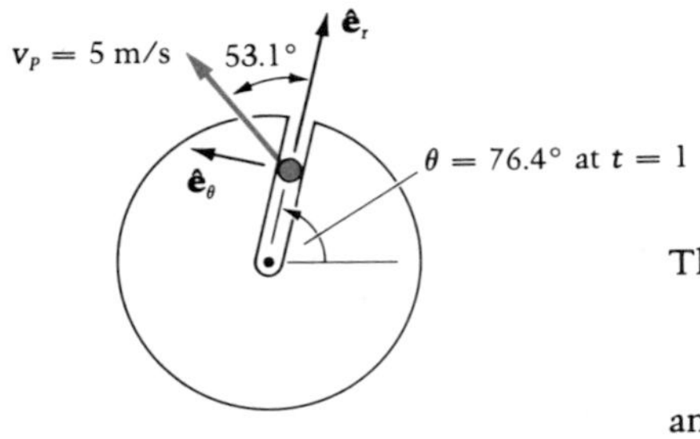

Velocity $\mathbf{v}_P$

Thus

$$\mathbf{v}_P\big|_{t=1} = 3\hat{\mathbf{e}}_r + 4\hat{\mathbf{e}}_\theta \text{ m/s}$$

and we note that the speed of P at $t = 1$ s is 5 m/s.

Continuing, from Equation (1.37) we get

$$\mathbf{a}_P = (\ddot{r} - r\dot{\theta}^2)\hat{\mathbf{e}}_r + (r\ddot{\theta} + 2\dot{r}\dot{\theta})\hat{\mathbf{e}}_\theta + \ddot{z}\hat{\mathbf{k}}$$

$$= \left[3 - \left(\frac{3}{2}t^2\right)\left(\frac{8}{3}t\right)^2\right]\hat{\mathbf{e}}_r + \left[\left(\frac{3}{2}t^2\right)\frac{8}{3} + 2\,(3t)\left(\frac{8}{3}t\right)\right]\hat{\mathbf{e}}_\theta + \mathbf{0}$$

$$= \left(3 - \frac{32}{3}t^4\right)\hat{\mathbf{e}}_r + 20t^2\hat{\mathbf{e}}_\theta$$

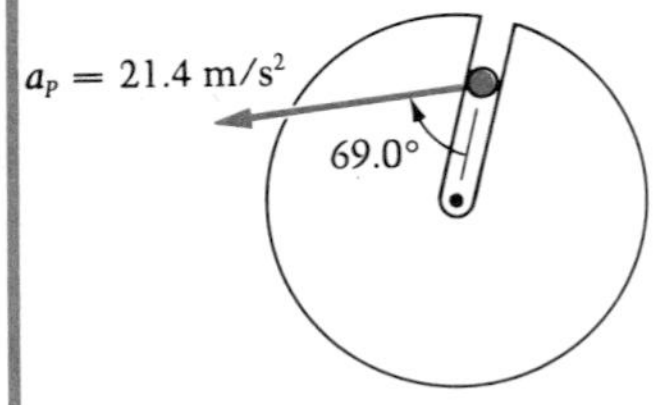

Acceleration $\mathbf{a}_P$

Thus

$$\mathbf{a}_P\big|_{t=1} = \frac{-23}{3}\hat{\mathbf{e}}_r + 20\hat{\mathbf{e}}_\theta \text{ m/s}^2$$

Since at $t = 1$ we have $r = 3/2$ m and $\theta = 4/3$ rad, we show the preceding results pictorially here.

(Continued)

Note that there is a time, $t = \sqrt[4]{9/32}$ s, when the $\ddot{r}$ and $-r\dot{\theta}^2$ parts of the radial component of $\mathbf{a}_P$ cancel each other, making this component zero at that instant of time. The reader is urged to compute and sketch $\mathbf{v}_P$ and $\mathbf{a}_P$ at another time, say $t = 2$ s.

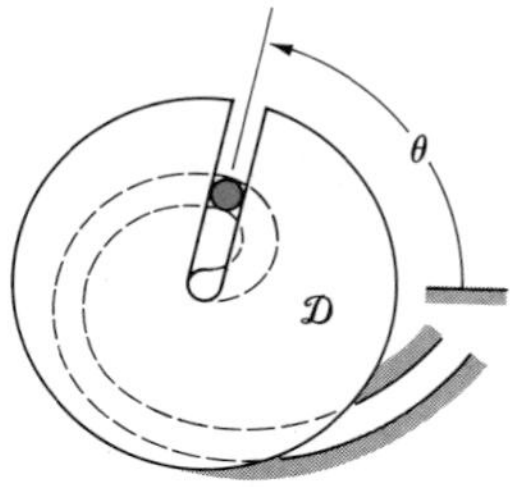

EXAMPLE 1.16

In the preceding example, discard the given r and θ. Suppose instead that $\dot{\theta}_{\mathcal{D}} =$ constant $= 0.3$ rad/s and that the pin slides not only in the slot of disk $\mathcal{D}$ but also in the spiral slot cut in the reference frame and defined by $r = 0.1\theta$ meters, with θ in radians. Find the velocity and acceleration of the pin when $\theta = \pi$ rad.

SOLUTION

From $r = 0.1\theta$, we have $\dot{r} = 0.1\dot{\theta}$ and $\ddot{r} = 0.1\ddot{\theta}$, which is zero since $\dot{\theta} =$ constant. Therefore

$$\begin{aligned}\mathbf{v}_P &= \dot{r}\hat{\mathbf{e}}_r + r\dot{\theta}\hat{\mathbf{e}}_\theta \\ &= 0.1\dot{\theta}\hat{\mathbf{e}}_r + 0.1\theta\dot{\theta}\hat{\mathbf{e}}_\theta \\ &= 0.0300\hat{\mathbf{e}}_r + 0.0942\hat{\mathbf{e}}_\theta \text{ m/s}\end{aligned}$$

Now for the acceleration:

$$\begin{aligned}\mathbf{a}_P &= (\ddot{r} - r\dot{\theta}^2)\hat{\mathbf{e}}_r + (r\ddot{\theta} + 2\dot{r}\dot{\theta})\hat{\mathbf{e}}_\theta \\ &= (0 - 0.1\theta\dot{\theta}^2)\hat{\mathbf{e}}_r + [0 + 2(0.1\dot{\theta})\dot{\theta}]\hat{\mathbf{e}}_\theta \\ &= -0.0283\hat{\mathbf{e}}_r + 0.0180\hat{\mathbf{e}}_\theta \text{ m/s}^2\end{aligned}$$

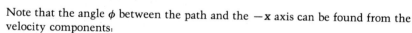

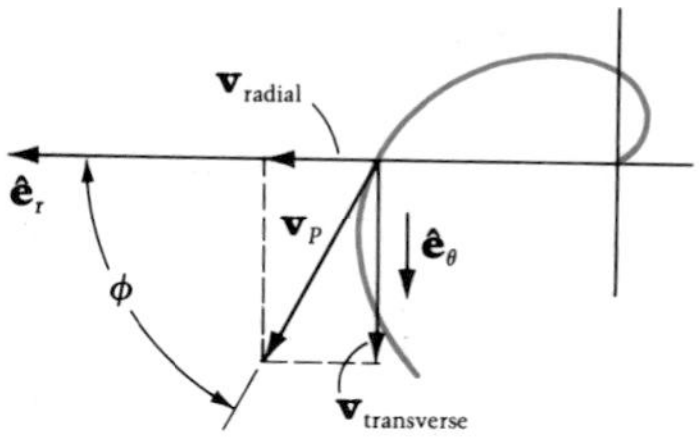

Note that the angle ϕ between the path and the $-\mathbf{x}$ axis can be found from the velocity components:

$$\phi = \tan^{-1}\left(\frac{0.0942}{0.0300}\right) = 72.3°$$

In our next examples, there is motion in the z direction as well as the radial (r) and transverse (θ) of the previous examples.

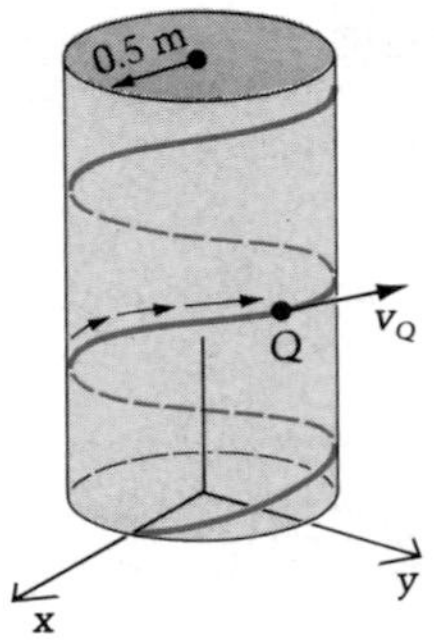

EXAMPLE 1.17

A point Q moves on a helix as shown in the figure. The pitch, p, of the helix is 0.2 m, and the point travels at constant speed $v_Q = 20$ m/s. Find the velocity of Q in terms of its cylindrical components.

SOLUTION

The meaning of the *pitch* of a helix is the (constant) advance of Q in the z direction for each revolution in θ. Therefore

$$\frac{p\theta}{2\pi} = z \tag{1}$$

so that

$$\dot{\theta} = \frac{2\pi\dot{z}}{p} \tag{2}$$

or, for this problem,

$$\dot{\theta} = 31.4\dot{z} \tag{3}$$

Noting that $\dot{r} = 0$ since Q travels on a cylinder (with r therefore constant), Equation (1.34) then gives the following for the point's velocity:

$$\mathbf{v}_Q = r\dot{\theta}\hat{\mathbf{e}}_\theta + \dot{z}\hat{\mathbf{k}} \tag{4}$$

$$= 0.5\dot{\theta}\hat{\mathbf{e}}_\theta + \dot{z}\hat{\mathbf{k}} \tag{5}$$

or, using (3),

$$\mathbf{v}_Q = 15.7\dot{z}\hat{\mathbf{e}}_\theta + \dot{z}\hat{\mathbf{k}} \tag{6}$$

The speed of Q is constant at 20 m/s; thus

$$\sqrt{(15.7\dot{z})^2 + \dot{z}^2} = 20 \tag{7}$$

$$\dot{z} = 1.27 \text{ m/s} \tag{8}$$

From Equation (3), we then get

$$\dot{\theta} = 31.4(1.27) = 39.9 \text{ rad/s} \tag{9}$$

Hence the velocity vector of Q is (substituting (9) and (8) into (5))

$$\mathbf{v}_Q = 20.0\hat{\mathbf{e}}_\theta + 1.27\hat{\mathbf{k}} \text{ m/s} \tag{10}$$

Note that $|\mathbf{v}_Q| = 20.0$ m/s, as it must be. Note also that a larger pitch will spread out the helix (see figure). The equations of this example then show that the $\hat{\mathbf{k}}$ component will become larger in comparison to the $\hat{\mathbf{e}}_\theta$ component for larger p.

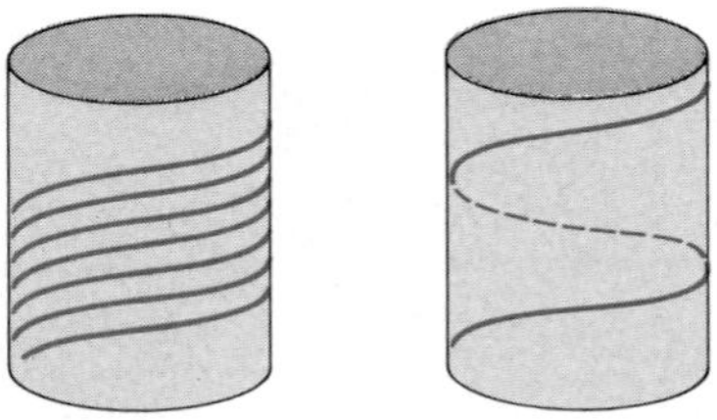

Small p Larger p

EXAMPLE 1.18

Find the acceleration vector of Q in the preceding example.

SOLUTION

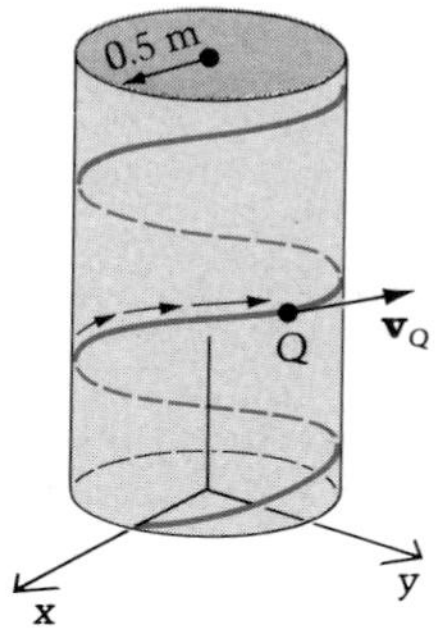

From Equation (1.37) we get

$$\mathbf{a}_Q = (\ddot{r} - r\dot{\theta}^2)\hat{\mathbf{e}}_r + (r\ddot{\theta} + 2\dot{r}\dot{\theta})\hat{\mathbf{e}}_\theta + \ddot{z}\hat{\mathbf{k}}$$

Because r is constant on the cylinder, this equation reduces to

$$\mathbf{a}_Q = -r\dot{\theta}^2\hat{\mathbf{e}}_r + r\ddot{\theta}\hat{\mathbf{e}}_\theta + \ddot{z}\hat{\mathbf{k}}$$

Furthermore, since $|\mathbf{v}_Q|$ is constant, Equations (8) and (3) of the previous example show that $\dot{z}$ and $\dot{\theta}$ are constants. Therefore

$$\mathbf{a}_Q = -r\dot{\theta}^2\hat{\mathbf{e}}_r = -0.5(39.9^2)\hat{\mathbf{e}}_r = -796\hat{\mathbf{e}}_r \text{ m/s}^2$$

Note that even though point Q ***never*** has a radial component of velocity (see figure), it has *only* a radial component of acceleration!

EXAMPLE 1.19

Find the velocity and acceleration vectors of point Q in Example 1.17 if, instead of the speed of Q being constant, we have its vertical position given as the function of time:

$$z = 0.08t^3 \text{ m}$$

SOLUTION

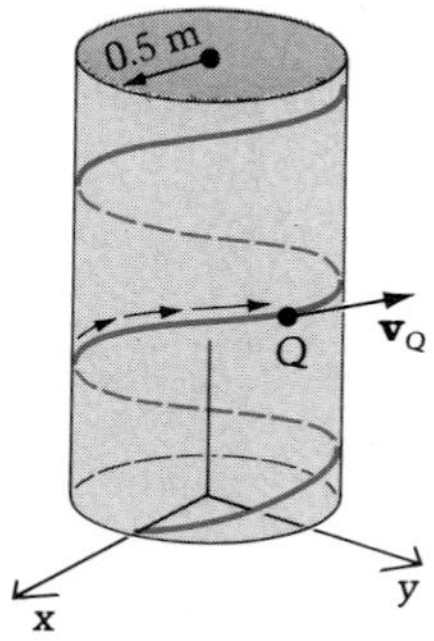

Referring to Example 1.17 (see the figure), we find

$$\dot{\theta} = 31.4\dot{z} = 31.4(0.24t^2)$$
$$= 7.54t^2 \text{ rad/s}$$

Therefore

$$\mathbf{v}_Q = 0.5(7.54t^2)\hat{\mathbf{e}}_\theta + 0.24t^2\hat{\mathbf{k}}$$
$$= 3.77t^2\hat{\mathbf{e}}_\theta + 0.24t^2\hat{\mathbf{k}} \text{ m/s}$$

This time, however, the velocity is seen to depend on the time; for example, at $t = 10$ s,

$$\mathbf{v}_Q|_{t=10} = 377\hat{\mathbf{e}}_\theta + 24.0\hat{\mathbf{k}} \text{ m/s}$$

(Continued)

For the acceleration, we note that $\dot{r}$ is still zero, so that

$$\mathbf{a}_Q = -r\dot{\theta}^2\hat{\mathbf{e}}_r + r\ddot{\theta}\hat{\mathbf{e}}_\theta + \ddot{z}\hat{\mathbf{k}}$$

This time, all three terms are nonzero. We have

$$\dot{z} = 0.24t^2 \text{ m/s} \qquad \dot{\theta} = 7.54t^2 \text{ rad/s}$$

$$\ddot{z} = 0.48t \text{ m/s}^2 \qquad \ddot{\theta} = 15.1t \text{ rad/s}^2$$

Thus

$$\begin{aligned}\mathbf{a}_Q &= -0.5(7.54t^2)^2\hat{\mathbf{e}}_r + 0.5(15.1t)\hat{\mathbf{e}}_\theta + (0.48t)\hat{\mathbf{k}} \\ &= -28.4t^4\hat{\mathbf{e}}_r + 7.55t\hat{\mathbf{e}}_\theta + 0.48t\hat{\mathbf{k}} \text{ m/s}^2\end{aligned}$$

In the following example we consider the case in which, in addition to the changing θ and z of the preceding three examples, the radius varies.

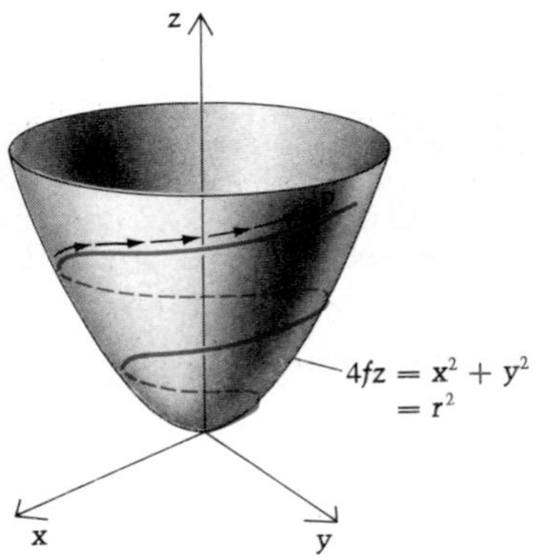

EXAMPLE 1.20

A point P moves on a spiraling path that winds around the paraboloid of revolution shown in the figure. The focal distance f is $\frac{1}{4}$ m, and the point P advances 4.0 m vertically with each revolution. If the speed of P is 0.7 m/s, a constant, determine the vertical component of the velocity vector of P as a function of r.

SOLUTION

From $z = r^2$, we obtain

$$\dot{z} = 2r\dot{r} \Rightarrow \dot{r} = \frac{\dot{z}}{2r} \text{ m/s} \tag{1}$$

And from the pitch relationship $p\theta/2\pi = z$, we get

$$4.0\dot{\theta} = 2\pi\dot{z} \Rightarrow \dot{\theta} = 1.57\dot{z} \quad \text{or} \quad \frac{\pi}{2}\dot{z} \text{ rad/s} \tag{2}$$

Therefore the speed of P may be expressed as

$$\begin{aligned}|\mathbf{v}_P| = v_\mathbf{P} = 0.7 &= \sqrt{\dot{r}^2 + (r\dot{\theta})^2 + \dot{z}^2} \\ &= \sqrt{\left(\frac{\dot{z}}{2r}\right)^2 + \left(\frac{\pi}{2}r\dot{z}\right)^2 + \dot{z}^2} \text{ m/s}\end{aligned}$$

Thus the answer is

$$\dot{z} = \frac{1.4r}{\sqrt{1 + 4r^2 + \pi^2 r^4}} \text{ m/s}$$

Let us extend the preceding example slightly. We can see that $\dot{z}$ varies with the radius r (distance from the z axis to P) and that it is zero initially and approaches zero again for large r. Its maximum may be determined from calculus:

$$\frac{d\dot{z}}{dr} = 0$$

or

$$0 = \frac{\sqrt{1 + 4r^2 + \pi^2 r^4}\,\dfrac{d(1.4r)}{dr} - 1.4r\,\dfrac{d\sqrt{1 + 4r^2 + \pi^2 r^4}}{dr}}{\left(\sqrt{1 + 4r^2 + \pi^2 r^4}\right)^2}$$

This yields the equation and result:

$$\pi^2 r^4 = 1 \Rightarrow r = 0.56 \text{ m}$$

at which

$$\dot{z}_{\max} = 0.44 \text{ m/s}$$

Note from Equation (2) in the example that at the preceding value of $\dot{z}$,

$$\dot{\theta} = 0.69 \text{ rad/s}$$

From Equation (1), we see that at the same time

$$\dot{r} = \frac{\dot{z}}{2r} = 0.39 \text{ m/s}$$

and therefore that when $\dot{z}$ is maximum, the speed is

$$\begin{aligned} v_P &= \sqrt{\dot{r}^2 + (r\dot{\theta})^2 + \dot{z}^2} \\ &= \sqrt{0.39^2 + (0.56 \times 0.69)^2 + 0.44^2} \\ &= 0.70 \text{ m/s} \end{aligned}$$

as it should be, since it does not change with time.

Question 1.5 By inspection (with little or no writing), what is the maximum value of the radial component of $\mathbf{v}_P$?

PROBLEMS / Section 1.6

1.104 An insect is asleep on a $33\frac{1}{3}$ rpm record, 6 in. from the spindle. When the record is turned on, the insect wakes up and dizzily heads toward the center, in a straight line relative to the disk, at 1 in./sec (Figure P1.104). If the bug can withstand a maximum acceleration magnitude of 100 in./sec^2, does it make it to the spindle (a) if it starts after the record is up to speed? (b) If it starts as soon as the record is turned on? Assume that the turntable accelerates linearly (with time) up to speed in one revolution.

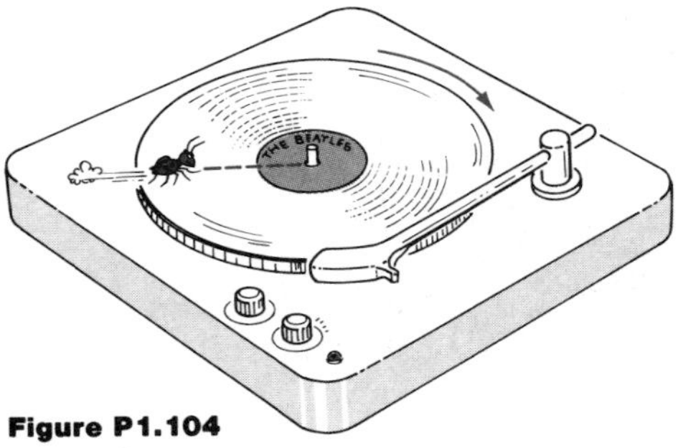

Figure P1.104

1.105 David throws a rock at Goliath with a sling. He whirls it around one revolution plus 135° more and releases it there, as shown in Figure P1.105, at 50 ft/sec. As he whirls the sling, the speed of the rock increases linearly with the time t; that is, $\dot{\theta} = kt$, where k is a constant. Find the acceleration of the rock just prior to release.

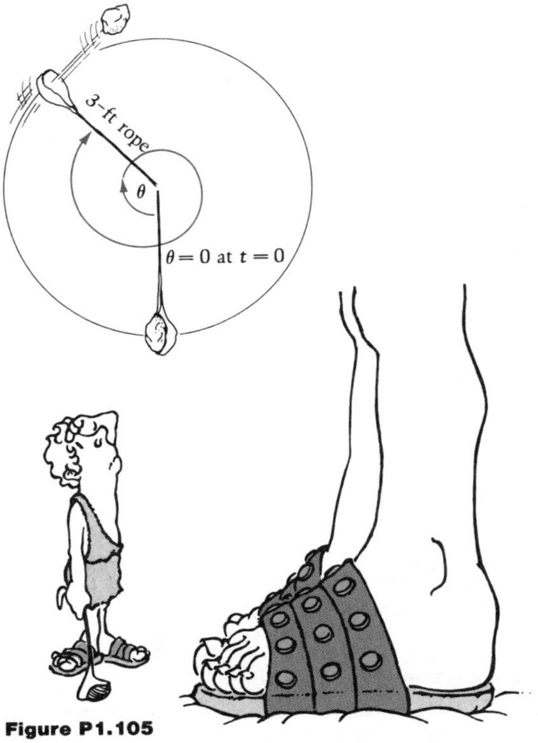

Figure P1.105

1.106 A ball bearing is moving radially outward in a slotted horizontal disk that is rotating about the vertical z axis. At the instant shown in Figure P1.106, the ball bearing is 3 in. from the center of the disk. It is traveling radially outward at a velocity of 4 in./sec relative to the disk and has a radial acceleration with respect to the disk of 5 in./sec^2 outward. What would $\dot{\theta}$ and $\ddot{\theta}$ have to be at the instant shown for the ball bearing to have a total acceleration of zero?

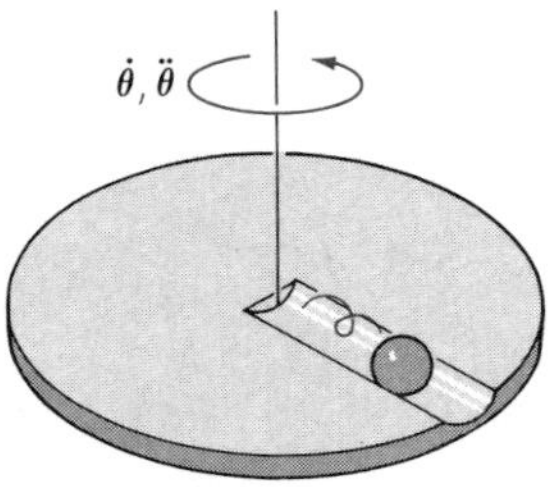

Figure P1.106

1.107 The disk shown in Figure P1.107 is horizontal and turns so that $\dot{\theta} = ct$ about the vertical. Forces cause a marble to move in a slot such that its radial distance from the center equals kt^2. Note that c and k are constants.

a. Find the acceleration of the marble.
b. At what time does the radial acceleration vanish?

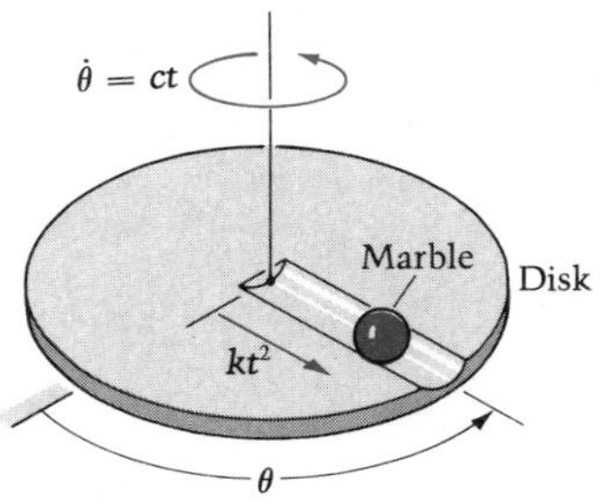

Figure P1.107

1.108 The point P in Figure P1.108 moves on the limaçon defined in polar coordinates by

$$r = 5 - 3\cos\theta \text{ m}$$

If the polar angle is quadratic in time according to $\theta = 10t^2$ rad, find the velocity of P when it is at its highest point.

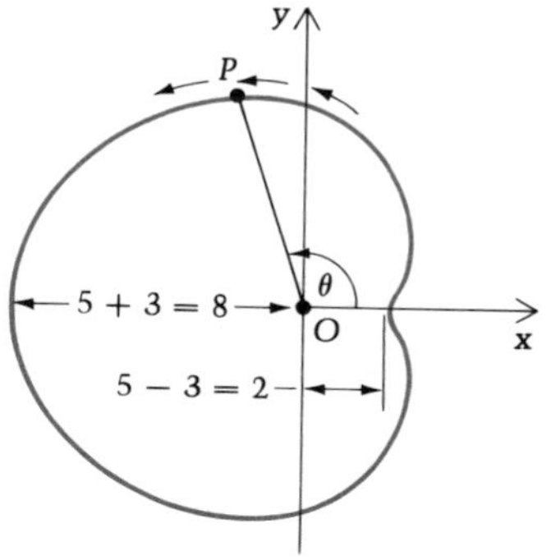

Figure P1.108

1.109 In the preceding problem, determine the acceleration of P at (a) the same highest point and (b) $\theta = \pi$ rad.

1.110 A point P starts at $t = 0$ at the origin and proceeds along a path on the paraboloid of revolution shown in Figure P1.110. The path is described (with time as a parameter) by

$$r = k_1 t$$

$$\theta = k_2 t^2$$

Find the position, velocity, and acceleration vectors of the point when it reaches the top edge of the paraboloid. (H, R, k_1, and k_2 are constants.)

1.111 A bead B slides down and around a cylindrical surface on a helical wire (Figure P1.111). The vertical drop of the bead as θ changes by 2π is called the pitch p of the helix; R is the radius of the helix.

a. Noting that θ (and therefore also z) is a function of time, write the equations for $\mathbf{r}_{OB}$, $\mathbf{v}_B$, and $\mathbf{a}_B$ in cylindrical coordinates.

b. For the values $R = 0.3$ m, $p = 0.2$ m, and $\theta = 0.6t$ rad/s, find and sketch the velocity and acceleration vectors of B when $t = 10$ s.

1.112 An ant travels up the banister of a spiral staircase (Figure P1.112) according to

$$\mathbf{r}_{OA} = 2\cos\frac{t}{50}\,\hat{\mathbf{i}} + 2\sin\frac{t}{50}\,\hat{\mathbf{j}} + \frac{t}{50}\,\hat{\mathbf{k}} \text{ m}$$

Find the position and velocity of the ant when $t = 30$ s.

1.113 Find the acceleration of the ant (again at $t = 30$ s) in the preceding problem.

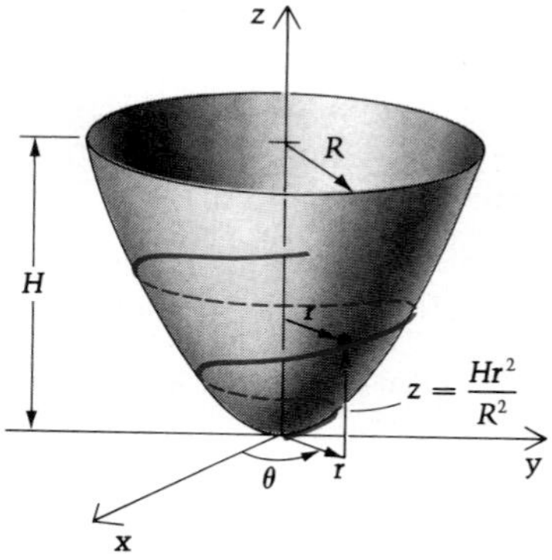

Figure P1.110

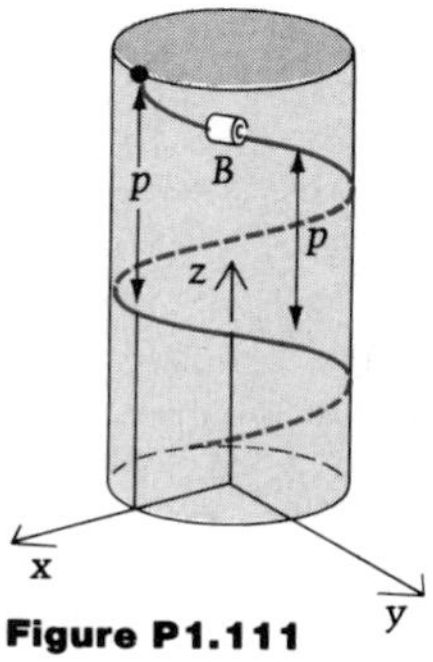

Figure P1.111

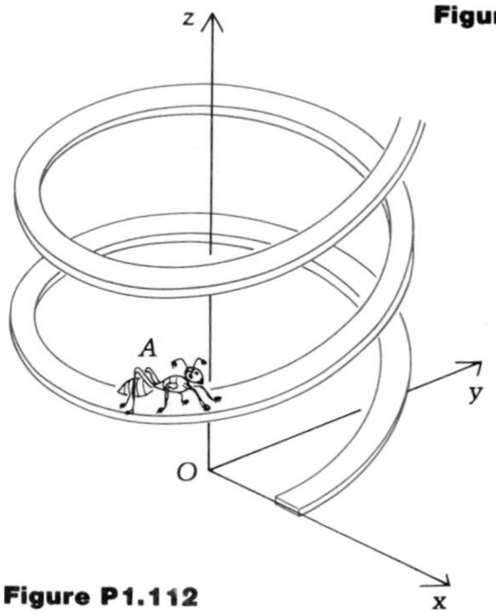

Figure P1.112

1.114 The mountain shown in Figure P1.114 is in the shape of the paraboloid of revolution $H - z = kr^2$, where H = height = 5000 ft, r is the radius at z, and k is a constant. The base radius is also 5000 ft. A car travels up the mountain on a spiraling path. Each time around, the car's altitude is 1000 ft higher. The car travels at the *constant speed* of 50 mph. Find the largest and smallest values of the radial component of velocity on the journey, and tell where the car is at these two times.

1.115 In the preceding problem, find the locations of the car (r, θ, z) for which the following velocity components are equal:

a. Radial ($\dot{r}$) and transverse ($r\dot{\theta}$)
b. Radial and vertical ($\dot{z}$)
c. Transverse and vertical

1.116 Find the differential equation whose solution would give $r(t)$ in Problem 1.114.

1.117 Determine whether any of the three cylindrical components of acceleration of the car in Problem 1.114 is ever zero on the trip up the mountain.

1.118 Show that the velocity of a point P in spherical coordinates (r, θ, ϕ) is given by

$$\mathbf{v}_P = \dot{r}\hat{\mathbf{e}}_r + r\dot{\theta}\hat{\mathbf{e}}_\theta + r\dot{\phi}\sin\theta\,\hat{\mathbf{e}}_\phi$$

See Figure P1.118. *Hint:* As intermediate steps, obtain the results

$$\dot{\hat{\mathbf{e}}}_r = \dot{\theta}\hat{\mathbf{e}}_\theta + \dot{\phi}\sin\theta\,\hat{\mathbf{e}}_\phi$$
$$\dot{\hat{\mathbf{e}}}_\theta = -\dot{\theta}\hat{\mathbf{e}}_r + \dot{\phi}\cos\theta\,\hat{\mathbf{e}}_\phi$$
$$\dot{\hat{\mathbf{e}}}_\phi = -\dot{\phi}(\sin\theta\,\hat{\mathbf{e}}_r + \cos\theta\,\hat{\mathbf{e}}_\theta)$$

Then differentiate the simple position vector $\mathbf{r}_{OP} = r\hat{\mathbf{e}}_r$.

1.119 Show by differentiating $\mathbf{v}_P$ in the preceding problem that the corresponding expression for the acceleration in spherical coordinates is

$$\begin{aligned}\mathbf{a}_P = {} & (\ddot{r} - r\dot{\theta}^2 - r\dot{\phi}^2\sin^2\theta)\hat{\mathbf{e}}_r \\ & + (2\dot{r}\dot{\theta} + r\ddot{\theta} - r\dot{\phi}^2\sin\theta\cos\theta)\hat{\mathbf{e}}_\theta \\ & + (2\dot{r}\dot{\phi}\sin\theta + 2r\dot{\theta}\dot{\phi}\cos\theta + r\ddot{\phi}\sin\theta)\hat{\mathbf{e}}_\phi\end{aligned}$$

1.120 In Problem 1.86 show that the velocity of the shingles may also be obtained by simply taking the component of the velocity of the bumper attachment point A *along the rope*. Using the cylindrical coordinate expression for velocity, explain why this procedure works.

1.121 In Problem 1.86 find the acceleration of the shingles by differentiating their velocity expression. Unlike the result for velocity, this answer is not equal to the radial component of acceleration ($\ddot{r} - r\dot{\theta}^2$) of the bumper A with origin at the pulley. Show this by using the velocity expression

$$\mathbf{v}_A = \frac{\mathbf{x}v_A}{\sqrt{19^2 + \mathbf{x}^2}}\hat{\mathbf{e}}_r + \frac{19v_A}{\sqrt{19^2 + \mathbf{x}^2}}\hat{\mathbf{e}}_\theta$$

which equals $\dot{r}\hat{\mathbf{e}}_r + r\dot{\theta}\hat{\mathbf{e}}_\theta$.

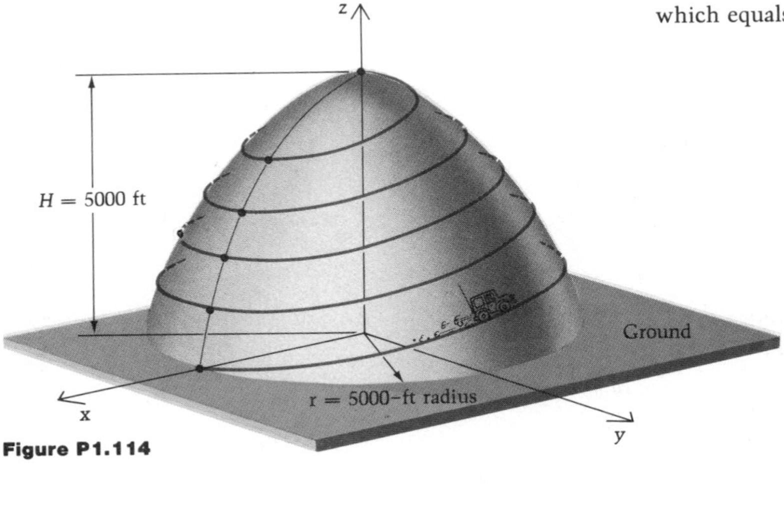

Figure P1.114

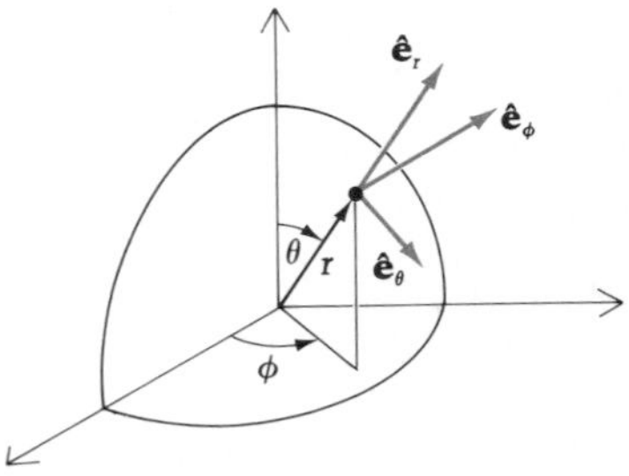

Figure P1.118

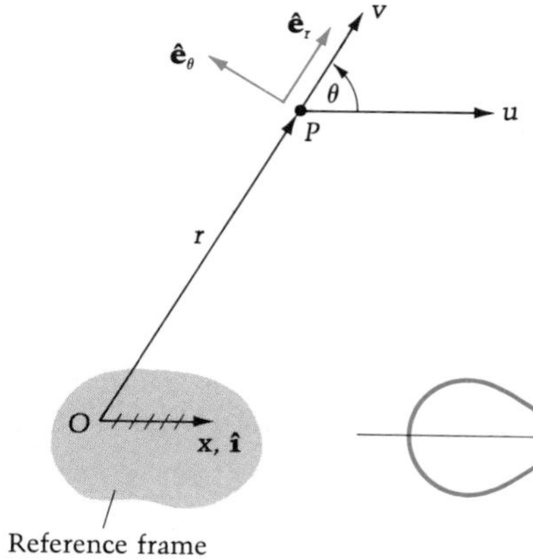

Figure P1.123

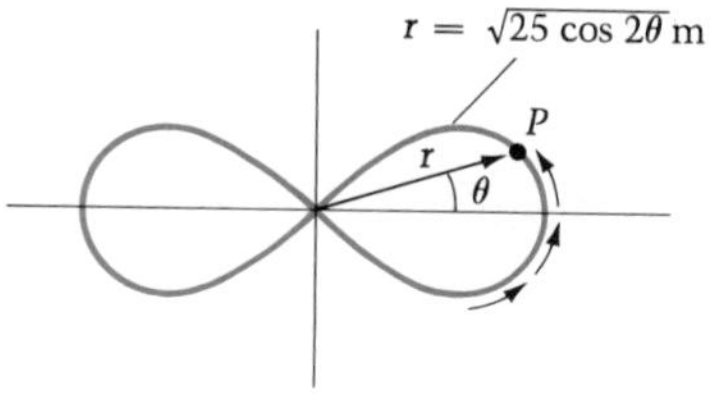

Figure P1.122

1.122 A point P moves on the figure eight in the indicated direction (Figure P1.122) at constant speed 2 m/s. Find the acceleration vector of P the next time its velocity is horizontal.

1.123 The velocity of a point P moving in a plane is the resultant of one part, $v\hat{\mathbf{e}}_r$, along the radius from a fixed point O to the point P, and another part, $u\hat{\mathbf{i}}$, which is always parallel to a fixed line. (See Figure P1.123.) Prove that the acceleration of P may be written as $a_v\hat{\mathbf{e}}_r + a_u\hat{\mathbf{i}}$, where

$$a_v = \frac{dv}{dt} - \frac{uv}{r}\cos\theta \qquad \text{and} \qquad a_u = \frac{du}{dt} + \frac{uv}{r}$$

where r is the length of the radius vector from O to P and θ is the angle it makes with the fixed direction.

1.7 Tangential and Normal Components

In this section we examine yet another means of expressing the velocity and acceleration of a point P. Instead of focusing on a specific coordinate system, this time we shall study the way in which the motion of P is related to its path. Consequently, the components of velocity and acceleration that result are sometimes called *intrinsic* or *natural.*

The path of point P, as mentioned in Section 1.3, is the locus of points of the reference frame $\mathcal{F}$ successively occupied by P as it moves. We begin, then, by defining some reference point on the path. From this arclength origin we then measure the arclength s along the path to the point P. Clearly the arclength coordinate depends on the time; that is, $s = s(t)$.

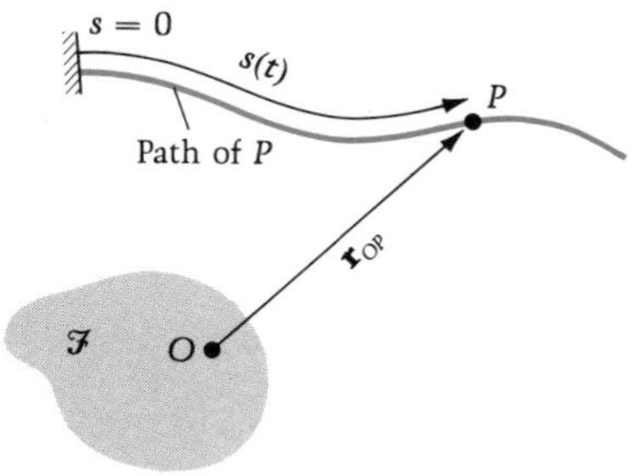

Figure 1.8 Arclength measurement of point P on its path.

In Figure 1.8 we see a position vector, $\mathbf{r}_{OP}$, for point P. This vector was seen in preceding sections to define the location of P, and thus it may be considered a function of the arclength s:

$$\mathbf{r}_{OP} = \mathbf{r}_{OP}(s) = \mathbf{r}_{OP}[s(t)] \tag{1.39}$$

Forming the velocity of P by differentiation (the definition is the same, regardless of how we choose to represent the vectors), we get

$$\mathbf{v}_P = \dot{\mathbf{r}}_{OP} = \frac{d\mathbf{r}_{OP}}{dt}$$

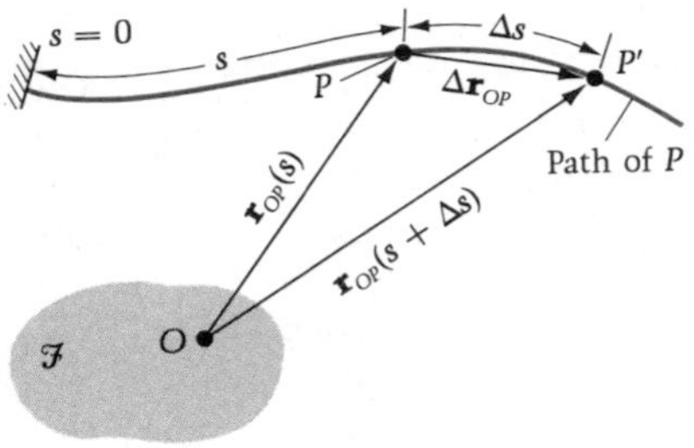

Figure 1.9 Change in $\mathbf{r}_{OP}$ as s changes.

and, by the chain rule,

$$\mathbf{v}_P = \frac{d\mathbf{r}_{OP}}{ds}\frac{ds}{dt}$$

$$= \dot{s} \lim_{\Delta s \to 0} \left[\frac{\mathbf{r}_{OP}(s + \Delta s) - \mathbf{r}_{OP}(s)}{\Delta s} \right]$$

$$= \dot{s} \lim_{\Delta s \to 0} \left(\frac{\Delta \mathbf{r}_{OP}}{\Delta s} \right) \tag{1.40}$$

Figure 1.9 shows the quantities $\Delta\mathbf{r}_{OP}$ and Δs.

We suggest that the reader sketch an arc on a large sheet of paper and then use a straightedge to draw the triangle OPP' (Figure 1.10). Then the limit in Equation (1.40) can be taken by, let us say, dividing Δs in half each time. After just a few more divisions of Δs on the large sheet, it will become clear that as Δs approaches zero—that is, as P' backs up toward P—two interesting things happen:

1. $\Delta\mathbf{r}_{OP}$ becomes tangent to the path of P at arclength s.
2. The magnitude of $\Delta\mathbf{r}_{OP}/\Delta s$ approaches $\Delta s/\Delta s = 1$.

These two results, taken together, prove that $d\mathbf{r}_{OP}/ds$ is always a unit vector that is tangent to the path and pointing in the direction of increasing s. It is for these reasons that this vector is called $\hat{\mathbf{e}}_t$, the **unit tangent.** Equation (1.40) may then be rewritten as

$$\mathbf{v}_P = \dot{s}\hat{\mathbf{e}}_t \tag{1.41}$$

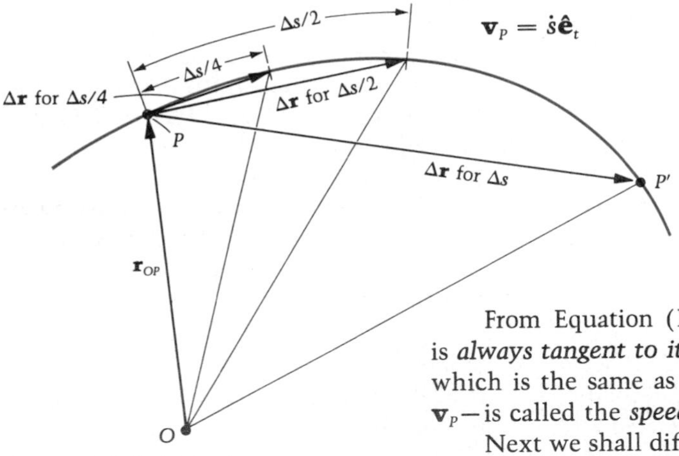

Figure 1.10 Shrinking Δs toward zero.

From Equation (1.41) we see that the velocity vector of point P is *always tangent to its path*. The absolute value $|\dot{s}|$ of the scalar part—which is the same as the magnitude $|\mathbf{v}_P|$, or v_P, of the velocity vector $\mathbf{v}_P$—is called the *speed* of P in $\mathcal{F}$, as we mentioned in Section 1.3.

Next we shall differentiate again in order to obtain the acceleration of P. Using Equation (1.41), we get

$$\mathbf{a}_P = \dot{\mathbf{v}}_P = \ddot{s}\hat{\mathbf{e}}_t + \dot{s}\frac{d\hat{\mathbf{e}}_t}{dt}$$

$$= \ddot{s}\hat{\mathbf{e}}_t + \dot{s}^2\frac{d\hat{\mathbf{e}}_t}{ds} \tag{1.42}$$

Since $\hat{\mathbf{e}}_t$ is a unit vector, $d\hat{\mathbf{e}}_t/ds$ is perpendicular to $\hat{\mathbf{e}}_t$ and hence perpendicular, or normal, to the path. Equation (1.42) shows an important separation of the acceleration into two parts, one tangent and the other normal to the path of P. The component tangent to the path, $\ddot{s}$, is (for $\dot{s} > 0$) the rate of change of the velocity *magnitude*, or speed, of P. The component normal to the path reflects the rate of change of the *direction* of the velocity vector.

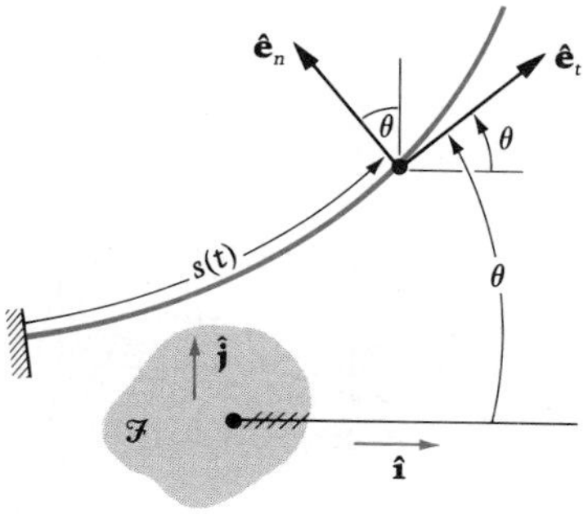

Figure 1.11 Tangent and normal to a plane curve.

Further examination of $d\hat{\mathbf{e}}_t/ds$ in Equation (1.42) is facilitated if we first restrict our attention to the case of a two-dimensional (plane) curve. To that end, let θ be the inclination of a tangent to the plane curve as shown in Figure 1.11. We can visualize that as s increases, $\hat{\mathbf{e}}_t$ turns in such a way that $d\hat{\mathbf{e}}_t/ds$ points toward the inside of the curve—that is, in the direction of $\hat{\mathbf{e}}_n$ shown in the figure. We can obtain this result analytically if we write

$$\frac{d\hat{\mathbf{e}}_t}{ds} = \frac{d\theta}{ds}\frac{d\hat{\mathbf{e}}_t}{d\theta} \tag{1.43}$$

Noting from Figure 1.11 that

$$\hat{\mathbf{e}}_t = \cos\theta\,\hat{\mathbf{i}} + \sin\theta\,\hat{\mathbf{j}} \tag{1.44}$$

we may differentiate Equation (1.44) to obtain

$$\frac{d\hat{\mathbf{e}}_t}{d\theta} = -\sin\theta\,\hat{\mathbf{i}} + \cos\theta\,\hat{\mathbf{j}} = \hat{\mathbf{e}}_n \tag{1.45}$$

From studies in calculus the reader probably recognizes the other factor in Equation (1.43)—namely $d\theta/ds$, whose magnitude is by definition the **curvature** of a plane curve. The reciprocal of the curvature is the **radius of curvature** ρ. The radius of curvature is the radius of the circle that provides the best local approximation to an infinitesimal segment of the curve. Equation (1.43) may thus be written

$$\frac{d\hat{\mathbf{e}}_t}{ds} = \frac{1}{\rho}\hat{\mathbf{e}}_n \tag{1.46}$$

In three dimensions the situation is more difficult to visualize. We cannot use the preceding development because $\hat{\mathbf{e}}_t$ cannot be expressed as a function of a single angle such as θ. Consequently, in the general case we adopt a definition of curvature that, in two dimensions, reduces to what we have just established. That is, we simply define the curvature $1/\rho$ to be the magnitude of the vector $d\hat{\mathbf{e}}_t/ds$. Then the unit vector $\hat{\mathbf{e}}_n$ as defined by

$$\hat{\mathbf{e}}_n = \frac{1}{|d\hat{\mathbf{e}}_t/ds|}\frac{d\hat{\mathbf{e}}_t}{ds} = \rho\frac{d\hat{\mathbf{e}}_t}{ds}$$

is called the **principal unit normal** to the curve. Upon substituting into Equation (1.42), we then obtain

$$\mathbf{a}_P = \ddot{s}\hat{\mathbf{e}}_t + \frac{(\dot{s})^2}{\rho}\hat{\mathbf{e}}_n \tag{1.47}$$

An alternative form in which the arclength parameter s is not explicitly involved follows if we choose the measurement of s so that at the instant of interest $\dot{s} > 0$. Hence $\dot{s} = |\mathbf{v}_P|$ and

$$\mathbf{a}_P = \left(\frac{d}{dt}|\mathbf{v}_P|\right)\hat{\mathbf{e}}_t + \frac{|\mathbf{v}_P|^2}{\rho}\hat{\mathbf{e}}_n \tag{1.48}$$

This expression more vividly depicts the natural decomposition of acceleration into parts related to rate of change of magnitude of velocity and rate of change of direction of velocity. We now consider some examples of the use of tangential and normal components.

EXAMPLE 1.21

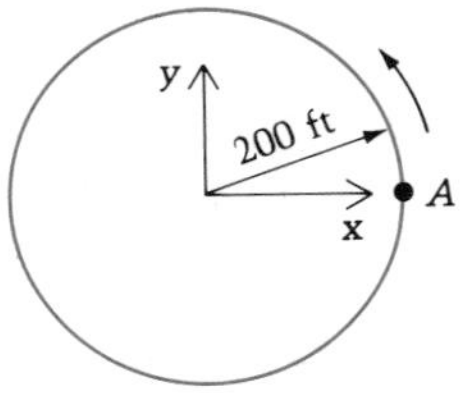

A car starts at rest at A and increases its speed around the track at 6 ft/sec², traveling counterclockwise. Determine the position and the time at which the car's acceleration magnitude reaches 20 ft/sec².

SOLUTION

$\ddot{s} = 6$ ft/sec²

$\dot{s} = 6t + \cancel{C_1}^{\,0}$ ft/sec (The constant is zero since $\dot{s} = 0$ at $t = 0$.)

$s = 3t^2 + \cancel{C_2}^{\,0}$ ft (The constant is zero since $s = 0$ at $t = 0$.)

The acceleration magnitude of Q is

$$|\mathbf{a}_Q| = a_Q = \sqrt{\ddot{s}^2 + (\dot{s}^2/\rho)^2}$$

$$= \sqrt{36 + (36t^2/200)^2} \text{ ft/sec}^2$$

When $a_Q = 20$ ft/sec², we obtain the equation

$$20^2 = 36 + (0.18t^2)^2 = 36 + 0.0324t^4$$

$$t = \sqrt[4]{\frac{364}{0.0324}} = 10.3 \text{ sec}$$

At $t = 10.3$ sec, $s = 318$ ft, which represents $318/(2\pi r) = 0.253$ of a revolution, or 91.1° counterclockwise from the **x** axis.

EXAMPLE 1.22

Verify the results of Example 1.13, at $\mathbf{x} = 2$ m, by using $\hat{\mathbf{e}}_t$ and $\hat{\mathbf{e}}_n$ components.

SOLUTION

We are given that $\dot{s} = |\mathbf{v}_P| = 0.2$ m/s. Since $\mathbf{v}_P$ is tangent to the path of P, we can calculate $\hat{\mathbf{e}}_t$:

$$\tan\theta = \frac{dy}{dx} = \mathbf{x} = 2 \qquad \text{(at the given point)}$$

$$\theta = \tan^{-1}(2) = 63.4°$$

$$\hat{\mathbf{e}}_t = \cos\theta\,\hat{\mathbf{i}} + \sin\theta\,\hat{\mathbf{j}}$$

$$= 0.448\,\hat{\mathbf{i}} + 0.894\,\hat{\mathbf{j}}$$

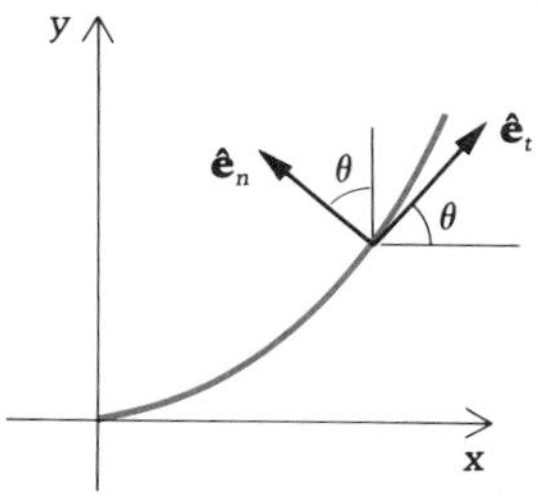

The radius of curvature comes from calculus:

$$\frac{1}{\rho} = \frac{y''}{(1 + y'^2)^{3/2}} = \frac{1}{(1 + \mathbf{x}^2)^{3/2}} = \frac{1}{5^{3/2}}$$

or

$$\rho = 11.2 \text{ m}$$

Thus with $\hat{\mathbf{e}}_n$ seen to be $-\sin\theta\,\hat{\mathbf{i}} + \cos\theta\,\hat{\mathbf{j}}$, we have

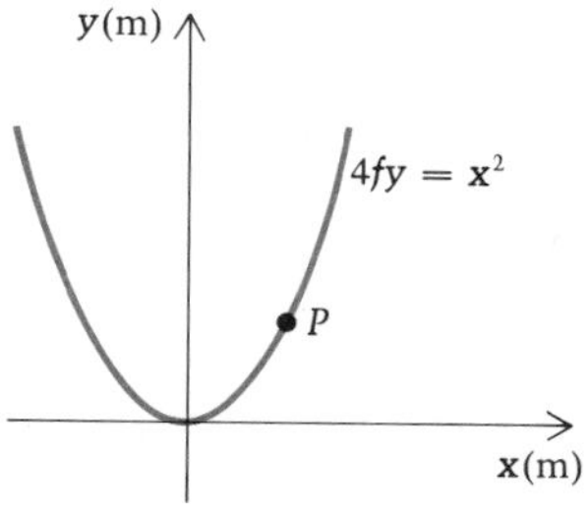

$$\mathbf{a}_P = \ddot{s}\hat{\mathbf{e}}_t + \frac{\dot{s}^2}{\rho}\hat{\mathbf{e}}_n$$

$$= \mathbf{0} + \frac{\dot{s}^2}{\rho}\hat{\mathbf{e}}_n \qquad \text{(since } \dot{s} = \text{constant)}$$

$$= \frac{0.04}{11.2}(-0.894\,\hat{\mathbf{i}} + 0.448\,\hat{\mathbf{j}})$$

$$= -0.0032\,\hat{\mathbf{i}} + 0.0016\,\hat{\mathbf{j}} \text{ m/s}^2 \qquad \text{(as before)}$$

EXAMPLE 1.23

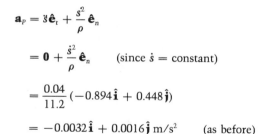

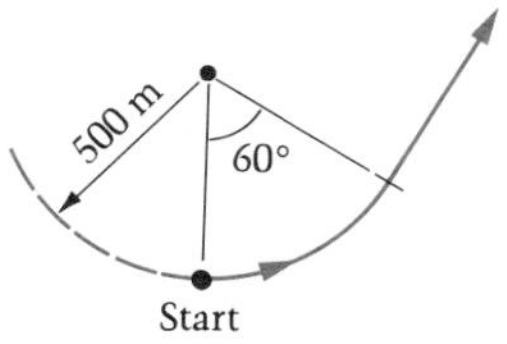

After picking up passengers, a train accelerates uniformly until its speed is 100 km/hr at $t = 300$ s. If during this time it is traveling on a planar path that approximates the circle and line shown in the figure, find the magnitude of the train's acceleration when (a) $t = 1$ min and (b) $t = 2$ min.

(Continued)

SOLUTION

First we determine the tangential component of the train's acceleration:*

$$\ddot{s} = \text{constant} = k \qquad \text{(This is the meaning of } \textit{uniform}.)$$

$$\dot{s} = kt + C_1 = kt \qquad (\text{since } \dot{s} = 0 \text{ at } t = 0)$$

$$s = \frac{kt^2}{2} + C_2 = \frac{kt^2}{2} \qquad (\text{since we may let } s = 0 \text{ at } t = 0)$$

At $t = 300$ s we have

$$\dot{s} = 100 \frac{\text{km}}{\text{hr}} \frac{1000 \text{ m}}{1 \text{ km}} \frac{1 \text{ hr}}{3600 \text{ s}} = 27.8 \text{ m/s}$$

Thus

$$k = \frac{27.8}{300} = 0.0927 \text{ m/s}^2$$

Substituting this acceleration value, we obtain the speed and position of the train at 1 min:

$$\dot{s} = kt = (0.0927)(60) = 5.56 \text{ m/s}$$

and

$$s = 0.0927 \frac{60^2}{2} = 167 \text{ m}$$

Note that $2\pi r = 2\pi(500) = 3140$ m, so that the train has covered only $(167/3140)360 = 19.1°$ of the circle and has therefore not yet reached the straight-line part of its path when $t = 60$ s. Therefore

$$|\mathbf{a}| = \sqrt{\ddot{s}^2 + \left(\frac{\dot{s}^2}{\rho}\right)^2} = \sqrt{(0.0927)^2 + \frac{5.56^4}{500^2}} \text{ m/s}^2$$

$$= 0.111 \text{ m/s}^2 \text{ at } t = 60 \text{ s}$$

At $t = 2$ min $= 120$ s, the arclength traversed by the train is

$$s = \frac{kt^2}{2} = \frac{0.0927(120^2)}{2} = 667 \text{ m}$$

corresponding to an angle of $(667/3140)360 = 76.5°$. Therefore since the train enters the straight-line portion of the path when $\theta = 60°$, the acceleration at $t = 120$ s has *no* normal component and

$$|\mathbf{a}| = \ddot{s} = 0.0927 \text{ m/s}^2 \text{ at } t = 120 \text{ s}$$

*We are treating the train as a particle in this example.

EXAMPLE 1.24

In Example 1.11 find the following for point P at $t = 1$ sec: tangential and normal components of acceleration, radius of curvature, and the principal unit normal.

SOLUTION

We obtained

$$\mathbf{r}_{OP} = 2t\hat{\mathbf{i}} + t^3\hat{\mathbf{j}} + 3t^2\hat{\mathbf{k}} \text{ ft}$$

$$\mathbf{v}_P = 2\hat{\mathbf{i}} + 3t^2\hat{\mathbf{j}} + 6t^2\hat{\mathbf{k}} \text{ ft/sec}$$

$$\mathbf{a}_P = 6t\hat{\mathbf{j}} + 6\hat{\mathbf{k}} \text{ ft/sec}^2$$

If we write the velocity $\mathbf{v}_P$ as a magnitude times a unit vector, we can determine $\dot{s}$ and $\hat{\mathbf{e}}_t$ for P:

$$\mathbf{v}_P = \sqrt{4 + 9t^4 + 36t^2}\left(\frac{2\hat{\mathbf{i}} + 3t^2\hat{\mathbf{j}} + 6t\hat{\mathbf{k}}}{\sqrt{4 + 9t^4 + 36t^2}}\right) \text{ ft/sec}$$

$$= \dot{s}\hat{\mathbf{e}}_t$$

where we note that $\dot{s} > 0$ since we are choosing the direction of increasing s to be that of the velocity.

Let us find the tangential and normal components of the acceleration of P at $t = 1$ sec:

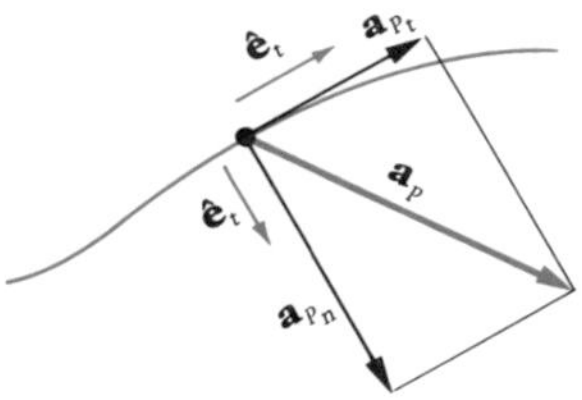

$$\mathbf{v}_P\big|_{t=1} = 7\left(\frac{2\hat{\mathbf{i}} + 3\hat{\mathbf{j}} + 6\hat{\mathbf{k}}}{7}\right) = |\mathbf{v}_P|\,\hat{\mathbf{e}}_t \text{ ft/sec}$$

Now that we have $\hat{\mathbf{e}}_t$, we can use it to split the acceleration $\mathbf{a}_P = 6\hat{\mathbf{j}} + 6\hat{\mathbf{k}}$ ft/sec^2 (at $t = 1$) into its tangential and normal components (see the figure). The tangential component of $\mathbf{a}_P$ (that is, the component parallel to $\hat{\mathbf{e}}_t$) is seen from the figure to be the dot product of $\mathbf{a}_P$ with $\hat{\mathbf{e}}_t$:

$$a_{P_t}\big|_{t=1} = \mathbf{a}_P\big|_{t=1} \cdot \hat{\mathbf{e}}_t\big|_{t=1}$$

$$= 6\sqrt{2}\left(\frac{\hat{\mathbf{j}} + \hat{\mathbf{k}}}{\sqrt{2}}\right) \cdot \left(\frac{2\hat{\mathbf{i}} + 3\hat{\mathbf{j}} + 6\hat{\mathbf{k}}}{7}\right)$$

$$= \frac{6}{7}(3 + 6) = \frac{54}{7} \text{ ft/sec}^2$$

Next we obtain the normal acceleration component by vectorially subtracting the component $\mathbf{a}_t$ $(= (54/7)\hat{\mathbf{e}}_t)$ from the total acceleration $\mathbf{a}_P$. That is, since

$$\mathbf{a}_P = \mathbf{a}_{P_t} + \mathbf{a}_{P_n}$$

(Continued)

we obtain

$$a_{P_n}\big|_{t=1} = \left|\mathbf{a}_P\big|_{t=1} - \mathbf{a}_{P_t}\big|_{t=1}\right|$$

$$= \left|6\hat{\mathbf{j}} + 6\hat{\mathbf{k}} - \frac{54}{7}\left(\frac{2\hat{\mathbf{i}} + 3\hat{\mathbf{j}} + 6\hat{\mathbf{k}}}{7}\right)\right| \text{ ft/sec}^2$$

$$= \left|\frac{-108\hat{\mathbf{i}} + 132\hat{\mathbf{j}} - 30\hat{\mathbf{k}}}{49}\right| = 3.53 \text{ ft/sec}^2$$

And since $a_{P_n} = \dot{s}^2/\rho = |\mathbf{v}_P|^2/\rho$, we obtain the radius of curvature:

$$\rho = \frac{7^2}{3.53} = 13.9 \text{ ft}$$

The unit vector $\hat{\mathbf{e}}_n$ follows from

$$\hat{\mathbf{e}}_n = \frac{\mathbf{a}_{P_n}}{a_{P_n}} = \frac{\mathbf{a}_P - \mathbf{a}_t}{a_{P_n}} = \frac{-108\hat{\mathbf{i}} + 132\hat{\mathbf{j}} - 30\hat{\mathbf{k}}}{49(3.53)}$$

$$= -0.624\hat{\mathbf{i}} + 0.763\hat{\mathbf{j}} - 0.173\hat{\mathbf{k}}$$

It is instructive to make a direct calculation of $d|\mathbf{v}_P|/dt$ since we here know $|\mathbf{v}_P|$ as a function of time:

$$|\mathbf{v}_P| = \sqrt{4 + 9t^4 + 36t^2} \text{ ft/sec}$$

$$\frac{d|\mathbf{v}_P|}{dt} = \frac{\frac{1}{2}(36t^3 + 72t)}{\sqrt{4 + 9t^4 + 36t^2}} \text{ ft/sec}^2$$

Thus

$$\left.\frac{d|\mathbf{v}_P|}{dt}\right|_{t=1} = \frac{\frac{1}{2}(108)}{\sqrt{49}} = \frac{54}{7} \text{ ft/sec}^2$$

which is, of course, the result we have already obtained by investigating the components of the acceleration vector.

Question 1.6 How would you find the position vector from the origin O to the center of curvature at $t = 1$ sec?

In closing this section, we remark that tangential and normal components of velocity and acceleration will be very useful to us later when we happen to know the path of a point (the center C of a wheel rolling on a curved track, for instance). We can then use Equations (1.41) and (1.47) to express $\mathbf{v}_P$ and $\mathbf{a}_P$.

PROBLEMS / Section 1.7

1.124 The point in Figure P1.124 moves on a circular path with its arclength given by $s = 2t^3 - 3t^2$ m.

a. What are its position, velocity, and acceleration vectors at $t = 3$ s?

b. How much distance has been traversed?

1.125 A particle moves in such a manner that its acceleration components in the x, y, and z directions are respectively $20t^3$, 0, and $4 \sin 2t$ m/s². The point's velocity at $t = 0$, when it is located at the origin, is $\mathbf{v} = \hat{\mathbf{i}} + 3\hat{\mathbf{j}} + (2 \cos 2)\hat{\mathbf{k}}$ m/s. Determine the particle's position and speed at $t = 1.7$ s, and find the radius of curvature of its path when $t = 5.6$ s.

1.126 At a particular instant a point has a velocity $3\hat{\mathbf{i}} + 4\hat{\mathbf{j}}$ in./sec and an acceleration $\hat{\mathbf{i}} + \hat{\mathbf{j}} - \hat{\mathbf{k}}$ in./sec². At this instant find: (a) the principal unit normal, (b) the curvature of the path, and (c) the time rate of change of the point's speed.

1.127 A point P has position vector $\mathbf{r}_{OP} = t^4\hat{\mathbf{i}} + t^3\hat{\mathbf{j}}$ meters. Find the tangential and normal components of the acceleration of P as functions of time. Use your answer to find the radius of curvature of the path of P when $t = 7$ s.

1.128 A car accelerates uniformly from rest and attains a speed of 60 mph in 1 min on the circular path shown in Figure P1.128. It then continues at 60 mph. Determine the car's location at the end of the 1-min interval, and find the time required for the car to reach the highest point T. Also find the car's acceleration when it reaches point T.

1.129 Use Example 1.22 to show that the center of curvature does not have to be on the y axis for a curve symmetric about y. *Hint:* Use $f = 1$ and $x = 2$, find ρ, and compare with the distance from (2, 1) to the y axis along the normal to the tangent at this point.

1.130 There is another formula for the radius of curvature ρ from the calculus; this one is in terms of a parameter such as time t:

$$\frac{1}{\rho} = \left| \frac{\dot{x}\ddot{y} - \dot{y}\ddot{x}}{(\dot{x}^2 + \dot{y}^2)^{3/2}} \right|$$

If the x and y coordinates of a moving point P are given by

$$x = 6 \sinh 0.02t \text{ ft} \qquad y = t^3 - 10t^2 + 7.2 \text{ ft}$$

find the radius of curvature of P's path at $t = 8$ sec.

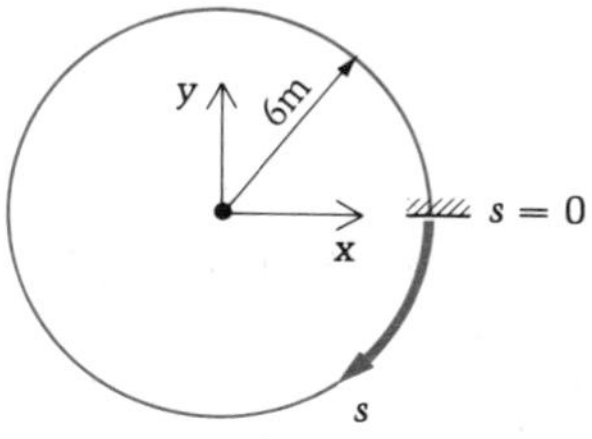

Figure P1.124

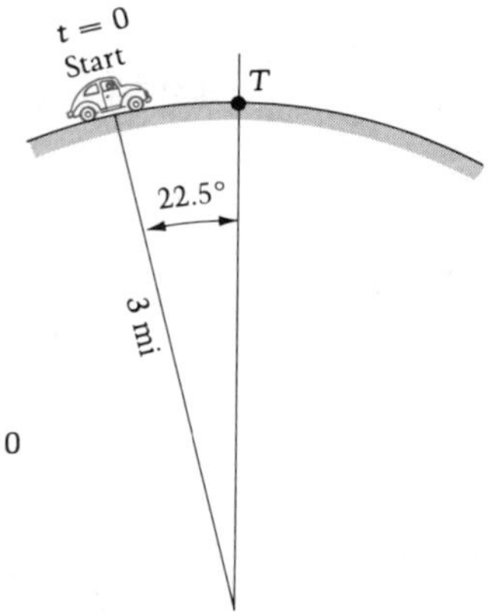

Figure P1.128

1.131 A point D moves along a curve in space with a speed given by $\dot{s} = 6t$ m/s, where t is measured from zero when D is at the arclength origin $s = 0$. If at a certain time t^* the acceleration magnitude of D is 12 m/s² and the radius of curvature is 3 m, determine t^*.

1.132 In Problem 1.67 find the tangential and normal components of the car's acceleration when $x = 2500$ ft. Check your result by also computing $\ddot{x}$ and $\ddot{y}$ there and showing that $\sqrt{\ddot{x}^2 + \ddot{y}^2} = \sqrt{a_t^2 + a_n^2}$.

1.133 Particle P moves on a circle (Figure P1.133) with an arclength given as a function of time as shown. Find the time(s) and the angle(s) θ when the tangential and normal acceleration components are equal.

1.134 In Example 1.23 calculate the distance the train has traveled on the straight-line portion of its path when its speed has reached 100 km/hr.

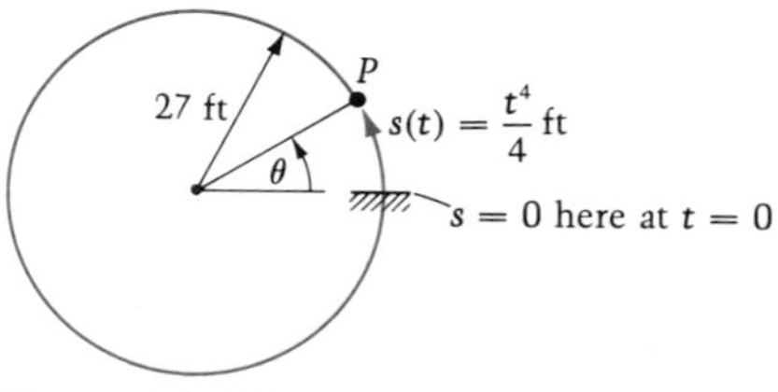

Figure P1.133

1.135 The position vector of a point is given as a function of time by $\mathbf{r}(t) = t^2\hat{\mathbf{i}} - t\hat{\mathbf{j}} + t^3\hat{\mathbf{k}}$ ft. Find the tangential and normal components of acceleration at $t = 1$ sec, and determine the radius of curvature at that time.

1.136 A particle P has the x, y, and z coordinates $(3t, 0, 4 \ln t)$ meters as functions of time. What is the vector from the origin to the center of curvature of the path of P at $t = 1$ s?

1.137 A point P has position vector $\mathbf{r}_{OP} = t^2\hat{\mathbf{i}} + t^3\hat{\mathbf{j}} - t^6\hat{\mathbf{k}}$ meters. Find the vector from the origin to the center of curvature of the path of P at $t = 1$ s. Find $d|\mathbf{v}_P|/dt$ at the same instant.

1.138 A point P moves on the 'Spiral of Archimedes' at constant speed 2 m/s. (See Figure P1.138.) The equation of the spiral is $r = 3\theta$. Find the acceleration of P when $\theta = 270°$.

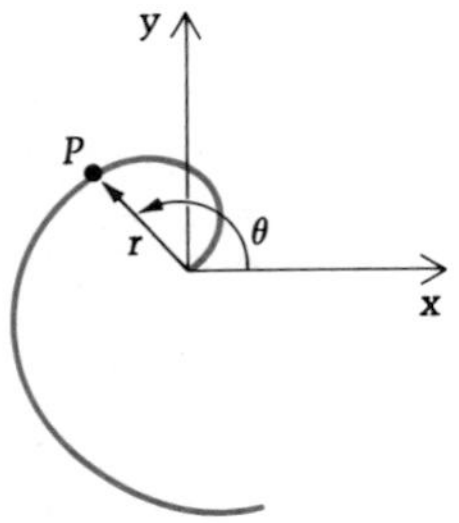

Figure P1.138

1.139 In the preceding problem, find the center of curvature of the path of P when $\theta = 270°$.

1.140 Find the radius of curvature of the 'Witch of Agnesi' curve at $x = 0$. (See Figure P1.140.)

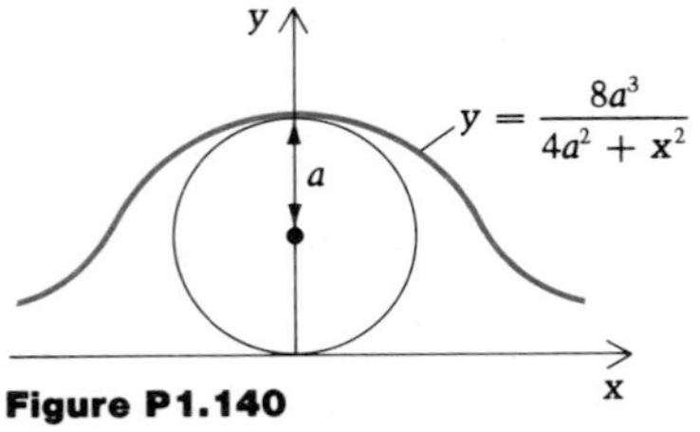

Figure P1.140

1.141 A point P moves from left to right along the curve defined in the preceding problem with a constant x component $(\dot{x}_0)$ of velocity. Find the acceleration of P when it reaches the point $(x, y) = (0, 2a)$.

1.142 There is a third unit vector associated with the motion of a point on its path. It is called the binormal $\hat{\mathbf{e}}_b$ and forms an orthogonal moving trihedral with $\hat{\mathbf{e}}_t$ and $\hat{\mathbf{e}}_n$, defined by $\hat{\mathbf{e}}_b = \hat{\mathbf{e}}_t \times \hat{\mathbf{e}}_n$.

a. Differentiate $\hat{\mathbf{e}}_b \cdot \hat{\mathbf{e}}_t = 0$ with respect to s. Then, using $d\hat{\mathbf{e}}_t/ds = \hat{\mathbf{e}}_n/\rho$,† prove that $d\hat{\mathbf{e}}_b/ds \cdot \hat{\mathbf{e}}_t = 0$ and therefore that $d\hat{\mathbf{e}}_b/ds$ is parallel to $\hat{\mathbf{e}}_n$.

b. Using part (a), let $d\hat{\mathbf{e}}_b/ds = \tau\hat{\mathbf{e}}_n$.† ($\tau$ is called the torsion of the path or curve.) Then differentiate $\hat{\mathbf{e}}_n = \hat{\mathbf{e}}_b \times \hat{\mathbf{e}}_t$ with respect to s and prove that

$$\frac{d\hat{\mathbf{e}}_n}{ds} = -\left(\frac{1}{\rho}\hat{\mathbf{e}}_t + \tau\hat{\mathbf{e}}_b\right)^{\dagger}$$

The three equations marked with daggers give the derivatives of the three unit vectors associated with a space curve and are called the Serret-Frenet formulas.

1.143 The derivative of acceleration is called the *jerk* and is studied in the dynamics of vehicle impact and in the kinematics of mechanisms involving cams and followers. Show that the jerk of a point has the following form in terms of its intrinsic components:

$$\mathbf{J}_P = \dot{\mathbf{a}}_P = \left(\dddot{s} - \frac{\dot{s}^3}{\rho^2}\right)\hat{\mathbf{e}}_t + \left(\frac{3\dot{s}\ddot{s}}{\rho} - \frac{\dot{\rho}\dot{s}^2}{\rho^2}\right)\hat{\mathbf{e}}_n - \frac{\dot{s}^3}{\rho}\tau\hat{\mathbf{e}}_b$$

1.144 Show by expressing the velocity and acceleration in tangential and normal components that

$$|\mathbf{v} \times \mathbf{a}| = va_n = \frac{|\dot{s}^3|}{\rho}$$

so that

$$\frac{1}{\rho} = \frac{|\mathbf{v} \times \mathbf{a}|}{|\mathbf{v}|^3} \quad \text{or} \quad \frac{|\mathbf{v} \times \mathbf{a}|}{v^3}$$

1.145 In Problem 1.92 determine the expression for $\dot{s}(t)$. Integrate, for a motion beginning at $t = 0$ at $(x, y) = (20, 0)$ m, and obtain $s(t)$. Evaluate the arclength at $t = 2$ s and show that the result, as it should be, is the circumference of the circle on which P travels.

1.146 Show that if a particle's position is defined by $x = 5 \cos t$ and $y = 5 \sin t$ ft, then the path of the particle is a circle. Find the equation for the arclength s as a function of time.

1.147 In Problem 1.83 find the arclength s as a function of time.

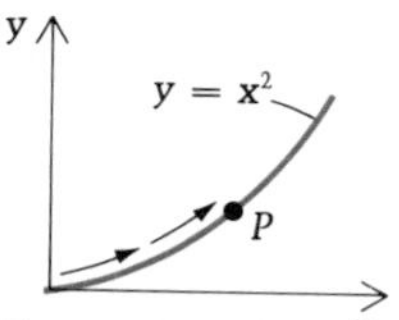

Figure P1.148

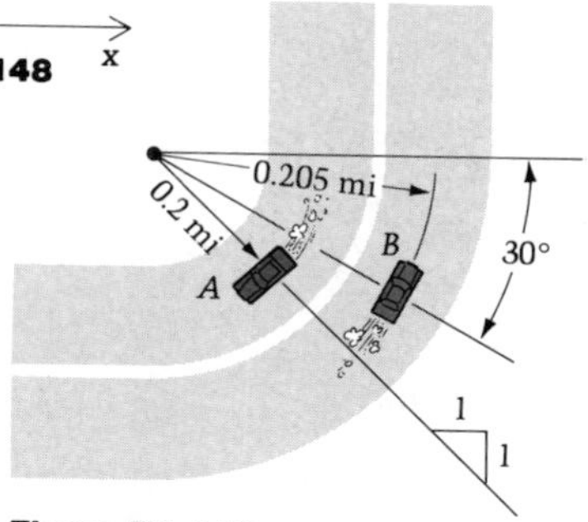

Figure P1.149

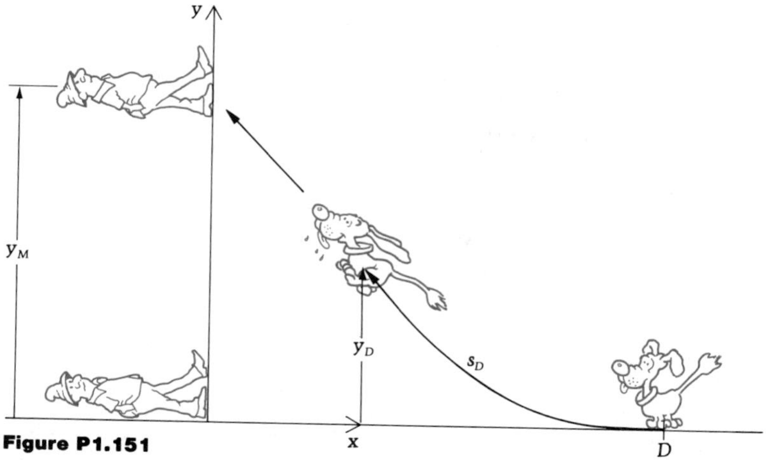

Figure P1.151

1.148 A particle P starts from rest at the origin and moves along the parabola shown in Figure P1.148. Its speed is given by $\dot{s} = 3s + 2$, where $\dot{s}$ is in meters per second when s is in meters. Determine the velocity of P when its x coordinate is 5 m. Also determine the elapsed time. *Hint:* $ds = \sqrt{dx^2 + dy^2} = \sqrt{1 + y'^2}\,dx$, so that $s = \int \sqrt{1 + y'^2}\,dx$. Substitute y' and use a table of integrals to get $s(x)$.

1.149 Find the difference between the velocities (and also the accelerations) of cars A and B in Figure P1.149 if, at the instant shown,

$$\dot{s}_A = 30 \text{ mph} \qquad \ddot{s}_A = 500 \text{ mph}^2$$

$$\dot{s}_B = 50 \text{ mph} \qquad \ddot{s}_B = -1000 \text{ mph}^2$$

1.150 A particle moves on the curve $(x - a)^2 + y^2 = a^2$, where a is a constant distance in meters. The first and second time derivatives of the arclength s are related by

$$\dot{s} = Ka\ddot{s}$$

in which the constant K has the value 1 second/meter. The distance s is measured counterclockwise on the curve from the point $(2a, 0)$ meters. When $t = 0$, the speed of the particle is $\dot{s} = 1$ meter/second and $s = a$. Find the normal and tangential components of the acceleration at time $t = 0$. Show these components on a diagram.

1.151 The following "pursuit" problem is very difficult, yet it illustrates exceptionally well the idea that the velocity vector is tangent to the path. Thus we include it along with a set of steps for the courageous student who wishes to "pursue" it. A dog begins at the point $(x, y) = (D, O)$ and runs toward his master at constant speed $2V_0$. (See Figure P1.151.) The dog's velocity direction is always toward his master, who starts at the same time at the origin and moves along the positive y direction at speed V_0. Find the man's position when his dog overtakes him, and determine how much time has elapsed. *Hints:* The man's y coordinate is y_M (which of course is V_0t). Show that:

1. $\dfrac{-dy_D}{dx} = \dfrac{y_M - y_D}{x}$, where (x, y_D) represent the dog's coordinates at any time.
2. $2V_0 = \dfrac{ds_D}{dt} = \dfrac{\sqrt{dx^2 + dy_D^2}}{dt} = \dfrac{-dx\sqrt{1 + y_D'^2}}{dt}$.
3. $V_0 = dy_M/dt$.
4. From dividing and rearranging steps 2 and 3, we get $2y_M' = -\sqrt{1 + y_D'^2}\ (y_M' = dy_M/dx)$.
5. From step 1, we have $y_M' = -xy_D''$.
6. From steps 4 and 5, we have $2xy_D'' = \sqrt{1 + y_D'^2}$.
7. Letting $y_D' = p$, from step 6 we get $dp/\sqrt{1 + p^2} = dx/2x$.
8. By integrating step 7 with a table of integrals, we get $$\ln(p + \sqrt{1 + p^2}) = \tfrac{1}{2}\ln(x) - \tfrac{1}{2}\ln(C_1) = \ln\left(\frac{x}{C_1}\right)^{1/2}$$
9. From step 8, we get $p + \sqrt{1 + p^2} = (x/C_1)^{1/2}$.
10. From step 9, we get $p - \sqrt{x/C_1} = -\sqrt{1 + p^2}$.
11. From step 10, squaring both sides and solving for p, $$2p = \sqrt{\frac{x}{C_1}} - \sqrt{\frac{C_1}{x}} = 2y_D'$$
12. $y_D' = 0$ when $x = D$ (initial condition).

13. From steps 11 and 12 we have $C_1 = D$, so that $2y'_D = \sqrt{x/D} - \sqrt{D/x}$.

14. Integrating step 13, we get

$$y_D = x^{3/2}/3\sqrt{D} - \sqrt{xD} + C_2.$$

15. $y_D = 0$ when $x = D$ (initial condition).

16. From steps 14 and 15 we have $C_2 = 2D/3$, so that

$$y_D = \frac{x^{3/2}}{3\sqrt{D}} - \sqrt{xD} + \tfrac{2}{3}D$$

17. $y_M = V_0 t$.

18. Finally, write the conditions relating to y_M, y_D, and x when the dog overtakes his master, and wrap it up!

Answers to Questions / Chapter 1

Q1.1 (a) Yes; we could simply define a frame in which P is fixed, and it would then have $\mathbf{v}_P = \mathbf{0} = \mathbf{a}_P$. (b) No; letting O' be a second origin in $\mathcal{F}$ and differentiating the relationship $\mathbf{r}_{OP} = \mathbf{r}_{OO'} + \mathbf{r}_{O'P}$ in $\mathcal{F}$ shows that: $\mathbf{v}_P$ (with origin O) $= \mathbf{v}_P$ (with origin O'). The derivative of $\mathbf{r}_{OO'}$ in $\mathcal{F}$ is of course zero! See Problem 1.17.

Q1.2 If the surface is flat, then a_D vanishes. If it is cubic, then a_D is linear in $\mathbf{x}$.

Q1.3 From Figure 1.5, we see that $\hat{\mathbf{e}}_\theta$ is perpendicular to $\mathbf{r}_{OP}$. Note, however, that implicit in the writing and use of Equation (1.30) is the polar angle θ.

Q1.4 The magnitude of velocity is correctly written as $|\mathbf{v}_P|$ or v_P or $|\dot{\mathbf{r}}_{OP}|$. The latter is not the same as $\dot{r}$, which is but one of its three components. As for $\ddot{r}$, this is but *a part* of one of three components of $\mathbf{a}_P$; the magnitude of the acceleration is $|\mathbf{a}_P|$ or a_P or $|\ddot{\mathbf{r}}_{OP}|$, but not $\ddot{r}$.

Q1.5 When $r = 0$, we have $\dot{z} = 0 = \dot{\theta}$. Thus $\dot{r}$, the radial component of $\mathbf{v}_P$, is maximum there at the value 0.7, which is the constant speed. (Note that $\dot{r}$ decreases continuously toward zero from there.)

Q1.6 If we call the center of curvature C, it is simply $\mathbf{r}_{OC} = \mathbf{r}_{OP} + \mathbf{r}_{PC} = \mathbf{r}_{OP} + \rho\hat{\mathbf{e}}_n$, with everything evaluated at the time of interest (in this case $t = 1$).

Review Questions / Chapter 1

True or False?

1. The velocity $\mathbf{v}_P$ of a point P is always tangent to its path.
2. $\mathbf{v}_P$ depends on the reference frame chosen to express the position of P.
3. $\mathbf{v}_P$ depends on the origin chosen in the reference frame.
4. The magnitude and direction in space of $\mathbf{v}_P$ depend on the choice of coordinates used to locate the point relative to the reference frame.
5. $\mathbf{a}_P$ always has a nonvanishing component normal to the path.
6. For any point P, $a_P = \sqrt{\ddot{s}^2 + \dot{s}^4/\rho^2}$.
7. A point can have $\ddot{r} = 0$ but still have a nonvanishing radial component of acceleration.
8. If a ball on a string is being whirled around in a horizontal circle at constant speed, its center has zero acceleration.

9. Studying the kinematics of a particle results in the same equations as studying the kinematics of a point.

10. In our study of the kinematics of a point, the following terms have not appeared in any of the equations: mass, force, moments, gravity, momentum, moment of momentum, inertia, or Newtonian (inertial) frames.

11. The acceleration vector of P, at the indicated point on the path shown in the figure, can lie in any of the four quadrants.

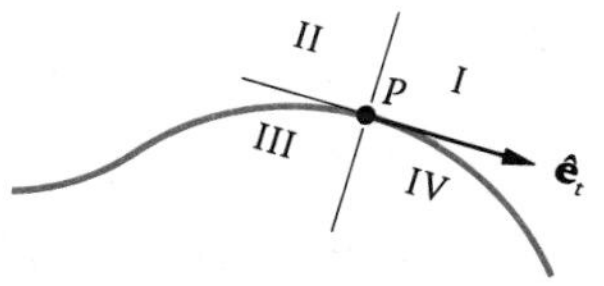

12. A particle moving in a plane, with constant values of $\dot{r}$ and $\dot{\theta}$, will at all times have zero acceleration.

Answers: T, T, F, F, F, T, T, F, T, T, F, F.

CHAPTER

2 Kinetics of Particles and of Mass Centers of Bodies

chapter outline ►

2.1 Introduction/Newton's Second Law and Euler's First Law

In this chapter we begin to consider the manner in which the motion of a body is related to external mechanical actions (forces and couples). Our kinematics notions of space and time must now be augmented by those of mass and force, which, like space and time, are **primitives** of the subject of mechanics. We simply have to agree in advance that some measures of quantity of matter (mass) and mechanical action (force) are basic ingredients in any attempt to analyze the motion of a body. We assume that the reader has a working knowledge, probably from a study of statics, of the characteristics of forces and moments and their vector descriptions. We use the term **body** to denote some material of fixed identity; we could think of a specific set of atoms, although the model we shall employ is based upon viewing material on a spatial scale such that mass is perceived to be distributed continuously. A body need not be rigid or even a solid, but, since our subject is classical dynamics (no relativistic effects), a body necessarily has constant mass.

The usual starting point for relating the external forces on a body to its motion is Newton's laws. These were proposed in the Englishman Isaac Newton's famous work the *Principia*, published in 1687, and are commonly expressed today as:

1. If the resultant force $\mathbf{F}$ on a particle is zero, then the particle has constant velocity.
2. If $\mathbf{F} \neq \mathbf{0}$, then $\mathbf{F}$ is proportional to the time derivative of the particle's momentum $m\mathbf{v}$ (product of mass and velocity).
3. The interaction of two particles is through a pair of self-equilibrating forces. That is, they have the same magnitude, opposite directions, and a common line of action.

Clearly the first law may be regarded a special case of the second, and one must add an assumption about the frame of reference, since a point may have its velocity constant in one frame of reference and varying in another. Frames of reference in which these laws are valid are variously called Newtonian, Galilean, or **inertial.** Furthermore, the constant of proportionality in the second law can be made unity by appropriate choices of units so that the law becomes

$$\mathbf{F} = \frac{d}{dt}(m\mathbf{v}) = m\mathbf{a}$$

where $\mathbf{a}$ is the acceleration of the particle.

As we mentioned briefly in Chapter 1, a particle is a piece of material sufficiently small that we need not make distinctions among its material points with respect to locations (or to velocities or accelerations). We also noted that this definition allows, for *some* purposes, a truck or a space vehicle or even a planet to be adequately modeled as a particle. In the *Principia,* Newton used heavenly bodies as the particles in his examples and treated them as moving points subject only to universal gravitation and their own inertia.* Newton did not extend his work to problems for which it is necessary to account for the actual sizes of the bodies and how their masses are distributed. It was to be over 60 years before the Swiss mathematician Leonhard Euler presented the first of the two principles that have come to be called **Euler's laws.** It is simply

$$\mathbf{F}_r = \frac{d\mathbf{L}}{dt} \tag{2.1}$$

in which $\mathbf{F}_r$ is the resultant of all the external forces acting upon the body, $\mathbf{L}$ is the momentum of the body, and the derivative is taken in an inertial frame.† This law, and a second one to follow in Chapter 4, is seen to avoid unnecessary questions of judgment about what is and what is not a particle. Euler's laws are valid for bodies (deformable, rigid, or even nonsolid) of any and all sizes.

We should mention the traditional way of obtaining governing equations of motion for a body. It is to imagine the body to be made up of a set of particles, invoke Newton's second law for each of the particles, and then add the corresponding equations, eliminating forces of interaction among the particles by the third law. There result two equations (analogous to the equilibrium equations of statics) relating both the sum of external forces and the sum of moments of external forces to the motion of the body.

Since nothing is lost by simply postulating Euler's laws at the outset, this will be our approach. In doing so, we shall more rapidly obtain the universally useful forms of the equations of motion. Moreover, there will be no sacrifice in logic, since we have the choice of postulating either the two vector equations of Newton (second and third laws) or the two vector equations constituting Euler's laws. (Appendix B demonstrates that Newton's third law follows from Euler's laws.)

We see from the statement of the first law (Equation 2.1) that we need an understanding of the vector $\mathbf{L}$ (the body's momentum) before proceeding. Thus we shall spend one section defining and studying it, and this discussion will take us to the very heart of the concept of the center of mass.

*See C. Truesdell, *Essays in the History of Mechanics* (Berlin: Springer-Verlag, 1968).

†Most experimental data correlate closely with the laws of motion of Newton and Euler. We should note, however, that there are dynamics problems in which these laws fall short. They are inadequate to explain the advance of the perihelion of the planet Mercury, for example. This was one of the problems used to test the general relativity theory of Einstein.

2.2 Momentum and the Center of Mass

The **momentum** $\mathbf{L}$ of a body $\mathcal{B}$ is defined as

$$\mathbf{L} = \int_{\mathcal{B}} \mathbf{v}\, dm \tag{2.2}$$

where $\mathbf{v}$ and dm are, respectively, the velocity vector and differential mass element associated with a typical (or generic) point in $\mathcal{B}$. Thus we are defining the momentum of a body as the sum of the momenta of all the elemental particles in the body. Note that $\mathbf{L}$ is intrinsically associated with the frame of reference in which $\mathbf{v}$ is the velocity field of the mass elements of $\mathcal{B}$. (The frame of reference is the rigid frame containing the origin from which the position vectors emanate; the velocity field of a body $\mathcal{B}$ is the set of velocity vectors of all its points.) Throughout this chapter a common frame of reference will be understood and henceforth will not be specifically denoted.

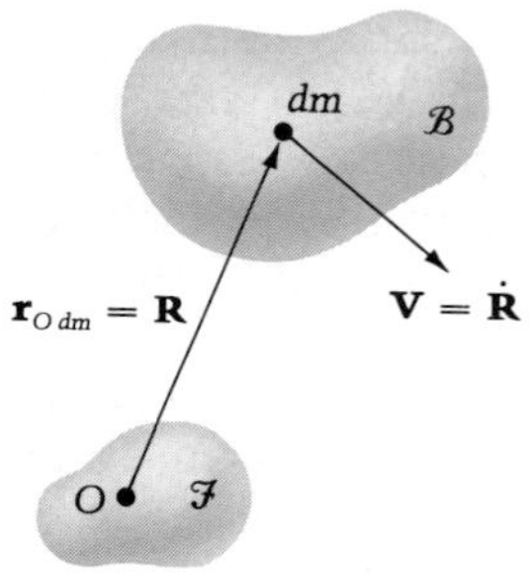

Figure 2.1

> **Question 2.1** What is the dimension of momentum?

If $\mathbf{R}$ is a position vector for the generic point (Figure 2.1), then

$$\mathbf{L} = \int_{\mathcal{B}} \frac{d\mathbf{R}}{dt}\, dm \tag{2.3}$$

or, since the integral is understood to be carried out over the body, which we recall is a well-defined quantity of material, independent of time,

$$\mathbf{L} = \frac{d}{dt}\int_{\mathcal{B}} \mathbf{R}\, dm \tag{2.4}$$

The integral $\int_{\mathcal{B}} \mathbf{R}\, dm$ has some interesting properties, and it is by way of this integral that the center of mass of $\mathcal{B}$ is defined. In particular, if m is the mass of $\mathcal{B}$ then

$$\frac{1}{m}\int_{\mathcal{B}} \mathbf{R}\, dm = \mathbf{r}_{OC} \tag{2.5}$$

defines the directed line segment $\mathbf{r}_{OC}$, which starts at the common origin (O) of the $\mathbf{R}$'s and ends at the mass center. In one of the problems at the end of the section you will be asked to show that the special point so identified is independent of the choice of the origin and therefore a unique point associated with $\mathcal{B}$.

In the remainder of the book, we shall usually designate the location of the mass center of a body by the letter C. We note that the centroid of any scalar quantity Q (such as arc length, area, volume) is defined in the same manner—that is, by the vector $(1/Q) \int \mathbf{R}\, dQ$. We now consider several examples in which we locate the mass center.

EXAMPLE 2.1

Using the definition of center of mass, find the mass center of a uniform right circular cone of density ρ.

SOLUTION

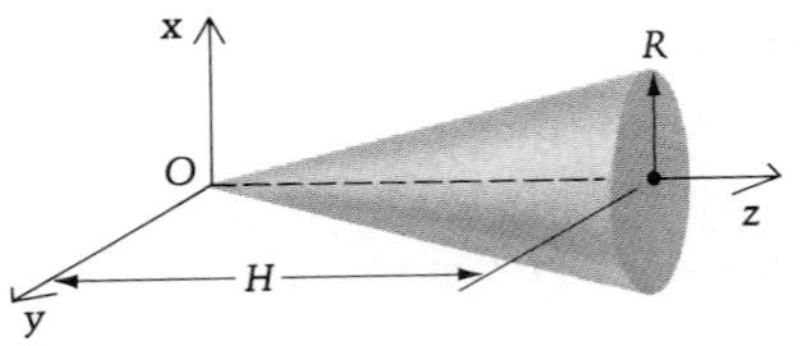

It is convenient to set up rectangular coordinates with their origin at the apex of the cone and to let one axis (z) coincide with the cone's axis, as shown in the accompanying diagram.

Then

$$m\mathbf{r}_{OC} = \int (x\hat{\mathbf{i}} + y\hat{\mathbf{j}} + z\hat{\mathbf{k}})\, dm$$

$$m(x_C\hat{\mathbf{i}} + y_C\hat{\mathbf{j}} + z_C\hat{\mathbf{k}}) = \hat{\mathbf{i}} \int x\, dm + \hat{\mathbf{j}} \int y\, dm + \hat{\mathbf{k}} \int z\, dm$$

The integrations are most easily accomplished by using cylindrical coordinates (r, θ, z); that is,

$$x = r\cos\theta \qquad \text{and} \qquad y = r\sin\theta$$

Thus

$$dm = \rho r\, dr\, d\theta\, dz \qquad (= \text{density} \times \text{differential volume})$$

and

$$mx_C = \int_0^H \int_0^{zR/H} \int_0^{2\pi} \rho r^2 \cos\theta\, d\theta\, dr\, dz = 0$$

Similarly, $y_C = 0$; that is, the mass center lies on the axis of the cone. The reader should note that this same result arises for a nonuniform cone as long as the density is independent of θ. That is, a body with an axially symmetric distribution of mass has its mass center on the symmetry axis.

The value of z_C is next determined from

$$mz_C = \int_0^H \int_0^{zR/H} \int_0^{2\pi} \rho z r\, d\theta\, dr\, dz = 2\pi\rho \int_0^H \left[\frac{r^2}{2}\right]_0^{zR/H} z\, dz$$

$$= \pi\rho \frac{R^2}{H^2} \int_0^H z^3\, dz = \pi\rho \frac{R^2 H^2}{4}$$

(Continued)

But

$$m = \int_0^H \int_0^{zR/H} \int_0^{2\pi} \rho r \, d\theta \, dr \, dz = \pi\rho \frac{R^2 H}{3}$$

Therefore

$$z_C = \frac{3}{4} H \qquad \text{and} \qquad \mathbf{r}_{OC} = \frac{3H}{4} \hat{\mathbf{k}}$$

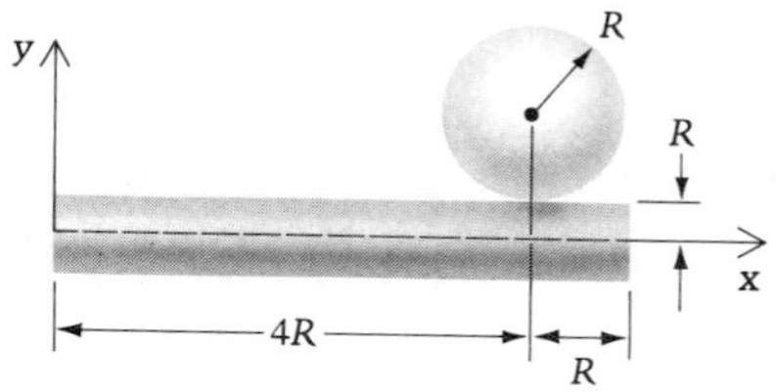

EXAMPLE 2.2

Consider a body that is a composite of a uniform sphere and a uniform cylinder, each of density ρ. Find the mass center of the body. (See the accompanying figure.)

SOLUTION

We begin by developing a useful property of mass centers. Suppose we have a body $\mathcal{B}$ divided into two parts, $\mathcal{B}_1$ and $\mathcal{B}_2$, of masses m_1 and m_2 and with centers of mass C_1 and C_2, respectively. If $m = m_1 + m_2$ is the mass of $\mathcal{B}$, whose center of mass is at C, then

$$m\mathbf{r}_{OC} = \int_{\mathcal{B}} \mathbf{R} \, dm = \int_{\mathcal{B}_1} \mathbf{R} \, dm + \int_{\mathcal{B}_2} \mathbf{R} \, dm$$

or

$$m\mathbf{r}_{OC} = m_1 \mathbf{r}_{OC_1} + m_2 \mathbf{r}_{OC_2}$$

Thus if we know the masses and mass centers of $\mathcal{B}_1$ and $\mathcal{B}_2$, we may easily compute the location of the mass center of the composite body $\mathcal{B}$.

Using the axis system shown (that is, the **x** axis coincides with the axis of the cylinder and the center of the sphere lies in the **xy** plane), we note that the coordinates of the mass center of the cylinder are $(2.5R, 0, 0)$ and those of the sphere are $(4R, 2R, 0)$.

Since the mass of the sphere is $(4/3)\pi R^3 \rho$ and the mass of the cylinder is $5R(\pi R^2)\rho = 5\pi R^3 \rho$, we have

$$\left(5\pi R^3 \rho + \frac{4}{3}\pi R^3 \rho\right) \mathbf{r}_{OC} = 5\pi R^3 \rho (2.5R\hat{\mathbf{i}}) + \frac{4}{3}\pi R^3 \rho (4R\hat{\mathbf{i}} + 2R\hat{\mathbf{j}})$$

Thus the solution is

$$\mathbf{r}_{OC} = 2.82R\hat{\mathbf{i}} + 0.421R\hat{\mathbf{j}}$$

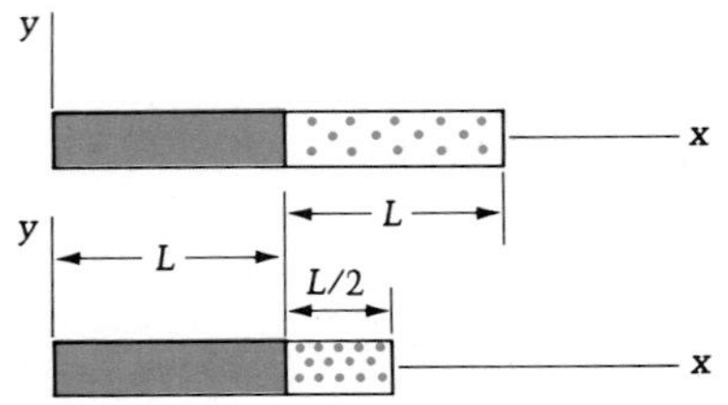

EXAMPLE 2.3

A uniform prismatic rod of density ρ and length $2L$ is deformed in such a way that the right half is uniformly compressed to length $L/2$ with no change in cross-sectional area A. (See the figure.) The left half of the rod is not altered. Letting the $\mathbf{x}$ axis be the locus of cross-sectional centroids, find the coordinates of the mass center in each configuration.

SOLUTION

In the first configuration the center-of-mass coordinates are $(L, 0, 0)$; that is, the center of mass is at the interface of the two segments. In the second configuration, however,

$$2\rho AL\mathrm{x}_C = \rho AL\left(\frac{L}{2}\right) + \rho AL\left(\frac{5}{4}L\right)$$

$$\mathrm{x}_C = \frac{L}{4} + \frac{5}{8}L = \frac{7}{8}L$$

Thus the mass center no longer lies in the interface. This example illustrates that the mass center of a deformable body does not in general coincide with the same material point in the body at different times.

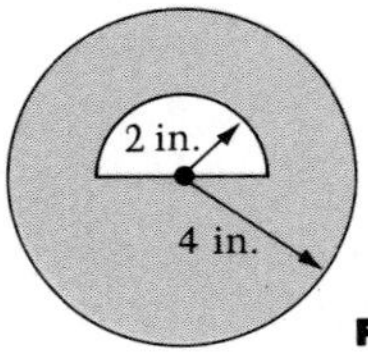

Figure 1

EXAMPLE 2.4

Find the mass center of a uniform cylinder with a smaller half-cylinder removed, having the cross section shown in Figure 1.

SOLUTION

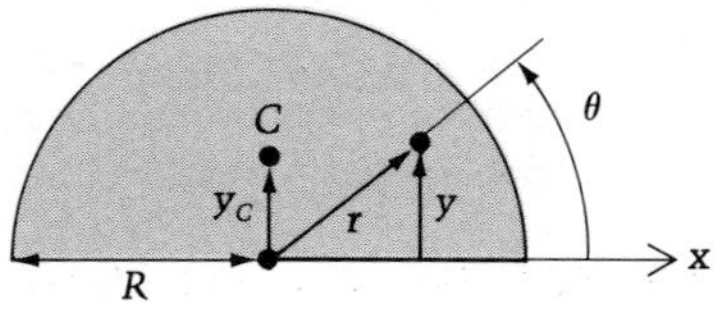

Figure 2

First we find the mass center of a half-cylinder. From Figure 2,

$$y_C = \frac{\displaystyle\int_{-L/2}^{L/2}\int_0^{\pi}\int_0^{R} (r\sin\theta)\rho\, r\, dr\, d\theta\, dz}{\rho\pi R^2 L/2}$$

$$= \frac{\rho L \displaystyle\int_0^{\pi} \left.\frac{r^3}{3}\right|_0^R \sin\theta\, d\theta}{\rho\pi R^2 L/2}$$

$$= \frac{2\rho L R^3(-\cos\theta)|_0^{\pi}}{3\rho\pi R^2 L}$$

$$= \frac{4R}{3\pi}$$

(Continued)

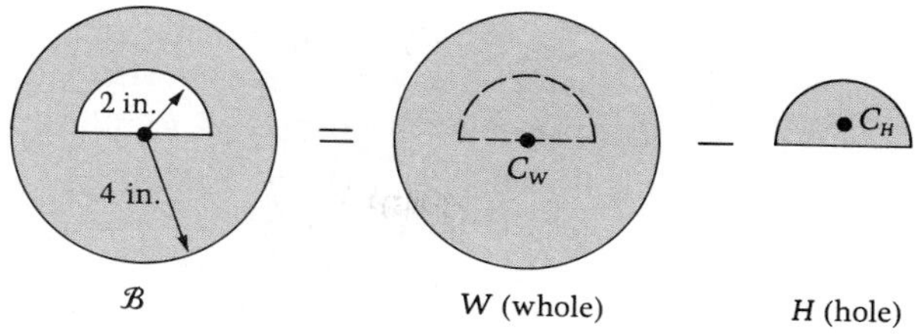

Figure 3

Now we do the body itself by using the idea of "whole" (W) minus "hole" (H). That is, we may calculate an integral by first including holes and then subtracting away their contributions, as suggested by Figure 3. The only thing that counts is that we end up having covered only the actually present mass elements exactly once. Therefore if W means the solid cylinder and H means the solid half-cylinder to be subtracted away, then

$$y_C = \frac{\int_{\mathcal{B}} y\,dm}{\int_{\mathcal{B}} dm} = \frac{\int_{W-H} y\,dm}{\int_{W-H} dm} = \frac{\int_W y\,dm - \int_H y\,dm}{\int_W dm - \int_H dm}$$

But for the individual bodies W and H, we know that

$$\int_W y\,dm = y_{C_W} m_W \qquad \text{and} \qquad \int_H y\,dm = y_{C_H} m_H$$

Thus

$$y_C = \frac{\overset{0}{\cancel{y_{C_W}}} m_W - y_{C_H} m_H}{m_W - m_H} = \frac{\rho L\left[0 - \frac{4(2)}{3\pi}\,\pi\,\frac{2^2}{2}\right]}{\rho L\left(\pi 4^2 - \pi\frac{2^2}{2}\right)}$$

$$= \frac{-8}{21\pi} = -0.121 \text{ in.}$$

Of course, $x_C = z_C = 0$ by symmetry.

Since O is a point fixed in the frame of reference, we may now write from Equations (2.4) and (2.5)

$$\mathbf{L} = \frac{d}{dt}(m\mathbf{r}_{OC}) = m\mathbf{v}_C \tag{2.6}$$

where $\mathbf{v}_C$ is the velocity of the mass center. Thus the momentum of *any* body can be reduced to the simple product of its mass times the velocity of its mass center.

Question 2.2 What happened to the $\dot{m}\mathbf{r}_{OC}$ term that results from the differentiation in Equation (2.6)?

We should note here that, for a rigid body, C coincides at every instant with a specific material point of the body or its rigid extension (for example, the center of a hollow sphere), but this is not the case for a deformable body, as we have seen in Example 2.3.

Finally, because of what we have learned about the mass center of a body, Euler's first law (Equation 2.1) is seen to take the various forms:*

$$\mathbf{F}_r = \dot{\mathbf{L}} = \frac{d\mathbf{L}}{dt}$$

$$= \frac{d}{dt}\int_{\mathcal{B}} \mathbf{v}\, dm \tag{2.7}$$

$$= \frac{d}{dt}(m\mathbf{v}_C) \tag{2.8}$$

$$\mathbf{F}_r = m\mathbf{a}_C \tag{2.9}$$

or

$$\mathbf{F}_r = m\ddot{\mathbf{r}}_{OC} \tag{2.10}$$

Thus we see that the resultant external force on the body is the product of the constant mass m of the body and the acceleration $\mathbf{a}_C$ of its mass center. Hence the motion of the mass center of a body is governed by an equation identical in form to Newton's second law for a particle.

Another point of contact with Newton's laws for particles arises by invoking Euler's first law in the form of Equation (2.7):

$$\mathbf{F}_r = \frac{d}{dt}\int_{\mathcal{B}} \mathbf{v}\, dm$$

If we interchange the differentiation with respect to time and integration over the body† in Equation (2.10) we obtain

$$\mathbf{F}_r = \int_{\mathcal{B}} \frac{d\mathbf{v}}{dt}\, dm$$

or

$$\mathbf{F}_r = \int_{\mathcal{B}} \mathbf{a}\, dm \tag{2.11}$$

which states that the sum of the external forces on the body equals the "sum of the $m\mathbf{a}$'s" of the particles making up the body.

*Whenever we invoke $\mathbf{F}_r = \dot{\mathbf{L}}$, as in obtaining Equations (2.7) to (2.11), the frame of reference is understood to be an inertial frame.

†The validity of this interchange depends on continuity of the velocities within the body. This is not the case for some mathematical models of shock waves, where discontinuous velocities (and hence "infinite" accelerations) are allowed. Euler's first law, $\mathbf{F}_r = \dot{\mathbf{L}}$, is still valid in such a circumstance, but Equation (2.11) is no longer a legitimate form of it. See Problem 2.18.

PROBLEMS / Section 2.2

2.1 Find the mass center of the composite body shown in Figure P2.1. Note that the three parts are composed of different materials.

2.2 Find the mass center of a uniform solid of revolution formed by revolving the curve around the x axis (Figure P2.2). The area is bounded by the curves $4y = x^2$, $y = 0$, and $x = 2$.

2.3 Find the mass center of a thin conical shell (Figure P2.3).

2.4 Fluid in a conical tank (Figure P2.4) contains sediment that results in a fluid density varying linearly with depth: $\rho_F = \rho_0 + Ky$. Find the mass center of the fluid.

2.5 Prove that a body may not lie entirely on one side of any plane through its center of mass.

2.6 Find y_C for the semielliptical prism shown in Figure P2.6. (Density $= \rho$; length normal to the plane of paper $= L$.)

2.7 Find x_C for the shaded plate in Figure P2.7; its density is ρ and its thickness is t.

2.8 Find y_C for the plate of the preceding problem. Note that these results, in terms of n, are quite general and apply to rectangles, triangles, parabolic curves, and so forth, for $n = 0, 1, 2$, and so on.

2.9 Show that the mass center of a solid, homogeneous, equilateral triangular plate is one-third the distance from any base to the opposite vertex. (This is true for *all* triangles!)

2.10 Find the mass center of a solid, homogeneous hemisphere.

2.11 Find the mass center of a thin, semicircular wire.

2.12 A thin wire is bent into the shape of an isosceles triangle (Figure P2.12). Find the mass center of the object, and show that it is at the same point as the centroid of a triangular plate of equal dimensions only if the triangle is equilateral. (Area of cross section $= A$ and mass density $= \rho$, both constant.)

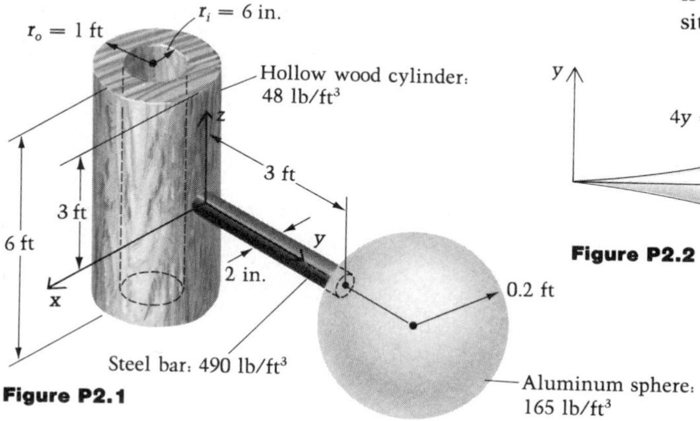

Figure P2.1

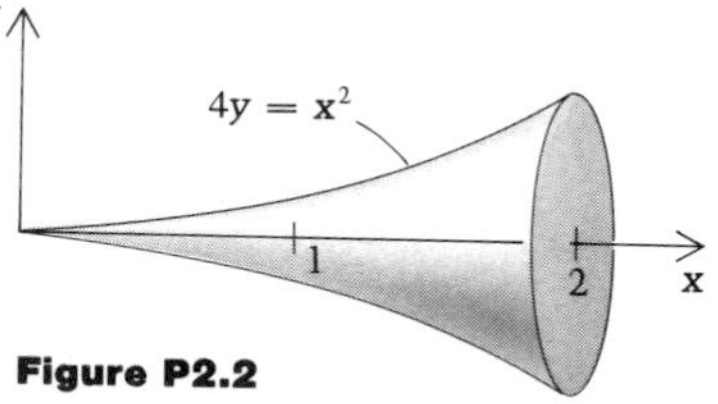

Figure P2.2

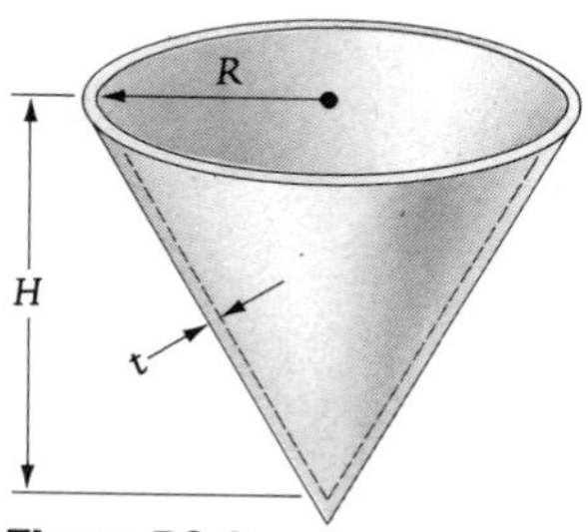

Figure P2.3

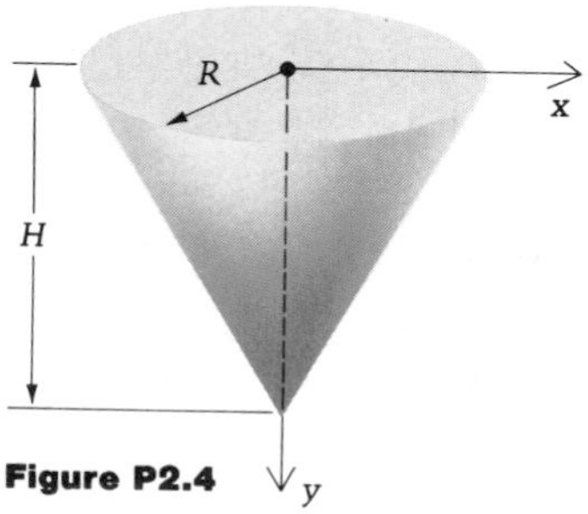

Figure P2.4

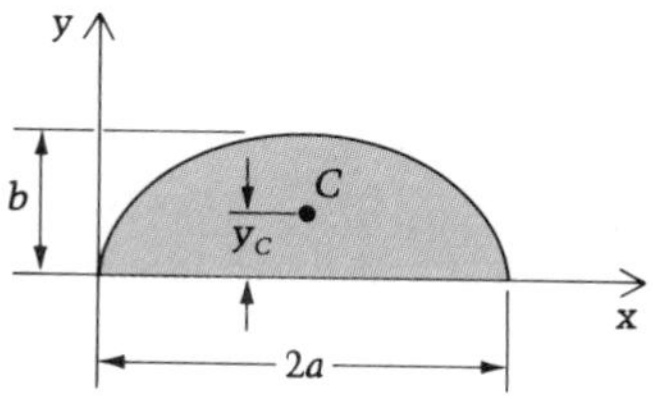

Figure P2.6

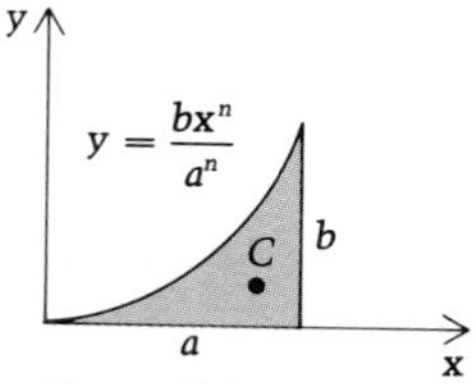

Figure P2.7

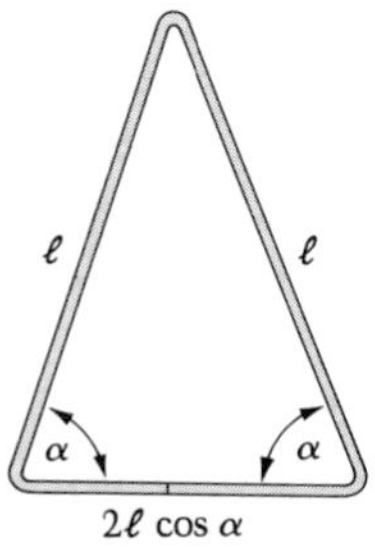

Figure P2.12

2.13 Find the doll's mass center (Figure P2.13) in terms of the densities ρ_1, ρ_2, and ρ_3. Then obtain numerical values for (x_C, y_C, z_C) if the three densities are equal.

2.14 The thin paraboloidal shell is an essential element of modern-day antenna structures. The parabola in Figure P2.14a has the equation $4fy = x^2$ and is revolved around the y axis to give the profile of the paraboloid. Show that the nondimensional distance to the centroid, in terms of the f/D ratio, is for the uniform shell

$$\frac{y_C}{y_T} = \frac{\frac{1}{5}\left\{\left[3 - 32\left(\frac{f}{D}\right)^2\right]\left[1 + 16\left(\frac{f}{D}\right)^2\right]^{3/2} + 2048\left(\frac{f}{D}\right)^5\right\}}{\left[1 + 16\left(\frac{f}{D}\right)^2\right]^{3/2} - 64\left(\frac{f}{D}\right)^3}$$

(Since the shell is thin, density ρ and thickness t may be factored from numerator and denominator and canceled, so that we are in effect computing the centroid of a nonplanar area in this problem.) *Hint:* The differential element of mass is (Figure P2.14b)

$$\begin{aligned} dm &= \rho d(\text{Vol}) = \rho t\, dA \\ &= \rho t x\, d\theta\, ds \\ &= \rho t x\, d\theta \sqrt{dx^2 + dy^2} \\ &= \rho t x\, d\theta \sqrt{1 + y'^2}\, dx \\ &= \rho t x \sqrt{1 + \frac{x^2}{4f^2}}\, d\theta\, dx \end{aligned}$$

and the limits are $\theta = 0$ to 2π and $x = 0$ to $D/2$. You will need integral tables!

2.15 Show in the preceding problem that the mass of the shell, nondimensionalized by the mass of a disk of the same density, thickness, and diameter, is given by

$$\frac{m}{\rho \pi D^2 t/4} = \frac{4}{3}\left\{\left[\frac{1}{4\left(\frac{f}{D}\right)^{2/3}} + 4\left(\frac{f}{D}\right)^{4/3}\right]^{3/2} - 8\left(\frac{f}{D}\right)^2\right\}$$

2.16 Show that the mass center C of a body $\mathcal{B}$ is unique—that is, regardless of the origin selected in $\mathcal{I}$, Equation (2.5) gives the same point for C. *Hint:* Consider the two mass centers C_1 and C_2 resulting from using Equation (2.5) with two origins O_1 and O_2, respectively:

$$\mathbf{r}_{O_1C_1} = \frac{1}{m}\int \mathbf{R}_1\, dm$$

$$\mathbf{r}_{O_2C_2} = \frac{1}{m}\int \mathbf{R}_2\, dm$$

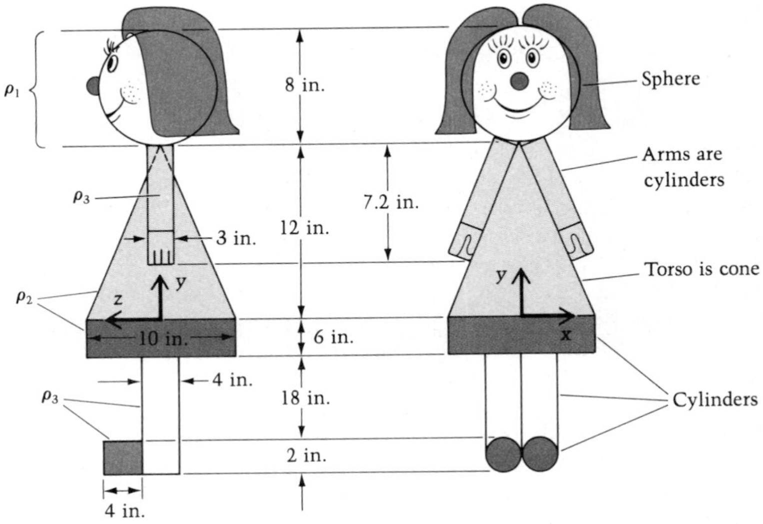

Figure P2.13

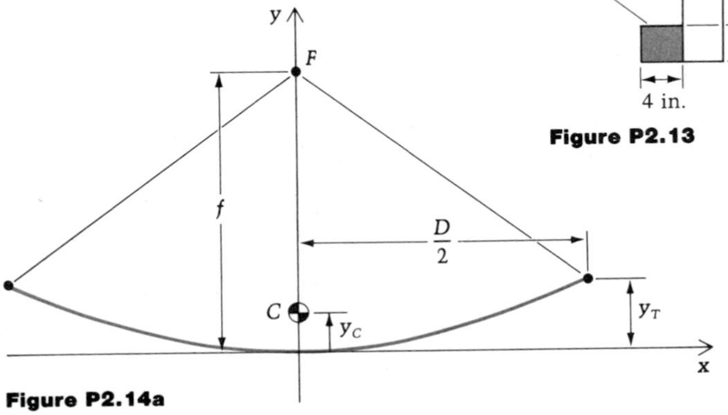

Figure P2.14a

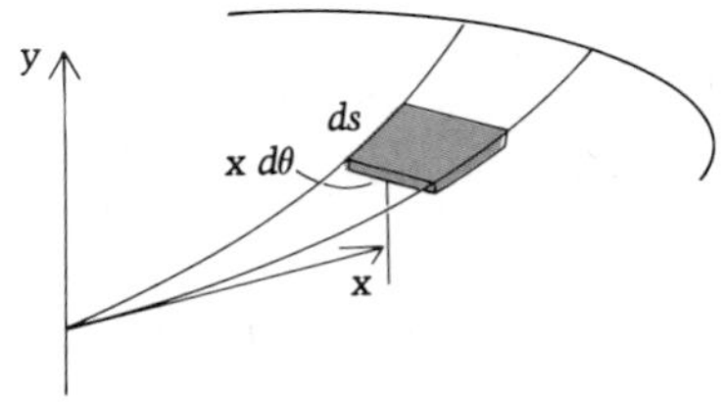

Figure P2.14b

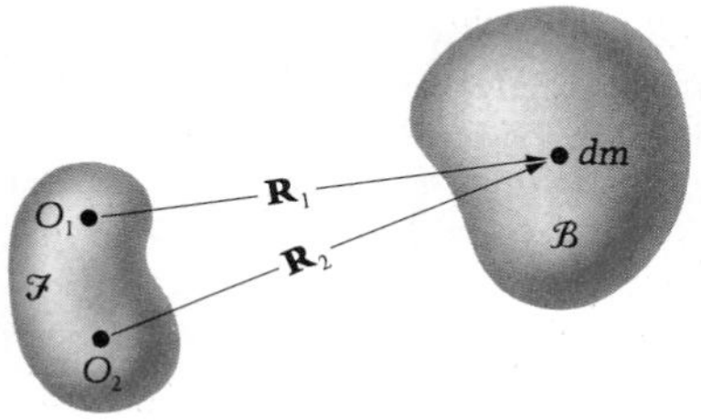

Figure P2.16

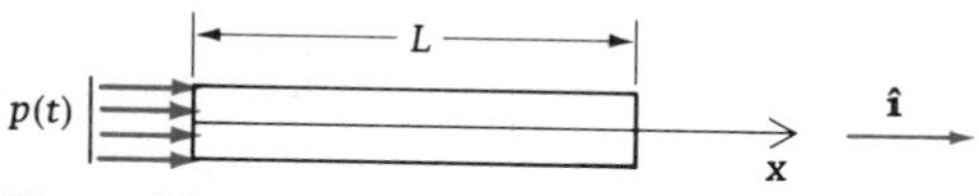

Figure P2.18a

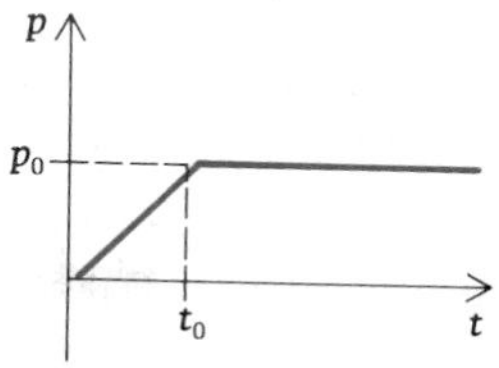

Figure P2.18b

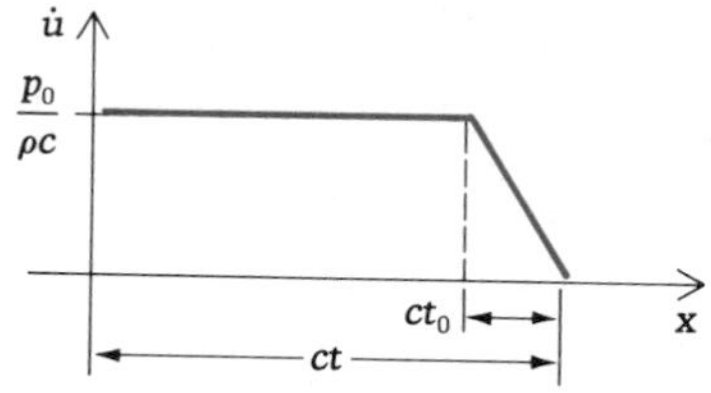

Figure P2.18c

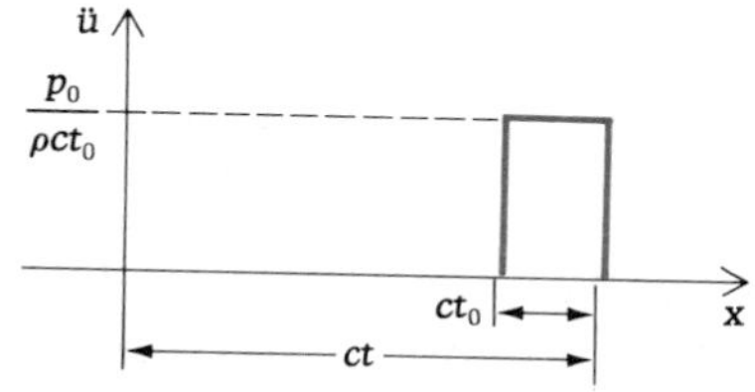

Figure P2.18d

and relate $\mathbf{R}_1$ to $\mathbf{R}_2$. (See Figure P2.16.) Using this relation, show that $\mathbf{r}_{O_1C_1} = \mathbf{r}_{O_1C_2}$, which means that C_1 and C_2 are the same point!

2.17 The laws of motion formulated by Newton and Euler are valid only in inertial frames; yet an inertial frame is one in which these laws hold true. What comes first: the chicken or the egg? Explain how you know you have an inertial frame in which to write, solve, and thereby utilize the equations of motion.

In the following extended problem, we compare the uses of the two forms (Equations 2.7 and 2.11) of Euler's first law.

2.18 An important problem in the dynamics of deformable solids is that of describing the motion which ensues when pressure is rapidly applied to the end of a slender, uniform, elastic bar. A useful approximate theory yields the one-dimensional wave equation as the governing equation of motion. This theory predicts that a pressure applied at one end of the bar creates a disturbance (wave) that propagates into the bar at a constant speed c. To be specific, suppose the bar shown in Figure P2.18a is at rest for $t < 0$ and is subjected to the uniform pressure (over the end of area A) shown in Figure P2.18b. If the disturbance has not reached the right end, that is if $t < L/c$, then for $t > t_0$ the particle velocities $\dot{u}\hat{\mathbf{i}}$ and accelerations $\ddot{u}\hat{\mathbf{i}}$, which vary only with $\mathbf{x}$ and t, are as shown in Figures P2.18c and d, where ρ is the density of the bar.

The first part of this problem is to evaluate the integral

$$\int_{\mathcal{B}} \mathbf{a}\, dm = \hat{\mathbf{i}} \int_0^L \int_A \rho \ddot{u}\, dA\, dx$$

The value that should be obtained is $p_0 A\hat{\mathbf{i}}$, and, since this equals the external force on the bar, Equation (2.11) is thus confirmed for this case. It is important to recognize that that only the interval from $\mathbf{x} = ct - ct_0$ to $\mathbf{x} = ct$ contributes to the value of the integral; that is, only the particles in that region are accelerating.

The second element of this problem is to evaluate the momentum

$$\mathbf{L} = \int_{\mathcal{B}} \mathbf{v}\, dm = \hat{\mathbf{i}} \int_0^L \int_A \rho \dot{u}\, dA\, dx$$

The result will be

$$\mathbf{L} = p_0 A\left[(t - t_0) + \frac{t_0}{2}\right]\hat{\mathbf{i}}$$

The second term in the brackets, a constant, is the contribution from integrating over the interval (ct_0) where the particles are accelerating. The time dependence of $\mathbf{L}$ appears through the increasing number of particles having velocity $(p_0/\rho c)\hat{\mathbf{i}}$. As expected, we see that $\dot{\mathbf{L}} = p_0 A\hat{\mathbf{i}}$.

In effect we have confirmed Euler's law, $\mathbf{F}_r = \dot{\mathbf{L}}$, in two forms. In the first,

$$\dot{\mathbf{L}} = \int \frac{d\mathbf{v}}{dt}\, dm = \int \mathbf{a}\, dm$$

In the second,

$$\dot{\mathbf{L}} = \frac{d}{dt}\int \mathbf{v}\, dm$$

For the case at hand there is no reason to express a preference for the order of differentiating with respect to time and integrating over the body. If the pressure were suddenly applied at full strength ($t_0 = 0$), however, there would be a discontinuity in particle velocity (shock wave) and a consequent undefined acceleration at the wavefront $x = ct$. Because of this undefined (or infinite) acceleration, $\int \mathbf{a}\, dm$ becomes meaningless and no longer provides $\dot{\mathbf{L}}$. There is no difficulty involved in evaluating $\mathbf{L}$, however, since the particle velocities are $p_0/\rho c$ for $x < ct$ and zero for $x > ct$. Thus

$$\mathbf{L} = \frac{p_0}{\rho c}(\rho Act)\hat{\mathbf{i}} = p_0 At\hat{\mathbf{i}}$$

and $\dot{\mathbf{L}} = p_0 A\hat{\mathbf{i}}$.

2.3 Motions of Particles and of Mass Centers of Bodies

Although the mass center of a body does not always coincide with a specific material point of the body, the mass center is nonetheless clearly an important point reflecting the distribution of the body's mass. Furthermore, there are a number of situations in which our objectives are satisfied if we can determine the motion of any material or characteristic point of the body. Clearly this is the case when we attempt to describe the orbits in which the planets move around the sun. Closer to home, a football coach is overjoyed if he finds a punter who can consistently kick the ball 60 yards in the air, regardless of whether the ball gets there end over end, spiraling, or floating like a "knuckleball." In such cases the material point upon which we focus our attention is unimportant. However, there is a strong computational advantage in focusing on the mass center: It is that the motion of that point is directly related to the external forces acting on the body.

We are more likely to think of the football as particle-like when exhibiting the knuckleball behavior than when it is rapidly spinning. Nonetheless, the *mass center's* motion in each case is governed by Euler's first law, although those motions might be quite different because of the different sets of external force induced by the differing interactions of the ball with the air.

If the external forces acting on the body are known functions of time, the mass center's motion can be calculated from Euler's first law:

$$\mathbf{F}_r = m\mathbf{a}_C \tag{2.12}$$

or, alternatively,

$$\mathbf{F}_r = m\frac{d^2\mathbf{r}_{OC}}{dt^2} \tag{2.13}$$

where $\mathbf{r}_{OC}$ is a position vector for the mass center. It is easily seen that two integrations of (2.13) with respect to time yield $\mathbf{r}_{OC}(t)$ provided that initial values of $\mathbf{r}_{OC}$ and $\dot{\mathbf{r}}_{OC}$ $(= \mathbf{v}_C)$ are known.

Figure 2.2

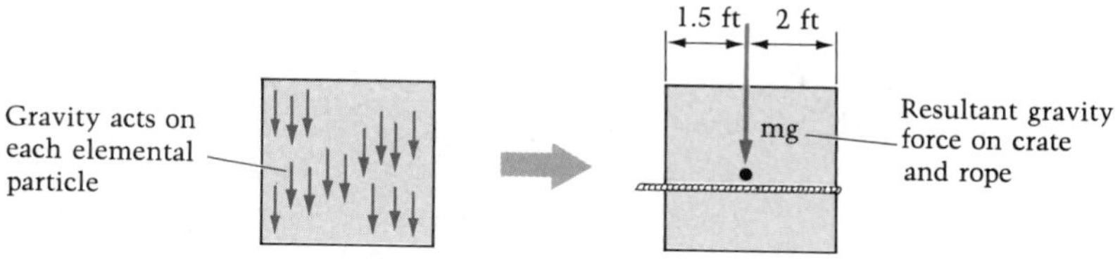

Figure 2.3

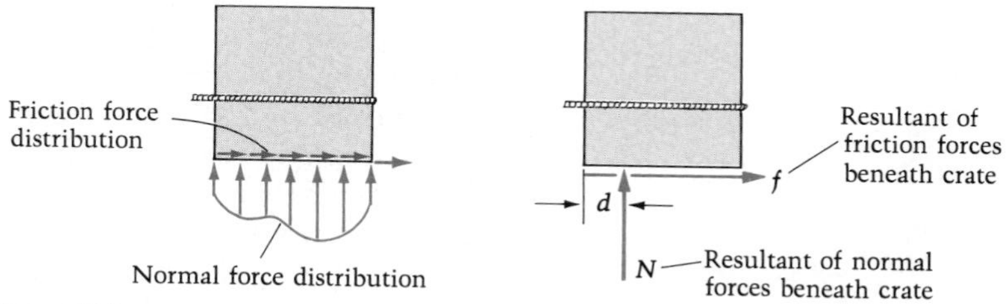

Figure 2.4

Only one thing remains to be done prior to studying several examples that make use of Euler's first law to analyze the motions of the mass centers of bodies. It is to review the concept of the **free-body diagram** which the reader should already have mastered in the study of statics. Without the ability to identify the external forces (and later the moments also), the student will not be able to write a correct set of equations of motion.

A free-body diagram is a sketch of a body in which all the external forces and couples acting upon it are carefully drawn with respect to location, direction, and magnitude. These forces might result from pushes or pulls, as the boy and girl are exerting on the crate and rope in Figure 2.2. Or the forces might result from gravity, such as the weight of the crate in Figure 2.3. (Note that the forces need not *touch* the body to be included in the free-body diagram; another such example is electromagnetic forces.) Or the forces might result from supports, such as the floor beneath the crate in Figure 2.4. If the crate/rope body is acted upon simultaneously by all the forces in these figures, its complete free-body diagram is as shown in Figure 2.5.

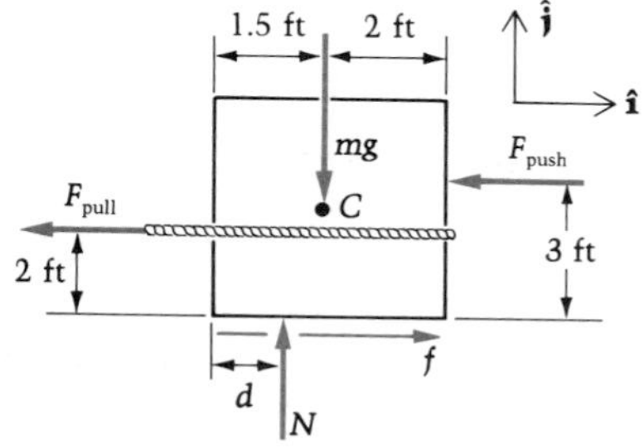

Figure 2.5

It is important to recognize that the free-body diagram:

1. Clearly identifies the body whose motion is to be analyzed.
2. Provides a catalog of all the *external* forces (and couples) *on* the body.

3. Allows us to express, in a compact way, what we know or can easily conclude about the lines of action of known and unknown forces. For example, we know that the pressure (distributed normal force) exerted by the floor on the bottom of the box has a resultant that is a force with a vertical line of action. The symbol N along with the arrow is a code for communicating the fact that we have decided to express that unknown (vector) force as $N\hat{\mathbf{j}}$. The fact that we do not know the location of the line of action of that force is displayed by the presence of the unknown length d.

In dynamics, as in statics, the only characteristics of a force that are manifest in the equations of motion are the vector describing the force and the location of its line of action; that is, we must sum up all the external forces, and we must also sum their moments about some point. Consequently, everything we need to know about the external forces is displayed on the free-body diagram, and we may readily check our work by glancing back and forth between our diagram and the equations we are writing.

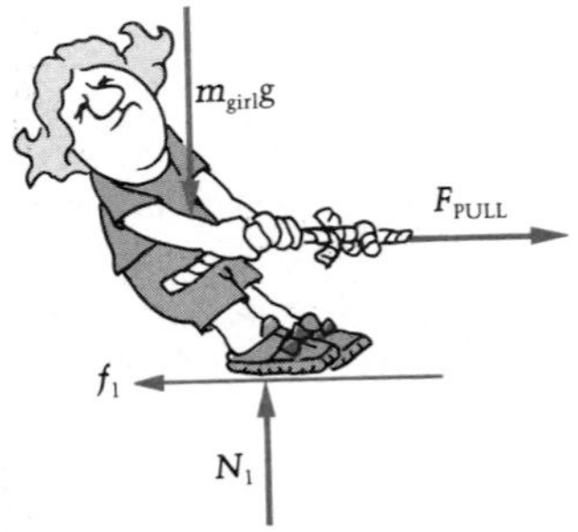

Figure 2.6

When we focus individually on two or more interacting bodies, the free-body diagrams provide an economical way to satisfy—and show that we have satisfied—the principle of action and reaction. The free-body diagram of the girl in our example is shown in Figure 2.6. Since we have already established by Figure 2.5 that the force exerted by the girl on the rope will be $F_{\text{pull}}\ (-\hat{\mathbf{i}})$, then, by the action-reaction principle, the force exerted by the rope on the girl must be $F_{\text{pull}}\ (+\hat{\mathbf{i}})$ as shown in Figure 2.6. In other words, consistent forces of interaction are expressed through the single scalar F_{pull} and the arrow code.

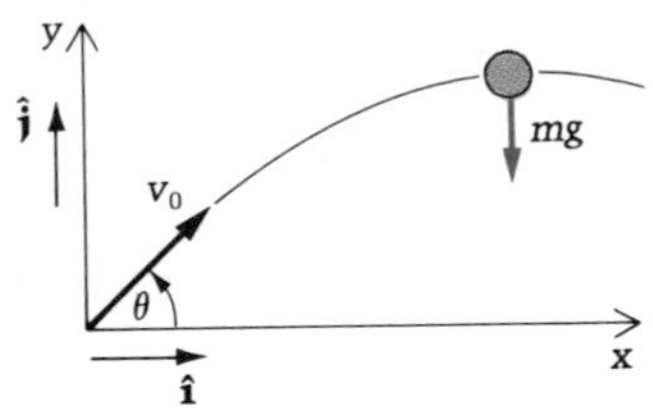

2.5

Ignoring air resistance, find the trajectory of a golf ball hit off a tee at speed v_0 and angle θ with the horizontal.

SOLUTION

It is convenient here to set up a rectangular coordinate system as shown in the diagram and let time $t = 0$ be the instant at which the ball leaves the club. With x, y, and z as the coordinates of the mass center of the ball and since the only external force on the ball is its weight, $-mg\hat{\mathbf{j}}$, we have from Equation (2.13):

$$-mg\hat{\mathbf{j}} = m(\ddot{x}\hat{\mathbf{i}} + \ddot{y}\hat{\mathbf{j}} + \ddot{z}\hat{\mathbf{k}})$$

(Continued)

Thus, collecting the coefficients of $\hat{\mathbf{i}}$, $\hat{\mathbf{j}}$, and $\hat{\mathbf{k}}$, we obtain

$$\ddot{x} = 0 \qquad \ddot{y} = -g \qquad \ddot{z} = 0$$

Integrating, we get

$$\dot{x} = C_1 \qquad \dot{y} = -gt + C_2 \qquad \dot{z} = C_3$$

Because of the way we have aligned the x and z axes,

$$\dot{x}(0) = v_0 \cos\theta \qquad \dot{z}(0) = 0 \qquad \dot{y}(0) = v_0 \sin\theta$$

Therefore

$$C_1 = v_0 \cos\theta \qquad C_2 = v_0 \sin\theta \qquad C_3 = 0$$

Integrating again, we get

$$x = v_0(\cos\theta)t + C_4$$

$$y = \frac{-gt^2}{2} + v_0(\sin\theta)t + C_5$$

$$z = C_6$$

Our location of the origin of the coordinate system at the 'launch' site yields $x(0) = y(0) = z(0) = 0$, so that $C_4 = C_5 = C_6 = 0$ and the trajectory of the mass center of the ball is given by

$$x = v_0(\cos\theta)t$$

$$y = -\frac{gt^2}{2} + v_0(\sin\theta)t$$

$$z = 0$$

which describes a parabola in the xy plane—that is, in the vertical plane defined by the launch point and the direction of the launch velocity.

Letting the time of maximum elevation be t_1, we find that $\dot{y}(t_1) = 0$ yields

$$0 = -gt_1 + v_0 \sin\theta$$

so that $t_1 = (v_0/g)\sin\theta$ and the maximum elevation is

$$y(t_1) = -\frac{v_0^2}{2g}\sin^2\theta + \frac{v_0^2}{g}\sin^2\theta = \frac{v_0^2}{2g}\sin^2\theta$$

If t_2 is the time the ball strikes the fairway (assumed level), then

$$y(t_2) = 0 = -\frac{gt_2^2}{2} + v_0 t_2 \sin\theta$$

$$t_2 = \frac{2v_0}{g}\sin\theta$$

(Continued)

which is, not surprisingly, twice the time (t_1) to reach maximum elevation. The length of the drive is

$$\mathbf{x}(t_2) = v_0(\cos\theta)\left(\frac{2v_0}{g}\sin\theta\right)$$

$$= \frac{2v_0^2}{g}\sin\theta\cos\theta = \frac{v_0^2}{g}\sin 2\theta$$

which, with v_0 fixed, is maximized by $\theta = 45°$. That is, for a given launch speed we get maximum range when the launch angle is 45°.

The results of this analysis apply to the unpowered flight of any projectile as long as the path is sufficiently limited that the gravitational force is constant (magnitude and direction) and we can ignore the medium (air) through which the body moves. Interaction with the air is responsible not only for the drag (retarding of motion) on a golf ball but also for the fact that its path is usually not planar (slice or hook!). On one of the Apollo moon landings in the early 1970s, astronaut Alan Shepard drove a golf ball a "country mile" on the moon because of the absence of air resistance and, more important, because the gravitational acceleration at the moon's surface is only about one-sixth that at the surface of the earth.

EXAMPLE 2.6

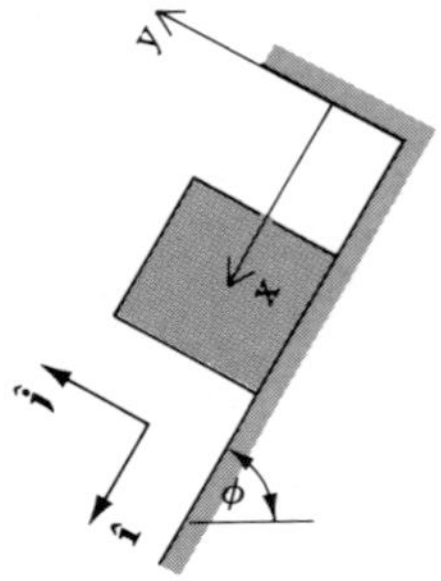

If the 20-kg block shown in the diagram is released from rest, find its speed after it has descended a distance $d = 5$ m down the plane. The angle $\varphi = 60°$ and the (Coulomb) coefficients of friction are

μ_s (static) = 0.30

μ_k (kinetic) = 0.25

SOLUTION

In the statement of the problem we are using some loose but common terminology in referring to the speed of the block. In fact we may only speak of the speed of a *point*, but here we are tacitly assuming that the block is *rigid* and *translating* so that every point in the block has the same velocity and the same acceleration. In contrast to the preceding example, note that here we do not know all the external forces on the body before we carry out the analysis, because the surface touching the block constrains its motion. That constraint is acknowledged by expressing the velocity of (the mass center of) the block by $\dot{x}\hat{\mathbf{i}}$ and its acceleration by $\ddot{x}\hat{\mathbf{i}}$.

(Continued)

Referring to the free-body diagram shown here,

$$\mathbf{F}_r = m\mathbf{a}_C$$

$$N\hat{\mathbf{j}} + mg(\sin\varphi\hat{\mathbf{i}} - \cos\varphi\hat{\mathbf{j}}) - f\hat{\mathbf{i}} = m\ddot{x}\hat{\mathbf{i}}$$

or

$$N = mg\cos\varphi$$

and

$$mg\sin\varphi - f = m\ddot{x}$$

First we must determine if in fact the block will move. For equilibrium, $\ddot{x} = 0$ and f is limited by $0 \le f \le f_{\max} = \mu_s N$. Hence

$$f = mg\sin\varphi \le \mu_s\, mg\cos\varphi = \mu_s N$$

or

$$\tan\varphi \le \mu_s$$

But

$$\tan 60° = 1.73 > 0.3 = \mu_s$$

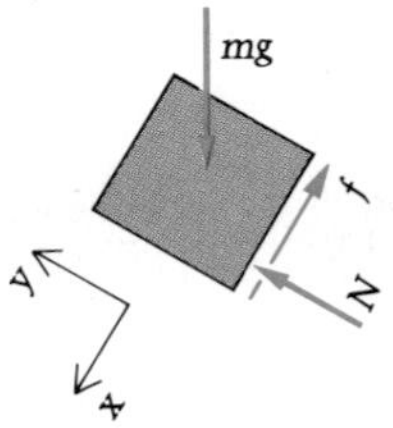

Thus the block moves (and, as it does, is acted on by $\mu_k N$ up the plane as shown in the diagram). We note that $\tan^{-1}(\mu_s)$ is sometimes called the ***angle of friction.*** Here $\tan^{-1}(\mu_s)$ is 16.7°, and this of course is the angle for which $\tan\varphi = \mu_s$; it means that any angle $\varphi > 16.7°$ (like our 60°) will result in sliding, or a loss of equilibrium.

Having checked the statics and briefly reviewed friction, we now solve the equation of motion for $\ddot{x}$:

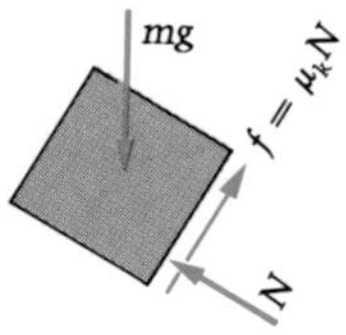

$$m\ddot{x} = mg\sin\varphi - \mu_k N$$

$$= mg\sin\varphi - \mu_k(mg\cos\varphi)$$

$$= mg(\sin\varphi - \mu_k\cos\varphi)$$

or

$$\ddot{x} = g(\sin\varphi - \mu_k\cos\varphi)$$

Thus

$$\dot{x} = g(\sin\varphi - \mu_k\cos\varphi)t + C_1$$

and $C_1 = 0$ since $\dot{x}(0) = 0$ if $t = 0$ is the instant at which the block is released. Hence

$$x = \frac{g}{2}(\sin\varphi - \mu_k\cos\varphi)t^2 + C_2$$

and $C_2 = 0$ if we choose the measurement of x so that $x(0) = 0$.

If we let t_1 be the time at which $x = d$, then

$$d = \frac{g}{2}(\sin\varphi - \mu_k\cos\varphi)t_1^2$$

(Continued)

For $\varphi = 60°$, $\mu_k = 0.25$, $d = 5$ m, and $g = 9.81$ m/s^2, we get

$$5 = \left(\frac{9.81}{2}\right)[0.866 - 0.25(0.5)]t_1^2$$

from which $t_1 = 1.17$ s.

Since the velocity is given by $\dot{x}\hat{\mathbf{i}}$, the speed at t_1 is merely the magnitude (or absolute value) of $\dot{x}(t_1)$ and

$$\dot{x}(t_1) = (9.81)[0.866 - 0.25(0.5)](1.17)$$

$$= 8.50 \text{ m/s}$$

Finally we should note that the plausibility of our numerical results can be verified from the fact that, owing to the steep angle and moderate coefficient of friction, they should be of the same orders of magnitude as those arising from a free vertical drop (acceleration g) for which

$$t_1 = \sqrt{\frac{2d}{g}} = \sqrt{\frac{2(5)}{9.81}} = 1.01 \text{ s}$$

and

$$\text{Speed} = \sqrt{2gd} = \sqrt{2(9.81)(5)} = 9.90 \text{ m/s}$$

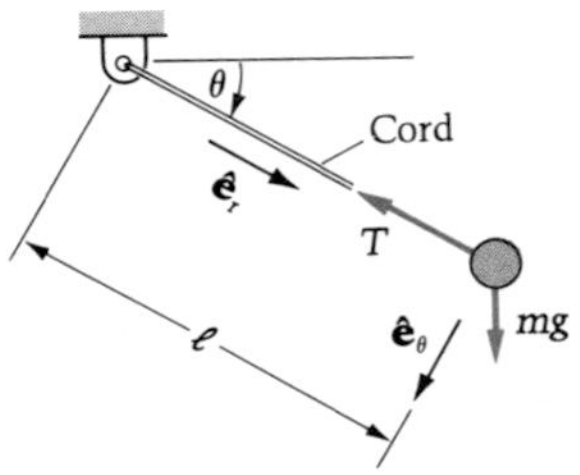

EXAMPLE 2.7

A ball of mass m (see the diagram) is released from rest with the cord taut and $\theta = 30°$. Find the tension in the cord during the ensuing motion.

SOLUTION

In this problem we make two basic assumptions:

1. The cord is inextensible.
2. The cord is attached to the ball at its mass center (or equivalently the ball is small enough to be treated as a particle). Either way the point whose motion is to be described has a path that is a circle. Thus the problem is similar to Example 2.6 in that the path of the mass center is known in advance (a circle here and a straight line there) and consequently among the external forces are unknowns caused by constraints (the tension in the cord here and the surface reaction in the preceding problem).

Using polar coordinates (Section 1.6), we may express the acceleration as

$$\mathbf{a} = (\ddot{r} - r\dot{\theta}^2)\hat{\mathbf{e}}_r + (r\ddot{\theta} + 2\dot{r}\dot{\theta})\hat{\mathbf{e}}_\theta$$

(Continued)

Since the polar coordinate r is the constant ℓ here, referring to the free-body diagram we have

$$-T\hat{\mathbf{e}}_r + mg(\hat{\mathbf{e}}_\theta \cos\theta + \hat{\mathbf{e}}_r \sin\theta) = m(-\ell\dot{\theta}^2\hat{\mathbf{e}}_r + \ell\ddot{\theta}\hat{\mathbf{e}}_\theta)$$

so that

$$T = mg\sin\theta + m\ell\dot{\theta}^2 \tag{2.14}$$

and

$$m\ell\ddot{\theta} - mg\cos\theta = 0 \tag{2.15}$$

The first of these component equations (Equation 2.14) yields the tension T if we know $\theta(t)$; the second (2.15) is the differential equation that we must integrate to obtain $\theta(t)$. In Example 2.6 the counterpart to Equation (2.15) is $\ddot{x} =$ constant, which of course was easily integrated.

Here not only do we have a nontrivial differential equation in that $\ddot{\theta}$ is a function of θ but we have the substantial complication that Equation (2.15) is nonlinear because $\cos\theta$ is a nonlinear function of θ. However, a partial integration of Equation (2.15) can be accomplished; to this end we write the equation in the standard form

$$\ddot{\theta} - \frac{g}{\ell}\cos\theta = 0$$

and then multiply by $\dot{\theta}$ to obtain

$$\dot{\theta}\ddot{\theta} - \frac{g}{\ell}\dot{\theta}\cos\theta = 0$$

which we recognize to be

$$\frac{d}{dt}\left(\frac{\dot{\theta}^2}{2} - \frac{g}{\ell}\sin\theta\right) = 0$$

or

$$\frac{\dot{\theta}^2}{2} - \frac{g}{\ell}\sin\theta = C_1 \qquad \text{(a constant)} \tag{2.16}$$

Equation (2.16) is called an **energy integral** of Equation (2.15) and is closely related to the "work and kinetic energy" principle that is introduced in the next section.

For the problem at hand the constant C_1 may be obtained from the fact that when $\theta = 30°$, then $\dot{\theta} = 0$; thus

$$0 - \frac{g}{\ell}\sin 30° = C_1$$

or

$$C_1 = \frac{-g}{2\ell}$$

(Continued)

Thus from Equation (2.16) we get

$$\dot{\theta}^2 = \frac{g}{\ell}(2\sin\theta - 1) \tag{2.17}$$

which we may substitute in Equation (2.14) to obtain

$$T = mg\sin\theta + mg(2\sin\theta - 1)$$

or

$$T = mg(3\sin\theta - 1) \tag{2.18}$$

Even though we have not obtained the time dependence of the tension,* the energy integral has enabled us to find the way in which the tension depends on the position of the ball. As we would anticipate intuitively, the maximum tension occurs when $\theta = 90°$, at which time $T = [3(1) - 1] = 2mg$.

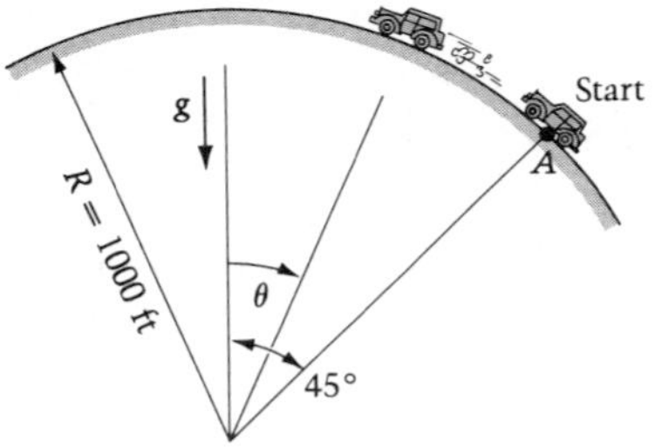

EXAMPLE 2.8

A car accelerates from rest, increasing its speed at the constant rate of $K = 6$ ft/sec^2. (See the accompanying diagram.) It travels on a circular path starting at point A. Find the time and the position of the car when it first leaves the surface due to excessive speed.

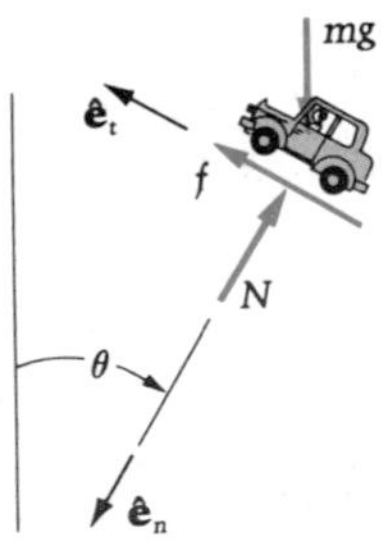

SOLUTION

Before the car (treated as a particle) leaves the surface, the free-body diagram is as shown in the figure. We shall work this problem in general (without substituting numbers until the end). The purpose is to illustrate the concept of nondimensional parameters. The equation of motion in the tangential ($\hat{\mathbf{e}}_t$) direction is

$$F_{r_t} = m\ddot{s}$$

$$f - mg\sin\theta = mK \tag{1}$$

Equation (1) shows that the friction exerted on the tires by the road is the external force which moves the car up the path. Note that after it passes the top of the circular hill, we have $\sin\theta < 0$ and then the gravity force ***adds*** to the friction in accelerating the car on the way down.

The following equation of motion is the one that will help us in this problem: It equates $\mathbf{F}_r$ and $m\mathbf{a}_C$ in the normal direction.

(Continued)

*This would require solving the differential equation (2.17) for $\theta(t)$ and substituting into Equation (2.18).

Question 2.3 Are the components of $\mathbf{F}_r$ and $m\mathbf{a}_C$ equal in *all* directions or just in coordinate directions?

$$F_{r_n} = m\frac{\dot{s}^2}{\rho}$$

$$mg\cos\theta - N = \frac{m(Kt)^2}{R} \qquad (2)$$

where $\dot{s} = \int \ddot{s}\,dt = Kt + C_1 = Kt$, since $\dot{s} = 0$ when $t = 0$.

We note that the car will lose contact with the road when N becomes zero. (The ground cannot pull down on the car for further increases of t, which would require $N < 0$!) Therefore, at the point of leaving the ground,

$$\not{m}g\cos\theta = \frac{\not{m}K^2t^2}{R} \qquad (3)$$

Question 2.4 What is the *meaning* of the fact that m cancels in Equation (3)?

Now θ is related to s according to

$$s = R\left(\frac{\pi}{4} - \theta\right) \qquad (4)$$

And from $\dot{s} = Kt$, we get another expression for s:

$$s = \frac{Kt^2}{2} + C_2 = \frac{Kt^2}{2} \qquad \text{(since } s = 0 \text{ when } t = 0\text{)} \qquad (5)$$

Hence from Equations (4) and (5) we get

$$\frac{Kt^2}{2} = R\left(\frac{\pi}{4} - \theta\right) \Rightarrow \theta = \frac{\pi}{4} - \frac{Kt^2}{2R} \qquad (6)$$

Substituting for θ from (6) into (3), we have

$$g\cos\left(\frac{\pi}{4} - \frac{Kt^2}{2R}\right) = \frac{K^2t^2}{R}$$

or

$$\cos\left(\frac{\pi}{4} - \frac{Kt^2}{2R}\right) = \frac{Kt^2}{2R} \cdot \frac{2K}{g} \qquad (7)$$

Equation (7) allows us to solve for the dimensionless parameter $q = (Kt^2/2R)$,

(Continued)

once we have selected a value of the car's dimensionless acceleration K/g. In this problem, for example,

$$\cos\left(\frac{\pi}{4} - q\right) = 2q\left(\frac{6}{32.2}\right) = 0.373q \tag{8}$$

The following table shows how (with a calculator)* we can quickly arrive at the value of q that solves Equation (8):

q	$\cos\left(\frac{\pi}{4} - q\right)$	$0.373q$
0.1	0.7742	0.0373
0.5	0.9595	0.1865
0.7854 (at the top)	1	0.2930
1.0	0.9771	0.3730
1.3	0.8705	0.4849
1.6	0.6862	0.5968
1.7	0.6101	0.6341
1.69	0.6180	0.6304
1.68	0.6258	0.6266

Thus at $q = Kt^2/2R \approx 1.68$, the car leaves the circular track due to excessive speed. Therefore for $K = 6$ ft/sec^2 and $R = 1000$ ft,

$$t = \sqrt{\frac{1.68 \times 2 \times 1000}{6}} = 23.7 \text{ sec}$$

The angle at loss of contact is given by Equation (6):

$$\theta = \frac{\pi}{4} - \frac{Kt^2}{2R} = \frac{\pi}{4} - q$$

$$= \frac{\pi}{4} - 1.68$$

$$= -0.895 \text{ rad}$$

$$= -51.3°$$

The speed at the point of loss of contact is $Kt = 6(23.7) = 142$ ft/sec $= 97.0$ mph. Note that for $K/g = 6/32.2$, the angle $-51.3°$ is the angle of leaving for many combinations of t and R (so long as $Kt^2/2R = 1.68$).

*See Appendix C for a numerical solution to this problem using the Newton–Raphson method with a programmable calculator.

PROBLEMS / Section 2.3

2.19 A cannonball is fired as shown in Figure P2.19. Neglecting air resistance, find the angle α that will result in the cannonball landing in the box.

2.20 There is a speed, called the *conical speed,* at which a ball on a string, in the absence of all friction, moves on a specific horizontal circle with the string sweeping out a conical surface with no radial or vertical component of velocity (Figure P2.20). If ℓ is the length of the string and $k\ell$ is the radius of the circle on which the ball moves, find the conical speed in terms of k, ℓ, and the acceleration of gravity.

2.21 Communications satellites are placed in *geosynchronous orbit,* an orbit in which the satellites are always located in the same position in the sky (Figure P2.21).

a. Give an argument why this orbit must lie in the equatorial plane. Why must it be circular?

b. If the satellites are to remain in orbit without expending energy, find the important ratio of the orbit radius r_s to the earth's radius r_e. *Hint:* Use Newton's law of universal gravitation

$$F = \frac{Gm_s m_e}{r_s^2}$$

together with the law of motion in the radial direction, and note that if the satellite were sitting on the earth's surface, the force would be

$$F = m_s g = \frac{Gm_s m_e}{r_e^2}$$

so that the product Gm_e may be rewritten as gr_e^2. Use $r_e = 3960$ mi.

2.22 Using the result of the preceding problem, show that a minimum of three satellites in geosynchronous orbit are required for continuous communications coverage over the whole earth except for small regions near the poles.

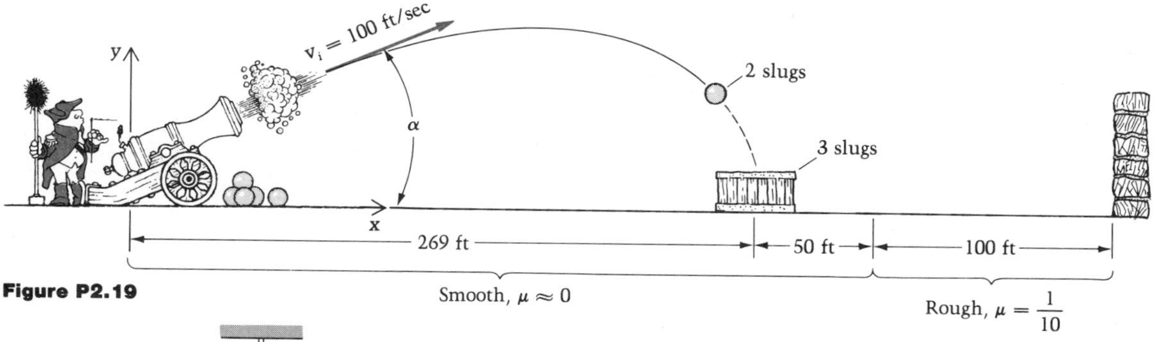

Figure P2.19

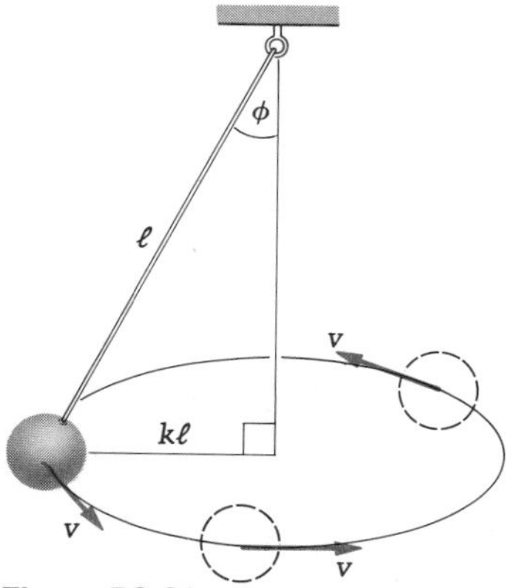

Figure P2.20

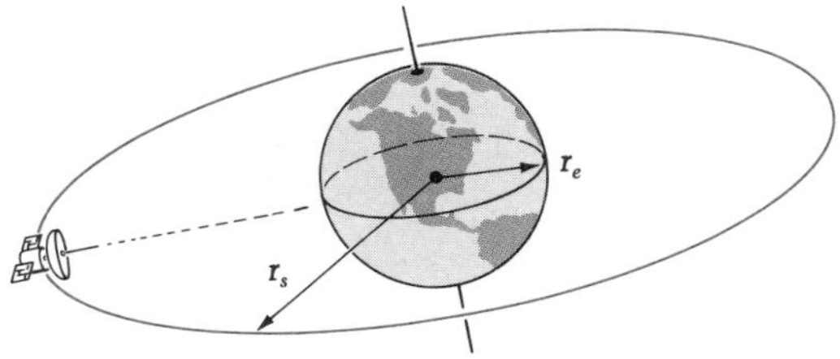

Figure P2.21

2.23 In terms of the parameters δ, R, μ, and g defined in Figure P2.23, find the minimum speed for which the motorcycle will not slip down the inside wall of the cylinder.

2.24 A ball of mass m on a string is swung at constant speed v_0 in a horizontal circle of radius R by a child. (See Figure P2.24.)

a. What holds up the ball?
b. What is the tension in the string?
c. If the child increases the speed of the ball, what provides the force in the forward direction needed to produce the $\mathfrak{a}$? Explain.

Figure P2.24

2.25 The acceleration of gravity varies with distance z above the earth's surface as

$$\ddot{z} = \frac{-gR^2}{(R+z)^2}$$

where g is the acceleration of gravity on the surface and R is the earth's radius. Find the minimum firing velocity v_i that a projectile must have in order to escape the earth if fired straight up (Figure P2.25). *Hint:* In one-dimensional motion, $dz/dt = v$ and $dv/dt = a$, so that $dz/v = dt = dv/a$. Thus $a\,dz = v\,dv$ (that is, $\ddot{z}\,dz = v\,dv$), and substitution of the preceding expression for $\ddot{z}$ allows us to integrate for v as a function of z. Not to return to earth thus requires the condition that $v \to 0$ as z gets large for the *minimum* possible v_i.

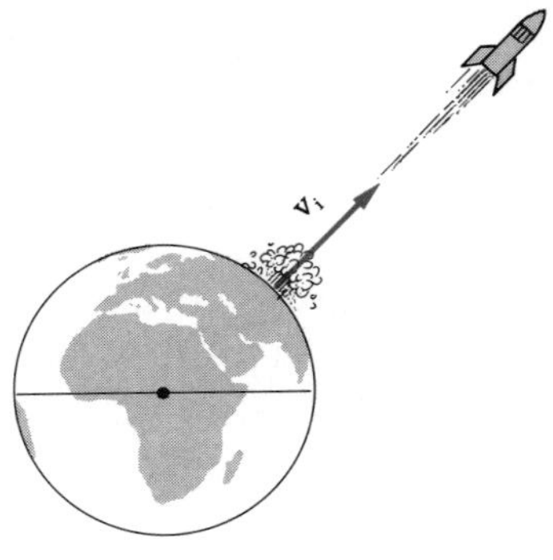

Figure P2.25

Figure P2.26

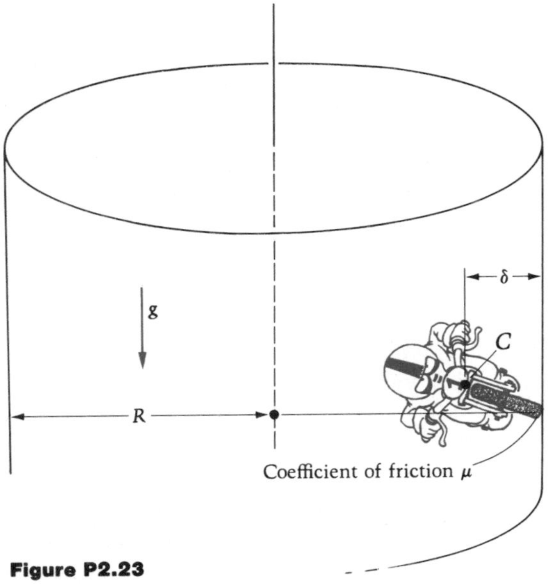

Figure P2.23

2.26 In the "spindle top" ride in an amusement park, people stand against a cylindrical wall and the cylinder is then spun up to a certain angular velocity ω_0. (See Figure P2.26.) The floor is then lowered, but the people remain against the wall at the same level. Use the equation $F_{r_n} = ma_{C_n} = mv^2/R$ to explain the phenomenon. Noting that each person is in equilibrium vertically, solve for the minimum ω_0 to prevent people from slipping if $R = 2$ m and the expected friction coefficient between the rough wall and the clothing is $\mu = 0.5$.

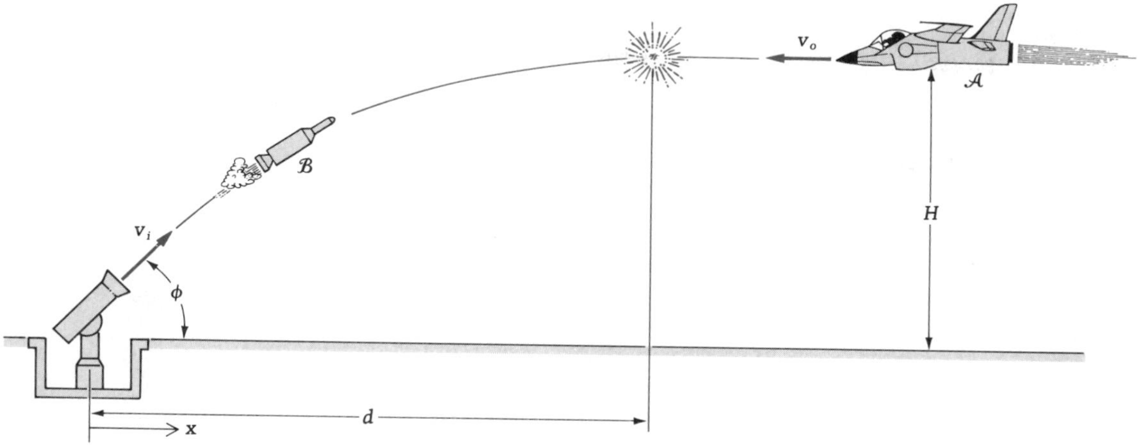

Figure P2.29

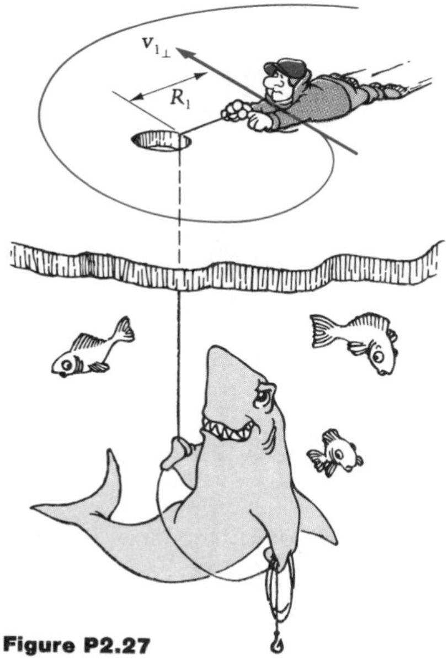

Figure P2.27

Figure P2.30

2.27 A wintertime fisherman of mass 70 kg is in trouble—*he* is being reeled in by Jaws on a lake of frozen ice. At the instant shown in Figure P2.27, the man has a velocity component, perpendicular to the radius r, of $v_{1_\perp} = 0.3$ m/s at an instant when $r = R_1 = 5$ m. If Jaws pulls in the line with a force of 100 N, find the value of $v_{2_\perp}$ when the radius is $R_2 = 1$ m. *Hint:*

$$r\ddot{\theta} + 2\dot{r}\dot{\theta} = \frac{(d/dt)(r^2\dot{\theta})}{r} \text{ and } F_{r_\theta} = ?$$

2.28 In the preceding problem, show that the differential equation of the man's radial motion is $\ddot{r} = (2.25/r^3) - (10/7)$. Use $\ddot{r}\,dr = \dot{r}\,d\dot{r}$ to integrate the equation, and, *if* $\dot{r} = 0$ when $r = 5$ m, show that the radial component ($\dot{r}$) of the man's velocity when $r = 1$ m is 3.04 m/s.

2.29 Find the angle ϕ, firing velocity v_i, and time t_f of intercept so that ballistic missile $\mathcal{B}$ shown in Figure P2.29 will intercept bomber $\mathcal{A}$ when $x = d$. The bomber, at $x = D$ when the missile is launched, travels horizontally at constant speed v_0 and altitude H. What has been neglected in your solution?

2.30 The pilot of an airplane flying at 300 km/hr wishes to release a package of mail at the right position so that it hits spot A. (See Figure P2.30.) What angle θ should his line of sight to the target make at the instant of release?

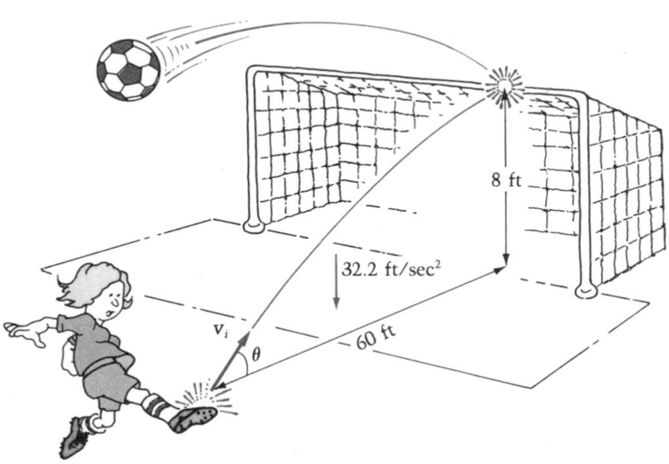

Figure P2.31

Figure P2.32

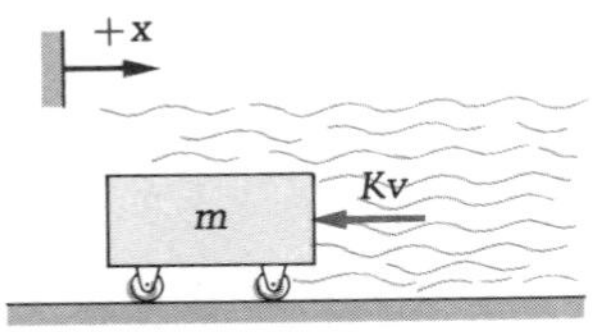

Figure P2.33

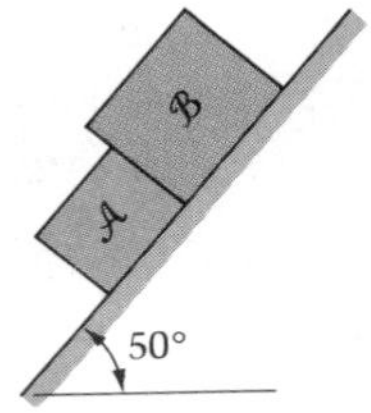

Figure P2.35

2.31 A soccer ball (Figure P2.31) is kicked toward the goal from 60 ft. It strikes the top of the goal at the highest point of its trajectory. Find the velocity and angle θ at which the ball was kicked, and determine the time of travel.

2.32 A child drops a rock into a well and hears it splash into the water at the bottom exactly 2 sec later. (See Figure P2.32.) If she is at a location where the speed of sound is $v_s = 1100$ ft/sec, determine the depth of the well with and without considering v_s. Compare the two results.

2.33 The mass m shown in Figure P2.33 is given an initial velocity of v_0 in the $\mathbf{x}$ direction. It moves in a medium that resists its motion with force proportional to its velocity, with proportionality constant K. By solving for $v(\mathbf{x})$, determine how far the mass travels before stopping. Then solve in a different manner for $v(t)$ if $v = v_0$ when $t = 0$.

2.34 Using the result of the preceding problem and expressing v as dx/dt, solve for $\mathbf{x}(t)$ if $\mathbf{x} = 0$ when $t = 0$.

2.35 The blocks in Figure P2.35 are in contact as they slide down the inclined plane. The masses of the blocks are $m_{\mathcal{B}} = 25$ kg and $m_{\mathcal{A}} = 20$ kg, and the friction coefficients between the blocks and the plane are 0.5 for $\mathcal{A}$ and 0.1 for $\mathcal{B}$. Determine the force between the blocks and find their common acceleration.

2.36 In the preceding problem, let μ be the coefficient of friction between $\mathcal{A}$ and the plane. Using the two motion equations of the blocks, find the range of values of μ for which the blocks will separate when released from rest.

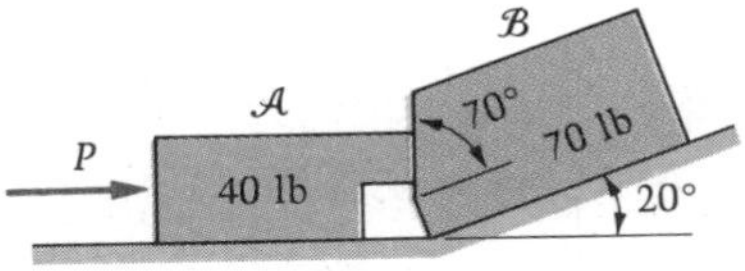

Figure P2.37

2.37 If all surfaces are smooth for the setup of blocks and planes in Figure P2.37, find the force P that will give block $\mathcal{B}$ an acceleration of 4 ft/sec² up the incline.

2.38 Work the preceding problem if the planes are still smooth but the friction coefficient between $\mathcal{A}$ and $\mathcal{B}$ is $\mu_s \approx \mu_k = 0.3$.

2.39 Work the preceding problem if the coefficient of friction is 0.3 for *all* contacting surfaces.

2.40 Find the largest force P for which $\mathcal{A}$ in Figure P2.40 will not slide on $\mathcal{B}$.

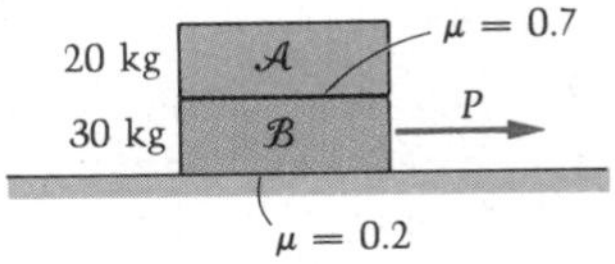

Figure P2.40

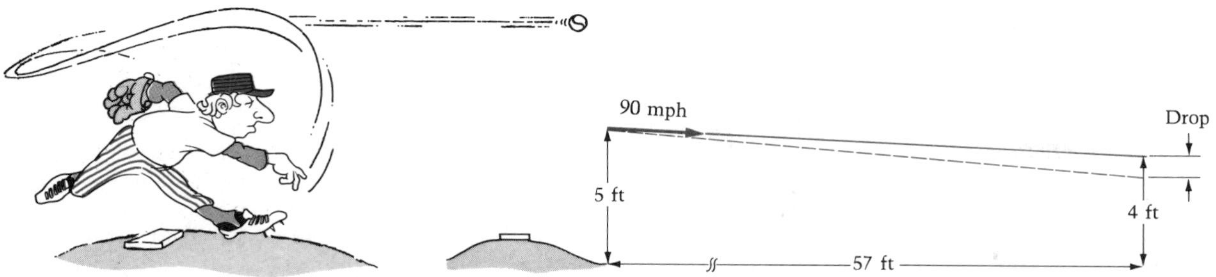

Figure P2.42

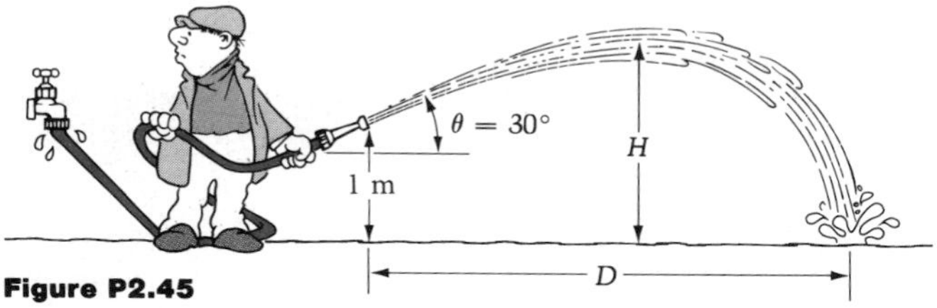

Figure P2.45

2.41 Work the preceding problem if P is applied to $\mathcal{A}$ instead of $\mathcal{B}$.

2.42 A baseball pitcher releases a 90-mph fastball 5 ft off the ground (Figure P2.42). If in the absence of gravity the ball would arrive at home plate 4 ft off the ground, find the drop in the actual path caused by gravity. Neglect air resistance.

2.43 In the preceding problem, find the radius of curvature of the path of the baseball's center at the instant it arrives at the plate.

2.44 In the preceding problem, the batter hits a pop-up that leaves the bat at a 45° angle with the ground. The shortstop loses the ball in the sun and it lands on second base, $90\sqrt{2}$ ft from home plate. What was the velocity of the baseball when it left the bat?

2.45 The garden hose shown in Figure P2.45 expels water at 13 m/s from a height of 1 m. Determine the maximum height H and horizontal distance D reached by the water.

2.46 In the preceding problem, use calculus to find the angle θ that will give maximum range D to the water.

2.47 A darts player releases a dart at the position indicated in Figure P2.47 with the initial velocity vector making a 10° angle with the horizontal. What must the dart's initial speed be if it lands at the center of the bullseye?

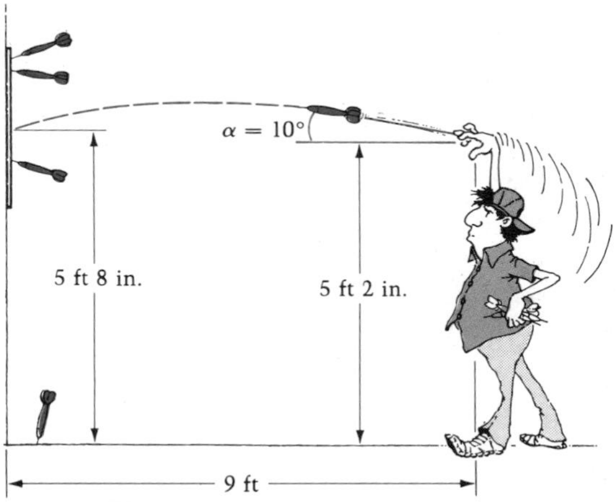

Figure P2.47

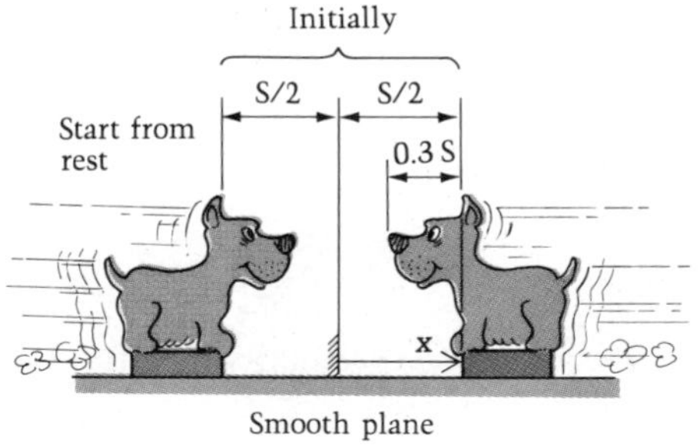

Figure P2.49

2.48 In the preceding problem, suppose the initial speed of the dart is 20 ft/sec. What must the angle α be if a bullseye is scored?

2.49 The identical plastic scottie dogs shown in Figure P2.49 are glued onto magnets and attract each other with a force $F = K/(2x)^2$, where K is a constant related to the strength of the magnets. Find the speeds at which the dogs collide if they are initially separated by the distance S.

2.50 A particle P moves along a curved surface $\mathcal{S}$ as shown in Figure P2.50. Show that P will remain in contact with $\mathcal{S}$ provided that, at all times, $v > \sqrt{\rho g \cos \alpha}$.

2.51 Find the condition for retention of contact if P moves along the *outside* of a surface defined by the same curve as in the preceding problem. (See Figure P2.51.)

2.52 At liftoff the space shuttle is powered upward by two solid rocket boosters of 11.8×10^6 N each and by the three Orbiter main liquid-rocket engines with thrusts of 2.1×10^6 N each. At liftoff the external tank holds 720,000 kg of liquid hydrogen and oxygen, and the payload (to be carried into orbit) has a mass of 29,500 kg. Assuming the fuel and payload to constitute nearly all the initial weight, determine the acceleration experienced by the crew members at liftoff. (This is only one-third the initial acceleration on earlier manned flights—demonstrate this by comparing with the Apollo moon rocket, which weighed 6×10^6 lb at liftoff and was powered by five engines each with a thrust of 1.5×10^6 lb.) Neglect the change in mass between ignition and liftoff.

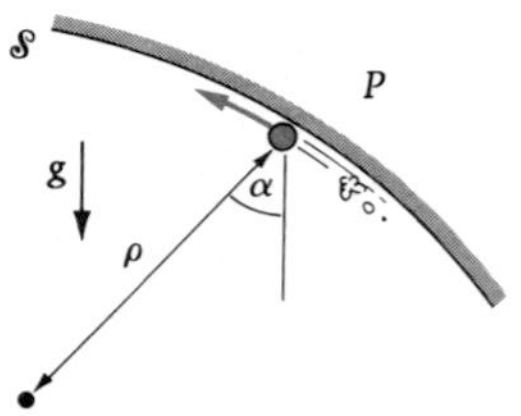

Figure P2.50

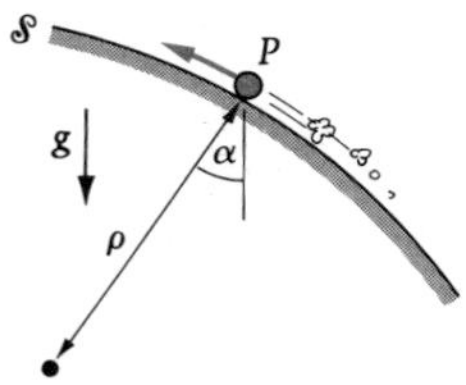

Figure P2.51

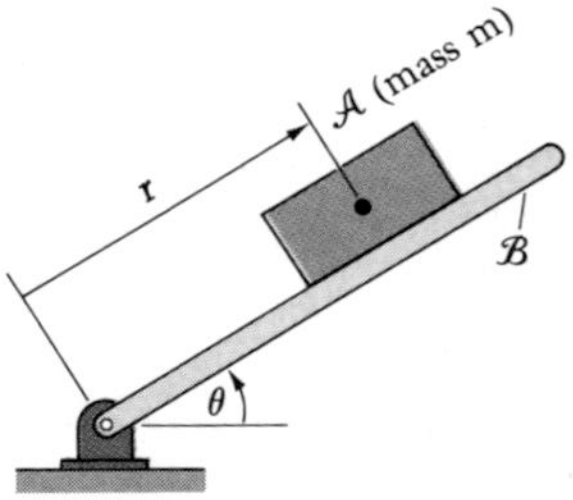

Figure P2.53

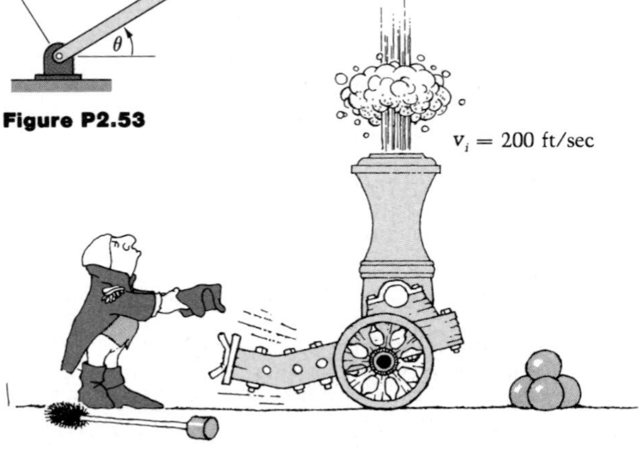

Figure P2.55

2.53 If bar $\mathcal{B}$ shown in Figure P2.53 were raised *slowly*, block $\mathcal{A}$ would start to slide at the angle $\theta = \tan^{-1}(\mu)$, which was seen in statics to be one way of determining the friction coefficient μ. Suppose now that the bar is suddenly rotated, starting from the position $\theta = 0$, at constant angular velocity ω_0 ↺. For $\mu = 0.5$ and $r\omega_0^2 = 0.1g$, compute the angle θ at which $\mathcal{A}$ slips downward on $\mathcal{B}$, and compare the result with $\tan^{-1}\mu = \tan^{-1}(0.5)$.

2.54 In the preceding problem, let μ remain at 0.5 but consider increasing the parameter $r\omega_0^2/g$. At what value of this parameter will $\mathcal{A}$ slide *outward* on $\mathcal{B}$? At what angle will this occur?

2.55 Over a certain range of velocities, the effect of air resistance on a projectile is proportional to the square of the object's speed. If the object can be regarded as a particle, the ***drag force*** is expressible as $F_D = \frac{1}{2}\rho A C_D v^2$, in which ρ is the density of the air, A is the projected area of the object onto a plane normal to the velocity vector, and C_D is a coefficient that depends on the object's shape. If $\rho A C_D = 0.0004$ lb-sec²/ft² for the 76-lb cannonball of Figure P2.55, find the maximum height it reaches. Compare your result with the answer neglecting air resistance.

2.56 In the preceding problem, find the velocity of the cannonball just before it hits the ground; again compare with the case of no air resistance.

2.57 Show that for a particle P moving in a viscous medium in which the air resistance is proportional to velocity (Figure P2.57), the differential equations of motion are

$$m\ddot{x} = -k\dot{x}$$

$$m\ddot{y} = -k\dot{y} - mg$$

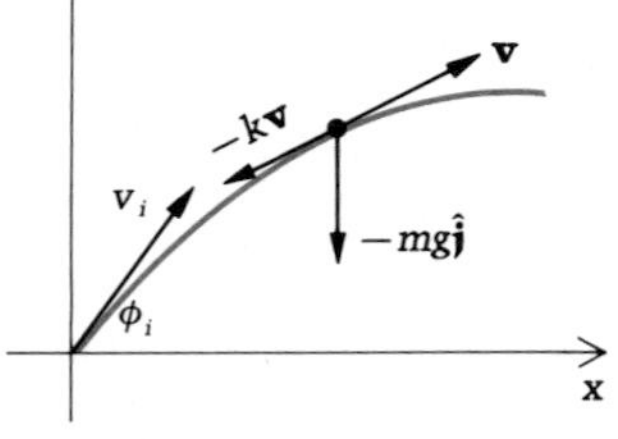

Figure P2.57

2.58 In the preceding problem, show by integration that the components of velocity of P are given by

$$\dot{x} = v_i \cos \phi_i \, e^{-kt/m}$$

$$\dot{y} = \frac{-mg}{k} + \left(v_i \sin \phi_i + \frac{mg}{k}\right) e^{-kt/m}$$

2.59 Continue the preceding exercise and show that the particle's position is given by the equations

$$x = \frac{mv_i \cos \phi_i}{k}(1 - e^{-kt/m})$$

$$y = \frac{m}{k}\left[\left(\frac{mg}{k} + v_i \sin \phi_i\right)(1 - e^{-kt/m}) - gt\right]$$

Show further that both x and $\dot{y}$ approach asymptotes as $t \to \infty$. (The limiting value of $\dot{y}$ is known as the terminal velocity of P, after which the weight is balanced by the viscous resistance so that the acceleration goes to zero.)

2.60 In Problem 1.105 what is the acceleration of the rock just *after* release?

2.61 In preparation for Problem 2.62, for the ellipse shown in Figure P2.61, the equation is

$$\frac{x^2}{a^2} + \frac{y^2}{b^2} = 1$$

Show that the radius of curvature ρ of the ellipse, as a function of x, is

$$\rho = \frac{[a^2(a^2 - x^2) + b^2x^2]^{3/2}}{a^4 b}$$

Hint: Recall from calculus that if $y = y(x)$, then

$$\rho = \left|\frac{(1 + y'^2)^{3/2}}{y''}\right|$$

2.62 In a certain amusement park, the tallest loop in a somersaulting ride (Figure P2.62) is 100 ft high and shaped approximately like an ellipse with a width of 95 ft. The ride advertises "five times the earth's pull at over 50 mph." Use the result of the preceding exercise to compute the radius of curvature at the bottom of the loop. Assuming that the normal force resultant is 5 mg, determine whether or not the maximum speed is over 50 mph. Treat the cars as a single particle.

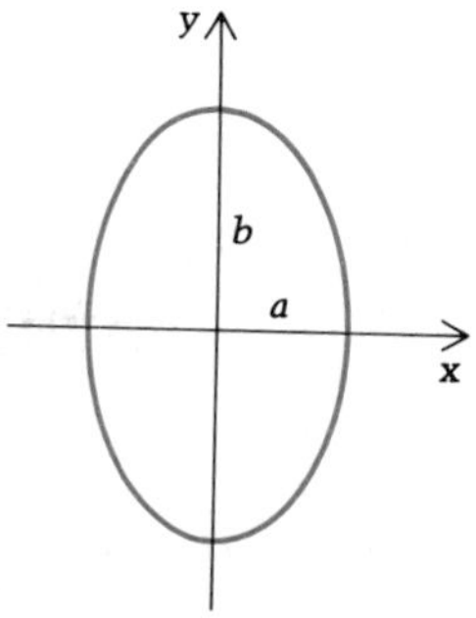

Figure P2.61

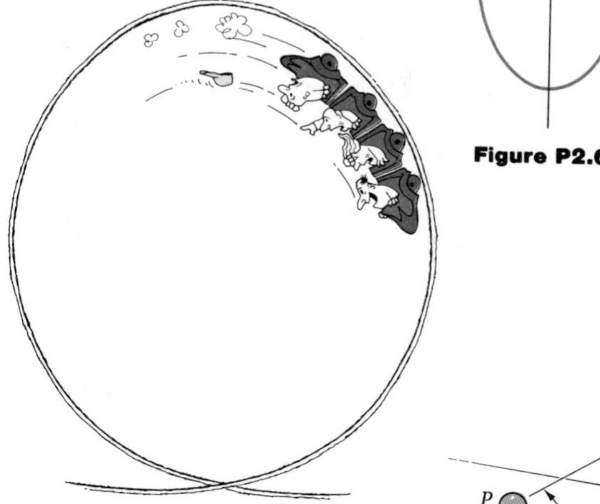

Figure P2.62

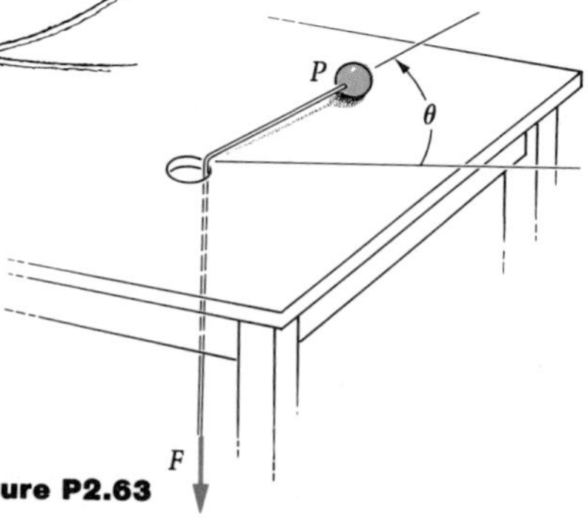

Figure P2.63

2.63 A particle P of mass m moves on a smooth, horizontal table and is attached to a light, inextensible cord that is being pulled downward by a force $F(t)$ as shown in Figure P2.63. Show that the differential equations of motion of P are

$$-F = m(\ddot{r} - r\dot{\theta}^2) \tag{1}$$

$$0 = r\ddot{\theta} + 2\dot{r}\dot{\theta} \tag{2}$$

Then show that Equation (2) implies that $r^2\dot{\theta} =$ constant.

2.64 In the preceding problem, let the particle be at $r = r_0$ at $t = 0$, and let the part of the cord beneath the table be descending at constant speed v_C. If the transverse component of velocity of P is $r\dot{\theta} = r_0\dot{\theta}_0$ at $t = 0$, find the tension in the cord as a function of time t.

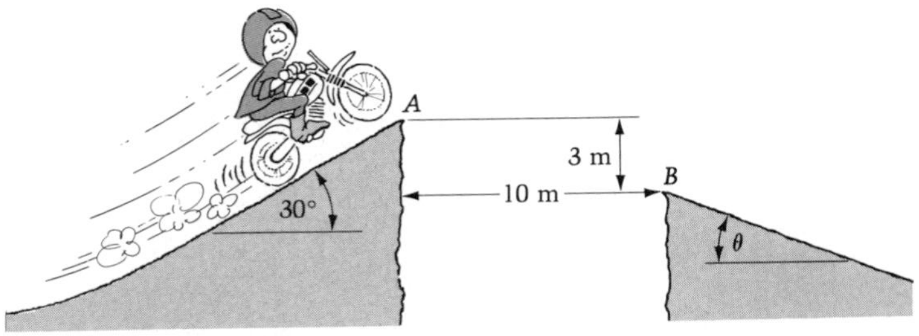

Figure P2.65

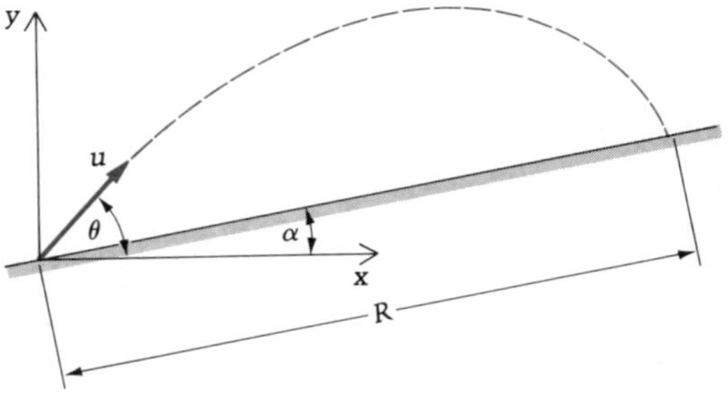

Figure P2.66

2.65 The motorcycle in Figure P2.65 is to be driven by a stunt man. Find the minimum takeoff velocity at A for which the motorcycle can clear the gap, and determine the corresponding angle θ for which the landing will be tangent to the road at B and hence smooth.

2.66 Find the range R for a projectile fired onto the inclined plane shown in Figure P2.66. Determine the maximum value of R for a given muzzle velocity u. (Angle $\alpha =$ constant.)

2.67 If a baseball player can throw a ball 90 m on the fly on earth, how far can he throw it on the moon where the gravitational acceleration is about one-sixth that on earth? Neglect the height of the player and the air resistance on earth.

2.68 The two particles in Figure P2.68 are at rest on a smooth horizontal table and connected by an inextensible string that passes through a small, smooth ring fixed to the table. The lighter particle (mass m) is then projected at right angles to the string with velocity v_0. Prove that the other particle will strike the ring with velocity $v_0\sqrt{3}/(2\sqrt{n+1})$. *Hint:* Use polar coordinates and note that $r^2\dot{\theta}$ is constant for each particle.

2.69 A particle moves on the inside of a fixed, smooth vertical hoop of radius a. It is projected from the lowest point A with velocity $\sqrt{7\,ga/2}$. Show that it will leave the hoop at a height $3a/2$ above A and meet the hoop again at A.

2.70 Particle P of mass m travels in a circle of radius a on the smooth table shown in Figure P2.70. Particle P is connected by an inextensible string to the stationary particle of mass M. Find the period of one revolution of P.

2.71 The block of mass m shown in Figure P2.71 is brought slowly down to the point of contact with the end of the spring, and then (at $t = 0$) the block is released. Write the differential equation governing the subsequent motion, clearly defining your choice of displacement parameter. What are the initial conditions? Find the maximum force induced in the spring and the first time at which it occurs.

2.72 In an emergency the driver of an automobile applies the brakes and locks all four wheels. Find the time and distance required to bring the car to rest in terms of the coefficient of sliding friction μ, the initial speed v, and the gravitational acceleration g.

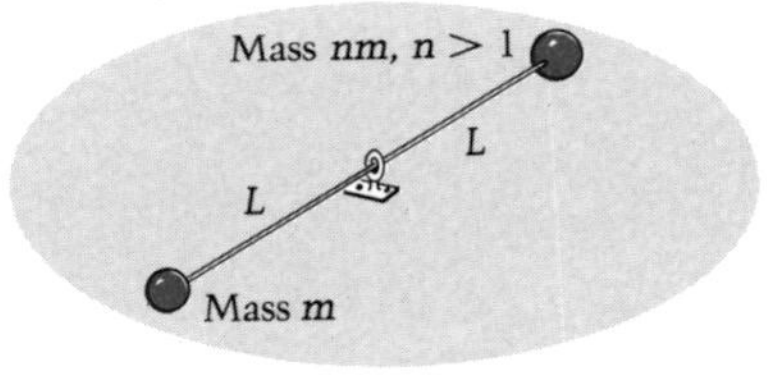

Figure P2.68

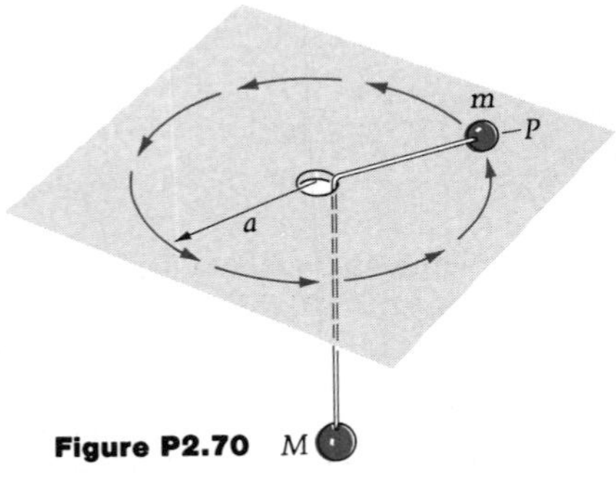

Figure P2.70

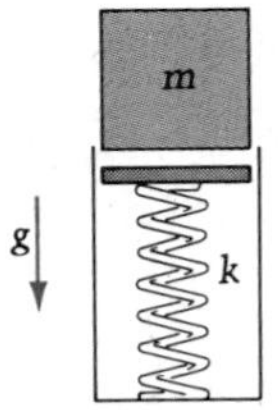

Figure P2.71

Figure P2.75

2.73 A box is placed in the rear of a pickup truck. Find the maximum acceleration of the truck for which the block does not slide on the truck bed. The coefficient of friction between the box and truck bed is μ.

2.74 A ball is dropped from the top of a tall building. The motion is resisted by the air, which exerts a drag force given by Dv^2; D is a constant and v is the speed of the ball. Find the ***terminal speed*** (the limiting speed of fall) if there is no limit on the drop height. What is the drop height for which the ball will strike the ground at 95 percent of the terminal speed?

2.75 The drag car of mass m shown in Figure P2.75, traveling at speed v_0, is to be initially slowed primarily by the deployment of a parachute. The parachute exerts a force F_d proportional to the square of the velocity of the car, $F_d = Cv^2$. Neglecting friction and the inertia of the wheels, determine the distance traveled by the car before its velocity is 40 percent of v_0. If the car and driver weigh 1000 lb and $C = 0.182$ lb-sec^2/ft^2, find the distance in feet.

2.76 In the preceding problem, suppose the drag car's speed at parachute release is 237 mph. Find the time to reach 40 percent speed.

2.77 A weight of 100 lb hangs freely from a light rope (Figure P2.77). It is pulled up by a force that is 150 lb at $t = 0$ but diminishes uniformly in magnitude at 1 lb per foot pulled up. Find the time required to pull the weight up to the platform from rest, and determine its velocity upon reaching the top.

2.78 What is the apparent weight, as perceived through pressure on the feet, of a passenger in an elevator accelerating at rate a upward (a) or downward (b)?

2.79 A 50-lb shell is fired from the cannon shown in Figure P2.79. The pressure of the expanding gases is inversely proportional to the volume behind the shell. Initially this pressure is 10 tons per square inch; just before exit, it is one-tenth this value. Find the exit velocity of the shell.

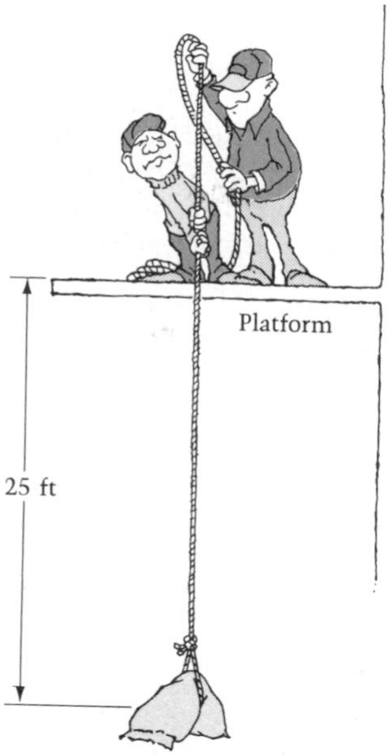

Figure P2.77

2.80 A baseball slugger connects with a pitch 4 ft above the ground. The ball heads toward the 10-ft-high center-field fence, 455 ft away. The ball leaves the bat with a velocity of 125 ft/sec and a slope of 3 vertical to 4 horizontal. Neglecting air resistance, determine whether the ball hits the fence (if it does, how high above the ground?) or whether it is a home run (if it is, by how much does it clear the fence?).

2.81 From a high vantage point in Yankee Stadium, a baseball fan observes a high-flying foul ball. Traveling vertically upward, the ball passes the level of the observer 1.5 sec after leaving the bat, and it passes this level again on its way down 4 sec after leaving the bat. Disregarding air friction, find the maximum height reached by the baseball and determine the ball's initial velocity as it leaves the bat (which is 3 ft above the ground at impact).

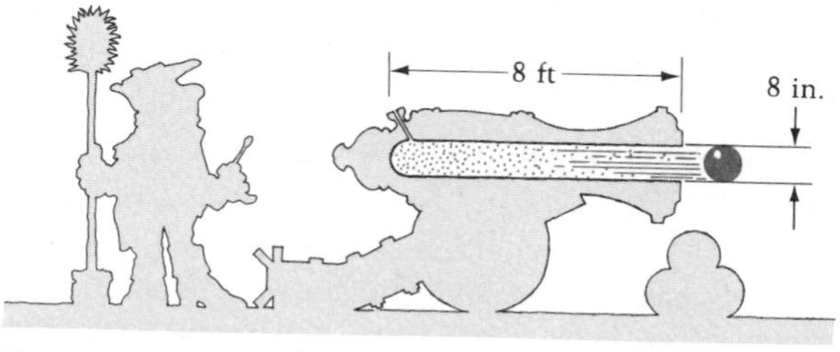

Figure P2.79

2.82 Rework Problem 2.77 but this time assume that the force *increases* by 1 lb per foot pulled up.

2.83 A 160-lb parachutist in the "free-fall spread-stable position" (Figure P2.83) reaches a velocity of 174 ft/sec in 12 sec after exiting a stationary blimp. Assuming velocity-squared air resistance, solve the differential equation of motion

$$\ddot{y} = -g - \frac{k}{m}\dot{y}^2$$

and determine the constant k.

Figure P2.83

2.84 In the preceding problem, suppose the parachutist opens his chute at a height of 1000 ft. If the value of k then becomes 0.63 lb/(ft/sec)^2, find the velocity at which the parachutist strikes the ground, if $v_i = 174$ ft/sec.

2.85 In the preceding problem, the drag force F_d can be calculated from

$$F_d = (\text{dynamic pressure}) \times (\text{projected area}) \times (\text{drag coefficient})$$

$$= (\tfrac{1}{2}\rho v^2)AC_D$$

where ρ = density of medium (0.00238 slug/ft^3 for the air) and C_D = drag coefficient = 1.35 for parachutes. Calculate the radius of the parachute.

2.86 When a man stands on a scale at one of the poles of the earth, the scale indicates weight W. Assuming the earth to be spherical (4000-mile radius) and assuming the earth to be an inertial frame, what will the scale read when the man stands on it at the equator?

2.87 Assuming the earth's orbit around the sun to be circular and supposing that a frame containing the earth's center and poles and the center of the sun is inertial, repeat Problem 2.86. Neglect the earth's tilt.

2.88 A chain of length L and mass per unit length β is held vertically above the platform scale shown in Figure P2.88 and is released from rest with its lower end just touching the platform. Assume that the links quickly come to rest as they stack up on the platform and that they do not interfere with the links still in free fall above the platform. Draw a free-body diagram of the entire chain and express the momentum as a function of the distance through which the upper end has fallen. Then determine the force read on the scale in terms of this distance.

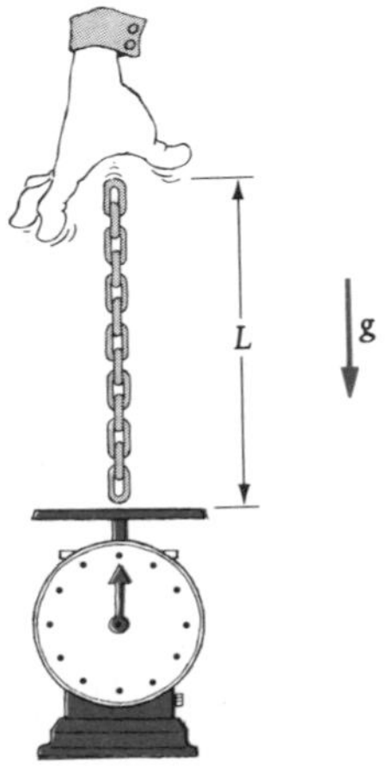

Figure P2.88

2.89 A block of mass mL, which can move on a smooth horizontal table, is attached to one end of a uniform chain of mass m per unit length. Initially the block and the chain are at rest, and the chain is completely coiled on the table. A constant horizontal force mLf is then applied to the block so that the chain begins to uncoil. Show that the length x uncoiled after time t is given by

$$(L + x)^2 = Lft^2 + L^2$$

until the chain is completely uncoiled. If the length of the chain is very large compared with L, show that the velocity of the block is approximately equal to $(Lf)^{1/2}$ at the moment when the chain is completely uncoiled.

2.4 Work and Kinetic Energy for Motions of Mass Centers

In Example 2.7 we were able to get useful information from an energy integral of the governing differential equation. The same result may be obtained in general by an integration of

$$\mathbf{F}_r = m\mathbf{a}_C$$

Forming the dot product of each side with the velocity $\mathbf{v}_C$ of the mass center, we have

$$\begin{aligned}\mathbf{F}_r \cdot \mathbf{v}_C &= m\mathbf{a}_C \cdot \mathbf{v}_C \\ &= m\frac{d\mathbf{v}_C}{dt} \cdot \mathbf{v}_C \\ &= \frac{m}{2}\frac{d}{dt}(\mathbf{v}_C \cdot \mathbf{v}_C) \\ &= \frac{d}{dt}\left(\frac{m}{2}\mathbf{v}_C \cdot \mathbf{v}_C\right) \\ &= \frac{d}{dt}\left(\frac{m}{2}|\mathbf{v}_C|^2\right)\end{aligned}$$

Integrating,* we get

$$\int_{t_1}^{t_2} \mathbf{F}_r \cdot \mathbf{v}_C\, dt = \frac{m}{2}[|\mathbf{v}_C(t_2)|^2 - |\mathbf{v}_C(t_1)|^2] \tag{2.19}$$

or, for a particle,

$$\int_{t_1}^{t_2} \mathbf{F}_r \cdot \mathbf{v}\, dt = \frac{m}{2}[|\mathbf{v}(t_2)|^2 - |\mathbf{v}(t_1)|^2] \tag{2.20}$$

For a particle, $\int_{t_1}^{t_2} \mathbf{F}_r \cdot \mathbf{v}\, dt$ is called the **work done on the particle** by the resultant of external forces.† We note that if there are N forces acting on the particle, then the resultant $\mathbf{F}_r$ is given by $\mathbf{F}_1 + \mathbf{F}_2 + \cdots + \mathbf{F}_N$ and

$$\begin{aligned}\mathbf{F}_r \cdot \mathbf{v} &= (\mathbf{F}_1 + \mathbf{F}_2 + \cdots + \mathbf{F}_N) \cdot \mathbf{v} \\ &= \mathbf{F}_1 \cdot \mathbf{v} + \mathbf{F}_2 \cdot \mathbf{v} + \cdots + \mathbf{F}_N \cdot \mathbf{v}\end{aligned}$$

Each term of this equation represents the **rate of work** of one of the forces. Thus the left side of Equation (2.20) may be read as the sum of the works of the individual forces acting on the particle. These statements are all consistent with the presentation to come in Chapter 5 in

*Sometimes (t_i, t_f), referring to 'initial' and 'final,' are used instead of (t_1, t_2).

†Thus the appropriate unit of work and of energy in SI is the joule (J), the joule being 1 N · m; in U.S. units the ft-lb is the unit of work and energy. The N · m and lb-ft are usually reserved for the moment of a force. Note that work, energy, and moment of force all have the same dimension.

which we define the rate of work done by a force $\mathbf{F}$ to be $\mathbf{F} \cdot \mathbf{v}$, where $\mathbf{v}$ is the velocity of the point in the body at which the force is applied. The left side of Equation (2.19) may then be interpreted as the work that would be done by the external forces if each had a line of action through the mass center.

For a *particle*, $(m/2)|\mathbf{v}|^2$ is called the **kinetic energy,** usually written T. Thus for the particle, Equation (2.20) is the work and kinetic energy principle:

$$\begin{matrix}\text{Work done on} \\ \text{the particle}\end{matrix} = \begin{matrix}\text{change in the particle's} \\ \text{kinetic energy}\end{matrix}$$

or

$$W = \Delta T \tag{2.21}$$

For a *body*, the kinetic energy is defined to be the sum of the kinetic energies of the particles constituting the body. If all the points in a body $\mathcal{B}$ have the same velocity (which is then $\mathbf{v}_C$), then $(m/2)|\mathbf{v}_C|^2$ is the *total* kinetic energy of $\mathcal{B}$. In general, however, the body is turning or deforming (or both) and this is not the case; the body then has *additional* kinetic energy due to its changes in orientation (that is, due to its angular motion) or due to the deformation. For a body $\mathcal{B}$, we shall also see in Chapter 5 that, in general, the left side of Equation (2.19) does not constitute the total work done on $\mathcal{B}$ by the external forces and couples. This is because, for a body, the forces do not have to be concurrent as they are for a particle. Equation (2.21) still turns out to be true for a rigid body, however, with the two sides of Equation (2.19) representing *parts* of W and ΔT.

Finally, with no restrictions on the size of the body, the energy integral (Equation 2.19) states that the work that would be done if the external forces acted at the mass center equals the change in what would be the kinetic energy if every point in the body had the velocity of the mass center. We could call this result the **"mass center work and kinetic energy principle."**

Before attempting to apply the work-energy principle to a specific problem it is helpful to determine the work done by two classes of forces. First, suppose $\mathbf{F}$ is a constant force and suppose we let $\mathbf{r}_{OC}$ be a position vector for the mass center C. Then

$$\begin{aligned}\int_{t_1}^{t_2} \mathbf{F} \cdot \mathbf{v}_C \, dt &= \mathbf{F} \cdot \int_{t_1}^{t_2} \mathbf{v}_C \, dt \\ &= \mathbf{F} \cdot \int_{t_1}^{t_2} \frac{d\mathbf{r}_{OC}}{dt} \, dt \\ &= \mathbf{F} \cdot [\mathbf{r}_{OC}(t_2) - \mathbf{r}_{OC}(t_1)] \end{aligned} \tag{2.22}$$

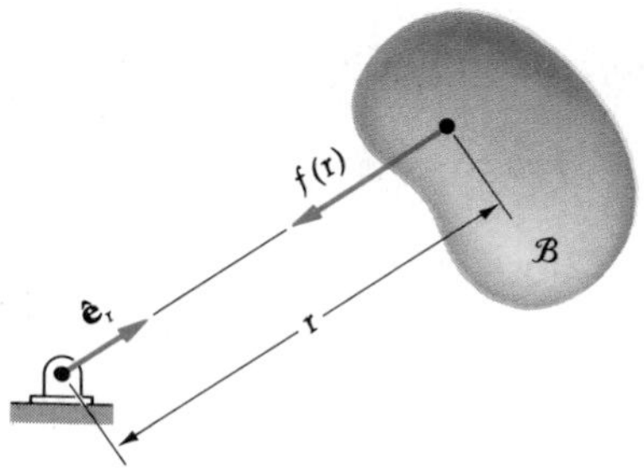

Figure 2.7

which states that the work done is the dot product of the force with the displacement of the mass center. We recall that this dot product can be expressed as the product of the force magnitude and the component of displacement in the direction of the force or as the product of the displacement magnitude and the component of force in the direction of the displacement.

The second case to which we give special attention is that of a **central force.** Such a force is defined to have a line of action always passing through the same fixed point in the frame of reference and a magnitude that depends only upon the distance r of the point of application from that fixed point, as shown in Figure 2.7.

The velocity of the point of application may be expressed as

$$\mathbf{v} = \frac{d}{dt}(r\hat{\mathbf{e}}_r) = \dot{r}\hat{\mathbf{e}}_r + r\dot{\hat{\mathbf{e}}}_r$$

$$\begin{aligned}\mathbf{F} \cdot \mathbf{v} &= [-f(r)\hat{\mathbf{e}}_r] \cdot [\dot{r}\hat{\mathbf{e}}_r + r\dot{\hat{\mathbf{e}}}_r] \\ &= -\dot{r}\,f(r)\end{aligned}$$

since by Equation (1.38) we know that $\hat{\mathbf{e}}_r \cdot \dot{\hat{\mathbf{e}}}_r = 0$. Thus the work done by $\mathbf{F} = -f(r)\hat{\mathbf{e}}_r$ is

$$W = \int_{t_1}^{t_2} -f(r)\frac{dr}{dt}\,dt = -\int_{r(t_1)}^{r(t_2)} f(r)\,dr \tag{2.23}$$

If φ is a function of r so that $f = d\varphi/dr$, then the work may be written as

$$W = -\int_{r(t_1)}^{r(t_2)} \frac{d\varphi}{dr}\,dr = -\varphi[r(t_2)] + \varphi[r(t_1)] \tag{2.24}$$

A special central force is that exerted by a spring on a body when the other end of the spring is fixed. In the case of a linear spring of instantaneous length r, we note that $f = k(r - L_0)$, where L_0 is its natural, or unstretched, length and k is called the **spring modulus** or **stiffness.** In this case, $\varphi = (k/2)(r - L_0)^2$ or, more simply, $\varphi = (k/2)\delta^2$, where δ is the spring stretch. Thus by Equation (2.24),

$$W = -\frac{k}{2}[\delta^2(t_2) - \delta^2(t_1)] \tag{2.25}$$

Question 2.5 What assumption about the mass of the spring is to be understood in the force-stretch relationship?

A second special case of a central force is the gravitational force exerted on a body by the earth. By Newton's law of universal gravitation,

$$f = \frac{mgr_e^2}{r^2} \tag{2.26}$$

where r_e is the radius of the earth, m is the mass of the attracted body, and g is the gravitational strength (or acceleration) at the surface of the earth.* By Equation (2.23) we have

$$W = -\int_{r(t_1)}^{r(t_2)} \frac{mgr_e^2}{r^2}\,dr = mgr_e^2\left[\frac{1}{r(t_2)} - \frac{1}{r(t_1)}\right]$$

We note that for this case the function φ is given by

$$\varphi = \frac{-mgr_e^2}{r} \tag{2.27}$$

If the motion is sufficiently near the surface of the earth,

$$\frac{r_e^2}{r^2} \approx 1$$

and so $f \approx mg$, a constant. In this case,

$$\begin{aligned} W &= \frac{mgr_e^2[r(t_1) - r(t_2)]}{r(t_1)r(t_2)} \\ &\approx mg[r(t_1) - r(t_2)] \\ &= (\text{weight}) \times (\text{decrease in altitude of mass center of body}) \end{aligned} \tag{2.28}$$

and the φ function becomes (if z_C is positive upward)

$$\varphi = mgz_C \tag{2.29}$$

In each case we have considered, the work has depended only on the initial and final positions of the point where the force is applied. Such a force whose work is independent of the path traveled by the point on which it acts is called **conservative.** Furthermore, the work may be expressed as the change in a scalar function of position; we saw this to be the case for the central force, and we may make the same statement for the constant force by defining φ to be $-\mathbf{F} \cdot \mathbf{r}_{OC}$.

*The force of gravity in fact results in infinitely many differential forces, each tugging on one of the body's particles. For nearly all applications on the planet earth, these forces may be thought of as equivalent to a single force through the mass center of the body. For applications in astronomy or in space vehicle dynamics, however, the gravity moment that accompanies the force at the mass center becomes important. In Skylab, for example, three huge control-moment gyros were present to 'take out' the angular momentum built up by a gravity moment of only a few lb-ft. And the gravity moment exerted on the earth by the sun and moon's gravitation causes the earth's axis to precess in the heavens once every 25,800 yr. The gravity moment vanishes if the body is a uniform sphere (which the earth is not, being bulged at the equator and having varying density). A further discussion of this *luni-solar precession* is presented in Chapter 7.

Question 2.6 Why the minus sign in front of $\mathbf{F} \cdot \mathbf{r}_{OC}$?

Thus if all forces acting are conservative and if φ is the sum of all their φ functions, then the work and kinetic energy equation (2.21) becomes

$$\varphi(t_1) - \varphi(t_2) = T(t_2) - T(t_1)$$

or

$$T(t_2) + \varphi(t_2) = T(t_1) + \varphi(t_1)$$

or

$$T + \varphi = \text{constant} \tag{2.30}$$

We call φ the **potential energy** and $T + \varphi$ the **total (mechanical) energy.** Thus Equation (2.30) is a statement of conservation of mechanical energy when all the forces are conservative (path-independent).

Question 2.7 How would Equation (2.30) read if instead of φ we had chosen to construct the scalar function ψ so that the work done by a force is the increase in its ψ?

In closing it is important to realize that not all forces are conservative. An example is the force of friction acting on a sliding block. That force does negative work regardless of the direction of the motion, and a potential function φ cannot be found for it.

EXAMPLE 2.9

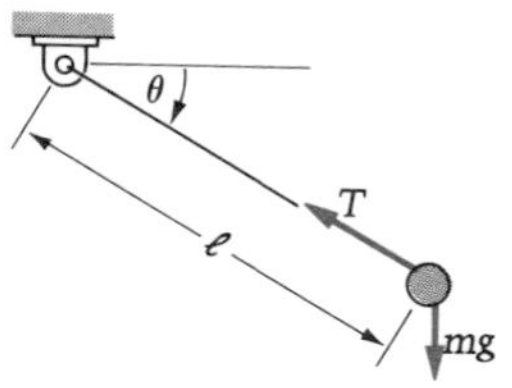

We repeat Example 2.7 (see the diagram): For a ball of mass m released from rest with the cord taut and $\theta = 30°$, we wish to find the tension in the cord as a function of θ.

SOLUTION

If, as before, we write the force-acceleration component equation in the radial direction, we have

$$T = mg \sin\theta + m\ell\dot{\theta}^2 \tag{1}$$

(Continued)

Now we apply Equation (2.19) by letting t_1 be the initial time at which $\theta = 30°$ and letting t_2 be the time at which we are applying (2.21). We note that

$$|\mathbf{v}_C(t_1)| = 0$$

$$|\mathbf{v}_C(t_2)|^2 = \ell^2\dot{\theta}^2$$

and the work done by the cord tension T is zero since that force is always perpendicular to the velocity of (the center of mass of) the ball. By Equation (2.21) the work of the weight is $mg[\ell \sin\theta - \ell \sin 30°]$. Thus Equation (2.19) yields

$$mg[\ell \sin\theta - \ell \sin 30°] = \tfrac{1}{2}m(\ell^2\dot{\theta}^2) - 0$$

or

$$m\ell\dot{\theta}^2 = 2mg(\sin\theta - \tfrac{1}{2})$$

Substituting in Equation (1) above, we get

$$T = mg\sin\theta + 2mg\sin\theta - mg$$

$$= mg(3\sin\theta - 1)$$

which is precisely the result obtained previously.

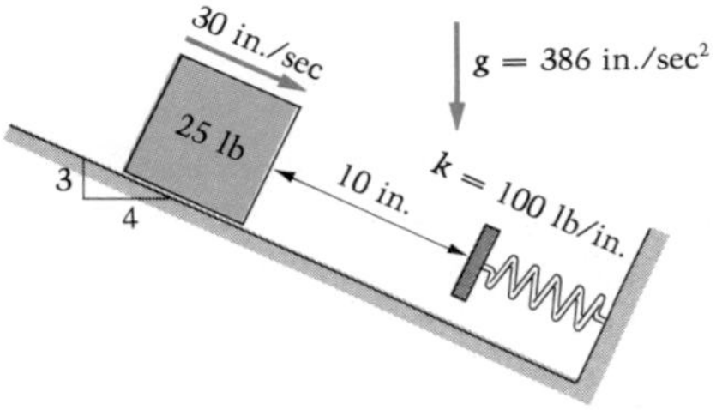

EXAMPLE 2.10

The block shown in the diagram slides on an inclined surface for which the coefficient of friction is $\mu = 0.3$. Find the maximum force induced in the spring if the motion begins under the conditions shown.

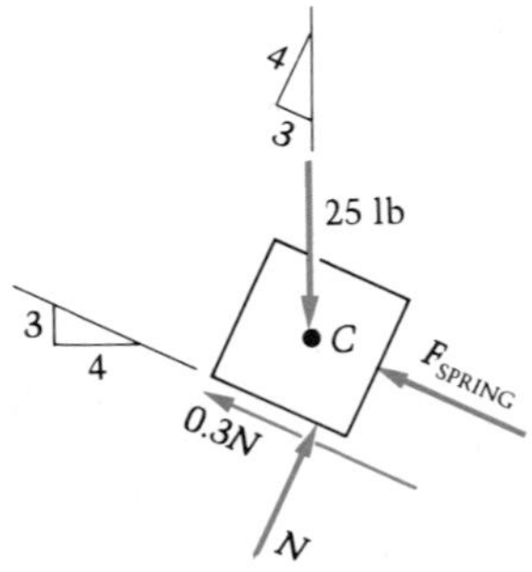

SOLUTION

We assume that the block can be treated as rigid; thus the end of the spring, once it contacts the block, will undergo the same displacements as the mass center (or for that matter any other point) of the block. To apply the mass-center work and kinetic energy principle we let t_1 be the initial time shown above and let t_2 be the time of maximum compression Δ of the spring. To catalog the external forces that do work we consider a free-body diagram at some arbitrary instant between t_1 and t_2. (See the figure.) Since the mass center has a path parallel to the inclined plane, there is no component of acceleration perpendicular to it and

$$N - \tfrac{4}{5}(25) = 0$$

or

$$N = 20 \text{ lb}$$

so that the friction force is $0.3(20) = 6$ lb.

(Continued)

Denoting the left side of Equation (2.19) by work(t_1, t_2) we have

$$\text{Work}(t_1, t_2) = \tfrac{1}{2}m\,[|\mathbf{v}_C(t_2)|^2 - |\mathbf{v}_C(t_1)|^2]$$

where

$$|\mathbf{v}_C(t_2)| = 0$$

$$|\mathbf{v}_C(t_1)| = 30 \text{ in./sec}$$

and the various works are:

1. For N, work(t_1, t_2) $= 0$ since the force is perpendicular to $\mathbf{v}_C$ at each instant.
2. For friction, work(t_1, t_2) $= -6(10 + \Delta)$ in.-lb
3. For the weight, work(t_1, t_2) $= [(\tfrac{3}{5})(25)](10 + \Delta)$ in.-lb
4. For the spring,

$$\text{Work}(t_1, t_2) = \text{work}(t_1, \text{contact}) + \text{work}(\text{contact}, t_2)$$

$$= 0 - \frac{100}{2}[(-\Delta)^2 - 0] \text{ in.-lb}$$

Thus, $W = \Delta T$ gives (with $g = 32.2 \times 12 = 386$ in./sec^2)

$$-6(10 + \Delta) + 15(10 + \Delta) - 50\Delta^2 = 0 - \frac{1}{2}\left(\frac{25}{386}\right)(30)^2$$

or

$$50\Delta^2 - 9\Delta - 119 = 0$$

$$\Delta^2 - 0.180\Delta - 2.38 = 0$$

From the quadratic formula,

$$\Delta = 0.09 \pm \sqrt{0.0081 + 2.38} \text{ in.}$$

from which only the positive root is meaningful:

$$\Delta = 0.09 + 1.55 = 1.64 \text{ in.}$$

The corresponding force is 100(1.64) = 164 lb.

EXAMPLE 2.11

In the preceding example, find the next position at which the block comes to rest.

SOLUTION

At time t_2 the spring force (164 lb) exceeds the sum of the component of weight along the plane (15 lb) and the maximum frictional resistance (6 lb), so we know that the block is not in equilibrium and must then begin to move back up the plane. Suppose we let t_3 be the time at which the block next comes to rest *(Continued)* and let d represent the corresponding compression of the spring. Then, since $|\mathbf{v}_C(t_2)| = |\mathbf{v}_C(t_3)| = 0$,

$$\text{Work}(t_2, t_3) = 0$$

For the spring, the work is $(-100/2)[d^2 - (-\Delta)^2]$; for the friction force, the work is $-6(\Delta - d)$; for the weight, the work is $-(\frac{3}{5})(25)(\Delta - d)$. Thus

$$(\Delta - d)\left[\frac{100}{2}(\Delta + d) - 6 - 15\right] = 0$$

or

$$d = \frac{21}{50} - 1.64 = -1.22 \text{ in.}$$

The negative sign here tells us that the spring must be stretched 1.22 in. when the block again comes to rest. If, as intended here, the spring does not become permanently attached to the block on first contact (that is, contact is maintained only in compression), our analysis only tells us that contact is broken before the block comes to rest. We therefore need to modify the expression for the work done by the spring, which we now see should have been $-(100/2)(0 - \Delta^2)$. It is convenient to let e be the distance from the end of the spring to the block (measured up the plane). Then

$$-\frac{100}{2}[0 - \Delta^2] - 6(\Delta + e) - \frac{3}{5}(25)(\Delta + e) = 0$$

$$21(\Delta + e) = 50\Delta^2$$

$$e = \frac{50}{21}\Delta^2 - \Delta$$

$$= \frac{50}{21}(1.64)^2 - 1.64$$

$$= 4.76 \text{ in.}$$

It is instructive to obtain this result by using the work and kinetic energy principle over the interval t_1 to t_3, for which the net work done by the spring is zero. Noting then that the mass center of the block drops $\frac{3}{5}(10 - e)$ in. vertically and that the distance traveled by the block on the plane is $(10 + 1.64 + 1.64 + e)$ in., we have

$$\text{Work}(t_1, t_3) = \frac{1}{2}m|\mathbf{v}_C(t_3)|^2 - \frac{1}{2}m|\mathbf{v}_C(t_1)|^2$$

$$25\left[\frac{3}{5}(10 - e)\right] - 6(13.3 + e) = 0 - \frac{1}{2}\left(\frac{25}{386}\right)(30)^2$$

$$150 - 15e - 79.8 - 6e = -29.2$$

$$e = 4.73 \text{ in.}$$

which is the same result we obtained before except for the third significant figure—a consequence of rounding off at an intermediate step.

EXAMPLE 2.12

A 120-lb person skis down a smooth parabolic incline. (See the diagram.) Find the force exerted on the bottom of the skis by the snow when the skier passes the lowest point (at the origin). Neglect friction.

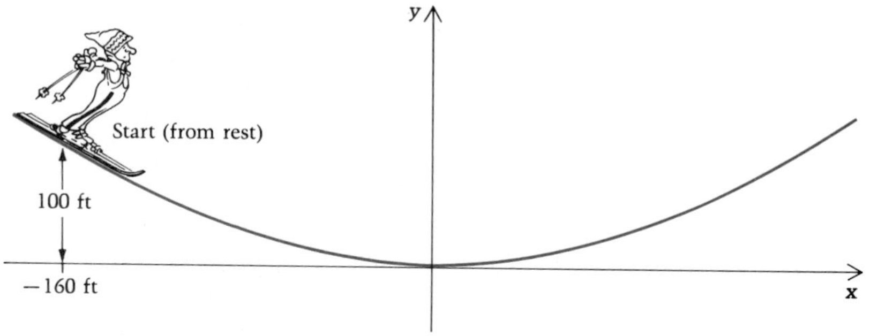

SOLUTION

This problem couples the solution to an equation of motion with an application of the principle of work and kinetic energy. First, using Equation (2.19) and noting that only the gravity force is doing work on the skier,

$$W = mgh = \tfrac{1}{2}m\,(v_2^2 - v_1^2)$$

$$32.2(100) = \tfrac{1}{2}(v_2^2 - 0^2)$$

$$v_2 = 80.3 \text{ ft/sec} \tag{1}$$

From the free-body diagram shown here, we can write the equation of motion in the normal direction at the origin:*

$$F_{r_n} = m\frac{v_2^2}{\rho}$$

or

$$N - 120 = \frac{120}{32.2}\frac{v_2^2}{\rho} \tag{2}$$

and we see that in order to proceed we need the radius of curvature at the origin. The parabola has the equation

$$4fy = x^2$$

(Continued)

*Note that the equations of motion are good for all times t, so that we can apply them at one instant if we wish.

and since $y = 100$ ft when $x = 160$ ft, the focal length $f = 64.0$ ft. Therefore the curvature formula gives

$$\kappa = \frac{1}{\rho} = \left|\frac{y''}{(1 + y'^2)^{3/2}}\right| = \frac{1/128}{[1 + (x^2/128^2)]^{3/2}}$$

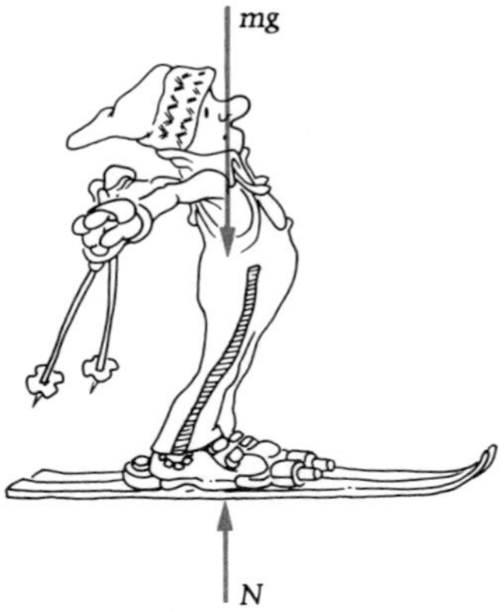

At $x = 0$, then, we have $\rho = 128$ ft. Substituting into Equation (2), we obtain

$$N - 120 = \frac{120}{32.2}\frac{v_2^2}{128} = 0.0291 v_2^2$$

Using Equation (1), we get

$$N = 120 + 0.0291(80.3^2)$$

$$= 308 \text{ lb}$$

The force N, the speed v_2, and the acceleration at the origin (nearly 1.6 g's) are all somewhat higher than one might realistically expect.

Question 2.8 Why?

EXAMPLE 2.13

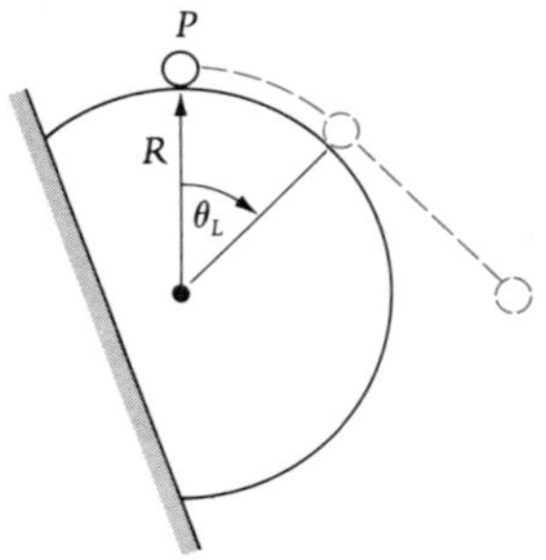

A particle P of mass m rests atop a smooth spherical surface. (See the diagram.) A slight nudge starts it sliding downward in a vertical plane. Find the angle θ_L at which the particle leaves the surface.

SOLUTION

Mechanical energy is conserved here because (1) there is no friction, (2) the normal force does no work since it is always normal to the velocity of P, and (3) the only other force is gravity. (See the diagram.) Therefore, using Equations (2.29) and (2.30),

$$T_1 + \varphi_1 = T_2 + \varphi_2$$

$$0 + mgR = \tfrac{1}{2} m v_2^2 + mgR \cos\theta_L$$

$$v_2^2 = 2gR(1 - \cos\theta_L) \qquad (1)$$

(Continued)

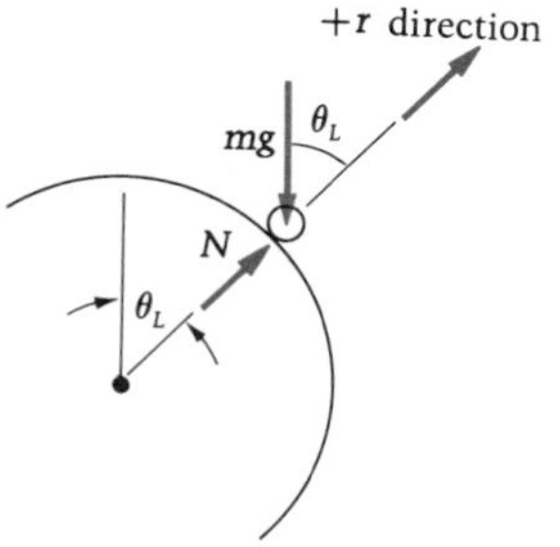

Equation (1) contains two unknowns; to eliminate the velocity v_2, we use the equation of motion in the radial direction:

$$F_r = ma_r$$

$$-mg\cos\theta_L + N = m\left(-\frac{v_2^2}{R}\right) \tag{2}$$

But N has just become zero when P is at the point of leaving. Therefore

$$v_2^2 = gR\cos\theta_L \tag{3}$$

Equating the right sides of Equations (1) and (3), we get

$$gR\cos\theta_L = 2gR(1-\cos\theta_L)$$

$$3\cos\theta_L = 2$$

$$\theta_L = \cos^{-1}(\tfrac{2}{3}) = 48.2°$$

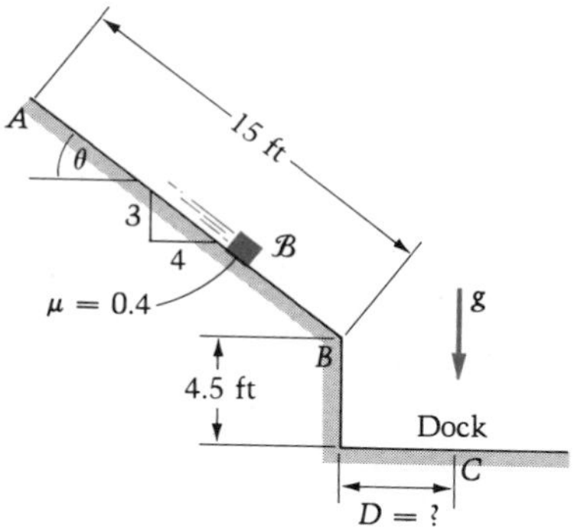

EXAMPLE 2.14

A small box $\mathcal{B}$ (see the diagram) slides from rest down a rough inclined plane from A to B and then falls onto the loading dock. The coefficient of sliding friction between the box and plane is $\mu = 0.4$. Find the distance D to the point C where the box strikes the dock.

SOLUTION

This problem has two parts. First we find the "leaving velocity" v_L at B via the principle of work and kinetic energy. Then we use this velocity as an initial condition for the second stage of the solution as the box falls onto the dock. From A to B,

$$W = \Delta T = T_f - T_i = T_f - 0$$

$$-15f + mg(\tfrac{3}{5}\times 15) = \tfrac{1}{2}mv_L^2 \quad (\mu N \text{ here})$$

$$-15(0.4\times\tfrac{4}{5}\times m\times 32.2) + 9m(32.2) = \tfrac{1}{2}mv_L^2 \tag{1}$$

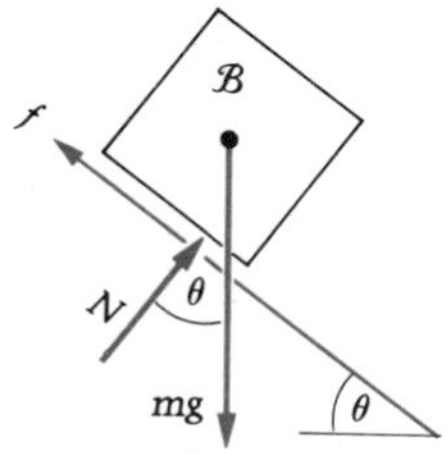

where the normal force, using the free-body diagram shown here and the fact that the box does not accelerate normal to the plane until leaving it, is $N = \frac{4}{5}mg$.

(Continued)

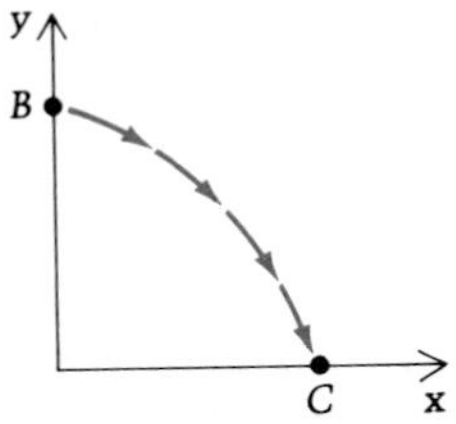

The mass cancels in Equation (1) and we obtain

$$v_L = 16.4 \text{ ft/sec}$$

Neglecting air resistance as the box then slides off the plane and falls (see the diagram), we get

$$F_{r_y} = m\ddot{y}$$

$$-mg = m\ddot{y}$$

$$\ddot{y} = -32.2 \text{ ft/sec}^2$$

$$\dot{y} = -32.2t + C_1 \text{ ft/sec}$$

If we start measuring time when the box is at B, then the initial condition on the vertical component of velocity is

$$\dot{y}(0) = -16.4(\tfrac{3}{5}) = -9.84 \text{ ft/sec}$$

so that

$$\dot{y} = -32.2t - 9.84 \text{ ft/sec}$$

Integrating again, we have

$$y = -16.1t^2 - 9.84t + C_2 \text{ ft}$$

But y is 4.5 ft when $t = 0$, so that

$$y(0) = 4.5 = -0 - 0 + C_2 \text{ ft}$$

The vertical position of the box is therefore given by

$$y = -16.1t^2 - 9.84t + 4.5 \text{ ft}$$

When $y = 0$, the box hits the dock (at a time t_f to be computed):

$$16.1t_f^2 + 9.84t_f - 4.5 = 0$$

The quadratic formula gives the roots:

$$t_f = \frac{-9.84 \pm \sqrt{9.84^2 - 4(16.1)(-4.5)}}{2(16.1)}$$

$$= 0.305 \text{ sec and } -0.916 \text{ sec}$$

The root $t_f = 0.305$ sec is the only physically meaningful striking time, since the box left the plane at $t = 0$. In the $\mathbf{x}$ direction, we have

$$F_{r_x} = m\ddot{x}$$

or, since the only force acting on the box is gravity,

$$\ddot{x} = 0$$

Integrating, we get

$$\dot{x} = C_3$$

(Continued)

The initial condition this time is

$$\dot{\mathbf{x}}(0) = 16.4(\tfrac{4}{5}) = 13.1 \text{ ft/sec}$$

Thus $C_3 = 13.1$ ft/sec and, integrating again, we have

$$\mathbf{x} = 13.1t + C_4 \text{ ft}$$

With $\mathbf{x} = 0$ at $t = 0$, we obtain $C_4 = 0$. Finally, when $t = t_f = 0.305$ sec, the distance D is calculated to be

$$\mathbf{x}\big|_{t=t_f} = D = 13.1(0.305) = 4.0 \text{ ft}$$

PROBLEMS / Section 2.4

2.90 The weight shown in Figure P2.90 is prevented from sliding down the inclined plane by a cable. An engineer wishes to lower the weight to the dashed position by inserting a spring and then cutting the cable. Find the modulus of a spring that will accomplish this task without allowing the block to move back up the incline after it stops. *Hint:* You are free to specify the initial stretch—try zero!

2.91 The block shown in Figure P2.91 is released from rest. Find how far it rebounds back up the plane after compressing the spring.

2.92 At the instant shown in Figure P2.92, the block is traveling to the left at 7 m/s and the spring is unstretched. Using $W = \Delta T$, find the velocity of the block (that is, the velocity of C) when it has moved 4 m to the left. *Hint:* You will need the integral

$$\int \frac{d\mathbf{x}}{\sqrt{a^2 + \mathbf{x}^2}} = \ln\left(\mathbf{x} + \sqrt{a^2 + \mathbf{x}^2}\right) + c$$

where a and c are constants. Also, $k = 80$ N/m.

2.93 Check the solutions to Problems 2.27 and 2.28 by using the principle of work and kinetic energy.

2.94 A particle of mass m slides down a frictionless chute and enters a circular loop of diameter d. (See Figure P2.94.) Find the minimum starting height h in order that the particle will make a complete circuit of the loop and exit normally (without having lost contact with the loop).

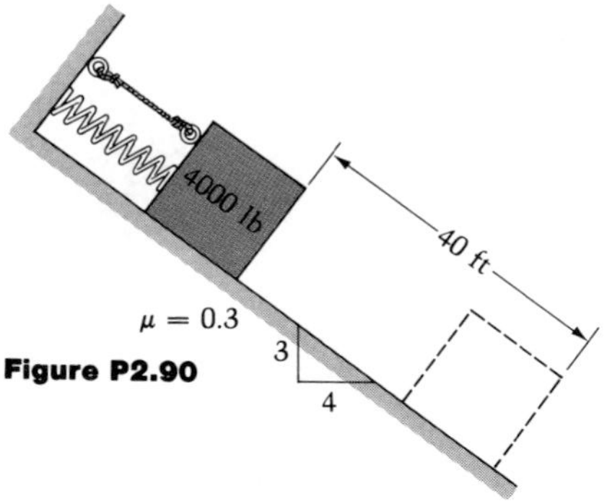

Figure P2.90

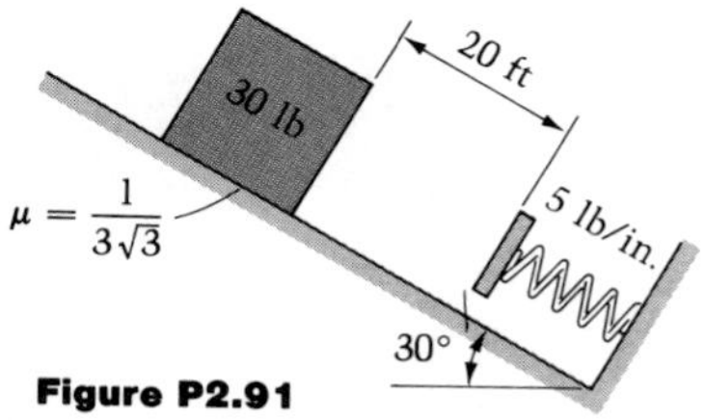

Figure P2.91

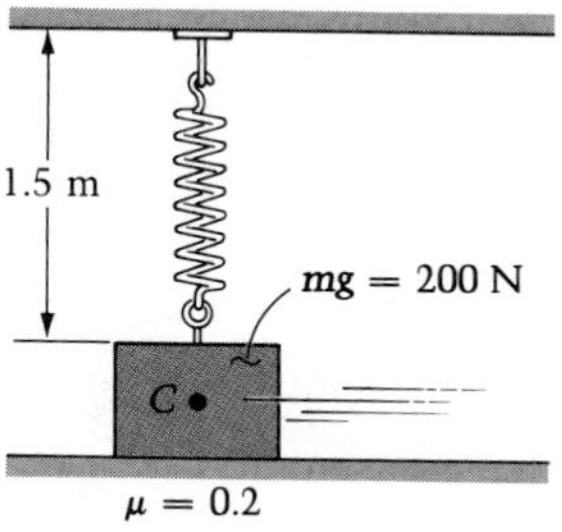

Figure P2.92

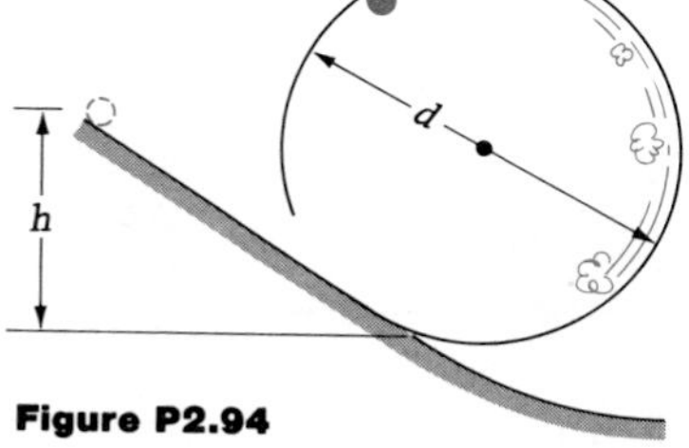

Figure P2.94

2.95 Show that if the surface in Example 2.13 is the parabola shown in Figure P2.95, the particle will *never* leave the surface. *Hint:* Show that

$$\rho = \left|\frac{(1 + y'^2)^{3/2}}{y''}\right| = \frac{(1 + 4x^2)^{3/2}}{2}$$

and use this in our equation:

$$F_{r_n} = m\frac{\dot{s}^2}{\rho}$$

together with $W = \Delta T$.

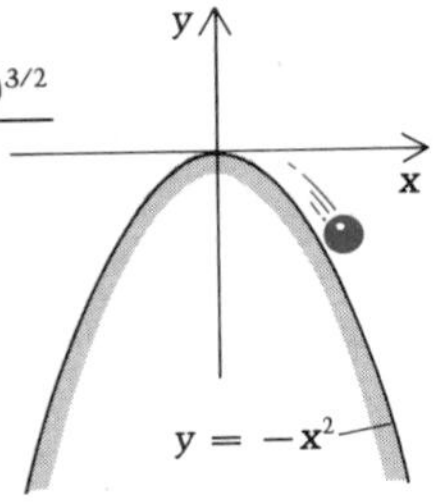

Figure P2.95

2.96 The block shown in Figure P2.96 weighs 100 lb and the spring's modulus is 10 lb/ft. The spring is unstretched when the block is released from rest. Find the minimum coefficient of friction μ such that the block will not start back up the plane after it stops.

2.97 Suppose the ends of a spring are attached to "particles" of mass m_1 and m_2. Show that the sum of the works of the spring forces on the particles is given by Equation (2.25).

2.98 The blocks in Figure P2.98 are released from rest. Determine where they are when they stop permanently. What is the spring force then? *Hint:* Write the work-energy equation for each block, add the two equations, and use the result of Problem 2.97. Also, think about the motion of the mass center.

2.99 Block $\mathcal{A}$ weighs 16.1 lb and translates along a smooth horizontal plane with a speed of 36 ft/sec. (See Figure P2.99.) The coefficient of friction between $\mathcal{A}$ and the inclined surface is $\mu = 0.5$, and the spring constant is 100 lb/ft. Determine the distance that $\mathcal{A}$ moves up the incline before coming to rest.

2.100 The Bernoulli brothers posed and then solved the "brachistochrone problem." (See Figure P2.100.) The problem was to determine on which single-valued, continuous, smooth path a particle would arrive at B in minimum time under uniform gravity, after beginning at rest at a higher point A. Their solution, beyond the scope of this book, was that this path of "quickest descent" is a cycloid. Show that, regardless of the path, the *speed on arrival* is as if the particle had been dropped freely through the same height H.

2.101 An 80-lb child rides a 10-lb wagon down an incline (Figure P2.101). Neglecting all frictional losses, find the "weight" of the child at A as indicated by a scale upon which she is sitting.

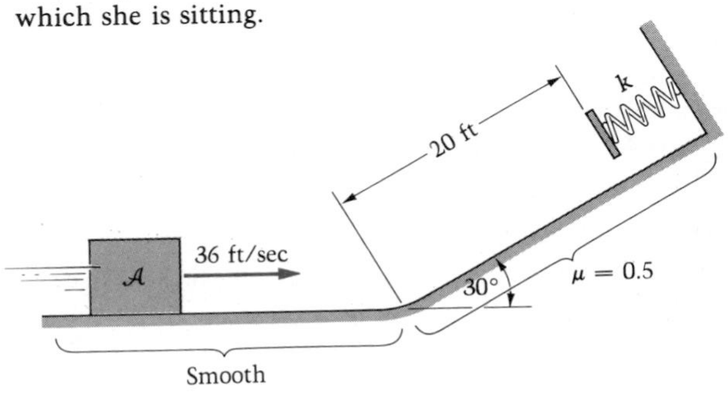

Figure P2.99

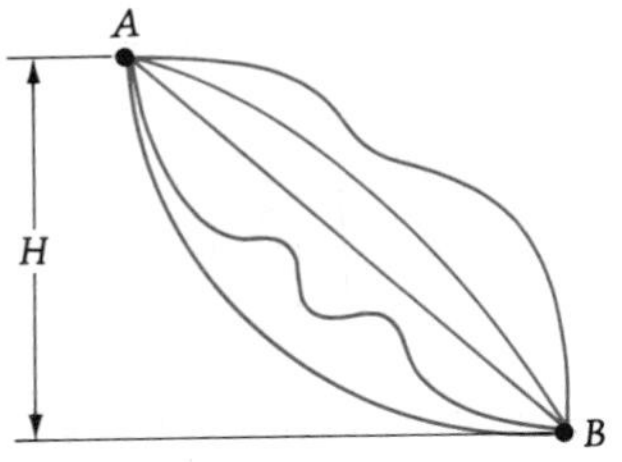

Figure P2.100

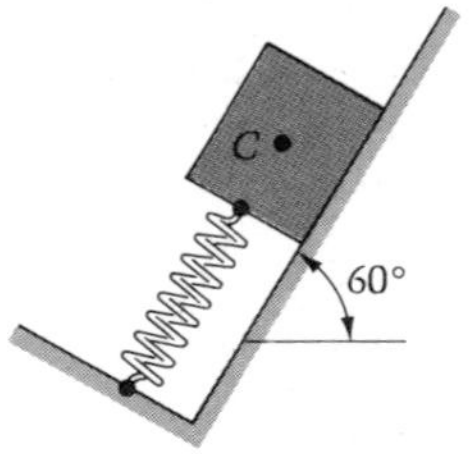

Figure P2.96

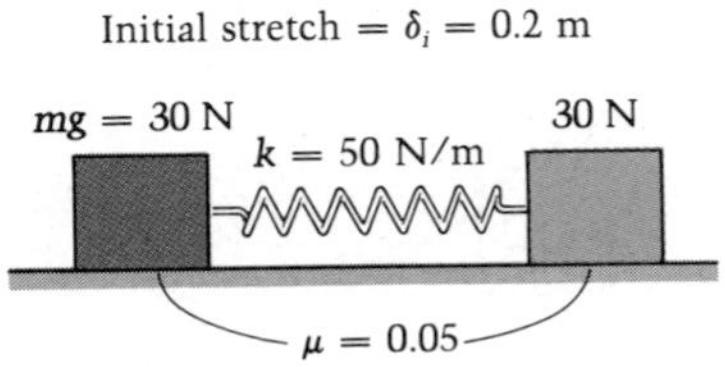

Figure P2.98

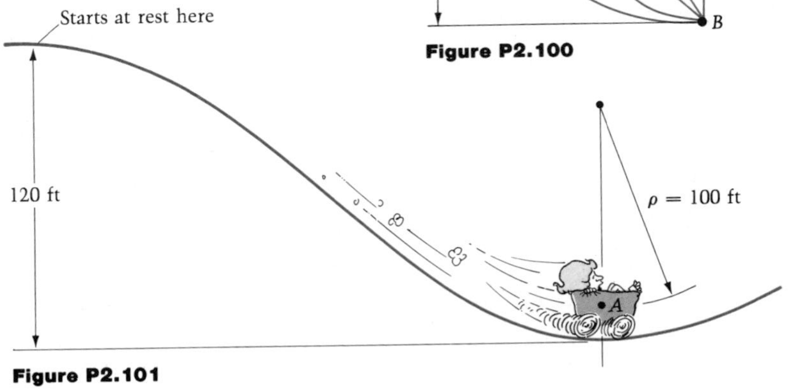

Figure P2.101

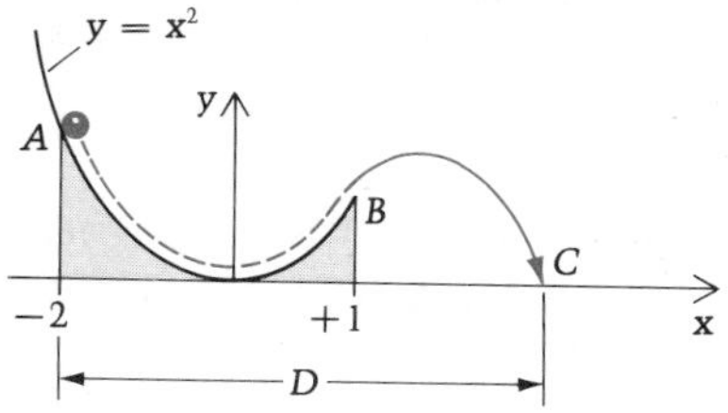

Figure P2.104

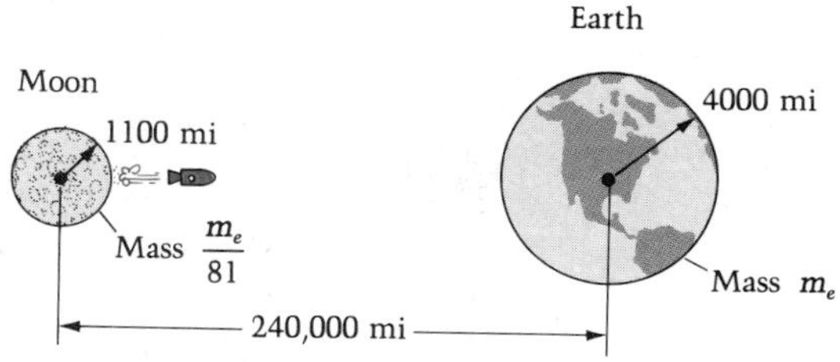

Figure P2.105

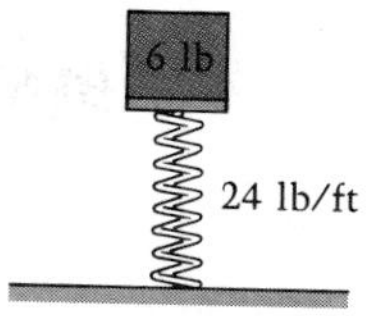

Figure P2.108

2.102 Show that for central gravitational force ($\mathbf{F} = [-GMm/r^2]\hat{\mathbf{e}}_r$), as distinct from the uniform gravity ($-mg\hat{\mathbf{k}}$) in the text, the potential is given by

$$\varphi = \frac{-GMm}{r}$$

where G is the universal gravitational constant and M and m are the masses of the two attracting bodies. Note that in view of Equation (2.27), one simply needs to show that $gr_e^2 = GM$.

2.103 Using the result of the preceding problem, calculate the work done by the earth's gravity on a satellite between the times of launch and insertion into a geosynchronous orbit with radius 6.61 times the radius of earth. (See Problem 2.21.)

2.104 A particle is released at rest A and slides on the smooth parabolic surface to B, where it flies off. (See Figure P2.104.) Find the total horizontal distance D that it travels before hitting the ground at C.

2.105 Find the least velocity with which a particle could be projected from the moon and reach the earth. (See Figure P2.105.) For this problem assume that the centers of the moon and earth are both fixed in an inertial frame.

2.106 Show that the equation $m\ell\dot{\theta}^2 = 2mg(\sin\theta - \frac{1}{2})$ can be obtained by conservation of mechanical energy ($T + \varphi =$ constant; Equation 2.30) in Example 2.9. Why can this principle not be used in Examples 2.10 and 2.11?

2.107 A smooth particle is projected along the inside of a vertical circle. At the lowest point its speed is $\sqrt{g(2a + 3H)}$, where a is the radius of the circle. If $0 < H < a$, show that the particle will leave the circle at some point on the upper half. If after leaving the circle at P the particle meets it again at Q, show that the time from P to Q is

$$4\left[\frac{H}{g}\left(1 - \frac{H^2}{a^2}\right)\right]^{1/2}$$

2.108 The 6-lb block shown in Figure P2.108 is released from rest when it just contacts the end of the unstretched spring. For the subsequent motion, find: (a) the maximum force in the spring; (b) the maximum speed of the block.

2.5 Impulse and Momentum/ Conservation of Momentum/Impact

A straightforward integration of the first law of motion (Equation 2.1) yields

$$\int_{t_1}^{t_2} \mathbf{F}_r\, dt = \mathbf{L}(t_2) - \mathbf{L}(t_1) = \text{change in momentum} = \Delta\mathbf{L} \qquad (2.31)$$

where the integral is usually called the **impulse** imparted to the body by the external forces;* note that the impulse is intrinsically associated with a specific time integral. If, during some time interval, the sum of the external forces vanishes, then $\dot{\mathbf{L}} = \mathbf{0}$ and hence the momentum is a constant, or is *conserved,* during that interval.

Since Equation (2.31) is a vector equation (unlike the scalar work and kinetic energy equation), we may use any or all of its component equations. For example:

$$\int_{t_1}^{t_2} F_{r_x}\,dt = L_x(t_2) - L_x(t_1) = m\dot{x}_C(t_2) - m\dot{x}_C(t_1) \tag{2.32}$$

and similarly for y and z. We note that we may have a planar situation, for example, in which $F_{r_x} = 0$ but $F_{r_y} \neq 0$ over an interval. If this is the case, momentum is conserved in the $\mathbf{x}$ direction but not in the y direction.†

Sometimes it is possible, by conservation of momentum, to obtain limited quantitative information about the motions of colliding bodies. As a rule this can be done when the bodies interact for a relatively brief interval—*before* and *after* which it is reasonable to treat their motions as rigid. While the analysis is best discussed with examples, we make the observation here that generally it makes little sense to treat the bodies as rigid during the collision. If we wish to describe the motion that ensues when a bullet is fired into a wooden block, for example, the block clearly cannot be regarded as rigid during the penetration process. On the other hand it may be quite plausible to assume that rigid motion of the block and embedded bullet occurs subsequent to permanent reorientation of material.

A key feature in the analysis of collision (or impact) problems is the fact that the momentum of a body made up of two parts is the sum of the momenta of the individual parts. This feature follows directly from the definition of a body's momentum as the integral

$$\begin{aligned}
\mathbf{L} &= \int_{\mathcal{B}} \mathbf{v}\,dm \\
&= \int_{\mathcal{B}_1} \mathbf{v}\,dm + \int_{\mathcal{B}_2} \mathbf{v}\,dm \\
&= \mathbf{L}_1 + \mathbf{L}_2
\end{aligned}$$

where the subscripts (1 and 2) identify the two constituent parts of the body.

*The impulse is sometimes called the *linear impulse.*
†Ballistics problems are of this type if air resistance is neglected.

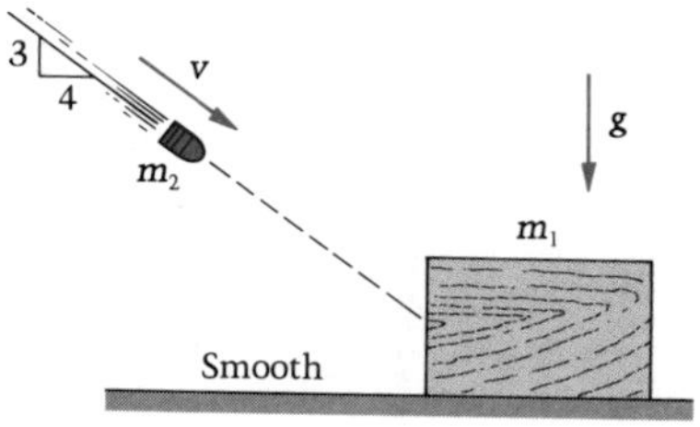

EXAMPLE 2.15

A wooden block of mass m_1 is at rest on a smooth horizontal surface when it is struck by a bullet of mass m_2 traveling at speed v as shown in the diagram. After the bullet becomes embedded in the block, the block slides to the right at speed V. Find the relationship between v and V.

SOLUTION

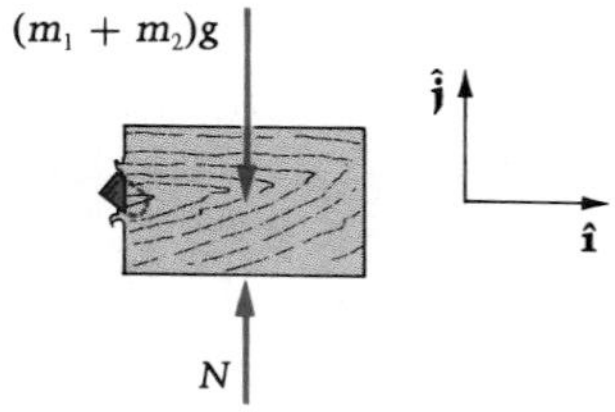

Let t_1 be the time at which the bullet first contacts the block and let t_2 be the time after which the bullet/block composite behaves as a rigid body in translation. For $t_1 < t < t_2$, a complex process of deformation and redistribution of mass is occurring within the block. If we isolate the block/bullet system during this interval (see the diagram), Equation (2.31) yields

$$\int_{t_1}^{t_2} [N(t) - m_1 g - m_2 g]\hat{\mathbf{j}}\, dt = \mathbf{L}(t_2) - \mathbf{L}(t_1)$$

But

$$\mathbf{L}(t_2) = (m_1 + m_2)V\hat{\mathbf{i}}$$

and

$$\mathbf{L}(t_1) = m_2(0.8v\hat{\mathbf{i}} - 0.6v\hat{\mathbf{j}})$$

since v is the speed of the mass center of the bullet and the block has no momentum at t_1. Therefore, equating the $\hat{\mathbf{i}}$ coefficients, we get

$$(m_1 + m_2)V = 0.8m_2 v$$

or

$$V = \frac{0.8m_2 v}{m_1 + m_2}$$

We note that, in the absence of an external force with a horizontal component, the horizontal component of momentum is conserved.

While we cannot calculate the reaction N during the collision, we *can* calculate its impulse:

$$\int_{t_1}^{t_2} [N - (m_1 + m_2)g]\, dt = 0.6m_2 v$$

or

$$\int_{t_1}^{t_2} N\, dt = (m_1 + m_2)g(t_2 - t_1) + 0.6m_2 v$$

(Continued)

Similarly, the impulse of the force **F** exerted on the bullet by the block can be calculated if we apply Equation (2.31) to the bullet:

$$\int_{t_1}^{t_2} (\mathbf{F} - m_2 g\hat{\mathbf{j}})\, dt = m_2 V\hat{\mathbf{i}} - m_2(0.8v\hat{\mathbf{i}} - 0.6v\hat{\mathbf{j}})$$

$$\int_{t_1}^{t_2} \mathbf{F}\, dt = m_2(V - 0.8v)\hat{\mathbf{i}} + m_2[g(t_2 - t_1) + 0.6v]\hat{\mathbf{j}}$$

$$= m_2\left(\frac{0.8m_2 v}{m_1 + m_2} - 0.8v\right)\hat{\mathbf{i}} + m_2[g(t_2 - t_1) + 0.6v]\hat{\mathbf{j}}$$

$$= -\frac{0.8m_1 m_2}{m_1 + m_2} v\hat{\mathbf{i}} + m_2[g(t_2 - t_1) + 0.6v]\hat{\mathbf{j}}$$

The reader should note that with a high-speed collision occurring in a short period of time, the impulses can be accurately estimated by neglecting the impulses of the weights of the bodies. In this example we would have $0.6v \gg g(t_2 - t_1)$. Other examples of impact problems are treated in Chapter 5.

EXAMPLE 2.16

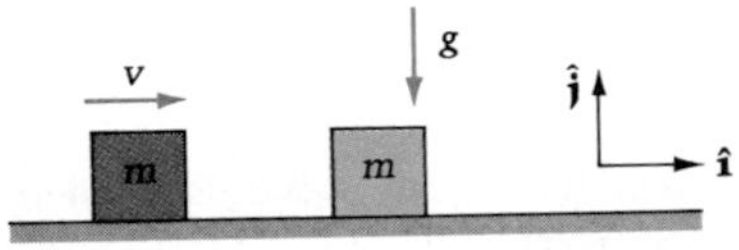

A block is at rest on a smooth horizontal surface before being struck by an identical block sliding at speed v. (See the diagram.) Find the velocities of the two blocks after the collision assuming (1) that they stick together or (2) that the system experiences no loss in kinetic energy.

SOLUTION

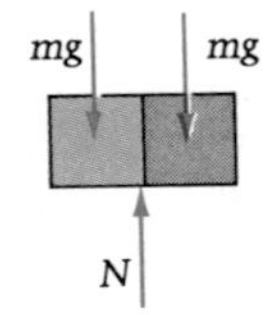

Let v_L and v_R be the speeds of the mass centers of the left and right blocks at the end of the collision; that is, $v_L\hat{\mathbf{i}}$ is the velocity of the left block. The free-body diagram of the system of two blocks during the collision (see the sketch) shows that there is no external force with a horizontal component. Thus the horizontal component (the only component not zero here) of momentum is conserved and

$$mv_L + mv_R = mv + m(0) \tag{1}$$

or

$$v_L + v_R = v \tag{2}$$

If the blocks remain attached after the collision is completed and they are behaving as rigid bodies, we have

$$v_R = v_L$$

so that

$$v_R = v_L = \frac{v}{2} \tag{3}$$

(Continued)

Question 2.9 What would be the common velocity if the right block had 100 times as much mass as the left block? If it had 10,000 times as much mass?

If, however, the blocks do not stick together, the conservation of momentum statement alone is not adequate to determine their subsequent velocities. What we need is some measure of their tendency to bounce off each other—or, to put it another way, a measure of how much energy is expended in permanent deformations or vibrations (or both) of the blocks. The parameter used to describe these effects (the coefficient of restitution) is discussed in Chapter 5. At this point we simply note that when the blocks stick together the kinetic energy of the system is less after the collision than before. That loss is

$$\frac{1}{2}mv^2 - \left[\frac{1}{2}m\left(\frac{v}{2}\right)^2 + \frac{1}{2}m\left(\frac{v}{2}\right)^2\right] = \frac{1}{4}mv^2 \tag{4}$$

which is to say that one-half of the mechanical energy was dissipated in the collision in this case.

The other extreme case is that in which *no* mechanical energy is expended during the collision process. In this case

$$\frac{1}{2}mv_L^2 + \frac{1}{2}mv_R^2 = \frac{1}{2}mv^2 \tag{5}$$

But since $v_R = v - v_L$ from Equation (2), we have

$$v_L^2 + (v - v_L)^2 = v^2 \tag{6}$$

$$v_L^2 + v^2 - 2vv_L + v_L^2 = v^2$$

or

$$2v_L(v_L - v) = 0 \tag{7}$$

Therefore either $v_L = 0$ and $v_R = v$ or $v_L = v$ and $v_R = 0$. The latter case must be rejected as physically meaningless since it would require the left block to pass through the stationary right block.

An extension of this result to the case of three blocks is shown below:

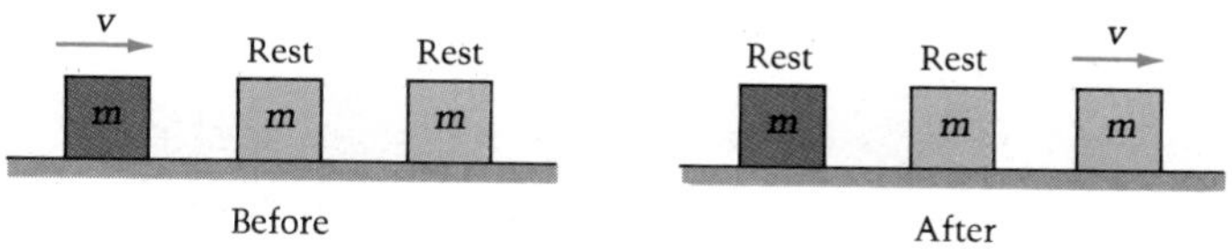

If we let the spacing between the blocks initially at rest approach zero and add more of them, then we have the mechanism for a popular adult toy:

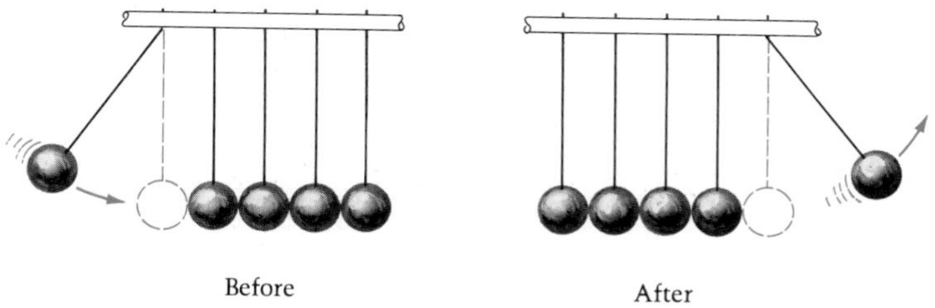

EXAMPLE 2.17

A 300-lb crate is being pulled up a rough incline with a winch and cable as shown in the diagram. Assume that the cable force P varies from $t = 0$ sec as $P = 300 + 2t$ pounds and suppose the coefficients of friction between the crate and plane are $\mu_s = 0.6$ and $\mu_k = 0.4$. Find the velocity of the crate at $t = 5$ sec.

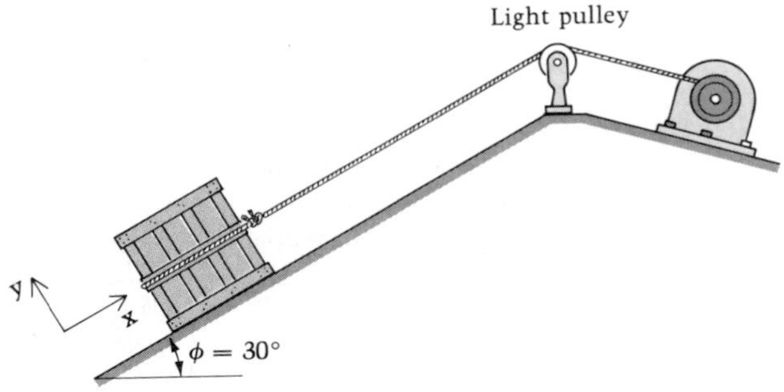

SOLUTION

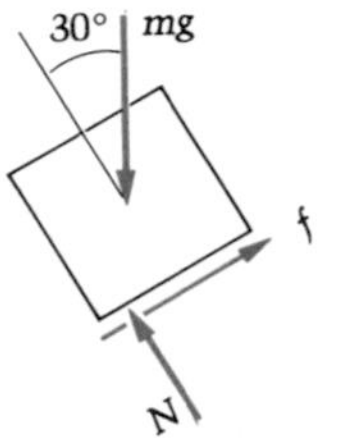

Prior to $t = 0$, there is sufficient friction to hold the crate in equilibrium without any cable force, since $\mu_s = 0.6 > \tan\phi = 0.577$. Therefore equilibrium exists prior to start, with (see the diagram)

$$N = mg\cos 30^\circ = 260 \text{ lb}$$

$$f = mg\sin 30^\circ = 150 \text{ lb}$$

$$f_{\max} = \mu_s N = 156 \text{ lb}$$

To move the cable up the plane, the cable force P must have grown from 300 lb (at $t = 0$) to the value shown below, which is the maximum resultant component down the plane of the friction plus gravity forces:

$$P = mg\sin 30^\circ + \mu_s(mg\cos 30^\circ) = 306 \text{ lb}$$

or

$$300 + 2t = 306$$

$$t = 3.0 \text{ sec}$$

Therefore the crate begins to move at $t = 3$ sec. At this time the friction coefficient drops from μ_s to μ_k, and the impulse-momentum principle in the $\mathbf{x}$ direction will give us the solution sought:

$$\int_3^5 F_{r_x}\,dt = m\dot{\mathbf{x}}_{C_f} - m\dot{\mathbf{x}}_{C_i}\overset{\nearrow 0}{}$$

$$\int_3^5 (300 + 2t - \underbrace{mg\sin 30^\circ}_{150} - \underbrace{f_{\text{sliding}}}_{0.4(260)})\,dt = \frac{300}{32.2}\dot{\mathbf{x}}_{C_f}$$

(Continued)

$$\int_3^5 (300 + 2t - 254)\, dt = 9.32\dot{\mathbf{x}}_{C_f}$$

$$46(5-3) + [t^2]_3^5 = 9.32\dot{\mathbf{x}}_{C_f}$$

$$\dot{\mathbf{x}}_{C_f} = 11.6 \text{ ft/sec}$$

This result assumes, of course, that the crate has not reached the top of the incline. Note from this example that the impulse-momentum principle can be applied in situations in which momentum is not conserved and no impact occurs.

PROBLEMS / Section 2.5

2.109 A $\frac{3}{4}$-oz bullet is fired with a speed of 1800 ft/sec into a 10-lb block. (See Figure P2.109.) If the coefficient of friction between block and plane is 0.3, find:

- **a.** The distance through which the block will slide
- **b.** The percentage of the bullet's loss of initial kinetic energy caused by sliding friction and the percentage caused by the collision
- **c.** How long it takes block and bullet to come to rest after the impact

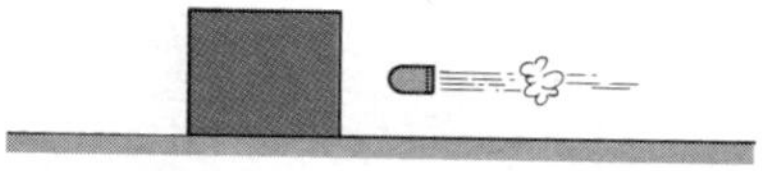

Figure P2.109

2.110 A 10-kg block swings down as shown in Figure P2.110 and strikes an identical block. Assume that the 6 m rope breaks during impact and the blocks stick together after colliding. How long will it be before they come to rest? How far will they have traveled?

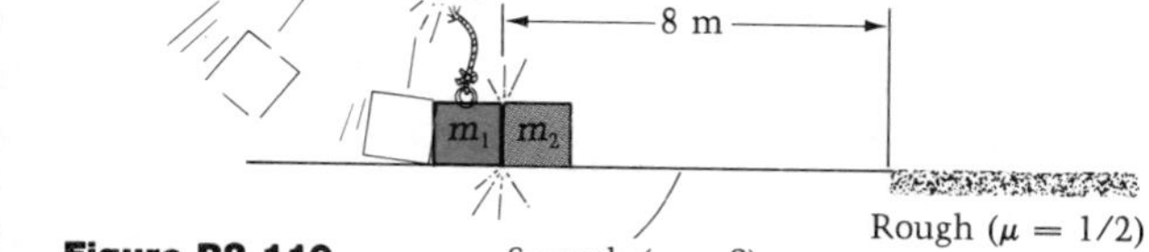

Figure P2.110

2.111 The astronaut in Figure P2.111 is finding it difficult to stop his forward momentum while jogging on the moon. Using a friction coefficient of $\mu = 0.3$ and a gravitational acceleration one-sixth that of earth's, illustrate the difficulty of stopping a forward momentum of $mv =$ (5 slugs)(12 ft/sec). Specifically, use the principle of impulse and momentum to find the time it takes to stop on earth versus on the moon.

Figure P2.111

2.112 The 16-kg body $\mathcal{A}$ and the 32-kg body $\mathcal{B}$ shown in Figure P2.112 are connected by a light spring of modulus 12,000 N/m. The unstretched length of the spring is 0.15 m. The blocks are pulled apart on the smooth horizontal plane until the distance between them is 0.3 m and then released from rest. Determine the velocity of each block when the distance between them has decreased to 0.22 m. *Hint:* As in Problem 2.98, form the sum of the work-energy equations for the two blocks.

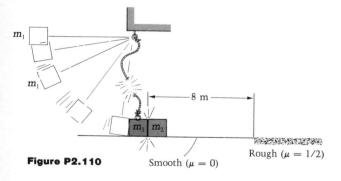

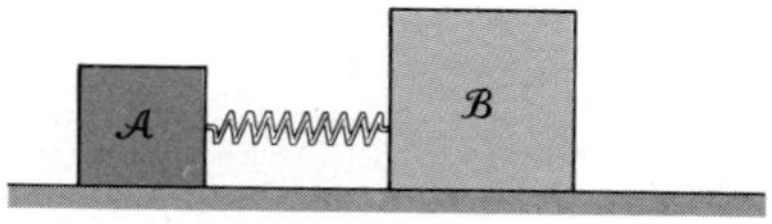

Figure P2.112

2.113 Two railroad cars are coupled by a collision occurring just after the instant shown in Figure P2.113. Neglecting the impulse caused by friction from the tracks, determine the final velocity of the two cars as they move together.

2.114 In the preceding problem, find the average impulsive force between the cars if the coupling requires 0.6 sec of contact.

2.115 An unattached 2.2-lb roofing shingle $\mathcal{S}$ slides downward and strikes a gutter $\mathcal{G}$. (See Figure P2.115.) The angle at which the shingle would be just on the verge of slipping is 20°. Determine the impulse imparted to the shingle by the gutter if there is no rebound. If the interval of impact is 0.1 sec, find the average force imparted to the gutter by the shingle.

2.116 A man of mass m and a boat of mass M are at rest as shown in Figure P2.116. If the man walks to the front of the boat, show that his distance from the pier is then $L\mathcal{M}/(1 + \mathcal{M})$, where $\mathcal{M} = m/M$ is the ratio of the masses of man and boat. Explain the answer in the limiting cases in which $m \ll M$ and $M \ll m$. Neglect the resistance of the water to the boat's motion.

2.117 In a rail yard a freight car moving at speed v strikes two identical cars at rest. (See Figure P2.117.) Neglecting any resistance to rolling, find the common velocity of the three-car system after the coupling has been completed and any associated vibrations have died out.

2.118 Two men each of mass m stand on a flatcar of mass M. The car is free to move on frictionless level tracks. All is at rest initially. One man runs to the right end of the car and jumps off horizontally, parallel to the tracks with a velocity U relative to the car. Then the other man runs to the left end of the car and jumps off horizontally, parallel to the tracks also with a velocity U relative to the car. Find the final velocity of the car and indicate clearly the direction of its motion.

2.119 Using the angle α that will land the cannonball of Problem 2.19 in the cart, find the maximum deflection of the spring. (See Figure P2.119.)

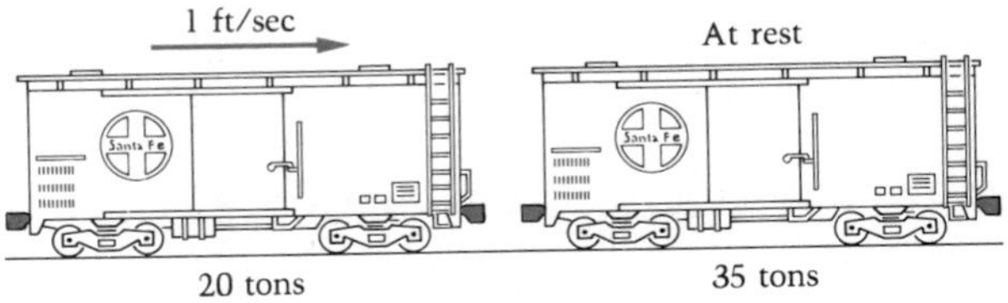

Figure P2.113

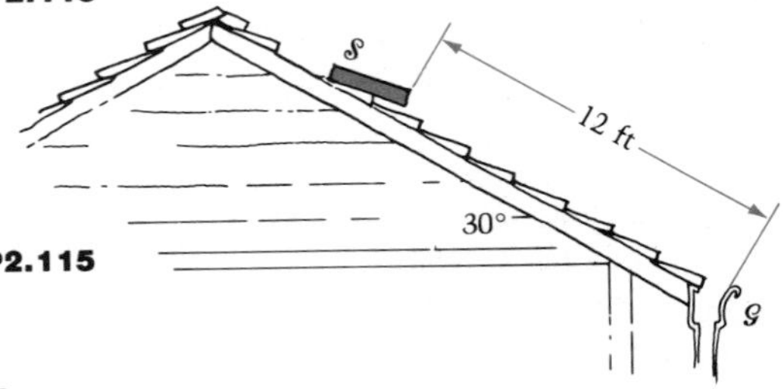

Figure P2.115

Figure P2.116

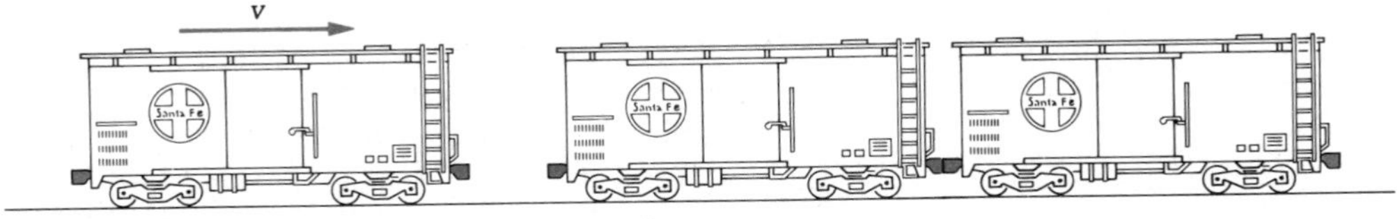

Figure P2.117

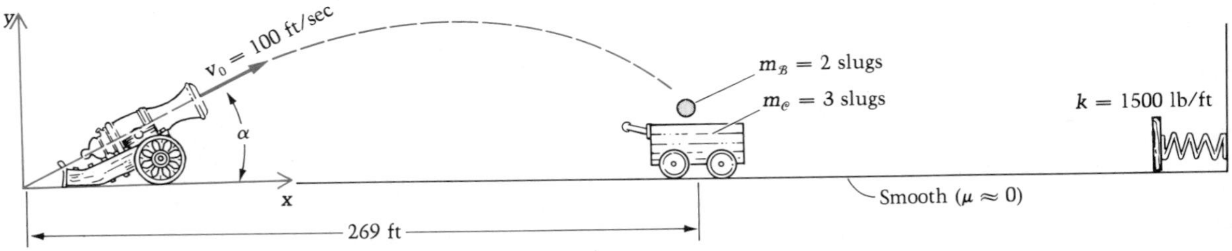

Figure P2.119

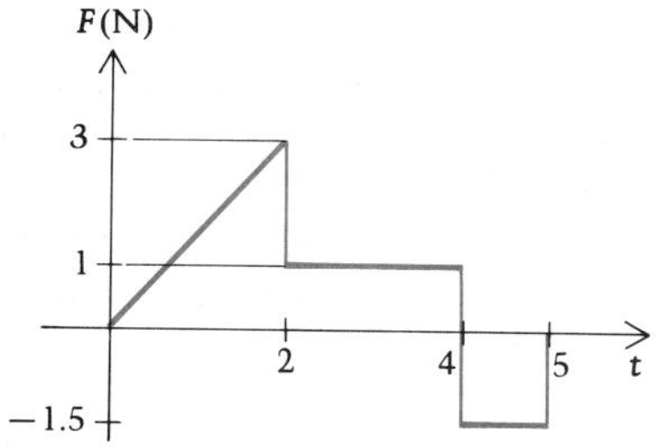

Figure P2.120

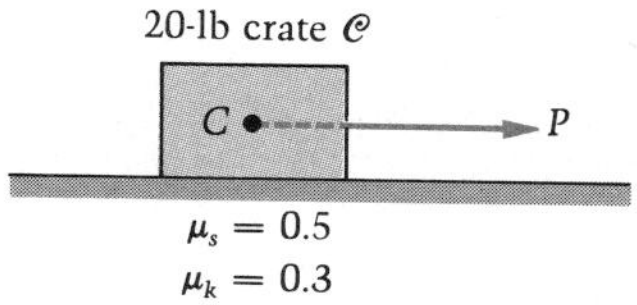

Figure P2.121

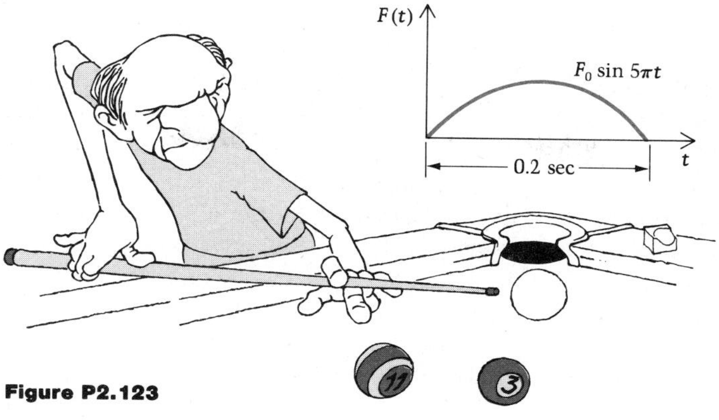

Figure P2.123

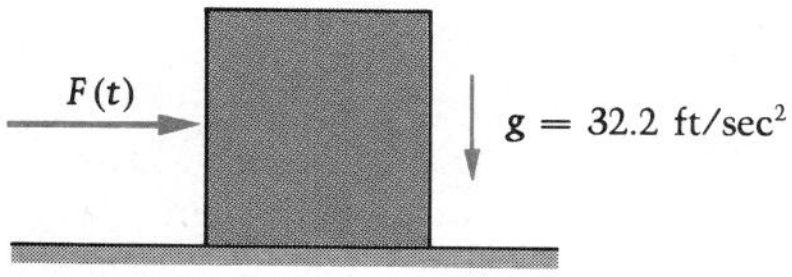

Figure P2.124

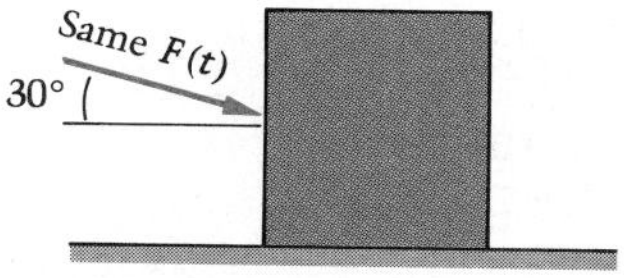

Figure P2.125

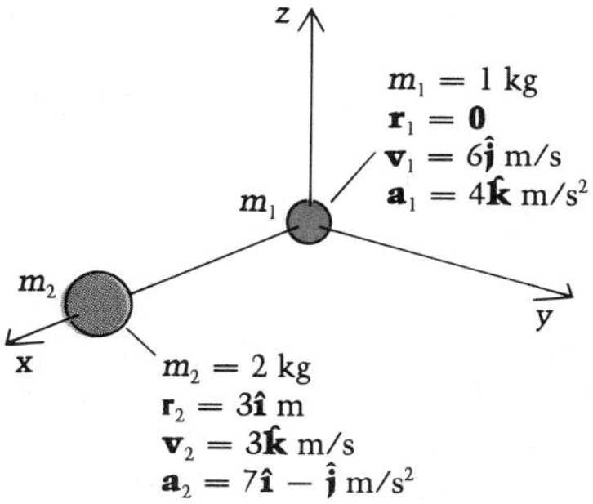

Figure P2.126

2.120 A 20-kg particle P is at rest at the origin in an inertial frame. At $t = 0$, it begins to be acted upon by a one-dimensional force that varies with time as shown in Figure P2.120. Find the velocity and the position of P at $t = 5$ s.

2.121 A force P applied to 𝒞 at $t = 0$ varies with time according to $P = 25 \sin(\pi t/60)$ lb, where t is in seconds. (See Figure P2.121.) How long will it take for 𝒞 to begin sliding? What will be the velocity of the point C at $t = 30$ sec?

2.122 Estimate the minimum impulse to be supplied by the earth (through the feet) for a high-jumper to clear a bar at 7 ft. After thinking about this problem, try to develop an explanation for the "arched-back" style of high-jumping.

2.123 A horizontal force $F(t)$ is applied for 0.2 sec to a cue ball (weighing 0.55 lb) by a cue stick; the form of the force is as shown in Figure P2.123. If the velocity of the center of the ball is 8 ft/sec after contact with the stick is broken, find the peak magnitude F_0 of the force. Neglect friction. Force F is measured in pounds.

2.124 The 50-lb box shown in Figure P2.124 is at rest before the force $F(t) = 5 + 2t$ pounds is applied at $t = 0$. Assume the box to be wide enough not to tip over and suppose the coefficient of friction between box and floor to be 0.2. Find the velocity of (the mass center of) the box at $t = 10$ sec.

2.125 Repeat the preceding problem for the case in which the force $F(t)$ has a vertical component as shown in Figure P2.125.

2.126 Figure P2.126 presents data pertaining to a system of two particles. At the instant shown find the:

- **a.** Position of the mass center
- **b.** Kinetic energy of the system
- **c.** Linear momentum of the system
- **d.** Velocity of the mass center
- **e.** Acceleration of the mass center

2.127 A cannonball is fired as shown in Figure P2.127 with an initial speed of 1600 ft/sec at 60°. Just after the cannon fires, it begins to recoil, and strikes a plate attached to a spring. Find the maximum spring deflection if the plane is smooth and the spring modulus is 500 lb/ft.

2.128 In Problem 2.19, find the total time for the cannonball and box to either stop or strike the wall, whichever comes first.

Figure P2.127

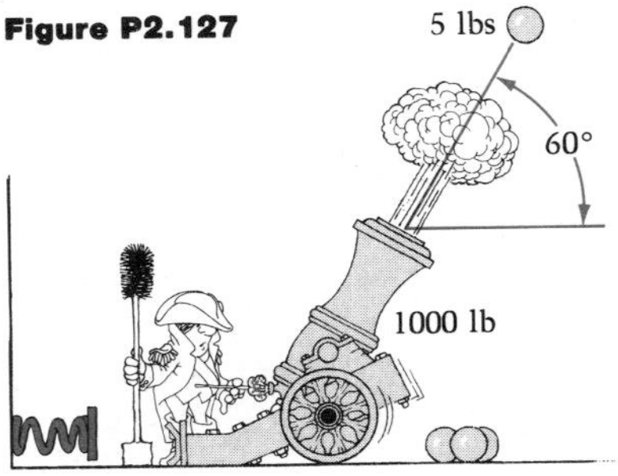

Answers to Questions / Chapter 2

Q2.1 Mass times length divided by time (or force times time).

Q2.2 $\dot{m}$ is zero since our definition of a body requires that its mass be constant.

Q2.3 The components of the two sides of a vector equation are equal in *any* direction.

Q2.4 It means the answer does not depend on the mass of the car.

Q2.5 $(k) \cdot$ (stretch) gives the equal in magnitude but opposite in direction forces acting at the ends of a spring in equilibrium. If particles in the spring are accelerating, as is generally the case in dynamics problems, there is no simple force-stretch law. If the spring is very light, however, so that its mass may be neglected (compared to the masses of other bodies in the problem), $\int_{\text{spring}} \mathbf{a}\, dm \approx \mathbf{0}$ and the forces on the spring are instantaneously related just as if the spring were in equilibrium. The hidden assumption is that the mass of the spring may be neglected.

Q2.6 It is needed so that the work equals the decrease in φ; that is, $\varphi[\mathbf{r}(t_1)] - \varphi[\mathbf{r}(t_2)]$.

Q2.7 $T - \psi =$ constant; we use the φ simply so that we may say that mechanical energy is the sum of its two parts.

Q2.8 Because the snow *does* exert a friction force on the skis that has been neglected. Friction *always* does negative work, and this work is seen in Equation (1) to reduce v_2; in turn, the calculated speed and acceleration magnitude at the origin will have been reduced by the effect of the friction from the snow. Air resistance against the skier's body will have the same slowing effect.

Q2.9 $v/101$; $v/10{,}001$.

Review Questions / Chapter 2

True or False?

1. At a given time, the mass center of a deformable body can be shown to be a unique point.

2. The momentum of any body (or system of bodies) in a frame $\mathcal{F}$ can be shown to be equal to the total mass times the velocity of the mass center in $\mathcal{F}$, even if $\mathcal{F}$ is not an inertial frame.

3. Euler's first law ($\mathbf{F}_r = \dot{\mathbf{L}}$) applies to deformable bodies whether solid, liquid, or gaseous, as well as to rigid bodies and particles.

4. Neither the laws of motion nor the inertial frame is of any value without the other.

5. The mass center of a body $\mathcal{B}$ has to be a physical, or material, point of $\mathcal{B}$.

6. The work done by a linear spring depends on the paths traversed by its end-points between the initial and final times.

7. The work done by the friction force upon a block sliding on a fixed plane depends on the path taken by the block.

8. The work done by gravity on a body $\mathcal{B}$ depends on the lateral as well as the vertical displacement of the mass center of $\mathcal{B}$.

9. Since no external work was done on the two bodies of Example 2.15 *during the impact*, their total kinetic energy is the same after the collision as it was before.

10. For all bodies of constant density, the centroid of volume and the center of mass coincide.

11. In studying the motion of the earth around the sun, it is acceptable to treat the earth as a particle; in studying the daily rotation of the earth on its axis, however, it would not make sense to consider the earth as a particle.

12. The external forces acting on a body $\mathcal{B}$, which together form the resultant $\mathbf{F}_r$, must each have a line of action passing through the mass center of $\mathcal{B}$ in order for Euler's first law to apply.

Answers: T, T, T, T, F, F, T, F, F, T, T, F

CHAPTER

3 Kinematics of Plane Motion of a Rigid Body

chapter outline ►

3.1 Introduction

In this chapter our goals are to develop the relationships between the velocities and accelerations of the points of a rigid body $\mathcal{B}$, and the angular velocities and angular accelerations of $\mathcal{B}$, as the body moves in plane motion in a reference frame $\mathcal{I}$. Before doing so, however, we shall first explain precisely what we mean by such terms as ***rigid body, plane motion, reference plane,*** and several other concepts we shall be needing in this chapter and those to follow.

A **rigid body** is taken to be a body in which the distance between each and every pair of its points remains the same throughout the motion.* There is, of course, no such thing as a truly rigid body (since all bodies do *some* deforming); however, the deformations of many bodies are sufficiently small during their motions to allow the bodies to be treated as though they were rigid with good results.

Plane motion is treated in this book as motion in the xy plane (fixed in $\mathcal{I}$) or in planes parallel to it. Let a point P be located originally at coordinates (x_P, y_P, z_P). To say that P has plane motion simply means that it stays in the plane $z = z_P$ throughout its motion. Extending this definition, we say that rigid body $\mathcal{B}$ has plane motion whenever *all* its points remain in the same planes (parallel to xy) in which they started.

Question 3.1 How few points of a rigid body must be in plane motion to ensure that they *all* are?

A third concept we need to understand in rigid-body kinematics is that of the **body extended,** also called **rigid extensions** of the body. This idea, briefly mentioned in Chapter 1, says that we sometimes need to imagine points (which are not physical or material points of $\mathcal{B}$) moving with $\mathcal{B}$ as though they were in fact attached to it. An example would be the points on the axis of a pipe that are in the space inside it but of course move rigidly with it. We shall imagine a "rigid extension" of the body to pick up such points whenever it is useful to do so. Note that *any* point Q may be considered a point of *any* body $\mathcal{B}$ extended, provided that Q moves with $\mathcal{B}$ as if it were rigidly attached to it.

With these three concepts in mind, we are now prepared to define the **reference plane.** First we must realize that we are faced with the problem of determining where all the points of a body are as functions of time t. These locations $(x(t), y(t))$ would take forever to find if we had to do so for each of the infinitely many points of $\mathcal{B}$. Fortunately, for a rigid body in plane motion, if we know the locations of all its

*We have already encountered this concept in Chapter 1, where it was seen to be synonymous with the concept of frame.

points in any one plane of $\mathcal{F}$ (which we shall call the reference plane), then we automatically know the locations of all its *other* points in all *other* planes. The reason for this is as follows. For each point B of $\mathcal{B}$ that does not lie in the reference plane, there is a "companion point" of $\mathcal{B}$ in the reference plane (suggested by A in Figure 3.1) that has the same (x, y) coordinates at the beginning of the motion. It then follows that the (x, y) coordinates of A and B *always* match *throughout* the motion of $\mathcal{B}$!

Question 3.2 Why is $x_A \equiv x_B$ and $y_A \equiv y_B$ as time passes?

Body $\mathcal{B}$ in Figure 3.1 is a cone pulley; we shall see an example of its industrial use in Section 3.6. But for now note from its varying cross-sectional diameter that a body need not have constant cross section to be in plane motion.

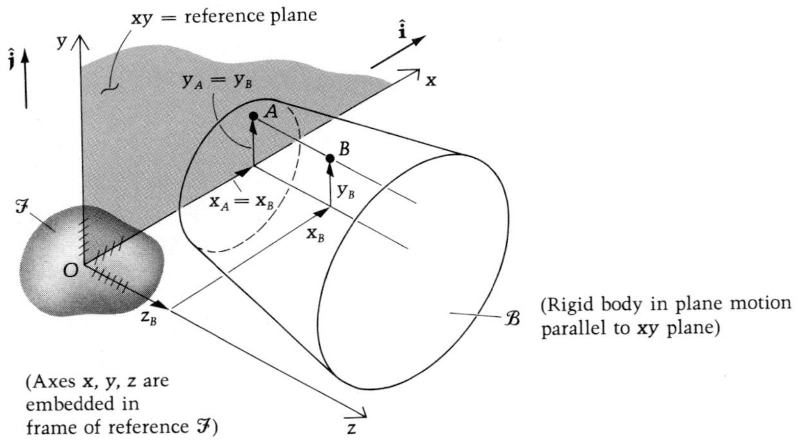

Figure 3.1

The reference plane is thus a very important concept, for it allows us to study the motion of an entire body by concerning ourselves only with those of its points that lie in this plane. We say we "know the motion" of $\mathcal{B}$ when we know where all its points are at all times. We have already reduced this task to knowing the locations of the points in the reference plane. (The rest of the body "goes along for the ride.") But in fact if we know the location of just *two* points (say P_1 and P_2) of the reference plane then we know the whereabouts of *all* points of this plane and thus of the whole body! This is because each point of the reference plane must maintain the same position relative to the points P_1 and P_2. This idea is illustrated in Figure 3.2. Note in the figure that if P_1 and P_2 are correctly located with respect to the reference frame $\mathcal{F}$, all other points of $\mathcal{B}$ are necessarily in their correct positions.

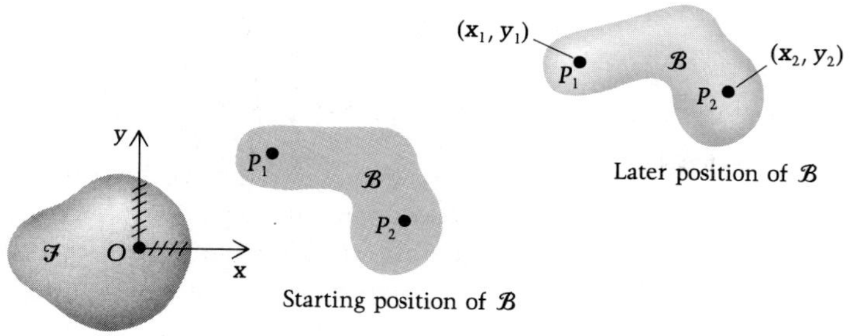

Figure 3.2

Question 3.3 Is knowledge of the locations of two points sufficient for us to know the motion of a body in general (three-dimensional) motion?

Instead of knowing the locations of two points of the body—(x_1, y_1) of P_1 and (x_2, y_2) of P_2—we may alternatively locate the body if we know where just ***one*** point, P_1, is located ***plus*** the value of the orientation angle θ (about an axis through P_1 and parallel to z); see Figure 3.3.

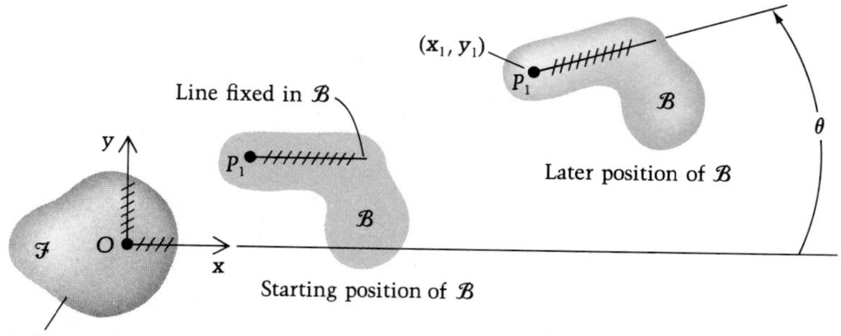

Figure 3.3

Question 3.4 Knowing the location of two points requires four variables (x_1, y_1, x_2, y_2), whereas one point plus the angle takes but three (x_1, y_1, θ). Why do these numbers of variables differ?

Having laid the necessary groundwork, we now let **xy** be our reference plane. From Figure 3.1 we see that

$$\mathbf{r}_{OA} = x_A\hat{\mathbf{i}} + y_A\hat{\mathbf{j}} \tag{3.1}$$

and we may differentiate this equation to obtain

$$\mathbf{v}_A = \dot{x}_A\hat{\mathbf{i}} + \dot{y}_A\hat{\mathbf{j}} = \dot{x}_B\hat{\mathbf{i}} + \dot{y}_B\hat{\mathbf{j}} = \mathbf{v}_B \tag{3.2}$$

$$\mathbf{a}_A = \ddot{x}_A\hat{\mathbf{i}} + \ddot{y}_A\hat{\mathbf{j}} = \ddot{x}_B\hat{\mathbf{i}} + \ddot{y}_B\hat{\mathbf{j}} = \mathbf{a}_B \tag{3.3}$$

In these equations we have used the facts that $x_B \equiv x_A$, $y_B \equiv y_A$, and $z_B =$ constant. Equations (3.2) and (3.3) show clearly that if we completely describe the velocities and accelerations in one reference plane, we then know them for *all* the points of the body.

PROBLEMS / Section 3.1

3.1 Which of the bodies $\mathcal{B}$ shown in the figures below are in plane motion in frame $\mathcal{F}$?

(a) A turkey being barbecued by slowly turning on a rotisserie.

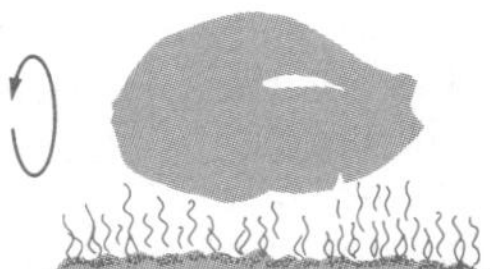

(b) A cone rolling on a tabletop.

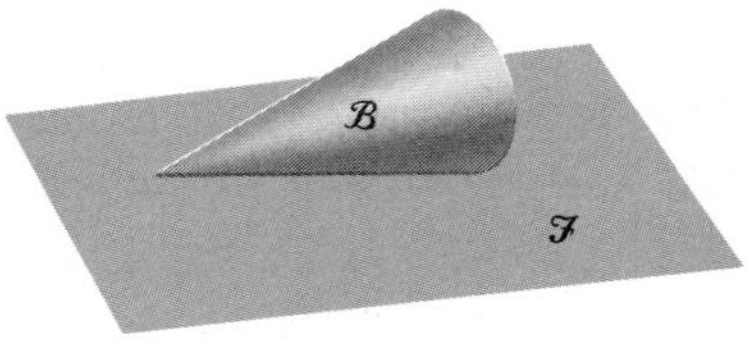

(c) A spinning coin if the base is fixed.

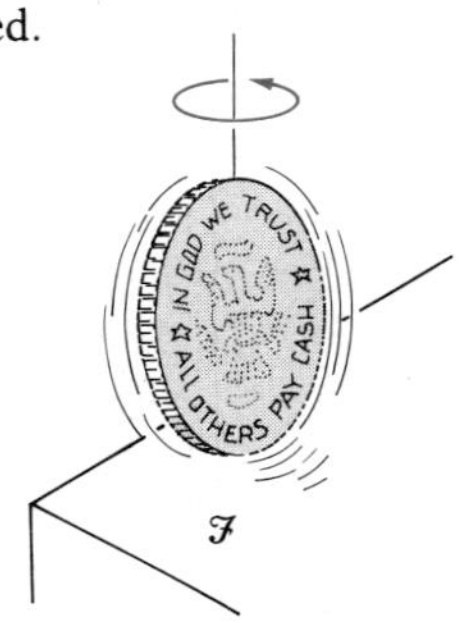

(d) A can rolling down an inclined plane.

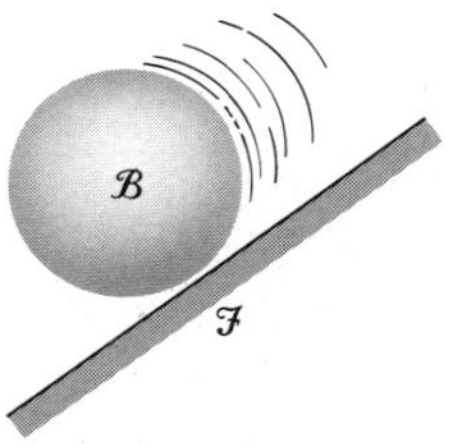

(e) The bevel gear $\mathcal{B}$, which meshes with another bevel gear $\mathcal{A}$.

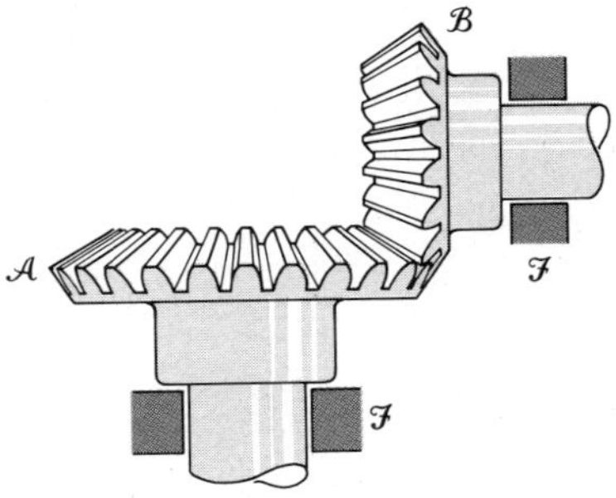

(f) The (shaded) crosspiece of a universal joint.

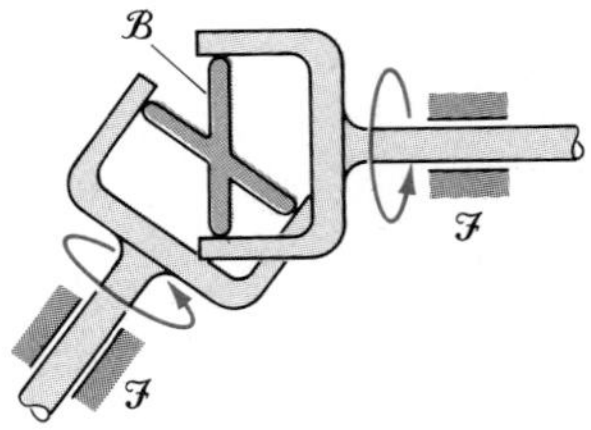

3.2 Give three examples of plane motion besides those in the previous problem. Then give three examples of motion that is not planar.

3.2 Velocity and Angular Velocity Relationship for Two Points of the Same Rigid Body

In this section we derive a very useful relationship between the velocities in $\mathcal{F}$ of any two points in the reference plane of a rigid body $\mathcal{B}$ in plane motion and the angular velocity vector of $\mathcal{B}$ in $\mathcal{F}$. Let P and Q denote these two points of $\mathcal{B}$, and let us embed the axes $(\mathbf{x}, \mathbf{y}, \mathbf{z})$ in reference frame $\mathcal{F}$ as shown in Figure 3.4.

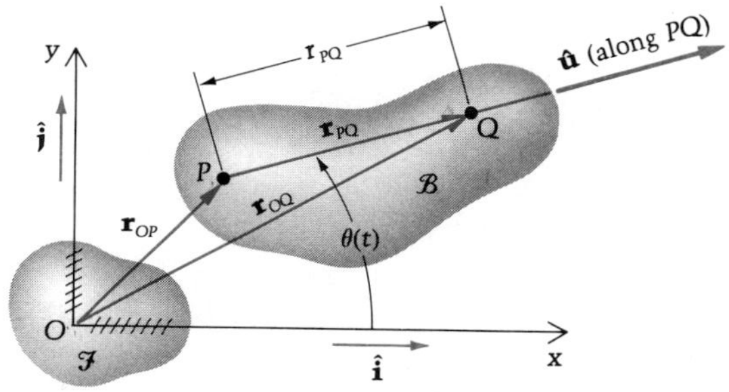

Figure 3.4

We are saying that even though $\mathcal{B}$ may move with respect to the reference frame $\mathcal{F}$, the $\mathbf{xy}$ plane of $\mathcal{F}$ always contains the points of interest P and Q of $\mathcal{B}$. A good example is found in the classroom; let the body $\mathcal{B}$ be a blackboard eraser. Letting the blackboard itself be the reference frame $\mathcal{F}$ (so that $\mathbf{x}$ and $\mathbf{y}$ are fixed in the plane of the blackboard), the eraser undergoes plane motion whenever the professor erases the board. Our points P and Q are any two points of the erasing surface of the eraser. Note how each point of the eraser remains the same $\mathbf{z}$ distance from the blackboard (where $z = 0$) during the erasing. The eraser is no longer in plane motion, however, once it *leaves* the surface of the board and its points move with $\mathbf{z}$ components of velocity.

Notice from Figure 3.4 that $\hat{\mathbf{u}}$ is a unit vector always directed from P toward Q so that $\mathbf{r}_{PQ} = r_{PQ}\hat{\mathbf{u}}$, where r_{PQ} is the distance PQ (that is, the magnitude of the vector $\mathbf{r}_{PQ}$). Note further that the orientation (angular rotation) of the body is described by the angle θ, measured between any line fixed in the reference frame $\mathcal{F}$ (we shall usually use the $\mathbf{x}$ axis) and any line fixed in the body (for the moment, we shall use the line segment from P to Q). For consistency in the vector operations, we always ensure that the axes $(\mathbf{x}, \mathbf{y}, \mathbf{z})$ form a right-handed Cartesian coordinate system; furthermore, θ is measured in radians of rotation about the $\mathbf{z}$ axis in accordance with the right-hand rule: When the fingers of the right hand curl in the positive θ direction, the right thumb points in the direction of the $\mathbf{z}$ axis.

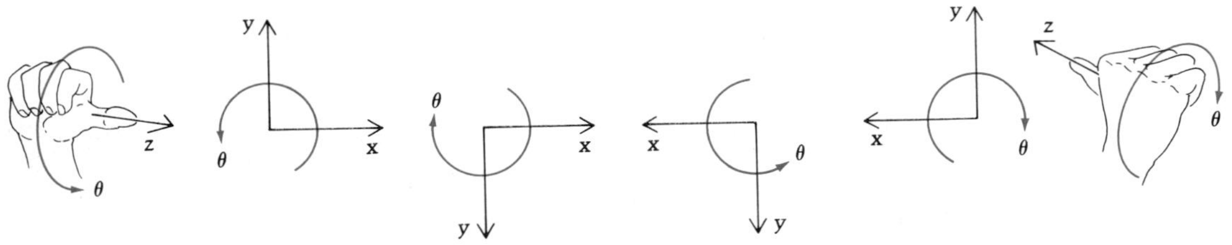

Figure 3.5

We may later find it useful to use the four plane-motion sign conventions shown in Figure 3.5, each of which is right-handed.* In terms of *base vectors*—that is, unit vectors ($\hat{\mathbf{i}}$, $\hat{\mathbf{j}}$, $\hat{\mathbf{k}}$) that are always respectively parallel to the coordinate axes (x, y, z)—we always have $\hat{\mathbf{i}} \times \hat{\mathbf{j}} = \hat{\mathbf{k}}$, $\hat{\mathbf{j}} \times \hat{\mathbf{k}} = \hat{\mathbf{i}}$, $\hat{\mathbf{k}} \times \hat{\mathbf{i}} = \hat{\mathbf{j}}$, $\hat{\mathbf{j}} \times \hat{\mathbf{i}} = -\hat{\mathbf{k}}$, $\hat{\mathbf{i}} \times \hat{\mathbf{k}} = -\hat{\mathbf{j}}$, and $\hat{\mathbf{k}} \times \hat{\mathbf{j}} = -\hat{\mathbf{i}}$ for each of the right-handed systems depicted in Figure 3.5.

We are now ready to develop the velocity and angular velocity relationship for rigid bodies. From Figure 3.4 we see that

$$\mathbf{r}_{OQ} = \mathbf{r}_{OP} + \mathbf{r}_{PQ} \tag{3.4}$$

so that we have, upon differentiation in frame $\mathcal{I}$,

$$\dot{\mathbf{r}}_{OQ} = \dot{\mathbf{r}}_{OP} + \dot{\mathbf{r}}_{PQ}$$

Recognizing the first two vectors as the definitions of the velocities of P and Q (in $\mathcal{I}$, where O is fixed), we may write

$$\mathbf{v}_Q = \mathbf{v}_P + \dot{\mathbf{r}}_{PQ} \tag{3.5}$$

In Equation (3.5), all derivatives are taken in $\mathcal{I}$, so that, for example, there is no need to write $\mathbf{v}_{Q/\mathcal{I}}$.

In order to write $\dot{\mathbf{r}}_{PQ}$ as a vector we can use, we express $\mathbf{r}_{PQ}$ as a magnitude times a unit vector. With the help of Figure 3.4 we get

$$\mathbf{r}_{PQ} = r_{PQ}\hat{\mathbf{u}} = r_{PQ}(\cos\theta\hat{\mathbf{i}} + \sin\theta\hat{\mathbf{j}}) \tag{3.6}$$

Differentiating this expression as in Section 1.6, we have

$$\begin{aligned}\dot{\mathbf{r}}_{PQ} &= r_{PQ}\dot{\theta}(-\sin\theta\hat{\mathbf{i}} + \cos\theta\hat{\mathbf{j}}) = r_{PQ}\dot{\theta}\hat{\mathbf{k}} \times (\cos\theta\hat{\mathbf{i}} + \sin\theta\hat{\mathbf{j}}) \\ &= r_{PQ}\dot{\theta}\hat{\mathbf{k}} \times \hat{\mathbf{u}} = \dot{\theta}\hat{\mathbf{k}} \times r_{PQ}\hat{\mathbf{u}}\end{aligned}$$

Therefore we have derived a useful expression for $\dot{\mathbf{r}}_{PQ}$:

$$\dot{\mathbf{r}}_{PQ} = \dot{\theta}\hat{\mathbf{k}} \times \mathbf{r}_{PQ} \tag{3.7}$$

Question 3.5 In the preceding development, why is $\dot{r}_{PQ} = 0$?

*The directions need not be left/right or top/bottom lines; it is often convenient to draw them along and normal to an inclined plane, for example.

Substituting Equation (3.7) into (3.5) yields

$$\mathbf{v}_Q = \mathbf{v}_P + \dot{\theta}\hat{\mathbf{k}} \times \mathbf{r}_{PQ} \tag{3.8}$$

which relates the velocities of the points P and Q and introduces the **angular velocity** of $\mathcal{B}$ in reference frame $\mathcal{I}$.* The Greek letter omega is usually used to denote this vector:

$$\boldsymbol{\omega}_{\mathcal{B}/\mathcal{I}} = \boldsymbol{\omega} = \dot{\theta}\hat{\mathbf{k}} \tag{3.9}$$

The magnitude $|\dot{\theta}|$ (or ω) of the angular velocity is called the **angular speed** of $\mathcal{B}$ in frame $\mathcal{I}$. Note that $\dot{\theta}$ itself can be negative. Note further that neither the angular velocity vector nor Equation (3.8) depends on which body-fixed line segment (such as PQ above) is chosen to measure θ. The proof of the preceding statement is not difficult and will be given later as an exercise.

We therefore have the result that the angular velocity vector is a property of the overall body $\mathcal{B}$, not a property of its individual *points.* This idea cannot be overemphasized. Remember:

1. A *point* has position, velocity, and acceleration.
2. A *body* has orientation, angular velocity, and angular acceleration.†

Remember too that a *point* does not have orientation, $\boldsymbol{\omega}$, or $\boldsymbol{\alpha}$,† and a finite-sized *body* does not have a unique $\mathbf{r}$, $\mathbf{v}$, and $\mathbf{a}$.‡

We now consider several examples of the use of our new equation (3.8); in each application of this equation, three rules must be followed *without exception:*

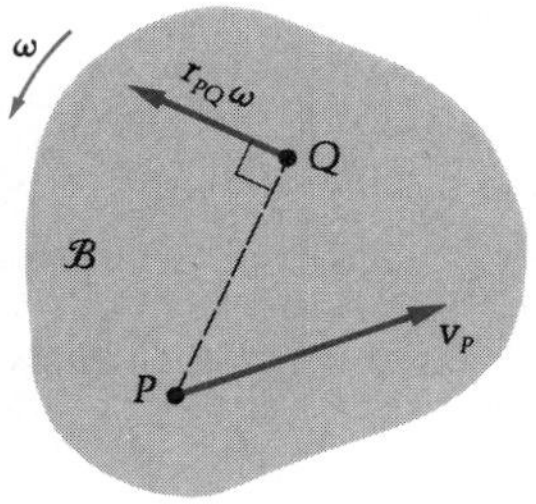

Figure 3.6

3. This vector extends *from* the point (P) on *this* (right) side of the equation *to* the point (Q) on the *other* (left) side.

$$\mathbf{v}_Q = \mathbf{v}_P + \dot{\theta}\hat{\mathbf{k}} \times \mathbf{r}_{PQ} \tag{3.10}$$

1. These two points are on the *same* rigid body $\mathcal{B}$.

2. This is the angular velocity vector of $\mathcal{B}$.

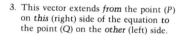

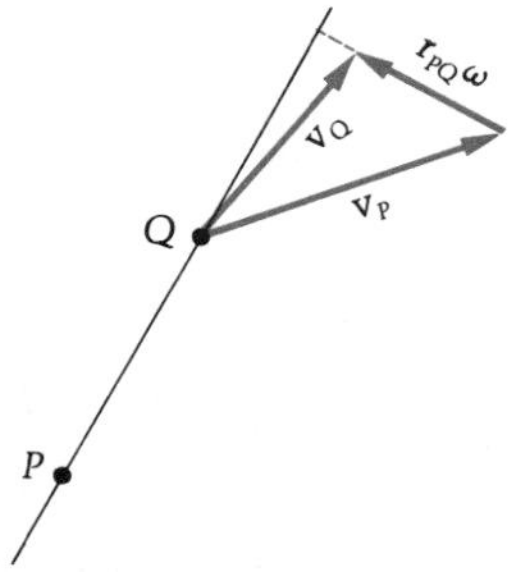

Figure 3.7

Also helpful in using Equation (3.8) is the kinematic diagram presented in Figure 3.6. The velocity of Q, from Equation (3.8), is the sum of the two vectors in Figure 3.6 (see Figure 3.7). Note that depending on the

*We remark that in general (three-dimensional) motion, such a simple relationship as $\dot{\theta}\hat{\mathbf{k}}$ between angular velocity and body orientation does not exist.

†Angular acceleration ($\boldsymbol{\alpha}$), the derivative of angular velocity, is discussed in Section 3.5.

‡Of course the *particle*, being treated as small enough that we need not distinguish between the velocities of its points, must be considered as having an $\mathbf{r}$, $\mathbf{v}$, and $\mathbf{a}$—and *not* an $\boldsymbol{\omega}$ or $\boldsymbol{\alpha}$.

relative sizes of $\mathbf{v}_P$ and $r_{PQ}\omega$, the velocity $\mathbf{v}_Q$ could lie on either side, or even along, line PQ. Note further that the difference between the velocities of Q and P—that is, $\mathbf{v}_Q - \mathbf{v}_P$—is simply $\dot{\theta}\hat{\mathbf{k}} \times \mathbf{r}_{PQ}$. This means that the only way in which the velocities of two points of a rigid body $\mathcal{B}$, in motion in frame $\mathcal{F}$, can differ is by the $r\omega$ term normal to the line joining them. We shall return to this idea following the first three examples of this section.

Incidentally, some books describe $\mathbf{v}_Q - \mathbf{v}_P$ as "the velocity of point Q relative to point P." We mention this only by way of explanation; our definition of $\mathbf{v}_P$ in Section 1.3 shows that points have velocities relative to frames, *not* relative to other points. If one uses the phrase "the velocity of point Q relative to point P," one means the velocity of Q in a reference frame in which P is fixed and which translates relative to $\mathcal{F}$.*

In the examples that follow, note the importance in each problem of selecting the sign convention to be used in the solution and showing it beside the figure for reference. We now illustrate the use of Equation (3.8).

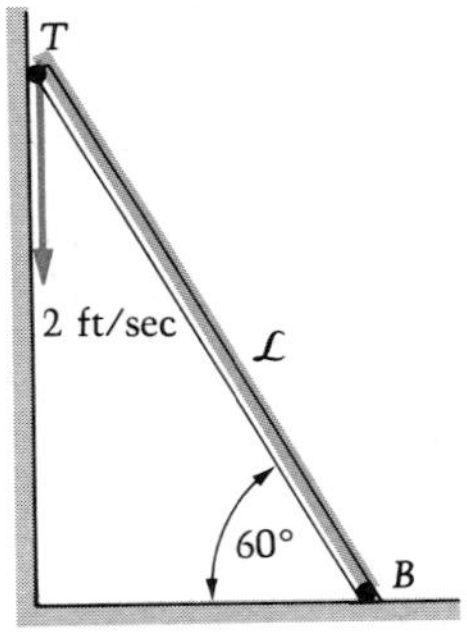

E X A M P L E 3.1

A 30-ft ladder $\mathcal{L}$ is slipping down in a warehouse with the upper contact point T moving downward on the wall at a speed of 2 ft/sec in the position shown in the diagram. Find the velocity of point B, which is sliding on the floor.

S O L U T I O N

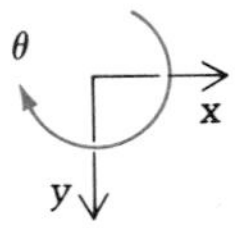

We relate $\mathbf{v}_T$ and $\mathbf{v}_B$ by using Equation (3.8):

$$\mathbf{v}_B = \mathbf{v}_T + \dot{\theta}\hat{\mathbf{k}} \times \mathbf{r}_{TB}$$

$$v_B\hat{\mathbf{i}} = +2\hat{\mathbf{j}} + \dot{\theta}\hat{\mathbf{k}} \times 30\left(\frac{1}{2}\hat{\mathbf{i}} + \frac{\sqrt{3}}{2}\hat{\mathbf{j}}\right)$$

$$= \left(-30\frac{\sqrt{3}}{2}\dot{\theta}\right)\hat{\mathbf{i}} + (2 + 15\dot{\theta})\hat{\mathbf{j}} \text{ ft/sec}$$

(Continued)

*"Translates" means that it moves in $\mathcal{F}$ without rotating. Translation is discussed in much more detail in Section 3.3.

Collecting the $\hat{\mathbf{j}}$ coefficients, we have

$$0 = 2 + 15\dot{\theta} \Rightarrow \dot{\theta} = -\frac{2}{15} = -0.133 \text{ rad/sec}$$

Collecting the $\hat{\mathbf{i}}$ coefficients, we have

$$v_B = -15\sqrt{3}\dot{\theta} = -15\sqrt{3}\left(-\frac{2}{15}\right) = 3.46 \text{ ft/sec}$$

The velocity of B is $\mathbf{v}_B = 3.46\hat{\mathbf{i}}$ ft/sec.

Note that the direction indicator must be attached to $\dot{\theta}$ in order to specify correctly the angular velocity vector of the ladder $\mathcal{L}$:

$$\boldsymbol{\omega}_{\mathcal{L}} = \dot{\theta}\hat{\mathbf{k}} = -0.133\hat{\mathbf{k}} \text{ rad/sec}$$

or

$$\boldsymbol{\omega}_{\mathcal{L}} = 0.133 \circlearrowright \text{ rad/sec}$$

Note also that the directions of $\mathbf{v}_B$ ($\longrightarrow$) and $\boldsymbol{\omega}_{\mathcal{L}}$ ($\circlearrowright$) make sense. Such visual checks on solutions should be made whenever possible.

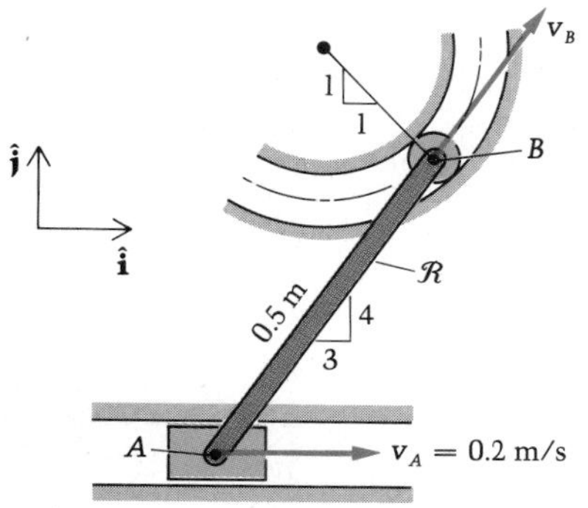

EXAMPLE 3.2

At the instant shown in the diagram, the velocity of point A is 0.2 m/s to the right. Find the angular velocity of rod $\mathcal{R}$, and determine the velocity of its other end (point B). The length of $\mathcal{R}$ is 0.5 m.

SOLUTION

We shall use Equation (3.8):

$$\mathbf{v}_B = \mathbf{v}_A + \dot{\theta}_{\mathcal{R}}\hat{\mathbf{k}} \times \mathbf{r}_{AB}$$

Noting that the velocity of B has a known direction (tangent to its path), we write $\mathbf{v}_B$ as an unknown scalar times a unit vector in this direction:

$$v_B\left(\frac{\hat{\mathbf{i}} + \hat{\mathbf{j}}}{\sqrt{2}}\right) = 0.2\hat{\mathbf{i}} + \dot{\theta}_{\mathcal{R}}\hat{\mathbf{k}} \times (0.3\hat{\mathbf{i}} + 0.4\hat{\mathbf{j}})$$

(Continued)

$\hat{\mathbf{i}}$ coefficients: $$\left(\frac{1}{\sqrt{2}}\right)v_B = 0.2 - 0.4\dot{\theta}_{\mathcal{R}} \qquad (1)$$

$\hat{\mathbf{j}}$ coefficients: $$\left(\frac{1}{\sqrt{2}}\right)v_B = 0.3\dot{\theta}_{\mathcal{R}} \qquad (2)$$

Solving Equations (1) and (2) gives

$$v_B = 0.121 \text{ m/s} \qquad \dot{\theta}_{\mathcal{R}} = 0.286 \text{ rad/s}$$

and therefore the answers (*vectors* are what are asked for!) are

$\mathbf{v}_B = 0.121$  m/s $\qquad \dot{\theta}_{\mathcal{R}}\hat{\mathbf{k}} = \boldsymbol{\omega}_{\mathcal{R}} = 0.286\hat{\mathbf{k}}$ rad/s

or $\qquad$ or

$\mathbf{v}_B = 0.0856\hat{\mathbf{i}} + 0.0856\hat{\mathbf{j}}$ m/s $\qquad \boldsymbol{\omega}_{\mathcal{R}} = 0.286$ ↺ rad/s

EXAMPLE 3.3

The crank arm $\mathcal{C}$ shown in the diagram turns about a horizontal z axis, through its pinned end O, with an angular velocity of 10 rad/sec clockwise at the given instant. Find the velocity of the piston pin B.

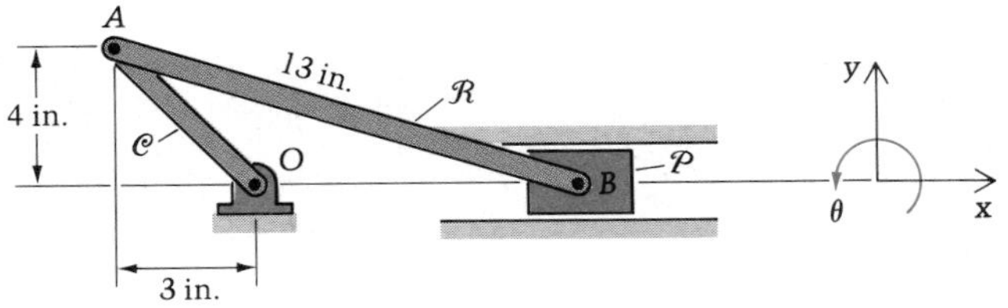

SOLUTION

We apply Equation (3.8) first to relate the velocities of A and O on body $\mathcal{C}$ and then to relate $\mathbf{v}_B$ to $\mathbf{v}_A$ on rod $\mathcal{R}$. Note that A is a "linking" point of both $\mathcal{C}$ and $\mathcal{R}$, since it belongs to *both* bodies. On body $\mathcal{C}$:

$$\begin{aligned}\mathbf{v}_A &= \mathbf{v}_O + \dot{\theta}_{\mathcal{C}}\hat{\mathbf{k}} \times \mathbf{r}_{OA} \\ &= \mathbf{0} + (-10\hat{\mathbf{k}}) \times (-3\hat{\mathbf{i}} + 4\hat{\mathbf{j}}) \\ &= 40\hat{\mathbf{i}} + 30\hat{\mathbf{j}} \text{ in./sec}\end{aligned}$$

We note from the diagram that

$$\text{Angle } ABO = \sin^{-1}\left(\frac{4}{13}\right) = 17.9° \Rightarrow \mathbf{r}_{AB} = 13(\cos 17.9°\,\hat{\mathbf{i}} - \sin 17.9°\,\hat{\mathbf{j}})$$

(Continued)

On body $\mathcal{R}$:

$$\mathbf{v}_B = \mathbf{v}_A + \dot{\theta}_{\mathcal{R}}\hat{\mathbf{k}} \times \mathbf{r}_{AB}$$

$$= (40\hat{\mathbf{i}} + 30\hat{\mathbf{j}}) + \dot{\theta}_{\mathcal{R}}\hat{\mathbf{k}} \times (12.4\hat{\mathbf{i}} - 4\hat{\mathbf{j}})$$

$$v_B\hat{\mathbf{i}} = (40 + 4\dot{\theta}_{\mathcal{R}})\hat{\mathbf{i}} + (30 + 12.4\dot{\theta}_{\mathcal{R}})\hat{\mathbf{j}}$$

Equating the $\hat{\mathbf{i}}$ coefficients:

$$v_B = 40 + 4\dot{\theta}_{\mathcal{R}} \tag{1}$$

Equating the $\hat{\mathbf{j}}$ coefficients:

$$0 = 30 + 12.4\dot{\theta}_{\mathcal{R}}$$

$$\dot{\theta}_{\mathcal{R}} = -2.42 \text{ rad/sec}$$

or

$$\boldsymbol{\omega}_{\mathcal{R}} = 2.42 \circlearrowright \text{ rad/sec} \tag{2}$$

Substituting $\dot{\theta}_{\mathcal{R}}$ into (1), we have $v_B = 30.3$ in./sec and $\mathbf{v}_B = 30.3\hat{\mathbf{i}}$ in./sec.

Note that we ***must*** use the kinematic restraint $\mathbf{v}_B = v_B\hat{\mathbf{i}}$, since each point of the piston is constrained to rectilinear motion in the **x** direction. Otherwise, our mathematics will not be working the correct problem but rather one in which end B of $\mathcal{R}$ is free to move in the y (as well as the **x**) direction.

It is often helpful in studying the kinematics of rigid bodies to make use of the following result, which is a corollary of Equation (3.8):

> Corollary: If P and Q are two points of a rigid body, their velocity components along the line joining them must be equal.

Intuitively, we see that the difference between these components is the rate of stretching of the line PQ, and this has to vanish. Also, we have seen that $\mathbf{v}_Q$ and $\mathbf{v}_P$ differ only by the term $\dot{\theta}\hat{\mathbf{k}} \times \mathbf{r}_{PQ}$, which is clearly ***normal*** to the line $\overline{PQ}$ joining the points. Mathematically, we can see this immediately by dotting Equation (3.8) with the unit vector parallel to $\mathbf{r}_{PQ}$, which is $\mathbf{r}_{PQ}/r_{PQ}$:

$$\underbrace{\frac{\mathbf{r}_{PQ}}{r_{PQ}} \cdot \mathbf{v}_Q}_{\substack{\text{component of} \\ \mathbf{v}_Q \text{ along } \overline{PQ}}} = \underbrace{\frac{\mathbf{r}_{PQ}}{r_{PQ}} \cdot \mathbf{v}_P}_{\substack{\text{component of} \\ \mathbf{v}_P \text{ along } \overline{PQ}}} + \underbrace{(\dot{\theta}\hat{\mathbf{k}} \times \mathbf{r}_{PQ}) \cdot \frac{\mathbf{r}_{PQ}}{r_{PQ}}}_{\substack{\text{zero (since } \dot{\theta}\hat{\mathbf{k}} \times \mathbf{r}_{PQ} \\ \text{is } \perp \mathbf{r}_{PQ})}} \tag{3.11}$$

We shall now make use of these ideas to rework the previous pair of examples.

EXAMPLE 3.4

Use the preceding corollary to rework Example 3.3.

SOLUTION

In Example 3.3 we had the geometry shown in the diagram. Since the velocities of A and B along line $\overline{AB}$ are equal, we have

$$50 \cos 54.8° = v_B \cos 17.9° \Rightarrow \mathbf{v}_B = 30.3 \rightarrow \text{in./sec}$$

as before. And since $\mathbf{v}_A$ and $\mathbf{v}_B$ differ by "$r\omega$," we obtain

$$\frac{50 \sin 54.8° - 30.3 \sin 17.9°}{13} = \omega_{\mathcal{R}} \Rightarrow \boldsymbol{\omega}_{\mathcal{R}} = 2.43 \circlearrowright \text{rad/sec}$$

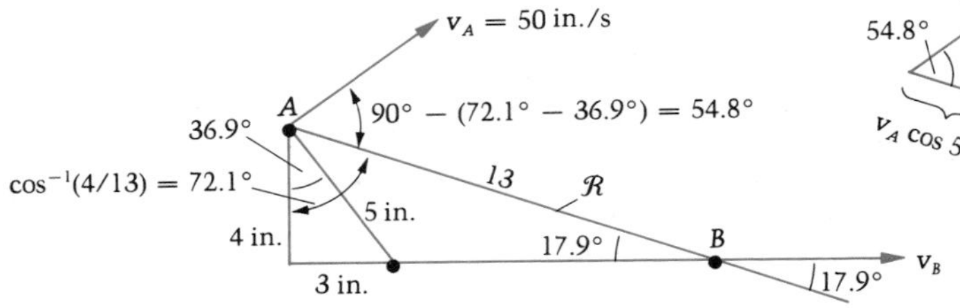

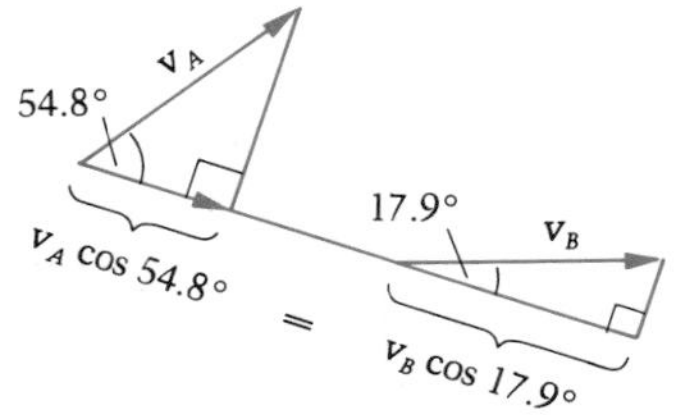

The chief value of these ideas is that we can often use them to solve for one unknown at a time, reducing the algebra, as in the next example.

EXAMPLE 3.5

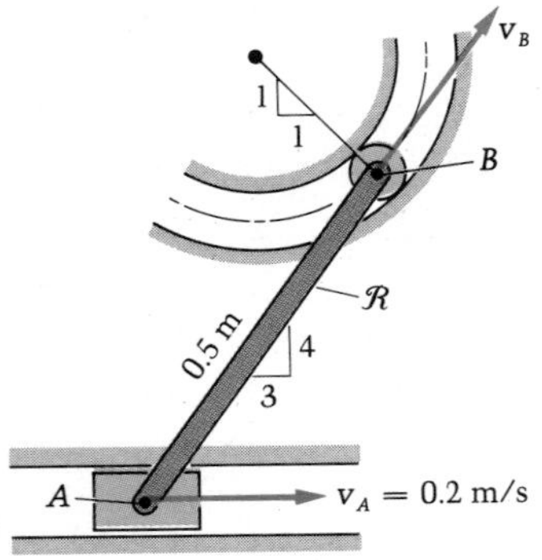

Use the corollary to rework Example 3.2. (See the diagram.) Find the angular velocity of rod $\mathcal{R}$, and determine the velocity of its other end, B.

SOLUTION

The component of $\mathbf{v}_A$ along $\mathcal{R}$ is seen from the diagram to be $0.2(\frac{3}{5})$ or 0.12 m/s directed toward B. Therefore we may write, using the corollary,

$$v_B \cos 8.1° = 0.120 \Rightarrow \mathbf{v}_B = 0.12 \ \angle_1^1 \ \text{m/s}$$

(Continued)

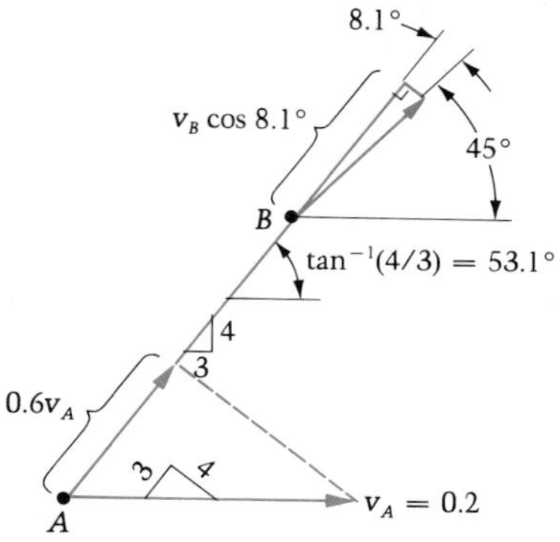

And since $\mathbf{v}_A - \mathbf{v}_B$ is the term $\dot{\theta}_{\mathcal{R}}\hat{\mathbf{k}} \times \mathbf{r}_{BA}$,

$$\dot{\theta}_{\mathcal{R}} = \frac{0.2(\frac{4}{5}) - 0.12 \sin 8.1°}{0.5}$$

$$\boldsymbol{\omega}_{\mathcal{R}} = 0.29 \circlearrowleft \text{rad/sec}$$

Note that we must keep up with the signs and directions ourselves when we do not use vectors; but we do gain in understanding for our trouble.

We now return to the vector formulation (Equation 3.8) for two final examples in this section.

EXAMPLE 3.6

In the linkage shown in the diagram, the velocities of A and C are given to be

$$\mathbf{v}_A = 2 \leftarrow \text{m/s}$$

$$\mathbf{v}_C = 3 \uparrow \text{m/s}$$

at the instant given. Find the velocity of point B at the same instant.

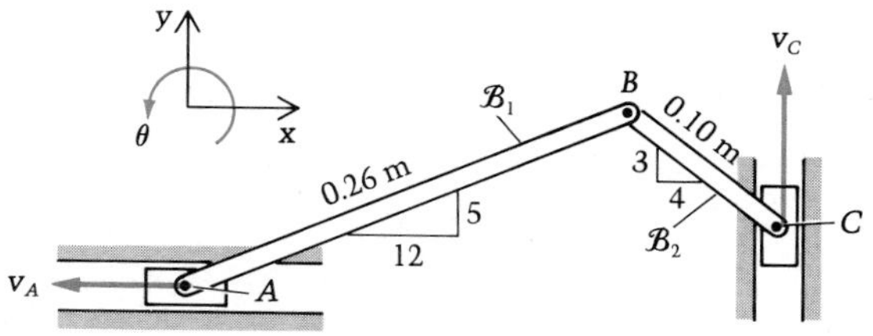

SOLUTION

On bar $\mathcal{B}_1$:

$$\mathbf{v}_B = \mathbf{v}_A + \dot{\theta}_{\mathcal{B}_1}\hat{\mathbf{k}} \times \mathbf{r}_{AB}$$

$$= -2\hat{\mathbf{i}} + \dot{\theta}_{\mathcal{B}_1}\hat{\mathbf{k}} \times (0.24\hat{\mathbf{i}} + 0.10\hat{\mathbf{j}})$$

$$= (-2 - 0.1\dot{\theta}_{\mathcal{B}_1})\hat{\mathbf{i}} + (0.24\dot{\theta}_{\mathcal{B}_1})\hat{\mathbf{j}} \text{ m/s} \quad (1)$$

(Continued)

On bar $\mathcal{B}_2$:

$$\mathbf{v}_B = \mathbf{v}_C + \dot{\theta}_{\mathcal{B}_2}\hat{\mathbf{k}} \times \mathbf{r}_{CB}$$
$$= 3\hat{\mathbf{j}} + \dot{\theta}_{\mathcal{B}_2}\hat{\mathbf{k}} \times (-0.08\hat{\mathbf{i}} + 0.06\hat{\mathbf{j}})$$
$$= (-0.06\dot{\theta}_{\mathcal{B}_2})\hat{\mathbf{i}} + (3 - 0.08\dot{\theta}_{\mathcal{B}_2})\hat{\mathbf{j}} \text{ m/s} \quad (2)$$

Equating the two vector expressions for $\mathbf{v}_B$, we get

$\hat{\mathbf{i}}$ coefficients: $-2 - 0.1\dot{\theta}_{\mathcal{B}_1} = -0.06\dot{\theta}_{\mathcal{B}_2}$

$\hat{\mathbf{j}}$ coefficients: $0.24\dot{\theta}_{\mathcal{B}_1} = 3 - 0.08\dot{\theta}_{\mathcal{B}_2}$

Solving, we obtain

$$\dot{\theta}_{\mathcal{B}_1} = 0.893 \qquad \text{and} \qquad \dot{\theta}_{\mathcal{B}_2} = 34.8 \text{ rad/s}$$

From Equation (1),

$$\mathbf{v}_B = -2.09\hat{\mathbf{i}} + 0.214\hat{\mathbf{j}} \text{ m/s}$$

and the same result follows from (2), as a check.

Question 3.6 If the velocities of A and C were given to be 2 m/s ← and 3 m/s ↑ for an *interval* of time, and not just at the instant shown, would the solution be any different at (a) the same instant? (b) some other instant?

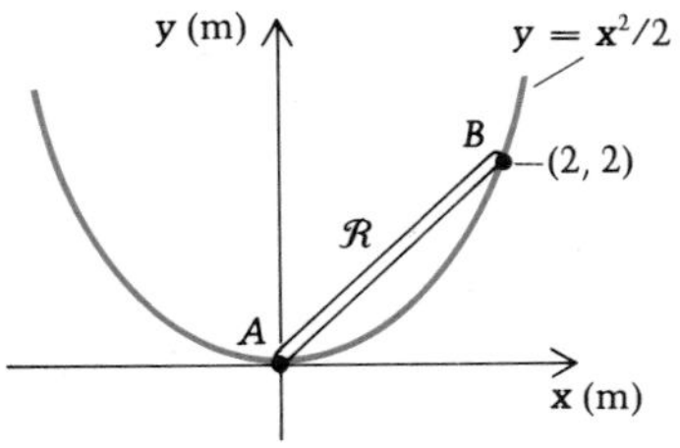

EXAMPLE 3.7

The end B of rod $\mathcal{R}$ travels up the right half of the parabolic incline in the diagram at the constant speed of 0.3 m/s. Find the angular velocity of $\mathcal{R}$ and the velocity of point A, which is at the origin at the given instant.

SOLUTION

The geometry shown in the accompanying diagram gives us the direction of $\mathbf{v}_B$, since we know it is tangent to its path along the parabola at all times:

$$\theta = \tan^{-1}(y') = \tan^{-1} 2 = 63.4°$$

(Continued)

Therefore

$$\hat{\mathbf{e}}_t = \cos\theta\,\hat{\mathbf{i}} + \sin\theta\,\hat{\mathbf{j}}$$
$$= 0.448\,\hat{\mathbf{i}} + 0.894\,\hat{\mathbf{j}}$$

And thus

$$\mathbf{v}_B = 0.3\hat{\mathbf{e}}_t = 0.134\,\hat{\mathbf{i}} + 0.268\,\hat{\mathbf{j}}\text{ m/s}$$

Since point A likewise has a velocity tangent to *its* path, we may write

$$\mathbf{v}_A = v_A\hat{\mathbf{i}}$$

and so Equation (3.8) gives

$$\mathbf{v}_B = \mathbf{v}_A + \boldsymbol{\omega}_{\mathcal{R}} \times \mathbf{r}_{AB}$$
$$0.134\,\hat{\mathbf{i}} + 0.268\,\hat{\mathbf{j}} = v_A\hat{\mathbf{i}} + \dot{\theta}_{\mathcal{R}}\hat{\mathbf{k}} \times (2\,\hat{\mathbf{i}} + 2\,\hat{\mathbf{j}})$$

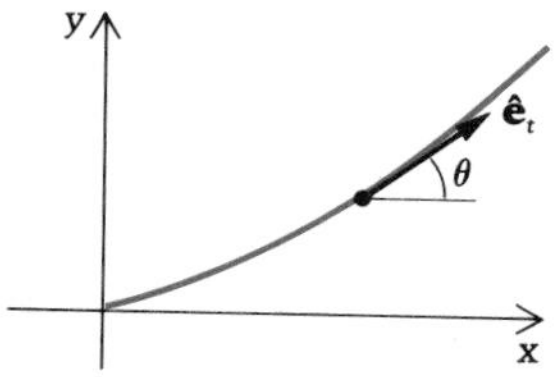

Collecting the coefficients of $\hat{\mathbf{i}}$ and $\hat{\mathbf{j}}$, we have

$\hat{\mathbf{j}}$ coefficients: $\quad 0.268 = 2\dot{\theta}_{\mathcal{R}} \Rightarrow \dot{\theta}_{\mathcal{R}} = 0.134$ rad/s

so that

$$\boldsymbol{\omega}_{\mathcal{R}} = \dot{\theta}_{\mathcal{R}}\hat{\mathbf{k}} = 0.134\,\hat{\mathbf{k}}\text{ rad/s or } 0.134 \circlearrowleft \text{ rad/s}$$

$\hat{\mathbf{i}}$ coefficients: $\quad 0.134 = v_A - 2\dot{\theta}_{\mathcal{R}}$

$$v_A = 0.402\text{ m/s}$$

so that

$$\mathbf{v}_A = 0.402\,\hat{\mathbf{i}}\text{ m/s}$$

Applications of Equation (3.8) to rolling bodies are presented in Section 3.6 after we have examined that topic in detail.

PROBLEMS / Section 3.2

3.3 The angular velocity of the bent bar $\mathcal{B}$ is indicated in Figure P3.3. Find the velocity of the endpoint B in this position.

3.4 Find the velocity of point B of rod $\mathcal{R}$ if end A has constant velocity 2 m/s to the right as shown in Figure P3.4. The rollers are small. Compare the use of Equation (3.8) with the procedure used to solve Problem 1.84.

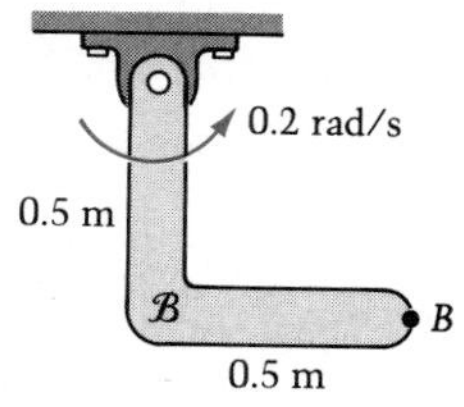

Figure P3.3

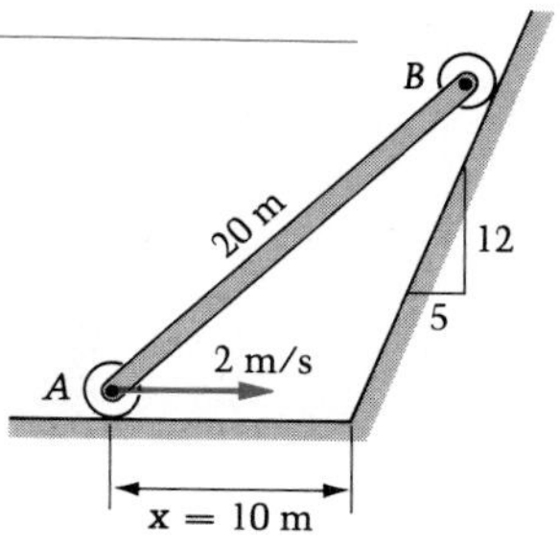

Figure P3.4

3.5 The velocities of the two endpoints A and B of a rigid bar in plane motion are shown in Figure P3.5. Find the velocity of the midpoint of the bar in the given position.

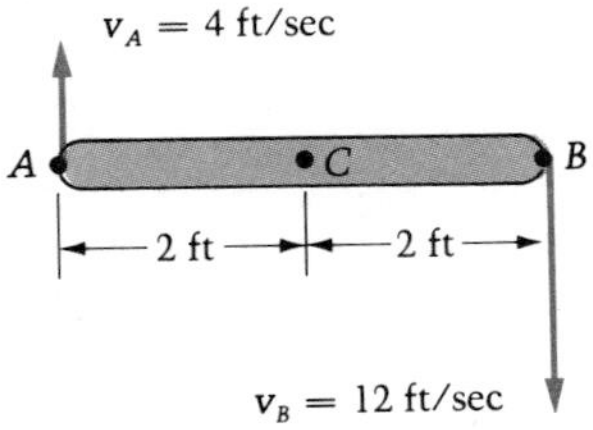

Figure P3.5

3.6 If $\mathbf{v}_A = 80\hat{\mathbf{i}}$ in./sec, find $\boldsymbol{\omega}_{\mathcal{B}_2}$ and $\boldsymbol{\omega}_{\mathcal{B}_3}$. See Figure P3.6.

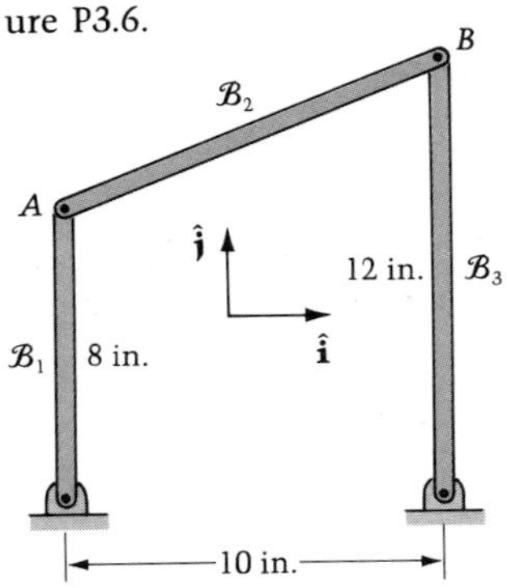

Figure P3.6

3.7 The wheel shown in Figure P3.7 turns and slips in such a manner that its angular velocity is 2 rad/s ↻ while the velocity of the center C is 0.3 m/s to the left. Determine the velocity of point A.

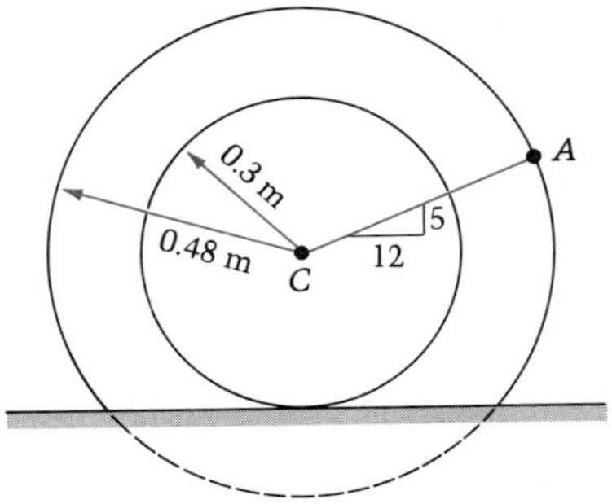

Figure P3.7

3.8–3.12 In the following five problems involving a "four-bar linkage" (the fourth bar in each case is the rigid ground length between fixed pins!), the angular velocity of one of the bars is indicated. Find the angular velocities of the other two bars.

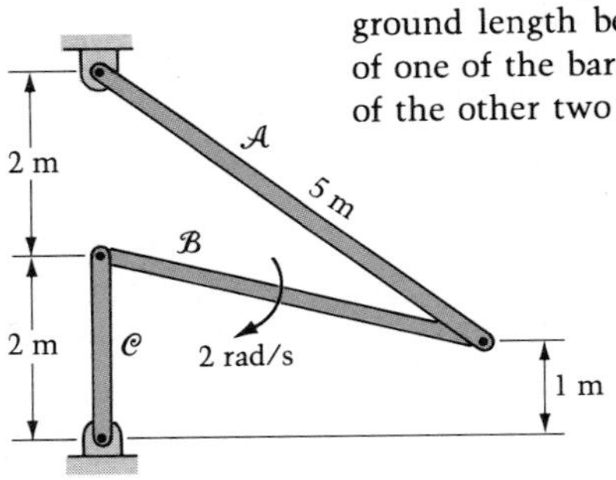

Figure P3.8

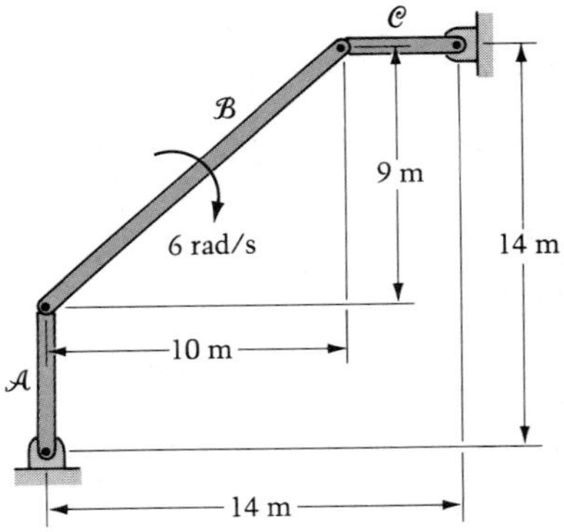

Figure P3.10

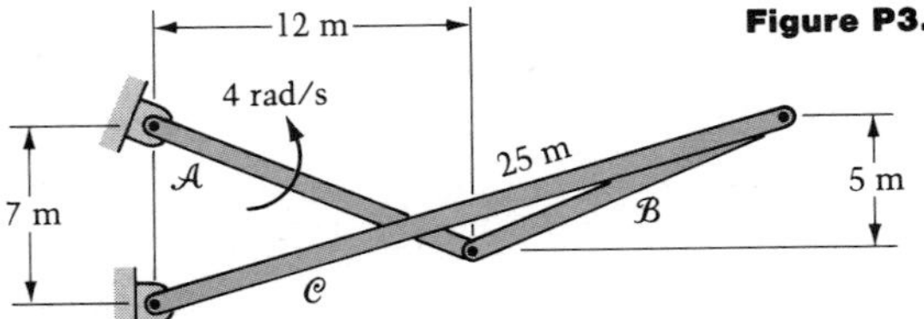

Figure P3.9

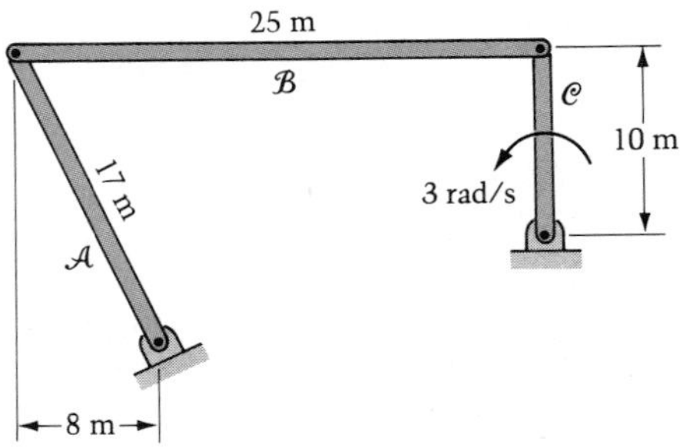

Figure P3.11

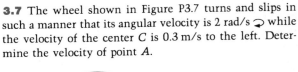

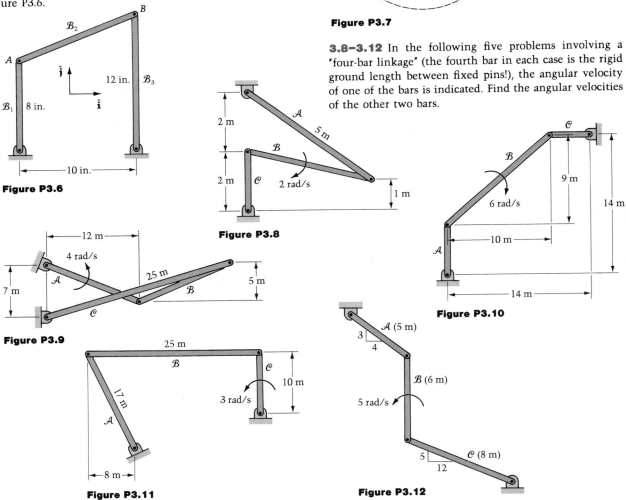

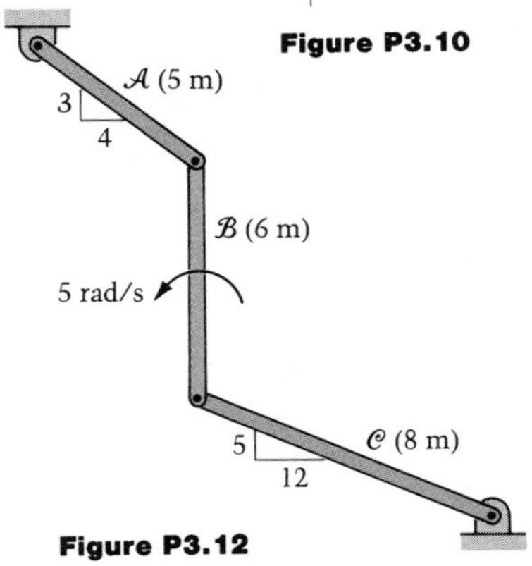

Figure P3.12

3.13 The equilateral triangular plate $\mathcal{T}$ shown in Figure P3.13 has three sides of length 0.3 m each. The bar $\mathcal{L}$ has an angular velocity $\omega_{\mathcal{L}} = 2$ rad/s counterclockwise and is pinned to $\mathcal{T}$ at A. Body $\mathcal{T}$ is also pinned to a block at B, which moves in the indicated slot. At the given time, find the angular velocity of $\mathcal{T}$.

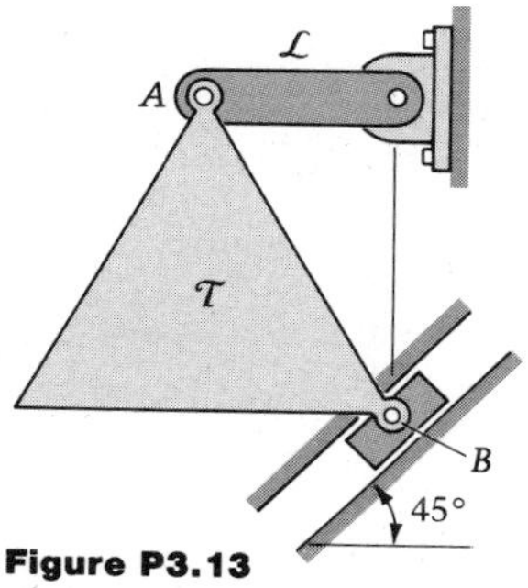

Figure P3.13

3.14 The four links shown in Figure P3.14 each have length 0.4 m, and two of their angular velocities are indicated. Find the velocity of point C and determine the angular velocities of $\mathcal{B}$ and $\mathcal{C}$ at the indicated instant.

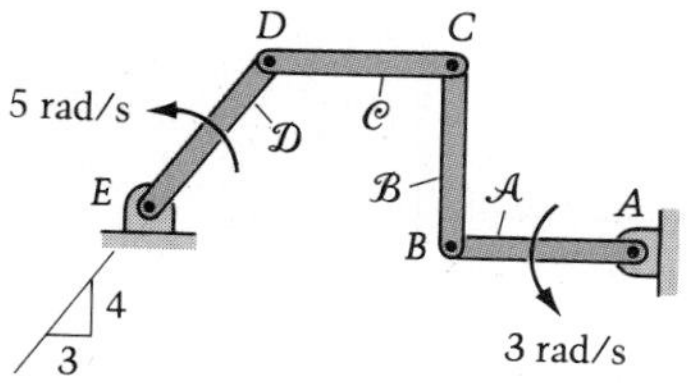

Figure P3.14

3.15 At a certain instant, the coordinates of two points A and B of a rigid body $\mathcal{B}$ in plane motion are given in Figure P3.15. Point A has $\mathbf{v}_A = 2\hat{\mathbf{i}}$ m/s, and the velocity of B is vertical. Find $\mathbf{v}_B$ and the angular velocity of $\mathcal{B}$.

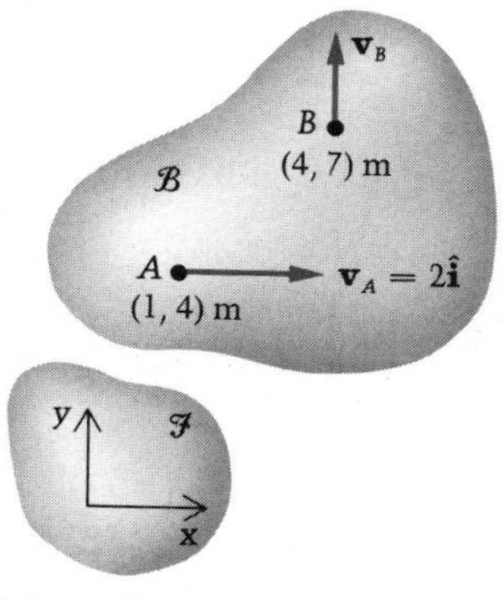

Figure P3.15

3.16 Crank arm $\mathcal{A}$ shown in Figure P3.16 turns counterclockwise at a constant rate of 1 rad/s. Rod $\mathcal{R}$ is pinned to $\mathcal{A}$ at A and to a roller at B that slides in a circular slot. Determine the velocity of B and the angular velocity of $\mathcal{R}$ at the given instant.

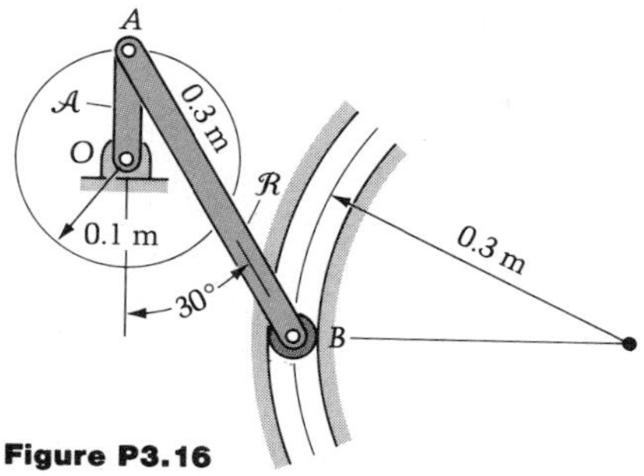

Figure P3.16

3.17 Wheel $\mathcal{A}$ in Figure P3.17 has a counterclockwise angular velocity of 6 rad/s. What is the velocity of point B at the instant shown?

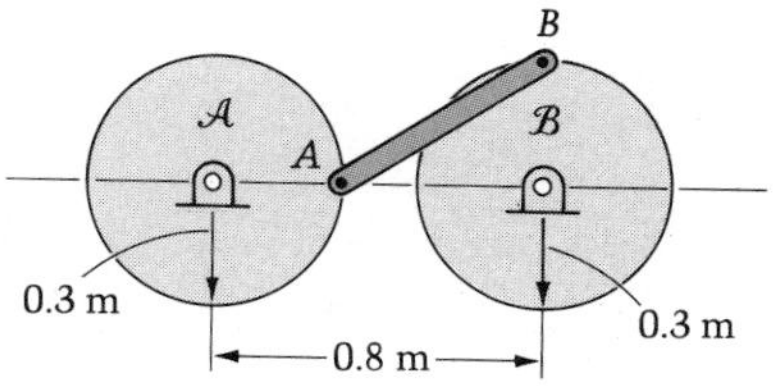

Figure P3.17

3.18 Point A of rod $\mathcal{R}$ slides along an inclined plane as in Figure P3.18, while the other end, B, slides on the horizontal plane. In the indicated position, $\boldsymbol{\omega} = 0.5\hat{\mathbf{k}}$ rad/sec. Find the velocity of the midpoint of $\mathcal{R}$ at this instant.

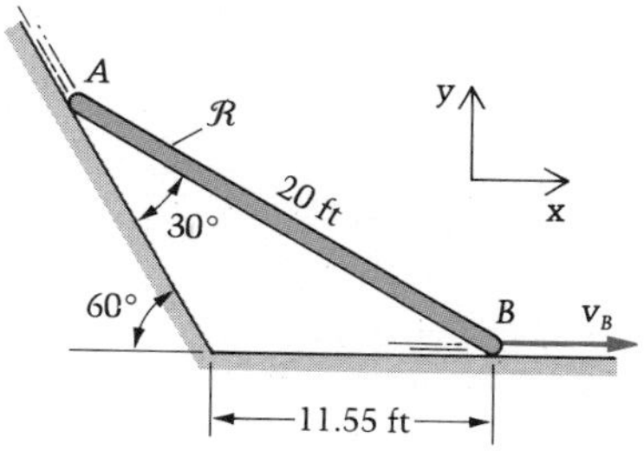

Figure P3.18

3.19 The speed of block $\mathcal{B}$ in Figure P3.19 has the value shown. Find the angular velocity of rod $\mathcal{R}$, and determine the velocity of pin A of block $\mathcal{A}$ when $\theta = 60°$.

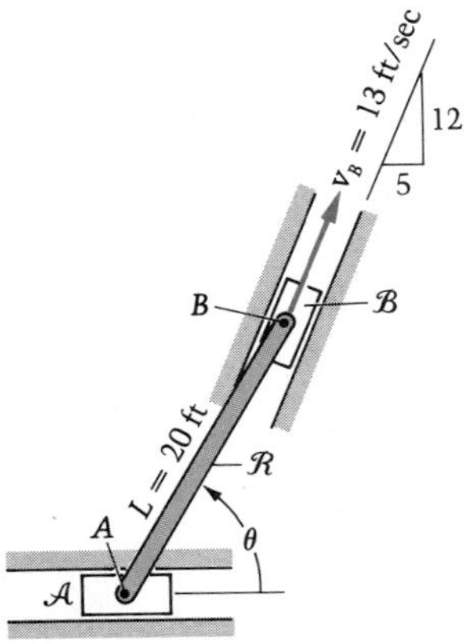

Figure P3.19

3.20 Wheel $\mathcal{C}$ (Figure P3.20) turns and slips in such a way that its angular velocity is 2 ↻ rad/s while the velocity of C is 0.4 m/s to the left. Determine the velocity of point B, which slides on the plane. Bar $\mathcal{B}$ is pinned to $\mathcal{C}$ at D.

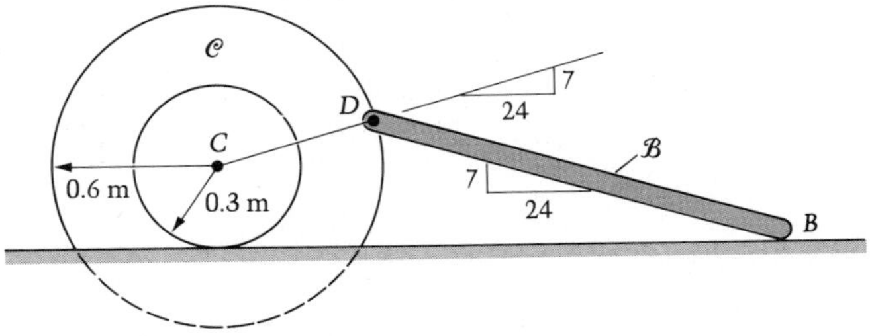

Figure P3.20

3.21 For the configuration shown in Figure P3.21, find the velocity of point P of the disk $\mathcal{B}_3$.

3.22 Block $\mathcal{B}$ has a controlled position in the slot given by $y = \sqrt{120}\,\sin(\pi t/10)$ in. for $0 \le t \le 10$ sec. (See Figure P3.22.) The time is $t = 0$ sec in the indicated position. Find the angular velocities of the rod $\mathcal{R}$ and the wheel $\mathcal{W}$ at (a) $t = 0$ sec and (b) $t = 5$ sec.

3.23 Block $\mathcal{B}$ in Figure P3.23 moves to the right according to the formula

$$x = 0.2 + 0.02t^2 \text{ m} \qquad \text{(with } t \text{ in seconds)}$$

Find the angular velocity of rod $\mathcal{R}$ when $x = 0.3$ m.

Hint:

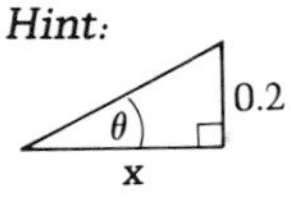

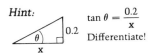

$\tan\theta = \dfrac{0.2}{x}$

Differentiate!

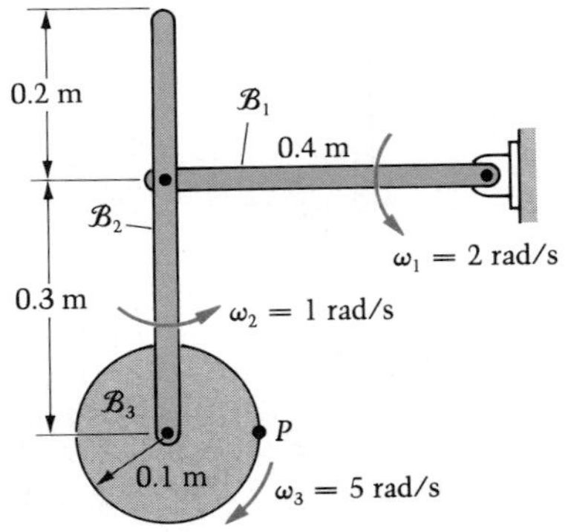

Figure P3.21

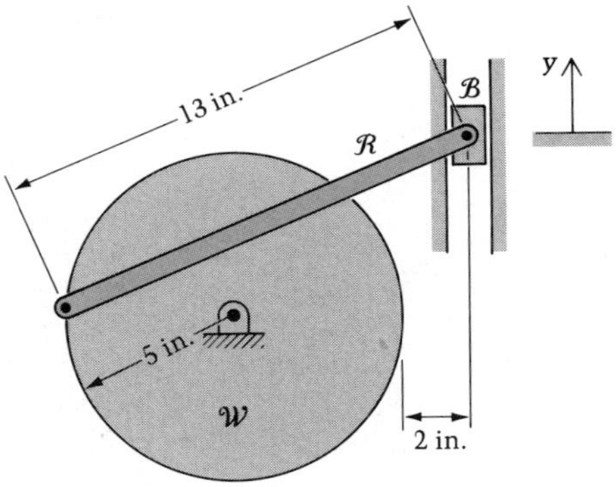

Figure P3.22

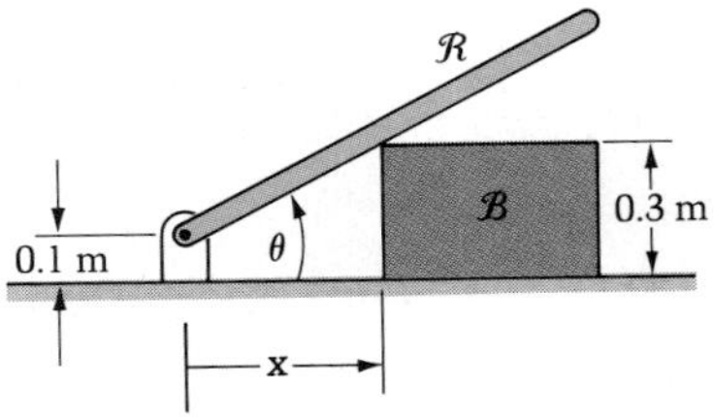

Figure P3.23

3.24 Referring to Section 3.2, show that neither the angular velocity vector nor Equation (3.8) depends on which body-fixed line segment (such as PQ in the text) is chosen to measure θ. Use two other points P' and Q' and their angle ϕ as suggested in Figure P3.24 for your proof.

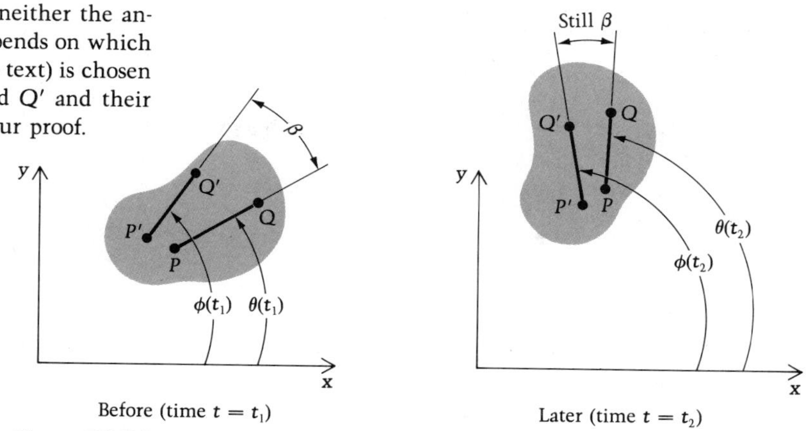

Figure P3.24

3.25 Block $\mathcal{A}$ in Figure P3.25, which slides in a vertical slot, is pinned to bars $\mathcal{B}$ and $\mathcal{C}$ at A. The other ends of $\mathcal{B}$ and $\mathcal{C}$ are pinned to blocks that slide in horizontal slots. Block $\mathcal{D}$ translates to the left at constant speed 0.2 m/s. Find the velocity of B: (a) at the given instant, (b) when C is at point D; (c) when C is at point E.

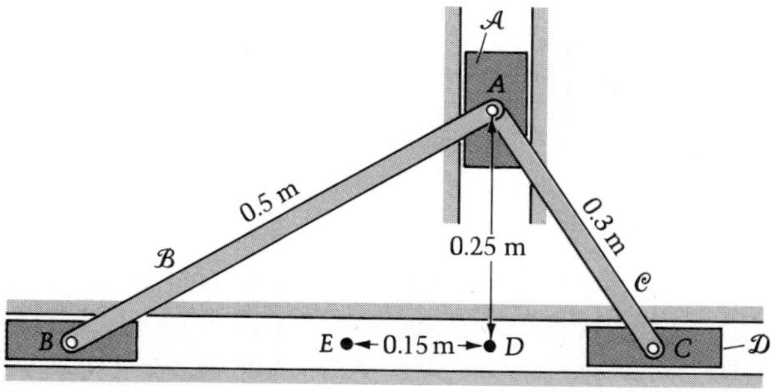

Figure P3.25

3.26 Crank $\mathcal{C}$ of the slider-crank mechanism shown in Figure P3.26 has a constant angular speed $\dot{\theta}$. Find the equation for the angular velocity $\dot{\phi}$ of the connecting rod $\mathcal{R}$ as a function of r, ℓ, θ, and $\dot{\theta}$.

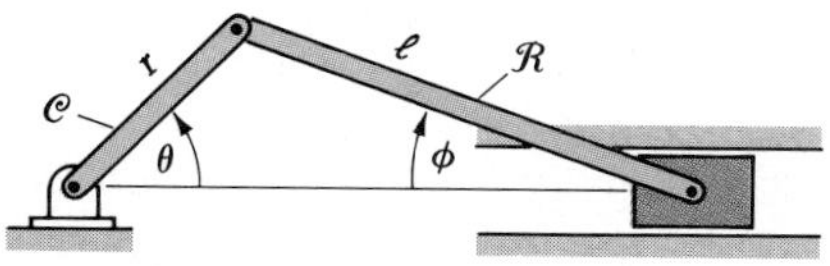

Figure P3.26

3.27 In the preceding problem, plot $\dot{\phi}/\dot{\theta}$ as a function of θ, from $\theta = 0$ to 2π, for $\ell/r = 1$, 2, and 5.

3.28 Show that xy = constant in Figure P3.28.

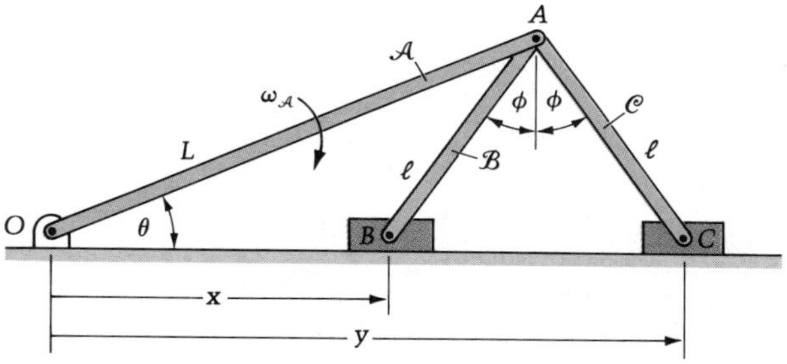

Figure P3.28

3.29 In the mechanism shown in Figure P3.29, the sleeve $\mathcal{S}$ is connected to the pivoted bar $\mathcal{A}$ by the 15-cm link $\mathcal{L}$. Over a certain range of motion of $\mathcal{A}$, the angle θ varies according to $\theta = 0.02t^2$ rad, starting at $t = 0$ with $\mathcal{A}$ and $\mathcal{L}$ horizontal. Find the velocity of pin S and the angular velocities of $\mathcal{A}$ and $\mathcal{L}$ when $\theta = 30°$.

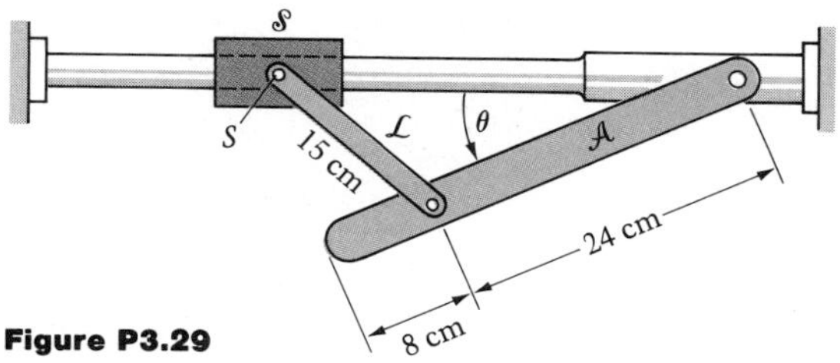

Figure P3.29

3.30 Rod $\mathcal{R}$ begins moving at $\theta = 0$ (see Figure P3.30) and is made to turn at the constant angular rate $\dot{\theta} = 0.2$ rad/s. The cord is attached to the end of $\mathcal{R}$ and passes around a pulley. The other end of the cord is tied to weight $\mathcal{B}$ at point B. Observe that $\mathcal{B}$ moves downward until $\theta = 90°$, when it reverses direction. Write an equation that gives the velocity of point B as a function of θ for $\pi \geq \theta \geq \pi/2$. *Hint:* Using trigonometry, write y as a function of θ and the length L of the cord. Then differentiate.

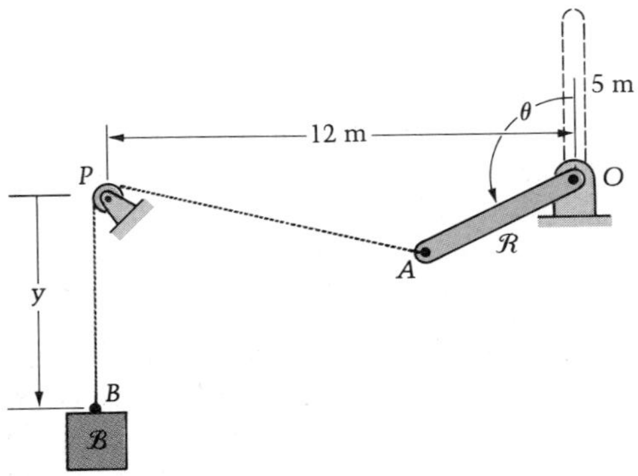

Figure P3.30

3.31 Repeat the preceding problem by using Equation (3.8) to obtain $\mathbf{v}_A$; then resolve $\mathbf{v}_A$ into two components: (a) one along PA that equals the magnitude of $\mathbf{v}_B$ and (b) the other normal to PA, which does not affect B. (These are sometimes called ***stretching*** and ***swinging*** components, respectively.)

3.32 Crank $\mathcal{C}$ in Figure P3.32 is driven at a constant angular speed $\omega_{\mathcal{C}}$ clockwise. Show that the speed of piston $\mathcal{P}$ is maximum when θ satisfies the equation

$$\cos\theta + \frac{\cos 2\theta}{\sqrt{(\ell_2/\ell_1)^2 - \sin^2\theta}} + \frac{\sin^2 2\theta}{4[(\ell_2/\ell_1)^2 - \sin^2\theta]^{3/2}} = 0$$

Solve with a calculator for the first root of this equation when the lengths of $\mathcal{C}$ and $\mathcal{R}$ are 8 cm and 20 cm, respectively. Note that the answer is independent of the value of $\omega_{\mathcal{C}}$. You may wish to read Appendix C and make use of the Newton–Raphson method described there, whether or not your calculator is programmable.

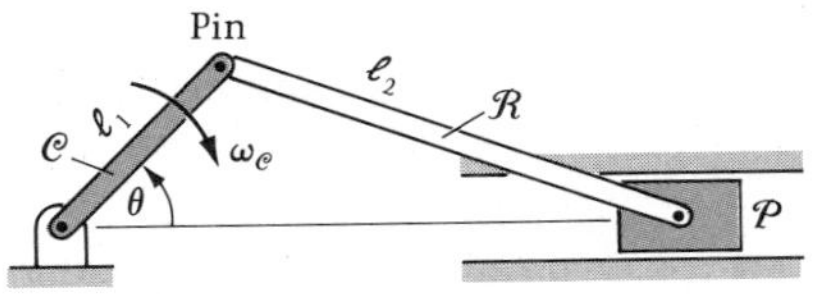

Figure P3.32

3.3 Translation

When a rigid body $\mathcal{B}$ moves during a certain time interval in such a way that its angular velocity vector remains identically zero, then the body is said to be **translating,** or to be in a state of **translational motion** during that interval. From Equation (3.8) we thus see that for translation

$$\mathbf{v}_Q = \mathbf{v}_P \tag{3.12}$$

That is, all points of the body have the same velocity vector. By differentiating Equation (3.12), we see that the accelerations of all points of $\mathcal{B}$ are also equal for translation. Note that if $\dot{\theta} = 0$ only at an *instant* (that is, at a single value of time rather than over an interval), then all points of the body have equal velocities at that instant but *need not have equal accelerations.*

Question 3.7 Why is this the case?

Translation can be either:

1. *Rectilinear:* Each point of $\mathcal{B}$ moves along a straight line in $\mathcal{I}$.
2. *Curvilinear:* Each point moves on a curved path in $\mathcal{I}$.

Examples of translation are shown in Figure 3.8. Part (a) shows an example of rectilinear translation: Body $\mathcal{B}$ is constrained to move in a straight slot. Part (b) shows an example of curvilinear translation: Body $\mathcal{B}$ is constrained by the identical links.

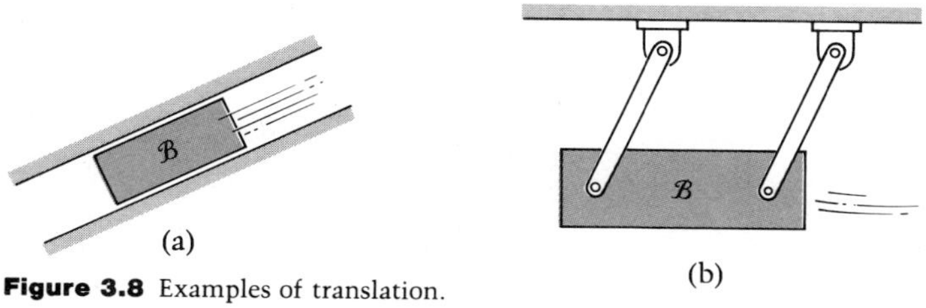

Figure 3.8 Examples of translation.

Perhaps an even better pair of examples is the blackboard eraser (Figure 3.9), which we used earlier to explain plane motion in Section 3.2. In part (a), the professor moves the eraser so that each of its points stays on a straight line; it is therefore in a state of rectilinear translation. In part (b), the professor moves the eraser on a curve; but if the word *eraser* is always horizontal during the erasing, then $\dot{\theta} \equiv 0$ and the eraser is in a state of curvilinear translation. Even though each of its points moves on a curve, all the velocities (and accelerations) are equal at all times. There is one notable exception to our earlier statement that "points, not bodies, have velocities and accelerations." In this present case of translation, since all the points have the *same* $\mathbf{v}$'s and $\mathbf{a}$'s, one could loosely refer to "the velocity of the eraser" without ambiguity.

There are no examples or problems in this section because rectilinear translation problems of rigid bodies are identical to examples of recti-

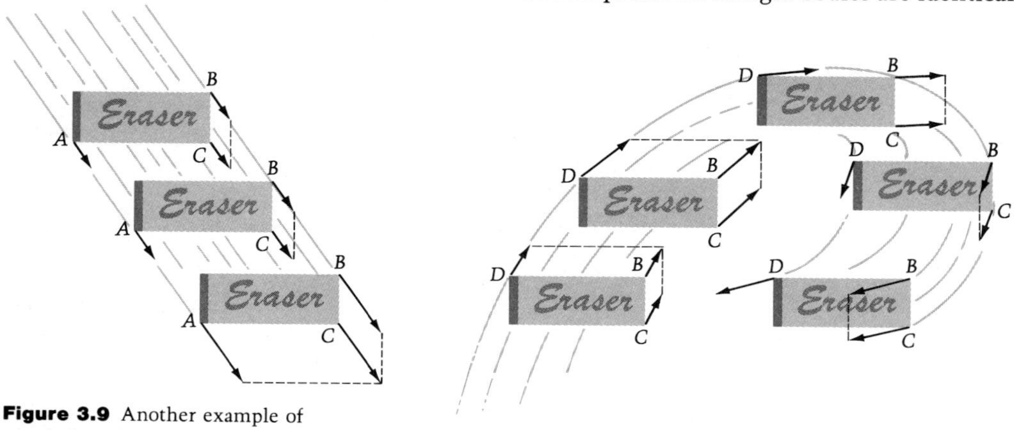

Figure 3.9 Another example of translation.

linear motion of particles, which we have already examined in Section 1.4. As for curvilinear translation, we shall give an example in Section 3.5 after we have defined angular acceleration.

Summarizing, when a body is translating (either rectilinearly or curvilinearly), its angular velocity $\dot{\theta}\hat{\mathbf{k}}$ is identically zero, and all its points have equal velocities (and accelerations). If $\dot{\theta} = 0$ only at an instant, then all the points of the body have the same velocity at that instant but need not have equal accelerations.

3.4 Instantaneous Center of Zero Velocity

If P is a point in the reference plane having zero velocity at some instant, then the velocity field of $\mathcal{B}$ is the same as if the body were constrained at that instant to rotate about an axis through P normal to the reference plane. This axis is called the **instantaneous axis of rotation,** and point P is called the **instantaneous center** (abbreviated Ⓘ) **of zero velocity*** of $\mathcal{B}$. Thus if Q is any *other* point of $\mathcal{B}$, then we have

$$\mathbf{v}_Q = \overset{\mathbf{0}}{\cancel{\mathbf{v}_{\text{Ⓘ}}}} + \dot{\theta}\hat{\mathbf{k}} \times \mathbf{r}_{\text{Ⓘ}Q} = \dot{\theta}\hat{\mathbf{k}} \times \mathbf{r}_{\text{Ⓘ}Q} \tag{3.13}$$

and thus each point moves with its velocity perpendicular to the line joining it to Ⓘ. This concept is illustrated in Figure 3.10 for a rolling wheel,[†] in which Ⓘ is the contact point.

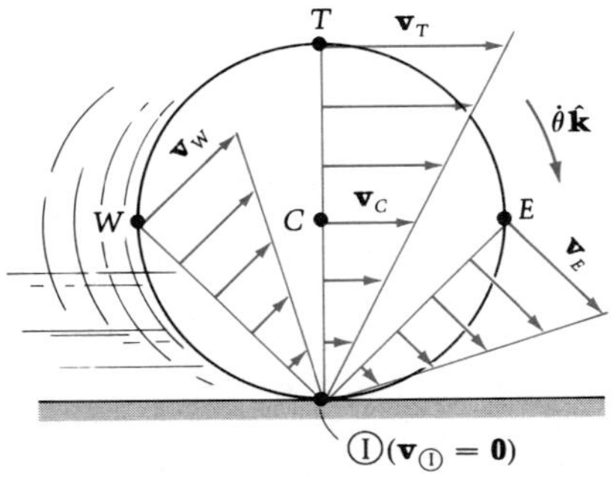

Figure 3.10 Instantaneous center of a rolling wheel.

We can show that if a body $\mathcal{B}$ has $\dot{\theta} \neq 0$ at a given instant, then it has an instantaneous center.

Question 3.8 Why can there be no point Ⓘ whenever $\dot{\theta}$ *is* zero?

To demonstrate the existence of Ⓘ, we shall use Equation (3.13) in conjunction with Figure 3.11. The vector $\mathbf{v}_Q$, being equal to $\boldsymbol{\omega} \times \mathbf{r}_{\text{Ⓘ}Q}$ for any point Ⓘ having $\mathbf{v}_{\text{Ⓘ}} = \mathbf{0}$, is therefore normal to both $\boldsymbol{\omega}$ and to $\mathbf{r}_{\text{Ⓘ}Q}$. Hence we have these results:

1. The vector $\mathbf{r}_{\text{Ⓘ}Q}$ lies in the reference plane and is normal to $\mathbf{v}_Q$. It thus lies along the line ℓ in Figure 3.11.
2. The point Ⓘ exists (and is unique) because $|\mathbf{r}_{\text{Ⓘ}Q}|$ is therefore seen to be $|\mathbf{v}_Q/\omega|$ in order that $\mathbf{v}_Q = \boldsymbol{\omega} \times \mathbf{r}_{\text{Ⓘ}Q}$.

*The phrase is admittedly redundant, but it is in common usage. "Instantaneous center of velocity" would perhaps be more concise, and "center of velocity" even more so. "Instantaneous center," however, is inadequate because of the possibility of confusion with points of zero acceleration.

[†]Rolling means no slipping, according to the definition we adopt in this book (see Section 3.6).

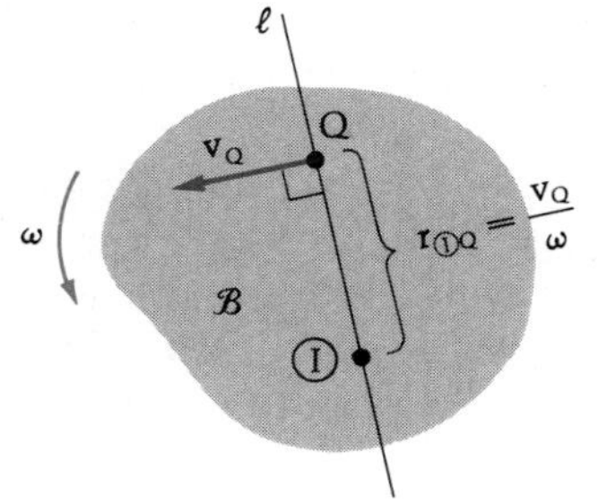

Figure 3.11

Question 3.9 Why is Ⓘ *below* Q in Figure 3.11 instead of being the same distance *above* Q?

We have thus verified the existence of the instantaneous center (unless $\dot{\theta} = 0$), because we know how to get *to* it from any arbitrary starting point Q of the body $\mathcal{B}$. We also note again that the velocity magnitude of every point of $\mathcal{B}$ (in the reference plane!) equals ω times the distance to the point from Ⓘ.

As time passes, Ⓘ may become ***different*** physical points of $\mathcal{B}$ (as in the rolling wheel example). In fact, the only time this does ***not*** happen is when the body rotates about a pivot fixed in both $\mathcal{B}$ and the frame of reference, in which case the motion is sometimes called **pure rotation.** Nor does the point Ⓘ have to be a material point of $\mathcal{B}$; it may be a point of the body extended, as discussed in Section 3.1. We now examine some examples involving the concept of the instantaneous center Ⓘ of velocity.

EXAMPLE 3.8

Rework Example 3.3 using instantaneous centers. (See the diagram.) The crank arm $\mathcal{C}$ turns about a horizontal z axis, through its pinned end O, with an angular velocity of 10 rad/sec clockwise at the given instant. Find the velocity of the piston pin B.

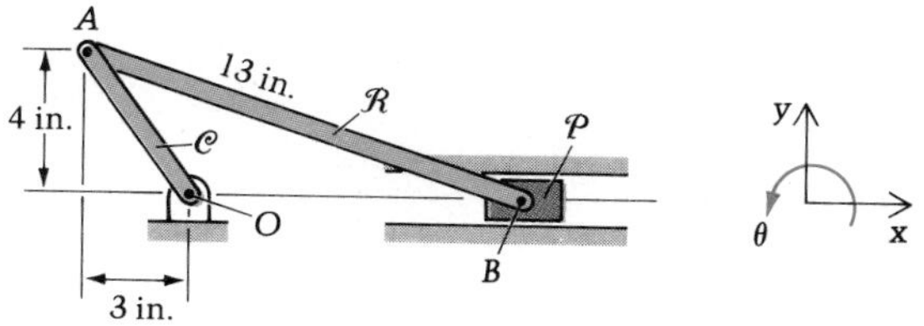

SOLUTION

Since O is the point Ⓘ for body $\mathcal{C}$, we have

$$\mathbf{v}_A = \boldsymbol{\omega}_{\mathcal{C}} \times \mathbf{r}_{OA} = -10\hat{\mathbf{k}} \times (-3\hat{\mathbf{i}} + 4\hat{\mathbf{j}}) \text{ in./sec}$$

$$= 40\hat{\mathbf{i}} + 30\hat{\mathbf{j}} = 50\left(\frac{4\hat{\mathbf{i}} + 3\hat{\mathbf{j}}}{5}\right) \text{ in./sec}$$

Next we find the Ⓘ of $\mathcal{R}$, using the fact that it is on lines perpendicular to the

(Continued)

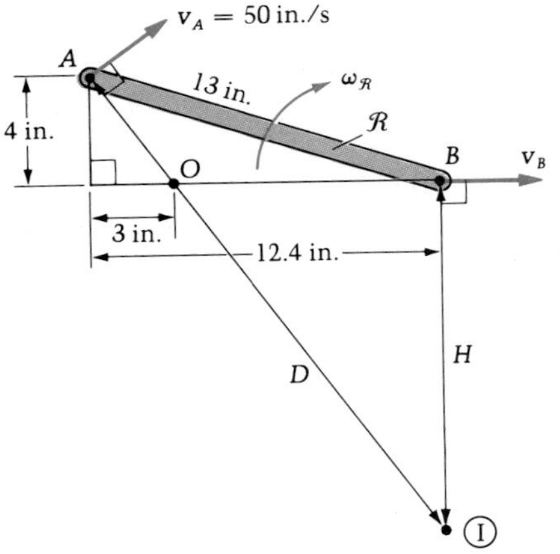

velocities of A and B as shown in the diagram. If we next find the distance D from Ⓘ to A, then $\omega_{\mathcal{R}}$ will be $v_A/D = 50/D$. By similar triangles:

$$\frac{D}{12.4} = \frac{5}{3} \Rightarrow D = 20.7 \text{ in.}$$

Then

$$\boldsymbol{\omega}_{\mathcal{R}} = 2.42 \circlearrowright \text{ rad/sec}$$

Again by similar triangles:

$$\frac{H+4}{12.4} = \frac{4}{3} \Rightarrow H = 12.5 \text{ in.}$$

so that $v_B = H\omega_{\mathcal{R}} = 12.5(2.42) = 30.3$ in./sec to the right, as we have seen twice before.

Note that when viewed from Ⓘ the senses of the velocity direction of any point and the angular velocity direction of the body must agree. For example, these are possible situations:

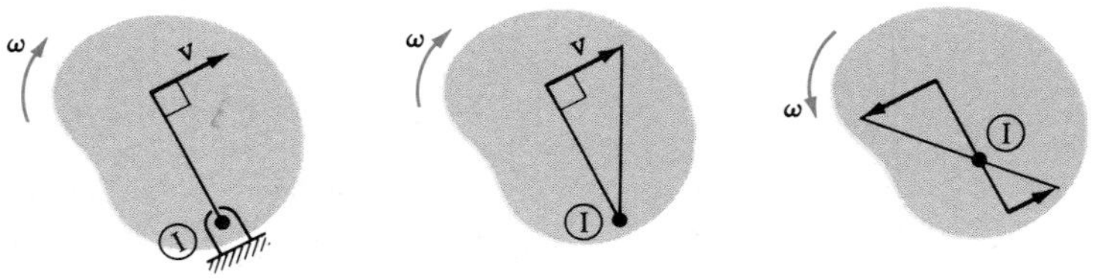

These are not:

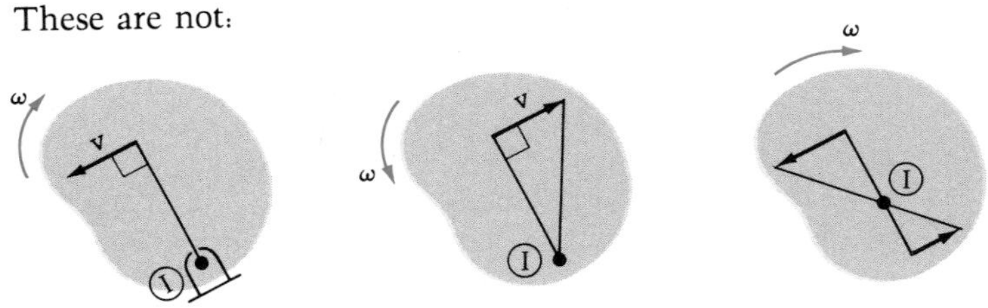

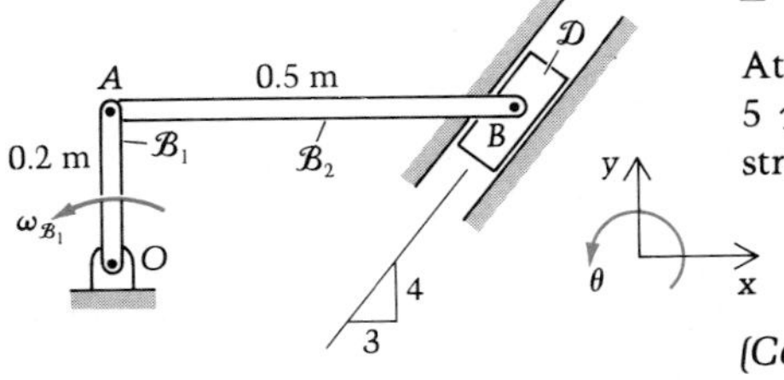

EXAMPLE 3.9

At the instant shown in the diagram, the angular velocity of bar $\mathcal{B}_1$ is $\boldsymbol{\omega}_{\mathcal{B}_1} = 5 \circlearrowright$ rad/sec. Find the velocity of pin B connecting bar $\mathcal{B}_2$ to block $\mathcal{D}$, constrained to slide in the slot.

(Continued)

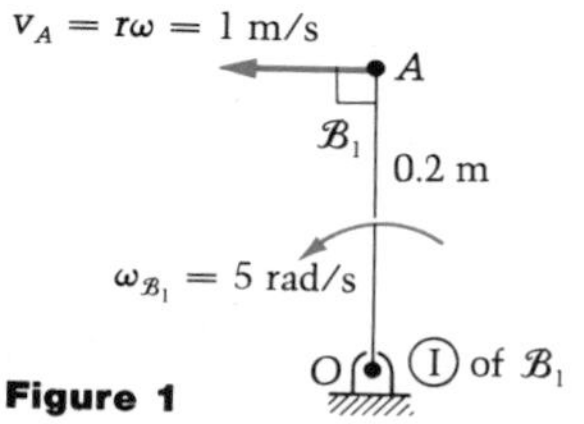

Figure 1

SOLUTION

As seen in Figure 1, the point Ⓘ for $\mathcal{B}_1$ is O, since it is pinned to the reference frame. The velocity of A is perpendicular to the line from Ⓘ to A (that is, from O to A) and has a direction in agreement with the angular velocity of $\mathcal{B}_1$ as the body turns about O. Its value is $r_{\text{Ⓘ}A}\,\omega_{\mathcal{B}_1}$, or 1 m/s ←.

Next we sketch body $\mathcal{B}_2$ and note the position of Ⓘ for $\mathcal{B}_2$, as explained in Figure 2. Similar triangles yield the height H of Ⓘ above A:

$$\frac{H}{0.5} = \frac{3}{4}$$

$$H = 0.375 \text{ m}$$

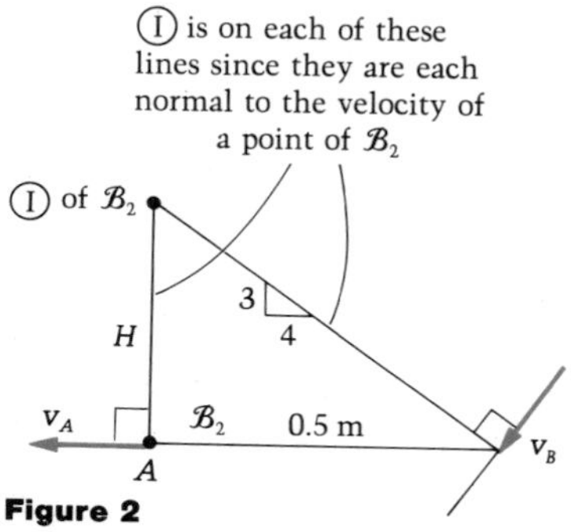

Figure 2

> **Question 3.10** Why does $\mathbf{v}_B$ have to be 'southwest' along the slot and not 'northeast'?

We may now use Ⓘ of $\mathcal{B}_2$ to get the angular velocity of $\mathcal{B}_2$:

$$\mathbf{v}_A = \omega_{\mathcal{B}_2}\hat{\mathbf{k}} \times \mathbf{r}_{\text{Ⓘ}A}$$

Substituting, we get

$$-1\hat{\mathbf{i}} = \omega_{\mathcal{B}_2}\hat{\mathbf{k}} \times (-0.375\hat{\mathbf{j}})$$

Solving gives

$$\omega_{\mathcal{B}_2} = -2.67 \Rightarrow \boldsymbol{\omega}_{\mathcal{B}_2} = -2.67\hat{\mathbf{k}} \text{ or } 2.67 \circlearrowright \text{ rad/s}$$

Note that when we write $\boldsymbol{\omega}_{\mathcal{B}_2} = \omega_{\mathcal{B}_2}\hat{\mathbf{k}}$, we are saying that $\omega_{\mathcal{B}_2}$ is counterclockwise in accordance with the sign convention adopted for the problem in the figure if its value turns out positive. Thus when its value is now found to be negative, we know that $\mathcal{B}_2$ is turning clockwise at the given instant. Of course, we do not have to use vectors on such a simple problem; we can often use what we know about the instantaneous center in scalar form to get a quick solution:

$$v_A = H|\boldsymbol{\omega}_{\mathcal{B}_2}| \Rightarrow |\boldsymbol{\omega}_{\mathcal{B}_2}| = \frac{v_A}{H} = \frac{1}{0.375} = 2.67 \Rightarrow \boldsymbol{\omega}_{\mathcal{B}_2} = 2.67 \circlearrowright \text{ rad/s}$$

where we assign the direction in accordance with the known velocity direction of A and the position of Ⓘ.

Next we use Ⓘ of $\mathcal{B}_2$ to obtain $\mathbf{v}_B$:

$$\begin{aligned} v_B &= r_{\text{Ⓘ}B}|\boldsymbol{\omega}_{\mathcal{B}_2}| \\ &= \sqrt{0.375^2 + 0.5^2}\,(|\boldsymbol{\omega}_{\mathcal{B}_2}|) \text{ m/s} \\ &= 0.625(2.67) = 1.67 \text{ m/s} \end{aligned}$$

The velocity of B is thus $\mathbf{v}_B = 1.67$ 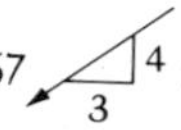 m/s.

Note that the arrow in this sketch is just as descriptive of the direction of the vector velocity of B as is the unit vector $-0.6\hat{\mathbf{i}} - 0.8\hat{\mathbf{j}}$.

For an instantaneous center Ⓘ that is more difficult to find, consider Figure 3.12. Point Ⓘ of bar $\mathcal{B}$ lies on perpendiculars to the velocities of A (which is horizontal) and B (the point of "$\mathcal{B}$ extended" that is passing over the pin). The velocity of this point B must be *along the slot* since the pin is rigidly fixed to the ground frame $\mathcal{F}$ and thus physically prevents $\mathbf{v}_B$ from having a component normal to the slot. The perpendiculars thus intersect at point Ⓘ as shown in the figure.

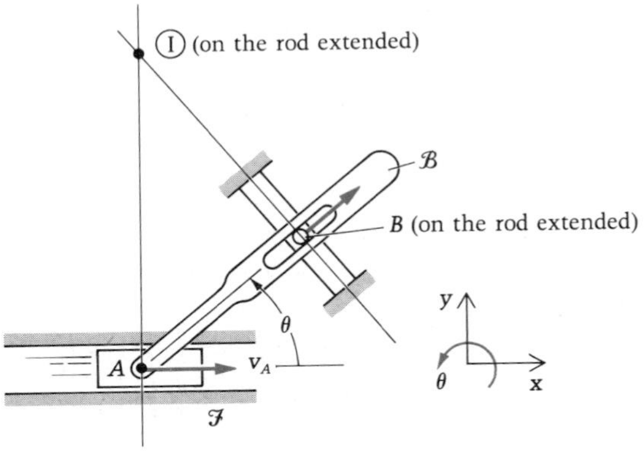

Figure 3.12 Instantaneous center as a point of the body extended.

EXAMPLE 3.10

Find the velocity of point B of $\mathcal{B}$-extended that is passing over the center of the pin in Figure 3.12 at an instant when:

$$\mathbf{v}_A = 2\hat{\mathbf{i}} \text{ in./sec} \qquad \overline{AB} = 9 \text{ in.} \qquad \theta = \tan^{-1}(3/4)$$

SOLUTION

See the accompanying diagram. Making use of Ⓘ, we have

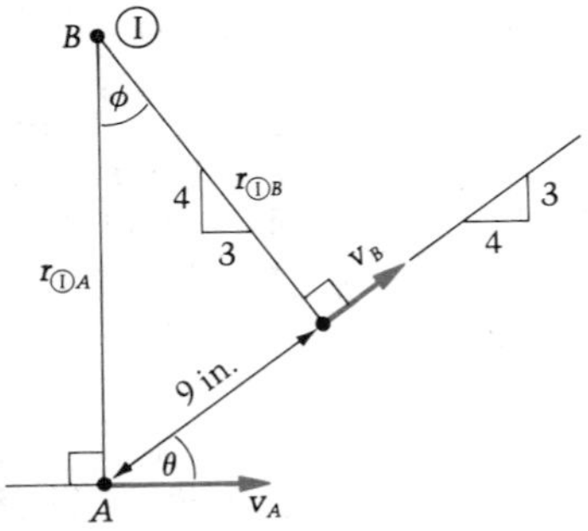

$$\mathbf{v}_A = \overset{\mathbf{0}}{\cancel{\mathbf{v}_{Ⓘ}}} + \dot{\theta}\hat{\mathbf{k}} \times r_{ⒾA}(-\hat{\mathbf{j}})$$

$$2\hat{\mathbf{i}} = r_{ⒾA}\dot{\theta}\hat{\mathbf{i}}$$

Next we calculate the distance from Ⓘ to A: $\phi = \tan^{-1}(3/4)$, so that

$$\sin\phi = \frac{3}{5} = \frac{9}{r_{ⒾA}}$$

(Continued)

Thus

$$r_{①A} = 15 \text{ in.}$$

So we have $2 = 15\dot{\theta} \Rightarrow \dot{\theta}\hat{\mathbf{k}} = 2/15$ ⟲ rad/sec. Continuing, we get

$$\tan\phi = \frac{3}{4} = \frac{9}{r_{①B}} \qquad r_{①B} = 12 \text{ in.}$$

Thus

$$\mathbf{v}_B = \frac{2}{15}\hat{\mathbf{k}} \times \mathbf{r}_{①B} = \frac{2}{15}\hat{\mathbf{k}} \times 12\left(\frac{3}{5}\hat{\mathbf{i}} - \frac{4}{5}\hat{\mathbf{j}}\right)$$

$$= \frac{32}{25}\hat{\mathbf{i}} + \frac{24}{25}\hat{\mathbf{j}} = \frac{8}{5}\left(\frac{4\hat{\mathbf{i}} + 3\hat{\mathbf{j}}}{5}\right) \text{ in./sec}$$

Note that $\mathbf{v}_B$ is parallel to the slot, a direction we already knew and used in finding ①. Alternatively, without using ①, we could have related $\mathbf{v}_B$ to $\mathbf{v}_A$:

$$\mathbf{v}_B = \mathbf{v}_A + \dot{\theta}\hat{\mathbf{k}} \times \mathbf{r}_{AB}$$

so that

$$v_B\left(\frac{4\hat{\mathbf{i}} + 3\hat{\mathbf{j}}}{5}\right) = 2\hat{\mathbf{i}} + \dot{\theta}\hat{\mathbf{k}} \times 9\left(\frac{4\hat{\mathbf{i}} + 3\hat{\mathbf{j}}}{5}\right)$$

$$= \left(2 - \frac{27}{5}\dot{\theta}\right)\hat{\mathbf{i}} + \frac{36}{5}\dot{\theta}\hat{\mathbf{j}}$$

$\hat{\mathbf{i}}$ coefficients: $$\frac{4}{5}v_B = 2 - \frac{27}{5}\dot{\theta} \tag{1}$$

$\hat{\mathbf{j}}$ coefficients: $$\frac{3}{5}v_B = \frac{36}{5}\dot{\theta} \tag{2}$$

From Equation (2) we get $v_B = 12\dot{\theta}$, which, substituted into (1), gives $\dot{\theta} = 2/15$ rad/sec so that $v_B = 8/5$ in./sec, the same answer as above. The velocity of B is thus

$$\mathbf{v}_B = \frac{8}{5}\left(\frac{4\hat{\mathbf{i}} + 3\hat{\mathbf{j}}}{5}\right) \text{ in./sec}$$

expressed as a magnitude times a unit vector. Alternatively, in component form,

$$\mathbf{v}_B = \frac{32}{25}\hat{\mathbf{i}} + \frac{24}{25}\hat{\mathbf{j}} \text{ in./sec}$$

Note that by using ① we avoided having to solve two coupled equations in two unknowns. Instead, we obtained $\dot{\theta}$ in one step and v_B in another.

Using a third approach—the idea that the components along a line of the

(Continued)

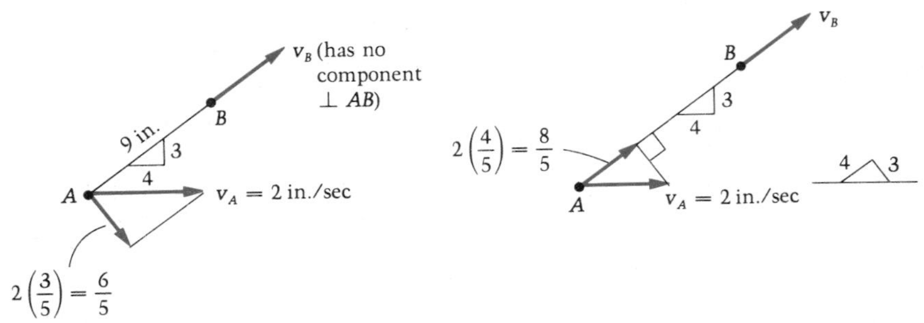

velocities of all points on that line are equal—we have the result that (from the left-hand diagram):

$$\frac{6}{5} - 0 = 9\dot{\theta}$$

or

$$\dot{\theta} = \frac{2}{15} \text{ rad/sec} \quad \text{(and we note that } \boldsymbol{\omega}_{\mathcal{B}} \text{ has direction} \circlearrowleft)$$

And from the right-hand sketch,

$$v_B = \text{component of } v_A \text{ along } AB$$

$$= 2\left(\frac{4}{5}\right) = \frac{8}{5} \text{ in./sec} \quad \text{(and we note that } \mathbf{v}_B \text{ has direction } \overset{3}{\underset{4}{\nearrow}})$$

EXAMPLE 3.11

The accompanying diagram shows a wheel of a large vehicle traveling at a constant speed of 60 mph ⟶. Find the velocity of the piston $\mathcal{P}$ when θ equals: (a) 0°, (b) 90°, (c) 180°, (d) 270°.

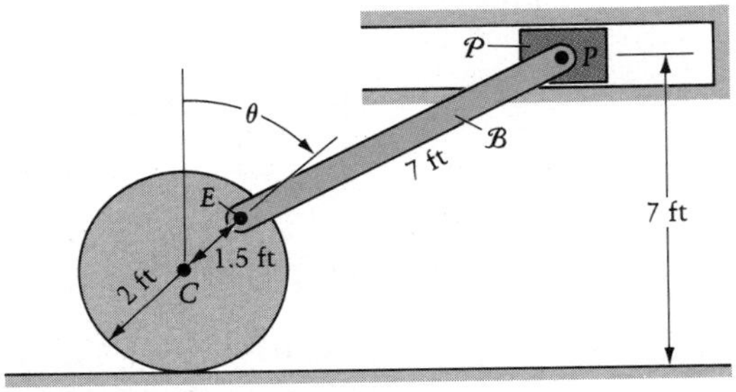

SOLUTION

First we solve for the velocities by using several approaches. In each case, the speed of the wheel's center is the same as the speed of the vehicle: 60 mph or

(Continued)

88 ft/sec. And since piston $\mathcal{P}$ translates, all its points have the same velocities and accelerations.

Case (a): On wheel $\mathcal{C}$, since velocities increase linearly with distance from the instantaneous center at the contact point,*

$$v_E = \frac{2 + 1.5}{2}(88) = 154 \text{ ft/sec}$$

If we draw lines at E and P perpendicular to $\mathbf{v}_E$ and $\mathbf{v}_P$ (P is constrained to move horizontally), they are parallel and thus will not intersect. Therefore $\omega_{\mathcal{B}} = 0$ at that instant. Thus $\mathbf{v}_P = \mathbf{v}_E = 154\hat{\mathbf{i}}$ ft/sec.

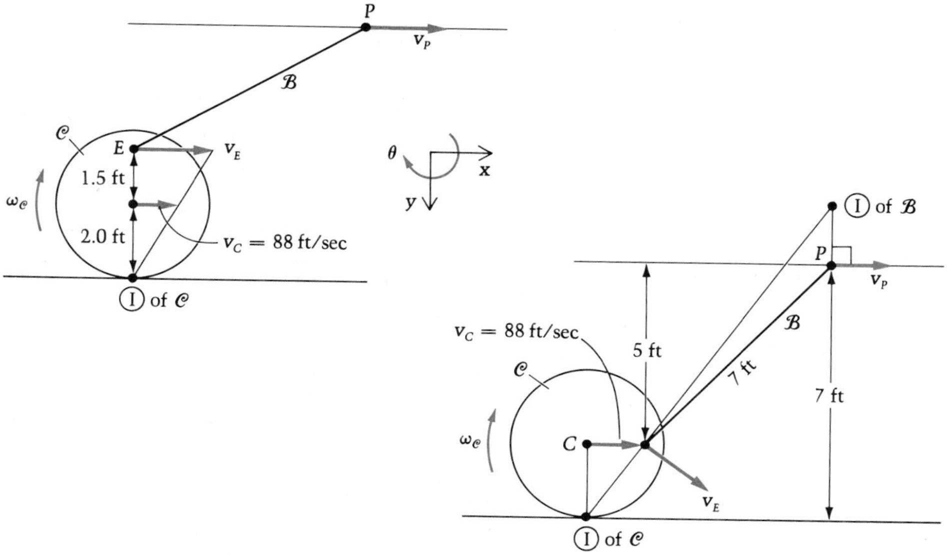

Case (b): This time we shall use vectors:

$$\mathbf{v}_C = \overset{\mathbf{0}}{\cancel{\mathbf{v}_{Ⓘ}}} + \omega_{\mathcal{C}}\hat{\mathbf{k}} \times \overset{-2\hat{\mathbf{j}}}{\cancel{\mathbf{r}_{ⓘC}}}$$

$$88\hat{\mathbf{i}} = 2.0\omega_{\mathcal{C}}\hat{\mathbf{i}} \Rightarrow \boldsymbol{\omega}_{\mathcal{C}} = 44.0 \circlearrowright \text{ rad/sec}$$

$$\mathbf{v}_E = 44.0\hat{\mathbf{k}} \times (1.5\hat{\mathbf{i}} - 2\hat{\mathbf{j}}) = 88.0\hat{\mathbf{i}} + 66.0\hat{\mathbf{j}}$$

or

$$\mathbf{v}_E = 110 \;\overset{4}{\searrow}3 \text{ ft/sec}$$

(Continued)

*Which we shall see more clearly in the discussion of rolling (Section 3.6). Our main purpose in this example is to use the various instantaneous centers of $\mathcal{B}$ in the four positions given.

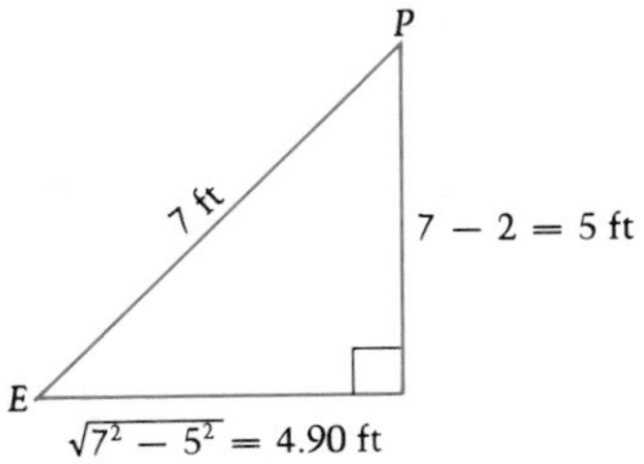

On $\mathcal{B}$:

$$\mathbf{v}_P = v_P\hat{\mathbf{i}} = \mathbf{v}_E + \omega_{\mathcal{B}}\hat{\mathbf{k}} \times \mathbf{r}_{EP}$$

$$= 88.0\hat{\mathbf{i}} + 66.0\hat{\mathbf{j}} + \omega_{\mathcal{B}}\hat{\mathbf{k}} \times (4.90\hat{\mathbf{i}} - 5.00\hat{\mathbf{j}})$$

Equating coefficients in the **x** and *y* directions, we get

x: $\quad v_P = 88.0 + 5.00\omega_{\mathcal{B}}$

y: $\quad 0 = 66.0 + 4.90\omega_{\mathcal{B}} \Rightarrow \omega_{\mathcal{B}} = -13.5$ rad/sec

Therefore

$$v_P = 88.0 + 5.00(-13.5) = 20.5 \text{ ft/sec}$$

$$\mathbf{v}_P = 20.5\hat{\mathbf{i}} \text{ ft/sec}$$

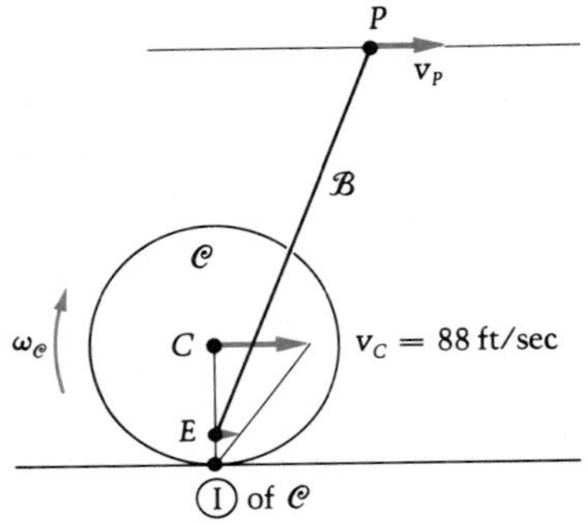

Case (c): In all four cases, $\boldsymbol{\omega}_{\mathcal{C}} = 44.0$ ↻ rad/sec. This time, then,

$$v_E = (2.0 - 1.5)\omega_{\mathcal{C}} = 22.0 \text{ ft/sec}$$

$$\mathbf{v}_E = 22.0 \rightarrow \text{ft/sec}$$

Again, as in Case (a), body $\mathcal{B}$ has $\boldsymbol{\omega}_{\mathcal{B}} = \mathbf{0}$ so that

$$\mathbf{v}_P = 22.0\hat{\mathbf{i}} \text{ ft/sec}$$

Case (d): $\qquad \mathbf{v}_C \; + \; \omega\hat{\mathbf{k}} \; \times \; \mathbf{r}_{CE}$

On $\mathcal{C}$: $\quad \mathbf{v}_E = 88\hat{\mathbf{i}} + 44.0\hat{\mathbf{k}} \times (-1.5\hat{\mathbf{i}})$

$$= 88.0\hat{\mathbf{i}} - 66.0\hat{\mathbf{j}}$$

This time let us use the instantaneous center of $\mathcal{B}$. From similar triangles:

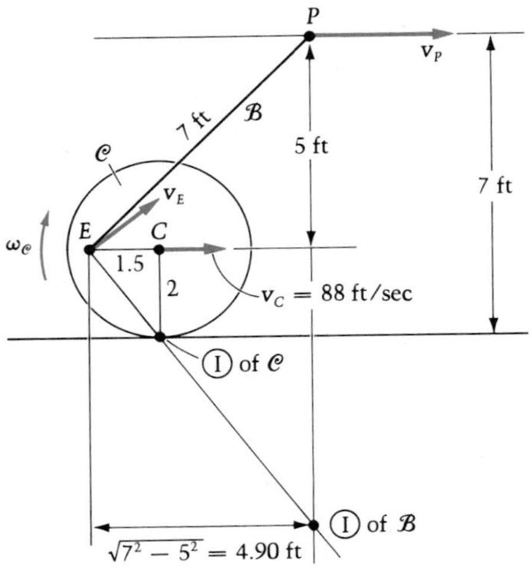

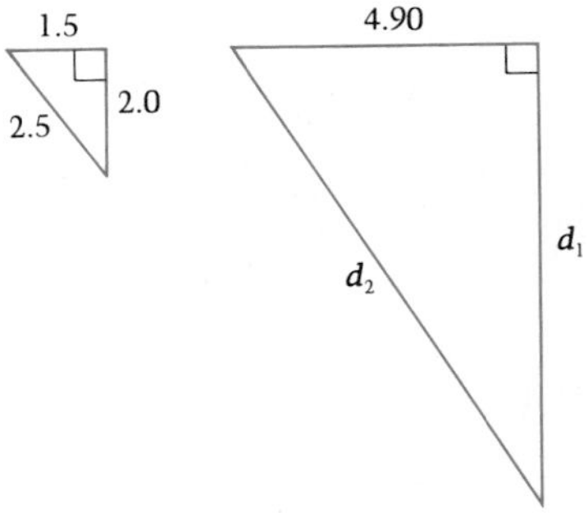

(Continued)

$$d_1 = \frac{2.0}{1.5}(4.90) = 6.53 \text{ ft}$$

$$d_2 = \frac{2.5}{1.5}(4.90) = 8.17 \text{ ft}$$

Therefore, on body $\mathcal{B}$,

$$\omega_{\mathcal{B}} = \frac{v_E}{8.17} = \frac{\sqrt{88.0^2 + 66.0^2}}{8.17} = \frac{110}{8.17} = 13.5 \text{ rad/sec}$$
$$\Rightarrow \boldsymbol{\omega}_{\mathcal{B}} = 13.5 \circlearrowright \text{ rad/sec}$$

and

$$v_P = (6.53 + 5)\omega_{\mathcal{B}} = 156 \text{ ft/sec} \Rightarrow \mathbf{v}_P = 156 \rightarrow \text{ft/sec}$$

Note from the four answers given above that the piston is moving faster during the parts of the wheel's revolution in which $\mathbf{v}_E$ makes small angles with rod $\mathcal{B}$; conversely, it moves slower when $\mathbf{v}_E$ is making a large angle with $\mathcal{B}$. This is because the components of $\mathbf{v}_E$ and $\mathbf{v}_P$ along $\mathcal{B}$ must always be the same, as we have shown earlier.

PROBLEMS / Section 3.4

3.33 The pin at B (Figure P3.33) has a constant speed of 51 cm/s and moves in a circle in the clockwise direction. Find the angular velocities of bars $\mathcal{B}_1$ and $\mathcal{B}_2$ in the given position.

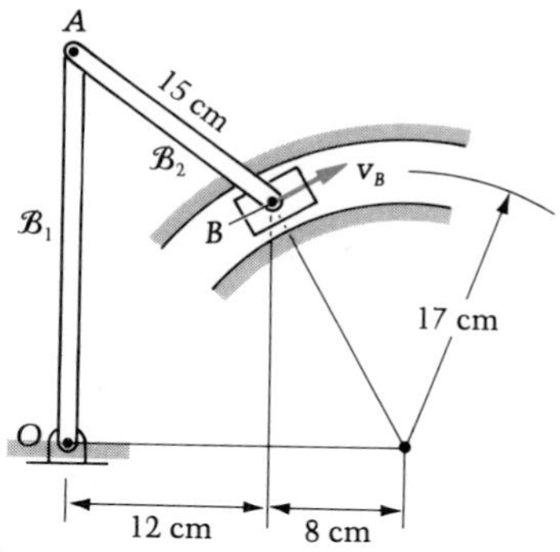

Figure P3.33

3.34 Rod $\mathcal{R}$ in Figure P3.34 was vertical at $t = 0$, at which time end B began a motion to the right at constant speed v_B. Find the velocity of the other end A as a function of v_B, t, and the length L of the rod by two methods: (a) differentiating $x^2 + y^2 = L^2$; (b) using the instantaneous center of the rod.

3.35 The linkage shown in Figure P3.35 is made up of rods $\mathcal{A}$, $\mathcal{B}$, and $\mathcal{C}$. Rod $\mathcal{A}$ has constant angular velocity $\boldsymbol{\omega}_{\mathcal{A}} = 0.6 \circlearrowleft$ rad/sec. Determine the angular velocities of $\mathcal{B}$ and $\mathcal{C}$ when the angle θ shown is equal to (a) 90° and (b) 180°.

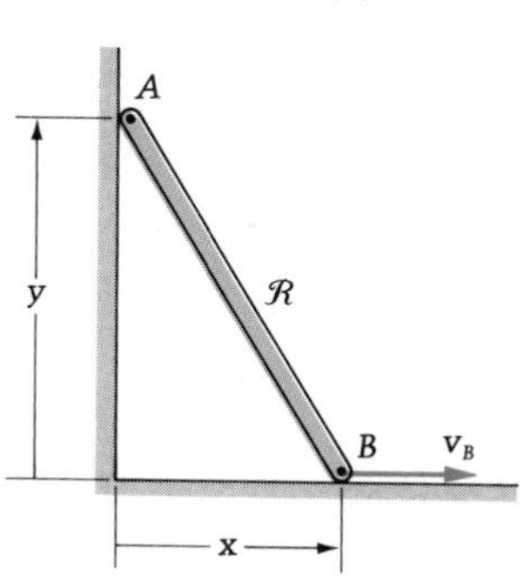

Figure P3.34

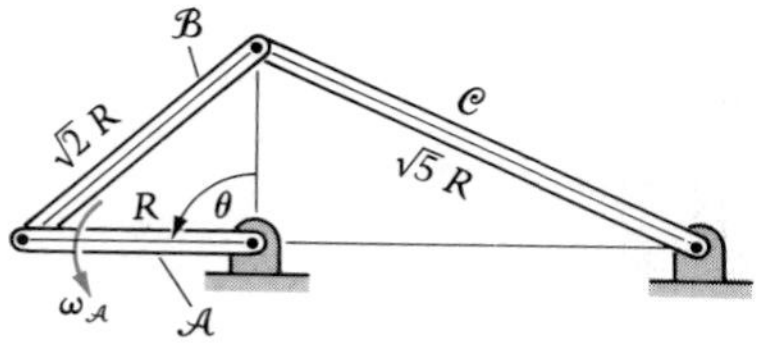

Figure P3.35

3.36 Bars $\mathcal{A}$ and $\mathcal{B}$ (see Figure P3.36) are pinned together at A. Find the angular velocity of bar $\mathcal{B}$ and the velocity of point B when bars $\mathcal{A}$ and $\mathcal{B}$ are next collinear.

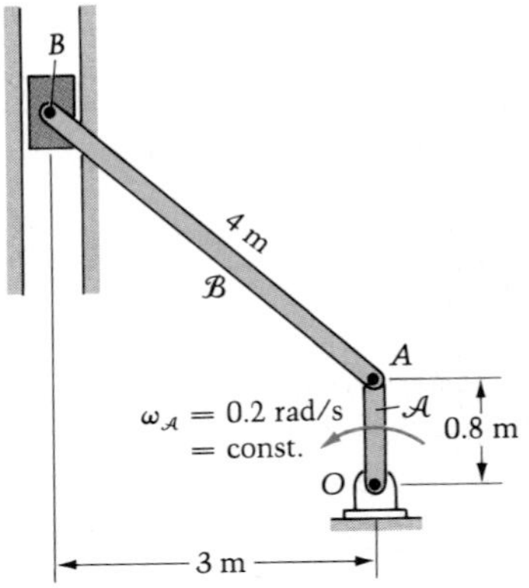

Figure P3.36

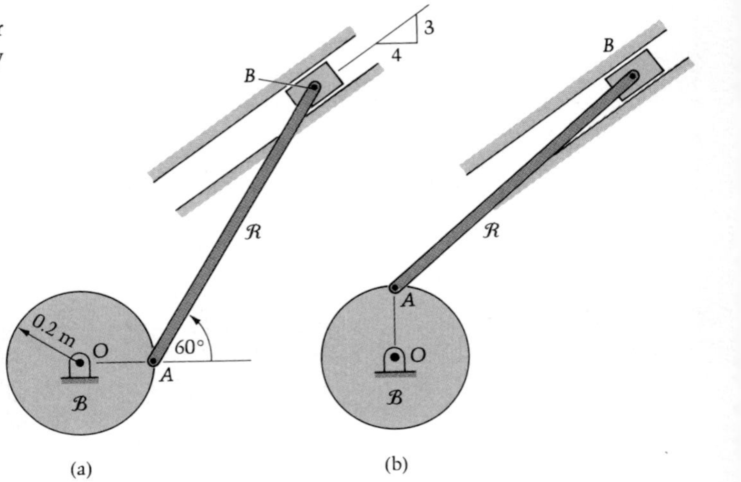

Figure P3.41

3.37 Repeat the preceding problem at the *second* instant of time when the bars are collinear.

3.38 Rods $\mathcal{A}$ and $\mathcal{B}$ are pinned at B and move in a vertical plane with the constant angular velocities shown in Figure P3.38. Locate the instantaneous center of $\mathcal{A}$ for the given position, and use it to find the velocity of point C. Then check by calculating $\mathbf{v}_C$ by relating it (on $\mathcal{A}$) to the velocity of B. Note that sometimes Ⓘ is more trouble to locate than it is worth!

3.39 The center of block $\mathcal{A}$ in Figure P3.39 travels at a constant speed of 30 mph to the right. Disk $\mathcal{D}$ is pinned to $\mathcal{A}$ at A and spins at 100 rpm counterclockwise. Find: (a) the velocity of P; (b) the instantaneous center Ⓘ of $\mathcal{D}$; and (c), using Ⓘ, the velocities of Q, S, and R.

3.40 See Figure P3.40. The angular velocity of rod $\mathcal{R}$ is a constant: $\boldsymbol{\omega}_{\mathcal{R}} = 0.3\hat{\mathbf{k}}$ rad/s. Determine the angular velocities of plate $\mathcal{P}$ and bar $\mathcal{B}$ in the indicated position.

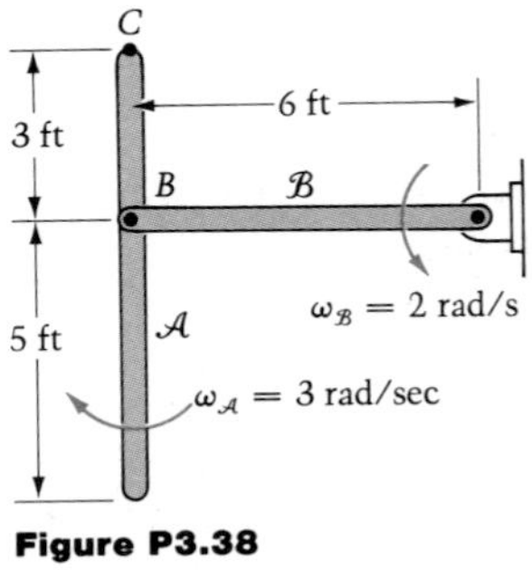

Figure P3.38

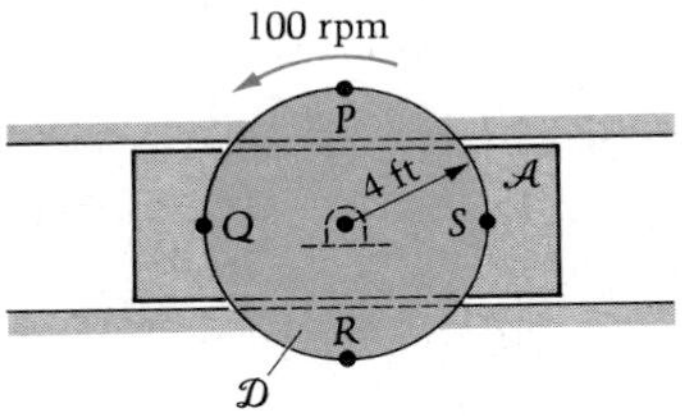

Figure P3.39

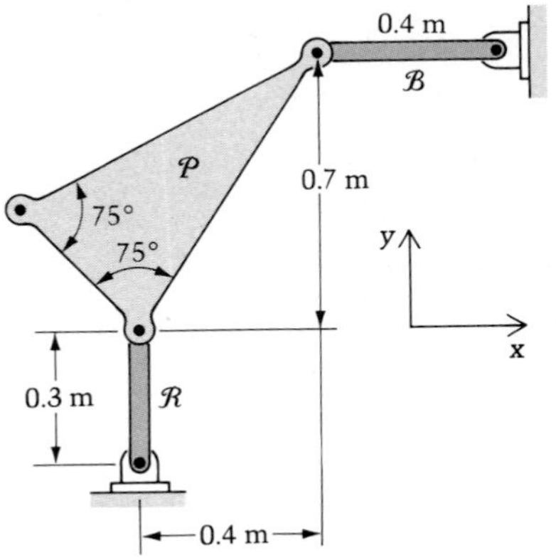

Figure P3.40

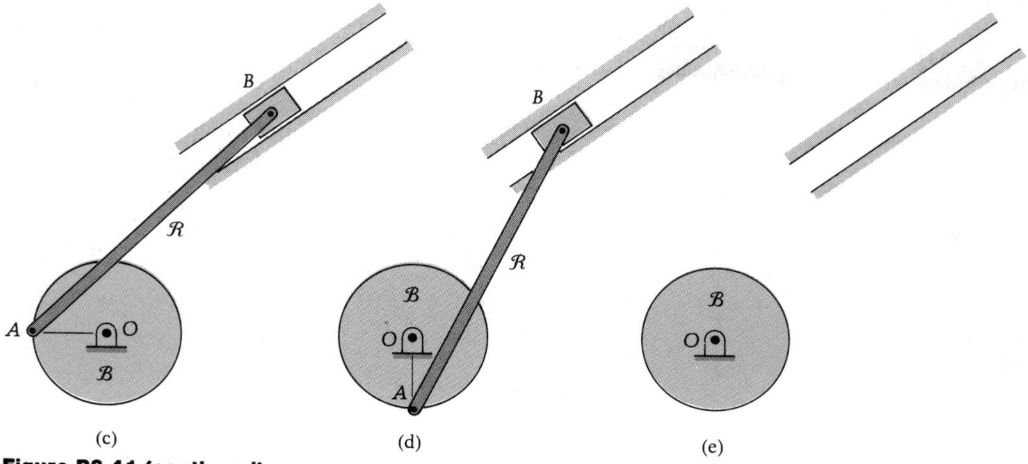

Figure P3.41 (continued)

3.41 The angular velocity of $\mathcal{B}$ in Figure P3.41 is 3 ⟲ rad/s = constant. Trace the five sketches and then show on parts (a) to (d) the position of Ⓘ for the rod $\mathcal{R}$. In part (e), using the proper length of $\mathcal{R}$, draw the positions of $\mathcal{R}$ at the two times when $\mathbf{v}_B = \mathbf{0}$. Rod $\mathcal{R}$ has length 0.9 m.

3.42 The piston rod of the hydraulic cylinder shown in Figure P3.42 moves outward at the constant speed of 0.13 m/s. Find the angular velocity of $\mathcal{B}_1$ at the instant shown.

3.43 In Figure P3.43 the crank arm $\mathcal{C}$ is 4 in. long and has constant angular velocity $\boldsymbol{\omega}_{\mathcal{C}} = 2$ ⟲ rad/sec. It is pinned to the triangular plate $\mathcal{P}$, which is also pinned to the block in the slot at D. Find the velocity of D at the instant shown.

3.44 A bar of length $2L$ moves with its ends in contact with the planes shown in Figure P3.44. Find the velocity and acceleration of point C, in terms of θ and its derivatives, by writing and then differentiating the position vector of C. Then check the velocity solution by using the instantaneous center.

3.45 Solve Problem 3.18 by using instantaneous centers.

3.46 Solve Problem 3.20 by using instantaneous centers.

3.47 Solve Problem 3.9 by using instantaneous centers.

3.48 Solve Problem 3.11 by using instantaneous centers.

3.49 Solve Problem 3.13 by using instantaneous centers.

3.50 Solve Problem 3.19 by using instantaneous centers.

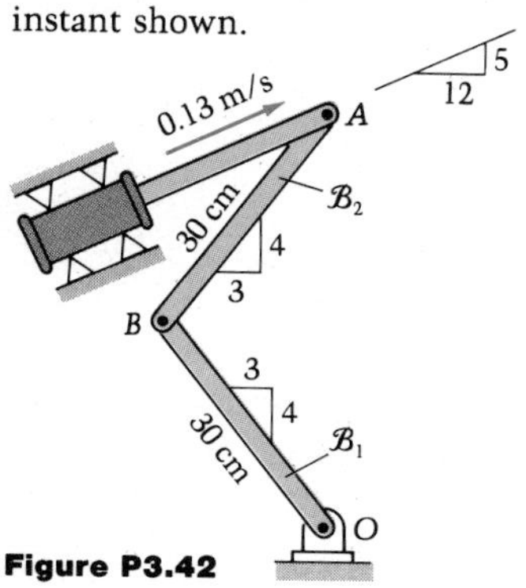

Figure P3.42

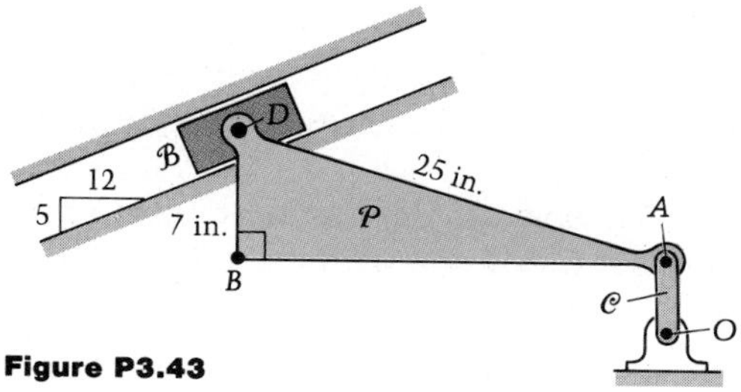

Figure P3.43

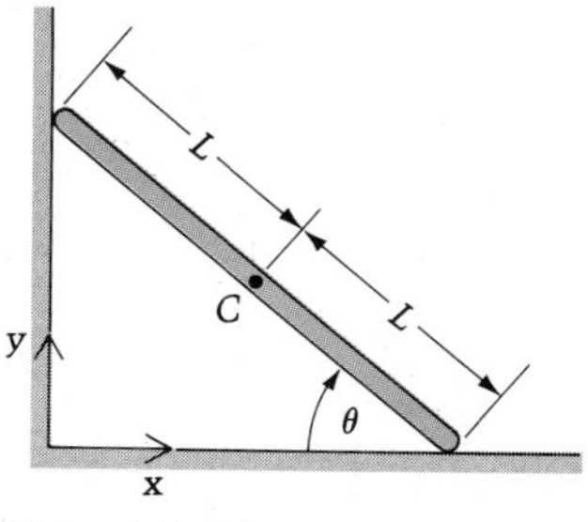

Figure P3.44

3.51 Solve Example 3.7 by using the instantaneous center of the rod.

3.52 Using the method of instantaneous centers, find the velocity of point B in Figure P3.52, which is constrained to move in the slot as shown. The angular velocity of $\mathcal{A}$ is 0.3 ⟲ rad/s at the indicated instant.

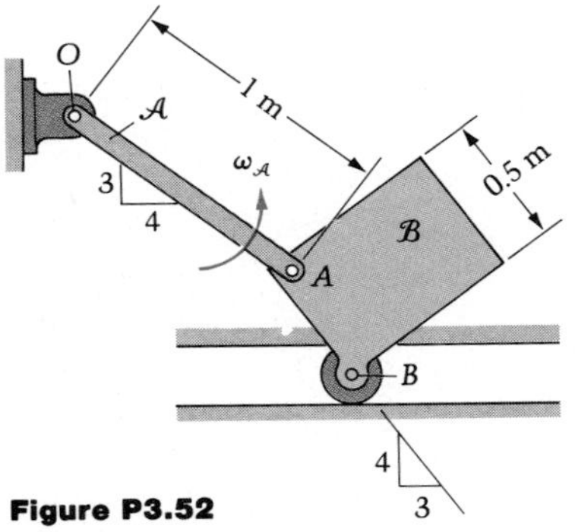

Figure P3.52

3.53 The four-bar linkage shown in Figure P3.53a is made up of bars $\mathcal{A}$, $\mathcal{B}$, $\mathcal{C}$, and a fourth "bar," which is the rigid distance OQ. Bar $\mathcal{A}$ has a constant angular velocity of 0.6 rad/s counterclockwise. Find the angular velocity of $\mathcal{C}$ when $\mathcal{A}$ is at the top of the circle on which it travels. *Hint:* Write equations relating the angles β and ϕ in Figure P3.53b, solve them with a calculator, and then make use of the instantaneous center of $\mathcal{B}$, which may then be found easily.

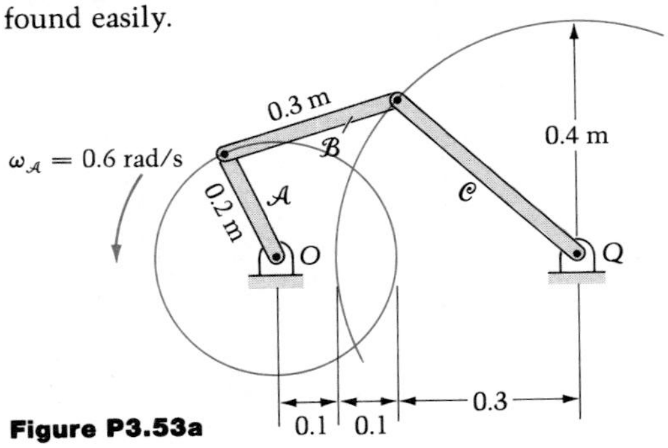

Figure P3.53a

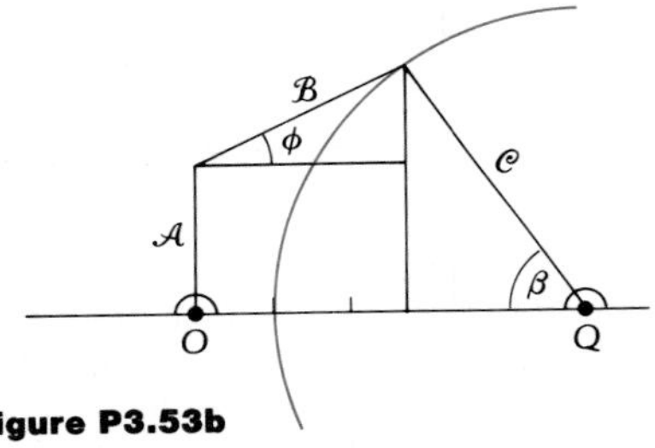

Figure P3.53b

3.5 Acceleration and Angular Acceleration Relationship for Two Points of the Same Rigid Body

The **angular acceleration** vector of a rigid body $\mathcal{B}$ in plane motion in a frame $\mathcal{I}$ is defined as the derivative in $\mathcal{I}$ of the angular velocity and is called $\boldsymbol{\alpha}$:

$$\boldsymbol{\alpha}_{\mathcal{B}/\mathcal{I}} = \boldsymbol{\alpha} = \frac{d\boldsymbol{\omega}_{\mathcal{B}/\mathcal{I}}}{dt} = \dot{\boldsymbol{\omega}} = \ddot{\theta}\hat{\mathbf{k}} \tag{3.14}$$

or

$$\boldsymbol{\alpha} = \ddot{\theta}\hat{\mathbf{k}} \tag{3.15}$$

where $\hat{\mathbf{k}}$ is a constant vector in both $\mathcal{B}$ and $\mathcal{I}$. Note that, as with $\boldsymbol{\omega}$, we delete the subscript when there is no confusion about the frame of reference being used.

We now develop the relationship between the accelerations of two points P and Q of a rigid body $\mathcal{B}$. Differentiating both sides of Equation (3.8) yields

$$\dot{\mathbf{v}}_Q = \mathbf{a}_Q = \mathbf{a}_P + \ddot{\theta}\hat{\mathbf{k}} \times \mathbf{r}_{PQ} + \dot{\theta}\hat{\mathbf{k}} \times \dot{\mathbf{r}}_{PQ} \tag{3.16}$$

Now using Equation (3.7),* we may rewrite the last term as

$$\dot{\theta}\hat{\mathbf{k}} \times \dot{\mathbf{r}}_{PQ} = \dot{\theta}\hat{\mathbf{k}} \times (\dot{\theta}\hat{\mathbf{k}} \times \mathbf{r}_{PQ})$$
$$= \dot{\theta}\hat{\mathbf{k}}(\dot{\theta}\hat{\mathbf{k}} \cdot \mathbf{r}_{PQ}) - \mathbf{r}_{PQ}(\dot{\theta}\hat{\mathbf{k}} \cdot \dot{\theta}\hat{\mathbf{k}}) \tag{3.17}$$

or

$$\dot{\theta}\hat{\mathbf{k}} \times \dot{\mathbf{r}}_{PQ} = -\dot{\theta}^2\mathbf{r}_{PQ} \tag{3.18}$$

> **Question 3.11** Why is the dot product $\dot{\theta}\hat{\mathbf{k}} \cdot \mathbf{r}_{PQ}$ in Equation (3.17) zero?

Equations (3.16) and (3.18) then yield the desired relation between the accelerations of P and Q:

$$\mathbf{a}_Q = \mathbf{a}_P + \ddot{\theta}\hat{\mathbf{k}} \times \mathbf{r}_{PQ} - \dot{\theta}^2\mathbf{r}_{PQ} \tag{3.19}$$

Note that, unlike velocities, the accelerations of P and Q do not generally have equal components along the line PQ joining them; these components differ by $r\dot{\theta}^2 = r\omega^2$. Likewise, the components *perpendicular* to PQ differ by $r\ddot{\theta} = r\alpha$ (in the same way as the velocity components normal to PQ differ by $r\dot{\theta} = r\omega$).

If the acceleration of point A of a body $\mathcal{B}$ is $\mathbf{a}_A$, for example, then the acceleration of any point B is the sum of the three vectors shown in Figure 3.13. If A is a pinned point, or pivot,† then $\mathbf{a}_B$ has two components: one along the line from B toward A and the other perpendicular to it and tangent to the circle (on which it necessarily travels when there is a pivot at A). Note that the direction of the tangential component depends on the direction of $\boldsymbol{\alpha}$, but the "radial" component is *always* inward toward the pivot. This case is shown in Figure 3.14. We now illustrate the use of Equation (3.19) with several examples.

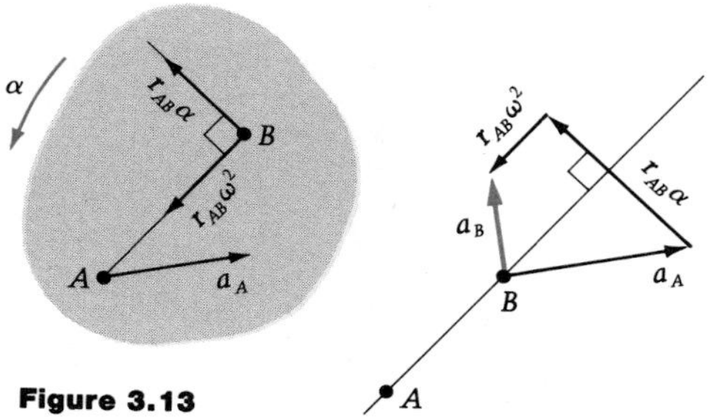

Figure 3.13

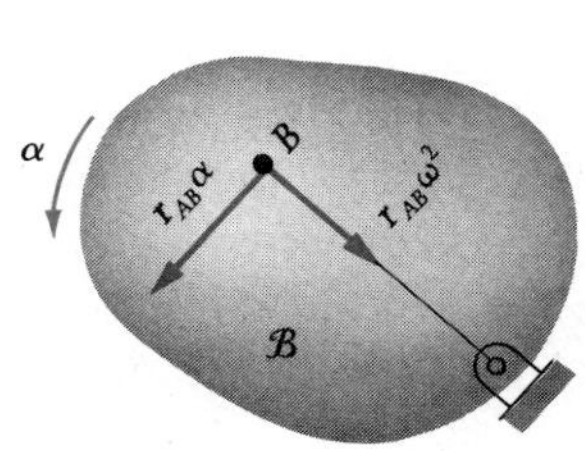

Figure 3.14

*And the vector identity $\mathbf{A} \times (\mathbf{B} \times \mathbf{C}) = \mathbf{B}(\mathbf{A} \cdot \mathbf{C}) - \mathbf{C}(\mathbf{A} \cdot \mathbf{B})$.

†Actually, *pivot* means a point of $\mathcal{B}$ that does not move throughout a motion. It includes, but is not limited to, a physical pin.

EXAMPLE 3.12

In Figure 3.8b, let the links have length 1 m and let them each have $\boldsymbol{\omega} = 2$ ↻ rad/s and $\boldsymbol{\alpha} = 3$ ↻ rad/s² at a time when they make an angle of 45° with the ceiling. Find the acceleration of block $\mathcal{B}$. (That is, find the acceleration of any of its points—they are all the same since $\mathcal{B}$ is translating.)

SOLUTION

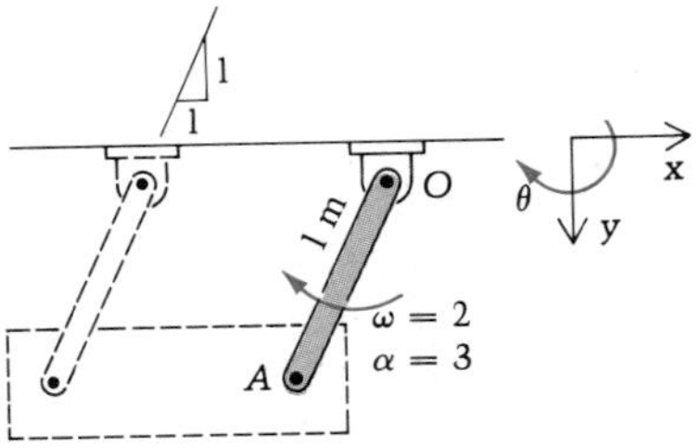

All points of $\mathcal{B}$ have the same $\mathbf{v}$ and $\mathbf{a}$ as point A. Using Equation (3.19), we get

$$\begin{aligned}\mathbf{a}_A &= \overset{\mathbf{0}}{\cancel{\mathbf{a}_O}} + \boldsymbol{\alpha} \times \mathbf{r}_{OA} - \omega^2 \mathbf{r}_{OA}\\ &= 3\hat{\mathbf{k}} \times (-0.707\hat{\mathbf{i}} + 0.707\hat{\mathbf{j}}) - 2^2(-0.707\hat{\mathbf{i}} + 0.707\hat{\mathbf{j}})\\ &= (-2.12 + 2.83)\hat{\mathbf{i}} + (-2.12 - 2.83)\hat{\mathbf{j}}\\ &= 0.71\hat{\mathbf{i}} - 4.95\hat{\mathbf{j}} \text{ m/s}^2\end{aligned}$$

EXAMPLE 3.13

In Example 3.3 find the acceleration of the translating piston $\mathcal{P}$ at the given instant if $\boldsymbol{\omega}_{\mathcal{C}} = 10$ ↻ rad/sec and $\boldsymbol{\alpha}_{\mathcal{C}} = 5$ ↺ rad/sec².

Question 3.12 What does it mean when $\boldsymbol{\alpha}$ is in the opposite direction from that of $\boldsymbol{\omega}$?

SOLUTION

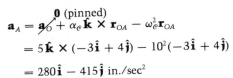

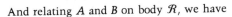

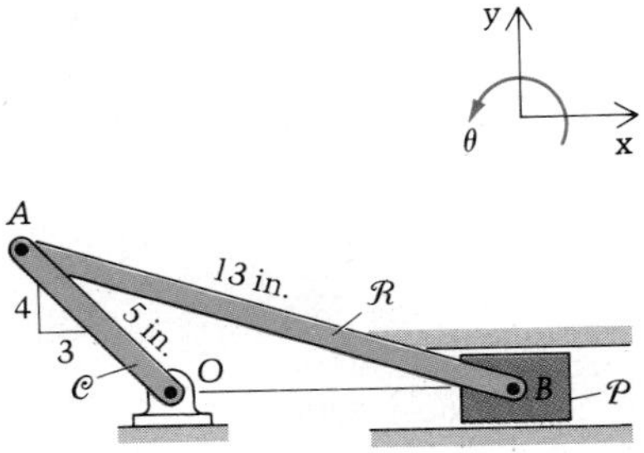

Relating O and A on body $\mathcal{C}$, we have

$$\begin{aligned}\mathbf{a}_A &= \overset{\mathbf{0}\text{ (pinned)}}{\cancel{\mathbf{a}_O}} + \alpha_{\mathcal{C}}\hat{\mathbf{k}} \times \mathbf{r}_{OA} - \omega_{\mathcal{C}}^2 \mathbf{r}_{OA}\\ &= 5\hat{\mathbf{k}} \times (-3\hat{\mathbf{i}} + 4\hat{\mathbf{j}}) - 10^2(-3\hat{\mathbf{i}} + 4\hat{\mathbf{j}})\\ &= 280\hat{\mathbf{i}} - 415\hat{\mathbf{j}} \text{ in./sec}^2\end{aligned}$$

And relating A and B on body $\mathcal{R}$, we have

$$\mathbf{a}_B = \mathbf{a}_A + \alpha_{\mathcal{R}}\hat{\mathbf{k}} \times \mathbf{r}_{AB} - \omega_{\mathcal{R}}^2 \mathbf{r}_{AB}$$

In the previous example we found

$$\mathbf{r}_{AB} = 12.4\hat{\mathbf{i}} - 4\hat{\mathbf{j}} \text{ in.} \qquad \text{and} \qquad \boldsymbol{\omega}_{\mathcal{R}} = 2.42 \circlearrowright \text{ rad/sec}$$

(Continued)

Noting that the piston $\mathcal{P}$ translates horizontally, we get

$$\begin{aligned}\mathbf{a}_B = a_B\hat{\mathbf{i}} &= 280\hat{\mathbf{i}} - 415\hat{\mathbf{j}} + 12.4\alpha_{AB}\hat{\mathbf{j}} + 4\alpha_{\mathcal{R}}\hat{\mathbf{i}} - 2.42^2(12.4\hat{\mathbf{i}} - 4\hat{\mathbf{j}}) \\ &= (207 + 4\alpha_{\mathcal{R}})\hat{\mathbf{i}} + (-392 + 12.4\,\alpha_{\mathcal{R}})\hat{\mathbf{j}}\end{aligned}$$

The y components (coefficients of $\hat{\mathbf{j}}$) yield

$$\alpha_{\mathcal{R}} = \frac{392}{12.4} = 31.6 \text{ rad/sec}^2$$

The **x** components (coefficients of $\hat{\mathbf{i}}$) then give our answer:

$$\mathbf{a}_B = a_B\hat{\mathbf{i}} = [207 + 4(31.6)]\hat{\mathbf{i}} = 333\hat{\mathbf{i}} \text{ in./sec}^2$$

EXAMPLE 3.14

Considering the instantaneous center Ⓘ of Figure 3.12 as a point of $\mathcal{B}$ extended, find its acceleration at the instant described in Example 3.10 if $\mathbf{v}_A = 2\hat{\mathbf{i}}$ in./sec *for all time* and not just at the instant shown.

SOLUTION

$$\mathbf{a}_{\textcircled{I}} = \mathbf{a}_A + \ddot{\theta}\hat{\mathbf{k}} \times \mathbf{r}_{A\textcircled{I}} - \dot{\theta}^2 \overset{15\hat{\mathbf{j}}}{\cancel{\mathbf{r}_{A\textcircled{I}}}}$$

Since $\mathbf{a}_A = \mathbf{0}$, we get

$$\mathbf{a}_{\textcircled{I}} = \mathbf{0} - 15\ddot{\theta}\hat{\mathbf{i}} - \left(\frac{2}{15}\right)^2 15\hat{\mathbf{j}}$$

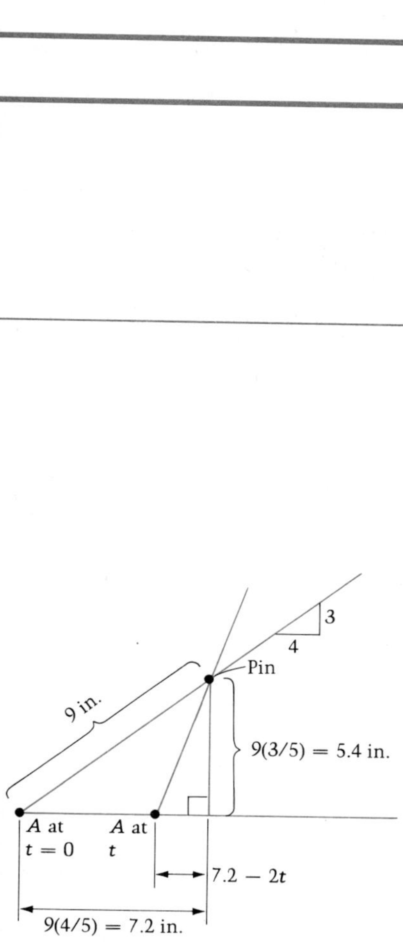

It thus remains to compute $\ddot{\theta}$ at the given instant. Note that $\theta = \tan^{-1}[5.4/(7.2 - 2t)]$ at all times. Thus, differentiating and simplifying, we get

$$\dot{\theta} = \frac{10.8}{(7.2 - 2t)^2 + 5.4^2}$$

and

$$\ddot{\theta} = \frac{43.2(7.2 - 2t)}{[(7.2 - 2t)^2 + (5.4)^2]^2}$$

So at $t = 0$ we have $\ddot{\theta} = 0.0474$ rad/sec². Hence

$$\mathbf{a}_{\textcircled{I}} = -0.711\hat{\mathbf{i}} - 0.267\hat{\mathbf{j}} \text{ in./sec}^2$$

From this example we note an important fact: ***A point of zero velocity does not generally have zero acceleration.*** The bottom point of the rolling wheel of Figure 3.10 is a simple example; even though this point has zero velocity when it is on the ground, it will be in motion during succeeding instants of time as new periphery points take their turns being Ⓘ. In the next section, when we study rolling in detail, we shall find that for a round body rolling on a flat surface, $\mathbf{a}_{Ⓘ}$ is $r\omega^2 \uparrow$.*

If P is the point of $\mathcal{B}$ passing over the pin in the preceding example, then

$$\begin{aligned}\mathbf{a}_{P_{\mathcal{B}}} &= \overset{\mathbf{0}}{\cancel{\mathbf{a}_A}} + \alpha_{\mathcal{B}}\hat{\mathbf{k}} \times \mathbf{r}_{AP_{\mathcal{B}}} - \omega_{\mathcal{B}}^2 \mathbf{r}_{AP_{\mathcal{B}}} \\ &= 0.0474\hat{\mathbf{k}} \times (7.2\hat{\mathbf{i}} + 5.4\hat{\mathbf{j}}) - 0.133^2(7.2\hat{\mathbf{i}} + 5.4\hat{\mathbf{j}}) \\ &= -0.384\hat{\mathbf{i}} + 0.245\hat{\mathbf{j}} \text{ in./sec}^2\end{aligned}$$

and we see that accelerations (unlike velocities) of slot points passing over pins *do not have to be along the slots.*

EXAMPLE 3.15

In Example 3.7 find the angular acceleration of rod $\mathcal{R}$ and the acceleration of its endpoint A.

SOLUTION

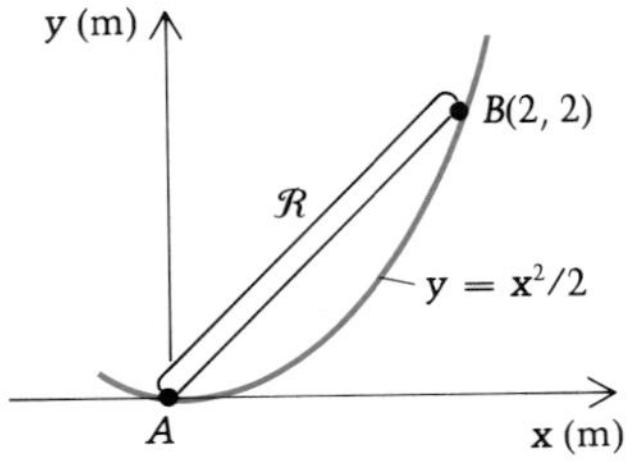

Figure 1

The curvature formula from calculus gives us the radii of curvature at points B and A (see Figure 1):

$$\frac{1}{\rho} = \frac{y''}{(1 + y'^2)^{3/2}} = \frac{1}{(1 + x^2)^{3/2}}$$

Therefore

$$\begin{aligned}\rho_B &= (1 + 2^2)^{3/2} \quad \text{and} \quad & \rho_A &= (1 + 0^2)^{3/2} \\ &= 5^{3/2} & &= 1 \text{ m} \\ &= 11.2 \text{ m}\end{aligned}$$

(Continued)

*Actually, in nontranslational cases there is a point of zero acceleration, but unless it is a pivot point, or a point of rolling contact at an instant when $\omega = 0$, it is more trouble to find than it is worth.

The acceleration of point B is then

$$\mathbf{a}_B = \ddot{s}\hat{\mathbf{e}}_t + \frac{\dot{s}^2}{\rho}\hat{\mathbf{e}}_n = \frac{\dot{s}^2}{\rho}\hat{\mathbf{e}}_n \qquad (\text{since } \dot{s}_B = \text{constant} = 0.3 \text{ m/s})$$

$$= \frac{(0.3)^2}{11.2}\hat{\mathbf{e}}_n = 0.00804\hat{\mathbf{e}}_n \text{ m/s}^2$$

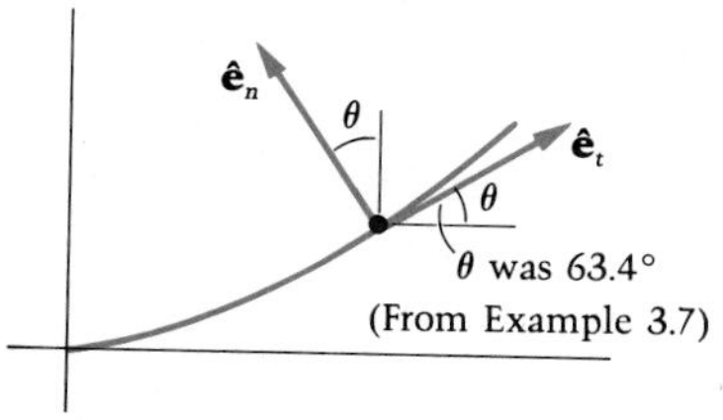

Figure 2

The unit vector $\hat{\mathbf{e}}_n$ is seen in Figure 2 to be

$$\hat{\mathbf{e}}_n = -\sin\theta\hat{\mathbf{i}} + \cos\theta\hat{\mathbf{j}}$$
$$= -0.894\hat{\mathbf{i}} + 0.448\hat{\mathbf{j}}$$

Therefore

$$\mathbf{a}_B = -0.00804\hat{\mathbf{e}}_n$$
$$= -0.00719\hat{\mathbf{i}} + 0.00360\hat{\mathbf{j}} \text{ m/s}^2$$

Now, relating $\mathbf{a}_B$ to $\mathbf{a}_A$ on rod $\mathcal{R}$, we have

$$\mathbf{a}_B = \mathbf{a}_A + \boldsymbol{\alpha}_{\mathcal{R}} \times \mathbf{r}_{AB} - \omega_{\mathcal{R}}^2\mathbf{r}_{AB}$$

$$-0.00719\hat{\mathbf{i}} + 0.00360\hat{\mathbf{j}} = \ddot{s}_A\hat{\mathbf{e}}_t + \frac{\dot{s}_A^2}{\rho_A}\hat{\mathbf{e}}_n$$
$$+ \ddot{\theta}_{\mathcal{R}}\hat{\mathbf{k}} \times (2\hat{\mathbf{i}} + 2\hat{\mathbf{j}}) - \omega_{\mathcal{R}}^2(2\hat{\mathbf{i}} + 2\hat{\mathbf{j}})$$

Substituting $\dot{s}_A = 0.403$ and $\omega_{\mathcal{R}} = 0.134$ from Example 3.7 (along with $\hat{\mathbf{e}}_t = \hat{\mathbf{i}}$ and $\hat{\mathbf{e}}_n = \hat{\mathbf{j}}$ at point A), we obtain the vector equation:

$$-0.00719\hat{\mathbf{i}} + 0.00360\hat{\mathbf{j}} = \ddot{s}_A\hat{\mathbf{i}} + \frac{(0.403)^2}{1}\hat{\mathbf{j}} - 2\ddot{\theta}_{\mathcal{R}}\hat{\mathbf{i}}$$
$$+ 2\ddot{\theta}_{\mathcal{R}}\hat{\mathbf{j}} - 2(0.134)^2(\hat{\mathbf{i}} + \hat{\mathbf{j}})$$

Writing the component equations, we have

$\hat{\mathbf{i}}$ coefficients: $\quad -0.00719 = \ddot{s}_A - 2\ddot{\theta}_{\mathcal{R}} - 2(0.134)^2$

$\hat{\mathbf{j}}$ coefficients: $\quad 0.00360 = \dfrac{(0.403)^2}{1} + 2\ddot{\theta}_{\mathcal{R}} - 2(0.134)^2$

The $\hat{\mathbf{j}}$ equation gives

$$\ddot{\theta} = -0.0612 \Rightarrow \boldsymbol{\alpha} = -0.0612\hat{\mathbf{k}} \text{ rad/s}^2$$

And the $\hat{\mathbf{i}}$ equation then yields

$$\ddot{s}_A = -0.0936 \text{ m/s}^2$$

Therefore

$$\mathbf{a}_A = -0.0936\hat{\mathbf{i}} + \frac{(0.403)^2}{1}\hat{\mathbf{j}}$$
$$= -0.0936\hat{\mathbf{i}} + 0.162\hat{\mathbf{j}} \text{ m/s}^2$$

Many more examples of the use of Equation (3.19) will be found in the next section, after we have discussed the topic of rolling.

PROBLEMS / Section 3.5

3.54 In Figure P3.54 the angular velocity of the bent bar $\mathcal{B}$ is 0.2 rad/s counterclockwise at an instant when its angular acceleration is 0.3 rad/s² clockwise. Find the acceleration of the endpoint B in the indicated position.

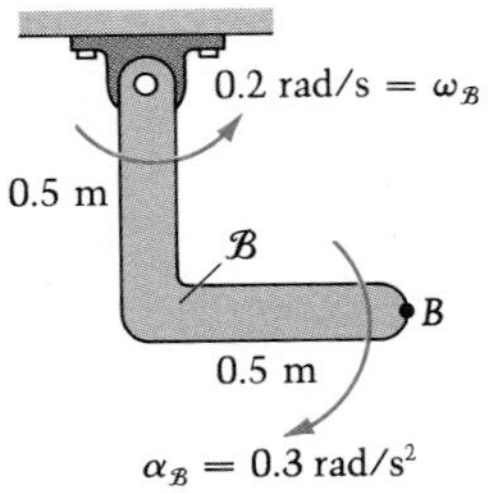

Figure P3.54

3.55 In the position indicated in Figure P3.55, the slider block $\mathcal{B}$ has the indicated velocity and acceleration. Find the angular acceleration of the wheel at this instant.

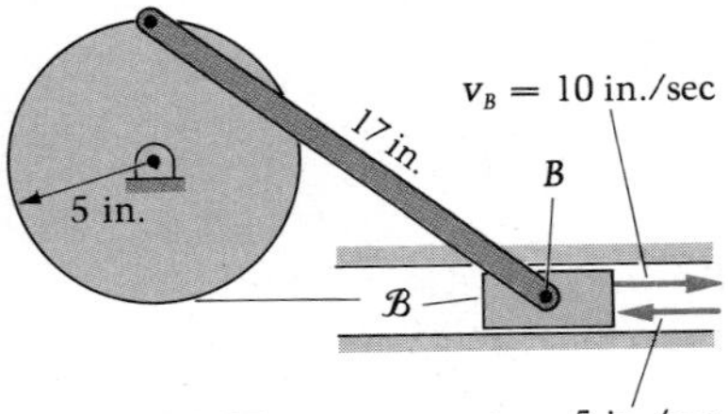

Figure P3.55

3.56 The velocities and accelerations of the two endpoints A and B of a rigid bar in plane motion are as shown in Figure P3.56. Find the acceleration of the midpoint of the bar in the given position.

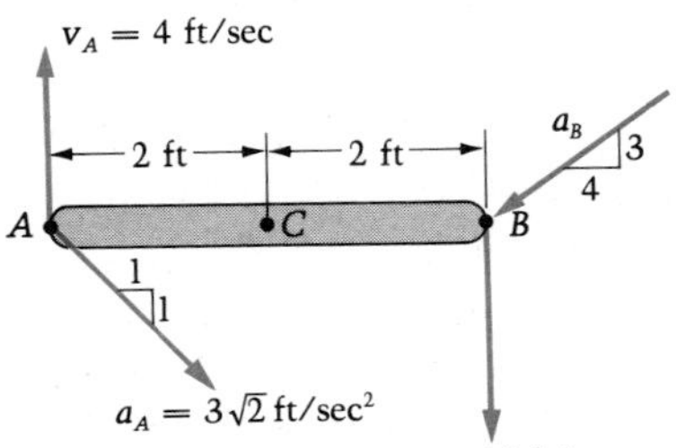

Figure P3.56 $v_B = 12$ ft/sec

3.57 In Problem 3.4, find the acceleration of B when $x = 10$ m.

3.58 The wheel $\mathcal{W}$ of diameter 2 m shown in Figure P3.58 turns about an axis through its center with an angular speed given by

$$\dot{\theta} = \omega = 5 - 2t$$

where ω is in radians per second and t is in seconds. Determine the angular acceleration of $\mathcal{W}$, and find the angular displacement of line OP during the time interval from $t = 0$ to $t = 5$ s.

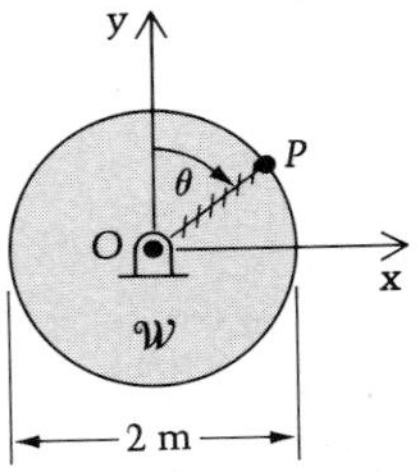

Figure P3.58

3.59 The angular velocity of $\mathcal{B}_1$ in Figure P3.59 is a constant 3 rad/sec clockwise. Find the velocity and acceleration of point C in the given configuration, and determine the acceleration of point C when $\mathbf{v}_C = \mathbf{0}$.

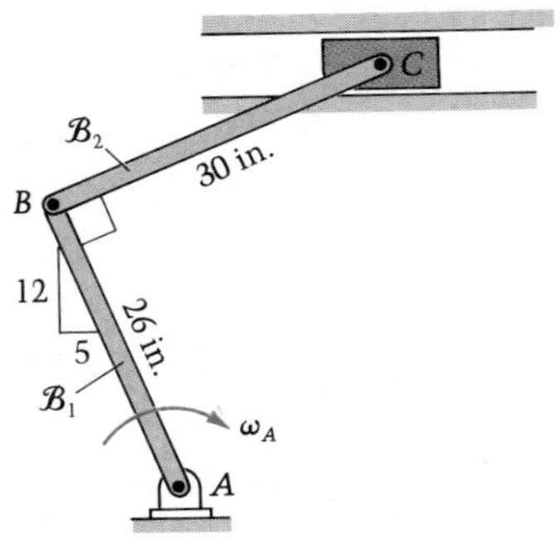

Figure P3.59

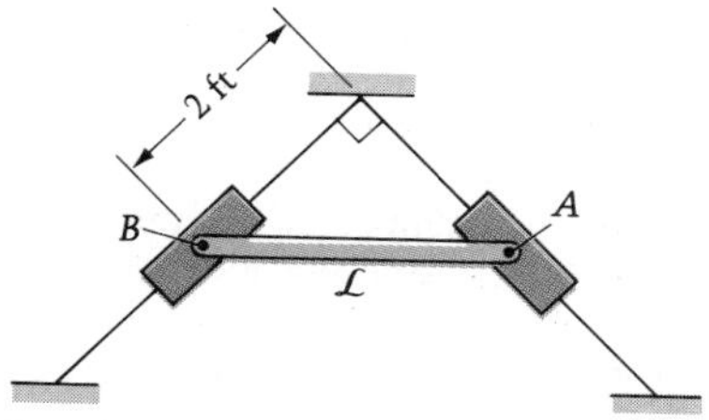

Figure P3.60

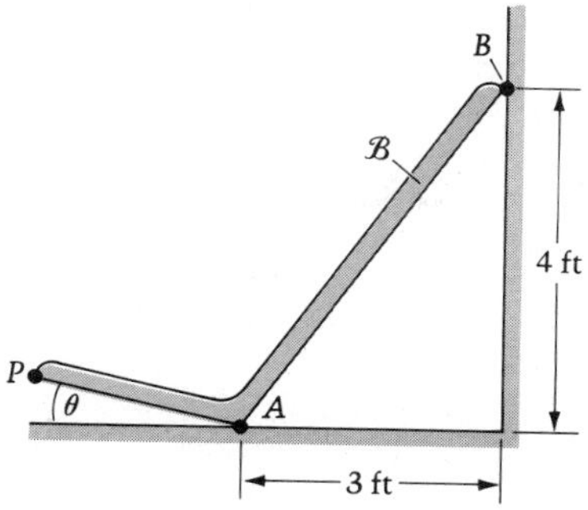

Figure P3.61

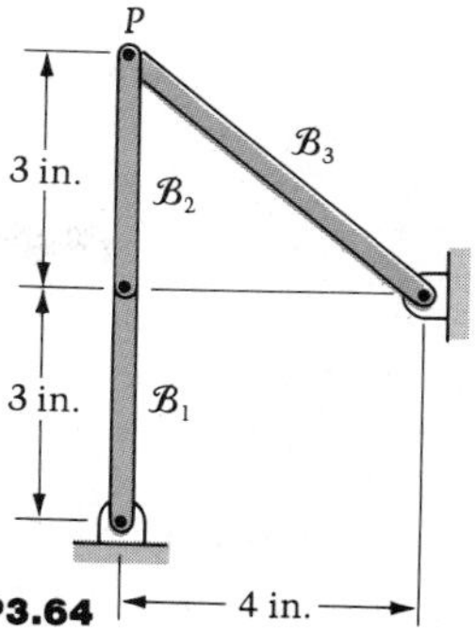

Figure P3.64

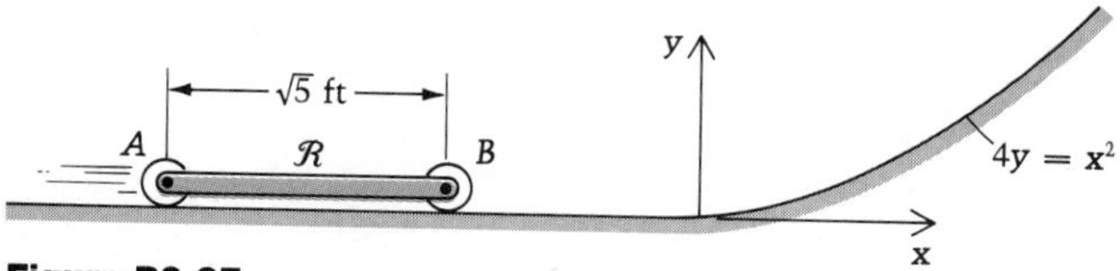

Figure P3.67

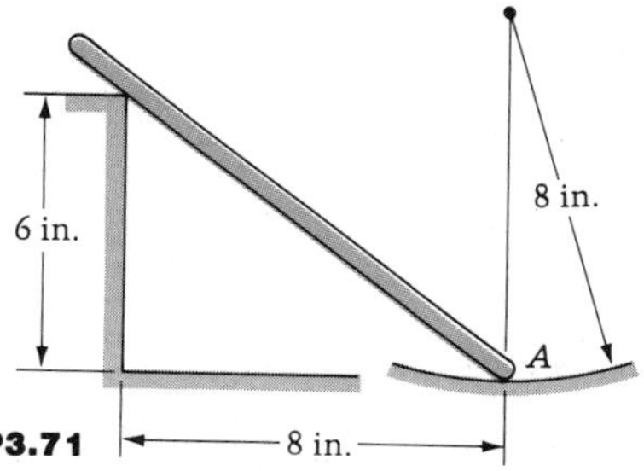

Figure P3.71

3.60 The acceleration of pin B in Figure P3.60 is 9.9 ft/sec^2 down and to the left, and its velocity is 4 ft/sec up and to the right, when $\mathcal{L}$ passes the horizontal. At this instant, find the angular acceleration of $\mathcal{L}$.

3.61 The bent bar $\mathcal{B}$ shown in Figure P3.61 slides on the vertical and horizontal surfaces. For the position shown, A has an acceleration of 4 ft/sec^2 to the left, while the bar has an angular velocity of 2 rad/sec clockwise and an angular acceleration of $\boldsymbol{\alpha}$.

a. Determine $\boldsymbol{\alpha}$ for the position shown.
b. Find, for the position shown, the angle θ and the distance PA such that point P has zero acceleration.

3.62 In Problem 3.29 find the acceleration of S and the angular accelerations of $\mathcal{A}$ and $\mathcal{L}$ at the instant when $\theta = 30°$.

3.63 In Problem 3.43 determine the angular acceleration of the plate and the acceleration of pin D in the indicated position.

3.64 Bar $\mathcal{B}_1$ rotates with a constant angular velocity of $2\hat{\mathbf{k}}$ rad/sec. Find the angular velocities and angular accelerations of $\mathcal{B}_2$ and $\mathcal{B}_3$ at the instant shown in Figure P3.64.

3.65 In Problem 3.42 determine the angular accelerations of $\mathcal{B}_1$ and $\mathcal{B}_2$ in the indicated position.

3.66 Find the acceleration of P in Problem 3.64 if at the same instant the body $\mathcal{B}_1$ has angular acceleration $\boldsymbol{\alpha}_{\mathcal{B}_1} = 3\hat{\mathbf{k}}$ rad/sec^2 instead of zero.

3.67 A rod $\mathcal{R}$ is pinned at A and B to the centers of two small rollers. (See Figure P3.67.) The speed of A is kept constant at v_0 even after B encounters the parabolic surface. Find the acceleration of B just after its roller begins to move on the parabola.

3.68 In the preceding problem, find $\mathbf{a}_B$ just after the *left* roller has begun to travel on the parabola.

3.69 In Problem 3.26 find the equation for the angular acceleration $\ddot{\phi}$ of the connecting rod $\mathcal{R}$ as a function of r, ℓ, θ, and $\dot{\theta}$.

3.70 In the preceding problem, plot $\ddot{\phi}/\dot{\theta}^2$ versus θ ($0 \le \theta \le 2\pi$) for $\ell/r = 1$, 2, and 5.

3.71 Point A in Figure P3.71 is constrained to move along the circular path; at the instant shown $\mathbf{v}_A = 4 \rightarrow$ in./sec and $d/dt\,(|\mathbf{v}_A|) = 3$ in./sec^2. If the bar is forced to stay in contact with the step, find the angular acceleration of the bar at the instant shown. *Hint:* Use polar coordinates.

3.72 In Figure P3.72 the velocity of point P (0.2 → m/s) and the angular velocity of bar $\mathcal{B}$ (0.4 ↻ rad/s) are constants. Find the acceleration of R and the angular acceleration of $\mathcal{C}$ in the given configuration.

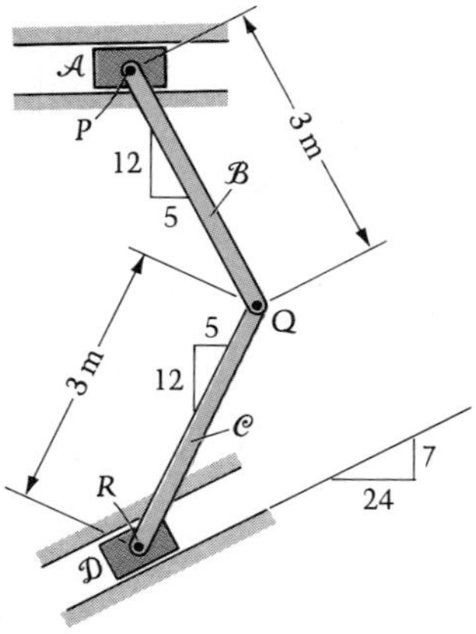

Figure P3.72

3.73 In Problem 3.38 find the acceleration of C in the position given in the figure.

3.74 Crank $\mathcal{C}$ in Figure P3.74 is pinned to rod $\mathcal{B}$; the other end of $\mathcal{B}$ slides on a parabolic incline and is at the origin in the position shown. The angular velocity of $\mathcal{C}$ is 3 ↻ rad/s = constant. Determine the acceleration of A and the angular acceleration of $\mathcal{B}$ at the given instant. *Hint:* The radius of curvature ρ of a plane curve $y = y(x)$ can be calculated from

$$\frac{1}{\rho} = \frac{y''}{(1 + y'^2)^{3/2}}$$

Use this result in computing the normal component of $\mathbf{a}_A$ as in Example 3.15.

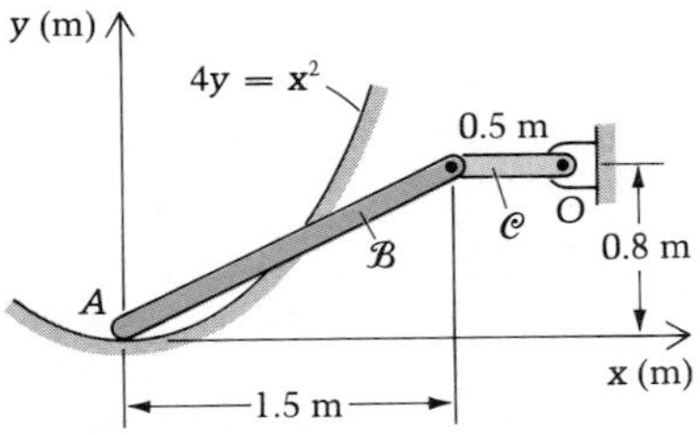

Figure P3.74

3.75 In Problem 3.41 find the acceleration of the upper end B of rod $\mathcal{R}$ in position (a).

3.76 Wheel $\mathcal{W}$ in Figure P3.76 has a constant clockwise angular velocity of 2 rad/sec. It is connected by link $\mathcal{L}$ to block $\mathcal{E}$. End B of rod $\mathcal{R}$ slides in a vertical slot in block $\mathcal{E}$. For the position shown, find the angular velocity and angular acceleration of rod $\mathcal{R}$ if block $\mathcal{E}$ translates.

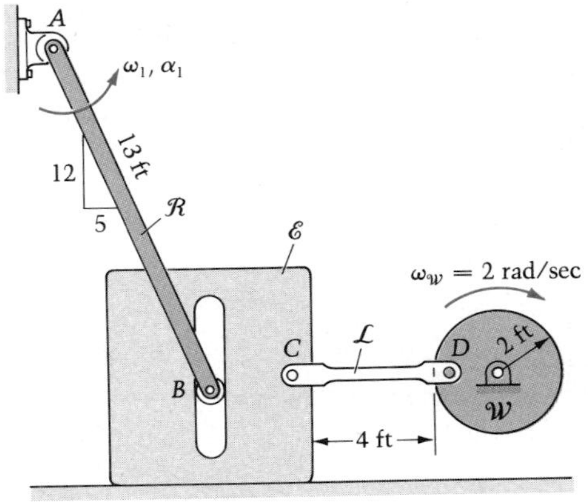

Figure P3.76

3.77 End B of the rod shown in Figure P3.77 has a constant velocity of 10 ft/sec down the plane. For the position shown (rod horizontal) determine, in terms of L, the velocity and acceleration of end A of the rod.

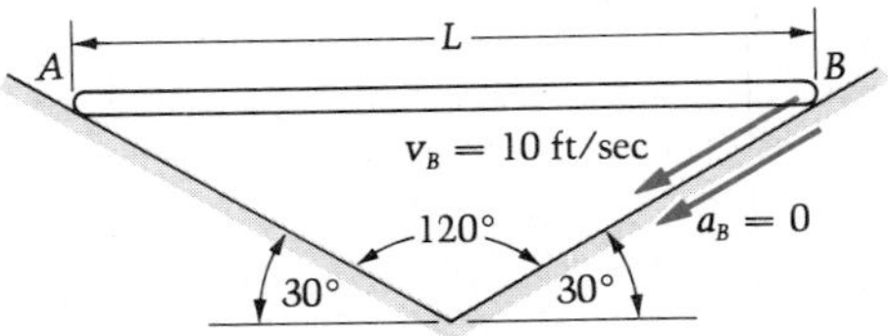

Figure P3.77

3.78 In Problem 3.36 find the acceleration of point B and the angular acceleration of body $\mathcal{B}$ at the described instant.

3.79 In Problem 3.37 find the acceleration of point B and the angular acceleration of body $\mathcal{B}$ at the described instant.

3.80 Rod $\mathcal{R}$ in Figure P3.80 is pinned to disk $\mathcal{D}$ at A and B. Disk $\mathcal{D}$ rotates about a fixed axis through O. The rod makes an angle θ radians with the line $\ell\ell$ as shown,

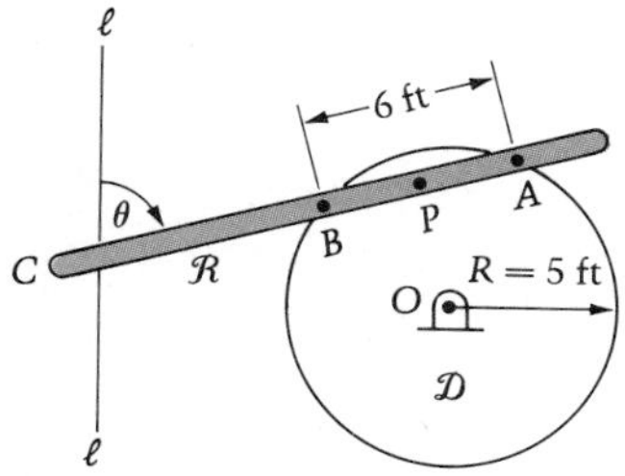

Figure P3.80

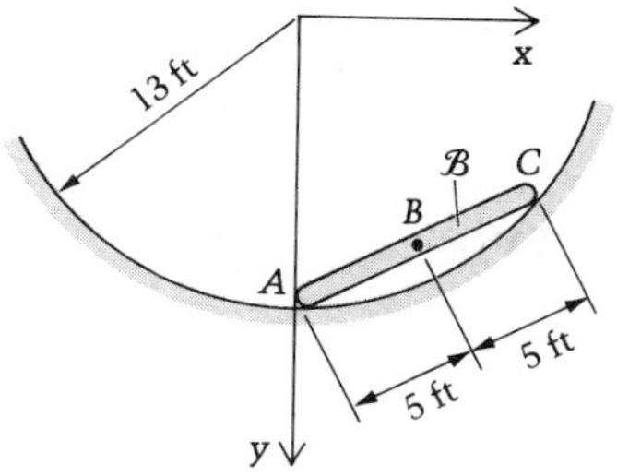

Figure P3.81

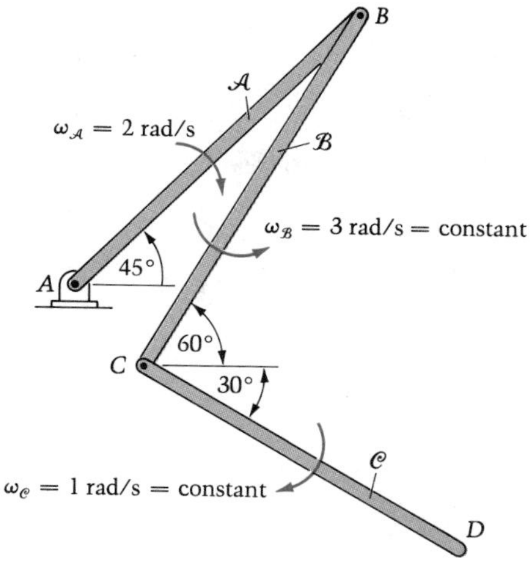

Figure P3.82

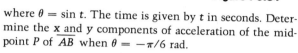

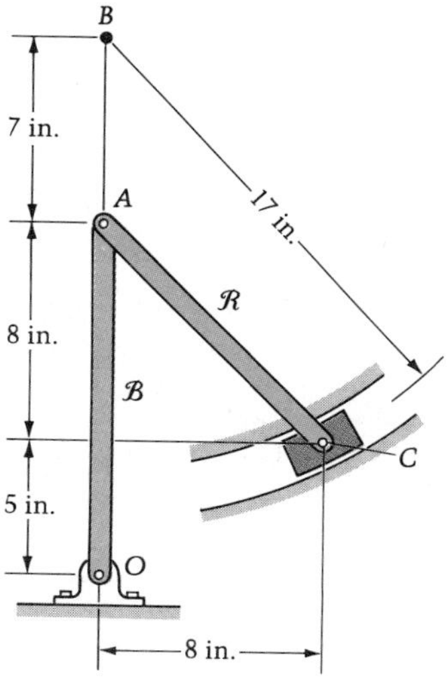

Figure P3.83

where $\theta = \sin t$. The time is given by t in seconds. Determine the $\mathbf{x}$ and $\mathbf{y}$ components of acceleration of the midpoint P of $\overline{AB}$ when $\theta = -\pi/6$ rad.

3.81 The 10-ft bar $\mathcal{B}$ in Figure P3.81 is sliding down the 13-ft-radius circle as shown. For the position shown, the bar has an angular velocity of 2 rad/sec and an angular acceleration of 3 rad/sec^2, both clockwise. Find the $\mathbf{x}$ and $\mathbf{y}$ components of acceleration of point B for this position.

3.82 Three bars each of length 1 m are pinned together as shown in Figure P3.82. Their angular velocities are given. Find:

a. The velocity of point D

b. The acceleration of D if the angular acceleration of $\mathcal{A}$ is 4 ↺ rad/sec^2

3.83 At the instant of time shown in Figure P3.83, the angular velocity and angular acceleration of rod $\mathcal{R}$ are 3 ↺ rad/sec and 2 ↻ rad/sec^2. At the same time, find the angular acceleration of bar $\mathcal{B}$.

3.84 The motion of a rotating element in a mechanism is controlled so that the rate of change of angular speed ω with angular displacement θ is a constant K. If the angular speed is ω_0 when both θ and the time t are zero, determine θ, ω, and the angular acceleration α as *functions of time*. *Hint:* Begin by showing that $\alpha\, d\theta = \omega\, d\omega$ and recall later that $d(\ln x) = dx/x$.

3.6 Rolling

Let $\mathcal{B}_1$ and $\mathcal{B}_2$ be two rigid bodies in motion. We define **rolling** to exist between $\mathcal{B}_1$ and $\mathcal{B}_2$ if during their motion:

1. A continuous sequence of points on the surface of $\mathcal{B}_1$ comes into one-to-one contact with a continuous sequence of points on the surface of $\mathcal{B}_2$.

2. At each instant during the interval of the motion, the contacting points have the same velocity vector.

Note that according to this definition there can be no slipping or sliding between the surfaces of $\mathcal{B}_1$ and $\mathcal{B}_2$ if rolling exists. Many authors, however, use the phrase "rolling without slipping" to describe the motion defined here. In their context, "rolling and slipping" would denote turning without the contact points having equal velocities; in our context, rolling *means* no slipping, so we shall say "turning and slipping" in cases of unequal contact-point velocities.

In this section we consider three classes of problems involving rolling contact:

1. Rolling of a wheel on a fixed straight line
2. Rolling of a wheel on a fixed plane curve
3. Gears

Rolling of a Wheel on a Fixed Straight Line

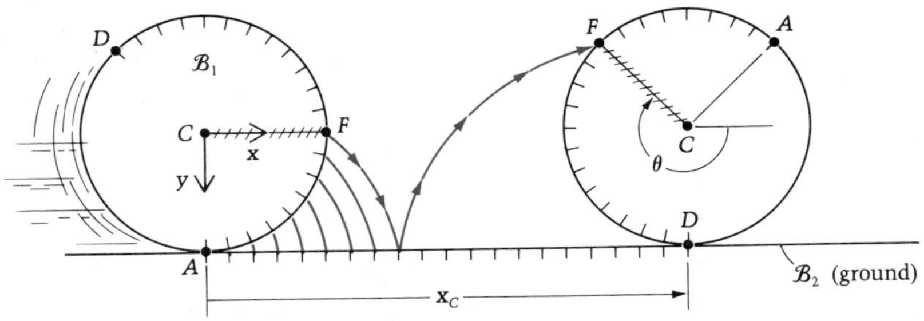

Figure 3.15 Rolling wheel, illustrating contacting sequences of points.

If the wheel ($\mathcal{B}_1$) shown in Figure 3.15 is rolling on the ground ($\mathcal{B}_2$, the reference frame in this case), then the continuous (shaded) sequences of points of $\mathcal{B}_1$ and $\mathcal{B}_2$ are in contact, one pair of points at a time. Since the points of $\mathcal{B}_2$ are all at rest, each point P_1 on the rim of $\mathcal{B}_1$ comes instantaneously to rest as it contacts a point P_2 of $\mathcal{B}_2$ (and is gripped for an instant by the ground). In this case, the velocities of P_1 and P_2 are each zero, although they need not vanish in general for rolling; all that is required is that $\mathbf{v}_{P_1} = \mathbf{v}_{P_2}$.*

Using Equation (3.8) and noting that the center point C moves only horizontally and that the contact point is always the instantaneous center of $\mathcal{B}_1$,

*If $\mathcal{B}_2$ is the flatbed of a truck, for example, itself in motion with respect to a ground reference frame $\mathcal{B}_3$, then $\mathbf{v}_{P_1} = \mathbf{v}_{P_2} \neq \mathbf{0}$, but $\mathcal{B}_1$ still rolls on $\mathcal{B}_2$.

$$\mathbf{v}_C = \mathbf{v}_{①} + \dot{\theta}\hat{\mathbf{k}} + \mathbf{r}_{①C} \tag{3.20}$$

$$\dot{x}_C\hat{\mathbf{i}} = \mathbf{0} + \dot{\theta}\hat{\mathbf{k}} \times (-r\hat{\mathbf{j}}) \tag{3.21}$$

$$\dot{x}_C\hat{\mathbf{i}} = r\dot{\theta}\hat{\mathbf{i}} \tag{3.22}$$

For the rolling wheel, therefore, the velocity of its center C and the body's angular velocity are related quite simply by

$$\dot{x}_C = r\dot{\theta} \text{ or } r\omega \tag{3.23}$$

We note that to express the rolling condition, the right side of Equation (3.23) would need a minus sign if the right-handed sign convention were established as either

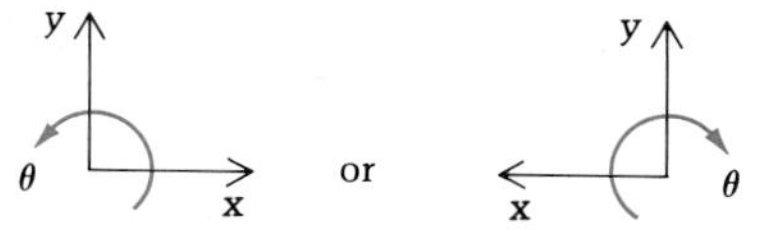

> **Question 3.13** Why will the signs not "take care of themselves" if we use $\dot{x}_C = r\dot{\theta}$ in these two cases?

The relation between the displacement of C and the rotation of $\mathcal{B}_1$ follows from integrating Equation (3.23):

$$x_C = r\theta + C_1 \tag{3.24}$$

where the integration constant is zero if we choose $x_C = 0$ when $\theta = 0$.

Another approach to rolling is to begin with a small displacement Δx_C while the base point ① grips the ground. If the angle of rotation produced is $\Delta\theta$, then

$$\Delta x_C = r\Delta\theta$$

if we envision the body turning about its instantaneous center. Dividing by a small time increment Δt and taking the limit, we get

$$\lim_{\Delta t \to 0}\left(\frac{\Delta x_C}{\Delta t}\right) = \dot{x}_C = \lim_{\Delta t \to 0}\left(\frac{r\Delta\theta}{\Delta t}\right) = r\dot{\theta} = r\omega$$

as was obtained in Equation (3.23).

Next we consider accelerations. From Equation (3.22):

$$\mathbf{a}_C = \dot{\mathbf{v}}_C = \ddot{x}_C\hat{\mathbf{i}} = r\ddot{\theta}\hat{\mathbf{i}} = r\alpha\hat{\mathbf{i}} \tag{3.25}$$

and the center point C accelerates parallel to the plane, as it must since it (alone) has rectilinear motion. Now let us compute the acceleration of the instantaneous center $\textcircled{I}$ of the wheel. Using Equation (3.19), we obtain

$$\begin{aligned}\mathbf{a}_{\textcircled{I}} &= \mathbf{a}_C + \ddot{\theta}\hat{\mathbf{k}} \times \mathbf{r}_{C\textcircled{I}} - \dot{\theta}^2\mathbf{r}_{C\textcircled{I}} \\ &= r\ddot{\theta}\hat{\mathbf{i}} + \ddot{\theta}\hat{\mathbf{k}} \times r\hat{\mathbf{j}} - \dot{\theta}^2 r\hat{\mathbf{j}} \\ &= -r\dot{\theta}^2\hat{\mathbf{j}} \text{ or } -r\omega^2\hat{\mathbf{j}} \text{ or } r\omega^2\uparrow \end{aligned} \tag{3.26}$$

Thus the contact point of a rolling wheel is accelerated toward its center with a magnitude $r\omega^2$. We see once again that a point of zero velocity need *not* be a point of zero acceleration, although of course it will be such if it is pinned to the reference frame. The point $\textcircled{I}$ in the example is at rest instantaneously, but it has an acceleration—which is why its velocity changes from zero as soon as it moves and a new $\textcircled{I}$ takes its place in the rolling. We now take up several examples,* each of which deals with a round object rolling on a flat surface.

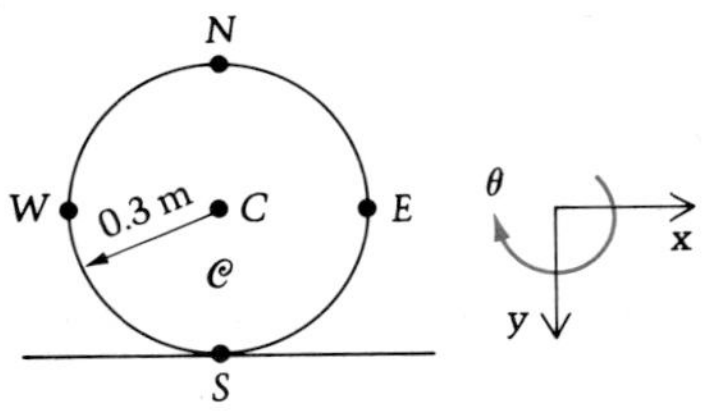

EXAMPLE 3.16

At a given instant, the rolling cylinder $\mathcal{C}$ in the accompanying diagram has $\boldsymbol{\omega}_{\mathcal{C}} = 2 \circlearrowright$ rad/s and $\boldsymbol{\alpha}_{\mathcal{C}} = 1.5 \circlearrowleft$ rad/s^2. Find the velocity and acceleration of points N and E.

SOLUTION

We have, because of the rolling,

$$\mathbf{v}_C = \dot{x}_C\hat{\mathbf{i}} = r\omega_{\mathcal{C}}\hat{\mathbf{i}} = 0.3(2)\hat{\mathbf{i}} = 0.6\hat{\mathbf{i}} \text{ m/s}$$

and

$$\mathbf{a}_C = \ddot{x}_C\hat{\mathbf{i}} = r\alpha_{\mathcal{C}}\hat{\mathbf{i}} = 0.3(-1.5)(\hat{\mathbf{i}}) = -0.45\hat{\mathbf{i}} \text{ m/s}^2$$

Therefore

$$\begin{aligned}\mathbf{v}_N &= \mathbf{v}_C + \boldsymbol{\omega}_{\mathcal{C}} \times \mathbf{r}_{CN} \\ &= 0.6\hat{\mathbf{i}} + 2\hat{\mathbf{k}} \times (-0.3\hat{\mathbf{j}}) \\ &= 1.2\hat{\mathbf{i}} \text{ m/s}\end{aligned}$$

(Continued)

*The examples of this section make continued use of Equations (3.8), (3.13), and (3.19), with the added feature that a rolling body is involved in each problem.

Note the agreement of this result with

$$
\begin{aligned}
\mathbf{v}_N &= \boldsymbol{\omega}_{\mathcal{C}} \times \mathbf{r}_{\textcircled{I}C} \\
&= 2\hat{\mathbf{k}} \times (-0.6\hat{\mathbf{j}}) \\
&= 1.2\hat{\mathbf{i}} \text{ m/s}
\end{aligned}
$$

Continuing, we get

$$
\begin{aligned}
\mathbf{a}_N &= \mathbf{a}_C + \boldsymbol{\alpha}_{\mathcal{C}} \times \mathbf{r}_{CN} - \omega_{\mathcal{C}}^2 \mathbf{r}_{CN} \\
&= -\,0.45\hat{\mathbf{i}} + (-1.5\hat{\mathbf{k}}) \times (-0.3\hat{\mathbf{j}}) - 2^2(-0.3\hat{\mathbf{j}}) \\
&= -\,0.90\hat{\mathbf{i}} + 1.2\hat{\mathbf{j}} \text{ m/s}^2
\end{aligned}
$$

For point E,

$$
\begin{aligned}
\mathbf{v}_E &= \mathbf{v}_C + \boldsymbol{\omega}_{\mathcal{C}} \times \mathbf{r}_{CE} \\
&= 0.6\hat{\mathbf{i}} + 2\hat{\mathbf{k}} \times 0.3\hat{\mathbf{i}} \\
&= 0.6\hat{\mathbf{i}} + 0.6\hat{\mathbf{j}} \text{ m/s}
\end{aligned}
$$

and

$$
\begin{aligned}
\mathbf{a}_E &= \mathbf{a}_C + \boldsymbol{\alpha}_{\mathcal{C}} \times \mathbf{r}_{CE} - \omega_{\mathcal{C}}^2 \mathbf{r}_{CE} \\
&= -0.45\hat{\mathbf{i}} + (-1.5\hat{\mathbf{k}}) \times 0.3\hat{\mathbf{i}} - 2^2(0.3\hat{\mathbf{i}}) \\
&= -\,1.65\hat{\mathbf{i}} - 0.45\hat{\mathbf{j}} \text{ m/s}^2
\end{aligned}
$$

The reader may wish to obtain the last result for $\mathbf{a}_E$ by relating it to $\mathbf{a}_{\textcircled{I}}$, which as we have seen is $r\omega^2\uparrow$, *not* zero.

EXAMPLE 3.17

In Example 3.11 find the piston acceleration when $\theta = 90°$.

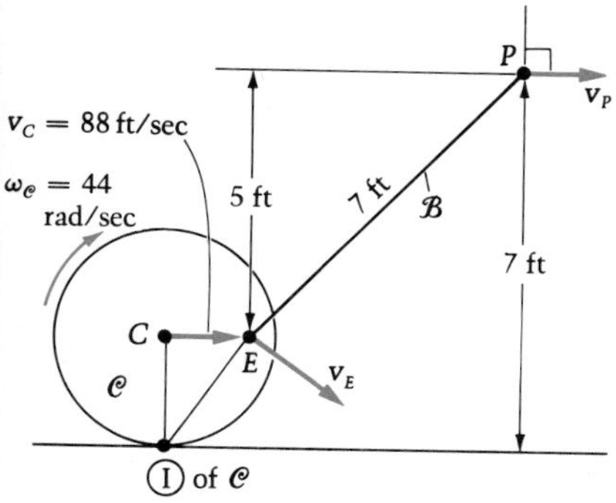

SOLUTION

In Case (b) of Example 3.11 we found $\boldsymbol{\omega}_{\mathcal{C}} = 44.0 \circlearrowright$ rad/sec, $\mathbf{v}_E = 88.0\hat{\mathbf{i}} + 66.0\hat{\mathbf{j}}$ ft/sec, $\boldsymbol{\omega}_{\mathcal{B}} = 13.5 \circlearrowleft$ rad/sec, and $\mathbf{v}_P = 20.5\hat{\mathbf{i}}$ ft/sec. On body $\mathcal{C}$, the velocity of C is constant $(88\hat{\mathbf{i}})$, so that $a_C = 0$. Also, $a_C = r\alpha_{\mathcal{C}}$ so that $\alpha_{\mathcal{C}} = 0$. Therefore

$$
\begin{aligned}
\mathbf{a}_E &= \mathbf{a}_C + \alpha_{\mathcal{C}}\hat{\mathbf{k}} \times \mathbf{r}_{CE} - \omega_{\mathcal{C}}^2 \mathbf{r}_{CE} \\
&= \mathbf{0} + \mathbf{0} - 44.0^2(2\hat{\mathbf{i}}) \\
&= -\,3870\hat{\mathbf{i}} \text{ ft/sec}^2
\end{aligned}
$$

(Continued)

On $\mathcal{B}$, we now relate $\mathbf{a}_E$ to the desired $\mathbf{a}_P$:

$$\underbrace{\mathbf{a}_P = a_P\hat{\mathbf{i}}}_{\text{kinematic constraint}} = \mathbf{a}_E + \alpha_{\mathcal{B}}\hat{\mathbf{k}} \times \mathbf{r}_{EP} - \omega_{\mathcal{B}}^2\mathbf{r}_{EP}$$

$$= -3870\hat{\mathbf{i}} + \underbrace{\alpha_{\mathcal{B}}\hat{\mathbf{k}} \times (4.90\hat{\mathbf{i}} - 5.00\hat{\mathbf{j}})} - \underbrace{13.5^2(4.90\hat{\mathbf{i}} - 5.00\hat{\mathbf{j}})}$$

Note that this is a cross product, but . . . *this* is not!

$$a_P\hat{\mathbf{i}} = (-3870 + 5.00\alpha_{\mathcal{B}} - 893)\hat{\mathbf{i}} + (4.90\alpha_{\mathcal{B}} + 911)\hat{\mathbf{j}}$$

Equating the $\hat{\mathbf{j}}$ coefficients first eliminates the unknown $\mathbf{a}_P$:

$$0 = 4.90\alpha_{\mathcal{B}} + 911 \Rightarrow \alpha_{\mathcal{B}} = -186 \text{ rad/sec}^2 \text{ or } \boldsymbol{\alpha}_{\mathcal{B}} = 186 \circlearrowleft \text{ rad/sec}^2$$

Then the coefficients of $\hat{\mathbf{i}}$ yield our answer:

$$a_P = -5690 \text{ ft/sec}^2 \qquad \text{or} \qquad \mathbf{a}_P = 5690 \leftarrow \text{ ft/sec}^2$$

EXAMPLE 3.18

The wheel $\mathcal{W}$ in the accompanying diagram rolls to the right on plane $\mathcal{P}$. At the instant shown in the figure, $\mathcal{W}$ has angular velocity $\boldsymbol{\omega}_{\mathcal{W}} = 1.2 \circlearrowright$ rad/s. Rod $\mathcal{R}$ is pinned to $\mathcal{W}$ at A, and the other end B of $\mathcal{R}$ slides along a plane $\mathcal{Q}$ parallel to $\mathcal{P}$. Determine the velocity of B and the angular velocity of $\mathcal{R}$ at the given instant.

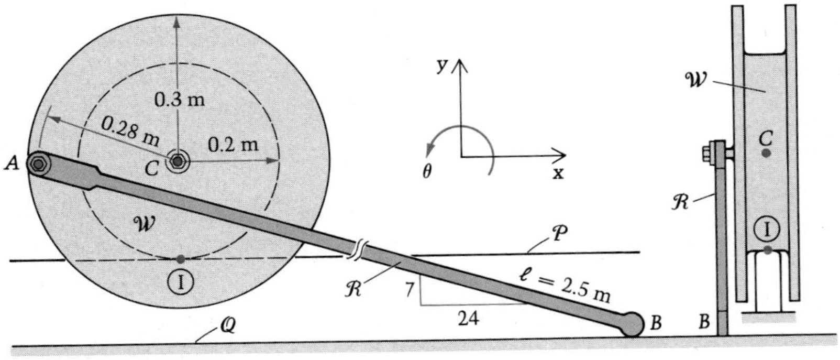

SOLUTION

We shall use Equation (3.8) together with the results we have derived for rolling. We first seek the velocity of A; when we have $\mathbf{v}_A$ we shall then relate it to $\mathbf{v}_B$ on rod $\mathcal{R}$. We may find $\mathbf{v}_A$ in either of two ways, each on wheel $\mathcal{W}$:

(Continued)

$$\mathbf{v}_A = \cancelto{0}{\mathbf{v}_{Ⓘ}} + \boldsymbol{\omega}_{\mathcal{W}} \times \mathbf{r}_{ⒾA}$$
$$= (-1.2\hat{\mathbf{k}}) \times (-0.28\hat{\mathbf{i}} + 0.2\hat{\mathbf{j}})$$
$$= 0.24\hat{\mathbf{i}} + 0.336\hat{\mathbf{j}} \text{ m/s}$$

$$\mathbf{v}_A = \mathbf{v}_C + \boldsymbol{\omega}_{\mathcal{W}} \times \mathbf{r}_{CA}$$
$$= 0.2(1.2)\hat{\mathbf{i}} + (-1.2\hat{\mathbf{k}}) \times (-0.28\hat{\mathbf{i}})$$
$$= 0.24\hat{\mathbf{i}} + 0.336\hat{\mathbf{j}} \text{ m/s}$$

We note for interest that when point A is beneath Ⓘ, its velocity is to the left—that is, it is going backwards!

Next, on $\mathcal{R}$,

$$\mathbf{v}_B = \mathbf{v}_A + \boldsymbol{\omega}_{\mathcal{R}} \times \mathbf{r}_{AB}$$

Point B is constrained to move horizontally; therefore

$$v_B\hat{\mathbf{i}} = 0.24\hat{\mathbf{i}} + 0.336\hat{\mathbf{j}} + \dot{\theta}_{\mathcal{R}}\hat{\mathbf{k}} \times (2.4\hat{\mathbf{i}} - 0.7\hat{\mathbf{j}})$$

or

$$v_B\hat{\mathbf{i}} = \hat{\mathbf{i}}(0.24 + 0.7\dot{\theta}_{\mathcal{R}}) + \hat{\mathbf{j}}(0.336 + 2.4\dot{\theta}_{\mathcal{R}})$$

$\hat{\mathbf{j}}$ coefficients: $\quad 0 = 0.336 + 2.4\dot{\theta}_{\mathcal{R}} \Rightarrow \dot{\theta}_{\mathcal{R}} = -0.140$

so that

$$\boldsymbol{\omega}_{\mathcal{R}} = -0.140\hat{\mathbf{k}} \text{ rad/s or } 0.140 \curvearrowright \text{ rad/s}$$

$\hat{\mathbf{i}}$ coefficients: $\quad v_B = 0.24 + 0.7(-0.140) = 0.142$

so that

$$\mathbf{v}_B = 0.142\hat{\mathbf{i}} \text{ m/s}$$

Mentally locate the Ⓘ point for $\mathcal{R}$ and from this deduce that the directions of $\boldsymbol{\omega}_{\mathcal{R}}$ and $\mathbf{v}_B$ are correct.

EXAMPLE 3.19

In the preceding example, at the same instant, $\boldsymbol{\alpha}_{\mathcal{W}} = 0.8 \curvearrowright \text{rad/s}^2$. Find the acceleration of point B and the angular acceleration of rod $\mathcal{R}$.

SOLUTION

As we did with the velocity of A, we can relate $\mathbf{a}_A$ to the acceleration of either Ⓘ or C:

$$\mathbf{a}_A = \mathbf{a}_{Ⓘ} + \boldsymbol{\alpha}_{\mathcal{W}} \times \mathbf{r}_{ⒾA} - \omega_{\mathcal{W}}^2\mathbf{r}_{ⒾA}$$

The terms are

$$\mathbf{a}_{Ⓘ} = r\omega_{\mathcal{W}}^2 \uparrow = 0.2(1.2^2)\hat{\mathbf{j}}$$
$$= 0.288\hat{\mathbf{j}} \quad \text{(note not zero!)}$$

$$\mathbf{a}_A = \mathbf{a}_C + \boldsymbol{\alpha}_{\mathcal{W}} \times \mathbf{r}_{CA} - \omega_{\mathcal{W}}^2\mathbf{r}_{CA}$$

The terms are

$$\mathbf{a}_C = r\alpha_{\mathcal{W}}\hat{\mathbf{i}} = 0.2(0.8)\hat{\mathbf{i}}$$
$$= 0.16\hat{\mathbf{i}}$$

(Continued)

$$\boldsymbol{\alpha}_{\mathcal{W}} \times \mathbf{r}_{①A} = -0.8\hat{\mathbf{k}} \times (-0.28\hat{\mathbf{i}} + 0.2\hat{\mathbf{j}})$$
$$= 0.16\hat{\mathbf{i}} + 0.224\hat{\mathbf{j}}$$
$$-\omega_{\mathcal{W}}^2 \mathbf{r}_{①A} = -1.2^2(-0.28\hat{\mathbf{i}} + 0.2\hat{\mathbf{j}})$$
$$= 0.403\hat{\mathbf{i}} - 0.288\hat{\mathbf{j}}$$

Adding the terms, we get

$$\mathbf{a}_A = 0.563\hat{\mathbf{i}} + 0.224\hat{\mathbf{j}} \text{ m/s}^2$$

$$\boldsymbol{\alpha}_{\mathcal{W}} \times \mathbf{r}_{CA} = (-0.8\hat{\mathbf{k}}) \times (-0.28\hat{\mathbf{i}})$$
$$= 0.224\hat{\mathbf{j}}$$
$$-\omega_{\mathcal{W}}^2 \mathbf{r}_{CA} = -1.2^2(-0.28\hat{\mathbf{i}})$$
$$= 0.403\hat{\mathbf{i}}$$

Again adding the terms, we get

$$\mathbf{a}_A = 0.563\hat{\mathbf{i}} + 0.224\hat{\mathbf{j}} \text{ m/s}^2$$

Now we relate $\mathbf{a}_A$ to $\mathbf{a}_B$ on the rod; note that the acceleration of B is constrained by the plane to be horizontal:

$$\mathbf{a}_B = a_B\hat{\mathbf{i}} = \mathbf{a}_A + \boldsymbol{\alpha}_{\mathcal{R}} \times \mathbf{r}_{AB} - \omega_{\mathcal{R}}^2 \mathbf{r}_{AB}$$
$$= 0.563\hat{\mathbf{i}} + 0.224\hat{\mathbf{j}} + \alpha_{\mathcal{R}}\hat{\mathbf{k}} \times (2.4\hat{\mathbf{i}} - 0.7\hat{\mathbf{j}})$$
$$-(-0.140)^2(2.4\hat{\mathbf{i}} - 0.7\hat{\mathbf{j}})$$

$\hat{\mathbf{j}}$ coefficients:

$$0 = 0.224 + 2.4\alpha_{\mathcal{R}} + 0.0137 \Rightarrow \alpha_{\mathcal{R}} = -0.0991 \text{ rad/s}^2$$

so that

$$\boldsymbol{\alpha}_{\mathcal{R}} = -0.0991\hat{\mathbf{k}} \text{ rad/s}^2$$

$\hat{\mathbf{i}}$ coefficients:

$$a_B = 0.563 + 0.7(-0.0991) - 0.0470 = 0.447$$

so that

$$\mathbf{a}_B = 0.447\hat{\mathbf{i}} \text{ m/s}^2$$

Rolling of a Wheel ($\mathcal{B}_1$) on a Fixed Plane Curve ($\mathcal{B}_2$)

In the second class of rolling problems to be considered in this section, the contact surface is curved. Let $\hat{\mathbf{e}}_t$ and $\hat{\mathbf{e}}_n$ be the principal unit tangent and normal vectors for the center point C of the wheel. (See Figure 3.16.) Then, since we are again defining a motion such that the contact point has zero velocity, Equation (3.8) yields

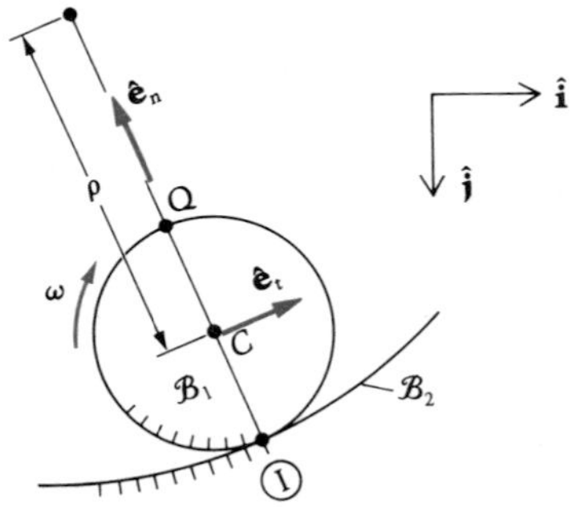

Figure 3.16 Wheel rolling on curved path concave upward.

$$\mathbf{v}_C = \mathbf{v}_① + \dot{\theta}\hat{\mathbf{k}} \times \mathbf{r}_{①C}$$
$$= \mathbf{0} + \dot{\theta}\hat{\mathbf{k}} \times r\hat{\mathbf{e}}_n$$
$$= r\dot{\theta}\hat{\mathbf{e}}_t \tag{3.27}$$

which gives us the velocity of C. Differentiating Equation (3.27), we get

$$\mathbf{a}_C = r\ddot{\theta}\hat{\mathbf{e}}_t + r\dot{\theta}\dot{\hat{\mathbf{e}}}_t$$
$$= r\ddot{\theta}\hat{\mathbf{e}}_t + r\dot{\theta}\frac{\dot{s}}{\rho}\hat{\mathbf{e}}_n \tag{3.28}$$

in which we have used Equations (1.42) and (1.46), where ρ is the instantaneous radius of curvature of the path on which C moves. If this path is a circle, as is often the case, then ρ = constant = radius of the circle.

Using $\dot{s} = v_C = r\dot{\theta}$ in Equation (3.28), we have

$$\mathbf{a}_C = r\ddot{\theta}\hat{\mathbf{e}}_t + \frac{(r\dot{\theta})^2}{\rho}\hat{\mathbf{e}}_n \tag{3.29}$$

The acceleration of Ⓘ is interesting, and it follows from Equation (3.19):

$$\mathbf{a}_{Ⓘ} = \mathbf{a}_C + \ddot{\theta}\hat{\mathbf{k}} \times \mathbf{r}_{CⓘI} - \dot{\theta}^2\mathbf{r}_{CⓘI} \tag{3.30}$$

or

$$\mathbf{a}_{Ⓘ} = r\ddot{\theta}\hat{\mathbf{e}}_t + \frac{(r\dot{\theta})^2}{\rho}\hat{\mathbf{e}}_n - r\ddot{\theta}\hat{\mathbf{e}}_t + r\dot{\theta}^2\hat{\mathbf{e}}_n = \left(1 + \frac{r}{\rho}\right)r\dot{\theta}^2\hat{\mathbf{e}}_n \tag{3.31}$$

Comparing the accelerations of the contact points Ⓘ of Figures 3.15 and 3.16, we observe from our results (Equations 3.26 and 3.31) that the point contacting the curved track has a greater acceleration than the one touching the flat track, due to the r/ρ term of Equation (3.31). This term represents the normal component of acceleration of C, which was zero on the flat track.

It is also interesting to examine the acceleration of point Q at the *top* of the wheel in Figure 3.16:

$$\begin{aligned}\mathbf{a}_Q &= \mathbf{a}_C + \ddot{\theta}\hat{\mathbf{k}} \times \mathbf{r}_{CQ} - \dot{\theta}^2\mathbf{r}_{CQ} \\ &= r\ddot{\theta}\hat{\mathbf{e}}_t + \frac{(r\dot{\theta})^2}{\rho}\hat{\mathbf{e}}_n + r\ddot{\theta}\hat{\mathbf{e}}_t - r\dot{\theta}^2\hat{\mathbf{e}}_n \\ &= 2r\ddot{\theta}\hat{\mathbf{e}}_t - r\dot{\theta}^2\left(1 - \frac{r}{\rho}\right)\hat{\mathbf{e}}_n \end{aligned} \tag{3.32}$$

Note that it is possible for the $\hat{\mathbf{e}}_n$ (normal) component of $\mathbf{a}_Q$ to be either away from or toward C, depending on whether $r > \rho$ or $r < \rho$, respectively (see Figure 3.17).

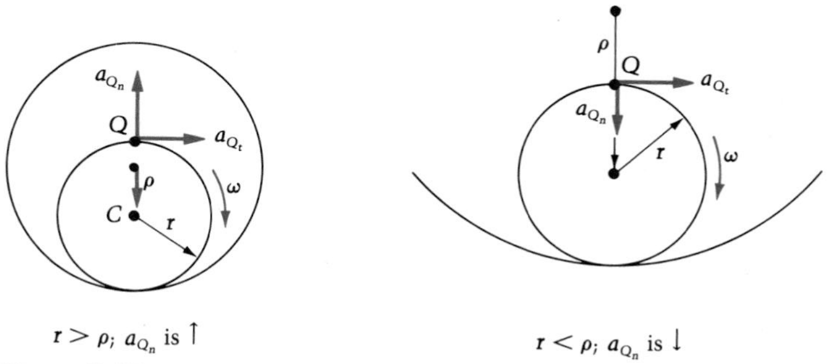
a_{Q_n}
Q
a_{Q_t}
ρ
C
r
ω
$r > \rho$; a_{Q_n} is ↑
ρ
Q
a_{Q_t}
a_{Q_n}
r
ω
$r < \rho$; a_{Q_n} is ↓

Figure 3.17

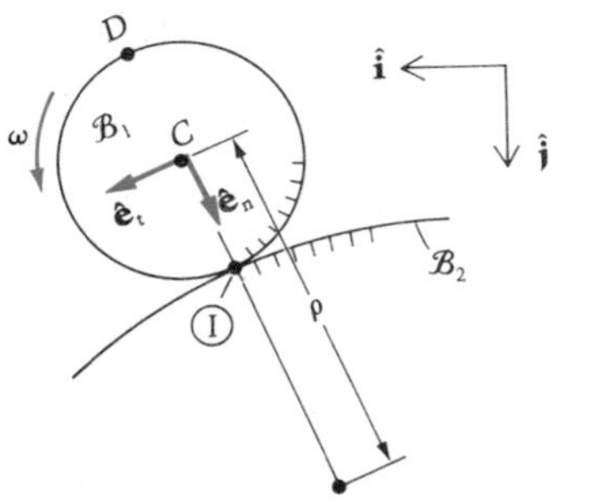

Figure 3.18 Wheel rolling on curved path concave downward.

If the track is concave ***downward*** as shown in Figure 3.18, similar results may be obtained for the accelerations of C, Ⓘ, and D. From Equation (3.29), which still holds:

$$\mathbf{a}_C = r\ddot{\theta}\hat{\mathbf{e}}_t + \frac{(r\dot{\theta})^2}{\rho}\hat{\mathbf{e}}_n \tag{3.33}$$

After relating $\mathbf{a}_{Ⓘ}$ and $\mathbf{a}_C$ on $\mathcal{B}_1$, we obtain:

$$\mathbf{a}_{Ⓘ} = \left(-1 + \frac{r}{\rho}\right) r\dot{\theta}^2\hat{\mathbf{e}}_n \tag{3.34}$$

And relating D to either C or Ⓘ gives

$$\mathbf{a}_D = 2r\ddot{\theta}\hat{\mathbf{e}}_t + \left(1 + \frac{r}{\rho}\right) r\dot{\theta}^2\hat{\mathbf{e}}_n \tag{3.35}$$

Note that this time the radius r cannot exceed ρ, so $\mathbf{a}_{Ⓘ}$ is always outward; the normal component of the acceleration of D is seen to be always inward in this case.

We have seen that in all the preceding "wheels on tracks" results, the acceleration of the contact point Ⓘ (the instantaneous center) lies along the line through the center C of the wheel. Therefore the reader may find the following expressions useful:

$$\mathbf{v}_C = \boldsymbol{\omega} \times \mathbf{r}_{ⒾC}$$

$$\mathbf{a}_C = \boldsymbol{\alpha} \times \mathbf{r}_{ⒾC} + \text{(normal part)}$$

If we knew either $\boldsymbol{\alpha}$ or the tangential component of $\mathbf{a}_C$, for example, we could get the other without concern for expressing the normal component of $\mathbf{a}_C$. We now consider several examples that feature a body rolling on a curved surface.

EXAMPLE 3.20

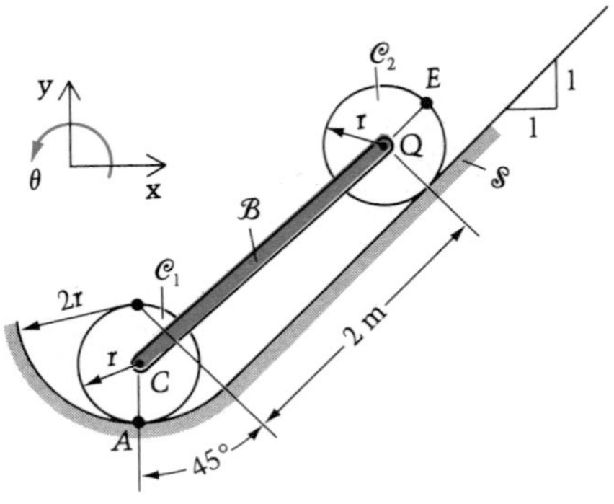

The two cylinders $\mathcal{C}_1$ and $\mathcal{C}_2$ shown in the diagram, connected by bar $\mathcal{B}$, are rolling on the surface $\mathcal{S}$ with $\omega_{\mathcal{C}_1} = 0.5$ ↻ rad/s at the given instant. The value of radius r is 0.4 m. Find the angular velocity of $\mathcal{B}$ and the velocity of point E on $\mathcal{C}_2$ at this time.

SOLUTION

Since A is instantaneously at rest (A is Ⓘ for $\mathcal{C}_1$), the velocity of C on $\mathcal{C}_1$ is

$$\mathbf{v}_C = \boldsymbol{\omega}_{\mathcal{C}_1} \times \mathbf{r}_{AC}$$
$$= -0.5\hat{\mathbf{k}} \times 0.4\hat{\mathbf{j}} = 0.2\hat{\mathbf{i}} \text{ m/s}$$

(Continued)

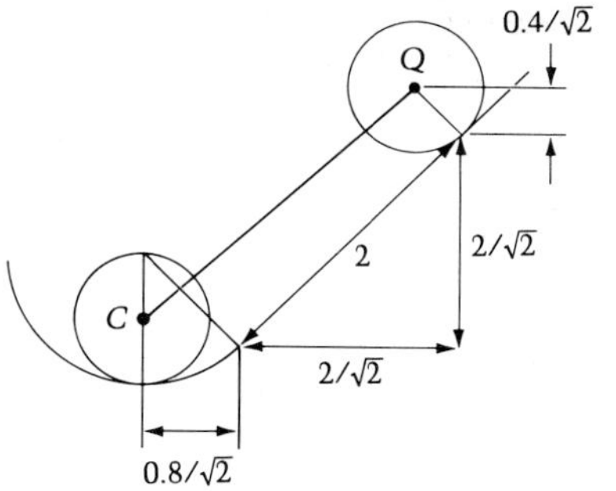

The diagram at the left shows that

$$\mathbf{r}_{CQ} = 1.70\hat{\mathbf{i}} + 1.53\hat{\mathbf{j}} \text{ m}$$

so that

$$\mathbf{v}_Q = \mathbf{v}_C + \boldsymbol{\omega}_{\mathcal{B}} \times \mathbf{r}_{CQ}$$

$$v_Q\left(\frac{\hat{\mathbf{i}}+\hat{\mathbf{j}}}{\sqrt{2}}\right) = 0.2\hat{\mathbf{i}} + \omega_{\mathcal{B}}\hat{\mathbf{k}} \times (1.70\hat{\mathbf{i}} + 1.53\hat{\mathbf{j}})$$

where $\mathbf{v}_Q$ must be parallel to the inclined plane. We solve for v_Q and $\omega_{\mathcal{B}}$ by equating the coefficients of the respective unit vectors:

$\hat{\mathbf{i}}$ coefficients: $$\frac{v_Q}{\sqrt{2}} = 0.2 - 1.53\omega_{\mathcal{B}}$$

$\hat{\mathbf{j}}$ coefficients: $$\frac{v_Q}{\sqrt{2}} = 1.70\omega_{\mathcal{B}}$$

Subtracting,

$$3.23\,\omega_{\mathcal{B}} = 0.2$$

$$\boldsymbol{\omega}_{\mathcal{B}} = 0.0619 \circlearrowleft \text{ rad/s}$$

Therefore

$$v_Q = \sqrt{2}(1.70)(0.0619) = 0.149 \text{ m/s}$$

Next, since B is at rest (as the rolling contact point of $\mathcal{C}_2$ with the fixed surface $\mathcal{S}$),

$$\mathbf{v}_Q = \boldsymbol{\omega}_{\mathcal{C}_2} \times \mathbf{r}_{BQ}$$

$$0.149 = \omega_{\mathcal{C}_2}(0.4) \Rightarrow \omega_{\mathcal{C}_2} = 0.373 \circlearrowright \text{ rad/s}$$

Therefore

$$\mathbf{v}_E = \overset{\mathbf{0}}{\cancel{\mathbf{v}_B}} + \boldsymbol{\omega}_{\mathcal{C}_2} \times \mathbf{r}_{BE}$$

$$= -0.373\hat{\mathbf{k}} \times [\sqrt{2}(0.4)\hat{\mathbf{j}}]$$

$$= 0.211\hat{\mathbf{i}} \text{ m/s}$$

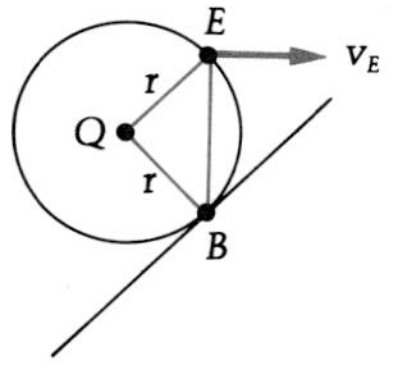

In the next two examples, the problems contain both "velocity parts" (as in the preceding example) and "acceleration parts."

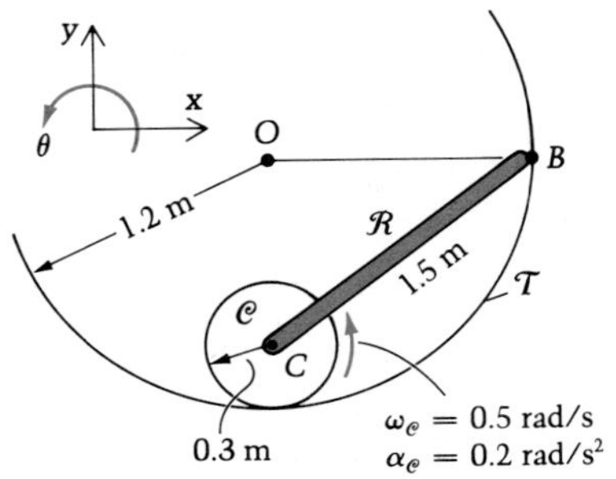

EXAMPLE 3.21

The cylinder $\mathcal{C}$ shown in the diagram is rolling on the track with the indicated angular velocity and acceleration when $\mathcal{C}$ is at the bottom of the track. Rod $\mathcal{R}$ is pinned to the center C of $\mathcal{C}$, and its other end, B, slides on track $\mathcal{T}$. Find the velocity and acceleration of B.

SOLUTION

Problems like this one have two parts. The "velocity part" must be solved before the "acceleration part" because, as will be seen, the velocities as well as angular velocities are needed in the expressions for acceleration. We shall use instantaneous centers to get $\mathbf{v}_B$; the steps are:

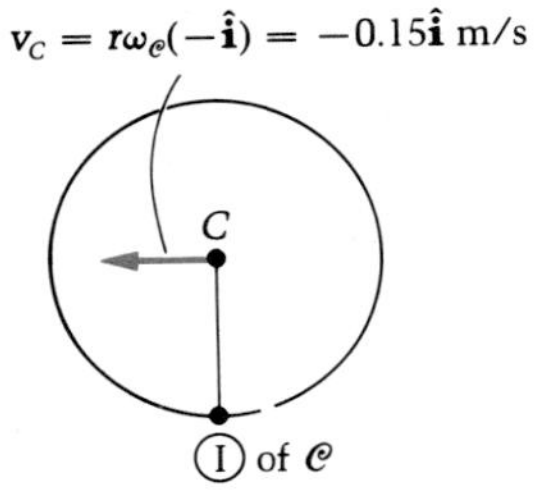

1. The contact point of $\mathcal{C}$ is its Ⓘ point, since the body is rolling.
2. $\mathbf{v}_C$ is determined as in the diagram from $r_{ⒾC}\,\omega_{\mathcal{C}}$.
3. The velocity of B is vertical (tangent to the path of the point).
4. Thus Ⓘ of $\mathcal{R}$ is at the intersection of the normals to $\mathbf{v}_C$ and $\mathbf{v}_B$.
5. Then $\omega_{\mathcal{R}} = v_C/d_1 = 0.15/0.9 = 0.167$ or $\boldsymbol{\omega}_{\mathcal{R}} = 0.167 \circlearrowleft$ rad/s
6. Finally, $v_B = d_2\omega_{\mathcal{R}} = 1.2(0.167) = 0.200$ or $\mathbf{v}_{\mathcal{B}} = 0.200 \downarrow$ m/s

From the preceding text, Equation (3.29) gives us the acceleration of C:

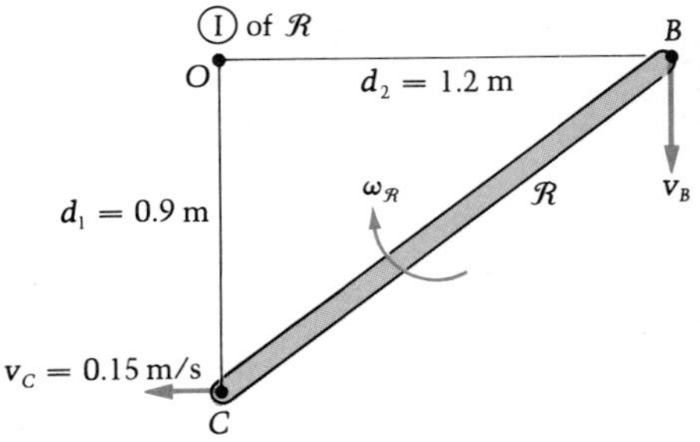

$$\mathbf{a}_C = r\alpha(-\hat{\mathbf{i}}) + \frac{(r\omega)^2}{\rho}\hat{\mathbf{j}}$$

$$= 0.3(0.2)(-\hat{\mathbf{i}}) + \frac{(0.3 \times 0.5)^2}{0.9}\hat{\mathbf{j}}$$

$$= -0.06\hat{\mathbf{i}} + 0.0250\hat{\mathbf{j}} \text{ m/s}^2$$

Relating $\mathbf{a}_C$ to $\mathbf{a}_B$, we get

$$\mathbf{a}_B = \mathbf{a}_C + \alpha_{\mathcal{R}}\hat{\mathbf{k}} \times \mathbf{r}_{CB} - \omega_{\mathcal{R}}^2\mathbf{r}_{CB}$$

$$\ddot{s}_B(-\hat{\mathbf{j}}) + \frac{\dot{s}_B^2}{\rho_B}(-\hat{\mathbf{i}}) = -0.06\hat{\mathbf{i}} + 0.0250\hat{\mathbf{j}} + \alpha_{\mathcal{R}}\hat{\mathbf{k}} \times (1.2\hat{\mathbf{i}} + 0.9\hat{\mathbf{j}})$$

$$-(0.167)^2(1.2\hat{\mathbf{i}} + 0.9\hat{\mathbf{j}})$$

(Continued)

In substituting for $\mathbf{a}_B$ we have used the facts that for B we have $\hat{\mathbf{e}}_t = -\hat{\mathbf{j}}$ (direction vector of $\mathbf{v}_B$) and $\hat{\mathbf{e}}_n = -\hat{\mathbf{i}}$ (always toward the center of curvature of the path of the point). Substituting $\dot{s}_B = |\mathbf{v}_B| = 0.2$ m/s and $\rho_B = 1.2$ m, we then have two scalar equations in the two unknowns $\ddot{s}_B$ and $\alpha_{\mathcal{R}}$:

$\hat{\mathbf{i}}$ coefficients:

$$-0.0333 = -0.06 - 0.9\alpha_{\mathcal{R}} - 0.0333$$

$$\alpha_{\mathcal{R}} = -0.0667 \text{ rad/s}^2$$

$\hat{\mathbf{j}}$ coefficients:

$$-\ddot{s}_B = 0.0250 + 1.2\alpha_{\mathcal{R}} - 0.0250 = -0.0800$$

$$\ddot{s}_B = 0.0800 \text{ m/s}^2$$

Therefore

$$\mathbf{a}_B = -0.0333\hat{\mathbf{i}} - 0.0800\hat{\mathbf{j}} \text{ m/s}^2$$

We wish to make a very important point regarding the preceding example. You may have noticed that the values of $\alpha_{\mathcal{R}}$ and $\ddot{s}_B$ could have been obtained more quickly from

$$|\boldsymbol{\alpha}_{\mathcal{R}}| = \frac{a_C}{r_{OC}} = \frac{0.06}{0.9} = 0.0667 \text{ rad/s}^2$$

and

$$\ddot{s}_B = r_{OB}\,|\boldsymbol{\alpha}_{\mathcal{R}}| = 1.2(0.0667) = 0.08 \text{ m/s}^2$$

This shortcut is very dangerous because it is not always valid! It is essential to understand when and why this procedure works. (It is *not* because O is the instantaneous center of zero velocity of $\mathcal{R}$!)

For a counterexample, consider the results of Example 3.14 together with the text following that example. If we calculate the component of $\mathbf{a}_B$ normal to the line ⒾB (see Figure 3.19), it is seen to be 0.160 in./sec². If we divide this result by $|\mathbf{r}_{ⒾB}|$, we obtain $0.160/12 = 0.0133$. Attaching the direction indicator ↻, we do *not* get the correct answer for $\boldsymbol{\alpha}$! It is not even in the correct direction, let alone the correct magnitude. The answer to the question of when the procedure is legitimate is covered by the following question.

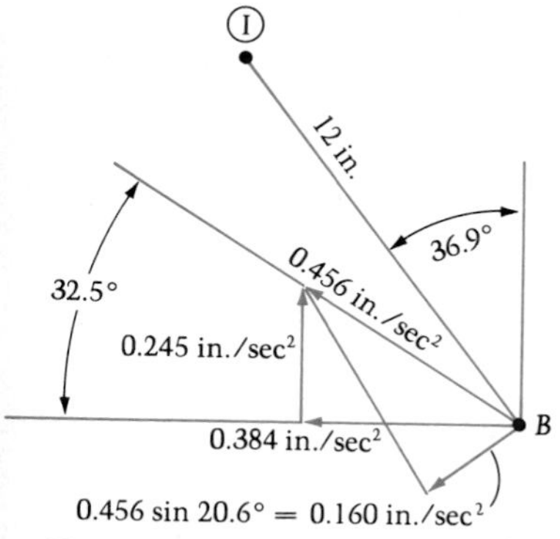

Figure 3.19

Question 3.14 When can we use $r_{ⒾP}\alpha$ to obtain the correct acceleration component of P normal to line ⒾP?

EXAMPLE 3.22

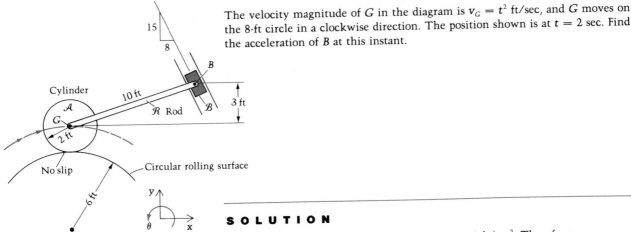

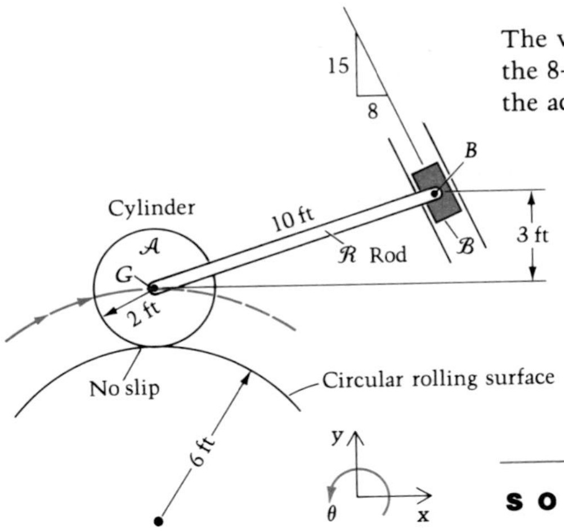

The velocity magnitude of G in the diagram is $v_G = t^2$ ft/sec, and G moves on the 8-ft circle in a clockwise direction. The position shown is at $t = 2$ sec. Find the acceleration of B at this instant.

SOLUTION

Since $v_G = \dot{s}_G = t^2$, we have $\ddot{s}_G = 2t = 2(2) = 4$ ft/sec². Therefore

$$\mathbf{a}_G = \ddot{s}_G\hat{\mathbf{e}}_t + \frac{\dot{s}_G^2}{\rho_G}\hat{\mathbf{e}}_n = 2t\hat{\mathbf{e}}_t + \frac{t^4}{8}\hat{\mathbf{e}}_n$$

$$= 4\hat{\mathbf{i}} - 2\hat{\mathbf{j}} \text{ ft/sec}^2 \qquad (\text{at } t = 2 \text{ sec})$$

On body $\mathcal{R}$:

$$\mathbf{a}_B = a_B\left(\frac{8\hat{\mathbf{i}} - 15\hat{\mathbf{j}}}{17}\right) = \overset{(4\hat{\mathbf{i}} - 2\hat{\mathbf{j}})}{\cancel{\mathbf{a}_G}} + \alpha_{\mathcal{R}}\hat{\mathbf{k}} \times \overset{(\sqrt{91}\,\hat{\mathbf{i}} + 3\hat{\mathbf{j}})}{\cancel{\mathbf{r}_{GB}}} - \omega_{\mathcal{R}}^2\mathbf{r}_{GB}$$

in which the acceleration of B is an unknown magnitude in a known direction as signified by the unit vector $(8\hat{\mathbf{i}} - 15\hat{\mathbf{j}})/17$ down the slot.

> **Question 3.15** If the direction of $\mathbf{a}_B$ turns out to be up the slot, will the solution be valid?

We have two equations ($\hat{\mathbf{i}}$ and $\hat{\mathbf{j}}$ coefficients) in the three unknowns a_B, $\alpha_{\mathcal{R}}$, and $\omega_{\mathcal{R}}$, but the angular speed $\omega_{\mathcal{R}}$ may be found from the instantaneous center of $\mathcal{R}$. From the geometry and similar triangles, we get

$$\frac{\sqrt{91}}{H} = \frac{15}{8} \Rightarrow H = 5.09 \text{ ft}$$

$$D = H - 3 = 2.09 \text{ ft}$$

Thus

$$\omega_{\mathcal{R}} = \frac{v_G}{D} = \frac{4}{2.09} = 1.91 \text{ rad/sec}$$

$$\boldsymbol{\omega}_R = 1.91 \circlearrowright \text{ rad/sec}$$

(Continued)

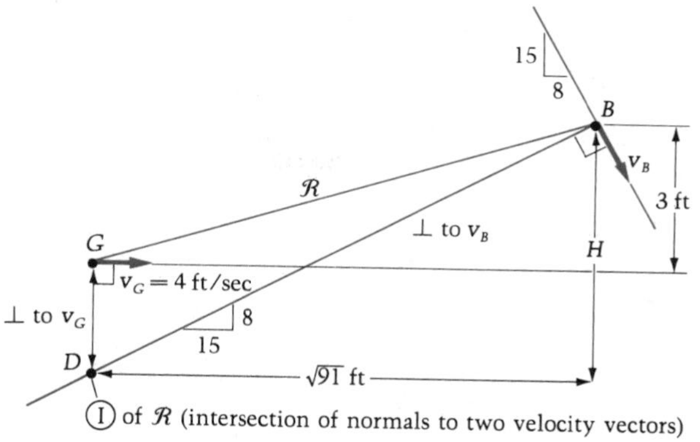

Note that until Ⓘ of $\mathcal{R}$ is located, we do not know whether $\mathbf{v}_B$ is up or down the slot; however, the *normal* to $\mathbf{v}_B$ is the same line in either case. Once Ⓘ is established, $\mathbf{v}_G$ to the right gives $\omega_{\mathcal{R}}$ clockwise—and then we know that $\mathbf{v}_B$ is *down* the slot. Substituting, we get

$$a_B\left(\frac{8\hat{\mathbf{i}} - 15\hat{\mathbf{j}}}{17}\right) = 4\hat{\mathbf{i}} - 2\hat{\mathbf{j}} + \sqrt{91}\,\alpha_{\mathcal{R}}\hat{\mathbf{j}} - 3\alpha_{\mathcal{R}}\hat{\mathbf{i}} - 1.91^2(\sqrt{91}\,\hat{\mathbf{i}} + 3\hat{\mathbf{j}})$$

$\hat{\mathbf{i}}$ coefficients: $\quad \dfrac{8}{17}a_B = 4 - 3\alpha_{\mathcal{R}} - 1.91^2\sqrt{91}$

$\hat{\mathbf{j}}$ coefficients: $\quad \dfrac{-15}{17}a_B = -2 + \sqrt{91}\,\alpha_{\mathcal{R}} - 1.91^2(3)$

Eliminating $\alpha_{\mathcal{R}}$ gives $a_B = -181$, so that

$$\mathbf{a}_B = 181 \;\text{(direction: 15–8 slope, up the slot)} \quad \text{or } (-85.2\hat{\mathbf{i}} + 160\hat{\mathbf{j}})\text{ ft/sec}^2$$

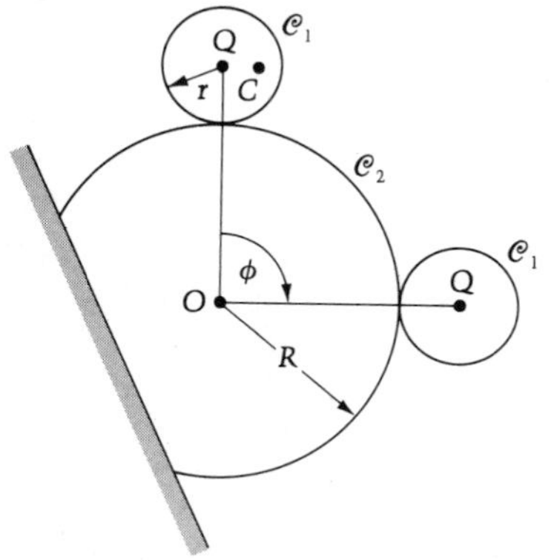

Figure 3.20

The final example in this part of Section 3.6 will be very helpful to us later in situations such as the one shown in Figure 3.20. Suppose we need to know the location of point C of cylinder $\mathcal{C}_1$ after $\mathcal{C}_1$ has rolled on $\mathcal{C}_2$ to the lower position.* The problem is to find the angle θ turned through by $\mathcal{C}_1$ for a given rotation ϕ of the line OQ. The procedure to be followed in solving this problem is illustrated by the next example.

*This need will arise, for example, in kinetics problems in which we seek the work done by gravity on $\mathcal{C}$ if its mass center is offset from its geometric center.

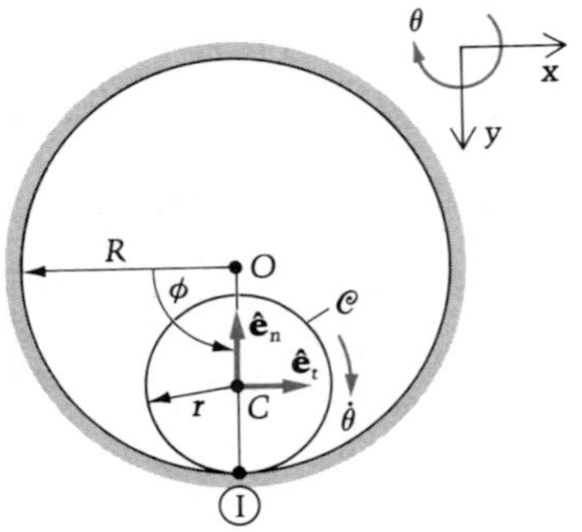

EXAMPLE 3.23

Find the relationship between the angle ϕ (locating the line OC) and the angle of rotation θ of the rolling cylinder $\mathscr{C}$.

SOLUTION

Treating C as a point whose path is a known circle, we get

$$\mathbf{v}_C = \dot{s}\hat{\mathbf{e}}_t = (R - r)\dot{\phi}\hat{\mathbf{i}}$$

Alternatively, we may also treat C as a point on the cylinder with instantaneous center at Ⓘ:

$$\mathbf{v}_C = \omega\hat{\mathbf{k}} \times \mathbf{r}_{\text{Ⓘ}C} = \dot{\theta}\hat{\mathbf{k}} \times (-r\hat{\mathbf{j}}) = r\dot{\theta}\hat{\mathbf{i}}$$

Thus, equating the two expressions for $\mathbf{v}_C$, we have

$$(R - r)\dot{\phi} = r\dot{\theta}$$

Integrating, and letting $\theta = 0$ when $\phi = 0$, we get

$$(R - r)\phi = r\theta + \overset{0}{\cancel{C_1}}$$

or

$$\theta = \left(\frac{R - r}{r}\right)\phi \tag{3.36}$$

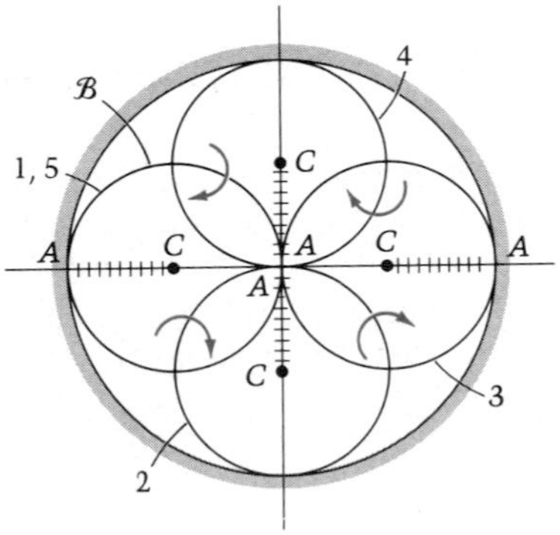

Figure 3.21

If we let $R = 2r$, then we see (Figure 3.21) that $\theta \equiv \phi$. Even though the circumferences of cylinder and track are $2\pi r$ and $4\pi r$ ($= 2\pi R$),* the curvature forces the angular velocities in space of the line OC and the cylinder to be the same. If the outer track were ***straight*** and of length $4\pi r$, the cylinder would turn ***two*** revolutions in space instead of just one in traversing it.

The line AC of $\mathscr{B}$ (and hence $\mathscr{B}$ itself) revolves once as C completes its circular path for the case $R = 2r$. When $R > 2r$, then $\theta > \phi$. This case is shown in Figure 3.22. If we now let $R = 7r$, then Equation (3.36) gives $\theta = 6\phi$ and $\mathscr{B}$ now revolves once in space for each 60° of turn of line OC.

*Which at first glance might lead one to believe that the cylinder would revolve twice per revolution of OC.

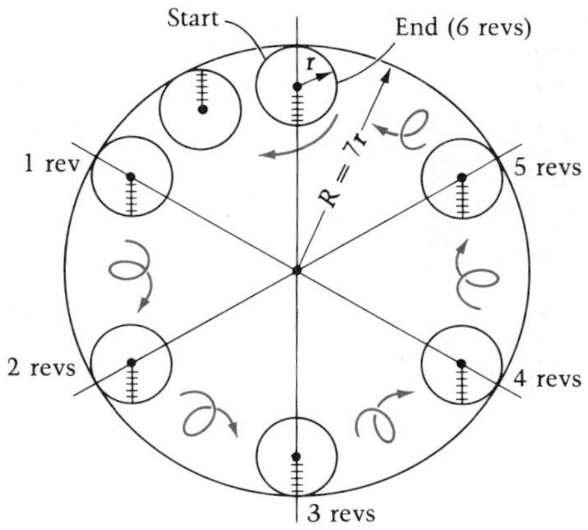

Figure 3.22

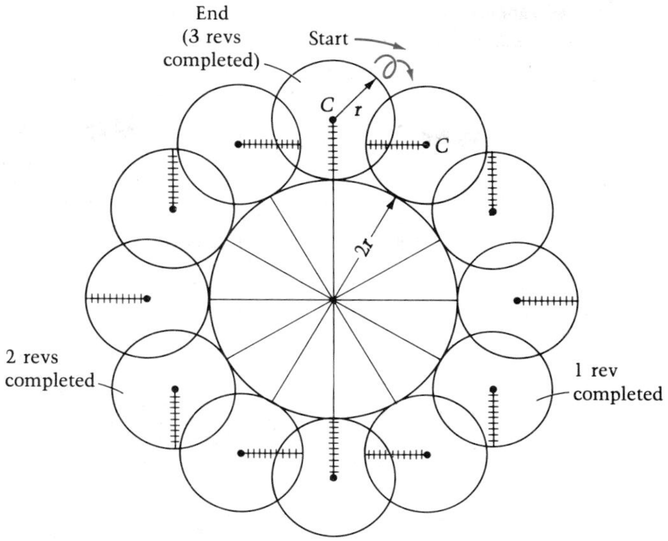

Figure 3.23

If the track were ***convex***, then for $R = 2r$:

$$\mathbf{v}_C = 3r\dot{\phi}\,\hat{\mathbf{e}}_t \qquad \text{and} \qquad \mathbf{v}_C = r\dot{\theta}\,\hat{\mathbf{e}}_t \Rightarrow 3\phi = \theta$$

and the wheel would turn in space ***three times*** as fast and as far as line OC (see Figure 3.23).

Gears

The final class of rolling problems is concerned with gears. Gears are used to transmit power. The teeth of the gears are cut so that they will give constant speed to the driven gear when the driving gear is itself turning at constant angular speed.

However, gears violate the rolling condition; there is necessarily some sliding since the contacting points do not have equal velocities (except at $\theta = 0$), as can be seen in Figure 3.24 for spur gears; nonetheless, the teeth are cut so that we may correctly treat the gears for dynamic purposes as if they were two cylinders rolling on each other at the pitch circles. Thus when the centers are pinned as in Figure 3.24, we may use the relation

$$r_1\omega_{\mathcal{g}} = r_2\omega_{\mathcal{p}} \tag{3.37}$$

where r_1 and r_2 are the respective pitch radii of the gear and the pinion. We may also use the derivative of Equation 3.37 (since it is valid for all t):

$$r_1\alpha_{\mathcal{g}} = r_2\alpha_{\mathcal{p}} \tag{3.38}$$

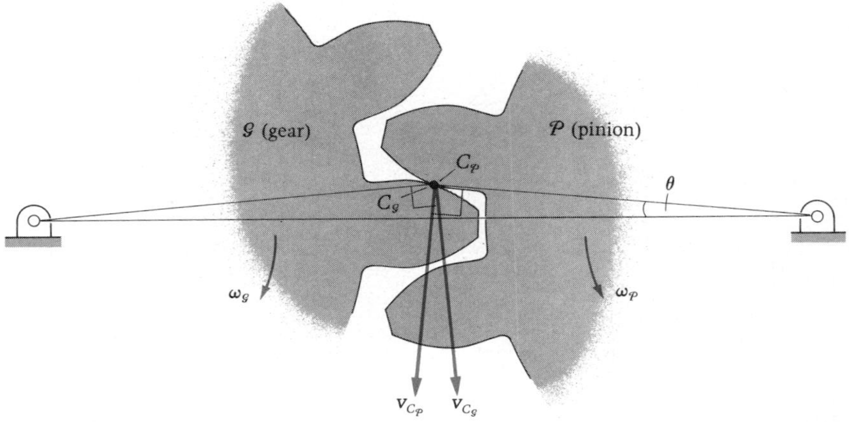

Figure 3.24 Spur gear teeth in contact.

We note that the radius ratio is inversely proportional to the ratio of angular speeds (and directly proportional to the ratio of numbers of gear teeth, since the shape and spacing of teeth must match). We now consider several examples in which there are gears in mesh.

EXAMPLE 3.24

Pin B of block $\mathcal{B}$ (see the diagram) has a velocity vector in the indicated position of 0.25 m/s upward. Rod $\mathcal{A}$ is pinned on its ends to $\mathcal{B}$ and to gear $\mathcal{G}$ as shown. The gear meshes with the rack $\mathcal{R}$, which slides on the floor $\mathcal{F}$ at 0.2 m/s to the right. Determine the angular velocities of $\mathcal{G}$ and $\mathcal{A}$, and find the velocity of point E.

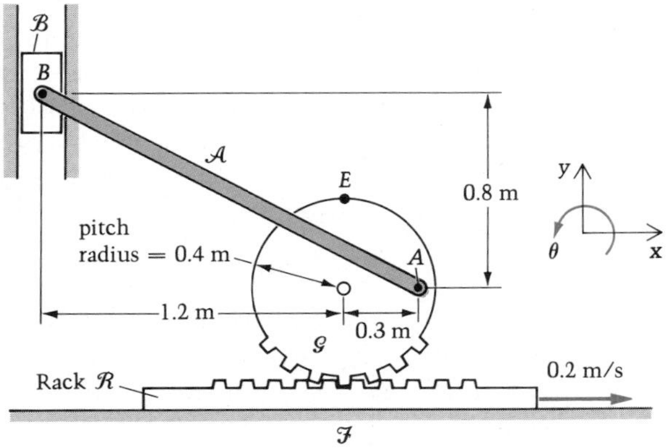

(Continued)

SOLUTION

As described in the foregoing text, we assume that the teeth of the gear and rack that are touching have rolling contact. This does not mean that these teeth have zero velocity (they do not!)—only that they are equal.

On rod $\mathcal{A}$ we relate the velocities of A and B:

$$\mathbf{v}_A = \overset{(v_B\hat{\mathbf{j}} = 0.25\hat{\mathbf{j}})}{\mathbf{v}_B} + \omega_{\mathcal{A}}\hat{\mathbf{k}} \times \overset{(1.5\hat{\mathbf{i}} - 0.8\hat{\mathbf{j}})}{\mathbf{r}_{BA}}$$

$$\mathbf{v}_A = +0.8\omega_{\mathcal{A}}\hat{\mathbf{i}} + (1.5\omega_{\mathcal{A}} + 0.25)\hat{\mathbf{j}} \tag{1}$$

Equation (1), which contains two scalar equations, contains three unknowns (v_{A_x}, v_{A_y}, and $\omega_{\mathcal{A}}$). Therefore we now relate $\mathbf{v}_A$ and $\mathbf{v}_{T_g}$ (where T_g is the tooth contact point of $\mathcal{G}$) on gear $\mathcal{G}$:

$$\mathbf{v}_A = \mathbf{v}_{T_g} + \omega_g\hat{\mathbf{k}} \times \overset{0.3\hat{\mathbf{i}} + 0.4\hat{\mathbf{j}}}{\mathbf{r}_{T_gA}}$$

Next we use the fact that since the teeth T_g and $T_{\mathcal{R}}$ are in rolling contact, their velocities are equal:

$$\mathbf{v}_A = \underset{\mathbf{v}_{T_{\mathcal{R}}} = +0.2\hat{\mathbf{i}}\text{ m/s}}{\mathbf{v}_{T_g}} + \omega_g(-0.4\hat{\mathbf{i}} + 0.3\hat{\mathbf{j}}) \tag{2}$$

Finally, we now equate Expressions (1) and (2) for $\mathbf{v}_A$:

$$0.8\omega_{\mathcal{A}}\hat{\mathbf{i}} + (1.5\omega_{\mathcal{A}} + 0.25)\hat{\mathbf{j}} = (0.2 - 0.4\omega_g)\hat{\mathbf{i}} + 0.3\omega_g\hat{\mathbf{j}}$$

$\hat{\mathbf{i}}$ coefficients: $\quad 0.8\omega_{\mathcal{A}} = 0.2 - 0.4\omega_g$

$\hat{\mathbf{j}}$ coefficients: $\quad 1.5\omega_{\mathcal{A}} + 0.25 = 0.3\omega_g$

The solution to these equations leads to

$$\boldsymbol{\omega}_{\mathcal{A}} = -0.0476\hat{\mathbf{k}}\text{ rad/s}$$

$$\boldsymbol{\omega}_g = 0.595\hat{\mathbf{k}}\text{ rad/s}$$

The velocity of E (the point at the top of $\mathcal{G}$) is, at this time,

$$\mathbf{v}_E = \mathbf{v}_A + \overset{0.595\hat{\mathbf{k}}}{\boldsymbol{\omega}_g} \times \overset{(-0.3\hat{\mathbf{i}} + 0.4\hat{\mathbf{j}})}{\mathbf{r}_{AE}} \tag{3}$$

From Equation (1),

$$\mathbf{v}_A = 0.8\omega_{\mathcal{A}}\hat{\mathbf{i}} + (1.5\omega_{\mathcal{A}} + 0.25)\hat{\mathbf{j}}$$

$$= -0.0381\hat{\mathbf{i}} + 0.179\hat{\mathbf{j}}$$

Substituting into Equation (3), we get

$$\mathbf{v}_E = -0.276\hat{\mathbf{i}}\text{ m/s}$$

Note that $\mathbf{v}_E$ could have been obtained much more easily by relating it to $\mathbf{v}_{T_g}$ on the gear $\mathcal{G}$:

$$\mathbf{v}_E = \mathbf{v}_{T_g} + \omega_g\hat{\mathbf{k}} \times \mathbf{r}_{T_gE} = 0.2\hat{\mathbf{i}} + 0.595\hat{\mathbf{k}} \times (0.8\hat{\mathbf{j}})$$

$$= -0.276\hat{\mathbf{i}}\text{ m/s} \qquad \text{(as above)}$$

EXAMPLE 3.25

Find the angular speed of the front sprocket (rigidly fixed to the pedal crank) in the accompanying diagram if the man is traveling at 10 mph. There are 26 teeth on the front sprocket and 9 on the rear sprocket (which turns rigidly with the rear wheel). The wheel diameters are 26 in.

SOLUTION

The velocities of A and B, the two ends of the straight upper length of chain, are equal. (As shown in the diagram, A is just leaving the rear sprocket $\mathcal{R}_1$; B is just about to enter the front sprocket $\mathcal{R}_2$.) To prove this, we note that the translating section AB of chain is behaving as if rigid, so that, calling this "body" $\mathcal{L}$,

$$\mathbf{v}_B = \mathbf{v}_A + \boldsymbol{\omega}_{\mathcal{L}} \times \mathbf{r}_{AB}$$

But $\boldsymbol{\omega}_{\mathcal{L}} = \mathbf{0}$, so that

$$\mathbf{v}_B = \mathbf{v}_A$$

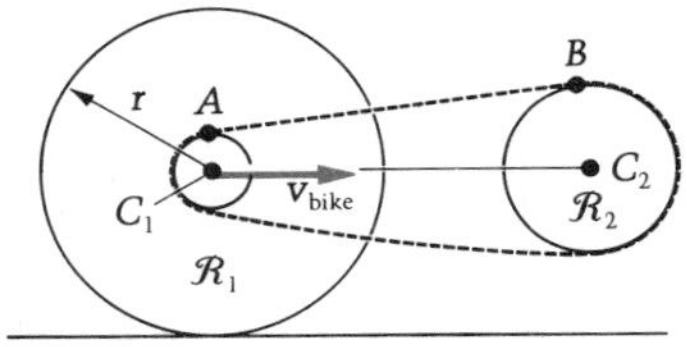

Next we relate the equal velocities of A and B to the respective centers of their sprockets $\mathcal{R}_1$ and $\mathcal{R}_2$:

$$\mathbf{v}_{C_1} + \boldsymbol{\omega}_{\mathcal{R}_1} \times \mathbf{r}_{C_1A} = \mathbf{v}_{C_2} + \boldsymbol{\omega}_{\mathcal{R}_2} \times \mathbf{r}_{C_2B}$$

Now the velocities of C_1 and C_2 are each equal to the "velocity of the bike," meaning the common velocity of all the points on the translating part of the bike, such as points of the frame and seat. Therefore $\mathbf{v}_{C_1}$ and $\mathbf{v}_{C_2}$ cancel, leaving

$$\boldsymbol{\omega}_{\mathcal{R}_1} \times \mathbf{r}_{C_1A} = \boldsymbol{\omega}_{\mathcal{R}_2} \times \mathbf{r}_{C_2B}$$

This says simply that (see the diagram):

$$r_1\omega_{\mathcal{R}_1} = r_2\omega_{\mathcal{R}_2}$$

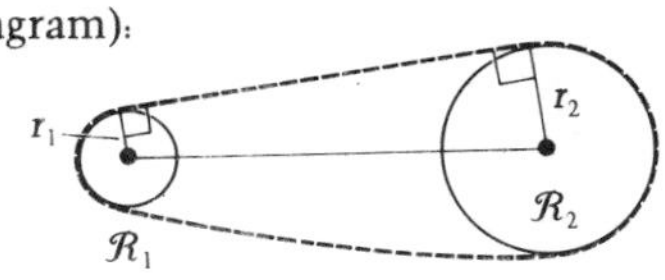

Now if the speed of the bike is to be 10 mph, we have

$$v_{C_1} = \overset{13\text{ in.}}{\cancel{r}}\omega_{\mathcal{R}_1} = 10\text{ mph}\left(\frac{88\text{ ft/sec}}{60\text{ mph}}\right)\frac{12\text{ in.}}{1\text{ ft}} = 176\text{ in./sec}$$

or

$$\omega_{\mathcal{R}_1} = 13.5 \circlearrowright \text{ rad/sec}$$

(Continued)

Thus

$$\omega_{\mathcal{R}_2} = \frac{r_1}{r_2}\,\omega_{\mathcal{R}_1} = \frac{9}{26}(13.5) \qquad \text{(Radii are proportional to number of teeth!)}$$

$$= 4.67 \text{ rad/sec}$$

So the rider must turn the pedal crank at $4.67/2\pi = 0.743$ rev/sec.

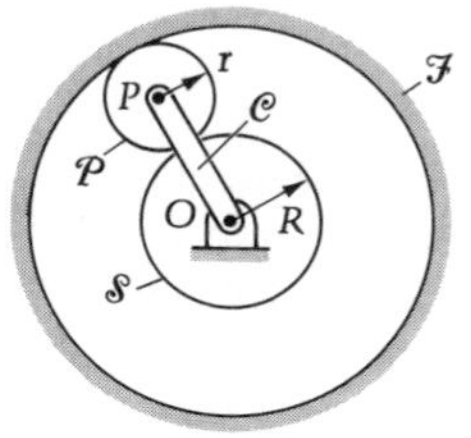

EXAMPLE 3.26

Frame $\mathcal{F}$ is a ring gear with internal teeth that mesh with those of the planetary gear $\mathcal{P}$. (See the diagram.) The teeth of $\mathcal{P}$ also mesh with those of the sun gear $\mathcal{S}$, which is pinned at its center point O to frame $\mathcal{F}$. The crank arm $\mathcal{C}$ is pinned at its ends to O and to the center point P of $\mathcal{P}$. The arm $\mathcal{C}$ has angular speed $\omega_{\mathcal{C}}(t)$ counterclockwise. Find the angular velocity of $\mathcal{S}$ in terms of R, r, and $\omega_{\mathcal{C}}$.

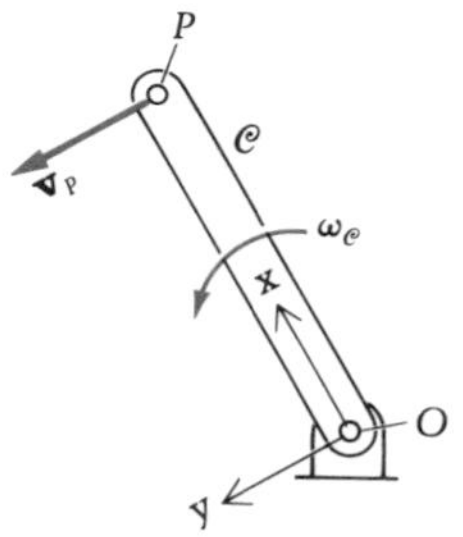

SOLUTION

We take $\mathcal{F}$ to be our reference frame, to which all motions are referred. We work first with the crank $\mathcal{C}$, since we know its angular velocity and the velocity of one of its points ($\mathbf{v}_O = \mathbf{0}$). From the following sketch of $\mathcal{C}$, we see that we can write

$$\mathbf{v}_P = \cancelto{\mathbf{0}}{\mathbf{v}_O} + \omega_{\mathcal{C}}\hat{\mathbf{k}} \times (R + r)\hat{\mathbf{i}}$$
$$= (R + r)\omega_{\mathcal{C}}\hat{\mathbf{j}} \tag{1}$$

(Note that we align **x** along OP for convenience; we need not always draw it to the right.)

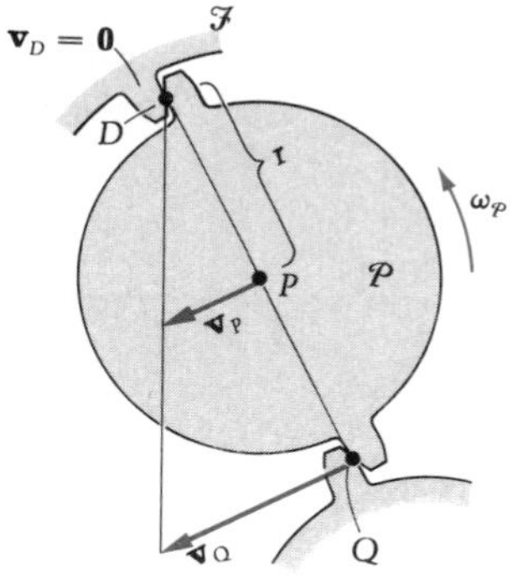

Next the points of $\mathcal{P}$ and $\mathcal{C}$ that are pinned together at P have the same velocity at all times. Furthermore, the points of $\mathcal{F}$ and $\mathcal{P}$ at D (see the following figure) are in contact and each has zero velocity* since D is fixed in $\mathcal{F}$. Thus

$$\mathbf{v}_P = \cancelto{\mathbf{0}}{\mathbf{v}_D} + \omega_{\mathcal{P}}\hat{\mathbf{k}} \times (-r\hat{\mathbf{i}})$$

or

$$(R + r)\omega_{\mathcal{C}}\hat{\mathbf{j}} = -r\omega_{\mathcal{P}}\hat{\mathbf{j}}$$

(Continued)

*As we have pointed out, the contact points of gear teeth necessarily have a very small radial velocity component. The points used in the analysis are actually not tooth points, however, but imaginary points on the pitch circles of the gears. Furthermore, the radii given in the examples and problems are the radii of these circles.

Solving for $\omega_{\mathcal{P}}$ gives

$$\omega_{\mathcal{P}} = \frac{-(R + r)\omega_{\mathcal{C}}}{r} \qquad \text{or} \qquad \boldsymbol{\omega}_{\mathcal{P}} = \frac{R + r}{r}\omega_{\mathcal{C}} \circlearrowright \tag{2}$$

We are now in a position to obtain the velocity of point Q of $\mathcal{P}$, the point in contact with the tooth of $\mathcal{S}$:

$$\mathbf{v}_Q = \cancelto{\mathbf{0}}{\mathbf{v}_D} + \omega_{\mathcal{P}}\hat{\mathbf{k}} \times (-2r\hat{\mathbf{i}})$$

Substituting for $\omega_{\mathcal{P}}$ in terms of $\omega_{\mathcal{C}}$ from Equation (2), we get

$$\mathbf{v}_Q = 2(R + r)\omega_{\mathcal{C}}\hat{\mathbf{j}} \tag{3}$$

Note that Q has twice the speed of P since it is twice as far from the instantaneous center D of $\mathcal{P}$ as is P.

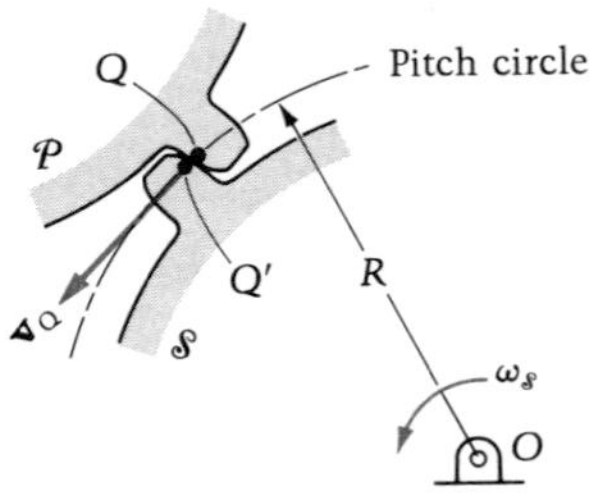

Finally, we come to the body $\mathcal{S}$ of interest (see the following diagram). Knowing that the points Q and Q′ (the tooth points in contact on $\mathcal{P}$ and $\mathcal{S}$, respectively) have equal velocities as they move together tangent to the pitch circle, we obtain

$$\mathbf{v}_{Q'} = \mathbf{v}_O + \omega_{\mathcal{S}}\hat{\mathbf{k}} \times R\hat{\mathbf{i}}$$

$$2(R + r)\omega_{\mathcal{C}}\hat{\mathbf{j}} = \mathbf{0} + R\omega_{\mathcal{S}}\hat{\mathbf{j}}$$

Thus the angular speed of $\mathcal{S}$ in $\mathcal{I}$ is

$$\omega_{\mathcal{S}} = \frac{2(R + r)\omega_{\mathcal{C}}}{R} \tag{4}$$

and the angular velocity of $\mathcal{S}$ in $\mathcal{I}$ is

$$\omega_{\mathcal{S}}\hat{\mathbf{k}} = \frac{2(R + r)\omega_{\mathcal{C}}}{R}\hat{\mathbf{k}}$$

Note that the point of $\mathcal{C}$ passing over point Q has velocity $R\omega_{\mathcal{C}}$, which is of course less than the velocity $\mathbf{v}_Q$ of the gear teeth in contact at Q. This velocity is $R\omega_{\mathcal{S}} = 2(R + r)\omega_{\mathcal{C}}$, which is more than twice as fast as $R\omega_{\mathcal{C}}$.

We remark that since the answers

$$\omega_{\mathcal{P}}(t) = \frac{R + r}{r}\omega_{\mathcal{C}}(t) \qquad \text{and} \qquad \omega_{\mathcal{S}}(t) = \frac{2(R + r)}{R}\omega_{\mathcal{C}}(t)$$

are completely general functions of time, the angular accelerations are obtainable immediately by differentiation, with $\alpha_{\mathcal{C}} = \dot{\omega}_{\mathcal{C}}$:

$$\alpha_{\mathcal{P}}(t) = \frac{R + r}{r}\alpha_{\mathcal{C}}(t) \qquad \text{and} \qquad \alpha_{\mathcal{S}}(t) = \frac{2(R + r)}{R}\alpha_{\mathcal{C}}(t)$$

Rather than differentiating, however, we shall obtain these two results in the following example in another manner: by repeated use of Equation (3.19). The purpose is to gain insight into its use in gearing situations involving several bodies. The procedure in the next example would have to be followed if the previous example had been worked using instantaneous values instead of generally (with symbols).

EXAMPLE 3.27

Find the angular accelerations $\alpha_{\mathcal{P}}$ and $\alpha_{\mathcal{S}}$ of the planetary and sun gears in the previous example in terms of R, r, and $\omega_{\mathcal{C}}$ and $\alpha_{\mathcal{C}}$, which are given functions of the time t.

SOLUTION

Relating the accelerations of P and O on body $\mathcal{C}$ gives (see the lower diagram)

$$\mathbf{a}_P = -(R + r)\omega_{\mathcal{C}}^2\hat{\mathbf{i}} + (R + r)\,\alpha_{\mathcal{C}}\hat{\mathbf{j}}$$

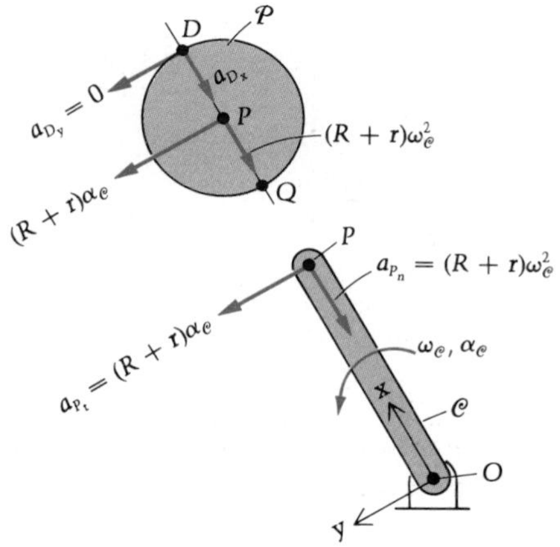

This acceleration is then carried over to the coincident point P on $\mathcal{P}$ (upper diagram). Relating D and P on the planetary gear $\mathcal{P}$, we have

$$\mathbf{a}_D = a_D\hat{\mathbf{i}} = \mathbf{a}_P + \alpha_{\mathcal{P}}\hat{\mathbf{k}} \times \overset{r\hat{\mathbf{i}}}{\mathbf{r}_{PD}} - \omega_{\mathcal{P}}^2 \overset{r\hat{\mathbf{i}}}{\mathbf{r}_{PD}}$$

$$= -(R + r)\omega_{\mathcal{C}}^2\hat{\mathbf{i}} + (R + r)\alpha_{\mathcal{C}}\hat{\mathbf{j}} + \alpha_{\mathcal{P}}\hat{\mathbf{k}} \times r\hat{\mathbf{i}} - \omega_{\mathcal{P}}^2 r\hat{\mathbf{i}}$$

Recalling that

$$\omega_{\mathcal{P}} = \frac{R + r}{r}\,\omega_{\mathcal{C}}$$

(Continued)

the coefficients of $\hat{\mathbf{i}}$ give

$$a_D = \frac{-(R + 2r)(R + r)}{r}\,\omega_{\mathcal{C}}^2$$

and the $\hat{\mathbf{j}}$ coefficients yield

$$\alpha_{\mathcal{P}} = \frac{-(R + r)\alpha_{\mathcal{C}}}{r} \qquad \text{or} \qquad \boldsymbol{\alpha}_{\mathcal{P}} = \frac{R + r}{r}\,\alpha_{\mathcal{C}} \curvearrowright$$

We now need $\mathbf{a}_Q$, where Q is again the tooth point in contact with the sun gear $\mathcal{S}$:

$$\mathbf{a}_Q = \mathbf{a}_D + \alpha_{\mathcal{P}}\hat{\mathbf{k}} \times \overset{-2r\hat{\mathbf{i}}}{\cancel{\mathbf{r}_{DQ}}} - \omega_{\mathcal{P}}^2\,\mathbf{r}_{DQ}$$

$$= \frac{-(R + 2r)(R + r)}{r}\,\omega_{\mathcal{C}}^2\,\hat{\mathbf{i}} + 2(R + r)\alpha_{\mathcal{C}}\,\hat{\mathbf{j}} + \frac{2r(R + r)^2}{r^2}\,\omega_{\mathcal{C}}^2\,\hat{\mathbf{i}}$$

$$= \frac{(R + r)R}{r}\,\omega_{\mathcal{C}}^2\,\hat{\mathbf{i}} + 2(R + r)\alpha_{\mathcal{C}}\,\hat{\mathbf{j}}$$

We now go to body $\mathcal{S}$ to complete the solution. Relating the tooth point Q' to O on $\mathcal{S}$ gives the components of $\mathbf{a}_{Q'}$ (see the following diagram). The tangential acceleration components of Q and Q' are equal* as the teeth contact and move together:

$$a_{Q'_t} = a_{Q'_y} = 2(R + r)\alpha_{\mathcal{C}} = R\alpha_{\mathcal{S}}$$

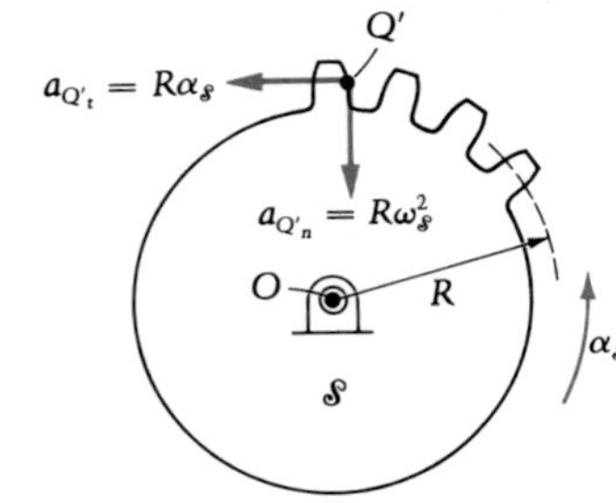

Thus

$$\boldsymbol{\alpha}_{\mathcal{S}} = \frac{2(R + r)}{R}\,\alpha_{\mathcal{C}} \circlearrowleft$$

Note that we may express the normal acceleration component of Q' in terms of $\omega_{\mathcal{C}}$ by using the result for $\omega_{\mathcal{S}}$ from the preceding problem:

$$a_{Q'_n} = -a_{Q'_x} = +R\omega_{\mathcal{S}}^2 = +R\left[\frac{2(R + r)}{R}\,\omega_{\mathcal{C}}\right]^2$$

$$= \frac{4(R + r)^2\,\omega_{\mathcal{C}}^2}{R}$$

(Continued)

*This is in fact true even when ***neither*** body's center is fixed and the geometry is irregular. As long as there is rolling, the acceleration components of the contacting points ***in the plane tangent to the two bodies*** are equal in plane motion at all times. See "Contact Point Accelerations in Rolling Problems," D. J. McGill, *Mechanics Research Communications*, **7**(3), 175–179, 1980.

Final notes: We could have alternatively obtained the accelerations of D and Q as points on rims of wheels rolling on curved tracks by using Formulas (3.31) and (3.32). Noting that ρ for P is $(R + r)$, we present these partial checks on our solution:

$$\mathbf{a}_D = \left(1 + \frac{r}{\rho}\right) r\dot{\theta}^2 \hat{\mathbf{e}}_n = \left(1 + \frac{r}{R + r}\right) r\omega_{\mathcal{P}}^2(-\hat{\mathbf{i}})$$

$$= -\left(\frac{R + 2r}{R + r}\right) r\omega_{\mathcal{P}}^2 \hat{\mathbf{i}} = -\left(\frac{R + 2r}{R + r}\right) r \left(\frac{R + r}{r}\omega_{\mathcal{C}}\right)^2 \hat{\mathbf{i}}$$

$$= \frac{-(R + 2r)(R + r)}{r}\omega_{\mathcal{C}}^2 \hat{\mathbf{i}} \qquad \text{(as above)}$$

$$\mathbf{a}_Q = 2r\ddot{\theta}\hat{\mathbf{e}}_t - r\dot{\theta}^2\left(1 - \frac{r}{\rho}\right)\hat{\mathbf{e}}_n$$

$$= 2r(-\alpha_{\mathcal{P}})(+\hat{\mathbf{j}}) - r\omega_{\mathcal{P}}^2\left(1 - \frac{r}{R + r}\right)(-\hat{\mathbf{i}})$$

$$= \frac{Rr}{R + r}\omega_{\mathcal{P}}^2 \hat{\mathbf{i}} - 2r\alpha_{\mathcal{P}}\hat{\mathbf{j}}$$

$$= \frac{Rr}{R + r}\left(\frac{R + r}{r}\omega_{\mathcal{C}}\right)^2 \hat{\mathbf{i}} - 2r\left[\frac{-(R + r)}{r}\alpha_{\mathcal{C}}\right]\hat{\mathbf{j}}$$

$$= \frac{(R + r)R}{r}\omega_{\mathcal{C}}^2 \hat{\mathbf{i}} + 2(R + r)\alpha_{\mathcal{C}}\hat{\mathbf{j}} \qquad \text{(as above)}$$

EXAMPLE 3.28

The accompanying diagram shows a dynamics application in the paper industry. A steam turbine, through a gearbox, drives the line shaft at 200 rpm. A major problem is that as the pulp and water dries into paper, it shrinks (a typical amount is 4 percent). Thus it must be possible to adjust the speed of the dryer

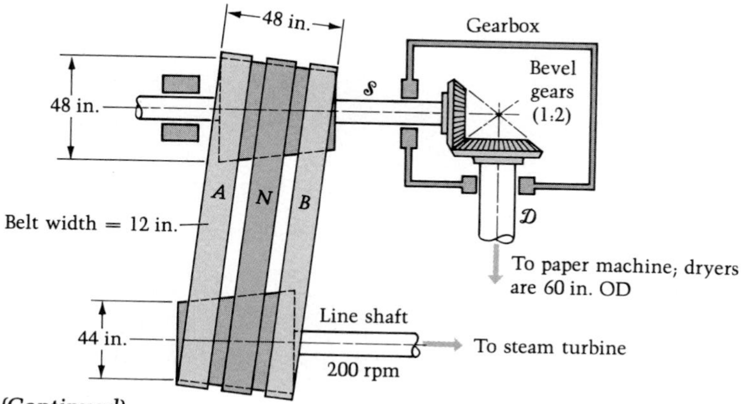

(Continued)

cylinders to prevent tearing. This is done by the two identical cone pulleys shown in the figure. Note that if the belt rides over the pulleys in the center position N (shaded), the shaft $\mathcal{S}$ turns at the same speed as the line shaft. In position A, however, the speed of $\mathcal{S}$ is less than 200 rpm, while in B it is greater than 200 rpm. In all cases, however, there is a rolling condition between the cone pulleys and the belt.

a. Calculate the nominal speed of the paper, which runs over dryer cylinders turning at the speed of the shaft $\mathcal{D}$ coming out of the gearbox $\mathcal{G}$. (In this calculation, the belt is in the center position.)

b. Compute the speed range made possible by altering the position of the belt on the cone pulleys.

SOLUTION

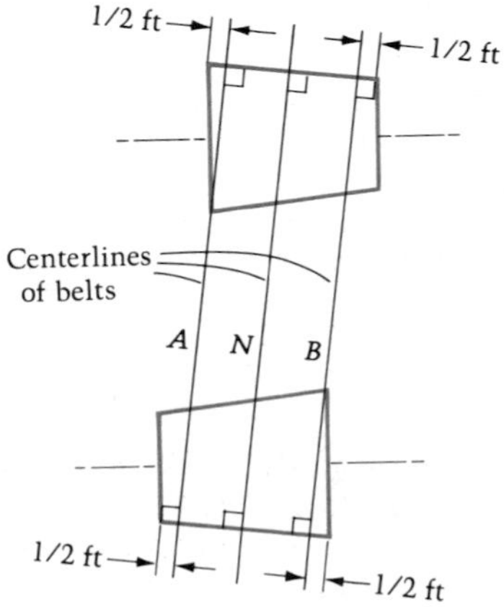

a. In the center position (see the diagram), the cone pulleys do not alter the speed (the ratio is 1:1), so the speed into the gearbox is 200 rpm. The bevel gears then give an output speed of shaft $\mathcal{D}$ of 100 rpm, or $100(2\pi) =$ 628 rad/min. Thus the nominal speed of the paper on the dryer cylinders is

$$v = r\omega = \frac{30}{12} \times 628 = 1570 \text{ ft/min}$$

b. The maximum speed decrease possible occurs when the pulley is in position A. The diameters change by 4 in. in 4 ft, or 1 in./ft. Thus if the 12-in.-wide belt is as far to the left as possible, its centerline is $\frac{1}{2}$ ft from the left edges of the pulleys and the diameters there have each changed by $\frac{1}{2}$ in. With the belt at A the paper speed is therefore

$$v = \underbrace{\underbrace{\underbrace{2\pi \times 200}_{\substack{\text{line shaft}\\\text{speed}\\\text{(rad/min)}}} \times \underbrace{\frac{44+\frac{1}{2}}{48-\frac{1}{2}}}_{\substack{\text{reduction}\\\text{through}\\\text{cone pulleys}}} \times \underbrace{\frac{1}{2}}_{\substack{\text{reduction}\\\text{through}\\\text{gearbox}}}}_{\omega\text{ of dryer cylinder}} \times \underbrace{\frac{30}{12}}_{\substack{\text{dryer cylinder}\\\text{radius (ft)}}}}_{\text{velocity of paper}}$$

$$v = 1470 \text{ ft/min}$$

The decrease in speed from nominal is

$$\left(\frac{1570 - 1470}{1570}\right)100 = 6.37\%$$

(Continued)

The speed *increase* possible, when the belt is in position B, is

$$v = 2\pi \times 200 \times \frac{48 - \frac{1}{2}}{44 + \frac{1}{2}} \times \frac{1}{2} \times \frac{30}{12}$$

$$= 1680 \text{ ft/min} \qquad \text{(a 7.01\% increase from nominal speed)}$$

The total percentage of increase in speed in going from belt position A to B is

$$\left(\frac{1680 - 1470}{1470}\right) 100 = 14.3\%$$

with a speed range of $1680 - 1470 = 210$ ft/min.

PROBLEMS / Section 3.6

3.85 The wheel in Figure P3.85 rolls on the plane with constant angular velocity 1 ↷ rad/sec. Find the velocity of point Q by using the instantaneous center Ⓘ of zero velocity of $\mathcal{C}$. Then check by using Equation (3.8) to relate $\mathbf{v}_Q$ to $\mathbf{v}_P$.

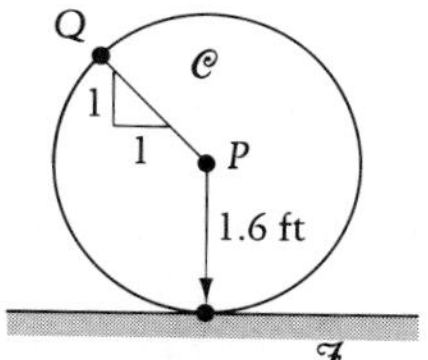

Figure P3.85

3.86 In the preceding problem, suppose that the plane $\mathcal{F}$ on which $\mathcal{C}$ rolls is not fixed to the reference frame but instead translates on it (this time the reference frame is $\mathcal{G}$) at constant velocity 3 ft/sec to the left. (See Figure P3.86.)

a. Find the instantaneous center of zero velocity Ⓘ for $\mathcal{C}$.

b. Find $\mathbf{v}_Q$ again.

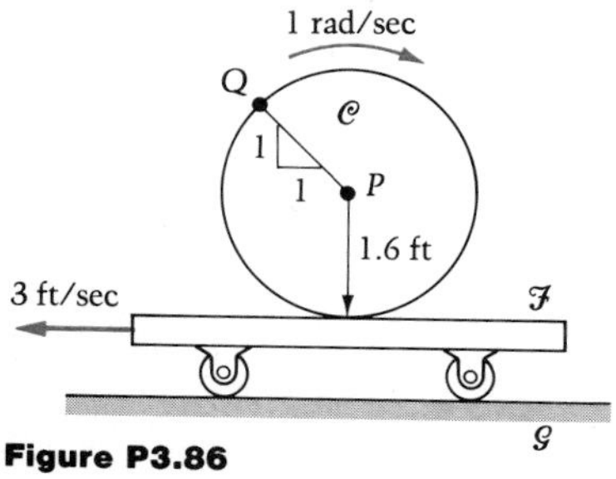

Figure P3.86

3.87 An inextensible string is wrapped around the cylinder $\mathcal{C}$ in Figure P3.87, fitting in a small slot near the rim. The center C of $\mathcal{C}$ is moving down the plane at a constant speed of 0.1 m/s. Find the velocities of points A, B, D, and E. *Hint:* The cylinder is not rolling on the plane, but it *is* rolling on ___?___.

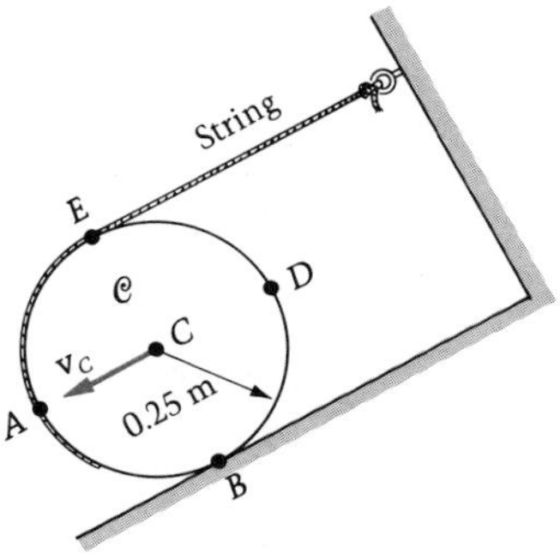

Figure P3.87

3.88 Body $\mathcal{B}$ in Figure P3.88 is one-quarter of a ring and is pinned at A near the periphery of the rolling cylinder $\mathcal{C}$. Its other end slides on the plane $\mathcal{F}$. Forces are applied to $\mathcal{C}$ that give it a constant angular velocity $\boldsymbol{\omega}_{\mathcal{C}} = 0.4$ ↶ rad/s. Find the velocity of the halfway point Q of $\mathcal{B}$ when A is:

a. In the position shown

b. 90° ahead of the position shown

c. 90° behind the position shown

d. At the bottom of $\mathcal{C}$

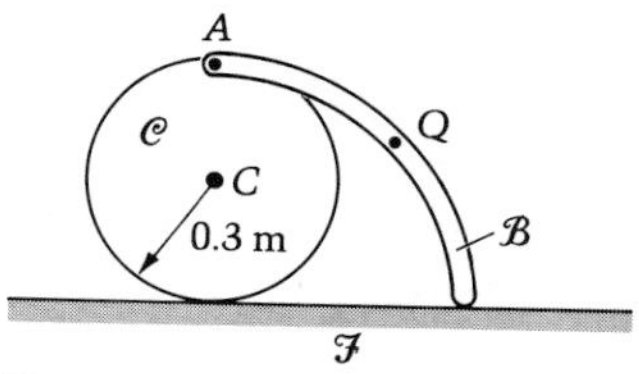

Figure P3.88

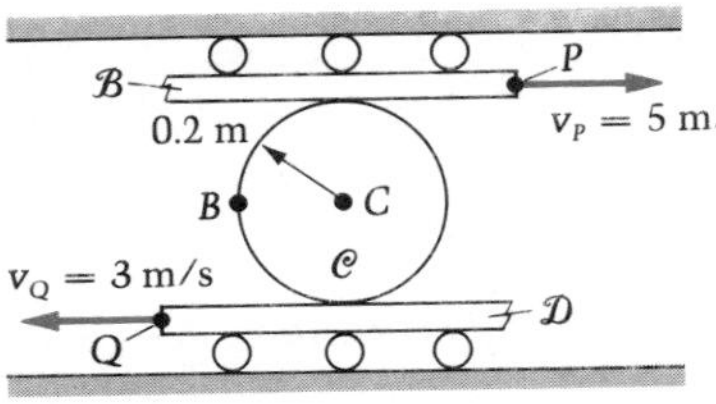

Figure P3.89

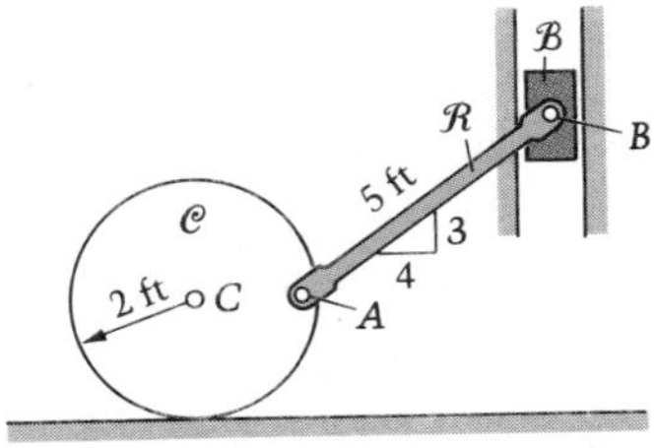

Figure P3.90

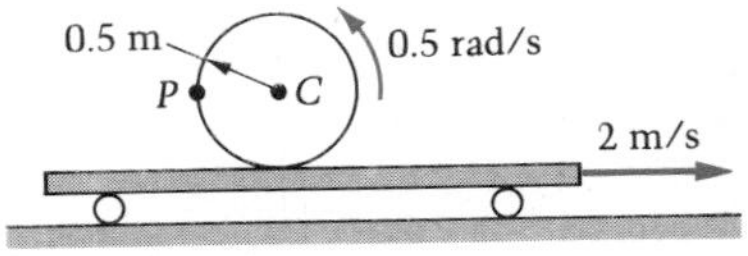

Figure P3.91

3.89 Find the velocities of points B and C in Figure P3.89 if the cylinder $\mathcal{C}$ does not slip on the translating bodies $\mathcal{B}$ and $\mathcal{D}$.

3.90 At the instant shown in Figure P3.90, point B of $\mathcal{B}$ (to which rod $\mathcal{R}$ is pinned) has $\mathbf{v}_B = 0.5 \downarrow$ ft/sec. Find the angular velocity of the rolling cylinder.

3.91 The wheel in Figure P3.91 rolls on the bar. If at a certain instant the bar has a velocity of 2 m/s to the right and the wheel has counterclockwise angular velocity of 0.5 rad/s, determine the velocity of (a) the center of the wheel and (b) point P.

3.92 Wheel $\mathcal{C}$ rolls on both $\mathcal{A}$ and $\mathcal{B}$. (See Figure P3.92.) The constant angular velocity of $\mathcal{C}$ in frame $\mathcal{F}$ is shown in the figure. Find:

a. The velocity of the points of $\mathcal{A}$ relative to $\mathcal{B}$

b. The constant velocity of C in $\mathcal{F}$ for which the velocities of T (on $\mathcal{A}$) and B (on $\mathcal{B}$) in $\mathcal{F}$ are equal in magnitude and opposite in direction.

3.93 Cylinder $\mathcal{C}$ in Figure P3.93 is rolling to the left with constant center velocity v_C. Stick $\mathcal{S}$ is pinned to $\mathcal{C}$ at B, and its other end A slides on the plane. Find the velocity of A when $\theta = 0°$, $90°$, $180°$, and $270°$.

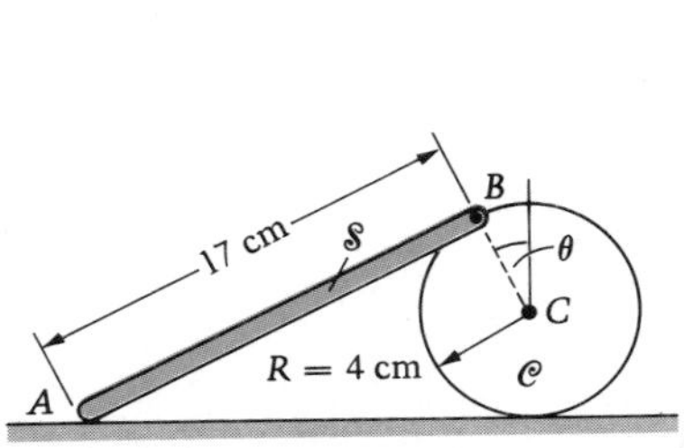

Figure P3.93

3.94 In the preceding problem, find the velocity of A at the time when the velocity of B is along $\mathcal{S}$.

3.95 Two men, a tall one and a short one, travel up identical inclines, pulling identical spools by means of ropes wrapped around the hubs. (See Figure P3.95.) The men travel at the same constant speed v_0, and the ropes are wrapped in the opposite directions indicated. If the spools do not slip on the plane, one of the men will be run over by his own spool. Prove which one it is, and show how long it will take, from the instant depicted, for the spool to make contact with him.

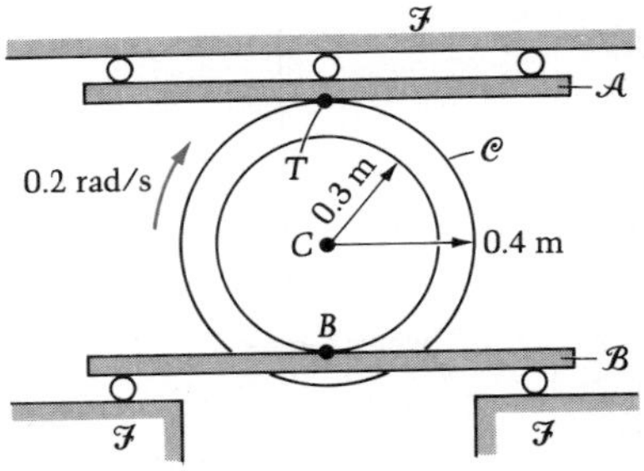

Figure P3.92

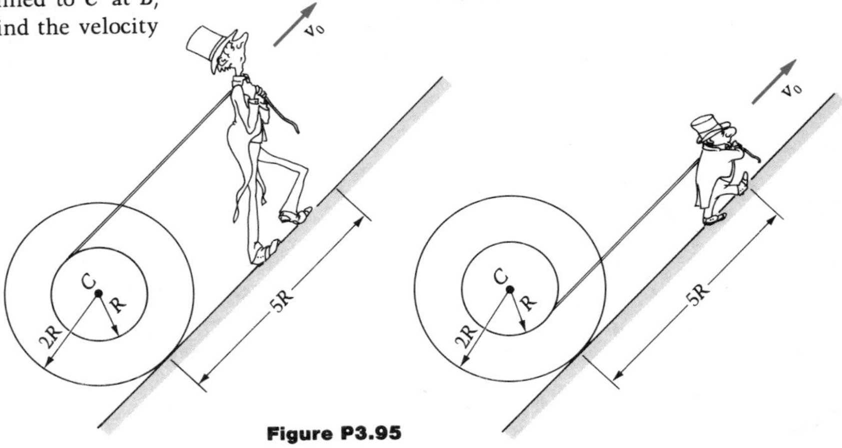

Figure P3.95

3.96 The constant angular velocities of the ring gear $\mathcal{R}$ and the spider arm $\mathcal{S}$ shown in Figure P3.96 are $2 \circlearrowleft$ rad/s and $10 \circlearrowright$ rad/s, respectively. Determine the angular velocity of gear $\mathcal{A}$ and the velocity of the point of $\mathcal{B}$ having maximum speed in the given position. The centers of $\mathcal{A}$ and $\mathcal{S}$ are pinned to the reference frame $\mathcal{I}$.

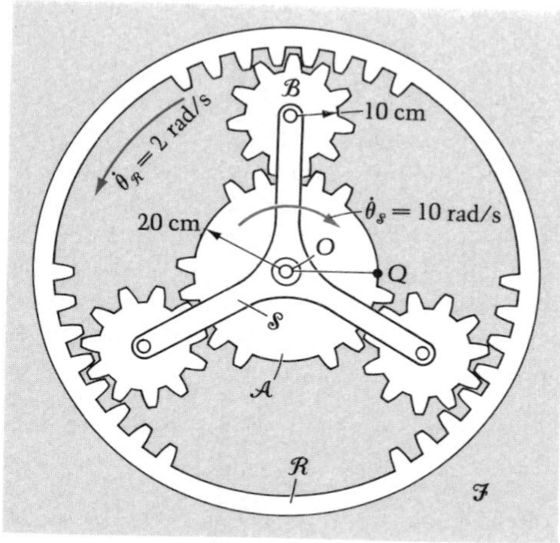

Figure P3.96

3.97 In the preceding problem let the ring gear $\mathcal{R}$ be turning at the constant counterclockwise rate of 10 rpm. Find the angular velocities of gear $\mathcal{B}$ and spider arm $\mathcal{S}$ that will make $\mathbf{v}_Q = \mathbf{0}$.

3.98 A cylinder $\mathcal{B}$ of radius r rolls without slipping over a circular arc of constant radius of curvature R (see Figure P3.98). What is the ratio of the angular speed of the cylinder to $\dot{\phi}$?

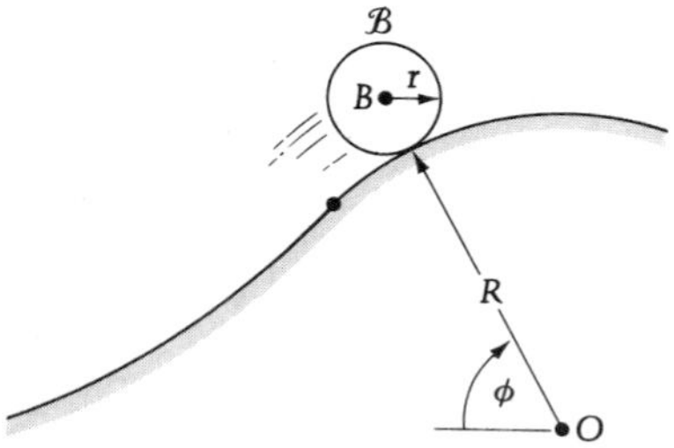

Figure P3.98

3.99 Figure P3.99a shows the manner in which a train wheel rests on the track. If the train travels at a constant speed of 80 mph and does not slip on the track, determine the velocities of points A, B, D, and E on the vertical line through the center C in Figure P3.99b. Which point is traveling backward? Why?

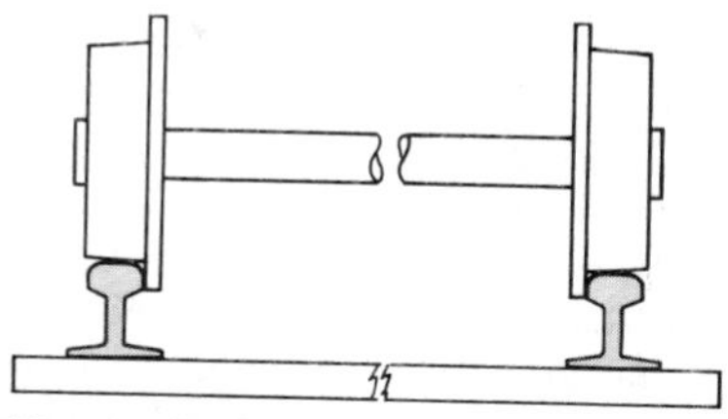

Figure P3.99a

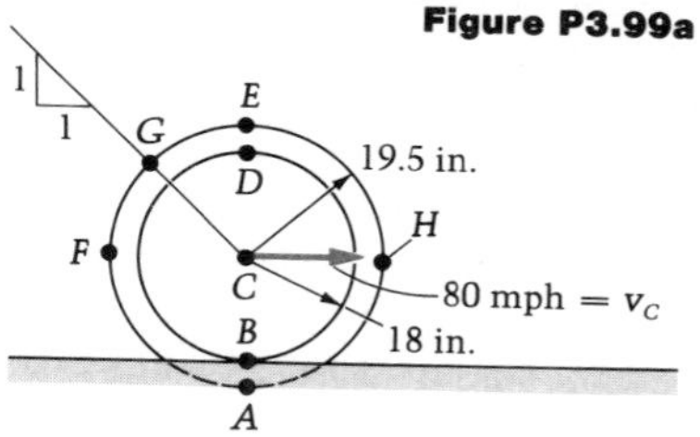

Figure P3.99b

3.100 In the preceding problem find the velocities of points F, G, and H.

3.101 Wheel $\mathcal{W}$ rolls up a parabolic incline having equation $4y = x^2$. (See Figure P3.101.) A bar $\mathcal{B}$ is pinned at C to the center of $\mathcal{W}$; the other end of $\mathcal{B}$ slides along the incline. The wheel has constant angular velocity $\boldsymbol{\omega}_{\mathcal{W}} = 3 \circlearrowright$ rad/s. At the instant shown, the contact point of $\mathcal{W}$ is at $x = 3$ m and A is at the origin. Determine the angular velocity of $\mathcal{B}$ and find the velocity of A at this time.

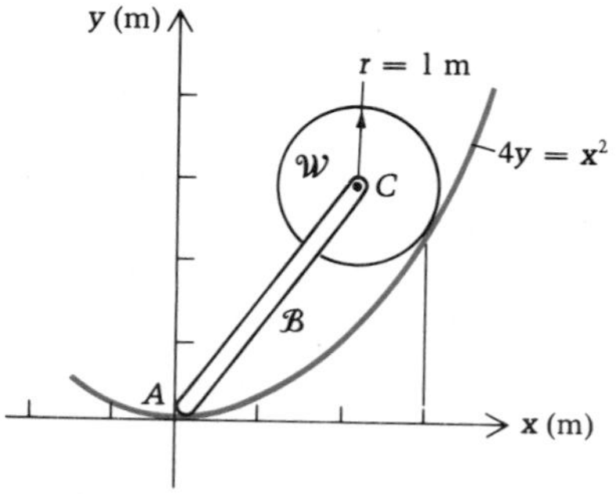

Figure P3.101

3.102 Rack $\mathcal{R}$ in Figure P3.102 moves to the right at a constant speed of 0.2 m/s. Bar $\mathcal{B}$ has length $\ell = 0.3$ m and is pinned at point D to the gear $\mathcal{G}$, 0.1 m from its center C. The other end of $\mathcal{B}$ is pinned at A to block $\mathcal{A}$, which slides in a horizontal slot. The hub (inner radius) of gear $\mathcal{G}$ rolls on plane $\mathcal{P}$.

a. Find $\dot{\theta}_{\mathcal{B}}$ in the given position and also when D is to the left, directly above, and to the right of C.

b. Find the velocity of the midpoint B of $\mathcal{B}$ when A is as far to the right of C as it goes.

3.103 Arm $\mathcal{A}_1$ in Figure P3.103a is being driven so that it turns at constant angular velocity ω_1 ↻ in the reference frame (ground) $\mathcal{F}$. Two bodies are attached to form a single rigid body $\mathcal{R}$: arm $\mathcal{A}_2$ and sprocket $\mathcal{S}_2$. Body $\mathcal{R}$ is then pinned to $\mathcal{A}_1$ at P. A chain is to ride over the sprockets $\mathcal{S}_2$ and $\mathcal{S}_1$ (which is ***fixed in*** $\mathcal{F}$) and force arm $\mathcal{A}_2$ to have an opposite angular velocity in $\mathcal{F}$ of ω_1 ↺. Find the required ratio r_2/r_1 of the radii of $\mathcal{S}_2$ and $\mathcal{S}_1$. *Hint:* Relate the velocities of P and Q in Figure P3.103b, and use the components normal to line PQ to find $\omega_{\mathcal{S}_2}$.

3.104 Extend the preceding problem as follows. (See Figure P3.104.) Sprocket $\mathcal{S}_3$ is fixed to $\mathcal{A}_1$, and $\mathcal{S}_4$ is pinned to the upper end Q of $\mathcal{A}_2$. A load $\mathcal{L}$ is also pinned to $\mathcal{A}_2$ at Q and is rigidly connected to $\mathcal{S}_4$. A second chain is to pass over $\mathcal{S}_3$ and $\mathcal{S}_4$ and result in body $\mathcal{L}$ having zero angular velocity in $\mathcal{F}$. Find the required ratio r_4/r_3 of the radii of $\mathcal{S}_4$ and $\mathcal{S}_3$. (This device has recently been proposed to simulate translational harmonic lateral accelerations of a shipboard antenna $\mathcal{L}$ produced by lateral sway of the ship's center of mass.)

3.105 The shaded arcs on $\mathcal{B}_1$ and $\mathcal{B}_2$ (Figure P3.105a) are always equal if the two bodies are in rolling contact; however, the converse is not necessarily true. Just because the contacting arclengths are equal does not mean that $\mathcal{B}_1$ rolls on $\mathcal{B}_2$. For the wheel $\mathcal{B}_1$ on the plane $\mathcal{B}_2$ shown in Figure P3.105b, give constant values of $\dot{x}_C$ and $\dot{\theta}$ for which the arclengths of contact are equal but the velocities of the contact points are not. *Hint:* Look at the shaded arcs on $\mathcal{B}_1$ and $\mathcal{B}_2$ in Figure P3.105b.

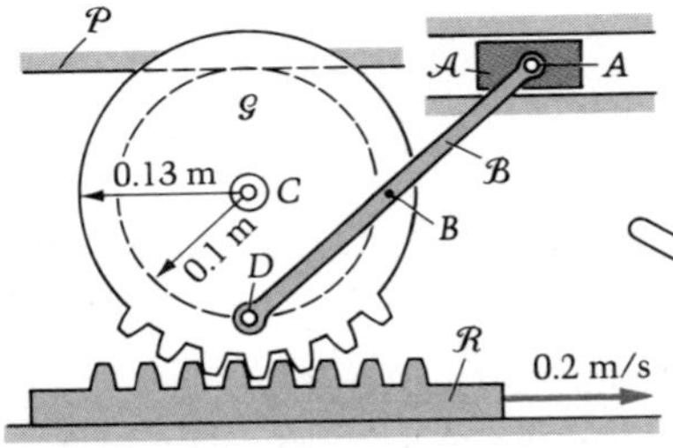

Figure P3.102

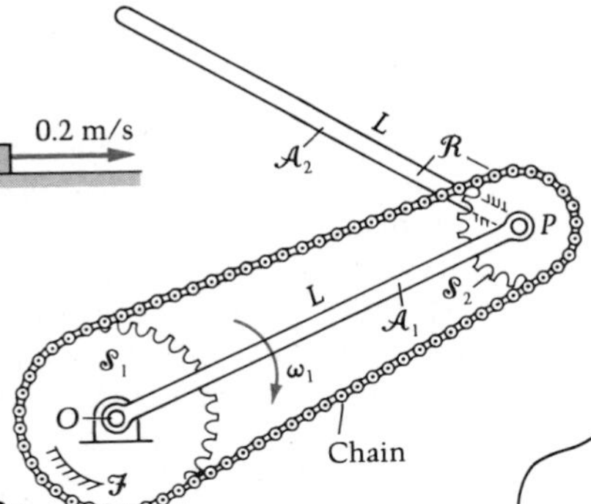

Figure P3.103a

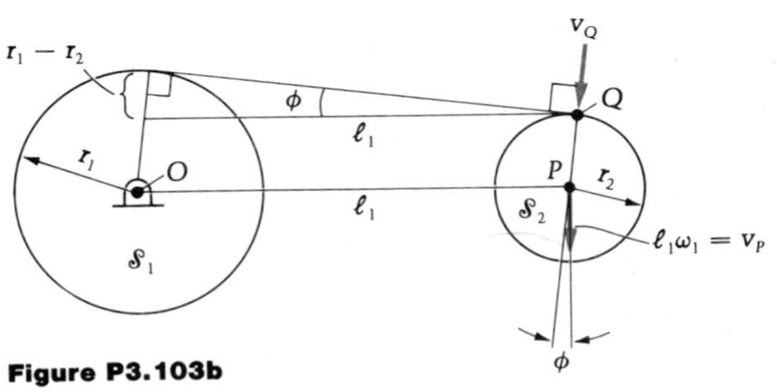

Figure P3.103b

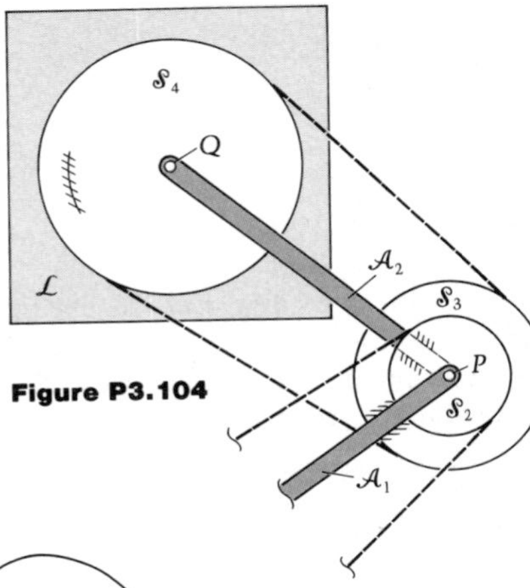

Figure P3.104

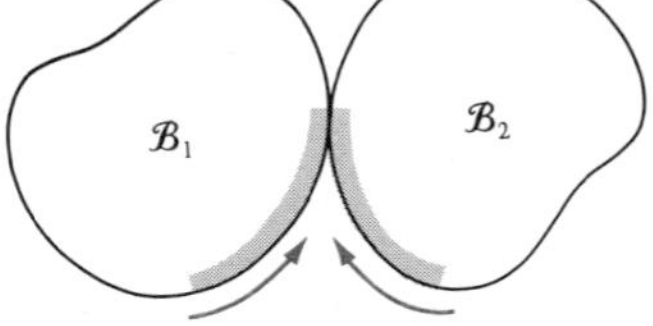

Figure P3.105a

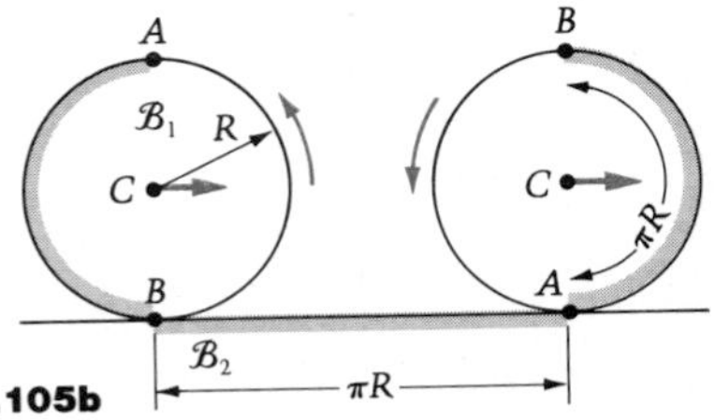

Figure P3.105b

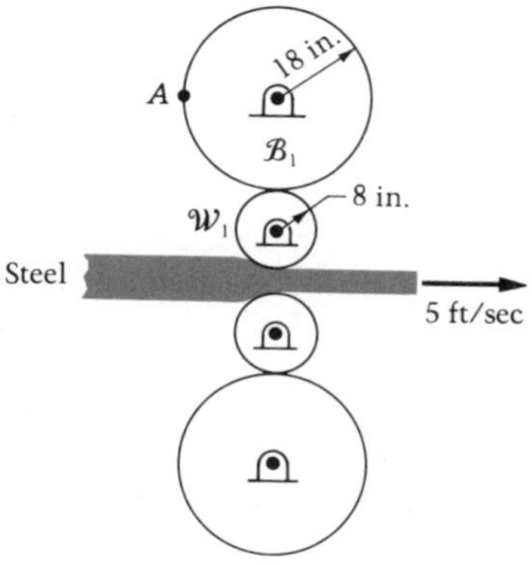

Figure P3.106

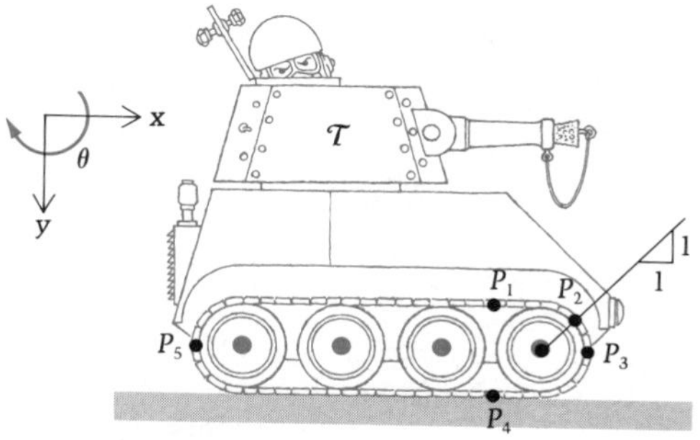

Figure P3.107

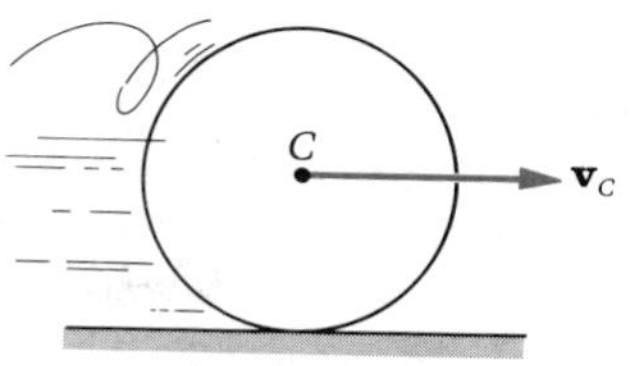

Figure P3.108

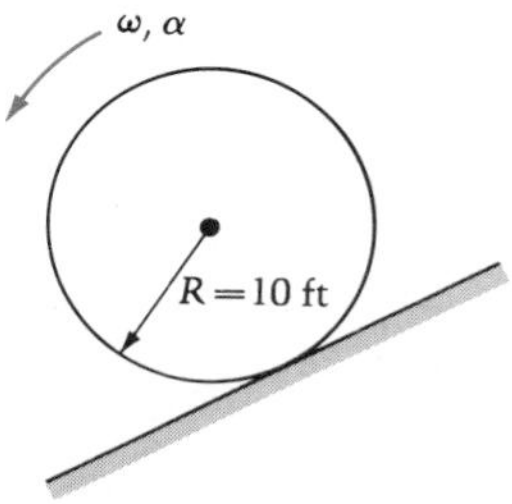

Figure P3.109

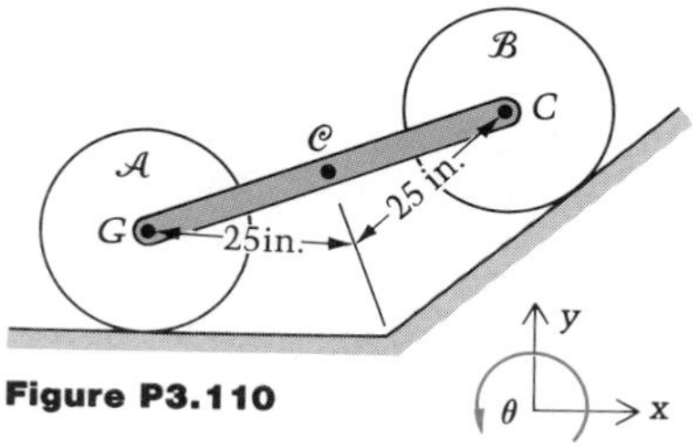

Figure P3.110

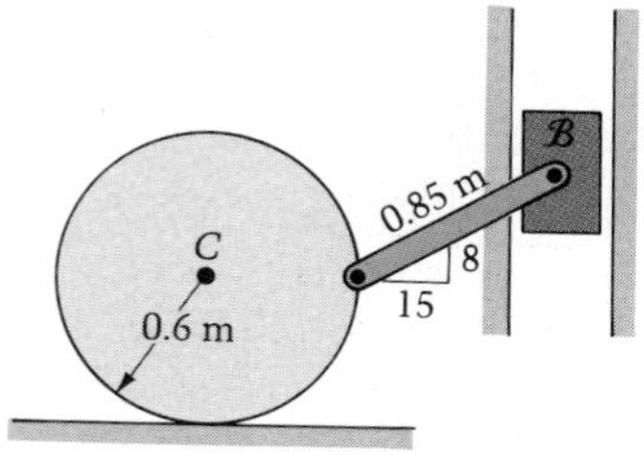

Figure P3.111

3.106 A sheet of steel, in the process of being cold-rolled, has the velocity indicated in Figure P3.106. Find the angular velocities of the upper "working roll" $\mathcal{W}_1$ and the upper "backing roll" $\mathcal{B}_1$. (The backing rolls greatly reduce the bending in the working rolls.) Also determine the velocity of point A on $\mathcal{B}_1$ at the given instant. Assume that no slipping occurs anywhere.

3.107 The tank $\mathcal{T}$ shown in Figure P3.107 is translating to the right, and at a certain instant it has velocity $v_0\hat{\mathbf{i}}$ and acceleration $a_0\hat{\mathbf{i}}$. (These values are $\mathbf{v}$ and $\mathbf{a}$ for all points in the body of the tank and for the centers of its wheels.) Find the velocities of the five points P_1, P_2, P_3, P_4, and P_5 if there is no slipping. The wheels have radius R.

3.108 A disk with diameter 1.2 m rolls along the plane as indicated in Figure P3.108. Its center point C has velocity

$$\mathbf{v}_C = (t^2 + 3t + 4)\hat{\mathbf{i}} \text{ m/s}$$

where t is the time in seconds. Find the acceleration of the point that lies 0.3 m directly below C when (a) $t = 2$ s and (b) $t = 5$ s.

3.109 Figure P3.109 shows a 10-ft-radius disk that rolls on a plane surface. It has, at the instant shown, an angular velocity of 2 rad/sec and an angular acceleration of 3 rad/sec^2, both counterclockwise. Find a point on the disk or the disk extended that has zero acceleration at this instant.

3.110 Cylinders $\mathcal{A}$ and $\mathcal{B}$ in Figure P3.110 have a radius of 10 in. each and roll on the respective planes. Bar $\mathcal{C}$ has length 48 in. and is pinned to the centers of the cylinders. The center G of $\mathcal{A}$ has velocity $\mathbf{v}_G = -10t\hat{\mathbf{i}}$ in./sec. If the time at the instant shown is $t = 5$ sec, find $\boldsymbol{\alpha}_{\mathcal{C}}$ and $\boldsymbol{\alpha}_{\mathcal{B}}$ at that instant.

3.111 See Figure P3.111. The velocity of the pin in block $\mathcal{B}$ is 0.2 ↓ m/s, and its acceleration is 0.5 ↑ m/s^2 in the given position. Find at this instant the angular velocity and angular acceleration of the cylinder if there is sufficient friction to prevent it from slipping.

3.112 During startup of the two friction wheels (see Figure P3.112), the angular velocity of $\mathcal{B}_1$ is $\boldsymbol{\omega}_1 = 5t^2$ ↺ rad/sec. Assuming rolling contact, compute the acceleration of the point T, which is at the top of $\mathcal{B}_2$ when $t =$ 3 sec.

3.113 The center point C of gear $\mathcal{G}_1$ in Figure P3.113 moves in a horizontal plane at constant speed v_0. The ring gear $\mathcal{F}$ is fixed in the reference frame, and the constant angular velocity of $\mathcal{G}_1$ is clockwise. Find the acceleration of point Q of $\mathcal{G}_1$.

3.114 If the ball in a ball bearing assembly (Figure P3.114) neither slips on the shaft nor on the fixed housing, find the velocity and acceleration of the center of the ball in terms of the angular velocity and angular acceleration of the shaft (ω and $\dot{\omega}$).

3.115 Find the acceleration of point B, the pin connecting rod $\mathcal{B}$ to the block in Figure P3.115, at $t = 1$ sec. The rod is horizontal at $t = 0$, and the velocity of C is $\mathbf{v}_C = 2t^2 \rightarrow$ ft/sec.

3.116 Bar $\mathcal{B}$ in Figure P3.116 is 25 cm long and is pinned to the rolling cylinder $\mathcal{C}$ at B. The other end of $\mathcal{B}$ is pinned to the roller at A as shown. The center of $\mathcal{C}$ has $v_C = 11.2$ cm/s and $a_C = 16.8$ cm/s^2 down the plane at the given instant; at this time line BⒾ is vertical and AⒾ is horizontal, and BC is parallel to the plane beneath it. Find the acceleration of point A and the angular acceleration of body $\mathcal{B}$ at the given instant.

3.117 A wheel rolls on a 10-cm-radius track. (See Figure P3.117.) At the instant shown the wheel has an angular velocity of $4\hat{\mathbf{k}}$ rad/s and an angular acceleration of $-3\hat{\mathbf{k}}$ rad/s^2. At the instant shown, find:

a. The velocity and acceleration of C
b. The velocity and acceleration of A
c. $(d/dt)|\mathbf{v}_A|$

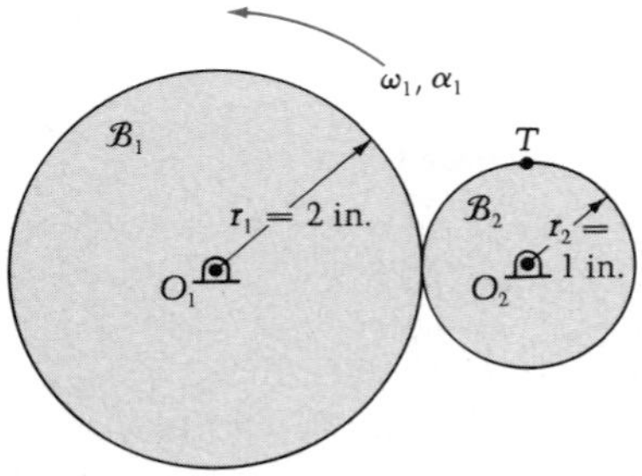

Figure P3.112

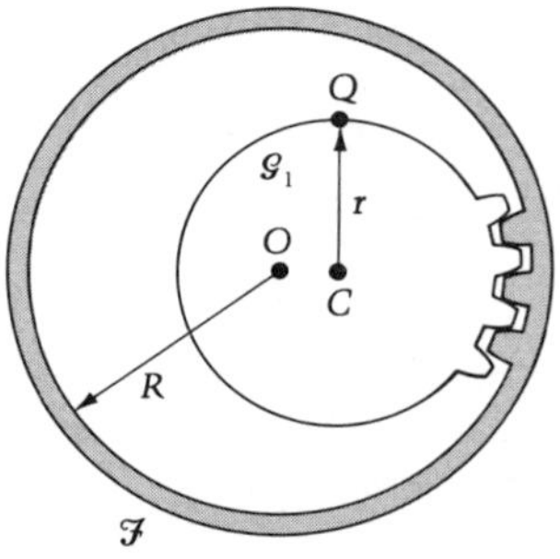

Figure P3.113

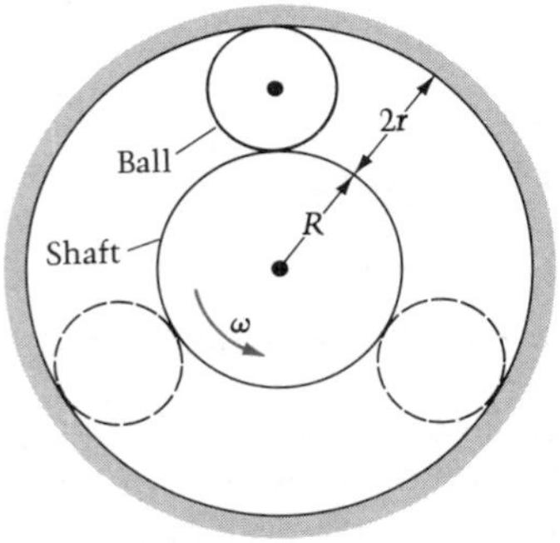

Figure P3.114

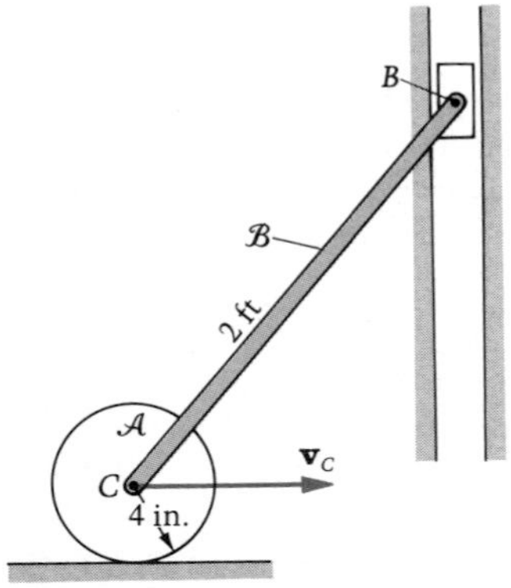

Figure P3.115

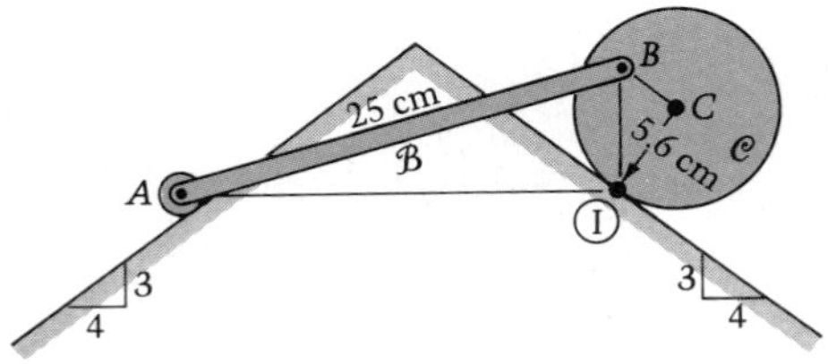

Figure P3.116

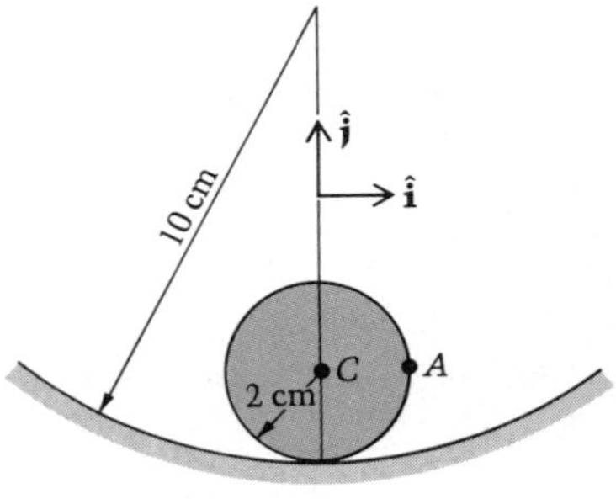

Figure P3.117

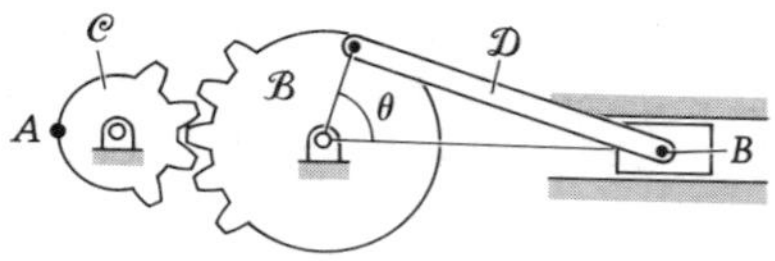

Figure P3.118

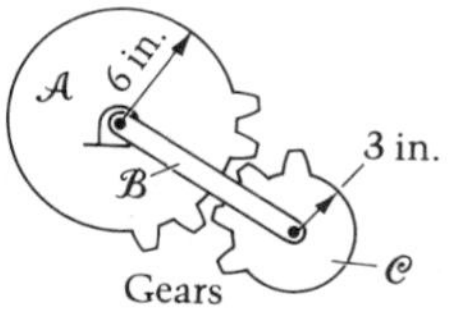

Figure P3.119

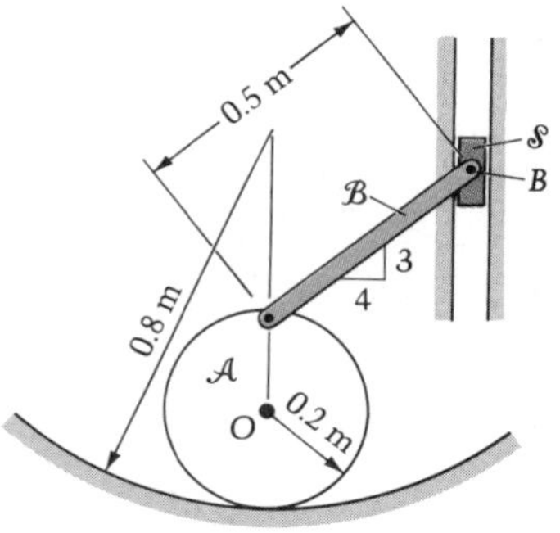

Figure P3.120

3.118 Gears $\mathcal{C}$ and $\mathcal{B}$ in Figure P3.118 have 25 and 50 teeth, respectively. Rod $\mathcal{D}$ is 2 ft long, and the radius of $\mathcal{B}$ is 1 ft. Determine the acceleration of point A when $t = 0$ if $\mathbf{x}_B = 0.2 \sin \pi t$ ft (positive to the left) with $\theta = 90°$ at $t = 0$.

3.119 At the instant shown in Figure P3.119, bar $\mathcal{B}$ has $\boldsymbol{\omega}_{\mathcal{B}} = \pi$ ↺ rad/sec and $\boldsymbol{\alpha}_{\mathcal{B}} = (\pi/3)$ ↻ rad/sec^2; and the gear $\mathcal{A}$ has $\boldsymbol{\omega}_{\mathcal{A}} = 2\pi$ rad/sec ↻ and $\boldsymbol{\alpha}_{\mathcal{A}} = \pi/2$ ↻ rad/sec^2. At this instant, determine the accelerations of each of the two contacting points.

3.120 A wheel $\mathcal{A}$ rolls along a curved surface. In the position shown in Figure P3.120, its angular velocity and angular acceleration are $\boldsymbol{\omega}_{\mathcal{A}} = 3$ ↻ rad/s^2 and $\boldsymbol{\alpha}_{\mathcal{A}} = 5$ ↻ rad/s^2. Determine at this instant the angular acceleration of bar $\mathcal{B}$ ($\boldsymbol{\alpha}_{\mathcal{B}}$) and the acceleration of pin B of slider block $\mathcal{S}$.

3.121 The angular velocity of crank $\mathcal{A}$ in Figure P3.121 is a constant 3 ↺ rad/s. In the given position, find the velocity of the center C of wheel $\mathcal{C}$ and determine the angular acceleration of $\mathcal{C}$, which rolls on the circular track.

3.122 The ball in Figure P3.122 rolls on the fixed surface and at the instant shown has angular velocity $\boldsymbol{\omega} = 3\hat{\mathbf{k}}$ rad/s and angular acceleration $\boldsymbol{\alpha} = -2\hat{\mathbf{k}}$ rad/s^2. At this instant find:

a. The velocities of A and B
b. The accelerations of A and B
c. $(d/dt)|\mathbf{v}_B|$

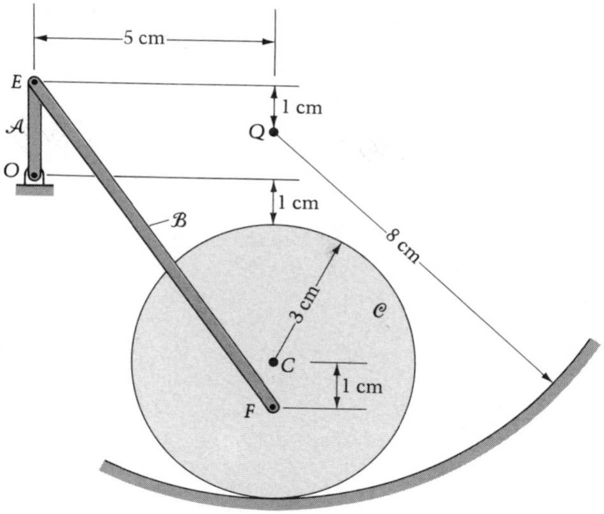

Figure P3.121

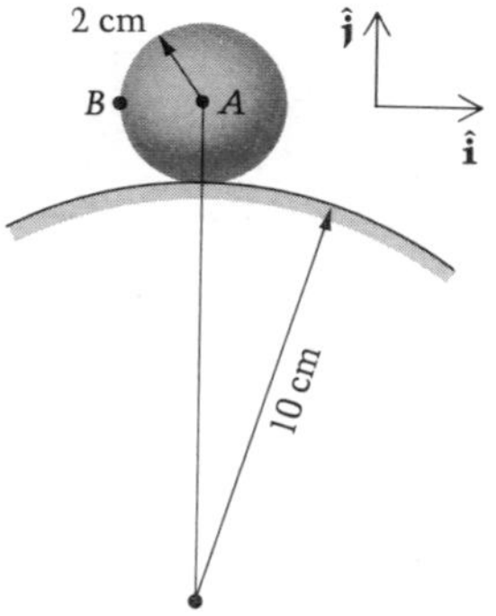

Figure P3.122

3.123 In Figure P3.123, gear $\mathcal{C}$ turns at constant angular velocity $\boldsymbol{\omega}_{\mathcal{C}} = 6$ ↺ rad/sec. Find the acceleration of point A and the angular acceleration of rod $\mathcal{R}$ when: (a) $\theta = 0°$; (b) $\theta = 90°$; (c) $\theta = 180°$; (d) $\theta = 270°$.

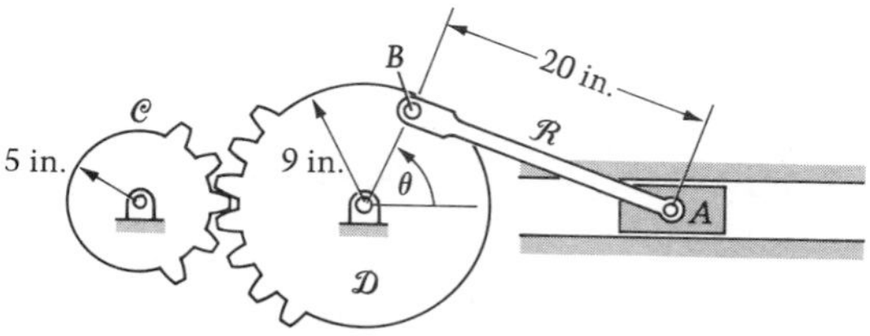

Figure P3.123

3.124 The outer gear $\mathcal{F}$ in Figure P3.124 is stationary. Crank $\mathcal{C}$ turns at the constant angular velocity of 10 rad/sec counterclockwise and is pinned at its ends to the centers of the sun gear $\mathcal{S}$ (at S) and the planetary gear $\mathcal{P}$ (at P). Find the accelerations of the points of $\mathcal{P}$ and $\mathcal{S}$ that are in contact with each other, if the radii of $\mathcal{P}$ and $\mathcal{S}$ are, respectively, 3 in. and 10 in.

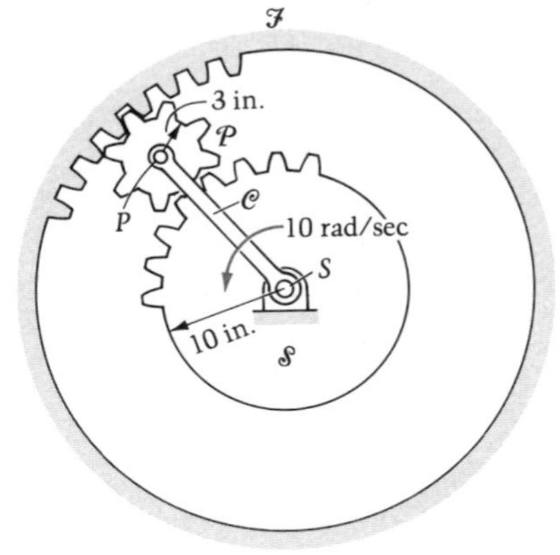

Figure P3.124

3.125 Point O is pinned to the reference frame $\mathcal{F}$. (See Figure P3.125.) The pitch radii of gears $\mathcal{P}$ and $\mathcal{C}$ are each 0.2 m. The angular velocities of $\mathcal{R}$ and $\mathcal{S}$ are 2 rad/s, clockwise for $\mathcal{R}$ and counterclockwise for $\mathcal{S}$, and both constant. Find the maximum acceleration magnitude experienced by any point of $\mathcal{P}$.

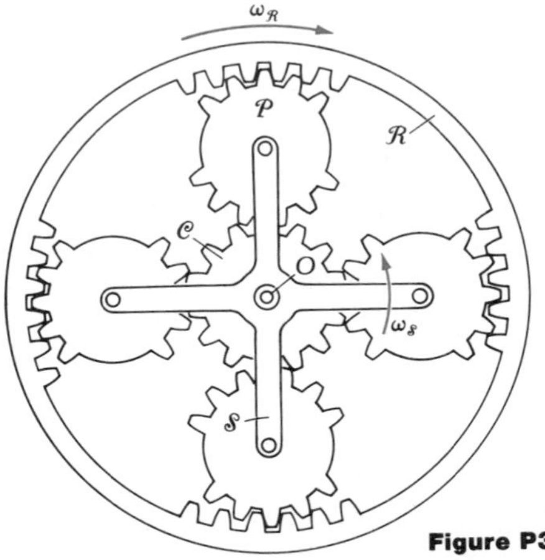

Figure P3.125

3.126 For the same data as in Problem 3.22, determine the angular accelerations of rod $\mathcal{R}$ and wheel $\mathcal{W}$ at the same two times.

3.127 The two bodies $\mathcal{B}_1$ and $\mathcal{B}_2$ in Figure P3.127 each have pinned centers and are in rolling contact; their angular velocities are constant. Show that the tangential components of the jerk vectors of the contact points P_1 and P_2 are equal if and only if the bodies have equal radii—that is, if $R_1 = R_2$. (See Problem 1.143.)

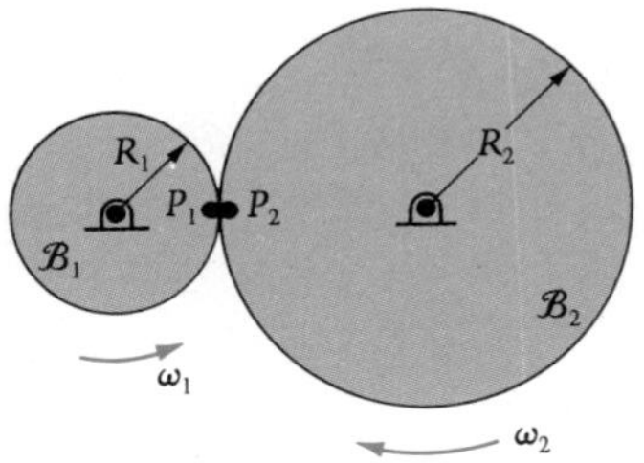

Figure P3.127

3.128 A moment applied to gear $\mathcal{C}$ in Figure P3.128 results in a constant angular acceleration $\boldsymbol{\alpha}_{\mathcal{C}} = 2 \circlearrowright$ rad/sec^2. The other gear, $\mathcal{B}$, is fixed in the reference frame. Determine:

a. The time required for C to return to its starting point after one revolution around $\mathcal{B}$ from rest

b. The number of revolutions turned through in space by $\mathcal{C}$ during the complete revolution

3.129 The two identical cylinders $\mathcal{A}$ and $\mathcal{B}$ (Figure P3.129) are connected by bar $\mathcal{D}$ (which is pinned to their centers), and they roll on the surface as shown. If the angular velocity of $\mathcal{A}$ is $\boldsymbol{\omega}_{\mathcal{A}} = \omega_0 \circlearrowright$ = constant, find the angular accelerations of both $\mathcal{D}$ and $\mathcal{B}$ at the given instant.

3.130 In Problem 3.89, let $\mathbf{a}_P = 0.2 \rightarrow$ m/s^2 and $\mathbf{a}_Q = 0.1 \leftarrow$ m/s^2 at the instant shown. Again assuming no slipping, find the accelerations of B and C.

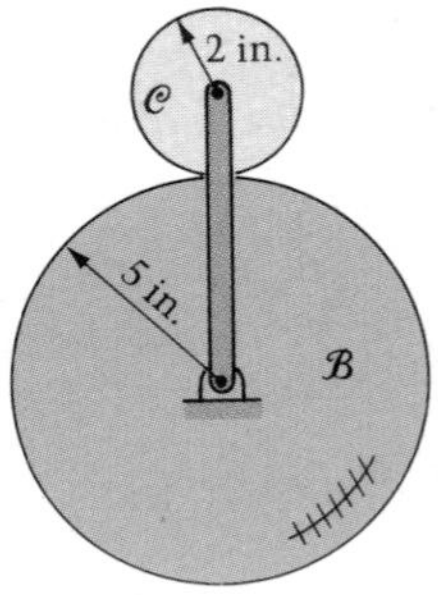

Figure P3.128

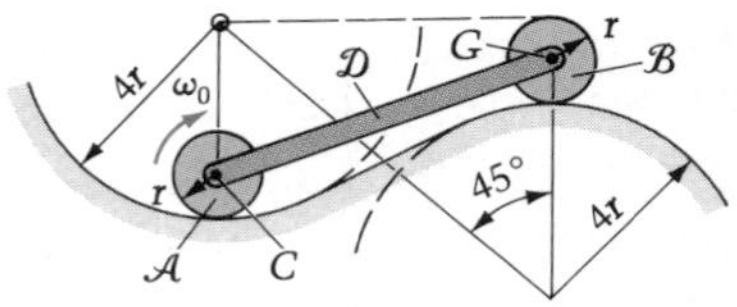

Figure P3.129

3.131 The center C of the small cylinder $\mathcal{C}$ in Figure P3.131 has a speed of $0.1t^2$ m/s as it moves clockwise on a circle. Body $\mathcal{C}$ rolls on the large cylinder $\mathcal{B}$. In the position given in the figure, $t = 10$ s. Find the acceleration of point B of the stick $\mathcal{S}$ that is in contact with $\mathcal{B}$ at the given instant.

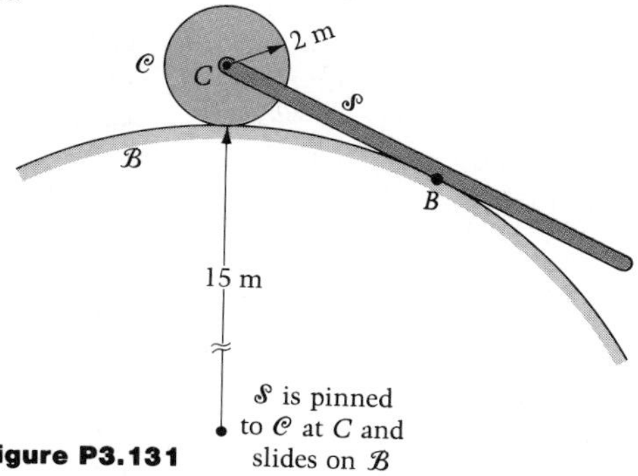

Figure P3.131

3.132 In Example 3.16 note that point S is the instantaneous center of $\mathcal{C}$, so that $\mathbf{v}_S = \mathbf{0}$ and $\mathbf{a}_S = r\omega^2 \uparrow = -1.2\hat{\mathbf{j}}$ (using the sign convention of that example). Use these results, instead of $\mathbf{v}_C$ and $\mathbf{a}_C$, to recalculate the velocity and acceleration of points N and E; compare your answers with those of the example.

3.133 Calculate the acceleration of the instantaneous center of rod $\mathcal{R}$ in Example 3.19.

3.134 Suppose rod $\mathcal{R}$ of Example 3.21 is pinned at point W, near the periphery as shown in Figure P3.134, instead of at C. Find $\mathbf{v}_B$ and $\mathbf{a}_B$. ($\mathcal{R}$ is now longer.)

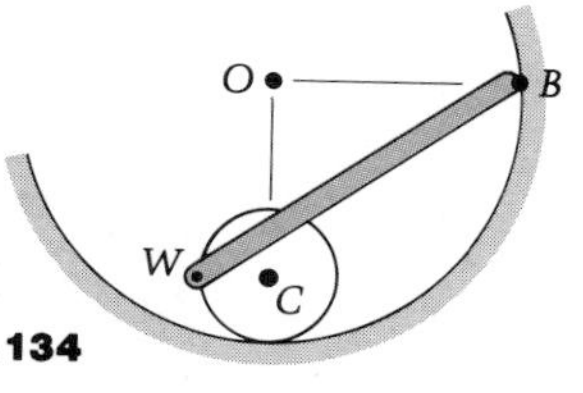

Figure P3.134

3.135 In Example 3.16 find the velocity and acceleration of point W.

3.136 In Example 3.17 find the piston acceleration when $\theta = 0°$, $180°$, and $270°$.

3.137 Pin A of block $\mathcal{A}$ has, at the instant shown in Figure P3.137, $\mathbf{v}_A = 2 \rightarrow$ m/s and $\mathbf{a}_A = 5 \leftarrow$ m/s². Wheel $\mathcal{C}$ rolls on the circular track. Find the angular acceleration of $\mathcal{C}$ at the instant shown.

3.138 Disk $\mathcal{D}$ rolls down a fixed circular arc as shown in Figure P3.138. The center C of the disk has a tangential component of acceleration given by $a_t = 21.5 \sin\phi$ ft/sec². If the disk starts at rest at $\phi = 0°$, find the radial component of acceleration of C at $\phi = 30°$. *Hint:* This is a tricky problem. To solve it, note that $a_t = \ddot{s} = (6+2)\ddot{\phi} = 21.5 \sin\phi$; also, $\ddot{\theta}\, d\theta = \dot{\theta}\, d\dot{\theta}$. Use these two ideas to integrate and obtain $\dot{\theta}$ (or ω) after you relate $\dot{\theta}$ and $\dot{\phi}$.

3.139 Point A of block $\mathcal{A}$ has, at the instant shown in Figure P3.139, $\mathbf{v}_A = 12 \uparrow$ in./sec and $\mathbf{a}_A = 6 \downarrow$ in./sec². Find the angular acceleration of bar $\mathcal{B}$.

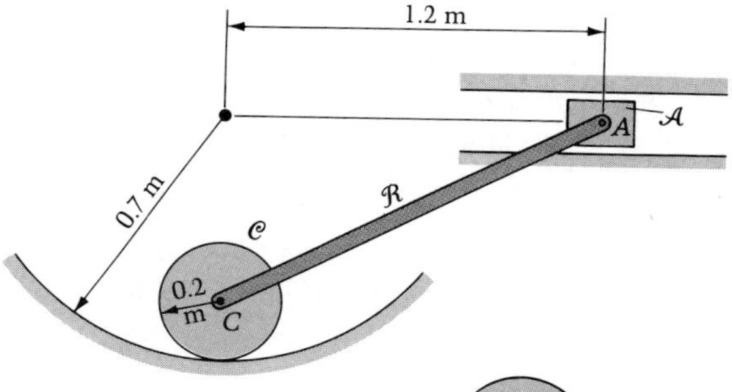

Figure P3.137

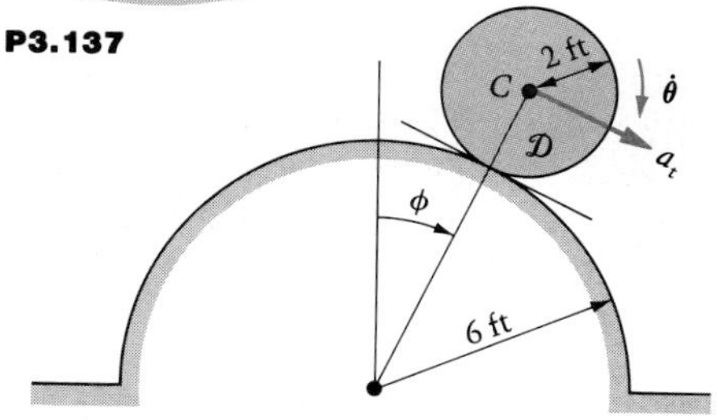

Figure P3.138

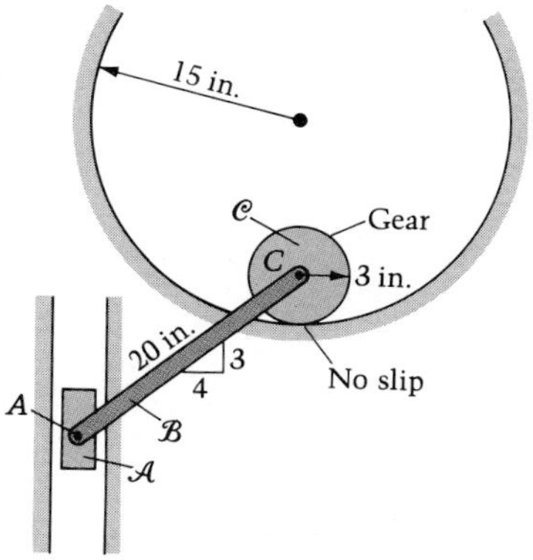

Figure P3.139

3.140 Two 5-in.-radius wheels roll on a plane surface. (See Figure P3.140.) A 13-in. bar $\mathcal{B}$ is pinned to the wheels at A and B as shown. If C has a constant velocity of 20 ft/sec to the right, find, for the position shown, the acceleration of A.

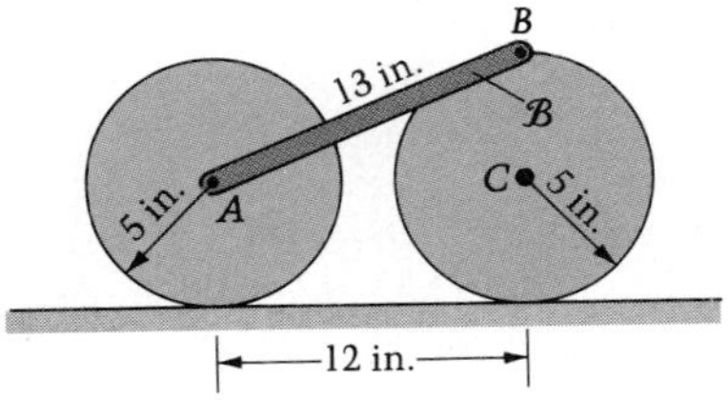

Figure P3.140

3.141 The ball in Figure P3.141 rolls on the fixed surface and at the instant shown has angular velocity $\boldsymbol{\omega} = 3\hat{\mathbf{k}}$ rad/sec and angular acceleration $\boldsymbol{\alpha} = -2\hat{\mathbf{k}}$ rad/sec^2. At this instant, find:

a. The acceleration of the center of the ball
b. The time derivative of the magnitude of the velocity of the ball's center.

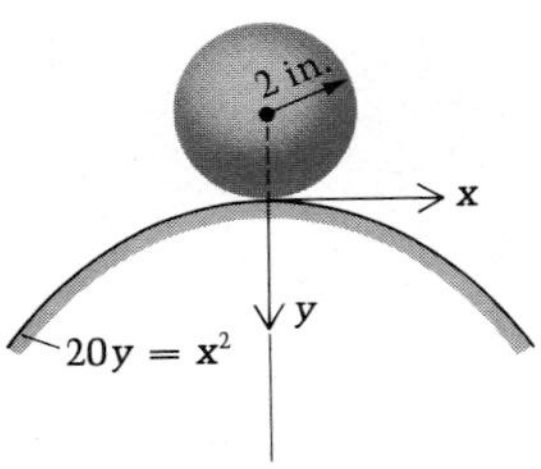

Figure P3.141

3.142 In Problem 3.107 find the accelerations of the same five points P_1, P_2, P_3, P_4, and P_5. (See Figure P3.142.)

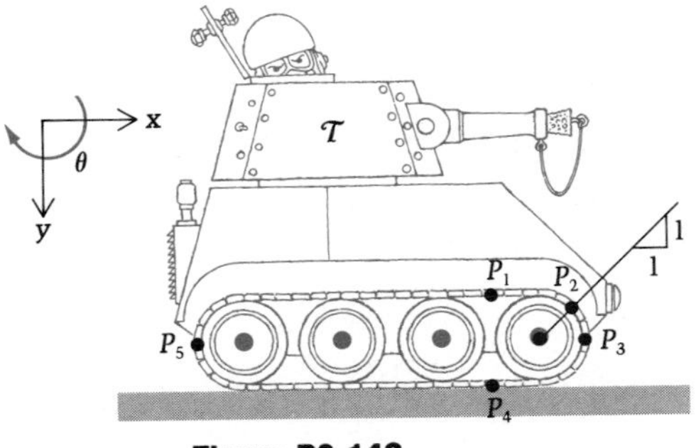

Figure P3.142

3.143 In Problem 3.99 find the accelerations of points A, B, C, D, and E.

3.144 In Problem 3.100 find the accelerations of points F, G, and H.

3.145 Rod $\mathcal{R}$ in Figure P3.145 slides on the plane surface at Q and is pinned to the disk at B. For the position shown, the disk has an angular velocity of 3 rad/sec clockwise and an angular acceleration of α rad/sec^2 clockwise; the disk does not slip. Find an expression for the acceleration of Q in terms of α.

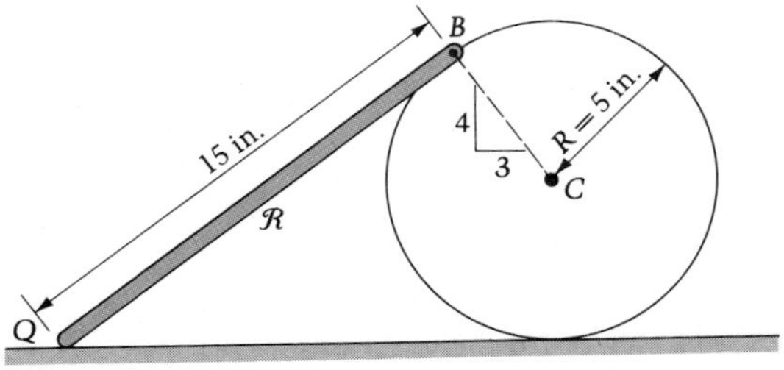

Figure P3.145

3.146 The wheel in Figure P3.146 rolls on the plane. Find the radius of curvature and the center of curvature of the path of point T at the given time in terms of r.

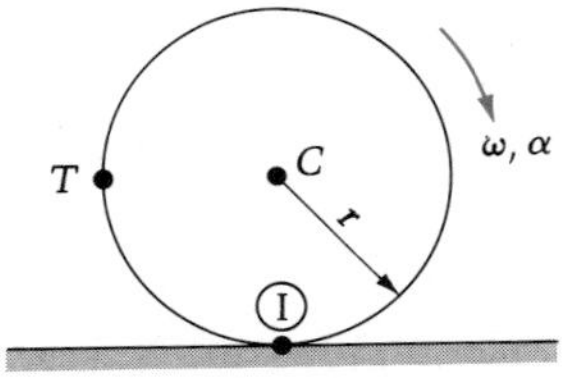

Figure P3.146

3.147 A *cycloid* is the curve traced out by a point on the rim of a rolling wheel. The equations for the rectangular coordinates of a point on the cycloid, in terms of the parameter φ (the angle shown in Figure P3.147), are

$$x = a(\varphi - \sin\varphi)$$

$$y = a(1 - \cos\varphi)$$

where a is the wheel's radius. Recall from calculus that the curvature of a plane curve is

$$\frac{1}{\rho} = \frac{y''}{(1 + y'^2)^{3/2}}$$

where ρ is the radius of curvature.

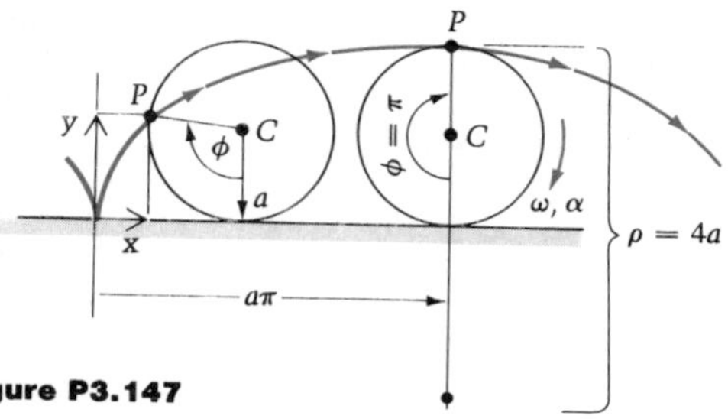

Figure P3.147

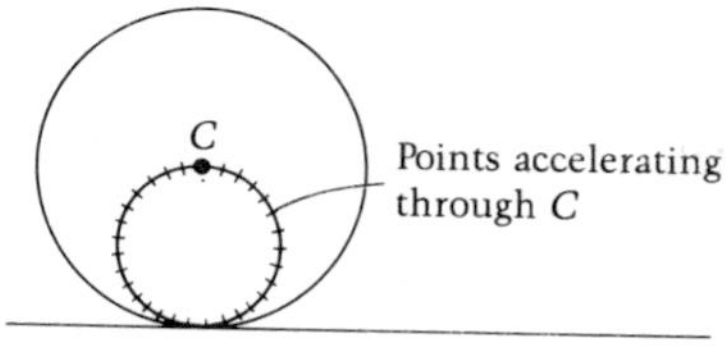

Figure P3.148

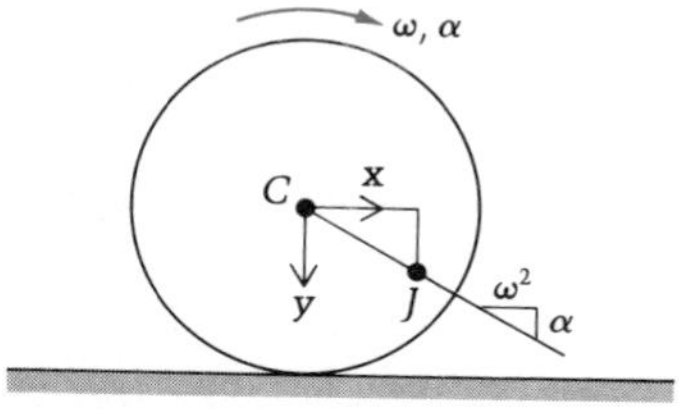

Figure P3.149

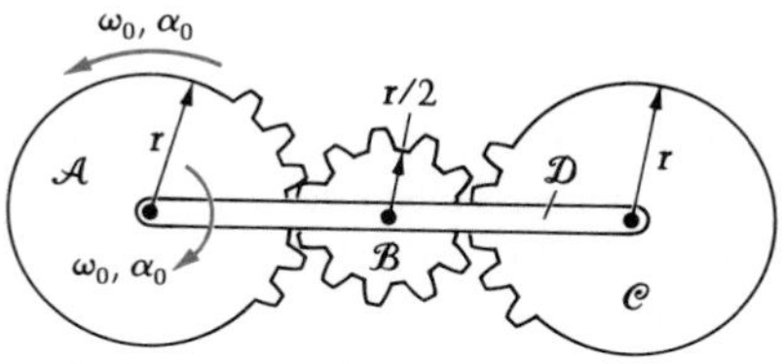

Figure P3.151

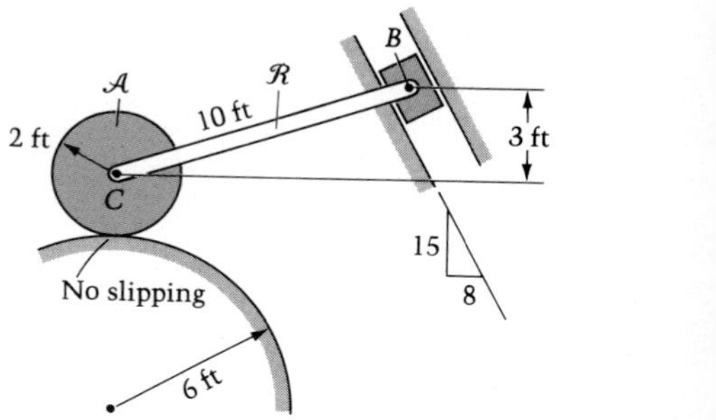

Figure P3.152

a. Use the chain rule

$$y' = \frac{dy}{dx} = \frac{dy}{d\varphi}\frac{d\varphi}{dx}$$

and show that, for the cycloid,

$$\rho = -2^{3/2}\, a\sqrt{1 - \cos\varphi}\,.$$

b. Explain what the minus sign in the expression for ρ means.

c. Observe that P is at its highest point when $\varphi = \pi$ and that $|\rho| = 4a$ there. In this configuration, show that the following two expressions for the acceleration of P agree:

$$\mathbf{a}_P = \underbrace{a\alpha\hat{\mathbf{i}}}_{\mathbf{a}_C} + \alpha\hat{\mathbf{k}} \times \mathbf{r}_{CP} - \omega^2\mathbf{r}_{CP}$$

$$\mathbf{a}_P = \ddot{s}\hat{\mathbf{e}}_t + \frac{\dot{s}^2}{\rho}\hat{\mathbf{e}}_n$$

3.148 Show that for a rigid body $\mathcal{B}$ in plane motion, as long as $\alpha \neq 0$ there is a circle of points P of $\mathcal{B}$ whose accelerations pass through any point C of $\mathcal{B}$. *Hint:* Write $\mathbf{a}_P = \mathbf{a}_C + \alpha\hat{\mathbf{k}} \times \mathbf{r}_{CP} - \omega^2\mathbf{r}_{CP}$ and dot both sides with the vector $\hat{\mathbf{k}} \times \mathbf{r}_{CP}$. Assume that $\mathbf{a}_P \parallel \mathbf{r}_{CP}$ and see if you can exhibit $\mathbf{r}_{CP}$. For the rolling wheel, show that the points are as shown in Figure P3.148.

3.149 Show that for the rolling uniform cylinder in Figure P3.149 there is a point of zero acceleration at the indicated position J if ω and α are in the given directions and are not both zero. You should find that the coordinates of J are $(x_J, y_J) = [r\alpha\omega^2/(\omega^4 + \alpha^2), r\alpha^2/(\omega^4 + \alpha^2)]$.

3.150 Show that for a rigid body $\mathcal{B}$ in plane motion, as long as ω and α are not *both* zero there is a point of $\mathcal{B}$ having zero acceleration. *Hint:* Let P be a reference point with acceleration $\mathbf{a}_P = a_{P_x}\hat{\mathbf{i}} + a_{P_y}\hat{\mathbf{j}}$. See if you can find a vector $\mathbf{r}_{PT} = x\hat{\mathbf{i}} + y\hat{\mathbf{j}}$ from P to a point T of zero acceleration. That is, solve

$$\mathbf{a}_T = \mathbf{0} = \mathbf{a}_P + \alpha\hat{\mathbf{k}} \times \mathbf{r}_{PT} - \omega^2\mathbf{r}_{PT}$$

for x and y.

3.151 Gear $\mathcal{A}$ and crank $\mathcal{D}$ have angular speeds ω_0 and angular acceleration magnitudes α_0 at the instant shown in Figure P3.151 in the indicated directions. Find the angular velocities and angular accelerations of gears $\mathcal{B}$ and $\mathcal{C}$ at the same time, if $\mathcal{D}$ is pinned to $\mathcal{A}$, $\mathcal{B}$, and $\mathcal{C}$.

3.152 In Figure P3.152 point C travels clockwise with a speed given by $v_C = 0.02t^2$ ft/sec and is in the given position at $t = 2$ sec. Find the acceleration of point B at this instant.

3.153 In Example 3.28 determine the minimum distance between the centers of the line shaft and shaft $\mathcal{S}$. *Hint:* The belt runs normal to the faces of the pulley.

3.154 In Example 3.28 suppose that a speed decrease of 7 percent from the nominal is needed to allow for shrinkage of a certain type of paper. Find the maximum belt width that will work in this application.

3.7 Relationship Between the Velocities of a Point *P* with Respect to Two Different Frames of Reference

While Equation (3.8) gives us the relationship between the velocities in $\mathcal{I}$ of two points of the *same* rigid body, we often need another equation relating the velocities of the ***same point*** relative to two ***different*** frames or bodies. This relationship (together with a companion equation for accelerations to be developed in the next section) will be essential in solving kinematics problems involving bodies moving in special ways relative to others (such as a pin of one body sliding in a slot of another).

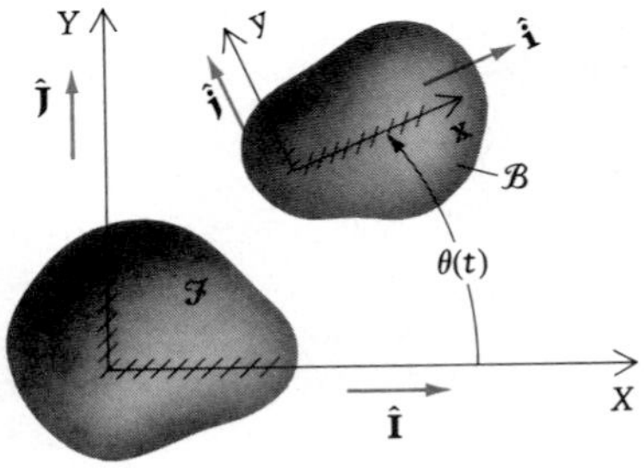

Figure 3.25

To develop this equation, we must first find the relationship between the derivatives of an arbitrary vector $\mathbf{T}$ in two frames $\mathcal{I}$ and $\mathcal{B}$. To do this, we begin by embedding axes X and Y in $\mathcal{I}$, and $\mathbf{x}$ and $\mathbf{y}$ in $\mathcal{B}$, as suggested by the hatch marks in Figure 3.25. Further, we let $(\hat{\mathbf{I}}, \hat{\mathbf{J}})$ and $(\hat{\mathbf{i}}, \hat{\mathbf{j}})$ be pairs of unit vectors, always respectively parallel to (X, Y) and to $(\mathbf{x}, \mathbf{y})$.

We note that if $\theta(t)$ again locates the $\mathbf{x}$ axis relative to X as shown, then

$$\hat{\mathbf{i}} = \cos\theta\hat{\mathbf{I}} + \sin\theta\hat{\mathbf{J}} \tag{3.39}$$

and

$$\hat{\mathbf{j}} = -\sin\theta\hat{\mathbf{I}} + \cos\theta\hat{\mathbf{J}} \tag{3.40}$$

Differentiating in frame $\mathcal{I}$ and noting that $\hat{\mathbf{I}}$ and $\hat{\mathbf{J}}$ are constants there, we obtain

$$\frac{d\hat{\mathbf{i}}}{dt} = \dot{\hat{\mathbf{i}}} = (-\sin\theta\hat{\mathbf{I}} + \cos\theta\hat{\mathbf{J}})\dot{\theta} = \dot{\theta}\hat{\mathbf{j}} \tag{3.41}$$

$$\frac{d\hat{\mathbf{j}}}{dt} = \dot{\hat{\mathbf{j}}} = -(\cos\theta\hat{\mathbf{I}} + \sin\theta\hat{\mathbf{J}})\dot{\theta} = -\dot{\theta}\hat{\mathbf{i}} \tag{3.42}$$

Now let the arbitrary vector $\mathbf{T}$ be written in frame $\mathcal{B}$ (meaning that $\mathbf{T}$ is expressed in terms of its components there):

$$\mathbf{T} = T_x\hat{\mathbf{i}} + T_y\hat{\mathbf{j}}$$

Differentiating this vector in $\mathcal{I}$, we get

$$^{\mathcal{I}}\dot{\mathbf{T}} = \dot{T}_x\hat{\mathbf{i}} + \dot{T}_y\hat{\mathbf{j}} + T_x\,^{\mathcal{I}}\dot{\hat{\mathbf{i}}} + T_y\,^{\mathcal{I}}\dot{\hat{\mathbf{j}}} \tag{3.43}$$

We now note that the first two terms on the right side of Equation (3.43) add up to the derivative of vector $\mathbf{T}$ in $\mathcal{B}$, because $\hat{\mathbf{i}}$ and $\hat{\mathbf{j}}$ do not change in magnitude *or* direction with time there. Thus

$$^{\mathcal{F}}\dot{\mathbf{T}} = {}^{\mathcal{B}}\dot{\mathbf{T}} + T_x\, {}^{\mathcal{F}}\dot{\hat{\mathbf{i}}} + T_y\, {}^{\mathcal{F}}\dot{\hat{\mathbf{j}}}$$

Substituting the derivatives of $\hat{\mathbf{i}}$ and $\hat{\mathbf{j}}$ in $\mathcal{F}$ from Equations (3.41) and (3.42) yields

$$\begin{aligned} ^{\mathcal{F}}\dot{\mathbf{T}} &= {}^{\mathcal{B}}\dot{\mathbf{T}} + \dot{\theta}(T_x\hat{\mathbf{j}} - T_y\hat{\mathbf{i}}) \\ &= {}^{\mathcal{B}}\dot{\mathbf{T}} + \dot{\theta}\hat{\mathbf{k}} \times (T_x\hat{\mathbf{i}} + T_y\hat{\mathbf{j}}) \end{aligned}$$

or

$$^{\mathcal{F}}\dot{\mathbf{T}} = {}^{\mathcal{B}}\dot{\mathbf{T}} + \dot{\theta}\hat{\mathbf{k}} \times \mathbf{T} \tag{3.44}$$

The angular velocity $\dot{\theta}\hat{\mathbf{k}}$ has reappeared, and Equation (3.44) shows us that this vector has a more general purpose than merely relating velocities in kinematics. It is in fact the link that allows us to relate the derivatives of any vector in two different frames. (This same result is in fact true in general three-dimensional motion, with three-dimensional vectors and a more general expression for angular velocity substituted, as will be seen in Chapter 6.)

We now use Equation (3.44) to relate the velocities of a point P in two frames $\mathcal{B}$ and $\mathcal{F}$. From Figure 3.26, the position vectors of P in these two frames are related by

$$\mathbf{r}_{OP} = \mathbf{r}_{OO'} + \mathbf{r}_{O'P} \tag{3.45}$$

Differentiating this equation in $\mathcal{F}$, we have

$$^{\mathcal{F}}\dot{\mathbf{r}}_{OP} = {}^{\mathcal{F}}\dot{\mathbf{r}}_{OO'} + {}^{\mathcal{F}}\dot{\mathbf{r}}_{O'P} \tag{3.46}$$

The first two vectors in Equation (3.46) are the velocities of P and O' in $\mathcal{F}$ by definition:

$$\mathbf{v}_{P/\mathcal{F}} = \mathbf{v}_{O'/\mathcal{F}} + {}^{\mathcal{F}}\dot{\mathbf{r}}_{O'P} \tag{3.47}$$

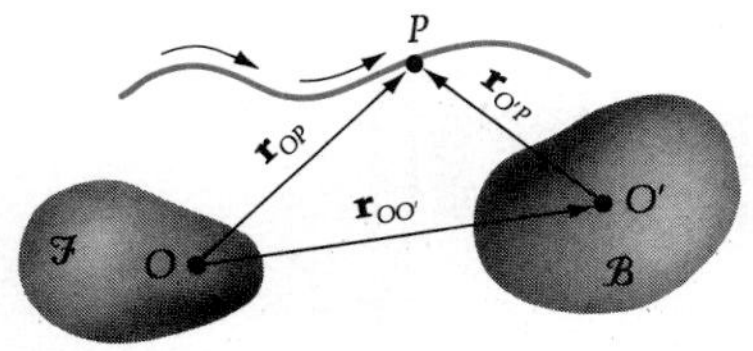

Figure 3.26

Question 3.16 (a) Why is the last vector in (3.47) not the velocity of P in $\mathcal{F}$? (b) Why is it not the velocity of P in $\mathcal{B}$?

To replace $^{\mathcal{F}}\dot{\mathbf{r}}_{O'P}$ by a vector that we can operate with, we use Equation (3.44), with $\mathbf{r}_{O'P}$ becoming the vector $\mathbf{T}$:

$$^{\mathcal{F}}\dot{\mathbf{r}}_{O'P} = {}^{\mathcal{B}}\dot{\mathbf{r}}_{O'P} + \dot{\theta}\hat{\mathbf{k}} \times \mathbf{r}_{O'P} \tag{3.48}$$

Therefore, recognizing that $^{\mathcal{B}}\dot{\mathbf{r}}_{O'P}$ is $\mathbf{v}_{P/\mathcal{B}}$ and substituting Equation (3.48) into (3.47), we obtain

$$\mathbf{v}_{P/\mathcal{F}} = \mathbf{v}_{P/\mathcal{B}} + \mathbf{v}_{O'/\mathcal{F}} + \dot{\theta}\hat{\mathbf{k}} \times \mathbf{r}_{O'P} \tag{3.49}$$

Another way of expressing Equation (3.49) is to consider frame $\mathcal{B}$ as a "moving frame" with respect to a "fixed" reference frame $\mathcal{F}$. Then the velocities of P can be written as simply $\mathbf{v}_P$ when the reference is $\mathcal{F}$ (thus $\mathbf{v}_P = \mathbf{v}_{P/\mathcal{F}}$) and as $\mathbf{v}_{\text{rel}}$ when the reference is the "moving frame" $\mathcal{B}$ (thus $\mathbf{v}_{\text{rel}} = \mathbf{v}_{P/\mathcal{B}}$). Hence we can write Equation (3.49) in abbreviated notation as

$$\mathbf{v}_P = \mathbf{v}_{\text{rel}} + \mathbf{v}_{O'} + \boldsymbol{\omega} \times \mathbf{r} \tag{3.50}$$

where $\mathbf{r} = \mathbf{r}_{O'P}$, the position of P in the moving frame. The reader may find this form of Equation (3.49) easier to use when there is just one "moving frame."

Now let us denote by $P_{\mathcal{B}}$ the point of $\mathcal{B}$ (or $\mathcal{B}$ extended) that is coincident with P. Then $\mathbf{r}_{O'P} = \mathbf{r}_{O'P_{\mathcal{B}}}$ and the last two terms of Equation (3.49) or (3.50) are seen (by Equation 3.8) to add to the velocity of $P_{\mathcal{B}}$ in $\mathcal{F}$:

$$\mathbf{v}_{P/\mathcal{F}} = \mathbf{v}_{P/\mathcal{B}} + \mathbf{v}_{P_{\mathcal{B}}/\mathcal{F}} \tag{3.51}$$

In words, Equation (3.51) says that at any time, the velocity of P in $\mathcal{F}$ is the sum of the velocity of P in $\mathcal{B}$ plus the velocity in $\mathcal{F}$ of the point of $\mathcal{B}$ coincident with P.*

As a preliminary example, consider Figure 3.27. Let P be the center of the pin, which is attached to frame $\mathcal{F}$. Then

$$\mathbf{v}_{P/\mathcal{F}} = \mathbf{0} = \mathbf{v}_{P/\mathcal{B}} + \mathbf{v}_{P_{\mathcal{B}}/\mathcal{F}} \Rightarrow \mathbf{v}_{P_{\mathcal{B}}/\mathcal{F}} = -\mathbf{v}_{P/\mathcal{B}}$$

Now the center of the pin's motion in $\mathcal{B}$ is along a straight line within the slot of $\mathcal{B}$. Therefore the velocity of $P_{\mathcal{B}}$ (the point of $\mathcal{B}$ extended coincident with P) is seen to be *also* parallel to the slot and in a direction (↗) opposite to that of $\mathbf{v}_{P/\mathcal{B}}$ (↙). We now consider several detailed examples of the use of Equation (3.51).

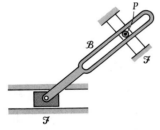

Figure 3.27

EXAMPLE 3.29

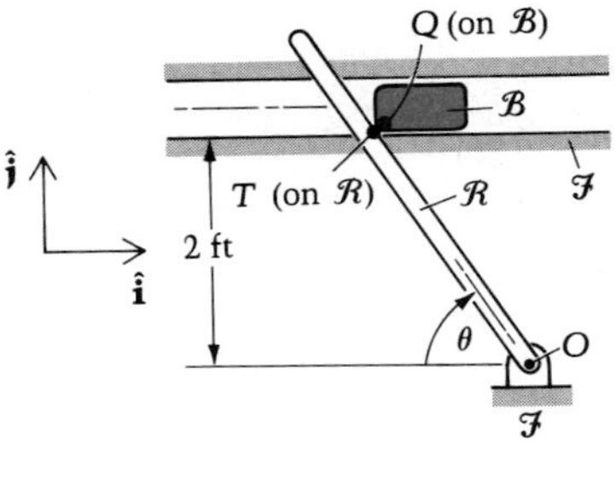

Block $\mathcal{B}$ translates in a horizontal slot (see the diagram) and is pushed along by a bar $\mathcal{R}$ that turns at angular velocity $\boldsymbol{\omega}_{\mathcal{R}} = 10$ ↻ rad/sec about the pin at point O. Find $\mathbf{v}_Q$, the velocity of the contact point of $\mathcal{B}$, when $\theta = 60°$.

(Continued)

*This latter term is sometimes called the *vehicle velocity* of P.

SOLUTION

Let the ground be the reference frame $\mathcal{F}$, and note that T is the point of $\mathcal{R}$ coincident with Q at the given instant (the point we have been calling $P_{\mathcal{B}}$ in the theory). Using Equation (3.51), we obtain

$$\mathbf{v}_{Q/\mathcal{F}} = \mathbf{v}_{Q/\mathcal{R}} + \mathbf{v}_{T/\mathcal{F}}$$

Therefore

$$\mathbf{v}_{Q/\mathcal{F}} = v_{Q/\mathcal{R}}\left(\frac{\hat{\mathbf{i}}}{2} - \frac{\sqrt{3}}{2}\hat{\mathbf{j}}\right) + (-10\hat{\mathbf{k}}) \times \left(\frac{-2}{\sqrt{3}}\hat{\mathbf{i}} + 2\hat{\mathbf{j}}\right) \tag{1}$$

Note that the velocity of Q relative to bar $\mathcal{R}$ has an unknown magnitude $v_{Q/\mathcal{R}}$ but a ***known*** direction (along $\mathcal{R}$). Now we also know the direction of $\mathbf{v}_{Q/\mathcal{F}}$, so that

$$\mathbf{v}_{Q/\mathcal{F}} = v_{Q/\mathcal{F}}\hat{\mathbf{i}} = \frac{20}{\sqrt{3}}\hat{\mathbf{j}} + 20\hat{\mathbf{i}} + \frac{v_{Q/\mathcal{R}}}{2}\hat{\mathbf{i}} - v_{Q/\mathcal{R}}\frac{\sqrt{3}}{2}\hat{\mathbf{j}} \tag{2}$$

Equating the $\mathbf{x}$ components of Equation (2), we get

$$v_{Q/\mathcal{F}} = 20 + \frac{v_{Q/\mathcal{R}}}{2} \tag{3}$$

And equating the y components:

$$0 = \frac{20}{\sqrt{3}} - v_{Q/\mathcal{R}}\frac{\sqrt{3}}{2} \tag{4}$$

Equation (4) gives $v_{Q/\mathcal{R}} = 13.3$ ft/sec, from which Equation (3) then yields

$$\mathbf{v}_{Q/\mathcal{F}} = v_{Q/\mathcal{F}}\hat{\mathbf{i}} = 26.7\hat{\mathbf{i}} \text{ ft/sec}$$

The correct triangle relating the velocities of Q and T is shown in the accompanying diagram. As a check, $v_Q \cos 30° = (26.7)\sqrt{3}/2 = 23.1$ and $v_T = \sqrt{(20/\sqrt{3})^2 + 20^2} = 23.1$ ft/sec.

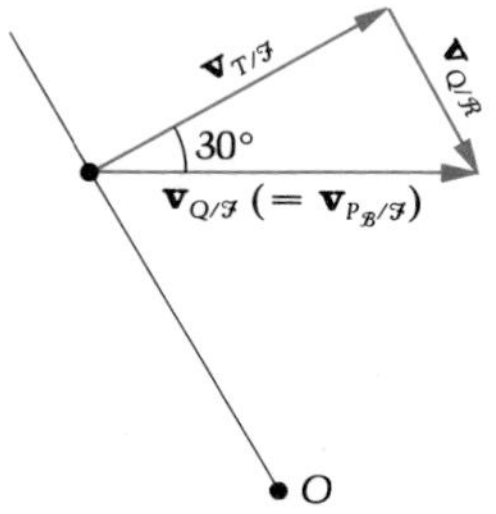

Question 3.17 Can the preceding example also be worked by using $\mathbf{v}_{T/\mathcal{F}} = \mathbf{v}_{T/\mathcal{B}} + \mathbf{v}_{Q/\mathcal{F}}$?

EXAMPLE 3.30

Collar $\mathcal{C}$ in the accompanying diagram is pinned to rod $\mathcal{R}$ at P and is free to slide along rod $\mathcal{B}$. The angular velocity of $\mathcal{R}$ is 0.2 rad/s ↻ at the instant shown. Find the angular velocity of $\mathcal{B}$ at this time, and determine the velocity of P relative to $\mathcal{B}$.

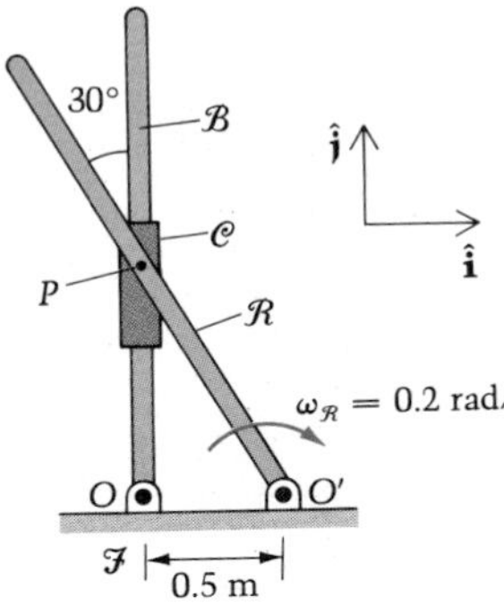

SOLUTION

We relate the velocities of P in $\mathcal{F}$ and in $\mathcal{B}$:

$$\mathbf{v}_{P/\mathcal{F}} = \mathbf{v}_{P/\mathcal{B}} + \mathbf{v}_{P_{\mathcal{B}}/\mathcal{F}}$$

$$\overbrace{\underbrace{\mathbf{v}_{O'/\mathcal{F}}}_{\text{zero}} + \boldsymbol{\omega}_{\mathcal{R}} \times \mathbf{r}_{O'P}}^{\text{on } \mathcal{R}} = v_{P/\mathcal{B}}\hat{\mathbf{j}} + \overbrace{(\underbrace{\mathbf{v}_{O/\mathcal{F}}}_{\text{zero}} + \boldsymbol{\omega}_{\mathcal{B}} \times \mathbf{r}_{OP_{\mathcal{B}}})}^{\text{on } \mathcal{B}}$$

$$-0.2\hat{\mathbf{k}} \times (-0.5\hat{\mathbf{i}} + 0.866\hat{\mathbf{j}}) = v_{P/\mathcal{B}}\hat{\mathbf{j}} + \omega_{\mathcal{B}}\hat{\mathbf{k}} \times 0.866\hat{\mathbf{j}}$$

$\hat{\mathbf{i}}$ coefficients: $\quad 0.173 = -0.866\omega_{\mathcal{B}}$

$$\omega_{\mathcal{B}} = -0.2 \Rightarrow \boldsymbol{\omega}_{\mathcal{B}} = -0.2\hat{\mathbf{k}} \text{ rad/s or } 0.2 \circlearrowright \text{ rad/s}$$

$\hat{\mathbf{j}}$ coefficients: $\quad 0.1 = v_{P/\mathcal{B}}$

$$\mathbf{v}_{P/\mathcal{B}} = 0.1\hat{\mathbf{j}} \text{ m/s}$$

Thus the pin is moving outward on $\mathcal{B}$, which is turning clockwise.

Question 3.18 Will $\mathcal{B}$ *always* have the same $\boldsymbol{\omega}$ as does $\mathcal{R}$?

EXAMPLE 3.31

Disk $\mathcal{C}$, with its attached pin P, has limited angular motion. (See the diagram.) After a 45° clockwise rotation from the depicted position, disk $\mathcal{C}$ has angular velocity $\boldsymbol{\omega}_{\mathcal{C}} = 2 \circlearrowright$ rad/sec. At this time, find $\boldsymbol{\omega}_{\mathcal{T}}$ of the slotted triangular body $\mathcal{T}$ and determine the velocity of pin P relative to $\mathcal{T}$.

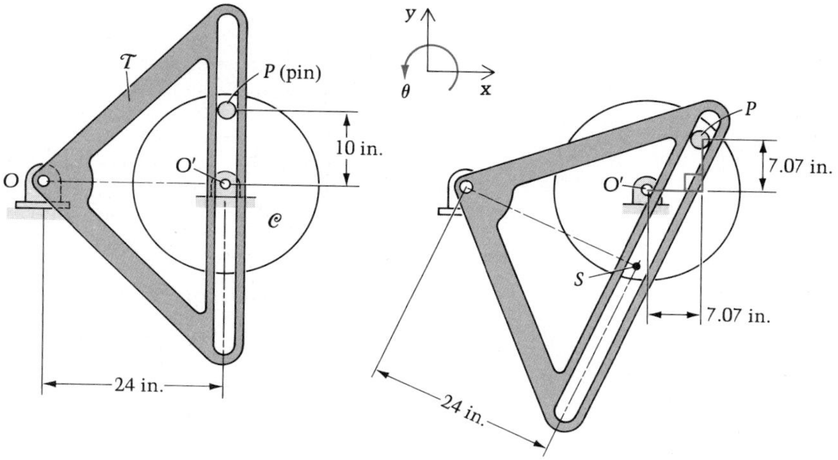

SOLUTION

At 45° $\circlearrowright$, the configuration is as shown on the right in the preceding diagram. Note that the slot center S moves on a circle about O and that a tangent to this circle at S must pass through P at all times, since P must stay within the slot. From the following diagram we get, from geometry and trigonometry,

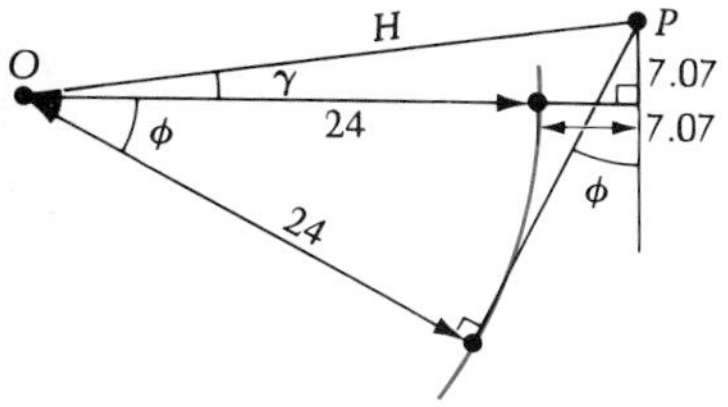

$$\gamma = \tan^{-1}\left(\frac{7.07}{31.07}\right) = 12.8°$$

$$H = \frac{7.07}{\sin \gamma} = 31.9 \text{ in.}$$

$$\varphi + \gamma = \cos^{-1}\left(\frac{24}{31.9}\right) = 41.2°$$

$$\varphi = 28.4°$$

These results will be used in what follows. Let us now call the ground frame $\mathcal{F}$ and then write

$$\mathbf{v}_{P/\mathcal{F}} = \mathbf{v}_{P/\mathcal{T}} + \mathbf{v}_{P_{\mathcal{T}}/\mathcal{F}}$$

(Continued)

But we may find $\mathbf{v}_{P/\mathcal{F}}$ by relating it to $\mathbf{v}_{O'/\mathcal{F}}$ (which vanishes):

$$\mathbf{v}_{P/\mathcal{F}} = \omega_{\mathcal{E}} \hat{\mathbf{k}} \times \mathbf{r}_{O'P} = -2\hat{\mathbf{k}} \times (7.07\hat{\mathbf{i}} + 7.07\hat{\mathbf{j}})$$

$$= 1.41\hat{\mathbf{i}} - 1.41\hat{\mathbf{j}} \text{ in./sec}$$

Therefore, substituting, we get

$$1.41\hat{\mathbf{i}} - 1.41\hat{\mathbf{j}} = v_{P/\mathcal{T}}\,(\sin 28.4^\circ\ \hat{\mathbf{i}} + \cos 28.4^\circ\ \hat{\mathbf{j}}) + \mathbf{v}_{P_{\mathcal{T}}/\mathcal{F}}$$

in which we have used the fact that we know the direction but not the magnitude of $\mathbf{v}_{P/\mathcal{T}}$. (It moves in the slot at the angle calculated earlier.) Further, relating the velocities of points $P_{\mathcal{T}}$ and O on $\mathcal{T}$, we have

$$\mathbf{v}_{P_{\mathcal{T}}/\mathcal{F}} = \overset{0}{\cancel{\mathbf{v}_{O/\mathcal{F}}}} + \omega_{\mathcal{T}} \hat{\mathbf{k}} \times \mathbf{r}_{OP_{\mathcal{T}}} = \omega_{\mathcal{T}} \hat{\mathbf{k}} \times (31.1\hat{\mathbf{i}} + 7.07\hat{\mathbf{j}})$$

or

$$\mathbf{v}_{P_{\mathcal{T}}/\mathcal{F}} = -7.07\omega_{\mathcal{T}}\hat{\mathbf{i}} + 31.1\omega_{\mathcal{T}}\hat{\mathbf{j}}$$

Substituting, and equating the coefficients of $\hat{\mathbf{i}}$ and of $\hat{\mathbf{j}}$, we get

$\hat{\mathbf{i}}$ coefficients: $\quad 0.476 v_{P/\mathcal{T}} - 7.07\omega_{\mathcal{T}} = 1.41$

$\hat{\mathbf{j}}$ coefficients: $\quad 0.880 v_{P/\mathcal{T}} + 31.1\omega_{\mathcal{T}} = -1.41$

Solving these equations gives

$$\mathbf{v}_{P/\mathcal{T}} = 1.61 \ \angle 28.4^\circ \text{ in./sec}$$

$$\omega_{\mathcal{T}} = -0.0909 \text{ rad/sec or } \boldsymbol{\omega}_{\mathcal{T}} = 0.0909 \circlearrowright \text{ rad/sec}$$

EXAMPLE 3.32

Crank $\mathcal{A}$ in Figure 1 rotates at constant angular velocity $\dot{\theta}\hat{\mathbf{k}}$. Plot the following two quantities as functions of θ for the case in which $D = 2\ell$:

a. The angle φ that locates slider $\mathcal{B}$
b. The ratio $\dot{\varphi}/\dot{\theta}$ of the angular speeds of $\mathcal{B}$ and $\mathcal{A}$

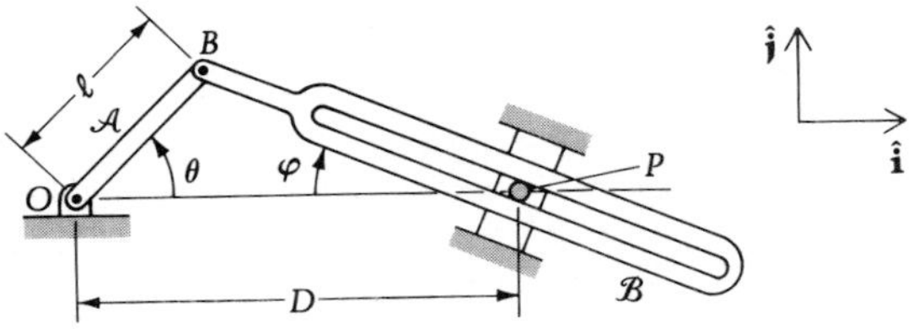

Figure 1

(Continued)

SOLUTION

We shall use the facts that P is fixed in the frame of reference $\mathcal{F}$ and that P has rectilinear motion relative to $\mathcal{B}$; thus Equation (3.51) yields

$$\mathbf{v}_{P/\mathcal{F}} = \mathbf{v}_{P/\mathcal{B}} + \mathbf{v}_{P_{\mathcal{B}}/\mathcal{F}}$$

$$\mathbf{0} = v_{P/\mathcal{B}}(\cos\varphi\,\hat{\mathbf{i}} - \sin\varphi\,\hat{\mathbf{j}}) + \mathbf{v}_{P_{\mathcal{B}}/\mathcal{F}} \qquad (1)$$

Now, since $P_{\mathcal{B}}$ and B are both points of the slider $\mathcal{B}$,

$$\mathbf{v}_{P_{\mathcal{B}}/\mathcal{F}} = \mathbf{v}_{B/\mathcal{F}} + (-\dot{\varphi}\hat{\mathbf{k}}) \times \overset{(D - \ell\cos\theta)\hat{\mathbf{i}} - \ell\sin\theta\,\hat{\mathbf{j}}}{\cancel{\mathbf{r}_{BP_{\mathcal{B}}}}}$$

But B is also a point of the crank $\mathcal{A}$, so that

$$\mathbf{v}_{B/\mathcal{F}} = \overset{\mathbf{0}}{\cancel{\mathbf{v}_{O/\mathcal{F}}}} + \dot{\theta}\hat{\mathbf{k}} \times \overset{\ell(\cos\theta\,\hat{\mathbf{i}} + \sin\theta\,\hat{\mathbf{j}})}{\cancel{\mathbf{r}_{OB}}}$$

$$= \ell\dot{\theta}(\cos\theta\,\hat{\mathbf{j}} - \sin\theta\,\hat{\mathbf{i}})$$

and we then obtain

$$\mathbf{v}_{P_{\mathcal{B}}/\mathcal{F}} = \ell\dot{\theta}(\cos\theta\,\hat{\mathbf{j}} - \sin\theta\,\hat{\mathbf{i}}) - \dot{\varphi}(D - \ell\cos\theta)\hat{\mathbf{j}} - \ell\dot{\varphi}\sin\theta\,\hat{\mathbf{i}} \qquad (2)$$

Eliminating $\mathbf{v}_{P_{\mathcal{B}}/\mathcal{F}}$ between (1) and (2) gives the following pair of scalar component equations:

$\hat{\mathbf{i}}$ coefficients: $\quad v_{P/\mathcal{B}}\cos\varphi - \ell\dot{\theta}\sin\theta - \ell\dot{\varphi}\sin\theta = 0$

$\hat{\mathbf{j}}$ coefficients: $\quad -v_{P/\mathcal{B}}\sin\varphi + \ell\dot{\theta}\cos\theta - \dot{\varphi}(D - \ell\cos\theta) = 0$

Eliminating $v_{P/\mathcal{B}}$, the magnitude of the velocity of the pin relative to $\mathcal{B}$, gives an equation that may be solved for the ratio of the variable $\dot{\varphi}$ to the constant $\dot{\theta}$:

$$\frac{\dot{\varphi}}{\dot{\theta}} = \frac{\ell\cos(\theta + \varphi)}{D\cos\varphi - \ell\cos(\theta + \varphi)} \qquad (3)$$

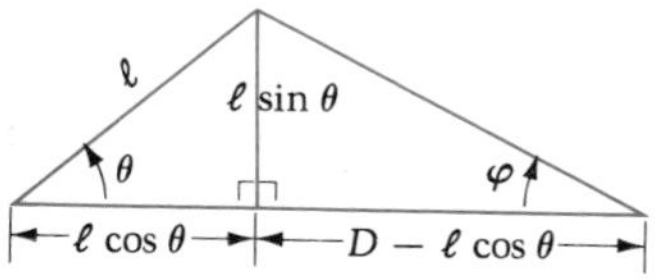

Figure 2

Figure 2 shows that

$$\tan\varphi = \frac{\ell\sin\theta}{D - \ell\cos\theta}$$

Therefore for $D = 2\ell$ we have

$$\frac{\dot{\varphi}}{\dot{\theta}} = \frac{\cos(\theta + \varphi)}{2\cos\varphi - \cos(\theta + \varphi)} \qquad \text{and} \qquad \varphi = \tan^{-1}\left(\frac{\sin\theta}{2 - \cos\theta}\right)$$

These relationships are graphed in Figure 3 for θ between 0 and 180°. About the line $\theta = 180°$, the φ versus θ curve is antisymmetric whereas the $\dot{\varphi}/\dot{\theta}$ versus θ curve is symmetric.

(Continued)

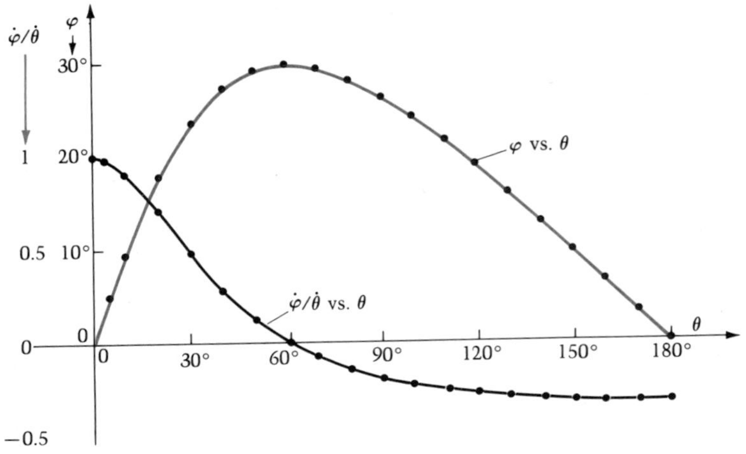

Figure 3

Question 3.19 Why is $\dot\varphi = 0$ at $\theta = 60°$?

Note that the equation (3) for $\dot\varphi$ may be checked by using the instantaneous center of $\mathcal{B}$ as shown in Figure 4. According to the law of sines:

$$\frac{H}{\sin(90° + \varphi)} = \frac{D}{\sin[90 - (\theta + \varphi)]}$$

$$H = \frac{D\cos\varphi}{\cos(\theta + \varphi)}$$

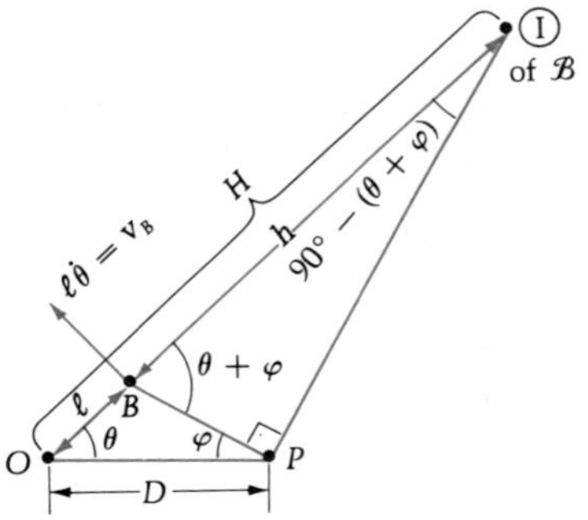

Figure 4

Therefore

$$h = \frac{D\cos\varphi}{\cos(\theta + \varphi)} - \ell$$

and

$$\dot\varphi = \frac{v_B}{h} = \frac{\ell\dot\theta\cos(\theta + \varphi)}{D\cos\varphi - \ell\cos(\theta + \varphi)}$$

In working out the following problems, the student is urged to begin by thinking carefully about the selection of a point whose velocities in two frames are to be related with Equation (3.51).

PROBLEMS / Section 3.7

3.155 Bar $\mathcal{B}$ slides through a collar in body $\mathcal{C}$ (see Figure P3.155) and is pinned at P to a second bar $\mathcal{R}$. Both $\mathcal{R}$ and $\mathcal{C}$ are pinned to the reference frame as shown, and $\mathcal{R}$ rotates with constant angular velocity $\dot{\theta}_{\mathcal{R}} = 1$ rad/s counterclockwise. Find the angular velocity of $\mathcal{C}$ when point P is at the top of the circle on which it travels.

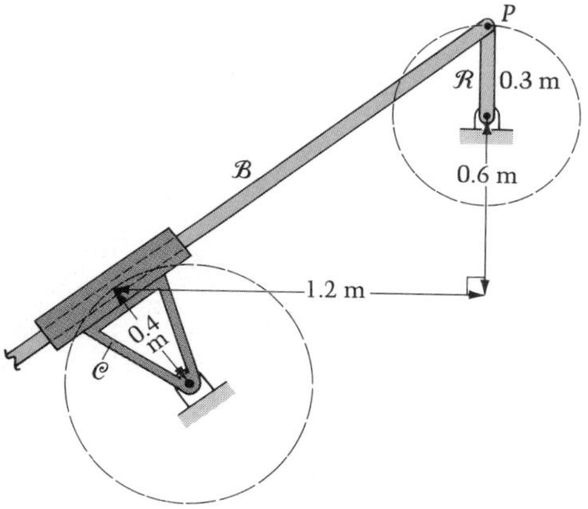

Figure P3.155

3.156 Rods $\mathcal{R}$ and $\mathcal{L}$ in Figure P3.156 are pinned at O and O′ to a reference frame $\mathcal{F}$. Rod $\mathcal{L}$ is also pinned to the slotted body $\mathcal{B}$ at B. The upper end of $\mathcal{R}$ is pinned at P to a roller that moves freely in the slot of $\mathcal{B}$. The angular velocities of rod $\mathcal{R}$ and link $\mathcal{L}$ are constants:

$$\boldsymbol{\omega}_{\mathcal{R}} = 0.2 \circlearrowright \text{ rad/s}$$

$$\boldsymbol{\omega}_{\mathcal{L}} = 0.4 \circlearrowright \text{ rad/s}$$

Determine the velocity of P in $\mathcal{B}$ and the angular velocity of $\mathcal{B}$ at the given instant.

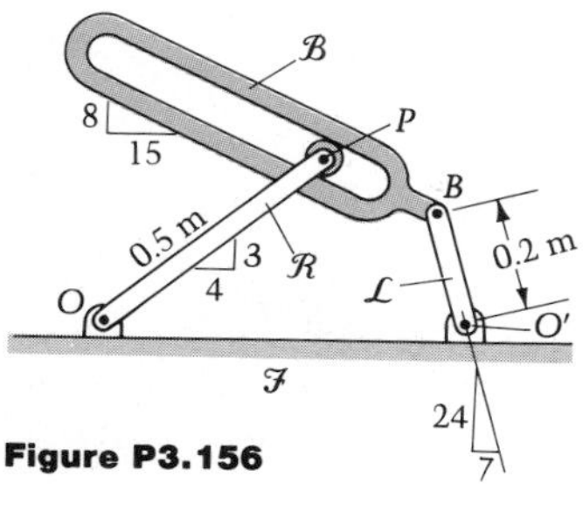

Figure P3.156

3.157 Figure P3.157 illustrates a "Geneva mechanism," in which disk $\mathcal{A}$ is driven with a constant counterclockwise angular speed and produces an intermittent (starting and stopping, waiting, then repeating) rotational motion of the slotted disk $\mathcal{B}$. Pin P is fixed to $\mathcal{A}$ and drives disk $\mathcal{B}$ by pressing on the surfaces of the slots. Show with the use of Equation (3.51) that disk $\mathcal{B}$ will have zero angular speed in the two positions shown, a varying angular speed in between these positions, and zero angular speed while P is returning to the $\theta = 135°$ position. Note that the operation of the mechanism requires that the distance between O and O′ be $\sqrt{2}R$.

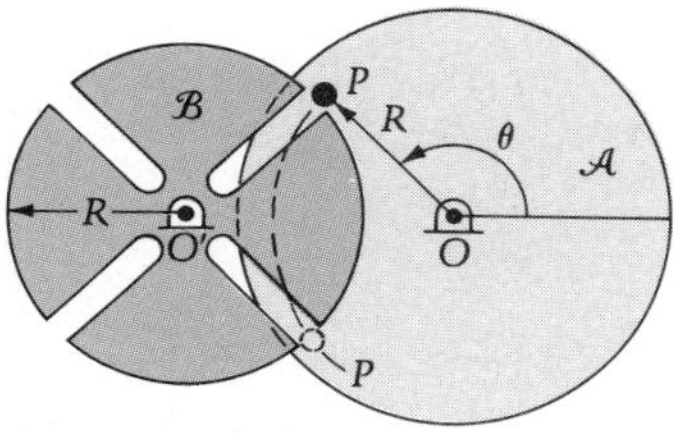

Figure P3.157

3.158 In the preceding problem let $R = 0.1$ m and $\dot{\theta}_{\mathcal{A}} = 5 \circlearrowleft$ rad/s = constant. Find the angular velocity of the slotted body $\mathcal{B}$ at the instant when $\theta = 160°$.

3.159 Cylinder $\mathcal{C}$ in Figure P3.159 rolls on a circular surface. When it is at the lowest point of the circle, its angular velocity and acceleration are $\boldsymbol{\omega}_{\mathcal{C}} = 0.2 \circlearrowleft$ rad/s and $\boldsymbol{\alpha}_{\mathcal{C}} = 0.02 \circlearrowleft$ rad/s^2. Rod $\mathcal{B}$ is pinned to $\mathcal{C}$ at E and is also pinned to a block at P that slides in the slot of $\mathcal{S}$. The constant angular velocity of $\mathcal{S}$ is $0.3 \circlearrowleft$ rad/s. Find the velocity of P in $\mathcal{S}$ and the angular velocity of $\mathcal{B}$ at the given instant.

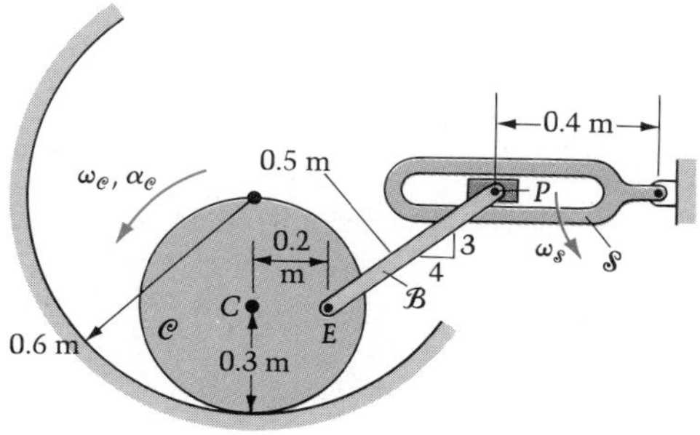

Figure P3.159

3.160 Show that the rigid-body velocity equation (3.8) can be derived from Equation (3.49). *Hint:* Fix P to $\mathcal{B}$ and the equation will relate the velocities in $\mathcal{F}$ of the two points P and O' of $\mathcal{B}$.

3.161 In Figure P3.161 collar $\mathcal{C}$ is fixed to arm $\mathcal{A}$ and slides along rod $\mathcal{R}$. Arm $\mathcal{A}$ is pinned to a second collar $\mathcal{H}$ at A; this collar slides on rod $\mathcal{L}$. At the given instant, $\boldsymbol{\omega}_{\mathcal{R}} = 0.2$ ↻ rad/s, $\boldsymbol{\omega}_{\mathcal{L}} = 0.1$ ↻ rad/s, and the velocity of all points of $\mathcal{C}$ relative to rod $\mathcal{R}$ is 0.3 m/s outward (along OC). Give the angular velocity of $\mathcal{A}$ by inspection, and find the velocity of the points of $\mathcal{H}$ relative to $\mathcal{L}$.

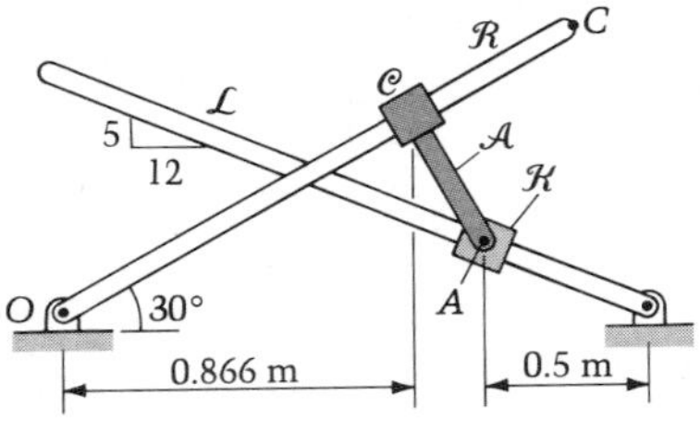

Figure P3.161

3.162 Referring to Example 3.29 for the meanings of the symbols, show that $\mathbf{v}_{Q/\mathcal{R}} = -\mathbf{v}_{T/\mathcal{B}}$.

3.163 Collars $\mathcal{A}$ and $\mathcal{C}$ in Figure P3.163 are pinned together at C, and they slide on rods $\mathcal{B}$ and $\mathcal{S}$, respectively. Rod $\mathcal{B}$ has a constant angular velocity $\dot{\theta}\hat{\mathbf{k}}$ for $10° \le \theta \le 45°$. Find the velocity of point C relative to $\mathcal{B}$, as a function of D, θ, and $\dot{\theta}$ in this range of angles.

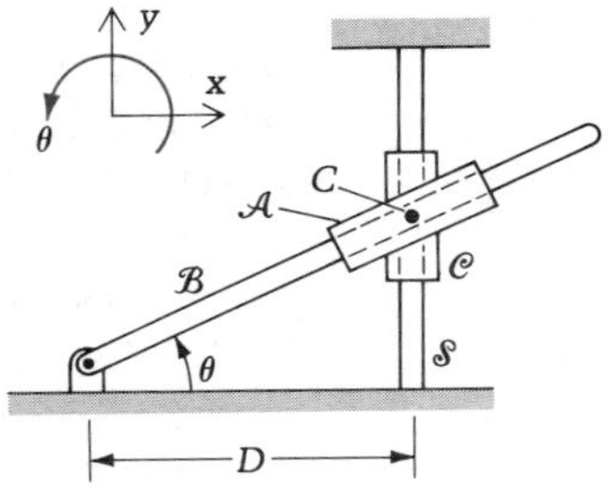

Figure P3.163

3.164 Bar $\mathcal{A}$ in Figure P3.164 is turning clockwise with angular speed 0.25 rad/s, pushing bar $\mathcal{B}$ as it goes. Find $\boldsymbol{\omega}_{\mathcal{B}}$ at the given instant.

3.165 Rod $\mathcal{R}$ is pinned to the ceiling at A and slides on wedge $\mathcal{W}$ at B. (See Figure P3.165.) The wedge moves to the right with constant velocity of 5 ft/sec. Find the angular velocity of the rod in the position shown.

3.166 A mechanism consists of crank $\mathcal{C}$ pinned to O, rocker $\mathcal{R}$ pinned to O′, and a small body $\mathcal{B}$ that is pinned to $\mathcal{C}$ and slides in the slot of $\mathcal{R}$. (See Figure P3.166.) The length of $\mathcal{C}$ is $\ell = D\sqrt{2}$, where D is the distance between O and O′. If $\mathcal{C}$ has constant angular velocity $\omega_{\mathcal{C}}$ ↻ over a range of its motion, find $\omega_{\mathcal{R}}$ when: (a) $\theta_{\mathcal{R}} = \tan^{-1}(1/2)$; (b) $\theta_{\mathcal{R}} = 90°$.

3.167 Plank $\mathcal{P}$ slides on the floor at A and on block $\mathcal{D}$ at Q. Block $\mathcal{D}$ moves to the right with a constant velocity of 6 ft/sec while end A moves to the left with a constant velocity of 4 ft/sec. For the position shown in Figure P3.167, find the angular velocity of the plank.

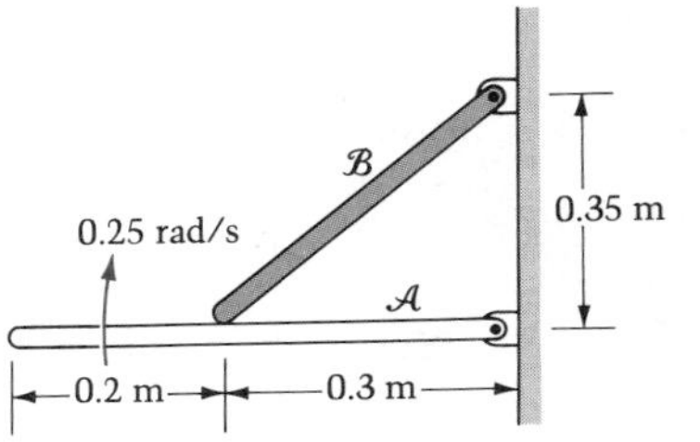

Figure P3.164

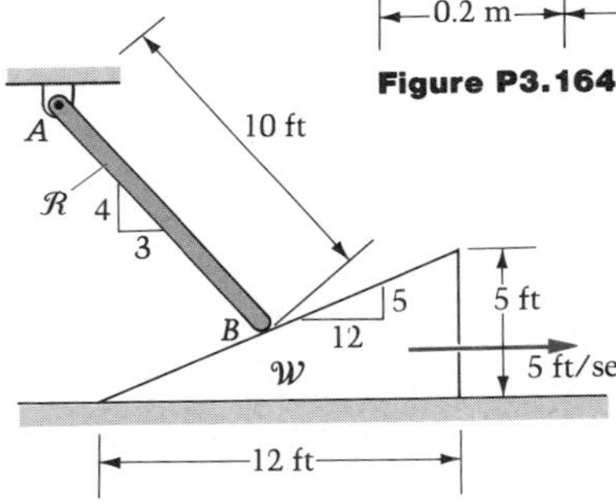

Figure P3.165

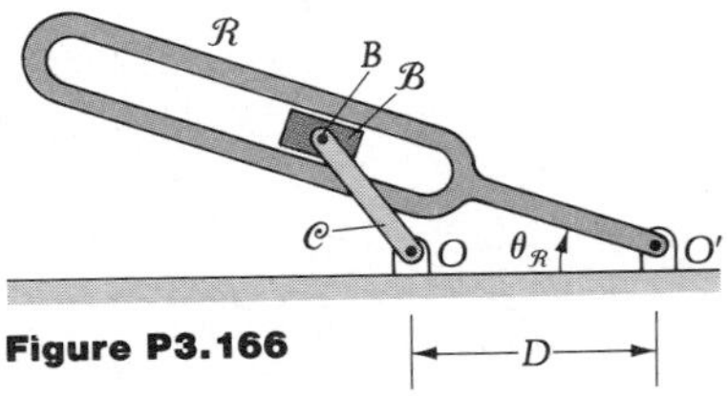

Figure P3.166

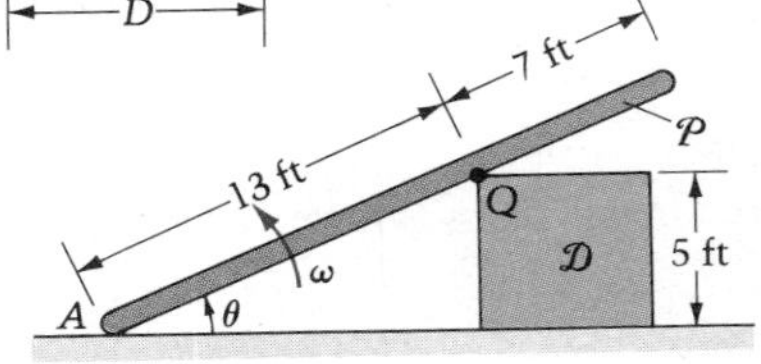

Figure P3.167

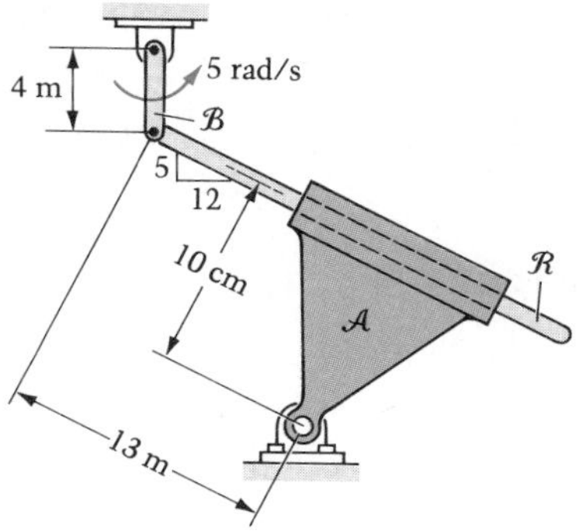

Figure P3.168

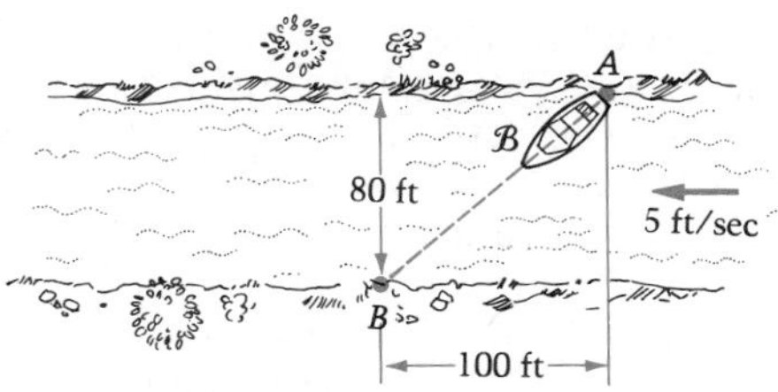

Figure P3.169

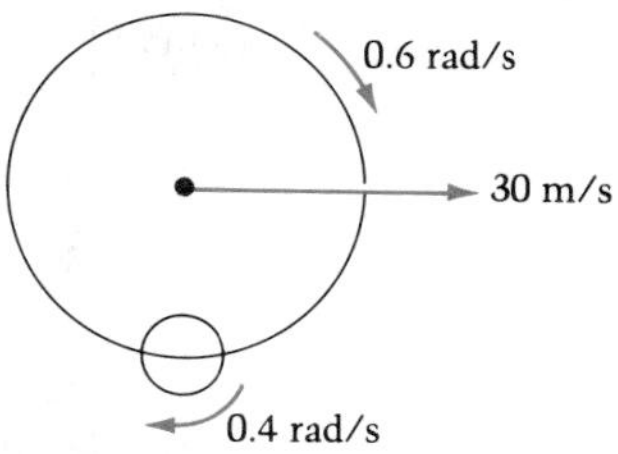

Figure P3.170

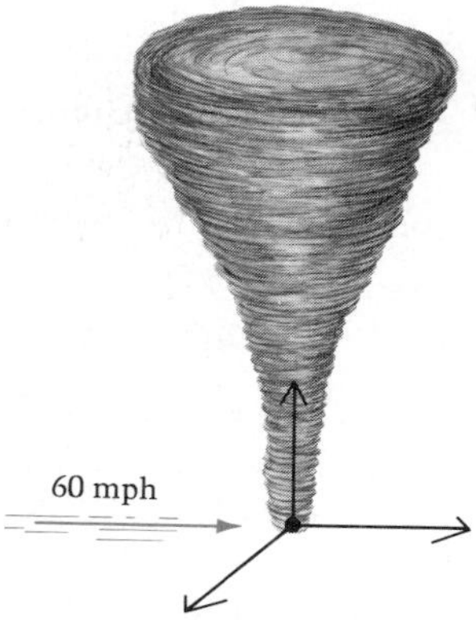

Figure P3.171

3.168 Rod $\mathcal{B}$ in Figure P3.168 has angular velocity 5 ⟲ rad/s. It is pinned to another rod $\mathcal{R}$, which passes through a slot in $\mathcal{A}$ as shown. At the given instant, find the angular velocity of body $\mathcal{A}$ and the velocity of any point of $\mathcal{R}$ relative to $\mathcal{A}$. *Hint:* All points of $\mathcal{R}$ translate in $\mathcal{A}$—what does this mean about the angular velocities of $\mathcal{R}$ and $\mathcal{A}$?

3.169 Boat $\mathcal{B}$ in Figure P3.169 departs from A and is supposed to arrive at point B some 100 ft downstream and on the other side of a river with a current of 5 ft/sec. If $\mathcal{B}$ can move at 10 ft/sec relative to the water, and if it travels on a straight line from A toward B, how long will it take?

3.170 In a large tornado there can be "suction vortices" circling about the tornado center near the (horizontal) circle of maximum wind speed. Suppose that a tornado is translating at 30 m/s and that particles in the circle of maximum wind speed have radii of curvature of 100 m in a frame translating with the tornado. In this frame (Figure P3.170), the circle turns at 0.6 rad/s. Moreover, a suction vortex rotates with its center fixed on this circle at a spin speed relative to the circle of 4 rad/s and with a radius of 10 m. Find the speed of the particle in the tornado that has the maximum velocity relative to the ground. Consider only horizontal velocities.

3.171 In analyzing tornados, it is sometimes assumed that in a frame translating along the ground with z representing the centerline of the tornado, the radial (r) velocity component of its particles is half the transverse (θ) and that the vertical component (z) varies from one-third to two-thirds of the transverse:

$$v_r \approx \frac{v_\theta}{2} \qquad v_z \approx \frac{v_\theta}{3} \text{ to } \frac{2v_\theta}{3}$$

Suppose a tornado is translating horizontally at 60 mph (Figure P3.171) and the particles on its circle of maximum transverse velocity are traveling at 120 mph. Find the velocity of the point with greatest speed in the tornado.

3.8 Relationship Between the Accelerations of a Point P with Respect to Two Different Frames of Reference

We shall now derive the companion equation to (3.51), this one relating the accelerations of P in $\mathcal{F}$ and $\mathcal{B}$. Differentiating Equation (3.49) in $\mathcal{F}$, we get

$$\mathbf{a}_{P/\mathcal{F}} = {}^{\mathcal{F}}\dot{\mathbf{v}}_{P/\mathcal{B}} + \mathbf{a}_{O'/\mathcal{F}} + \ddot{\theta}\hat{\mathbf{k}} \times \mathbf{r}_{O'P} + \dot{\theta}\hat{\mathbf{k}} \times {}^{\mathcal{F}}\dot{\mathbf{r}}_{O'P} \tag{3.52}$$

Using Equation (3.44) to "move the derivative" in the first and last terms on the right side of Equation (3.52) gives

$$\mathbf{a}_{P/\mathcal{I}} = ({}^{\mathcal{B}}\dot{\mathbf{v}}_{P/\mathcal{B}} + \dot{\theta}\hat{\mathbf{k}} \times \mathbf{v}_{P/\mathcal{B}}) + \mathbf{a}_{O'/\mathcal{I}} + \ddot{\theta}\hat{\mathbf{k}} \times \mathbf{r}_{O'P}$$
$$+ \dot{\theta}\hat{\mathbf{k}} \times ({}^{\mathcal{B}}\dot{\mathbf{r}}_{O'P} + \dot{\theta}\hat{\mathbf{k}} \times \mathbf{r}_{O'P}) \tag{3.53}$$

Recognizing that ${}^{\mathcal{B}}\dot{\mathbf{r}}_{O'P} = \mathbf{v}_{P/\mathcal{B}}$ and that ${}^{\mathcal{B}}\dot{\mathbf{v}}_{P/\mathcal{B}} = \mathbf{a}_{P/\mathcal{B}}$, and rearranging terms, from Equation (3.53) we obtain

$$\mathbf{a}_{P/\mathcal{I}} = \mathbf{a}_{P/\mathcal{B}} + (\mathbf{a}_{O'/\mathcal{I}} + \ddot{\theta}\hat{\mathbf{k}} \times \mathbf{r}_{O'P} - \dot{\theta}^2\mathbf{r}_{O'P}) + 2\dot{\theta}\hat{\mathbf{k}} \times \mathbf{v}_{P/\mathcal{B}} \tag{3.54}$$

in which $\dot{\theta}\hat{\mathbf{k}} \times (\dot{\theta}\hat{\mathbf{k}} \times \mathbf{r}_{O'P}) = -\dot{\theta}^2\mathbf{r}_{O'P}$ as we have already seen in Section 3.5.

The parenthesized term in Equation (3.54) is seen from Equation (3.19) to be $\mathbf{a}_{P_{\mathcal{B}}/\mathcal{I}}$, where, as before, $P_{\mathcal{B}}$ is the point of $\mathcal{B}$ (or $\mathcal{B}$ extended) that is coincident with P. Therefore we have the following for our result:

$$\mathbf{a}_{P/\mathcal{I}} = \mathbf{a}_{P/\mathcal{B}} + \mathbf{a}_{P_{\mathcal{B}}/\mathcal{I}} + 2\dot{\theta}\hat{\mathbf{k}} \times \mathbf{v}_{P/\mathcal{B}} \tag{3.55}$$

In words: The acceleration of P in $\mathcal{I}$ equals its acceleration in $\mathcal{B}$, plus the acceleration in $\mathcal{I}$ of the point of $\mathcal{B}$ coincident with P, *plus* the **Coriolis acceleration,** $2\dot{\theta}\hat{\mathbf{k}} \times \mathbf{v}_{P/\mathcal{B}}$. The Coriolis acceleration is seen to provide an unexpected but essential difference between the forms of Equations (3.51) and (3.55).

Question 3.20 If we differentiate Equation (3.51) instead of (3.49), we might (*erroneously!*) obtain

$$\begin{aligned}\mathbf{a}_{P/\mathcal{I}} &= {}^{\mathcal{I}}\dot{\mathbf{v}}_{P/\mathcal{B}} + \mathbf{a}_{P_{\mathcal{B}}/\mathcal{I}} \\ &= {}^{\mathcal{B}}\dot{\mathbf{v}}_{P/\mathcal{B}} + \boldsymbol{\omega}_{\mathcal{B}/\mathcal{I}} \times \mathbf{v}_{P/\mathcal{B}} + \mathbf{a}_{P_{\mathcal{B}}/\mathcal{I}} \\ &= \mathbf{a}_{P/\mathcal{B}} + \mathbf{a}_{P_{\mathcal{B}}/\mathcal{I}} + \boldsymbol{\omega}_{\mathcal{B}/\mathcal{I}} \times \mathbf{v}_{P/\mathcal{B}}\end{aligned}$$

and we come up (incorrectly) short by half on the Coriolis term. What is wrong with this approach?

In the same procedure we used for velocities, we can simplify the notation of Equation (3.54) if there is but one "moving frame" ($\mathcal{B}$) involved, which is in motion relative to the reference frame ($\mathcal{I}$):

$$\mathbf{a}_P = \mathbf{a}_{\text{rel}} + \mathbf{a}_{O'} + \boldsymbol{\alpha} \times \mathbf{r} - \omega^2\mathbf{r} + 2\boldsymbol{\omega} \times \mathbf{v}_{\text{rel}}$$

In this equation, $\mathbf{a}_P$ and $\mathbf{a}_{\text{rel}}$ are the respective accelerations of P in $\mathcal{I}$ and in $\mathcal{B}$. The vectors $\boldsymbol{\omega}$ and $\boldsymbol{\alpha}$ are the angular velocity and angular acceleration of $\mathcal{B}$ in $\mathcal{I}$ (equal to $\dot{\theta}\hat{\mathbf{k}}$ and $\ddot{\theta}\hat{\mathbf{k}}$) and $\mathbf{r} = \mathbf{r}_{O'P}$ (the position vector of P in the "moving frame"). We now consider several examples showing the use of Equation (3.55).

EXAMPLE 3.33

In Example 3.29, suppose that at the given instant all the data are the same and in addition $\boldsymbol{\alpha}_{\mathcal{R}} = 30$ rad/sec^2 ↺. Find the acceleration of block $\mathcal{B}$ at $\theta = 60°$ (see the accompanying diagram).

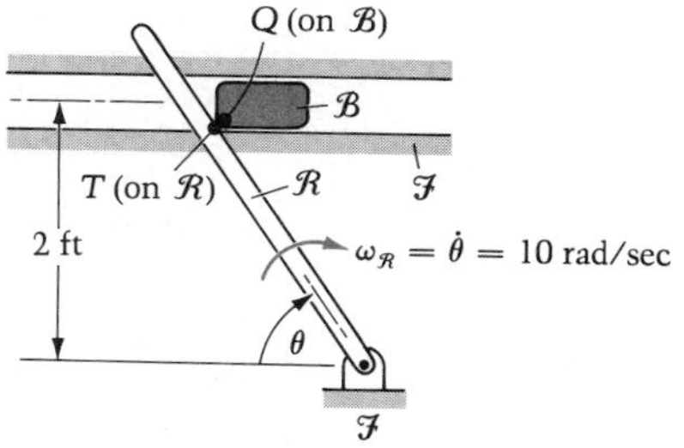

SOLUTION

Note that if we again use Q as our point that is moving relative to two bodies ($\mathcal{R}$ and $\mathcal{F}$), we know the direction of $\mathbf{a}_{Q/\mathcal{R}}$. (It is *along* $\mathcal{R}$ since point Q moves on a *straight line* in $\mathcal{R}$. We do not, however, know the motion in $\mathcal{B}$ of point T.

Equation (3.54) thus gives

$$\mathbf{a}_{Q/\mathcal{F}} = \mathbf{a}_{Q/\mathcal{R}} + \mathbf{a}_{Q_{\mathcal{R}}/\mathcal{F}} + 2\omega_{\mathcal{R}}\hat{\mathbf{k}} \times \mathbf{v}_{Q/\mathcal{R}}$$

Noting that point $Q_{\mathcal{R}}$ is point T in Example 3.29, we obtain

$$a_{Q/\mathcal{F}}\hat{\mathbf{i}} = a_{Q/\mathcal{R}}\left(-\frac{1}{2}\hat{\mathbf{i}} + \frac{\sqrt{3}}{2}\hat{\mathbf{j}}\right) + \Big(\overset{\mathbf{0}}{\cancel{\mathbf{a}_O}} + \overset{30}{\cancel{\alpha_{\mathcal{R}}}}\hat{\mathbf{k}} \times \overset{\left(\frac{-2}{\sqrt{3}}\hat{\mathbf{i}} + 2\hat{\mathbf{j}}\right)}{\cancel{\mathbf{r}_{OP}}} - \overset{10^2}{\cancel{\omega^2}}\mathbf{r}_{OP}\Big)$$

$$+ 2(-10\hat{\mathbf{k}}) \times \left(\frac{40}{3}\right)\left(-\frac{1}{2}\hat{\mathbf{i}} + \frac{\sqrt{3}}{2}\hat{\mathbf{j}}\right)$$

This result yields the following scalar component equations:

$\hat{\mathbf{j}}$ coefficients:
$$0 = a_{Q/\mathcal{R}}\frac{\sqrt{3}}{2} - \frac{60}{\sqrt{3}} - 200 + \frac{800}{6}$$

$$a_{Q/\mathcal{R}} = 117 \text{ in./sec}^2$$

$\hat{\mathbf{i}}$ coefficients:
$$a_{Q/\mathcal{F}} = 117\left(-\frac{1}{2}\right) - 60 + \frac{200}{\sqrt{3}} + \frac{800}{6}\sqrt{3}$$

$$\mathbf{a}_{Q/\mathcal{F}} = a_{Q/\mathcal{F}}\hat{\mathbf{i}} = 228\,\hat{\mathbf{i}} \text{ in./sec}^2 \qquad \text{(the acceleration of } \mathcal{B})$$

EXAMPLE 3.34

If in Example 3.30 we have the additional data that $\boldsymbol{\alpha}_{\mathcal{R}} = 0.25\hat{\mathbf{k}}$ rad/s^2 at the given time, find $\boldsymbol{\alpha}_{\mathcal{B}}$ and $\mathbf{a}_{P/\mathcal{B}}$. (See the diagram.)

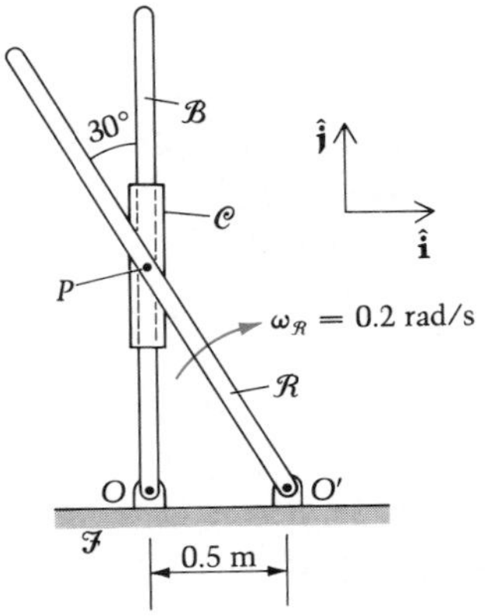

SOLUTION

Relating the accelerations of P in $\mathcal{B}$ and $\mathcal{F}$, we have

$$\mathbf{a}_{P/\mathcal{F}} = \mathbf{a}_{P/\mathcal{B}} + \mathbf{a}_{P_{\mathcal{B}}/\mathcal{F}} + 2\boldsymbol{\omega}_{\mathcal{B}/\mathcal{F}} \times \mathbf{v}_{P/\mathcal{B}}$$

$$\overbrace{\underbrace{\mathbf{a}_{O'/\mathcal{F}}}_{\text{zero}} + \boldsymbol{\alpha}_{\mathcal{R}} \times \mathbf{r}_{O'P} - \omega_{\mathcal{R}}^2 \mathbf{r}_{O'P}}^{\text{on } \mathcal{R}} = a_{P/\mathcal{B}}\hat{\mathbf{j}} + \overbrace{\underbrace{\mathbf{a}_{O/\mathcal{F}}}_{\text{zero}} + \boldsymbol{\alpha}_{\mathcal{B}} \times \mathbf{r}_{OP_{\mathcal{B}}} - \omega_{\mathcal{B}}^2 \mathbf{r}_{OP_{\mathcal{B}}}}^{\text{on } \mathcal{B}}$$

$$+ 2(-0.2\hat{\mathbf{k}}) \times (0.1\hat{\mathbf{j}})$$

in which the answers to Example 3.30 are used in the Coriolis term. Filling in the variables, we get

$$0.25\hat{\mathbf{k}} \times (-0.5\hat{\mathbf{i}} + 0.866\hat{\mathbf{j}}) - 0.04(-0.5\hat{\mathbf{i}} + 0.866\hat{\mathbf{j}})$$
$$= a_{P/\mathcal{B}}\hat{\mathbf{j}} + \alpha_{\mathcal{B}}\hat{\mathbf{k}} \times 0.866\hat{\mathbf{j}} - 0.04(0.866\hat{\mathbf{j}}) + 0.04\hat{\mathbf{i}}$$

$\hat{\mathbf{i}}$ coefficients: $\quad -0.217 + 0.02 = -0.866\alpha_{\mathcal{B}} + 0.04$

$$\alpha_{\mathcal{B}} = 0.274 \Rightarrow \boldsymbol{\alpha}_{\mathcal{B}} = 0.274\hat{\mathbf{k}} \text{ rad/s}^2$$

$\hat{\mathbf{j}}$ coefficients: $\quad -0.125 - 0.0346 = a_{P/\mathcal{B}} - 0.0346$

$$\mathbf{a}_{P/\mathcal{B}} = -0.125\hat{\mathbf{j}} \text{ m/s}^2$$

Thus point P is slowing down as it moves outward along $\mathcal{B}$, and $\mathcal{B}$ is slowing down as it rotates clockwise.

EXAMPLE 3.35

If in Example 3.31 we add $\boldsymbol{\alpha}_{\mathcal{C}} = 10 \circlearrowleft$ rad/sec² to the data, find the angular acceleration of $\mathcal{T}$ and the acceleration of P relative to $\mathcal{T}$. (See the accompanying diagram.)

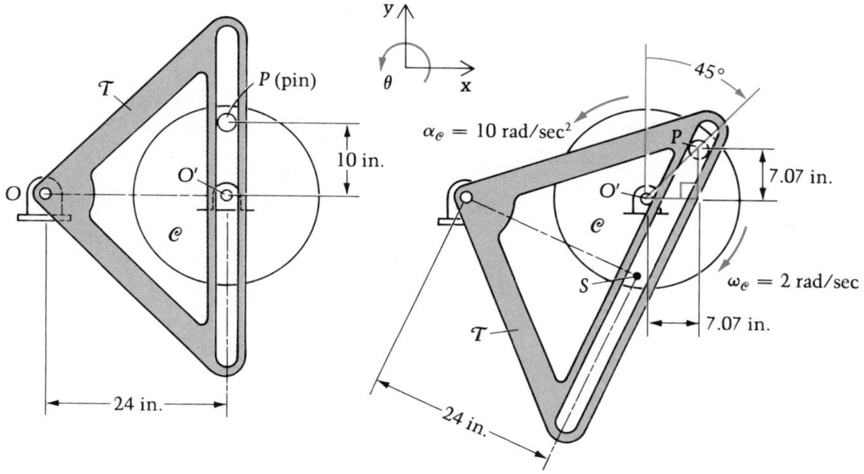

SOLUTION

Again we apply Equation (3.55):

$$\mathbf{a}_{P/\mathcal{F}} = \mathbf{a}_{P/\mathcal{T}} + \mathbf{a}_{P_{\mathcal{T}}/\mathcal{F}} + 2\omega_{\mathcal{T}}\hat{\mathbf{k}} \times \mathbf{v}_{P/\mathcal{T}}$$

$$\overset{\mathbf{0}}{\mathbf{a}_{O'/\mathcal{F}}} + \overset{10}{\alpha_{\mathcal{C}}}\hat{\mathbf{k}} \times \overset{(7.07\,\hat{\mathbf{i}} + 7.07\,\hat{\mathbf{j}})}{\mathbf{r}_{O'P}} - \overset{2^2}{\omega_{\mathcal{C}}^2}\mathbf{r}_{O'P} = a_{P/\mathcal{T}}(\sin 28.4^\circ\ \hat{\mathbf{i}} + \cos 28.4^\circ\ \hat{\mathbf{j}}) + \overset{\mathbf{0}}{\mathbf{a}_{O/\mathcal{F}}}$$

$$+ \alpha_{\mathcal{T}}\hat{\mathbf{k}} \times (31.1\,\hat{\mathbf{i}} + 7.07\,\hat{\mathbf{j}}) - \overset{0.091^2}{\omega_{\mathcal{T}}^2}(31.1\,\hat{\mathbf{i}} + 7.07\,\hat{\mathbf{j}})$$

$$+ 2(-0.091\,\hat{\mathbf{k}}) \times 1.62(\sin 28.4^\circ\ \hat{\mathbf{i}} + \cos 28.4^\circ\ \hat{\mathbf{j}})$$

and again we equate the $\hat{\mathbf{i}}$ coefficients, and then the $\hat{\mathbf{j}}$ coefficients, to obtain the scalar component equations:

$\hat{\mathbf{i}}$ coefficients: $\quad -99.0 = 0.476a_{P/\mathcal{T}} - 7.07\alpha_{\mathcal{T}} - 0.258 + 0.259$

$\hat{\mathbf{j}}$ coefficients: $\quad 42.4 = a_{P/\mathcal{T}}(0.880) + 31.1\alpha_{\mathcal{T}} - 0.0586 - 0.140$

Solving these equations results in

$$\boldsymbol{\alpha}_{\mathcal{T}} = 5.12 \text{ rad/sec}^2 \circlearrowleft$$

$$a_{P/\mathcal{T}} = -132 \text{ in./sec}^2 \text{ or } \mathbf{a}_{P/\mathcal{T}} = 132 \measuredangle 28.4^\circ \text{ in./sec}^2$$

EXAMPLE 3.36

Pin P in the accompanying diagram is attached to cart $\mathcal{B}$ and slides in the smooth slot cut in wheel $\mathcal{C}$. The wheel rolls on the rough plane $\mathcal{I}$. The cart's position is given by $\mathbf{x}_B = 0.3t^2$, with $\mathbf{x}_B$ in meters when t is in seconds. Find $\boldsymbol{\omega}_{\mathcal{C}}$ and $\boldsymbol{\alpha}_{\mathcal{C}}$ at the given instant (which is at $t = 3$ s), and determine the acceleration of P in the slot at this time.

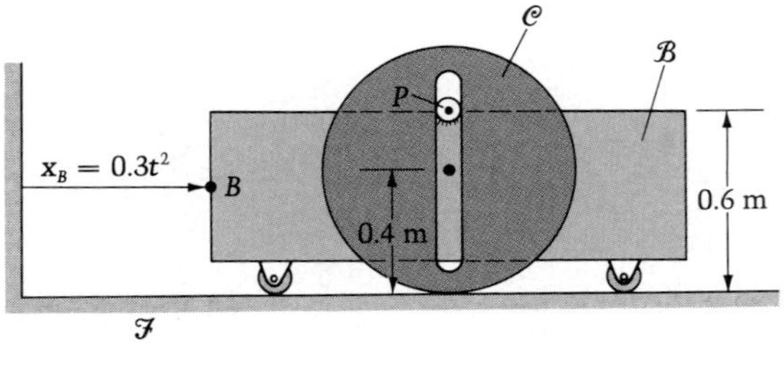

SOLUTION

Since $\mathbf{x}_B = 0.3t^2$, we have $\mathbf{v}_B = 0.6t\hat{\mathbf{i}}$ and $\mathbf{a}_B = 0.6\hat{\mathbf{i}}$; these are the velocity and acceleration vectors of all points of the cart $\mathcal{B}$. At $t = 3$, we have $\mathbf{v}_B = 1.8\hat{\mathbf{i}}$ m/s and $\mathbf{a}_B = 0.6\hat{\mathbf{i}}$ m/s².

Relating the velocities of P in frames $\mathcal{C}$ and $\mathcal{I}$, we obtain

$$\mathbf{v}_{P/\mathcal{I}} = \mathbf{v}_{P/\mathcal{C}} + \mathbf{v}_{P_{\mathcal{C}}/\mathcal{I}}$$

$$v_B\hat{\mathbf{i}} = v_{P/\mathcal{C}}\hat{\mathbf{j}} + 0.6\omega_{\mathcal{C}}\hat{\mathbf{i}}$$

$\hat{\mathbf{i}}$ coefficients: $\quad v_B = 1.8 = 0.6\omega_{\mathcal{C}} \Rightarrow \boldsymbol{\omega}_{\mathcal{C}} = 3 \circlearrowright$ rad/s

$\hat{\mathbf{j}}$ coefficients: $\quad v_{P/\mathcal{C}} = 0$

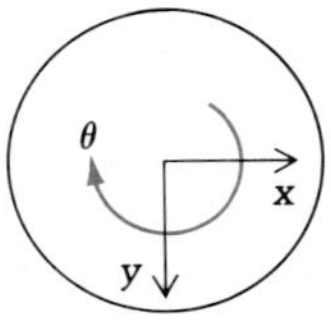

Relating the accelerations using Equation (3.54), we get

$$\mathbf{a}_{P/\mathcal{I}} = \mathbf{a}_{P/\mathcal{C}} + \underbrace{\mathbf{a}_{P_{\mathcal{C}}/\mathcal{I}}} + 2\boldsymbol{\omega}_{\mathcal{C}/\mathcal{I}} \times \mathbf{v}_{P/\mathcal{C}}$$

$$a_B\hat{\mathbf{i}} = a_{P/\mathcal{C}}\hat{\mathbf{j}} + \overbrace{0.6\alpha_{\mathcal{C}}\hat{\mathbf{i}} + 0.2\omega_{\mathcal{C}}^2\hat{\mathbf{j}}} + \mathbf{0}$$

Here we have related $P_{\mathcal{C}}$ (the point of $\mathcal{C}$ extended coincident with P) and the center of $\mathcal{C}$ (call it C) to get

$$\mathbf{a}_{P_{\mathcal{C}}/\mathcal{I}} = \mathbf{a}_C + \alpha_{\mathcal{C}}\hat{\mathbf{k}} \times \cancelto{-0.2\hat{\mathbf{j}}}{\mathbf{r}_{CP_{\mathcal{C}}}} - \omega_{\mathcal{C}}^2\mathbf{r}_{CP_{\mathcal{C}}}$$

$$= \cancelto{0.4}{R}\,\alpha_{\mathcal{C}}\hat{\mathbf{i}} + 0.2\alpha_{\mathcal{C}}\hat{\mathbf{i}} + 0.2\omega_{\mathcal{C}}^2\hat{\mathbf{j}}$$

Solving, we obtain

$\hat{\mathbf{i}}$ coefficients: $\quad a_B = 0.6 = 0.6\alpha_{\mathcal{C}} \Rightarrow \boldsymbol{\alpha}_{\mathcal{C}} = 1 \circlearrowright$ rad/s²

$\hat{\mathbf{j}}$ coefficients: $\quad a_{P/\mathcal{C}} = -0.2\omega_{\mathcal{C}}^2 = -0.2(3^2) = -1.8$ m/s²

(Continued)

Thus the acceleration of P in the slot (which is its acceleration in $\mathcal{C}$) is $1.8 \uparrow \text{m/s}^2$; it is upward because we assumed it to be in the positive y direction ($\downarrow$) and got a negative answer. Note that although P is momentarily stopped in the slot ($v_{P/\mathcal{C}} = 0$), it has to be "getting ready" to move outward since $\mathcal{B}$ is translating to the right and $\mathcal{C}$ is rolling that way. This is what is indicated by $\mathbf{a}_{P/\mathcal{C}} = 1.8 \uparrow \text{m/s}^2$.

PROBLEMS / Section 3.8

3.172 The mechanism shown in Figure P3.172 is used to raise and lower hammer $\mathcal{H}$. The 26-cm crank $\mathcal{C}$ turns clockwise at the constant rate of 30 rpm. It is pinned to block $\mathcal{B}$, which slides in a slot in $\mathcal{H}$. If at $t = 0$ point A is directly above O, find the velocity and acceleration of $\mathcal{H}$ as a function of time. (The block and hammer are slightly offset from $\mathcal{C}$ so they do not interfere with the pin at O.)

3.173 Rods $\mathcal{A}$ and $\mathcal{B}$ (see Figure P3.173) pass smoothly through the short collars, which can turn relative to each other by virtue of the ball and socket connection. If bars $\mathcal{A}$ and $\mathcal{B}$ turn with constant angular velocities 0.4 ↺ rad/sec and 0.2 ↻ rad/sec, respectively, find the velocity and acceleration of the ball and socket connection with respect to $\mathcal{A}$ and to $\mathcal{B}$ in the indicated position.

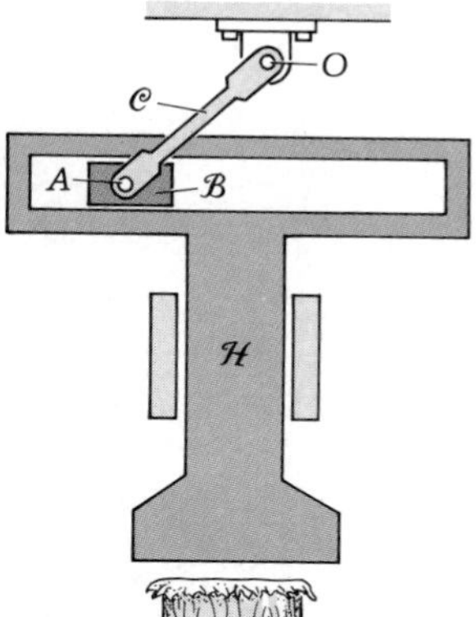

Figure P3.172

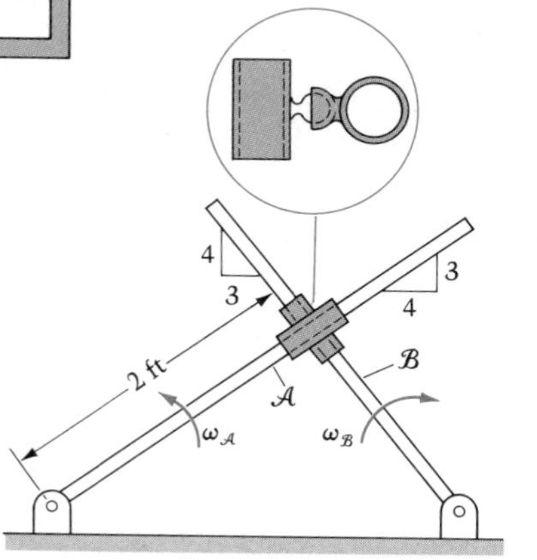

Figure P3.173

3.174 In Problem 3.159 find the acceleration of P in $\mathcal{S}$ and the angular acceleration of $\mathcal{B}$.

3.175 In Problem 3.156 find, at the same instant of time, the acceleration of P in $\mathcal{B}$ and the angular acceleration of $\mathcal{B}$.

3.176 In Problem 3.157 let $R = 0.1$ m and $\dot{\theta}_{\mathcal{A}} = 5$ rad/s as in Problem 3.158. This time, again at $\theta = 160°$, find: (a) the angular acceleration of $\mathcal{B}$; (b) the acceleration of P in $\mathcal{B}$ ($= \mathbf{a}_{P/\mathcal{B}}$).

3.177 Show that the rigid-body acceleration equation (3.19) can be ***derived*** from Equation (3.54). *Hint:* If you fix P to $\mathcal{B}$, the equation will then relate the accelerations in $\mathcal{I}$ of the two points P and O' of $\mathcal{B}$.

3.178 Referring to Example 3.33 for the meanings of the symbols, show that $\mathbf{a}_{Q/\mathcal{R}} \neq -\mathbf{a}_{T/\mathcal{B}}$.

3.179 Bar $\mathcal{A}$ in Figure P3.179 has angular velocity 0.25 ↻ rad/s and angular acceleration 0.15 ↺ rad/s² at the given instant. Find the angular acceleration of $\mathcal{B}$ at this time. (See Problem 3.164.)

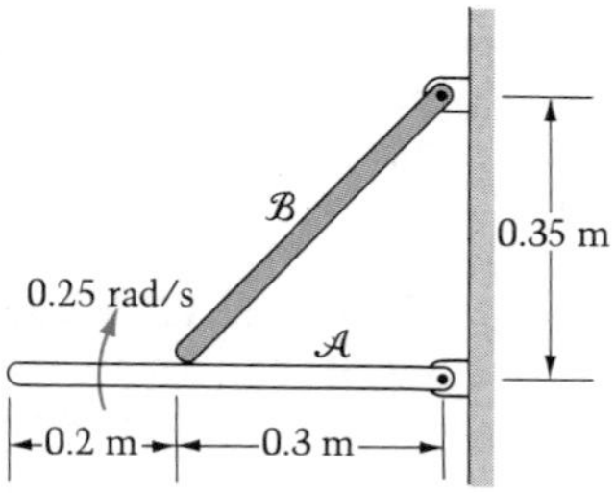

Figure P3.179

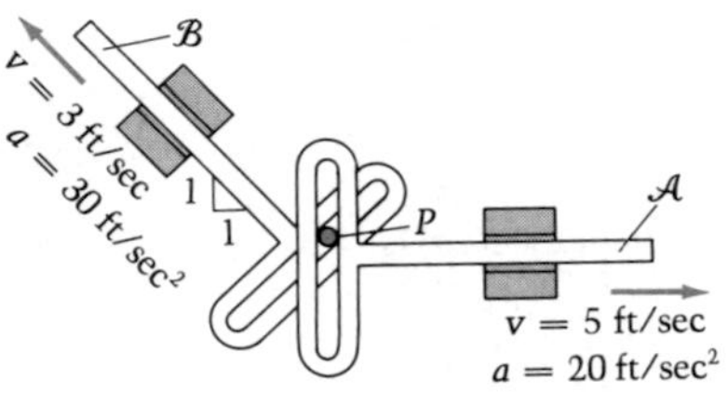

Figure P3.180

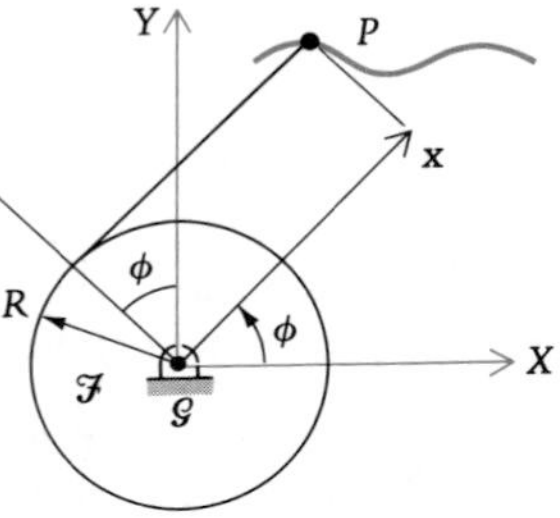

Figure P3.181

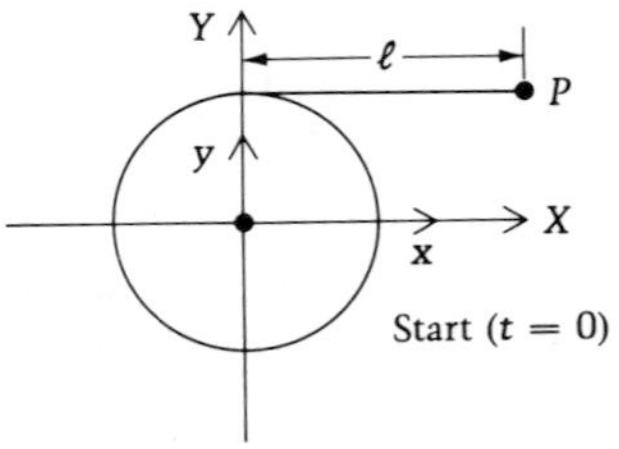

Figure P3.182

3.180 Pin P in Figure P3.180 moves along a curved path and is controlled by the motions of the slotted links $\mathcal{A}$ and $\mathcal{B}$. At the instant shown, each point of $\mathcal{A}$ has a velocity of 5 ft/sec and an acceleration of 20 ft/sec^2, both to the right, while each point of $\mathcal{B}$ has a velocity of 3 ft/sec and an acceleration of 30 ft/sec^2 in the direction shown in the figure. Find the radius of curvature of the path of P in this position.

3.181 Suppose that a point P moves in the XY plane, where X and Y are fixed in a reference frame $\mathcal{G}$. (See Figure P3.181.) Let the frame $\mathcal{F}$ rotate at $\boldsymbol{\omega}_{\mathcal{F}/\mathcal{G}} = \dot{\phi}\hat{\mathbf{k}}$ with respect to $\mathcal{G}$, where ϕ is measured as shown so that the tangent to the circle that touches the y axis always passes through P. Find the velocity and acceleration of P in $\mathcal{G}$, expressed in terms of their components in $\mathcal{F}$.

3.182 In the preceding problem let P be a particle on the end of a string that is wrapped around the circle of radius a. Show that at any time $\mathbf{x} = \ell + a\phi$, where ℓ is the unwrapped length at the start, when $\phi = 0$, as shown in Figure P3.182. Then specify the previous problem's answers for the velocity and acceleration of P in $\mathcal{G}$ for this special case.

3.183 The 26-ft rod $\mathcal{R}$ in Figure P3.183 slides on a plane surface at A and on the fixed half-cylinder at Q as shown below. If end A is moved at a constant velocity of 4 ft/sec to the right along the plane, determine the vertical component of the acceleration of B for the position shown.

3.184 Extending Problem 3.168, suppose that the angular velocity of rod $\mathcal{B}$, 5 ↺ rad/s, is constant in time. (See Figure P3.184.) Find, at the given instant, the angular acceleration of $\mathcal{A}$ and the acceleration of any point of $\mathcal{R}$ relative to $\mathcal{A}$.

3.185 In Problem 3.163 determine the acceleration of C relative to $\mathcal{B}$ as a function of D, θ, and $\dot{\theta}$.

3.186 A circular turntable $\mathcal{T}$ (see Figure P3.186) rotates about a vertical axis through O (normal to the plane of the paper) with θ changing at a constant rate ω_0. A block $\mathcal{B}$ rests in a groove cut in the turntable. If the cable is reeled in at a constant velocity v relative to $\mathcal{T}$, find expressions for the radial and transverse components of the block's acceleration.

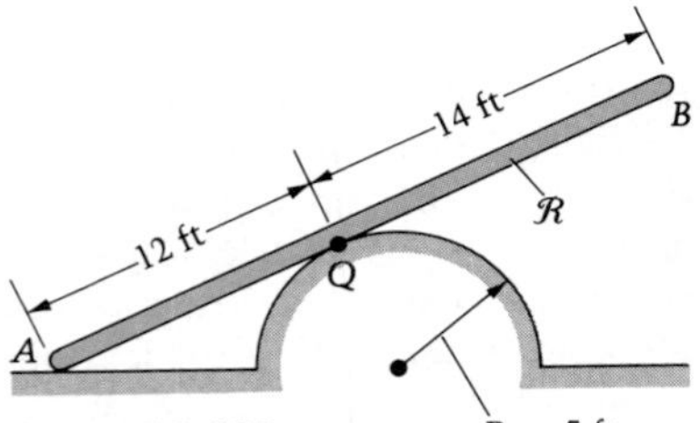

Figure P3.183

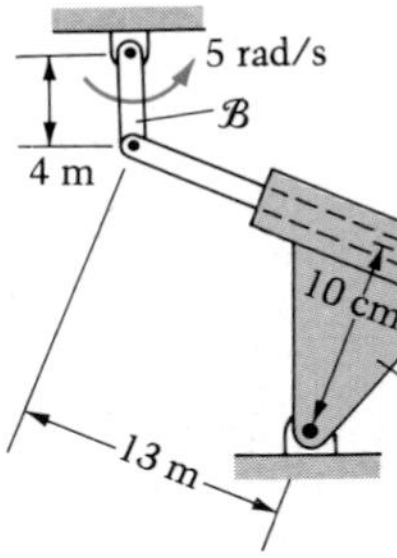

Figure P3.184

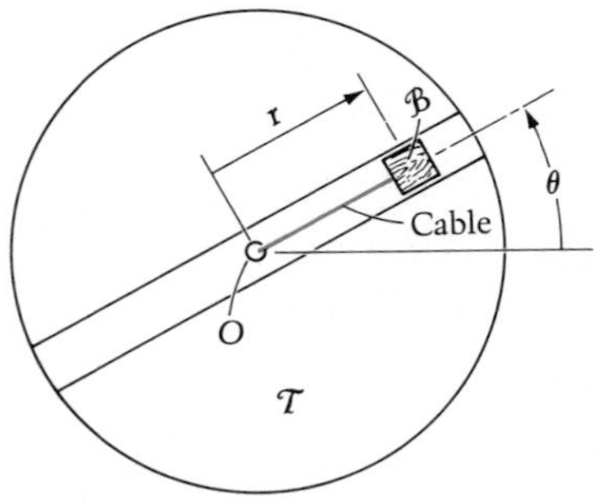

Figure P3.186

3.187 A bug $\mathcal{B}$ is crawling at a uniform speed of 3 ft/sec outward along the rotating arm $\mathcal{A}$. In the position shown in Figure P3.187, $\omega = 2$ rad/sec and $\alpha = 4$ rad/sec^2, both counterclockwise. What is the Coriolis acceleration of the bug? Indicate the direction in a sketch.

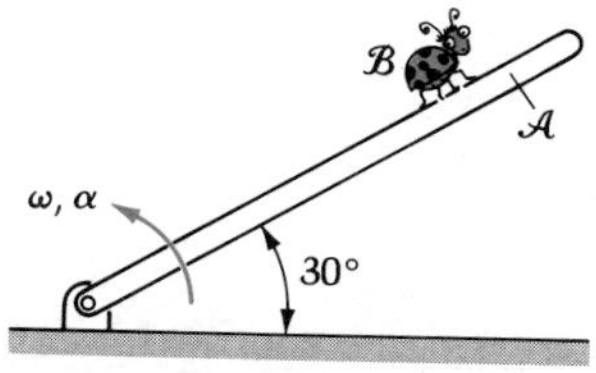

Figure P3.187

Answers to Questions / Chapter 3

Q3.1 Two points on a line originally parallel to the z axis are sufficient. Otherwise, three noncollinear points are needed.

Q3.2 If ever $x_A \neq x_B$ or $y_A \neq y_B$ (or both), either the rigid body or the plane motion assumptions (or both) will have been violated.

Q3.3 No; the body could rotate around the line joining the two points. In three dimensions it takes three points, not all on the same line!

Q3.4 The $2 \times 2 = 4$ coordinates of P_1 and P_2 are not independent. The distance between the points is a constant, so

$$\sqrt{(x_1 - x_2)^2 + (y_1 - y_2)^2} = \text{constant}$$

can be used to find any one of x_1, y_1, x_2, y_2 in terms of the other three.

Q3.5 Because $\mathcal{B}$ is rigid, r_{PQ} is the *constant* distance between points P and Q.

Q3.6 (a) No. (b) Yes, because the geometry would be different.

Q3.7 The derivative of $\dot{\theta}\hat{\mathbf{k}} \times \mathbf{r}_{PQ}$ is not zero merely because $\dot{\theta}$ happens to be zero at one instant of time. To be able to differentiate $\mathbf{v}_Q = \mathbf{v}_P$, it must be true for *all* values of t and not just one!

Q3.8 If $\dot{\theta} = 0$, Equation (3.8) says that $\mathbf{v}_Q = \mathbf{v}_P$; that is, all points of $\mathcal{B}$ have the *same* velocity vector. This common velocity vector is then zero only if the body is at rest. Some think of $\mathcal{B}$ as having an instantaneous center Ⓘ at infinity when $\dot{\theta} = 0$.

Q3.9 If point Ⓘ were above Q, then $\boldsymbol{\omega} \times \mathbf{r}_{ⒾQ}$ would give an incorrect direction for the velocity $\mathbf{v}_Q$—it would be opposite to the actual direction. The senses of $\mathbf{v}_Q$ and $\boldsymbol{\omega}$ when viewed from Ⓘ must agree, as we point out later in the section.

Q3.10 The known velocity direction of A dictates that $\mathcal{B}_2$ is turning clockwise around Ⓘ, so $\mathbf{v}_B$ has to be "southwest" for this to be the case.

Q3.11 Because $\mathbf{r}_{PQ}$ lies in the (xy) reference plane and is therefore perpendicular to $\hat{\mathbf{k}}$.

Q3.12 It means that the angular speed of the body is decreasing.

Q3.13 If $\dot{x}_C = r\dot{\theta}$ for either

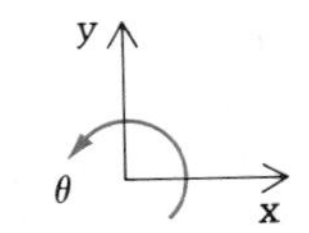

or

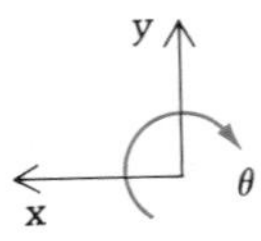

,

then the point of zero velocity will be at the *top* of the wheel!

Q3.14 From Figure 3.13 we can see exactly when the component of $\mathbf{a}_B$ normal to line AB is given by $r_{AB}\alpha$: It is when $\mathbf{a}_A$ has no component normal to AB. (This result does not require that A be the point Ⓘ, incidentally.) The reader should show that in Example 3.14 the component of $\mathbf{a}_{Ⓘ}$ normal to ⒾB is *not* zero, while in Example 3.21 the components of $\mathbf{a}_O$ normal to both OC and OB are zero since O is a pivot point of $\mathcal{R}$.

Q3.15 Yes; this will be manifested by a_B turning out negative. Note that (negative a_B) $[(8\hat{\mathbf{i}} - 15\hat{\mathbf{j}})/17]$ is the same as (positive a_B) $[(-8\hat{\mathbf{i}} + 15\hat{\mathbf{j}})/17]$.

Q3.16 (a) It is not $\mathbf{v}_{P/\mathcal{I}}$ because the origin of the position vector is not fixed in $\mathcal{I}$. (b) And it is not $\mathbf{v}_{P/\mathcal{B}}$ because the derivative of $\mathbf{r}_{O'P}$ is not taken in $\mathcal{B}$.

Q3.17 Yes, provided we recognize that the direction of $\mathbf{v}_{T/\mathcal{B}}$ is along the axis of the rod (see Problem 3.162). That is, $\mathbf{v}_{T/\mathcal{B}}$ must be tangent to the surface at which $\mathcal{R}$ and $\mathcal{B}$ touch. It is important to realize, however, that while the path of Q in $\mathcal{R}$ is a straight line, the path of T in $\mathcal{B}$ is not. Thus $\mathbf{a}_{T/\mathcal{B}}$ would *not* be in the direction of the axis of $\mathcal{R}$.

Q3.18 Definitely not! This happened because of the geometry at the given instant.

Q3.19 Because at that time, the bars form a right angle and there is no rotation about the pin.

Q3.20 The error is that the derivative ${}^{\mathcal{I}}\dot{\mathbf{v}}_{P_{\mathcal{B}}/\mathcal{I}}$ is not equal to $\mathbf{a}_{P_{\mathcal{B}}/\mathcal{I}}$, because $P_{\mathcal{B}}$ denotes a *succession of points* of $\mathcal{B}$ which are at each instant coincident with P.

Review Questions / Chapter 3

True or False?

In these questions, P and Q are points in the reference plane of a rigid body $\mathcal{B}$ in plane motion.

1. For a rigid body in plane motion, $\dot{\theta}$ depends on a certain choice of points whose velocities are to be related.

2. For a rigid body in plane motion, there is always an instantaneous center Ⓘ of zero velocity located a finite distance from the body.

3. $\mathbf{a}_{Ⓘ}$ is not necessarily zero.

4. $\mathbf{v}_P = \dot{\theta}\hat{\mathbf{k}} \times \mathbf{r}_{ⒾP}$ if the rigid body in plane motion is not translating.

5. For a rigid body $\mathcal{B}$ in rectilinear translation, the velocities of all points of $\mathcal{B}$ are equal, and so are the accelerations.

6. For a rigid body $\mathcal{B}$ in curvilinear translation, the velocities of all points of $\mathcal{B}$ are equal, but the accelerations are not.

7. At a given instant, any point can be considered to lie on a rigid extension of any body.

8. The smallest number of variables required to locate a rigid body in plane motion is four.

9. For a rigid body $\mathcal{B}$ in plane motion,

$$\mathbf{v}_P = \mathbf{v}_Q + \dot{\theta}_{\mathcal{B}}\hat{\mathbf{k}} \times \mathbf{r}_{QP}$$

10. A point P can have an angular velocity.

11. The components of $\mathbf{v}_P$ and $\mathbf{v}_Q$ along the line PQ are not always equal.

12. The components of $\mathbf{a}_P$ and $\mathbf{a}_Q$ along the line PQ are not always equal.

Answers: F, F, T, T, T, F, T, F, T, F, F, T

CHAPTER

4

Kinetics of a Rigid Body in Plane Motion/ Development and Solution of the Differential Equations Governing the Motion

chapter outline ►

4.1 Introduction

In this chapter we develop the differential equations governing the plane motion of rigid bodies and then illustrate their solution. Motion of the mass center of a rigid body is described by the solution to the general equation discussed in Section 2.1. The rotational motion of the body, however, is governed by the *second* law of Euler. In general, this law requires us to grapple with a new concept (the moment of momentum), and this in turn leads to yet another new topic for the particular case of the rigid body: moments and products of inertia. Thus we shall study these two concepts in Sections 4.2 and 4.3 before discussing the laws governing the plane motion of a rigid body. Euler's second law is discussed in Section 4.4; it is applied to the rigid body in plane motion in Section 4.5. Together with the mass center equations, the resulting rotational equation will then be used to solve kinetics problems of the rigid body in plane motion.

In the next chapter, we continue our study of plane-motion kinetics of rigid bodies by investigating the use of three special solutions (which can be obtained in general) to the differential equations of motion. These special integrals are known as the principles of work and kinetic energy; linear impulse and momentum; and angular impulse and angular momentum.

It is possible to obtain all the results of this chapter on plane motion of rigid bodies from the general three-dimensional results developed in Chapter 7. It is not necessary to travel this complex route in order to learn plane motion, however, and in this chapter we take a simpler path. It is worth noting that while the planar case covers a restricted class of motions, it does in fact contain a large number of problems with important engineering applications.

4.2 Moment of Momentum (Angular Momentum)/Reduction to the Rigid Body in Plane Motion

We have seen in Chapter 3 that the kinematics of a rigid body $\mathcal{B}$ in plane motion can be described very simply. If we know the velocity of just one point and the angular velocity of the body, then we know the velocity of *every point* of $\mathcal{B}$—quite a bargain. Because of this simplicity, we shall see that compact and yet completely general expressions can be written for the moment of momentum (or angular momentum, as it is often called for rigid bodies). First, though, we shall define the moment of momentum for the general motion of *any* body (or system of bodies) and develop an equation that will prove useful.

The **moment of momentum** of a body with respect to a point P (not necessarily fixed in our reference frame $\mathcal{F}$) is defined* by

$$\mathbf{H}_P = \int_{\mathcal{B}} \mathbf{R} \times \mathbf{v}\, dm \tag{4.1}$$

where in this section $\mathbf{R}$ is the vector from P to the element of mass dm. (See Figure 4.1.)

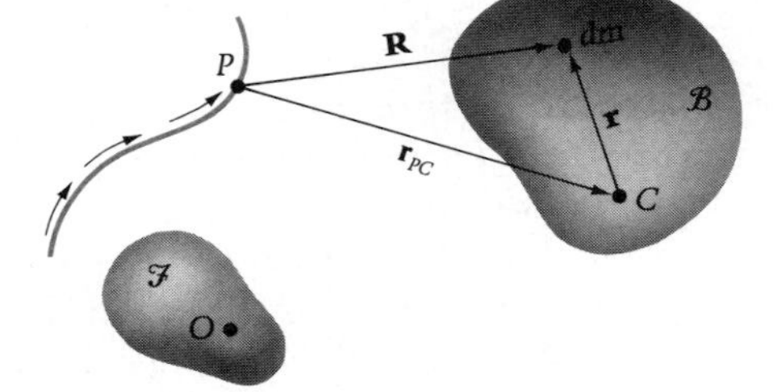

Figure 4.1

We note that $\mathbf{H}_P$ depends on the location of P as well as the distributions of mass and velocities in the body; it is seen to be the sum of the moments of momenta of all the mass elements of $\mathcal{B}$. If P is now taken to be the mass center C, then (see the meaning of the vector $\mathbf{r}$ in Figure 4.1) the moment of momentum there is

$$\mathbf{H}_C = \int \mathbf{r} \times \mathbf{v}\, dm \tag{4.2}$$

and, since $\mathbf{R} = \mathbf{r} + \mathbf{r}_{PC}$, we obtain the following from Equation (4.1):

$$\mathbf{H}_P = \int (\mathbf{r} + \mathbf{r}_{PC}) \times \mathbf{v}\, dm \tag{4.3}$$

$$= \mathbf{H}_C + \mathbf{r}_{PC} \times \int \mathbf{v}\, dm \tag{4.4}$$

$$= \mathbf{H}_C + \mathbf{r}_{PC} \times \mathbf{L} \tag{4.5}$$

Thus the moment of momentum of a body about any point P in space is the sum of the moment of momentum about its mass center and the moment of its linear momentum $\mathbf{L}$ about P, where $\mathbf{L}$ is given a "line of action" through C.

Suppose (see Figure 4.2) a planet $\mathcal{P}$ is spinning about an axis z_C normal to its orbital plane. Then Equation (4.5) tells us that its moment of momentum with respect to an axis z_P through P fixed in the sun (taken to be our frame $\mathcal{F}$) is made up of two parts:

1. The moment of momentum $\mathbf{H}_C$—its angular momentum relative to its *own* mass center C due to the spin, which is associated with motion of its points relative to C
2. The moment of the planet's linear momentum ($m\mathbf{v}_C$) about the axis z_P

*We note that $\mathbf{v}$ is the derivative of the position vector from O to dm, not the derivative of any of the three vectors in Figure 4.1.

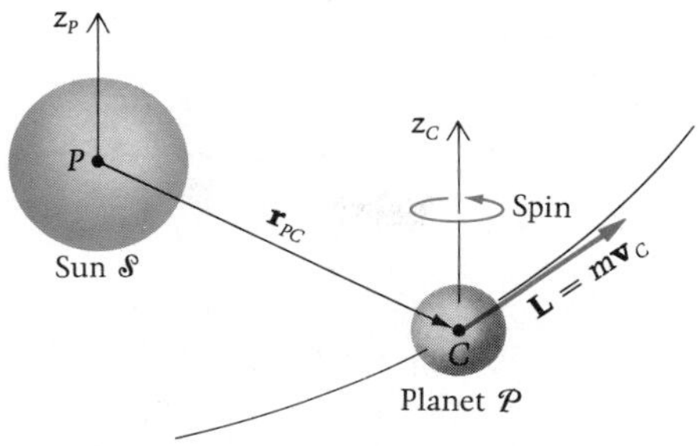

Figure 4.2 $\mathbf{H}_P$ equals $\mathbf{H}_C$ plus the moment, about P, of $\mathbf{L}$.

For the earth, the vectors are not parallel because of the $23\frac{1}{2}°$ tilt, but nonetheless, in magnitude, the term $\mathbf{r}_{PC} \times \mathbf{L}$ dwarfs the $\mathbf{H}_C$ term by a factor of about 3.7×10^6. In other examples, the $\mathbf{H}_C$ term might predominate, while in still others the two terms might be of the same order of magnitude. We shall next learn in this section how to express and calculate $\mathbf{H}_C$ for rigid bodies.

We now let our general point P in Figure 4.1 become any point of rigid body $\mathcal{B}$ that lies in the reference plane (xy) of motion. We also choose the reference plane to contain the mass center C. The rectangular axes (x_P, y_P, z_P) are fixed in $\mathcal{B}$ with origin at P as shown in Figure 4.3.

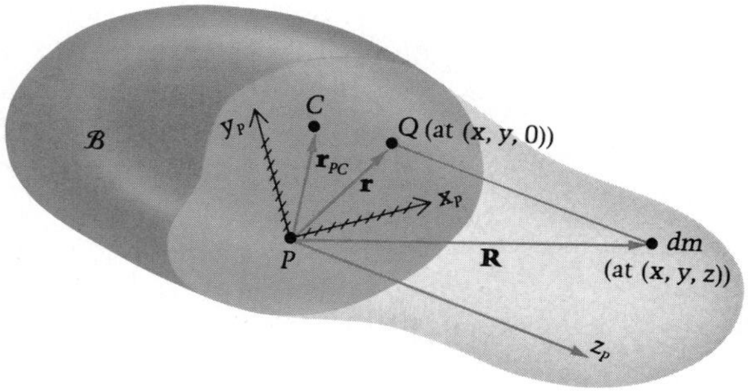

Figure 4.3

Noting that the velocity $\mathbf{v}$ of dm is the same as its companion point Q in the reference plane,* we proceed to express $\mathbf{H}_P$ in terms useful for our study of the rigid body in plane motion:

$$\mathbf{v} = \mathbf{v}_Q = \mathbf{v}_P + \omega\hat{\mathbf{k}} \times \mathbf{r}_{PQ}$$

*Recall from Chapter 3 that each point of the body has a "companion" in the reference plane that always has the same x and y as the point and hence the same $\mathbf{v}$ and $\mathbf{a}$.

Hence*

$$\mathbf{H}_P = \int \underbrace{(x_P\hat{\mathbf{i}} + y_P\hat{\mathbf{j}} + z_P\hat{\mathbf{k}})}_{\mathbf{R}} \times [\mathbf{v}_P + \omega\hat{\mathbf{k}} \times \underbrace{(x_P\hat{\mathbf{i}} + y_P\hat{\mathbf{j}})}_{\mathbf{r}_{PQ}}]\, dm$$

or

$$\mathbf{H}_P = \left(\int \mathbf{R}\, dm\right) \times \mathbf{v}_P + \int (x_P\hat{\mathbf{i}} + y_P\hat{\mathbf{j}} + z_P\hat{\mathbf{k}}) \times [\omega\hat{\mathbf{k}} \times (x_P\hat{\mathbf{i}} + y_P\hat{\mathbf{j}})]\, dm$$

The integral $\int \mathbf{R}\, dm$ in the first term above is equal to $m\mathbf{r}_{PC}$ by the definition of the mass center. Making this substitution, and also carrying out the cross products in the second term, gives the moment of momentum vector in terms of $\mathbf{v}_P$, ω, and certain mass distribution integrals:

$$\mathbf{H}_P = \mathbf{r}_{PC} \times m\mathbf{v}_P + \hat{\mathbf{k}}\omega \int (x_P^2 + y_P^2)\, dm - \hat{\mathbf{i}}\omega \int x_P z_P\, dm - \hat{\mathbf{j}}\omega \int y_P z_P\, dm \quad (4.6)$$

We call the integrals in (4.6) inertia properties. Specifically:

$$\int (x_P^2 + y_P^2)\, dm = I_{z_P} = \text{moment of inertia of mass of } \mathcal{B} \text{ about } z_P \text{ axis through } P \quad (4.7a)$$

$$-\int x_P z_P\, dm = I_{xz_P} = \text{product of inertia of mass of } \mathcal{B} \text{ with respect to } x_P \text{ and } z_P \text{ axes through } P^{\dagger} \quad (4.7b)$$

$$-\int y_P z_P\, dm = I_{yz_P} = \text{product of inertia of mass of } \mathcal{B} \text{ with respect to } y_P \text{ and } z_P \text{ axes through } P \quad (4.7c)$$

We are now at the point in our development where the inertia properties, like the mass center in Chapter 2, have arisen naturally. We shall spend the next section studying the moments and products of inertia; readers already familiar with inertia properties may wish to skip Section 4.3.

Before leaving Equation (4.6), we note for future reference that its first term, $\mathbf{r}_{PC} \times m\mathbf{v}_P$, vanishes if P is the mass center or has zero velocity.‡ In both these cases, which will prove valuable to us, $\mathbf{H}_P$ takes the form

$$\mathbf{H}_P = I_{xz_P}\omega\hat{\mathbf{i}} + I_{yz_P}\omega\hat{\mathbf{j}} + I_{z_P}\omega\hat{\mathbf{k}} \quad (4.8)$$

*The reader may have noticed that we have previously let x_P, y_P, z_P denote the coordinates of a *point* P in a rectangular coordinate system whose origin is implied or understood. In this section and the next, we shall deviate from that convention. Here, x_P, y_P, z_P denote the coordinates of a *general point* in a rectangular coordinate system with *origin at* P. We have done this to emphasize the importance of that origin in our development of the inertia properties.

†If the products of inertia are defined *with* the minus sign, then and only then will the inertia properties transform as a tensor.

‡Or if $\mathbf{r}_{PC}$ is parallel to $\mathbf{v}_P$, a case we need not consider here.

Thus in these cases the moment of momentum can be expressed in terms of the angular velocity of $\mathcal{B}$ (hence its other name: **angular momentum**), along with three measures of its mass distribution.

PROBLEMS / Section 4.2

4.1 Show that for *any* two points P and Q, the angular momenta of a body with respect to these points are related by

$$\mathbf{H}_P = \mathbf{H}_Q + \mathbf{r}_{PQ} \times \mathbf{L}$$

Thus Q need not be the mass center for Equation (4.5) to apply!

4.2 Show that if $\mathbf{r}$ is the vector in Figure P4.2 and $\mathbf{v}$ is the velocity of the element dm in the reference frame $\mathcal{I}$, then

$$\frac{d}{dt}\int_{\mathcal{B}} \mathbf{r} \times \mathbf{v}\, dm = \int_{\mathcal{B}} \mathbf{r} \times \mathbf{a}\, dm$$

where the derivative is taken in $\mathcal{I}$.

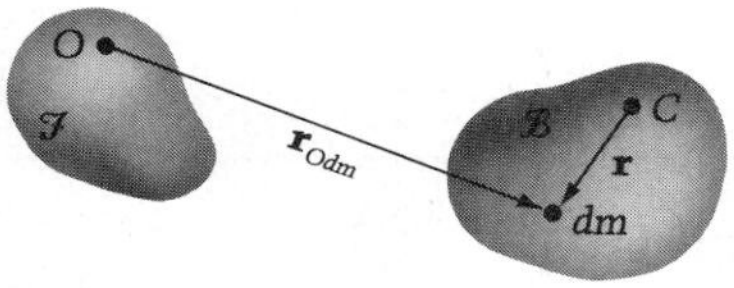

Figure P4.2

4.3 Referring to Figure 4.1, the *relative* angular momentum of $\mathcal{B}$ with respect to P is defined by

$$\mathbf{H}_{P_{\text{rel}}} = \int \mathbf{R} \times \dot{\mathbf{R}}\, dm$$

The difference between $\mathbf{H}_P$ and $\mathbf{H}_{P_{\text{rel}}}$ is that in the latter we do not use $\mathbf{v}$, but rather $\dot{\mathbf{R}}$, where

$$\dot{\mathbf{R}} = \mathbf{v} - \mathbf{v}_P$$

Show that $\mathbf{H}_P$ equals $\mathbf{H}_{P_{\text{rel}}}$ if and only if

$$\mathbf{r}_{PC} \times \mathbf{v}_P = \mathbf{0}$$

and observe that this is always so for the mass center—that is, if $P = C$. We have not used $\mathbf{H}_{P_{\text{rel}}}$ anywhere else in the theory of the text; this problem is included only for completeness and comparison.

4.3 Moments and Products of Inertia/the Parallel-Axis Theorems

Examples of Moments of Inertia

From the definition of moment of inertia,

$$I_{z_P} = \int (x_P^2 + y_P^2)\, dm$$

we see that I_{z_P} is a measure of "how much mass is located how far" from the z_P axis. In cylindrical coordinates we have $I_{z_P} = \int r^2\, dm$, and thus I_{z_P} measures the sum total of mass times distance squared over the body's volume. The quantity I_{z_P} is thus seen to be always positive. We now compute the centroidal moments of inertia of a number of common shapes. In Examples 4.1–4.9, we are seeking I_{z_C}:

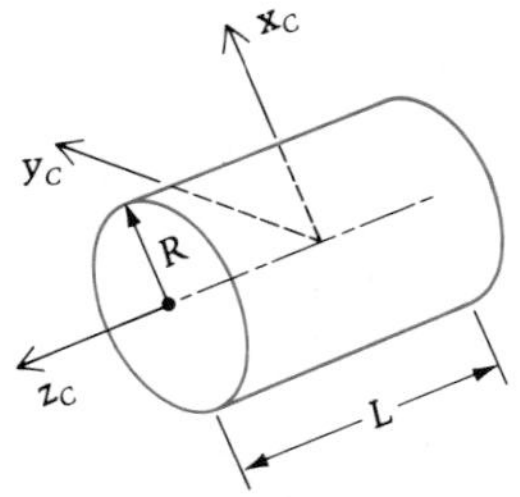

EXAMPLE 4.1

Homogeneous solid cylinder if z_C is its axis (see the diagram).

SOLUTION

Noting that $dm = \rho\, dV$, where $\rho =$ mass density,

$$I_{z_C} = \int_{\text{vol}} (x_C^2 + y_C^2)\, \overbrace{\rho\, dV}^{dm} = \int_{z=-L/2}^{L/2} \int_{\theta=0}^{2\pi} \int_{r=0}^{R} r^2 \rho\, (r\, dr\, d\theta\, dz_C)$$

$$I_{z_C} = \rho \frac{r^4}{4}\Bigg]_0^R \theta\Bigg]_0^{2\pi} z_C\Bigg]_{-L/2}^{L/2} = (\rho\pi R^2 L)\frac{R^2}{2} = \boxed{\frac{mR^2}{2}} \tag{4.9}$$

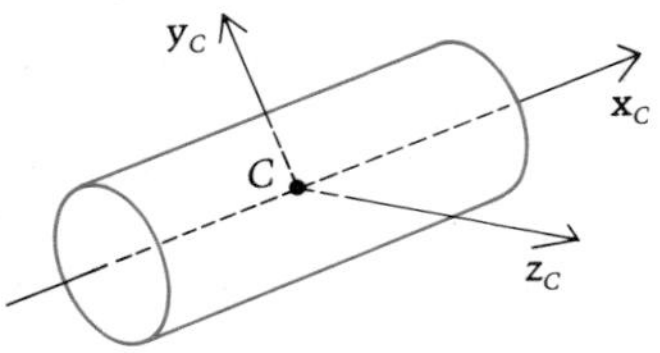

EXAMPLE 4.2

Homogeneous solid cylinder if z_C is an axis *normal* to the axis of the cylinder (see the diagram).

SOLUTION

$$I_{z_C} = \int_{\text{vol}} (x_C^2 + y_C^2)\, \rho\, dV = \int_{x_C=-L/2}^{L/2} \int_{\theta=0}^{2\pi} \int_{r=0}^{R} [\underbrace{(r \sin\theta)^2}_{y_C = r\sin\theta} + x_C^2]\, \rho r\, dr\, d\theta\, dx_C$$

$$= \int_{x=-L/2}^{L/2} \int_{\theta=0}^{2\pi} \left(\frac{r^4}{4} \underbrace{\sin^2\theta}_{\frac{1-\cos 2\theta}{2}} + \frac{x_C^2 r^2}{2}\right)\Bigg]_0^R \rho\, d\theta\, dx_C$$

(Continued)

$$= \int_{-L/2}^{L/2} \left[\frac{R^4}{4} \left(\frac{\theta}{2} - \frac{\sin 2\theta}{4} \right) \Bigg|_0^{2\pi} + \frac{x_C^2 R^2 \theta}{2} \Bigg|_0^{2\pi} \right] \rho \, dx_C$$

$$= \rho \frac{\pi R^4}{4} x_C \Bigg|_{-L/2}^{L/2} + \frac{\rho R^2 \pi x_C^3}{3} \Bigg|_{-L/2}^{L/2} = (\rho \pi R^2 L)\left(\frac{R^2}{4} + \frac{L^2}{12} \right)$$

$$= \frac{mR^2}{4} + \frac{mL^2}{12} \tag{4.10}$$

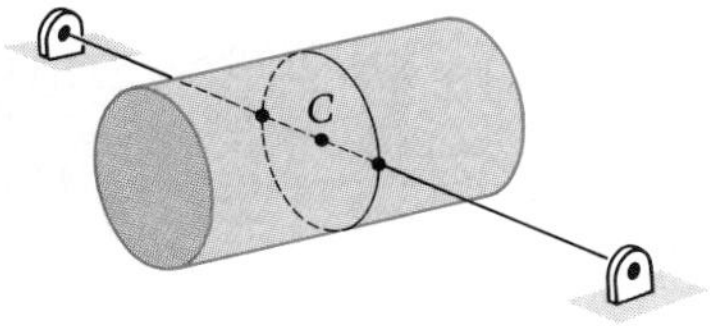

It may be confusing at first to know which moment of inertia to use for a cylinder in a plane kinetics problem; the answer is that it is always the value associated with the axis normal to the **xy** plane of the motion. If the problem is a rolling cylinder, $I_{z_C} = mR^2/2$. If we have a cylinder turning around a diametral axis (see the figure at the left), then $I_{z_C} = mR^2/4 + mL^2/12$.

EXAMPLE 4.3

Two special cases of Example 4.2: Slender rods and disks.

SOLUTION

1. If the body is "pencillike"—that is, $L \gg R$—the moment of inertia for a lateral axis through C is approximated as $mL^2/12$. This is also a correct result even if the cross section is not circular but has a maximum dimension within the cross section much less than L. Such a body is called a *slender bar* or *rod*.
2. If the body is a disk, however, we have $R \gg L$ and the moment of inertia is approximately $mR^2/4$.

For a cylinder with the dimensions of a pencil, for example, with $R = \frac{1}{8}$ in. and $L = 7$ in., we see that (see Figure 1)

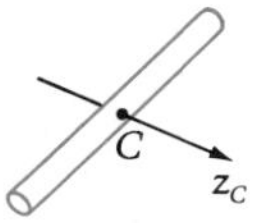

Figure 1

$$I_{z_C} = \frac{mL^2}{12} + \frac{mR^2}{4} = \frac{mL^2}{12}\left[1 + 3\left(\frac{R}{L}\right)^2\right]$$

$$= \frac{mL^2}{12}\left[1 + 3\left(\frac{1/8}{7}\right)^2\right] \tag{4.11a}$$

$$\approx \frac{mL^2}{12}$$

where the second term is less than 0.1 of 1 percent of the retained $mL^2/12$

(Continued)

term. For a typical coin, on the other hand, with $R \approx \frac{15}{16}$ in. and $L \approx \frac{1}{16}$ in., we obtain (see Figure 2)

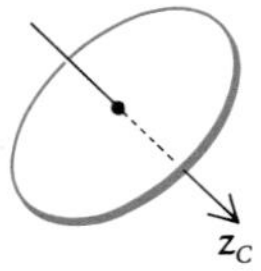

Figure 2

$$
\begin{aligned}
I_{z_C} &= \frac{mR^2}{4} + \frac{mL^2}{12} = \frac{mR^2}{4}\left[1 + \frac{1}{3}\left(\frac{L}{R}\right)^2\right] \\
&= \frac{mR^2}{4}\left[1 + \frac{(1/16)^2}{3(15/16)^2}\right] \qquad (4.11b) \\
&\approx \frac{mR^2}{4}
\end{aligned}
$$

This time it is the $mL^2/12$ term that is negligible; it is less than 0.15 of 1 percent of the $mR^2/4$ term. We emphasize, however, that with respect to the ***axis*** of any solid homogeneous cylinder (disk, rod, or anything in between), the moment of inertia is $mR^2/2$.

EXAMPLE 4.4

A uniform rectangular solid.

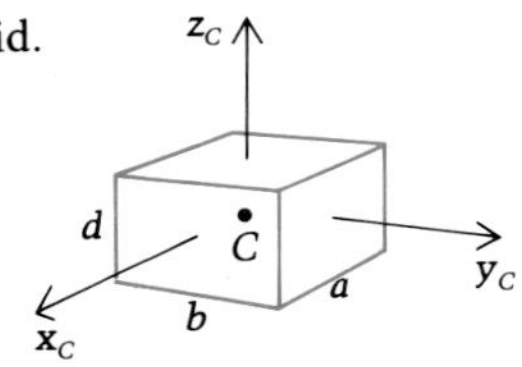

SOLUTION

$$I_{z_C} = \int_{-d/2}^{d/2}\int_{-b/2}^{b/2}\int_{-a/2}^{a/2} (x_C^2 + y_C^2)\,\rho\, dx_C\, dy_C\, dz_C$$

This integration yields

$$I_{z_C} = (\rho\, abd)\frac{a^2 + b^2}{12} = \boxed{\frac{m}{12}(a^2 + b^2)} \qquad (4.12)$$

EXAMPLE 4.5

Special case of Example 4.4: Rectangular Plate.

(Continued)

SOLUTION

If the rectangular solid is a plate—that is, it has one edge much smaller than the other two dimensions—then:

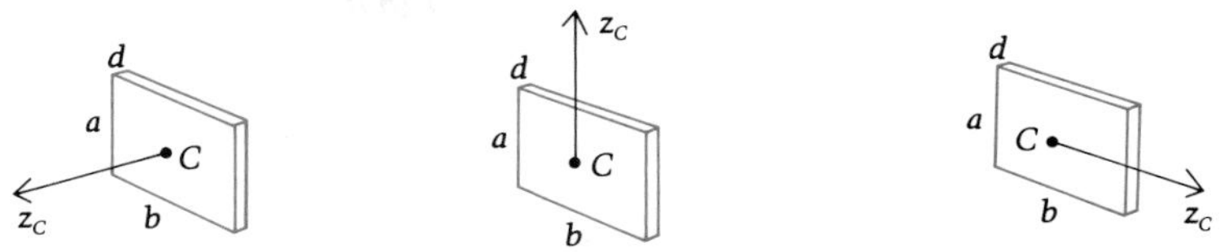

$$I_{z_C} = \frac{m}{12}(a^2 + b^2) \qquad I_{z_C} = \frac{m}{12}(b^2 + d^2) \approx \frac{mb^2}{12} \qquad I_{z_C} = \frac{m}{12}(a^2 + d^2) \approx \frac{ma^2}{12} \tag{4.13}$$

Again it depends on how the body's plane motion is set up as to which axis is z_C (normal to the plane of the motion) and hence which formula to use.

EXAMPLE 4.6

Solid, homogeneous, right circular cone about its axis.

SOLUTION

Here we encounter a variable limit, since $\mathscr{r} = \mathscr{r}(z)$. Noting that z_O (originating from O) and z_C are the same line, so that $I_{z_O} = I_{z_C}$, then

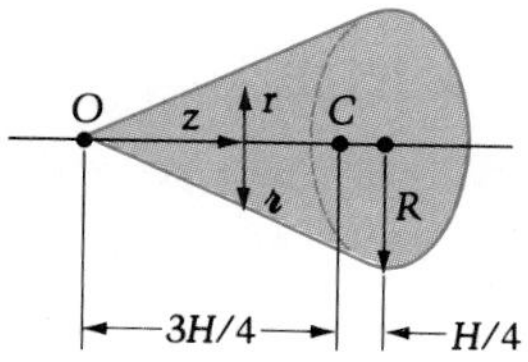

$$I_{z_C} = I_{z_O} = \int (x_C^2 + y_C^2)\, dm = \int_{z=0}^{H} \int_{\theta=0}^{2\pi} \int_{r=0}^{\mathscr{r}(z)} r^2 \rho r \, dr\, d\theta\, dz$$

From similar triangles,

$$\frac{\mathscr{r}}{z} = \frac{R}{H} \Rightarrow \mathscr{r} = \frac{Rz}{H} = \mathscr{r}(z)$$

which gives the varying radius in terms of z. Then

$$I_{z_C} = \int_{z=0}^{H} \int_{\theta=0}^{2\pi} \left.\frac{\rho r^4}{4}\right]_0^{Rz/H} \rho\, d\theta\, dz = \int_{z=0}^{H} \left.\frac{\rho R^4 z^4}{4H^4}\theta\right]_0^{2\pi} dz$$

$$= \left.\frac{\rho\pi R^4}{2H^4}\frac{z^5}{5}\right]_0^H = \left(\frac{\rho\pi R^2 H}{3}\right)\frac{3R^2}{10} = \boxed{\frac{3}{10}mR^2} \tag{4.14}$$

EXAMPLE 4.7

Hollow, homogeneous cylinder (about its axis).

SOLUTION

$$I_{z_C} = \int_{-L/2}^{L/2} \int_0^{2\pi} \int_{R_i}^{R_o} r^2 \rho r \, dr \, d\theta \, dz_C = \frac{\rho\pi(R_o^4 - R_i^4)L}{2}$$

$$= \underbrace{[\rho\pi(R_o^2 - R_i^2)L]}_{m} \frac{(R_o^2 + R_i^2)}{2} = \boxed{\frac{m(R_o^2 + R_i^2)}{2}} \tag{4.15a}$$

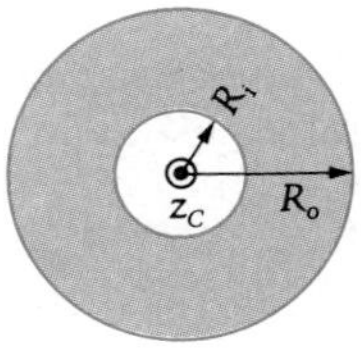

The same result can be obtained by subtracting the moment of inertia of the "hole" (H) from that of the "whole" (W). The basis for this procedure is that we may integrate over ***more*** than the required region provided we subtract away the integral over the part that is not to be included:

$$I_{z_C} = \frac{m_W R_o^2}{2} - \frac{m_H R_i^2}{2} = \frac{\rho\pi R_o^2 L R_o^2}{2} - \frac{\rho\pi R_i^2 L R_i^2}{2}$$

$$= \underbrace{\rho\pi(R_o^2 - R_i^2)L}_{m} \frac{(R_o^2 + R_i^2)}{2} = \frac{m(R_o^2 + R_i^2)}{2}$$

Note that if the wall thickness is small, we have a cylindrical shell (or a hoop if the length is small) for which $R_o \approx R_i$ and

$$I_{z_C} \approx \frac{m(2R^2)}{2} = \boxed{mR^2} \tag{4.15b}$$

(It is obvious that if all the mass is the same distance R from the axis z_C, we should indeed get mR^2.)

EXAMPLE 4.8

A uniform solid sphere about any diameter.

SOLUTION

$$I_{z_C} = \int (x_C^2 + y_C^2) \, dm$$

(Continued)

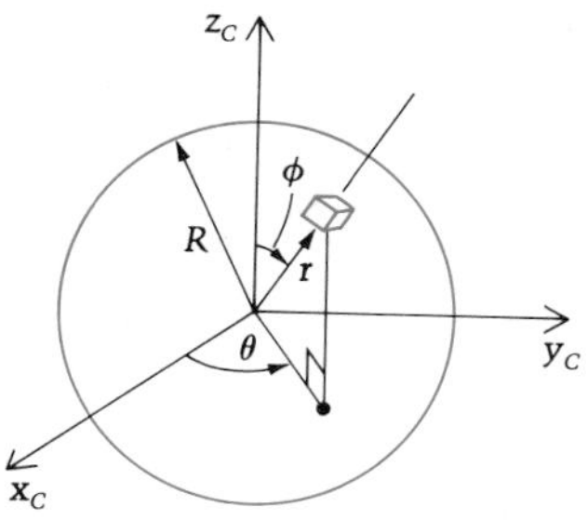

Also:

$$I_{x_C} = \int (y_C^2 + z_C^2)\, dm$$

and

$$I_{y_C} = \int (z_C^2 + x_C^2)\, dm$$

Adding:

$$\underbrace{I_{x_C} + I_{y_C} + I_{z_C}}_{\substack{= 3I \text{ since they are all} \\ \text{equal by symmetry}}} = \int 2\underbrace{(x_C^2 + y_C^2 + z_C^2)}_{r^2}\, dm$$

Thus

$$3I = 2\int r^2 \rho_0\, dV = 2\rho_0 \int_{\theta=0}^{2\pi} \int_{\phi=0}^{\pi} \int_{r=0}^{R} r^2 \underbrace{(r^2 \sin\phi\, dr\, d\phi\, d\theta)}_{dV \text{ in spherical coordinates}}$$

$$I = \frac{2}{3}\rho_0 \frac{r^5}{5}\Big]_0^R (-\cos\phi)\Big]_0^{\pi} \theta\Big]_0^{2\pi} = \frac{8\pi\rho_0 R^5}{15}$$

$$= \underbrace{\left(\frac{4}{3}\pi R^3 \rho_0\right)}_{m} \frac{2}{5} R^2 = \boxed{\frac{2}{5} mR^2} \qquad (4.16)$$

A less tricky way to do the sphere is to use spherical coordinates directly; the integral is

$$I_{z_C} = \int_{\theta=0}^{2\pi} \int_{\phi=0}^{\pi} \int_{r=0}^{R} \rho_0 (r \sin\phi)^2\, dr\, (r\, d\phi)\, (r \sin\phi\, d\theta)$$

which yields the same result of $(2/5)mR^2$, as the reader may show by carrying out the integration.

EXAMPLE 4.9

An example in which the density is not constant. Sometimes a body's density varies; if it does, it must stay inside the integral when we calculate the inertia properties. An example is the earth; we now know that the density of the solid central core of the earth is about four times that of the outermost part of its crust and, moreover, that this central density is nearly twice that of steel!

(Continued)

Let us imagine a sphere with the same mass and radius as in the preceding example but with a density that varies linearly and is twice as high at $r = 0$ as at $r = R$. We shall find I about any diameter. (See the figure on the left.)

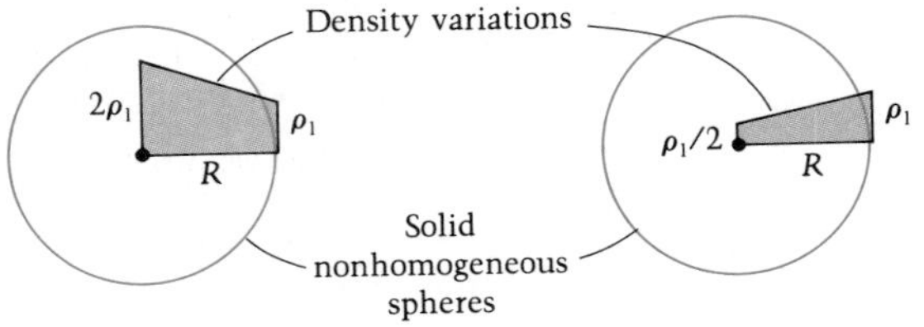

SOLUTION

The mass of the body is

$$m = \int \rho \, dv = \int_0^{2\pi} \int_0^{\pi} \int_0^{R} \rho r^2 \sin \phi \, dr \, d\phi \, d\theta$$

Letting the density at R be ρ_1, then

$$\rho = \frac{-\rho_1}{R} r + 2\rho_1$$

Substituting and integrating with the same limits as before, and then equating the new mass to the old, gives

$$\rho_1 = \tfrac{4}{5} \rho_0$$

Thus to have the same mass as the uniform sphere, the density varies from $(8/5)\rho_0$ to $(4/5)\rho_0$ going outward from the center. Then, integrating to find I,

$$3I = 2 \int_0^{2\pi} \int_0^{\pi} \int_0^{R} r^2 \rho(r) (r^2 \sin \phi \, dr \, d\phi \, d\theta)$$

which gives

$$I = \frac{28}{75} mR^2 = 0.373 mR^2 \qquad \text{(slightly less than } \tfrac{2}{5} mR^2\text{)}$$

Alternatively, if the density varies linearly but is only half as much ($\rho_1/2$) at the center as at R, then the results, if m and R are again the same as in the uniform case, are (see the figure on the right above)

$$\rho_1 = \frac{8}{7} \rho_0 \qquad \text{and} \qquad I = \frac{44}{105} mR^2 = 0.419 mR^2 \qquad \text{(slightly more than } \tfrac{2}{5} mR^2\text{)}$$

The Parallel-Axis Theorem for Moments of Inertia

In many applications the body consists of a number of different smaller bodies of familiar shapes. In such cases, there is fortunately no need to integrate in order to find the inertia of each part with respect to a common axis of interest, thanks to what is called the **parallel-axis theorem** or **transfer theorem.** If we know the moment of inertia about an axis through the mass center C of any body $\mathcal{B}$, we can then easily find it about any axis parallel to C by a simple calculation. The theorem states that the moment of inertia of the mass of $\mathcal{B}$ about *any* line z_P is the sum of the moment of inertia about an axis z_C through C parallel to z_P, plus the mass of $\mathcal{B}$ times the square of the distance between the two axes:

$$I_{z_P} = I_{z_C} + md^2 \tag{4.17}$$

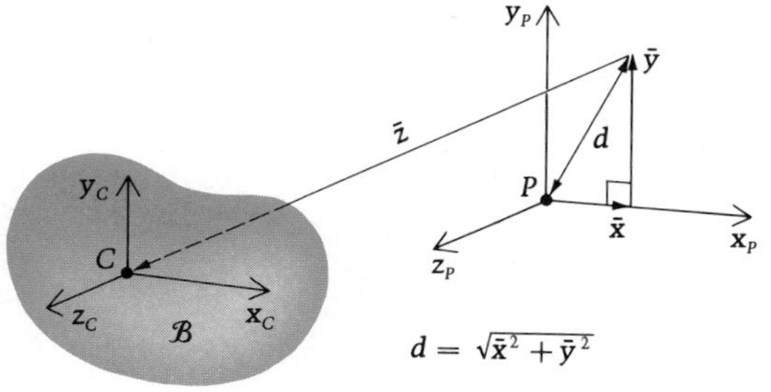

Figure 4.4

To prove this theorem, we let (x_P, y_P, z_P) and (x_C, y_C, z_C) be rectangular cartesian coordinate axes through P and C with the corresponding axes respectively parallel, as shown in Figure 4.4. Then we have, by definition,

$$I_{z_C} = \int (x_C^2 + y_C^2)\, dm \qquad \text{and} \qquad I_{z_P} = \int (x_P^2 + y_P^2)\, dm \tag{4.18}$$

Since it is seen that

$$x_P = x_C + \bar{x} \qquad \text{and} \qquad y_P = y_C + \bar{y}$$

substitution gives

$$\begin{aligned} I_{z_P} &= \int [(x_C + \bar{x})^2 + (y_C + \bar{y})^2]\, dm \\ &= \int (x_C^2 + y_C^2)\, dm + (\bar{x}^2 + \bar{y}^2) \int dm \\ &\qquad + 2\bar{x} \int x_C\, dm + 2\bar{y} \int y_C\, dm \end{aligned} \tag{4.19}$$

or

$$I_{z_P} = I_{z_C} + md^2 \tag{4.20}$$

in which $\bar{x}^2 + \bar{y}^2 = d^2$, the square of the distance between axes z_P and z_C.

Question 4.1 Why are the last two integrals in Equation (4.19) zero?

Equation (4.20) is the parallel-axis theorem for moments of inertia. But note: *We can only transfer from the mass center C and not from any other point A about which we may happen to know* I_{z_A}.

EXAMPLE 4.10

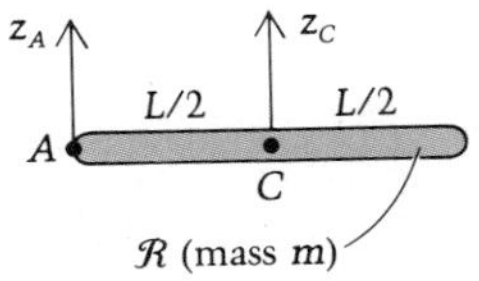

For the uniform slender rod ℛ shown in the diagram find I_{z_A}, the moment of inertia of the mass of ℛ with respect to a lateral axis through one end. (This exercise will be useful in pendulumlike applications in which a rod is pinned at one end.)

SOLUTION

$$I_{z_A} = I_{z_C} + md^2 = \frac{mL^2}{12} + m\left(\frac{L}{2}\right)^2 = \frac{mL^2}{3} \tag{4.21}$$

We next consider an example of the buildup of the moment of inertia for a composite body.

EXAMPLE 4.11

Find I_{z_O} for the body shown in the diagrams. The mass densities each $= \rho =$ constant, so that the respective masses are:

$$m_{\text{sph}} = m_1 = \rho \frac{4}{3}\pi R^3$$

$$m_{\text{bar}} = m_2 = \rho \frac{\pi\, d^2 L}{4}$$

and

$$m_{\text{cyl}} = m_3 = \rho \frac{\pi r^2 \ell}{2}$$

(Continued)

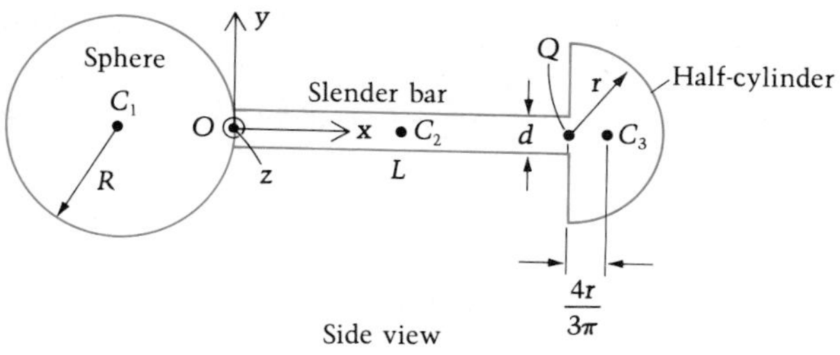

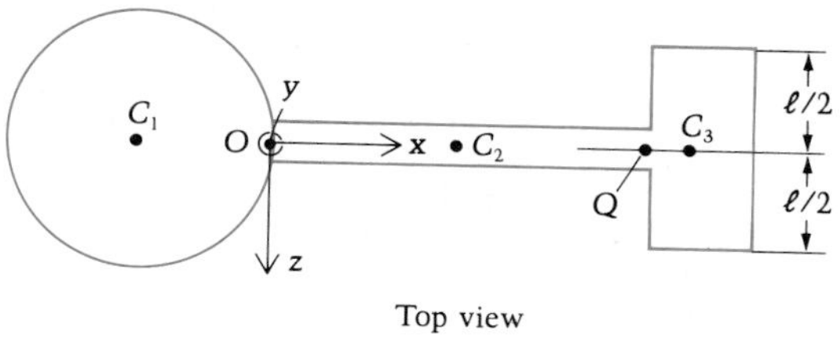

SOLUTION

First we observe that $I_{z_O} = I_{z_O}^{\text{sph}} + I_{z_O}^{\text{bar}} + I_{z_O}^{\text{cyl}}$, since the inertia integral may be carried out over the bodies separately, so long as we cover all the elemental masses of the total body. Filling in the separate integrals, we get

$$I_{z_O} = \left(\frac{2}{5} m_1 R^2 + m_1 R^2\right) + \underbrace{\left[\frac{m_2 L^2}{12} + m_2\left(\frac{L}{2}\right)^2\right]}_{\text{or } m_2 L^2/3}$$

$$+ \left\{\left[\frac{m_3 r^2}{2} - m_3\left(\frac{4r}{3\pi}\right)^2\right] + m_3\left(L + \frac{4r}{3\pi}\right)^2\right\}$$

in which

$$I_{z_{C_3}} = I_{z_Q} - m_3\left(\frac{4r}{3\pi}\right)^2 = \frac{m_3 r^2}{2} - m_3\left(\frac{4r}{3\pi}\right)^2$$

Note that we cannot correctly transfer the inertia of the semicylinder from Q to O; it must be done from the mass center C_3. So first we go "through the back door" to *find* $I_{z_{C_3}}$ (since we know the moment of inertia already with respect to Q) and only ***then*** may we transfer to O.

EXAMPLE 4.12

A closed, empty wooden box is 5 ft × 3 ft × 2 ft and weighs 124 lb.

a. Find its moment of inertia about an axis through C parallel to the 2 ft dimension.

b. If the box is then filled with homogeneous material weighing 240 lb (excluding the box), how much does the moment of inertia about the axis of part (a) increase?

(Continued)

SOLUTION

a. The masses of the various sides of the box are proportional to their areas (thickness and density assumed constant):

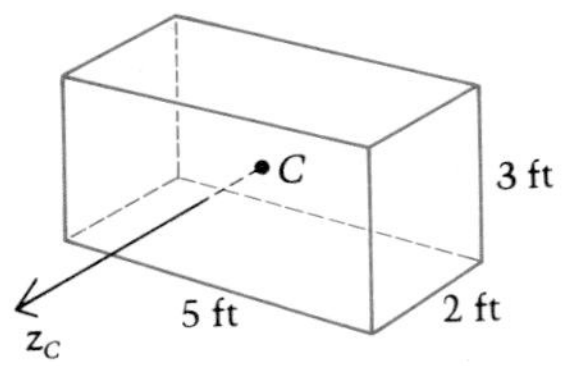

1. 3×5's: $2 \times 15 = 30 \text{ ft}^2$
2. 2×3's: $2 \times 6 = 12 \text{ ft}^2$
3. 2×5's: $2 \times 10 = 20 \text{ ft}^2$

total area $= 62 \text{ ft}^2$

$$W_1 = \frac{30}{62}(124) = 60 \text{ lb for two } 3 \times 5\text{'s}$$

$$W_2 = \frac{12}{62}(124) = 24 \text{ lb for two } 2 \times 3\text{'s}$$

$$W_3 = \frac{20}{62}(124) = 40 \text{ lb for two } 2 \times 5\text{'s}$$

Therefore, taking the contributions from the three pairs of sides,

$$I_{z_C} = I_{z_{C_1}} + I_{z_{C_2}} + I_{z_{C_3}}$$

$$= \frac{1}{32.2}\left[\frac{2(30)(3^2 + 5^2)}{12} + 2(12)\left(\frac{3^2}{12} + 2.5^2\right) + 2(20)\left(\frac{5^2}{12} + 1.5^2\right)\right]$$

$$= 15.9 \text{ slug-ft}^2 \qquad (\text{or lb-ft-sec}^2)$$

b. $$I_{z_{C_{\text{contents}}}} = \frac{240(5^2 + 3^2)}{32.2(12)} = 21.1 \text{ slug-ft}^2$$

Even though the box weighs only about half as much as the contents, the position of its mass makes its moment of inertia over three-fourths that of the contents. The *total* moment of inertia is 37.0 slug-ft^2.

There is a distance called the **radius of gyration** that is often used in connection with moments of inertia. The radius of gyration of the mass of a body about a line z_P (through a point P) is called k_{z_P}, or just k_P if the axis is understood to be z, and is defined by the equation

$$I_{z_P} = mk_P^2 \tag{4.22}$$

If one insists on a physical interpretation of k_P, it may be thought of as the distance from P in any direction perpendicular to z_P at which a point mass, with the same mass as the body, would have the same resulting moment of inertia that the body itself has about axis z_P. For example, a solid homogeneous cylinder has a radius of gyration with respect to its axis of $R/\sqrt{2}$, since $I_{z_C} = mk_C^2 = \frac{1}{2}mR^2$. The usefulness of k_C is seen here, since regardless of the mass of a cylinder (and hence of its density) k_C will be the same for all homogeneous cylinders of equal radii.

Note further that (using the parallel-axis theorem)

$$mk_P^2 = I_{z_P} = I_{z_C} + md^2 = m(k_C^2 + d^2)$$

Thus

$$k_P^2 = k_C^2 + d^2 \tag{4.23}$$

and we see from Equation (4.23) that the radius of gyration, like the moment of inertia itself, is a minimum at C.

Products of Inertia

We now turn to the other two measures of mass distribution that have arisen in our study of plane motion of a rigid body—namely I_{yz_P} and I_{xz_P}, taken here to be with respect to body-fixed axes $(\mathbf{x}, \mathbf{y}, \mathbf{z})$ through any point P in the reference plane (which contains the mass center C).*

Our first step is to gain insight into the meaning of products of inertia as we show that they in fact vanish for two large classes of commonly occurring symmetry. These classes are defined by the two conditions (with ρ constant in both): (1) $\mathbf{z}_P$ is an axis of symmetry and (2) $\mathbf{x}_P\mathbf{y}_P$ is a plane of symmetry. Let us examine why the two products of inertia I_{xz_P} and I_{yz_P} are zero in these cases. We recall that their definitions are

$$I_{xz_P} = -\int xz\,dm \qquad I_{yz_P} = -\int yz\,dm \tag{4.24}$$

Class 1: $\mathbf{z}_P$ Is an Axis of Symmetry. For each dV at (x, y, z) there is a corresponding dV at $(-x, -y, z)$. Thus the contributions of these two elements cancel in both the I_{xz_P} and I_{yz_P} integrals. Since ***each*** point of $\mathcal{B}$ has a "canceling point" reflected through the z axis, I_{xz_P} and I_{yz_P} are each zero for this class of bodies. (See Figure 4.5.)

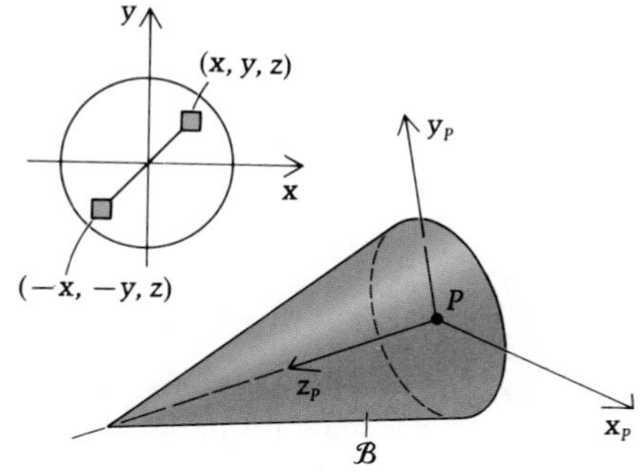

Figure 4.5

*In general (three-dimensional) motion, there arise six distinct inertia properties: three moments of inertia and three products of inertia.

Class 2: $x_P y_P$ Is a Plane of Symmetry. In this case each differential volume dV at (x, y, z) necessarily has a mirror image at $(x, y, -z)$. Thus the contributions of these two elements cancel in both integrals, and taken over the whole of $\mathcal{B}$ we see again that I_{xz_P} and I_{yz_P} are each zero. (See Figure 4.6.)

Just because a body does not belong to either of these two classes does not mean it cannot have zero products of inertia. However, these are simply common cases worthy of note.

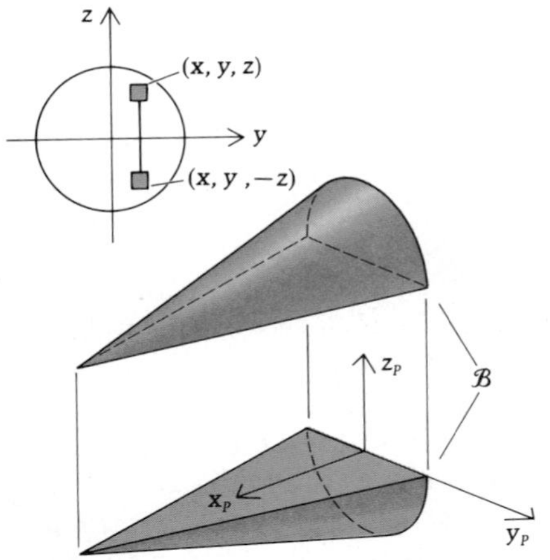

Figure 4.6

Question 4.2 Think of a rigid body for which both products of inertia are zero, but which does not fall into either of the two categories.

There is a transfer theorem for products of inertia, just as there is one for moments of inertia. To derive it, we write from Figure 4.4:

$$I_{xz_P} = -\int x_P z_P \, dm = -\int (x_C + \bar{x})(z_C + \bar{z}) \, dm$$
$$= -\int x_C z_C \, dm - \overline{xz} \int dm - \bar{x} \int z_C \, dm - \bar{z} \int x_C \, dm$$

The last two terms vanish by virtue of the definition of the mass center (for example, $\int z_C \, dm = m$ times the z distance from C to C, which is zero). Thus

$$I_{xz_P} = I_{xz_C} - m\overline{xz} \tag{4.25a}$$

We note that if $(\bar{x}, \bar{y}, \bar{z})$ were measured from C to P instead of from P to C, the result would be identical since both $\bar{x}$ and $\bar{z}$ would change signs. Therefore we need not memorize whether $(\bar{x}, \bar{y}, \bar{z})$ measure from C to P or from P to C. Similarly,

$$I_{yz_P} = I_{yz_C} - m\overline{yz} \tag{4.25b}$$

EXAMPLE 4.13

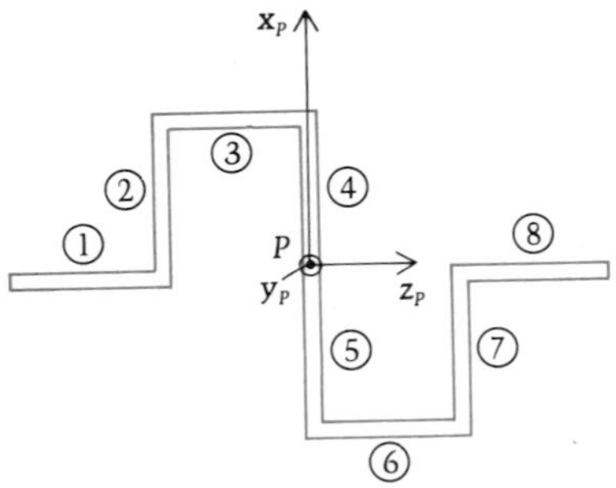

Find I_{xz_P} and I_{yz_P} for the body shown in the accompanying diagram; it is composed of eight identical uniform slender rods, each of mass m and length ℓ.

(Continued)

SOLUTION

We have $I_{yz_P} = 0$ since $\mathbf{x}_P\mathbf{z}_P$ is a plane of symmetry. Recall that when this happens, the two products of inertia containing (as a subscript) the coordinate normal to the plane are zero.* With superscripts identifying the various rods, we then have the following for the other product of inertia:

$$I_{xz_P} = I_{xz_P}^{①} + I_{xz_P}^{②} + I_{xz_P}^{③} + I_{xz_P}^{④} + I_{xz_P}^{⑤} + I_{xz_P}^{⑥} + I_{xz_P}^{⑦} + I_{xz_P}^{⑧}$$

Note that by symmetry each rod has zero I_{xz} about axes through its *own* center of mass parallel to $\mathbf{x}_P$ and $\mathbf{z}_P$. Therefore the eight terms listed above will consist only of transfer terms in this problem.

Furthermore, since $\bar{z}$ is zero for rods 4 and 5, and since $\bar{x}$ is zero for rods 1 and 8, only four rods contribute to the overall I_{xz_P}:

$$\begin{aligned} I_{xz_P} &= I_{xz_P}^{②} + I_{xz_P}^{③} + I_{xz_P}^{⑥} + I_{xz_P}^{⑦} \\ &= -m\left(\frac{\ell}{2}\right)(-\ell) - m(\ell)\left(-\frac{\ell}{2}\right) - m(-\ell)\left(\frac{\ell}{2}\right) - m\left(-\frac{\ell}{2}\right)\ell \qquad (4.26) \\ &= 4\left(\frac{m\ell^2}{2}\right) = 2m\ell^2 \end{aligned}$$

Note that the "unbalanced" masses lie in the second and fourth quadrants in this example; hence the sign of I_{xz_P} is positive since its definition carries a minus sign outside the integral. We shall return to this example later in the chapter and examine the reactions caused by the nonzero value of I_{xz_P} when the body is spun up in bearings about the z_P axis.

*Thus I_{xy_P} is also zero in this example, but I_{xy_P} does not appear anywhere in the plane-motion equations, as we have seen.

Now that we have developed a feel for the inertia properties appearing in the angular momentum vector $\mathbf{H}_P$ (Equation 4.6), in the next section we examine the other law of motion, in which $\mathbf{H}_P$ plays a central role.

PROBLEMS / Section 4.3

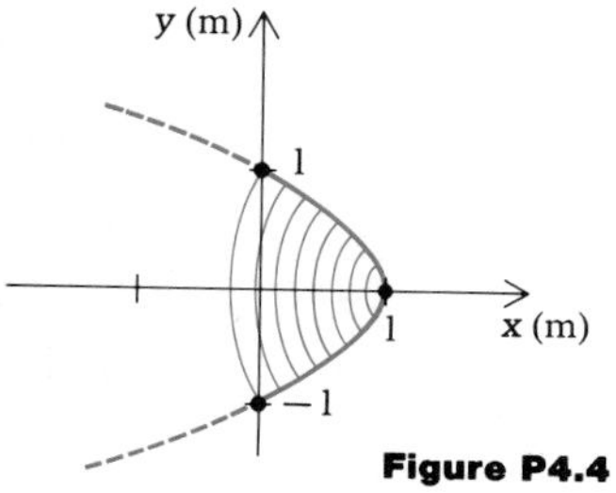

Figure P4.4

4.4 In Figure P4.4, the area bounded by the x and y axes and the parabola $y^2 = 1 - x$ is rotated about the x axis to form a solid of revolution. The density is $\rho = c(1 - x)^8$ kg/m³, where c is a constant. Find the moment of inertia of the solid mass about: (a) the x axis; (b) the line $y = -2$.

4.5 An ellipsoid of revolution is formed by rotating the ellipse about the **x** axis as in Figure P4.5. Find the moment of inertia of this solid body of density 15 slug/ft³ about the **x** axis.

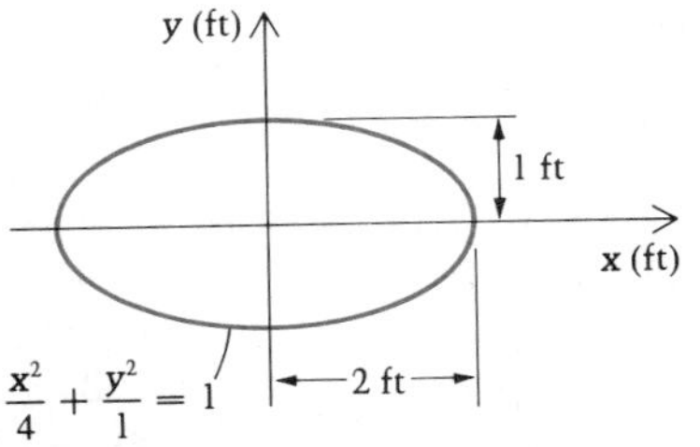

Figure P4.5

4.6 Find the moment of inertia of the mass of $\mathcal{B}$ abo axis z_B if $I_{z_A} = 40$ kg · m². (See Figure P4.6.)

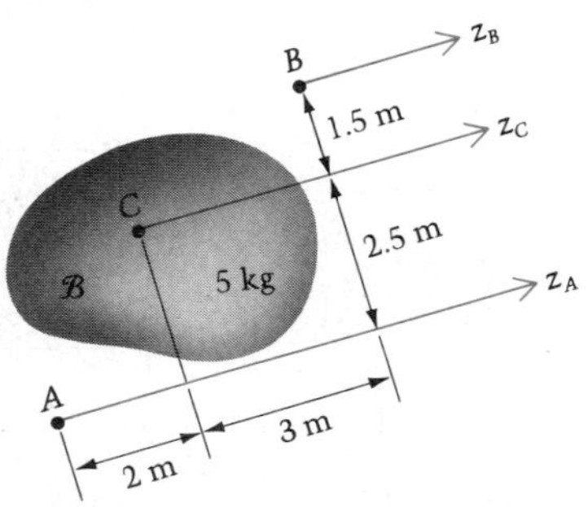

Figure P4.6

4.7 The rod in Figure P4.7 has a mass density given by

$$\rho = \rho_1\left(\frac{x}{L}\right)^2 + \rho_0$$

in which ρ_0 and ρ_1 are constants. The rod has length L and radius R. Find its moment of inertia about the line defined by $x = 0$, $y = L/2$.

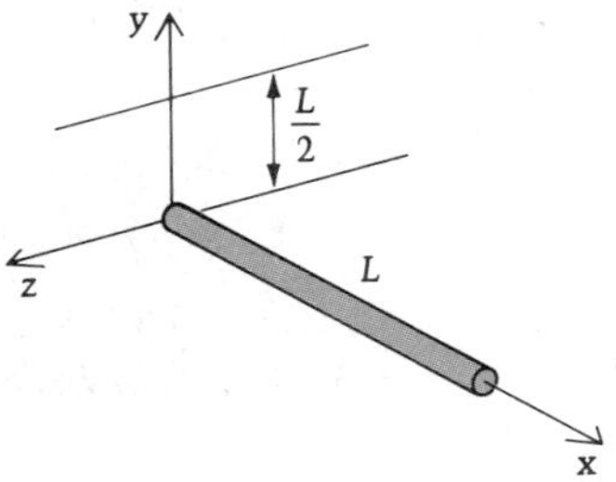

Figure P4.7

4.8 A rod $\mathcal{R}$ of length 1 m is welded on its ends to a disk $\mathcal{D}$ and a sphere $\mathcal{S}$. (See Figure P4.8.) The uniform bodies have masses $m_{\mathcal{R}} = 10$ kg, $m_{\mathcal{D}} = 5$ kg, and $m_{\mathcal{S}} = 15$ kg. The radii of $\mathcal{D}$ and $\mathcal{S}$ are 0.3 m and 0.1 m, respectively. Find I_z.

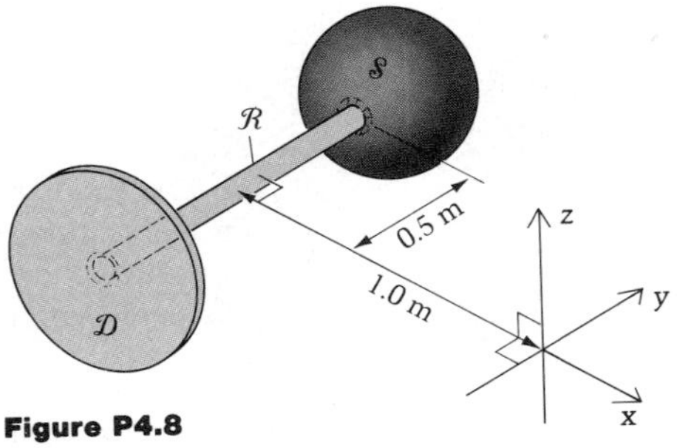

Figure P4.8

4.9 The three bodies shown in Figure P4.9, welded together to form a single body $\mathcal{B}$, have masses of 64 (rectangular plate), 56 (rod), and 48 (disk), all in kilograms. Find the moment of inertia of $\mathcal{B}$ with respect to the z_Q axis.

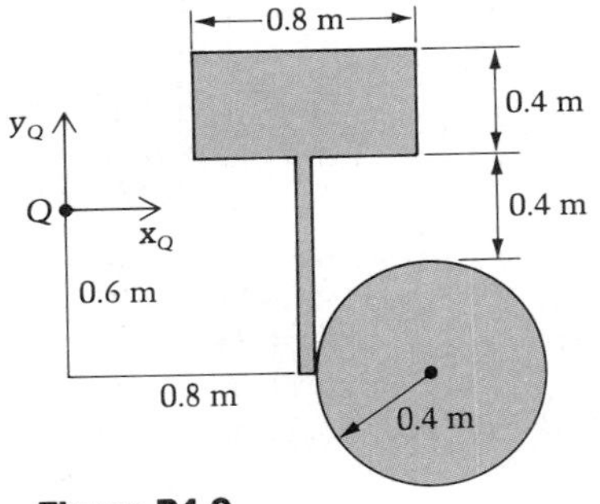

Figure P4.9

4.10 The body shown in Figure P4.10 is composed of a slender uniform bar ($m = 4$ slugs) and a uniform sphere ($m = 5$ slugs). Find I_{z_C} for the body, where z_C is normal to the figure.

4.11 Two bars, each weighing 5 lb per foot, are welded together as shown in Figure P4.11. (a) Locate the center of mass of the body. (b) Find I_{z_C}.

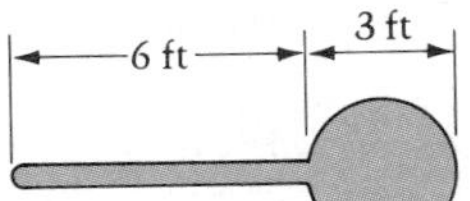

Figure P4.10

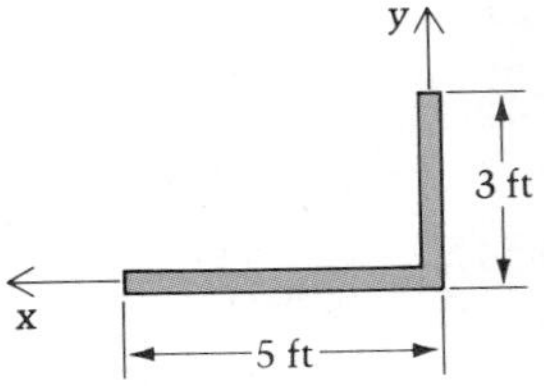

Figure P4.11

4.12 Find I_{z_C} for the semicircular ring $\mathcal{B}$ in Figure P4.12. *Hint:* If the dashed portion were present, I_{z_O} would be $(2m)R^2$; by symmetry, our semicircular ring contributes half of this, so that for $\mathcal{B}$ we have

$$I_{z_O} = mR^2$$

Now use the transfer theorem to complete the solution without integration.

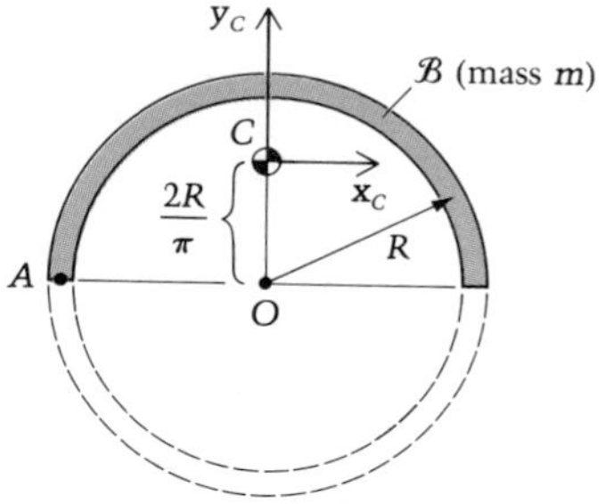

Figure P4.12

4.13 In Figure P4.13 the moment of inertia I_{z_G} of the uniform rectangular plate $\mathcal{P}$, made up of two identical triangular plates $\mathcal{A}$ and $\mathcal{B}$ is

$$I_{z_G} = \frac{m}{12}(b^2 + H^2)$$

as in Example 4.5. Use the parallel-axis theorem as suggested below to derive the moment of inertia $I_{z_C}^{\mathcal{B}}$ of the triangular plate $\mathcal{B}$:

$$I_{z_G}^{\mathcal{P}} = I_{z_G}^{\mathcal{A}} + I_{z_G}^{\mathcal{B}} = 2I_{z_G}^{\mathcal{B}} = 2[I_{z_C}^{\mathcal{B}} + m_{\mathcal{B}}d^2]$$

4.14 Find the moments of inertia of the pendulum about axes x_O, y_O, and z_O. (See Figure P4.14.) Axes x_O and y_O are in the plane of the pendulum, $\mathcal{R}$ is a slender rod, and $\mathcal{D}$ is a semicircular disk, each of constant density.

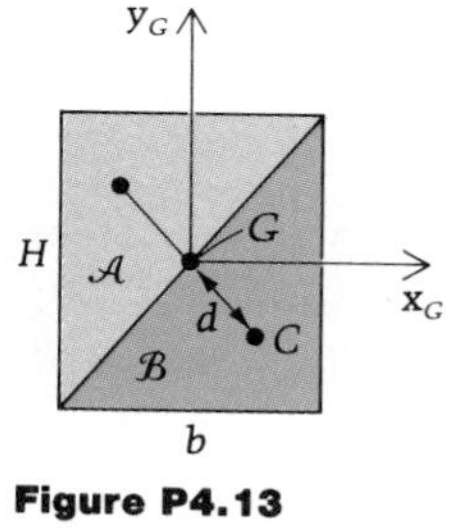

Figure P4.13

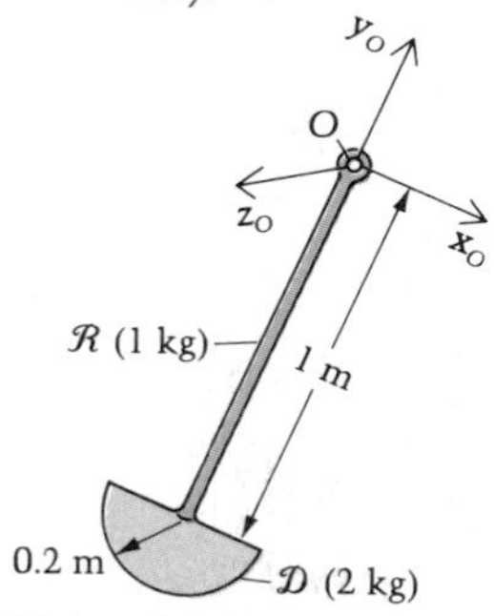

Figure P4.14

4.15 In Problem 2.14, show that the nondimensional axial moment of inertia, defined by $I_{y_C}/[m(D/2)^2]$, has the same equation as y_C/y_T in that problem.

4.16 Derive the moment of inertia of a thin conical shell of mass m about its axis z_C. (See Figure P4.16.)

4.17 See Figure P4.17. (a) Show that the moment of inertia (I_{x_O}) for a solid homogeneous cone about a lateral axis through the base is $I_{x_O} = (m/20)(3R^2 + 2H^2)$. (b) Using the transfer theorem, find the expression for I_{x_C}.

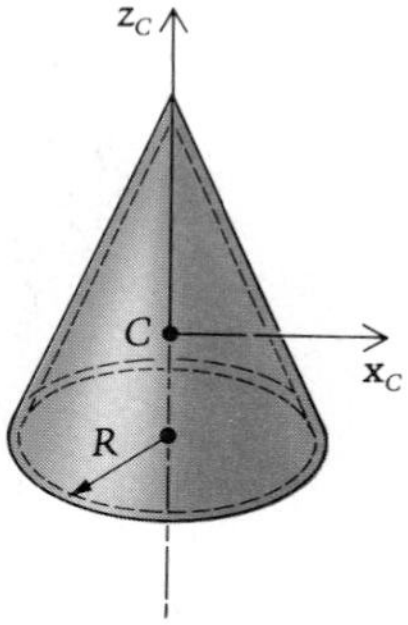

Figure P4.16

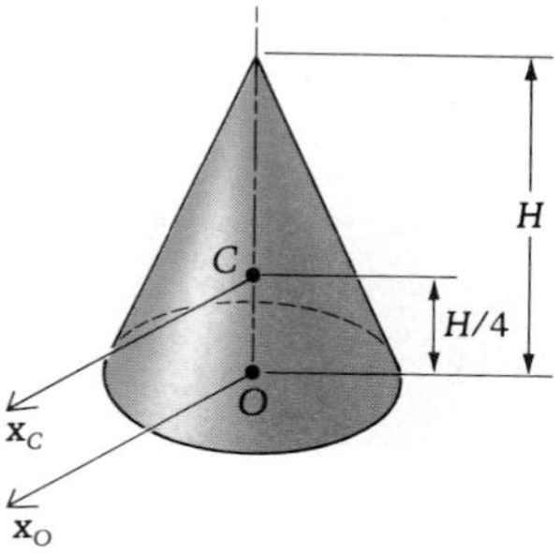

Figure P4.17

4.18 In the previous problem, find I_{x_O} for the body in the figure if the part above $z = H/2$ is cut away to form a truncated cone.

4.19 Find I_{x_C} (which equals I_{y_C}) for a thin plate (density ρ, thickness t) in the shape of a quarter-circle. (See Figure P4.19.)

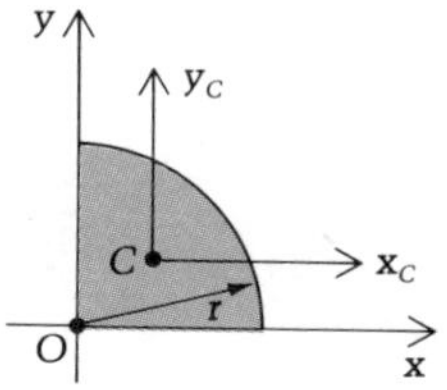

Figure P4.19

4.20 Find I_{x_C} for the semielliptical prism shown in Figure P4.20 (density ρ, length normal to plane of paper $= L$).

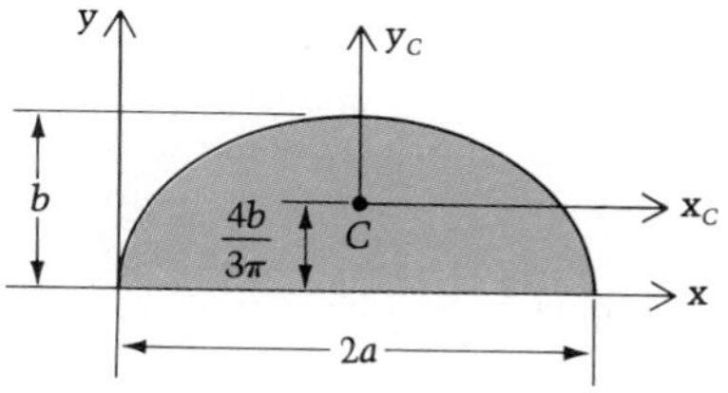

Figure P4.20

4.21 In the preceding problem, determine I_{y_C}.

4.22 Find I_x in Problem 4.11.

4.23 Use the result of the preceding problem, together with the transfer (parallel axis) theorem, to find I_{x_C}.

4.24 Find I_y in Problem 4.11.

4.25 Use the result of the preceding problem, together with the transfer theorem, to determine I_{y_C}.

4.26 Find the moment of inertia of a uniform hemispherical solid about the lateral axis $\mathbf{x}_C$ through its mass center. (See Figure P4.26.)

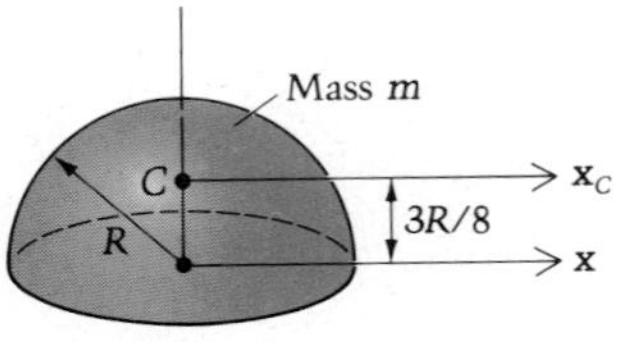

Figure P4.26

4.27 The surface area of a solid of revolution is formed by rotating the curve $y = x^2$ (for $0 \le x \le 2$ m) about the x axis. (See Figure P4.27.) The density of the material varies according to the equation $\rho = 20x$, where ρ is in kg/m^3 when x is in meters. Find I_{x_O} and tell why your answer is also I_{x_C}.

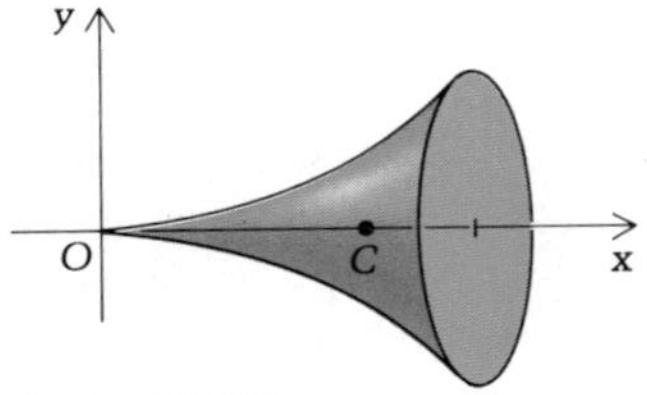

Figure P4.27

4.28 In the preceding problem, find I_{y_O} and I_{y_C}.

4.29 In supporting an antenna subreflector, either three or four spars are normally used to connect it to the main reflector, as suggested in Figure P4.29a. Sometimes blockage (interference with the signal caused by the spars' projected area) is a serious problem. One solution is to make the spars in an ogival shape as shown in Figure P4.29b. The cross section is composed of sections of two circular tubes welded at their intersection. Find the moment of inertia of the ogival spar about the $\mathbf{x}_C$ axis. (This information is of use in studying and minimizing vibrations of the subreflector/spar assembly caused by the wind.)

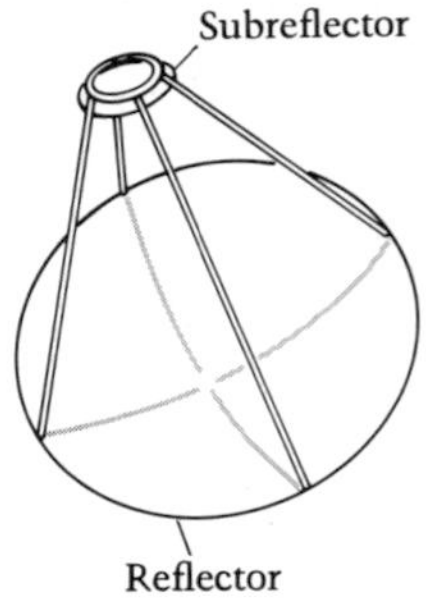

Figure P4.29a

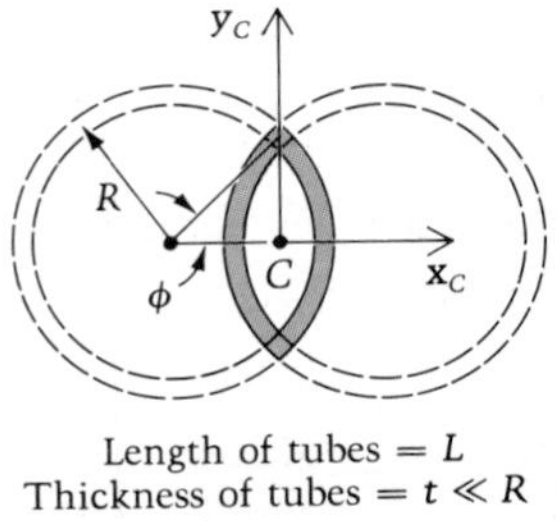

Figure P4.29b

4.30 Find I_{y_C} in the preceding problem.

4.31 Find I_{x_C} for a uniform thin plate in the form of a pie-shaped circular sector as shown in Figure P4.31.

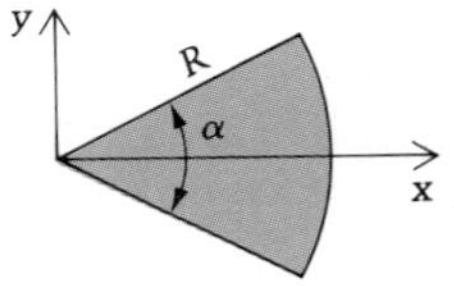

Figure P4.31

4.32 Find I_{y_C} for the plate in the preceding problem.

4.33 The moment of inertia of a uniform, equilateral triangular plate $\mathcal{P}$ about the axis z_C normal to the plane of the plate is $mS^2/12$, where S is the length of one side. Show this by using the "S by $(\sqrt{3}/2)\,S$" rectangular plate shown in Figure P4.33 and subtracting the contribution of the pair of (shaded) right triangles. (Refer to Problem 4.13.)

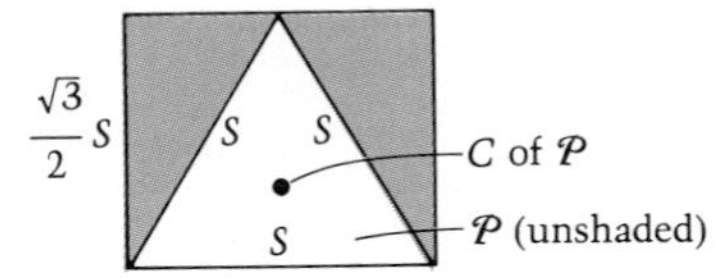

Figure P4.33

4.34 Find the product of inertia I_{xy_A} for the ring of Problem 4.12, where $\mathbf{x}_A$ and $\mathbf{y}_A$ have origin at A and are parallel to $\mathbf{x}_C$ and $\mathbf{y}_C$.

4.35 Find I_{xy_Q} for the welded body $\mathcal{B}$ of Problem 4.9.

4.36 Determine I_{xy} in Problem 4.8. The **xy** plane contains the centers of $\mathcal{D}$, $\mathcal{R}$, and $\mathcal{S}$.

4.37 Show that for a symmetric rigid body (see Section 4.3) in plane motion, the angular momentum about any point Q of the body lying in the reference plane containing C is given by

$$\mathbf{H}_Q = I_Q \omega \hat{\mathbf{k}} + \mathbf{r}_{QC} \times m\mathbf{v}_Q$$

4.38 Refer to Problem 4.3. Let P be a point of a rigid body in plane motion, and let the body be such that $I_{xz_P} = I_{yz_P} = 0$. Show that $\mathbf{H}_{P_{\text{rel}}} = I_{z_P}\omega\hat{\mathbf{k}}$.

4.39 Refer to Problem 4.15. Show that

$$\frac{I_{x_v}}{mD^2/4} = \frac{\left[3 + 96\left(\frac{f}{D}\right)^2 - 1024\left(\frac{f}{D}\right)^4\right]\left[1 + 16\left(\frac{f}{D}\right)^2\right]^{3/2} + 65{,}536\left(\frac{f}{D}\right)^7}{28\left\{16\left(\frac{f}{D}\right)^2\left[1 + 16\left(\frac{f}{D}\right)^2\right]^{3/2} - 1024\left(\frac{f}{D}\right)^5\right\}}$$

4.40 The antenna $\mathcal{A}$ in Figure P4.40 has a moment of inertia about z_C of I_C, and the counterweights $\mathcal{C}$ have a collective moment of inertia about z_G of I_G. Points C and G are the respective mass centers of the antenna and its counterweights. The purpose of the counterweights is to place the combined mass center at O to reduce stresses. Thus $MD = md$, where we neglect the mass of the connecting frame for this problem. Compute the values of M and D that will minimize the total moment of inertia I_O (I of $\mathcal{C}$ about O plus I of $\mathcal{A}$ about O). Use $I_G = Mk_G^2$, where k_G is a constant.

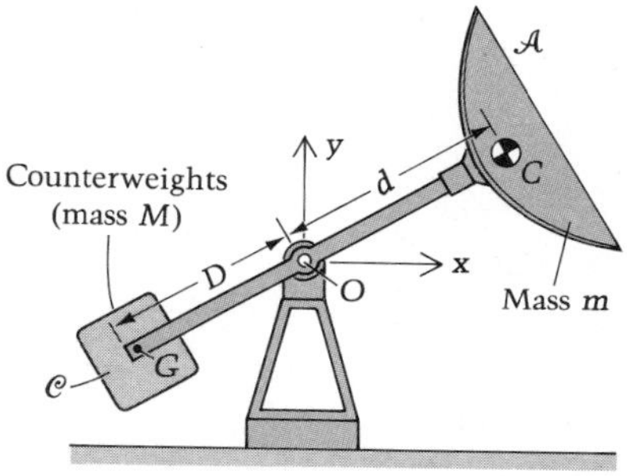

Figure P4.40

4.4 Euler's Second Law

Newton's laws, as described in Chapter 2, were written for, and apply to, the particle. We have also seen that Euler's extension of Newton's work allows us to find the motion of the mass center of any system, whether or not it can properly be treated as a particle.

Neither Newton's second law nor Euler's first law, however, is able to tell us anything about the orientation or the angular motion of the finite-sized body. Euler spent 50 years, off and on, wrestling with this problem of how we may completely solve for a body's position in space—not only where a representative point such as the mass center is located but also how we must then rotate the body about a line through the point in order to establish its orientation also.

His discovery, completed in 1775, was to hypothesize additionally that in the same (inertial, Newtonian, Galilean) frames in which $\mathbf{F}_r = \dot{\mathbf{L}}$, it is also true that

$$\mathbf{M}_{r_O} = \dot{\mathbf{H}}_O \qquad (4.27)$$

in which $\mathbf{M}_{r_O}$ is the resultant moment of all external forces and couples about any point O *fixed in the inertial frame.** It seems simple 200 years

*If O is fixed in the frame, then it never moves in the frame—that is, $\mathbf{v}_O = \mathbf{0}$, $\mathbf{a}_O = \mathbf{0}$, and all further derivatives are zero too in that frame.

after the fact that if *force* produces change in *linear* momentum, then *moment* should likewise produce change in *angular* momentum. But then hindsight is always twenty-twenty.

It is worth noting that in many textbooks Newton's second and third laws for a particle and a pair of particles, respectively, are adopted as the basic postulates; Euler's laws are derived by adding equations appropriate to the motion of each of the many particles of which the body is composed. In our approach an action-reaction principle is not postulated but may in fact be deduced from Equations (4.27) and (2.1). The result, derived in Appendix B, states that the resultant (force and couple) exerted on body 2 by body 1 is the opposite (negative) of the resultant exerted on body 1 by body 2. This, of course, is the same action-reaction principle the student has used many times in analyzing problems in statics.

We shall now derive several useful alternative forms of **Euler's second law** (the moment equation of motion, also known as the **principle of angular momentum**). If we let the (arbitrary) point P in Equation (4.5) be a fixed point O in our inertial frame, then

$$\mathbf{H}_O = \mathbf{H}_C + \mathbf{r}_{OC} \times \mathbf{L} \tag{4.28}$$

This equation relates the two moment of momentum vectors with respect to O and C. Differentiating and using (4.27), we get

$$\mathbf{M}_{r_O} = \dot{\mathbf{H}}_O = \dot{\mathbf{H}}_C + \dot{\mathbf{r}}_{OC} \times \mathbf{L} + \mathbf{r}_{OC} \times \dot{\mathbf{L}}$$

But

$$\dot{\mathbf{r}}_{OC} = \mathbf{v}_C \qquad \text{and} \qquad \mathbf{L} = m\mathbf{v}_C$$

Therefore

$$\dot{\mathbf{r}}_{OC} \times \mathbf{L} = \mathbf{0}$$

and so

$$\mathbf{M}_{r_O} = \dot{\mathbf{H}}_C + \mathbf{r}_{OC} \times \dot{\mathbf{L}}$$

Now since $\dot{\mathbf{L}} = \mathbf{F}_r$ by Euler's first law (Equation 2.1),

$$\mathbf{M}_{r_O} = \dot{\mathbf{H}}_C + \mathbf{r}_{OC} \times \mathbf{F}_r \tag{4.29}$$

However, we know from the definition of resultants that $\mathbf{M}_{r_O} = \mathbf{M}_{r_C} + \mathbf{r}_{OC} \times \mathbf{F}_r$, so that

$$\mathbf{M}_{r_C} + \mathbf{r}_{OC} \times \mathbf{F}_r = \dot{\mathbf{H}}_C + \mathbf{r}_{OC} \times \mathbf{F}_r$$

and thus

$$\mathbf{M}_{r_C} = \dot{\mathbf{H}}_C \tag{4.30}$$

The steps leading to Equation (4.30) reverse themselves back to $\mathbf{M}_{r_O} = \dot{\mathbf{H}}_O$, so that either equation may be taken as Euler's second law.

We shall also find it helpful to have an expression for moments taken with respect to an *arbitrary* point P, not necessarily fixed in the inertial frame of reference. Again using the definition of resultants, we obtain

$$\mathbf{M}_{r_P} = \mathbf{M}_{r_C} + \mathbf{r}_{PC} \times \mathbf{F}_r$$

Therefore, using (4.30) and (2.1),

$$\mathbf{M}_{r_P} = \dot{\mathbf{H}}_C + \mathbf{r}_{PC} \times \dot{\mathbf{L}}$$

But $\dot{\mathbf{L}} = m\mathbf{a}_C$ so that

$$\mathbf{M}_{r_P} = \dot{\mathbf{H}}_C + \mathbf{r}_{PC} \times (m\mathbf{a}_C) \qquad (4.31)$$

This equation allows us to take moments about any point we wish, a feature that is often very handy. It is important to note that satisfaction of the force-momentum equation (2.1) and one of the three moment-moment of momentum equations (4.27, 4.30, or 4.31) guarantees satisfaction of the other two moment equations, so they are not all independent.

In the remainder of the chapter, we shall use these equations to solve problems of the rigid body in plane motion. In the next section we solve them by direct use of the differential equations; in Chapter 5 we use certain general integrals of the equations that produce the principles of work and kinetic energy as well as (linear and angular) impulse and momentum.

In summary, then, Euler's laws are

First law: $\mathbf{F}_r = \dot{\mathbf{L}}$ $\qquad (4.32)$

$(= m\mathbf{a}_C)$

and

Second law: $\mathbf{M}_{r_O} = \dot{\mathbf{H}}_O$ or $\mathbf{M}_{r_C} = \dot{\mathbf{H}}_C$ $\qquad (4.33)$

or $\mathbf{M}_{r_P} = \dot{\mathbf{H}}_C + \mathbf{r}_{PC} \times m\mathbf{a}_C$

If the forces and couples acting on the body $\mathcal{B}$ are not functions of the motion quantities (linear and angular positions, velocities, and accelerations) making up $\mathbf{L}$ and $\mathbf{H}_C$, then (4.32) and (4.33) are not coupled; in this case, solving (4.32) independently locates the mass center of $\mathcal{B}$ and solving (4.33) tells us the orientation of $\mathcal{B}$ in space, as we shall see in the remainder of the chapter.

PROBLEM / Section 4.4

4.41 Show that if Euler's laws

$$\mathbf{F}_r = \dot{\mathbf{L}}$$

$$\mathbf{M}_{r_O} = \dot{\mathbf{H}}_O$$

hold in an inertial frame $\mathcal{I}$ in which O is a fixed point, then it follows that $\mathbf{M}_{r_{O'}} = \dot{\mathbf{H}}_{O'}$, where O' is any other fixed point of $\mathcal{I}$.

4.5 Equations of Motion

Derivation, Discussion, and Solution of the Equations

We have seen in Chapter 2 that the first of Euler's two laws gives us a vector equation governing the mass center motion of any system of bodies. This equation is in every respect analogous to Newton's second law for a particle; it is more general, however, because its use is not restricted to particles. Hence for the purposes of this chapter (rigid bodies in plane motion), the motion of the mass center C of rigid body $\mathcal{B}$ will be governed by the $\mathbf{x}$ and y components of Equation (2.1):

$$F_{r_x} = m\ddot{x}_C \tag{4.34}$$

$$F_{r_y} = m\ddot{y}_C \tag{4.35}$$

where x_C and y_C are coordinates of the mass center in a rectangular coordinate system fixed in the inertial frame.* Note also that since $\dot{z}_C$ vanishes identically in plane motion, the third equation from (2.1) is a statics equation: $F_{r_z} = m\ddot{z}_C = 0$.

We also need to specialize Euler's second law for plane motion of the rigid body $\mathcal{B}$. This step will allow us to solve for the ***orientation*** of the body. Then, with the mass center located by the solutions to Equations (4.34) and (4.35), the motion, or exact location of all points of $\mathcal{B}$ at all times, will be known.

Euler's second law may be equivalently expressed as either

$$\mathbf{M}_{r_C} = \dot{\mathbf{H}}_C \qquad \text{or} \qquad \mathbf{M}_{r_O} = \dot{\mathbf{H}}_O$$

in which the derivative is taken in an inertial frame $\mathcal{I}$ and where O is any given point of $\mathcal{I}$, and C is the mass center of $\mathcal{B}$.

*Which component equations are used and the specific forms they take depend, of course, on what type of coordinate system is used to describe the motion of C. A rectangular coordinate system is the natural choice for most of the problems taken up in this chapter. A polar coordinate system is the natural choice, however, for problems of orbital mechanics (see Section 8.3).

Using the mass center form presented above and substituting Equation (4.8) for $\mathbf{H}_C$ gives

$$\mathbf{M}_{r_C} = \frac{d}{dt}\mathbf{H}_C = \frac{d}{dt}(I_{xz_C}\omega\hat{\mathbf{i}} + I_{yz_C}\omega\hat{\mathbf{j}} + I_{z_C}\omega\hat{\mathbf{k}}) \tag{4.36}$$

At this stage we must make a decision with regard to how the **x** and **y** axes of Equation (4.36), which have ***their*** origin at *C*, will be allowed to change relative to the frame of reference. Note that the direction of the z axis has already been fixed perpendicular to the reference plane. If we fix the directions of **x** and **y** relative to the inertial frame, then $\hat{\mathbf{i}}$ and $\hat{\mathbf{j}}$ (as well as $\hat{\mathbf{k}}$) are constant relative to that frame but I_{xz_C} and I_{yz_C} are in general time-dependent.

Question 4.3 Why will I_{z_C} not change in this case?

A more convenient choice is to let the axes **x**, **y**, and **z** all be fixed in the ***body*** so that the moments and products of inertia are all constant. Now $\hat{\mathbf{i}}$ and $\hat{\mathbf{j}}$ are time-dependent relative to the inertial frame $\mathcal{I}$, however, and their derivatives, from Equations (3.41) and (3.42), are

$$\frac{d\hat{\mathbf{i}}}{dt} = (-\sin\theta\hat{\mathbf{I}} + \cos\theta\hat{\mathbf{J}})\dot{\theta} = \dot{\theta}\hat{\mathbf{j}} \text{ or } \omega\hat{\mathbf{j}} \tag{4.37a}$$

$$\frac{d\hat{\mathbf{j}}}{dt} = -(\cos\theta\hat{\mathbf{I}} + \sin\theta\hat{\mathbf{J}})\dot{\theta} = -\dot{\theta}\hat{\mathbf{i}} \text{ or } -\omega\hat{\mathbf{i}} \tag{4.37b}$$

Figure 4.7 shows the unit vectors $\hat{\mathbf{i}}$ and $\hat{\mathbf{j}}$ (fixed to $\mathcal{B}$) expressed in terms of $\hat{\mathbf{I}}$ and $\hat{\mathbf{J}}$ (which are fixed in the inertial reference frame $\mathcal{I}$):

$$\hat{\mathbf{i}} = 1(\cos\theta\hat{\mathbf{I}} + \sin\theta\hat{\mathbf{J}})$$

$$\hat{\mathbf{j}} = 1(-\sin\theta\hat{\mathbf{I}} + \cos\theta\hat{\mathbf{J}})$$

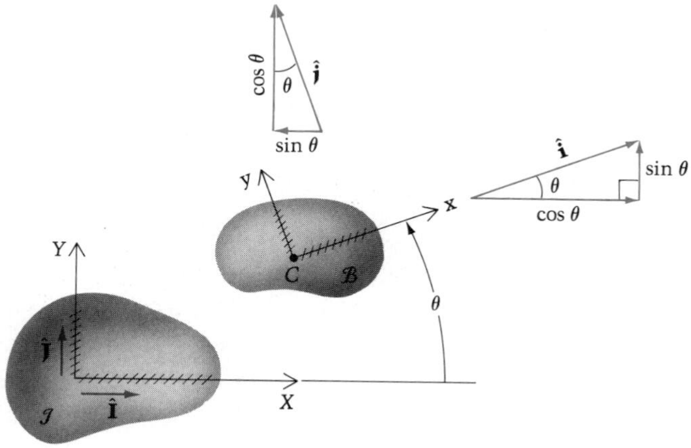

Figure 4.7

Therefore, carrying out the differentiations in Equation (4.36), we have

$$\mathbf{M}_{r_C} = I_{xz_C}\dot{\omega}\hat{\mathbf{i}} + I_{yz_C}\dot{\omega}\hat{\mathbf{j}} + I_{z_C}\dot{\omega}\hat{\mathbf{k}} + I_{xz_C}\omega(\omega\hat{\mathbf{j}}) + I_{yz_C}\omega(-\omega\hat{\mathbf{i}})$$

or

$$\mathbf{M}_{r_C} = (I_{xz_C}\alpha - I_{yz_C}\omega^2)\hat{\mathbf{i}} + (I_{yz_C}\alpha + I_{xz_C}\omega^2)\hat{\mathbf{j}} + I_{z_C}\alpha\hat{\mathbf{k}} \tag{4.38}$$

This expression represents three scalar equations:

$$M_{r_{C_x}} = I_{xz_C}\alpha - I_{yz_C}\omega^2 \tag{4.39a}$$

$$M_{r_{C_y}} = I_{yz_C}\alpha + I_{xz_C}\omega^2 \tag{4.39b}$$

$$M_{r_{C_z}} = I_{z_C}\alpha \tag{4.39c}$$

Equations (4.39a,b), along with $F_{r_z} = 0$, tell us about the nature of reactions necessary to maintain the plane motion. If I_{xz_C} and I_{yz_C} are both zero,* then $M_{r_{C_x}} = 0 = M_{r_{C_y}}$ and the system of external forces (loads plus reactions) has a planar resultant. Thus, with a coplanar system of external loads, the resultant reactions must likewise be coplanar. This is the basis for "two-dimensionalizing" the analysis of problems for which these two products of inertia vanish; we shall first work with symmetrical bodies for which this is the case. Later in this section we examine some problems in which I_{xz_C} and I_{yz_C} are *not* both zero. In any case, the rotational motion of the body is governed by the simple kinetics equation

$$M_{r_{C_z}} = I_{z_C}\alpha \qquad (= I_{z_C}\ddot{\theta}) \tag{4.40}$$

We see that while force produces acceleration with the "resistance" being the mass, it is also true that moment produces angular acceleration with the "resistance" being the body's moment of inertia. Note also that—whether or not I_{yz_C} and I_{xz_C} are zero—the moment of the external forces about the z axis through the mass center is $\mathbf{M}_{r_C} \cdot \hat{\mathbf{k}}$, and we have $\mathbf{M}_{r_C} \cdot \hat{\mathbf{k}} = I_{z_C}\alpha$. Thus the resultant moment about this axis equals the moment of inertia about the axis multiplied by the angular acceleration of the body, regardless of whether the products of inertia vanish. And so we find that the following three differential equations *always* apply to the plane motion of a rigid body:

$$F_{r_x} = m\ddot{x}_C \tag{4.41a}$$

$$F_{r_y} = m\ddot{y}_C \tag{4.41b}$$

and, dropping the z subscript in Equation (4.40) with the understanding that we are to use the z component of $\mathbf{M}_{r_C}$,

*When I_{xz_C} or I_{yz_C} is not zero, there have to be couples $M_{r_{C_x}}$ and $M_{r_{C_y}}$ present to maintain the motion; these are usually formed by forces (such as bearing reactions) at different positions along the z axis.

$$M_{r_C} = I_C\alpha \tag{4.41c}$$

In Equations (4.41) the coordinates (x_C, y_C) that locate the mass center C are with respect to axes that are fixed, as they must be, in the inertial frame.

At this point we briefly discuss what we are doing in kinetics. There is, as we have observed, a "force side" $(\mathbf{F}_r, \mathbf{M}_{r_C})$ of the equations of motion and likewise a "motion side" $(\dot{\mathbf{L}}, \dot{\mathbf{H}}_C)$. Sometimes we know the motions, either through observation or design requirements, and desire to find the forces and moments that produce them. For example:

1. Motions of heavenly bodies, when observed and run through the equations of motion, yield the forces of gravitational attraction that lead to estimates of the masses of the bodies.
2. Perhaps certain speeds and accelerations are desired in a piston problem in order to generate a certain amount of power by an engine. Using the equations to solve for the forces will allow us to compute stresses in the piston, pins, connecting rod, and so forth and ensure that the parts are strong enough to take the loads.

At other times we know the forces and are looking for the motions. A simple example is the computation of the trajectory of a falling body under a known gravity force. When Skylab prematurely fell to earth in 1979, engineers and their computers were working back and forth between *both* sides of the equations, trying to orient the body in such a way as to produce the drag forces required to bring the space station down over uninhabited areas of the earth.

There are other useful, alternative forms of the rotational equation $M_{r_C} = I_C\alpha$ that we will discuss later in this section. First, however, we must study some examples of the solution of Equations (4.41), which always apply without restriction. We recommend the following procedure:

1. Draw a complete free-body diagram of each body in the problem. As in statics, a free-body diagram is a sketch of the body in isolation showing all the external forces and couples that act upon it.
2. Establish a sign convention for x, y, and θ. Use a right-handed system and do not feel obliged always to establish the coordinate directions in the usual way:

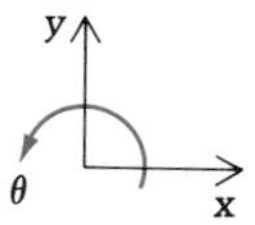

For inclined planes, for example, it is usually helpful to take x parallel to the plane and y normal to it as shown in Figure 4.8. The equations will then be much easier to write and solve. We have to separate the gravity force into components, but all else ($\ddot{x}_C$, $\ddot{y}_C$, and the friction and normal forces) will already be in either the x or

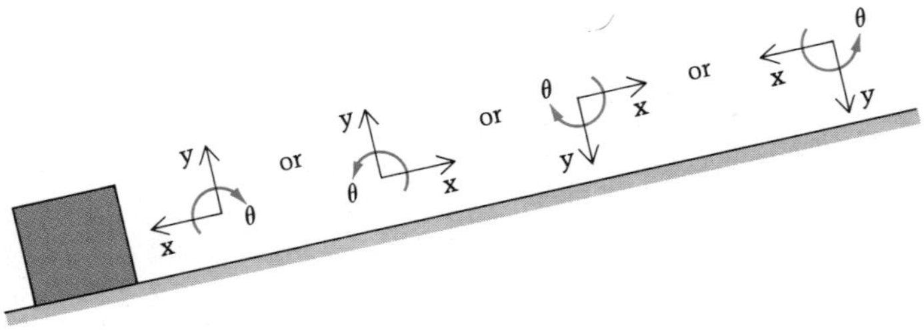

Figure 4.8

the y direction. Moreover, $\ddot{y}$ is identically zero if y is taken normal to the plane. In all cases: *Choose coordinate axes that make your work easier!*

3. Write the governing equations of motion, remembering that the sign convention chosen establishes the positive directions of not only x, y, and θ but also the directions of positive velocity and acceleration components and the directions of positive angular velocity and acceleration. Components of forces and moments, as in statics, are given positive signs if they are assumed in positive directions on the free-body diagram (and minus signs if they are not).
4. Supplement the equations with given or implied information. Three common examples of this step are:
   a. *Constraints:* For example, $\dot{y}_C \equiv \ddot{y}_C \equiv 0$ for the sliding block depicted in Figure 4.8.
   b. *Friction:* If we *know* the body is slipping, for example, then the friction force f is at its maximum (μN) and opposes the velocity direction of the contact points. If, however, we are sure there is no slipping, then $|f|$ is somewhere *between* zero and μN, and *both* f and N are unknowns. Finally, if we do not know whether the body is slipping or not, then we can assume no slipping, find f as part of our solution, and then compare it with f_{max}, or μN, where N is also found in the solution. If then $|f| \le \mu N$, our assumption is correct and the solution based on it is valid. If not, then $|f| > \mu N$, the maximum of f. This cannot be true, so we must "return to the drawing board" with $|f| = \mu N$ and solve again, abandoning any no-slip conditions that were originally employed.
   c. *Kinematics:* Actually, constraint (a) is a kinematic constraint, but here we are also speaking of velocity conditions, such as no-slip or rolling conditions. For example, the velocity of the center of a wheel rolling on a flat, fixed surface is related to the angular velocity by $\dot{x}_C = R\dot{\theta}$.
5. Solve the resulting equations after the number of equations and unknowns agree. Eliminate the unknowns that do not matter and solve for those that do.
6. Check any assumptions you have made, and rework the solution if necessary.

7. Check your units (especially important in these days of transition from U.S. to SI units), and also check to see if your solution seems reasonable.

In the first two examples that follow, the bodies of interest have zero angular velocity and angular acceleration. Such motions are called ***translation***, and Equations (4.30) and (4.31) become, in this case,

$$\mathbf{M}_{I_C} = \mathbf{0} \tag{4.42a}$$

and

$$\mathbf{M}_{I_P} = \mathbf{r}_{PC} \times m\mathbf{a}_C \tag{4.42b}$$

What we have, then, in translation problems, speaking loosely, is "translational dynamics" and "rotational statics."* However, we shall not have to use the moment equation (4.42a or b) at all in these first two examples. We withhold its use until later in the section where it is more appropriately emphasized. The two examples that follow use only mass center equations and are thus in no way different from the problems of Chapter 2, in which we studied the motions of particles and those of mass centers of bodies without having to write any moment equations of motion. Rectilinear translation is reviewed in the first example and curvilinear translation in the second.

EXAMPLE 4.14

Find the accelerations of the blocks shown in the diagram when released from rest. Then repeat the problem with the friction coefficients reversed.

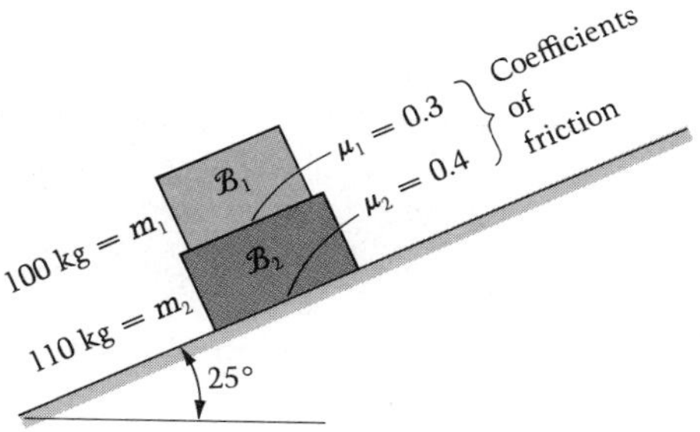

(Continued)

*Except that, unlike in statics, we cannot take moments anywhere we wish and set the result to zero. Equation (4.42b) shows that for translation, $\mathbf{M}_{I_P} = \mathbf{0}$ in only three instances: (1) if $\mathbf{a}_C = \mathbf{0}$, which is the trivial case of equilibrium; (2) if $\mathbf{a}_P$ passes through C (meaning that the vector $\mathbf{a}_P$, drawn or placed at P, lies on the line through points P and C), which case is not very useful, since the points about which we wish to sum moments in translation are usually not accelerating through C; (3) if P is the mass center. Therefore, in translation problems we must set the moment resultant equal to the moment of the $m\mathbf{a}_C$ vector in accordance with Equation (4.42b) if we desire to sum moments about a point other than C.

SOLUTION

We know from statics that *if* the two blocks move as a unit, their motion will occur when

$$\tan 25° > \mu_2$$

that is, when

$$0.466 > 0.4$$

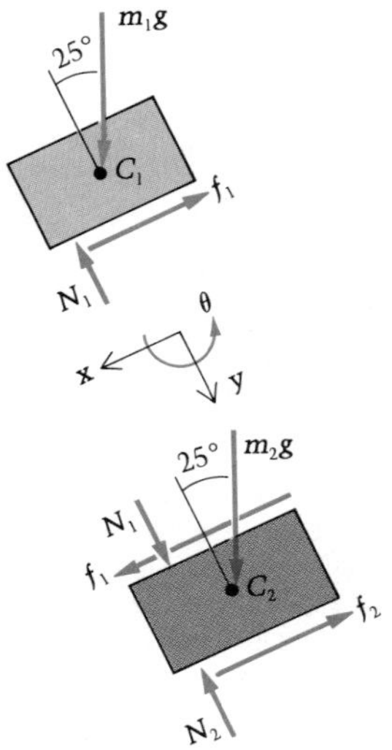

which is the case here. But before our solution is complete we must determine whether either block moves without the other. We consider the free-body diagrams of each translating block and write the equations of motion:*

$$F_{r_{x_1}} = m_1\ddot{x}_{C_1} \Rightarrow 100(9.81)\sin 25° - f_1 = 100\ddot{x}_{C_1} \tag{1}$$

$$F_{r_{y_1}} = m_1\ddot{y}_{C_1} \Rightarrow 100(9.81)\cos 25° - N_1 = 100\cancelto{0}{\ddot{y}_{C_1}} \tag{2}$$

$$F_{r_{x_2}} = m_2\ddot{x}_{C_2} \Rightarrow 110(9.81)\sin 25° + f_1 - f_2 = 110\ddot{x}_{C_2} \tag{3}$$

$$F_{r_{y_2}} = m_2\ddot{y}_{C_2} \Rightarrow 110(9.81)\cos 25° + N_1 - N_2 = 110\cancelto{0}{\ddot{y}_{C_2}} \tag{4}$$

We mention that the sum of Equations (1) and (3) gives the "**x** equation" of the overall system; the sum of (2) and (4) yields the "**y** equation" (C is the mass center of the combined blocks):

$$\begin{aligned} F_{r_x} &= (100 + 110)9.81 \sin 25° - f_2 \\ &= 100\ddot{x}_{C_1} + 110\ddot{x}_{C_2} \\ &= (100 + 110)\ddot{x}_C \end{aligned} \tag{5}$$

$$\begin{aligned} F_{r_y} &= (100 + 110)9.81 \cos 25° - N_2 \\ &= (100 + 110)\cancelto{0}{\ddot{y}_C} = 0 \end{aligned} \tag{6}$$

Note that f_1 and N_1 disappear in (5) and (6), as they become internal forces on the combined system.

Equation (6) tells us that $N_2 = 1870$ N, regardless of which motion takes place. The equation for the **x** motion (Equation 5) shows again that if $f_{2_{max}} <$ $mg(\sin 25°)$ then one or both of the blocks must slide:

$$f_{2_{max}} = \mu_2 N_2 = 0.4(1870) = 748 < 210(9.81)(0.423) = 871 \text{ N}$$

and so $\ddot{x}_C$ cannot be zero. Assuming first that the blocks *both* move, then f_2 is at its maximum:

$$f_2 = f_{2_{max}} = 748 \text{ N}$$

(Continued)

*Note that in problems of this type the moment equation would only tell us where the normal force resultant N_1 or N_2 acts along the bottom of the block.

If they move *together* as one body, then Equation (5) gives us

$$\ddot{x}_C (= \ddot{x}_{C_1} = \ddot{x}_{C_2}) = (mg \sin 25° - f_{2_{max}}) \div m$$

$$= (871 - 748) \div 210$$

$$= 0.586 \text{ m/s}^2$$

Substituting this acceleration into Equation (1), we can check to see if body $\mathcal{B}_1$ additionally slides relative to $\mathcal{B}_2$:

$$f_1 = -100(0.586) + 415 = 356 \text{ N}$$

But the maximum value that f_1 can have is given by

$$f_{1_{max}} = \mu_1 N_1 = 0.3(889) = 267 \text{ N}$$

Hence block $\mathcal{B}_1$ slides on $\mathcal{B}_2$ and the blocks do not move together; our assumption was incorrect. We than substitute $f_1 = \mu_1 N_1$ into Equation (1) and proceed:

$$\ddot{x}_{C_1} = (415 - 267) \div 100 = 1.48 \text{ m/s}^2$$

This is then the acceleration of the top block. Substituting f_1 into Equation (3) gives

$$456 + 267 - f_2 = 110\ddot{x}_{C_2}$$

$$723 - f_2 = 110\ddot{x}_{C_2}$$

For no motion of the bottom block, $f_{2_{max}}$ clearly needs to be at least 723 N. Since it is in fact 748 N, the bottom block does *not* move for this combination of parameters, and $\ddot{x}_{C_2} = 0$.

If the friction coefficients are now swapped, nothing changes until we begin to analyze the six equations. We have

$$f_{2_{max}} = \mu_2 N_2 = 0.3(1870) = 561 < mg \sin 25° = 871 \text{ N} \qquad \text{(as before)}$$

Again, then, $\ddot{x}_C$ cannot be zero. Assuming again that the blocks both move, f_2 is its maximum and Equation (5) gives

$$\ddot{x}_C = (871 - 561) \div 210 = 1.48 \text{ m/s}^2$$

Substituting this acceleration into Equation (1), we get

$$f_1 = 415 - 100(1.48) = 267 < \mu_1 N_1 = 0.4(889) = 356 \text{ N}$$

This time we have more friction than we need in order to prevent $\mathcal{B}_1$ from slipping on $\mathcal{B}_2$. Thus both $\ddot{x}_{C_1}$ and $\ddot{x}_{C_2}$ are 1.48 m/s^2.

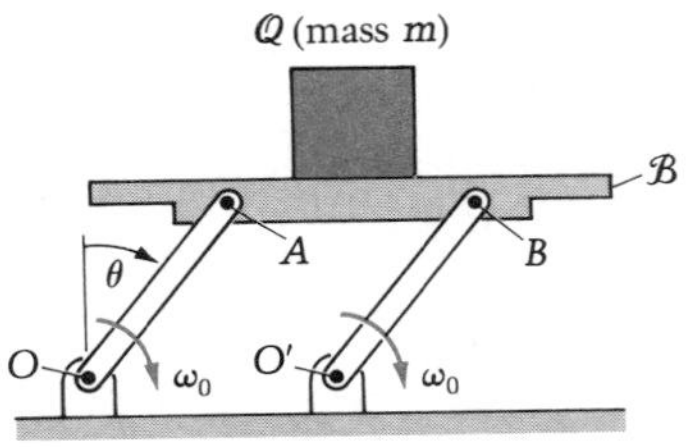

EXAMPLE 4.15

The link bars OA and $O'B$ of body $\mathcal{B}$ in the accompanying diagram are turning clockwise at the constant rate of ω_0 rad/sec. The bars have length ℓ, and the coefficient of static friction between $\mathcal{B}$ and block $\mathcal{Q}$ is μ, sufficient to prevent $\mathcal{Q}$ from sliding on $\mathcal{B}$. Show that if $\omega_0 \le \sqrt{g/\ell}$ the block will stay with the body $\mathcal{B}$, meaning that $\mathcal{B}$ will not fall faster than $\mathcal{Q}$ and leave it behind. Then find the condition under which $\mathcal{Q}$ will not *slide* on $\mathcal{B}$.

SOLUTION

The acceleration of any point P on the translating body $\mathcal{B}$ is

$$\mathbf{a}_P = \mathbf{a}_A = \cancelto{\mathbf{0}}{\mathbf{a}_O} + \cancelto{0}{\alpha_{OA}}\hat{\mathbf{k}} \times \mathbf{r}_{OA} - \omega_0^2 \mathbf{r}_{OA}$$

$$= \ell\omega_0^2$$

Question 4.4 Why is N offset from the centerline of $\mathcal{Q}$ in the free-body diagram?

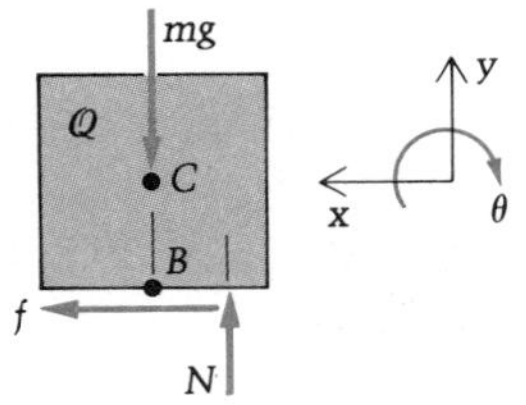

Thus, using the free-body diagram shown here, we may write the mass center equations of motion for body $\mathcal{Q}$:

$$F_{r_x} = m\ddot{x}_C \Rightarrow f = m\ell\omega_0^2 \sin\theta \qquad (1)$$

$$F_{r_y} = m\ddot{y}_C \Rightarrow N - mg = -m\ell\omega_0^2 \cos\theta \qquad (2)$$

Note that the acceleration of C is the same as that of ***all*** the points of $\mathcal{B}$ if there is no motion of $\mathcal{Q}$ relative to $\mathcal{B}$ ***and*** if the bodies are translating. Equation (2) may now be solved for the normal force N:

$$N = mg - m\ell\omega_0^2 \cos\theta$$

Therefore, enforcing the inequality $N \ge 0$, which must be true if the bodies remain in contact, will lead to our answer:

$$N = mg - m\ell\omega_0^2 \cos\theta \ge 0 \Rightarrow \frac{g}{\ell\cos\theta} \ge \omega_0^2 \qquad (\theta \ne 90°, 270°)$$

Since we wish the largest value of ω_0 for which the bodies stay together for any and all θ's, we must set $\cos\theta = 1$, and we find the condition

$$\omega_0 \le \sqrt{\frac{g}{\ell}}$$

Note that for $90° \le \theta \le 270°$, N is positive regardless of the parameters. Thus, the only positions where there is a danger of $\mathcal{B}$ "coming out from under" the block are those ***above*** the line 00′.

(Continued)

Next we use Equations (1) and (2) to examine the sliding possibility. If no slipping is to occur, we must have

$$f \leq \mu N$$

Therefore,

$$m\ell\omega_0^2 \sin\theta \leq \mu(mg - m\ell\omega_0^2 \cos\theta)$$

or

$$\omega_0^2 \leq \frac{g\mu}{\ell(\sin\theta + \mu\cos\theta)}$$

Using calculus, the largest value of the function of θ in the denominator, $\sin\theta + \mu\cos\theta$, is $\sqrt{1+\mu^2}$ at $\theta = \tan^{-1}(1/\mu)$. Thus, if

$$\omega_0 \leq \sqrt{\frac{g\mu}{\ell\sqrt{1+\mu^2}}}$$

then $\mathcal{Q}$ will not slip on $\mathcal{B}$. This is a slower angular speed than the value needed to prevent separation of $\mathcal{Q}$ and $\mathcal{B}$.

The following examples involve bodies in plane motion which are *not* translating and which therefore require the additional use of the rotational equation $M_{r_C} = I_C\alpha$ for their solution.

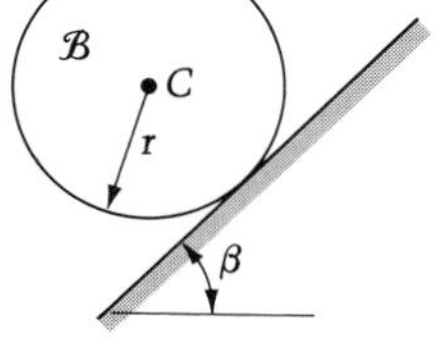

E X A M P 4.16

Cylinder $\mathcal{B}$ (mass m, radius r) is released from rest on the inclined plane shown in the diagram. The coefficient of friction between cylinder and plane is μ. Determine the motion of $\mathcal{B}$.

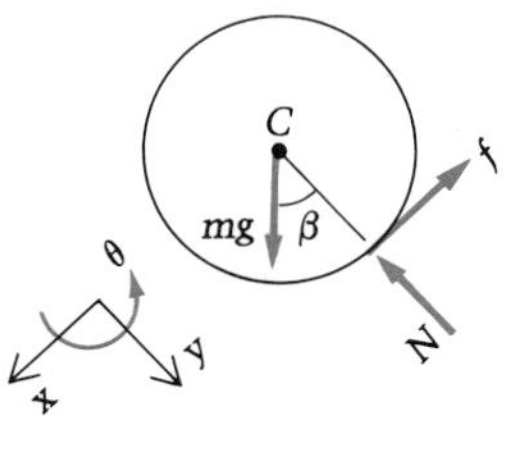

SOLUTION

As in statics, a good first step is to draw a free-body diagram (see the following figure). We shall assume that the cylinder rolls. In this case the friction force f is an unknown and has a value satisfying

$$0 \leq f \leq f_{max}$$

where $f_{max} = \mu N$ from the study of Coulomb friction in statics. We shall also use a kinematic equation expressing the rolling. After solving for f, we then determine whether it is in fact less than or equal to μN. If it is not, then we must rework the problem *with* $f = \mu N$ and *without* the kinematic equation of rolling.

(Continued)

We choose $\mathbf{x}$, y, and θ to be in the indicated directions, motivated by the fact that C will move down the plane and $\mathcal{B}$ will turn counterclockwise. The equations of motion are

$$F_{r_x} = m\ddot{x}_C \Rightarrow mg\sin\beta - f = m\ddot{x}_C \tag{1}$$

$$F_{r_y} = m\ddot{y}_C \Rightarrow mg\cos\beta - N = m\ddot{y}_C = 0 \tag{2}$$

(Note that, kinematically, y_C is constant so that $\ddot{y}_C$ vanishes.)

$$M_{r_C} = I_C\alpha \Rightarrow fr = \tfrac{1}{2}mr^2\ddot{\theta} \tag{3}$$

We can solve (2), getting $N = mg\cos\beta$. There remain two equations in the three unknowns f, $\ddot{x}_C$, and $\ddot{\theta}$. We must therefore supplement our equations of motion with the remaining kinematics result, which comes from the rolling condition:

$$\ddot{x}_C = r\ddot{\theta} \tag{4}$$

Solving Equations (1), (3), and (4) gives

$$\ddot{x}_C = \tfrac{2}{3}g\sin\beta \qquad \ddot{\theta} = \frac{2g\sin\beta}{3r} \qquad f = \frac{mg\sin\beta}{3}$$

Integrating twice, and noting that the integration constants vanish, we get

$$x_C = \frac{gt^2}{3}\sin\beta \qquad \theta = \frac{gt^2}{3r}\sin\beta$$

And if

$$f \le f_{\max} = \mu N$$

that is, if

$$\frac{mg\sin\beta}{3} \le \mu mg\cos\beta$$

or

$$\tan\beta \le 3\mu$$

then our assumption and solution are valid and the cylinder will roll as assumed. If $\mu = 0.5$ and $\beta = 30°$, for example, then

$$\tan\beta = 0.577 \le 3\mu = 1.5$$

and the cylinder rolls. But if $\mu = 0.2$ and $\beta = 60°$, then

$$\tan\beta = 1.73 \not\le 3\mu = 0.6$$

and we must "go back to the drawing board." In such a case,

$$f = \mu N = \mu mg\cos\beta$$

since N still equals $mg\cos\beta$. And since Equation (3) still holds,

$$fr = \mu mgr\cos\beta = \tfrac{1}{2}mr^2\ddot{\theta}$$

(Continued)

or

$$\ddot{\theta} = \frac{2g\mu \cos \beta}{r}$$

Therefore, integrating twice, we get

$$\theta = \frac{\mu g t^2 \cos \beta}{r}$$

where the integration constants are zero since $\theta = \dot{\theta} = 0$ at $t = 0$. Equation (1) now yields, with $f = \mu N$,

$$mg \sin \beta - \mu mg \cos \beta = m\ddot{x}_C$$

or

$$\ddot{x}_C = g(\sin \beta - \mu \cos \beta)$$

Thus

$$x_C = \frac{gt^2}{2}(\sin \beta - \mu \cos \beta)$$

and the solutions for the motion ($x_C(t)$ and $\theta(t)$) are indeed quite different when the cylinder turns and slips than they are when it rolls. Note that if we wish to distinguish between static and kinetic coefficients of friction (μ_s and μ_k), then the rolling assumption would be correct if $\tan \beta \leq 3\mu_s$. But if $\tan \beta > 3\mu_s$, we would then use $f = \mu_k N$ in the remainder of the solution, and the μ in the answers for x_C and θ would become μ_k.

EXAMPLE 4.17

At the instant when the system consisting of the spool $\mathcal{C}$ and weight $\mathcal{B}$ is released from rest (see the diagram), find the accelerations of C and B (the mass centers of $\mathcal{C}$ and $\mathcal{B}$). Also find how long it takes for $\mathcal{C}$ to roll off the beam, and determine how much rope is wrapped or unwrapped (tell which) at that time. Assume enough friction to prevent slipping.

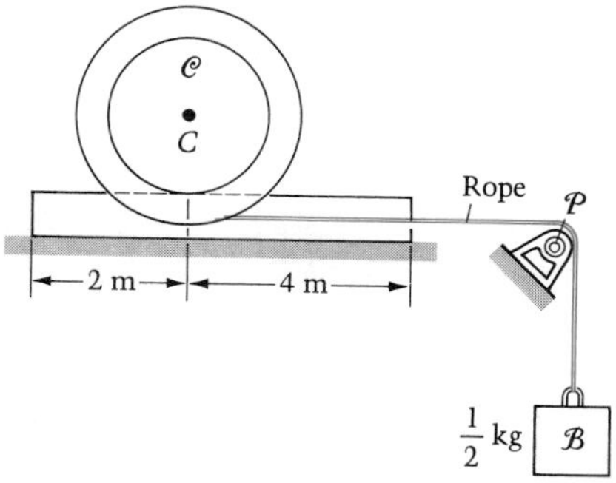

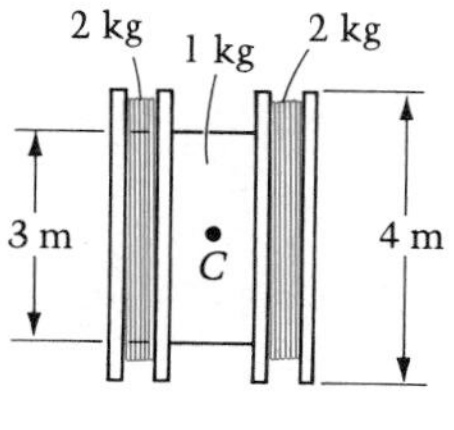

Body 𝒞

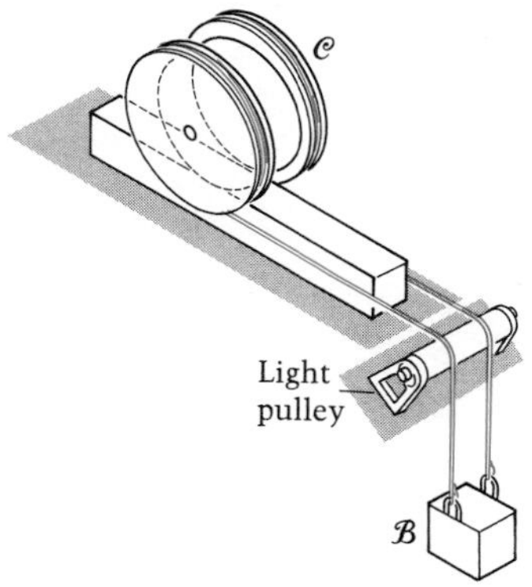

(Continued)

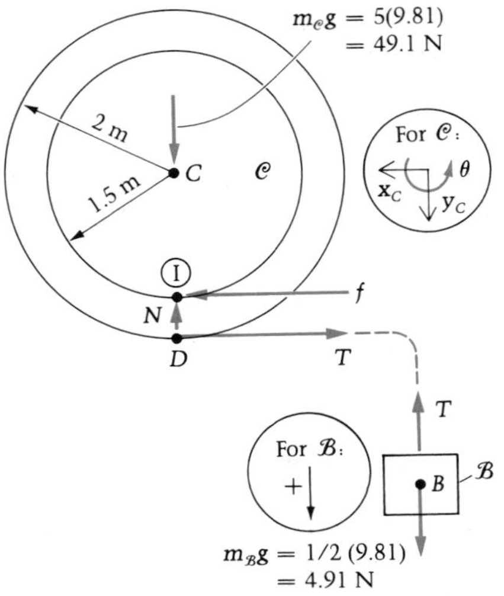

SOLUTION

We draw the free-body diagrams and write the equations of motion. On $\mathcal{C}$:

$$F_{r_x} = f - T = m\ddot{x}_C = 5\ddot{x}_C \qquad (1)$$

$$F_{r_y} = 49.1 - N = m\ddot{y}_C = 0 \qquad (2)$$

$$M_{r_C} = 2T - 1.5f = I_C\alpha = 9.13\alpha \qquad (3)$$

where $\mathcal{C}$ is composed of three cylinders, so that

$$I_C = 2\left[\frac{1}{2}\,2(2^2)\right] + \frac{1}{2}\,1(1.5^2)$$

$$= 9.13 \text{ kg-m}^2$$

Adding 1.5 times Equation (1) to Equation (3) eliminates the friction force:

$$0.5T = 7.5\ddot{x}_C + 9.13\alpha$$

Since the spool is rolling, $\ddot{x}_C = r\alpha = 1.5\alpha$, so that

$$0.5T = 7.5(1.5\alpha) + 9.13\alpha = 20.4\alpha \qquad (4)$$

From Equation (2), we get $N = 49.1$ N, and Equations (1) and (3) have produced Equation (4). So we have the one equation (4) remaining in the two unknowns T and α. Thus we go to the *other* free-body diagram, that of body $\mathcal{B}$, and write its equation of vertical motion. In doing so, we note that the tension is approximately the same on both sides of the small pulley $\mathcal{P}$, because if its mass is negligible then $I_{C_{\mathcal{P}}} \approx 0$ so that $M_{r_{C_{\mathcal{P}}}} = I_{C_{\mathcal{P}}}\alpha \approx 0$. (If it weighs nothing, it takes no moment to turn it.) Therefore

$$F_{r_y} = m_{\mathcal{B}}\ddot{y}_B$$

$$4.91 - T = \tfrac{1}{2}\ddot{y}_B \qquad (5)$$

Equations (4) and (5) are now two equations in three unknowns; we resort to kinematics to relate $\ddot{y}_B$ and α. The (vertical) acceleration of B (or any other point of $\mathcal{B}$ since it is translating) has the same magnitude as that of any point of the horizontal or vertical sections of the rope; furthermore the point of the rope just leaving the body at its lowest point D has the same acceleration as the tangential (horizontal) part of $\mathbf{a}_D$:

$$\ddot{y}_B = (2 - 1.5)\alpha = 0.5\alpha \text{ m/s}^2 \qquad (6)$$

Substituting (6) into (5) then gives

$$4.91 - T = \tfrac{1}{2}(0.5\alpha) = 0.25\alpha \qquad (7)$$

Equating T (from Equation 4) to T (from Equation 7) allows us to compute α:

$$40.8\alpha = 4.91 - 0.25\alpha \Rightarrow \alpha = \frac{4.91}{41.1} = 0.119 \text{ rad/s}^2$$

Therefore we have the accelerations of B and C:

(Continued)

$$\ddot{y}_B = 0.5\alpha = 0.0595 \text{ m/s}^2$$

$$\ddot{x}_C = 1.5\alpha = 0.179 \text{ m/s}^2$$

So that, integrating, we get

$$\dot{x}_C = 0.179t + \cancelto{0 \text{ since } \dot{x}_C = 0 \text{ at } t=0}{C_1} \text{ m/s}$$

$$x_C = 0.0895t^2 + \cancelto{0 \text{ since } x_C = 0 \text{ at } t=0}{C_2} \text{ m}$$

Body $\mathcal{C}$ will roll off the (left) end of the beam when x_C becomes 2 m:

$$2 = 0.0895t_{\text{leave}}^2$$

$$t_{\text{leave}} = 4.73 \text{ s}$$

As the point C moves to the left, rope is *unwrapped* as the bodies move. We can see this in either of two ways:

First Approach

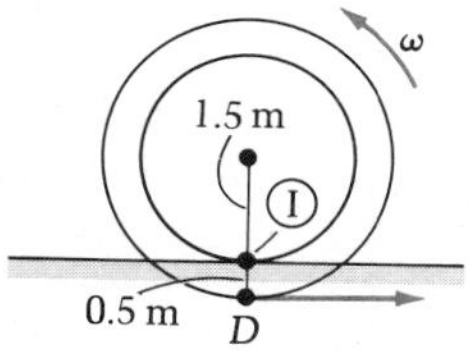

As the body $\mathcal{C}$ turns, points D (at the bottom of the body) successively move to the right as the body turns about Ⓘ. This means that $\mathcal{B}$ falls in space by the amount of $0.5\theta = 0.5(2/1.5) = 2/3$ m. In addition, an amount of rope is unwrapped from $\mathcal{C}$ equal to the displacement to the left of the locus of contact points, which is 2 m. Thus an 8/3-m length of rope that was originally wrapped around $\mathcal{C}$ is now in the space between $\mathcal{C}$ and $\mathcal{B}$.

Second Approach

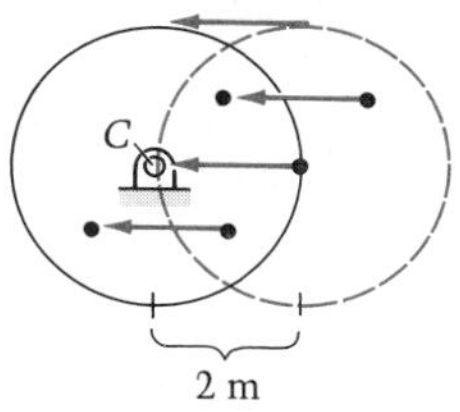

First let $\mathcal{C}$ *translate* 2 m to the left. This *raises* $\mathcal{B}$ by 2 m but does not wrap *or* unwrap any rope. Then if we turn $\mathcal{C}$(↺) about C until the correct point is on the plane,* the angle we need is $\theta = 2/1.5 = 4/3$ rad. This lets out $2\theta = 8/3$ m of rope, and $\mathcal{B}$ drops 8/3 m. The *net* movement of $\mathcal{B}$ is therefore to fall $8/3 - 2 = 2/3$ m, which gets accomplished by 8/3 m of rope unwrapping from $\mathcal{C}$.

We conclude this example with a few instructional points. First we note that the choice of sign convention should, whenever possible, have the bodies moving consistently with respect to one another. If y_B were taken positive ↑ in this example, then positive $\omega_{\mathcal{C}}$ and y_B at the same time would mean that the rope

(Continued)

*The "correct point" means the point that would be contacting the plane if the body had *rolled* to the left as it actually does.

goes slack; to avoid this, one would have to use $\dot{y}_B = -0.5\omega_{\mathcal{C}}$, which gives $\ddot{y}_B = -0.5\alpha$. It is better to use the sign convention of y_B positive *down*, for then $\ddot{y}_B = +0.5\alpha$ and $\ddot{y}_B$ is positive whenever α is (and vice versa) and the motions of $\mathcal{C}$ and $\mathcal{B}$ are consistent with each other. We emphasize again that one need not use the same sign convention for each body in a problem.

A second point is that, if desired, the condition $\ddot{y}_B = 0.5\alpha$ can be obtained via our vector kinematic equation from Chapter 3 relating accelerations:

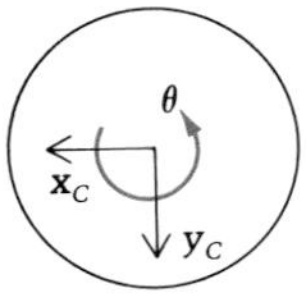

$$\begin{aligned}\mathbf{a}_D &= \mathbf{a}_C + \alpha\hat{\mathbf{k}} \times \overset{2\hat{\mathbf{j}}}{\mathbf{r}_{CD}} - \omega^2\mathbf{r}_{CD}\\ &= \ddot{x}_C\hat{\mathbf{i}} - 2\alpha\hat{\mathbf{i}} - 2\omega^2\hat{\mathbf{j}}\\ &= 1.5\alpha\hat{\mathbf{i}} - 2\alpha\hat{\mathbf{i}} - 2\omega^2\hat{\mathbf{j}}\\ &= -0.5\alpha\hat{\mathbf{i}} - 2\omega^2\hat{\mathbf{j}} \text{ m/s}^2\end{aligned}$$

Thus $a_{D_x} = 0.5\alpha$ to the right, which is equated to $\ddot{y}_B$ since the rope remains inextensible and taut. Also $a_{D_y} = -2\omega^2$, which is zero initially since the bodies are released from rest.

EXAMPLE 4.18

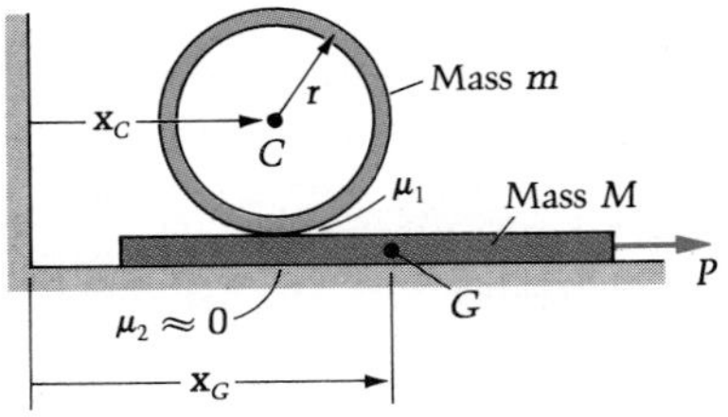

Force P is applied to a plate that rests on a smooth surface. (See the diagram.) Find the largest force P for which the pipe will not slip on the plate.

SOLUTION

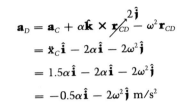

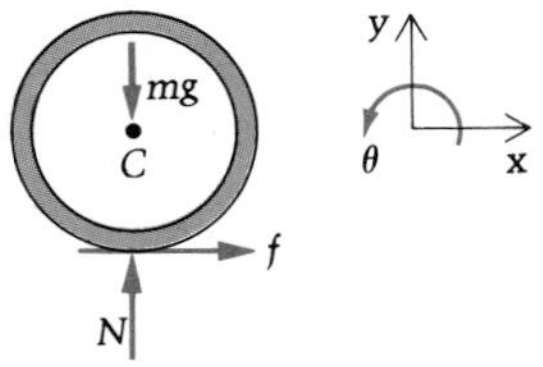

Figure 1

For the pipe (Figure 1):

$$F_{r_x} = f = m\ddot{x}_C \tag{1}$$

$$F_{r_y} = N - mg = m\ddot{y}_C = 0 \Rightarrow N = mg \tag{2}$$

$$M_{r_C} = fr = I_C\alpha = mr^2\alpha \tag{3}$$

(Note that $I_C = (m/2)(r_o^2 + r_i^2) \approx mr^2$ if $r_o \approx r_i$. If the thickness $(r_o - r_i)$ is not given, assume it is small.)

For the plate (Figure 2), we note that only the **x** equation of motion is of help; $F_{r_y} = m\ddot{y}_G = 0$ gives $N_2 = N + Mg = (m + M)g$ as expected, and dimen-

(Continued)

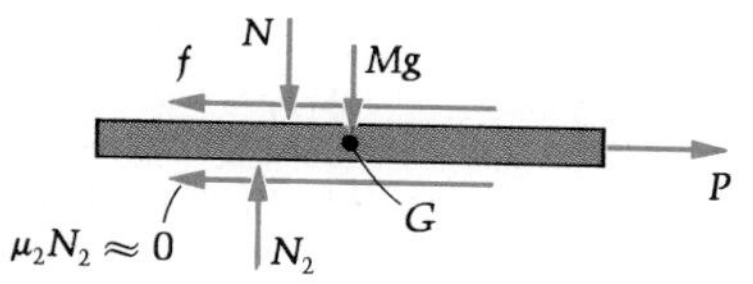

Figure 2

sions are not given so moments cannot be taken. (The moment equation would only give us the location of N_2, anyway.) Therefore

$$F_{r_x} = P - f = M\ddot{x}_G{}^* \tag{4}$$

Eliminating f between (1) and (3) gives

$$m\ddot{x}_C = mr\alpha \tag{5}$$

And between (1) and (4) gives

$$P = m\ddot{x}_C + M\ddot{x}_G \tag{6}$$

We note that if $m_T = m + M$ and C_T is the mass center of pipe ***plus*** plate, then Equation (6) could have been written immediately from $F_{r_x} = m_T\ddot{x}_{C_T}$ for the combined system. Here $F_{r_x} = P$; the right side follows from two derivatives of the definition of the mass center ($m_T x_{C_T} = m x_C + M x_G$).

The kinematics equation is tricky here. It is a rolling condition, but we must remember that x_G and x_C are necessarily measured relative to an inertial frame, here assumed to be fixed in the ground. Thus it is $\ddot{x}_C - \ddot{x}_G$ that is related to α:**

$$\ddot{x}_C - \ddot{x}_G = -r\alpha \tag{7}$$

The minus sign on the right side is caused by the fact that if the left side is positive then C will move to the right, necessitating a negative α if there is to be no slipping.

Substituting $\alpha = \ddot{x}_C/r$ from (5) into (7) relates the accelerations of the two mass centers:

$$2\ddot{x}_C = \ddot{x}_G \tag{8}$$

Then (6) and (8) may be combined to give

$$P = (m + 2M)\ddot{x}_C \tag{9}$$

And combining (9) and (1) gives us the relationship between P and f:

$$P = \frac{m + 2M}{m} f \tag{10}$$

Since $f \leq \mu_1 N$ for no slip, (10) gives:

$$\frac{m}{m + 2M} P \leq \mu_1 mg \Rightarrow P \leq (m + 2M)g\mu_1$$

Any larger P than $(m + 2M)g\mu_1$ will cause the pipe to slip on the plate.

*Sometimes G is used to designate a mass center.
**This difference is just the acceleration of C in the frame consisting of the plate.

PROBLEMS / Section 4.5

4.42 Generalizing Example 4.14, let the blocks, friction coefficients, and angle of the plane be as shown in Figure P4.42. Show that:

a. If $\tan\varphi > \mu_2$, motion will occur, and if so:
b. If $\mu_2 \le \mu_1$, the blocks move together
c. If $\mu_2 > \mu_1$, then $\mathcal{B}_1$ slides on $\mathcal{B}_2$. In this case, the lower block does not move if

$$\tan\varphi \le \mu_2 + (\mu_2 - \mu_1)\frac{m_1}{m_2}$$

d. If $\tan\varphi \le \mu_2$, then the lower block will not move. In this case, the upper block slides on it if and only if $\tan\varphi > \mu_1$.

4.43 In Figure P4.43 the masses of $\mathcal{A}$, $\mathcal{B}$, and $\mathcal{C}$ are 10, 60, and 50 kg, respectively. The coefficient of friction between $\mathcal{B}$ and the plane is $\mu = 0.35$, and the pulleys have negligible mass and friction. Find the tensions in each cord, and the acceleration of B, upon release from rest.

4.44 A 400-lb cabinet is to be transported on a truck as shown in Figure P4.44. Assuming sufficient friction so that the cabinet will not slide on the truck, what is the maximum forward acceleration for which the cabinet will not tip over?

4.45 A nonuniform block rests on a flatcar as shown in Figure P4.45. If the coefficient of friction between car and block is 0.40, for what range of accelerations of the car will the block neither tip nor slide?

4.46 The force P causes the rectangular box of weight W in Figure P4.46 to slide. Find the range of values of H for which the box will not tip (as it slides) about either the front *or* rear lower corner.

4.47 If block $\mathcal{Q}$ in Example 4.15 is homogeneous with height $2H$ and base $2b$, determine the largest value of ω_0 for which it will not tip over on $\mathcal{B}$.

4.48 The cords in Figure P4.48 have a tensile strength of 12 N. Cart $\mathcal{C}$ has a mass of 35 kg exclusive of the 10-kg and 1.2-m vertical rod $\mathcal{R}$, which is pinned to it at A. Find the maximum value of P that can be exerted without breaking either cord if: (a) P acts to the right as shown; (b) P acts to the left. Neglect friction.

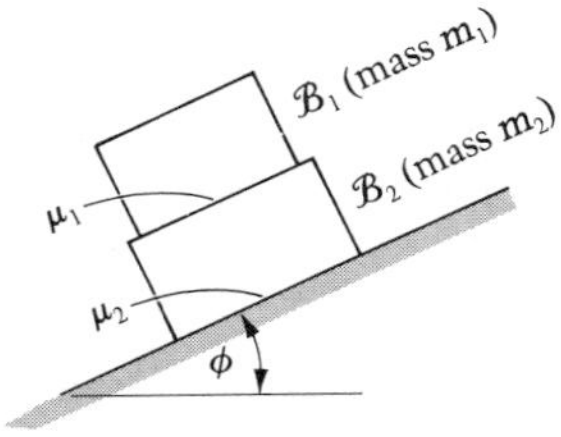

Figure P4.42

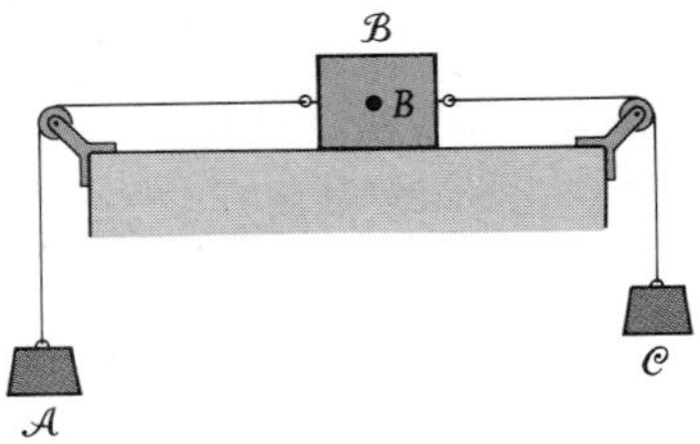

Figure P4.43

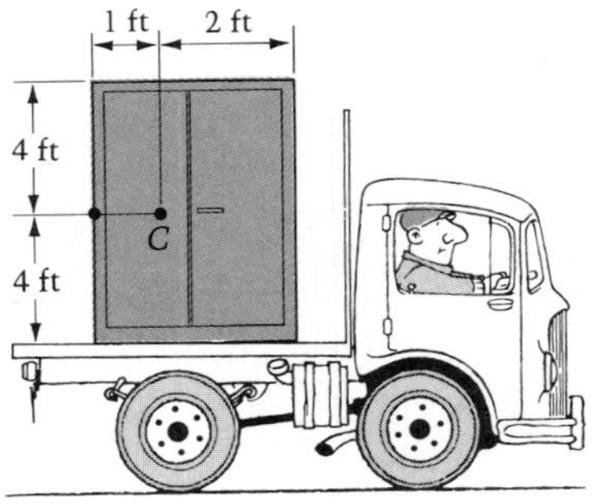

Figure P4.44

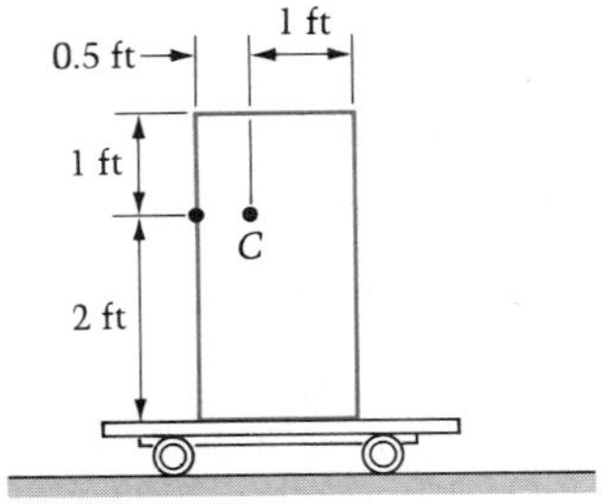

Figure P4.45

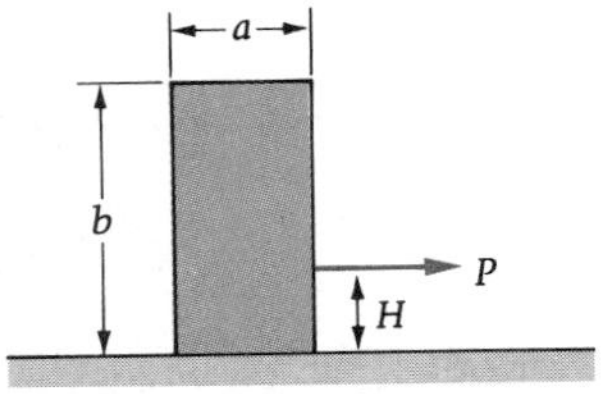

Figure P4.46

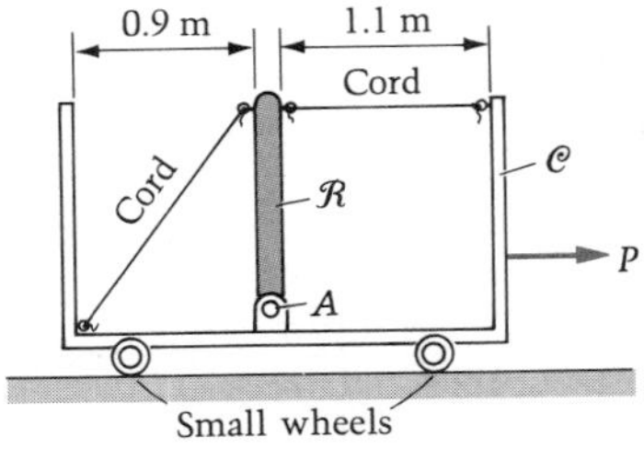

Figure P4.48

4.49 The coefficient of friction at both ends of the bar in Figure P4.49 is 0.5. Find the maximum forward acceleration that the truck may have without the bar slipping.

Figure P4.49

4.50 Find the value of F for which one of the wheels of the garage door in Figure P4.50 lifts out of its track. Which one? Assume negligible friction.

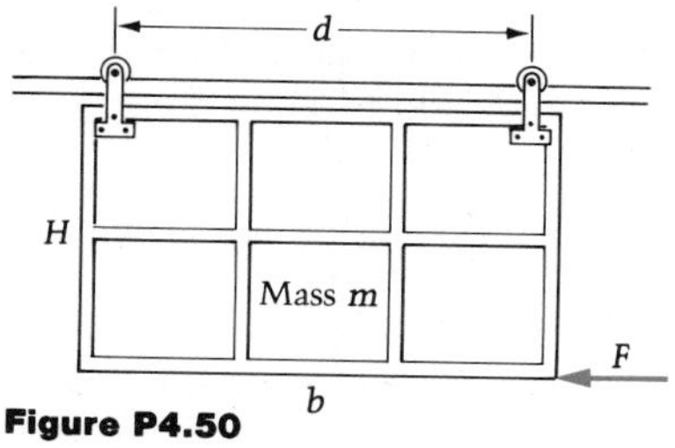

Figure P4.50

4.51 A 100-lb cabinet, rolling on small wheels, is subjected to a 40-lb force as shown in Figure P4.51. Neglecting friction, find (a) the acceleration of the cabinet; (b) the reactions of the floor on the wheels.

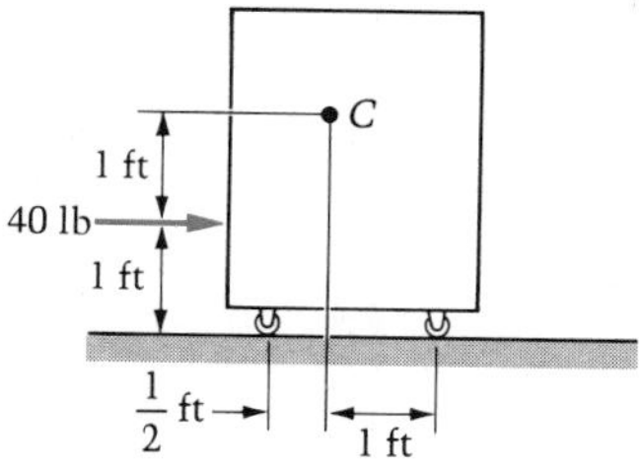

Figure P4.51

4.52 The monorail car in Figure P4.52 is driven through its front wheel and moves forward from left to right. If the coefficient of friction between wheels and track is $\mu = 0.55$, determine the maximum acceleration possible for the car.

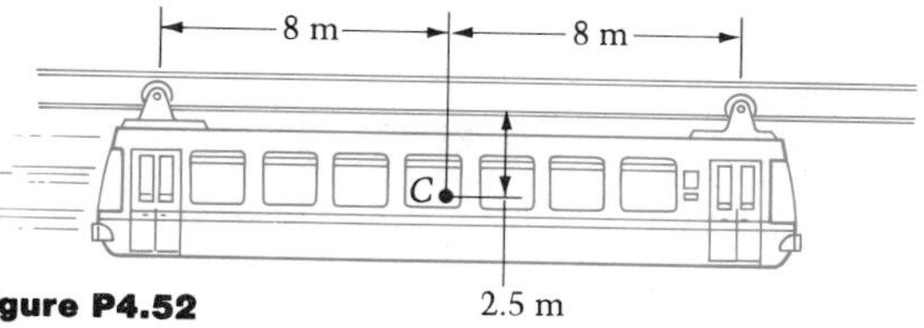

Figure P4.52

4.53 A child notices that sometimes the ball $\mathcal{M}$ does not roll down the inclined surface of toy $\mathcal{T}$ when she pushes it along the floor. (See Figure P4.53.) What is the minimum acceleration $a_{\min}$ of $\mathcal{T}$ to prevent this rolling?

Figure P4.53

4.54 In the preceding problem, suppose the acceleration of $\mathcal{T}$ is $2a_{\min}$. What is the normal force between the smooth vertical surface of $\mathcal{T}$ and the ball? The ball's weight is 0.06 lb.

4.55 The uniform bar $\mathcal{B}$ in Figure P4.55 weighs 60 lb and is pinned at A (and fastened by the cable DB) to the frame $\mathcal{F}$. If the frame is given an acceleration $a = 32.2$ ft/sec^2 as shown, determine the tension T in the cable and the force exerted by the pin at A on the bar.

4.56 The truck in Figure P4.56 is traveling at 45 mph. Find the minimum stopping distance such that the 250-lb crate will neither slide nor tip over.

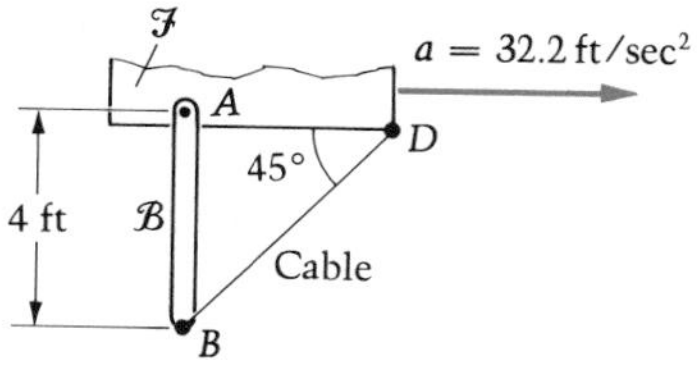

Figure P4.55

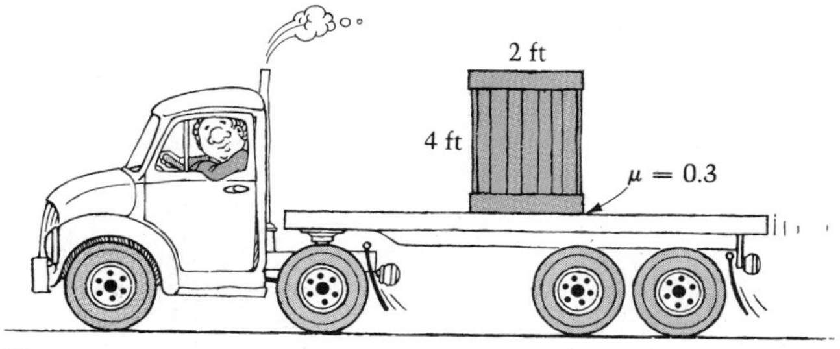

Figure P4.56

4.57 Find the range of accelerations of $\mathcal{W}$ that F can produce without $\mathcal{B}$ moving in any manner relative to $\mathcal{W}$. (See Figure P4.57.) Note carefully the position of the mass center of $\mathcal{B}$.

4.58 A can $\mathcal{C}$ that may be considered a uniform solid cylinder (see Figure P4.58) is pushed along a surface $\mathcal{B}$ by a moving arm $\mathcal{A}$. If it is observed that $\mathcal{C}$ *translates* to the right with $\ddot{x}_C = g/10$, what must be the minimum coefficient of friction between $\mathcal{A}$ and $\mathcal{C}$? The coefficient of friction between $\mathcal{C}$ and $\mathcal{B}$ is μ.

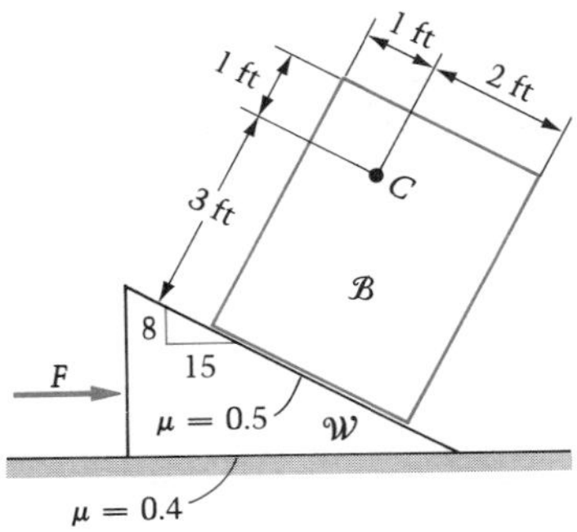

Figure P4.57

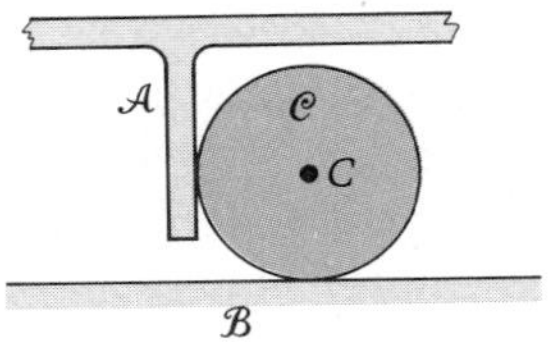

Figure P4.58

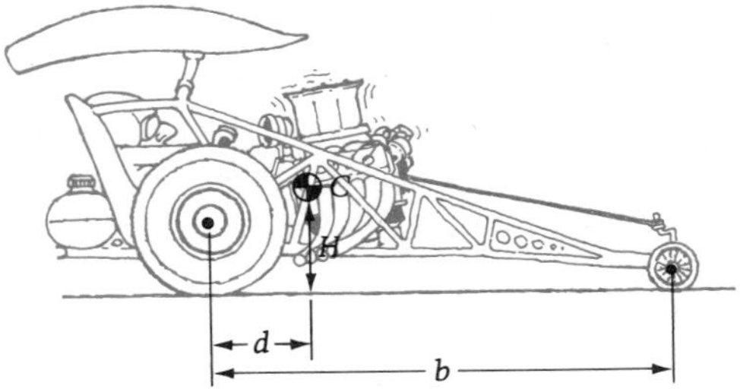

Figure P4.59

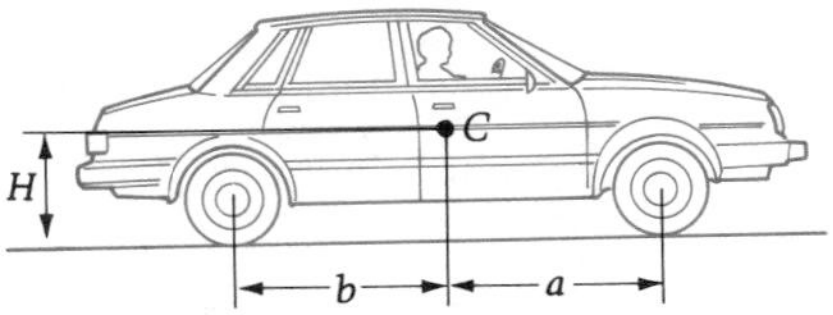

Figure P4.61

4.59 A dragster is all set for the annual neighborhood race. (See Figure P4.59.)

a. In terms of the dimensions b, H, and d and the coefficient of friction μ, find the maximum possible acceleration of the car. Neglect the rotational inertia of the tires. (From a free-body diagram of the driven rear wheel, show that including its inertia will reduce the maximum acceleration.)

b. How would you adjust the four parameters b, H, d, and μ to further increase the driver's acceleration?

4.60 Rework the preceding problem for a car with (a) front-wheel drive and (b) four-wheel drive.

4.61 In an emergency the driver of an automobile applies his brakes; the front brakes fail and the rear wheels are locked. Find the time and distance required to bring the car to rest. Neglect the masses of the wheels, and express the results in terms of the coefficient of sliding friction μ, the initial speed v, the gravitational acceleration g, and the dimensions shown in Figure P4.61.

4.62 If the radius of gyration of each driven wheel in Problem 4.59 is k_C, derive a more accurate formula for the maximum acceleration of the car. Let m be the mass of each driven rear wheel and let M be the mass of the car without these wheels. Give the largest value of the driving couple from the differential onto the rear axle for which the driven wheels will not slip.

4.63 A force F, alternating in direction, causes carriage $\mathcal{A}$ to move with rectilinear horizontal motion defined by the equation $x = 2 \sin \pi t$ ft, where x is the displacement in feet and t is the time in seconds. (See Figure P4.63.) A slender homogeneous rod $\mathcal{R}$ of weight 32.2 lb and length 6 ft is welded to the carriage at B and projects vertically upward. Find, in magnitude and direction, the bending moment that the carriage exerts on the rod at B when $t = \frac{1}{2}$ sec.

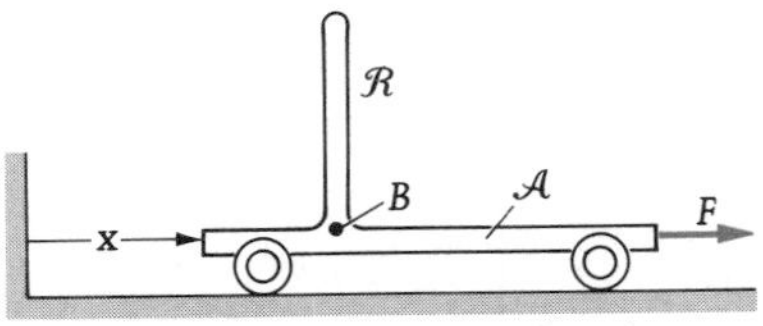

Figure P4.63

4.64 Body $\mathcal{A}$ in Figure P4.64 weighs 223 N and body $\mathcal{B}$ weighs 133 N. Neglect the weight of the rigid member connecting $\mathcal{A}$ and $\mathcal{B}$. The coefficient of friction is 0.3 between all surfaces. Determine the accelerations of $\mathcal{A}$ and $\mathcal{B}$ just after the cord is cut.

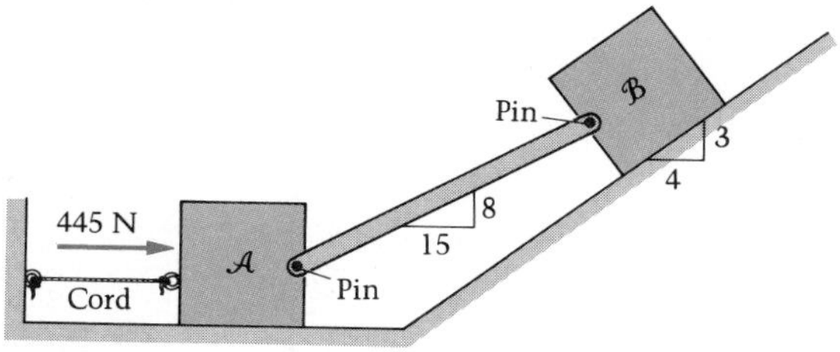

Figure P4.64

4.65 A man pushes a cart that weighs 16 lb without the wheels; there is a load of dirt weighing 20 lb inside. (See Figure P4.65.) Each of the four wheels may be treated as a 2-lb disk with radius 6 in. If the man pushes with a constant force F along the 30° angle of the handle, find F to give a cart speed of 3 ft/sec after the cart moves 6 ft from rest. Assume that there is enough friction to prevent slipping.

Figure P4.65

4.66 Cylinder $\mathcal{C}$ in Figure P4.66 weighs 100 lb; it is rolling on the plane and is pinned at its center C to the 10-lb rod $\mathcal{R}$. If v_C is initially 10 ft/sec to the left, and if the coefficient of kinetic friction between the plane and each body is $\mu = 0.4$, determine how long it will take the system to come to rest.

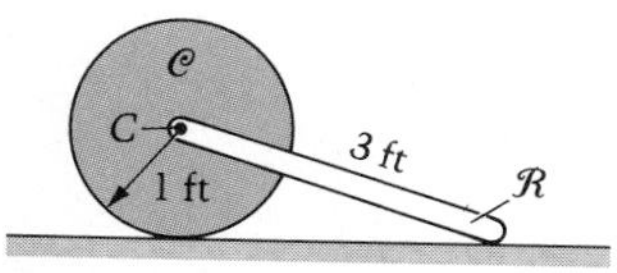

Figure P4.66

4.67 In Figure P4.67, find how far down the incline C travels in 5 s if the 20-kg cylinder $\mathcal{C}$ is released from rest.

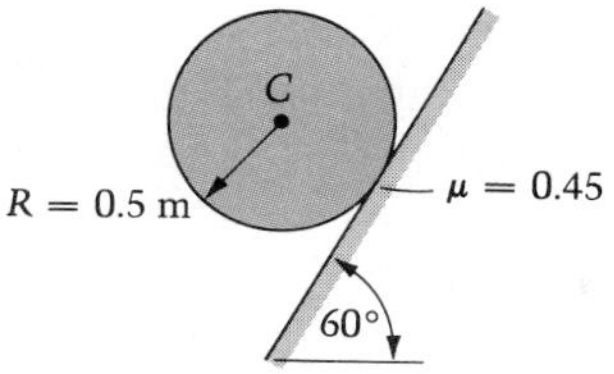

Figure P4.67

4.68 The system shown in Figure P4.68 is initially at rest. A force P is then applied that varies with time according to $P = 7t^2$, where P is in newtons and t in seconds. If the coefficient of friction between cylinder and cart is $\mu = 0.5$, find how much time elapses before the cylinder starts to slip on the cart.

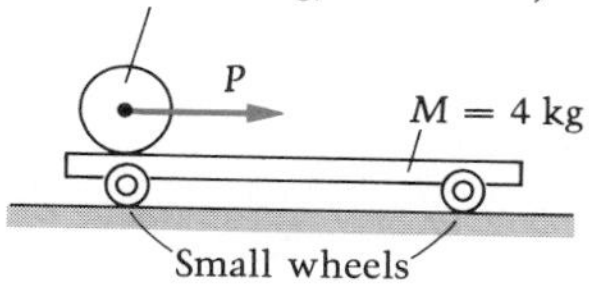

Figure P4.68

4.69 In the previous problem, determine how much time passes (from $t = 0$) before the cylinder leaves the surface of the cart. Initially, the center of the cylinder is 2 m from the right end of the cart.

4.70 Two drums of radius 4.5 in. are mounted on each end of a cylinder of radius 6 in. to form a 40-lb rigid body $\mathcal{B}$ with radius of gyration $k_{z_C} = 5$ in. (See Figure P4.70.) Ropes are wrapped around the drum and tied to a horizontal bar to which a 3-lb force is applied. As $\mathcal{B}$ rolls from rest, tell (a) the number of inches of rope wound or unwound (tell which) in three seconds and (b) the minimum friction coefficient needed for the rolling to take place.

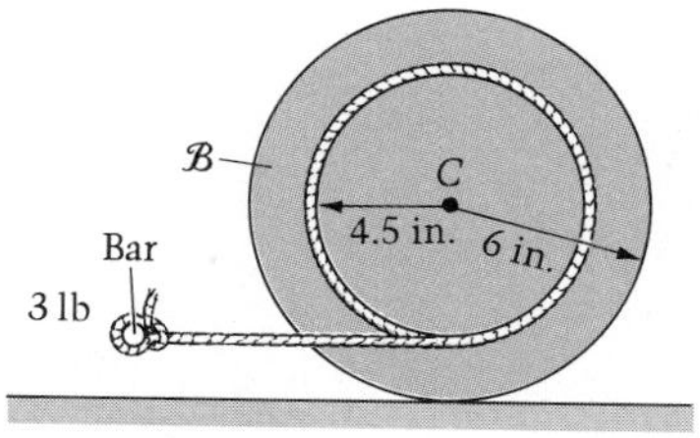

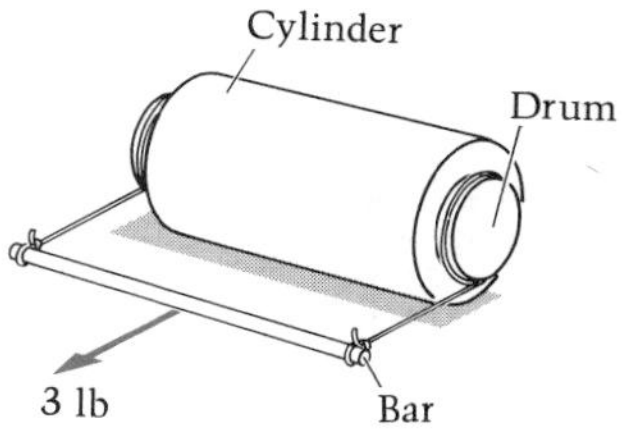

Figure P4.70

4.71 The bowling ball in Figure P4.71 is released with $v_C = 22$ ft/sec and $\omega = 0$ as it contacts the surface of the alley. Neglecting the effect of the three finger holes, and using a coefficient of friction of 0.3, find the distance traveled by the center of the ball before slipping stops.

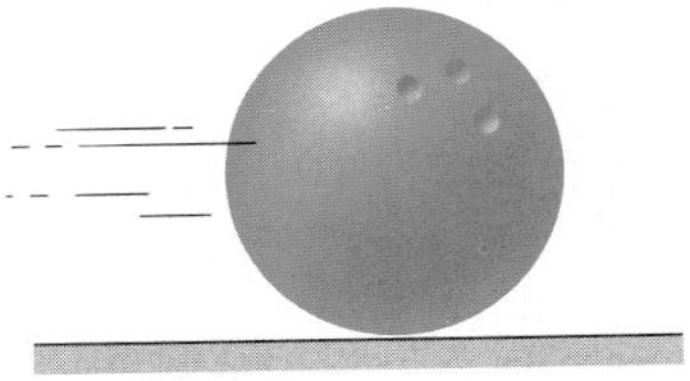

Figure P4.71

4.72 The wheel in Figure P4.72 has a mass of 10 kg and a radius of gyration with respect to z_C (normal to the page) of 0.3 m. Determine the acceleration of C, and find how far C moves in 5 s.

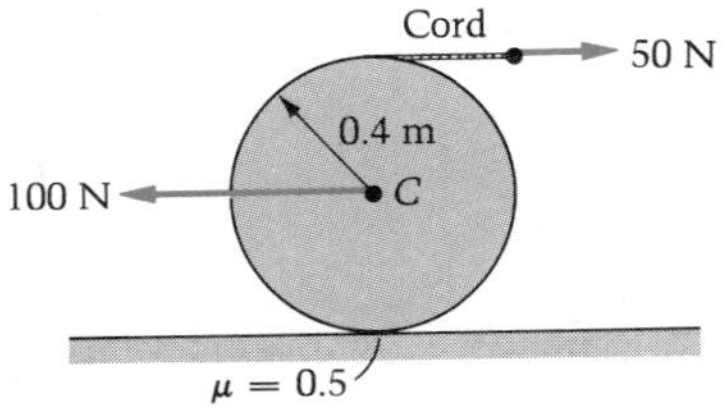

Figure P4.72

4.73 The semicylinder in Figure P4.73a is released from rest, and there is enough friction to prevent slipping throughout the ensuing motion (Figure P4.73b).

a. Find I_{z_C}.
b. Write the three differential equations of motion of the body (good at any angle θ).
c. Find the two equations relating $\ddot{x}_C$ and $\ddot{y}_C$ to $\ddot{\theta}$, $\dot{\theta}$, and θ.
d. Eliminate f, N, $\ddot{x}_C$, and $\ddot{y}_C$ and obtain the single differential equation in the variable $\theta(t)$. Note the complexity of the equation!

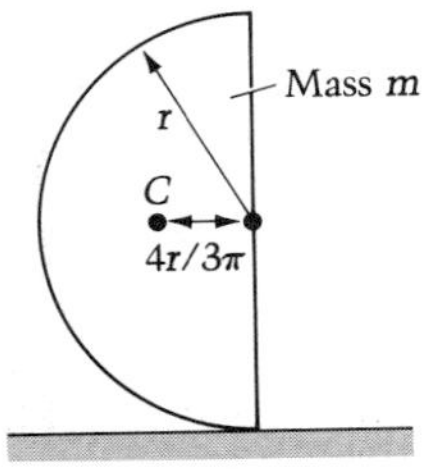

Figure P4.73a

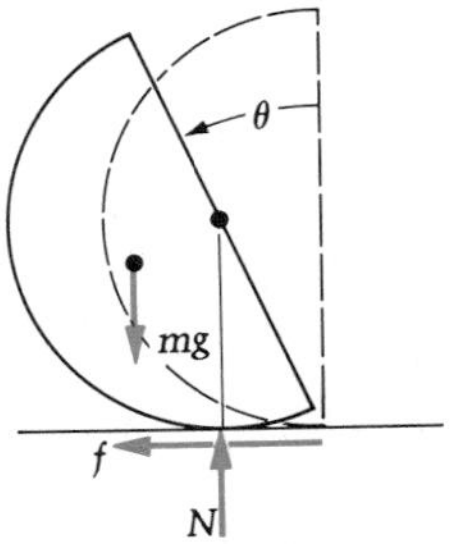

Figure P4.73b

4.74 The homogeneous block $\mathcal{B}$ shown in Figure P4.74 rests on the horizontal turntable at a 24-in. radius to its center. The coefficient of friction between the block and the turntable is $\mu = 0.7$. If the angular speed of the turntable is slowly increased from zero, will the block tip first or will it slide? At what value of ω will this occur?

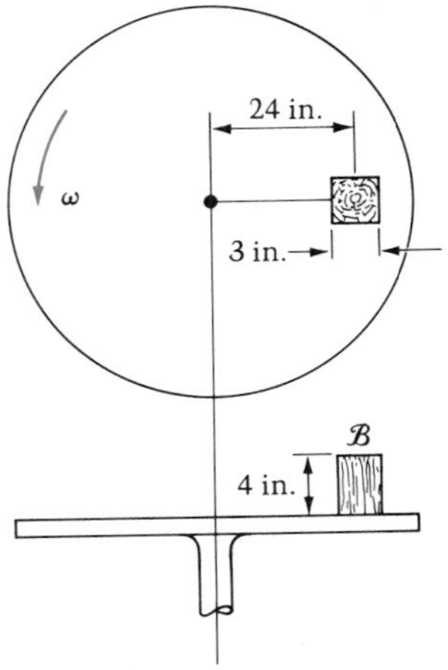

Figure P4.74

4.75 A uniform rod $\mathcal{A}$ of length L and weight W is connected to smooth hinges at E and D by the light members $\mathcal{B}$ and $\mathcal{C}$, each of length L. In the position shown in Figure P4.75, $\mathcal{B}$ has an angular velocity of ω rad/sec clockwise. Find the forces in members $\mathcal{B}$ and $\mathcal{C}$, and determine the acceleration of center C of rod $\mathcal{A}$ in terms of the given variables.

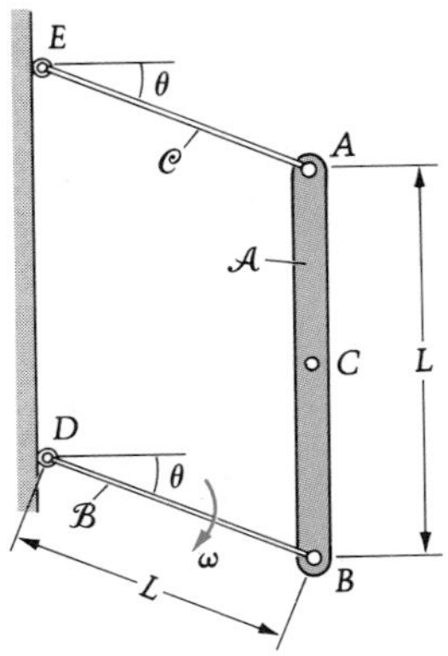

Figure P4.75

4.76 A hula hoop $\mathcal{H}$ (mass m, radius r) is thrown forward with backspin; $v_C = v_0$ to the right and $\omega = \omega_0$ counterclockwise. (See Figure P4.76.)

a. How long and how far does the mass center move before $\mathcal{H}$ stops slipping?
b. Find the relationship between v_0 and ω_0 such that when $\mathcal{H}$ stops slipping: (i) it rolls right; (ii) it rolls left; (iii) it stops.

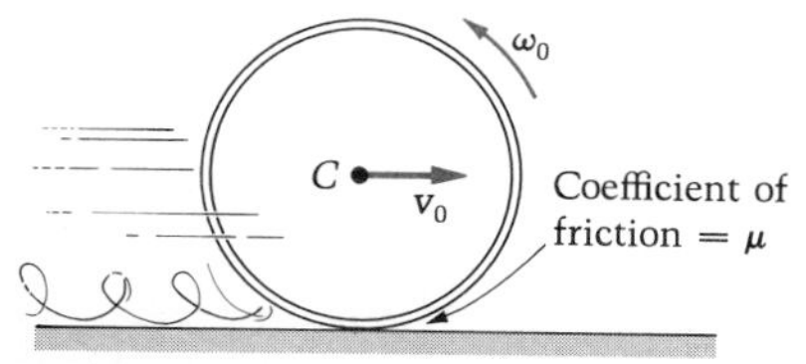

Figure P4.76

4.77 The uniform sphere (mass = 1 slug, radius = 1 ft) and the slab (mass = 2 slugs) shown in Figure P4.77 are at rest before the force $P = 24$ lb is suddenly applied to the slab. The coefficient of friction is 0.2 between the sphere and slab and between the slab and horizontal plane. (a) Does the sphere slip on the slab? (b) What is the acceleration of the center of the sphere?

Figure P4.77

4.78 The stick shown in Figure P4.78, originally at rest with $\theta = 0$, is disturbed slightly and begins to slide on a smooth wall and floor. Derive the differential equation of motion of the stick (valid until contact with the wall is lost). Your equation should be in terms only of θ and its derivatives, g, and ℓ.

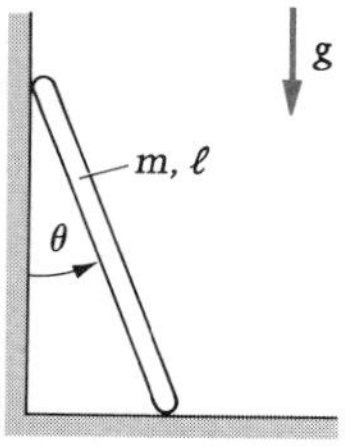

Figure P4.78

4.79 Disks $\mathcal{A}$ and $\mathcal{B}$ each weigh 64.4 lb and are rigidly attached to the light shaft $\mathcal{S}$ that joins their centers. (See Figure P4.79.) A 96.6-lb cylinder $\mathcal{C}$ has a hole drilled along its axis, through which $\mathcal{S}$ passes. A force of 20 lb is applied horizontally to an inextensible string wrapped around $\mathcal{C}$. If friction is negligible between $\mathcal{S}$ and $\mathcal{C}$, and if $\mathcal{A}$ and $\mathcal{B}$ roll on the plane, find:

a. The angular acceleration of the cylinder
b. The angular acceleration of the disks
c. The minimum coefficient of friction between disks and plane for no slipping

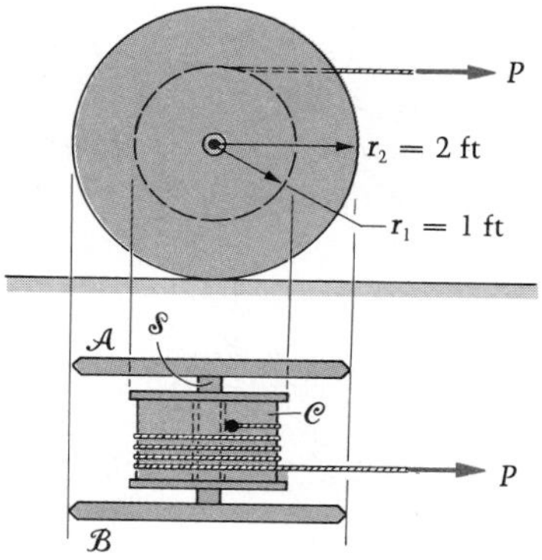

Figure P4.79

4.80 Rework the preceding problem, but this time assume that the string is wrapped so that it comes off the *bottom* of $\mathcal{C}$.

4.81 The strong, flexible cable shown in Figure P4.81 is wrapped around a light hub attached to the 130-lb cylinder $\mathcal{C}$. Find the angular acceleration of $\mathcal{C}$ upon release from rest. Note that it is impossible for the wheel to roll down the plane (meaning without slipping); to do so the cord would have to break.

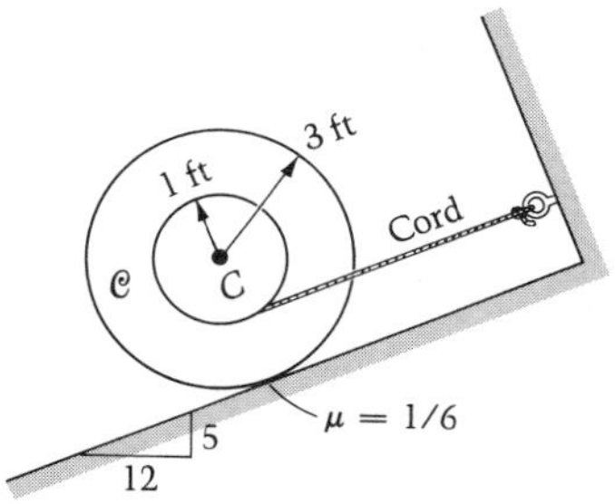

Figure P4.81

4.82 Repeat the previous problem for $\mu = 0.25$.

4.83 The 15-lb carriage shown in Figure P4.83 is supported by two uniform rollers each of weight 10 lb and radius 3 in. The rollers roll on the ground and on the carriage. Determine the acceleration of the carriage when the 5-lb force is applied to it.

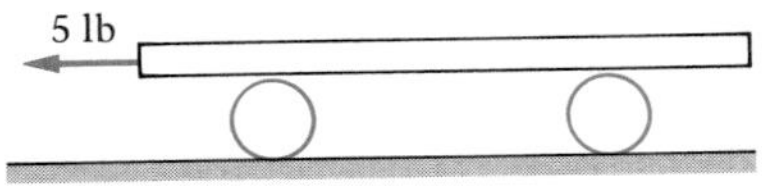

Figure P4.83

4.84 A cylinder has been cut into two equal parts by slicing it along its length (Figure P4.84b). One part is released in the position shown in Figures P4.84a, b. Find the minimum μ to prevent slipping at the instant of release.

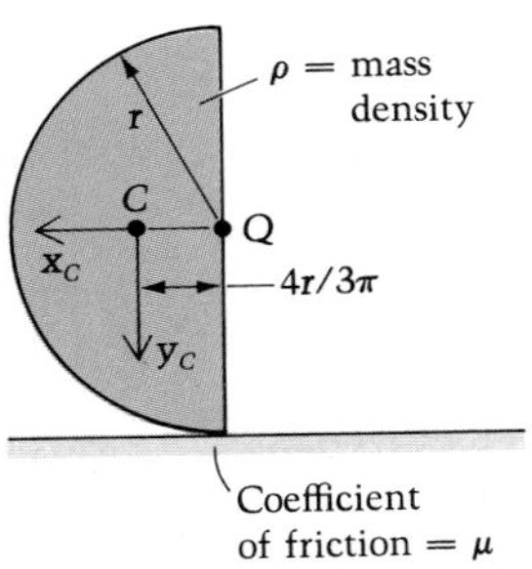

Figure P4.84a

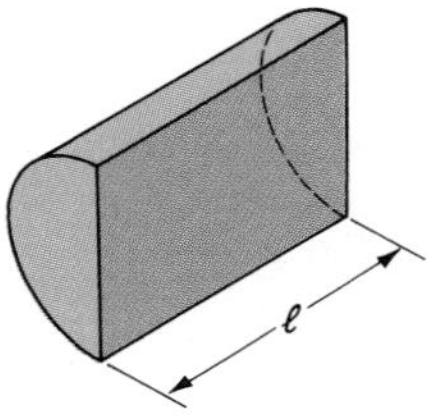

Figure P4.84b

4.85 The cylinder shown in Figure P4.85 is made of two halves of different densities. The left half is steel, with mass density $\rho_1 = 15.2\ \text{slug/ft}^3$; the right half is wood with $\rho_2 = 1.31\ \text{slug/ft}^3$. Recalling that the mass center of each half is located $4r/3\pi$ from the geometric center Q, find the acceleration of Q when the cylinder is released from rest. Assume enough friction to prevent slipping.

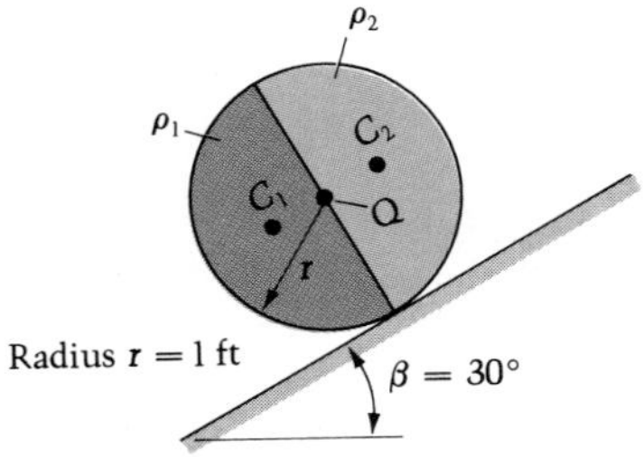

Figure P4.85

4.86 A uniform rod $\mathcal{R}$ is supported by two cords as shown in Figure P4.86. If the right-hand cord suddenly breaks, determine the initial tension in the left cord AD. ("Initial" means before the rod has had time to move and before it has had time to generate any velocities.)

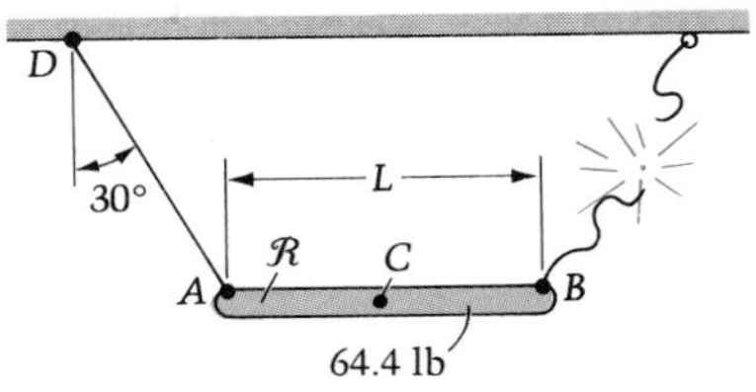

Figure P4.86

4.87 Two uniform bars $\mathcal{A}$ and $\mathcal{B}$ are released from rest in the position shown in Figure P4.87. Each bar is 2 ft long and weighs 10 lb. Determine the angular acceleration of each bar and the reactions at A and D immediately after release. The rollers are light and the pins smooth.

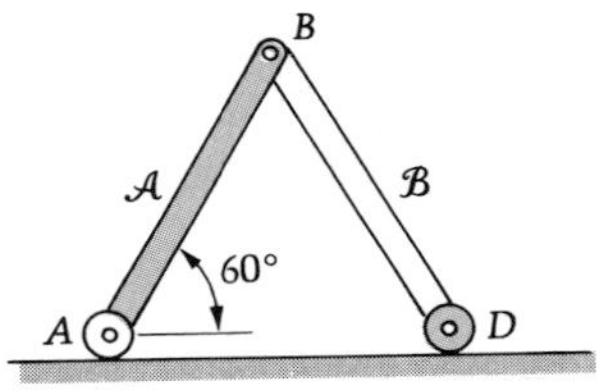

Figure P4.87

4.88 A thin rod AB of length ℓ and mass m is released from rest in the position shown in Figure P4.88. Point A of the rod is in contact with a surface whose coefficient of friction is μ.

a. Determine the minimum value of μ, say $\mu = \mu_{\text{min}}$, required to prevent end A from slipping upon release.

b. Find the acceleration of the mass center of the rod immediately after release for $\mu \geq \mu_{\text{min}}$ and for $\mu < \mu_{\text{min}}$.

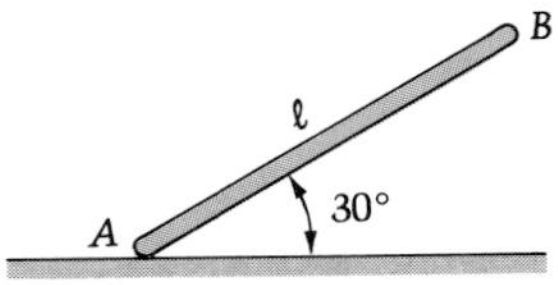

Figure P4.88

4.89 A uniform slender rod, 10 ft long and weighing 90 lb, is supported by wires attached to its ends. (See Figure P4.89.) Find the tension in the right wire just after the left wire is cut. Take the wires to be inextensible.

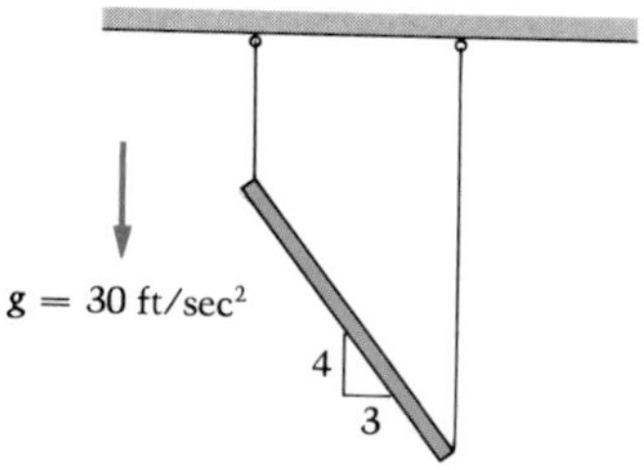

Figure P4.89

4.90 The disk shown in Figure P4.90 has mass m and radius r. Show that at the instant the right-hand string is cut, the tension in the other string changes to $\frac{2}{5}mg$, so that the acceleration of the mass center is $\frac{3}{5}g\downarrow$.

4.91 A uniform half-cylinder of radius r and mass m is held in the position shown in Figure P4.91 by the string tied to B. Find the reaction of the floor just after the string is cut. There is sufficient friction to prevent slipping.

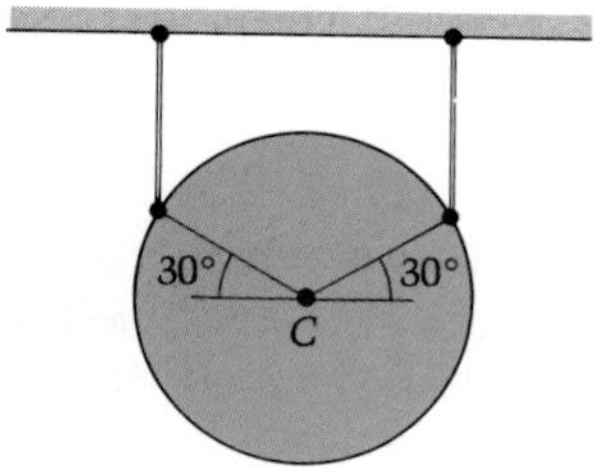

Figure P4.90

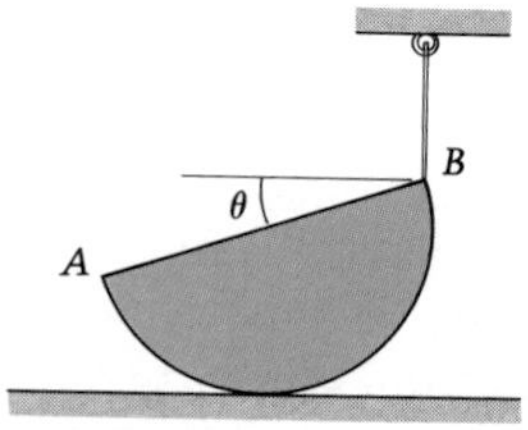

Figure P4.91

4.92 The uniform 10-lb bar in Figure P4.92 is suspended by two inextensible cables. At the instant shown, when each point in the bar has a velocity of $10\hat{\mathbf{i}}$ ft/sec, the right cable breaks. Find the force in the left cable immediately after the break.

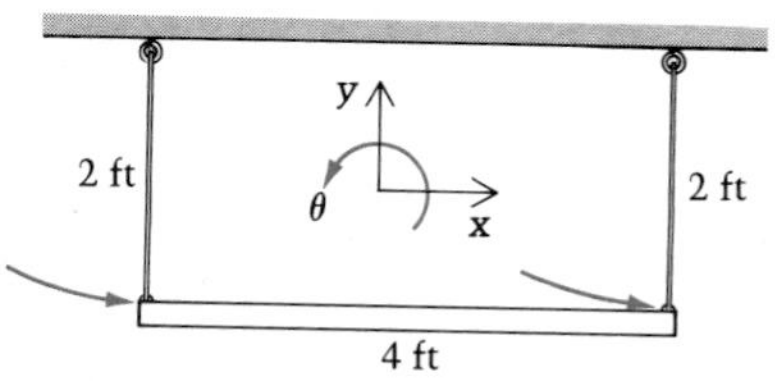

Figure P4.92

4.93 A slender homogeneous bar $\mathcal{B}$ weighing 193 lb has an angular velocity of 2 rad/sec clockwise and an angular acceleration of 8 rad/sec^2 clockwise when in the position shown in Figure P4.93. The wall at B is smooth; the coefficient of sliding friction at A is 0.10. Find the reactions at A and B on $\mathcal{B}$ in this position. *Hint:* The force P can be found.

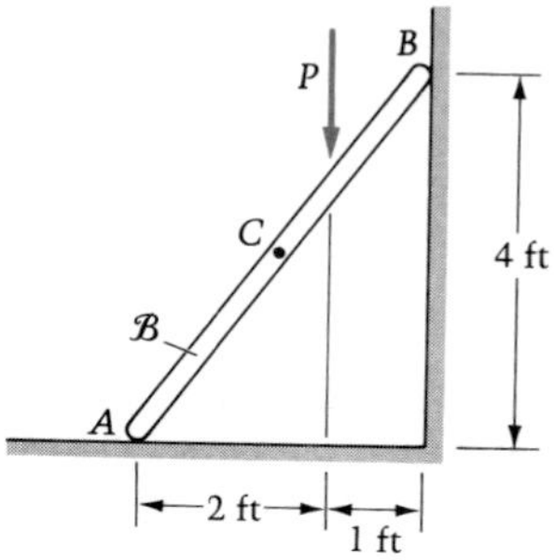

Figure P4.93

4.94 See Figure P4.94. (a) Find the initial angular acceleration of the cylinder, with one-quarter cut away as shown, when released from rest in the given position. Assume no slipping. (b) What is the minimum friction coefficient required to prevent slipping in part (a)?

4.95 A slender bar $\mathcal{B}$ weighing 64.4 lb is attached by massless cables to a fixed pivot A as shown in Figure P4.95. The system is swinging about A as a pendulum. At $\theta = 0$ the angular velocity is 2 ⟲ rad/sec and cable AD breaks. Find the tension in cable AB just after the break.

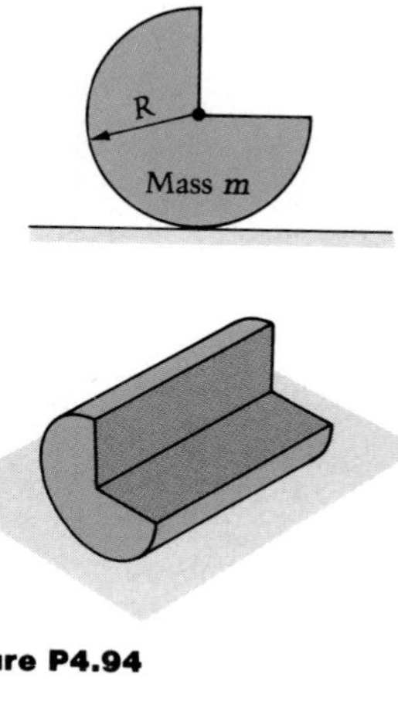

Figure P4.94

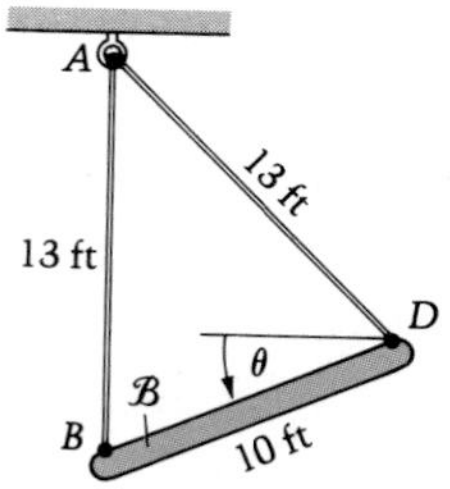

Figure P4.95

4.96 The system shown in Figure P4.96 is released from rest (angular velocity zero). Determine the initial tension in the cable. (The pulley is light and its pin is smooth, so that the tension is the same on both sides of it.)

4.97 A beam of length L and weight W per unit length is supported by two cables at A and B. (See Figure P4.97.) If the cable at B should break, find the shear force V and bending moment M at section xx just after the cable breaks. *Hint:* Euler's laws apply to every part of the body.

4.98 In Figure P4.98, pin P is welded to the 2-kg block $\mathcal{A}$ and extends through the slot in bar $\mathcal{B}$. The bar is driven so that its angular velocity is $\boldsymbol{\omega}_{\mathcal{B}} = 0.2$ ↻ rad/s = constant.

a. Find the force exerted on the pin by the slot in $\mathcal{B}$ when $\theta = 60°$.

b. Find the resultant force exerted on the block $\mathcal{A}$, by the horizontal slot in which it moves, at $\theta = 60°$.

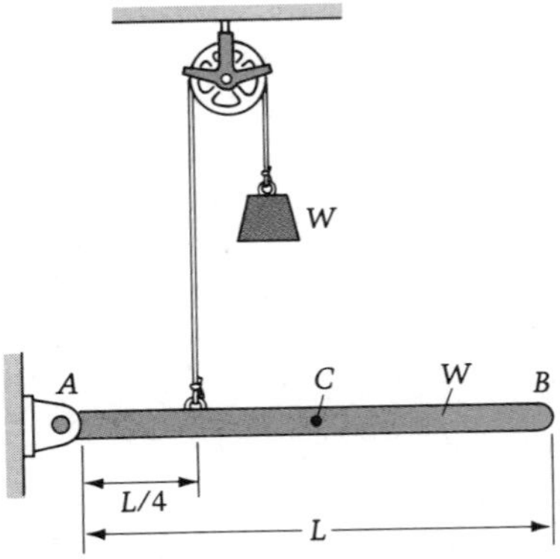

Figure P4.96

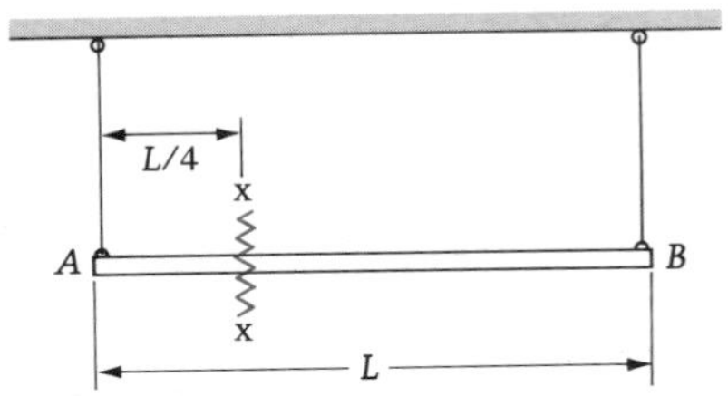

Figure P4.97

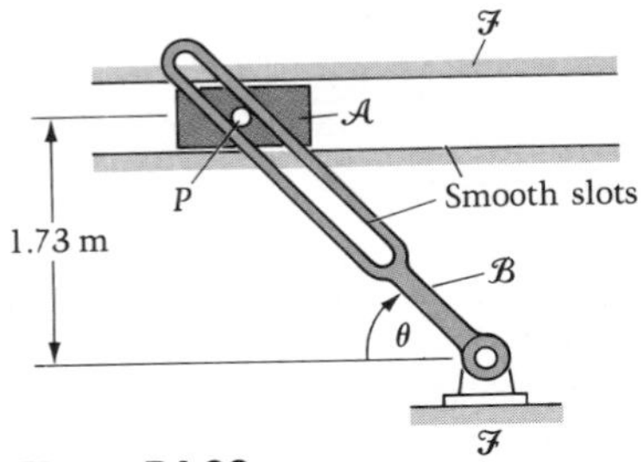

Figure P4.98

Other Useful Forms of the Rotational Equation

First Useful Additional Form. There are two other forms of the rotational equation of motion that are commonly used because they simplify matters considerably in certain cases of practical interest. The first one applies when body $\mathcal{B}$ has a pivot point O that is defined to be a point fixed in both $\mathcal{B}$ and the inertial reference frame $\mathcal{I}$ throughout the motion. (A pinned-down point is of course a pivot, but the body may be pivoting around a point *not* physically pinned.)

Recalling that Equation (4.8) applies equally well to the mass center and to points of zero velocity such as O, we may write

$$\mathbf{M}_{r_O} = \frac{d}{dt}\mathbf{H}_O = \frac{d}{dt}(I_{xz_O}\omega\hat{\mathbf{i}} + I_{yz_O}\omega\hat{\mathbf{j}} + I_{z_O}\omega\hat{\mathbf{k}}) \tag{4.43}$$

Equation (4.43) is exactly like (4.36) if C is replaced by O. Therefore the same steps lead to three alternative equations having the same form as Equations (4.39) whenever $\mathcal{B}$ has a pivot:

$$M_{r_{O_x}} = I_{xz_O}\alpha - I_{yz_O}\omega^2 \tag{4.44a}$$

$$M_{r_{O_y}} = I_{yz_O}\alpha + I_{xz_O}\omega^2 \tag{4.44b}$$

$$M_{r_{O_z}} = I_{z_O}\alpha \tag{4.44c}$$

Note that the only difference between Equations (4.39) and (4.44) is that the pivot point O replaces C. We emphasize that the C equations are *always* valid; the O equations are only of value to us if the body happens to have a pivot.

Question 4.5 Since Equation (4.8) applies for any point P having zero velocity, why can we not use equations such as (4.44) for the instantaneous center Ⓘ of $\mathcal{B}$ when Ⓘ is *not* a pivot?

When the rigid body $\mathcal{B}$ in plane motion does have a pivot point O, the body is forced to rotate about O and this motion is called **pure rotation.** This is, incidentally, the *only* time it makes any sense to say that the body continuously rotates about a point.* Each point of $\mathcal{B}$ moves on a circle centered at O, and to know the motion of $\mathcal{B}$ (that is, to "pin down" its position at any t) we need only know the angle $\theta = \theta(t)$, since we already know where one point (O itself) is. Hence we generally need only the equation $M_{r_O} = I_O\alpha$ (we may omit the z subscript) to find the motion. We may have to use the x and y force equations of motion, however, to determine unknowns of interest such as pin reactions.

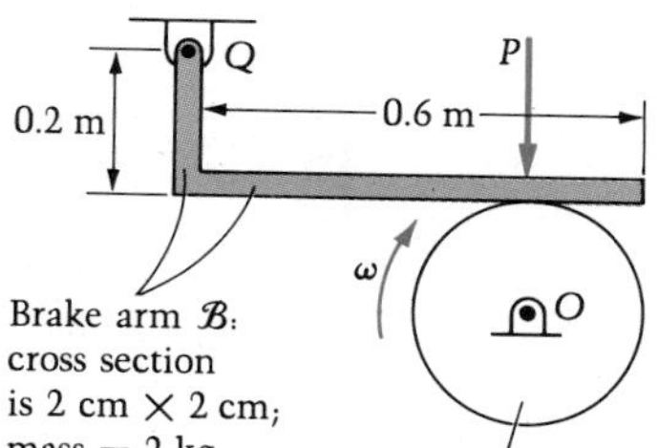

Brake arm $\mathcal{B}$: cross section is 2 cm × 2 cm; mass = 2 kg

Cylinder $\mathcal{A}$: mass = 10 kg; radius = 0.2 m

EXAMPLE 4.19

Just after the brake arm in the accompanying diagram contacts the top of the cylinder $\mathcal{A}$, the cylinder is turning at 1000 rpm ↻. The coefficient of kinetic friction between $\mathcal{A}$ and $\mathcal{B}$ is $\mu = 0.3$. Find how long it takes for $\mathcal{A}$ to come to rest under the constant force $P = 40$ N.

SOLUTION

Since body $\mathcal{B}$ is in equilibrium, we may find the normal force between it and the cylinder by statics. The bar's weight is proportioned between its horizontal and vertical parts as shown in the following diagram. Note that equilibrium requires that $f = Q_x$, where $f = f_{max} = \mu N$ since slipping is taking place.

(Continued)

**Instantaneously*, of course, it always rotates, velocity-wise, about its instantaneous center Ⓘ, whether or not it has a pivot.

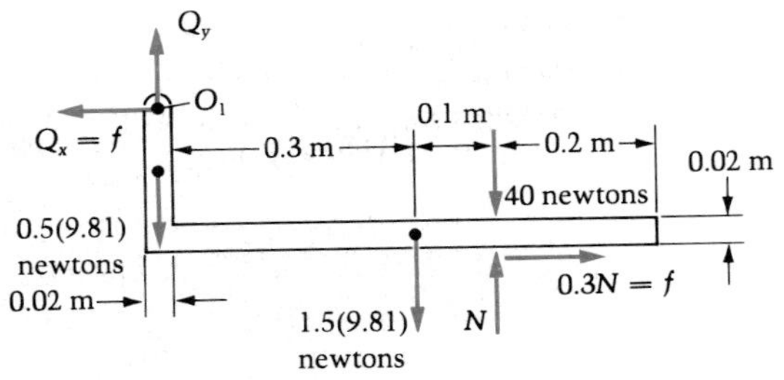

Summing moments about Q, we have

$$0.401N + 0.2(0.3N) - 40(0.401) - 1.5(9.81)(0.301) = 0$$

from which we get

$$N = \frac{20.5}{0.461} = 44.6 \text{ newtons}$$

and

$$\mu N = 0.3(44.6) = 13.4 \text{ newtons}$$

Question 4.6 Would the normal force N be any different if ω was counterclockwise?

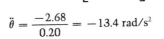

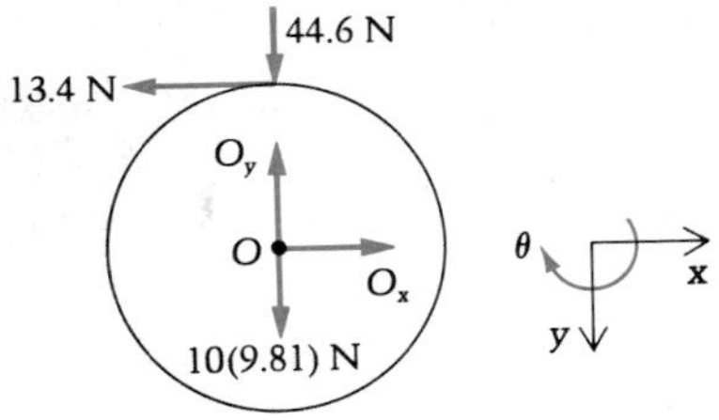

The motion of body $\mathcal{A}$ is one of pure rotation. Its free-body diagram is shown at the left.

$$M_{r_O} = I_O\ddot{\theta}$$

$$-13.4(0.2) = \left[\frac{1}{2}\,10(0.2)^2\right]\ddot{\theta}$$

$$\ddot{\theta} = \frac{-2.68}{0.20} = -13.4 \text{ rad/s}^2$$

(The minus sign indicates a *de*celeration, as must be the case.)

Integrating, we get

$$\dot{\theta} = -13.4t + C_1$$

$$= -13.4t + 1000\left(\frac{2\pi}{60}\right)$$

where the initial condition on ω allows us to calculate the integration constant C_1.

Body $\mathcal{A}$ stops when $\dot{\theta} = 0$ at a time t_s that we are now in a position to calculate:

$$0 = -13.4t_s + 105$$

$$t_s = 7.84 \text{ s}$$

(Continued)

Note that the mass center $O = C$ of $\mathcal{A}$ is fixed in the inertial frame, so that the pin reactions follow from the mass center equations:

$$F_{r_x} = -13.4 + O_x = m\ddot{x}_C = 0 \Rightarrow O_x = 13.4 \text{ newtons}$$

$$F_{r_y} = +44.6 + 98.1 - O_y = m\ddot{y}_C = 0 \Rightarrow O_y = 143 \text{ newtons}$$

We note that if the body is pinned at its mass center C, then Equations (4.39) and (4.44) are the same. If the body has a pivot O and $O \neq C$, however, we nearly always use Equation (4.44c) because the pin reactions are eliminated. This case is covered in the following example, which also involves translation of a second body.

EXAMPLE 4.20

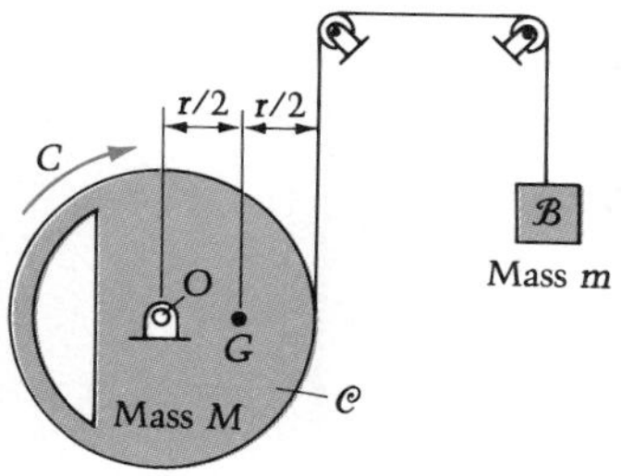

($\theta = 0$ at $t = 0$)
Mass center $= G$

Figure 1

How must the applied couple C in Figure 1 vary with time in order to turn the unbalanced (but round) wheel at constant angular velocity ω_0 ↻? The moment of inertia of the mass of $\mathcal{C}$ with respect to its axis of rotation is I_O, and the mass center is located at G.

SOLUTION

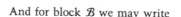

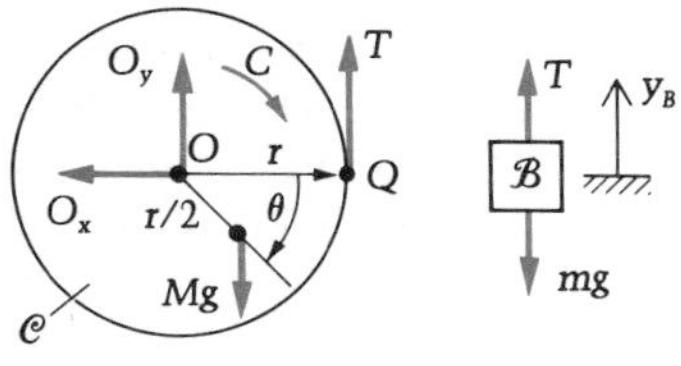

Figure 2

Since $M_{r_O} = I_O\alpha$ for body $\mathcal{C}$,

$$-Tr + C + Mg\frac{r}{2}\cos\theta = I_O\ddot{\theta} \tag{1}$$

And for block $\mathcal{B}$ we may write

$$F_{r_y} = m\ddot{y}_B$$

$$T - mg = m\ddot{y}_B \tag{2}$$

The point Q of $\mathcal{C}$ located where the rope leaves the rim has the same velocity ($r\dot{\theta}\downarrow$) and tangential acceleration component ($r\ddot{\theta}\downarrow$) as does the rope itself at that point. Since the rope is assumed inextensible, this acceleration has the same magnitude as $\ddot{y}_B$. Therefore our kinematics gives us an additional equation:

$$\ddot{y}_B = r\ddot{\theta} \tag{3}$$

(Continued)

Substituting (3) into (2) gives

$$T = mg + mr\ddot{\theta}$$

And substituting T into the moment equation (1) for $\mathscr{C}$ then yields

$$C = mgr - Mg\frac{r}{2}\cos\theta + (I_O + mr^2)\ddot{\theta}$$

Since $\dot{\theta} = \omega_0 =$ constant, we have $\theta = \omega_0 t$ and $\ddot{\theta} = 0$, so that

$$C = mgr - \frac{Mgr}{2}\cos\omega_0 t$$

and the required couple varies harmonically.

If there is *no* couple C and the system is released from rest in the position shown, then the equations are all still valid and

$$(I_O + mr^2)\ddot{\theta} = Mg\frac{r}{2}\cos\theta - mgr$$

The initial angular acceleration of $\mathscr{C}$ (with $\theta = 0$) is thus seen to be

$$\ddot{\theta}_0 = \frac{(Mgr/2) - mgr}{I_O + mr^2}$$

which is positive (↻) if $M > 2m$.

Question 4.7 What happens if $M < 2m$? If $M = 2m$?

It is interesting to write the equations of motion in their usual form for this problem:

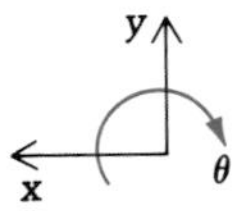

$$F_{r_x} = O_x = M\ddot{x}_G \tag{4}$$

$$F_{r_y} = O_y + T - Mg = M\ddot{y}_G \tag{5}$$

$$M_{r_G} = O_y\frac{r}{2}\cos\theta - O_x\frac{r}{2}\sin\theta - T\left(r - \frac{r}{2}\cos\theta\right) = I_G\ddot{\theta} \tag{6}$$

Equations (4) and (5) are useful if the pin reactions are desired,* but Equation (6) is nowhere near as handy to use as $M_{r_O} = I_O\alpha$, which we have used earlier in the example since the body has a pin. The student may wish to eliminate O_x, O_y, and T from (6) by using (4), (5), and the previous equation for $\mathscr{B}$ ($T = mg + mr\ddot{\theta}$) and to show that the same result is obtained (after a good deal more work than in the example) for $\ddot{\theta}$. (Kinematics must also be used to relate $\ddot{x}_G$ and $\ddot{y}_G$ to θ, $\dot{\theta}$, and $\ddot{\theta}$!)

*For instance, the pins must be designed strong enough to take the forces caused by the accelerations.

EXAMPLE 4.21

The purpose of this example is to show how to write a single differential equation for the rotational motion of a load $\mathcal{L}$ driven by a motor through a series of parallel meshing spur gears. In the accompanying figure, a motor armature $\mathcal{M}$ with inertia I_m delivers a torque T_m through shaft $\mathcal{S}_1$, which turns pinion $\mathcal{P}_1$ (inertia $I_{\mathcal{P}_1}$). The pinion teeth then engage and drive gear $\mathcal{G}_1$ (inertia $I_{\mathcal{G}_1}$), which is connected via shaft $\mathcal{S}_2$ to a second pinion $\mathcal{P}_2$ ($I_{\mathcal{P}_2}$). This pinion drives gear $\mathcal{G}_2$ ($I_{\mathcal{G}_2}$), which is rigidly connected to the load $\mathcal{L}$ ($I_{\mathcal{L}}$) by a third shaft $\mathcal{S}_3$. Find the relation between T_m and $\alpha_{\mathcal{L}}$.

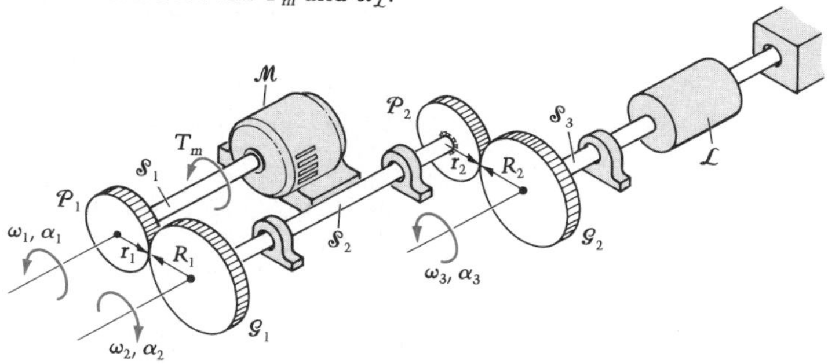

SOLUTION

From kinematics we know that the equality of velocities of the gear teeth implies that angular velocities (and angular accelerations) through a gear train are changed by the ratio of the pitch radii. Therefore

$$\alpha_2 = \frac{r_1}{R_1}\alpha_1 \qquad \text{and} \qquad \alpha_3 = \frac{r_2}{R_2}\alpha_2$$

Or, letting $\alpha_m = \alpha_1$ = angular acceleration of the motor shaft, $\alpha_{\mathcal{L}} = \alpha_3$ = angular acceleration of the load shaft, and $n_i = R_i/r_i$ = gear ratio ($i = 1, 2$), we have

$$\alpha_2 = \frac{\alpha_m}{n_1} \qquad \text{and} \qquad \alpha_{\mathcal{L}} = \frac{\alpha_2}{n_2} \tag{1}$$

Next we study the pinions and gears that are in mesh with each other (see the diagrams). Note that the moments of the forces at the gear teeth about the centers of the gears ($F_1 r_1$ and $F_1 R_1$, for example) grow in proportion to the gear ratios just as the angular speeds decrease. Thus we give up speed through gear drives if we wish to produce large torques; conversely, if the gear ratios defined above (such as $n_1 = R_1/r_1$) are *smaller* than unity, the angular speeds *increase* while the torques *decrease*.

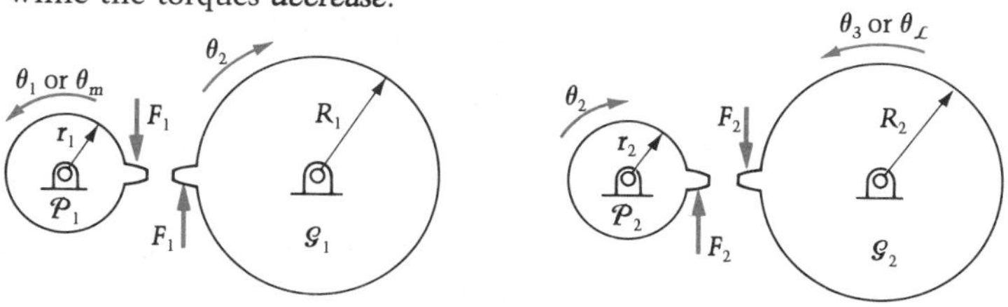

(Continued)

Writing the rotational equation of motion for the three (assumed rigid) bodies ($\mathcal{B}_1 = \mathcal{M} + \mathcal{S}_1 + \mathcal{P}_1$; $\mathcal{B}_2 = \mathcal{G}_1 + \mathcal{S}_2 + \mathcal{P}_2$; $\mathcal{B}_3 = \mathcal{G}_2 + \mathcal{S}_3 + \mathcal{L}$), noting the positive directions in the preceding sketches, gives

$$T_m - F_1 r_1 = (I_{\mathcal{P}_1} + I_m)\alpha_m \tag{2}$$

$$F_1 R_1 - F_2 r_2 = (I_{\mathcal{G}_1} + I_{\mathcal{P}_2})\alpha_2 \tag{3}$$

$$F_2 R_2 = (I_{\mathcal{G}_2} + I_{\mathcal{L}})\alpha_{\mathcal{L}} \tag{4}$$

in which the inertias of the shafts $\mathcal{S}_1$, $\mathcal{S}_2$, $\mathcal{S}_3$ about their axes are grouped (if they are not small enough to be neglected) together with $I_{\mathcal{P}_1}$, $I_{\mathcal{P}_2}$, and $I_{\mathcal{L}}$, respectively.

Eliminating the gear tooth forces F_1 and F_2 leaves an equation that may be put into the form

$$T_m n_1 n_2 = (I_{\mathcal{L}} + I_{\mathcal{G}_2})\alpha_{\mathcal{L}} + (I_{\mathcal{G}_1} + I_{\mathcal{P}_2})\alpha_2 n_2 + (I_{\mathcal{P}_1} + I_m)\alpha_m n_1 n_2 \tag{5}$$

Using Equation (1), we may replace α_2 by $n_2\alpha_{\mathcal{L}}$ and α_m by $n_1\alpha_2 = n_1 n_2 \alpha_{\mathcal{L}}$. Substituting these results into Equation (5) gives

$$T_m n_1 n_2 = [(I_{\mathcal{L}} + I_{\mathcal{G}_2}) + (I_{\mathcal{G}_1} + I_{\mathcal{P}_2})n_2^2 + (I_{\mathcal{P}_1} + I_m)n_1^2 n_2^2]\alpha_{\mathcal{L}} \tag{6}$$

Equation (6) allows us to conclude that the strength of the motor torque *at the load* is multiplied by the product of all gear ratios between the input and output. Also, each inertia, reflected through the gear train to the load, is multiplied by the square of all gear ratios lying between it and the load.

The same procedure for N gear passes, instead of two, would result in

$$\begin{aligned} T_m n_1 n_2 \cdots n_N = {} & [(I_{\mathcal{L}} + I_{\mathcal{G}_N}) + (I_{\mathcal{G}_{N-1}} + I_{\mathcal{P}_N})n_N^2 \\ & + (I_{\mathcal{G}_{N-2}} + I_{\mathcal{P}_{N-1}})n_N^2 n_{N-1}^2 + \cdots \\ & + (I_{\mathcal{P}_2} + I_{\mathcal{G}_1})n_N^2 n_{N-1}^2 \cdots n_2^2 + (I_{\mathcal{P}_1} + I_m)n_N^2 \cdots n_1^2]\alpha_{\mathcal{L}} \end{aligned} \tag{7}$$

as the student may wish to verify. Thus the preceding results about motor torque and inertias are true for any number of gear passes.

PROBLEMS / Section 4.5 (continued)

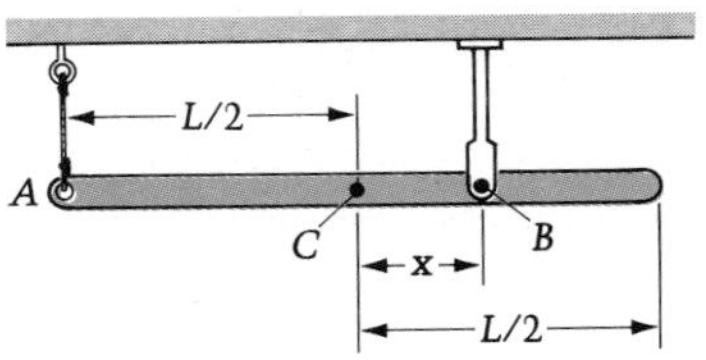

Figure P4.99

4.99 The slender, homogeneous rod in Figure P4.99 is supported by a cord at A and a horizontal pin at B. The cord is cut. Determine, at that instant, the location of pin B that will result in the maximum initial angular acceleration of the rod.

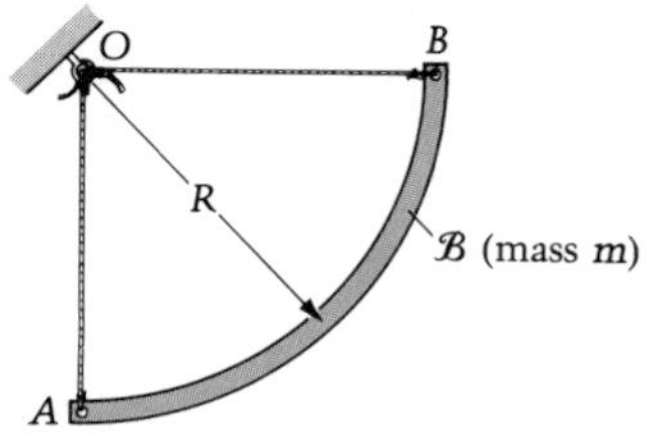

Figure P4.100

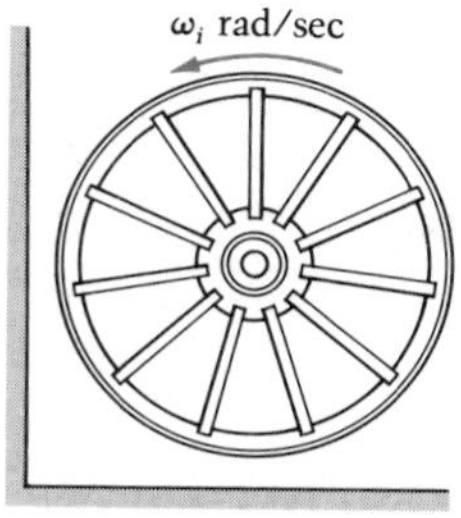

Figure P4.101

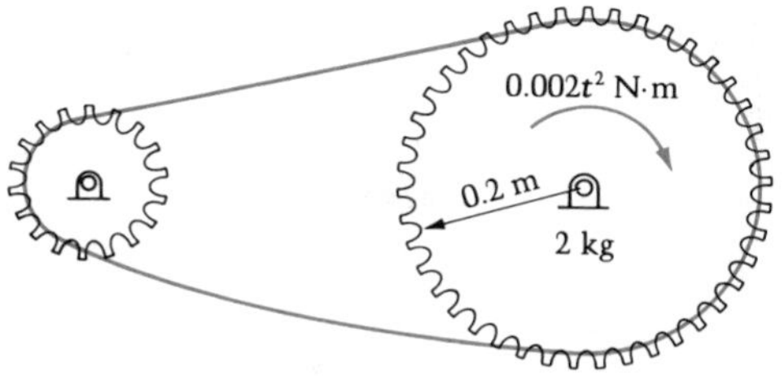

Figure P4.102

4.100 Body $\mathcal{B}$ is a slender bar bent into the shape of a quarter-circle (Figure P4.100). Find the tensions in strings OA and OB when the system is released from rest.

4.101 A wagon wheel spinning counterclockwise is placed in a corner and contacts the wall and floor. (See Figure P4.101.)

a. Show with a free-body diagram that the wheel cannot climb the wall.

b. Show with a free-body diagram that the wheel cannot move to the right along the floor either.

c. Therefore the wheel stays in the corner. Treat it as a ring of mass m and radius R, with friction coefficient μ at both surfaces of contact. Determine how long it takes for the wheel to stop, and find how many radians it has turned through since first contacting the surfaces.

4.102 The chain drive in Figure P4.102 may be considered as two disks with equal density and thickness. The larger sprocket has a mass of 2 kg and a radius of 0.2 m. If the couple is applied starting from rest at $t = 0$, find the angular speed of the smaller sprocket at $t = 10$ s. *Hint:* What does a dentist do?

4.103 Figure P4.103 shows a scene from Edgar Allen Poe's "The Pit and the Pendulum." Find the reaction of the pin onto the bar if the pendulum is instantaneously at rest in a horizontal position.

4.104 Find the angular acceleration of cylinder $\mathcal{C}$ in Figure P4.104. The rope passes over it without slipping and ties to $\mathcal{A}$ and $\mathcal{B}$ as shown.

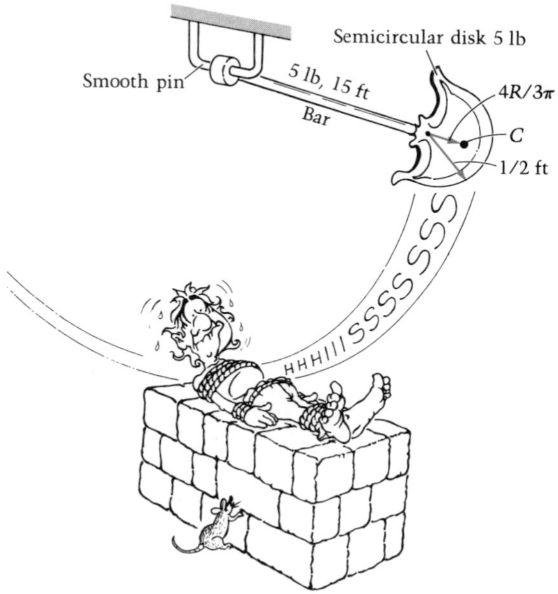

Figure P4.103

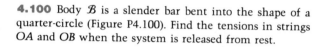

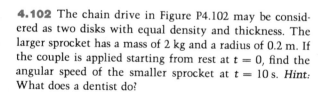

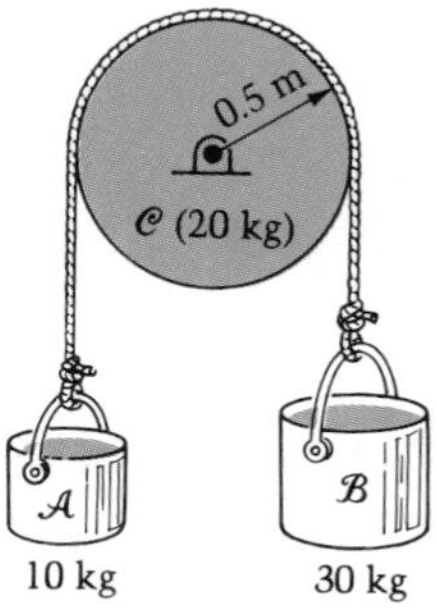

Figure P4.104

4.105 If the man in Figure P4.105 wishes to use a chain hoist to lift the engine, he will find that a force of $F_s = W(R - r)/(2R)$ will hold it in equilibrium. Thus for a force slightly greater than F_s the chain (which cannot slip) will move and the engine will start to rise. Show that the speed of the vertically translating engine is given by

$$\frac{(R - r)v_0}{2R}$$

where v_0 is the constant downward speed of the point of pulling. Then compute the force F_s which will accelerate the engine at $g/8$ if its mass is m and the pulley masses are negligible. Take $r = 3R/4$.

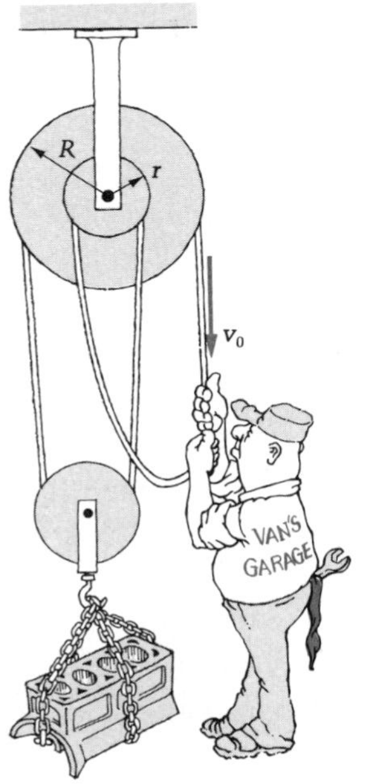

Figure P4.105

4.106 In the preceding exercise, note that as r is constructed closer and closer to R, then F_s decreases, suggesting that the man could theoretically lift *very* large weights. What is the practical limitation on this idea?

4.107 Cylinder $\mathcal{C}$ in Figure P4.107 with four cutouts is rotating at 200 rpm initially. A uniform 100-lb cylinder $\mathcal{D}$ is placed in the position shown, and the friction produces a braking moment that will stop $\mathcal{C}$. The friction coefficient is $\mu = \frac{1}{3}$, and before the four holes were drilled the uniform body $\mathcal{C}$ weighed 200 lb. For whichever rotation direction of $\mathcal{C}$ results in a quicker stop, find the stopping time.

4.108 The cylinder in Figure P4.108 has a mass of 30 kg and rotates about an axis normal to the clevis at O. At the instant shown, $\omega = 5$ ↻ rad/s and $\dot{\omega} = 10$ ↺ rad/s^2. Find the force P that acts on the cylinder, and determine the reactions exerted by the pin onto the clevis at O, all at the given instant.

4.109 In Figure P4.109 the moments of inertia of the motor armature, gear 1, gear 2, and the symmetric antenna load are I_M, I_1, I_2, and I_L; the gear ratio is given by $R/r = n$. Show that the torque T_M developed by the motor is related to the angular acceleration α_M of its output shaft by

$$T_M = \left(\frac{I_L + I_2}{n^2} + I_1 + I_M\right)\alpha_M$$

Thus when inertias are reflected *back* through gears, their effective values at the motor are *divided* by the square of the gear ratio for each mesh through which they are reflected.

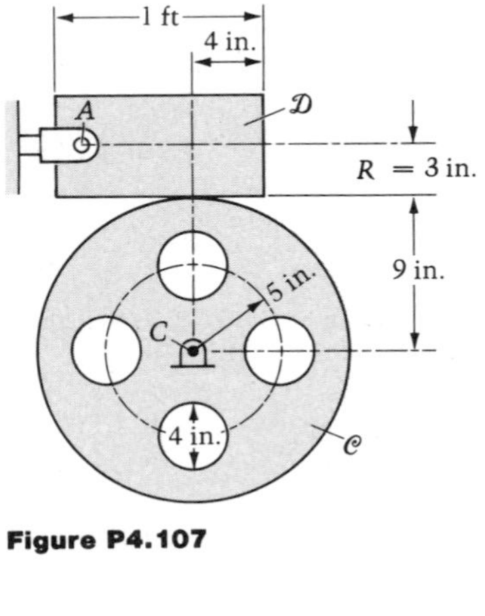

Figure P4.107

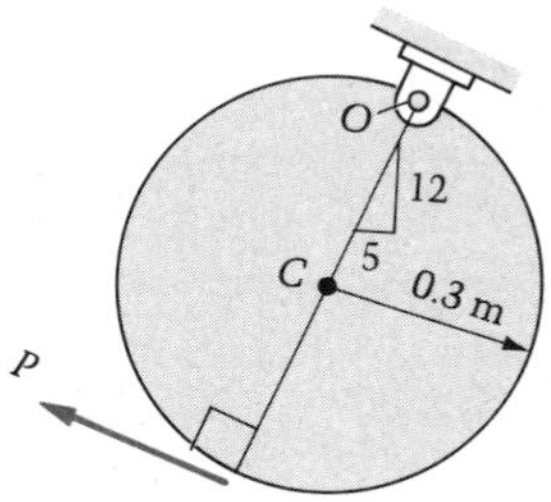

Figure P4.108

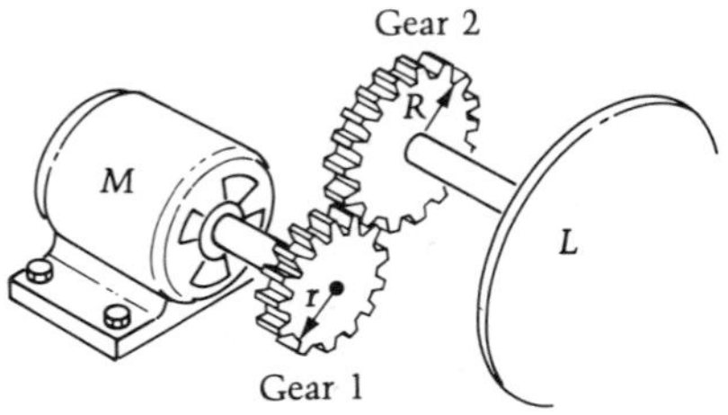

Figure P4.109

4.110 The metacenter of a ship $\mathcal{S}$ is the point M on its symmetry axis through which the resultant buoyant force passes. (See Figure P4.110.) Discuss whether:

a. Point M is a fixed point of $\mathcal{S}$. (Does it change as the ship rolls?)

b. Point M is above, at, or below the mass center C. Relate this answer to the rolling of the ship in the sea.

Figure P4.110

4.111 The slender homogeneous rod in Figure P4.111 is 12 ft long and weighs 5 lb; it is connected by a rusty hinge to a support at A. Because of friction in the hinge, the hinge exerts a couple of 9 ft-lb on the rod when it rotates. If the rod is released from rest with $\theta = 30°$, find: (a) the angular acceleration of the rod when $\theta = 30°$, $60°$ and $90°$; (b) the angle θ at which the angular acceleration of the rod is zero.

4.112 The uniform slender bar of weight W and length L in Figure P4.112 is released from rest at $\theta = 0$ and pivots on its square end about corner O.

a. If the bar is observed to slip at $\theta = 30°$, find the coefficient of limiting static friction μ_s.

b. If the end of the bar is notched so that it cannot slip, find the angle θ at which contact between bar and corner ceases. *Hint:* Write the moment equation of motion about the pivot O, multiply it by $\dot{\theta}$, and integrate, obtaining $\dot{\theta}$ as a function of θ. Use this relation together with the component equation of $\mathbf{F}_r = m\mathbf{a}_C$ in the $\hat{\mathbf{e}}_n$ direction.

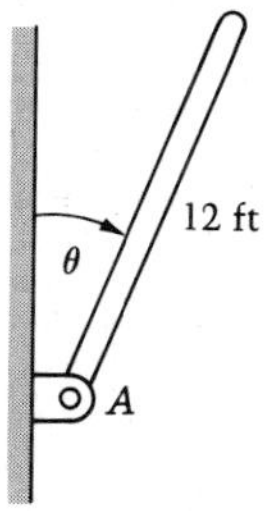

Figure P4.111

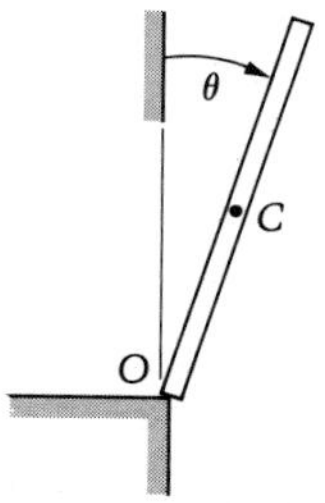

Figure P4.112

Second Useful Additional Form. The second useful alternative form of the rotational equation is one which, just like $M_{r_C} = I_C\alpha$, has no restrictions whatsoever on its use in plane motion. If P is any point (P could even be moving and not fixed in the body $\mathcal{B}$), Equation (4.31) allows us to use P as a moment center:

$$\mathbf{M}_{r_P} = \dot{\mathbf{H}}_C + \mathbf{r}_{PC} \times m\mathbf{a}_C \tag{4.45}$$

And for plane motion of a rigid body $\mathcal{B}$ this becomes, using Equations (4.36) and (4.38) and with axes fixed in $\mathcal{B}$ at C,

$$\mathbf{M}_{r_P} = (I_{xz_C}\alpha - I_{yz_C}\omega^2)\hat{\mathbf{i}} + (I_{yz_C}\alpha + I_{xz_C}\omega^2)\hat{\mathbf{j}} + I_{z_C}\alpha\hat{\mathbf{k}} + \mathbf{r}_{PC} \times m\mathbf{a}_C \tag{4.46}$$

Taking P to be in the reference plane containing C (but otherwise still arbitrary), we therefore have the following useful scalar component equations for the body's rotational motion:

$$M_{r_{P_x}} = I_{xz_C}\alpha - I_{yz_C}\omega^2 \tag{4.47a}$$

$$M_{r_{P_y}} = I_{yz_C}\alpha + I_{xz_C}\omega^2 \tag{4.47b}$$

$$M_{r_{P_z}} = I_{z_C}\alpha + (\mathbf{r}_{PC} \times m\mathbf{a}_C)_z \tag{4.48}$$

in which $(\mathbf{r}_{PC} \times m\mathbf{a}_C)_z$ means the indicated cross product without its unit vector (which is $\hat{\mathbf{k}}$, since $\mathbf{r}_{PC}$ and $\mathbf{a}_C$ both lie *in* the reference plane). Note that $(\mathbf{r}_{PC} \times m\mathbf{a}_C)_z$ may or may not be the magnitude of the cross product since it could be negative.

We note that since P and C are in the same **xy** plane, by Equations (4.25) we have

$$I_{xz_P} = I_{xz_C} - m\overline{x}\overline{z} = I_{xz_C} \qquad (\text{since } \overline{z} = 0)$$

$$I_{yz_P} = I_{yz_C} - m\overline{y}\overline{z} = I_{yz_C} \qquad (\text{since } \overline{z} = 0)$$

Also, it is always true that

$$\mathbf{M}_{r_P} = \mathbf{M}_{r_C} + \mathbf{r}_{PC} \times \mathbf{F}_r$$
$$= \mathbf{M}_{r_C} + \mathbf{r}_{PC} \times m\mathbf{a}_C$$

Since $\mathbf{r}_{PC}$ and $\mathbf{a}_C$ both have only **x** and **y** components, the cross product term is in the **z** direction. Thus

$$M_{r_{P_x}} = M_{r_{C_x}} \tag{4.49a}$$

and

$$M_{r_{P_y}} = M_{r_{C_y}} \tag{4.49b}$$

Therefore the moments in the **x** and **y** moment equations (4.47a and b) may be taken with respect to *any* point in the plane (moving or not, fixed to $\mathcal{B}$ or not) and the product of inertia terms I_{xz} and I_{yz} may likewise be calculated at *any* point of the reference plane. We now concentrate on Equation (4.48) and ignore the other two equations until later in this section (realizing that they become statics equations for the symmetrical bodies described earlier).

EXAMPLE 4.22

The forklift truck $\mathcal{T}$ weighs 2000 lb and carries a crate $\mathcal{B}$ of uniform density that weighs 1500 lb. (See the diagram.) The coefficients of friction are $\mu_1 = 0.3$ between crate and truck and $\mu_2 = 0.5$ between tires and floor. If the truck is

(Continued)

moving at 15 ft/sec, find the minimum stopping distance if none of the following is to occur upon application of the brakes.

a. The truck slides.
b. The box slides relative to the truck.
c. The box tips relative to the truck.
d. The truck and crate tip as a unit.

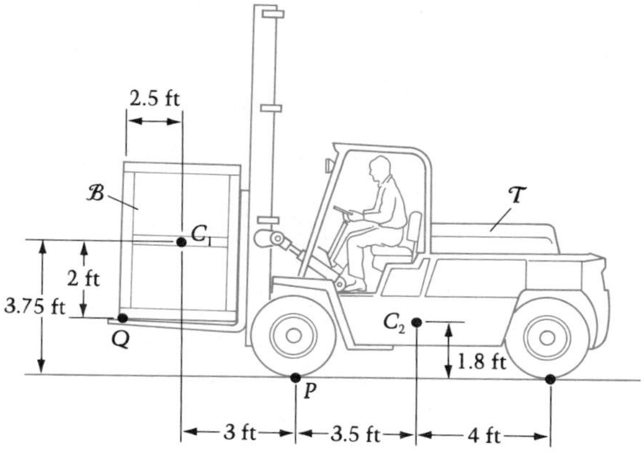

SOLUTION

In a problem such as this one in which many different things could happen, we assume the onset of one case, solve the problem, and then check the others, repeating until we find the solution for the governing situation (the one that occurs first). We begin by assuming that the first thing that happens if the brakes are increasingly but rapidly applied is Case (d)—the truck and crate tip together as a unit about the contact point P of the front wheel with the ground.

The free-body diagram for Case (d) is shown at left; in this case, we are at a "best of both worlds" point:

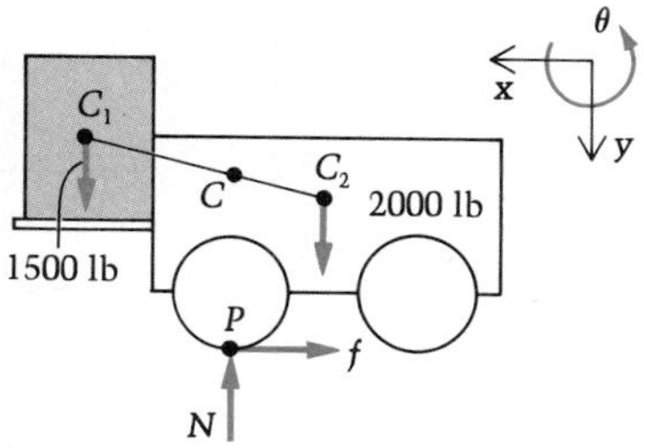

Case (d)

1. The rear wheel force has gone to zero.
2. But at the same time the angular acceleration is still zero. (Any tiny further brake application will cause the truck to rotate about P.)

Thus using Equation (4.48), where C is the combined mass center of the crate (body 1) and truck (body 2), which move as a combined body,

$$M_{r_{P_z}} = 0 + (\mathbf{r}_{PC} \times m\mathbf{a}_C)_z$$

Because of the definition of C, we know that

$$\mathbf{r}_{PC} = \frac{m_1 \mathbf{r}_{PC_1} + m_2 \mathbf{r}_{PC_2}}{m}$$

(Continued)

And since the acceleration $\mathbf{a}_C = a_C\hat{\mathbf{i}}$ is the same for $\mathcal{B}$ and $\mathcal{T}$ (if our assumption of tipping turns out to be correct), we have*

$$1500(3) - 2000(3.5) = \left[3.75\left(\frac{1500}{32.2}\right) + 1.8\left(\frac{2000}{32.2}\right)\right]a_C$$

so that

$$a_C = \frac{-2500}{286} = -8.74 \text{ ft/sec}^2$$

To check Case (a), note that on the overall free-body diagram

$$F_{r_x} = m\ddot{x}_C = ma_C$$

$$-f = \frac{2000 + 1500}{32.2}a_C = (109 \text{ slugs})(-8.74 \text{ ft/sec}^2)$$

$$f = 953 \text{ lb}$$

Note that a plus sign on the answer for an unknown's value means that the direction is *as assumed*, and not necessarily in a positive coordinate direction. Next we have

$$F_{r_y} = m\ddot{y}_C = 0$$

$$N - 1500 - 2000 = 0$$

$$N = 3500 \text{ lb}$$

And since

$$f_{\max} = \mu N = 0.5(3500) = 1750 \text{ lb}$$

We see that $f_{\max} > f$, so we have more than enough friction to prevent slipping of the tires.

Next we check Case (b). We note two items in conjunction with the free-body diagram of the crate shown here:

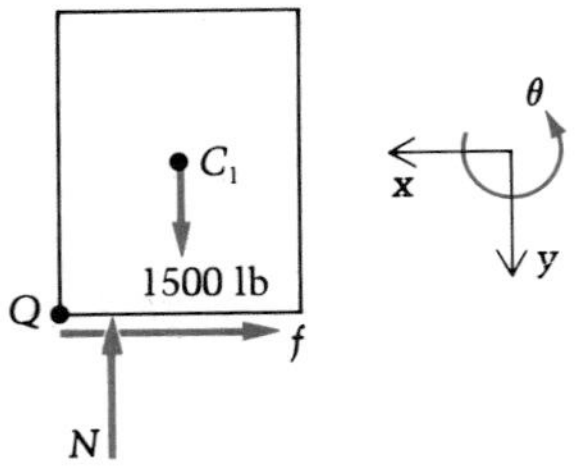

Case (b)

1. If sliding motion of the crate impends, the forces exerted on its vertical right face by the truck have gone to zero.
2. To prevent tipping, N is somewhere *between* the corner Q and the projection of C_1 (in order that $M_{r_{C_1}} = 0$!).

Clearly

$$F_{r_y} = m\ddot{y}_{C_1} = 0$$

$$1500 - N = 0 \Rightarrow N = 1500 \text{ lb}$$

and

$$f_{\max} = \mu N = 0.3(1500) = 450 \text{ lb}$$

(Continued)

*Note that the combined mass center C need not be found!

But we can determine the *actual* force f, assuming that $\mathcal{B}$ moves with $\mathcal{T}$:

$$F_{r_x} = m_1 \ddot{x}_{C_1}$$

$$-f = \left(\frac{1500}{32.2}\text{ slugs}\right)(-8.74\text{ ft/sec}^2)$$

$$f = 407\text{ lb} < f_{\max}$$

Thus the crate will not slide before the truck tips.

Finally we check Case (c). Again we use Equation (4.48), this time with the reference point Q on the lower left corner of the crate (see the diagram). Here we must see whether the normal force N can act on the bottom of the crate without its tipping. If $\delta < 0$ (see the figure) for $\alpha = 0$, then it will *already* have tipped at the value of a_C found for Case (d):

$$M_{r_Q} = 0 + (\mathbf{r}_{QC_1} \times m\mathbf{a}_{C_1})_z$$

$$-2.5(1500) + N\delta = -2\left(\frac{1500}{32.2}\right)8.74\text{ ft-lb}$$

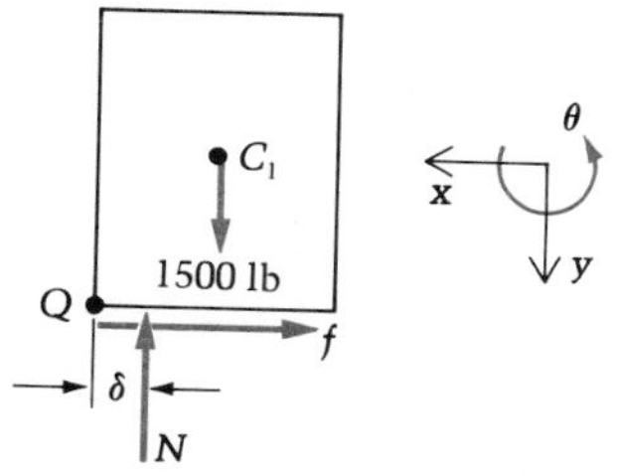

Case (c)

And since we are assuming $\ddot{y}_{C_1} = 0$, we find that $F_{r_y} = 0$ has given $N = 1500$ lb. Thus

$$\delta = \frac{-814 + 3750}{1500} = 1.96\text{ ft} \qquad \text{(which is safe)}$$

If Cases (a), (b), or (c) had happened to govern the solution instead of (d), we would simply have solved again, in that particular case, for the a_C needed to prevent the particular case from occurring.* The reader may wish to show that the (higher) decelerations required to produce Cases (a), (b), and (c) are respectively 16.1, 9.66, and 40.3 ft/sec².

Continuing, we have the following for the governing Case (d):

$$a_C = \ddot{x}_C = -8.74\text{ ft/sec}^2$$

Integrating gives

$$\dot{x}_C = -8.74t + K_1\text{ ft/sec}$$

And since our initial condition on velocity is $\dot{x}_C = +15$ ft/sec at $t = 0$,

$$\dot{x}_C = -8.74t + 15\text{ ft/sec}$$

Integrating again:

$$x_C = -4.37t^2 + 15t + K_2\text{ ft}$$

The initial condition on the position of the truck is that $x_C = 0$ at $t = 0$; thus $K_2 = 0$.

(Continued)

*And then we would have to check only the cases we had not yet compared!

Finally, the truck stops when its velocity $\dot{x}_C$ goes to zero:

$$-8.74t_{\text{stop}} + 15 = 0$$

$$t_{\text{stop}} = 1.72 \text{ sec}$$

Thus the minimum stopping distance to prevent all four calamities is

$$\begin{aligned} x_{\text{stop}} &= -4.37(1.72^2) + 15(1.72) \\ &= 12.9 \text{ ft} \end{aligned}$$

EXAMPLE 4.23

There is an angle γ at which the rod $\mathcal{R}$ of length ℓ can translate as the block $\mathcal{B}$, to which it is pinned at P, slides down the rough plane. (See Figure 1.) Find γ.

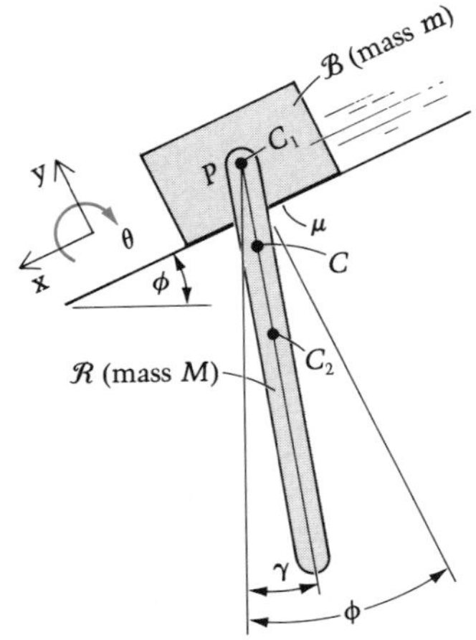

Figure 1

SOLUTION

Since the bodies are to translate together, all points of each have identical accelerations and $\ddot{x}_{C_1} = \ddot{x}_{C_2} = \ddot{x}_C$. Using the free-body diagram of the combined body (Figure 2):

$$F_{r_x} = (m + M)g \sin\phi - f = (m + M)\ddot{x}_C \tag{1}$$

And since the bar does not swing relative to the block, we have $\ddot{y}_{C_2} = 0$ so that $\ddot{y}_C = 0$. Thus

$$F_{r_y} = N - (m + M)g\cos\phi = (m + M)\ddot{y}_C = 0$$

which results in

$$N = (m + M)g\cos\phi \tag{2}$$

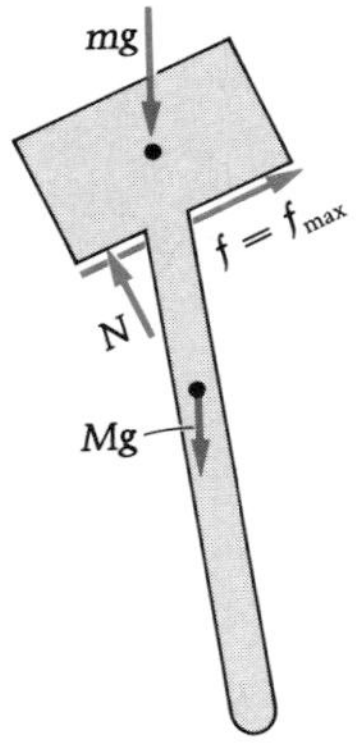

Figure 2

(Continued)

Since the block is sliding, we have $f = f_{\max} = \mu N$; substituting (2) into this relation and the results into (1) yields the acceleration down the plane:

$$\ddot{x}_C = g(\sin\phi - \mu\cos\phi) \tag{3}$$

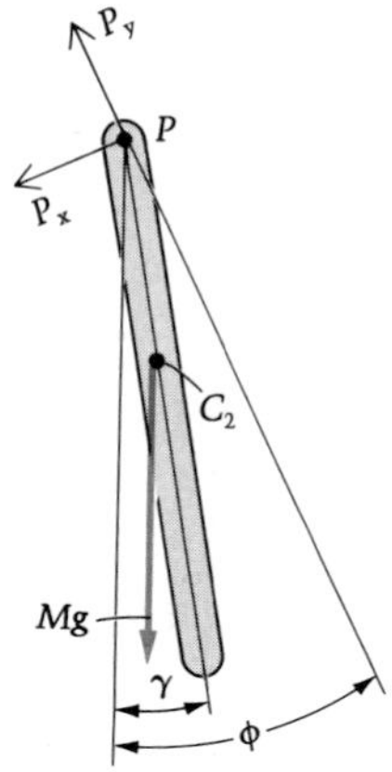

Figure 3

We next use the free-body diagram of $\mathcal{R}$ alone (Figure 3) and sum moments about $P = C_1$ using Equation (4.48):

$$M_{r_P} = \overset{0}{\cancel{I_C}}\alpha + (\mathbf{r}_{PC_2} \times M\mathbf{a}_{C_2})_z \tag{4}$$

The vectors needed for the cross product in (4) are

$$\mathbf{r}_{PC_2} = \frac{\ell}{2}[\sin(\phi - \gamma)\hat{\mathbf{i}} - \cos(\phi - \gamma)\hat{\mathbf{j}}] \tag{5}$$

and

$$\mathbf{a}_{C_2} = \ddot{x}_C\hat{\mathbf{i}} = g(\sin\phi - \mu\cos\phi)\hat{\mathbf{i}} \tag{6}$$

Noting from Figure 3 that the moment M_{r_P} is $Mg(\ell/2)\sin\gamma$ and substituting this and (5) and (6) into (4) gives

$$Mg\frac{\ell}{2}\sin\gamma = Mg(\sin\phi - \mu\cos\phi)\frac{\ell}{2}\cos(\phi - \gamma)$$

or

$$\sin\gamma = (\sin\phi - \mu\cos\phi)(\cos\phi\cos\gamma + \sin\phi\sin\gamma)$$

from which we find γ:

$$\gamma = \tan^{-1}\left(\frac{\sin\phi - \mu\cos\phi}{\cos\phi + \mu\sin\phi}\right)$$

We see that the angle γ between the rod and the vertical depends on the angle of the plane ϕ and the coefficient of friction. If the plane is smooth, $\mu \approx 0$ and $\tan\gamma = \tan\phi$ so that $\gamma = \phi$ and the bar assumes a position normal to the plane. A physical interpretation for this may be seen from the free-body diagram of the block alone (Figure 4).

The normal force N, in the case $\mu = 0$, must pass through C_1 in order that $M_{r_{C_1}} = 0$. Then, examining the lower free-body diagram of *both* bodies translating together as one, we see that M_{r_C} (C is the combined mass center) can equal zero (as it must since the bodies are translating) only if the bar is aligned normal to the inclined plane. Otherwise the normal force beneath the block would have an unbalanced moment about C.

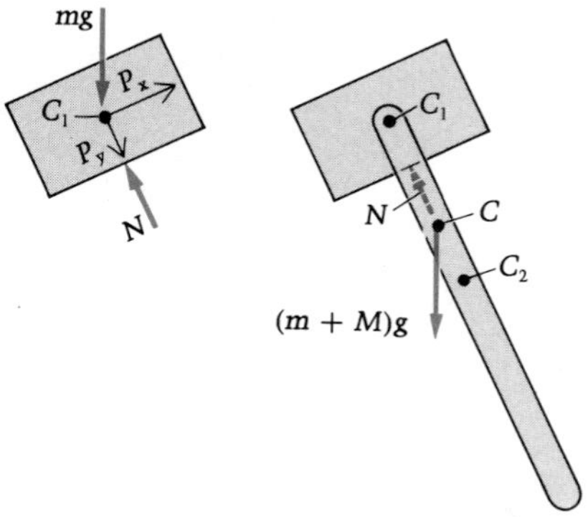

Figure 4

EXAMPLE 4.24

Ring $\mathcal{R}$ in Figure 1 has a mass of 10 kg and radius 0.5 m and is pinned to the 30-kg frame $\mathcal{F}$ at P. Force Q pushes the bodies up the smooth plane. Find the largest force Q for which the ring will not turn with respect to $\mathcal{F}$.

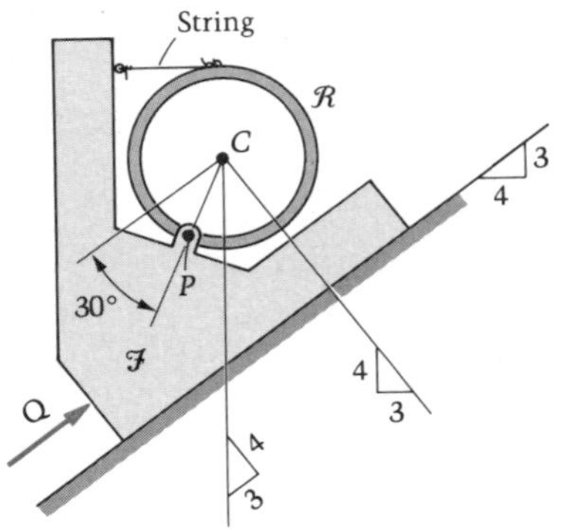

Figure 1

SOLUTION

Ring $\mathcal{R}$ is translating parallel to the plane, so that $\mathbf{a}_C = \ddot{x}_C\hat{\mathbf{i}}$. If we use Equation (4.48) to sum moments about P on the free-body diagram of $\mathcal{R}$ (Figure 2), we eliminate the two pin reactions at P. Furthermore, at the largest Q the string tension has gone to zero. Thus the only contributor to $\mathbf{M}_{r_P}$ is the gravity force at C. We have, then,

$$M_{r_{P_z}} = I_{z_C}\overset{0 \text{ (since } \mathcal{R} \text{ is translating)}}{\cancel{\alpha}} + (\mathbf{r}_{PC} \times m\mathbf{a}_C)_z$$

$$mg(R \sin 23.1°) = [R(\cos 30°\ \hat{\mathbf{i}} - \sin 30°\ \hat{\mathbf{j}}) \times m\ddot{x}_C\hat{\mathbf{i}}]_z$$

$$= mR\ddot{x}_C \sin 30°$$

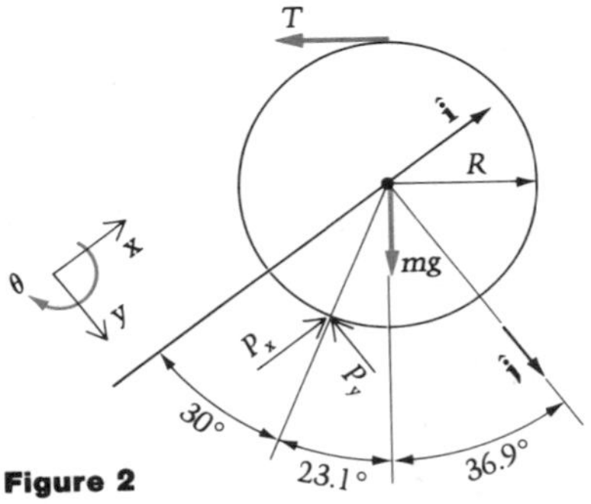

Figure 2

Thus

$$\ddot{x}_C = 2g \sin 23.1° = 7.70 \text{ m/s}^2 \tag{1}$$

Next we consider a free-body diagram of the combined body of $\mathcal{F}$ and $\mathcal{R}$ (Figure 3). Note that there is no friction force beneath the frame, because the plane is smooth. Note also that the accelerations of C, G, and the combined mass center* C_T are all equal to $\ddot{x}_C$, since the bodies translate together so long as Q does not exceed the Q_{max} we are seeking.

We shall use the equation

$$F_{r_x} = m_T\ddot{x}_{C_T} \tag{2}$$

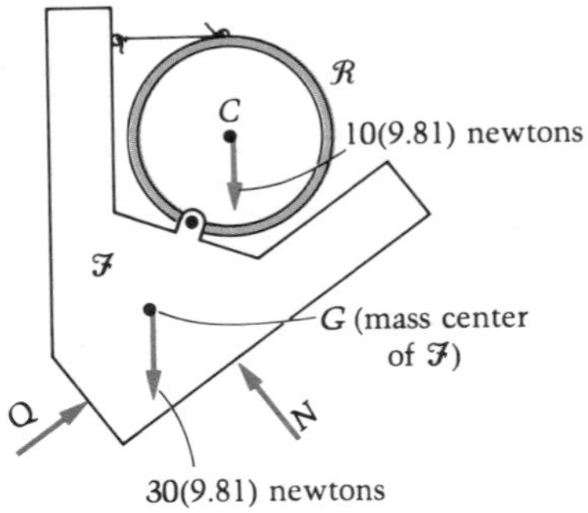

Figure 3

in which m_T is the total mass. This equation is the x-component equation of $\mathbf{F}_r = \dot{\mathbf{L}}$, and it is valid for *any* system of bodies or particles (or both); it would

(Continued)

*Located one-fourth of the way from G to C!

even apply if $\mathcal{R}$ and $\mathcal{J}$ were not translating as one body. By the definition of the mass center, we can avoid finding and using C_T by rewriting the right side of Equation (2) in terms of the masses and mass center accelerations of the two component bodies:

$$F_{r_x} = m_{\mathcal{R}}\ddot{x}_C + m_{\mathcal{J}}\ddot{x}_G = (m_{\mathcal{R}} + m_{\mathcal{J}})\ddot{x}_C$$

Thus

$$Q_{\max} - 10(9.81)\tfrac{3}{5} - 30(9.81)\tfrac{3}{5} = (10 + 30)\,\text{kg}\,(7.71)\,\text{m/s}^2$$

and

$$Q_{\max} = 235 + 308$$

$$= 543 \text{ newtons}$$

Note that by using Equation (4.48) and then the combined body, we avoid having to find the pin reactions at P.

PROBLEMS / Section 4.5 (continued)

4.113 The 30-kg sphere and 15-kg rod in Figure P4.113 are welded together to form a single rigid body. Determine the angular acceleration of the body immediately after the right-hand string is cut.

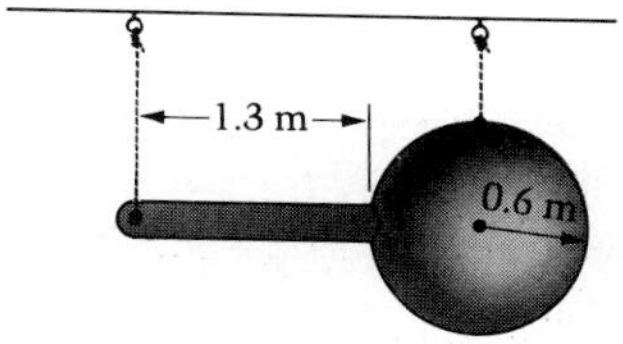

Figure P4.113

4.114 If the right-hand string in Figure P4.114 is cut, find the initial tension in the left string. The slender rod has mass m and length L. *Hint:* The acceleration of A has no vertical component ($v_A^2/\ell_{\text{string}} = 0$); thus, at most, $\mathbf{a}_A$ has only a component along the direction from A toward C. Actually, in this problem you will find that $\mathbf{a}_A = \mathbf{0}$.

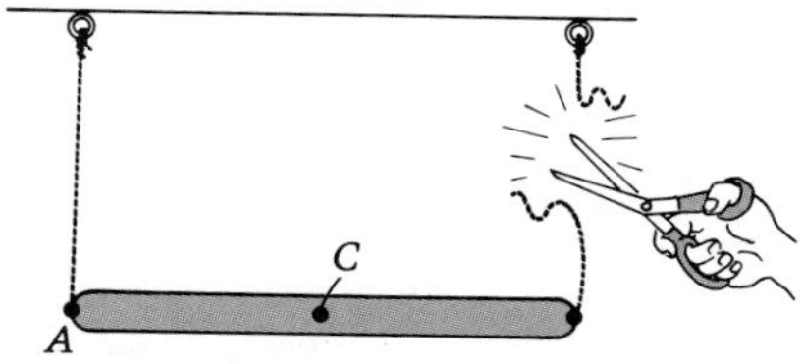

Figure P4.114

4.115 Repeat the preceding problem if the rod is replaced by a rectangular plate suspended from the two upper corners. The width (between strings) is B and the height is H.

4.116 Rods $\mathcal{B}_1$ and $\mathcal{B}_2$ have respective masses m_1 and m_2. (See Figure P4.116.) Upon release from rest in the horizontal position indicated, find the reactions at O, and at A, onto $\mathcal{B}_1$.

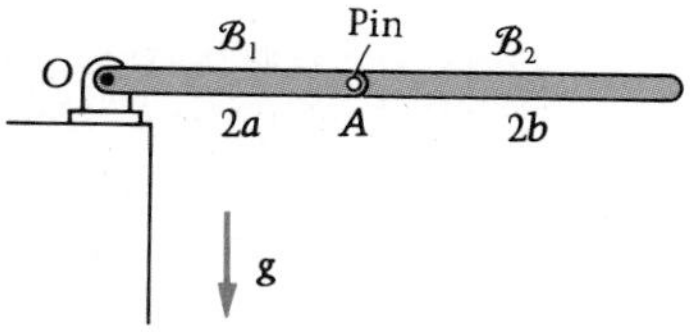

Figure P4.116

4.117–4.122 The six equilateral triangular plates (Figures P4.117–4.122) are each supported on their rightmost corner B by a string; each has a different support condition at the left corner A. At the instant when the string at B is cut, find $\mathbf{a}_C$ and $\boldsymbol{\alpha}$ in each case. The length of each side is s.

4.123 The rectangular door of a railroad car has mass m (Figure P4.123); it is of uniform width 2ℓ and has its hinges on the side of the doorway closest to the engine. Initially the door makes an angle β with the train, which begins to move forward from rest at constant acceleration a_0. Find the initial resultant horizontal reaction component that the hinges exert on the door. *Hint:* Use Equation (4.48).

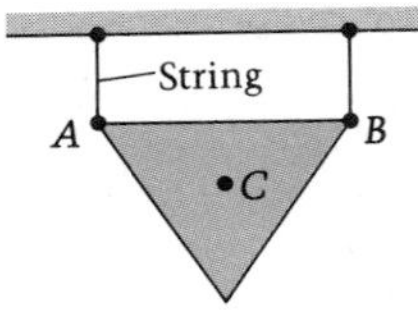

Figure P4.117

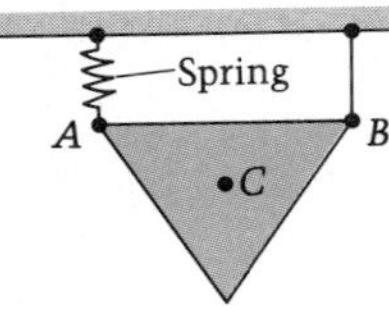

Figure P4.118

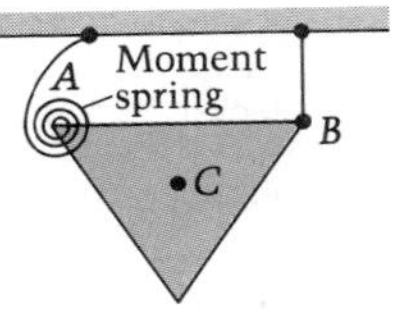

Figure P4.119

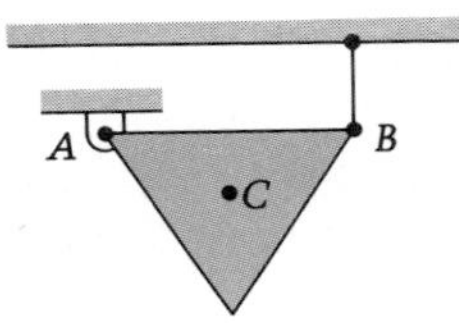

Figure P4.120

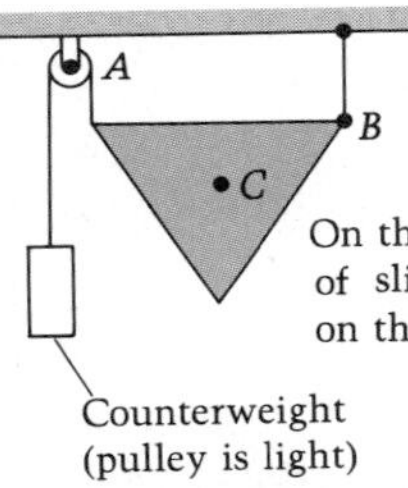

Figure P4.121

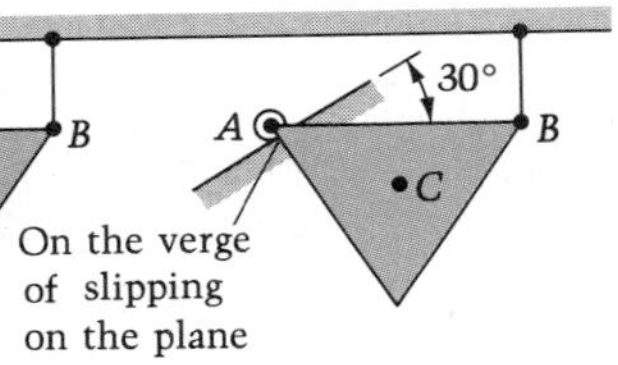

Figure P4.122

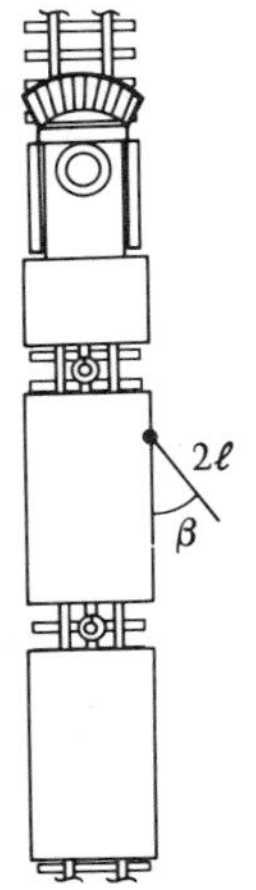

Figure P4.123

A Note About the Rotational Equation. What remains in this section is not essential to an understanding of plane motion. It is concerned with the remaining special cases of the rotational equation of motion and is more important in preventing ***misuse*** of the equation than in offering a positive alternative to $M_{r_C} = I_C\alpha$. Though not essential, it is often quite useful.

We have seen that the equation $M_{r_C} = I_C\alpha$ is always valid for a rigid body $\mathcal{B}$ in plane motion and that the same equation applies for a pivot O ($M_{r_O} = I_O\alpha$) *if* $\mathcal{B}$ has one. A natural question to ask is: Are there any ***other*** special points P of body $\mathcal{B}$ for which $M_{r_P} = I_P\alpha$? In fact there are two others; with few exceptions, however, they are difficult to use. Nevertheless we shall briefly derive the result for the reader to prevent the use of *invalid* points P in the equation. We start with Equation (4.46), which is valid for any point P:*

$$\mathbf{M}_{r_P} = M_{r_P}\hat{\mathbf{k}} = I_{z_C}\alpha\hat{\mathbf{k}} + \mathbf{r}_{PC} \times m\mathbf{a}_C \tag{4.50}$$

Next we specialize P to be any point of $\mathcal{B}$ in the reference plane containing C. In this case, Equation (3.19) may be used to rewrite $\mathbf{a}_C$ in terms of $\mathbf{a}_P$:

$$\mathbf{M}_{r_P} = I_{z_C}\alpha\hat{\mathbf{k}} + \mathbf{r}_{PC} \times m(\mathbf{a}_P + \alpha\hat{\mathbf{k}} \times \mathbf{r}_{PC} - \omega^2\mathbf{r}_{PC}) \tag{4.51}$$

Noting that $\mathbf{r}_{PC} \times (-m\omega^2)\mathbf{r}_{PC} = \mathbf{0}$ and that $\mathbf{r}_{PC} \times (m\alpha\hat{\mathbf{k}} \times \mathbf{r}_{PC}) = m\alpha r_{PC}^2\hat{\mathbf{k}}$, we obtain

$$\mathbf{M}_{r_P} = (I_{z_C} + mr_{PC}^2)\alpha\hat{\mathbf{k}} + (\mathbf{r}_{PC} \times m\mathbf{a}_P)$$

Question 4.8 Why does the cross-product term in this equation have only a z component?

Using the parallel-axis theorem,

$$\mathbf{M}_{r_P} = I_{z_P}\alpha\hat{\mathbf{k}} + \mathbf{r}_{PC} \times m\mathbf{a}_P \tag{4.52}$$

We now see that we may legitimately use $M_{r_P} = I_P\alpha$ under, and *only* under, the three conditions for which the cross product $\mathbf{r}_{PC} \times m\mathbf{a}_P$ vanishes:

1. $\mathbf{r}_{PC} = \mathbf{0}$, which means that P is the mass center C, a case we have already discovered.
2. $\mathbf{a}_P = \mathbf{0}$, meaning that P has zero acceleration. This case includes the pivot case (which is valid for all t) but is in fact more general. It allows us to use points of zero acceleration as instantaneous moment centers for problems taking place at one specific time. Note that this does *not* say zero *velocity*; just because a point may have zero velocity at some instant does *not* mean that its acceleration vanishes then (and vice versa).
3. $\mathbf{r}_{PC}$ is parallel to $\mathbf{a}_P$, meaning that the acceleration vector of P passes through C. (That is, it has the same direction as the line through P and C.)

We do not ordinarily use conditions 2 or 3, but there are two exceptions. The first, as we have already discovered, is if the point of condition 2 is a pivot. Otherwise it is usually more troublesome to *find* the

*We are still assuming $I_{xz_C} = I_{yz_C} = 0$, although the points P to be discovered for which $M_{r_P} = I_P\alpha$ are valid for nonzero products of inertia if the equation reads $\mathbf{M}_{r_P} \cdot \hat{\mathbf{k}} = M_{r_{P_z}} = I_{z_P}\alpha$.

point of zero acceleration than simply to go ahead and work the problem with C as the moment center.*

The other exception, relating to condition 3, is very tricky and therefore dangerous. First we remark that it is just as much a waste of time to search for a point whose acceleration passes through C as it is to seek a point of zero acceleration. There is, however, one point whose acceleration passes through C that we mention because students who fail to understand the rules for its use are bound to make serious errors. The point is the contact point of a round body whose mass center coincides with its geometric center, which rolls on a surface fixed in the inertial reference frame. The acceleration of such a point, call it Q, is normal to the rolling surface's tangent and therefore passes directly through C *at all times*. For these conditions, $M_{r_Q} = I_Q\alpha$.†

We call this application "dangerous" because students may erroneously be led to think that they can sum moments about the instantaneous center Ⓘ of zero velocity and equate the result to $I_{Ⓘ}\,\alpha$. This procedure is absolutely invalid, but nonetheless it has occasionally found its way into the literature of dynamics.

We summarize this discussion with some warning sketches. In Figures 4.9 to 4.11, let C be the mass center of the round body $\mathcal{B}$; G is the geometric center of the circular periphery of $\mathcal{B}$. In Figure 4.9, $M_{r_Q} = I_Q\alpha$ since $\mathbf{a}_Q$ passes through C. In Figure 4.10, $M_{r_Q} \neq I_Q\alpha$ because $\mathbf{a}_Q$, still vertical, does *not* pass through C (except at the two isolated instants when C is above or below G). In Figure 4.11, $M_{r_Q} \neq I_Q\alpha$ because slipping results in the contact point having a horizontal acceleration component that, together with $r\omega^2 \uparrow$, prevents $\mathbf{a}_Q$ from passing through C.

In the following example we first solve the problem with Equations (4.41). We then illustrate that the same result is obtained by using a point accelerating through the mass center C, a point of zero acceleration, Equation (4.50), and Equation (4.52).

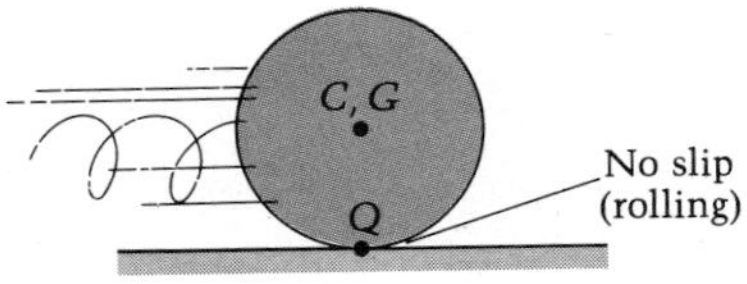

Figure 4.9 $M_{r_Q} = I_Q\alpha$

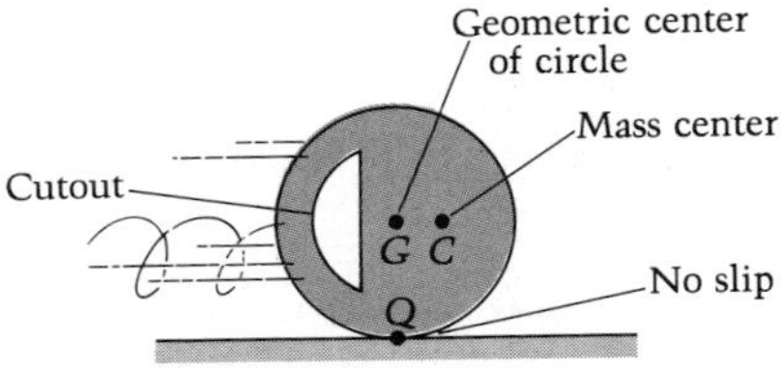

Figure 4.10 $M_{r_Q} \neq I_Q\alpha$

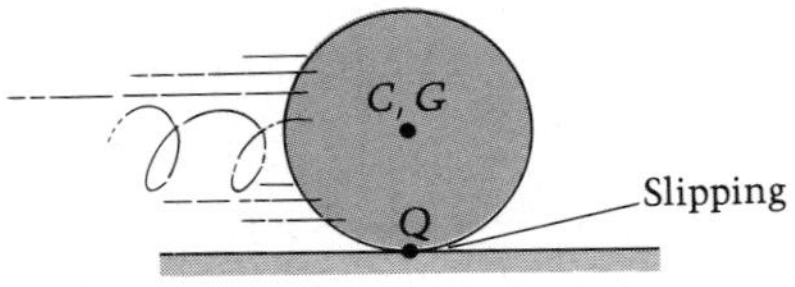

Figure 4.11 $M_{r_Q} \neq I_Q\alpha$

*When $\omega = 0$, as at a release from rest in a rolling problem, we can often locate a point of zero acceleration by inspection and make use of this case. If a wheel is rolling on a fixed surface, then when $\omega = 0$ the contact point has zero acceleration.

†If the mass center and geometric center do not coincide, the equation is valid at the isolated instants when those two points are on-line with the contact point. This assumes that the body is still round, although asymmetric with regard to mass distribution.

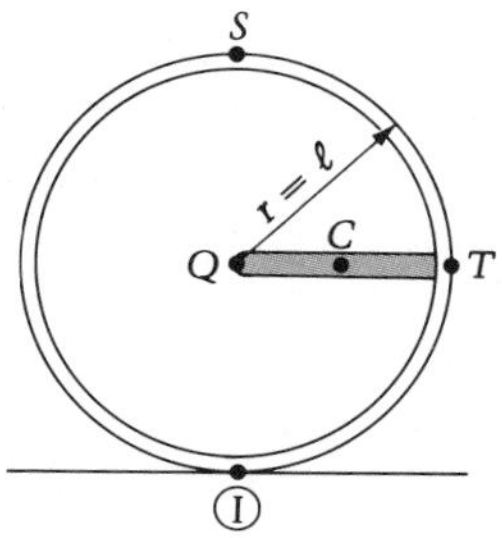

EXAMPLE 4.25

The rigid body $\mathcal{B}$ in the diagram consists of a heavy bar of mass m welded to a light hoop; the radius of the hoop thus equals the length of the bar. Find the minimum coefficient of friction between the hoop and the ground for which the body will roll when released from rest in the given position.

SOLUTION

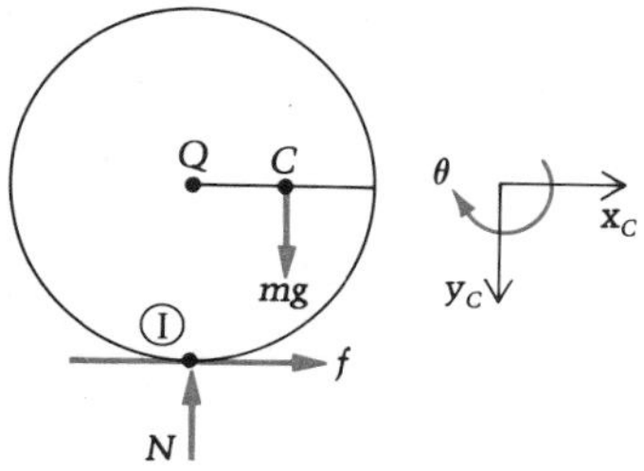

The free-body diagram is shown here along with the sign convention adopted for the problem.

> **Question 4.9** Why is this a good sign convention to use in this problem?

Note that the gravity force resultant acts through the center of the bar since we are neglecting the weight of the hoop.

Next we write the three differential equations of motion:

$$F_{r_x} = m\ddot{x}_C \Rightarrow f = m\ddot{x}_C \tag{1}$$

$$F_{r_y} = m\ddot{y}_C \Rightarrow mg - N = m\ddot{y}_C \tag{2}$$

$$M_{r_C} = I_{z_C}\alpha \Rightarrow \frac{N\ell}{2} - f\ell = \frac{m\ell^2}{12}\alpha \tag{3}$$

These equations contain the unknowns f, N, $\ddot{x}_C$, $\ddot{y}_C$, and α. We draw upon kinematics for two more equations. We know that the acceleration of the geometric center Q of the hoop is $\mathbf{a}_Q = r\alpha\hat{\mathbf{i}}$, so that

$$\mathbf{a}_C = \underbrace{\mathbf{a}_Q}_{a_Q\hat{\mathbf{i}} = r\alpha\hat{\mathbf{i}}} + \alpha\hat{\mathbf{k}} \times \mathbf{r}_{QC} - \overbrace{\omega^2}^{0}\,\mathbf{r}_{QC}$$

or

$$\ddot{x}_C\hat{\mathbf{i}} + \ddot{y}_C\hat{\mathbf{j}} = r\alpha\hat{\mathbf{i}} + \alpha\hat{\mathbf{k}} \times \left(\frac{\ell}{2}\,\hat{\mathbf{i}}\right)$$

Therefore, equating the $\hat{\mathbf{i}}$ coefficients and then the $\hat{\mathbf{j}}$ coefficients, we find

$$\ddot{x}_C = \ell\alpha \tag{4}$$

$$\ddot{y}_C = \frac{\ell\alpha}{2} \tag{5}$$

(Continued)

The student may wish to verify that (4) and (5) also result from relating $\mathbf{a}_C$ to $\mathbf{a}_{Ⓘ} = r\omega^2\uparrow$, true for any round body rolling on a flat, fixed plane. In this problem, $\mathbf{a}_{Ⓘ}$ is then zero at release because, until time passes, ω is still zero.

Solving Equations (1) to (5) for the five unknowns gives the results:

$$f = \frac{3}{8}mg \qquad N = \frac{13}{16}mg \qquad \ddot{x}_C = \frac{3g}{8} \qquad \ddot{y}_C = \frac{3g}{16} \qquad \alpha = \frac{3g}{8\ell}$$

To complete the solution, we must get the coefficient of friction μ into the picture. We know that for any friction force f,

$$0 \le f \le f_{\text{max}} = \mu N$$

Therefore, in our problem,

$$\frac{3}{8}mg \le \mu\frac{13}{16}mg$$

so that

$$\mu \ge \frac{6}{13}$$

This means that for the body to roll, a friction coefficient of at least $\mu = 6/13$ is required; this is then the desired minimum.

We emphasize that students should always make "eyeball checks" of their answers—glancing over the results to be sure they make sense physically. In this problem, for instance, note that:

1. $N < mg$ as expected, for otherwise the mass center could not begin to move downward as the body rolls.
2. $\ddot{x}_C$, $\ddot{y}_C$, and α are all positive and therefore in the expected directions.

We now mention several alternatives available to us in writing the rotational equation of motion for this problem. First we note that Q accelerates through C at the instant of interest, so that we may legitimately use

$$M_{r_Q} = I_Q\alpha$$

And indeed we can easily verify that the moment M_{r_Q} on the left has the same value as the $I_Q\alpha$ term on the right:

$$M_{r_Q} = mg\frac{\ell}{2} - \overset{\mu N \text{ here}}{\cancel{f}}\ell = mg\ell\left(\frac{1}{2} - \frac{6}{13}\cdot\frac{13}{16}\right) = \frac{mg\ell}{8}$$

which equals

$$I_Q\alpha = \frac{m\ell^2}{3}\cdot\frac{3g}{8\ell} = \frac{mg\ell}{8}$$

Point Ⓘ is a valid moment center also, but only because it is a point H with $\mathbf{a}_H = \mathbf{0}$ and *not* because it has zero velocity. As soon as C moves slightly, the new point Ⓘ would then give incorrect results since then $\mathbf{a}_{Ⓘ} = r\omega^2\uparrow \neq \mathbf{0}$

(Continued)

even though $\mathbf{v}_{①}$ still equals zero. Thus

$$M_{r_H} = \frac{mg\ell}{2}$$

which is equal to

$$I_H\alpha = \Bigg\{\overbrace{\frac{m\ell^2}{12}}^{I_C} + \overbrace{m\left[\left(\frac{\ell}{2}\right)^2 + \ell^2\right]}^{md^2}\Bigg\}\left(\frac{3g}{8\ell}\right) = \frac{mg\ell}{2}$$

Finally, we note that point T (refer to the figure), for example, is an *invalid* moment center at the given instant, since it is not the mass center, nor does it have zero acceleration, nor does it accelerate through C. If we compute the resultant moment of the external forces about T,

$$M_{r_T} = N\ell - f\ell - \frac{mg\ell}{2} = mg\ell\left(\frac{13}{16} - \frac{3}{8} - \frac{1}{2}\right)$$

$$= \frac{-mg\ell}{16}$$

we see that this is *not* equal to

$$I_T\alpha = \frac{m\ell^2}{3} \cdot \frac{3g}{8\ell} = \frac{mg\ell}{8}$$

For a point such as T, there are only two legitimate ways to treat it as a moment center. One way is to use Equation (4.50):

$$M_{r_T} = I_C\alpha + (\mathbf{r}_{TC} \times m\mathbf{a}_C)_z$$

$$= \left(\frac{m\ell^2}{12}\right)\left(\frac{3g}{8\ell}\right) + \left[-\frac{\ell}{2}\hat{\mathbf{i}} \times m\left(\ell\alpha\hat{\mathbf{i}} + \frac{\ell\alpha}{2}\hat{\mathbf{j}}\right)\right]_z$$

$$= \frac{-mg\ell}{16}$$

which equals the moment vector $\mathbf{M}_{r_T}$ calculated above. Or, alternatively, Equation (4.52) gives

$$M_{r_T} = I_T\alpha + (\mathbf{r}_{TC} \times m\mathbf{a}_T)_z$$

$$= \frac{m\ell^2}{3}\left(\frac{3g}{8\ell}\right) + \Big[-\frac{\ell}{2}\hat{\mathbf{i}} \times m(\underbrace{\mathbf{a}_Q}_{\ell\alpha\hat{\mathbf{i}}} + \alpha\hat{\mathbf{k}} \times \underbrace{\mathbf{r}_{QT}}_{\ell\hat{\mathbf{i}}} - \underbrace{\omega^2\mathbf{r}_{QT}}_{0\text{ at release}})\Big]_z$$

$$= \frac{mg\ell}{8} - \frac{m\ell^2}{2}\left(\frac{3g}{8\ell}\right) = \frac{-mg\ell}{16}$$

We suggest that the student show by calculation that $M_{r_S} \neq I_S\alpha$. (S is the highest point of the body as shown in the figure.) Then show that the correct result is forthcoming from both Equations (4.50) and (4.52), as was done above for point T.

PROBLEMS / Section 4.5 (continued)

4.124 A homogeneous spool of weight W rolls on an inclined plane; a string tension of amount $4W$ acts up the plane as shown. With I_C given approximately by $WR^2/2g$, find the acceleration of C. Assume unlimited friction.

4.125 Force T is given to be small enough, and the friction coefficient large enough, that both wheels in Figure P4.125 will roll on the plane.

a. Give arguments why one wheel rolls left and the other right.

b. Find the ratio of r to R for which the accelerations of C are equal in magnitude.

4.126 The uniform sphere in Figure P4.126, of mass m and radius r, is subjected to a couple of moment M_0. If there is no slip, find the acceleration of the center of the sphere.

4.127 Assuming that sufficient friction is present to prevent slipping between $\mathcal{C}$ and the plane, find the angular accelerations of $\mathcal{C}$ and $\mathcal{R}$ just after force P is applied to the bodies at rest. They are connected by a smooth pin.

4.128 The cylinder in Figure P4.128 is observed to be rolling on the planes in such a way that the particle $\mathcal{A}$, to which it is attached by a cord, makes a constant angle γ with the vertical. Find the angle γ and the tension in the cord.

4.129 The 128.8-lb homogeneous plank shown in Figure P4.129 is placed on two homogeneous cylindrical rollers, each of weight 32.2 lb. The system is released from rest. Determine the initial acceleration of the plank if no slipping occurs. Is this the acceleration for later times as well?

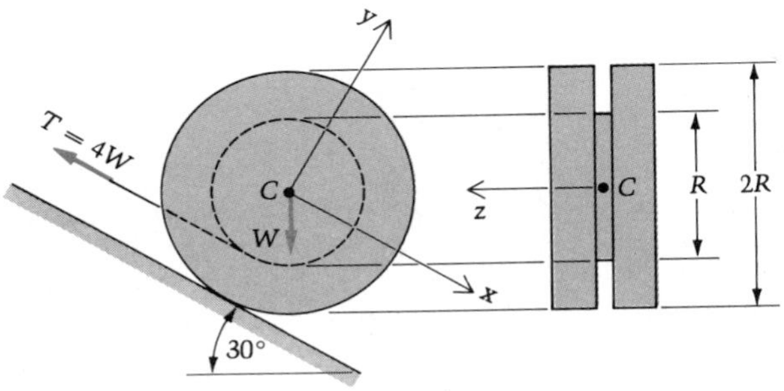

Figure P4.124

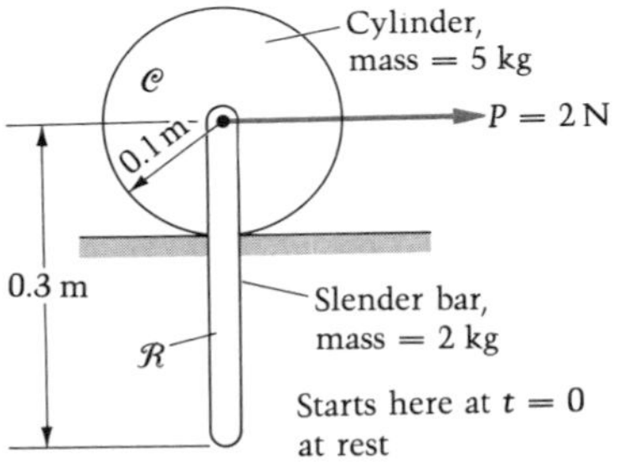

Figure P4.127

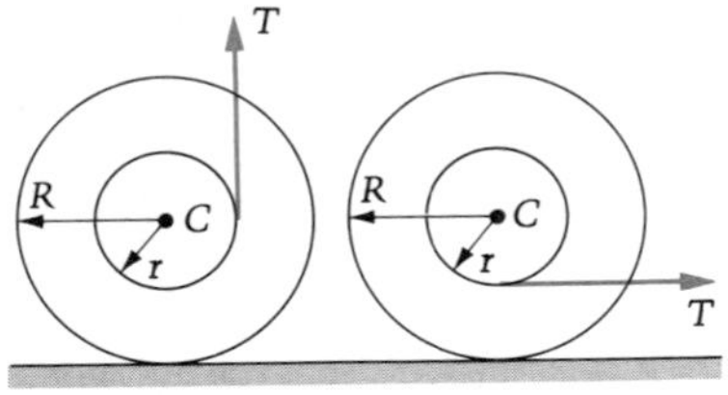

Figure P4.125

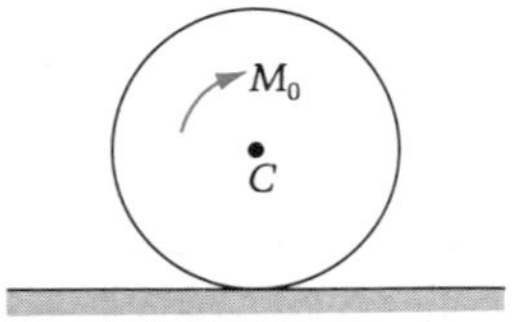

Figure P4.126

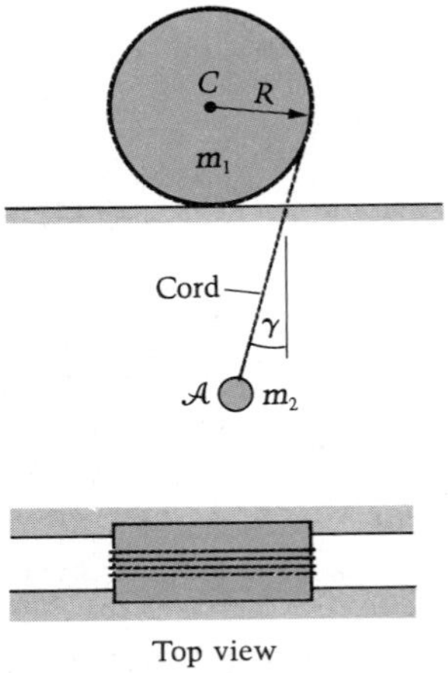

Figure P4.128

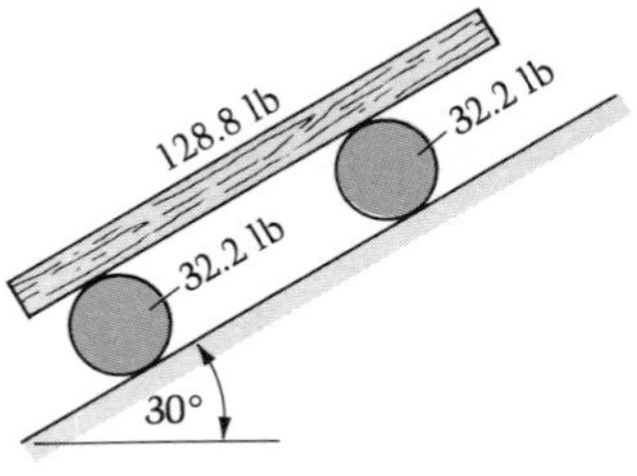

Figure P4.129

Rotation of Unbalanced Bodies (Nonzero I_{xz} or I_{yz})

We have seen earlier in this section that

$$M_{r_{xp}} = I_{xz_P}\alpha - I_{yz_P}\omega^2 \tag{4.53a}$$

$$M_{r_{yp}} = I_{yz_P}\alpha + I_{xz_P}\omega^2 \tag{4.53b}$$

in which any single point P in the reference plane containing C may be chosen as the reference point for the four quantities M_{r_x}, M_{r_y}, I_{xz}, and I_{yz}.

Let us now focus our attention on the case in which the body $\mathcal{B}$ is constrained to rotate about the z-axis, with point P specialized to lie on the axis as shown in Figure 4.12.

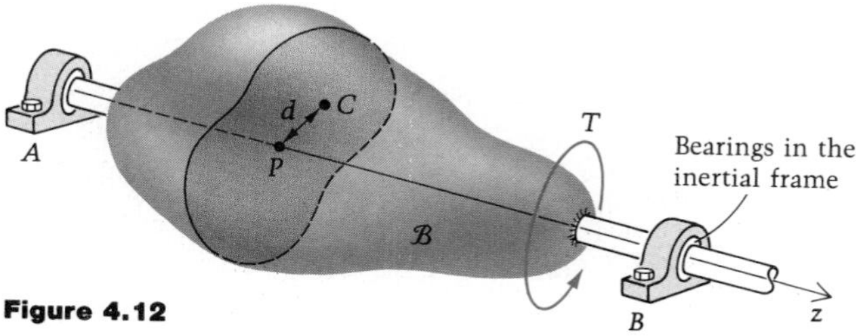

Figure 4.12

Since P is now a pivot of $\mathcal{B}$, Equation (4.44c) applies:

$$M_{r_{P_z}} = T = I_{z_P}\alpha \tag{4.54}$$

We see again that regardless of the vanishing or nonvanishing of the two products of inertia in Equations (4.53), the angular acceleration of the body is still given by the net torque about the axis of rotation divided by the moment of inertia about that axis.

If now the mass center is off-axis (that is, C not at P), then using the free-body diagram in Figure 4.13, the "force-equations" of motion yield

$$F_{r_x} = A_x + B_x = m(-d\,\alpha \sin\theta - d\omega^2 \cos\theta) \tag{4.55a}$$

$$F_{r_y} = A_y + B_y = m(d\,\alpha \cos\theta - d\omega^2 \sin\theta) \tag{4.55b}$$

$$F_{r_z} = A_z = m(0) \tag{4.55c}$$

For simplicity, gravity (weight) has been ignored here. To account for its influence on bearing reactions, we need only add the reactions considered in this section to the reactions that would be necessary to support the weight of the body if it were in equilibrium.

We will make one additional simplification in this section—to place the mass center C on the axis of rotation (i.e., C at P and $d = 0$). This will result in what is called ***static balance***. This terminology is derived from the fact that if the body, free to turn about the bearing axis, is released from rest at any θ, it will remain at rest. Note now that with $d = 0$, Equations (4.55) show that the external forces acting on $\mathcal{B}$ are in

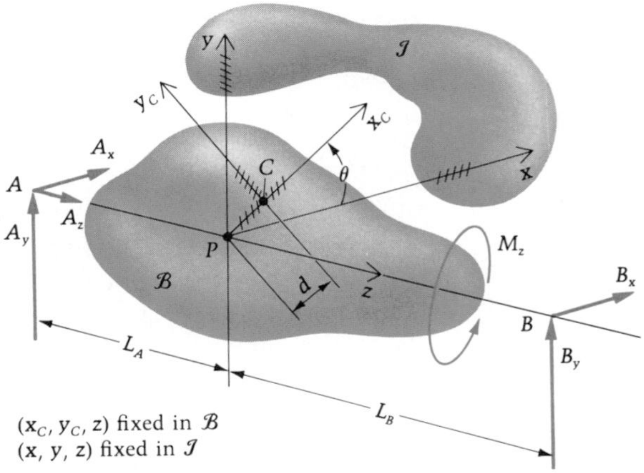

Figure 4.13

balance; this is evidenced by the fact that with C at P, the mass center is stationary in the inertial frame. Also, if the body turns at constant angular speed, then $\alpha = 0$ and Equation (4.54) shows that the resultant moment about the rotation axis (applied torque less friction moments) is zero. Therefore, what remains is the study of the forces which produce the moments in Equations (4.53).

Returning to Equations (4.53), with C at P on the z-axis, we see that even though $\mathbf{a}_C = \mathbf{0}$, we still need moments about the lateral axes $\mathbf{x}_C$ and $\mathbf{y}_C$ to sustain the motion unless I_{xz_C} and I_{yz_C} are both zero. These two equations (4.53) will reveal these moments to us.

Since Equations (4.55a,b) tell us that $A_x = -B_x$ and $A_y = -B_y$, the "bearing reactions" are equal in magnitude and opposites in direction, and thus they constitute a couple. It is convenient to refer components of these "dynamic bearing reactions" to the body-fixed (in $\mathcal{B}$) axes ($\mathbf{x}_C$, $\mathbf{y}_C$) rather than to ($\mathbf{x}$, $\mathbf{y}$) fixed in $\mathcal{I}$; z, of course, is fixed in both $\mathcal{B}$ *and* $\mathcal{I}$. We now illustrate how to find the bearing reactions in several examples.

EXAMPLE 4.26

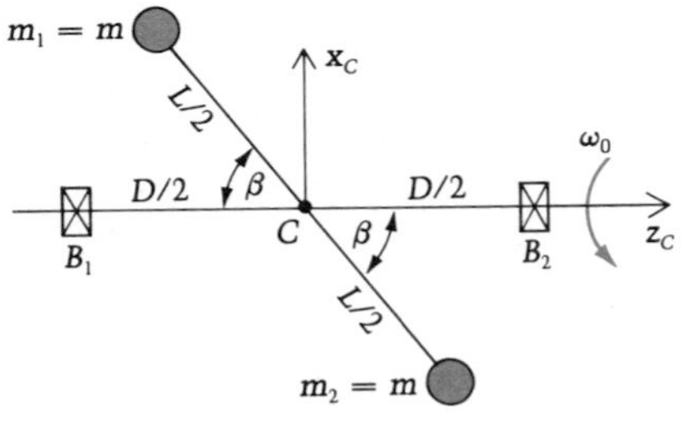

Determine the dynamic reactions of the bearings B_1 and B_2 on the horizontal shaft, which is welded to the slanted shaft of length L. (See the diagram.) Both shafts are light compared to the heavy concentrated masses m, and the shaft turns at constant angular speed ω_0.

(Continued)

SOLUTION

Computing the products of inertia, we get

$$I_{yz_C} = 0 \qquad (\mathbf{x}_C\mathbf{z}_C \text{ is a plane of symmetry})$$

$$I_{xz_C} = -\int xz\,dm = -\sum_{i=1}^{2} x_i z_i m_i \qquad \text{(here)}$$

$$= -m\left(\frac{L}{2}\sin\beta\right)\left(-\frac{L}{2}\cos\beta\right) - m\left(-\frac{L}{2}\sin\beta\right)\left(\frac{L}{2}\cos\beta\right)$$

$$= \frac{mL^2}{2}\sin\beta\cos\beta$$

Therefore

$$M_{r_{x_C}} = I_{xz_C}\alpha - I_{yz_C}\omega^2 = 0$$

$$M_{r_{y_C}} = I_{yz_C}\alpha + I_{xz_C}\omega^2 = \frac{mL^2\omega_0^2\sin\beta\cos\beta}{2}$$

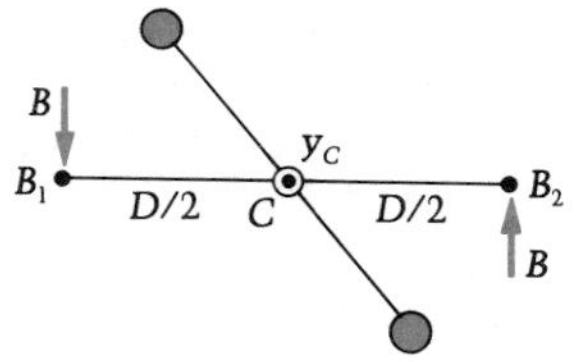

This moment $M_{r_{y_C}}$ is provided by the equal and opposite bearing reactions B shown in the accompanying diagram. Thus

$$BD = \frac{mL^2\omega_0^2\sin\beta\cos\beta}{2}$$

and the bearing forces are each

$$B = \frac{mL^2\omega_0^2\sin\beta\cos\beta}{2D}$$

We note that if shaft B_1B_2 also has an angular acceleration α, then our equations give a couple of additional bearing reactions in the y_Cz_C plane, for then

$$M_{r_{x_C}} = I_{xz_C}\alpha = \frac{mL^2\alpha\sin\beta\cos\beta}{2}$$

Each pair of bearing reactions, producing $M_{r_{x_C}}$ and $M_{r_{y_C}}$, *rotates* with the shaft. The reactions onto the bearings are of course equal and opposite to the forces B at B_1 and B_2. This example is simple enough to be solved by considering *centripetal acceleration* and *centrifugal force*. Thus we shall take this opportunity to explain these two terms as we rework the problem.

We again let $\omega = \omega_0 =$ constant and note that the masses m_1 and m_2 each move on a vertical circle of radius R at constant speed. Consider

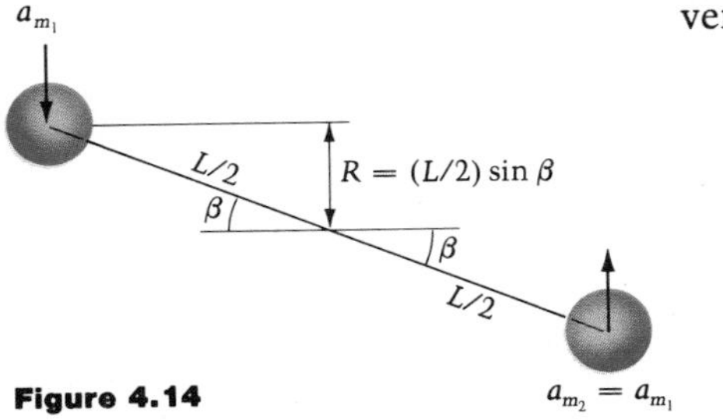

Figure 4.14

Figure 4.14, which shows the masses detached from the slanted bar. Their accelerations (obtained from knowing that each mass moves on a vertical circle at constant speed) are

$$a_{m_1} = \frac{\dot{s}^2}{\rho} = \frac{v^2}{R}$$

$$= \frac{(R\omega)^2}{R} = R\omega^2$$

$$= \frac{L\omega_0^2}{2} \sin \beta$$

By Newton's second law for the particle, there are forces on each mass ***in the same direction*** as these accelerations, which are equal to $m\mathbf{a}_{m_1}$ and $m\mathbf{a}_{m_2}$, respectively, as shown in Figure 4.15. The forces come from the slanted shaft.

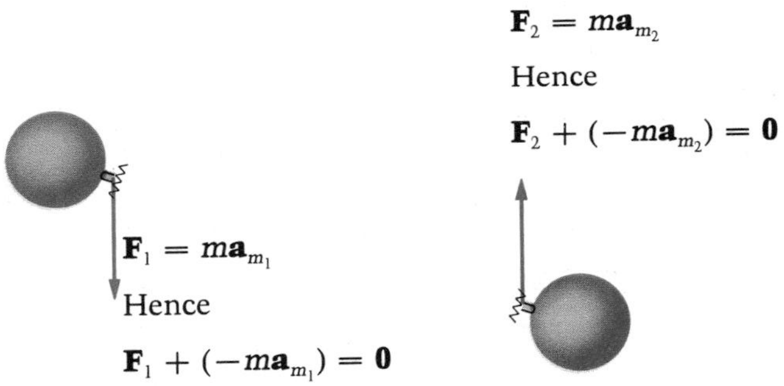

Figure 4.15

The forces $\mathbf{F}_1$ and $\mathbf{F}_2$ in a problem like this one, and the inward accelerations they produce, are called ***centripetal,*** meaning "toward the center" (of rotation). As can be seen, one can subtract the right side $m\mathbf{a}_C$ vector from the left-side $\mathbf{F}_r$ and get zero. The problem may then be treated as a statics problem if $-m\mathbf{a}$ is regarded as a force. If it is, we call it ***centrifugal force.*** It is, of course, ***not*** a real external force arising from a source of external mechanical action; rather (with the plus sign) it is the resulting $m\mathbf{a}_C$ produced by the actual (or real) external forces.* The idea of these ***reversed effective forces*** is attributed to d'Alembert, and we do not recommend it since we believe that one should not confuse statics and dynamics problems. We think it important to keep the force and couple terms of the equations (the left side) equal to, but ***separated*** from, the inertia term (the right side).

*Similarly, there is an *inertia torque* (reversed effective moment) that is used in the rotational motion of systems in the same way.

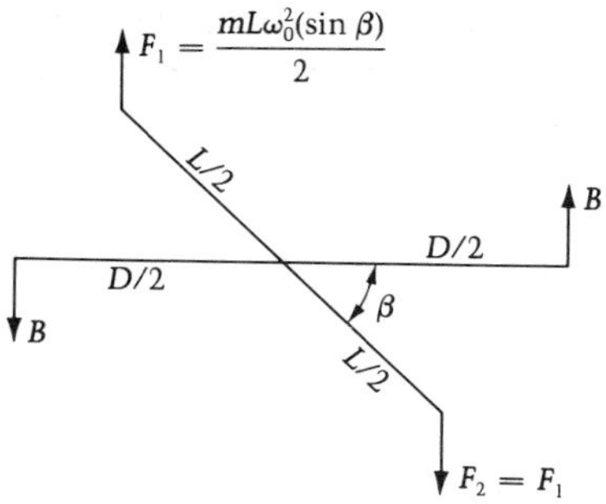

Figure 4.16

Continuing, we find that the equal and opposite forces to F_1 and F_2 act on the rigid shafts (Figure 4.16). Since the shafts are light, they have negligible inertia. Thus for the body consisting of the shafts we have

$$M_{r_{C_y}} = I_{yz_C}\alpha + I_{xz_C}\omega^2 \approx 0$$

so that

$$BD - \frac{mL\omega_0^2 \sin\beta}{2}\left(\frac{L}{2}\cos\beta\right)2 = 0$$

or

$$B = \frac{mL^2\omega_0^2 \sin\beta\cos\beta}{2D} \qquad \text{(as before)}$$

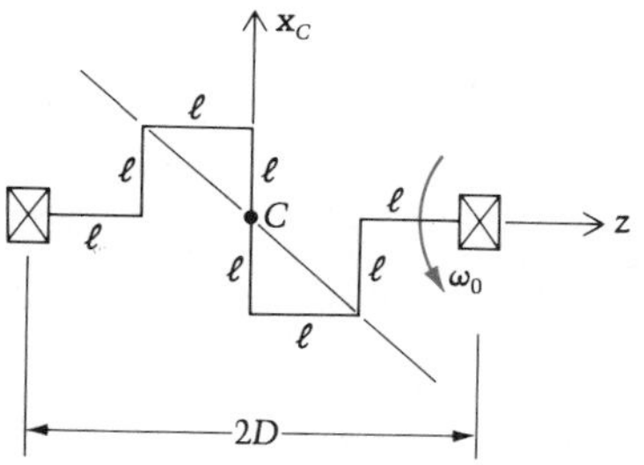

EXAMPLE 4.27

For the body of Example 4.13 (see the diagram), find the dynamic reactions of the bearings on the shaft if $\omega = \omega_0 =$ constant. Let M be the mass of each of the eight segments. Recall that I_{xz_C} is $2M\ell^2$.

SOLUTION

As in the preceding example, $I_{yz_C} = 0$. Thus

$$M_{r_{x_C}} = I_{xz_C}\overset{0}{\cancel{\alpha}} - \overset{0}{\cancel{I_{yz_C}}}\omega^2 = 0$$

$$M_{r_{y_C}} = I_{yz_C}\overset{0}{\cancel{\alpha}} + I_{xz_C}\omega^2 = 2M\ell^2\omega_0^2$$

Note that if in Example 4.26 we let $\beta = 45°$, $L/2 = 3\sqrt{2}\ell/4$, and $m = 2M$, the result for the bearing couple is

$$M_{r_{y_C}} = \frac{2M[(3\sqrt{2}\ell)/2]^2\omega_0^2(1/\sqrt{2})(1/\sqrt{2})}{2}$$

$$= 2.25M\ell^2\omega_0^2$$

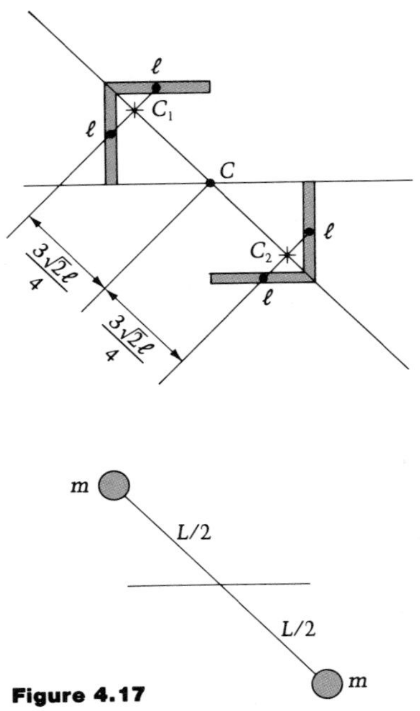

Figure 4.17

Here we have lumped the same amount of mass at the mass center of the pairs of unbalanced bars (see Figure 4.17). It gives a reasonable approximation to

$$M_{r_{y_C}} = 2M\ell^2\omega_0^2 \qquad \text{(11.1\% difference)}$$

but it is not exact because the mass in the eight-bar problem is actually distributed. The bearing reactions B (in the same directions as in the preceding example) are then

$$B = \frac{M_{r_{y_C}}}{2D} = \frac{M\ell^2\omega_0^2}{D}$$

We usually wish to eliminate dynamic bearing reactions because we want things to run quietly and smoothly and last as long as possible. Only when vibrations are desired do we purposely have imbalance. We now examine an example of ***elimination*** of imbalance (making I_{xz} and $I_{yz} = 0$), which, as we have seen, automatically eliminates dynamic bearing reactions in problems with a single shaft on two supports.

EXAMPLE 4.28

Determine where, and of what diameter, holes may be drilled in the identical disks in the diagram to eliminate the bearing reactions. The shaft is welded to the disks with the indicated offsets.

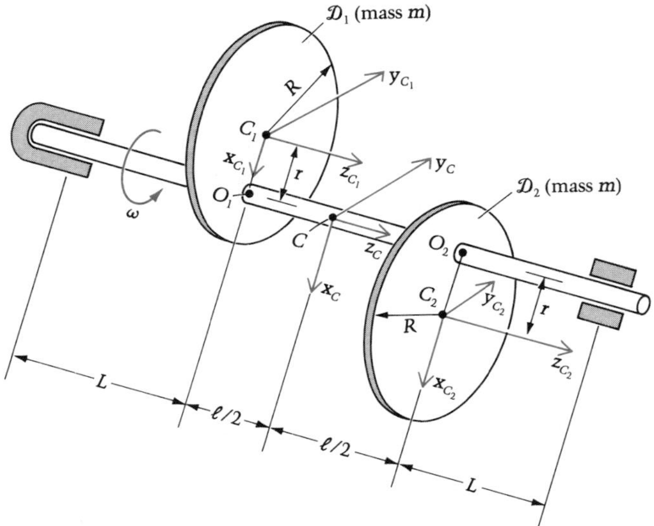

(Continued)

SOLUTION

Since $\mathbf{x}_C\mathbf{z}_C$ is a plane of symmetry, $I_{yz_C} = 0$. Thus

$$M_{r_{x_C}} = I_{xz_C}\alpha$$

$$M_{r_{y_C}} = I_{xz_C}\omega^2$$

and we see that there will be neither dynamic couple if $I_{xz_C} = 0$. Now

$$I_{xz_C} = I_{xz_C}^{\mathcal{D}_1} + I_{xz_C}^{\mathcal{D}_2}$$

$$= I_{xz_{C_1}} - m\bar{x}_1\bar{z}_1 + I_{xz_{C_2}} - m\bar{x}_2\bar{z}_2$$

The first and third terms vanish since $y_{C_1}z_{C_1}$ and $y_{C_2}z_{C_2}$ are planes of symmetry.* Thus

$$I_{xz_C} = -m(-r)\left(\frac{-\ell}{2}\right) - m(r)\left(\frac{\ell}{2}\right) = -mr\ell$$

If now there are holes H_1 and H_2 drilled (on the lines O_1C_1 and O_2C_2), then I_{xz_C} becomes

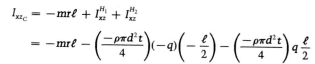

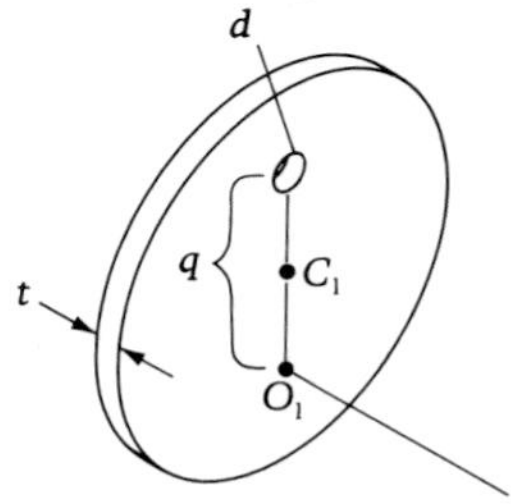

$$I_{xz_C} = -mr\ell + I_{xz}^{H_1} + I_{xz}^{H_2}$$

$$= -mr\ell - \left(\frac{-\rho\pi d^2 t}{4}\right)(-q)\left(-\frac{\ell}{2}\right) - \left(\frac{-\rho\pi d^2 t}{4}\right)q\frac{\ell}{2}$$

Setting $I_{xz_C} = 0$ gives

$$qd^2 = \frac{4mr\ell}{\rho\pi t\ell} = \frac{4\rho\pi R^2 tr\ell}{\rho\pi t\ell}$$

$$= 4R^2 r$$

Hence if $R = 0.2$ m and the offset is $r = 0.001$ m, for example, then for a 0.01-m diameter hole we have

$$q = \frac{4(0.2^2)(0.001)}{(0.01)^2} = 1.6 \text{ m}$$

which is unrealistic (off the disk!). For a 0.03-m diameter hole,

$$q = \frac{4(0.2^2)(0.001)}{(0.03)^2} \approx 0.18 \text{ m} \qquad \text{(which fits)}$$

Alternatively, weights could be added to the two disks on the *other* side of the shaft to balance the system. This of course is how automobile wheels are balanced.

*Recall that if yz is a plane of symmetry, the two products of inertia involving $\mathbf{x}$ (I_{xy} and I_{xz}) vanish.

PROBLEMS / Section 4.5 (continued)

4.130 Body $\mathcal{C}$ of mass m is attached to a shaft that turns in two bearings mounted in the reference frame as shown along the z_O axis. (See Figure P4.130.) The mass center C of $\mathcal{C}$ is not on the axis of rotation; it has coordinates $(x_O, y_O, z_O) = (a, b, c)$. Explain carefully how the following four equations will guarantee static *and* dynamic balance of the assembly if two masses m_A and m_B are respectively placed on symmetric disks $\mathcal{A}$ and $\mathcal{B}$ at (x_A, y_A) in the plane $z_O = z_A$ and at (x_B, y_B) in the plane $z_O = z_B$ (< 0). The centroidal products of inertia of $\mathcal{B}$ of interest to us are $I_{x_C z_C}$ and $I_{y_C z_C}$.

$$m_A x_A + m_B x_B + ma = 0$$

$$m_A y_A + m_B y_B + mb = 0$$

$$-m_A x_A z_A - m_B x_B z_B + I_{x_C z_C} - mac = 0$$

$$-m_A y_A z_A - m_B y_B z_B + I_{y_C z_C} - mbc = 0$$

What are the four unknowns?

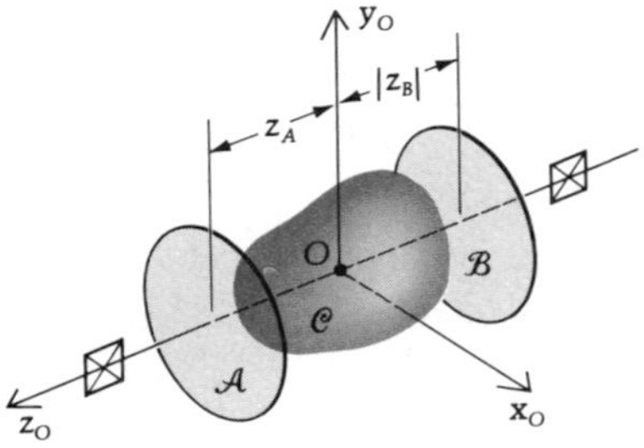

Figure P4.130

4.131 Rotor $\mathcal{R}$ in Figure P4.131 has a mass of 2 slugs, and its mass center C is at a 5-in. offset from its shaft as shown ($x = 3$ in., $y = 4$ in., and $z = 10$ in. in the coordinate system fixed in the shaft at the point B). The products of inertia of $\mathcal{R}$ with respect to the center of mass axes (x_C, y_C, z_C) are $I_{x_C z_C} = -20$ slug-in.^{2} and $I_{y_C z_C} = -10$ slug-in.^{2}. By adding a $\frac{1}{7}$-slug mass in each of the two correction planes $\mathcal{A}$ and $\mathcal{B}$, balance the rotor. That is, determine the x and y coordinates of each of the two added masses by ensuring that the mass center of the final system is on the shaft and that the products of inertia vanish.

4.132 The *S*-shaped shaft in Figure P4.132a is made of two half-rings, each of radius R and mass $m/2$. Find the dynamic bearing reactions for the instant given. *Hint:* For a half-ring, the mass center is located as shown in Figure P4.132b.

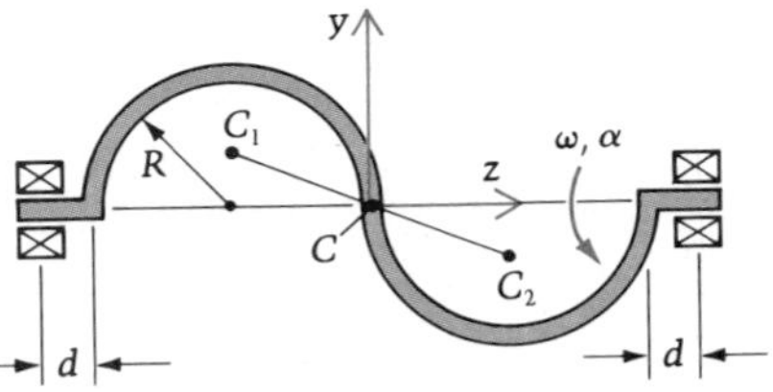

Figure P4.132a

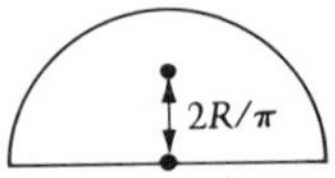

Figure P4.132b

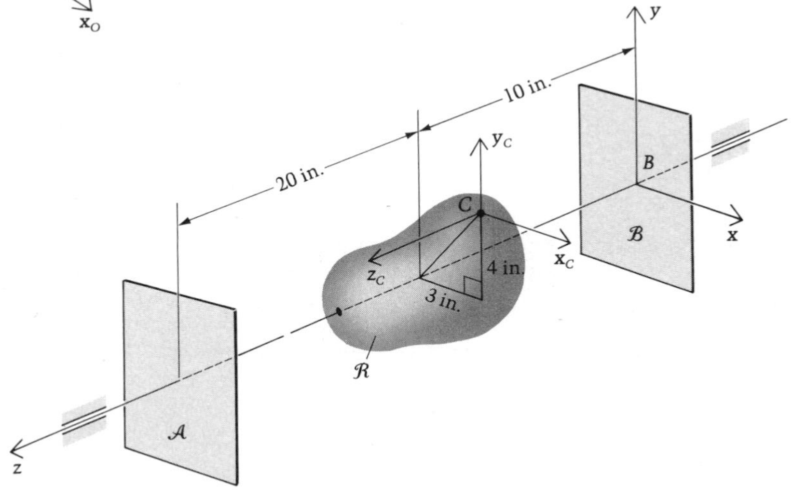

Figure P4.131

4.133 Two plates, each weighing 32.2 lb, are welded to a light shaft as shown in Figure P4.133. A torque T of 10 ft-lb is applied about the z axis until the assembly is turning at angular speed ω_0; then T is removed. If the bearings can hold a force perpendicular to the shaft of no more than 320 lb, find the maximum value that ω_0 can be without failure. Note that xz is the plane of the plates and (x, y, z) are fixed to the assembly.

4.134 The circular disk in Figure P4.134 is offset by the amount δ from the shaft to which it is attached.

a. Find the dynamic bearing reactions at A and B in terms of the system parameters shown in the figure.
b. If $\delta = r/20$, find the radius of a hole (in terms of r) at Q that will eliminate these bearing reactions.

4.135 A light rod of length ℓ, with a concentrated end mass M, is welded to a vertical shaft turning at 10 rad/sec. (See Figure P4.135.) Find the force and moment exerted by the rod onto the shaft.

4.136 The shaft in Figure P4.136 turns at constant angular velocity 10 rad/sec. If the bars are light compared with the two weights, determine the bending moment exerted on $\mathcal{S}_2$ (length 2ℓ) by $\mathcal{S}_1$ at the point where they are welded together. Sketch the way the shaft will deform in reality under the action of this couple. Ignore gravity.

4.137 Balance the shaft in Figure P4.137 by adding a mass of 3 sl in plane A and a mass of 4 sl in plane B.

4.138 In Example 4.28 suppose that $\mathcal{D}_1$ is centered on the shaft while $\mathcal{D}_2$ has the indicated offset. Calculate where a weight W must be bolted in order to balance the system dynamically.

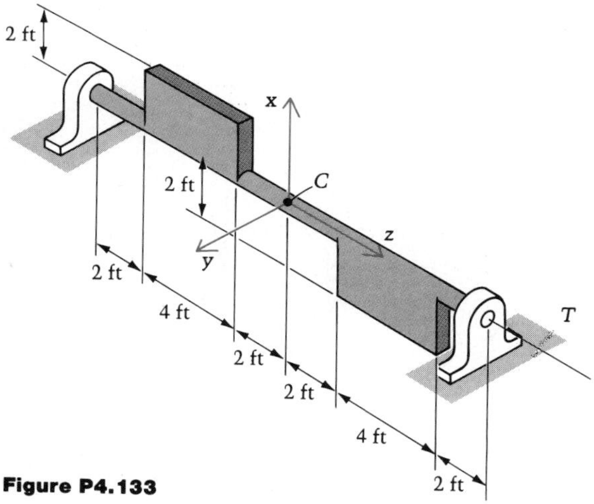

Figure P4.133

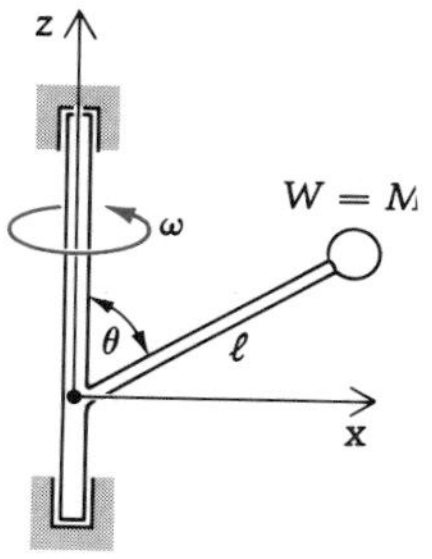

Figure P4.135

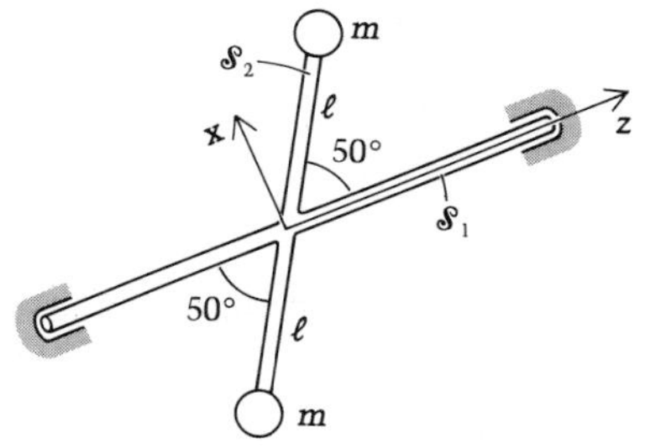

Figure P4.136

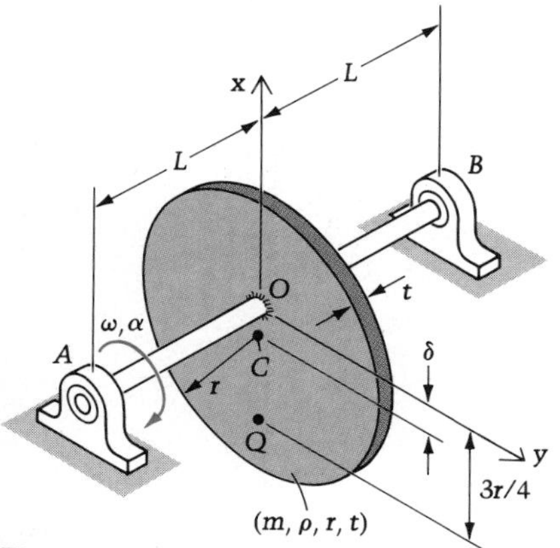

Figure P4.134

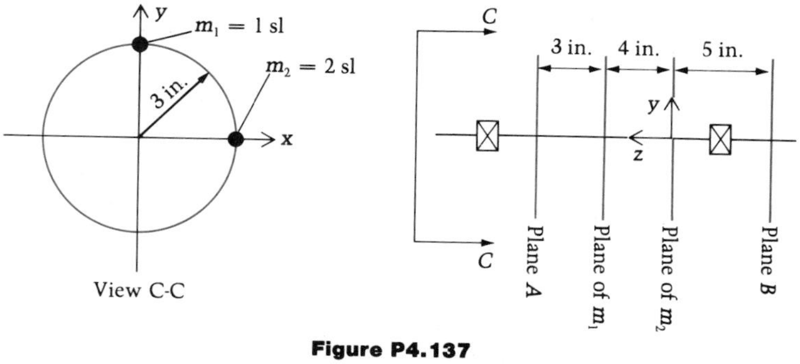

Figure P4.137

4.139 The shaft in Figure P4.139 supports the eccentrically located weights W_1 (1 lb) and W_2 (2 lb) as shown. It is desired to add a 3-lb weight in plane A and a 4-lb weight in plane B to balance the shaft dynamically. Determine the **x** and *y* coordinates of the added weights.

4.140 Two thin disks are mounted on a shaft, each midway between the center and one of the bearings, as indicated in Figure P4.140. The disks are each mounted off center by the amount $\delta = 0.05$ in. as shown. Determine the **x** and *y* locations of two small 4-oz magnetic weights (one for each disk), which when stuck to the disks will balance the shaft. Neglect the thicknesses of the disks, and treat the weights as particles.

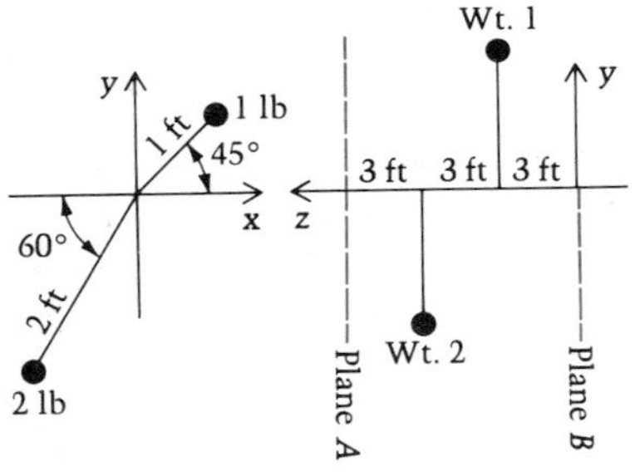

Figure P4.139

4.141 In Figure P4.141, $\mathcal{A}$ is the axle of a bicycle, mounted in bearings $2d$ apart. The cranks $\mathcal{C}$ are rigidly connected to the axle and also to the pedal shafts $\mathcal{P}_1$ and $\mathcal{P}_2$. If the rigid body consisting of axle $\mathcal{A}$, the cranks $\mathcal{C}$, and the pedal shafts is turning freely about axis z_C at constant angular speed ω, find the forces exerted on the bearings in the given configuration.

4.142 A solid cylinder (of mass m, radius r, and length $\ell = 4r$) and a light rod are welded together at angle ψ as shown in Figure P4.142. The rigid assembly is spun up to ω_0 rad/sec and then maintained at that speed. Find the dynamic bearing reactions after $\omega = \omega_0$.

4.143 Repeat the preceding problem if the lower half (shaded) of the cylinder is missing. (The mass is now $m/2$.) See Figure P4.143.

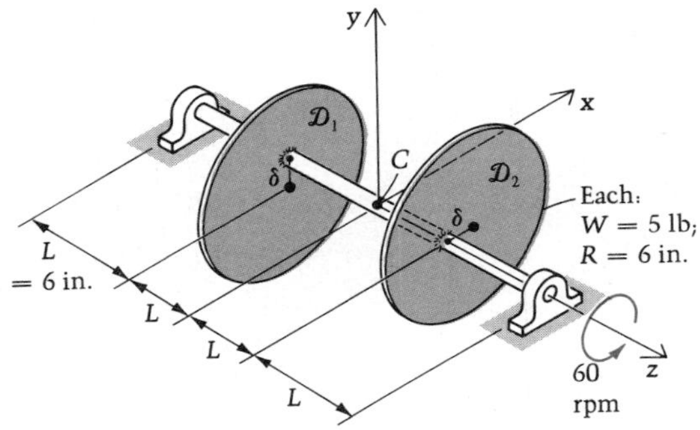

Figure P4.140

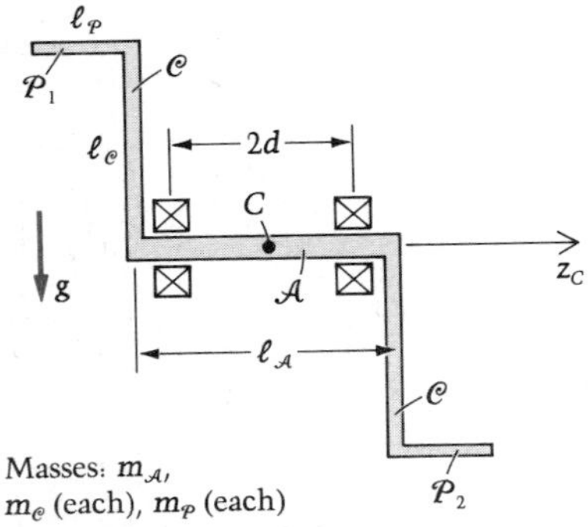

Figure P4.141

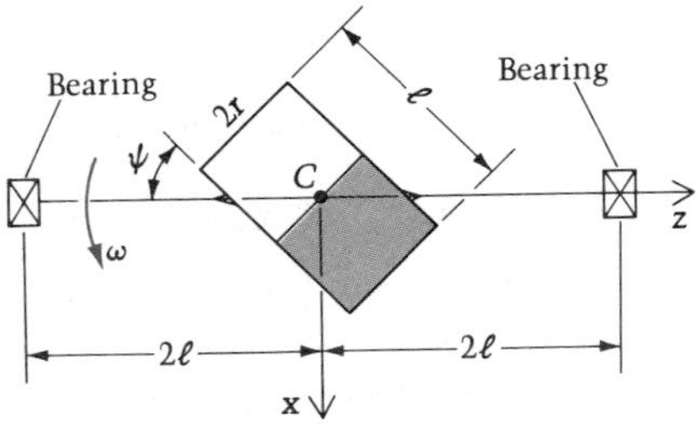

Figure P4.142

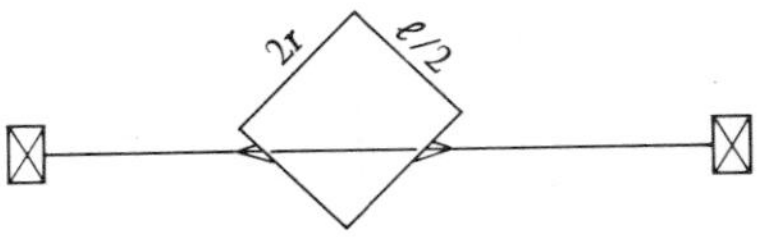

Figure P4.143

Summary of Equations of Motion

In this section we summarize the general equations of plane motion of rigid bodies. We have seen that for each rigid body in the problem being considered we may *always* write:

$$F_{r_x} = m\ddot{x}_C \tag{4.56a}$$

$$F_{r_y} = m\ddot{y}_C \tag{4.56b}$$

$$M_{r_{C_z}} = I_{z_C}\alpha \tag{4.56c}$$

Moreover, the plane motion requires that

$$F_{r_z} = m\ddot{z}_C = 0 \tag{4.57}$$

which simply states that all forces in the z direction on the body must equilibrate. We also have

$$M_{r_{C_x}} = I_{xz_C}\alpha - I_{yz_C}\omega^2 \tag{4.58}$$

$$M_{r_{C_y}} = I_{yz_C}\alpha + I_{xz_C}\omega^2 \tag{4.59}$$

The right-hand sides of these two equations vanish if the body is either symmetrical or balanced. (We have seen that C may be replaced in Equations (4.58) and (4.59) by any point in the reference plane containing the mass center.) If either or both products of inertia in Equations (4.58) and (4.59) do *not* vanish, however, these two equations give us the non-zero moments (usually caused by bearing reactions) that must be present about the x_C and y_C axes to sustain the plane motion. If these moments are not physically available, body $\mathcal{B}$ cannot remain in a state of plane motion.

We have also seen that we have some leeway in the use of the rotational equation (4.56c). Specifically, we may use any of its various forms (in which all moments and moments of inertia are taken with respect to z axes through the respective named points):

$M_{r_C} = I_C\alpha$ (*always* valid for each rigid body) (4.60a)

$M_{r_O} = I_O\alpha$ (if $\mathcal{B}$ has a pivot, such as a pinned point O) (4.60b)

$M_{r_H} = I_H\alpha$ (if H is *any* point of zero acceleration whether or not it is a pivot; this includes the preceding case) (4.60c)

$M_{r_Q} = I_Q\alpha$ (if $\mathbf{a}_Q$ passes through C) (4.60d)

$M_{r_P} = I_C\alpha + (\mathbf{r}_{PC} \times m\mathbf{a}_C)_z$ (P is *any* point) (4.60e)

$M_{r_P} = I_P\alpha + (\mathbf{r}_{PC} \times m\mathbf{a}_P)_z$ (P is a point of $\mathcal{B}$ in the reference plane) (4.60f)

Three of these forms (4.60a, b, and e) are normally used in working problems; the others are given for completeness. There are *no other* points than those mentioned above (4.60a–d) for which $M_{r_P} = I_P\alpha$. It is important to note one final time that, in general, $M_{r_{Ⓘ}} \neq I_{Ⓘ}\alpha$.

From Equations (4.56a), (4.56b), and (4.60a, e) we can see the special reduction to *translation:*

$$F_{r_x} = m\ddot{x}_C \tag{4.61a}$$

$$F_{r_y} = m\ddot{y}_C \tag{4.61b}$$

$$M_{r_{C_z}} = 0 \quad \text{or} \quad M_{r_{P_z}} = (\mathbf{r}_{PC} \times m\mathbf{a}_C)_z \tag{4.61c}$$

And from (4.56a), (4.56b), and (4.60b) the special case of *pure rotation* results:

$$F_{r_x} = m\ddot{x}_C \qquad (= \text{zero if } C \text{ is the pivot}) \tag{4.62a}$$

$$F_{r_y} = m\ddot{y}_C \qquad (= \text{zero if } C \text{ is the pivot}) \tag{4.62b}$$

$$M_{r_O} = I_O\alpha \qquad (O \text{ is the pivot}) \tag{4.62c}$$

In Equations (4.60a, b), if the mass center is *not* the pivot point, then $\ddot{x}_C$ and $\ddot{y}_C$ may be obtained in terms of α and ω from the component equations of

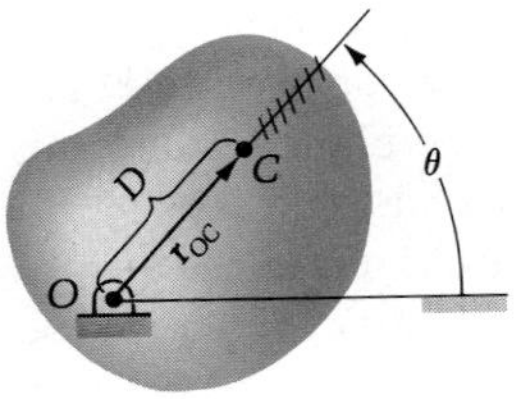

Figure 4.18

$$\mathbf{a}_C = \ddot{x}_C\hat{\mathbf{i}} + \ddot{y}_C\hat{\mathbf{j}} = \overset{\mathbf{0}}{\cancel{\mathbf{a}_O}} + \boldsymbol{\alpha} \times \mathbf{r}_{OC} - \omega^2\mathbf{r}_{OC},$$

which yields (see Figure 4.18):

$$\ddot{x}_C = -D\alpha \sin\theta - D\omega^2 \cos\theta \tag{4.63a}$$

$$\ddot{y}_C = D\alpha \cos\theta - D\omega^2 \sin\theta \tag{4.63b}$$

PROBLEMS / Section 4.5 (continued)

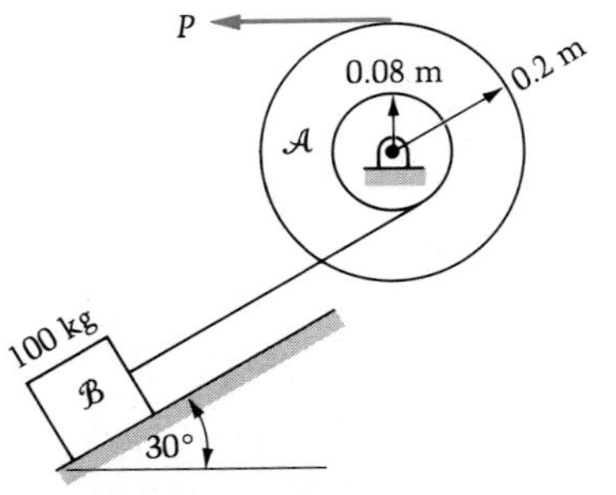

Figure P4.144

4.144 The 30-kg drum $\mathcal{A}$ shown in Figure P4.144 has a radius of gyration $k_C = 0.17$ m with respect to its axis. The force P varies as $3t^3$ with time (P in newtons, t in seconds).

a. If the 100-kg crate $\mathcal{B}$ begins to slide up the plane at $t = 5$ s, determine the coefficient of static friction.

b. If the velocity of $\mathcal{B}$ is 1.5 m/s at $t = 6$ s, find the coefficient of kinetic friction.

4.145 The radius of gyration of the 20-kg wheel in Figure P4.145 with respect to its axis is 0.3 m. Motion starts from rest. Find the acceleration of the mass center C, and determine how far C moves in 5 s.

4.146 The two wheels are identical 16.1-lb cylinders with smooth axles at their centers. (See Figure P4.146.) The carriage weighs 32.2 lb and has its mass center at C. The cylinders do not slip on the inclined plane.

a. Find the acceleration of point Q.

b. If the front cylinder is replaced by one with equal density and length but half the radius (6 in.), find the acceleration of Q. Assume no interference.

4.147 The pipe in Figure P4.147 has a mass of 500 kg and rests on the flatbed of the truck. The coefficient of friction between the pipe and truck bed is $\mu = 0.4$. The truck starts from rest with a constant acceleration a_0.

a. How large can a_0 be without the pipe slipping at any time?

b. For the value of a_0 in part (a), how far has the truck moved when the pipe rolls off the back?

4.148 Cylinder $\mathcal{C}$ in Figure P4.148 has a mass of 4 slugs, and the effect of the hub on its moment of inertia is negligible. It is connected by means of a cord to the 1-slug block $\mathcal{B}$. The mass of the pulley is negligible. The coefficient of friction between $\mathcal{C}$ and the plane is $\mu = 0.5$, and the radii of $\mathcal{C}$ are given in the figure. If the system is released from rest, determine the time that will elapse before $\mathcal{B}$ hits the ground.

4.149 Bicycles looked something like the one in Figure P4.149 prior to the use of chains and sprockets. Explain why the front wheel was made so large, and compute the distance moved forward for each complete revolution of the pedal cranks.

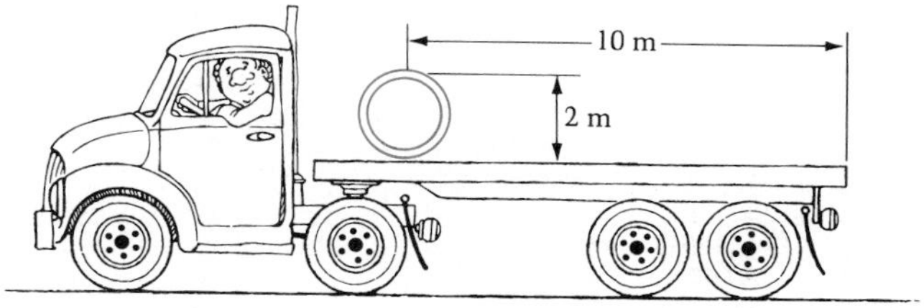

Figure P4.147

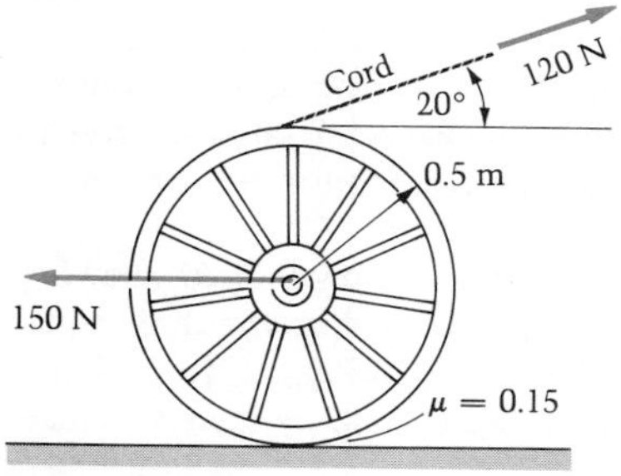

Figure P4.145

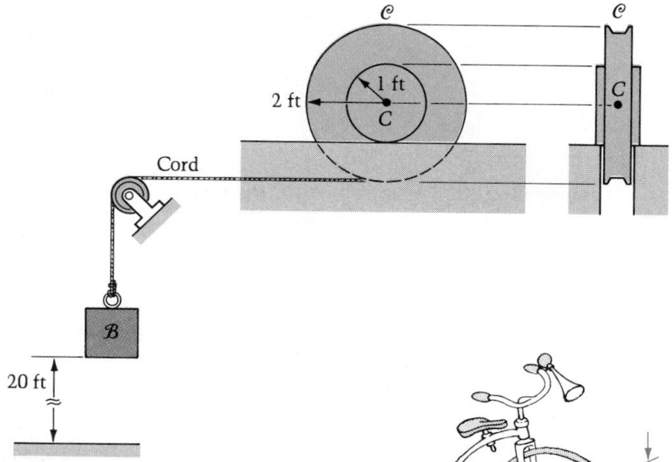

Figure P4.148

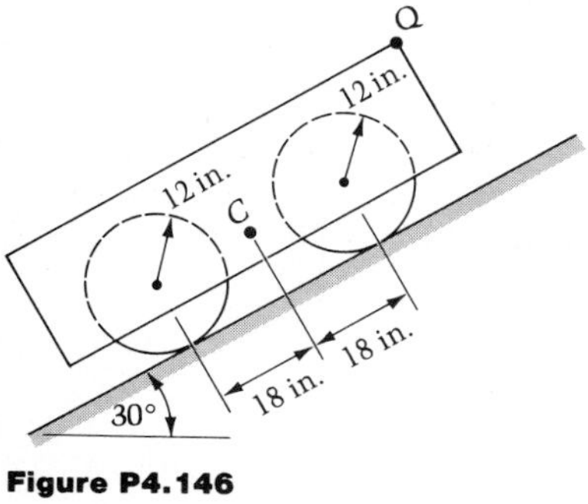

Figure P4.146

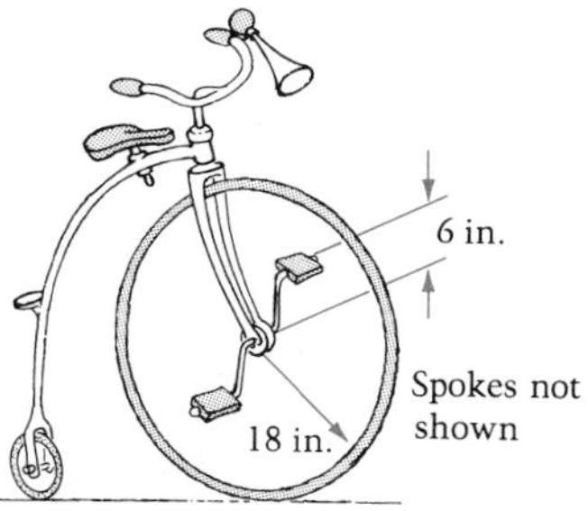

Figure P4.149

4.150 A man rides the early bicycle of Problem 4.149. (a) If he exerts an average moment of 18 ft-lb onto the front wheel with his feet, what is the acceleration of the bicycle? Neglect air and rolling resistance and the inertia of the wheels. Man plus bicycle weigh 150 lb. (b) Repeat the problem considering the inertia of the front wheel; treat it as a ring weighing 18 lb.

4.151 A 70-lb child is riding a 15-lb unicycle. (See Figure P4.151.) If the coefficient of friction between the tire and the street is $\mu = 0.3$, find the maximum possible acceleration of the center of the wheel (when $f = f_{max}$) and the actual acceleration. Neglect the rotational inertia of the wheel and legs for the rough calculation.

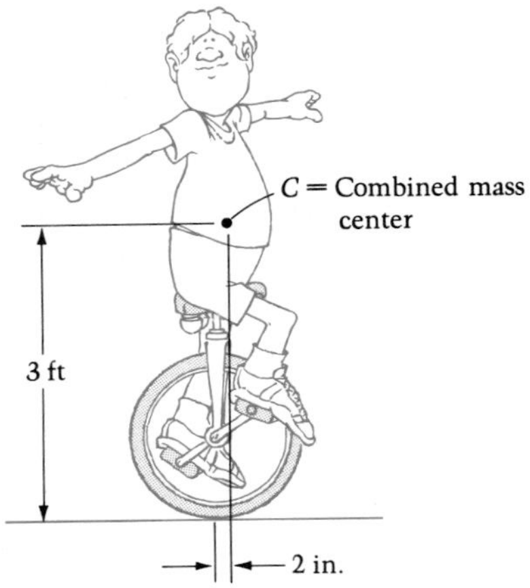

Figure P4.151

4.152 A string is wrapped around the hub of the spool shown in Figure P4.152. There are four indicated string directions. For the direction that will result in the largest displacement of C in 3 s, find this displacement. Assume sufficient friction to prevent slipping. The spool has a mass of 12 kg and a radius of gyration about z_C of 0.6 m. Each force equals 10 N, and the spool starts from rest.

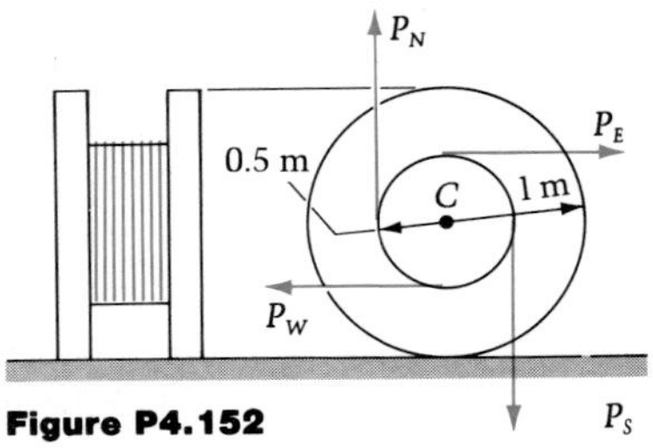

Figure P4.152

4.153 A constant torque T_0 is applied to the crank arm $\mathcal{C}$ of the planetary mechanism shown in Figure P4.153. The axes of the identical gears $\mathcal{S}$ and $\mathcal{P}$ are vertical, and the ends of the crank are pinned to the centers of $\mathcal{S}$ and $\mathcal{P}$. Determine the angular acceleration of $\mathcal{C}$ if $\mathcal{S}$ is fixed in the inertial frame of reference. Treat the gears as uniform disks. The plane of the page is horizontal.

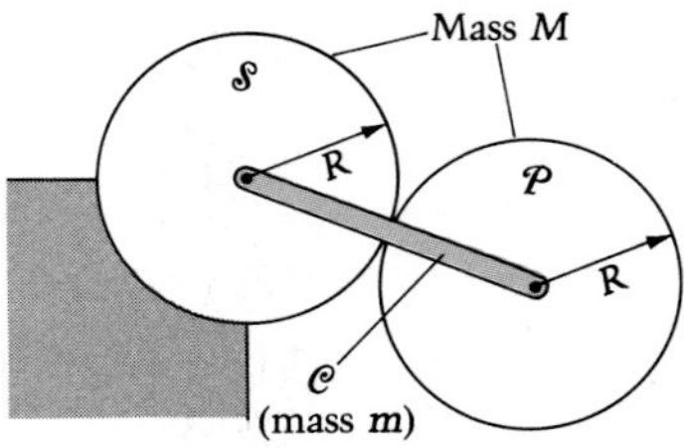

Figure P4.153

4.154 Figure P4.154 shows a hoop $\mathcal{H}$ of inside radius R supported on two roller bearings. The disk $\mathcal{D}$ is free to roll around on the inside of the hoop. The hoop is driven to rotate about its center O in such a manner that the disk's center C remains in the indicated position in the (inertial) frame $\mathcal{I}$. What must be the angular velocity and angular acceleration of the hoop? *Hint:* Kinematics: Relate the velocities of P and C; do the same for accelerations. Remember that C is at rest and that P is a point in rolling contact with the hoop.

Kinetics: Draw a free-body diagram of $\mathcal{D}$ showing friction, normal, and gravity forces. Write the three equations of motion, and solve for $\alpha_{\mathcal{D}}$. Integrate it to obtain $\omega_{\mathcal{D}}$, and then relate these to $\alpha_{\mathcal{H}}$ and $\omega_{\mathcal{H}}$ by your kinematics results.

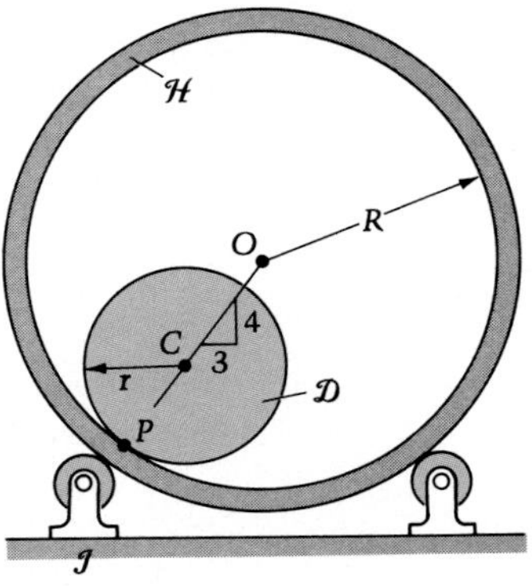

Figure P4.154

4.155 The angular speed of the turntable in Figure P4.155 is slowly increased from zero. Two blocks rest on it along the same radial line, and they are connected by a cord as shown. The radii of the masses are 0.4 m and 0.6 m, and the mass of the inner block is 30 kg.

a. Find the mass of the outer block for which the blocks will never slip.

b. Suppose the mass of the outer block is 10 kg, the coefficient of friction is 0.5 at all surfaces, and the rope breaks at a tension of 20 N. Find out whether the rope breaks or the blocks slide first.

4.156 The homogeneous cylinder $\mathcal{A}$ in Figure P4.156 weighs 64.4 lb and rolls on the 96.6-lb truck $\mathcal{B}$. The mass of the truck rollers may be neglected. Find the force P such that C does not move relative to the plane.

4.157 A uniform circular disk $\mathcal{D}$ of radius r has a uniform bar $\mathcal{B}$ of length L and weight W welded to it at A such that L is perpendicular to AO. (See Figure P4.157.) The disk and bar rotate about O as shown in such a manner that $\theta = \sin \beta t$, where θ is in radians and t is time. Find the reaction of the disk on the bar at A when $\theta = -1$ rad.

4.158 Determine the largest value of x such that the rope supporting block $\mathcal{B}$ in Figure P4.158 will not go slack when the other rope is cut. If $x = L/4$, find the tension in the rope supporting $\mathcal{B}$ just after the other rope is cut.

4.159 An unbalanced cylinder $\mathcal{C}$ rolls on an inclined plane, starting from rest at $t = 0$ in the position indicated in Figure P4.159. Write the three equations of motion for $\mathcal{C}$ in the arbitrary later position at time t. Use them plus kinematics to obtain a single differential equation involving θ, $\dot{\theta}$, $\ddot{\theta}$, the constants g, r, R, β, and the radius of gyration k_c.

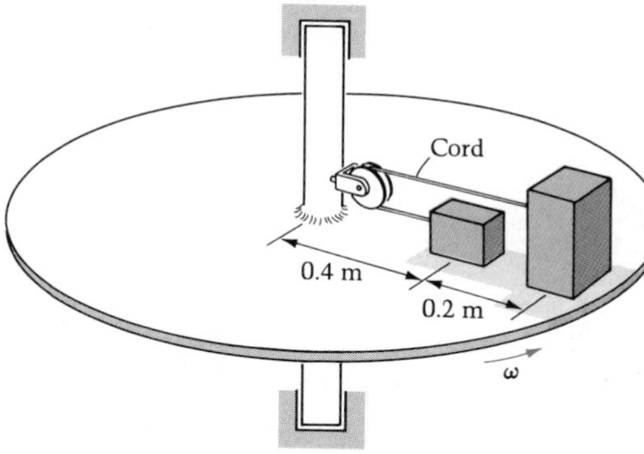

Figure P4.155

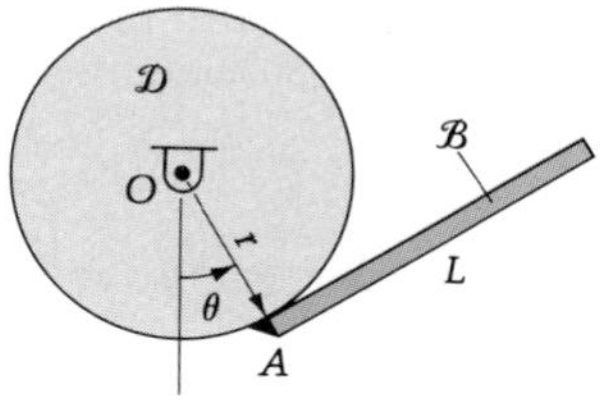

Figure P4.157

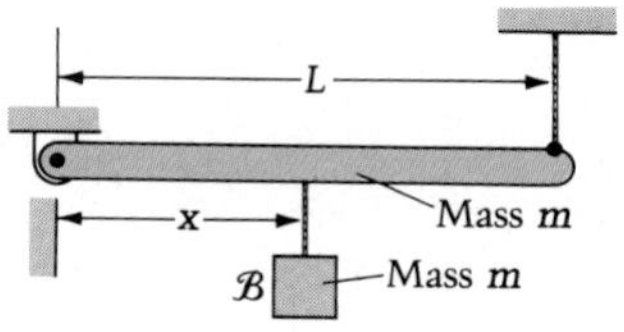

Figure P4.158

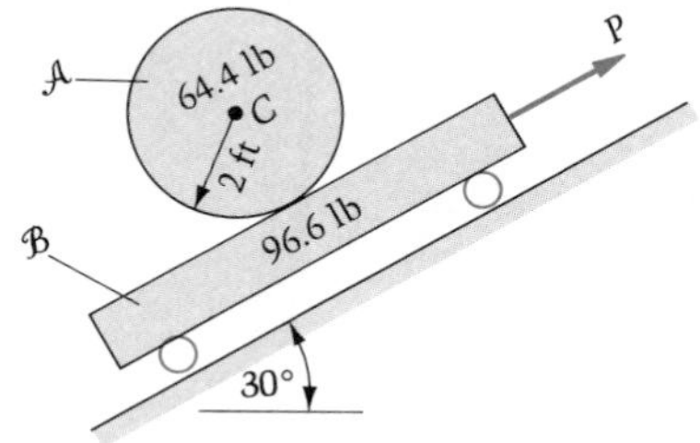

Figure P4.156

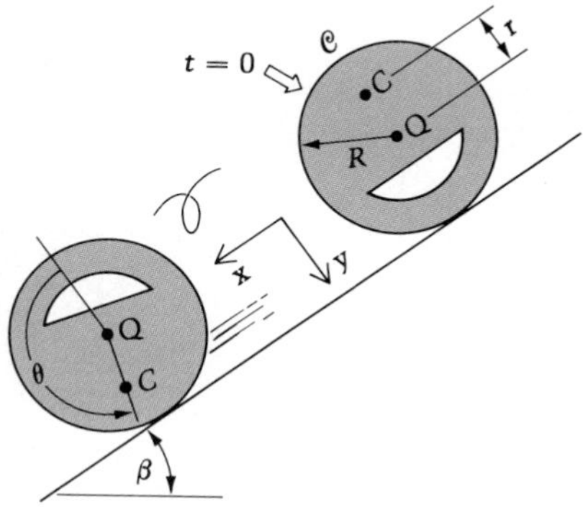

Figure P4.159

4.160 Sally Sphere, Carolyn Cylinder, Harry Hoop, and Wally Wheel each have mass m and radius R. Wally's spokes and rim are very light compared to his hub. (See Figure P4.160.) They are going to have a race by rolling down a rough plane. Give (a) the order in which they finish and (b) the times.

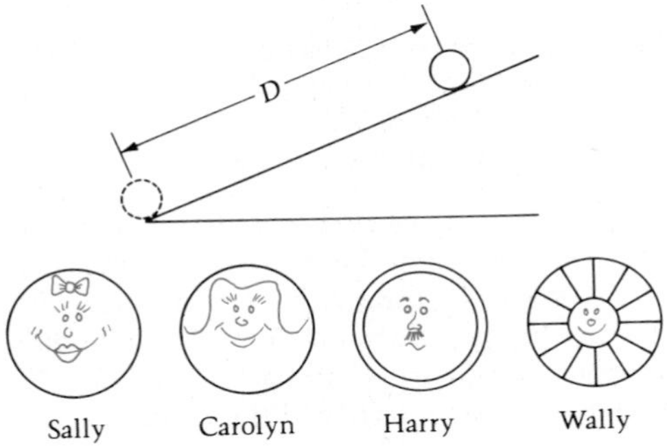

Figure P4.160

4.161 In the preceding problem, Wally and Carolyn are connected by a bar of negligible mass and released from rest on the same incline. (See Figure P4.161.) Determine the force in the bar.

4.162 Repeat the preceding problem, but suppose that Wally and Carolyn switch places.

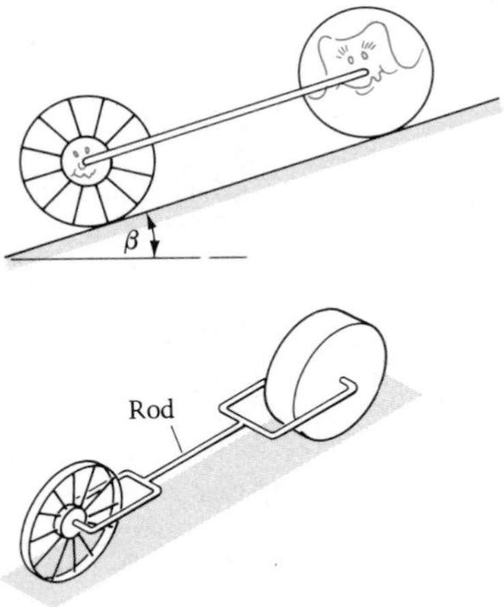

Figure P4.161

4.163 The cylinder of mass m in Figure P4.163 is projected up the plane with backspin; initially $\mathbf{v}_C = v_0$  and $\boldsymbol{\omega} = \omega_0$ ↺.

a. Determine how high up the plane (H_1) the cylinder travels before slipping stops, and find how long (t_1) it takes.

b. Find the additional distance (H_2) traveled by the cylinder before it reaches its highest point, and determine the additional time ($t_2 - t_1$) elapsed after the slipping stops.

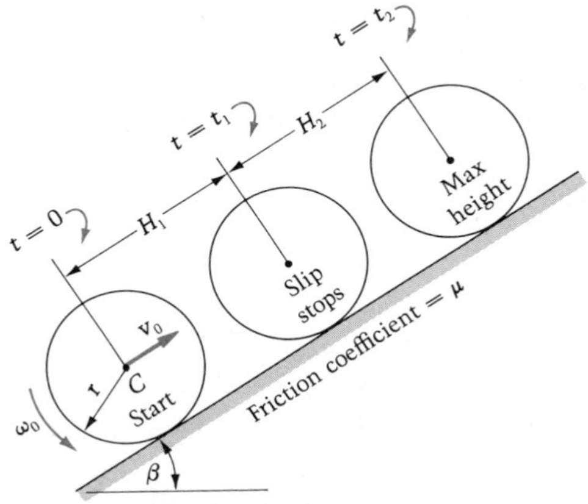

Figure P4.163

4.164 Cylinder $\mathcal{B}$ is released from rest on wedge $\mathcal{W}$ in the position shown in Figure P4.164. Find how long it takes for the cylinder to roll off the wedge. Use the equations of motion and note that: (1) $\mathcal{W}$ translates, so $\dot{x}_C = \dot{x}_A$; (2) unlike x_A, x_G in the figure is *not* measured from an inertial origin.

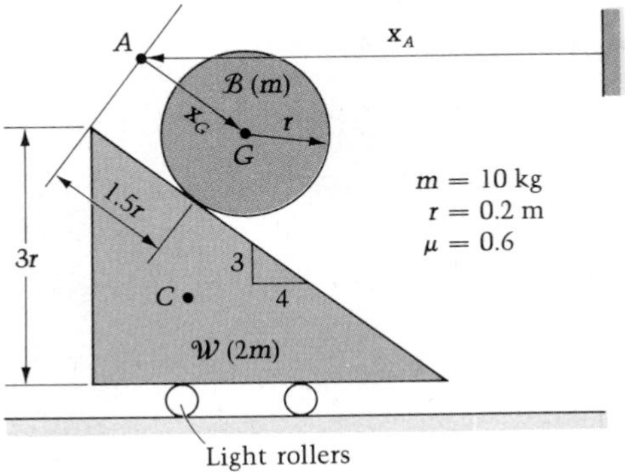

Figure P4.164

4.165 An airplane lands on a level strip at 200 mph. (See Figure P4.165.) Initially, just before the wheels touch the runway, the wheels are not turning. After they touch the runway they will skid for some distance and then roll free. If during this skidding the plane has a constant velocity of 200 mph and the normal force between the wheel and the runway is 10 times the wheel weight, find the length of the skid mark. (The coefficient of friction is $\frac{1}{2}$; the radius of gyration of the wheel is three-fourths of its radius.)

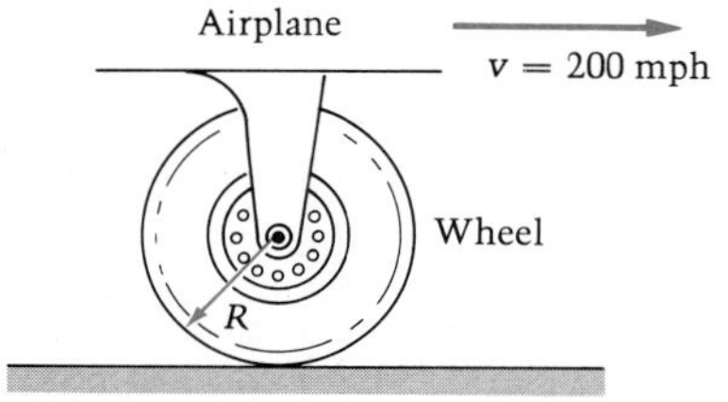

Figure P4.165

4.166 The homogeneous cylinder $\mathcal{C}$ in Figure P4.166 weighs 64.4 lb. The acceleration of the 96.6-lb cart $\mathcal{D}$ is 10 ft/sec^2 to the right.

a. Determine the acceleration of the center C of the cylinder and the friction force exerted on $\mathcal{C}$ by $\mathcal{D}$ if there is sufficient friction to prevent slipping.

b. How large does the friction coefficient μ have to be for this to occur?

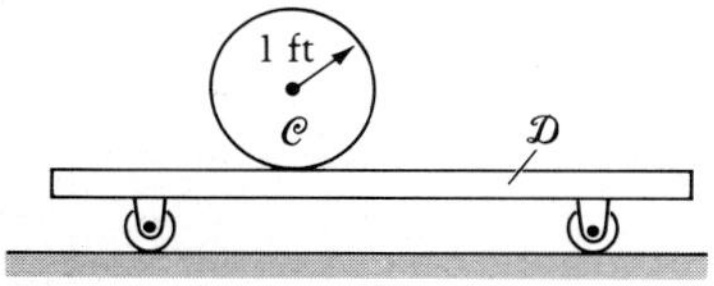

Figure P4.166

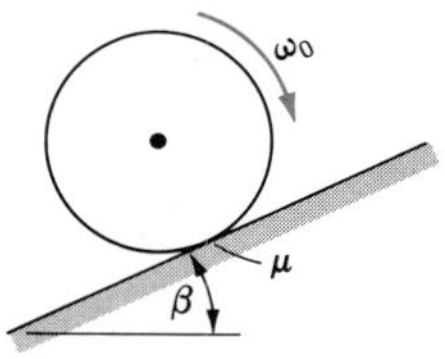

Figure P4.167

4.167 A cylinder spinning at angular speed ω_0 rad/sec clockwise is placed on an inclined plane. (See Figure P4.167.) Show that the cylinder center will begin moving up the plane if $\mu > \tan \beta$. Why does this result have nothing to do with the size of ω_0?

4.168 The homogeneous cylinder $\mathcal{C}$ in Figure P4.168 is at rest on the conveyor belt when the latter is started up with a constant acceleration of 3 ft/sec^2 to the right. If the cylinder rolls on the belt, find the elapsed time when the cylinder reaches the end A.

4.169 Referring to Problem 4.47 and Example 4.15, find the relationship between b, μ, and H for which the block tips and slides simultaneously.

4.170 Find how long it takes for $\mathcal{A}$ to roll off the plane in Figure P4.170, assuming sufficient friction to prevent slipping. The system is released from rest.

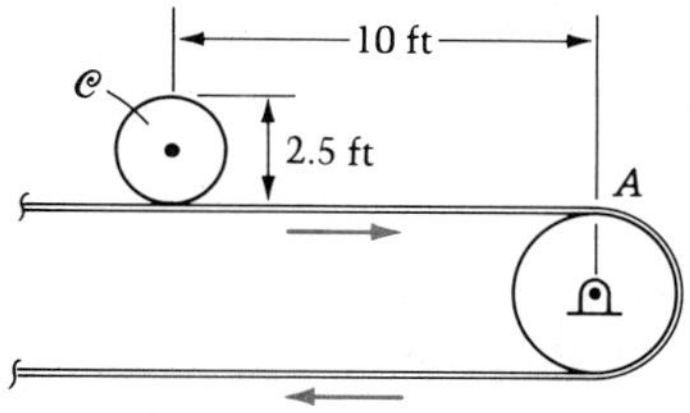

Figure P4.168

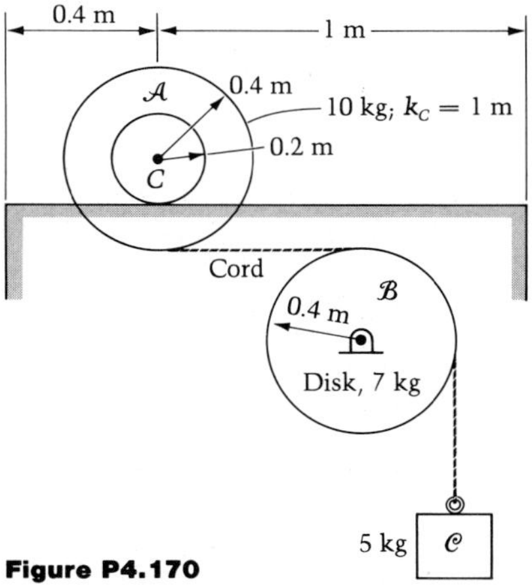

Figure P4.170

4.171 The force $P = 60$ N is applied as shown in Figure P4.171 to the 10-kg cylinder $\mathcal{C}$, originally at rest beneath the mass center of the thin, 5-kg rectangular plate $\mathcal{P}$. The coefficient of friction between $\mathcal{C}$ and $\mathcal{P}$ is 0.5, and the plane beneath $\mathcal{C}$ is smooth. Determine: (a) the initial acceleration of C; (b) the value of $\mathbf{x}$ when $\mathcal{C}$ is slipping on both surfaces. The length of $\mathcal{P}$ is 2 m.

4.172 The three cylinders $\mathcal{A}$, $\mathcal{B}$, and $\mathcal{C}$ in Figures P4.172a and b are identical except that $\mathcal{B}$ has a narrow slot cut in it to accommodate a cord at a smaller radius. The cord is wrapped around $\mathcal{C}$ as shown in the two figures. Neglecting the effect of the slot on the moment of inertia of $\mathcal{B}$, find the accelerations of the mass centers of $\mathcal{B}$ and $\mathcal{C}$.

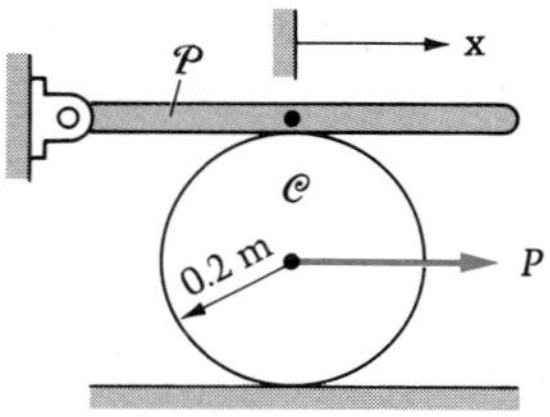

Figure P4.171

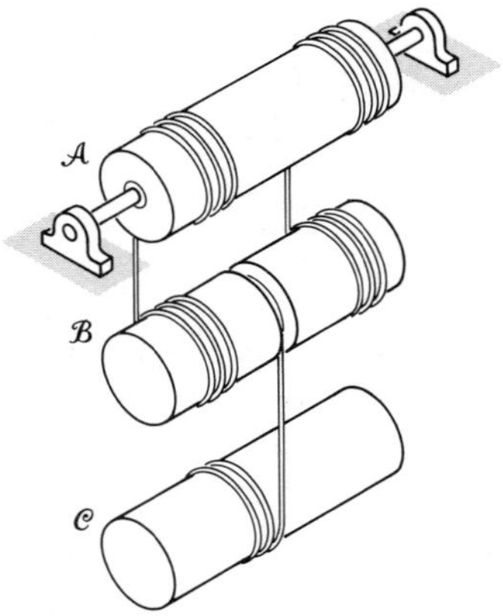

Figure P4.172a

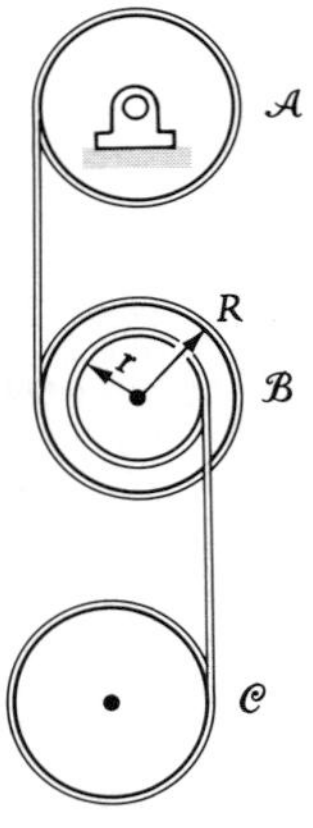

Figure P4.172b

4.173 A symmetric body $\mathcal{B}$ has mass m and radius R; a cord is wrapped around it as shown in Figure P4.173. Compute the downward acceleration of the center C if $\mathcal{B}$ is (a) a cylinder; (b) a sphere (with a small slot to accommodate the cord); (c) a thin ring. *Hint:* Work the problem just once with radius of gyration k_C; then substitute the three values $R/\sqrt{2}$, $\sqrt{\frac{2}{5}}\,R$, and $1R$ for k_C.

4.174 Assume that enough friction is available to prevent the cylinder in Figure P4.174 from slipping.

a. Show that:
(i) $\mathcal{A}$ rolls to the right if $\theta < \cos^{-1}(r/R)$.
(ii) $\mathcal{A}$ rolls to the left if $\theta > \cos^{-1}(r/R)$.
(iii) $\mathcal{A}$ is in equilibrium if $\theta = \cos^{-1}(r/R)$ (and will translate if P increases enough to overcome friction).

b. Find $\ddot{x}_C$ and α if $r = 0.2$ m, $R = 0.4$ m, $P = 20$ N, $mg = 40$ N, and $\theta = 45°$.

4.175 The body $\mathcal{B}$ of mass m shown in Figure P4.175 is in equilibrium, and the sensor at B reads that the force it exerts on $\mathcal{B}$ is fmg, where f is the fraction of the weight supported by the sensor. The cord at A is then cut, and the force at B immediately changes to $f'mg$. Find the moment of inertia of the mass of $\mathcal{B}$ with respect to its mass center C in terms of the measurable (assumed known) quantities m, d, f, and f'.

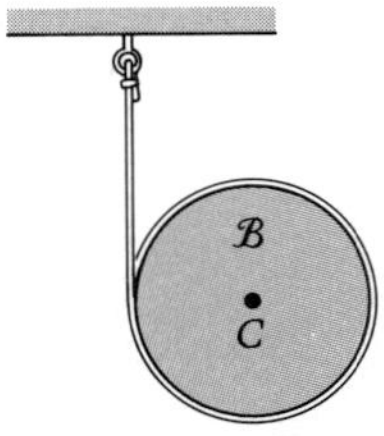

Figure P4.173

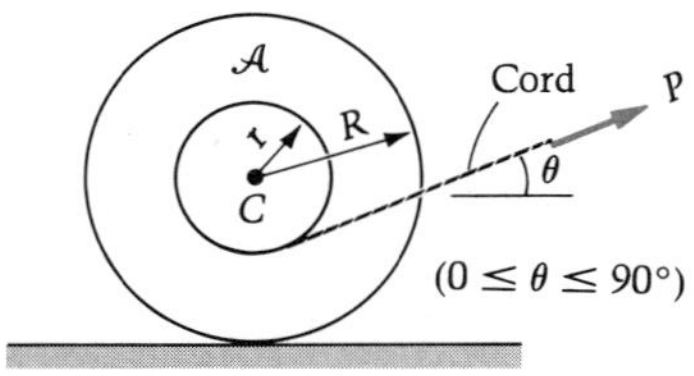

Figure P4.174

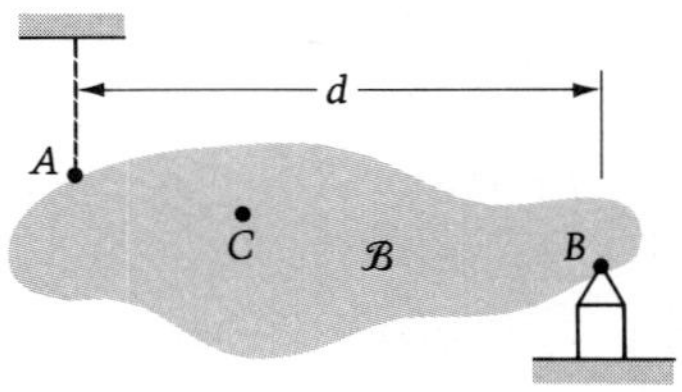

Figure P4.175

4.176 Cylinder $\mathcal{A}$ in Figure P4.176 rolls down a wedge that can slide without friction on a smooth floor. Show that the acceleration of wedge $\mathcal{D}$ is a constant given by the equation

$$a_G = \frac{m_1 g \sin 2\beta}{3(m_1 + m_2) - 2m_1 \cos^2 \beta}$$

4.177 Body $\mathcal{P}$ in Figure P4.177 is a rigid plate of mass M, resting on a number n of cylinders each of mass m and radius R. Force F is constant and starts the system moving from the position shown. If there is no slipping at any surface, find: (a) the acceleration of the plate and (b) its position x_C as a function of M, m, F, n, and time t.

4.178 A rope is wound around a cylinder that is released at the instant shown on a 30° incline. (See Figure P4.178.) Find how long it takes for the cylinder to get to the bottom if $\mu_s = 0.25$ and $\mu_k = 0.2$.

4.179 In Figure P4.179 the masses of blocks $\mathcal{A}$, $\mathcal{B}$, and $\mathcal{C}$ are 50, 20, and 30 kg. Find the accelerations of each if the table is removed. Which block will hit the floor first? How long will it take?

4.180 Two weights W_1 and W_2 in Figure P4.180 are connected by an inextensible cord that passes over a pulley. The pulley weighs W and its mass is concentrated at the rim of radius R. Show that if the system is released and the cord does not slip on the pulley, the acceleration magnitude of W_1 and W_2 is

$$\left| \frac{W_2 - W_1}{W_1 + W_2 + W} g \right|$$

4.181 Wheel $\mathcal{C}$ is made up of the solid disk $\mathcal{A}$, rim $\mathcal{R}$, and four spokes $\mathcal{S}$. Masses and radii are given in Figure P4.181 and the table.

a. Compute I_{z_C} for the wheel.

b. The coefficient of friction between $\mathcal{C}$ and the plane is $\mu = 0.3$. If a cord is wrapped around the disk and connected to the 50-kg body $\mathcal{B}$, determine the acceleration of the mass center C of $\mathcal{C}$.

Part	Mass (kg)
Disk $\mathcal{A}$	20
Spokes $\mathcal{S}$	5 (each)
Rim $\mathcal{R}$	10

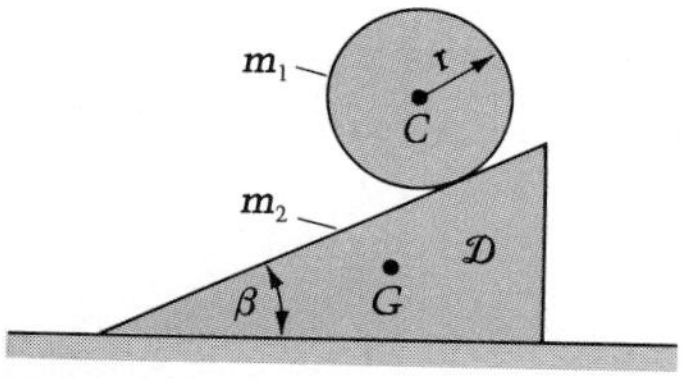

Figure P4.176

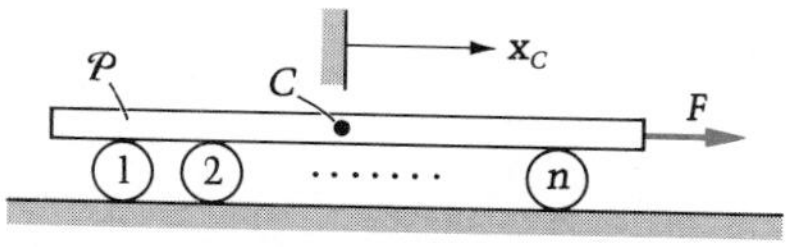

Figure P4.177

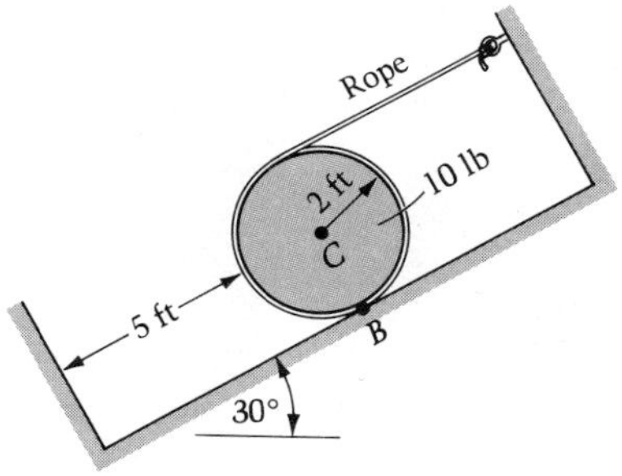

Figure P4.178

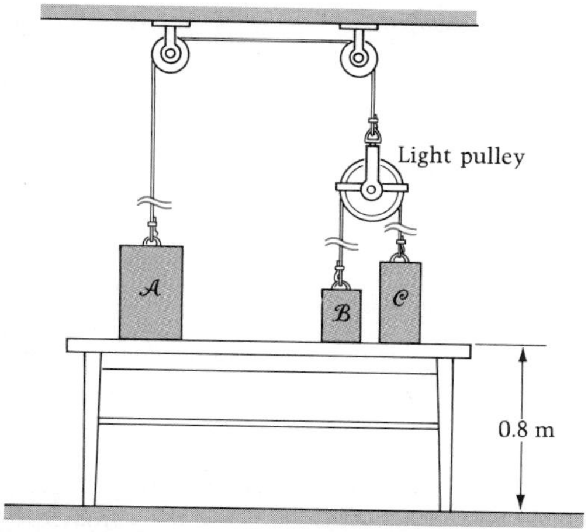

Figure P4.179

Figure P4.180

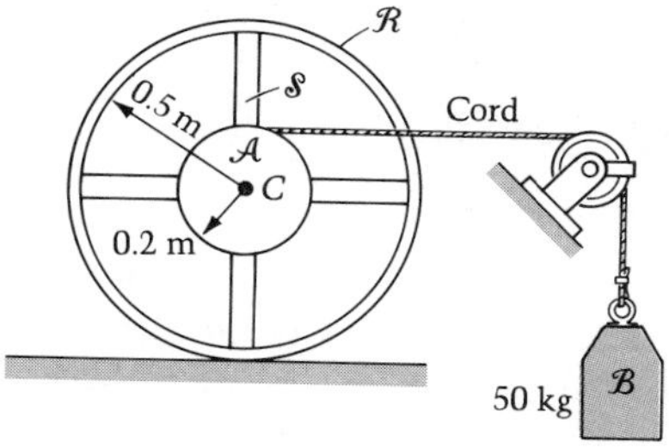

Figure P4.181

4.182 In the preceding problem, allow both μ and the mass of $\mathcal{B}$ ($m_{\mathcal{B}}$) to be general.

a. Show that with the cord on the disk, wheel $\mathcal{C}$ will *never* slip to the left but its base point B will slip to the *right* if $\mu < 1/49$ at values of $m_{\mathcal{B}}$ given by

$$m_{\mathcal{B}} \geq \frac{1800\mu}{1 - 49\mu} \text{ kg}$$

b. Show that with the cord in a slot near the rim, the wheel will not slip to the *right* but its base point B will slip *left* for $\mu < 7/50$ at masses given by

$$m_{\mathcal{B}} \geq \frac{900\mu}{7 - 50\mu} \text{ kg}$$

4.183 The constant force F_0 is applied to the cylinder, initially at rest, as shown in Figure P4.183. Show in the following two ways that the cylinder will slip provided that

$$\frac{F_0}{mg} > 3\mu$$

a. Assume rolling; then obtain the inequality from $f > \mu N$ after solving for f and N.

b. Assume slipping; then integrate $\ddot{x}_C$ and $\ddot{\theta}$ to obtain $\dot{x}_C$ and $\dot{\theta}$, and then find the velocity of the contact point B; if it is to the right (that is, positive), this is consistent with $f = \mu N$ to the left and we have slipping.

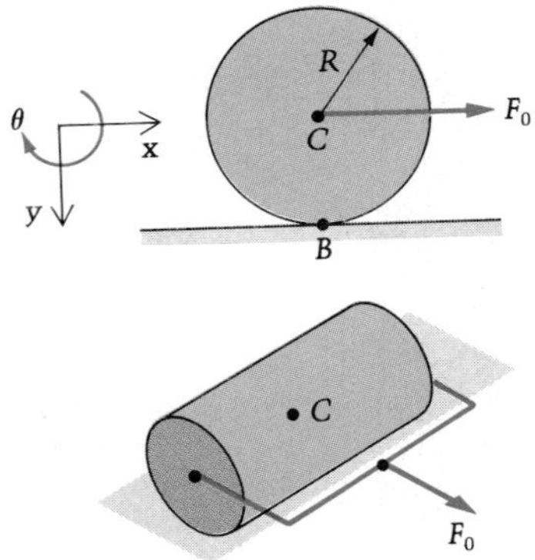

Figure P4.183

4.184 In Example 4.17 note that the acceleration of the contact point (of $\mathcal{C}$ with the plane) passes through C. Calling the point Q, it is thus correct to write $M_{r_Q} = I_Q\alpha$. Show that Equation (4) of that example follows immediately from this equation.

4.185 Two identical drums $\mathcal{A}$ and $\mathcal{B}$ rotate about fixed horizontal axes as shown in Figure P4.185 with constant angular speeds of 40 rpm. The coefficient of kinetic friction between the 20-lb homogeneous plank $\mathcal{P}$ and the drums is 0.3. The plank is displaced 3 in. to the right from its position of equilibrium and released from rest. Find the velocity of C when C first passes the equilibrium position, and determine the minimum wheel speed to keep the friction forces always directed toward line ℓ.

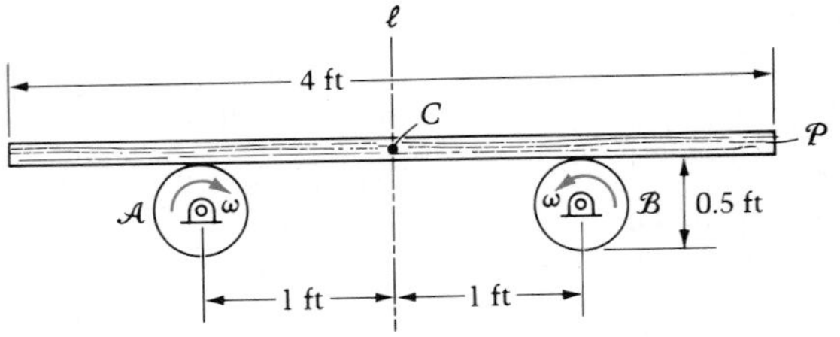

Figure P4.185

4.186 A slender homogeneous rod weighing 64.4 lb and 20 ft long is supported as shown in Figure P4.186. Bars $\mathcal{A}$ and $\mathcal{C}$ are of negligible mass and have frictionless pins at each end. The system is released from rest with $\theta = 0$.

a. Derive expressions for the angular velocity and acceleration of bars $\mathcal{A}$ and $\mathcal{C}$ as functions of θ.

b. Derive expressions for the axial force in bars $\mathcal{A}$ and $\mathcal{C}$ as a function of θ.

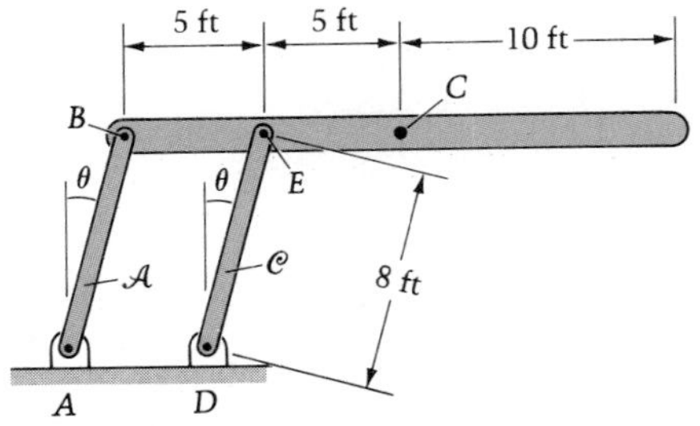

Figure P4.186

4.187 A 6-ft gymnast makes a somersault dive into a net by standing stiff and erect on the edge of a platform and allowing himself to overbalance. He loses foothold (without having slipped) when the platform's reaction on his feet becomes zero; he preserves his rigidity during his fall. Show that he falls flat on his back if the drop from the platform to the net is about 40 ft.

4.188 Find the range of possible values of the couple M_0 for which the cylinder in Figure P4.188 will not slip in either direction when released from rest on the incline. The mass is 15 kg; the radius is 0.2 m; and the coefficient of friction is $\mu = 0.2$.

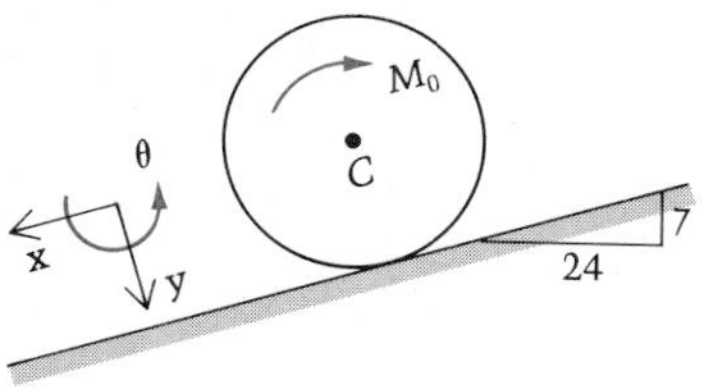

Figure P4.188

4.189 A uniform thin hollow cylinder of radius a rolls inside a rough fixed cylinder of radius b; the angular velocity at the lowest point is ω_0. (See Figure P4.189.) Show that slipping will not occur if

$$a^2\omega_0^2 < g(b - a)(1 - \cos k)$$

where $k = \tan^{-1} 2\mu$; μ is the coefficient of friction between the cylinders.

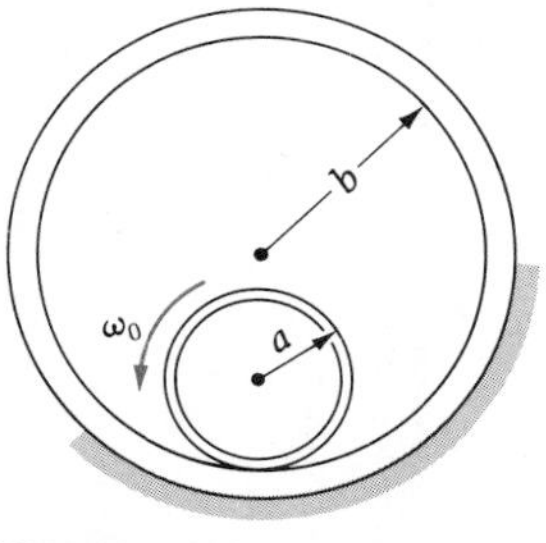

Figure P4.189

4.190 The cylinder of weight W and radius r shown in Figure P4.190 has an angular velocity of 100 rad/s clockwise. It is lowered onto the rough incline. If its center C is observed to remain momentarily at rest, determine the coefficient of sliding friction. Find how long the center C remains at rest.

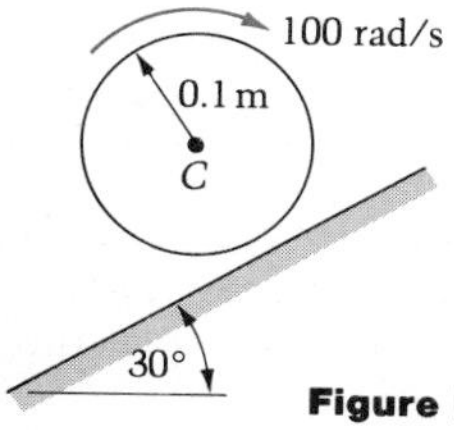

Figure P4.190

4.191 The 32.2-lb body $\mathscr{C}$ in Figure P4.191 is a spool having a radius of gyration $k_c = 5$ in. about its axis. Cords are wrapped around the peripheries; one is connected to a ceiling, the others to the 48.3-lb block $\mathscr{B}$. Find the accelerations of the centers C (of $\mathscr{C}$) and B (of $\mathscr{B}$).

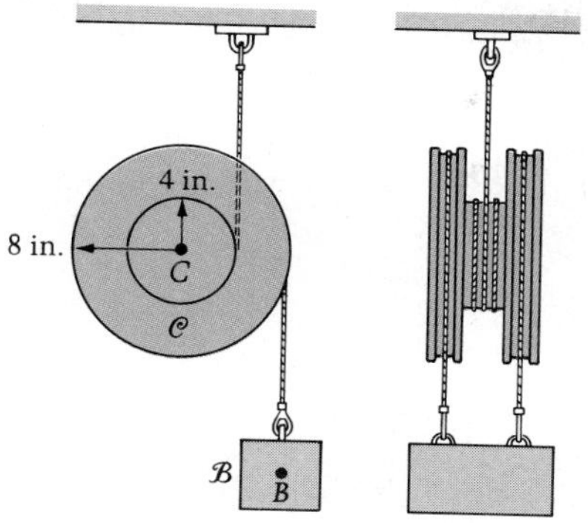

Figure P4.191

4.192 The uniform solid sphere in Figure P4.192, rotating at angular velocity Ω about a horizontal diameter, is gently placed on a rough horizontal plane with coefficient of friction μ. Prove that the center C of the sphere moves a distance $2\Omega^2 a^2/(49\,\mu g)$ before slipping stops. Find the angular speed at that instant.

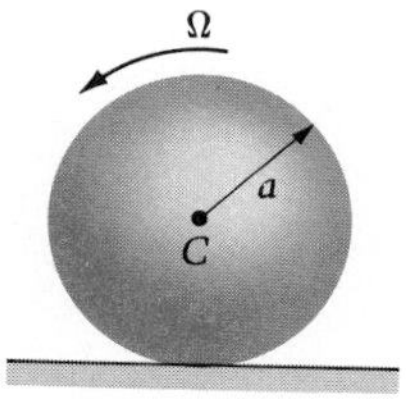

Figure P4.192

4.193 After release from a slightly displaced position, the rod in Figure P4.193 will remain in contact with the floor throughout its fall. Describe the path of C and find the reaction onto the floor just before the rod becomes horizontal.

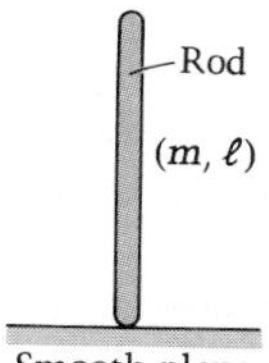

Figure P4.193

4.194 Figure P4.194 shows a solid circular disk of weight 64.4 lb and radius 4 ft connected to a 161-lb block by a massless link $\mathcal{L}$. The coefficient of friction between disk and plane, and also block and plane, is $\frac{1}{2}$. The block is given an initial velocity of 5 ft/sec up the plane. For motion up the plane, find: (a) the acceleration of the block; (b) the magnitudes and directions of the friction forces on the disk and block.

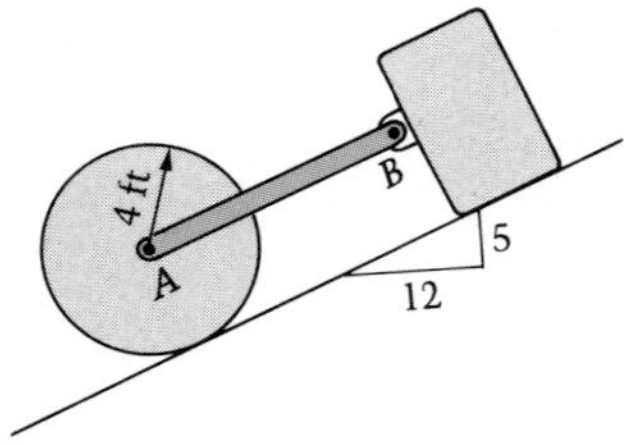

Figure P4.194

4.195 In Example 4.28 determine q and d so that the equation $4R^2r = qd^2$ still holds but, in addition, the moment of inertia of the total mass about axis z_C is maximized for a given offset r.

4.196 Explain why the thin body in Figure P4.196, asymmetric though it is, is nonetheless dynamically balanced.

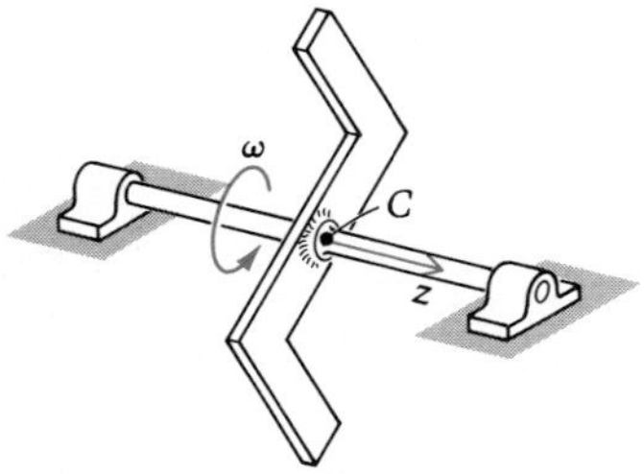

Figure P4.196

4.197 Review the Problems 2.86 and 2.87. Then refer to Figure P4.197, and make the following simplifying assumptions:

1. $|\mathbf{r}_{GC}| = 93 \times 10^6$ mi.
2. $r_{\text{earth}} = 4000$ mi.
3. Earth's axis tilt is zero.
4. Earth's orbit is circular.
5. Earth is round.
6. Angular speed of line GC is 2π rad/yr in an inertial frame.
7. Angular speed of earth is 2π rad/day in the inertial frame.
8. Gravitational acceleration at the poles is 32.2 ft/sec^2.

(a) A man (5 slugs) stands on a scale at the equator at midnight as shown in Figure P4.197. Let the inertial frame have its origin fixed at the mass center of the sun; also, assume the earth's orbital curve to be fixed in this frame. Calculate the reading of the man's weight and the percent difference of this result from 161 lb. (b) Repeat the problem if the man weighs himself at noon.

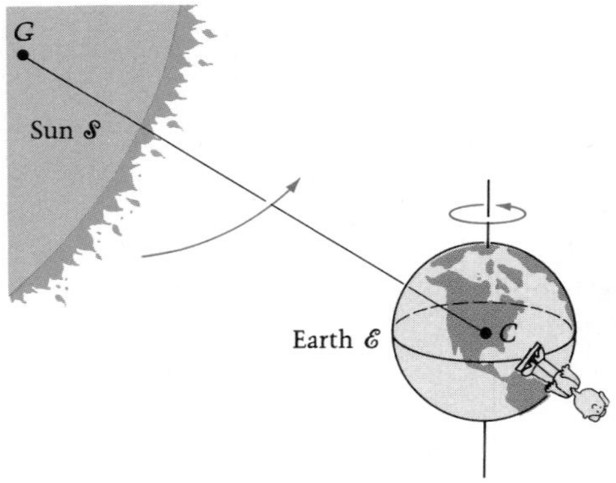

Figure P4.197

Answers to Questions / Chapter 4

Q4.1 Any integral such as $\int_{\mathcal{B}} \mathbf{x}\, dm$, where $\mathbf{x}$ is measured from an origin at, say, Q is equal to $m\mathbf{x}_{QC}$; this *is* the definition of the mass center. So if the origin is C, then $\mathbf{x}_{CC} = 0$ and $\int_{\mathcal{B}} \mathbf{x}\, dm = 0$.

Q4.2 Both products of inertia are zero for the body shown in the diagram. Three bars of any lengths lying along the ($\mathbf{x}$, $\mathbf{y}$, $\mathbf{z}$) axes are joined at the origin to form a rigid body. Neither the $\mathbf{z}$ axis nor the $\mathbf{xy}$ plane is one of symmetry, yet I_{xz_O} and I_{yz_O} are zero.

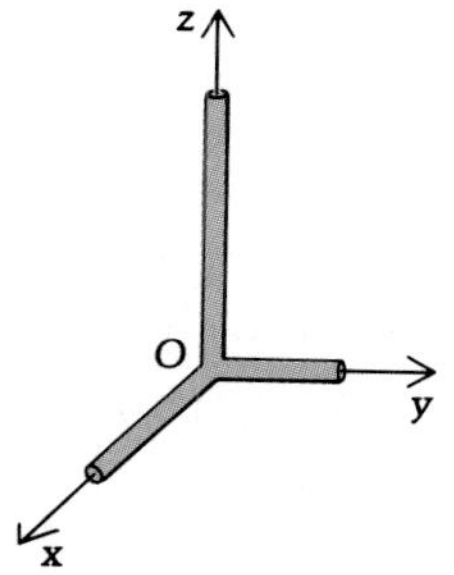

Q4.3 Let ($\mathbf{x}$, $\mathbf{y}$, $\mathbf{z}$) be fixed in the reference frame $\mathcal{J}$. As the body rotates with respect to frame $\mathcal{J}$, its mass is then distributed differently, at different times, with respect to $\mathbf{x}$ and $\mathbf{y}$. But in plane motion the $\mathbf{z}$ axis is fixed in direction in *both* the body and in space. Thus since the square of the distance from the $\mathbf{z}$ axis is always the same r^2 regardless of the orientation of $\mathbf{x}$ and $\mathbf{y}$, we see that I_{z_C} does not change as the body turns.

Q4.4 So that the summation of moments about C can be zero. Note that $M_{z_C} = I_C\ddot{\theta} = 0$ for translation.

Q4.5 Although $\mathbf{H}_O$ may always be written as $I_{xz_O}\omega\hat{\mathbf{i}} + I_{yz_O}\omega\hat{\mathbf{j}} + I_{z_O}\omega\hat{\mathbf{k}}$ whenever $\mathbf{v}_O$ is zero, its derivative is only equal to $\mathbf{M}_{z_O}$ when $\mathbf{v}_O$ is *identically* zero—in other words, *zero all the time.*

Q4.6 Yes, for then the friction force would be in the opposite direction and ΣM_{O_1} would give a larger N. The normal force N would then have to balance the moments about O_1 of all three of P, f, and the weight.

Q4.7 If $M = 2m$, then (when the couple is not present) $T = mg$ and $\ddot{\theta} = 0$ are solutions to the problem and there is *no* motion. If $M < 2m$, the block moves *downward* and $\ddot{\theta}$ is negative (↺).

Q4.8 When both P and C are in the reference plane, $\mathbf{r}_{PC}$ has only $\mathbf{x}$ and $\mathbf{y}$ components. The same is true of $\mathbf{a}_P$ because of the plane motion. Thus $\mathbf{r}_{PC} \times \mathbf{a}_P$ has to be perpendicular to the $\mathbf{xy}$ plane—that is, in the $\mathbf{z}$ direction.

Q4.9 Because the initial acceleration of the mass center C will be to the right and downward (↘) and the initial angular acceleration will be clockwise (↻). It is a good idea to set the coordinates in the direction of the motion if it is known.

Review Questions / Chapter 4

True or False?

These questions all refer to rigid bodies in plane motion.

1. Euler's second law enables us to study the rotational motion of rigid bodies.
2. The moment of inertia is always positive, whereas the products of inertia can have either sign.
3. The formula $m\ell^2/12$ gives the exact value of the moment of inertia of a slender rod about a lateral axis through its mass center.
4. Euler's second law, $\mathbf{M}_{r_O} = \dot{\mathbf{H}}_O$, is valid only in an inertial frame (meaning that the position vectors and velocities inherent in $\mathbf{H}_O$, the origin O, and the time derivative are all taken in an inertial frame).
5. In $I_{z_P} = I_{z_C} + md^2$, the quantity d is the distance between the points P and C. (C is in the reference plane, whereas P is *any* point of the body.)
6. Euler's second law, $\mathbf{M}_{r_O} = \dot{\mathbf{H}}_O$, applies to deformable bodies, liquids, and gases, as well as to rigid bodies.
7. If Ⓘ represents the instantaneous center of zero velocity, then $M_{r_{Ⓘ}} \neq I_{Ⓘ}\alpha$ in general.
8. Products of inertia are not found in the equations of plane motion.
9. $\mathbf{M}_{r_C} = \dot{\mathbf{H}}_C$ is just as general as $\mathbf{M}_{r_O} = \dot{\mathbf{H}}_O$, where O is fixed in an inertial frame.
10. In translation problems, the moments of external forces and couples taken about any point add to zero.
11. Suppose you buy a new set of automobile tires and a dynamic balance is performed on each wheel by adding weights in two planes (inner and outer rims). The products of inertia I_{xz} and I_{yz} have thus been eliminated, which otherwise would have caused bearing reactions and vibration.
12. $M_{r_C} = I_C\alpha$ applies to deformable as well as to rigid bodies, as long as they are in plane motion.

Answers: T, T, F, T, F, T, T, F, T, F, T, F

CHAPTER

5

Special Integrals of the Equations of Plane Motion of Rigid Bodies: Work-Energy and Impulse-Momentum Methods

chapter outline ►

5.1 Introduction

In this chapter we shall still be concerned with the kinetics of rigid bodies in plane motion, but the methods we employ will result from special solutions of the equations of Chapter 4. It turns out that we can effect certain special integrations of these equations in general, thereby proving three principles that are often useful in solving kinetics problems: the principles of work and kinetic energy, impulse and momentum, and angular impulse and angular momentum. The three principles will be seen to be nothing more than various first integrals of the differential equations of motion (which, as we have seen, are themselves second-order equations in time). Thus the principles of this chapter allow us to begin our solutions halfway between accelerations and positions. They therefore involve velocities but not accelerations.

5.2 The Principle(s) of Work and Kinetic Energy

Kinetic Energy of a Rigid Body in Plane Motion

There is a principle, derived from the equations of motion, that will help us to solve for unknowns of interest in kinetics problems. In this section we shall see that this principle arises from first deriving and then differentiating the kinetic energy of the body.

Kinetic energy is usually denoted by the letter T; for any body or system of bodies, it is defined as the summation of $\frac{1}{2}(dm)v^2$ over all its elements of mass:

$$T = \frac{1}{2}\int (\mathbf{v} \cdot \mathbf{v})\, dm \tag{5.1}$$

In this section we need to specialize Definition (5.1) for a rigid body $\mathcal{B}$ in plane motion. To this end we kinematically relate the velocity $\mathbf{v}$ of the differential mass to the velocity of the mass center (Figure 5.1):

$$\mathbf{v} = \mathbf{v}_C + \omega\hat{\mathbf{k}} \times (\mathrm{x}\hat{\mathbf{i}} + \mathrm{y}\hat{\mathbf{j}}) = \mathbf{v}_C + \omega(-\mathrm{y}\hat{\mathbf{i}} + \mathrm{x}\hat{\mathbf{j}})$$

We note that the x and y axes are fixed in $\mathcal{B}$ with their origin at C. Forming v^2, that is, $\mathbf{v} \cdot \mathbf{v}$, we have

$$\mathbf{v} \cdot \mathbf{v} = \mathbf{v}_C \cdot \mathbf{v}_C + \omega^2(\mathrm{x}^2 + \mathrm{y}^2) + 2\omega\mathbf{v}_C \cdot (\mathrm{x}\hat{\mathbf{j}} - \mathrm{y}\hat{\mathbf{i}}) \tag{5.2}$$

Thus the kinetic energy becomes, substituting (5.2) into (5.1),

$$T = \frac{1}{2}\mathbf{v}_C \cdot \mathbf{v}_C \int dm + \frac{\omega^2}{2}\int (\mathrm{x}^2 + \mathrm{y}^2)\, dm + \omega(\mathbf{v}_C \cdot \hat{\mathbf{j}})\int \mathrm{x}\, dm - \omega(\mathbf{v}_C \cdot \hat{\mathbf{i}})\int \mathrm{y}\, dm$$

Recognizing the moment of inertia integral in the second term, we obtain

$$T = \frac{1}{2}mv_C^2 + \frac{1}{2}I_{z_C}\omega^2 \tag{5.3}$$

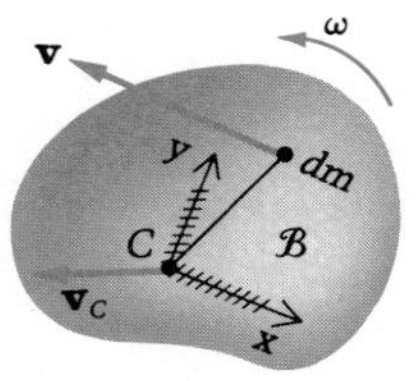

Figure 5.1

in which $\mathbf{v}_C \cdot \mathbf{v}_C = |\mathbf{v}_C|^2 = v_C^2$, the square of the magnitude of the velocity of C.

> **Question 5.1** Why do $\int x\,dm$ and $\int y\,dm$ both vanish?

We note that the (scalar) kinetic energy has two identifiable ***parts*** (*not* components!): one relating to the motion of the mass center C ($T_v = \frac{1}{2}mv_C^2$) and the other to the motion of the body relative to C ($T_\omega = \frac{1}{2}I_{z_C}\omega^2$). This clear division of T even exists in general motion of rigid bodies (that is, in three dimensions), though there are more terms in T_ω then.

EXAMPLE 5.1

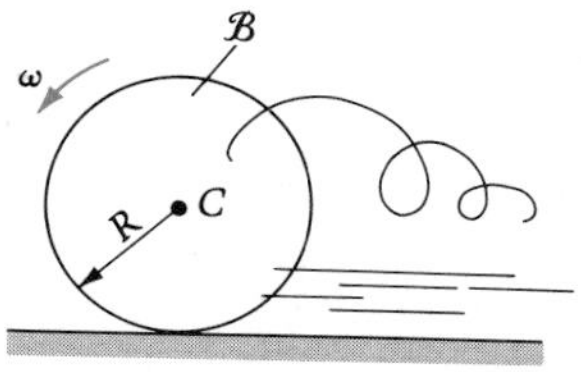

Calculate the kinetic energy of a round rolling body $\mathcal{B}$ having mass m, radius R, and radius of gyration k_C with respect to the z_C axis. The mass center C (see the diagram) lies at the geometric center.

SOLUTION

$$T = \frac{1}{2}mv_C^2 + \frac{1}{2}I_{z_C}\omega^2$$

$$= \frac{1}{2}m[(R\omega)^2 + k_C^2\omega^2]$$

$$= \frac{mR^2\omega^2}{2}\left[1 + \left(\frac{k_C}{R}\right)^2\right]$$

Note that if $\mathcal{B}$ is a solid cylinder, then $k_C = R/\sqrt{2}$ and

$$T = \frac{mR^2\omega^2}{2}\left(1 + \frac{1}{2}\right)$$

In this case, two-thirds of the kinetic energy rests in the translation term of T ($\frac{1}{2}mv_C^2$).

If $\mathcal{B}$ is a ring (or hoop), however, $I_{z_C} = mR^2$ so that $k_C = R$ and

$$T = \frac{mR^2\omega^2}{2}(1 + 1)$$

and this time half the kinetic energy is in each of the translational and rotational terms.

EXAMPLE 5.2

Compare the translational and rotational parts of the kinetic energy of the earth (neglecting the axis tilt and assuming the orbit to be circular).

SOLUTION

Let

$$R_o = \text{average orbital radius} \approx 93(10^6)\text{ mi}$$

$$R_e = \text{earth's radius} \approx 4000\text{ mi}$$

$$\omega_d = \text{angular velocity of spin} = 2\pi\text{ rad/day}$$

$$\omega_o = \text{angular velocity in orbit around sun} = 2\pi\text{ rad/year}$$

Then

$$\begin{aligned}
T &= \frac{1}{2}mv_C^2 + \frac{1}{2}I_{z_C}\omega^2 \\
&= \frac{1}{2}m(R_o\omega_o)^2 + \frac{1}{2}\left(\frac{2}{5}mR_e^2\right)(\omega_d + \omega_o)^2 \\
&= \frac{1}{2}mR_o^2\omega_o^2\left[1 + \frac{2}{5}\left(\frac{R_e}{R_o}\right)^2\left(1 + \frac{\omega_d}{\omega_o}\right)^2\right] \\
&= \frac{1}{2}mR_o^2\omega_o^2\left[1 + \frac{2}{5}\left(\frac{4}{93{,}000}\right)^2(1 + 365)^2\right] \\
&= \frac{1}{2}mR_o^2\omega_o^2(1 + 0.000099)
\end{aligned}$$

We have assumed the earth to be a solid sphere having uniform density. (Its average density, incidentally, is 5.51 times that of water, but it is neither spherical nor uniform nor completely solid!) It is seen that in this case the spin (rotational) term is tiny compared to $\frac{1}{2}mv_C^2$.

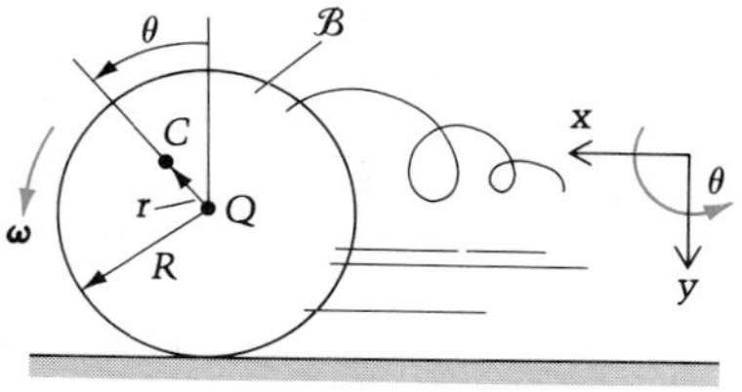

EXAMPLE 5.3

Work Example 5.1 for the case when the mass center C is offset by a distance r from the geometric center Q of a round rolling body $\mathcal{B}$. (See the diagram.)

SOLUTION

In order to use our equation for kinetic energy,

$$T = \frac{1}{2} m v_C^2 + \frac{1}{2} I_C \omega^2$$

we must first calculate v_C^2:

$$\mathbf{v}_C = \mathbf{v}_Q + \dot{\theta}\hat{\mathbf{k}} \times \underbrace{\mathbf{r}_{QC}}_{r \sin\theta\hat{\mathbf{i}} - r\cos\theta\hat{\mathbf{j}}}$$

$$= (R\dot{\theta} + r\dot{\theta}\cos\theta)\hat{\mathbf{i}} + r\dot{\theta}\sin\theta\hat{\mathbf{j}}$$

Therefore

$$v_C^2 = R^2\dot{\theta}^2 + r^2\dot{\theta}^2 + 2Rr\dot{\theta}^2\cos\theta$$

Substituting, we get

$$T = \frac{mR^2\dot{\theta}^2}{2}\left[1 + \frac{2r}{R}\cos\theta + \left(\frac{r}{R}\right)^2 + \left(\frac{k_C}{R}\right)^2\right]$$

Note that if $r = 0$, the answer agrees as it should with Example 5.1.

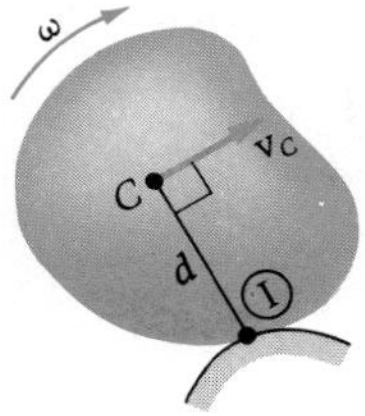

Figure 5.2

There is an alternative means of writing the kinetic energy T of a rigid body in plane motion by making use of the instantaneous center of zero velocity Ⓘ (Figure 5.2):

$$T = \overbrace{\frac{1}{2} m v_C^2}^{T_v} + \overbrace{\frac{1}{2} I_C \omega^2}^{T_\omega}$$

$$= \frac{1}{2} m (d\omega)^2 + \frac{1}{2} I_C \omega^2$$

$$= \frac{1}{2}(I_C + md^2)\omega^2$$

Thus by using the parallel-axis theorem we obtain

$$T = \frac{1}{2} I_{Ⓘ} \omega^2 \qquad (5.4)$$

The translational (T_v) and rotational (T_ω) terms composing the scalar T are thus seen to collapse into the one term $\frac{1}{2}I_{\textcircled{I}}\omega^2$ if we choose to work with Ⓘ instead of C.

As an example, we consider a rolling cylinder again (see Figure 5.3):

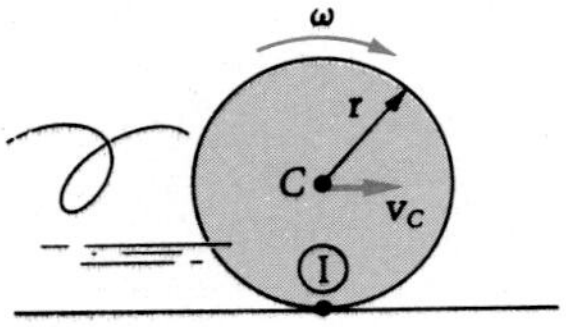

Figure 5.3

$$T = T_v + T_\omega = \frac{1}{2}mv_C^2 + \frac{1}{2}I_C\omega^2$$

$$= \frac{1}{2}m(r\omega)^2 + \frac{1}{2}\left(\frac{1}{2}mr^2\right)\omega^2$$

$$= \frac{1}{2}mr^2\omega^2\left(1 + \frac{1}{2}\right) = \frac{3}{4}mr^2\omega^2$$

We have noted that two-thirds of the cylinder's kinetic energy is associated with the "translational part" of T and one-third with the "rotational part." If we now use Ⓘ, we get *all* of T at once:

$$T = \frac{1}{2}I_{\textcircled{I}}\omega^2 = \frac{1}{2}\underbrace{\left[\underbrace{\frac{1}{2}mr^2}_{I_C} + \underbrace{mr^2}_{\text{transfer term}}\right]}_{I_{\textcircled{I}}}\omega^2 = \frac{3}{4}mr^2\omega^2 \qquad \text{(as above)}$$

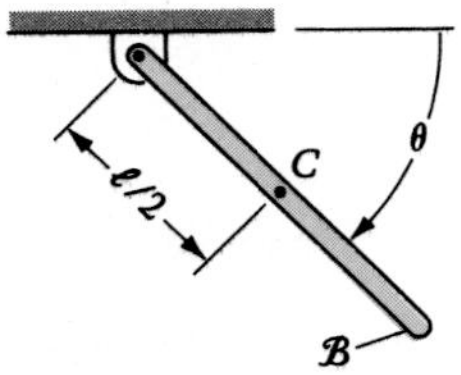

Figure 5.4

As a second example of the use of Equation (5.4), consider the slender rod $\mathcal{B}$ swinging about a pivot at A as shown in Figure 5.4. The kinetic energy of $\mathcal{B}$ may be found in either of two ways:

$$T = \frac{1}{2}I_{\textcircled{I}}\omega^2 \qquad\qquad T = \frac{1}{2}mv_C^2 + \frac{1}{2}I_C\omega^2$$

$$= \frac{1}{2}\left[\frac{1}{3}m\ell^2\right]\omega^2 \qquad\qquad = \frac{1}{2}m\left(\frac{\ell}{2}\omega\right)^2 + \frac{1}{2}\left(\frac{1}{12}m\ell^2\right)\omega^2$$

$$= \frac{m\ell^2\omega^2}{6} \qquad\qquad = m\ell^2\omega^2\left(\frac{1}{8} + \frac{1}{24}\right)$$

$$= \frac{m\ell^2\omega^2}{6}$$

Derivation of the Principle $W = \Delta T$
Work and Power of Systems of Forces and Couples

Returning now to the derivation of our principle, we next compute the rate of change of kinetic energy:

$$\frac{dT}{dt} = \frac{d}{dt}\left(\frac{1}{2}m\mathbf{v}_C \cdot \mathbf{v}_C + \frac{1}{2}I_{z_C}\omega\hat{\mathbf{k}} \cdot \omega\hat{\mathbf{k}}\right)$$

$$= \frac{1}{2}m\underbrace{(\mathbf{a}_C \cdot \mathbf{v}_C + \mathbf{v}_C \cdot \mathbf{a}_C)}_{2\mathbf{a}_C \cdot \mathbf{v}_C} + \frac{1}{2}I_{z_C}\underbrace{(\alpha\hat{\mathbf{k}} \cdot \omega\hat{\mathbf{k}} + \omega\hat{\mathbf{k}} \cdot \alpha\hat{\mathbf{k}})}_{2\alpha\hat{\mathbf{k}} \cdot \omega\hat{\mathbf{k}}}$$

Therefore

$$\frac{dT}{dt} = m\mathbf{a}_C \cdot \mathbf{v}_C + (I_{z_C}\alpha\hat{\mathbf{k}}) \cdot \omega\hat{\mathbf{k}} \tag{5.5}$$

Recalling that $\mathbf{F}_r = m\mathbf{a}_C$ and that the z component of $\mathbf{M}_{r_C}$ is $I_{z_C}\alpha$ for rigid bodies in plane motion, we may write

$$\frac{dT}{dt} = \mathbf{F}_r \cdot \mathbf{v}_C + \mathbf{M}_{r_C} \cdot \omega\hat{\mathbf{k}} \tag{5.6}$$

Question 5.2 Since $\mathbf{M}_{r_C}$ can contain x_C and y_C components (see Equations 4.39) why may we substitute the *total* vector $\mathbf{M}_{r_C}$ for just the z component $I_{z_C}\alpha\hat{\mathbf{k}}$ in Equation 5.5?

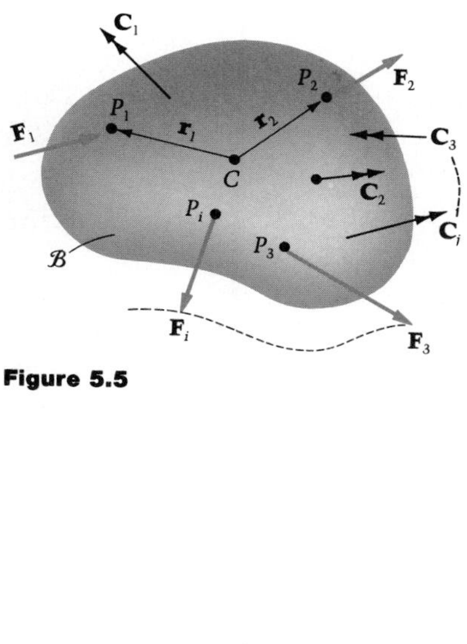

Figure 5.5

Our next goal is to get the individual external forces and couples acting on the body $\mathcal{B}$ into the equation. (See Figure 5.5.) Note the abbreviations $\mathbf{r}_{CP_1} = \mathbf{r}_1$, $\mathbf{r}_{CP_2} = \mathbf{r}_2$, and so on, of the vectors to the points of application of $\mathbf{F}_1$, $\mathbf{F}_2$, and so on. We assume that the external mechanical actions on the body arise from a system of forces ($\mathbf{F}_1$, $\mathbf{F}_2$, . . .) and couples with moment vectors ($\mathbf{C}_1$, $\mathbf{C}_2$, . . .), as shown in Figure 5.5. Further, we let ($\mathbf{v}_1$, $\mathbf{v}_2$, . . .) be the velocities of the material points (P_1, P_2, . . .) on which the forces act instantaneously.

Clearly, then, the resultant of the external forces is

$$\mathbf{F}_r = \mathbf{F}_1 + \mathbf{F}_2 + \cdots \tag{5.7}$$

and the moment of the $\mathbf{F}_i$'s and $\mathbf{C}_i$'s about C is

$$\mathbf{M}_{r_C} = \mathbf{r}_1 \times \mathbf{F}_1 + \mathbf{r}_2 \times \mathbf{F}_2 + \cdots + \mathbf{C}_1 + \mathbf{C}_2 + \cdots \tag{5.8}$$

Substituting Equations (5.7) and (5.8) into (5.6) gives

$$\begin{aligned}\frac{dT}{dt} &= (\mathbf{F}_1 + \mathbf{F}_2 + \cdots) \cdot \mathbf{v}_C + (\mathbf{r}_1 \times \mathbf{F}_1 + \mathbf{r}_2 \times \mathbf{F}_2 + \cdots) \cdot \omega\hat{\mathbf{k}} \\ &\quad + (\mathbf{C}_1 + \mathbf{C}_2 + \cdots) \cdot \omega\hat{\mathbf{k}} \\ &= (\mathbf{F}_1 + \mathbf{F}_2 + \cdots) \cdot \mathbf{v}_C + \omega\hat{\mathbf{k}} \cdot (\mathbf{r}_1 \times \mathbf{F}_1 + \mathbf{r}_2 \times \mathbf{F}_2 + \cdots) \\ &\quad + (\mathbf{C}_1 + \mathbf{C}_2 + \cdots) \cdot \omega\hat{\mathbf{k}} \end{aligned} \tag{5.9}$$

But since the dot and cross may be interchanged without altering the value of a scalar triple product,

$$\begin{aligned}\frac{dT}{dt} &= (\mathbf{F}_1 + \mathbf{F}_2 + \cdots) \cdot \mathbf{v}_C + (\omega\hat{\mathbf{k}} \times \mathbf{r}_1) \cdot \mathbf{F}_1 + (\omega\hat{\mathbf{k}} \times \mathbf{r}_2) \cdot \mathbf{F}_2 \\ &\quad + \cdots + (\mathbf{C}_1 + \mathbf{C}_2 + \cdots) \cdot \omega\hat{\mathbf{k}} \end{aligned} \tag{5.10}$$

But the velocities of P_1 and C are related:

$$\mathbf{v}_{P_1} = \mathbf{v}_1 = \mathbf{v}_C + \omega\hat{\mathbf{k}} \times \mathbf{r}_1$$

so that

$$\frac{dT}{dt} = \mathbf{F}_1 \cdot \mathbf{v}_1 + \mathbf{F}_2 \cdot \mathbf{v}_2 + \cdots + (\mathbf{C}_1 + \mathbf{C}_2 + \cdots) \cdot \omega\hat{\mathbf{k}} \tag{5.11}$$

The right-hand side of Equation (5.11) is called the **power,** or **rate of work,** of the external system of forces and couples acting on the body. The power of a force is its dot product with the velocity of the point on which it acts; the power of a couple is its dot product with the angular velocity of the body on which it acts:

$$\text{Rate of work of force } \mathbf{F}_1 = \mathbf{F}_1 \cdot \mathbf{v}_1 = \text{power of } \mathbf{F}_1 \tag{5.12}$$

$$\text{Rate of work of couple } \mathbf{C}_1 = \mathbf{C}_1 \cdot \omega\hat{\mathbf{k}} = \text{power of } \mathbf{C}_1 \tag{5.13}$$

Hence one form of the principle of this section is

$$\text{Power} = \frac{dT}{dt} \tag{5.14}$$

or

$$P = \dot{T}$$

Integrating, we obtain another principle:*

$$\int_{t_1}^{t_2} P\,dt = T(t_2) - T(t_1) = T_2 - T_1$$

or

$$W = \Delta T = \left(\frac{1}{2} m v_C^2 + \frac{1}{2} I_{z_C} \omega^2\right)\Bigg]_{t_1}^{t_2} \tag{5.15}$$

where the integral of the power is called the **work** W of the external forces and couples. It is the work done by the $\mathbf{F}_i$'s and $\mathbf{C}_i$'s on the body between the two times t_1 and t_2. Hence we have a principle that can be stated in words:

$$\begin{matrix}\text{Work done by external forces} \\ \text{and couples on } \mathcal{B}\end{matrix} = \begin{matrix}\text{change of kinetic} \\ \text{energy of } \mathcal{B}\end{matrix}$$

It is essential to recognize that our derivation of the principle of work and kinetic energy depends crucially on the body being rigid. In fact the work of external forces on a deformable body is *not* in general equal to the change in its kinetic energy. That is the case even when the "deformable" body is composed of several individually rigid parts. However, there are a number of special circumstances, usually easy to recognize, for which the principle is valid for such a system of rigid bodies.

*Sometimes (t_i, t_f) is used to denote the time interval, rather than (t_1, t_2); the subscripts stand for "initial" and "final" values.

To give an example for which this is true, suppose we have two rigid bodies, $\mathcal{B}_1$ and $\mathcal{B}_2$, making up the system, and suppose the bodies are connected by a pin (or hinge) with negligible friction as shown in Figure 5.6.

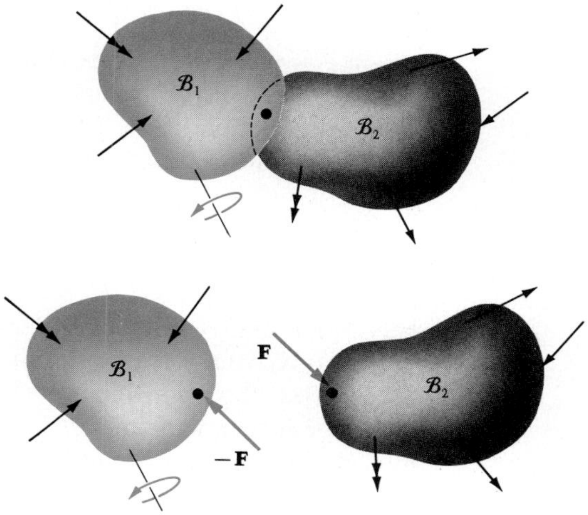

Figure 5.6

Let $\mathbf{F}$ be the force exerted by $\mathcal{B}_1$ on $\mathcal{B}_2$ at the pin, and consequently $-\mathbf{F}$ is the force exerted by $\mathcal{B}_2$ on $\mathcal{B}_1$. Furthermore, let

$\mathbf{v}$ = common velocity of attachment points in the two bodies

P_{E_1} = power (rate of work) of forces acting on $\mathcal{B}_1$ that are also external to system

P_{E_2} = power of forces acting on $\mathcal{B}_2$ that are also external to system

$T_{\mathcal{B}_1}$ = kinetic energy of $\mathcal{B}_1$

$T_{\mathcal{B}_2}$ = kinetic energy of $\mathcal{B}_2$

Now if we apply Equations (5.11, 14) to each of the bodies, we obtain

$$P_{E_1} + (-\mathbf{F}) \cdot \mathbf{v} = \frac{dT_{\mathcal{B}_1}}{dt}$$

and

$$P_{E_2} + \mathbf{F} \cdot \mathbf{v} = \frac{dT_{\mathcal{B}_2}}{dt}$$

which may be added to yield

$$P_{E_1} + P_{E_2} = \frac{d}{dt}(T_{\mathcal{B}_1} + T_{\mathcal{B}_2})$$

or

$$P = \frac{dT}{dt}$$

where P is the power of the external forces on the system and T is the kinetic energy of the system.

With friction in the pin we also would have interactive couples $\mathbf{C}$ and $-\mathbf{C}$, and the sum of their work rates would be

$$\mathbf{C} \cdot (\omega_{\mathcal{B}_2} - \omega_{\mathcal{B}_1})\hat{\mathbf{k}}$$

which in general would *not* vanish.* This net rate of work of friction couples would be negative, reflecting the fact that the friction will reduce the kinetic energy of the system. We can expect the principle of work (of external forces) and kinetic energy to be valid for a system of rigid bodies whenever the interaction of the bodies leads neither to dissipation of mechanical energy by friction nor to a storing of energy as in a spring. When in doubt, follow the procedure we have just been through—that is, apply Equation (5.14) to each of the bodies, add the equations, and see whether the rates of work of interactive forces cancel out.

Computing the Work Done by Various Types of Forces and Moments

Before we can put Equation (5.15) to use, it is essential to demonstrate how to compute the work W done on $\mathcal{B}$ by five common types of forces and moments.

Type 1: $\mathbf{F}_1$ is constant. In this case,

$$W = \int \mathbf{F}_1 \cdot \mathbf{v}_1 \, dt = \mathbf{F}_1 \cdot \int \mathbf{v}_1 \, dt \tag{5.16}$$

Type 2: $\mathbf{F}_1$ acts on the same point P_1 of $\mathcal{B}$ throughout its motion. In this case,

$$W = \int_{t_i}^{t_f} \mathbf{F}_1 \cdot \mathbf{v}_1 \, dt = \int_{t_i}^{t_f} \mathbf{F}_1 \cdot \frac{d\mathbf{r}_1}{dt} \, dt = \int_{\mathbf{r}_i(t_i)}^{\mathbf{r}_f(t_f)} \mathbf{F}_1 \cdot d\mathbf{r}_1 \tag{5.17}$$

where $\mathbf{r}_{OP_1} = \mathbf{r}_1$ and i and f denote starting (initial) and ending (final) times and positions. It is true, of course, that the velocity $\mathbf{v}_1$, which combines with $\mathbf{F}_1$ to produce its power, is at each instant the derivative

*It would vanish, of course, if the friction were enough to prevent relative rotation so that $\omega_{\mathcal{B}_1} = \omega_{\mathcal{B}_2}$; then the system would behave as a single rigid body!

of *some* position vector. If the force acts on ***different*** material points of $\mathcal{B}$ at different times throughout a motion (such as friction from a brake), however, the path integral $\int \mathbf{F}_1 \cdot d\mathbf{r}_{OP_1}$ has no real functional utility and the general $\int \mathbf{F}_1 \cdot \mathbf{v}_1\, dt$ must be used.

Type 3: $\mathbf{F}_1$ is due to gravity. This is an example of both Types 1 *and* 2. Thus, letting z be positive downward, we get*

$$W = \int mg\hat{\mathbf{k}} \cdot d\mathbf{r}_{OC} = mg\hat{\mathbf{k}} \cdot \int d\mathbf{r}_{OC}$$

Expressing the differential of the position vector in terms of rectangular cartesian coordinates, we get

$$d\mathbf{r}_{OC} = dx_C\,\hat{\mathbf{i}} + dy_C\,\hat{\mathbf{j}} + dz_C\,\hat{\mathbf{k}}$$

and substituting we obtain a simple result for the work of gravity:

$$W = mg \int_{z_{C_1}}^{z_{C_2}} dz_C = mg(z_{C_2} - z_{C_1}) = mgh \tag{5.18}$$

Note that gravity does positive work if the body moves downward. (Indeed, a good rule of thumb to remember is that a force does positive work if it "gets to move in the direction it wants to"—that is, it has a component in the direction of the motion of the point on which it acts. If it does not, it does negative work during the motion of that point.)

Type 4: $\mathbf{F}_1$ is the force in a linear spring connected to the same two points P and Q of bodies $\mathcal{B}$ and $\mathcal{R}$ during an interval of their motions. (See Figure 5.7 on the next page.) We denote:

k = spring modulus (which when multiplied by the stretch yields the force in the linear spring)

ℓ_u = unstretched length

δ = stretch ($\delta < 0$ if compressed)

$\hat{\mathbf{u}}$ = unit vector along spring toward body $\mathcal{B}$

We first note that the work of spring S on body $\mathcal{B}$ is

$$W_{s \text{ on } \mathcal{B}} = \int_{t_1}^{t_2} \mathbf{F} \cdot \mathbf{v}_P\, dt = \int_{t_1}^{t_2} -k\delta\,\hat{\mathbf{u}} \cdot \mathbf{v}_P\, dt$$

Using

$$\mathbf{r}_{OP} = \mathbf{r}_{OQ} + (\ell_u + \delta)\hat{\mathbf{u}}$$

*The work done by *any* constant force $\mathbf{F}$ always acting on the same point with position vector $\mathbf{r}$ is thus $\mathbf{F} \cdot [\mathbf{r}(t_f) - \mathbf{r}(t_i)]$.

we may differentiate and obtain

$$\mathbf{v}_P = \mathbf{v}_Q + \dot{\delta}\hat{\mathbf{u}} + (\ell_u + \delta)\dot{\hat{\mathbf{u}}}$$

Therefore, substituting for $\mathbf{v}_P$, we get

$$W_{s \text{ on } \mathcal{B}} = \int_{t_1}^{t_2} -k\,\delta\hat{\mathbf{u}} \cdot [\mathbf{v}_Q + \dot{\delta}\hat{\mathbf{u}} + (\ell_u + \delta)\dot{\hat{\mathbf{u}}}]\,dt$$

$$= -\int_{t_1}^{t_2} k\,\delta\hat{\mathbf{u}} \cdot \mathbf{v}_Q\,dt - k\int_{t_1}^{t_2} \delta\dot{\delta}\,dt - k\int_{t_1}^{t_2} \delta(\ell_u + \delta)\hat{\mathbf{u}} \cdot \dot{\hat{\mathbf{u}}}\,dt$$

Since the derivative of a unit vector is perpendicular to the unit vector, the last integral vanishes and we obtain

$$W_{s \text{ on } \mathcal{B}} = -W_{s \text{ on } \mathcal{R}} - k\int_{\delta_1}^{\delta_2} \delta d\delta$$

Thus

$$W_{s \text{ on } \mathcal{B}} + W_{s \text{ on } \mathcal{R}} = W_{s \text{ on system of } (\mathcal{B} + \mathcal{R})} = \frac{k}{2}(\delta_1^2 - \delta_2^2) \qquad (5.19)$$

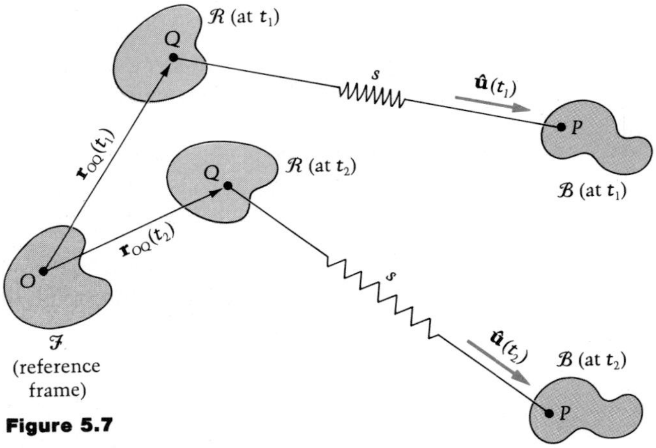

Figure 5.7

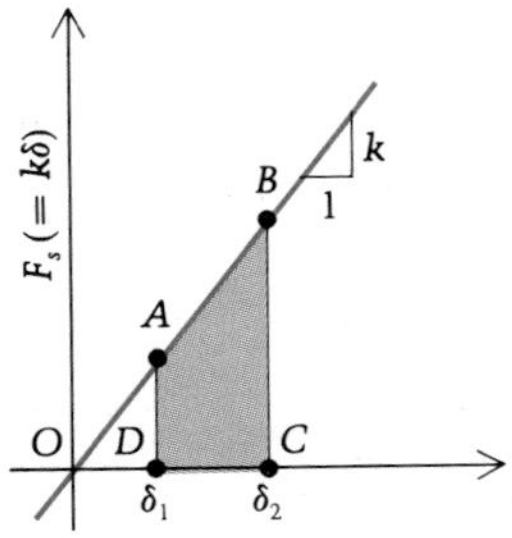

Figure 5.8

If Q is fixed in the reference frame $\mathcal{F}$, the work of S on $\mathcal{B}$ alone is given by the right side of (5.19);* if Q moves, however, we can only say that the total work on *both* bodies by S is given by $(k/2)(\delta_1^2 - \delta_2^2)$.

We note from the spring's force-stretch diagram (Figure 5.8) that the work done by the spring is in fact the negative of the change in energy E stored in it; namely, in stretching from δ_1 to δ_2,

*And its work on $\mathcal{R}$ is of course then zero.

$$E = (\text{area of triangle } OCB) - (\text{area of } ODA)$$
$$= \frac{k}{2}(\delta_2^2 - \delta_1^2)$$

Type 5: We have a couple **C**. In this case, the work of the couple in plane motion is given by

$$W = \int_{t_1}^{t_2} \mathbf{C} \cdot \boldsymbol{\omega}\, dt = \int_{t_1}^{t_2} C\hat{\mathbf{k}} \cdot \dot{\theta}\hat{\mathbf{k}}\, dt$$
$$= \int_{t_1}^{t_2} C\dot{\theta}\, dt \quad \text{or} \quad \int_{\theta_1}^{\theta_2} C\, d\theta \tag{5.20}$$

Thus if C is constant, the work of the couple is given by

$$W = C(\theta_2 - \theta_1) \tag{5.21}$$

That is, the work of C is the strength of the couple times the angle through which the body turns. As with the work of forces, the couple's work is positive if it "gets to move" in the direction in which it acts (or turns, in this case).

Examples Solved by the Principle $W = \Delta T$

We are now in a position to solve some problems by using the principle of work and kinetic energy. A number of examples follow.

EXAMPLE 5.4

This example illustrates the work done by forces and couples belonging to Types 1, 2, and 5 on the preceding pages. The force $F = 52$ lb is applied to the uniform cylinder $\mathcal{C}$ at rest in Figure 1. (This type of force might be applied by a cord on a hub, as is suggested by Figure 2.) If force F continues to act with the same magnitude and direction as the cylinder rolls, find:

a. The work done by F during transit to the dotted position

b. The velocity of C and the angular velocity of the cylinder in the dotted position

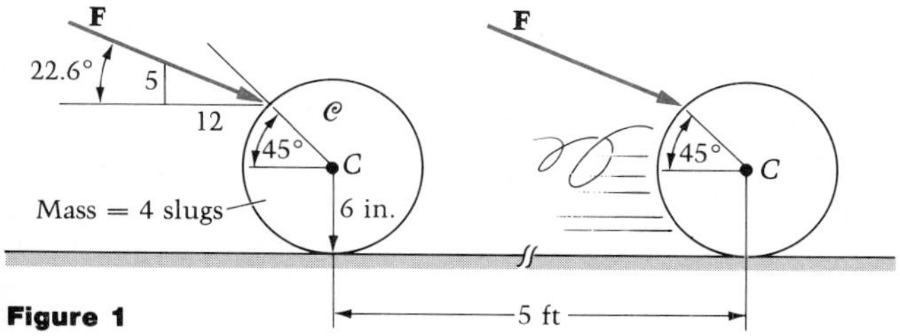

Figure 1

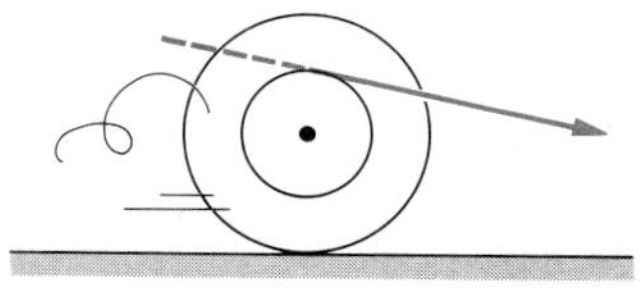

Figure 2

SOLUTION

We shall work part (a) in two ways. First, the definition of the work of **F** is

$$W = \int \mathbf{F} \cdot \mathbf{v}_Q \, dt$$

where Q is the point of $\mathscr{C}$ in contact with **F** at any time. The geometry in Figure 3 gives an angle of 45.1° between **F** and $\mathbf{v}_Q$, since

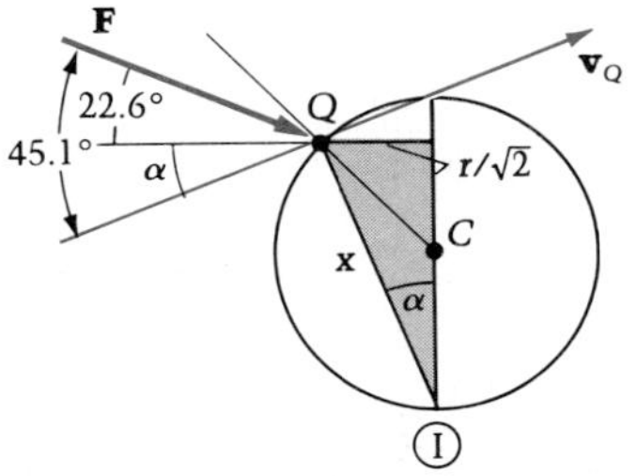

Figure 3

$$\alpha = \tan^{-1}\left(\frac{r/\sqrt{2}}{r + r/\sqrt{2}}\right) = 22.5°$$

Also,

$$r_{ⒾQ} = \sqrt{\left(\frac{1}{\sqrt{2}}\right)^2 + \left(1 + \frac{1}{\sqrt{2}}\right)^2}\, r = 1.85r$$

Therefore

$$\mathbf{F} \cdot \mathbf{v}_Q = F(r_{ⒾQ}\dot{\theta}) \cos 45.1°$$

and

$$W = \int (52 \cos 45.1°) 1.85\, r \frac{d\theta}{dt}\, dt$$

$$= (52 \cos 45.1°)(1.85) \frac{\theta}{2}\Big]_0^{5/0.5}$$

$$= 340 \text{ ft-lb}$$

A second and simpler approach is to note that **F** at Q may be moved to C, using the idea of resultants, as in Figure 4. The force at Q is replaced by the force and couple at C which produce the same effect on the rigid body.

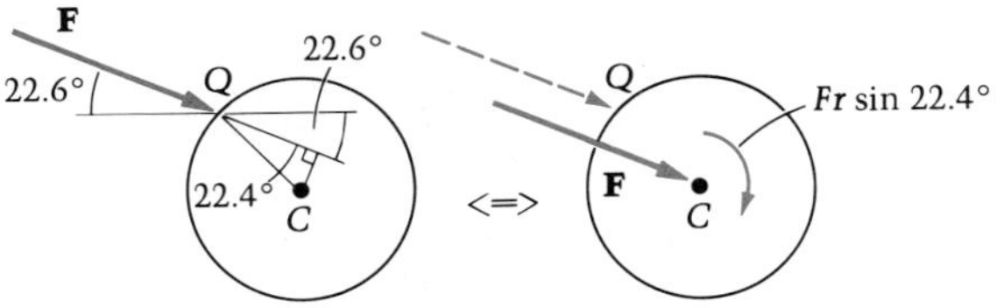

Figure 4

Since, at C, force **F** always acts on the *same point of the body* (it didn't at Q!) we may write

$$W = \text{work of } F \text{ at Q}$$

$$= (\text{work of } F \text{ at } C) + (\text{work of couple on } \mathscr{C})$$

$$= (F \cos 22.6°)x_C + (Fr \sin 22.4°)\theta$$

$$= (0.923F)5 + (0.191F)10$$

$$= 6.53F = 340 \text{ ft-lb} \qquad \text{(as before)}$$

(Continued)

Note that the work of a constant couple in plane motion is simply the moment of the couple times the angle through which the body turns.

For part (b) we equate the work to the change in the kinetic energy of $\mathscr{C}$:

$$W = \Delta T = \frac{1}{2}mv_{C_f}^2 + \frac{1}{2}I_G\omega_f^2 - 0 \qquad \text{(initial } T = 0)$$

$$340 = \frac{1}{2}4v_{C_f}^2 + \frac{1}{2}\left(\frac{1}{2}\cdot 4 \cdot \frac{1}{4}\right)\left(\frac{v_{C_f}}{1/2}\right)^2$$

$$v_{C_f} = \sqrt{\frac{340}{2+1}} = 10.6 \text{ ft/sec}$$

Note that the gravity, friction, and normal forces do no work in this problem. There are two reasons for this:

1. $(-mg\hat{\mathbf{j}}) \cdot (v_C\hat{\mathbf{i}}) = 0.$
2. The normal and friction forces act on a point of zero velocity.

Question 5.3 If the cylinder slips, do the normal and friction forces *then* do work?

EXAMPLE 5.5

This example illustrates the work done by gravity, the third item described in the preceding pages. In Example 4.20 suppose that wheel $\mathscr{C}$ is released from rest (with the couple $C = 0$). Find the angle θ at which $\mathscr{C}$ first stops if $M = 4m$.

SOLUTION

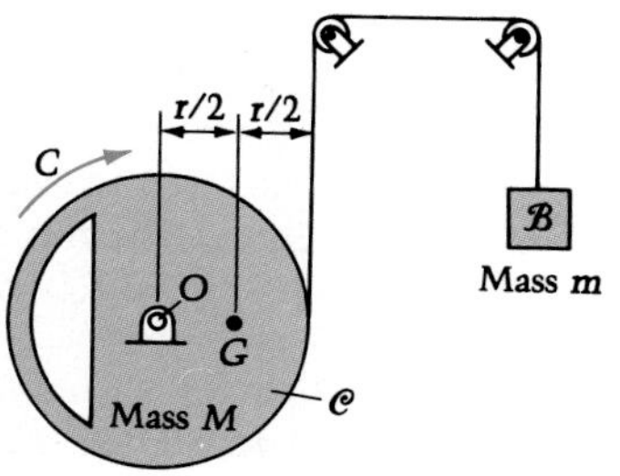

($\theta = 0$ at $t = 0$)
Mass center = G

Only gravity does work, and $T_i = T_f = 0$. Thus

$$W = \Delta T = T_f - T_i = 0 - 0 = 0$$

$$Mg\frac{r}{2}\sin\theta - mgr\theta = 0$$

$$\sin\theta = \frac{2m}{M}\theta$$

We see that the solution is the angle θ (in radians) where the two functions $f_1 = \sin\theta$ and $f_2 = (2m/M)\theta$ are equal. (See the diagram.)

(Continued)

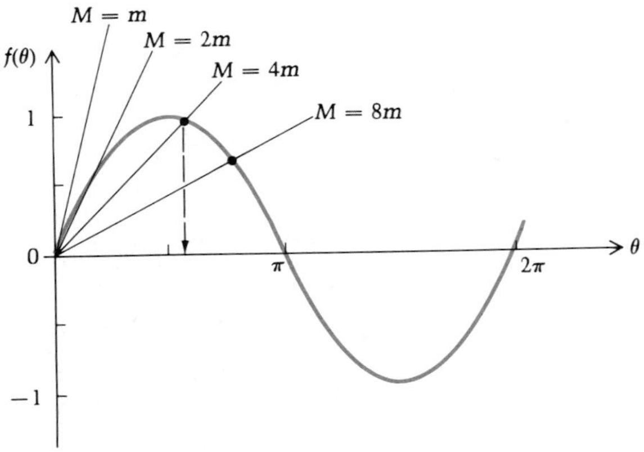

Note that if $2m/M > 1$, there is no solution—that is, no intersection of f_1 and f_2, meaning that $\sin\theta$ cannot equal $(2m/M)\theta$. This is because if $m > M/2$, then upon release block $\mathcal{B}$ moves ***downward*** (M_{r_O} is then counterclockwise on $\mathcal{C}$) and will ***never*** stop. If $m = M/4$, for instance, then we have a solution where $\sin\theta = \frac{1}{2}\theta$, which is, as a few minutes on a calculator will show, $\theta \approx 1.90$ rad or 109°:*

θ (rad)	$\sin\theta - \theta/2$
1	0.3415
1.2	0.3320
1.4	0.2855
1.6	0.1996
1.8	0.07385
1.9	−0.003700
1.89	0.004486
1.895	0.000405

EXAMPLE 5.6

The work done by a linear spring is illustrated in this example. The system (see Figure 1) consists of a cylinder $\mathcal{C}$ (100 kg) and (equilateral) triangular plate $\mathcal{P}$ (20 kg) pinned together at the mass center G_1 of the cylinder. The other two vertices of the plate are connected to springs, the left one of which (S_1) remains vertical in the slot. (Spring S_1 is shown only in its initial position.) The initial stretches of the two springs (in the position shown) are 0.2 m for S_1 and 0.4 m for S_2. The moduli are 40 N/m for S_1 and 10 N/m for S_2. If the system is released from rest in the given position, find the velocity of G_1 when vertex B reaches its lowest point in the slot. Assume sufficient friction to prevent $\mathcal{C}$ from slipping on the plane.

(Continued)

*See Appendix C for a numerical solution to this problem using the Newton–Raphson method with a programmable calculator.

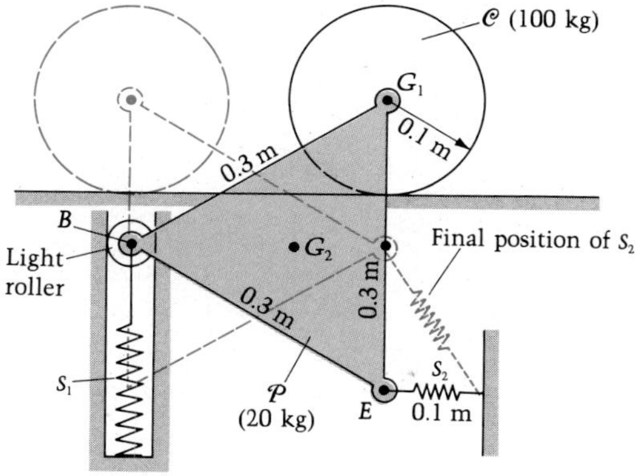

Figure 1

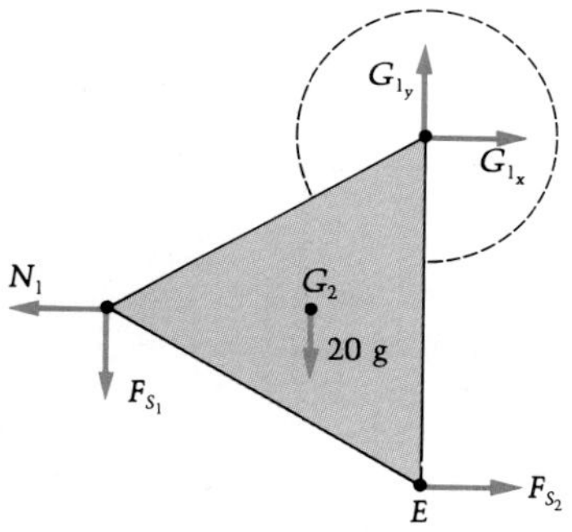

Figure 2

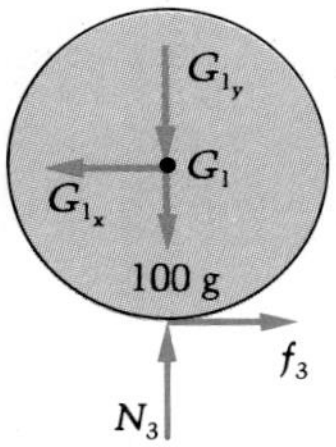

Figure 3

SOLUTION

It is always a good idea to make sure that there *is* a problem—that is, that the system cannot remain in equilibrium upon release. In the free-body diagram of the cylinder (Figure 2) we note from $M_{z_{G_1}} = I_{G_1}\alpha$ that $\mathcal{C}$ will roll unless $f_3 = 0$. If $f_3 = 0$, though, G_{1_x} must also vanish if $\mathcal{C}$ is to be in equilibrium. If G_{1_x} is zero, then on the free-body diagram of the plate $\mathcal{P}$ (Figure 3) N_1 has to be to the left to balance F_{s_2} (the tensile force in the spring S_2). But then the summation of moments about E cannot be zero! Thus the bodies necessarily begin to move.

We now use $W = \Delta T$ to solve the problem. Let the system be the two bodies, $\mathcal{C}$ plus $\mathcal{P}$:

$$W_g + W_{S_1} + W_{S_2} = T_f - \overset{0}{\cancel{T_i}}$$

Both mass centers start and end at the same level, so the work of gravity W_g vanishes. For the springs, we have

$$W_{S_1} = \frac{k}{2}(\delta_i^2 - \delta_f^2)$$

$$= 20[0.2^2 - (0.2 - 0.15)^2] = 0.750 \text{ J}$$

$$W_{S_2} = \frac{k}{2}(\delta_i^2 - \delta_f^2)$$

$$= 5[0.4^2 - (0.4 + 0.080)^2] = -0.352 \text{ J}$$

Note that S_1 has ***shortened*** from its initial stretch by 0.15 m whereas S_2 has ***additionally*** stretched by the amount $0.1803 - 0.1 = 0.0803$ m.

For the kinetic energy, we note that B is Ⓘ for $\mathcal{P}$ in the final configuration (Figure 4); point B is changing direction and hence is stopped. Therefore

$$T_{\mathcal{P}_f} = \frac{1}{2} I_{\textcircled{I}}\omega_{\mathcal{P}_f}^2 = \frac{1}{2}(0.750)\left(\frac{v_{G_{1_f}}}{0.3}\right)^2 = 4.17v_{G_{1_f}}^2$$

(Continued)

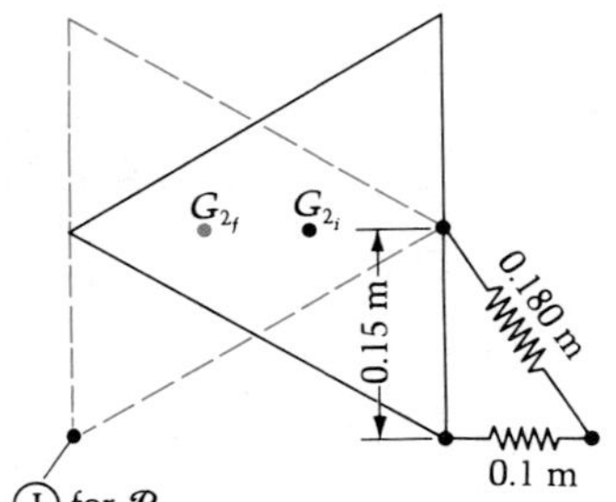

① for $\mathcal{P}$ in final position

Figure 4

where

$$\mathbf{v}_{G_1} = \cancelto{\mathbf{0}}{\mathbf{v}_{①}} + \omega_{\mathcal{P}_f}\,\hat{\mathbf{k}} \times \mathbf{r}_{①G_1} \Rightarrow \frac{v_{G_{1_f}}}{0.3} = \omega_{\mathcal{P}_f}$$

and

$$\begin{aligned} I_{①} &= I_{G_2} + md^2 \\ &= \frac{m(0.3)^{2*}}{12} + m\left[0.15^2 + \left(\frac{0.3\sqrt{3}/2}{3}\right)^2\right] \\ &= 20(0.00750 + 0.0300) = 0.750 \text{ kg} \cdot \text{m}^2 \end{aligned}$$

For the cylinder,

$$\begin{aligned} T_{\mathcal{C}_f} &= \frac{1}{2}mv_{G_{1_f}}^2 + \frac{1}{2}I_{G_1}\omega_{\mathcal{C}_f}^2 \\ &= \frac{1}{2}100v_{G_{1_f}}^2 + \frac{1}{2}\left[\frac{1}{2}100(0.1^2)\right]\left[\frac{v_{G_{1_f}}}{0.1}\right]^2 \\ &= 75.0v_{G_{1_f}}^2 \end{aligned}$$

Therefore $W = \Delta T$ gives

$$0.750 - 0.352 = 4.17v_{G_{1_f}}^2 + 75.0v_{G_{1_f}}^2$$

$$v_{G_{1_f}} = 0.0709 \text{ m/s}$$

Note that strengthening S_1 (increasing k_1) or weakening S_2 (decreasing k_2) will both increase $v_{G_{1_f}}$. But if, for example, we swap k_1 and k_2, then the negative work done by S_2 overcomes the positive work of S_1 and the system will ***not even reach*** the bottom configuration! This would be manifested mathematically by $v_{G_{1_f}}^2$ being negative, as the reader may wish to show.

EXAMPLE 5.7

The unstretched length of the spring in Figure 1 is $\ell_u = 0.3$ m. The initial angular velocity of body $\mathcal{A}$ in the top position is $\boldsymbol{\omega}_i = 2.5$ ↻ rad/s. There is enough friction to prevent slipping of $\mathcal{A}$ on $\mathcal{B}$ at all times. Determine the modulus of the spring that will cause $\mathcal{A}$ to stop in the $\varphi = 90°$ position. Will it ***remain*** there?

(Continued)

*For an equilateral triangle, $I_{z_C} = m(\text{side})^2/12$.

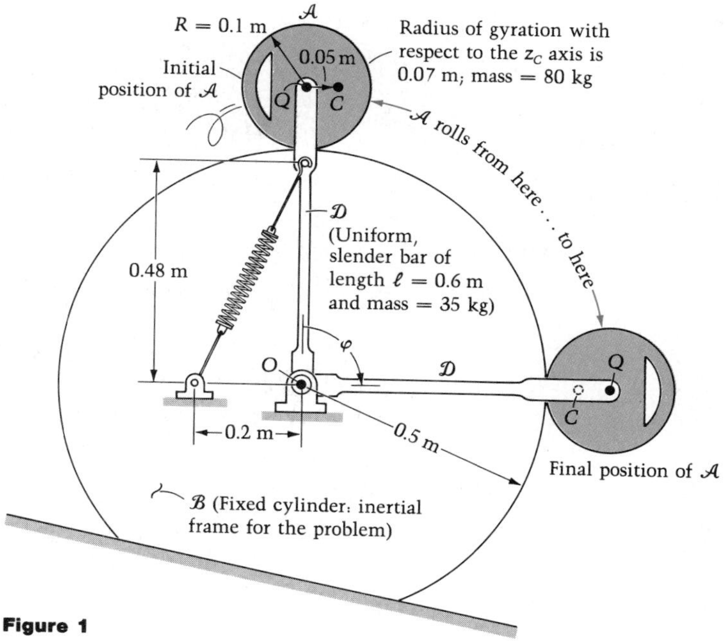

Figure 1

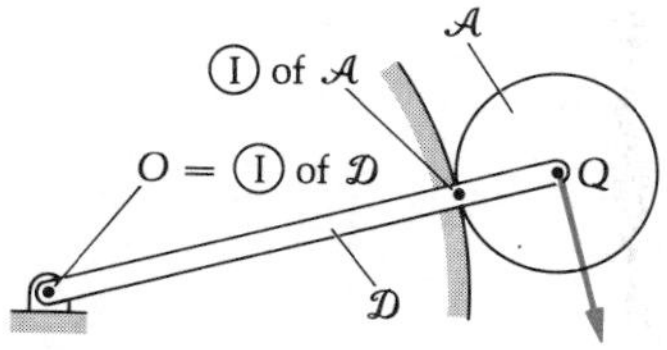

Figure 2

SOLUTION

Part of the problem is determining where the mass center C is located when $\mathcal{A}$ reaches the final position. Reviewing the kinematics, we find that the velocity of the geometric center Q of $\mathcal{A}$ is expressible in two ways (see Figure 2):

1. As a point of $\mathcal{A}$, $\mathbf{v}_Q = \cancelto{0}{\mathbf{v}_{\text{Ⓘ}}} + R\dot{\theta}\hat{\mathbf{e}}_t$.

2. As a point of $\mathcal{D}$, $\mathbf{v}_Q = \cancelto{0}{\mathbf{v}_O} + \ell\dot{\varphi}\hat{\mathbf{e}}_t$.

Thus we see that $R\dot{\theta} = \ell\dot{\varphi}$. Integrating, we get

$$R\theta = \ell\varphi$$

in which the constant of integration is zero if we select $\theta = 0$ when $\varphi = 0$. Therefore when $\varphi = \pi/2$, we may find the orientation of body $\mathcal{A}$:

θ = angle that body $\mathcal{A}$ turns through in reference frame (angle seen by *stationary* observer in body $\mathcal{B}$)

$$= \frac{\ell}{R}\varphi = \frac{0.6}{0.1} \times \frac{\pi}{2} = 3\pi$$

And so the final position of C is to the left of Q (see Figure 3). We can now write the work of gravity W_g because we now know the h moved through by C:

$$W_g = (mgh)_{\mathcal{D}} + (mgh)_{\mathcal{A}} = 35(9.81)(0.3) + 80(9.81)(0.6) = 574 \text{ J}$$

(Continued)

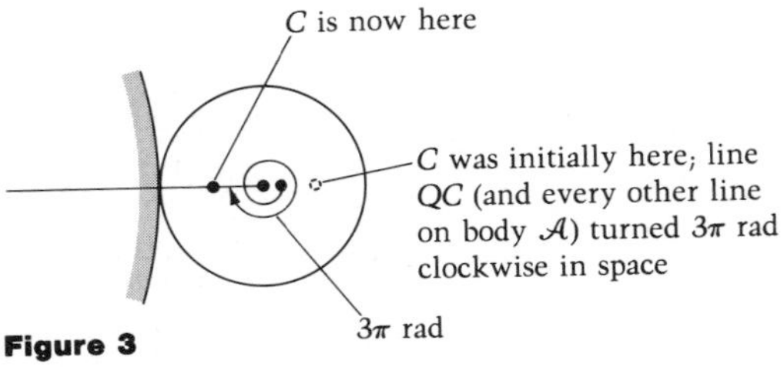

C is now here
C was initially here; line QC (and every other line on body 𝒜) turned 3π rad clockwise in space
3π rad

Figure 3

The work done by the linear spring is *always* given by $(k/2)(\delta_i^2 - \delta_f^2)$:

$$W_s = \frac{k}{2}(\delta_i^2 - \delta_f^2) = \frac{k}{2}(0.220^2 - 0.380^2)$$

$$= -0.0480k \text{ J}$$

where k is our unknown and the initial and final stretches are computed as follows:

$$\ell_i = \ell_u + \delta_i \qquad \text{(unstretched length plus initial stretch = initial length of spring)}$$

and

$$\ell_f = \ell_u + \delta_f$$

so that the stretches are

$$\delta_i = \sqrt{0.2^2 + 0.48^2} - 0.3 = 0.520 - 0.300 = 0.220 \text{ m}$$

$$\delta_f = (0.48 + 0.2) - 0.3 = 0.680 - 0.300 = 0.380 \text{ m}$$

For the kinetic energy side of $W = \Delta T$, we need the moments of inertia; first we consider body $\mathcal{A}$:

$$I_C = mk_C^2 = 80(0.07^2) = 0.392 \text{ kg-m}^2$$

We shall use the "short form" of T—namely $T = \frac{1}{2}I_{\textcircled{I}}\omega^2$ (always valid whenever $\omega \neq 0$ in plane motion). Thus we need $I_{\textcircled{I}_i}$ and $I_{\textcircled{I}_f}$.* Note that when Ⓘ is a different point of a body in the initial and final positions, the value of $I_{\textcircled{I}}$ is generally different in the two configurations, as in this problem.

At $\varphi = 0°$:

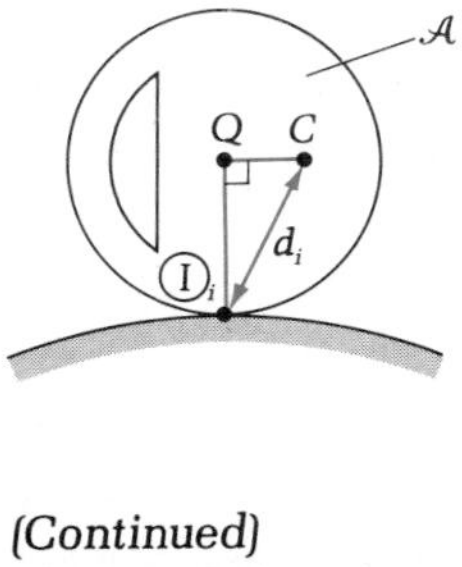

𝒜
Q
C
d_i
$\textcircled{I}_i$

At $\varphi = \pi/2$:

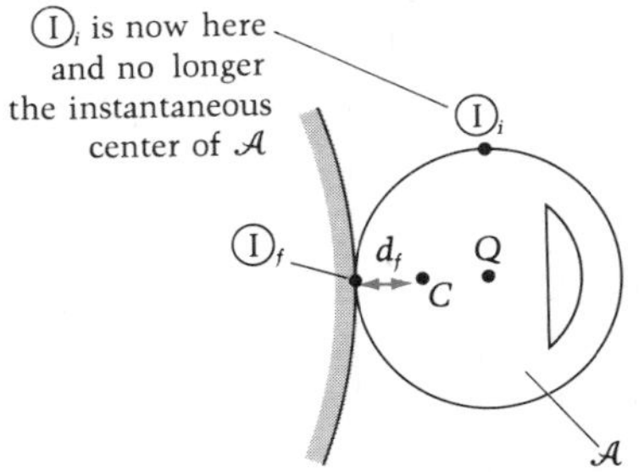

$\textcircled{I}_i$ is now here and no longer the instantaneous center of 𝒜
$\textcircled{I}_i$
$\textcircled{I}_f$
d_f
C
Q
𝒜

(Continued)

*Since $\omega_f = 0$, we do not have to calculate $I_{\textcircled{I}_f}$ here, but we do so to illustrate the procedure in general.

$$I_{①_i} = I_C + md_i^2 \qquad I_{①_f} = I_C + md_f^2$$

$$= 0.392 + 80(0.1^2 + 0.05^2) \qquad = 0.392 + 80(0.1 - 0.05)^2$$

$$= 1.39 \text{ kg} \cdot \text{m}^2 \qquad = 0.592 \text{ kg} \cdot \text{m}^2$$

Therefore the kinetic energies of $\mathcal{A}$ we need are

$$T_i^{\mathcal{A}} = \frac{1}{2} I_{①_i}\omega_i^2 = \frac{1}{2}(1.39)2.5^2 = 4.34 \text{ J}$$

$$T_f^{\mathcal{A}} = \frac{1}{2} I_{①_f}\omega_f^2 = \frac{1}{2}(0.592)0^2 = 0 \text{ J} \qquad \text{(since the final angular speed is to be zero)}$$

For the bar $\mathcal{D}$, $I_{①}$ is the same in any position since ① is point O, which is pinned to the reference frame. Therefore, using $v_Q = R\omega_{\mathcal{A}}$, we have

$$T_i^{\mathcal{D}} = \frac{1}{2} I_{①_i}\omega_i^2 = \frac{1}{2}\left(\frac{m\ell^2}{3}\right)\left(\frac{v_{Q_i}}{r_{①Q}}\right)^2 = \frac{1}{2}\left(\frac{35 \times 0.6^2}{3}\right)\left(\frac{0.1 \times 2.5}{0.6}\right)^2$$

$$= 0.365 \text{ J}$$

$$T_f^{\mathcal{D}} = \frac{1}{2} I_{①_f}\omega_f^2 = \frac{1}{2}\left(\frac{m\ell^2}{3}\right)\left(\frac{v_{Q_f}}{r_{①Q}}\right)^2 = 0 \qquad (\text{since } v_{Q_f} = 0.1\omega_{\mathcal{A}_f} = 0)$$

Applying the work and kinetic energy principle, we get

$$W = \Delta T$$

$$W_g + W_s = T_f - T_i = -T_i$$

$$574 - 0.0480k = 0 - (4.34 + 0.365)$$

$$k = 12{,}100 \text{ N/m}$$

This is equivalent to 829 lb/ft of stiffness in the U.S. system of units, since 1 lb/ft is the same stiffness as 14.6 N/m.

It is instructive to consider whether motion will continue at $\varphi = 90°$ after the stop—that is, whether the stop is permanent or merely instantaneous. The reaction R_y on the two free-body diagrams (Figure 4) furnishes the key to the answer. Clearly on $\mathcal{D}$, $M_{r_O} = I_O\alpha$; and since $\mathcal{A}$ does not slip, we also know that $M_{r_J} = I_J\alpha$.* Therefore, if the resultant moment on $\mathcal{A}$ about J is clockwise, motion will continue.

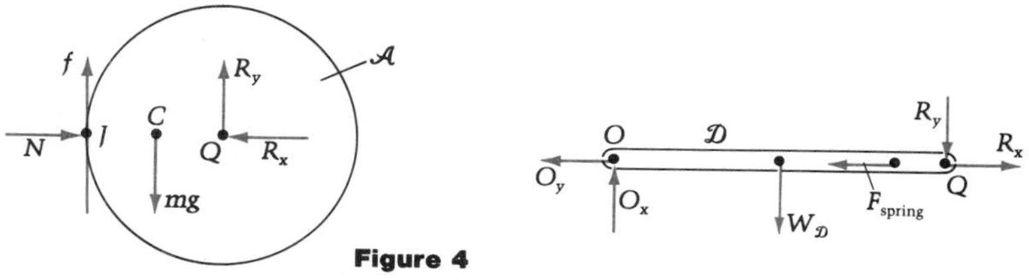

Figure 4

(Continued)

*At this instant, the acceleration $\mathbf{a}_J$ passes through C!

Now if the reaction at the pin at Q in the vertical direction is as shown (note equal and opposite on the two diagrams), then the sign of M_{r_I} on $\mathcal{A}$ is uncertain; but M_{r_O} on $\mathcal{D}$ is clockwise, and further downward motion of $\mathcal{D}$ will ensue. If the R_y's are reversed, then the free-body diagram of $\mathcal{D}$ is this time inconclusive; but M_{r_I} on $\mathcal{A}$ is clockwise, and again motion continues. If $R_y = 0$, both free-body diagrams yield the same result of continued motion.

The momentary stopping described here is not the same as a condition of unstable equilibrium that is sometimes encountered. In fact, we have just shown that the bodies will continue moving because moments do *not* add to zero as they do in equilibrium.

Question 5.4 What happens if k is larger than the calculated value? What happens if it is smaller?

EXAMPLE 5.8

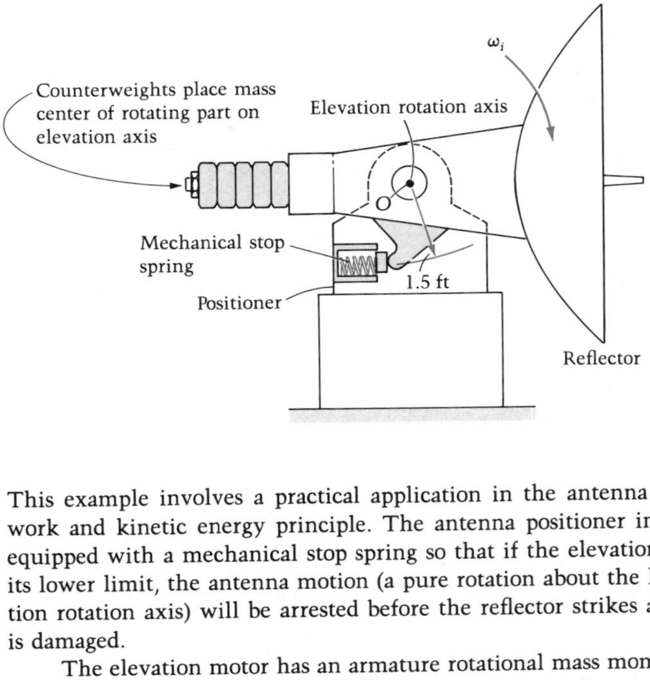

This example involves a practical application in the antenna industry of the work and kinetic energy principle. The antenna positioner in the diagram is equipped with a mechanical stop spring so that if the elevation drive overruns its lower limit, the antenna motion (a pure rotation about the horizontal elevation rotation axis) will be arrested before the reflector strikes another part and is damaged.

The elevation motor has an armature rotational mass moment of inertia of 0.01 lb-ft-sec^2 (or slug-ft^2) and drives the reflector through a gear reducer with a 700:1 gear ratio. The combined moment of inertia of the reflector, its counterweights, and the supporting structure is 12,000 sl-ft^2 = I_O.

(Continued)

It is desired to arrest a rotational speed of 30°/sec during a rotation from contact to full stop of 3°. The radius from the elevation rotation axis to the stop spring is 1.5 ft. The spring is unstretched at initial contact and may be assumed to have linear load-deflection behavior. It is further assumed that the motor is switched off but remains mechanically coupled while the rotation is being arrested. Find:

a. The required stiffness of the spring.
b. The maximum force induced in it.
c. The rotational position when it sustains its maximum force.
d. The angular accelerations of the reflector and motor armature at the position of maximum force. (Are these the maximum accelerations?)

SOLUTION

Since the spring is linear, its greatest force is the spring stiffness times the maximum deflection. This is also the position for which motion is completely arrested. At this position the kinetic energy has been brought to zero with the stop spring storing the energy; the principle $W = \Delta T$ gives

$$W = \frac{1}{2}k(\overset{0}{\cancel{\delta_i^2}} - \delta_f^2) = \Delta T = \frac{1}{2}I_O\overset{0}{\cancel{\omega_f^2}} - \frac{1}{2}I_O\omega_i^2$$

Note that point O is Ⓘ for the rotating body and that gravity does no work between contact and stop.

Question 5.5 Why does gravity do no work?

The values of δ_f, I_O, and ω_i needed in the equation are calculated as follows:

$$I_O = \text{total moment of inertia at axis of rotation}$$

$$= I_{\text{motor}_O} + I_{(\text{reflector, counterweights, structure})_O}$$

$$= 0.01 \times 700^{2*} + 12{,}000 = 16{,}900 \text{ slug-ft}^2$$

$$\omega_i = 30 \times \frac{\pi}{180} = 0.524 \text{ rad/sec}$$

$$\delta_f = 3 \times \frac{\pi}{180} \times 1.5 = 0.0785 \text{ ft}$$

(Note that over the very small angle of 3° the spring compression is approximately the arclength $R\theta$.)

Solving for the spring's stiffness, we get

$$k = \frac{I_O\omega_i^2}{\delta_f^2} = \frac{16{,}900 \times 0.524^2}{0.0785^2}$$

$$= 753{,}000 \text{ lb/ft}$$

(Continued)

*In Example 4.21 it was proved that moments of inertia reflect through gear trains from input to output with the gear ratio squared as a factor; also, the torque increases (while the speed decreases) with the gear ratio as the multiplying factor.

The maximum spring force $= k\delta_f = 753{,}000 \times 0.0785 = 59{,}100$ lb. The rotational position is 3° beyond contact—that is, the position at full stop. The angular acceleration of the reflector is

$$\alpha = \frac{M_{r_O}}{I_O} = \frac{1.5 \times 59{,}100}{16{,}900} = 5.25 \text{ rad/sec}^2$$

and that of the motor armature is $5.25 \times 700^* = 3680$ rad/sec^2.

These are the maximum accelerations, since here the force (and torque) are greatest. In closing, we note that motor torque and friction, omitted in this problem for simplicity, limit the rebound in the actual case.

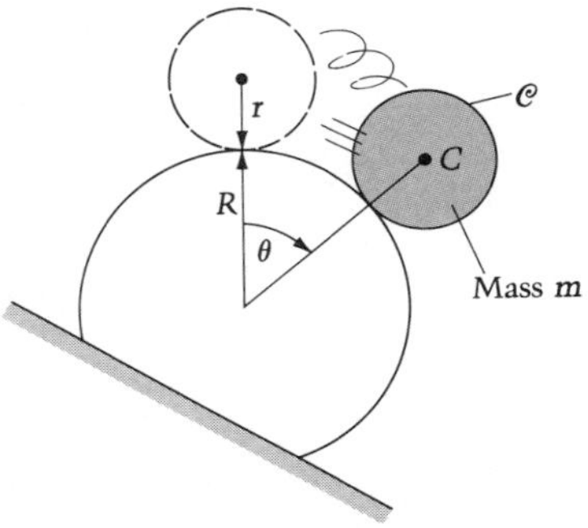

EXAMPLE 5.9

Sometimes it is useful to combine the work and kinetic energy principle with one or more of the differential equations of motion in order to obtain a desired solution. This example involves such a combination. The small cylinder $\mathscr{C}$ starts from rest at $\theta = 0$ in the dotted position (see the diagram) and begins to roll down the large cylinder. Find the angle θ_s at which slipping starts, and show that the small cylinder will *always* slip before it leaves the surface for a finite coefficient of friction.

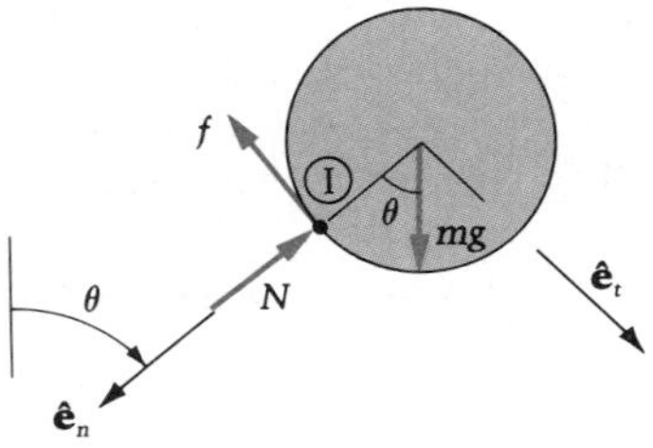

SOLUTION

Using the free-body diagram shown here, the equations of motion are

$$F_{r_n} = mg\cos\theta - N = ma_{C_n} = \frac{mv_C^2}{R+r} = \frac{mr^2\omega^2}{R+r} \tag{1}$$

$$F_{r_t} = mg\sin\theta - f = ma_{C_t} = m\ddot{s}_C = mr\alpha \tag{2}$$

$$M_{r_C} = fr = I_C\alpha = \frac{mr^2}{2}\alpha \tag{3}$$

Just prior to slipping, the friction force $f \approx \mu N$ while a_C is still equal to $r\alpha$ and v_C is still equal to $r\omega$. Therefore the equations can be rewritten as

$$mg\cos\theta_s - N = \frac{mr^2\omega_s^2}{R+r} \tag{1a}$$

$$mg\sin\theta_s - \mu N = mr\alpha_s \tag{2a}$$

$$\mu N r = \frac{mR^2}{2}\alpha_s \tag{3a}$$

(Continued)

*See footnote on page 341.

These equations may be supplemented with the work and kinetic energy equation for body $\mathscr{C}$, written between $\theta = 0$ and $\theta = \theta_s$:

$$W_g = mg(R + r)(1 - \cos\theta_s) = \frac{1}{2} I_{①}\omega_s^2 = \frac{1}{2}\left(\frac{3}{2}mr^2\right)\omega_s^2 \qquad (4)$$

Equations (1) to (4) may now be treated as four equations in the unknowns N, θ_s, ω_s^2, and α_s, where the last three quantities are the angle, angular velocity, and angular acceleration at slip. Solving them for θ_s yields the equation

$$7\mu \cos\theta_s - 4\mu = \sin\theta_s$$

Writing $\sqrt{1 - \cos^2\theta_s}$ for $\sin\theta_s$, and then squaring and solving the resulting quadratic for $\cos\theta_s$, gives

$$\theta_s = \cos^{-1}\left(\frac{28\mu^2 + \sqrt{33\mu^2 + 1}}{1 + 49\mu^2}\right)$$

which plots as shown in the diagram.

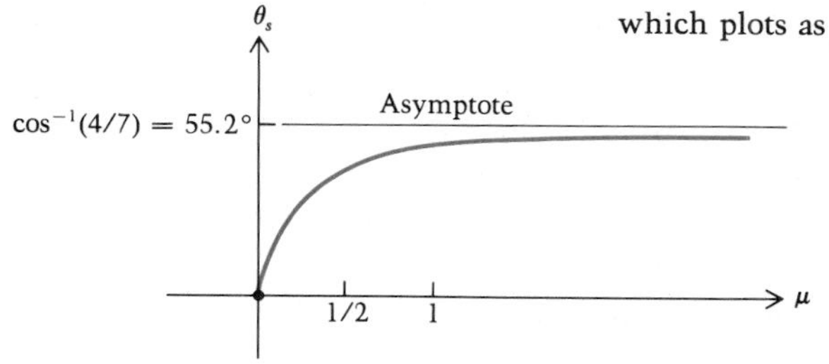

The curve in the diagram gives the slipping angle as a function of the friction coefficient; this is *not* the angle at which body $\mathscr{C}$ leaves the surface. We note that if we were to look for the angle at which the small cylinder leaves the surface of the large cylinder, assuming no slipping has occurred, we would be trying to solve a problem with no solution if the friction coefficient is finite. The curve clearly shows that for $\mathscr{C}$ to reach the angle $\cos^{-1}(4/7)$, an infinite coefficient of friction is required. Since the solution to the 'leaving without slip' problem is precisely $\cos^{-1}(4/7)$, shown below, then regardless of the friction coefficient (so long as it is finite) $\mathscr{C}$ will have to slip before it leaves.

Assuming now that the cylinder $\mathscr{C}$ leaves without having slipped, we obtain the (simpler) solution:

$$\text{Equation (1)} \Rightarrow mg - \overset{\text{0 at leaving!}}{\cancel{N}} = \frac{mr^2\omega_L^2}{R + r}$$

$$\text{Equation (4)} \Rightarrow mg(R + r)(1 - \cos\theta_L) = \frac{1}{2}\left(\frac{3}{2}mr^2\right)\omega_L^2$$

Eliminating ω_L gives

$$\theta_L = \cos^{-1}(4/7) = 55.2°$$

As we have noted, this solution is valid only for an infinite coefficient of friction between the cylinders. If $\mathscr{C}$ were a *particle* (no rotational kinetic energy) with a smooth surface, we would obtain (Example 2.13) $\theta_L = \cos^{-1}(2/3) = 48.2°$. Note the differences between these solutions.

Two Subcases of the Work and Kinetic Energy Principle

There is an important subcase of the principle of work and kinetic energy that we have already seen in Chapter 2. Using

$$\mathbf{F}_r = \frac{d}{dt}(m\mathbf{v}_C)$$

we obtained

$$\int_{t_1}^{t_2} \mathbf{F}_r \cdot \mathbf{v}_C\, dt = m \int_{t_1}^{t_2} \dot{\mathbf{v}}_C \cdot \mathbf{v}_C\, dt = \frac{1}{2} m v_C^2 \Bigg]_{v_C(t_1)}^{v_C(t_2)} \tag{5.22}$$

This principle states again that the work done by the external force resultant, when considered to act on the mass center, equals the change in the translational part of the kinetic energy:

$$\int \mathbf{F}_r \cdot \mathbf{v}_C\, dt = \Delta T_v \tag{5.23}$$

The integral of Equation (5.6) is

$$\int \mathbf{F}_r \cdot \mathbf{v}_C\, dt + \int \mathbf{M}_{r_C} \cdot \omega\hat{\mathbf{k}}\, dt = \int \frac{dT}{dt}\, dt = \Delta T = \Delta T_v + \Delta T_\omega \tag{5.24}$$

If we subtract (5.23) from (5.24), we obtain yet another result:

$$\int \mathbf{M}_{r_C} \cdot \omega\hat{\mathbf{k}}\, dt = \Delta T_\omega = \Delta\left(\frac{1}{2} I_C \omega^2\right)$$

or

$$\int M_{r_{C_z}}\, d\theta = \Delta T_\omega \tag{5.25}$$

This second subprinciple says that the work done by the external moments (about C) on the body, as it turns in the inertial frame, is equal to the change in the rotational part of the kinetic energy. We may use the "total" $W = \Delta T$ principle or either of its two "subparts" (Figure 5.9).

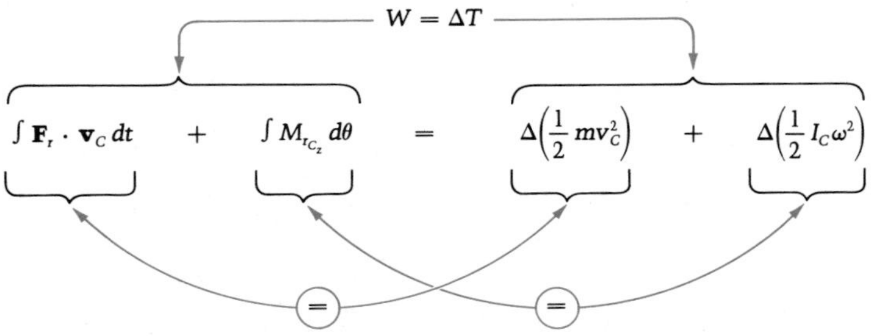

Figure 5.9

EXAMPLE 5.10

As an illustration of the two subcases of the principle of work and kinetic energy, we consider the cylinder of mass m rolling down the inclined plane shown in the diagram. If the cylinder is released from rest, find the velocity v_C of its mass center as a function of the distance $\mathbf{x}_C$ traveled by C.

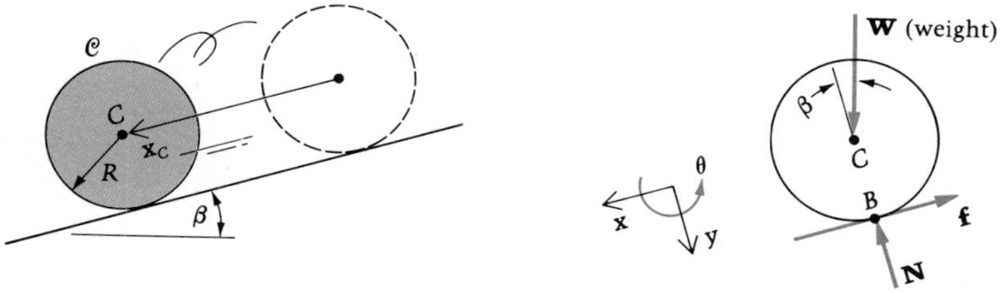

SOLUTION

Referring to the free-body diagram we see that the normal and friction forces do no work because, as the cylinder rolls on the incline, they always act on a point at rest. That is,

$$\int \mathbf{N} \cdot \mathbf{v}_B \, dt = 0 \qquad \text{and} \qquad \int \mathbf{f} \cdot \mathbf{v}_B \, dt = 0$$

Applying the principle that work $= \Delta T$, only the component of the gravity force $\mathbf{W}$ that acts parallel to the plane does any work:

$$\int (mg \sin \beta \hat{\mathbf{i}}) \cdot v_C \hat{\mathbf{i}} \, dt = \frac{1}{2} m v_C^2 + \frac{1}{2} I_{z_C} \omega^2 \tag{1}$$

Since $\mathbf{W}$ always acts on the same point (C) of $\mathcal{C}$, and since $dx_C/dt = v_C = R\omega$,

$$\int mg \sin \beta \, dx_C = \frac{1}{2} m v_C^2 + \frac{1}{2}\left(\frac{1}{2} mR^2\right)\frac{v_C^2}{R^2} \tag{2}$$

The mass center's velocity is therefore

$$\mathbf{v}_C = v_C \hat{\mathbf{i}} = \sqrt{\frac{4g \sin \beta \, x_C}{3}} \, \hat{\mathbf{i}} \tag{3}$$

Now suppose we apply Equation (5.23):

$$\int \mathbf{F}_r \cdot \mathbf{v}_C \, dt = \Delta\left(\frac{1}{2} m v_C^2\right) \tag{4}$$

In this problem the resultant force acting on $\mathcal{C}$ is

$$\mathbf{F}_r = \mathbf{W} + \mathbf{f} + \mathbf{N} = mg(\sin \beta \hat{\mathbf{i}} + \cos \beta \hat{\mathbf{j}}) - f\hat{\mathbf{i}} - N\hat{\mathbf{j}}$$

and Equation (4) becomes

$$\int mg \sin \beta \, dx_C - \int f v_C \, dt = \frac{1}{2} m v_C^2 \tag{5}$$

(Continued)

We see that, as expected, the friction force (though it does no net work) retards the motion of the mass center C while *turning* the cylinder, as can be seen from the *other* subcase of $W = \Delta T$:

$$\int \underbrace{\mathbf{M}_{r_C} \cdot \boldsymbol{\omega}\, dt}_{M_{r_C}\, d\theta} = \Delta \frac{1}{2} I_{z_C} \omega^2 \tag{6}$$

$$\int fR\, d\theta = \frac{1}{2}\left(\frac{1}{2} mR^2\right)\omega^2 \tag{7}$$

or

$$\int f \frac{d(R\theta)}{dt}\, dt = \int f v_C\, dt = \frac{1}{2}\left(\frac{1}{2} mR^2\right)\frac{v_C^2}{R^2} \tag{8}$$

And the sum of Equations (5) and (8) indeed gives Equation (2): the total $W = \Delta T$ equation!

Potential Energy, Conservative Forces, and Conservation of Mechanical Energy

In Section 2.4 we introduced the concept of **potential energy,** or the potential of a force. When the work done by a force on a body is independent of the path taken as the body moves from one configuration to another, the force is said to be **conservative** and the work is expressible as the decrease in a scalar function φ, the potential (energy). Thus as a body moves from a configuration at time t_1 to a second configuration at time t_2, the work done by an external conservative force is

$$W = \varphi(t_1) - \varphi(t_2)$$

or simply

$$W = \varphi_1 - \varphi_2$$

If all the external forces that do work on a rigid body are conservative and φ is now the *sum* of the potentials of those forces, Equation (5.15) yields

$$\varphi_1 - \varphi_2 = W = \Delta T = T_2 - T_1$$

or

$$T_2 + \varphi_2 = T_1 + \varphi_1$$

or

$$T + \varphi = \text{constant}$$

which expresses the **conservation of mechanical energy.**

From Chapter 2 and earlier in this section we can easily identify two common conservative forces: (1) the constant force acting always on the same material point in the body and (2) the force exerted on a body by a linear spring attached at one end to the body and at the other to a point fixed in the inertial frame of reference.

In the case of the constant force, a potential is $\varphi = -\mathbf{F} \cdot \mathbf{r}$, where $\mathbf{r}$ is a position vector for the point of application. When the force is that exerted by gravity (weight) on a body near the surface of the earth,

$$\varphi = mgh$$

where h is the altitude of the mass center of the body.

For the linear spring, we recall that $\varphi = (k/2)\delta^2$, where k is the spring modulus, or stiffness, and δ is the stretch. It is important to recognize that when a spring is attached to, or between, two bodies that are both moving (relative to the inertial frame), then $(k/2)\delta^2$ is a potential for the two spring forces ***taken together*** (see Equation 5.19). That is, while neither of the forces acting on the bodies can be judged by itself to be conservative, the net work done on the two bodies by the two forces is expressible as a decrease in the potential, $\varphi = (k/2)\delta^2$. This is helpful in the analysis of problems in which we have two or more interacting rigid bodies. We have already noted earlier in this section that the work of the ***external*** forces on a system of rigid bodies is not in general equal to the change in kinetic energy of the system; this is because there may be net work done on the rigid bodies by the equal and opposite forces of interaction. Suppose now that our system is made up of two bodies joined by a spring, and suppose the spring forces are the ***only*** internal ones that produce net work on the system. We may then write $W = \Delta T$ for each rigid body. Upon adding these equations there results

(Work of forces external to system) + (work of pair of spring forces)

= (change in kinetic energy of system)

If the forces external to the system that do work are conservative, we may add the various potential energies associated with them to that for the pair of spring forces and conclude that

$$T + \varphi = \text{constant}$$

That is to say, in this case the mechanical energy of the ***system*** is conserved.

An example of a ***nonconservative*** force is sliding friction. A potential cannot be found for friction, since the (negative) work it does depends on the path taken by the body on which it acts. In this case, $W = \Delta T$ must be used, and it is seen to be more general than the principle of conservation of mechanical energy.

EXAMPLE 5.11

Show that the same equation for the spring modulus in Example 5.7 is obtained by conservation of mechanical energy. (See the diagram.)

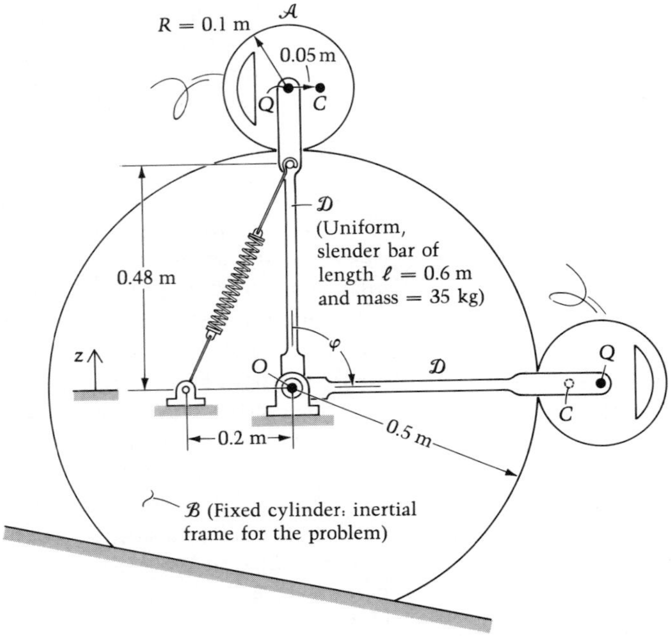

SOLUTION

The potentials for gravity and for the spring are

$$\phi_g = +m_{\mathcal{A}} g z_{C_{\mathcal{A}}} + m_{\mathcal{D}} g z_{C_{\mathcal{D}}} \qquad \varphi_{\text{spr}} = \frac{k\delta^2}{2}$$

Therefore, measuring z_C from O, we have

$$\varphi_{g_i} = +80(9.81)(+0.6) + 35(9.81)(+0.3) \qquad \varphi_{g_f} = -mg(0)$$

$$= +471 + 103$$

$$= +574 \text{ J} \qquad\qquad = 0$$

For the spring, using i for the initial and f for the final configuration, we have

$$\varphi_{\text{spr}_i} = \frac{k(0.22)^2}{2} = 0.0242k \text{ J} \qquad \varphi_{\text{spr}_f} = \frac{k(0.38)^2}{2} = 0.0722k \text{ J}$$

Thus, adding the potentials ($\varphi = \varphi_g + \varphi_{\text{spr}}$), we get

$$\varphi_i = 574 + 0.0242k \text{ J} \qquad \varphi_f = 0 + 0.0722k \text{ J}$$

(Continued)

The kinetic energies were $T_i = 4.34 + 0.365 = 4.71$ J and $T_f = 0$. Therefore

$$\varphi_i + T_i = \varphi_f + T_f$$

$$574 + 0.0242k + 4.71 = 0.0722k + 0$$

or, rearranging,

$$574 - 0.0480k = -4.71$$

This is the same final equation that resulted from $W = \Delta T$ in the earlier Example 5.7.

PROBLEMS / Section 5.2

5.1 Find the kinetic energy of the system of bodies $\mathcal{A}$, $\mathcal{B}$, and $\mathcal{C}$ at an instant when the speed of $\mathcal{A}$ is 4 ft/sec. (See Figure P5.1.)

5.2 See Figure P5.2. (a) Explain why the friction force f does no work on the rolling cylinder $\mathcal{C}$ if the plane $\mathcal{F}$ *is* the reference frame. (b) If, however, $\mathcal{F}$ is the top surface of a moving block (dotted lines) and the reference frame is now the ground $\mathcal{G}$, does f then do work on $\mathcal{C}$?

5.3 One end of the linear spring in Figure P5.3 is attached to an inextensible cord that is lightly wrapped around a groove in the wheel (mass = 1 slug, radius of gyration about center = 1.5 ft). If the wheel rolls and starts from rest when the spring is stretched 1 ft, find the velocity of the center of the wheel when the center has moved 2 ft. The mass center of the wheel coincides with the geometric center. Note that the spring is not attached to a specific material point on the wheel. However, the moving end of the spring has, at every instant the spring is taut, the same velocity as the point of force application on the wheel. Thus Equation (5.19) gives the work done on the wheel by the spring.

5.4 The three rods shown in Figure P5.4 are pinned together with one vertex also pinned to the ground. The length of the bar labeled $\mathcal{B}$ is given by $2b = 0.4$ m, and the density of the material of all bars is 7850 kg/m^3. Their cross-sectional area is 0.002 m^2. Find the angular velocity of the combined body after it swings 90° from rest if: (a) $H = 2b$; (b) $H = \sqrt{3}b$; (c) $H = b$.

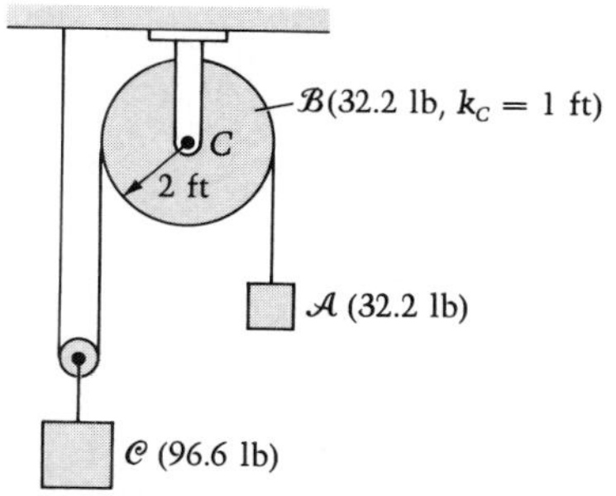

Figure P5.1

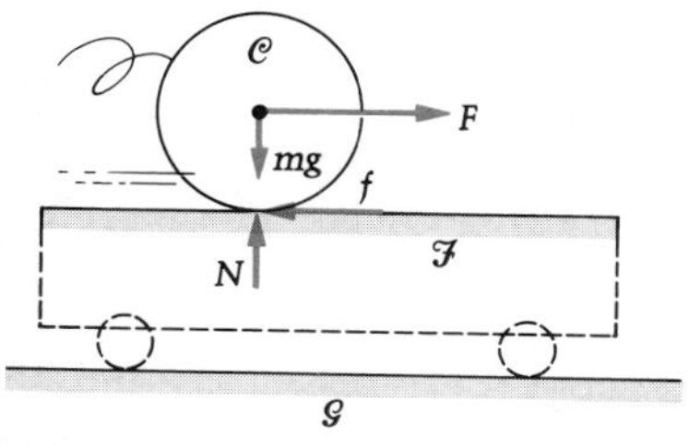

Figure P5.2

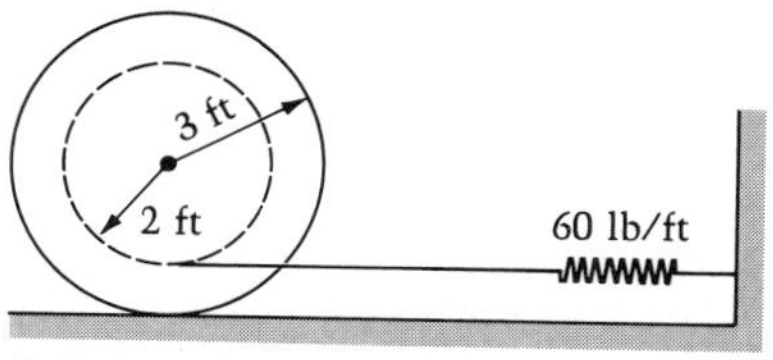

Figure P5.3

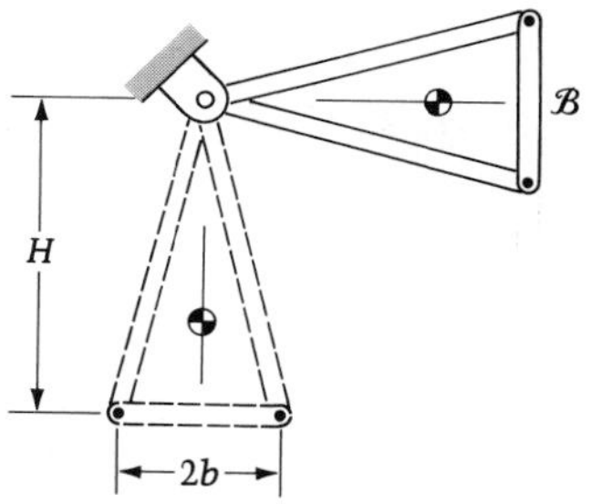

Figure P5.4

5.5 The radius of gyration of the wheel and hub in Figure P5.5, with respect to its axis of symmetry through C, is $k_C = 2.5$ m. The springs are unstretched at an initial position of rest, when the 50-N force is applied.

a. Find how far to the right the mass center moves in the ensuing motion, assuming sufficient friction to prevent slipping.

b. When $\mathscr{C}$ stops instantaneously at its farthest right point, what increase in the 50-N force and what minimum friction coefficient are needed to keep it there?

5.6 The spring in Figure P5.6 has an unstretched length of 0.8 m and a modulus of 60 N/m. The 20-kg wheel $\mathscr{G}$ is released from rest in the upper position. Find the angular velocity of $\mathscr{G}$ when it passes through the lower (dashed) position if its radius of gyration is $k_{z_C} = 0.2$ m.

5.7 Upon application of the 10-N force F to the cord in Figure P5.7, the cylinder begins to roll to the right. After C has moved 5 m, how much work has been done by F?

5.8 Pulley $\mathscr{A}$ weighs 100 lb and has a centroidal radius of gyration $k_C = 7$ in. (See Figure P5.8.) The disk pulley $\mathscr{B}$ weighs 20 lb. Find the velocity of weight $\mathscr{D}$ (50 lb) after it falls 2 ft from rest. (Assume that the rope does not slip on the pulleys.)

5.9 The uniform slender rod in Figure P5.9 (mass = 5 slugs, length = 10 ft) is released from rest in the position shown. Neglecting friction, find the force that the floor exerts on the lower end of the rod when the upper end is 6 ft above the floor. *Hint:* First use a free-body diagram and the equations of motion to deduce the path of the mass center.

5.10 A vertical rod $\mathscr{R}$ is resting in unstable equilibrium when it begins to fall over. (See Figure P5.10.) End A slides along a smooth floor. Find the velocity of the mass center C of $\mathscr{R}$ as a function of L, g, and its height H above the floor.

5.11 The system depicted in Figure P5.11 is released from rest with 2 ft of initial stretch in the spring. There is sufficient friction to prevent slipping at all times. Determine whether $\mathscr{C}$ will leave the horizontal surface during the subsequent motion. Note that the string $\mathscr{S}$ goes slack if the stretch tries to become negative.

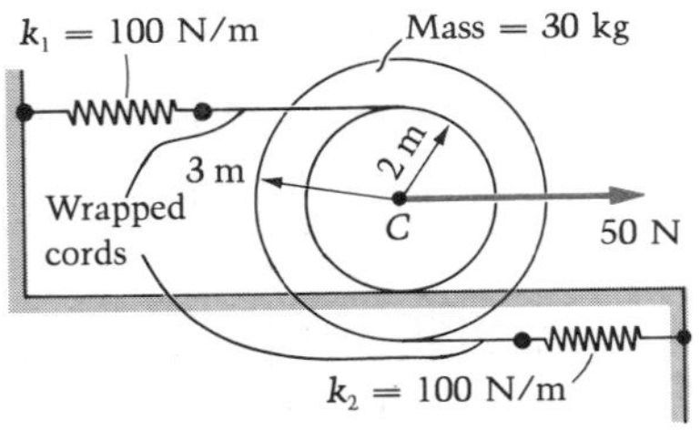

Figure P5.5

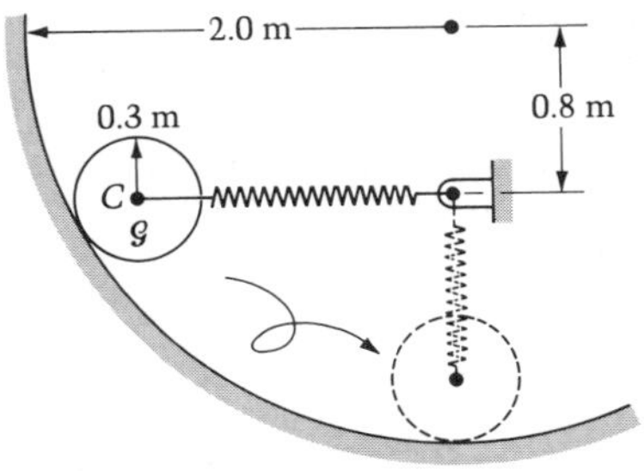

Figure P5.6

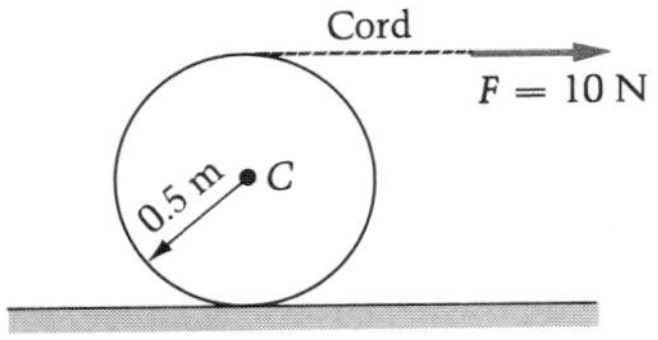

Figure P5.7

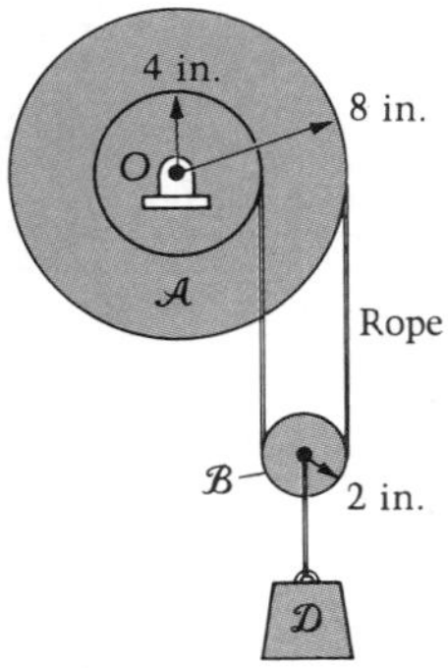

Figure P5.8

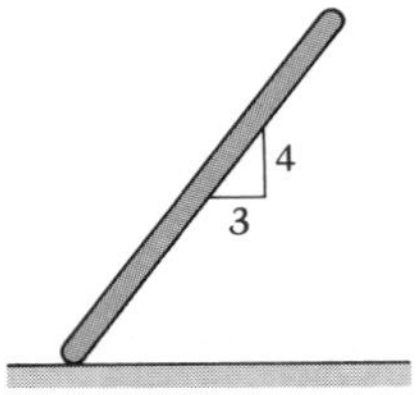

Figure P5.9

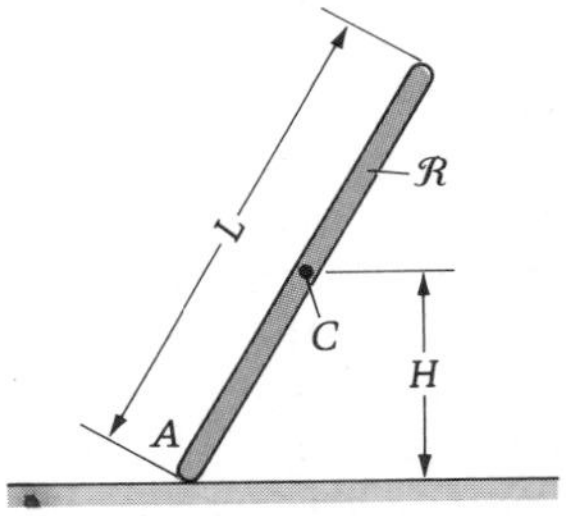

Figure P5.10

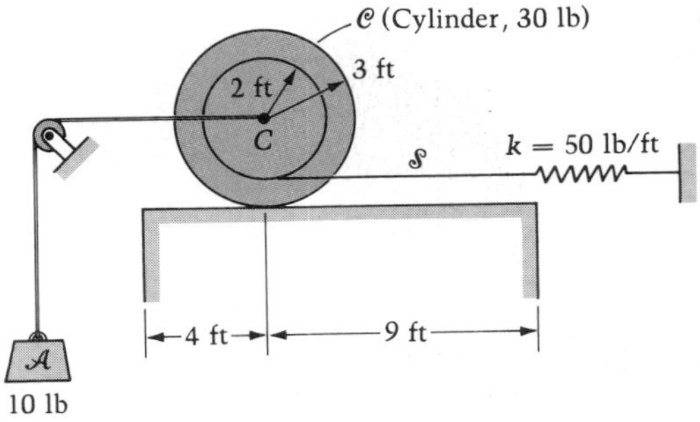

Figure P5.11

5.12 A slender bar (mass m, length ℓ) is originally at rest against a vertical wall. (See Figure P5.12.) Suppose both surfaces are smooth and the lower end of the bar is disturbed very slightly and released.

a. Show with a free-body diagram that the bar will begin sliding with its ends against the wall and floor.
b. Find the angle θ at which the top of the bar leaves the wall.

5.13 In the preceding problem, let the ends of the bar be constrained to vertical and horizontal paths by the smooth rollers in the slots shown in Figure P5.13. Find the reactions onto the bar at A and B just before the bar becomes horizontal.

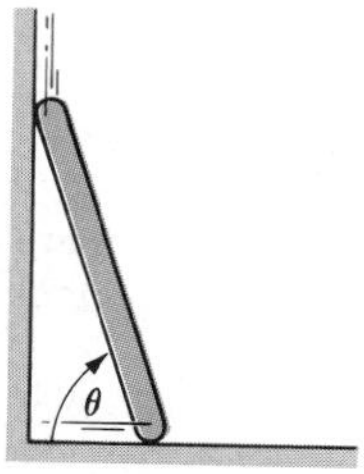

Figure P5.12

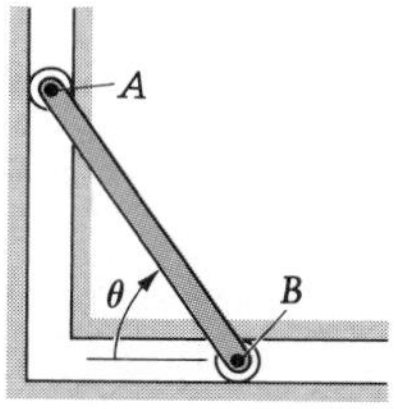

Figure P5.13

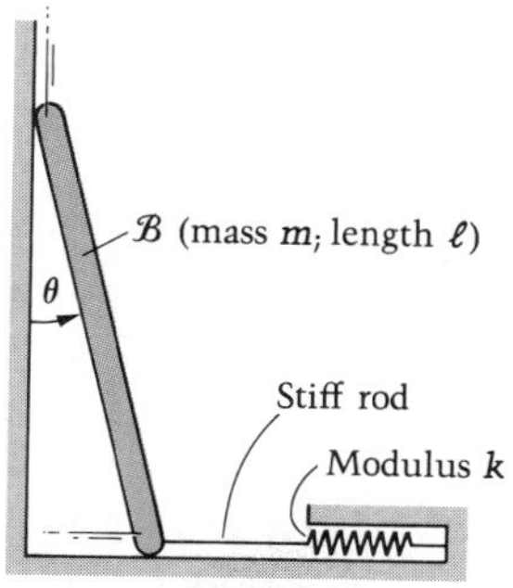

Figure P5.14

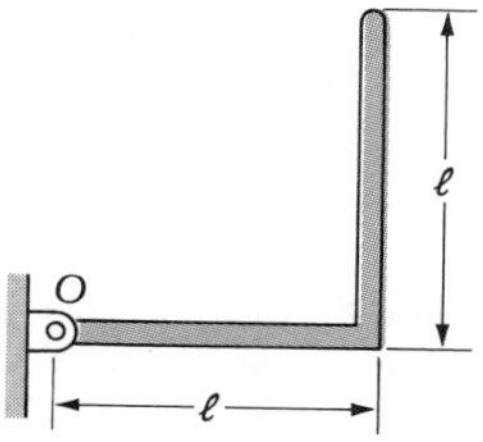

Figure P5.15

5.14 Bar $\mathcal{B}$ in Figure P5.14 is initially at rest in the vertical position, where the spring is unstretched. The wall and floor are smooth. Point B is then given a very slight displacement to the right, opening up a small angle $\Delta\theta$.

a. Draw a free-body diagram of the slightly displaced bar and use it to show that the bar will start to slide downward if $k < mg/2\ell$.
b. Find the angular velocity of $\mathcal{B}$ as a function of θ for such a spring.

5.15 The body of mass $2m$ in Figure P5.15 is composed of two identical uniform slender rods welded together. If friction in the bearing at O is neglected and the body is released from rest in the position shown, find the ***magnitude*** of the force exerted on the rod by the bearing after the body has rotated through 90°.

5.16 Solve Problem 4.65 by $W = \Delta T$.

5.17 Solve Problem 4.112 with the help of $W = \Delta T$. Ignore the hint.

5.18 Solve Problem 4.160(a) by $W = \Delta T$.

5.19 Solve Problem 4.177(b) by $W = \Delta T$.

5.20 In Problem 4.111, find the angular velocity of the rod when $\theta = 90°$.

5.21 The 12-ft, 32.2-lb homogeneous rod $\mathcal{R}$ shown in Figure P5.21 is free to move on the smooth horizontal and vertical guides as shown. The modulus of the spring is 8 lb/ft and the spring is unstressed when in the position shown. Rod $\mathcal{R}$ is released from rest with $\theta = \pi/2$ and nudged to the right to begin motion. (a) Determine the angular velocity of the rod when it becomes horizontal. (b) What is the angular acceleration of the rod in this position ($\theta = 0$)?

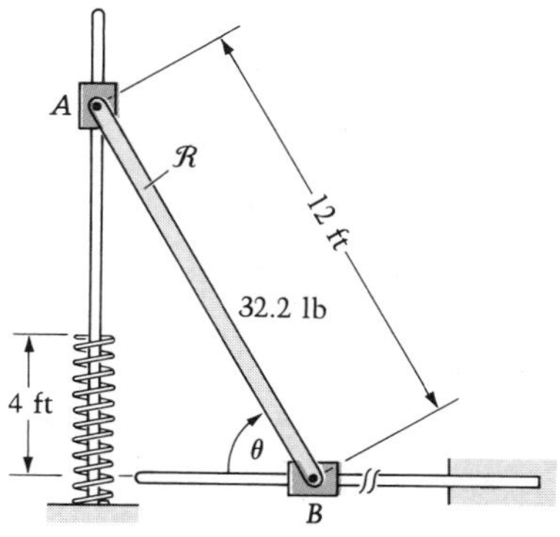

Figure P5.21

5.22 The bodies in Figure P5.22 have masses $m_{\mathcal{B}} = 0.3$ slug, $m_{\mathcal{C}} = 0.5$ slug, and $m_{\mathcal{A}}$ negligible. A spring is attached to A that is stretched 25 in. in the dotted position when everything is at rest. Find the spring modulus if $\omega_{\mathcal{B}} = 2 \circlearrowright$ rad/sec when $\mathcal{B}$ is horizontal as shown.

5.23 The mass center C of a rolling 2-kg wheel of radius $R = 15$ cm is located 5 cm from its geometric center Q. (See Figure P5.23.) The spring is attached at C and is not shown in position 2; its unstretched length is 0.3 m, and its modulus is 3 N/m. The radius of gyration is $k_C = 0.09$ m. Find the angular speed in position 2 (one-quarter turn from position 1).

5.24 The 20-lb wheel $\mathcal{A}$ in Figure P5.24 has a radius of gyration of 4 in. with respect to its (z_C) axis. A cable wrapped around its inner radius passes under and over two small pulleys and is then tied to the 50-lb block $\mathcal{B}$. The spring has a modulus of 3 lb/ft and is constrained to remain horizontal. There is sufficient friction to prevent $\mathcal{A}$ from slipping on the plane. (a) If the system is released from rest, find the angular speed of $\mathcal{A}$ after the block then falls 1 ft. (b) Would the answer be different if block $\mathcal{B}$ were replaced by a device that keeps the cable force constant at 50 lb? Why or why not?

5.25 Body $\mathcal{A}$ in Figure P5.25 rolls to the right along the plane and has a radius of gyration with respect to its axis of symmetry of $k_c = 0.5$ m. The corresponding radius of gyration for $\mathcal{C}$ is 0.12 m. The spring is stretched 0.6 m at an instant when $\omega_{\mathcal{A}} = 5 \circlearrowright$ rad/s. Find $\omega_{\mathcal{A}}$ after C has traveled 1 m to the right. (C_0 is an externally applied couple acting on $\mathcal{A}$.)

5.26 The cylinder in Figure P5.26 is rolling at $\omega = 2 \circlearrowleft$ rad/sec in the initial (i) position, where the spring is unstretched. Other data are:

$$m = 2 \text{ slugs}$$

$$r = 3 \text{ ft}$$

$$k = 3 \text{ lb/ft}$$

$$\mu = 0.2$$

$$\ell_u = \text{unstretched spring length} = 9 \text{ ft}$$

Find the final position of $C(x_C)$ at which either the cylinder has stopped (for an instant) or started to slip, whichever comes first. *Hint:* Try one, check the other!

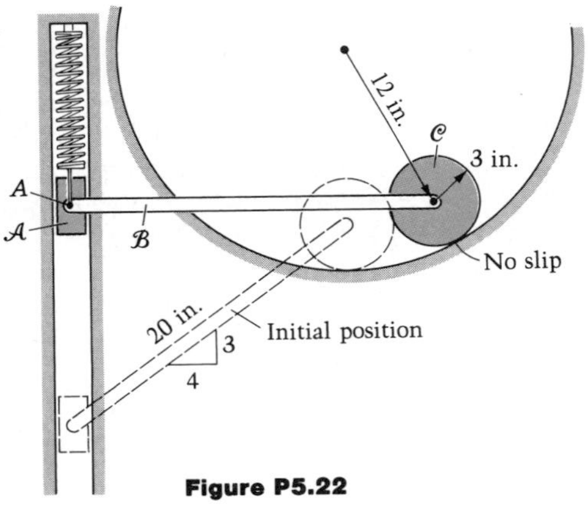

Figure P5.22

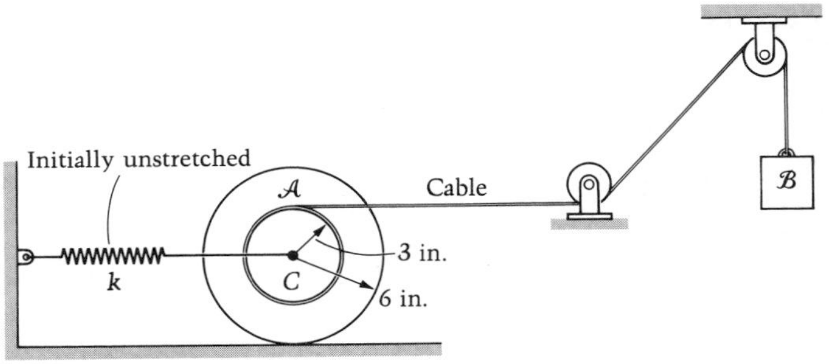

Figure P5.24

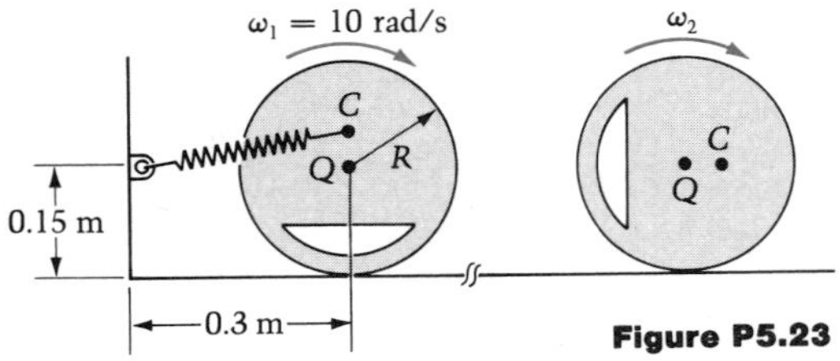

Figure P5.23

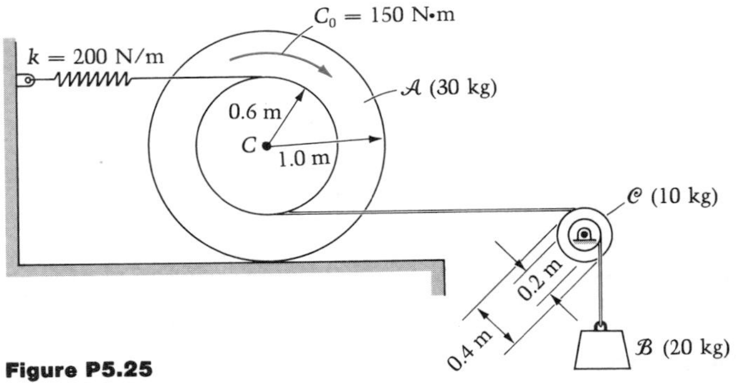

Figure P5.25

5.27 The 20-kg bar in Figure P5.27 has an angular velocity of 3 rad/s clockwise in the horizontal configuration shown. In that position the tensile force in the spring is 30 N. After a 90° clockwise rotation the angular velocity has increased to 4 rad/s. Determine the spring modulus k.

5.28 Block $\mathcal{A}$ in Figure P5.28 is moving downward at 5 ft/sec at a certain time when the spring is compressed 6 in. The coefficient of friction between block $\mathcal{B}$ and the plane is 0.2, and the radius of cylinder $\mathcal{C}$ is 0.5 ft. Weights of $\mathcal{A}$, $\mathcal{B}$, and $\mathcal{C}$ are 161, 193, and 322 lb, respectively.

a. Find the distance that $\mathcal{A}$ falls from its initial position before coming to zero speed.

b. Determine whether or not body $\mathcal{A}$ will start to move back upward.

5.29 A uniform 50-lb sphere $\mathcal{S}$ (radius = 1 ft) is released from rest in the position shown in Figure P5.29. If the sphere rolls (no slip), find its maximum angular speed.

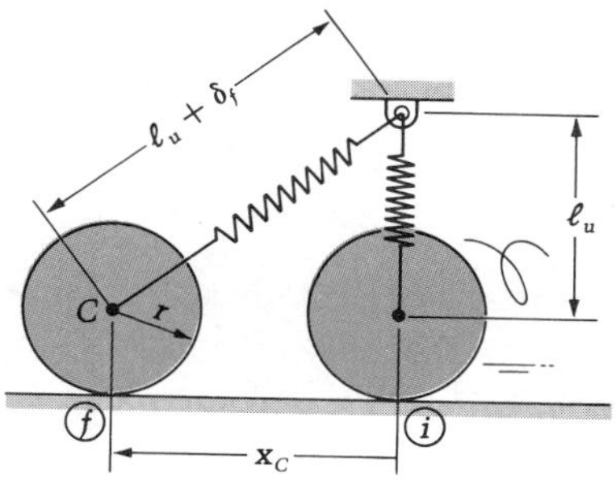

Figure P5.26

5.30 Show that if the rolling body in Example 5.9 is a sphere instead of a cylinder, it will slip at the angle θ_s satisfying the equation

$$\mu = \frac{2 \sin \theta_s}{17 \cos \theta_s - 10}$$

5.31 Two quarter-rings are pinned together at P and released from rest in the indicated position (Figure P5.31) on a smooth plane. Find the angular velocities of the rings when their mass centers are passing through their lowest points. *Hint:* By symmetry, point P always has only a vertical velocity component; this means that no work is done on either ring by the other, because (again by symmetry) the force between the rings has only a horizontal component normal to the velocity of P. More generally, as long as the pin is smooth the work done by two pinned bodies in motion on each other will be the negatives of each other because the velocities will be equal whereas the forces will be opposites.

5.32 A flatcar $\mathcal{C}$ of weight $W_{\mathcal{C}} = Mg$, plus a mast $\mathcal{M}$ of weight $W_{\mathcal{M}} = mg$ and length L, is moving with acceleration a_0 when the cable BD breaks and the mast falls. (See Figure P5.32.) Suppose the force P on the car is varied so that the acceleration of the car remains constant during the motion of the mast.

a. Determine the angular velocity of the mast when it becomes horizontal.

b. Write an expression for the force P in terms of m, M, and a_0 at the instant the mast becomes horizontal.

The car wheels may be considered light and frictionless.

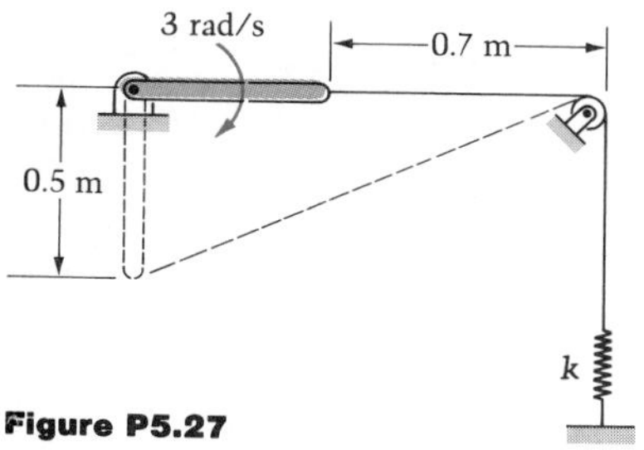

Figure P5.27

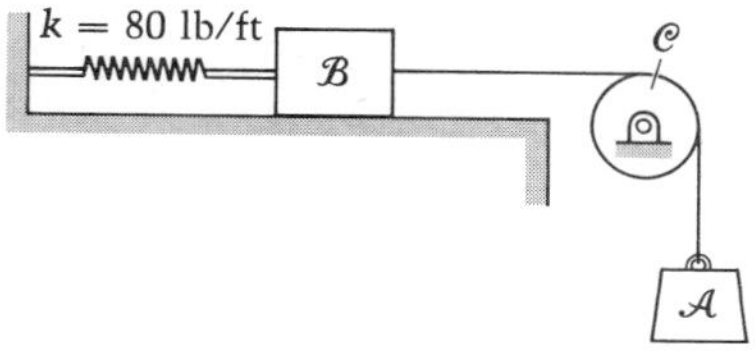

Figure P5.28

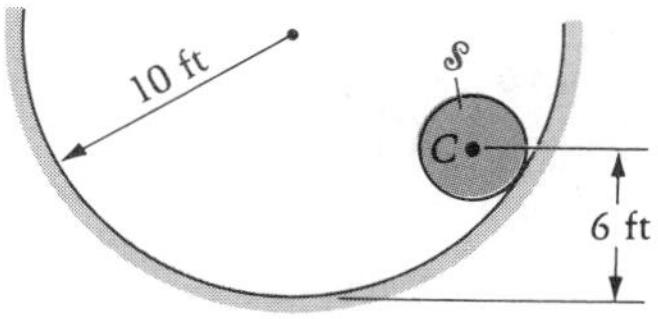

Figure P5.29

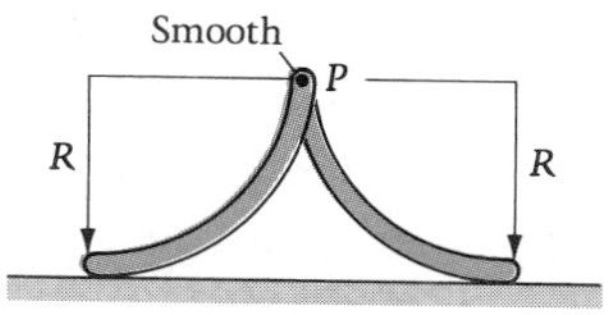

Figure P5.31

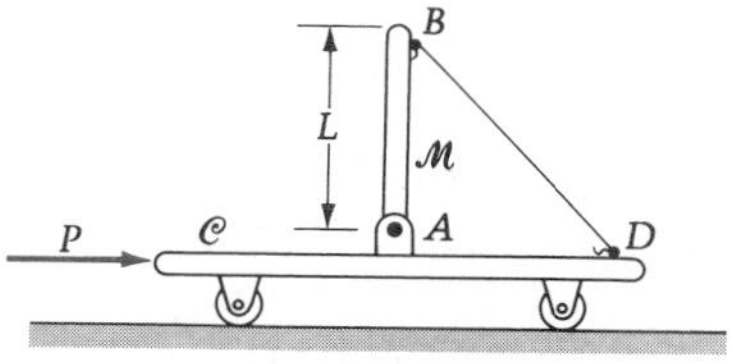

Figure P5.32

5.33 Use $W = \Delta T$ in Problem 4.178 to find the velocity of C as a function of its position ($\mathbf{x}_C$). Then substitute $\mathbf{x}_C(t)$ from that problem and verify the solution for $v_C(t)$. In computing the work of friction W_f, note that $v_B = 2v_C$ since the cylinder *rolls clockwise on the rope*. Thus

$$W_f = \int (-\mu N\hat{\mathbf{i}}) \cdot \underbrace{[(v_C + r\omega)\hat{\mathbf{i}}]}_{2v_C = 2\frac{dx_C}{dt}} dt = -\int 2\mu N \, dx_C$$

and the work computes like $\int \mathbf{F} \cdot d\mathbf{r}$ if written in this form. It is important, however, to note that the final integral is not the same as $-\int \mu N \, dx_B$, which makes no sense since B doesn't stay on the floor!

5.34 In Problem 4.86 use the principle of work and energy to obtain an upper bound on the rod's angular speed in its subsequent motion after the right-hand string is cut.

5.35 For the data of Problem 4.177 use $W = \Delta T$ to find the speed $\dot{\mathbf{x}}_C$ of the plate as a function of the distance $\mathbf{x}_C$ it has traveled to the right. Use the $\mathbf{x}_C = \mathbf{x}_C(t)$ result to check your answer; differentiate and eliminate t to produce the same $\dot{\mathbf{x}}_C = \dot{\mathbf{x}}_C(\mathbf{x}_C)$ result.

5.36 The 10-lb wheel shown in Figure P5.36 is attached at its center to a spring of modulus 20 lb/in. The radius of gyration of the wheel about the center is 2.5 in. The wheel rolls (no slip) after being released from rest with the spring stretched 1 in. Find: (a) the maximum magnitude of force in the spring; (b) the maximum speed of the center of the wheel during the ensuing motion.

5.37 A slender rod is placed on a table as shown in Figure P5.37. It will begin to pivot about the edge E and, at some angle θ_s, it will begin to slip. Find this angle, which will depend on the coefficient of friction μ and on k. *Hint:* Use all three equations of motion together with $W = \Delta T$. Eliminate α and ω^2, obtaining expressions for f and N. Setting $f = \mu N$ then permits a solution for θ_s. Solve the resulting equation when $\mu = 0.3$ and $k = 0.25$.

5.38 Rod $\mathcal{B}$ and disk $\mathcal{G}$ in Figure P5.38 have weights $W_{\mathcal{B}} = 5$ lb and $W_{\mathcal{G}} = 6$ lb. The rod's length is 8 in., the disk's radius is 4 in., the mass center offset of the disk is 2 in. from Q, and the radius of gyration of the mass of $\mathcal{G}$ with respect to the z axis through C is 3 in. It is desired to attach a spring between point Q and a fixed point so that the disk and rod come to a stop (in the dotted position) after $\mathcal{B}$ turns 90° clockwise from rest. The spring has a modulus of 25.5 lb/ft and an unstretched length of 4 in.; it is to be unstretched initially. Find the final spring stretch, and from this result determine where to attach the fixed end of the spring. (There are two possible points!)

5.39 Determine the spring modulus that will allow the 2-kg bar in Figure P5.39 to arrive at the position $\theta = 90°$ at zero angular velocity if it passes through the vertical (where the spring is stretched 0.05 m) at 2 rad/s ↻.

5.40 Find the spring modulus k that will result in the system momentarily stopping at $\theta = 0$ after being released from rest at $\theta = 50°$ if: (a) the initial stretch δ_i in the spring is zero; (b) the initial stretch is 0.2 m. (See Figure P5.40.) *Hint:* Use symmetry!

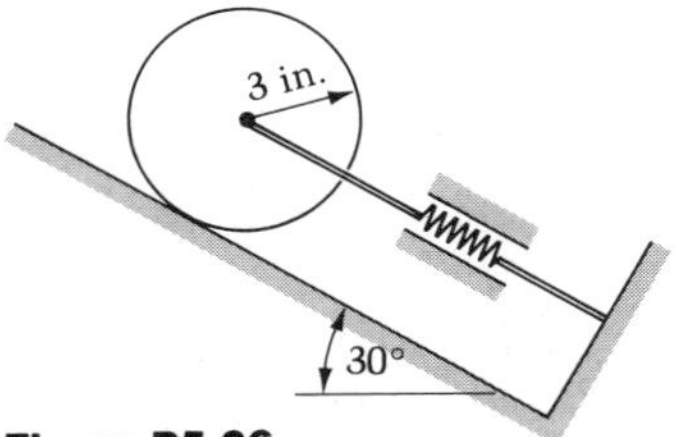

Figure P5.36

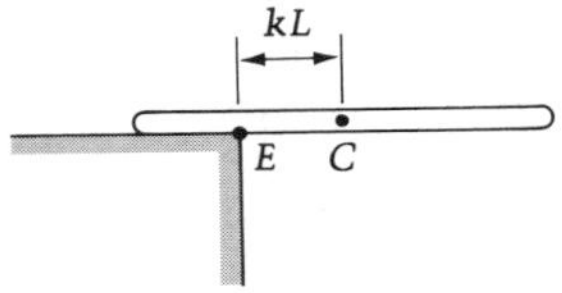

Figure P5.37

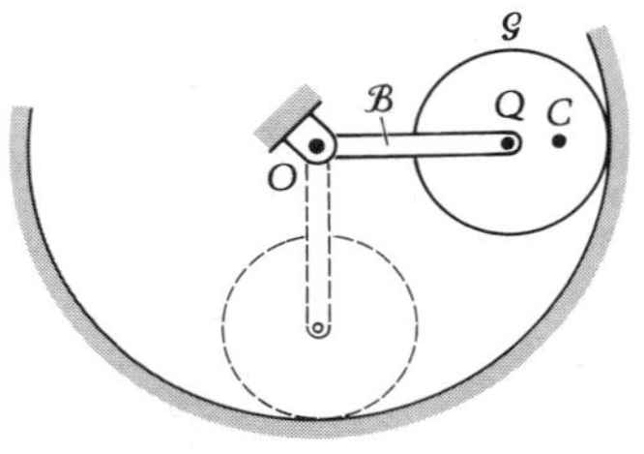

Figure P5.38

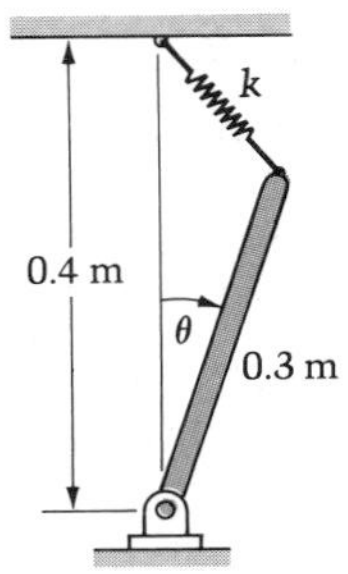

Figure P5.39

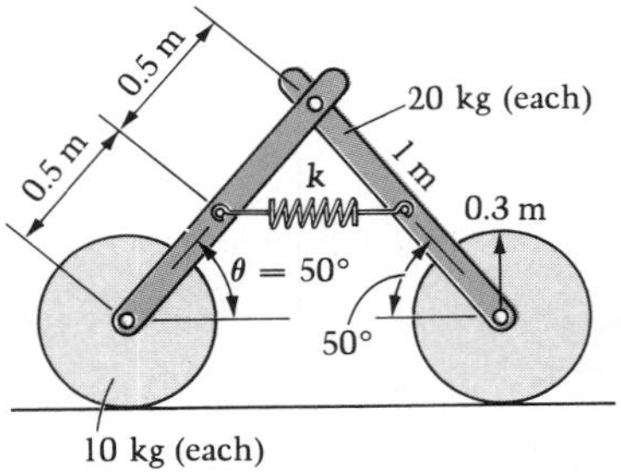

Figure P5.40

5.41 Body $\mathcal{D}$ translates in the slot without friction. (See Figure P5.41.) Disk $\mathcal{A}$ (radius R) is pinned to block $\mathcal{D}$ through their mass centers at G. Body $\mathcal{A}$ and body $\mathcal{D}$ *each* has mass m; body $\mathcal{B}$ has mass $2m$. The system is released from rest a distance D above the floor. Find: (a) the starting accelerations of $\mathcal{B}$ and $\mathcal{G}$; (b) the velocity of $\mathcal{B}$ when it hits the floor, using $W = \Delta T$.

5.42 The uniform cylinder $\mathcal{A}$ weighs 20 lb and is connected to the 10-lb body $\mathcal{B}$ by a cord wrapped around a slot near its rim which has a negligible effect on its moment of inertia. The spring has a modulus of 4 lb/ft and is unstretched when the system is released from rest. It is observed that the spring first stretches, then returns to its natural length at an instant when $\mathcal{B}$ has fallen 12.1 ft. If $\mathcal{A}$ does not slip on the plane, find the velocity of $\mathcal{B}$ at this instant. (See Figure P5.42.)

5.43 A 5-lb cylinder is raised from rest by a force $P =$ 20 lb. (See Figure P5.43.) Find the modulus of the spring that will cause the cylinder to stop after its center has been raised 2 ft. Will it then start back down? The spring is initially unstretched.

5.44 The 50-kg wheel in Figure P5.44 is to be treated as a cylinder of radius $R = 0.2$ m. If it is rolling to the left with $v_C = 0.07$ m/s at an initial instant when the spring is unstretched, find: (a) the distance moved by C before v_C is instantaneously zero; (b) the minimum coefficient of friction that will prevent slip.

5.45 Cylinder $\mathcal{C}$ in Figure P5.45 is moving up the plane with $v_C = 0.3$ m/s at an initial instant when the spring is stretched 0.2 m. If $\mathcal{C}$ does not slip at any time, determine how far *down* the plane the point C will move in the subsequent motion. *Note:* The spring, connected to the cord, cannot be in compression.

5.46 In Problem 4.91 determine the velocity of corner B of the half-cylinder when the diameter AB becomes horizontal for the first time.

5.47 The cylinder in Figure P5.47 rolls on the incline. If the velocity of the mass center C is 5 ft/sec down the plane in the upper (starting) position, find v_C in the bottom position.

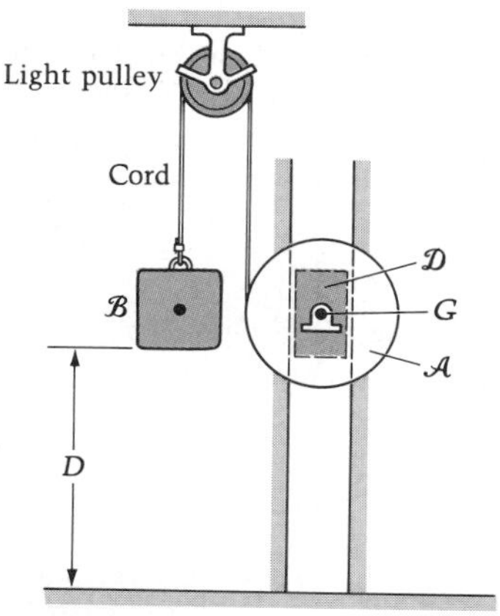

Figure P5.41

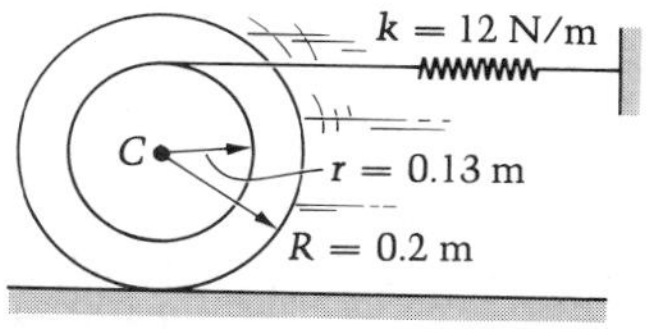

Figure P5.44

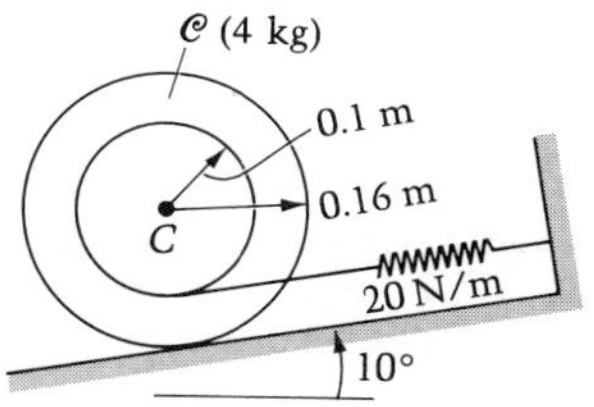

Figure P5.45

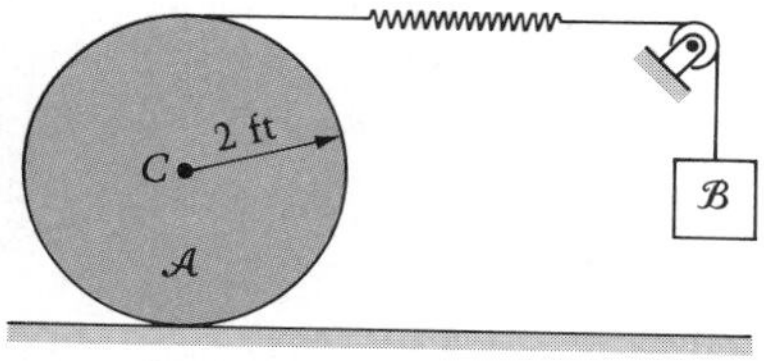

Figure P5.42

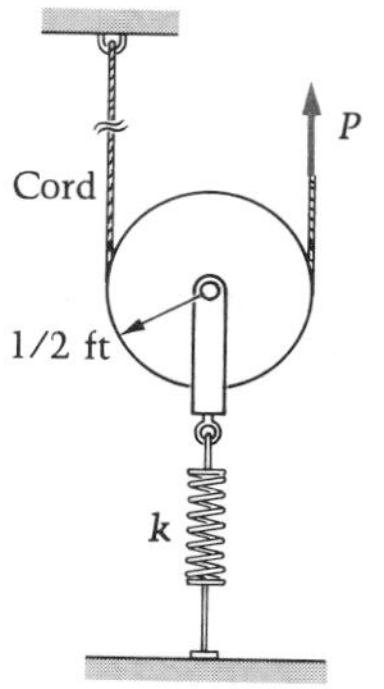

Figure P5.43

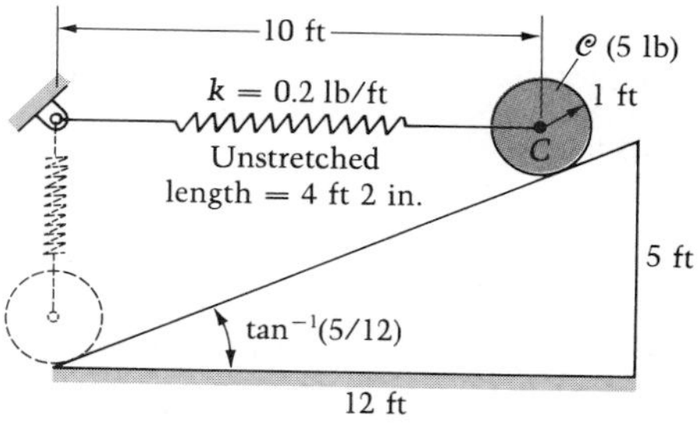

Figure P5.47

5.48 Cylinders $\mathcal{A}$ and $\mathcal{B}$ in Figure P5.48 are released from rest and turn without slip at the contact point. A cord is wrapped around an attached hub of each, which has negligible effect on the moment of inertia. There is enough friction to prevent the rope from slipping on the pulley. Find the velocity of the mass center of the pulley $\mathcal{C}$ after body $\mathcal{D}$ has fallen 10 ft.

5.49 The bar in Figure P5.49 weighs the same (W) as the hoop to which it is welded. The combined body is released from rest on the incline in the position shown. If there is no slipping, determine the velocity of Q after one revolution of the hoop.

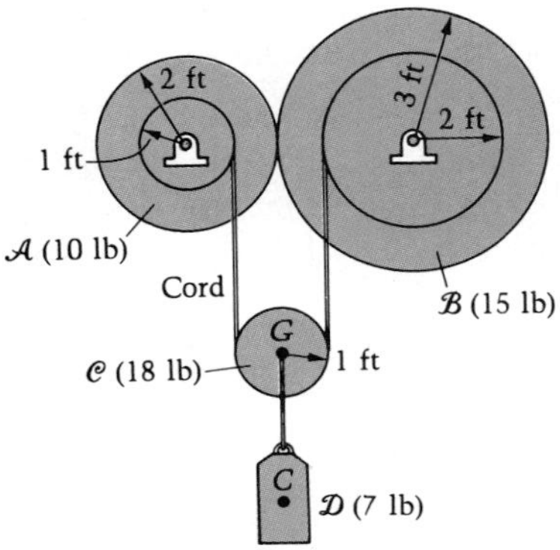

Figure P5.48

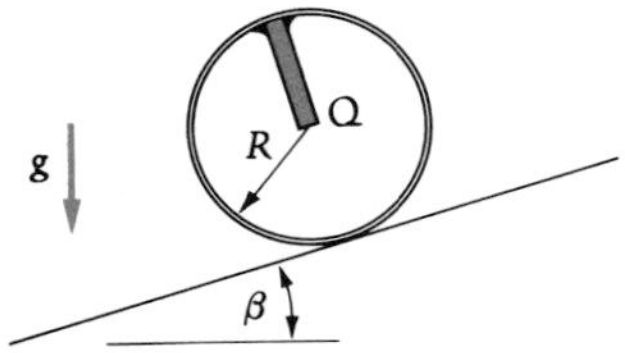

Figure P5.49

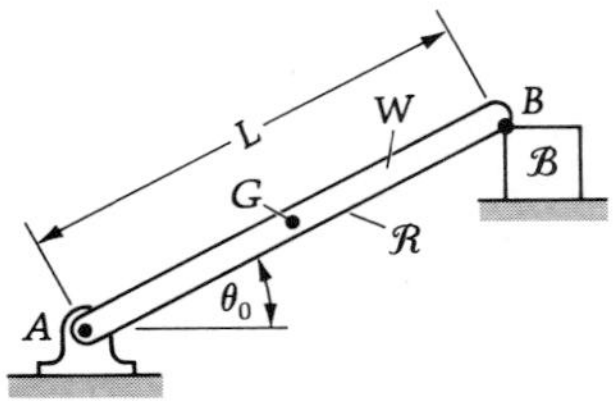

Figure P5.50

5.50 A slender uniform rod $\mathcal{R}$ of length L and weight W is smoothly hinged to a fixed support at A and rests on a block at B. (See Figure P5.50.) Block $\mathcal{B}$ is suddenly removed. Find: (a) the initial angular acceleration and components of reaction at A; (b) the components of reaction at A when the rod becomes horizontal.

5.51 Show that the kinetic energy of a uniform straight rod moving in any manner can be written as

$$\tfrac{1}{6}M(u^2 + v^2 + uv\cos\alpha)$$

where M is the mass of the rod, $\mathbf{u}$ and $\mathbf{v}$ are the velocities of its ends, and α is the angle between the directions of these velocities. (See Figure P5.51.) *Hints:* (1) Components of $\mathbf{u}$ and $\mathbf{v}$ along AB are equal. This gives $\tan\phi = [(v\cos\alpha - u)/v\sin\alpha]$ so that ϕ is not unknown. (2) Velocity components of C are $u\cos\phi$ along AB and $[u\sin\phi + v\sin(\alpha + \phi)]/2$ normal to it.

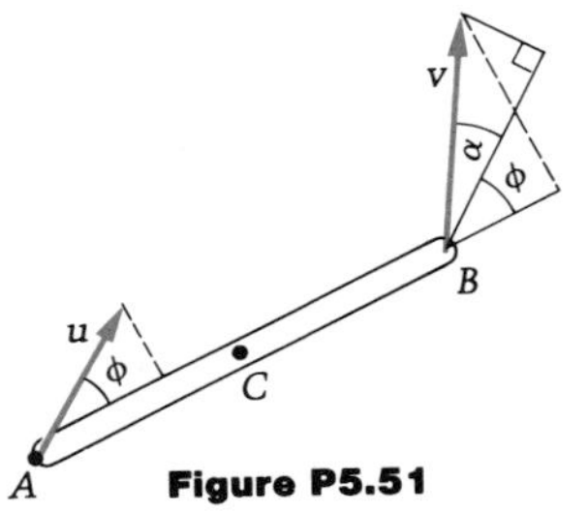

Figure P5.51

5.52 The slender nonuniform bar in Figure P5.52 (the mass is m and the radius of gyration with respect to the mass center C is $L/2$) is supported by two inextensible wires. If the bar is released from rest with $\theta = 0$, find the tension in each wire as a function of θ.

5.53 A truck body weighing 4000 lb is carried by four solid disk wheels that roll on the sloping surface. (See Figure P5.53.) Each wheel weighs 322 lb and is 3 ft in diameter. The truck has a velocity of 5 ft/sec in the position shown. Determine the modulus of the spring if the truck is brought to rest by compressing the spring 6 in.

5.54 A string is released from rest in the position shown in Figure P5.54. Find an upper limit on its kinetic energy at any time following release.

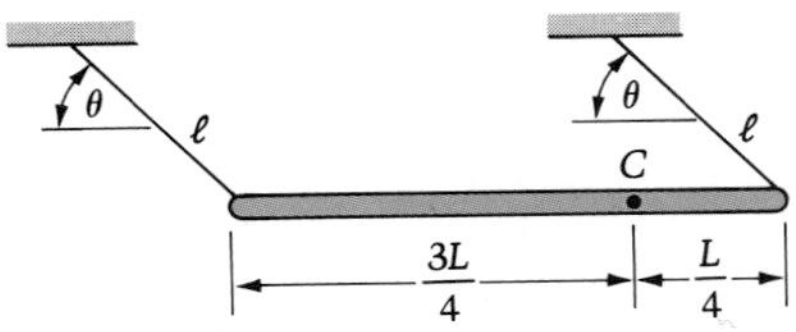

Figure P5.52

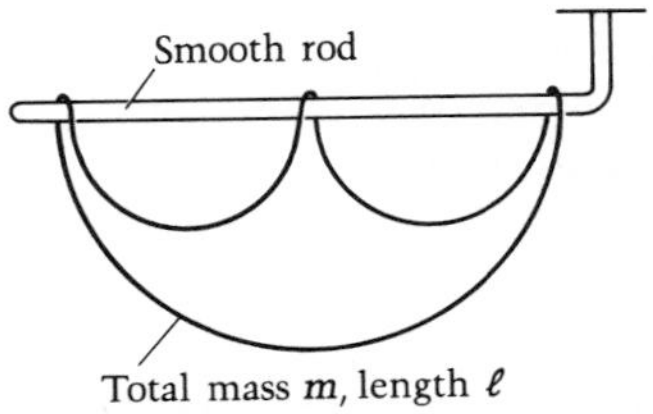

Figure P5.54

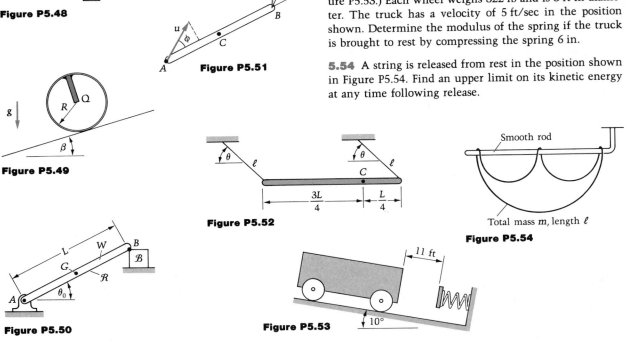

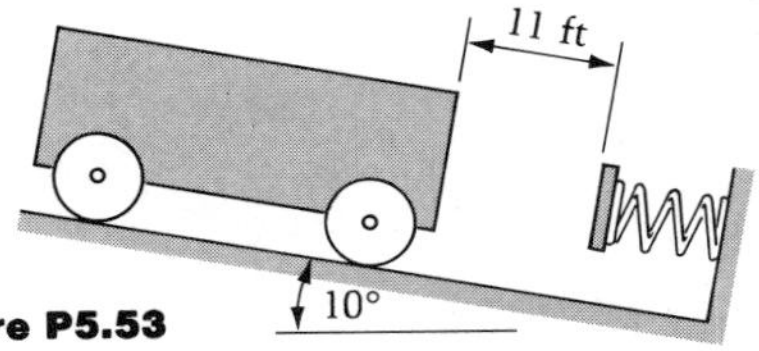

Figure P5.53

5.55 Find the angular velocity of the elliptical cylinder in the position shown in Figure P5.55 following release from rest in the position at the left. *Hint:* Use the equation of the ellipse ($x^2/a^2 + y^2/b^2 = 1$) to find the slope of the tangent line at A, and then use this result to help determine distance d. Assume no slipping, and that $b > a$.

5.56 The prehistoric car shown in Figure P5.56 is powered by the falling rock m, connected to the main wheel (a cylinder of mass M) by a vine as shown. If the weights of the frame, pulley, and front wheel are small compared with Mg, find the velocity v_C of the car as a function of y if there is no slipping and it starts from rest with $y = 0$ at $t = 0$. Assume that m moves only vertically relative to the car's frame.

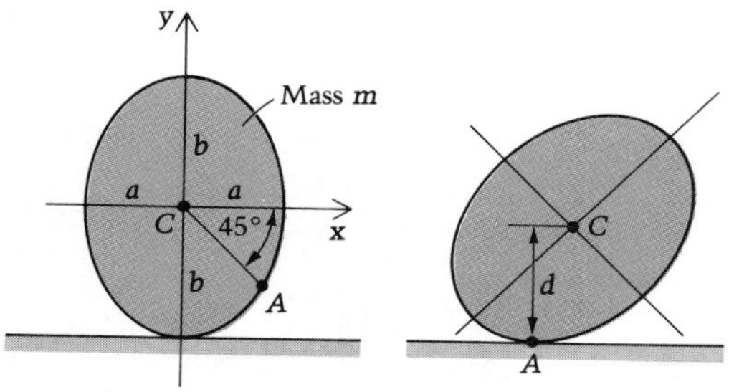

Figure P5.55

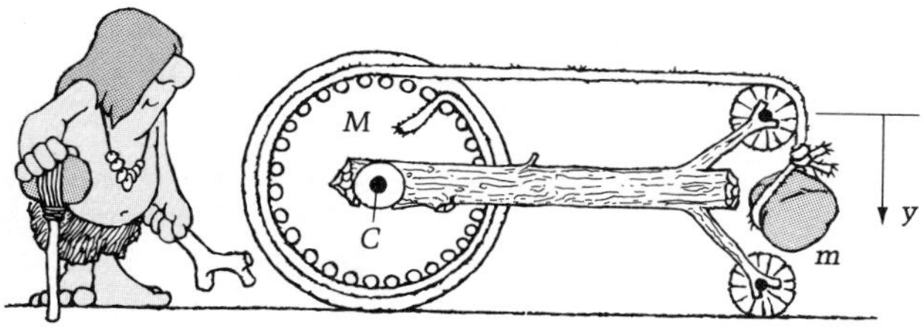

Figure P5.56

5.57 The suspended log shown in Figure P5.57 is to be used as a battering ram. At what angle θ should the ruffian release the log from rest so that it strikes the door at $\theta = 0$ with a velocity of 25 ft/sec?

5.58 Bar $\mathcal{A}$ is smoothly pinned to the support at A and smoothly pin-jointed to $\mathcal{B}$ at B. (See Figure P5.58.) End D slides on a smooth horizontal surface. If D starts from rest at $\theta = \theta_0$, determine the angular velocities of the rods just before they become horizontal.

5.59 Link $\mathcal{L}$ weighs 10 lb and may be treated as a uniform slender rod (Figure P5.59). The 15-lb wheel is a circular disk with sufficient friction on the horizontal surface to prevent slipping. The spring is unstretched as shown. Link $\mathcal{L}$ is released from rest, and the light block $\mathcal{A}$ slides down the smooth slot. Neglecting friction in the pins, determine: (a) the angular velocity of the link as A strikes the spring with $\mathcal{L}$ horizontal; (b) the maximum deflection of the spring. (The modulus k of the spring is 10 lb/in.)

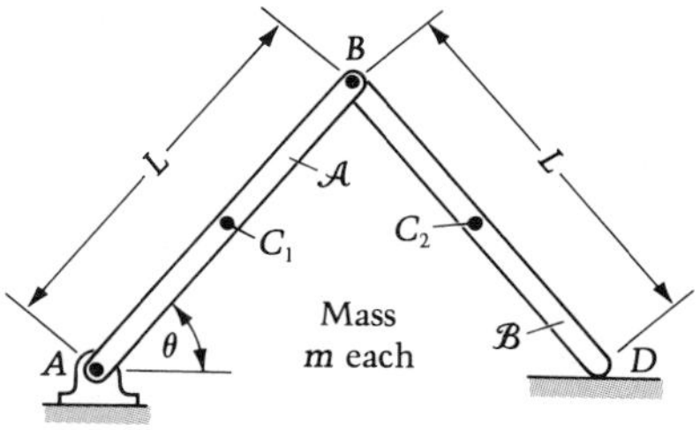

Figure P5.58

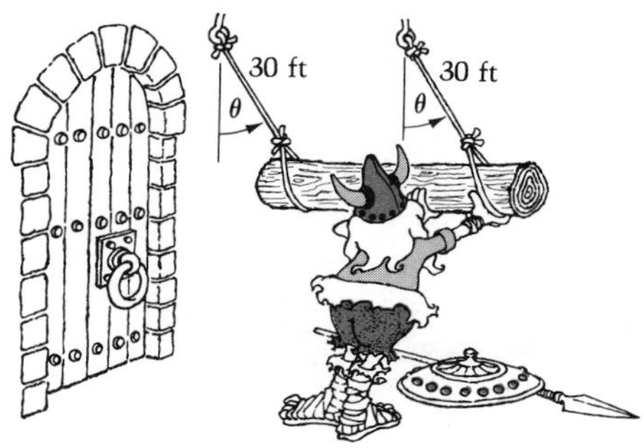

Figure P5.57

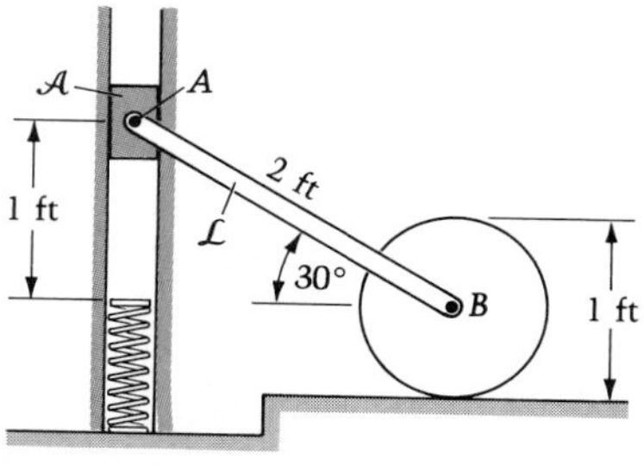

Figure P5.59

5.60 The center of mass of a uniform triangular plate is two-thirds of the distance from any vertex to the opposite side. The moment of inertia of an equilateral triangular plate is $ms^2/12$ with respect to the z_C axis through C. For the plate shown in Figure P5.60, with mass 30 kg and side 2 m, find its angular velocity when it reaches the dotted position where C is beneath O. The spring has unstretched length 0.5 m and modulus 20 N/m, and the plate is released from rest.

5.61 The unbalanced wheel of radius 2 ft and weight 64.4 lb shown in Figure P5.61 has a centroidal moment of inertia of 6 slug-ft². In position 1, with C above O, the wheel has a clockwise angular velocity of 2 rad/sec. The wheel then rolls to position 2, where OC is horizontal. Determine the angular velocity of the wheel in position 2.

5.62 Figure P5.62 shows a spring pushing on a massless plunger that in turn bears on a disk of weight 32.2 lb. There is sufficient friction between the disk and floor to prevent slipping at their point of contact. The coefficient of friction between all other surfaces is $\mu = \frac{1}{2}$. If the system is released at rest in the position shown with an initial compressive spring force of 150 lb, find the maximum velocity the center of the disk will attain.

5.63 Figure P5.63 shows a fire door on the roof of a building. The door $\mathcal{D}$, 4 ft wide, 6 ft long, and 4 in. thick, is wooden (at 30 lb/ft³) and can rotate about a frictionless hinge at O. A cantilever arm $\mathcal{A}$ of negligible weight is rigidly fastened to the door and carries a 150-lb weight at its free end. During a fire the link $\mathcal{L}$ melts and the door swings open 45°. Find the angular velocity of the door just before the 150-lb weight hits the roof: (a) with no snow on the roof; (b) with snow at 1 lb/ft² on the roof.

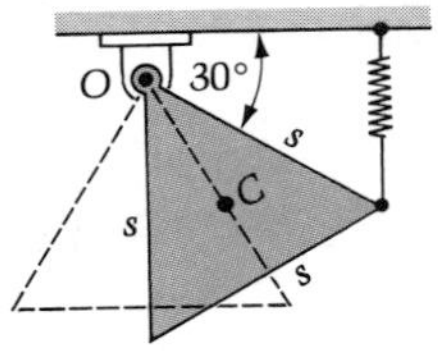

Figure P5.60

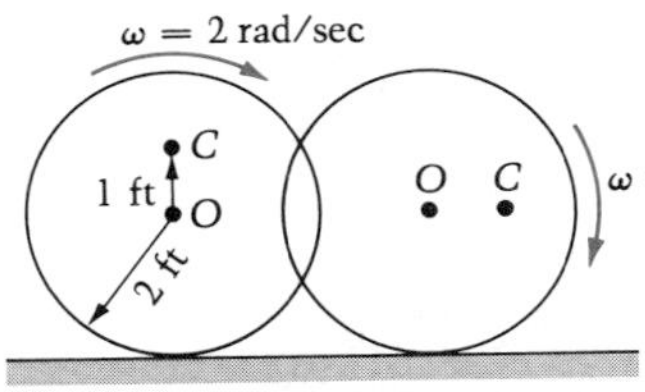

Figure P5.61

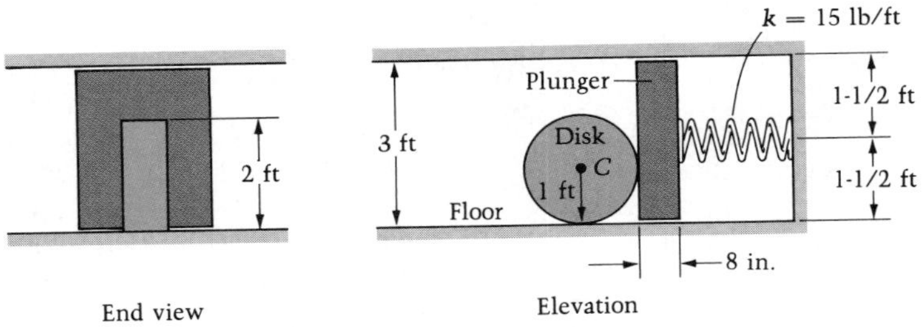

Figure P5.62

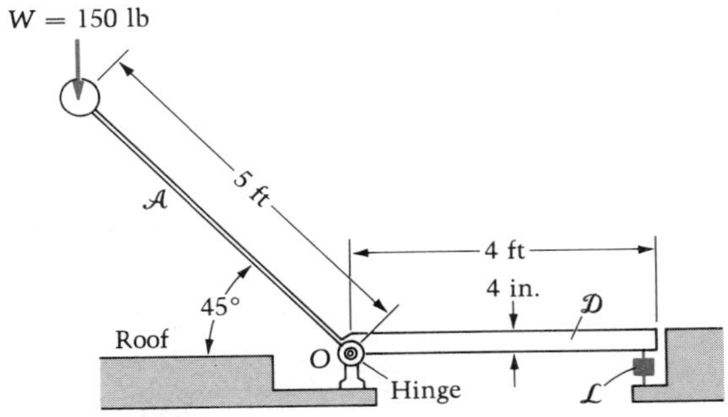

Figure P5.63

5.3 The Principles of Impulse and Momentum

The Equations of Impulse and Momentum for the Rigid Body in Plane Motion

The principle of work and kinetic energy is very helpful when the problem is posed in terms of positions and velocities. When time, rather than position, is the main concern, we often draw on a pair of principles concerned with impulse and momentum vectors. Like $W = \Delta T$, these principles too are obtained by general integrations of the equations of motion, but now the integration is directly with respect to time, without first dotting the equations with velocity. Thus they leave us with a set of vector equations instead of a single scalar result.

We have encountered one of the principles in Section 2.5 in our study of mass center motion: The impulse of the external forces imparted to any system equals its change of momentum over the same time interval. In Chapter 2 the system was general, so this principle holds for the rigid bodies we are now studying. From Equation (2.1), $\mathbf{F}_r = \dot{\mathbf{L}}$ so that

$$\int \mathbf{F}_r \, dt = \int \dot{\mathbf{L}} \, dt = \int d\mathbf{L} = \mathbf{L}\Big]_i^f = \mathbf{L}_f - \mathbf{L}_i = \Delta\mathbf{L} = m\mathbf{v}_{C_f} - m\mathbf{v}_{C_i} \quad (5.26)$$

Reviewing, we note that the integral $\int \mathbf{F}_r \, dt$ is called the **impulse** (or **linear impulse**) imparted to the system by external forces. The vector $\Delta\mathbf{L}$ is the change in the system's momentum (or linear momentum) from the initial to the final time.

We need only the $\mathbf{x}$ and y components of Equation (5.26) for the rigid body in plane motion:

$$F_{r_x} = \Delta L_x = \Delta(m\dot{x}_C) = m\dot{x}_{C_f} - m\dot{x}_{C_i} \quad (5.27)$$

$$F_{r_y} = \Delta L_y = \Delta(m\dot{y}_C) = m\dot{y}_{C_f} - m\dot{y}_{C_i} \quad (5.28)$$

There is also a corresponding principle of *angular* impulse and momentum. From Equation (4.33),

$$\mathbf{M}_{r_C} = \dot{\mathbf{H}}_C$$

so that

$$\int \mathbf{M}_{r_C} \, dt = \int \dot{\mathbf{H}}_C \, dt = \int d\mathbf{H}_C = \mathbf{H}_C\Big]_i^f = \mathbf{H}_{C_f} - \mathbf{H}_{C_i} = \Delta\mathbf{H}_C \quad (5.29)$$

This equation may be put into a convenient form for rigid bodies in plane motion by recalling Equation (4.8) for the angular momentum:

$$\mathbf{H}_C = I_{xz_C}\omega\hat{\mathbf{i}} + I_{yz_C}\omega\hat{\mathbf{j}} + I_{z_C}\omega\hat{\mathbf{k}}$$

For symmetric bodies in which the products of inertia vanish and $\mathbf{M}_{r_C} = M_{r_C}\hat{\mathbf{k}}$, this equation becomes

$$\mathbf{H}_C = I_{z_C}\omega\hat{\mathbf{k}}$$

Therefore

$$\int M_{r_C}\,dt = \Delta \mathbf{H}_C = \Delta(I_{z_C}\omega\hat{\mathbf{k}})$$

or

$$\int M_{r_C}\,dt = \Delta(I_{z_C}\omega) = I_{z_{C_f}}\omega_f - I_{z_{C_i}}\omega_i \tag{5.30}$$

The integral $\int M_{r_C}\,dt$ is called the **angular impulse** imparted to the system by the external forces and couples, and the quantity $\Delta(I_{z_C}\omega)$ is the change in angular momentum, both taken about C.

A subtle but important point regarding Equation (5.30) must be understood here. We note from Equation (5.29) that angular impulse equals the change in angular momentum for any body (deformable as well as rigid); therefore the use of Equation (5.30) only requires that the body of interest behave rigidly at the start (t_i) and end (t_f) of the time interval (t_i, t_f). At those times the moment of momentum is $\mathbf{H}_C = I_{z_C}\omega\hat{\mathbf{k}}$, even though this simple expression for $\mathbf{H}_C$ may not apply ***between*** t_i and t_f. A good example is an ice skater drawing in her arms to increase angular speed, as we shall see later in Example 5.15.

In summary, for the rigid body in plane motion we have the following two principles at our disposal:

1. Linear impulse and momentum:

$$\int \mathbf{F}_r\,dt = \Delta(m\mathbf{v}_C)$$

from which we get

$\hat{\mathbf{i}}$ coefficients:
$$\int_{t_i}^{t_f} F_{r_x}\,dt = m(\dot{x}_{C_f} - \dot{x}_{C_i}) \tag{5.31}$$

$\hat{\mathbf{j}}$ coefficients:
$$\int_{t_i}^{t_f} F_{r_y}\,dt = m(\dot{y}_{C_f} - \dot{y}_{C_i}) \tag{5.32}$$

2. Angular impulse and momentum:

$$\int_{t_i}^{t_f} M_{r_C}\,dt = I_{z_{C_f}}\omega_f - I_{z_{C_i}}\omega_i \tag{5.33}$$

We note also that if the products of inertia are not zero, we have

$$\int M_{r_{C_x}}\,dt = I_{xz_{C_f}}\omega_f - I_{xz_{C_i}}\omega_i \tag{5.34}$$

and

$$\int M_{r_{C_y}}\,dt = I_{yz_{C_f}}\omega_f - I_{yz_{C_i}}\omega_i \tag{5.35}$$

where the directions of **x**, **y**, and **z** are fixed in the inertial frame.

Question 5.6 Are the coordinate axes associated with Equations (5.31) and (5.32) the same as those of (5.34) and (5.35)?

In the remainder of this section we treat only problems of symmetric bodies, for which Equations (5.31) to (5.33) are the impulse and momentum equations.

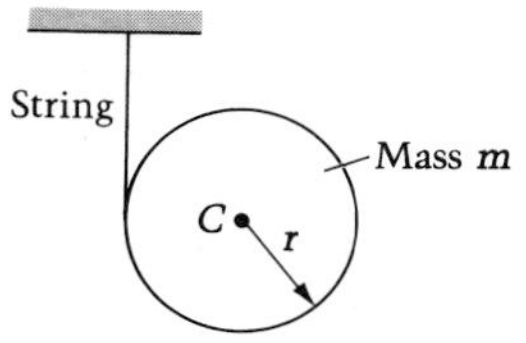

EXAMPLE 5.12

A cylinder has a string wrapped around it (see the diagram) and is released from rest. Determine the velocity of C as a function of time.

M824

SOLUTION

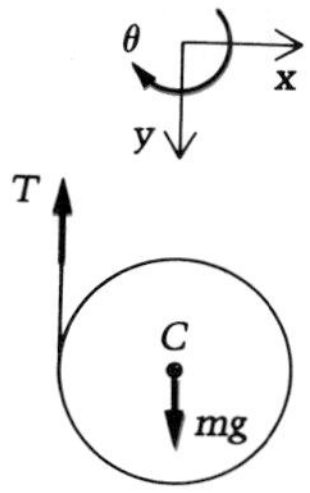

We choose the sign convention to be as shown in the following diagram, since the cylinder turns clockwise as C moves downward. Applying the impulse and momentum equations in the y and θ directions (note that $F_{r_x} = 0$ means $\ddot{x}_C = 0$ so that $\dot{x}_C = \text{constant} = 0$) gives

$$\int_0^t F_{r_y}\,dt = m(\dot{y}_{C_f} - \dot{y}_{C_i})$$

$$\int_0^t (mg - T)\,dt = m(\dot{y}_C - 0) \tag{1}$$

$$mgt - \int_0^t T\,dt = m\dot{y}_C$$

and

$$\int_0^t M_{r_C}\,dt = I_C(\omega_f - \omega_i)$$

$$\int_0^t Tr\,dt = \frac{1}{2}mr^2\omega_f \tag{2}$$

$$\int_0^t T\,dt = \frac{1}{2}mr\omega_f$$

We note that $\int T\,dt$ is itself an unknown and should be treated as such. (In this problem use of the equations of motion would separately tell us that $T = mg/3$, so that the integral is in fact Tt. But sometimes T is time-dependent, in which case Tt would be incorrect for the value of the integral.)

(Continued)

Eliminating $\int_0^t T\,dt$ gives

$$mgt - m\dot{y}_C = \frac{1}{2} mr\omega_f$$

However, $\dot{y}_C = r\omega_f$ because the cylinder rolls on the rope, so that

$$gt = \left(1 + \frac{1}{2}\right)\dot{y}_C$$

which gives our result:

$$\dot{y}_C = \frac{2}{3} gt$$

The cylinder falls with "$\frac{2}{3}$ of a g" because of the retarding force of the rope.

We recall that if the rigid body in plane motion has a point Q associated with it that always accelerates through the mass center C, we may write

$$M_{r_Q} = I_Q \alpha$$

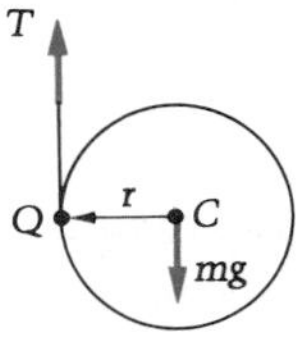

The contact point of a round rolling body whose geometric and mass centers coincide is just such a point Q.* We can integrate this equation and obtain

$$\int M_{r_Q}\,dt = I_Q(\omega_f - \omega_i)$$

In the preceding example, the cylinder rolls down the vertical string so that

$$\int_0^t M_{r_Q}\,dt = I_Q(\omega_f - \omega_i)$$

$$mgrt = \left(\frac{1}{2} mr^2 + mr^2\right)\omega_f$$

$$= \frac{3}{2} mr^2 \frac{\dot{y}_c}{r}$$

$$\dot{y}_C = \frac{2}{3} gt$$

and we have obtained the same result for the velocity of C without having to deal with the unknown $\int T\,dt$.

*We are considering Q to be a point moving in space along the fixed contact line, Q always being the contact point. Note that I_Q is constant for this case even though the moving point Q is not a point of the cylinder!

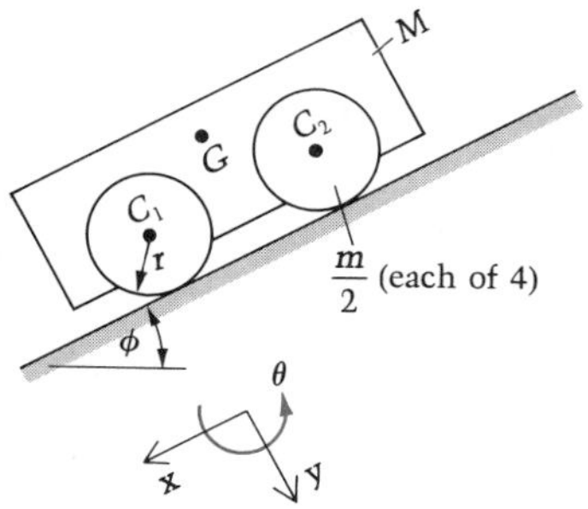

Figure 1

EXAMPLE 5.13

The cart shown in Figure 1 has mass M exclusive of its four wheels, each of which is a disk of mass $m/2$. The front wheels and their axle are rigidly connected, and the same is true for the rear wheels. If the axles are smooth, find the acceleration of G (the cart's mass center) as a function of time. The system starts from rest. Assume that there is enough friction to prevent the wheels from slipping.

SOLUTION

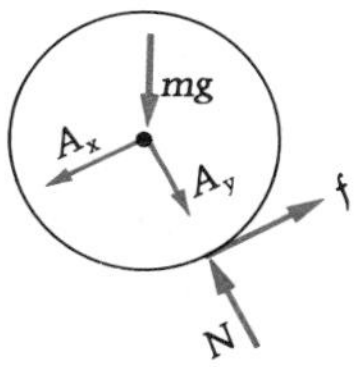

Figure 2

We first consider the free-body diagram (Figure 2) of a wheel pair (either front or back). Since the front and rear wheels are constrained to have identical angular velocities at all times, the fr's (front and back) must produce identical $I\alpha$'s; hence the friction force is the same for the rear wheels as for the front. And since the wheels' mass centers must always have identical velocities, the forces acting down the plane on each pair (front and back) must also be equal. These resultants are $A_x + mg\sin\phi - f$, so the reaction A_x is also the same on each pair of wheels.

Question 5.7 Are A_y and N also the same for front and rear wheels?

From the free-body diagram, we may write the following linear and angular equations of impulse and momentum:

$$\int_0^t F_{r_x}\,dt = m(\dot{x}_{C_f} - \dot{x}_{C_i})$$

$$\int_0^t (A_x + mg\sin\phi - f)\,dt = m\dot{x}_{C_{1\text{ or }2}} \tag{1}$$

$$\int_0^t M_{r_C}\,dt = I_C(\omega_f - \omega_i) = I_C(\omega - 0)$$

$$\int_0^t fr\,dt = \frac{1}{2}mr^2\omega \tag{2}$$

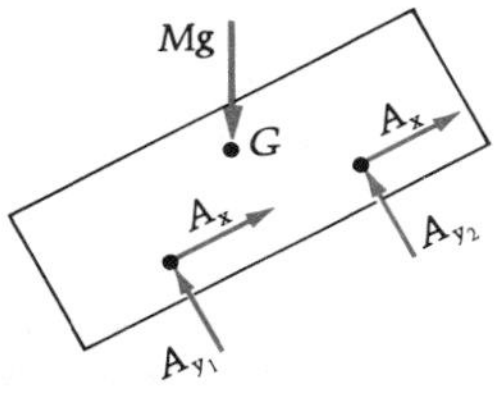

Figure 3

We may also isolate a free-body diagram of the translating cart (Figure 3) and write its equation of impulse and momentum in the **x** direction from the plane:

$$\int_0^t F_{r_x}\,dt = M(\dot{x}_{G_f} - \dot{x}_{G_i})$$

$$\int_0^t (-2A_x + Mg\sin\phi)\,dt = M(\dot{x}_G - 0) \tag{3}$$

(Continued)

Next we note that C_1 and C_2, the mass centers of the front and back wheel pairs, are also points of the cart; thus $\dot{\mathbf{x}}_{C_1} \equiv \dot{\mathbf{x}}_{C_2} \equiv \dot{\mathbf{x}}_G$. Using this relation, and adding Equation (3) to twice Equation (1), eliminates the unknown impulse of the reaction A_x:

$$\int_0^t [(M + 2m)g \sin\phi - 2f]\, dt = (M + 2m)\dot{\mathbf{x}}_G{}^* \tag{4}$$

Similarly, adding Equation (4) to twice Equation (2) results in an equation free of the unknown friction force:

$$\int_0^t (M + 2m)g \sin\phi\, dt = (M + 2m)\dot{\mathbf{x}}_G + m\overbrace{(r\omega)}^{\dot{\mathbf{x}}_G}$$

Carrying out the integration and solving for $\dot{\mathbf{x}}_G$, we get

$$\dot{\mathbf{x}}_G = \frac{(M + 2m)gt \sin\phi}{M + 3m}$$

We note from this result that if the wheels are very light compared to the weight of the cart ($m \ll M$), then $\dot{\mathbf{x}}_G = gt \sin\phi$, which is the answer for the problem of Figure 4. Thus light wheels on smooth axles make the cart move as if it were on a smooth plane, as expected. If the cart is light compared to ***heavy*** wheels ($M \ll m$), then each rolls as it would ***alone***, with $\dot{\mathbf{x}}_G = \frac{2}{3}gt \sin\phi$, because (Figure 5)

$$\int M_{z_Q}\, dt = I_Q(\omega_f - \omega_i) \qquad \text{(Note that } \mathbf{a}_Q \text{ passes through } G!)$$

$$\int_0^t mg \sin\phi\, dt = \left(\frac{1}{2}mr^2 + mr^2\right)\overset{\dot{\mathbf{x}}_C/R}{\cancel{\phi}}$$

$$\dot{\mathbf{x}}_{C_1} = \dot{\mathbf{x}}_{C_2} = \frac{2}{3}gt \sin\phi$$

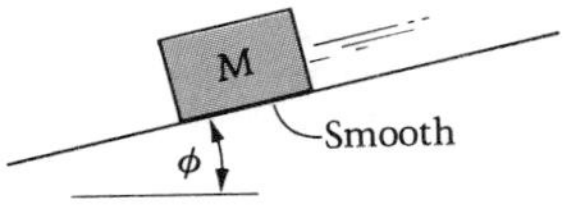

Figure 4

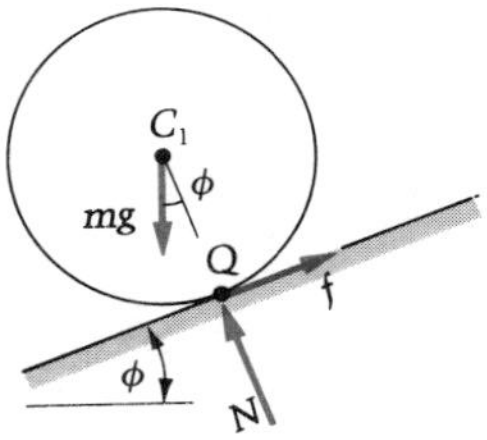

Figure 5

*This is the equation of linear impulse and momentum for the total (nonrigid) system of cart plus wheels.

EXAMPLE 5.14

Force P acts on the rolling cylinder $\mathcal{C}$ beginning at $t = 0$ with $\mathcal{C}$ at rest. (See Figure 1.) Force P varies with the time t in seconds according to

$$P = 5 \sin \frac{\pi t}{10} \text{ Newtons} \qquad \text{(positive to the left as shown)}$$

Cylinder $\mathcal{C}$ and weight $\mathcal{B}$ respectively weigh 100 and 40 N. Find the velocity of G (the mass center of $\mathcal{B}$) when $t = 10$ s. Neglect the effect of the hubs in Figure 2 (and the drilled hole to accommodate force P) on the moment of inertia of $\mathcal{C}$.

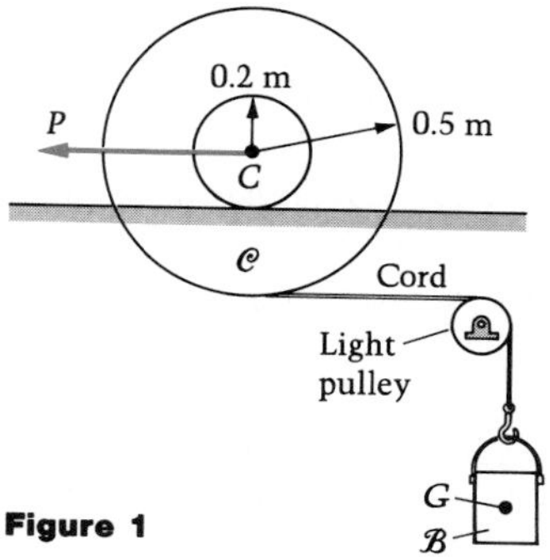

Figure 1

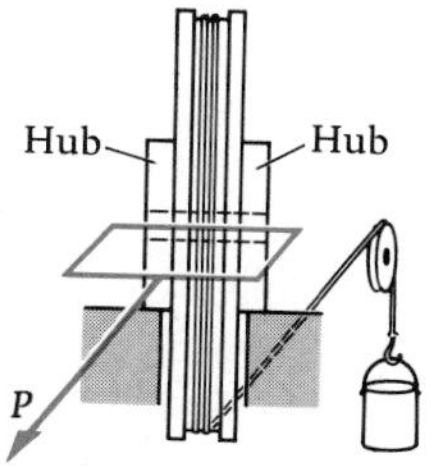

Figure 2

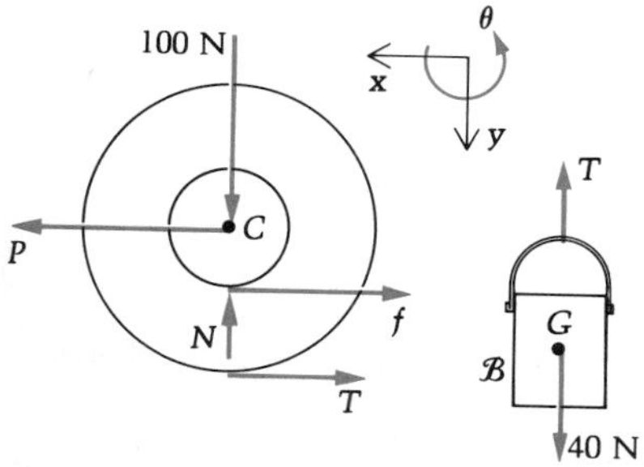

Figure 3

SOLUTION

Using the free-body diagrams (Figure 3), we may write the equations of impulse and momentum. On $\mathcal{C}$:

$$\int F_{r_x}\, dt = m\dot{x}_{C_f} - m\dot{x}_{C_i}^{\nearrow 0}$$

or

$$\int_0^{10} \left(5 \sin \frac{\pi t}{10} - f - T\right) dt = \frac{100}{9.81}\dot{x}_{C_f} = 10.2\dot{x}_{C_f} \tag{1}$$

Also on $\mathcal{C}$:

$$\int M_{r_C}\, dt = I_C\omega_f - I_C\omega_i^{\nearrow 0}$$

or

$$\int_0^{10} (0.2f + 0.5T)\, dt = \left[\frac{1}{2}\frac{100}{9.81}(0.5)^2\right]\omega_f = 1.27\omega_f \tag{2}$$

On $\mathcal{B}$:

$$\int F_{r_y}\, dt = m\dot{y}_{G_f} - m\dot{y}_{G_i}^{\nearrow 0}$$

(Continued)

or

$$\int_0^{10} (40 - T)\, dt = \frac{40}{9.81} \dot{y}_{G_f} = 4.08\dot{y}_{G_f} \tag{3}$$

Subtracting Equation (3) from (1), after integrating the sine function, we obtain

$$\frac{100}{\pi} - \int_0^{10} f\, dt - 40(10) = 10.2\dot{x}_{C_f} - 4.08\dot{y}_{G_f} \tag{4}$$

To obtain a second, independent equation that is also free of the integral of the unknown tension T, we add Equation (1) to twice Equation (2):

$$\frac{100}{\pi} - 0.6 \int_0^{10} f\, dt = 10.2\dot{x}_{C_f} + 2.54\omega_f \tag{5}$$

Multiplying Equation (4) by 0.6 and subtracting from Equation (5) gives

$$\frac{40}{\pi} + 240 = 4.08\dot{x}_{C_f} + 2.54\omega_f + 2.45\dot{y}_{G_f} \tag{6}$$

Kinematics now relates $\dot{x}_{C_f}$, $\dot{y}_{G_f}$, and ω_f; the lowest point of $\mathcal{C}$ has the same velocity magnitude as does G because of the inextensibility of the cord (Figure 4):

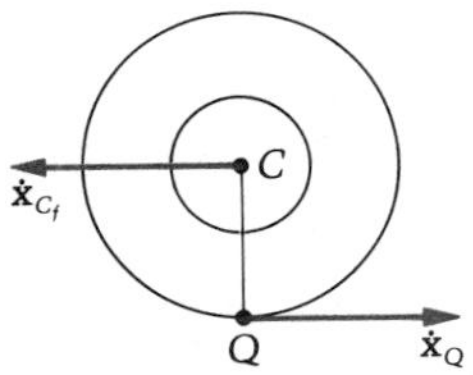

Figure 4

Kinematic conditions:
$$\left.\begin{aligned} \dot{x}_{C_f} &= 0.2\omega_f \\ \dot{x}_Q &= 0.3\omega_f \\ \dot{x}_Q &= \dot{y}_{G_f} \end{aligned}\right\} \Rightarrow \begin{cases} \omega_f = 3.33\dot{y}_{G_f} \\ \dot{x}_{C_f} = 0.2\omega_f = 0.667\dot{y}_{G_f} \end{cases}$$

Substituting these expressions for $\dot{x}_{C_f}$ and ω_f into Equation (6) gives

$$253 = \dot{y}_{G_f}[4.08(0.667) + 2.54(3.33) + 2.45]$$

$$\dot{y}_{G_f} = \frac{253}{13.6} = 18.6 \text{ m/s}$$

Hence the velocity of the mass center of $\mathcal{B}$ at $t = 10$ s (when the force changes direction) is

$$\mathbf{v}_G = 18.6 \downarrow \text{m/s}$$

We note that in this case one could legitimately take moments about the contact point of $\mathcal{C}$ and thereby eliminate the need for one of the three impulse-momentum equations.

Question 5.8 Which one?

We have chosen not to do this because the problem, as presented, gives applications of impulse and momentum in both the $\mathbf{x}$ and y directions, as well as illustrating the principle of angular impulse and angular momentum. We must emphasize that Equations (1) to (3) are simply first integrals of the three Equations (4.56a,b,c) we studied in detail in Chapter 4.

PROBLEMS / Section 5.3

5.64 Drum $\mathcal{A}$ has a radius of gyration of mass with respect to a horizontal axis through O of 1 m and a mass of 800 kg. Body $\mathcal{C}$ has a mass of 600 kg and a velocity of 20 m/s upward when in the position shown in Figure P5.64. Find the velocity of $\mathcal{C}$ 3 s later.

5.65 The hollow drum shown in Figure P5.65 weighs 161 lb and rotates about a fixed horizontal axis through O. The diameter of the drum is 2.4 ft, and the radius of gyration of the mass with respect to the axis through O is 0.8 ft. The angular speed changes from 30 rpm ↻ to 90 rpm ↺ during a certain time interval. Find the time interval.

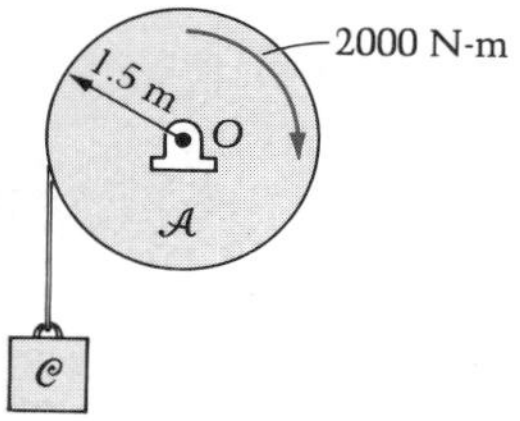

Figure P5.64

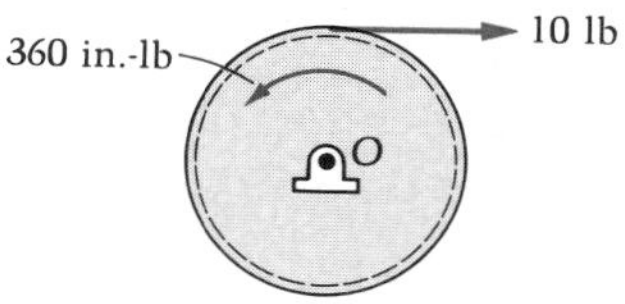

Figure P5.65

5.66 In Figure P5.66 the force P is applied to the cord at $t = 0$, when the 25-N cylinder is at rest. Find the velocity of the mass center and the angular velocity of the cylinder when $t = 6$ s.

5.67 A body $\mathcal{B}$ weighing 805 lb with radius of gyration 0.8 ft about its z_C axis (see Figure P5.67) is pinned at its mass center. A clockwise couple of magnitude e^t lb-ft is applied to $\mathcal{B}$ starting at $t = 0$. Find the angle through which $\mathcal{B}$ has turned during the interval $0 \leq t \leq 3$ sec. *Hint:* Use impulse and momentum principles from 0 to t and then integrate again.

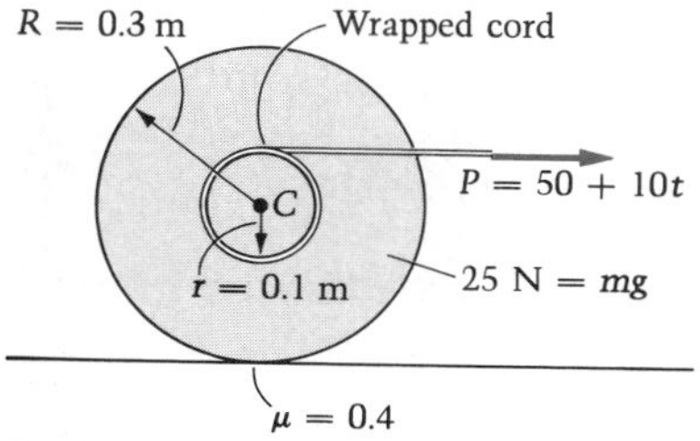

Figure P5.66

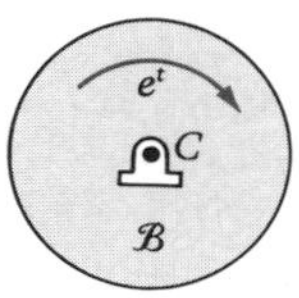

Figure P5.67

5.68 A pipe rolls (from rest) down an incline (Figure P5.68). Using the ***equations of motion***, find:

a. $\dot{x}_C$ at time t
b. $\dot{x}_C$ after C moves the distance x_C

Then use work and energy to verify the answer to part (b) and impulse and momentum to verify part (a). Finally, give the minimum μ to prevent slipping.

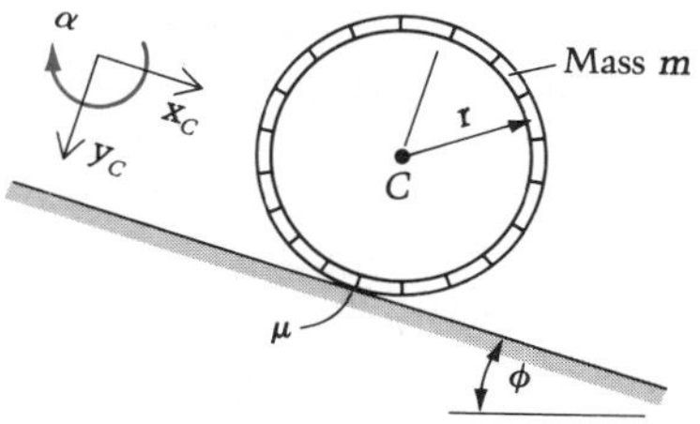

Figure P5.68

5.69 The sinusoidal force P is applied to the string in Figure P5.69 for a half-cycle. If the cylinder (initially at rest) does not slip, find its angular velocity at the end of the load application (at $t = t_0$).

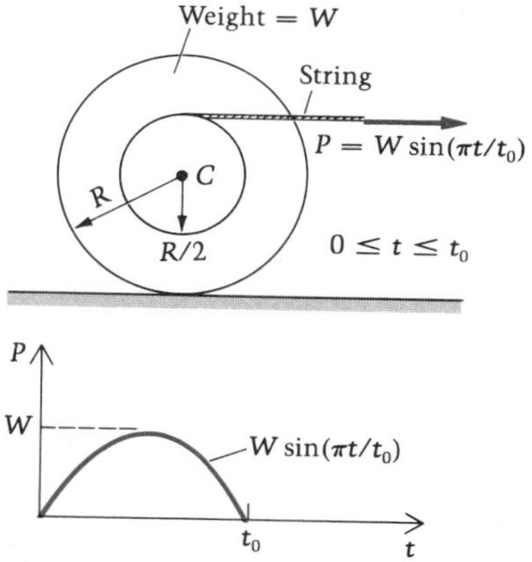

Figure P5.69

5.70 Force F in Figure P5.70 varies with time according to $F = 0.02t^2$ newtons, where t is measured in seconds. If there is enough friction to prevent slipping of the cylinder $\mathcal{C}$ on the plane, find the velocity of C at: (a) $t = 3$ s; (b) $t = 10$ s. $\mathcal{C}$ starts from rest at $t = 0$.

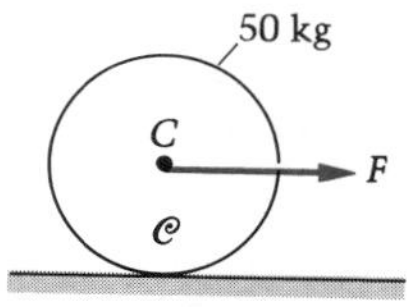

Figure P5.70

In problems 5.71–5.78, see the hint in Problem 5.67:

5.71 Given that the slot (for the cord) in the cylinder in Figure P5.71 (mass 10 kg) has a negligible effect on I_C, find:

a. The largest θ for which no motion down the plane will occur

b. The time required for C to move 3 m down the incline if $\theta = 60°$

5.72 A uniform sphere (radius r, mass m) rolls on the plane in Figure P5.72. If the sphere is released from rest at $t = 0$ when $x = L$, find $x(t)$.

5.73 The cord in Figure P5.73 is wrapped around the cylinder, which is released from rest on the 60° incline shown. Find the velocity and position of C as a function of time t.

5.74 The 50-lb body $\mathscr{C}$ in Figure P5.74 may be treated as a solid cylinder of radius 2 ft. The coefficient of friction between $\mathscr{C}$ and the plane is $\mu = 0.2$, and a force $P = 10$ lb is applied vertically to a cord wrapped around the hub. Find the position of the center C 10 sec after starting from rest.

5.75 A child pulls on an old wheel with a force of 5 lb by means of a rope looped through the hub of the wheel. (See Figure P5.75.) The friction coefficient between wheel and ground is $\mu = 0.2$. Find I_C for the wheel, and use it to determine the location of C after 3 sec.

5.76 Two cables are wrapped around the hub of the 10-kg spool shown in Figure P5.76, which has a radius of gyration of 500 mm with respect to its axis. A constant 40-N force is applied to the upper cable as shown. Find the mass center location 5 s after starting from rest if: (a) $\mu = 0.2$; (b) $\mu = 0.5$.

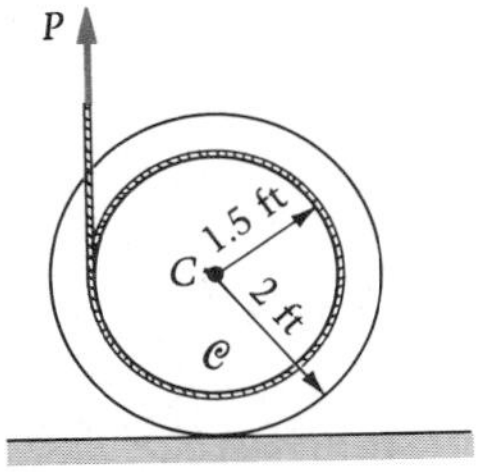

Figure P5.74

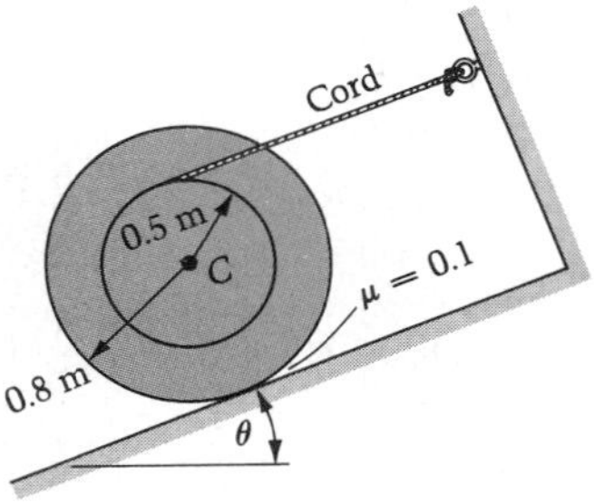

Figure P5.71

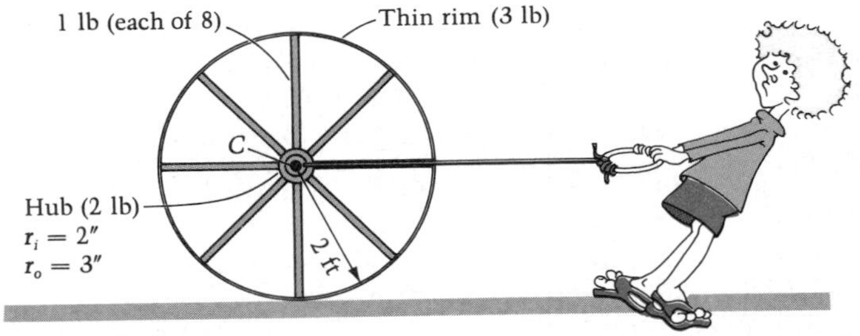

Figure P5.75

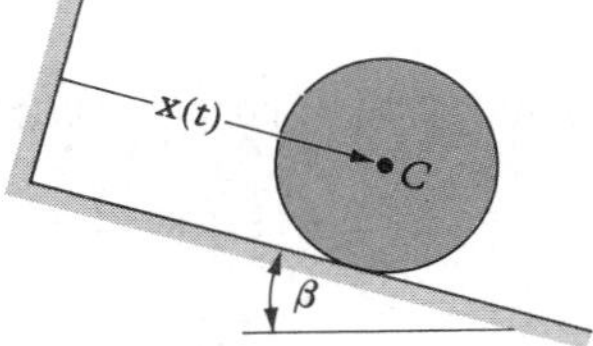

Figure P5.72

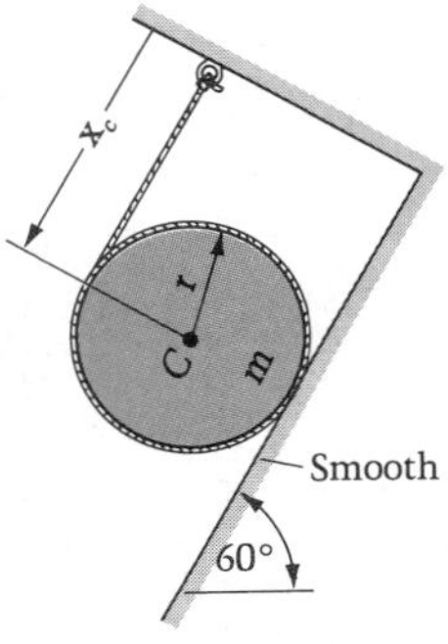

Figure P5.73

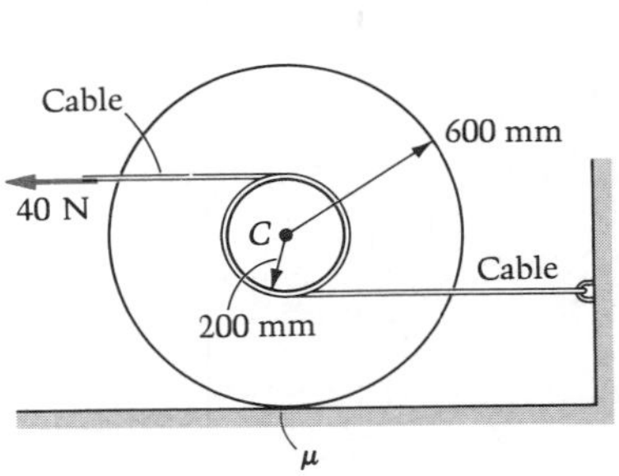

Figure P5.76

5.77 In Figure P5.77 the force $F = 20$ N is applied to the 2-kg block $\mathcal{A}$. Block $\mathcal{B}$ has mass 1.5 kg and is connected to $\mathcal{A}$ by means of a cord that wraps around the light pulleys $\mathcal{P}_1$ and $\mathcal{P}_2$; the center of $\mathcal{P}_2$ is pinned to $\mathcal{A}$. Find the locations of $\mathcal{A}$ and $\mathcal{B}$ (as measured by **x** and y in the figure) as functions of time.

5.78 A sphere of radius $\frac{1}{2}$ ft and weight 16.1 lb is projected onto a horizontal plane (Figure P5.78). Its center has initial velocity $\mathbf{v}_0$ at $t = 0$ and the sphere has initial angular velocity ω_0, defined as shown. If the coefficient of sliding friction between the sphere and the plane is 0.15, plot graphs of distance gone ($\mathbf{x}_C$) against time t up to $t = 3$ sec for the following cases:

a. $v_0 = 10$ ft/sec; $\omega_0 = 100$ rad/sec
b. $v_0 = 10$ ft/sec; $\omega_0 = 50$ rad/sec
c. $v_0 = 10$ ft/sec; $\omega_0 = 30$ rad/sec

5.79 Acting on the gear $\mathcal{G}$ is a couple C with a time-dependent strength given by $C = (6 + 0.8t)$ N-m, where t is measured in seconds. (See Figure P5.79.) If the system is released from rest at $t = 0$, find the velocity of block $\mathcal{B}$ when (a) $t = 3$ s; (b) $t = 10$ s. The centroidal radius of gyration of $\mathcal{G}$ is 0.25 m.

5.80 A uniform cylinder of mass m and radius R is projected onto a horizontal plane with its center at speed v_0. The angular velocity of the wheel as it makes contact with the ground is zero and the coefficient of friction between the wheel and ground is μ. Using the principles of impulse and momentum, find the time from the instant of contact until the wheel rolls.

Solve the following problems by the impulse and momentum, and/or angular impulse and angular momentum method.

5.81 Problem 4.66

5.82 Problem 4.67

5.83 Problem 4.101(c)

5.84 Problem 4.102

5.85 Problem 4.107

5.86 Problem 4.148

5.87 Problem 4.152

5.88 Problem 4.178

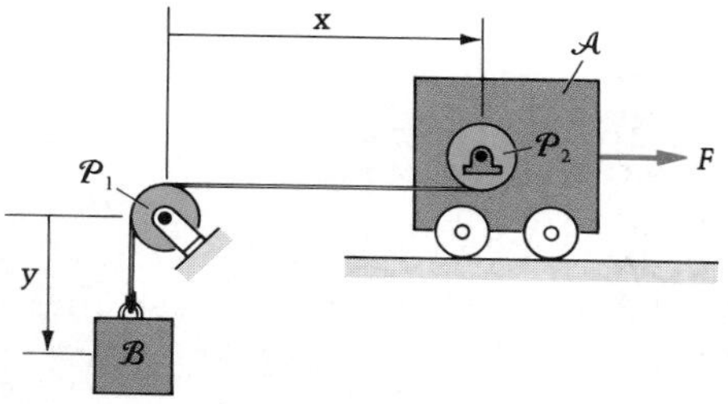

Figure P5.77

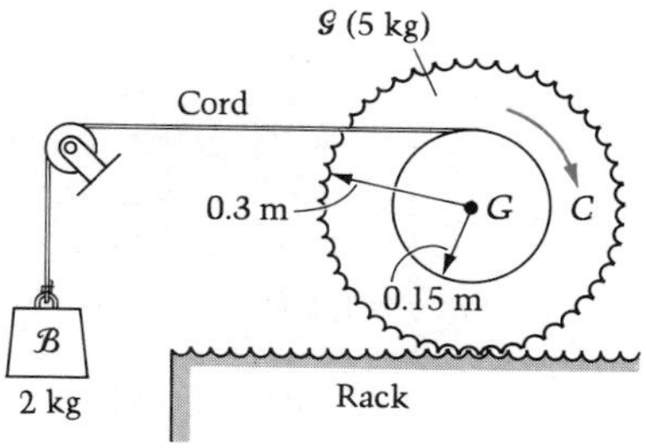

Figure P5.79

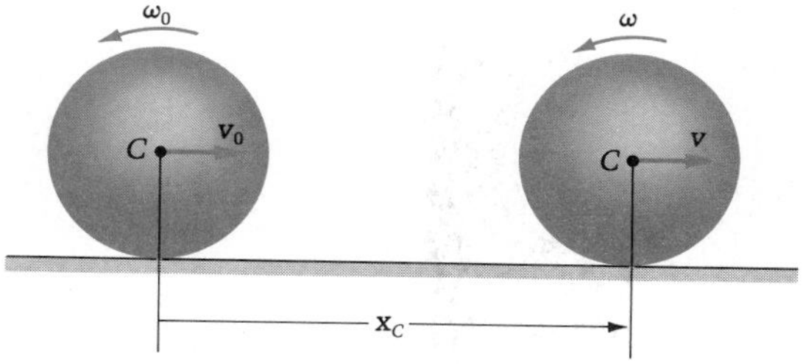

Figure P5.78

Conservation of Momentum

If the force in any direction (let us use $\mathbf{x}$, for example) vanishes over a time interval, then the impulse in that direction vanishes also:

$$\int_{t_i}^{t_f} F_{r_x}\, dt = 0$$

Since this impulse equals the change in momentum in the $\mathbf{x}$ direction, we have zero change when $F_{r_x} \equiv 0$ and thus the momentum is *conserved* in that direction between t_i and t_f:*

$$0 = m(\dot{x}_{C_f} - \dot{x}_{C_i})$$

or

$$m\dot{x}_{C_i} = m\dot{x}_{C_f} \tag{5.36}$$

We would of course also have **conservation of momentum** in the y (or any other) direction in which the force resultant vanished.

Finally, if the z component of $\mathbf{M}_{r_C}$ is zero between t_i and t_f, then the *angular* impulse vanishes and we have **conservation of *angular* momentum:**

$$\int_{t_i}^{t_f} M_{r_{C_z}}\, dt = 0 = H_{C_{z_f}} - H_{C_{z_i}}$$

or

$$H_{C_{z_i}} = H_{C_{z_f}} \tag{5.37}$$

For plane motion of symmetric bodies, we have $\mathbf{H}_C = I_{z_C}\omega\hat{\mathbf{k}}$; there is then no need for the z subscript, and Equation (5.37) may be rewritten

$$(I_C\omega)_i = (I_C\omega)_f \tag{5.38}$$

We now consider a well-known example of conservation of angular momentum.

*Actually the force need not vanish; more generally, only its time integral need be zero to have conservation of momentum.

EXAMPLE 5.15

A skater spinning about a point on the ice (see the diagram) draws in her arms and her angular speed increases.

a. Is angular momentum conserved?
b. Is kinetic energy conserved?
c. Account for any gains or losses if either answer is no.

SOLUTION

We begin with part (a). ***Before*** the skater draws in her arms, we may treat her as a rigid body and thus $H_{C_i} = I_1\omega_1$. The same is true ***after*** the arms are drawn in, so that $H_{C_f} = I_2\omega_2$. If we neglect the small friction couple at the skates and the small drag moments caused by air resistance, then the answer to part (a) is yes because $\mathbf{M}_{I_C}$ is then zero. Thus

$$I_1\omega_1 = I_2\omega_2$$

Therefore

$$\omega_2 = \frac{I_1}{I_2}\omega_1 > \omega_1 \qquad \text{(showing the angular speed increase since } I_1 > I_2\text{)}$$

For part (b) the kinetic energies are

$$T_1 = \frac{1}{2}I_1\omega_1^2$$

$$T_2 = \frac{1}{2}I_2\omega_2^2 = \frac{1}{2}I_1\omega_1^2\left(\frac{I_1}{I_2}\right) > T_1$$

Thus kinetic energy is ***not*** conserved.

For part (c) the change in kinetic energy is seen to be positive:

$$\Delta T = T_2 - T_1 = \frac{1}{2}I_1\omega_1^2\underbrace{\left(\frac{I_1}{I_2} - 1\right)}_{>0 \text{ since } I_1 > I_2}$$

Since there is no work done by the external forces and couples,* it is clear that this kinetic energy increase is accompanied by an internal energy decrease within the skater's body as her muscles do (nonexternal) work on her (nonrigid) arms in drawing them inward. Since ***total*** energy is always conserved (first law of thermodynamics), the skater has lost ***internal*** energy in the process.

*Except possibly for an incidental amount of work done by gravity if C changes slightly.

We now consider an example in which, uniquely, both angular momentum *and* kinetic energy are conserved.

EXAMPLE 5.16

The 2-kg collar $\mathcal{C}$ in the diagram turns along with the smooth rod $\mathcal{R}$, which is 1 m long, has a mass of 3 kg, and is mounted in bearings with negligible friction. The angular speed is increased until the cord breaks (its tensile strength is 60 N), and at that instant the external moment is removed. Determine the angular velocity of $\mathcal{R}$ and the velocity of C (the mass center of $\mathcal{C}$) when the collar leaves the rod.

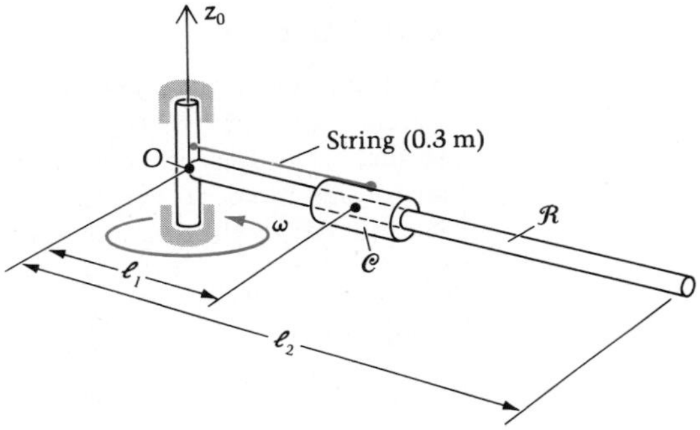

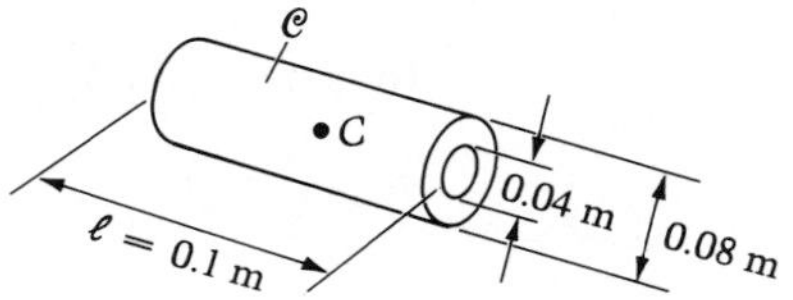

SOLUTION

The string provides the force causing the centripetal (inward) acceleration until it breaks. At that instant, we may solve for the angular velocity of $\mathcal{C}$:

$$F_{r_n} = ma_{C_n}$$

$$T = m\ell_1\omega_1^2$$

$$60 = 2(0.3)\omega_1^2$$

$$\omega_1 = 10 \text{ rad/s}$$

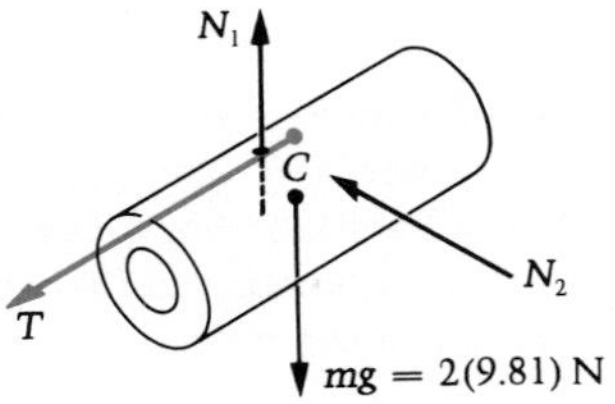

In the accompanying free-body diagram, N_1 and N_2 are the vertical and horizontal resultants of the pressures on the inside wall of $\mathcal{C}$ exerted by $\mathcal{R}$.

After the rope breaks at time t_1, collar $\mathcal{C}$ moves outward in addition to turning with $\mathcal{R}$; this is because there is no longer any inward force to keep it

(Continued)

from "flying off on a tangent." Between times t_1 and t_2 (when it leaves $\mathcal{R}$), we have the following for the system ($\mathcal{C}$ plus $\mathcal{R}$):

1. Conservation of angular momentum H_{O_z} about z_O (because the external forces have no moment about z_O).
2. Conservation of kinetic energy T (since no net work is done on the system). Note that the normal forces between rod and collar, being equal in magnitude but opposite in direction, act on points with equal velocity components in the direction of either force; hence their net work vanishes.

Condition 1 gives

$$H_{z_{O_i}} = H_{z_{O_f}}$$

$$\underbrace{I_{z_O}^{\mathcal{C}}\omega_i + I_{z_O}^{\mathcal{R}}\omega_i}_{\substack{\text{Until the string breaks,}\\ O \text{ is a point of both bodies!}}} = I_{z_O}^{\mathcal{R}}\omega_f + \overbrace{I_{z_C}^{\mathcal{C}}\omega_f + \left[\left(\ell_2 + \frac{\ell}{2}\right)\hat{\mathbf{i}} \times m_{\mathcal{C}}\mathbf{v}_C\right]_z}^{\mathbf{H}_{z_{O_f}}^{\mathcal{C}} = \mathbf{H}_{z_{C_f}}^{\mathcal{C}} + \mathbf{r}_{OC} \times m_{\mathcal{C}}\mathbf{v}_{C_f}}$$

Thus

$$\underbrace{\left[m_{\mathcal{C}}\left(\frac{r_i^2 + r_o^2}{4} + \frac{\ell^2}{12}\right) + m_{\mathcal{C}}\,\ell_1^2\right]}_{I_{z_C}^{\mathcal{C}} = 0.00267}10 + \frac{m_{\mathcal{R}}\,\ell_2^2}{3}\,10 = \frac{m_{\mathcal{R}}\,\ell_2^2}{3}\,\omega_f + I_{z_C}^{\mathcal{C}}\omega_f + \left(\ell_2 + \frac{\ell}{2}\right)^2\omega_f m_{\mathcal{C}}$$

$$(0.00267 + 0.180)10 + \frac{3(1^2)}{3}\,10 = \omega_f + 0.00267\omega_f + (1.05)^2 2\omega_f$$

$$1.83 + 10 = \omega_f(1 + 0.00267 + 2.21)$$

$$\omega_f = 3.69 \text{ rad/s}$$

The component of $\mathbf{v}_C$ perpendicular to the rod $\mathcal{R}$ is thus $v_C = 1.05\omega_f = 3.87$ m/s. We can now obtain the radial component by conservation of T (condition 2):

$$\underbrace{\frac{1}{2}I_{z_O}\omega_i^2}_{\substack{O \text{ is (I) for both } \mathcal{C}\\ \text{and } \mathcal{R} \text{ initially}}} = \frac{1}{2}I_{z_O}^{\mathcal{R}}\omega_f^2 + \frac{1}{2}m_{\mathcal{C}}(\underbrace{v_{C_\parallel}^2 + v_{C_\perp}^2}_{\substack{\text{components of } \mathbf{v}_C \text{ parallel}\\ \text{and perpendicular to } \mathcal{R}}}) + \frac{1}{2}I_{z_C}^{\mathcal{C}}\omega_f^2$$

$$\tfrac{1}{2}(0.183 + 1)10^2 = \tfrac{1}{2}(1)3.69^2 + \tfrac{1}{2}(2)(v_{C_\parallel}^2 + 3.87^2) + \tfrac{1}{2}(0.00267)3.69^2$$

$$59.2 = 6.81 + v_{C_\parallel}^2 + 15.0 + 0.0182$$

$$v_{C_\parallel} = 6.11 \text{ m/s}$$

(Continued)

Thus since the initial kinetic energy was 59.2 J and since

$$\frac{\frac{1}{2}(2)(6.11^2)}{59.2} = 0.631$$

we see that 63 percent of the original energy has gone into the outward motion of the collar.

PROBLEMS / Section 5.3 (continued)

5.89 A massless rope hanging over a massless, frictionless pulley supports two monkeys (one of mass M, the other of mass $2M$). The system is released at rest at $t = 0$, as shown in Figure P5.89. During the following 2 sec, monkey B travels down 15 ft of rope to obtain a massless peanut at end P. Monkey A holds tightly to the rope during these 2 sec. Find the displacement of A during the time interval.

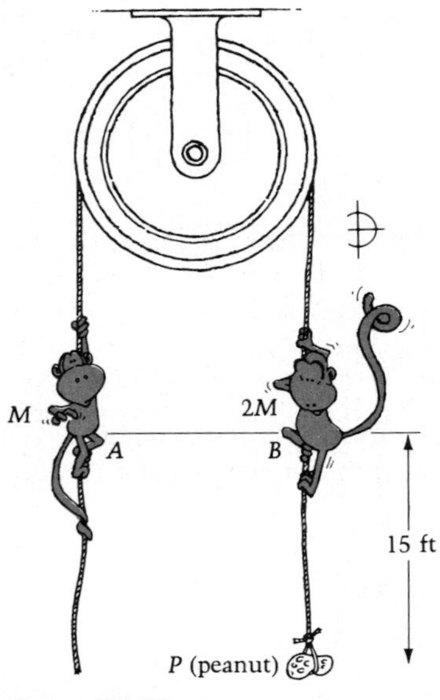

Figure P5.89

5.90 A starving monkey of mass m spies a bunch of delicious bananas of the same mass. (See Figure P5.90.) He climbs at constant speed v_0 relative to the (light) rope. Determine whether the monkey reaches the bananas before they sail over the pulley if:

a. The pulley's mass is negligible ($\ll m$).

b. The pulley's mass is fm, where $f > 0$ and the radius of gyration of the pulley with respect to its axis is k.

If either answer is yes, give the relationship between d and H in the figure for which overtaking the bananas is possible.

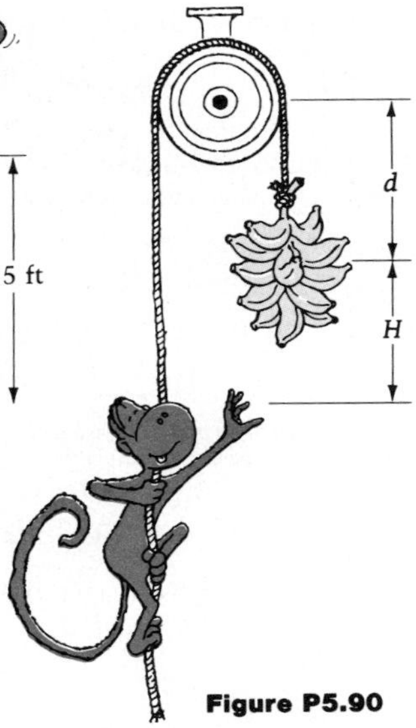

Figure P5.90

5.91 Two gymnasts A and B, each of weight W, hold onto the left side of a rope that passes over a cylindrical pulley (weight W, radius r) to a counterweight C of weight $2W$. (See Figure P5.91.) Initially the gymnast A is at depth d below B. He climbs the rope to join gymnast B. Determine the displacement of the counterweight C at the end of the climb.

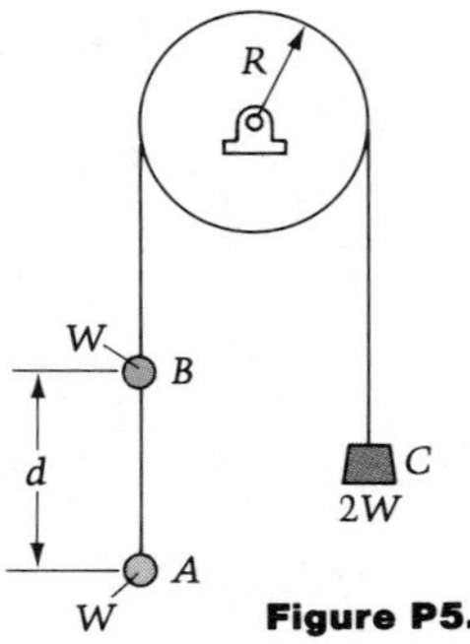

Figure P5.91

5.92 The masses of four bodies are shown in Figure P5.92. The radius of gyration of wheel $\mathcal{C}$ with respect to its axis is $k_c = 0.4$ m. Initially there is 0.6 m of slack in the cord between $\mathcal{A}$ and the linear spring. (Modulus $k =$ 1000 N/m, and the spring is initially unstretched.) Determine how far downward body $\mathcal{B}$ will move.

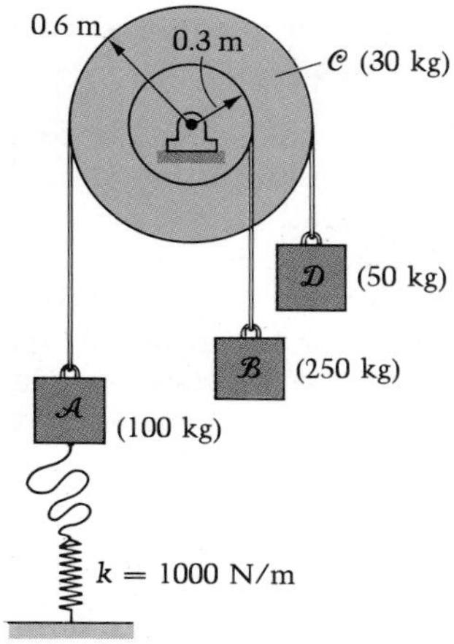

Figure P5.92

5.93 A bird of mass m, flying horizontally at speed v_0 perpendicular to a stick, lands on the stick and holds fast to it. (See Figure P5.93.) The stick (mass M, length ℓ) is lying on a frozen pond. Find the velocity of the stick's mass center G and its angular velocity as bird and stick start to move together. (Answer in terms of m, M, ℓ, and v_0.) Assume that the bird lands on the end of the stick.

Figure P5.93

5.94 In Figure P5.94 the man of mass m stands at end A of a 20-ft plank of mass $3m$ that is held at rest on the smooth inclined plane by the cord. The man cuts the cord and runs down to end B of the plank. When he gets there, end B is in the same position on the plane as it was originally. Find the time it takes the man to run down from A to B.

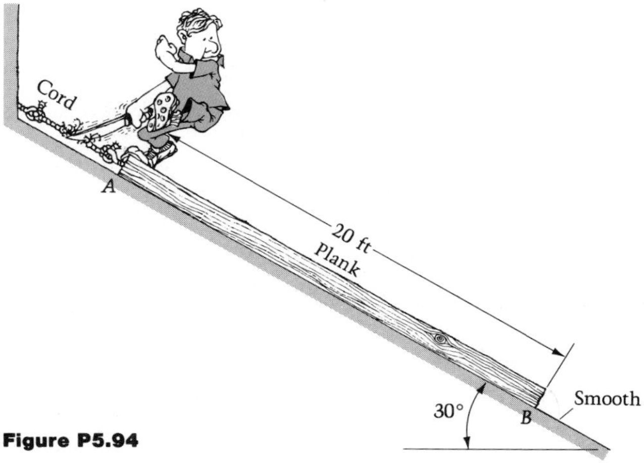

Figure P5.94

5.95 Define the angular momentum of a particle about a fixed axis and state the conditions under which the angular momentum remains constant. A man (to be regarded as a particle) stands on a swing. His distance from the smooth horizontal axis of the swing is L when he crouches and $L - H$ when he stands. As the swing falls he crouches; as it rises he stands—the changeover is assumed instantaneous. If the swing falls through an angle α and then rises through an angle β, show that

$$\sin\frac{\beta}{2} = \left(\frac{L}{L - H}\right)^{3/2} \sin\frac{\alpha}{2}$$

5.96 A light rigid rod of length $2a$ has a particle A of mass m fixed to one end and lies on a smooth horizontal table. (See Figure P5.96.) A bead B, also of mass m, is placed over the rod at its midpoint. Suppose a blow at right angles to the rod at A causes A to move off with velocity v_0 relative to the table.

a. Prove that if x is the distance between the particles and θ is the angle turned through in time t, then (neglecting friction)

$$x^2\dot{\theta} = v_0 a \qquad \dot{x}^2 + x^2\dot{\theta}^2 = v_0^2$$

b. Prove that time $\sqrt{3}a/v_0$ elapses before B separates from the rod.

5.97 A hemispherical block of mass M and radius a whose surfaces are smooth rests with its plane face in contact with a smooth horizontal table. A particle of mass m is placed at the highest point of the block and is slightly disturbed. Show that as long as the particle remains in contact with the block, the radius to the particle makes an angle θ with the upward vertical where

$$a\dot{\theta}^2(M + m\sin^2\theta) = 2g(M + m)(1 - \cos\theta)$$

5.98 Two equal uniform rods AB and BC are smoothly jointed at B and move freely in a horizontal plane about a smooth fixed pivot at point A. (See Figure P5.98.) Initially AB and BC lie in the same straight line, and BC has angular velocity ω_0 with AB at rest. If θ and ϕ are the angles made respectively by AB and BC with their initial directions, show that

$$\frac{16}{3}\dot{\theta}^2 + 4\dot{\theta}\dot{\phi}\cos(\theta - \phi) + \frac{4}{3}\dot{\phi}^2 = \frac{4}{3}\omega_0^2$$

and

$$\frac{16}{3}\dot{\theta} + 2(\dot{\theta} + \dot{\phi})\cos(\theta - \phi) + \frac{4}{3}\dot{\phi} = \frac{10}{3}\omega_0$$

Hint: Both energy and angular momentum about axis z_A are conserved.

5.99 In the preceding problem show that the greatest value of the angle between AB and BC is $\cos^{-1}(5/12)$.

5.100 A block of mass M rests with one face on a smooth horizontal plane. The block contains a smooth-surfaced spherical hole of radius a whose center is the center of mass of the block; a particle of mass m is at rest at the lowest point of the hole. The block is now projected with velocity v_0 along the plane. By considering momentum and energy, show that the particle just reaches the height of the center of the hole if

$$Mv_0^2 = 2ga(M + m)$$

Find the velocity of the block at this instant.

5.101 A circular disk of mass M rotates without friction about O. (See Figure P5.101.) A string passed over the disk (and not slipping on it) carries a mass M at each end. The system is released at rest as shown with the right-hand mass carrying a washer of mass M. As the system moves, the left-hand weight picks up a washer of mass M at the same instant the right mass deposits its washer. Find the velocity of the right-hand weight just after this exchange of washers.

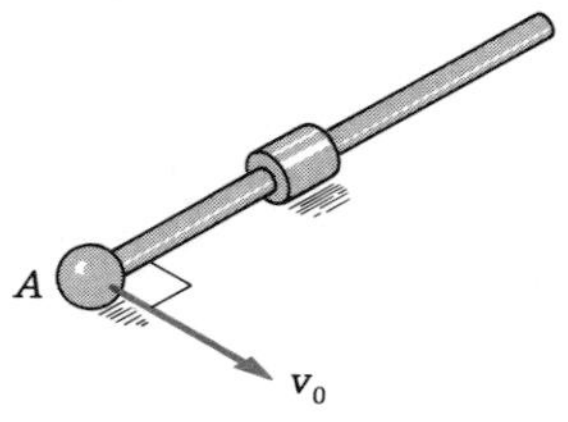

Figure P5.96

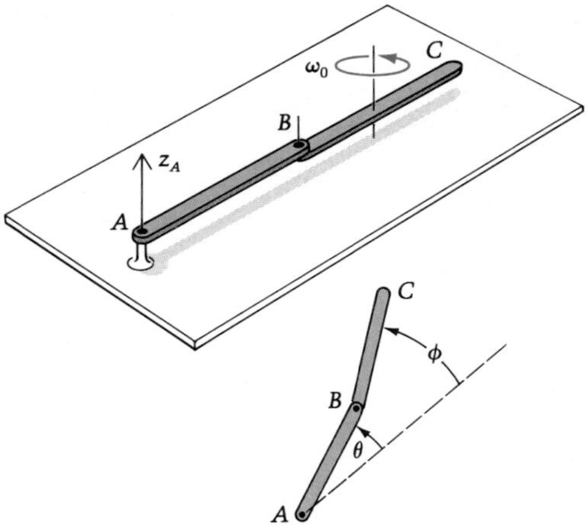

Figure P5.98

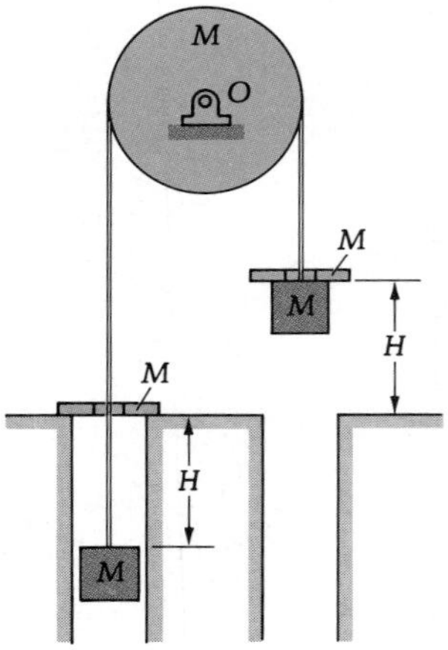

Figure P5.101

5.102 The system in Figure P5.102 consists of the 12-lb body $\mathcal{A}$, the 6-lb (disk) pulley $\mathcal{B}$, the 8-lb "rider" $\mathcal{C}$, and the 10-lb body $\mathcal{D}$. Everything is released from rest in the given position. Body $\mathcal{D}$ then falls through a hole in bracket $\mathcal{E}$, which stops body $\mathcal{C}$. Find how far $\mathcal{D}$ descends from its original position.

5.103 Figure P5.103a shows a rough guess at a skater's mass distribution. Calculate the percentage increase in his angular speed about the vertical if he draws in his arms as shown in Figure P5.103b. Assume that his arms are wrapped around the 6-in. radius circle of his upper body.

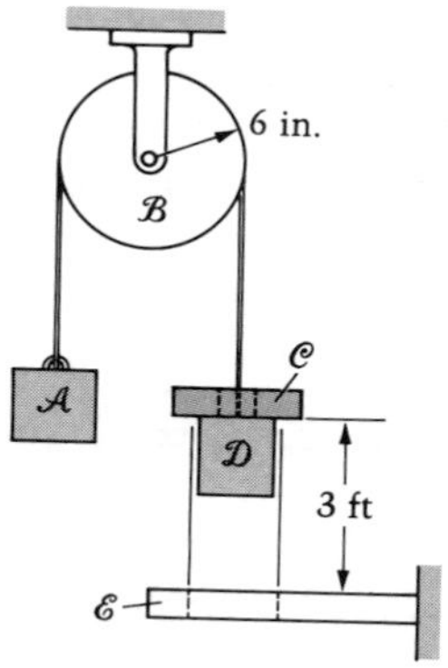

Figure P5.102

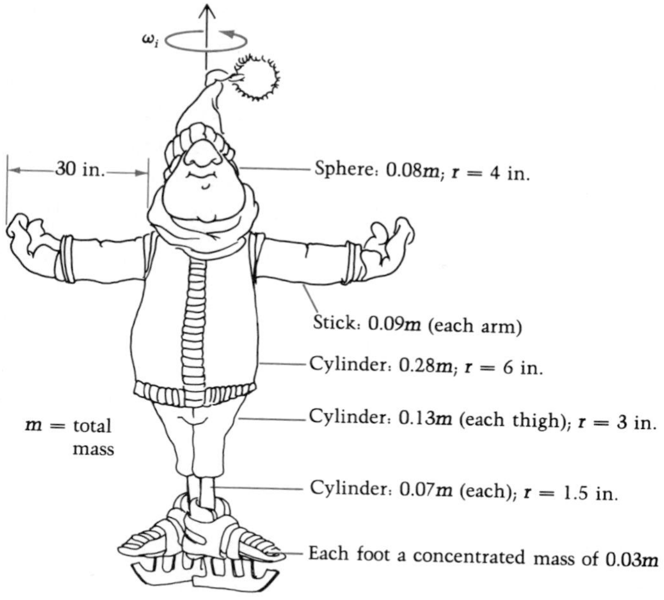

Figure P5.103a

5.104 Two disks are spinning in the directions shown in Figure P5.104. The upper disk is lowered until it contacts the bottom disk (around the rim). (a) Find how long it takes for the two disks to reach a common angular velocity, and (b) determine its value. (c) Finally, determine the energy lost. Show that if $I_1 = I_2$ and $\omega_1 = -\omega_2$, your solution predicts that 100 percent of the energy is lost (as it should). Determine which of the three answers (time, ω_f, energy loss) are the same if the two disks are instantaneously locked together instead of slipping.

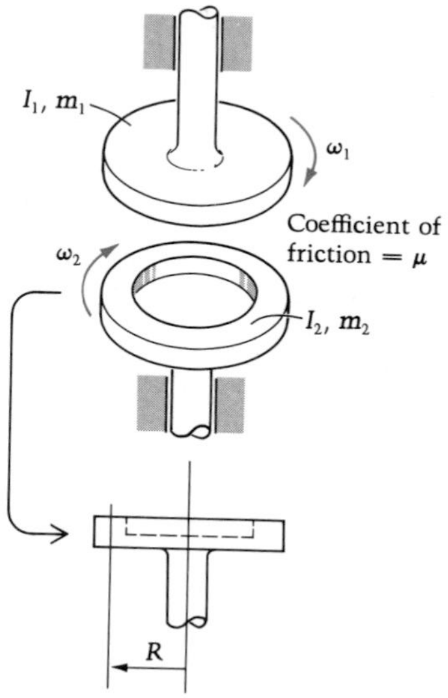

Figure P5.104

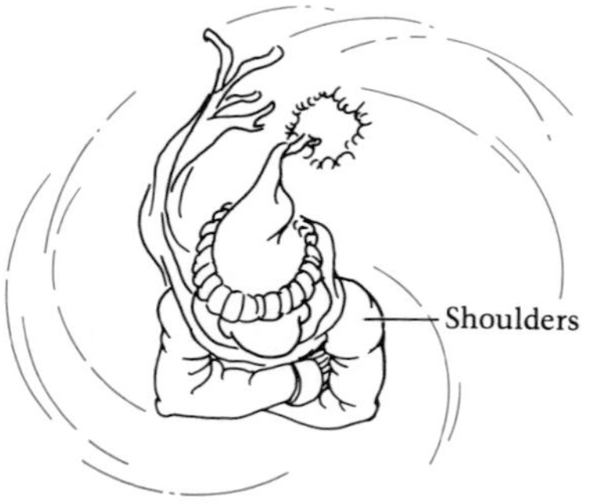

Figure P5.103b

5.105 Disk $\mathcal{A}$ and the light shaft in Figure P5.105 rotate freely at 40 rpm. Disk $\mathcal{B}$ (initially not turning) slides down the shaft and strikes $\mathcal{A}$; after a brief period of slipping, they move together. Find the average frictional moment exerted on $\mathcal{A}$ by $\mathcal{B}$ if the slipping lasted for 3 sec.

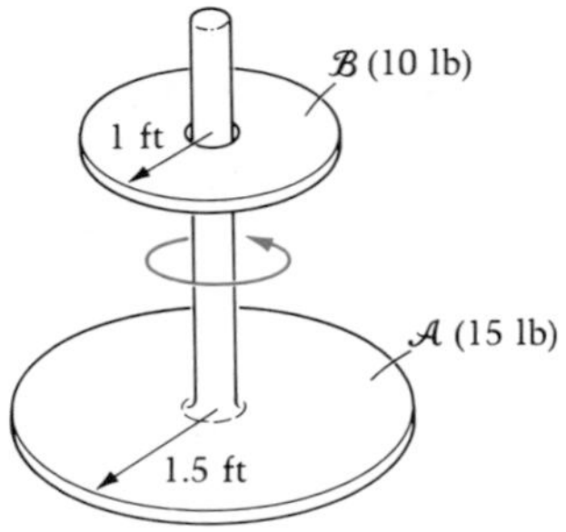

Figure P5.105

Impact

The large forces occurring during an impact between two bodies $\mathcal{B}_1$ and $\mathcal{B}_2$ obviously deform the bodies. Because of vibrations and permanent deformations that are produced, some of the mechanical energy will be dissipated in the collision. However, it is often possible to treat a body as rigid *before*, and then again *after*, the impact in order to gain information of value. In impact problems we assume that:

1. Velocities and angular velocities may change greatly over the short impact interval Δt.
2. Positions of the bodies do not change appreciably.
3. Forces (and moments) that do not grow large over the interval Δt are neglected (such as gravity and spring forces). Such forces are called **nonimpulsive;** the large contact forces are called **impulsive.** It is the impulsive forces and moments that produce the sudden changes in velocities and angular velocities.

When two bodies $\mathcal{B}_1$ and $\mathcal{B}_2$ collide by contacting at points along a line joining their mass centers, the impact is called **central;** otherwise, it is called **eccentric.** And if the initial velocities of the mass centers $\mathcal{B}_1$ and $\mathcal{B}_2$ are parallel, the impact is also called **direct;** if they are not, it is labeled **oblique.** Thus there are four types of impact (Figure 5.10).

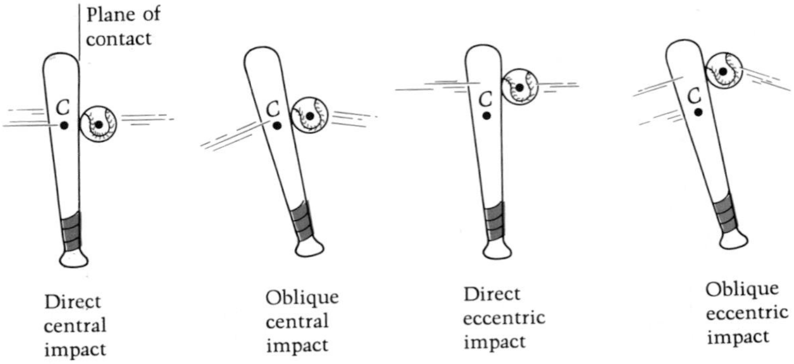

Figure 5.10

There is a quantity defined in mechanics that gives a measure of the rebound, or resilience, following a collision—the **coefficient of restitution*** (usually labeled e). It is related to the impulses and velocity components normal to the plane of contact. This plane, illustrated in Figure 5.11, is simply the plane tangent to the two bodies at the point of initial impact.

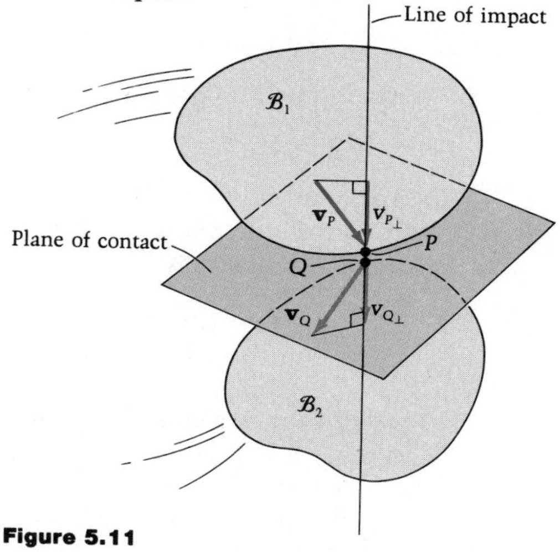

Figure 5.11

The coefficient of restitution is defined to be

$$e = -\left[\frac{v_{P_\perp}(t_f) - v_{Q_\perp}(t_f)}{v_{P_\perp}(t_i) - v_{Q_\perp}(t_i)}\right] \tag{5.39}$$

where t_i and t_f are the initial and final times of the collision interval and $v_{P_\perp}$ and $v_{Q_\perp}$ denote the components of velocities of the contacting points perpendicular to the plane of impact as shown in Figure 5.11. Thus the coefficient of restitution—the negative of the ratio of the "relative velocity of separation" to the "relative velocity of approach"—is positive.

Question 5.9 Why does the minus sign appear in Equation (5.39)?

The limiting cases of collision are given by $e = 0$ and $e = 1$. When $e = 0$ the bodies stick together; when $e = 1$ the mechanical energy is conserved.

The coefficient of restitution is not the best of physical properties to measure; it turns out to depend on the sizes and shapes of $\mathcal{B}_1$ and $\mathcal{B}_2$ and even on the velocity of impact. Nonetheless, its definition does pro-

*See the book *Impact* by W. Goldsmith, Edward Arnold Publishers, Ltd., London, 1960.

vide an approximate, much-needed equation that allows us to solve many problems of impact. We now consider two forms of the angular impulse and angular momentum equation that are applicable at the beginning and end of impacts involving the plane motion of bodies that may be regarded as rigid except during the collision phase of the motion.

If the body has a pivot O, we recall that

$$\mathbf{H}_O = I_{xz_O}\omega\hat{\mathbf{i}} + I_{yz_O}\omega\hat{\mathbf{j}} + I_{z_O}\omega\hat{\mathbf{k}}$$

With O fixed in the inertial frame we have

$$\mathbf{M}_{r_O} = \dot{\mathbf{H}}_O$$

Thus we may replace the C by an O in the angular impulse and momentum equation (5.33) for such pivot cases. The resulting equation about the axis of rotation is

$$\int_{t_i}^{t_f} M_{r_{O_z}}\,dt = H_{O_f} - H_{O_i} = I_{z_O}(\omega_f - \omega_i)^* \tag{5.40}$$

This formula is of considerable value in impact problems because impulsive pivot reactions have no moment about O and thus do not appear in the equation. Note that if $O = C$, then Equation (5.40) is the same as our previous Equation (5.33) written about the mass center.

Another of our useful equations follows from

Another of our useful equations follows from

$$\mathbf{M}_{r_P} = \dot{\mathbf{H}}_C + (\mathbf{r}_{PC} \times m\mathbf{a}_C) \tag{5.41}$$

This equation is

$$M_{r_{P_z}} = I_{z_C}\alpha + (\mathbf{r}_{PC} \times m\mathbf{a}_C)_z$$

in which P is an arbitrary point. If we state that P is now a fixed point O of the inertial frame $\mathcal{I}$, we may integrate this equation, getting

$$\begin{aligned}\int_{t_i}^{t_f} M_{r_{O_z}}\,dt &= (H_{C_f} - H_{C_i}) + (\mathbf{r}_{OC} \times m\mathbf{v}_C)_z]_i^f \\ &= \{I_{z_C}\omega^* + (\mathbf{r}_{OC} \times m\mathbf{v}_C)_z\}]_i^f\end{aligned} \tag{5.42}$$

Question 5.10 Why is the right-hand side not the integral of the right side of Equation 5.41 if O is moving?

Equation (5.42) is especially useful in impact problems in which a point of body $\mathcal{B}$ collides with a point O of the reference frame $\mathcal{I}$. Using

*We emphasize again that the angular momentum $\mathbf{H}_O$ (or $\mathbf{H}_C$) is not equal to its rigid body form $I_{z_O}\omega\hat{\mathbf{k}}$ (or $I_{z_C}\omega\hat{\mathbf{k}}$) during the impact, but these substitutions may be made at t_i before the collision and at t_f afterward.

Equation (5.42), we may then sum moments about O and lose the impulsive forces of collision. Equation (5.42) is ***more general*** for this purpose than Equation (5.40); the latter requires O to be a fixed point of both the body ***and*** $\mathcal{J}$, which is the case in impacts only if the striking point sticks and does not rebound (that is, if the coefficient of restitution $e = 0$). We now consider an impact application of the impulse and momentum equations.

EXAMPLE 5.17

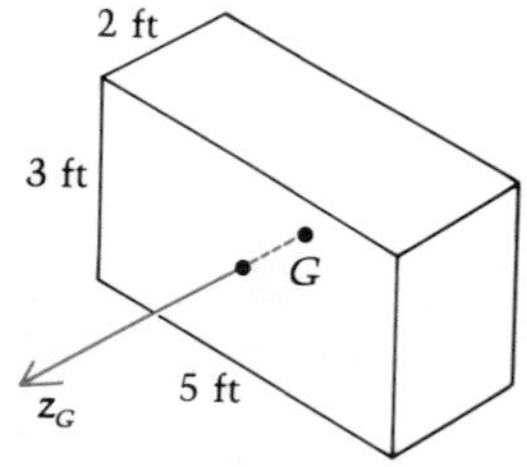

A CARE package consists of the box plus contents described in Example 4.12. At impact the crate has $\mathbf{v}_{G_i} = 50 \downarrow$ ft/sec and is translating. The coefficient of restitution is $e = 0.2$. Find the angular velocity of the box and the velocity of its mass center G just after the impact. (See the diagram.)

SOLUTION

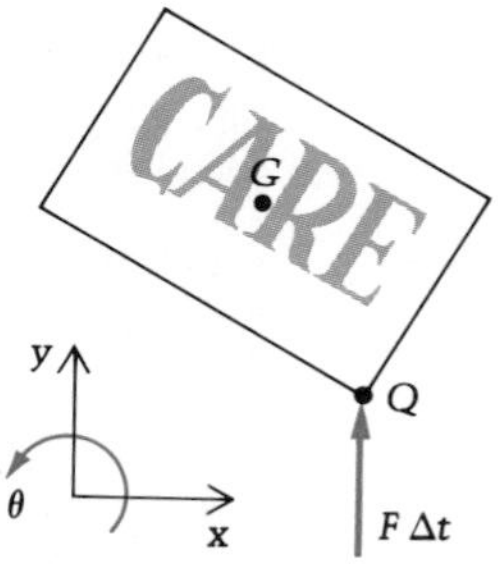

The mass of the box plus contents is 11.3 slugs; the moment of inertia for z_G is 37.0 slug-ft^2. Neglecting friction, there is no horizontal force during the impact; thus, since the **x** component of the linear impulse vanishes, the linear momentum in that direction is conserved. Hence $\dot{x}_{G_f} = \dot{x}_{G_i} = 0$. Next we write the impulse and momentum equations in the y and θ directions, letting $F\,\Delta t$ be the impulse imparted to the box by the ground. This means that F represents the average normal force over the short time interval Δt, so that $F\,\Delta t = \int N\,dt$:

$$y: \quad F\,\Delta t = 11.3[v_{G_{f_y}} - (-50)]$$

$$\theta: \quad F\,\Delta t\,(2.5 \cos 30° - 1.5 \sin 30°) = 37\,(\omega_f - 0)$$

Eliminating $F\,\Delta t$ from the two equations gives the equation

$$11.3 v_{G_{f_y}} + 565 = 26.1\omega_f \qquad (1)$$

(Continued)

The definition of the coefficient of restitution gives

$$0.2 = \frac{v_{Q_{f_y}} - (0)}{0 - (-50)} \Rightarrow v_{Q_{f_y}} = 10$$

We can now get another equation relating $v_{G_{f_y}}$ and ω_f by relating the velocities of Q and G after the impact:

$$\mathbf{v}_{Q_f} = \mathbf{v}_{G_f} + \boldsymbol{\omega}_f \times \mathbf{r}_{GQ}$$

$$v_{Q_{f_x}}\hat{\mathbf{i}} + 10\hat{\mathbf{j}} = v_{G_{f_y}}\hat{\mathbf{j}} + \omega_f\hat{\mathbf{k}} \times \underbrace{\mathbf{r}_{GQ}}_{1.42\hat{\mathbf{i}} - 2.55\hat{\mathbf{j}}}$$

The $\hat{\mathbf{j}}$ coefficients yield

$$10 = v_{G_{f_y}} + 1.42\omega_f \tag{2}$$

Solving (1) and (2), we obtain our answers:

$\mathbf{v}_{G_{f_y}} = 12.8 \downarrow$ ft/sec and $\boldsymbol{\omega}_f = 16.1 \circlearrowleft$ rad/sec

We note two points for emphasis:

1. To use the change of angular momentum in the form $I_{z_C}(\omega_f - \omega_i)$, it is only necessary for the body to be behaving rigidly just before and again just after the impact with the ground.
2. During the impact, the body is *not* behaving rigidly, because it has lost the following percentage of its initial mechanical energy:

$$\left[\frac{E_i - E_f}{E_i}\right] \times 100 = \left[1 - \frac{\frac{1}{2}(11.3)12.8^2 + \frac{1}{2}(37.0)16.1^2}{\frac{1}{2}(11.3)50^2}\right] \times 100$$

$$= 59.5\%$$

Question 5.11 What is the meaning of the cases $e = 0$ and $e = 1$?

(Courtesy American Rigging Co.)

EXAMPLE 5.18

A 770-ton steel nuclear reactor vessel is being transported down a 6.5 percent grade using a specially designed suspended hauling platform together with crawler transporters. (See the photograph.) Determine the maximum velocity at which the reactor can be transported without tipping over if it should strike, and pivot about, a rigid obstacle at the front edge of the vessel's base ring.

SOLUTION

The reactor vessel $\mathcal{R}$ will tip over if there is any kinetic energy left after it pivots about the front edge at O (see the diagram) and the mass center C reaches its highest point B, directly above O. Thus we solve for the velocity that will cause C to reach B; any higher velocity will cause overturning.

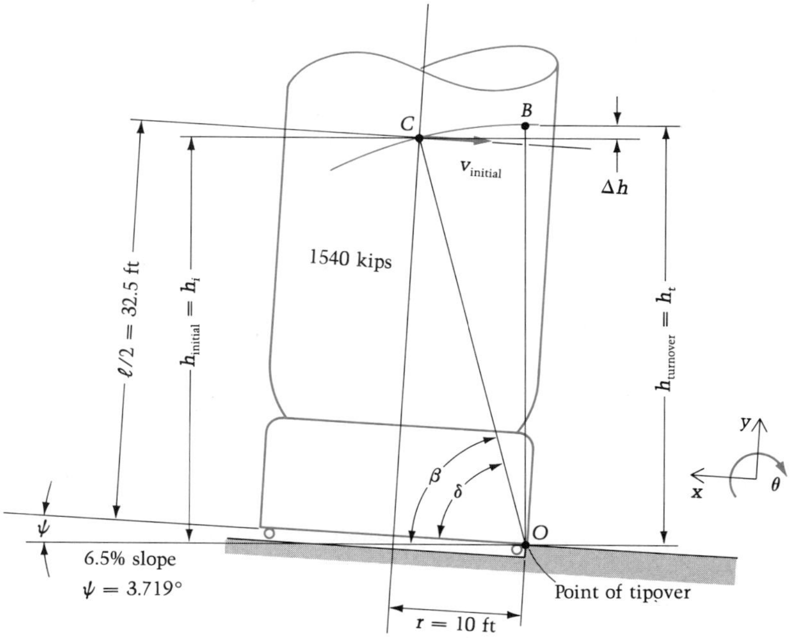

C = position of mass center at instant of contact at O

B = position of mass center at point of no return

(Continued)

We begin the solution with some preliminary geometric and trigonometric calculations based on the diagram. The 6.5 percent slope means $\psi = \tan^{-1}(0.065) = 3.719°$.* The turnover height h_t is given by the distance OC:

$$h_t = OC = OB = \sqrt{10^2 + 32.5^2} = 34.00 \text{ ft}$$

Also

$$\delta = \tan^{-1}\left(\frac{32.5}{10}\right) = 72.90°$$

and

$$\beta = \delta + \psi = 72.90° + 3.72° = 76.62°$$

The initial height h_i of C, above the horizontal line through O, is

$$\begin{aligned} h_i &= (OC) \sin \beta \\ &= (34.00) \sin 76.62° = 33.08 \text{ ft} \end{aligned}$$

and thus the vertical distance through which C will move in reaching B is

$$\begin{aligned} \Delta h &= h_t - h_i = 34.00 - 33.08 \\ &= 0.92 \text{ ft} \end{aligned}$$

(If we had adhered to three significant digits, the subtraction would have reduced us to just one good digit.)

There now remain two separate main parts to the solution of this problem. We first have to consider that mechanical energy is lost during the impact of $\mathcal{R}$ with the obstacle at O. Thus we are prevented from using the principle of work and kinetic energy over the short period of impact. What does apply, however, is conservation of angular momentum about O. This is because the impulsive forces (in both the **x** and **y** directions!) causing the sudden changes in the mass center velocity $\mathbf{v}_C$ and in the angular velocity $\boldsymbol{\omega}_{\mathcal{R}}$ are acting at O, so that $\mathbf{M}_{r_O} = \mathbf{0}$. Therefore

$$\mathbf{M}_{r_O} = \mathbf{0} = {}^{\mathcal{I}}\dot{\mathbf{H}}_O \Rightarrow \mathbf{H}_O = \text{constant vector}$$

or

$$\mathbf{H}_{O_i} = \mathbf{H}_{O_f} \tag{1}$$

We shall use Equation (4.5) to express $\mathbf{H}_{O_i}$; this is the best formula for $\mathbf{H}_{O_i}$ for translation problems because $\mathbf{H}_{C_i} = 0$ in that case. For $\mathbf{H}_{O_f}$, however, we have a nonzero $\omega_{\mathcal{R}}$. Thus we draw on the fact that since the vessel does not bounce at O, we may consider O a fixed point of both $\mathcal{R}$ and the inertial frame during and following the short period of impact. This in turn means that $\mathbf{H}_{O_f}$ is simply $I_O \omega_f \hat{\mathbf{k}}$ after impact. Therefore Equation (1) becomes

$$\overset{\mathbf{0}}{\cancel{\mathbf{H}_{C_i}}} + \mathbf{r}_{OC} \times m\mathbf{v}_{C_i} = I_O \omega_f \hat{\mathbf{k}} \tag{2}$$

Now since

(Continued)

*We use four significant digits in this example.

$$\mathbf{r}_{OC} = -r\hat{\mathbf{i}} - \frac{\ell}{2}\hat{\mathbf{j}} \qquad \text{and} \qquad \mathbf{v}_{C_i} = v_{C_i}\hat{\mathbf{i}}$$

the left side of Equation (2) is simply

$$\frac{\ell}{2} m v_{C_i} \hat{\mathbf{k}} \tag{3}$$

By the parallel-axis (transfer) theorem for moments of inertia we have

$$I_O = I_C + md^2 = m\left[k_C^2 + \left(\frac{\ell}{2}\right)^2 + r^2\right]$$

The vessel is essentially a thick shell; considering then that

$$I_C = mk_C^2 \approx \frac{mr^2}{2} + \frac{m\ell^2}{12} \tag{4}$$

we see that

$$I_O = \frac{m\ell^2}{3} + \frac{3}{2}mr^2 \tag{5}$$

By substituting Equations (3) and (5) into (2), we thus obtain

$$m\frac{\ell}{2}v_{C_i} = m\left(\frac{\ell^2}{3} + \frac{3}{2}r^2\right)\omega_f$$

and the angular velocity after impact is then

$$\boldsymbol{\omega}_f = \frac{3\ell v_{C_i}}{2\ell^2 + 9r^2} \circlearrowleft \tag{6}$$

We are now ready to proceed to the second part of the solution.

Between the start of pivoting (immediately following impact) and the arrival at point B, the system is easily analyzed by work and kinetic energy:

$$W = \Delta T = T_f - T_i$$

To obtain the least possible value of v_{C_i} for no overturning, we set $T_f = 0$. The only work done in this phase of the vessel's motion is by gravity, so that

$$W = -mg\Delta h = 0 - T_i = -\frac{1}{2}I_O\omega_f^2$$

where ω_f is now an *initial* angular speed for this final stage of the problem and O is still a pivot point for $\mathcal{R}$. Thus

$$-mg\Delta h = \frac{1}{2}m\left(\frac{\ell^2}{3} + \frac{3}{2}r^2\right)\left(\frac{3\ell v_{C_i}}{2\ell^2 + 9r^2}\right)^2 \tag{7}$$

or

$$v_{C_i} = \sqrt{\frac{4g\Delta h(2\ell^2 + 9r^2)}{3\ell^2}} = \sqrt{\frac{4(32.17)(0.92)(2 \times 65^2 + 9 \times 10^2)}{3 \times 65^2}}$$

$$= 9.4 \text{ ft/sec}$$

(Continued)

There are several important remarks yet to be made about this problem. The first is that it can be shown (with a coefficient of restitution analysis) that more energy is lost with no bounce at O than if rebounding takes place. This energy loss is

$$\Delta E = T_i - T_f$$

where i and f refer to the instants just before and after the impact. Substituting, we get

$$\begin{aligned}\Delta E &= \frac{1}{2} m v_{C_i}^2 - \frac{1}{2} I_O \omega_f^2 \\ &= \frac{1}{2} m \left[v_{C_i}^2 - \left(\frac{\ell^2}{3} + \frac{3}{2} r^2 \right) \left(\frac{3\ell v_{C_i}}{2\ell^2 + 9r^2} \right)^2 \right] \\ &= \frac{1}{2} m v_{C_i}^2 \left(1 - \frac{9\ell^2/6}{2\ell^2 + 9r^2} \right)\end{aligned}$$

For $\ell = 65$ ft and $r = 10$ ft, we obtain

$$\Delta E = \frac{1}{2} m v_{C_i}^2 (0.322)$$

Thus 32.2%, or nearly a third of the original mechanical energy, is lost during impact if the lower front corner of $\mathcal{R}$ sticks to, and pivots about, point O. Of great importance here is the fact that we do not *know* how much rebounding would actually occur in the physical situation and hence how much energy would be lost. This means that 9.4 ft/sec may not be a conservative engineering answer for the safe speed. If the plane is flat ($\psi = 0$), for example, it can be shown that:

1. The speed corresponding to pivoting as in this example is 11.9 ft/sec. (It has further to pivot so it can be going faster prior to impact.)
2. At this speed initially, and with a no-energy-lost rebound, the vessel will easily overturn even though the striking corner backs up.

A conservative safe speed for the vessel in the inclined plane case can be obtained by assuming that all the vessel's initial kinetic energy goes into tilting it up about O. This approach gives

$$-mg\Delta h = -\frac{1}{2} m v_{C_i}^2$$

$$v_{C_i} = \sqrt{2g\Delta h} = 7.7 \text{ ft/sec}$$

In practice, the engineers in this case decided not to exceed 3 ft/sec, in view of the importance of the work and the danger involved.

Besides the mass center C (but of much lesser importance), there is another special point of interest of a rigid body $\mathcal{B}$ in plane motion, a point that differs from C in that it depends not only on the mass distribution of $\mathcal{B}$ but also on the motion of the body. This point lies along the resultant of the **ma** vectors of all the body's mass elements. The point is called the **center of percussion,** and it has value in certain applications such as impact testing.

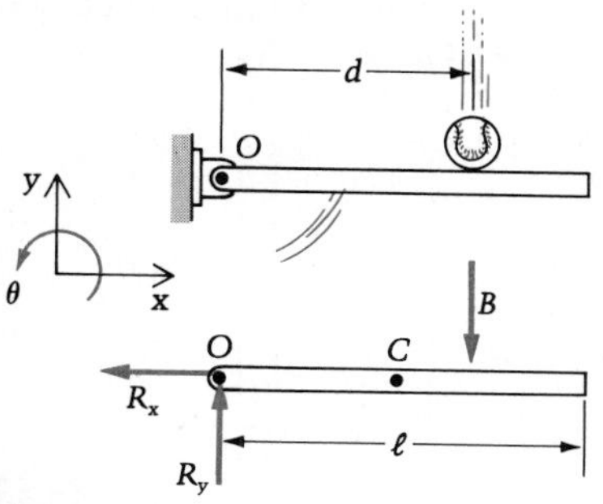

Figure 5.12

Before getting into the theory behind the center of percussion, we first illustrate its existence and demonstrate its value by means of an example. If a youngster hits a baseball with a stick and does not translate his hands too much, we may model the situation as shown in Figure 5.12 and ask where the ball should hit the stick in order to eliminate the "sting" (transverse reaction R_y of the stick onto the boy's hands). If the boy hits the ball at just the right place, called the "sweet spot," he hits it a long way while hardly feeling it and is said to have gotten "good wood" on the ball. Assuming the stick to be rigid at the beginning ($t = 0$) and end ($t = \Delta t$) of the short impact interval, the two principles of impulse and momentum are used as follows.

The impulse-momentum equation for the stick in the y direction is

$$\int_0^{\Delta t} (R_y - B)\, dt = m(\dot{y}_{C_f} - \dot{y}_{C_i}) \tag{1}$$

The angular impulse–angular momentum equation is

$$\int_0^{\Delta t} -Bd\, dt = I_O(\omega_f - \omega_i) = \frac{m\ell^2}{3}(\omega_f - \omega_i) \tag{2}$$

Using $R_y = 0$ for "no sting," setting $y_C = (\ell/2)\omega$ by kinematics at $t = 0$ and Δt, and multiplying Equation (1) by d gives

$$-\int_0^{\Delta t} Bd\, dt = \frac{md\ell}{2}(\omega_f - \omega_i) \tag{3}$$

Dividing Equation (3) by (2) gives

$$1 = \frac{d}{2} \div \frac{\ell}{3} \Rightarrow d = \frac{2}{3}\ell$$

Thus if the ball is struck two-thirds of the way from O to the end of the stick, the transverse reaction R_y will be zero.

In this example the point at which the ball is struck ($d = 2\ell/3$) is the center of percussion of the stick. To show this, we first recall that in Chapter 4 we proved that for *any* point P (moving or not, fixed to $\mathcal{B}$ or not), we can always write

$$\mathbf{M}_{r_P} = \int \mathbf{R} \times \mathbf{a}\, dm$$

in which $\mathbf{R}$ is the position vector from point P to a generic differential mass element. Therefore, since the integral of the vectors ($\mathbf{R} \times \mathbf{a}\, dm$) over the body in fact represents the resultant moment about P of the **ma** vectors over the mass of $\mathcal{B}$, this integral vanishes for all points P^* on the line along which the resultant of the **ma** vectors lies and hence $\mathbf{M}_{r_{P^*}} = \mathbf{0}$ for these points.

Armed with the fact that $\mathbf{M}_{r_{P^*}} = \mathbf{0}$ for the center of percussion, we can now derive the general equation for the distance from a pivot O to the center of percussion P^* in plane motion (Figure 5.13):

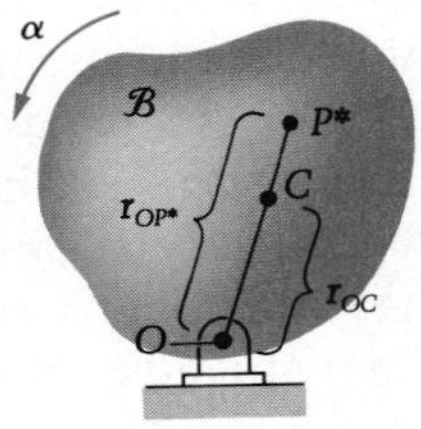

Figure 5.13

$$M_{r_{O_z}} = \overset{0}{\cancel{M_{r_{P^*_z}}}} + (\mathbf{r}_{OP^*} \times \overset{m\mathbf{a}_C}{\cancel{\mathbf{F}_r}})_z$$

$$\alpha(I_C + mr_{OC}^2) = r_{OP^*}m(r_{OC}\alpha)$$

$$mk_C^2 + mr_{OC}^2 = mr_{OP^*}r_{OC}$$

Therefore

$$r_{OP^*} = r_{OC} + \frac{k_C^2}{r_{OC}} \tag{5.43}$$

and we see that, for the stick,

$$r_{OP^*} = \frac{\ell}{2} + \frac{\ell^2/12}{\ell/2} = \frac{2}{3}\ell \qquad \text{(as before)}$$

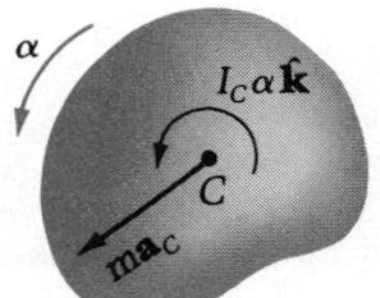

Figure 5.14

Note from Equation (5.43) that the center of percussion is always farther from the pivot than is the mass center. We make one final remark about the center of percussion. If we treat the **a** dm's of $\mathcal{B}$ as a collection of vectors, its resultant may be expressed (for a rigid body in plane motion) at the mass center (Figure 5.14), where

$$\int \mathbf{a}\, dm = m\mathbf{a}_C$$

$$\int \mathbf{r} \times \mathbf{a}\, dm = \mathbf{M}_{r_C} = I_C\alpha\hat{\mathbf{k}}$$

In a manner identical to reducing a force and couple to its simplest form, we may reduce this resultant of the **ma** vectors as shown in Figure 5.15, where the distance D is $I_C\alpha/(ma_C)$. We note that there is in fact a *line* $\ell\ell$ of points along the resultant of the **ma** vectors, making the location of a single point P^* ambiguous. However, the concept of the center of percussion is usually used in conjunction with problems in which the body has a pivot O (as in the previous example). In these problems P^* is the well-defined single point at the intersection of lines $\ell\ell$ and OC as in Figure 5.15.

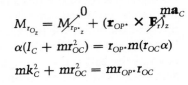

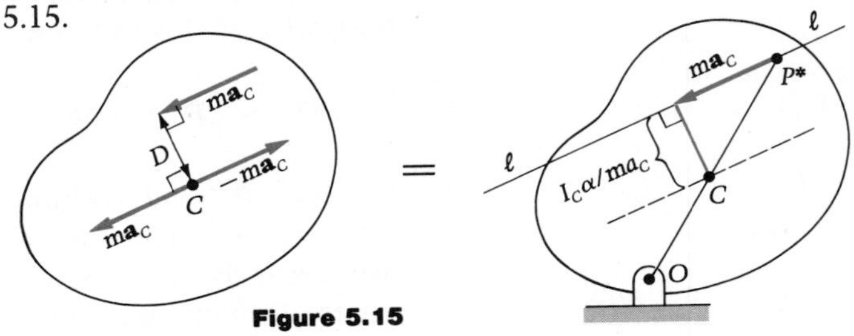

Figure 5.15

EXAMPLE 5.19

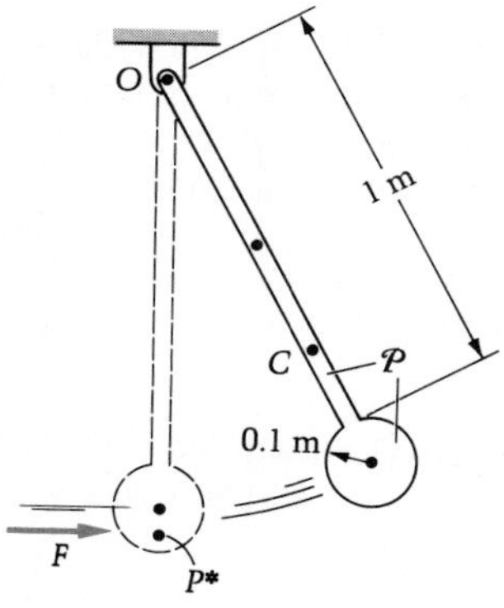

Find the center of percussion P^* for a pendulum $\mathcal{P}$ consisting of a rod plus disk, each of which has equal mass m. (See the diagram.)

SOLUTION

The mass center of $\mathcal{P}$ is located at a distance r_{OC} from O given by

$$m(0.5) + m(1.1) = 2mr_{OC}$$

$$r_{OC} = 0.8 \text{ m}$$

(Note that with equal masses C lies halfway between the mass centers of the rod and disk.) The radius of gyration with respect to C is calculated next:

$$I_{z_C} = \frac{m(1^2)}{12} + m(0.3^2) + \frac{m(0.1^2)}{2} + m(0.3^2)$$

$$= 0.268m = m(0.518)^2 = mk_C^2$$

Thus $k_C = 0.518$ m and Equation (5.43) then gives us the location of P^*:

$$r_{OP^*} = r_{OC} + \frac{k_C^2}{r_{OC}}$$

$$= 0.8 + \frac{(0.518)^2}{0.8}$$

$$= 1.14 \text{ m}$$

Striking the pendulum at P^* eliminates the horizontal pin reaction at O, as we have seen.

PROBLEMS / Section 5.3 (continued)

5.106 The bar in Figure P5.106 is welded to the end of the cylinder, which is traveling downward in translation. The bar strikes the tables at speed v_0, and the cylinder begins to rotate about the bar without rebound.

a. Find the angular velocity of the cylinder when C is at its lowest point.

b. Find the percentage of energy lost during the impact; that is,

$$\left(1 - \frac{\text{new energy}}{\text{old energy}}\right) \times 100$$

5.107 Block $\mathcal{A}$ in Figure P5.107 weighs 16.1 lb and is traveling to the right on the smooth plane at 50 ft/sec. Block $\mathcal{B}$ weighs 8.05 lb and is in equilibrium with the spring barely preventing it from sliding down the rough section of plane. Body $\mathcal{A}$ impacts $\mathcal{B}$; the coefficient of restitution $e = \frac{1}{2}$. Find the maximum spring deflection.

5.108 In soccer, a goal is scored only when the *entire* ball is over the *entirety* of the 4-in.-wide goal line. (See Figure P5.108.) Assuming a coefficient of restitution of 0.4 and neglecting friction between ball and post, determine whether a goal is scored if the velocity of the (translating) ball's center C makes an angle with the horizontal of 15°. Neglect the deviation caused by gravity on the trajectory between post and ground.

5.109 In the preceding problem determine the maximum coefficient of restitution for which a goal will be scored; all other data is the same.

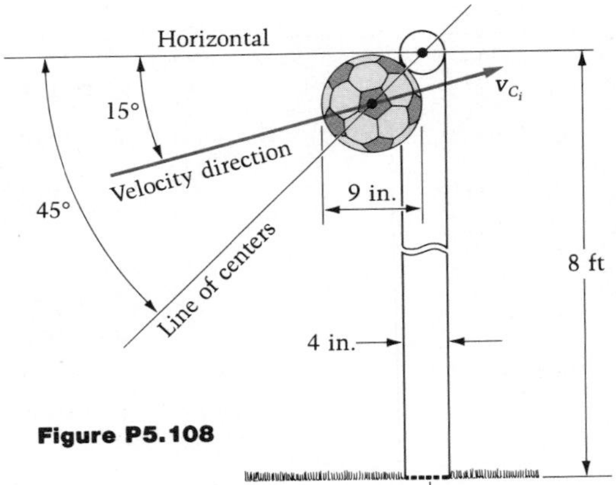

Figure P5.108

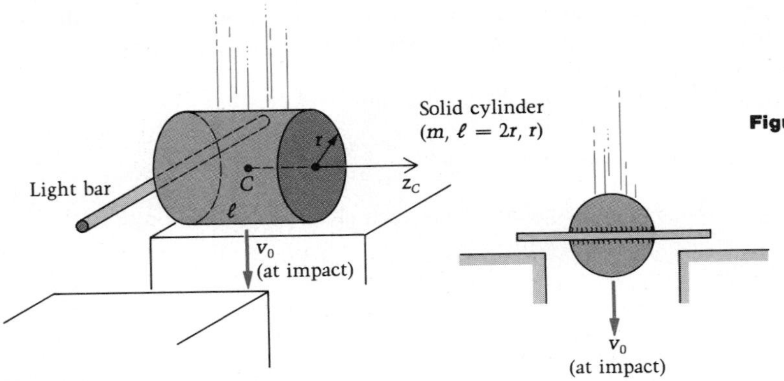

Figure P5.106

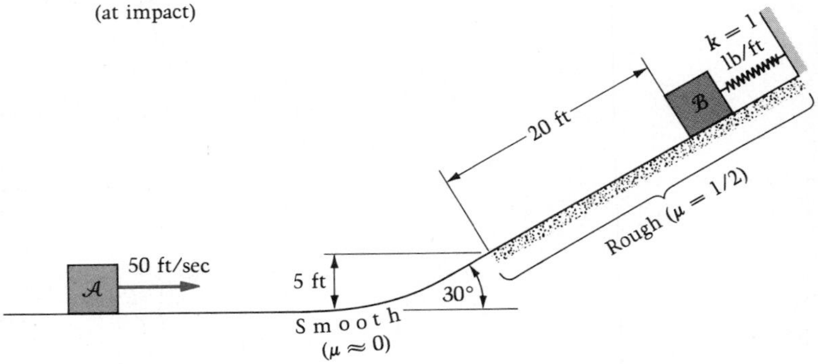

Figure P5.107

5.110 A uniform bar AB of length L and mass M is moving on a smooth horizontal plane with $\mathbf{v}_C = \hat{\mathbf{i}}v_0$ and $\boldsymbol{\omega} = \hat{\mathbf{k}}\omega_0$, when end B strikes a peg P (see Figure P5.110). If $\omega_0 = 2v_0/L$ and the coefficient of restitution $e = \frac{1}{6}$, find the loss of kinetic energy.

5.111 A homogeneous cube of side a and mass M slides on a level, frictionless table with velocity v_0. (See Figure P5.111.) It strikes a small lip on the table at A of negligible height. Find the velocity of the center of mass just after impact if the coefficient of restitution is unity. (The centroidal moment of inertia of a cube about an axis parallel to an edge is $Ma^2/6$.)

5.112 Show that the cylinder in Figure P5.112, following release from rest, will reach the lower wall. Find the velocity with which C will rebound up the plane following impact, and determine the amount of energy lost. All data are shown on the figure.

5.113 An arrow of length L traveling with speed v_0 strikes a smooth hard wall obliquely as shown in Figure P5.113. End A does not penetrate but slides downward along the wall without friction or rebound. Find the angular velocity of the arrow after impact.

5.114 A sphere rolling with speed v_C on a horizontal surface strikes an obstacle of height H. What is the largest value that H can have if the sphere is able to make it over the obstacle? Consider the coefficient of restitution to be zero during the sphere's impact with the corner point O. The answer will be a function of g, r, and v_C—in fact, H/r may be solved for as a function of the single nondimensional parameter gr/v_C^2. (See Figure P5.114.)

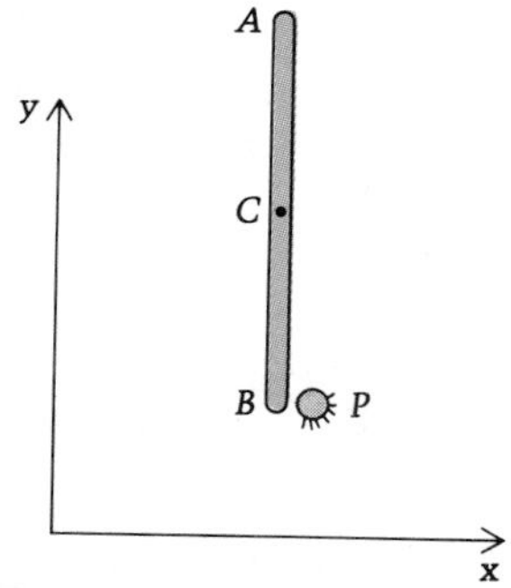

Figure P5.110

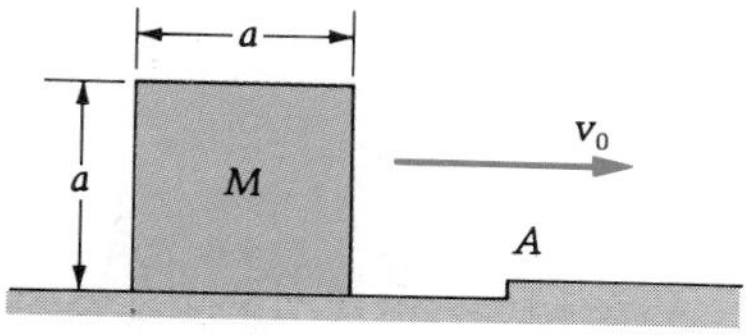

Figure P5.111

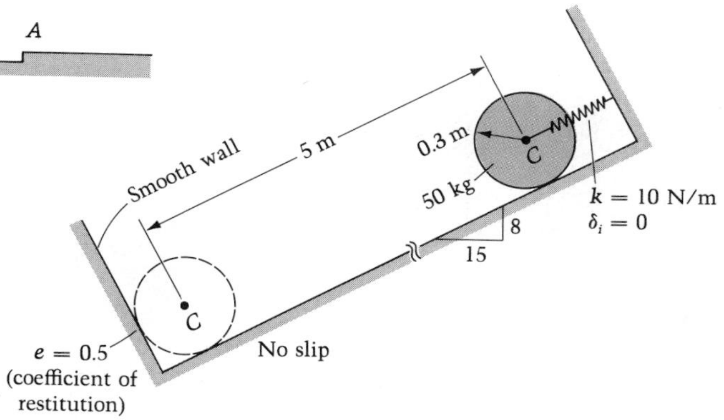

Figure P5.112

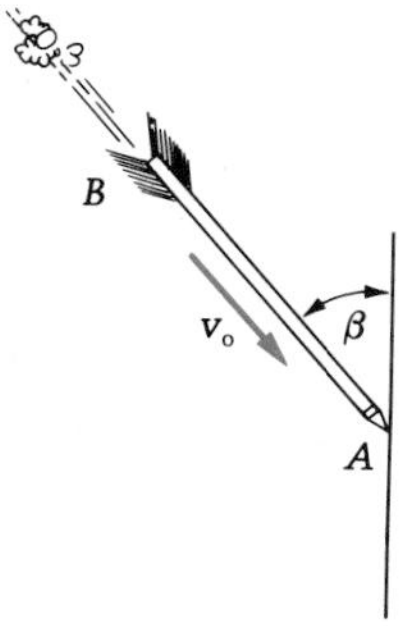

Figure P5.113

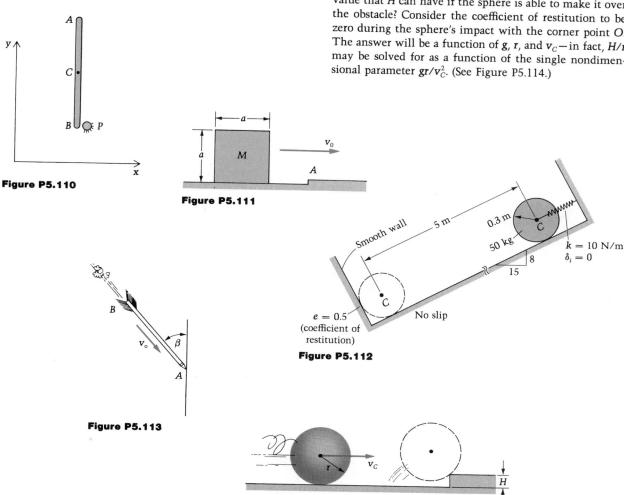

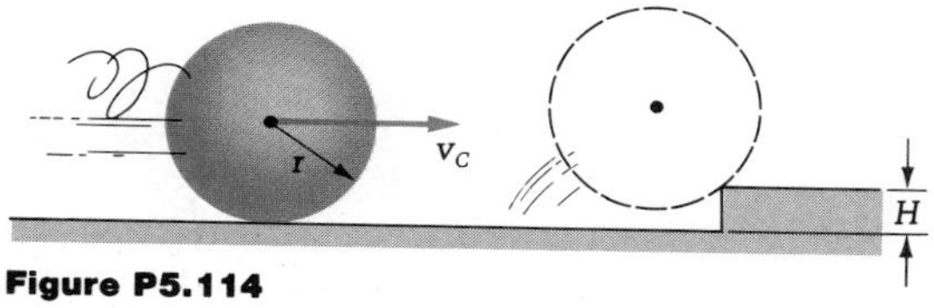

Figure P5.114

5.115 A block slides to the right and strikes a small obstruction at a speed of 20 ft/sec. (See Figure P5.115.)

a. If the coefficient of restitution is zero, find the energy loss caused by the impact.
b. What is the minimum striking velocity required to overturn the block after collision?

5.116 Two identical elastic balls $\mathcal{A}$ and $\mathcal{B}$ move toward each other. Find the approach velocity ratio v_{A_i}/v_{B_i} that will result in $\mathcal{A}$ coming to rest following the collision. The coefficient of restitution is e. (See Figure P5.116.)

5.117 Use the two equations

$$m_{\mathcal{A}}v_{A_i} + m_{\mathcal{B}}v_{B_i} = m_{\mathcal{A}}v_{A_f} + m_{\mathcal{B}}v_{B_f}$$

and

$$e = \frac{v_{B_f} - v_{A_f}}{v_{A_i} - v_{B_i}} \qquad (= \text{coefficient of restitution})$$

to prove that the loss in kinetic energy as the bodies $\mathcal{A}$ and $\mathcal{B}$ collide (Figure P5.117) is

$$\Delta T = \frac{m_{\mathcal{A}}m_{\mathcal{B}}(1 - e^2)(v_{A_i} - v_{B_i})^2}{2(m_{\mathcal{A}} + m_{\mathcal{B}})}$$

Deduce from this result that $e \le 1$!

5.118 Use the result of the preceding problem to show that for a head-on collision at equal speeds v and equal masses m,

$$\Delta T = 2\frac{mv^2}{2}(1 - e^2)$$

so that if $e = 0$ then *all* of the initial T is lost and if $e = 1$ then *none* of T is lost. Is this true for differing speeds and masses?

5.119 Sphere $\mathcal{A}$ has mass m and radius r, and it rolls with mass center velocity $v_0 \rightarrow$ on a horizontal plane. (See Figure P5.119.) It hits squarely an identical sphere $\mathcal{B}$ that is at rest. The coefficient of friction between a sphere and the plane is μ, and between spheres it is negligible. The impact is nearly elastic ($e \approx 1$).

a. Find v_{G_f} and ω_f of each sphere right after impact.
b. Find v_G of each sphere after it has started rolling uniformly.
c. Discuss the special case when $\mu = 0$.

5.120 A ball is dropped from a height H and bounces until it comes to rest. (See Figure P5.120.) If the coefficient of restitution is e, find the total distance traveled by the ball.

5.121 A 4-lb sphere is released from rest in the position shown in Figure P5.121 and two observations are then made: (1) The sphere comes immediately to rest after the

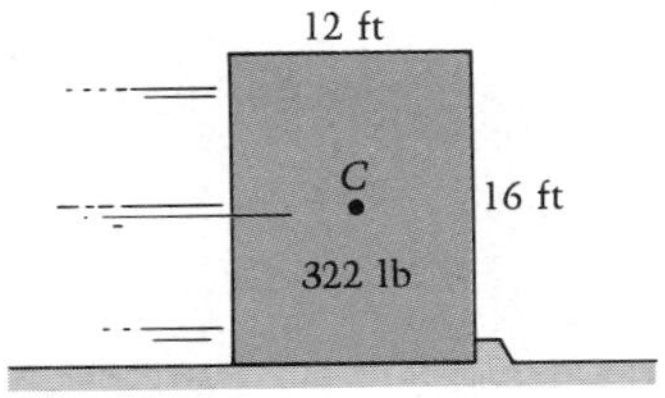

Figure P5.115

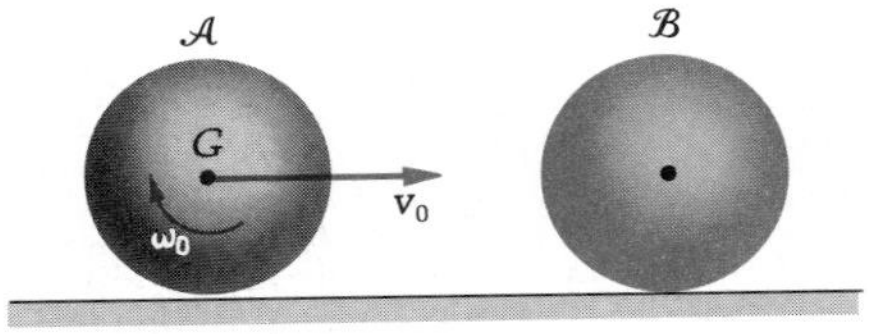

Figure P5.119

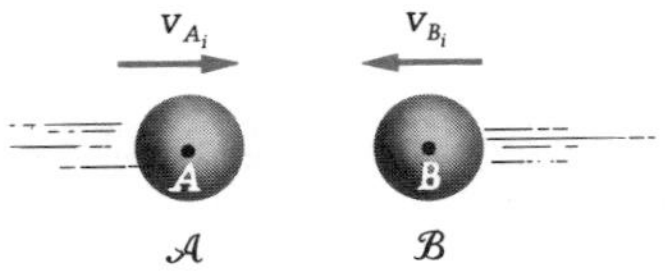

Figure P5.116

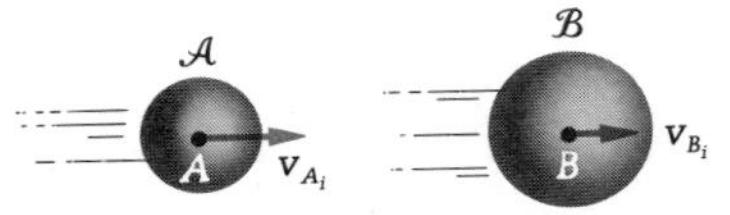

Figure P5.117

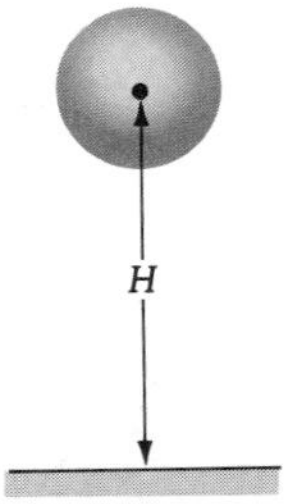

Figure P5.120

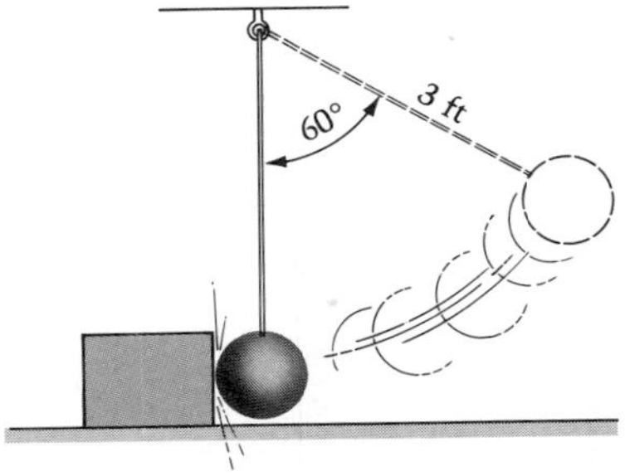

Figure P5.121

impact; (2) the 5-lb block slides 3 ft before coming to rest. Using these observations, find the coefficients of restitution (between sphere and block) and friction (between block and floor).

5.122 There is only one height H above a pool table at which a cue ball may be struck by the stick without the ball slipping for a while after the impact. (See Figure P5.122.) Find this value of H, in terms of R, for which the ball immediately rolls.

5.123 The assembly in Figure P5.123 is turning at $\omega_i = 2$ rad/sec when the collar is released. All surfaces are smooth, and the disk is fixed to the bar; the light vertical shaft ends in bearings and turns freely. The data are:

$m_{\mathcal{D}} = \frac{1}{4}$ slug $\quad \ell_{\mathcal{C}} = 2$ ft

$m_{\mathcal{B}} = 1$ slug $\quad \ell_{\mathcal{B}} = 4$ ft

$m_{\mathcal{C}} = \frac{1}{4}$ slug

The collar moves outward and impacts the disk without rebound. Find: (a) the angular speed of the bar just before and just after impact; (b) the percentage of energy lost during impact. The radii of $\mathcal{B}$ and $\mathcal{C}$ are small compared to their lengths. Treat $\mathcal{D}$ as a particle.

5.124 Using a density of wood of 673 kg/m^3, find the mass center C of the baseball bat in Figure P5.124 and then determine its moment of inertia with respect to the z_O axis, perpendicular to the axis of symmetry of the bat. Use the parallel-axis theorem to obtain I_{z_C}, and find the bat's center of percussion if it is swinging about a fixed point O.

5.125 The boy in Figure P5.125 flings his yo-yo with $\mathbf{v}_{G_i} = 3 \downarrow$ ft/sec and $\boldsymbol{\omega} = 36$ ↶ rad/sec. It rolls down the (double) string and "sleeps" when it hits bottom. (a) Find the impulse exerted by the loop of the string on the yo-yo when it hits bottom. (b) Assuming no change in angular velocity during this short time and assuming the string provides a subsequent resistive couple to the yo-yo's rotation of $0.4(2m)gR$, find how long it "sleeps" before stopping.

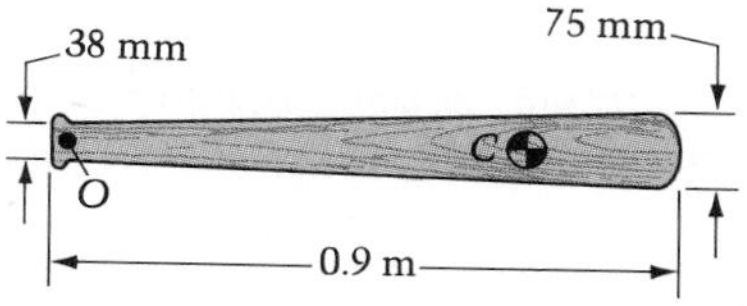

Figure P5.124

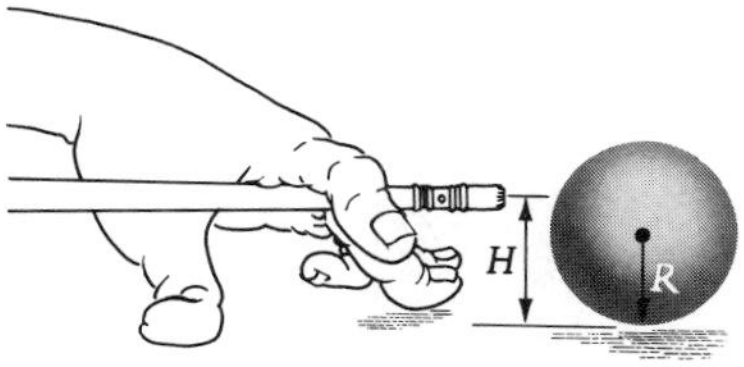

Figure P5.122

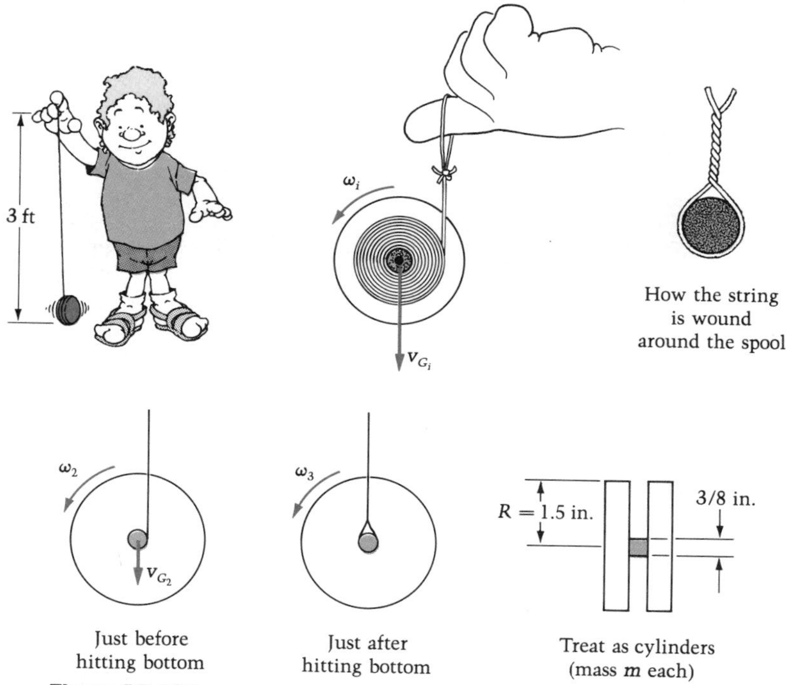

Figure P5.125

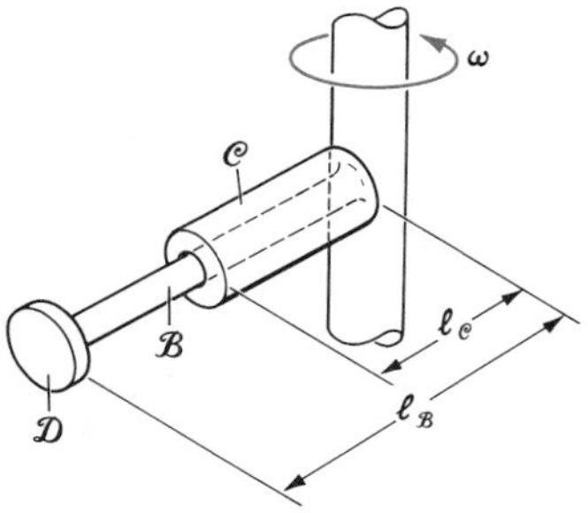

Figure P5.123

5.126 In Example 5.17 use Equation (5.42) instead of (5.33) to solve the problem. Use point Q as the moment center, and the impulse $F\,\Delta t$ will be eliminated. (You will still need Equation (2), however.)

5.127 The rod-sphere rigid body in Figure P5.127 is released from rest in the horizontal position. It swings down and strikes the box at its lowest point. Find how far the box slides before coming to rest if the coefficient of restitution is $e = 0.5$. The data are:

1. Rod: length = 1 m; mass = 3 kg
2. Sphere: radius = 0.2 m; mass = 10 kg
3. Block: $b = 0.3$ m; $H = 0.35$ m; mass = 5 kg
4. Coefficient of friction between block and plane = 0.3

Assume the sphere hits the block just once.

5.128 The 30-kg bent bar $\mathcal{B}$ in Figure P5.128 falls from the dashed position onto the spinning cylinder $\mathcal{C}$, which was initially turning at 3000 rad/s ↻. If the bar does not bounce (coefficient of restitution is zero), find the stopping time for $\mathcal{C}$ following the impact.

5.129 Weight W_1 falls from rest through a distance H; it lands on another weight W_2, which was in equilibrium atop a spring of modulus k. (See Figure P5.129.) If the coefficient of restitution is zero, find the spring compression when the weights are at their lowest point.

5.130 Specialize Example 5.17 to the limiting case of a perfectly elastic collision ($e = 1$). Compute the percentage of kinetic energy lost in the impact.

5.131 Repeat the preceding problem for the perfectly plastic case ($e = 0$).

5.132 The bar in Figure P5.132 swings downward from the dotted horizontal position and strikes mass m. The bar has mass M and length ℓ. The collision takes place with a coefficient of restitution of zero. If the coefficient of friction between m and the plane is μ, find the distance moved by m before stopping. Treat m as a particle.

5.133 A tennis ball ($I_C = \frac{2}{3}mr^2$) hits the court with $v_C = v_0$ and $\omega = \omega_0$ as shown in Figure P5.133. Assuming the court grips the ball during the impact (no skidding), find the relationship between v_0 and ω_0 so that the ball will rebound back toward the net (to the left).

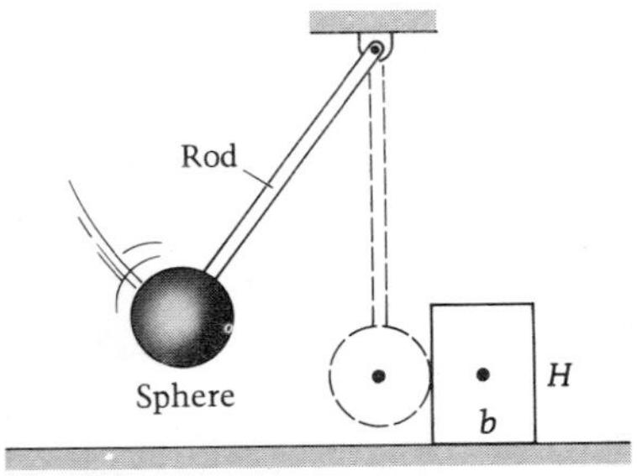

Figure P5.127

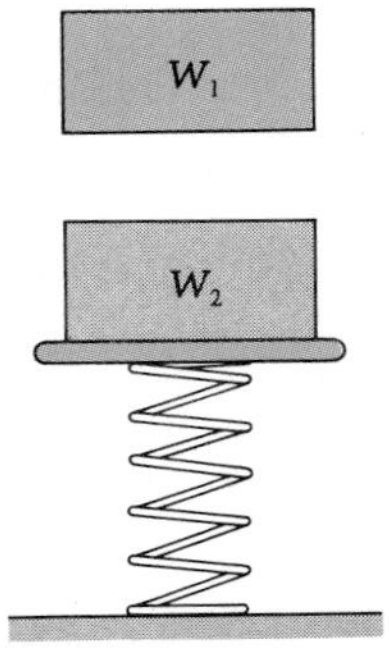

Figure P5.129

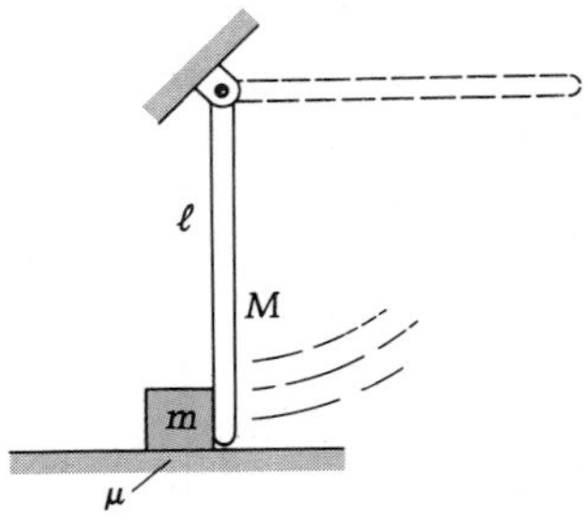

Figure P5.132

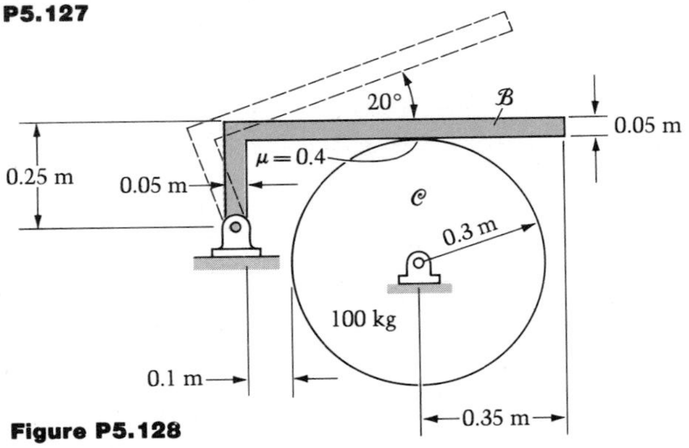

Figure P5.128

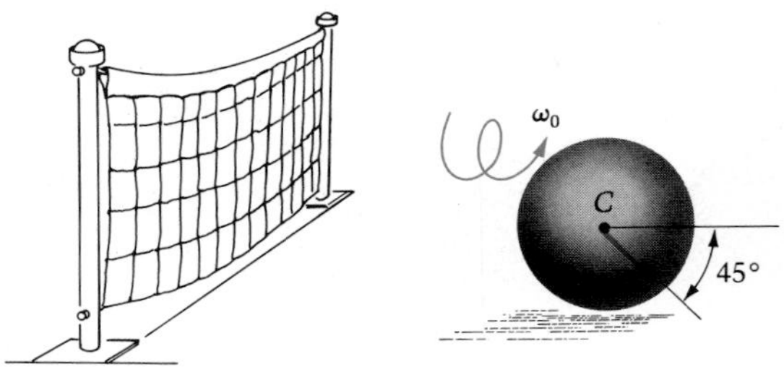

Figure P5.133

5.134 A uniform rod $\mathcal{R}$ of length L is dropped and translates downward at an angle θ with the vertical as shown in Figure P5.134. If end A of $\mathcal{R}$ does not rebound after striking the ground at speed v_0, find: (a) the energy lost during the impact of A with the ground; (b) the speed at which the other end B then hits the ground.

5.135 A wooden sphere weighing 0.644 lb swings down from a position where the rod is horizontal, and it impacts the block. (See Figure P5.135.) The coefficient of restitution is $e = 0.6$. The block weighs 3.22 lb and is initially at rest. Find the position of the block when it comes *permanently* to rest. (Assume that the sphere is removed from the problem after impact and that the spring cannot rebound past its original unstretched position.)

5.136 Compute the error in I_{z_C} in Example 5.18 that was incurred in assuming the vessel to be a shell (so that I_{z_C} was $m\ell^2/12 + mr^2/2$). Use the weight, height, outer radius, and density to compute the thickness of the vessel; then calculate a more accurate I_{z_C} and compare.

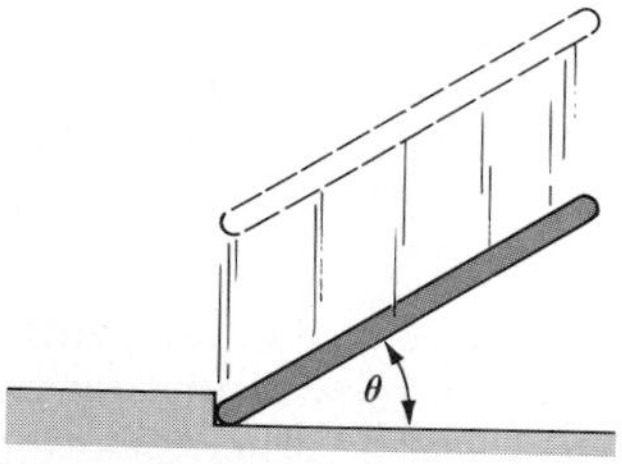

Figure P5.134

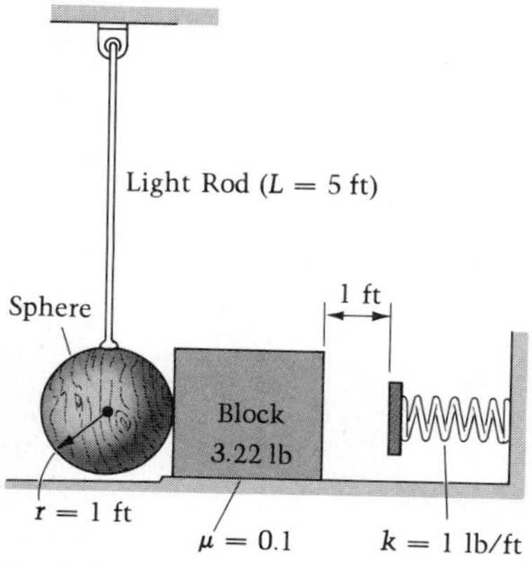

Figure P5.135

5.137 In Example 5.18 assume that the lower front striking corner of the vessel $\mathcal{R}$ rebounds back up the plane with velocity ev_{C_i}, where e is the coefficient of restitution ($0 < e \leq 1$). Use the equations of impulse and momentum in the $\mathbf{x}$ and $\mathbf{y}$ directions, and the equation of angular impulse and angular momentum, to find the two mass-center velocity components and the angular velocity of $\mathcal{R}$ after impact. Compare the results of $\dot{x}_{C_f}$, $\dot{y}_{C_f}$, and $\dot{\theta}_f$ for $e = 1$ with those at $e = 0$. Show that no energy is lost when $e = 1$; that is, show

$$\frac{1}{2}mv_{C_i}^2 = \frac{1}{2}m(\dot{x}_{C_f}^2 + \dot{y}_{C_f}^2) + \frac{1}{2}I_C\dot{\theta}_f^2$$

5.138 *After* the impact in the preceding problem, show that the equations of motion of the vessel are

$$m\ddot{x}_C = -mg\sin\psi \tag{1}$$

$$m\ddot{y}_C = N - mg\cos\psi \tag{2}$$

$$I_C\ddot{\theta} = \frac{\ell}{2}N\sin\theta - rN\cos\theta \tag{3}$$

where N is the normal reaction at the corner Q. (See Figure P5.138.) Observe that until there is another impact, the mass center has a constant $\mathbf{x}$ component of acceleration. Use kinematics to prove that

$$\ddot{y}_C = \left(r\cos\theta - \frac{\ell}{2}\sin\theta\right)\ddot{\theta} - \dot{\theta}^2\left(r\sin\theta + \frac{\ell}{2}\cos\theta\right) \tag{4}$$

Use Equations (2) and (4) to eliminate N from (3), thus obtaining a single differential equation in θ governing the rotational motion of $\mathcal{R}$, and note its complexity.

5.139 Prove statement 1 near the end of Example 5.18 for the case when the plane is level ($\psi = 0$) and $e = 0$. *Hint:* Note carefully that the angle of the plane does not affect Equation (6), so you only need to alter the Δh in Equation (7) to obtain the new result.

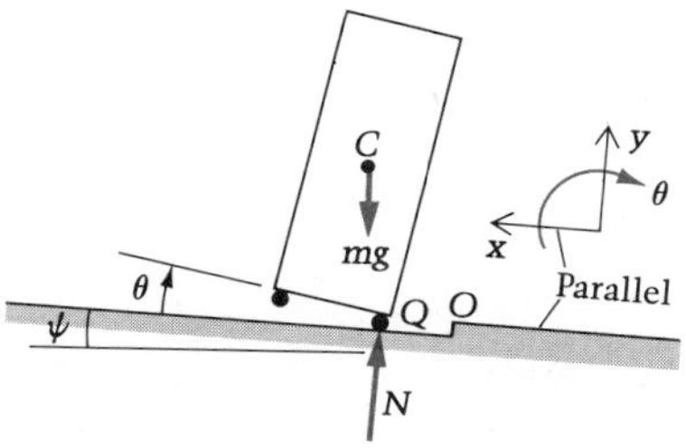

Figure P5.138

5.140 Prove statement 2 near the end of Example 5.18. Again the plane is to be level in this problem, but now $e = 1$. *Hint:* The $\mathbf{x}_C$ component of velocity is constant after impact, since with $\psi = 0$ all external forces (mg and N) are vertical. To find this velocity $\dot{x}_{C_f}$, use ω_f and the velocity of the striking corner Q just after impact (ω_f is the same as with $\psi = 3.719$ in Problem 5.137 with $e = 1$; v_Q is v_{C_i} back to the left). Then use

$$W = \Delta T = \frac{1}{2} m\dot{x}_{C_f}^2 + \frac{1}{2} m\dot{y}_{C_f}^2 + \frac{1}{2} I_C\omega_f^2 - \frac{1}{2} mv_{C_i}^2$$

to show that C reaches the top with energy to spare.

5.141 Three equal uniform rods AB, BC, CD, each of mass m and length $2a$, are smoothly jointed together at B and C and placed on a smooth horizontal table. As Figure P5.141 shows, all three lie in a straight line. An impulsive couple of moment M_0 is applied to AB, in the plane of the table, for a short time Δt. Find the initial motion of the rods.

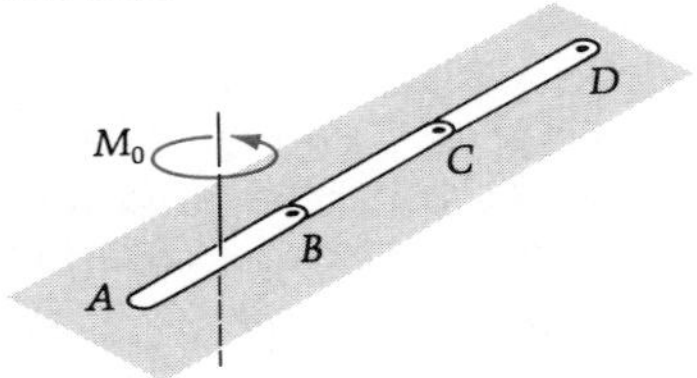

Figure P5.141

5.142 In the previous problem show that the total kinetic energy generated by the angular impluse $M_0\,\Delta t$ is $9(M_0\,\Delta t)^2/(10ma^2)$.

5.143 Rod $\mathcal{R}$ in Figure P5.143 is freely falling in a vertical plane. At a certain instant it is horizontal with its ends A and B having the velocities shown. If end A is suddenly fixed, prove that the rod will start to rise around end A provided that $v_1 < 2v_2$.

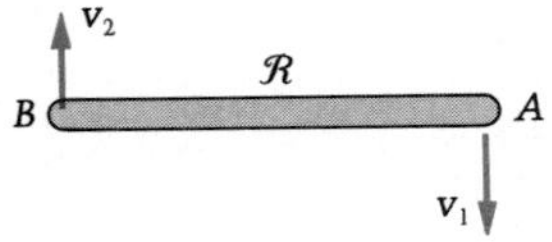

Figure P5.143

5.144 In the preceding problem show that the energy loss in instantaneously stopping point A is independent of v_2.

5.145 Four identical rods are pinned together as shown in Figure P5.145; they rest on a smooth horizontal desk. If the system is struck a blow F at A, imparting the impulse $I\,(= F\,\Delta t)$, find the velocities of the mass centers and the angular velocities of the rods just after impact. Note that, due to symmetry, only AB and BC need be found.

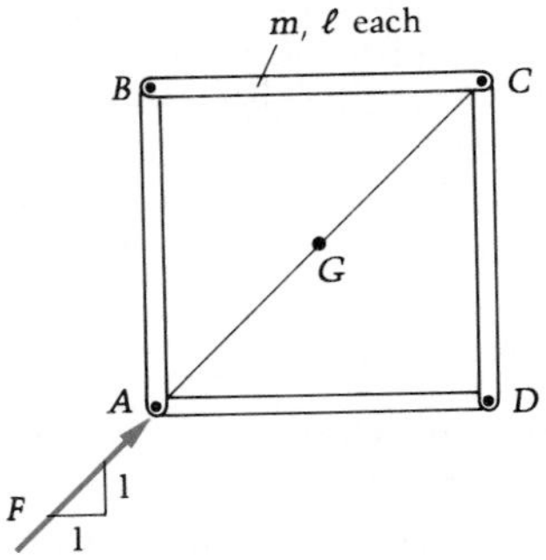

Figure P5.145

5.146 If the rods in the previous problem are welded into a single rigid body, the final velocity of G is $I/4m$ and $\omega_f = 0$. Thus the kinetic energy after impact is $\frac{1}{2}(4m)(I/4m)^2 = I^2/8m$. Show that this is less than the kinetic energy produced in the preceding problem by the amount $3\,I^2/16m$.

5.147 The plate in Figure P5.147, supported by ball joints at its top corners, is suddenly struck as shown with a force that produces the impulse I normal to the plate. Find the kinetic energy produced by the impact.

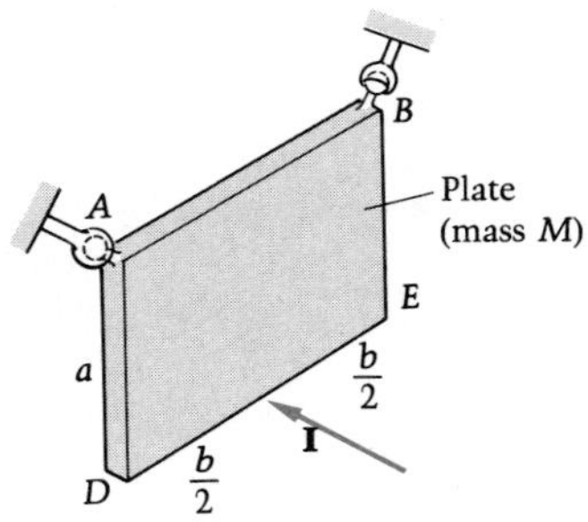

Figure P5.147

5.148 Work the preceding problem but assume that the plate is initially free. If you work both problems, show that the difference in the energies is $I^2/(2M)$.

5.149 The hammer in Figure P5.149 strikes the sphere $\mathcal{S}$ and imparts a horizontal impulse I to it. Determine the initial angular velocity of $\mathcal{S}$.

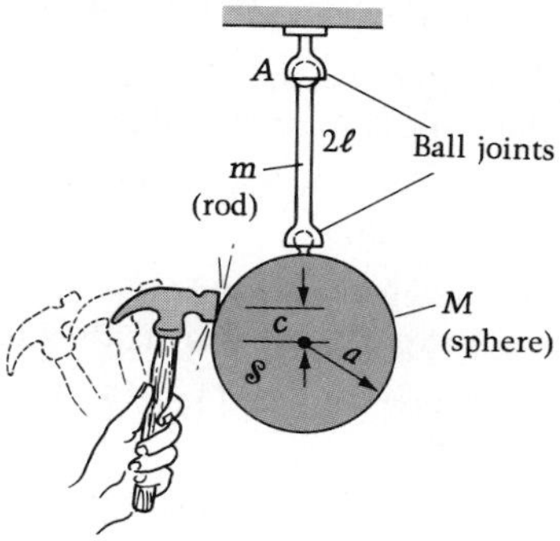

Figure P5.149

5.150 Repeat the preceding problem but suppose that the sphere and rod are welded together to form one rigid body.

5.151 Two toothed gear wheels, which may be treated as uniform disks of radii a and b and masses M and m, respectively, are rotating in the same plane. They are not quite in contact and have angular velocities ω_1 and ω_2 about fixed axes through their centers. Their axes are then slightly moved so that the wheels engage. Prove that the loss of energy is

$$\frac{Mm}{4(M+m)}(a\omega_1 + b\omega_2)^2$$

Answers to Questions / Chapter 5

Q5.1 By the definition of the mass center.

Q5.2 Because $(M_{r_{C_x}}\hat{\mathbf{i}} + M_{r_{C_y}}\hat{\mathbf{j}}) \cdot \omega\hat{\mathbf{k}}$ is zero!

Q5.3 Normal force: no; it is perpendicular to the horizontal velocity of the contact point B. Friction force: yes; $W_f = \int f\hat{\mathbf{i}} \cdot v_B\hat{\mathbf{i}}\,dt$.

Q5.4 Larger: the rod will not reach the horizontal position; smaller: the system passes through this position without stopping.

Q5.5 Since the counterweights place the mass center on the elevation axis, the mass center does not move.

Q5.6 No. In Equations (5.31) and (5.32) they are rectangular axes fixed in the inertial frame $\mathcal{I}$; in (5.34) and (5.35), however, they are fixed in body $\mathcal{B}$.

Q5.7 No, provided that G is equidistant from C_1 and C_2 and lies above the line C_1C_2. Note that $M_{r_C} = I_C\alpha = 0$ for the cart, and $M_{r_C} = 0$ requires that A_y be larger in the front. It then follows from $F_{r_y} = 0$ for the wheels that N is also larger in the front.

Q5.8 The $\mathbf{x}$ equation; summing moments about the bottom point has eliminated the impulse of the friction force.

Q5.9 The minus sign makes e positive, since the difference in velocity components at approach will have a sign opposite to the difference in the components at separation.

Q5.10 If O is not fixed in $\mathcal{I}$, then $(d/dt)(\mathbf{r}_{OC} \times m\mathbf{v}_C) = \mathbf{r}_{OC} \times m\mathbf{a}_C - \mathbf{v}_0 \times m\mathbf{v}_C$ and the second term is not zero then!

Q5.11 The value $e = 0$ produces a *plastic collision,* in which the separation velocities are equal—that is, the bodies stick together. If $e = 1$, however, the relative velocity of separation equals that of approach. This represents the upper limit of e, called *purely elastic,* because when $e = 1$ it can be shown for impacting particles that no energy is lost during the collision.

Review Questions / Chapter 5

True or False?

These questions all refer to rigid bodies in plane motion.

1. If you raise a 2-lb object 3 ft from rest and stop it there, gravity has done −6 ft-lb of work and you have done +6 ft-lb on the object.
2. The work of a couple $C\hat{\mathbf{k}}$ on body $\mathcal{B}$ is always $C\hat{\mathbf{k}} \cdot \theta\hat{\mathbf{k}} = C\theta$, where θ is the angle through which the body turns.
3. The work done by a linear spring is always $k(\delta_i^2 - \delta_f^2)/2$, where δ_i and δ_f are the amounts of initial and final stretch. (If negative, they represent compression.)
4. There are actually three separate work and kinetic energy principles; two of the equations add to give the third.
5. The principles of work and kinetic energy and (linear and angular) impulse and momentum result from general integrations of the equations of motion, and thus they are free of accelerations.
6. Not all forces acting on a body have to do work on it in general.
7. The friction force beneath a rolling wheel does work on it if the surface of contact is curved and fixed.
8. The normal force exerted on a rolling wheel by a surface, whether fixed or in motion, never does work on the wheel.
9. The principle $\int \mathbf{F}_r\, dt = \int \dot{\mathbf{L}}\, dt = m\mathbf{v}_{C_f} - m\mathbf{v}_{C_i}$ is valid for deformable bodies.
10. The principle $\int M_{r_C}\, dt = I_{z_C}\omega_f - I_{z_C}\omega_i$ is valid for deformable bodies.
11. Any problem that can be solved by $W = \Delta T$ can likewise by solved by using "kinetic + potential energy = constant."
12. The formula $T = \frac{1}{2}I_{\textcircled{1}}\omega^2$ gives *all* the kinetic energy of the rigid body in plane motion, assuming the body is not translating.

Answers: T, F, T, T, T, T, F, F, T, F, F, T

CHAPTER

6

Kinematics of a Rigid Body in Three-Dimensional Motion

chapter outline ►

6.1 Introduction

In this chapter we study the kinematics of a rigid body in general motion. We begin by defining the angular velocity vector $\boldsymbol{\omega}$ in three dimensions and then studying its important properties. We shall see that this vector no longer has the simple $\dot{\theta}\hat{\mathbf{k}}$ form it had in plane motion and that consequently there is no natural extension of this special case into general motion. We shall also study the derivative of $\boldsymbol{\omega}$, which is the angular acceleration vector.

There follows a study of what is often called velocity and acceleration equations in "moving frames of reference." It will be seen in Section 6.6 that the rigid-body velocity and acceleration equations may in fact be obtained from these general equations. In Section 6.7 we shall study the relationship between angular velocity and the angles of orientation of the body. We conclude the chapter with a study of rotation matrices.

6.2 Relation Between Derivatives/the Angular Velocity Vector

In this section we consider the relationship between the derivatives of a vector taken in two different frames. In the process we shall arrive at a concise and useful definition of angular velocity. The reader is strongly encouraged to persevere until this section and the next are fully understood. Even though the angular velocity vector in three dimensions is a difficult subject at first, it *must* be comprehended before we can consider the kinematics and kinetics of general rigid-body motion. The angular velocity vector is the key to the subject. It will either make life easier for students of three-dimensional motion (if they work hard at understanding it) or much more difficult (if they do not).

Let $\mathbf{Q}$ be an arbitrary vector. We may express $\mathbf{Q}$ in terms of its components (Q_1, Q_2, Q_3) in a rectangular Cartesian coordinate system embedded in frame $\mathcal{B}$ by

$$\begin{aligned}\mathbf{Q} &= (\mathbf{Q} \cdot \hat{\mathbf{b}}_1)\hat{\mathbf{b}}_1 + (\mathbf{Q} \cdot \hat{\mathbf{b}}_2)\hat{\mathbf{b}}_2 + (\mathbf{Q} \cdot \hat{\mathbf{b}}_3)\hat{\mathbf{b}}_3 \\ &= Q_1\hat{\mathbf{b}}_1 + Q_2\hat{\mathbf{b}}_2 + Q_3\hat{\mathbf{b}}_3 \qquad (6.1)\end{aligned}$$

in which the unit vectors $(\hat{\mathbf{b}}_1, \hat{\mathbf{b}}_2, \hat{\mathbf{b}}_3)$ are parallel at all times to the respective axes of the coordinate system fixed in $\mathcal{B}$. Now consider another reference frame $\mathcal{A}$, in which we wish to differentiate vector $\mathbf{Q}$ (see Figure 6.1). As an example, we may wish to find the velocity of a point in frame $\mathcal{A}$ even though the point's location may be defined in $\mathcal{B}$ (say by the vector $\mathbf{Q}$). In this case, part of the solution will require that we be able to differentiate $\mathbf{Q}$ in $\mathcal{A}$ even though it is expressed in terms of its components in $\mathcal{B}$.

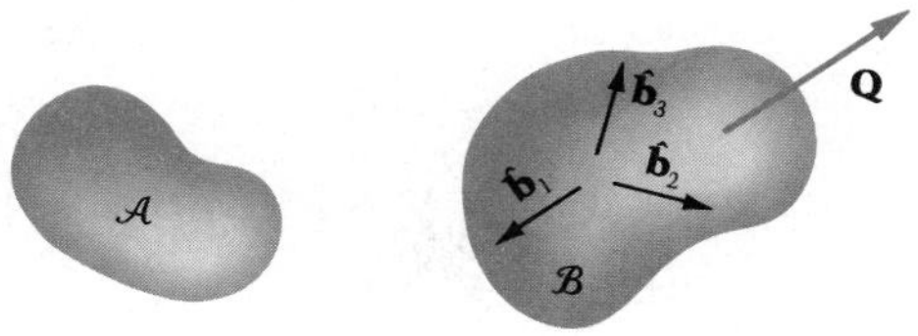

Figure 6.1 Vector **Q** and frames $\mathcal{A}$ and $\mathcal{B}$.

Therefore it is now time to learn how to relate derivatives of a vector taken in two different frames. We emphasize at the outset that these vectors are completely arbitrary—they need not even be related to dynamics! Nor does the derivative have to be with respect to time, although this is the independent variable of interest to us in dynamics and thus the one we shall use in the development to follow.

Letting ${}^{\mathcal{A}}\dot{\mathbf{Q}}$ represent (see Equation 1.8) the derivative of **Q** with respect to time in $\mathcal{A}$, we have

$$ {}^{\mathcal{A}}\dot{\mathbf{Q}} = \dot{Q}_1\mathbf{b}_1 + \dot{Q}_2\mathbf{b}_2 + \dot{Q}_3\mathbf{b}_3 + Q_1{}^{\mathcal{A}}\dot{\mathbf{b}}_1 + Q_2{}^{\mathcal{A}}\dot{\mathbf{b}}_2 + Q_3{}^{\mathcal{A}}\dot{\mathbf{b}}_3 \qquad (6.2) $$

Recognizing the first three terms on the right of Equation (6.2) as the derivative of **Q** in $\mathcal{B}$, we have

$$ {}^{\mathcal{A}}\dot{\mathbf{Q}} = {}^{\mathcal{B}}\dot{\mathbf{Q}} + (Q_1{}^{\mathcal{A}}\dot{\mathbf{b}}_1 + Q_2{}^{\mathcal{A}}\dot{\mathbf{b}}_2 + Q_3{}^{\mathcal{A}}\dot{\mathbf{b}}_3) \qquad (6.3) $$

Clearly the last three (parenthesized) terms in Equation (6.3) represent a vector depending upon both **Q** *and* the change of orientation of frame $\mathcal{B}$ with respect to $\mathcal{A}$. We now proceed to obtain a useful and compact expression for this vector; in the process, the angular velocity vector will arise.

Since for $i = 1, 2,$ or 3, we have

$$ \mathbf{b}_i \cdot \mathbf{b}_i = 1 \qquad (6.4) $$

it follows that

$$ \mathbf{b}_i \cdot {}^{\mathcal{A}}\dot{\mathbf{b}}_i = 0 \qquad (6.5) $$

so that the three derivatives of the $\mathbf{b}_i$ in Equation (6.3) are each perpendicular to the respective $\mathbf{b}_i$ themselves.*

Question 6.1 Will this be true for *any* vector of constant magnitude (not necessarily a unit vector)?

*This assumes that the $\mathbf{b}_i$ are not constant in frame $\mathcal{A}$. If two of the $\mathbf{b}_i$ are constant in $\mathcal{A}$, then all three are and the angular velocity vanishes; if only *one* is constant in $\mathcal{A}$, we have a simple special case to be considered later.

This means that there are three vectors $\boldsymbol{\alpha}$, $\boldsymbol{\beta}$, and $\boldsymbol{\gamma}$ for which

$$\begin{aligned} {}^{\mathcal{A}}\dot{\hat{\mathbf{b}}}_1 &= \boldsymbol{\alpha} \times \hat{\mathbf{b}}_1 \\ {}^{\mathcal{A}}\dot{\hat{\mathbf{b}}}_2 &= \boldsymbol{\beta} \times \hat{\mathbf{b}}_2 \\ {}^{\mathcal{A}}\dot{\hat{\mathbf{b}}}_3 &= \boldsymbol{\gamma} \times \hat{\mathbf{b}}_3 \end{aligned} \tag{6.6}$$

The cross products ensure that the $\hat{\mathbf{b}}_i$ are perpendicular to their derivatives ($\hat{\mathbf{b}}_1 \perp {}^{\mathcal{A}}\dot{\hat{\mathbf{b}}}_1$ and so on) and the magnitudes of $\boldsymbol{\alpha}$, $\boldsymbol{\beta}$, and $\boldsymbol{\gamma}$ give to ${}^{\mathcal{A}}\dot{\hat{\mathbf{b}}}_1$, ${}^{\mathcal{A}}\dot{\hat{\mathbf{b}}}_2$, and ${}^{\mathcal{A}}\dot{\hat{\mathbf{b}}}_3$ their correct magnitudes.

In terms of their components in $\mathcal{B}$, we can write $\boldsymbol{\alpha}$, $\boldsymbol{\beta}$, and $\boldsymbol{\gamma}$ as

$$\begin{aligned} \boldsymbol{\alpha} &= \alpha_1\hat{\mathbf{b}}_1 + \alpha_2\hat{\mathbf{b}}_2 + \alpha_3\hat{\mathbf{b}}_3 \\ \boldsymbol{\beta} &= \beta_1\hat{\mathbf{b}}_1 + \beta_2\hat{\mathbf{b}}_2 + \beta_3\hat{\mathbf{b}}_3 \\ \boldsymbol{\gamma} &= \gamma_1\hat{\mathbf{b}}_1 + \gamma_2\hat{\mathbf{b}}_2 + \gamma_3\hat{\mathbf{b}}_3 \end{aligned} \tag{6.7}$$

Substituting these component expressions into Equations (6.6) gives

$$\begin{aligned} {}^{\mathcal{A}}\dot{\hat{\mathbf{b}}}_1 &= \alpha_3\hat{\mathbf{b}}_2 - \alpha_2\hat{\mathbf{b}}_3 \\ {}^{\mathcal{A}}\dot{\hat{\mathbf{b}}}_2 &= \beta_1\hat{\mathbf{b}}_3 - \beta_3\hat{\mathbf{b}}_1 \\ {}^{\mathcal{A}}\dot{\hat{\mathbf{b}}}_3 &= \gamma_2\hat{\mathbf{b}}_1 - \gamma_1\hat{\mathbf{b}}_2 \end{aligned} \tag{6.8}$$

and we see that α_1, β_2, and γ_3, at this point, remain arbitrary.

Question 6.2 Why do they remain arbitrary?

Here we are seeking to relate the components of the vectors $\boldsymbol{\alpha}$, $\boldsymbol{\beta}$, and $\boldsymbol{\gamma}$ in the hope of finding a way to express the last three terms of Equation (6.3). To this end we note that, for all time t,

$$\hat{\mathbf{b}}_1 \cdot \hat{\mathbf{b}}_2 = 0$$

from which differentiation yields

$${}^{\mathcal{A}}\dot{\hat{\mathbf{b}}}_1 \cdot \hat{\mathbf{b}}_2 + {}^{\mathcal{A}}\dot{\hat{\mathbf{b}}}_2 \cdot \hat{\mathbf{b}}_1 = 0 \tag{6.9}$$

Substitution of the first two equations of (6.6) into (6.9) yields

$$(\boldsymbol{\alpha} \times \hat{\mathbf{b}}_1) \cdot \hat{\mathbf{b}}_2 + (\boldsymbol{\beta} \times \hat{\mathbf{b}}_2) \cdot \hat{\mathbf{b}}_1 = 0 \tag{6.10}$$

Interchanging the dot and cross in each term (which leaves the scalar triple product unchanged) results in

$$\boldsymbol{\alpha} \cdot \hat{\mathbf{b}}_3 - \boldsymbol{\beta} \cdot \hat{\mathbf{b}}_3 = 0 \tag{6.11}$$

so that

$$\alpha_3 = \beta_3 \tag{6.12}$$

Similarly from

$$\hat{\mathbf{b}}_2 \cdot \hat{\mathbf{b}}_3 = 0 \qquad \text{and} \qquad \hat{\mathbf{b}}_3 \cdot \hat{\mathbf{b}}_1 = 0 \tag{6.13}$$

we respectively obtain (as the student should verify)

$$\beta_1 = \gamma_1 \qquad \text{and} \qquad \gamma_2 = \alpha_2 \tag{6.14}$$

The only components not involved in Equations (6.12) and (6.14) are α_1, β_2, and γ_3, which were arbitrary. If we now select them as follows,

$$\begin{aligned} \alpha_1 &= \beta_1 = \gamma_1 \\ \beta_2 &= \gamma_2 = \alpha_2 \\ \gamma_3 &= \alpha_3 = \beta_3 \end{aligned} \tag{6.15}$$

then all three vectors are *identical*, and we call the resulting common vector $\boldsymbol{\omega}_{\mathcal{B}/\mathcal{A}}$:

$$\boldsymbol{\alpha} = \boldsymbol{\beta} = \boldsymbol{\gamma} = \boldsymbol{\omega}_{\mathcal{B}/\mathcal{A}} \tag{6.16}$$

If we now dot the three equations (6.8) respectively with $\hat{\mathbf{b}}_2$, $\hat{\mathbf{b}}_3$, and $\hat{\mathbf{b}}_1$, we get the three components of $\boldsymbol{\omega}_{\mathcal{B}/\mathcal{A}}$:

$$\begin{aligned} {}^{\mathcal{A}}\dot{\hat{\mathbf{b}}}_1 \cdot \hat{\mathbf{b}}_2 &= \alpha_3 = \omega_{\mathcal{B}/\mathcal{A}_3} \\ {}^{\mathcal{A}}\dot{\hat{\mathbf{b}}}_2 \cdot \hat{\mathbf{b}}_3 &= \beta_1 = \omega_{\mathcal{B}/\mathcal{A}_1} \\ {}^{\mathcal{A}}\dot{\hat{\mathbf{b}}}_3 \cdot \hat{\mathbf{b}}_1 &= \gamma_2 = \omega_{\mathcal{B}/\mathcal{A}_2} \end{aligned} \tag{6.17}$$

Thus the vector $\boldsymbol{\omega}_{\mathcal{B}/\mathcal{A}}$ may be expressed as*

$$\boldsymbol{\omega}_{\mathcal{B}/\mathcal{A}} = ({}^{\mathcal{A}}\dot{\hat{\mathbf{b}}}_2 \cdot \hat{\mathbf{b}}_3)\hat{\mathbf{b}}_1 + ({}^{\mathcal{A}}\dot{\hat{\mathbf{b}}}_3 \cdot \hat{\mathbf{b}}_1)\hat{\mathbf{b}}_2 + ({}^{\mathcal{A}}\dot{\hat{\mathbf{b}}}_1 \cdot \hat{\mathbf{b}}_2)\hat{\mathbf{b}}_3 \tag{6.18}$$

We call the vector $\boldsymbol{\omega}_{\mathcal{B}/\mathcal{A}}$ defined by Equation (6.18) the **angular velocity of frame $\mathcal{B}$ with respect to frame $\mathcal{A}$**, or more briefly the **angular velocity of $\mathcal{B}$ in $\mathcal{A}$**. It is clear that the angular velocity vector depends intimately on the way frame $\mathcal{B}$ is changing its orientation with respect to $\mathcal{A}$. In the next section we examine some special properties of this vector. We shall see that $\boldsymbol{\omega}_{\mathcal{B}/\mathcal{A}}$ is ***unique***, which means that we lost no generality when we let $\alpha_1 = \beta_2 = \gamma_3$ in our development of angular velocity.

6.3 Properties of Angular Velocity

We now return to Equation (6.3). Substituting from Equations (6.6) for the ${}^{\mathcal{A}}\dot{\hat{\mathbf{b}}}_i$ and using Equation (6.16) to replace $\boldsymbol{\alpha}$, $\boldsymbol{\beta}$, and $\boldsymbol{\gamma}$ by $\boldsymbol{\omega}_{\mathcal{B}/\mathcal{A}}$, we obtain

$$ {}^{\mathcal{A}}\dot{\mathbf{Q}} = {}^{\mathcal{B}}\dot{\mathbf{Q}} + Q_1(\boldsymbol{\omega}_{\mathcal{B}/\mathcal{A}} \times \hat{\mathbf{b}}_1) + Q_2(\boldsymbol{\omega}_{\mathcal{B}/\mathcal{A}} \times \hat{\mathbf{b}}_2) + Q_3(\boldsymbol{\omega}_{\mathcal{B}/\mathcal{A}} \times \hat{\mathbf{b}}_3) \tag{6.19}$$

*This is the definition of angular velocity set forth by the dynamicist T. R. Kane. See his book *Dynamics* (New York: Holt, Rinehart and Winston, 1968), p. 21.

which may be expressed, using Equation (6.1), as

$$^{\mathcal{A}}\dot{\mathbf{Q}} = {}^{\mathcal{B}}\dot{\mathbf{Q}} + \boldsymbol{\omega}_{\mathcal{B}/\mathcal{A}} \times \mathbf{Q} \tag{6.20}$$

Equation (6.20) will turn out to be of vital importance in this chapter and, moreover, to be equally invaluable in our later study of the kinetics of rigid bodies in general motion in Chapter 7. It permits us easily to calculate the derivative of a vector in one frame if it is expressed in another; the only price we have to pay is to add the cross product $\boldsymbol{\omega}_{\mathcal{B}/\mathcal{A}} \times \mathbf{Q}$. Thus the first property of $\boldsymbol{\omega}_{\mathcal{B}/\mathcal{A}}$ is that it allows us to relate (by Equation 6.20) the derivatives of any vector in two different frames.

There remains the nagging question of whether there might be more than one vector satisfying Equation (6.20); remember that we arbitrarily selected the components α_1, β_2, and γ_3 in the preceding section in order to make $\boldsymbol{\alpha} = \boldsymbol{\beta} = \boldsymbol{\gamma} = \boldsymbol{\omega}_{\mathcal{B}/\mathcal{A}}$. We now proceed to show that the angular velocity vector is unique. We do so by postulating that *two* vectors $\boldsymbol{\omega}_{\mathcal{B}/\mathcal{A}_1}$ and $\boldsymbol{\omega}_{\mathcal{B}/\mathcal{A}_2}$ *both* satisfy Equation (6.20) and then showing that they are necessarily equal.* We have

$$^{\mathcal{A}}\dot{\mathbf{Q}} = {}^{\mathcal{B}}\dot{\mathbf{Q}} + \boldsymbol{\omega}_{\mathcal{B}/\mathcal{A}_1} \times \mathbf{Q} \tag{6.21}$$

$$^{\mathcal{A}}\dot{\mathbf{Q}} = {}^{\mathcal{B}}\dot{\mathbf{Q}} + \boldsymbol{\omega}_{\mathcal{B}/\mathcal{A}_2} \times \mathbf{Q} \tag{6.22}$$

so that, subtracting, we get

$$(\boldsymbol{\omega}_{\mathcal{B}/\mathcal{A}_1} - \boldsymbol{\omega}_{\mathcal{B}/\mathcal{A}_2}) \times \mathbf{Q} = \mathbf{0} \tag{6.23}$$

Finally, since $\mathbf{Q}$ is arbitrary, the parenthesized expression of Equation (6.23) must vanish, and thus the angular velocity vector has been shown to be unique as the two $\boldsymbol{\omega}$'s are one and the same.

Nothing has yet been said about dynamics in this section or in the preceding one; thus it is clear that angular velocity is a far more general vector than one that is simply useful in describing rotational motions of rigid bodies. We have seen that angular velocity is in fact the vector that may be used in relating the derivatives in two frames of any arbitrary vector. Furthermore, even though we have used time as the independent variable, these derivatives may be taken with respect to ***any*** scalar variable. Finally, we note from the defining equation (6.18) for $\boldsymbol{\omega}_{\mathcal{B}/\mathcal{A}}$ that angular velocity is a vector relating two ***frames;*** thus it is meaningless to talk about the angular velocity of a point.

Now let us consider several additional properties of $\boldsymbol{\omega}_{\mathcal{B}/\mathcal{A}}$ that will prove useful in what is to follow. First we note for emphasis that if two frames $\mathcal{A}$ and $\mathcal{B}$ maintain a constant orientation (even if they are each in motion in a third frame $\mathcal{C}$), then $\boldsymbol{\omega}_{\mathcal{B}/\mathcal{A}} \equiv \mathbf{0}$.† The proof is simply to

*Let $\boldsymbol{\omega}_{\mathcal{B}/\mathcal{A}_1}$ be calculated with $\hat{\mathbf{b}}_i$'s as described in Section 6.2, for example, and let $\boldsymbol{\omega}_{\mathcal{B}/\mathcal{A}_2}$ be computed with another triad of unit vectors (say $\hat{\mathbf{d}}_i$'s) fixed in $\mathcal{B}$. The question of uniqueness is whether the resulting $\boldsymbol{\omega}_{\mathcal{B}/\mathcal{A}}$'s are the same.

†*Constant orientation* means that $\mathcal{A}$ and $\mathcal{B}$ move as if they were rigidly attached except for a possible translation of one with respect to the other.

observe that no unit vector $\hat{\mathbf{b}}_i$ fixed in direction in $\mathcal{B}$ can change with time in $\mathcal{A}$ if there is no change in orientation between $\mathcal{A}$ and $\mathcal{B}$. Thus from Equation (6.18), $\boldsymbol{\omega}_{\mathcal{B}/\mathcal{A}} \equiv \mathbf{0}$.

Next we shall prove that the angular velocity of $\mathcal{B}$ in $\mathcal{A}$ is the negative of the angular velocity of $\mathcal{A}$ in $\mathcal{B}$. If we add Equation (6.20) and

$$^{\mathcal{B}}\dot{\mathbf{Q}} = {}^{\mathcal{A}}\dot{\mathbf{Q}} + \boldsymbol{\omega}_{\mathcal{A}/\mathcal{B}} \times \mathbf{Q} \tag{6.24}$$

we obtain

$$(\boldsymbol{\omega}_{\mathcal{B}/\mathcal{A}} + \boldsymbol{\omega}_{\mathcal{A}/\mathcal{B}}) \times \mathbf{Q} = \mathbf{0} \tag{6.25}$$

Again, since $\mathbf{Q}$ is arbitrary, we have the expected result:

$$\boldsymbol{\omega}_{\mathcal{B}/\mathcal{A}} = -\boldsymbol{\omega}_{\mathcal{A}/\mathcal{B}} \tag{6.26}$$

We now prove the **addition theorem,** which states that

$$\boldsymbol{\omega}_{\mathcal{C}/\mathcal{A}} = \boldsymbol{\omega}_{\mathcal{C}/\mathcal{B}} + \boldsymbol{\omega}_{\mathcal{B}/\mathcal{A}} \tag{6.27}$$

For the proof, we know from the first property of $\boldsymbol{\omega}_{\mathcal{B}/\mathcal{A}}$ that

$$^{\mathcal{A}}\dot{\mathbf{Q}} = {}^{\mathcal{C}}\dot{\mathbf{Q}} + \boldsymbol{\omega}_{\mathcal{C}/\mathcal{A}} \times \mathbf{Q} \tag{6.28}$$

$$^{\mathcal{B}}\dot{\mathbf{Q}} = {}^{\mathcal{C}}\dot{\mathbf{Q}} + \boldsymbol{\omega}_{\mathcal{C}/\mathcal{B}} \times \mathbf{Q} \tag{6.29}$$

$$^{\mathcal{A}}\dot{\mathbf{Q}} = {}^{\mathcal{B}}\dot{\mathbf{Q}} + \boldsymbol{\omega}_{\mathcal{B}/\mathcal{A}} \times \mathbf{Q} \tag{6.30}$$

Adding Equations (6.29) and (6.30) yields

$$^{\mathcal{A}}\dot{\mathbf{Q}} = {}^{\mathcal{C}}\dot{\mathbf{Q}} + (\boldsymbol{\omega}_{\mathcal{C}/\mathcal{B}} + \boldsymbol{\omega}_{\mathcal{B}/\mathcal{A}}) \times \mathbf{Q} \tag{6.31}$$

and subtracting Equation (6.31) from (6.28) gives

$$(\boldsymbol{\omega}_{\mathcal{C}/\mathcal{A}} - \boldsymbol{\omega}_{\mathcal{C}/\mathcal{B}} - \boldsymbol{\omega}_{\mathcal{B}/\mathcal{A}}) \times \mathbf{Q} = \mathbf{0} \tag{6.32}$$

Therefore, again since $\mathbf{Q}$ is arbitrary,

$$\boldsymbol{\omega}_{\mathcal{C}/\mathcal{A}} = \boldsymbol{\omega}_{\mathcal{C}/\mathcal{B}} + \boldsymbol{\omega}_{\mathcal{B}/\mathcal{A}} \tag{6.33}$$

and the theorem is proved. It may seem intuitively obvious to the reader that Equation (6.33) is true, but later in this section we show that such a relationship *does not exist* for angular acceleration!

The addition theorem is an extremely powerful result. With it we are able to build up the angular velocity, one pair of frames at a time, of a body turning in complicated ways relative to a reference frame. This theorem makes it possible for us to avoid using the definition (6.18), which has served us well but in practice is normally supplanted by the properties described in this section.

We note for emphasis that if $\mathcal{A}$ and $\mathcal{B}$ both move in $\mathcal{C}$ and maintain constant orientation with respect to each other, then $\boldsymbol{\omega}_{\mathcal{B}/\mathcal{A}} = \mathbf{0}$. Thus, by the addition theorem,

$$\boldsymbol{\omega}_{\mathcal{B}/\mathcal{C}} = \boldsymbol{\omega}_{\mathcal{B}/\mathcal{A}} + \boldsymbol{\omega}_{\mathcal{A}/\mathcal{C}} = \boldsymbol{\omega}_{\mathcal{A}/\mathcal{C}}$$

so that, as expected, the angular velocities in $\mathcal{C}$ of two frames $\mathcal{A}$ and $\mathcal{B}$ maintaining constant orientation with each other are identical.

Conversely, if two frames $\mathcal{A}$ and $\mathcal{B}$ have equal angular velocities in $\mathcal{C}$, we may show that their relative orientation is constant. Using the addition theorem, $\boldsymbol{\omega}_{\mathcal{B}/\mathcal{A}} = \mathbf{0}$; thus from Equations (6.6) and (6.16)

$$^{\mathcal{A}}\dot{\hat{\mathbf{b}}}_1 = \boldsymbol{\omega}_{\mathcal{B}/\mathcal{A}} \times \hat{\mathbf{b}}_1 = \mathbf{0}$$

$$^{\mathcal{A}}\dot{\hat{\mathbf{b}}}_2 = \boldsymbol{\omega}_{\mathcal{B}/\mathcal{A}} \times \hat{\mathbf{b}}_2 = \mathbf{0}$$

$$^{\mathcal{A}}\dot{\hat{\mathbf{b}}}_3 = \boldsymbol{\omega}_{\mathcal{B}/\mathcal{A}} \times \hat{\mathbf{b}}_3 = \mathbf{0}$$

Therefore the $\hat{\mathbf{b}}_i$'s are constant in $\mathcal{A}$, so the orientation of $\mathcal{B}$ in $\mathcal{A}$ is constant.

Thus, for two frames $\mathcal{A}$ and $\mathcal{B}$, the descriptions "constant orientation" and "$\boldsymbol{\omega}_{\mathcal{B}/\mathcal{A}} \equiv \mathbf{0}$" are completely equivalent. Note also that the addition theorem can be extended to *any number of frames* by repetition of the following procedure, two frames at a time:

$$\boldsymbol{\omega}_{\mathcal{A}/\mathcal{D}} = \boldsymbol{\omega}_{\mathcal{A}/\mathcal{B}} + \boldsymbol{\omega}_{\mathcal{B}/\mathcal{D}} = \boldsymbol{\omega}_{\mathcal{A}/\mathcal{B}} + \boldsymbol{\omega}_{\mathcal{B}/\mathcal{C}} + \boldsymbol{\omega}_{\mathcal{C}/\mathcal{D}}$$

Next we show that when there exists a unit vector $\hat{\mathbf{k}}$ whose time derivative in each of two frames $\mathcal{A}$ and $\mathcal{B}$ vanishes (that is, $^{\mathcal{A}}\dot{\hat{\mathbf{k}}} = {}^{\mathcal{B}}\dot{\hat{\mathbf{k}}} = \mathbf{0}$), then

$$\boldsymbol{\omega}_{\mathcal{B}/\mathcal{A}} = \dot{\theta}\hat{\mathbf{k}} \tag{6.34}$$

in which θ is the angle between a pair of directed line segments $\ell_{\mathcal{A}}$ and $\ell_{\mathcal{B}}$ fixed respectively in $\mathcal{A}$ and $\mathcal{B}$, each perpendicular to $\hat{\mathbf{k}}$. The angle is measured in a reference plane containing projections of the two lines intersecting at point P as shown in Figure 6.2. The sign of the angle θ is given by the right-hand rule: If the right thumb is placed in the positive $\hat{\mathbf{k}}$ direction at P, then the direction of positive θ is that of the right hand's fingers when they curl from $\ell_{\mathcal{A}}$ into $\ell_{\mathcal{B}}$ as shown.

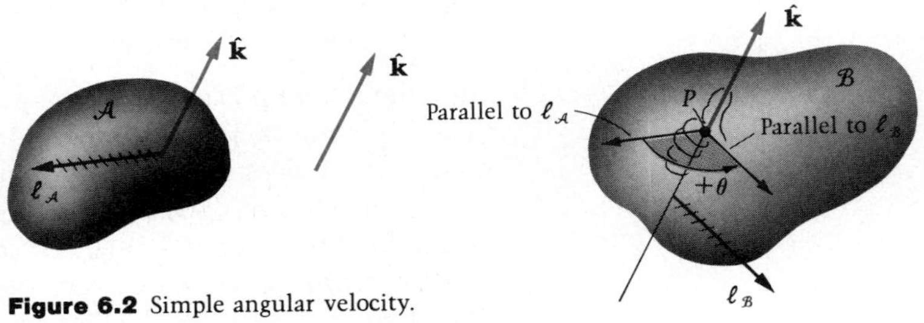

Figure 6.2 Simple angular velocity.

The type of rotational motion given by Equation (6.34) is called **simple angular velocity.** One case in which Equation (6.34) holds is that of plane motion; note, however, that there are more general cases of simple angular velocity in which the body $\mathcal{B}$ may also have a translational motion in $\mathcal{A}$ parallel to $\hat{\mathbf{k}}$ that would prevent the plane motion designation.

To prove Equation (6.34), we make use of Figure 6.3. (The reference plane is the plane of the paper, and the unit vector that is constant in $\mathcal{A}$ and in $\mathcal{B}$ is $\hat{\mathbf{k}}$, perpendicular to the paper.) From this figure we may write

$$\begin{aligned} \hat{\mathbf{b}}_1 &= \hat{\mathbf{a}}_1 \cos\theta + \hat{\mathbf{a}}_2 \sin\theta \\ \hat{\mathbf{b}}_2 &= -\hat{\mathbf{a}}_1 \sin\theta + \hat{\mathbf{a}}_2 \cos\theta \\ \hat{\mathbf{b}}_3 &= \hat{\mathbf{a}}_3 = \hat{\mathbf{k}} \end{aligned} \tag{6.35}$$

Therefore, differentiating Equations (6.35), we get

$$\begin{aligned} {}^{\mathcal{A}}\dot{\hat{\mathbf{b}}}_1 &= (-\hat{\mathbf{a}}_1 \sin\theta + \hat{\mathbf{a}}_2 \cos\theta)\dot{\theta} = \dot{\theta}\hat{\mathbf{b}}_2 \\ {}^{\mathcal{A}}\dot{\hat{\mathbf{b}}}_2 &= (-\hat{\mathbf{a}}_1 \cos\theta - \hat{\mathbf{a}}_2 \sin\theta)\dot{\theta} = -\dot{\theta}\hat{\mathbf{b}}_1 \\ {}^{\mathcal{A}}\dot{\hat{\mathbf{b}}}_3 &= \mathbf{0} \end{aligned} \tag{6.36}$$

and Equation (6.18) yields, upon direct substitution of Equation (6.36),

$$\boldsymbol{\omega}_{\mathcal{B}/\mathcal{A}} = 0\hat{\mathbf{b}}_1 + 0\hat{\mathbf{b}}_2 + \dot{\theta}\hat{\mathbf{b}}_3 = \dot{\theta}\hat{\mathbf{a}}_3 = \dot{\theta}\hat{\mathbf{k}} \tag{6.37}$$

which is the desired result.

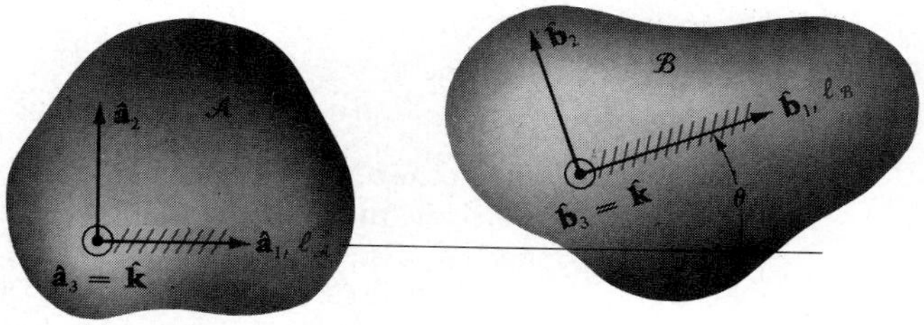

Figure 6.3 Unit vectors drawn in the reference plane for simple angular velocity.

One interesting final property of $\boldsymbol{\omega}_{\mathcal{B}/\mathcal{A}}$ is that its derivative is the same whether computed in $\mathcal{A}$ or in $\mathcal{B}$. Using Equation (6.20) and letting $\mathbf{Q}$ be $\boldsymbol{\omega}_{\mathcal{B}/\mathcal{A}}$ itself, we get

$${}^{\mathcal{A}}\dot{\boldsymbol{\omega}}_{\mathcal{B}/\mathcal{A}} = {}^{\mathcal{B}}\dot{\boldsymbol{\omega}}_{\mathcal{B}/\mathcal{A}} + \boldsymbol{\omega}_{\mathcal{B}/\mathcal{A}} \times \boldsymbol{\omega}_{\mathcal{B}/\mathcal{A}} = {}^{\mathcal{B}}\dot{\boldsymbol{\omega}}_{\mathcal{B}/\mathcal{A}}$$

This result is not true for any other nonvanishing vector, unless it happens to be parallel to $\boldsymbol{\omega}_{\mathcal{B}/\mathcal{A}}$.

The properties of $\boldsymbol{\omega}_{\mathcal{B}/\mathcal{A}}$ that we have examined are summarized here:

1. It is a unique vector that satisfies

$${}^{\mathcal{A}}\dot{\mathbf{Q}} = {}^{\mathcal{B}}\dot{\mathbf{Q}} + \boldsymbol{\omega}_{\mathcal{B}/\mathcal{A}} \times \mathbf{Q}$$

2. $\boldsymbol{\omega}_{\mathcal{B}/\mathcal{A}} \equiv \mathbf{0}$ is synonymous with "the orientations of $\mathcal{B}$ and $\mathcal{A}$ do not change." And if $\mathcal{A}$ and $\mathcal{B}$ maintain constant orientation, their angular velocities in any third frame are equal.
3. $\boldsymbol{\omega}_{\mathcal{B}/\mathcal{A}} = -\boldsymbol{\omega}_{\mathcal{A}/\mathcal{B}}$.
4. $\boldsymbol{\omega}_{\mathcal{D}/\mathcal{A}} = \boldsymbol{\omega}_{\mathcal{D}/\mathcal{B}} + \boldsymbol{\omega}_{\mathcal{B}/\mathcal{A}} = (\boldsymbol{\omega}_{\mathcal{D}/\mathcal{C}} + \boldsymbol{\omega}_{\mathcal{C}/\mathcal{B}}) + \boldsymbol{\omega}_{\mathcal{B}/\mathcal{A}}$, which can be further extended to any number of frames.
5. If $\hat{\mathbf{k}}$ is constant in both $\mathcal{A}$ and $\mathcal{B}$, then

$$\boldsymbol{\omega}_{\mathcal{B}/\mathcal{A}} = \dot{\theta}\hat{\mathbf{k}}$$

where θ was defined earlier.
6. ${}^{\mathcal{A}}\dot{\boldsymbol{\omega}}_{\mathcal{B}/\mathcal{A}} = {}^{\mathcal{B}}\dot{\boldsymbol{\omega}}_{\mathcal{B}/\mathcal{A}}$.

We now present an extended practical example of the use of the $\boldsymbol{\omega}$ properties. In this example three separate bodies are in motion in a reference frame, and their angular velocities are related by using the simple angular velocity and addition theorem properties.

EXAMPLE 6.1

The *Hooke's joint,* or universal joint, is a device used to transmit power between two shafts that are not collinear. Figure 1 shows a Hooke's joint in which the shafts $\mathcal{S}_1$ and $\mathcal{S}_2$ are out of alignment by the angle α.

Each shaft is mounted in a bearing fixed to the reference frame $\mathcal{F}$. The shafts, whose axes intersect at point A, are rigidly attached to the yokes $\mathcal{Y}_1$ and $\mathcal{Y}_2$. A rigid cross $\mathcal{C}$ is the connecting body between the yokes. One leg of the cross (indicated by the unit vector $\hat{\mathbf{u}}_1$) turns in bearings fixed in $\mathcal{Y}_1$ at D_1 and E_1, while the other leg (unit vector $\hat{\mathbf{u}}_2$) turns in bearings fixed in $\mathcal{Y}_2$ at D_2 and E_2. The arms of cross $\mathcal{C}$ are identical; they form a right angle with each other, and each is perpendicular to its respective shaft.

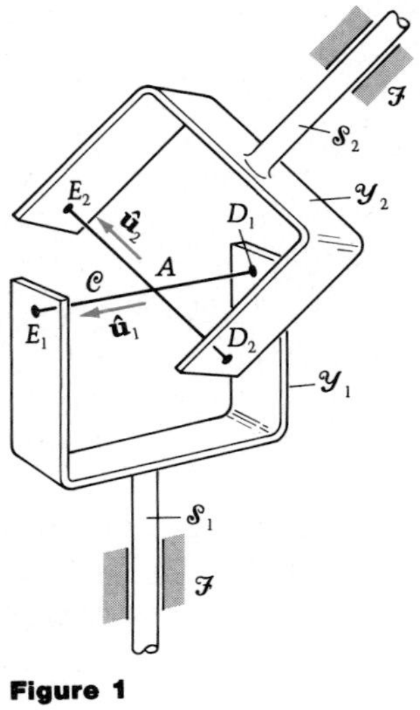

Figure 1

(Continued)

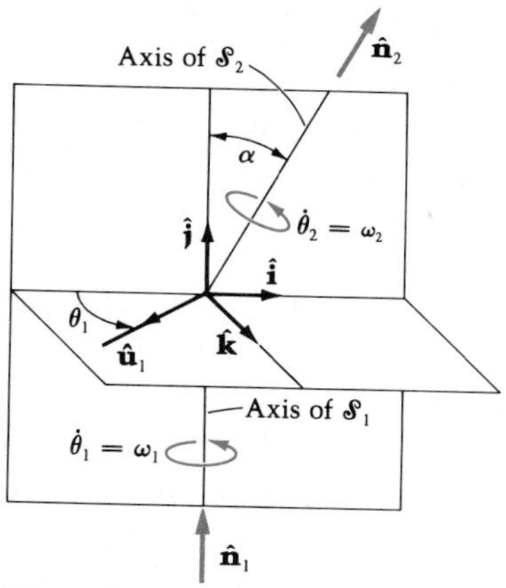

Figure 2

Figure 2 shows that θ_1 measures the angular position of $\mathcal{S}_1$ in $\mathcal{F}$. If $\mathcal{S}_1$ (considered to be the driven shaft) has angular velocity $\boldsymbol{\omega}_{\mathcal{S}_1/\mathcal{F}} = \omega_1 \hat{\mathbf{n}}_1$ and the resulting angular velocity of $\mathcal{S}_2$ is $\boldsymbol{\omega}_{\mathcal{S}_2/\mathcal{F}} = \omega_2 \hat{\mathbf{n}}_2$, find the ratio of ω_2 to ω_1 in terms of θ_1 and α, and plot ω_2/ω_1 versus θ_1 for $\alpha = 0$, 20, 40, 60, and 80°. Letting θ_2 be the rotation angle of $\mathcal{S}_2$, further investigate θ_2 versus θ_1 for the same five α values.

Question 6.3 In Figure 2 note that $\hat{\mathbf{n}}_2$ has no $\hat{\mathbf{k}}$ component. Why does this not represent a loss of generality?

SOLUTION

Using the addition theorem, we may relate the angular velocities of the four rigid bodies $\mathcal{B}_2$ (shaft $\mathcal{S}_2$ plus its yoke $\mathcal{Y}_2$), $\mathcal{C}$, $\mathcal{B}_1$ (shaft $\mathcal{S}_1$ plus its yoke $\mathcal{Y}_1$), and $\mathcal{F}$:

$$\boldsymbol{\omega}_{\mathcal{B}_2/\mathcal{F}} = \boldsymbol{\omega}_{\mathcal{B}_2/\mathcal{C}} + \boldsymbol{\omega}_{\mathcal{C}/\mathcal{B}_1} + \boldsymbol{\omega}_{\mathcal{B}_1/\mathcal{F}} \tag{1}$$

Since $\mathcal{B}_1$ and $\mathcal{B}_2$ both have simple angular velocity in $\mathcal{F}$, we may write

$$\boldsymbol{\omega}_{\mathcal{B}_1/\mathcal{F}} = \omega_1 \hat{\mathbf{n}}_1 \text{ or } \dot{\theta}_1 \hat{\mathbf{n}}_1 \qquad \boldsymbol{\omega}_{\mathcal{B}_2/\mathcal{F}} = \omega_2 \hat{\mathbf{n}}_2 \text{ or } \dot{\theta}_2 \hat{\mathbf{n}}_2 \tag{2}$$

We also know from Figure 1 that the cross $\mathcal{C}$ has simple angular velocity in *each* of $\mathcal{B}_1$ and $\mathcal{B}_2$. For example, the only motion that $\mathcal{C}$ can have with respect to $\mathcal{B}_1$ is a rotation about D_1E_1, the line fixed in *both* bodies. The same is true for the motion of $\mathcal{C}$ in $\mathcal{B}_2$. Thus

$$\boldsymbol{\omega}_{\mathcal{C}/\mathcal{B}_1} = \omega_{\mathcal{C}/\mathcal{B}_1} \hat{\mathbf{u}}_1 \qquad \boldsymbol{\omega}_{\mathcal{C}/\mathcal{B}_2} = -\boldsymbol{\omega}_{\mathcal{B}_2/\mathcal{C}} = -\omega_{\mathcal{B}_2/\mathcal{C}} \hat{\mathbf{u}}_2 \tag{3}$$

in which $\omega_{\mathcal{C}/\mathcal{B}_1}$ and $\omega_{\mathcal{B}_2/\mathcal{C}}$ are the unknown magnitudes of the respective vectors.

Next we must express all of $\hat{\mathbf{n}}_1$, $\hat{\mathbf{n}}_2$, $\hat{\mathbf{u}}_1$, and $\hat{\mathbf{u}}_2$ in terms of a common set of unit vectors. Then we shall be able to obtain three scalar equations from (1) and hence solve for ω_2 in terms of ω_1. From Figure 2, three of the unit vectors are obvious:

$$\begin{aligned} \hat{\mathbf{n}}_1 &= \hat{\mathbf{j}} \\ \hat{\mathbf{n}}_2 &= \sin\alpha\,\hat{\mathbf{i}} + \cos\alpha\,\hat{\mathbf{j}} \\ \hat{\mathbf{u}}_1 &= -\cos\theta_1\,\hat{\mathbf{i}} + \sin\theta_1\,\hat{\mathbf{k}} \end{aligned} \tag{4}$$

To obtain $\hat{\mathbf{u}}_2$, we note that it is perpendicular to both $\hat{\mathbf{n}}_2$ and $\hat{\mathbf{u}}_1$. Crossing $\hat{\mathbf{u}}_1$ into $\hat{\mathbf{n}}_2$ then gives the assigned direction of $\hat{\mathbf{u}}_2$ (note that $\hat{\mathbf{n}}_2 \times \hat{\mathbf{u}}_1$ is opposite!); $\hat{\mathbf{u}}_1 \times \hat{\mathbf{n}}_2$ is not generally a unit vector, however, so to get $\hat{\mathbf{u}}_2$ we divide this vector by its magnitude:

(Continued)

$$\hat{\mathbf{u}}_2 = \frac{\hat{\mathbf{u}}_1 \times \hat{\mathbf{n}}_2}{|\hat{\mathbf{u}}_1 \times \hat{\mathbf{n}}_2|} = \frac{-\cos\alpha\sin\theta_1\hat{\mathbf{i}} + \sin\alpha\sin\theta_1\hat{\mathbf{j}} - \cos\alpha\cos\theta_1\hat{\mathbf{k}}}{\sqrt{\cos^2\alpha + \sin^2\alpha\sin^2\theta_1}} \tag{5}$$

Substituting Equations (4) and (5) for the four unit vectors into Equations (2) and (3), and then substituting the resulting angular velocity expressions into Equation (1), we get a vector equation that has the following three scalar component equations:

$\hat{\mathbf{i}}$ coefficients: $$\omega_2 \sin\alpha = R\omega_{\mathcal{B}_2/\mathcal{C}}(-\cos\alpha\sin\theta_1) + \omega_{\mathcal{C}/\mathcal{B}_1}(-\cos\theta_1) \tag{6}$$

$\hat{\mathbf{j}}$ coefficients: $$\omega_2 \cos\alpha = R\omega_{\mathcal{B}_2/\mathcal{C}}(\sin\alpha\sin\theta_1) + \omega_1 \tag{7}$$

$\hat{\mathbf{k}}$ coefficients: $$0 = R\omega_{\mathcal{B}_2/\mathcal{C}}(-\cos\alpha\cos\theta_1) + \omega_{\mathcal{C}/\mathcal{B}_1}(\sin\theta_1) \tag{8}$$

in which $R = 1/\sqrt{\cos^2\alpha + \sin^2\alpha\sin^2\theta_1}$. Eliminating $\omega_{\mathcal{C}/\mathcal{B}_1}$ between (6) and (8) gives

$$\omega_{\mathcal{B}_2/\mathcal{C}} = \frac{-\omega_2\tan\alpha\sin\theta_1}{R} \tag{9}$$

and substitution of (9) into (7) yields

$$\omega_2 = \left(\frac{\cos\alpha}{\cos^2\alpha + \sin^2\alpha\sin^2\theta_1}\right)\omega_1$$

so that

$$\frac{\omega_2}{\omega_1} = \frac{\cos\alpha}{1 - \sin^2\alpha\cos^2\theta_1} \tag{10}$$

A plot of this expression (Figure 3) shows the manner in which ω_2 changes over a quarter-turn of $\mathcal{S}_1$ in space. Note that since $\cos\theta_1$ is squared, the curves reflect around the vertical line at $\theta_1 = 90°$ for $90° \le \theta_1 \le 180°$; between 180° and 360° we again have a mirror image, this time of the curves between 0° and 180°. Note that for large misalignment angles α, shaft $\mathcal{S}_2$ must turn very rapidly at and near $\theta_1 = 0$; in fact, when $\alpha = 90°$ the bodies reach a configuration in which they cannot turn at all. This is called ***gimbal lock***. Note further that a misalignment of as much as 10° results in an output speed variation (ω_2) over a revolution of only about 3 percent.

Next we examine the angles of rotation θ_1 (of $\mathcal{S}_1$) and θ_2 (of $\mathcal{S}_2$). Since $\mathcal{S}_1$ and $\mathcal{S}_2$ both have simple angular velocity in $\mathcal{I}$, we have $\omega_1 = \dot{\theta}_1$ and $\omega_2 = \dot{\theta}_2$, so that (from Equation 10)

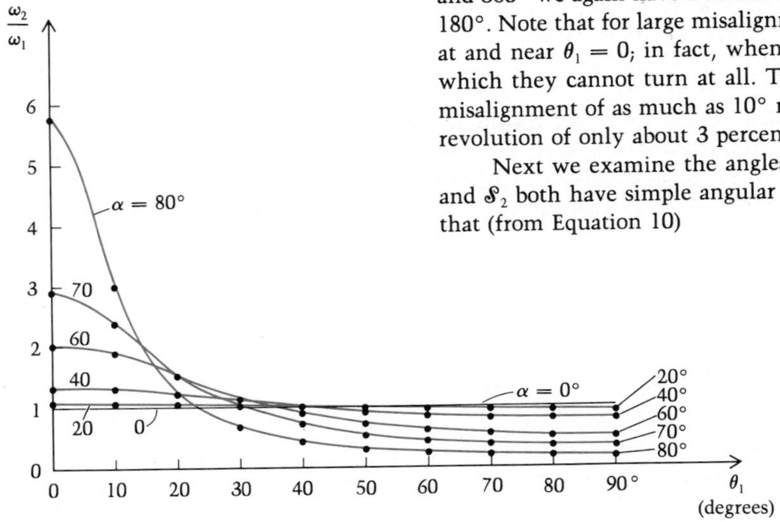

Figure 3

$$\dot{\theta}_2 = \frac{\dot{\theta}_1 \cos \alpha}{1 - \sin^2 \alpha \cos^2 \theta_1} \tag{11}$$

Integrating (11) gives

$$\theta_2 + \text{constant} = \int \frac{\cos \alpha \, d\theta_1}{1 - \sin^2 \alpha \cos^2 \theta_1}$$

$$= \tan^{-1}\left(\frac{\tan \theta_1}{\cos \alpha}\right) \tag{12}$$

The constant of integration is zero if we define $\theta_2 = 0$ when $\theta_1 = 0$. Then θ_2 may be plotted as a function of θ_1 for the same representative values of α (see Figure 4). If $\alpha = 0$, then $\theta_1 \equiv \theta_2$ and the curve is a 45° line. If there is misalignment ($\alpha \neq 0$), then note from the curves that for $0 \leq \theta_1 \leq 90°$, shaft $\mathcal{S}_2$ is always turned ***more*** than $\mathcal{S}_1$. Then $\mathcal{S}_1$ catches up at $\theta = 90°$, and from $\theta_1 = 90°$ to 180° the angle θ_2 of $\mathcal{S}_2$ lags ***behind*** θ_1.

Question 6.4 Explain why this is so by using Equation (12).

From 180° to 360°, the cycle repeats and everything returns to the same starting position ($\theta_1 = \theta_2 = 360°$) at the same time.

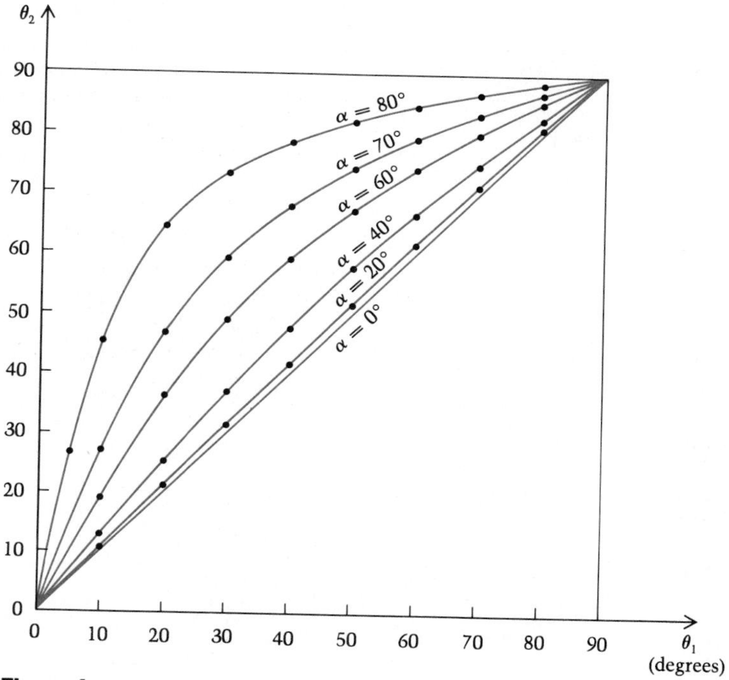

Figure 4

PROBLEMS / Section 6.3

6.1 Verify, in Example 1.1, that ${}^{\mathcal{F}}\dot{\mathbf{A}}$ is indeed $\boldsymbol{\omega}_{\mathcal{B}/\mathcal{F}} \times \mathbf{A}$, where $\boldsymbol{\omega}_{\mathcal{B}/\mathcal{F}} = \dot{\theta}(-\hat{\mathbf{k}})$. (See Figure P6.1.)

6.2 A vector $\mathbf{v}$ is given as a function of time t by $\mathbf{v} = t^3\hat{\mathbf{i}} + t^2\hat{\mathbf{j}} + t\hat{\mathbf{k}}$ m/s, where $(\hat{\mathbf{i}}, \hat{\mathbf{j}}, \hat{\mathbf{k}})$ are unit vectors whose directions are fixed in a frame $\mathcal{H}$. The angular velocity of $\mathcal{H}$ in frame $\mathcal{R}$ is $\boldsymbol{\omega}_{\mathcal{H}/\mathcal{R}} = t\hat{\mathbf{i}} + t^2\hat{\mathbf{j}} + t^3\hat{\mathbf{k}}$ rad/s. Find the derivative of $\mathbf{v}$ in frame $\mathcal{R}$, that is, ${}^{\mathcal{R}}\dot{\mathbf{v}}$: (a) as a function of t; (b) at $t = 1$ s; (c) at $t = 2$ s.

6.3 Show that the output angle θ_2 of the Hooke's joint in Example 6.1 can be alternatively obtained by dotting $\hat{\mathbf{k}}$ with $\hat{\mathbf{u}}_2$.

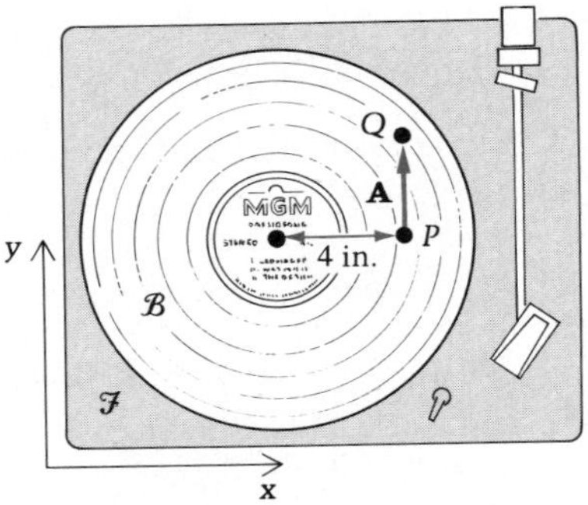

Figure P6.1

6.4 The outer cone $\mathcal{B}$ in Figure P6.4 has the following prescribed motion with respect to the fixed inner cone $\mathcal{C}$:

1. The vertices remain together.
2. The line AB (a base radius fixed in $\mathcal{B}$) always lies in some vertical plane parallel to XY.
3. Cone $\mathcal{B}$ slides on $\mathcal{C}$; that is, there is always a line of contact between O and a point of the base circle of $\mathcal{C}$.
4. Point A of $\mathcal{B}$ revolves around the $\mathbf{x}$ axis in a vertical circle at constant speed $H\dot{\theta}$.

Use the addition theorem to show that the angular velocity of $\mathcal{B}$ in $\mathcal{C}$ is given by

$$\boldsymbol{\omega}_{\mathcal{B}/\mathcal{C}} = \frac{\dot{\theta}(\tan^2\gamma\cos^2\theta\,\hat{\mathbf{i}} - \tan\gamma\cos\theta\,\hat{\mathbf{j}} - \sin\theta\tan\gamma\,\hat{\mathbf{k}})}{1 + \tan^2\gamma\cos^2\theta}$$

6.5 In the preceding problem find $\boldsymbol{\omega}_{\mathcal{B}/\mathcal{C}}$ if the projection of AB into the YZ plane through A is always aligned with the radius. (See Figure P6.5.)

6.6 Find $\boldsymbol{\omega}_{\mathcal{B}/\mathcal{C}}$ in the preceding problem if the outer cone *rolls* on the fixed inner cone. (See Figure P6.6.)

A popular method of stabilizing shipboard antennas is by means of pendulous masses together with the gyroscopic effect of spinning flywheels. In Figure P6.7 the ship (frame $\mathcal{S}$) pitches (about $\mathbf{x}$), rolls (about y), and yaws (about z) in the sea (frame $\mathcal{E}$). Frame $\mathcal{F}$, just above a Hooke's joint, is to form a stable platform on which the

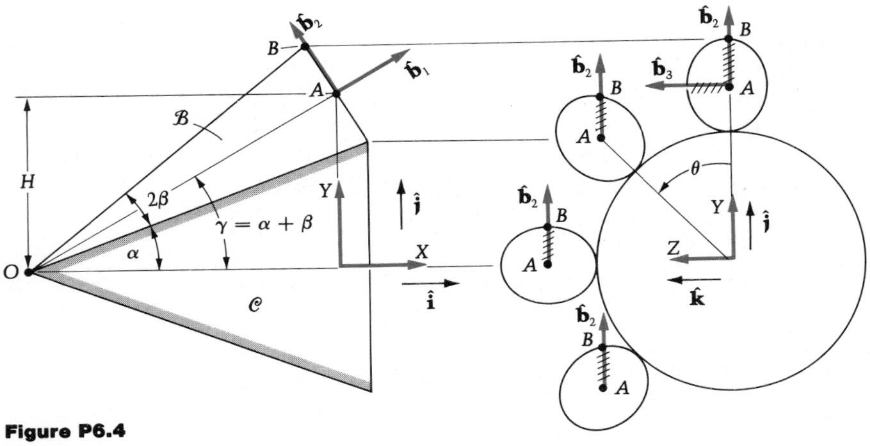

Figure P6.4

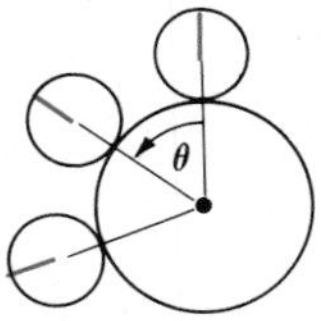

Figure P6.5

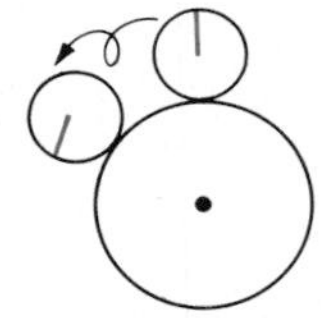

Figure P6.6

antenna can then be easily positioned in azimuth (angle A) and elevation (angle E). Precessing flywheels (not shown) are suspended from $\mathcal{F}$; (x, y, z) are attached to $\mathcal{S}$. The mass center of $\mathcal{F}$ plus the antenna and flywheels is below point Q. The gyroscopic effect of the flywheels causes $\mathcal{F}$ to remain level by the "depitching" rotation P above the "derolling" rotation R. The following three problems are based on this system.

The INMARSAT communications satellite system requires that shipboard antenna systems remain operational up to the following oscillatory limits:

Pitch:	$\pm 10°$ in 6 sec
Roll:	$\pm 30°$ in 8 sec
Yaw:	$\pm 8°$ in 50 sec

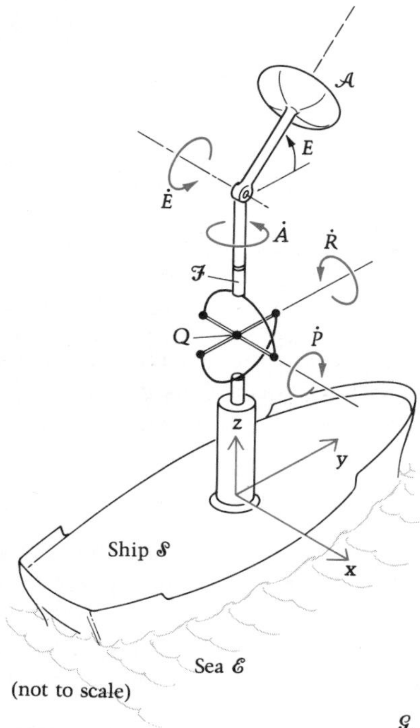

Figure P6.7

6.7 Assume sine waves for each of these three motions and assume yaw over roll over pitch—that is, the assumed order of ship rotations is (1) pitch, from frame $\mathcal{E}$ to an intermediate frame $\mathcal{F}_1$; (2) roll, from $\mathcal{F}_1$ to a second intermediate frame $\mathcal{F}_2$; and (3) yaw, from $\mathcal{F}_2$ to frame $\mathcal{S}$. Write the angular velocity of the ship $\mathcal{S}$ in the sea (earth-fixed frame $\mathcal{E}$), expressed in the ship-fixed axes (x, y, z).

6.8 Write the angular velocity of $\mathcal{F}$ in $\mathcal{S}$, expressed in the axes (x, y, z).

6.9 Write the angular velocity of $\mathcal{F}$ in $\mathcal{E}$, using the results of the preceding two problems together with the addition theorem.

6.10 A robot manufactured by the Heath Company has the mechanical arm shown in Figure P6.10 and accompanying photograph. Its shoulder $\mathcal{S}$ extends from the head $\mathcal{H}$, which can itself rotate 350° about z in 30 sec relative to the reference frame $\mathcal{F}$. The arm is able to travel 150° about axis y in 26 sec. (The axes (x, y, z) are fixed in the head $\mathcal{H}$.) The part of the arm to the left of point E is able to extend (and retract) up to 5 in. in 30 sec. The wrist $\mathcal{W}$ has two motions: It is able to pivot up to 180° about y′ in 10 sec and to rotate (about x′) up to 350° in 4 sec. Axes (x′, y′, z′) are fixed in $\mathcal{W}$. Finally, the gripper $\mathcal{G}$ is able to open (and close) 3.5 in. in 3 sec, but is assumed here to be a closed circle with a 2.5-inch diameter. Approximate dimensions are shown in the figure.

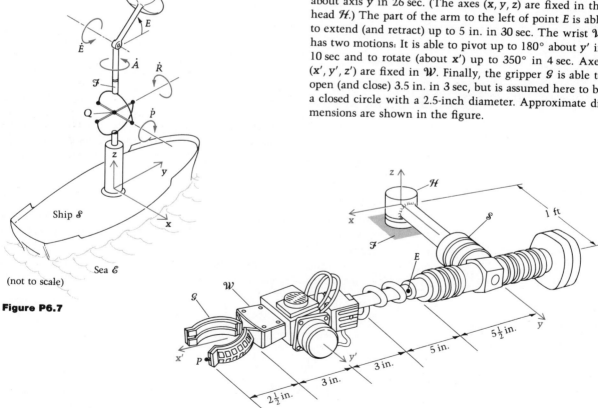

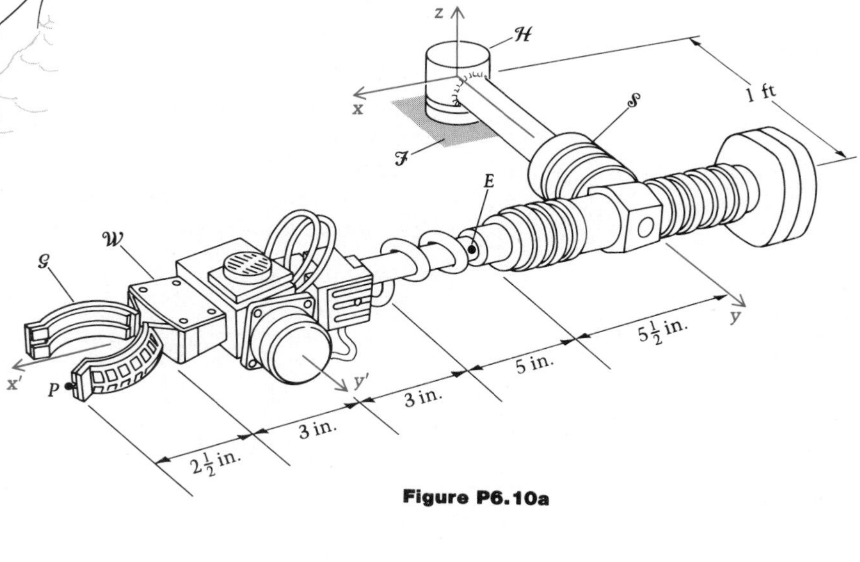

Figure P6.10a

Figure P6.10b *(Courtesy of the Heath Company.)*

For this problem, assume that all the robots motions (except the gripper opening) are occurring simultaneously about positive axes with their respective average speeds. Find the angular velocity of the gripper $\mathcal{G}$, relative to $\mathcal{F}$ and expressed in terms of unit vectors in $\mathcal{H}$, at an instant when:

1. The shoulder rotation angle θ is $-60°$:

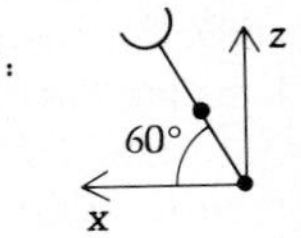

2. The arm is extended 3 in.
3. The wrist is pivoted 30°:

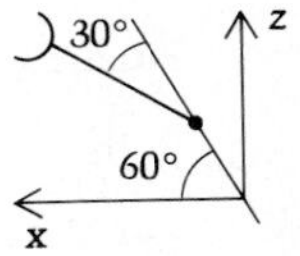

4. The wrist is rotated 90°:

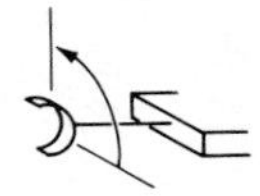

6.4 The Angular Acceleration Vector

In applying what we have learned about angular velocity to the kinematics of rigid bodies, we also need to understand its derivative. The **angular acceleration** of frame $\mathcal{B}$ relative to frame $\mathcal{A}$ is defined to be

$$\boldsymbol{\alpha}_{\mathcal{B}/\mathcal{A}} = {}^{\mathcal{A}}\dot{\boldsymbol{\omega}}_{\mathcal{B}/\mathcal{A}} \tag{6.38}$$

(Note from the last property of $\boldsymbol{\omega}_{\mathcal{B}/\mathcal{A}}$ in the previous section that the derivative could equally well be taken in $\mathcal{B}$, but generally not in any other frame.)

It is important to note that the addition theorem (Equation 6.27) *does not hold* for angular acceleration. Watch:

$$\begin{aligned}
\boldsymbol{\alpha}_{\mathcal{C}/\mathcal{A}} &= {}^{\mathcal{A}}\dot{\boldsymbol{\omega}}_{\mathcal{C}/\mathcal{A}} \\
&= {}^{\mathcal{A}}\dot{\overline{\boldsymbol{\omega}_{\mathcal{C}/\mathcal{B}} + \boldsymbol{\omega}_{\mathcal{B}/\mathcal{A}}}} \\
&= {}^{\mathcal{A}}\dot{\boldsymbol{\omega}}_{\mathcal{C}/\mathcal{B}} + {}^{\mathcal{A}}\dot{\boldsymbol{\omega}}_{\mathcal{B}/\mathcal{A}} \\
&= ({}^{\mathcal{B}}\dot{\boldsymbol{\omega}}_{\mathcal{C}/\mathcal{B}} + \boldsymbol{\omega}_{\mathcal{B}/\mathcal{A}} \times \boldsymbol{\omega}_{\mathcal{C}/\mathcal{B}}) + \boldsymbol{\alpha}_{\mathcal{B}/\mathcal{A}}
\end{aligned}$$

$$\boldsymbol{\alpha}_{\mathcal{C}/\mathcal{A}} = \boldsymbol{\alpha}_{\mathcal{C}/\mathcal{B}} + \boldsymbol{\alpha}_{\mathcal{B}/\mathcal{A}} + \boldsymbol{\omega}_{\mathcal{B}/\mathcal{A}} \times \boldsymbol{\omega}_{\mathcal{C}/\mathcal{B}} \tag{6.39}$$

We see that there is an extra term (the cross product of two angular velocity vectors) that prevents the simple theorem we have derived for $\boldsymbol{\omega}$'s from working for $\boldsymbol{\alpha}$'s. This term is sometimes called a *gyroscopic term;* note that it vanishes for plane motion, in which case we do have an addition theorem for the $\boldsymbol{\alpha}$'s (which are then of the form $\ddot{\theta}\hat{\mathbf{k}}$).

EXAMPLE 6.2

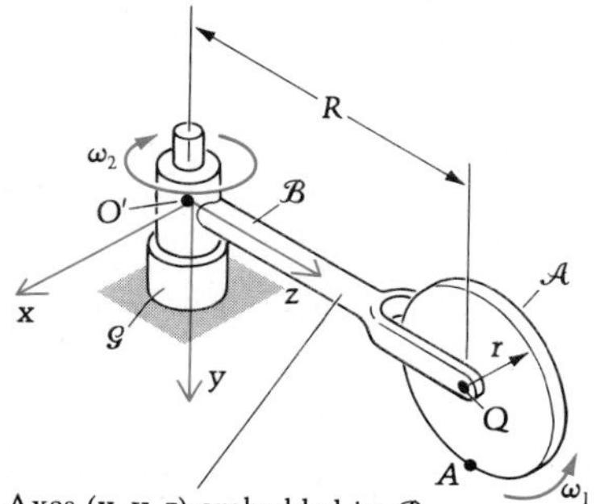

Axes (x, y, z) embedded in $\mathcal{B}$

Body $\mathcal{B}$ in the diagram rotates in frame $\mathcal{G}$ about the vertical at constant angular speed ω_2; in $\mathcal{B}$, disk $\mathcal{A}$ rotates about its pinned axis at constant angular speed ω_1. (The directions of rotation are as shown.) Determine the angular acceleration of $\mathcal{A}$ in $\mathcal{G}$.

SOLUTION

The coordinate axes shown are fixed in $\mathcal{B}$. By the addition theorem,*

$$\boldsymbol{\omega}_{\mathcal{A}/\mathcal{G}} = \boldsymbol{\omega}_{\mathcal{A}/\mathcal{B}} + \boldsymbol{\omega}_{\mathcal{B}/\mathcal{G}}$$
$$= \omega_1\hat{\mathbf{i}} + \omega_2\hat{\mathbf{j}}$$

Next, using Equation (6.20) and noting that $\boldsymbol{\omega}_{\mathcal{A}/\mathcal{G}}$ is expressed in terms of axes embedded in $\mathcal{B}$, we move the derivative and obtain†

$$\boldsymbol{\alpha}_{\mathcal{A}/\mathcal{G}} = {}^{\mathcal{G}}\dot{\boldsymbol{\omega}}_{\mathcal{A}/\mathcal{G}}$$
$$= {}^{\mathcal{B}}\dot{\boldsymbol{\omega}}_{\mathcal{A}/\mathcal{G}} + \boldsymbol{\omega}_{\mathcal{B}/\mathcal{G}} \times \boldsymbol{\omega}_{\mathcal{A}/\mathcal{G}}$$
$$= \mathbf{0} + \omega_2\hat{\mathbf{j}} \times (\omega_1\hat{\mathbf{i}} + \omega_2\hat{\mathbf{j}})$$
$$= -\omega_1\omega_2\hat{\mathbf{k}}$$

Note that the same result is obtained by using Equation (6.39) with frames $\mathcal{C}$, $\mathcal{B}$, and $\mathcal{A}$ replaced by $\mathcal{A}$, $\mathcal{B}$, and $\mathcal{G}$, respectively.

*While the defining equation (6.18) is always available for directly computing the angular velocity, it is usually easier to build up the $\boldsymbol{\omega}$ vector by using the addition theorem.

†By "moving the derivative," we mean shifting it *from* a frame in which it is inconvenient to differentiate *to* a frame in which we desire to differentiate.

We note that up to this point in Chapter 6 we have not mentioned velocities or accelerations. As long as the angular velocities from one body to another are all simple, we can do a considerable amount of angular velocity and angular acceleration computation merely by using the definitions and properties of $\boldsymbol{\omega}$ and $\boldsymbol{\alpha}$.

PROBLEMS / Section 6.4

6.11 The antenna $\mathcal{A}$ in Figure P6.11 is oriented with the following three rotations:

1. Azimuth, about y fixed in $\mathcal{G}$, at the rate $\omega_{\mathcal{F}_1/\mathcal{G}} = 3t^2$ rad/sec
2. Elevation, about $\mathbf{z}_1$ fixed in a first intermediate frame $\mathcal{F}_1$, at the constant rate $\omega_{\mathcal{F}_2/\mathcal{F}_1} = 1$ rad/sec
3. A polarization rotation about the antenna axis $\mathbf{x}_2$ (fixed in both the second intermediate frame $\mathcal{F}_2$ and in $\mathcal{A}$) at $\omega_{\mathcal{A}/\mathcal{F}_2} = 4t$ rad/sec

If the structure is in the $\phi = 0$ position at $t = 0$, find (a) $\boldsymbol{\omega}_{\mathcal{A}/\mathcal{G}}$ and (b) $\boldsymbol{\alpha}_{\mathcal{A}/\mathcal{G}}$ at the time $t = \pi/2$ sec. Use unit vectors fixed in direction in $\mathcal{F}_1$.

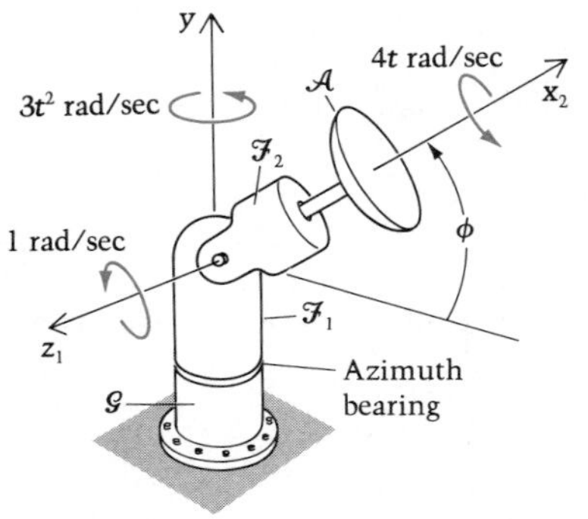

Figure P6.11

6.12 Let the angular velocity and angular acceleration vectors $\boldsymbol{\omega}_{\mathcal{C}/\mathcal{A}}$ and $\boldsymbol{\alpha}_{\mathcal{C}/\mathcal{A}}$ be expressed in terms of their components in a third frame $\mathcal{B}$:

$$\left.\begin{aligned}\boldsymbol{\omega}_{\mathcal{C}/\mathcal{A}} &= \omega_1 \hat{\mathbf{b}}_1 + \omega_2 \hat{\mathbf{b}}_2 + \omega_3 \hat{\mathbf{b}}_3 \\ \boldsymbol{\alpha}_{\mathcal{C}/\mathcal{A}} &= \alpha_1 \hat{\mathbf{b}}_1 + \alpha_2 \hat{\mathbf{b}}_2 + \alpha_3 \hat{\mathbf{b}}_3\end{aligned}\right\} \quad \text{(The } \hat{\mathbf{b}}_i\text{'s are fixed in } \mathcal{B}\text{)}$$

Find the restriction on frame $\mathcal{B}$ for which $\alpha_i = \dot{\omega}_i$ $(i = 1, 2, 3)$.

6.13 In Example 6.1 find the angular acceleration of shaft $\mathcal{S}_2$ in the reference frame $\mathcal{F}$.

6.14 In Problem 6.2 find ${}^{\mathcal{R}}\ddot{\mathbf{v}}$.

6.15 In Problem 6.7, find the angular acceleration of $\mathcal{S}$ in $\mathcal{E}$ at $t = 24$ sec.

6.16 We know that one of the properties of the angular velocity vector is that ${}^{\mathcal{F}}\dot{\boldsymbol{\omega}}_{\mathcal{B}/\mathcal{F}} = {}^{\mathcal{B}}\dot{\boldsymbol{\omega}}_{\mathcal{B}/\mathcal{F}}$. Show that this is *not* a property of the angular acceleration vector $\boldsymbol{\alpha}_{\mathcal{B}/\mathcal{F}}$.

6.17 The components of two angular velocity vectors are shown in the following table as functions of time. The orthogonal unit vectors $(\hat{\mathbf{b}}_1, \hat{\mathbf{b}}_2, \hat{\mathbf{b}}_3)$ are fixed in direction in frame $\mathcal{B}$. Find the angular acceleration of $\mathcal{C}$ in $\mathcal{A}$: (a) as a function of time; (b) at $t = 0$ sec; (c) at $t = 0.5$ sec. (See Figure P6.17.)

	$\hat{\mathbf{b}}_1$	$\hat{\mathbf{b}}_2$	$\hat{\mathbf{b}}_3$
$\boldsymbol{\omega}_{\mathcal{C}/\mathcal{B}}$	$4t^2$	$2t$	6
$\boldsymbol{\omega}_{\mathcal{B}/\mathcal{A}}$	$\sin t$	$\cos t$	$7t$

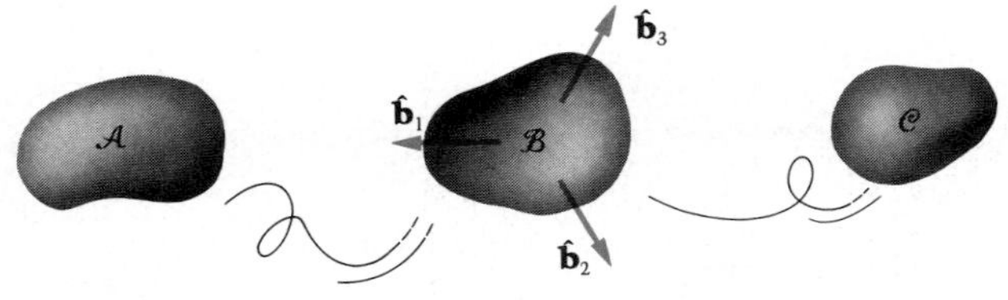

Figure P6.17

6.18 In Problem 6.10 find the angular acceleration of the gripper $\mathcal{G}$ in $\mathcal{F}$ at the same instant.

6.5 Velocity and Acceleration in Moving Frames of Reference

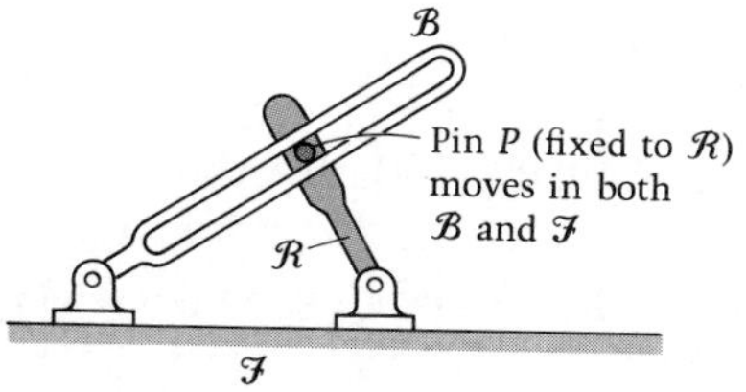

Figure 6.4 Example of a point moving relative to two frames.

In certain practical situations a point is moving relative to *two* frames (or bodies) of interest. For example, a pin P may be sliding in a slot of a body $\mathcal{B}$ that is itself in motion in another frame $\mathcal{F}$ (see Figure 6.4). In problems such as these, we are often interested in the relationship between the velocities (and also the accelerations) of P in the two frames $\mathcal{B}$ and $\mathcal{F}$. We shall arbitrarily choose $\mathcal{F}$ as a reference frame for the moving body $\mathcal{B}$, but we emphasize that both are frames and both are bodies; as long as they are considered rigid, the terms mean the same. Figure 6.5 shows the general picture.

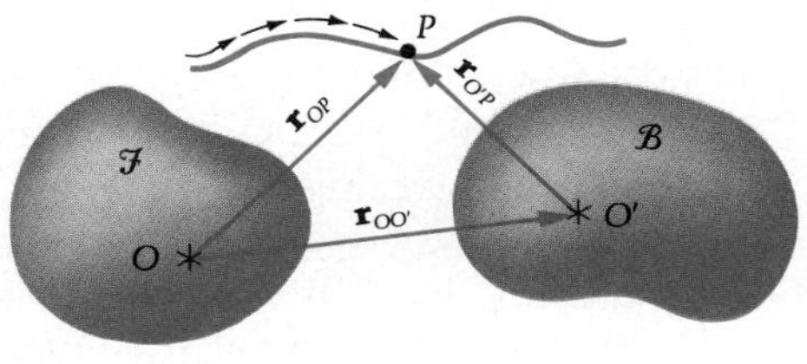

Figure 6.5 Point P moving with respect to frames $\mathcal{B}$ and $\mathcal{F}$.

Letting O and O' be fixed points of $\mathcal{F}$ and $\mathcal{B}$, respectively, we differentiate the connecting relation

$$\mathbf{r}_{OP} = \mathbf{r}_{OO'} + \mathbf{r}_{O'P} \tag{6.40}$$

in $\mathcal{F}$ and obtain*

$${}^{\mathcal{F}}\dot{\mathbf{r}}_{OP} = {}^{\mathcal{F}}\dot{\mathbf{r}}_{OO'} + {}^{\mathcal{F}}\dot{\mathbf{r}}_{O'P} \tag{6.41}$$

Because O is fixed in $\mathcal{F}$, the first two of the three vectors in Equation (6.41) are the velocities in $\mathcal{F}$ of P and O':

$$\mathbf{v}_{P/\mathcal{F}} = \mathbf{v}_{O'/\mathcal{F}} + {}^{\mathcal{F}}\dot{\mathbf{r}}_{O'P} \tag{6.42}$$

The last vector in Equation (6.42) presents a problem. It is not the velocity of P in $\mathcal{F}$, because the point O' is fixed in frame $\mathcal{B}$; nor is it the velocity in $\mathcal{B}$ either, because the derivative is not taken there. To over-

*The superscripts in Equation (6.41) are now necessary to denote the frame in which the derivative is taken.

come this dilemma, we shall rewrite the term by moving the derivative from $\mathcal{F}$ to $\mathcal{B}$ using Equation (6.20):

$$\mathbf{v}_{P/\mathcal{F}} = \mathbf{v}_{O'/\mathcal{F}} + ({}^{\mathcal{B}}\dot{\mathbf{r}}_{O'P} + \boldsymbol{\omega}_{\mathcal{B}/\mathcal{F}} \times \mathbf{r}_{O'P}) \tag{6.43}$$

$$= \mathbf{v}_{O'/\mathcal{F}} + \mathbf{v}_{P/\mathcal{B}} + \boldsymbol{\omega}_{\mathcal{B}/\mathcal{F}} \times \mathbf{r}_{O'P}$$

$$= \mathbf{v}_{P/\mathcal{B}} + (\mathbf{v}_{O'/\mathcal{F}} + \boldsymbol{\omega}_{\mathcal{B}/\mathcal{F}} \times \mathbf{r}_{O'P}) \tag{6.44}$$

We are now in a position to derive the equation relating the velocities of two points of the same rigid body from the (therefore more general) equation (6.44). Temporarily let P be a fixed point of $\mathcal{B}$; then $\mathbf{v}_{P/\mathcal{B}} = \mathbf{0}$ and

$$\mathbf{v}_{P/\mathcal{F}} = \mathbf{v}_{O'/\mathcal{F}} + \boldsymbol{\omega}_{\mathcal{B}/\mathcal{F}} \times \mathbf{r}_{O'P} \tag{6.45}$$

or

$$\mathbf{v}_{P} = \mathbf{v}_{O'} + \boldsymbol{\omega} \times \mathbf{r}_{O'P} \tag{6.46}$$

which is the same as the plane motion equation (3.5), except that now the $\mathbf{r}$ and $\mathbf{v}$ vectors may also have z components and the $\boldsymbol{\omega}$ vector can have x and y components in addition to z.

Returning to the general case in which P is not necessarily attached to either $\mathcal{B}$ or $\mathcal{F}$, let us denote the point of $\mathcal{B}$ (or $\mathcal{B}$ extended) coincident with P by $P_{\mathcal{B}}$. Then Equation (6.44) becomes

$$\mathbf{v}_{P/\mathcal{F}} = \mathbf{v}_{P/\mathcal{B}} + \mathbf{v}_{P_{\mathcal{B}}/\mathcal{F}} \tag{6.47}$$

In words, we may restate Equation (6.47) as follows:

$$\begin{bmatrix}\text{Velocity of}\\ P \text{ in } \mathcal{F}\end{bmatrix} = \begin{bmatrix}\text{velocity of}\\ P \text{ in } \mathcal{B}\end{bmatrix} + \begin{bmatrix}\text{velocity in } \mathcal{F} \text{ of}\\ \text{the fixed point of}\\ \mathcal{B} \text{ coincident with } P\end{bmatrix}$$

$$\mathbf{v}_{P/\mathcal{F}} \quad = \quad \mathbf{v}_{P/\mathcal{B}} \quad + \quad \mathbf{v}_{P_{\mathcal{B}}/\mathcal{F}}$$

Equation (6.47) has the virtue of compactness. However, it is a less convenient form than is (6.44) for differentiation to produce a corresponding relationship of accelerations.

Question 6.5 Why?

A common alternative means of stating this equation is to view body $\mathcal{B}$ as a moving frame—that is, as a body moving relative to another frame $\mathcal{F}$. (See Figure 6.6.) We may then rewrite Equation (6.44) as

$$\mathbf{v}_{P} = \dot{\mathbf{R}} + \mathbf{v}_{\text{rel}} + \boldsymbol{\omega} \times \mathbf{r} \tag{6.48}$$

in which

$$\mathbf{v}_P = \text{velocity of } P \text{ in reference frame } \mathcal{F}$$

$$\dot{\mathbf{R}} = \text{velocity of moving origin} = \mathbf{v}_{O'/\mathcal{F}}$$

$$\mathbf{v}_{\text{rel}} = \text{velocity of } P \text{ in moving frame} = \mathbf{v}_{P/\mathcal{B}}$$

$$\boldsymbol{\omega} = \text{angular velocity of moving frame} = \boldsymbol{\omega}_{\mathcal{B}/\mathcal{F}}$$

$$\mathbf{r} = \text{position vector of } P \text{ in moving frame} = \mathbf{r}_{O'P}$$

We now consider several examples involving the use of the velocity equation for moving frames (Equation 6.44, 6.47, or 6.48).

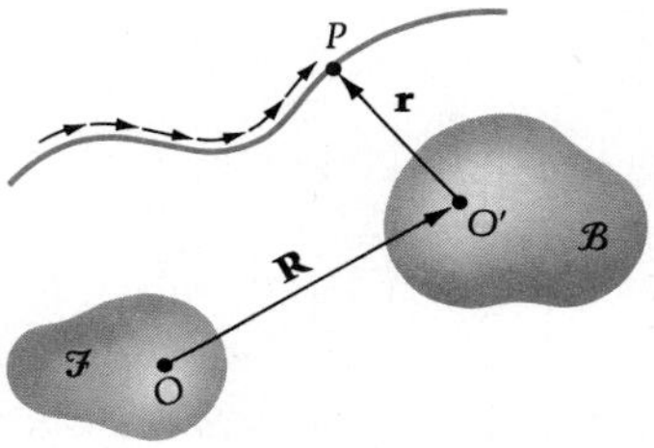

Figure 6.6

EXAMPLE 6.3

Find the velocity in $\mathcal{G}$ of point A at the bottom of the disk in Example 6.2. (Refer to the sketch in that example.)

SOLUTION

We select $\mathcal{B}$ as the moving frame, and Equation (6.44) gives the following (where $\mathcal{G}$ is $\mathcal{F}$ and A is P):

$$\mathbf{v}_{A/\mathcal{G}} = \mathbf{v}_{A/\mathcal{B}} + \mathbf{v}_{O'/\mathcal{G}} + \boldsymbol{\omega}_{\mathcal{B}/\mathcal{G}} \times \mathbf{r}_{O'A}$$

In this case $\mathbf{v}_{O'/\mathcal{G}} = \mathbf{0}$; also $\boldsymbol{\omega}_{\mathcal{B}/\mathcal{G}} = \omega_2\hat{\mathbf{j}}$ from the previous example. The velocity of A in $\mathcal{B}$ is given by

$$\mathbf{v}_{A/\mathcal{B}} = \cancelto{\mathbf{0}}{\mathbf{v}_{Q/\mathcal{B}}} + \omega_1\hat{\mathbf{i}} \times r\hat{\mathbf{j}} = r\omega_1\hat{\mathbf{k}}$$

Therefore

$$\begin{aligned}\mathbf{v}_{A/\mathcal{G}} &= r\omega_1\hat{\mathbf{k}} + \omega_2\hat{\mathbf{j}} \times (r\hat{\mathbf{j}} + R\hat{\mathbf{k}}) \\ &= R\omega_2\hat{\mathbf{i}} + r\omega_1\hat{\mathbf{k}}\end{aligned}$$

EXAMPLE 6.4

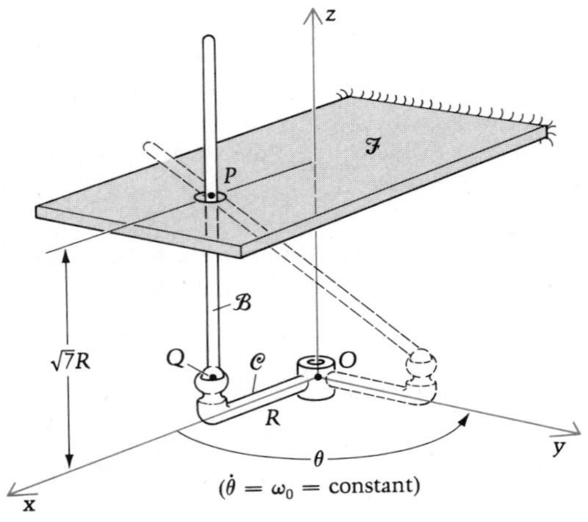

Crank $\mathcal{C}$ in the diagram rotates about axis z through point O. Its other end, Q, is attached to a ball and socket joint as shown. The ball forms the end of rod $\mathcal{B}$, which passes through a hole in the ceiling $\mathcal{F}$. Find the velocity of point P of the bar, which is passing through the hole when $\theta = 90°$.

SOLUTION

We denote by H the point of $\mathcal{F}$ at the center of the hole; knowing that the motion of H in $\mathcal{B}$ must be *along* the axis of $\mathcal{B}$, we may write

$$\cancelto{\mathbf{0}}{\mathbf{v}_{H/\mathcal{F}}} = \mathbf{v}_{H/\mathcal{B}} + \mathbf{v}_{P/\mathcal{F}} \tag{1}$$

in which P is the point of $\mathcal{B}$ coincident with point H. Continuing, we have

$$\mathbf{0} = v_{H/\mathcal{B}}\left(\frac{R\hat{\mathbf{i}} - R\hat{\mathbf{j}} + \sqrt{7}R\hat{\mathbf{k}}}{3R}\right) + (\overbrace{\mathbf{v}_{Q/\mathcal{F}} + \boldsymbol{\omega}_{\mathcal{B}/\mathcal{F}} \times \mathbf{r}_{QP}}^{\mathbf{v}_{P/\mathcal{F}}}) \tag{2}$$

or

$$\mathbf{0} = v_{H/\mathcal{B}}\left(\frac{\hat{\mathbf{i}} - \hat{\mathbf{j}} + \sqrt{7}\hat{\mathbf{k}}}{3}\right)$$
$$+ [-R\omega_0\hat{\mathbf{i}} + (\omega_x\hat{\mathbf{i}} + \omega_y\hat{\mathbf{j}} + \omega_z\hat{\mathbf{k}}) \times (R\hat{\mathbf{i}} - R\hat{\mathbf{j}} + \sqrt{7}R\hat{\mathbf{k}})] \tag{3}$$

Collecting the coefficients of $\hat{\mathbf{i}}$, $\hat{\mathbf{j}}$, and $\hat{\mathbf{k}}$, respectively, we find

$$\frac{1}{3}v_{H/\mathcal{B}} + \sqrt{7}R\omega_y + R\omega_z = R\omega_0 \tag{4}$$

(Continued)

$$-\frac{1}{3}v_{H/\mathcal{B}} + R\omega_z - \sqrt{7}R\omega_x = 0 \tag{5}$$

$$\frac{\sqrt{7}}{3}v_{H/\mathcal{B}} - R\omega_x - R\omega_y = 0 \tag{6}$$

These three equations (4–6) obviously cannot be solved for unique values of all four unknowns ω_x, ω_y, ω_z, $v_{H/\mathcal{B}}$. However, an answer for $v_{H/\mathcal{B}}$ is obtainable by subtracting Equation (5) from (4) and then adding $\sqrt{7}$ times Equation (6). The result is

$$v_{H/\mathcal{B}} = \frac{R\omega_0}{3}$$

Substituting this result back into Equations (4–6) gives three equations in ω_x, ω_y, and ω_z whose coefficient matrix is singular (has a zero determinant). Thus they cannot be solved for the angular velocity components. A more practical reason for this is that the component of $\boldsymbol{\omega}_{\mathcal{B}/\mathcal{I}}$ along bar $\mathcal{B}$ cannot affect the answer for $\mathbf{v}_{P/\mathcal{I}}$ because $\mathcal{B}$ can turn freely in its socket about its axis without altering $\mathbf{v}_{P/\mathcal{I}}$. Mathematically this is manifested by this "axial" component of $\boldsymbol{\omega}_{\mathcal{B}/\mathcal{I}}$ being parallel to $\mathbf{r}_{QP}$ and thus canceling out of Equation (2). These $\boldsymbol{\omega}$ components are not needed for a solution, however, because from Equation (1) we can obtain our desired result:

$$\mathbf{v}_{P/\mathcal{I}} = -\mathbf{v}_{H/\mathcal{B}} = \frac{-R\omega_0}{3}\left(\frac{\hat{\mathbf{i}} - \hat{\mathbf{j}} + \sqrt{7}\hat{\mathbf{k}}}{3}\right)$$

Although we cannot solve for the component of $\boldsymbol{\omega}_{\mathcal{B}/\mathcal{I}}$ along $\mathcal{B}$ with the given information, we *are* able to calculate the component of $\boldsymbol{\omega}_{\mathcal{B}/\mathcal{I}}$ normal to $\mathcal{B}$. It will be made up of values ω_x', ω_y', and ω_z' that enforce the relationship

$$\boldsymbol{\omega}' \cdot \mathbf{r}_{QP} = 0$$

That is,

$$\cancel{R}(\omega_x' - \omega_y' + \sqrt{7}\omega_z') = 0 \tag{7}$$

This equation states that these $\boldsymbol{\omega}'$ components form a vector normal to the line OP; again, this vector is the only part of $\boldsymbol{\omega}$ that can affect $\mathbf{v}_{P/\mathcal{I}}$.

After adding primes to ω_x, ω_y, ω_z in Equations (4–6), the solution of Equations (4–7), as the reader may verify, is

$$\omega_x' = 0 \qquad \omega_y' = \frac{\sqrt{7}}{9}\omega_0 \qquad \omega_z' = \frac{\omega_0}{9} \qquad v_{H/\mathcal{B}} = \frac{R\omega_0}{3}$$

And we may now calculate the velocity of P from Equation (1) as before or from Equation (2) as follows:

$$\begin{aligned}\mathbf{v}_{P/\mathcal{I}} &= \mathbf{v}_{Q/\mathcal{I}} + \boldsymbol{\omega}_{\mathcal{B}/\mathcal{I}} \times \mathbf{r}_{QP} \\ &= -R\omega_0\hat{\mathbf{i}} + \omega_0\left(\frac{\sqrt{7}\hat{\mathbf{j}} + \hat{\mathbf{k}}}{9}\right) \times R(\hat{\mathbf{i}} - \hat{\mathbf{j}} + \sqrt{7}\hat{\mathbf{k}}) \\ &= \frac{R\omega_0}{9}(-\hat{\mathbf{i}} + \hat{\mathbf{j}} - \sqrt{7}\hat{\mathbf{k}}) \qquad \text{(as before)}\end{aligned}$$

After these examples we are now ready to derive the corresponding relationship between the *accelerations* of P in two frames $\mathcal{B}$ and $\mathcal{I}$. Differentiating Equation (6.44) gives

$$^{\mathcal{I}}\dot{\mathbf{v}}_{P/\mathcal{I}} = {^{\mathcal{I}}}\dot{\mathbf{v}}_{O'/\mathcal{I}} + {^{\mathcal{I}}}\dot{\boldsymbol{\omega}}_{\mathcal{B}/\mathcal{I}} \times \mathbf{r}_{O'P} + \boldsymbol{\omega}_{\mathcal{B}/\mathcal{I}} \times {^{\mathcal{I}}}\dot{\mathbf{r}}_{O'P} + {^{\mathcal{I}}}(\dot{\mathbf{v}}_{P/\mathcal{B}}) \tag{6.49}$$

or, again using Equation (6.20) (once in each of the last two terms), we get

$$\mathbf{a}_{P/\mathcal{I}} = \mathbf{a}_{O'/\mathcal{I}} + \boldsymbol{\alpha}_{\mathcal{B}/\mathcal{I}} \times \mathbf{r}_{O'P} + \boldsymbol{\omega}_{\mathcal{B}/\mathcal{I}} \times ({^{\mathcal{B}}}\dot{\mathbf{r}}_{O'P} + \boldsymbol{\omega}_{\mathcal{B}/\mathcal{I}} \times \mathbf{r}_{O'P}) + ({^{\mathcal{B}}}\dot{\mathbf{v}}_{P/\mathcal{B}} + \boldsymbol{\omega}_{\mathcal{B}/\mathcal{I}} \times {^{\mathcal{B}}}\dot{\mathbf{r}}_{O'P}) \tag{6.50}$$

Rearranging the terms, we have

$$\mathbf{a}_{P/\mathcal{I}} = \underbrace{{^{\mathcal{B}}}\dot{\mathbf{v}}_{P/\mathcal{B}}} + \underbrace{\mathbf{a}_{O'/\mathcal{I}} + \boldsymbol{\alpha}_{\mathcal{B}/\mathcal{I}} \times \mathbf{r}_{O'P} + \boldsymbol{\omega}_{\mathcal{B}/\mathcal{I}} \times (\boldsymbol{\omega}_{\mathcal{B}/\mathcal{I}} \times \mathbf{r}_{O'P})} + \underbrace{2\boldsymbol{\omega}_{\mathcal{B}/\mathcal{I}} \times {^{\mathcal{B}}}\dot{\mathbf{r}}_{O'P}} \tag{6.51}$$

$$\mathbf{a}_{P/\mathcal{I}} = \mathbf{a}_{P/\mathcal{B}} + \mathbf{a}_{P_{\mathcal{B}}/\mathcal{I}} + 2\boldsymbol{\omega}_{\mathcal{B}/\mathcal{I}} \times \mathbf{v}_{P/\mathcal{B}} \tag{6.52}$$

The middle three terms on the right side of Equation (6.51) make up the acceleration of the point $P_{\mathcal{B}}$ of $\mathcal{B}$ (or $\mathcal{B}$ extended) coincident with P. (The proof is brief: If P is *fixed* to $\mathcal{B}$ at point $P_{\mathcal{B}}$, then the other two terms vanish since $\mathbf{r}_{O'P}$ becomes a constant vector in $\mathcal{B}$, and what remains is necessarily $\mathbf{a}_{P_{\mathcal{B}}/\mathcal{I}}$.) The term ${^{\mathcal{B}}}\dot{\mathbf{v}}_{P/\mathcal{B}}$ (which is ${^{\mathcal{B}}}\ddot{\mathbf{r}}_{O'P}$, with both derivatives taken in $\mathcal{B}$) is clearly the acceleration of P in $\mathcal{B}$. The last term, $2\boldsymbol{\omega}_{\mathcal{B}/\mathcal{I}} \times {^{\mathcal{B}}}\dot{\mathbf{r}}_{O'P}$, is called the **Coriolis acceleration** of P. Note that due to the presence of the Coriolis acceleration it is *not true* that the acceleration of P in frame $\mathcal{I}$ is its acceleration in $\mathcal{B}$ plus the acceleration of the point of $\mathcal{B}$ with which it is coincident (as was in fact the case with the velocity $\mathbf{v}_P$ of P). This result is interestingly analogous to the fact that the addition theorem for angular velocity is not true for angular acceleration.

As we did for the velocity equation, we now restate Equation (6.52) in words:

$$\begin{bmatrix}\text{Acceleration}\\ \text{of } P \text{ in } \mathcal{I}\end{bmatrix} = \begin{bmatrix}\text{acceleration}\\ \text{of } P \text{ in } \mathcal{B}\end{bmatrix} + \begin{bmatrix}\text{acceleration in } \mathcal{I} \text{ of}\\ \text{the fixed point of } \mathcal{B}\\ \text{coincident with } P\end{bmatrix} + \begin{bmatrix}\text{Coriolis}\\ \text{acceleration}\end{bmatrix}$$

$$\mathbf{a}_{P/\mathcal{I}} = \mathbf{a}_{P/\mathcal{B}} + \mathbf{a}_{P_{\mathcal{B}}/\mathcal{I}} + 2\boldsymbol{\omega}_{\mathcal{B}/\mathcal{I}} \times \mathbf{v}_{P/\mathcal{B}}$$

In the abbreviated notation of Equation (6.48) we have

$$\mathbf{a}_P = \ddot{\mathbf{R}} + \boldsymbol{\alpha} \times \mathbf{r} + \boldsymbol{\omega} \times (\boldsymbol{\omega} \times \mathbf{r}) + 2\boldsymbol{\omega} \times \mathbf{v}_{\text{rel}} + \mathbf{a}_{\text{rel}} \tag{6.53}$$

in which

$$\mathbf{a}_P = \text{acceleration of } P \text{ in reference frame } \mathcal{F}$$

$$\ddot{\mathbf{R}} = \text{acceleration of moving origin} = \mathbf{a}_{O'/\mathcal{F}}$$

$$\boldsymbol{\alpha} = \text{angular acceleration of moving frame} = \boldsymbol{\alpha}_{\mathcal{B}/\mathcal{F}}$$

$$\mathbf{a}_{\text{rel}} = \text{acceleration of } P \text{ in moving frame} = \mathbf{a}_{P/\mathcal{B}}$$

and in which all other terms in Equation (6.53) are defined directly after Equation (6.48).

EXAMPLE 6.5

Compute the acceleration in $\mathcal{G}$ of point A in Examples 6.2 and 6.3.

SOLUTION

We shall use Equation (6.51); the reference frame $\mathcal{F}$ is $\mathcal{G}$ and the moving point P is A:

$$\mathbf{a}_{A/\mathcal{G}} = \mathbf{a}_{A/\mathcal{B}} + \mathbf{a}_{O'/\mathcal{G}} + \boldsymbol{\alpha}_{\mathcal{B}/\mathcal{G}} \times \mathbf{r}_{O'A} + \boldsymbol{\omega}_{\mathcal{B}/\mathcal{G}} \times (\boldsymbol{\omega}_{\mathcal{B}/\mathcal{G}} \times \mathbf{r}_{O'A}) + 2\boldsymbol{\omega}_{\mathcal{B}/\mathcal{G}} \times \mathbf{v}_{A/\mathcal{B}}$$

The various terms on the right side are calculated as follows:

1. $\mathbf{a}_{A/\mathcal{B}} = \cancelto{\mathbf{0}}{\mathbf{a}_{Q/\mathcal{B}}} + \cancelto{\mathbf{0}}{\boldsymbol{\alpha}_{\mathcal{A}/\mathcal{B}}} \times \mathbf{r}_{QA} - \underbrace{\omega_1^2 \mathbf{r}_{QA}}$

   Note that $\mathcal{A}$ is in plane motion relative to $\mathcal{B}$, so that this short form is acceptable.

   $= -r\omega_1^2 \hat{\mathbf{j}}$ (A moves on a circle at constant speed in $\mathcal{B}$)
2. $\mathbf{a}_{O'/\mathcal{G}} = \mathbf{0}$. Note that O' is fixed in $\mathcal{G}$ in this example.
3. $\boldsymbol{\alpha}_{\mathcal{B}/\mathcal{G}} \times \mathbf{r}_{O'A} = \mathbf{0}$. Note that $\boldsymbol{\alpha}_{\mathcal{B}/\mathcal{G}} = {}^{\mathcal{G}}(d/dt)(\omega_2 \hat{\mathbf{j}}) = \mathbf{0}$ since ω_2 is a constant, and $\hat{\mathbf{j}}$ does not change in direction in $\mathcal{G}$.
4. $\boldsymbol{\omega}_{\mathcal{B}/\mathcal{G}} \times (\boldsymbol{\omega}_{\mathcal{B}/\mathcal{G}} \times \mathbf{r}_{O'A}) = \omega_2 \hat{\mathbf{j}} \times [\omega_2 \hat{\mathbf{j}} \times (r\hat{\mathbf{j}} + R\hat{\mathbf{k}})] = -R\omega_2^2 \hat{\mathbf{k}}$.
5. $2\boldsymbol{\omega}_{\mathcal{B}/\mathcal{G}} \times \mathbf{v}_{A/\mathcal{B}} = 2\omega_2 \hat{\mathbf{j}} \times (r\omega_1 \hat{\mathbf{k}}) = 2r\omega_1\omega_2 \hat{\mathbf{i}}$.

Thus the answer for the acceleration of point A is

$$\mathbf{a}_{A/\mathcal{G}} = 2r\omega_1\omega_2 \hat{\mathbf{i}} - r\omega_1^2 \hat{\mathbf{j}} - R\omega_2^2 \hat{\mathbf{k}}$$

PROBLEMS / Section 6.5

6.19 The crane $\mathcal{C}$ in Figure P6.19 turns about the vertical at $\omega_v = 0.2$ rad/sec = constant, and simultaneously its boom $\mathcal{B}$ is being elevated at the increasing rate $\omega_H = 0.1t$ rad/sec. The $(\mathbf{x}, \mathbf{y}, \mathbf{z})$ axes are fixed to the crane $\mathcal{C}$ at O, and the boom was along the $\mathbf{y}$ axis when t was zero. When the boom makes a 60° angle with the horizontal, find: (a) $\boldsymbol{\omega}_{\mathcal{B}/\mathcal{G}}$; (b) $\boldsymbol{\alpha}_{\mathcal{B}/\mathcal{G}}$; (c) $\mathbf{v}_{P/\mathcal{G}}$; (d) $\mathbf{a}_{P/\mathcal{G}}$.

6.20 Disk $\mathcal{D}$ rotates about its axis at the constant angular speed $\omega_0 = 0.1$ rad/s. The round wire is rigidly affixed to $\mathcal{D}$ at points A and B as shown in Figure P6.20. A bug crawls around the wire from A to B; its speed relative to the wire (initially zero) is always increasing at the constant rate of 0.001 m/s^2. Find the velocity and acceleration of the bug, relative to the reference frame $\mathcal{F}$ in which the disk turns, when it arrives at B.

6.21 A platform translates past a turntable at 6 mph. (See Figure P6.21.) People step onto the turntable and walk straight toward the center where they exit onto stairs. There is rolling contact between platform and turntable. Suppose the people walk at the constant rate of approximately 3 mph relative to the turntable. If it is desired that they do not experience more than 3 ft/sec^2 of lateral acceleration, find the required turntable radius.

6.22 A bird flies horizontally past a man's head in a straight line at constant speed v_0 toward the axis of a turntable on which the man is standing. The axes $\mathbf{xy}$ have origin at the man's feet; the $\mathbf{z}$ axis is vertical and $\mathbf{x}$ always points toward the center of the turntable, which has radius R and angular velocity $\boldsymbol{\omega} = \omega_0\hat{\mathbf{k}}$. Derive $\mathbf{x}(t)$ and $\mathbf{y}(t)$ of the bird in terms of R, v_0, ω_0, and the time t. (See Figure P6.22.)

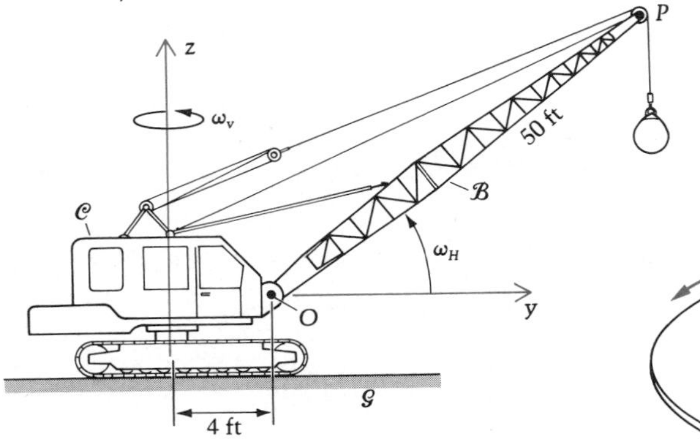

Figure P6.19

Figure P6.21

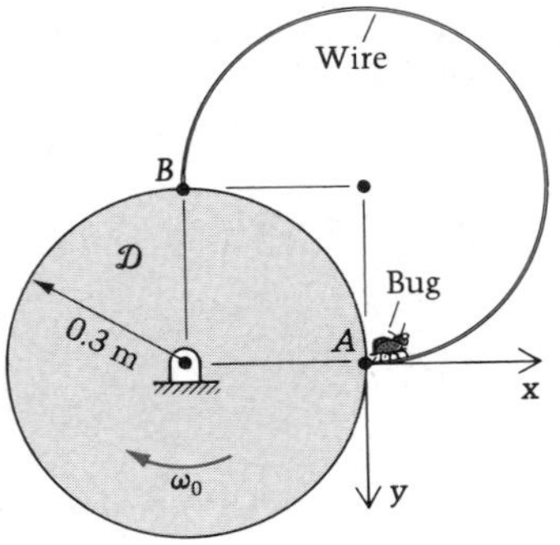

Figure P6.20

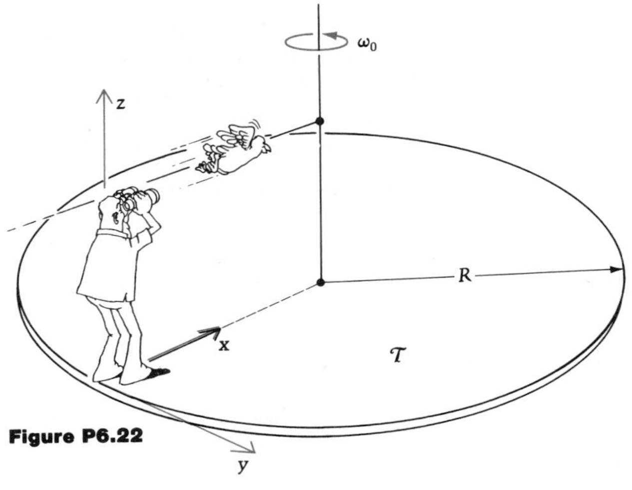

Figure P6.22

6.23 A car travels south along a meridian at a certain time; its speed relative to the earth is 60 mph, increasing at the rate of 2 ft/sec^2. (See Figure P6.23.) Find the acceleration of the car in a frame with origin always at the center of the earth and z axis along the polar axis of rotation but not rotating about the axis with the earth.

6.24 If in the preceding problem the car is traveling from west to east at 45°N latitude instead of along a meridian, find the car's acceleration in the same frame. (See Figure P6.24.)

6.25 The bent bar $\mathcal{B}$ in Figure P6.25 revolves about the vertical at $2\pi\hat{\mathbf{k}}$ rad/sec. The center C of the collar $\mathcal{C}$ has velocity and acceleration relative to $\mathcal{B}$ of $20\hat{\mathbf{e}}$ in./sec and $-10\hat{\mathbf{e}}$ in./sec^2, respectively, where $\hat{\mathbf{e}}$ is in the direction of the velocity of C in $\mathcal{B}$. At the given instant, find the velocity and acceleration of C in the frame $\mathcal{F}$ in which $\mathcal{B}$ turns.

6.26 Shaft $\mathcal{S}$ in Figure P6.26 turns in the clevis $\mathcal{F}$ at 2 rad/sec in the direction shown. The wheel simultaneously rotates at 3 rad/sec about its axis as indicated. Both rates are constants. The bug is crawling outward on a spoke at 0.2 in./sec with an acceleration of 0.1 in./sec^2 both relative to the spoke. At the instant shown, find: (a) the angular velocity of the wheel; (b) the velocity of the bug.

6.27 Find the angular acceleration of the wheel and the acceleration of the bug in Problem 6.26.

6.28 The large disk $\mathcal{A}$ in Figure P6.28 rotates at 10 rad/sec counterclockwise (looking down on its horizontal surface). A smaller disk $\mathcal{B}$ rolls radially outward along a radius OD of $\mathcal{A}$. At the instant shown, the center C of $\mathcal{B}$ is 4 ft from the axis of rotation of $\mathcal{A}$, and this distance is increasing at the constant rate of 2 ft/sec. Determine the velocity and acceleration of point E, which is at the top of $\mathcal{B}$ at the given instant.

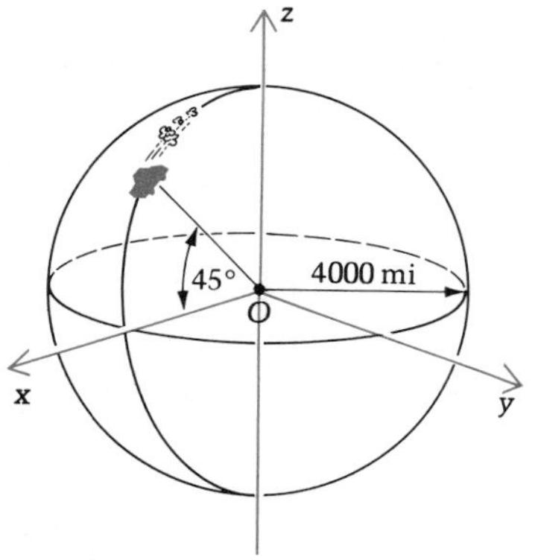

Figure P6.23

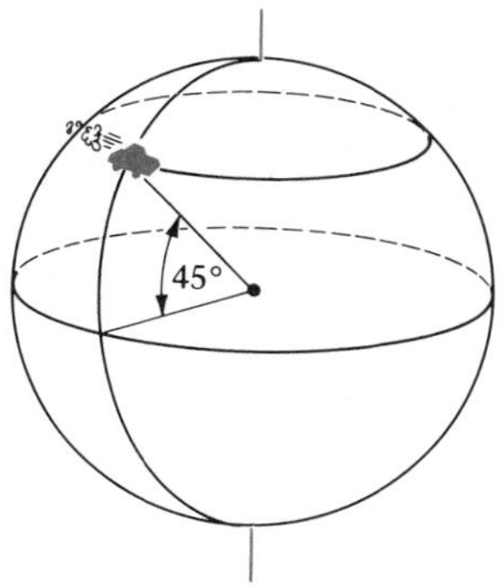

Figure P6.24

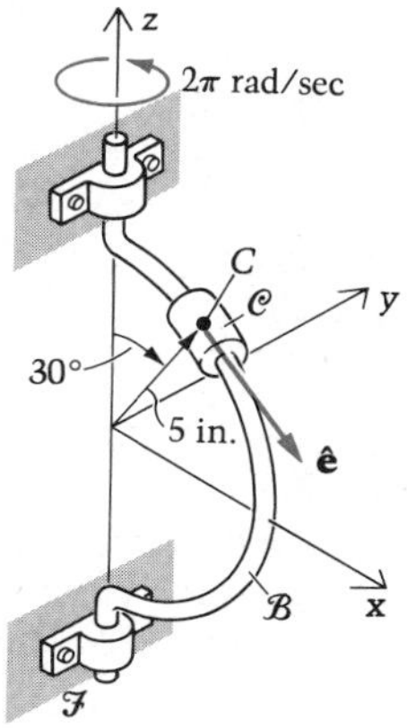

Figure P6.25

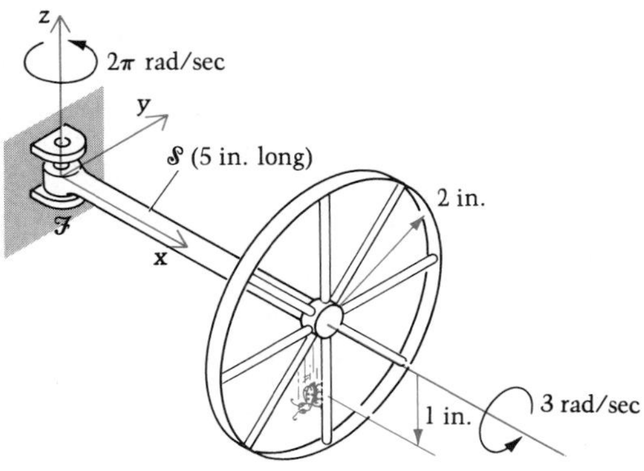

Figure P6.26

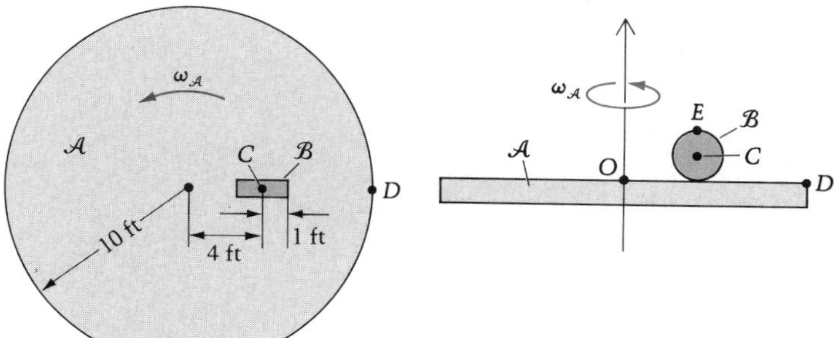

Figure P6.28

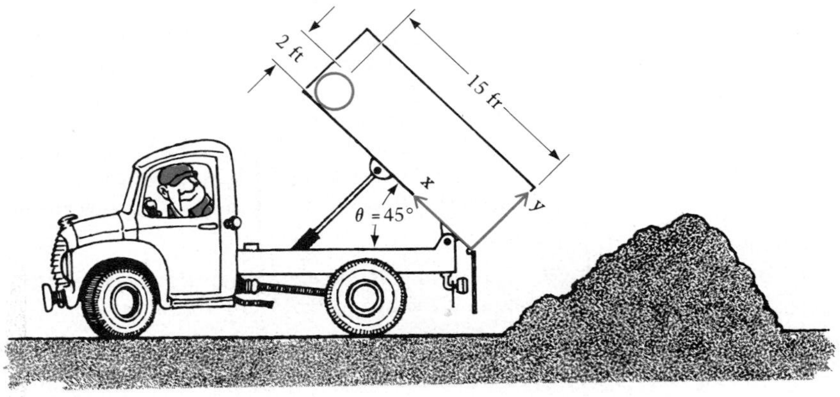

Figure P6.29

6.29 The truck in Figure P6.29 moves to the left at a constant speed of 7.07 ft/sec. At the instant shown, the loading compartment has an angular speed $\dot{\theta} = \frac{1}{3}$ rad/sec and an angular acceleration of $\ddot{\theta} = -\frac{1}{15}$ rad/sec^2. The (dotted) cylinder on the truck bed comes loose and rolls toward the ground at an angular velocity, relative to the compartment, of 1 ↻ rad/sec at this instant; it is speeding up at a rate of $\frac{1}{2}$ rad/sec^2, also relative to the compartment. Find the velocity and acceleration of the center of the cylinder relative to the ground. Use the rotating reference frame shown.

6.30 A man walks dizzily outward along a sine wave fixed to a merry-go-round that is 40 ft in diameter and turns at 10 rpm. (See Figure P6.30.) If the man's speed relative to the turntable is a constant 2 ft/sec, what is the magnitude of his acceleration when he is 10 ft from the center?

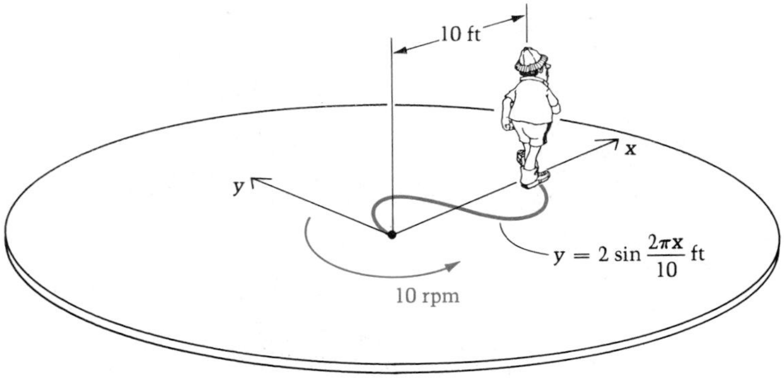

Figure P6.30

6.31 A device for simulating conditions in space allows rotations about orthogonal axes as shown in Figure P6.31. Determine the angular velocity in frame $\mathcal{G}$ of the capsule $\mathcal{C}$ containing the astronaut. Express the result in terms of unit vectors $(\hat{\mathbf{i}}, \hat{\mathbf{j}}, \hat{\mathbf{k}})$ fixed in the beam $\mathcal{B}$. Note that the ω_y rotation is about an axis fixed in $\mathcal{C}$ and in $\mathcal{A}$ but *not* in $\mathcal{B}$; this axis is parallel to y at $t = 0$, and ω_x is a constant.

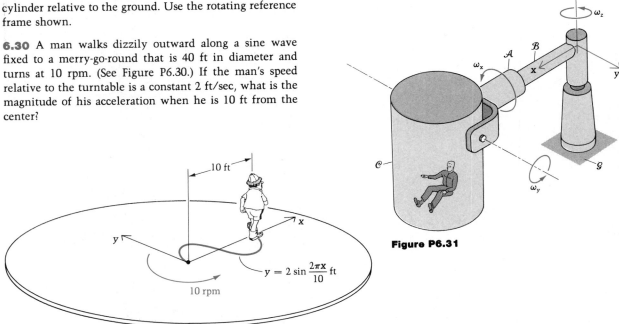

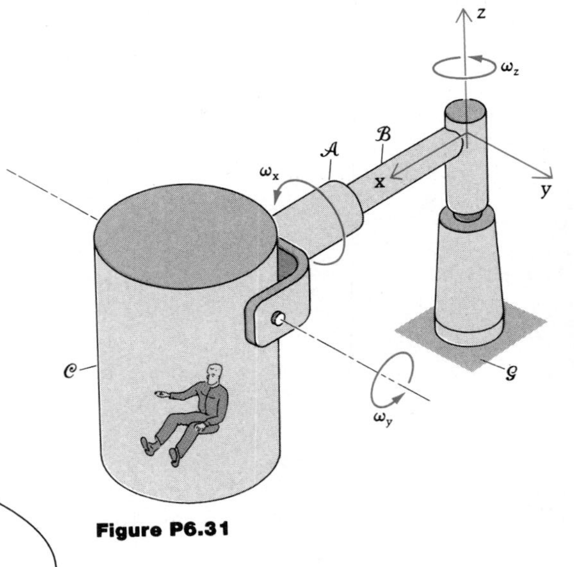

Figure P6.31

6.32 In Problem 6.31 find the angular acceleration of the capsule $\mathcal{C}$ in $\mathcal{G}$. Take ω_z and ω_x to be constants.

6.33 In Problem 6.32 let $\omega_z = 0.5$ rad/sec and $\omega_x = 0.7$ rad/sec in the directions indicated. With the dimensions given in Figure P6.33, find the maximum value of ω_y for which the acceleration magnitude of the astronaut's head will not exceed 5g at the given instant. Let $\omega_y =$ constant.

6.34 A man lifts a drink in a lounge that revolves around the local vertical at 1°/sec. The glass moves about his fixed elbow at 3 ft/sec. (See Figure P6.34.) Calculate the Coriolis acceleration of the center of the glass at the instant shown.

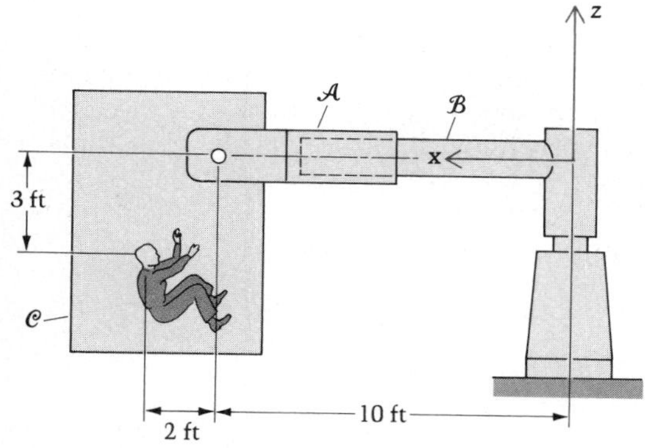

Figure P6.33

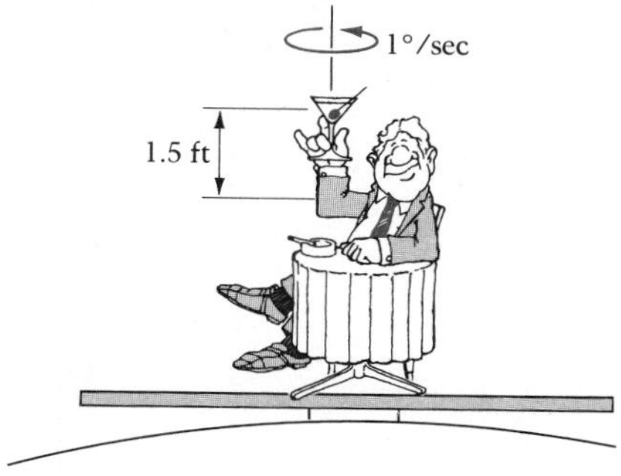

Figure P6.34

6.35 A centrifugal pump $\mathcal{P}$ turns at 500 rpm in $\mathcal{F}$, and the water particles have respective tangential velocity and acceleration components ***relative to the blades*** of 120 ft/sec and 80 ft/sec^2 outward when they reach the outermost point of their blades. (See Figure P6.35.) At the instant before exit determine the velocity and acceleration vectors of the water particle at P with respect to the ground $\mathcal{F}$.

6.36 In Problem 6.10 find the velocity and acceleration of point P at the tip of the gripper $\mathcal{G}$ at the instant given. (See also Problem 6.18.)

6.37 A bug travels outward from the center of the disk in Figure P6.37 with distance from O given by $r = ct^2$, where $c =$ constant and the path relative to the disk is a straight line. The disk turns with the angle $\theta = t^2/(2\sqrt{8})$ rad. Find all times at which the bug's radial acceleration component is zero.

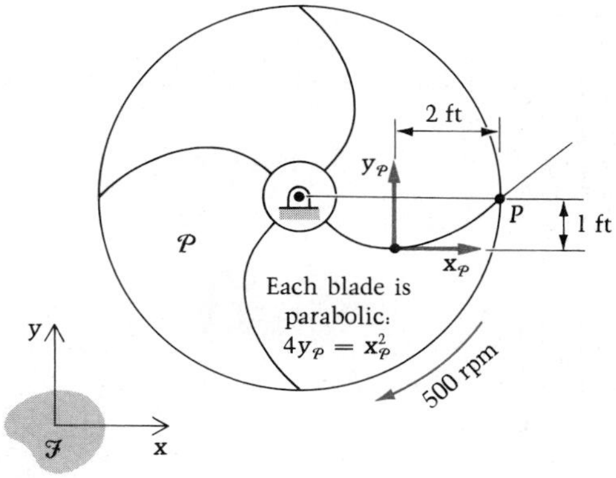

Figure P6.35

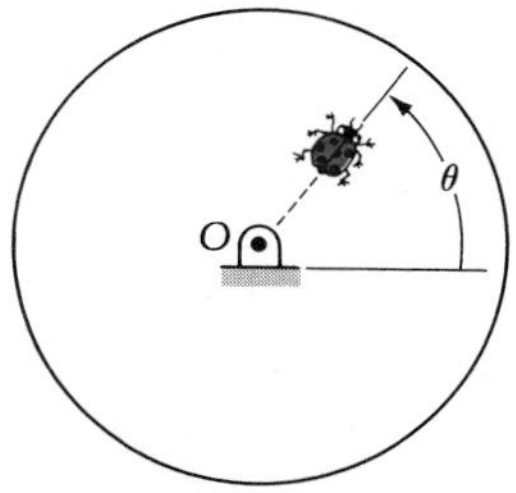

Figure P6.37

6.38 In Figure P6.38 the axes **x** and *y*, and the origin O, are fixed on the deck of a ship. The ship $\mathcal{S}$ has an angular velocity relative to the earth $\mathcal{E}$ of

$$\boldsymbol{\omega}_{\mathcal{S}/\mathcal{E}} = \omega_r\hat{\mathbf{i}} + \omega_P\hat{\mathbf{j}}$$

where **x** and *y* are fore-and-aft and athwartships axes, respectively; thus ω_r is a rolling component and ω_P a pitching component of angular velocity. Point T is a target fixed relative to earth, such as a geosynchronous satellite. Find the angular rates $\dot{\theta}$ and $\dot{\phi}$, in terms of θ, ϕ, ω_r, and ω_P, that are required to track point T.

6.39 Differentiate the results of the preceding problem and obtain $\ddot{\theta}$ and $\ddot{\phi}$ as functions of the same four parameters plus α_r and α_P.

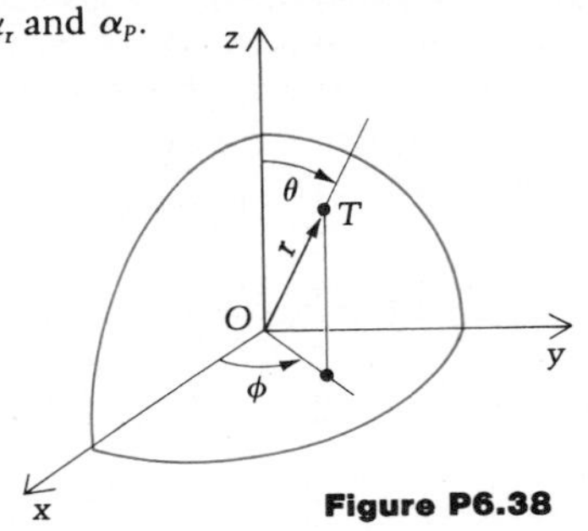

Figure P6.38

6.40 Extend the preceding problem. Consider the case where roll and pitch are sinusoidal, and specialize $\ddot{\theta}$ and $\ddot{\phi}$ for:

a. Roll and pitch in phase; $\alpha_r = \alpha_P = 0$; ω_r and ω_P at maximum values
b. Roll and pitch in phase; $\omega_r = \omega_P = 0$; α_r and α_P at maximum values
c. Roll and pitch out of phase; $\omega_r = \alpha_P = 0$; α_r and ω_P at maximum values
d. Roll and pitch out of phase; $\omega_P = \alpha_r = 0$; α_P and ω_r at maximum values

6.41 Prove that if two bodies are in rolling contact, the shaded arcs in Figure P6.41, representing the loci of former contact points, are equal in length. (Note from Problem 3.105 that the converse is not true.)

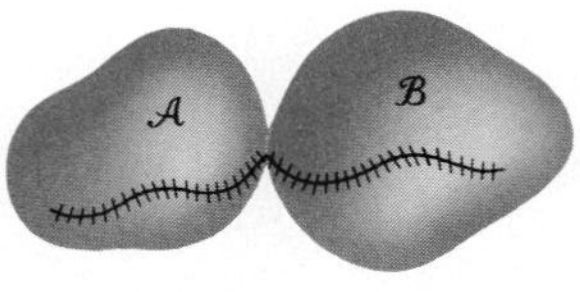

Figure P6.41

6.6 Velocity and Acceleration Equations for Two Points of the Same Rigid Body

We next apply the concepts of position, velocity, acceleration, angular velocity, and angular acceleration (developed in Chapter 1 and Sections 6.2 to 6.5) to the kinematics of a rigid body $\mathcal{B}$ in general motion in a frame $\mathcal{F}$. The equations relating the velocities in $\mathcal{F}$ of two points of a rigid body $\mathcal{B}$ to its angular velocity $\boldsymbol{\omega}_{\mathcal{B}/\mathcal{F}}$ are special cases of Equations (6.44) and (6.51). We have seen in the text following Equations (6.44) and (6.51) that if P joins O′ as a fixed point of $\mathcal{B}$ (see Figure 6.7), the relationship between the velocities of these two points is given by the equation

$$\mathbf{v}_{P/\mathcal{F}} = \mathbf{v}_{O'/\mathcal{F}} + \boldsymbol{\omega}_{\mathcal{B}/\mathcal{F}} \times \mathbf{r}_{O'P} \tag{6.54}$$

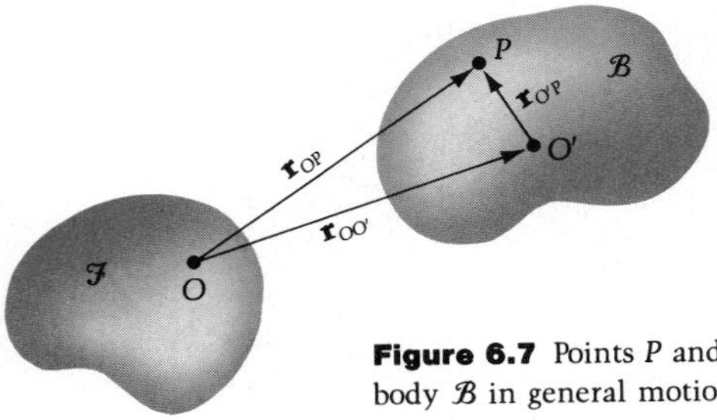

Figure 6.7 Points P and O′ of a rigid body $\mathcal{B}$ in general motion.

and the relationship between the accelerations of these two points of $\mathcal{B}$ is

$$\mathbf{a}_{P/\mathcal{J}} = \mathbf{a}_{O'/\mathcal{J}} + \boldsymbol{\alpha}_{\mathcal{B}/\mathcal{J}} \times \mathbf{r}_{O'P} + \boldsymbol{\omega}_{\mathcal{B}/\mathcal{J}} \times (\boldsymbol{\omega}_{\mathcal{B}/\mathcal{J}} \times \mathbf{r}_{O'P}) \tag{6.55}$$

We remark again that Equations (6.54) and (6.55) follow from the general equations (6.44) and (6.51) when and if $\mathbf{r}_{O'P}$ is a constant in $\mathcal{B}$; in that case, $\mathbf{v}_{P/\mathcal{B}} = {}^{\mathcal{B}}\dot{\mathbf{r}}_{O'P} = \mathbf{0}$ and $\mathbf{a}_{P/\mathcal{B}} = {}^{\mathcal{B}}\ddot{\mathbf{r}}_{O'P} = \mathbf{0}$. Furthermore, the two points P and O' in Equations (6.54) and (6.55) may be replaced by *any* pair of points fixed to $\mathcal{B}$, since being fixed to $\mathcal{B}$ is the only restriction on either of them. Speaking loosely, with Equations (6.54, 55) we are interested in *two points* on *one body*, whereas with Equations (6.44, 51) we were studying *one point* in motion relative to *two bodies*. We now consider examples involving the use of the two rigid-body equations (6.54) and (6.55).

EXAMPLE 6.6

Rework Example 6.5 by treating point A as a point of body $\mathcal{A}$ instead of as a point moving with a known motion in $\mathcal{B}$.

SOLUTION

From Equation (6.54) and the sketch in Example 6.2, we have

$$\mathbf{v}_{A/\mathcal{G}} = \mathbf{v}_{Q/\mathcal{G}} + \boldsymbol{\omega}_{\mathcal{A}/\mathcal{G}} \times \mathbf{r}_{QA}$$

Now recognizing that Q is also a point of $\mathcal{B}$, we obtain

$$\begin{aligned}\mathbf{v}_{A/\mathcal{G}} &= (\mathbf{v}_{O'/\mathcal{G}} + \boldsymbol{\omega}_{\mathcal{B}/\mathcal{G}} \times \mathbf{r}_{O'Q}) + \boldsymbol{\omega}_{A/\mathcal{G}} \times \mathbf{r}_{QA} \\ &= \mathbf{0} + \omega_2\hat{\mathbf{j}} \times R\hat{\mathbf{k}} + (\omega_1\hat{\mathbf{i}} + \omega_2\hat{\mathbf{j}}) \times r\hat{\mathbf{j}} \\ &= R\omega_2\hat{\mathbf{i}} + r\omega_1\hat{\mathbf{k}} \qquad \text{(as before)}\end{aligned}$$

Next we relate the accelerations of A and Q with Equation (6.55):

$$\mathbf{a}_{A/\mathcal{G}} = \mathbf{a}_{Q/\mathcal{G}} + \boldsymbol{\alpha}_{\mathcal{A}/\mathcal{G}} \times \mathbf{r}_{QA} + \boldsymbol{\omega}_{\mathcal{A}/\mathcal{G}} \times (\boldsymbol{\omega}_{\mathcal{A}/\mathcal{G}} \times \mathbf{r}_{QA})$$

The first term on the right side, using Q and O' as points of body $\mathcal{B}$, is

$$\begin{aligned}\mathbf{a}_{Q/\mathcal{G}} &= \mathbf{a}_{O'/\mathcal{G}} + \boldsymbol{\alpha}_{\mathcal{B}/\mathcal{G}} \times \mathbf{r}_{O'Q} + \boldsymbol{\omega}_{\mathcal{B}/\mathcal{G}} \times (\boldsymbol{\omega}_{\mathcal{B}/\mathcal{G}} \times \mathbf{r}_{O'Q}) \\ &= \mathbf{0} + \mathbf{0} + \omega_2\hat{\mathbf{j}} \times (R\omega_2\hat{\mathbf{i}}) \\ &= -R\omega_2^2\hat{\mathbf{k}}\end{aligned}$$

Thus

$$\mathbf{a}_{A/\mathcal{G}} = -R\omega_2^2\hat{\mathbf{k}} + (-\omega_1\omega_2\hat{\mathbf{k}}) \times (r\hat{\mathbf{j}}) + (\omega_1\hat{\mathbf{i}} + \omega_2\hat{\mathbf{j}}) \times (r\omega_1\hat{\mathbf{k}})$$

or

$$\mathbf{a}_{A/\mathcal{G}} = 2r\omega_1\omega_2\hat{\mathbf{i}} - r\omega_1^2\hat{\mathbf{j}} - R\omega_2^2\hat{\mathbf{k}}$$

which we previously obtained in Example 6.5 by another approach.

EXAMPLE 6.7

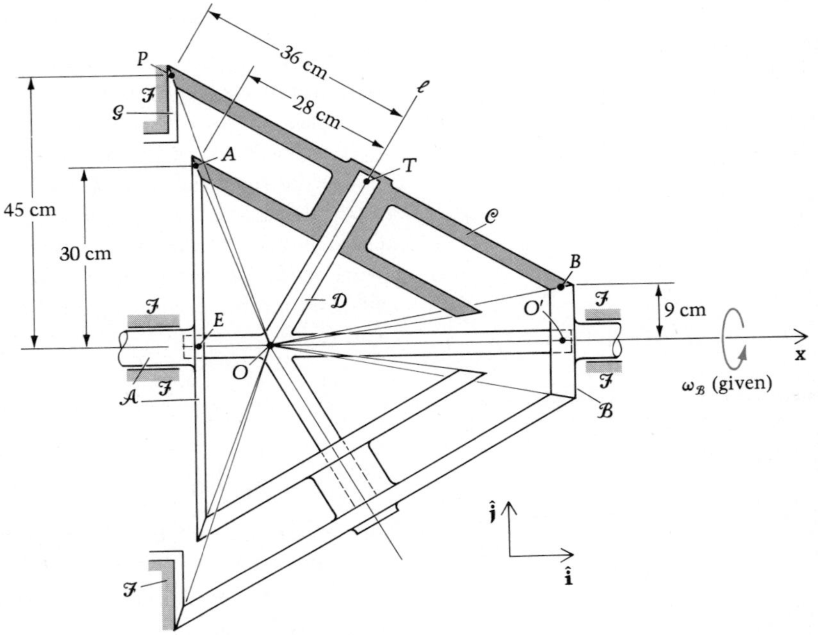

Very large alterations in speed along a given direction may be obtained by using the gear arrangement shown in the diagram. Gears $\mathcal{A}$, $\mathcal{B}$, and $\mathcal{D}$ all rotate about the **x** axis in $\mathcal{F}$, but $\mathcal{C}$ has a more complicated motion:

1. It rolls on the fixed (to $\mathcal{F}$) gear $\mathcal{G}$, currently contacting it at P.
2. It rolls on $\mathcal{A}$. (The contacting teeth are at A in the figure.)
3. It rolls on $\mathcal{B}$ (at B in the figure).
4. It turns with respect to $\mathcal{D}$ about the line ℓ, which is fixed in both $\mathcal{C}$ and $\mathcal{D}$.

Considering $\mathcal{B}$ to be the driven gear, find the ratio of $\omega_{\mathcal{A}}$ to $\omega_{\mathcal{B}}$.

SOLUTION

The velocity of the contacting points of $\mathcal{B}$ and $\mathcal{C}$ is, using body $\mathcal{B}$,

$$\mathbf{v}_B = \overset{\mathbf{0}}{\cancel{\mathbf{v}_{O'}}} + \boldsymbol{\omega}_{\mathcal{B}/\mathcal{F}} \times \mathbf{r}_{O'B}$$

$$\mathbf{v}_B = \omega_{\mathcal{B}}\hat{\mathbf{i}} \times 9\hat{\mathbf{j}} = 9\omega_{\mathcal{B}}\hat{\mathbf{k}} \text{ cm/s} \tag{1}$$

in which we use just one subscript on $\boldsymbol{\omega}$ when it is the angular velocity of a body with respect to $\mathcal{F}$.

Next we find another expression for $\mathbf{v}_B$, this time by relating the tooth point of $\mathcal{C}$ at B to the point of $\mathcal{C}$ that contacts the reference frame ($\mathcal{G}$ is fixed to $\mathcal{F}$) at P:

$$\mathbf{v}_B = \overset{\mathbf{0}}{\cancel{\mathbf{v}_P}} + \boldsymbol{\omega}_{\mathcal{C}} \times \mathbf{r}_{PB} \tag{2}$$

(Continued)

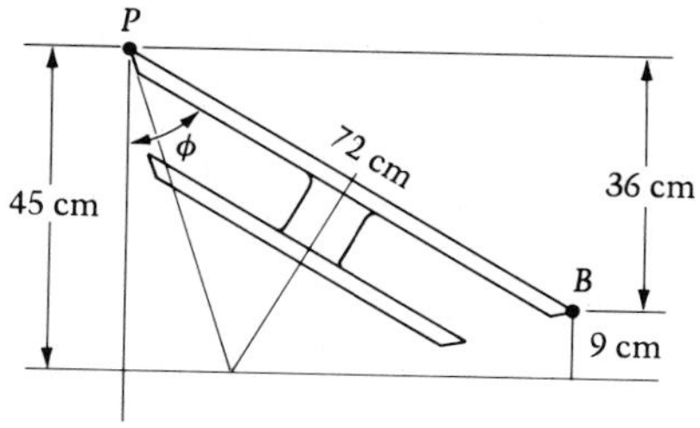

To obtain $\mathbf{r}_{PB}$, we use the accompanying sketch and see that

$$\phi = \cos^{-1}\left(\frac{36}{72}\right) = 60°$$

Therefore

$$\mathbf{r}_{PB} = 72\left(\frac{\sqrt{3}}{2}\hat{\mathbf{i}} - \frac{1}{2}\hat{\mathbf{j}}\right) \text{ cm} \tag{3}$$

Also, by the addition theorem,

$$\begin{aligned} \boldsymbol{\omega}_{\mathcal{C}} = \boldsymbol{\omega}_{\mathcal{C}/\mathcal{I}} &= \boldsymbol{\omega}_{\mathcal{C}/\mathcal{D}} + \boldsymbol{\omega}_{\mathcal{D}/\mathcal{I}} \\ &= \omega_{\mathcal{C}/\mathcal{D}}\left(\frac{1}{2}\hat{\mathbf{i}} + \frac{\sqrt{3}}{2}\hat{\mathbf{j}}\right) + \omega_{\mathcal{D}}\hat{\mathbf{i}} \end{aligned} \tag{4}$$

in which we have used the fact that we know the directions (but not the magnitudes yet!) of $\boldsymbol{\omega}_{\mathcal{C}/\mathcal{D}}$ and $\boldsymbol{\omega}_{\mathcal{D}/\mathcal{I}}$.

Substituting Equations (3) and (4) into (2) and substituting the result into Equation (1) gives us

$$\begin{aligned} 9\omega_{\mathcal{B}}\hat{\mathbf{k}} &= \left[\omega_{\mathcal{C}/\mathcal{D}}\left(\frac{1}{2}\hat{\mathbf{i}} + \frac{\sqrt{3}}{2}\hat{\mathbf{j}}\right) + \omega_{\mathcal{D}}\hat{\mathbf{i}}\right] \times 72\left(\frac{\sqrt{3}}{2}\hat{\mathbf{i}} - \frac{1}{2}\hat{\mathbf{j}}\right) \\ &= \left\{\omega_{\mathcal{C}/\mathcal{D}}\left[72\left(-\frac{1}{4} - \frac{3}{4}\right)\right] + \omega_{\mathcal{D}}\left[72\left(-\frac{1}{2}\right)\right]\right\}\hat{\mathbf{k}} \end{aligned}$$

or, simplifying,

$$\omega_{\mathcal{B}} = -8\omega_{\mathcal{C}/\mathcal{D}} - 4\omega_{\mathcal{D}} \tag{5}$$

To get another equation in these variables, we shall use the point T, which belongs to both $\mathcal{C}$ and $\mathcal{D}$ and is shown in the diagram at the left with some essential geometry. First, as a point of $\mathcal{D}$, we have

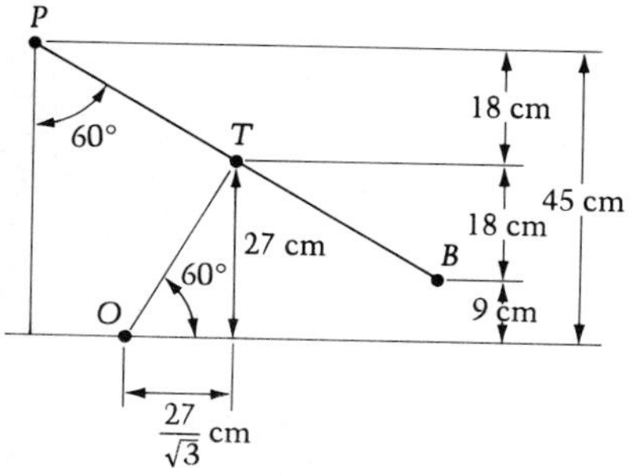

$$\begin{aligned} \mathbf{v}_T &= \cancelto{\mathbf{0}}{\mathbf{v}_O} + \boldsymbol{\omega}_{\mathcal{D}} \times \mathbf{r}_{OT} \\ &= \omega_{\mathcal{D}}\hat{\mathbf{i}} \times \left(\frac{27}{\sqrt{3}}\hat{\mathbf{i}} + 27\hat{\mathbf{j}}\right) \\ &= 27\omega_{\mathcal{D}}\hat{\mathbf{k}} \text{ cm/s} \end{aligned} \tag{6}$$

Next, as a point of $\mathcal{C}$,

$$\mathbf{v}_T = \cancelto{\mathbf{0}}{\mathbf{v}_P} + \boldsymbol{\omega}_{\mathcal{C}} \times \mathbf{r}_{PT} \tag{7}$$

The cross product in Equation (7) is one-half the cross product in Equation (2).

Question 6.6 Why is this so?

(Continued)

Therefore

$$\mathbf{v}_T = (-36\omega_{\mathcal{C}/\mathcal{D}} - 18\omega_{\mathcal{D}})\hat{\mathbf{k}} \tag{8}$$

Equating the right sides of Equations (6) and (8) gives

$$27\omega_{\mathcal{D}} = -36\omega_{\mathcal{C}/\mathcal{D}} - 18\omega_{\mathcal{D}}$$

or

$$5\omega_{\mathcal{D}} = -4\omega_{\mathcal{C}/\mathcal{D}} \tag{9}$$

Substituting Equation (9) into (5) leads to

$$\omega_{\mathcal{D}} = \frac{\omega_{\mathcal{B}}}{6} \text{ rad/s} \tag{10}$$

and

$$\omega_{\mathcal{C}/\mathcal{D}} = -\frac{5}{24}\omega_{\mathcal{B}} \text{ rad/s} \tag{11}$$

Substituting Equations (10) and (11) into (4) gives us $\boldsymbol{\omega}_{\mathcal{C}}$, which we shall need in the last step of the problem. The result is

$$\boldsymbol{\omega}_{\mathcal{C}} = \left(-\frac{5}{24}\omega_{\mathcal{B}}\right)\left(\frac{1}{2}\hat{\mathbf{i}} + \frac{\sqrt{3}}{2}\hat{\mathbf{j}}\right) + \left(\frac{\omega_{\mathcal{B}}}{6}\right)\hat{\mathbf{i}}$$

$$\boldsymbol{\omega}_{\mathcal{C}} = \frac{\omega_{\mathcal{B}}}{16}\hat{\mathbf{i}} - \frac{5\sqrt{3}}{48}\omega_{\mathcal{B}}\hat{\mathbf{j}} \tag{12}$$

We can now relate the velocities of the contacting points of bodies $\mathcal{C}$ and $\mathcal{A}$ at point A; first, using points A and E of $\mathcal{A}$, we get

$$\mathbf{v}_A = \overset{\mathbf{0}}{\cancel{\mathbf{v}_E}} + \omega_{\mathcal{A}}\hat{\mathbf{i}} \times 30\hat{\mathbf{j}}$$

$$\mathbf{v}_A = 30\omega_{\mathcal{A}}\hat{\mathbf{k}} \text{ cm/s} \tag{13}$$

To get another expression for $\mathbf{v}_A$, we relate the velocities of the two points A and P on body $\mathcal{C}$:

$$\mathbf{v}_A = \overset{\mathbf{0}}{\cancel{\mathbf{v}_P}} + \boldsymbol{\omega}_{\mathcal{C}} \times \mathbf{r}_{PA} \tag{14}$$

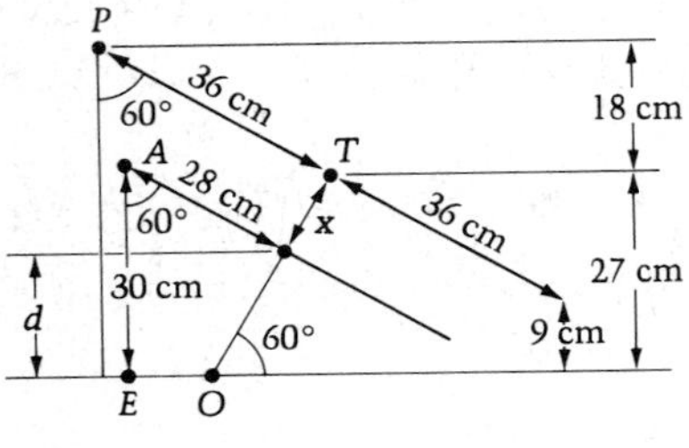

To obtain the position vector $\mathbf{r}_{PA}$, we use the geometry shown in the diagram. The distance $\mathbf{x}$, needed in forming $\mathbf{r}_{PA}$, is equal to $(27 - d)/\sin 60°$:

$$\mathbf{x} = \frac{27 - d}{\sqrt{3}/2} = \frac{27 - [30 - 28(\frac{1}{2})]}{\sqrt{3}/2} = \frac{22}{\sqrt{3}} \tag{15}$$

Therefore

$$\mathbf{r}_{PA} = (36 - 28)\left(\frac{\sqrt{3}}{2}\hat{\mathbf{i}} - \frac{1}{2}\hat{\mathbf{j}}\right) + \frac{22}{\sqrt{3}}\left(-\frac{1}{2}\hat{\mathbf{i}} - \frac{\sqrt{3}}{2}\hat{\mathbf{j}}\right)$$

$$= 0.577\hat{\mathbf{i}} - 15.0\hat{\mathbf{j}} \text{ cm} \tag{16}$$

(Continued)

Substituting Equations (12) and (16) into (14), we get

$$\mathbf{v}_A = \omega_{\mathcal{B}}\left(\frac{1}{16}\hat{\mathbf{i}} - \frac{5\sqrt{3}}{48}\hat{\mathbf{j}}\right) \times (0.577\hat{\mathbf{i}} - 15.0\hat{\mathbf{j}})$$

$$\mathbf{v}_A = -0.833\omega_{\mathcal{B}}\hat{\mathbf{k}} \text{ cm/s} \tag{17}$$

Equating the two expressions for $\mathbf{v}_A$ in Equations (13) and (17) gives

$$\omega_{\mathcal{A}} = -0.0278\omega_{\mathcal{B}} \text{ rad/s} \tag{18}$$

It is seen that the angular speed of gear $\mathcal{B}$ is 36 times that of $\mathcal{A}$, and in the opposite direction.

Question 6.7 Give an argument why $\mathcal{A}$ has to be moving in the opposite direction from that of $\mathcal{B}$. (*Hint:* Use the original figure and focus your attention on points P and O of $\mathcal{C}$.)

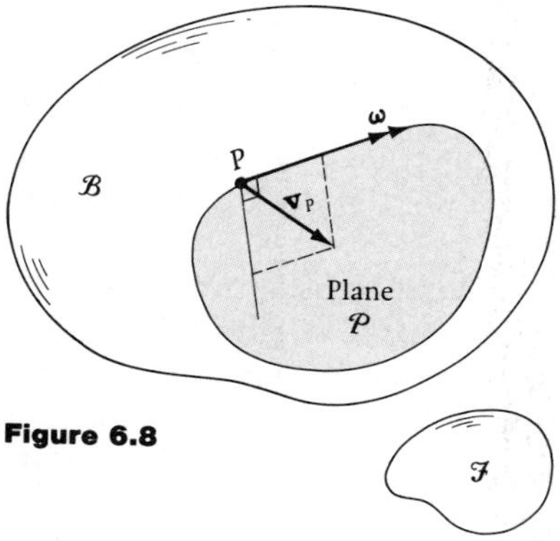

Figure 6.8

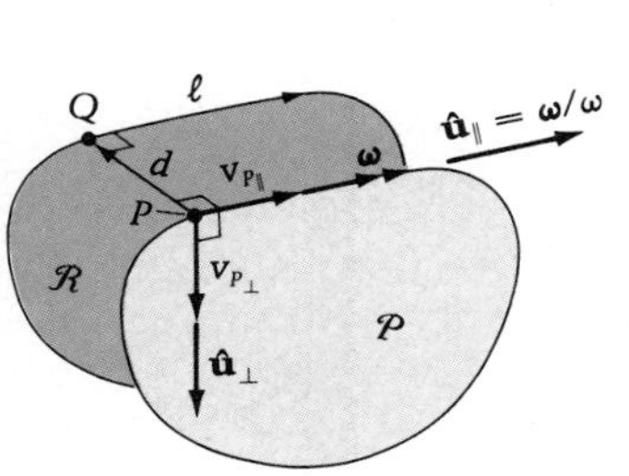

Figure 6.9

We recall that in Chapter 3 we were able to show that in plane motion a rigid body $\mathcal{B}$, except when its angular velocity vanishes, always has a point of zero velocity (the instantaneous center Ⓘ). We now show that in general (three-dimensional) motion, this is not the case. We start with an arbitrary point P with velocity $\mathbf{v}_P$, and sketch its velocity along with the angular velocity vector $\boldsymbol{\omega}_{\mathcal{B}/\mathcal{F}} = \boldsymbol{\omega}$ of $\mathcal{B}$ in the reference frame $\mathcal{F}$. (See Figure 6.8.)

Note that in Figure 6.8 there is a plane $\mathcal{P}$ defined by the vectors $\mathbf{v}_P$ and $\boldsymbol{\omega}$ drawn through P, unless the two vectors are parallel. If they are, then the motion of $\mathcal{B}$ in $\mathcal{F}$ is like that of a screwdriver—the body turns around a line that translates along its axis. The general case ($\mathbf{v}_P$ not parallel to $\boldsymbol{\omega}$) may also be reduced to a screwdriver motion as follows. First we replace $\mathbf{v}_P$ by its components parallel ($v_{P_\parallel}$) and perpendicular ($v_{P_\perp}$) to $\boldsymbol{\omega}$. (See Figure 6.9.) Next we consider a plane $\mathcal{R}$ parallel to $\mathcal{P}$ and separated from $\mathcal{P}$ by the distance d as shown in Figure 6.9. Point Q is the projection of P into the plane $\mathcal{R}$, and we may write its velocity in terms of P by Equation (6.46):

$$\mathbf{v}_Q = \mathbf{v}_P + \boldsymbol{\omega} \times \mathbf{r}_{PQ}$$

$$= v_{P_\parallel}\hat{\mathbf{u}}_\parallel + v_{P_\perp}\hat{\mathbf{u}}_\perp + \boldsymbol{\omega} \times d\underbrace{(\hat{\mathbf{u}}_\parallel \times \hat{\mathbf{u}}_\perp)}$$

Note from Figure 6.9 that this is the unit vector directed from P to Q.

The vector triple product is equal to

$$d\hat{\mathbf{u}}_{\parallel}(\boldsymbol{\omega}\cdot\hat{\mathbf{u}}_{\perp}) - d\hat{\mathbf{u}}_{\perp}(\boldsymbol{\omega}\cdot\hat{\mathbf{u}}_{\parallel}) = \mathbf{0} - d\hat{\mathbf{u}}_{\perp}(\omega)$$

so that

$$\mathbf{v}_Q = v_{P_{\parallel}}\hat{\mathbf{u}}_{\parallel} + \hat{\mathbf{u}}_{\perp}(v_{P_{\perp}} - d\omega)$$

Therefore if d is chosen equal to $v_{P_{\perp}}/\omega$, the line ℓ will be a "screwdriver line" and the motion of $\mathcal{B}$ will be as shown in Figure 6.10.

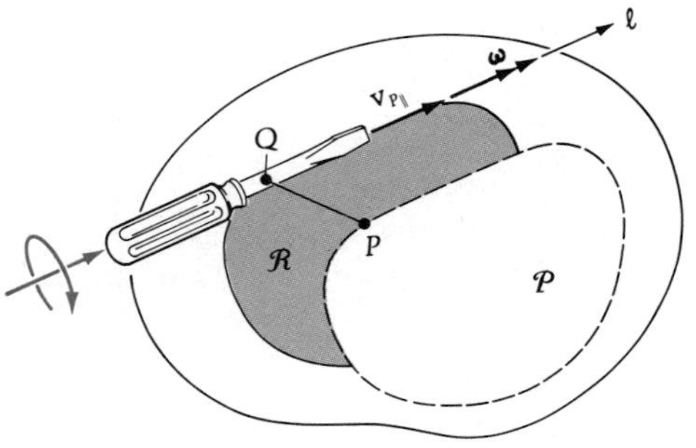

Figure 6.10

All points *on* ℓ have velocities *along* ℓ at the given instant, while those off the line have the same velocity component parallel to ℓ; in addition they rotate around it. This is the simplest reduction possible for the motion of $\mathcal{B}$, and it is clear that unless $v_{P_{\parallel}} = 0$, *no* points can have zero velocity. Thus in three dimensions we are no longer assured of having an instantaneous center Ⓘ as we were in dealing with plane motion.

There are, however, special cases in three dimensions in which Ⓘ exists; a good example is a cone rolling on a plane (Figure 6.11). Note that in this case the *entire line* of contact is at each instant at rest on the plane. Since the angular velocity of the cone is parallel to this line, *all* the contact points have $v_{P_{\parallel}} = 0$, which we have shown must be true if Ⓘ is to exist in three dimensions.

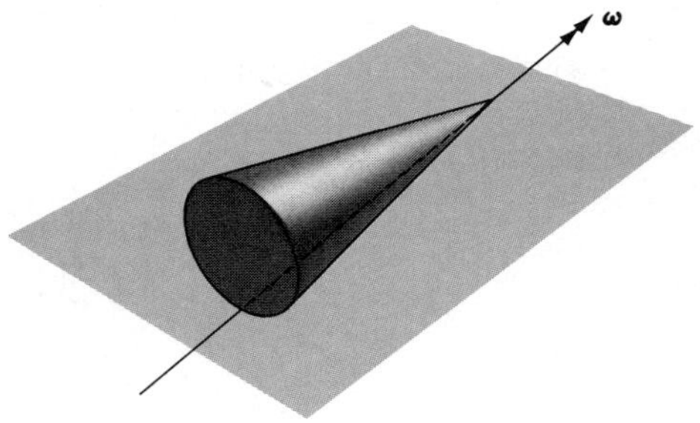

Figure 6.11

PROBLEMS / Section 6.6

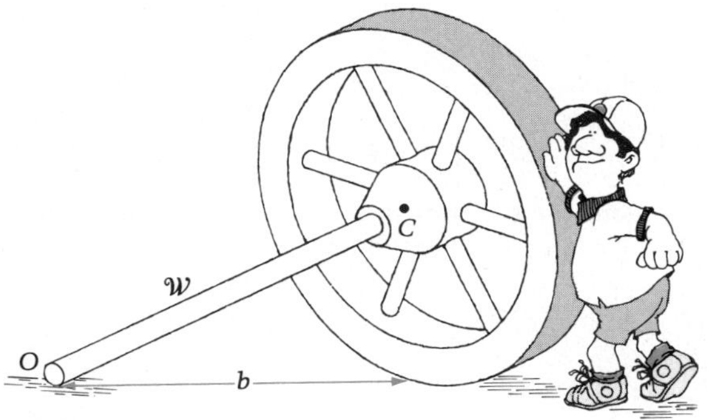

Figure P6.42a

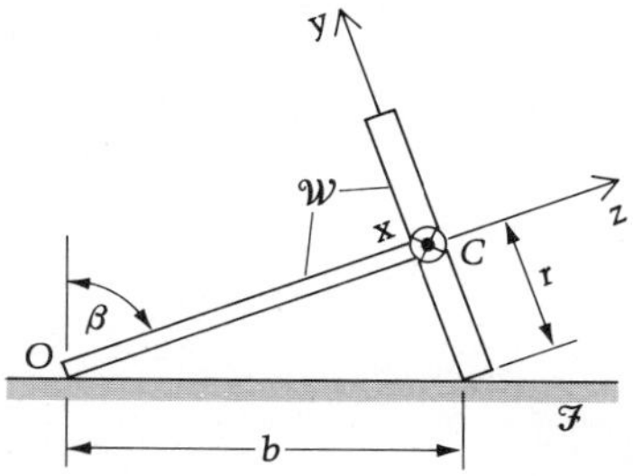

Figure P6.42b

6.42 A youngster finds an old wagon wheel $\mathcal{W}$ and pushes it around with the center C moving at constant speed in a horizontal circle. (See Figure P6.42a.) If the end O of the axle stays fixed while C returns to its starting point in T seconds, compute the angular velocity vector of the wheel $\boldsymbol{\omega}_{\mathcal{W}/\mathcal{F}}$, where $\mathcal{F}$ is the ground frame (Figure P6.42b). Give the result in terms of b, β, and T.

6.43 The bevel gears $\mathcal{B}_1$ and $\mathcal{B}_2$ in Figure P6.43 support the turning shaft $\mathcal{S}$, whose angular velocity is given by $\boldsymbol{\omega}_{\mathcal{S}/\mathcal{F}} = 30\hat{\mathbf{k}}$ rad/sec, in which $\hat{\mathbf{k}}$ is parallel to the z direction in both frames $\mathcal{F}$ and $\mathcal{S}$. (Gears $\mathcal{A}$ and $\mathcal{C}$ are seen to be part of $\mathcal{F}$.) Find $\boldsymbol{\omega}_{\mathcal{B}_1/\mathcal{F}}$ and $\boldsymbol{\omega}_{\mathcal{B}_2/\mathcal{F}}$.

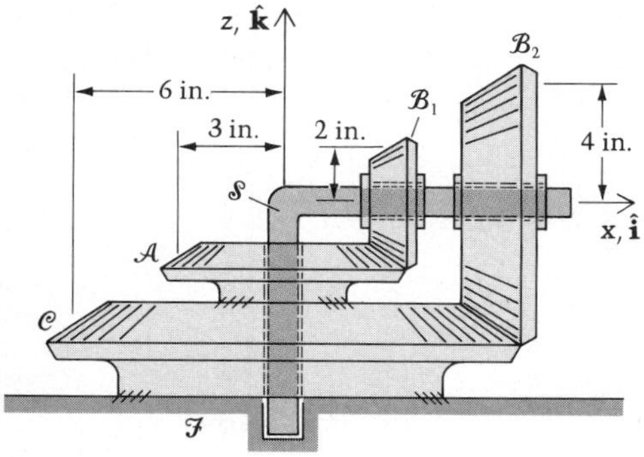

Figure P6.43

It is to be assumed in the following three problems that not all points of body $\mathcal{B}$ have zero acceleration.

6.44 Show that for a rigid body $\mathcal{B}$ in general motion, there is a point Q of zero acceleration if $\boldsymbol{\omega} \neq \mathbf{0}$, $\boldsymbol{\alpha} \neq \mathbf{0}$, and $\boldsymbol{\alpha}$ is not parallel to $\boldsymbol{\omega}$. *Hint:* Let P be an arbitrary point, and let $\boldsymbol{\omega} = \omega\hat{\mathbf{i}}$, $\boldsymbol{\alpha} = \alpha_1\hat{\mathbf{i}} + \alpha_2\hat{\mathbf{j}}$, and $\mathbf{a}_P = a_{P_1}\hat{\mathbf{i}} + a_{P_2}\hat{\mathbf{j}} + a_{P_3}\hat{\mathbf{k}}$, noting that there is no loss in generality in these assumptions. Set $\mathbf{a}_Q = \mathbf{0} = \mathbf{a}_P + \boldsymbol{\alpha} \times \mathbf{r}_{PQ} + \boldsymbol{\omega} \times (\boldsymbol{\omega} \times \mathbf{r}_{PQ})$, set $\mathbf{r}_{PQ} = x\hat{\mathbf{i}} + y\hat{\mathbf{j}} + z\hat{\mathbf{k}}$, and solve for x, y, and z.

6.45 (a) Following up the previous problem, show that if $\boldsymbol{\omega} = \mathbf{0}$ and $\boldsymbol{\alpha} \neq \mathbf{0}$ at a given instant then at this instant there is a point Q of zero acceleration if and only if the accelerations of all points of $\mathcal{B}$ are perpendicular to $\boldsymbol{\alpha}$. (b) Investigate the case $\boldsymbol{\omega} \neq \mathbf{0}$ and $\boldsymbol{\alpha} = \mathbf{0}$.

6.46 Here is another follow-up on Problem 6.44: Show that at any instant when the two vectors $\boldsymbol{\omega}$ and $\boldsymbol{\alpha}$ are parallel, then there is a point of $\mathcal{B}$ with zero acceleration if and only if the accelerations of all its points are perpendicular to $\boldsymbol{\omega}$ and $\boldsymbol{\alpha}$.

6.47 Find the angular acceleration of the wagon wheel of Problem 6.42.

6.48 In Problem 6.43 find $\boldsymbol{\alpha}_{\mathcal{B}_1/\mathcal{F}}$ and $\boldsymbol{\alpha}_{\mathcal{B}_2/\mathcal{F}}$.

6.49 Cone $\mathcal{C}_1$ rolls on cone $\mathcal{C}_2$ so that its axis of symmetry (x) moves in a horizontal plane through O, turning about z at rate Ω_2 rad/sec. (See Figure P6.49.) Cone $\mathcal{C}_2$ is rotating about (−z) at Ω_1 rad/sec. Find, with respect to the frame $\mathcal{F}$ in which $\mathcal{C}_2$ turns, the angular velocity and angular acceleration of $\mathcal{C}_1$, and the acceleration of point A. (Axes (x, y, z) are fixed in frame $\mathcal{B}$, which turns so that **x** is always along the axis of symmetry of $\mathcal{C}_1$ and z is always vertical. Also, Ω_1, and Ω_2 are constants.)

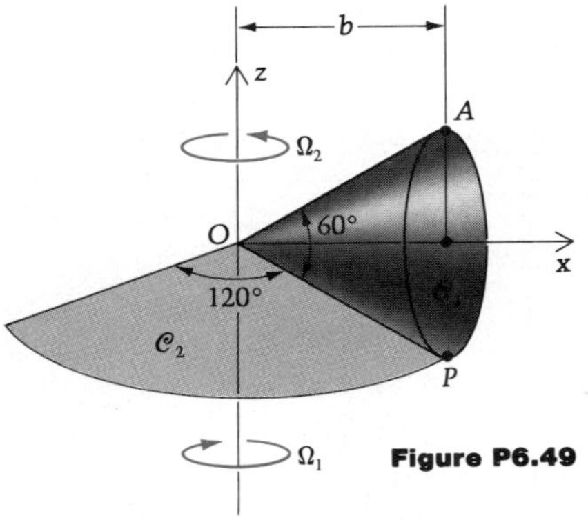

Figure P6.49

6.50 Find the angular acceleration of gear $\mathcal{C}$ in reference frame $\mathcal{F}$ in Example 6.7 if $\omega_{\mathcal{B}}$ = constant.

6.51 A wheel of radius r turns on an axle that rotates with angular velocity $\omega_1\hat{\mathbf{k}}$ about a vertical axis (z) fixed relative to ground (Figure P6.51). If the wheel rolls on the horizontal plane and ω_1 is constant, find:

- **a.** The angular velocity and angular acceleration of the wheel relative to the ground
- **b.** The acceleration, relative to the ground, of the point on the wheel in contact with the horizontal plane

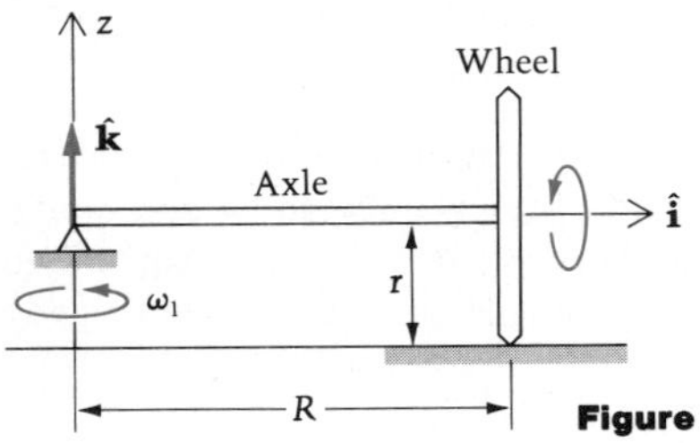

Figure P6.51

6.52 Rework Example 6.7, but this time suppose that the radius of gear $\mathcal{A}$ is 44 cm instead of 30 cm (and that its center is at the same point of $\mathcal{F}$). Explain why gears $\mathcal{A}$ and $\mathcal{B}$ are now moving in the same direction.

6.53 In Example 6.7 find the radius of gear $\mathcal{A}$ for which it will remain stationary in $\mathcal{F}$ as $\mathcal{B}$, $\mathcal{C}$, and $\mathcal{D}$ turn.

6.54 In Example 6.7 label the radius of gear $\mathcal{A}$ as H and call the radius of the 28-cm gear R. Show that the relationship between $\omega_{\mathcal{A}}$ and $\omega_{\mathcal{B}}$ is given by $96\omega_{\mathcal{A}}H = (16H - 20R)\omega_{\mathcal{B}}$. (The center of $\mathcal{A}$ is fixed in $\mathcal{F}$.)

6.55 Collars $\mathcal{C}_1$ and $\mathcal{C}_2$ in Figure P6.55 are attached at C_1 and C_2 to rod $\mathcal{R}$ by ball and socket joints. Point C_2 has a motion along the **x** axis given by $\mathbf{x}_2 = -0.012t^3$ mm. Find the velocity of C_1 as a function of time.

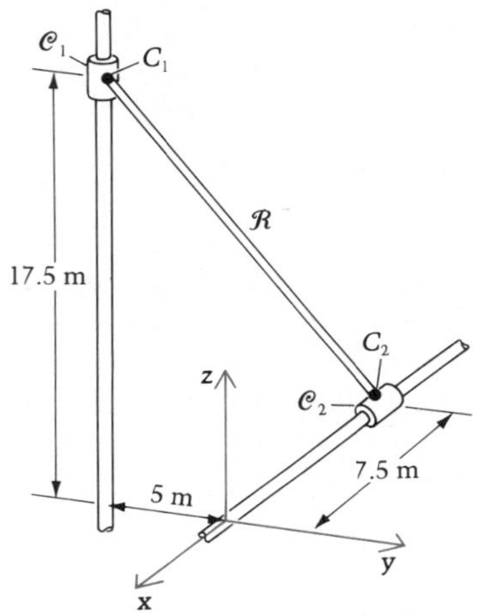

Figure P6.55

6.56 Plate $\mathcal{P}$ in Figure P6.56 has the following motion:

1. Corner A moves on the **x** axis.
2. Corner B moves on the **y** axis with constant velocity $6\hat{\mathbf{j}}$ in./sec.
3. Some point of the top edge of $\mathcal{P}$ (point Q at the instant shown) is always in contact with the z axis.

Find the angular velocity vector of the plate when $\mathbf{x}_A$ = 3 in.

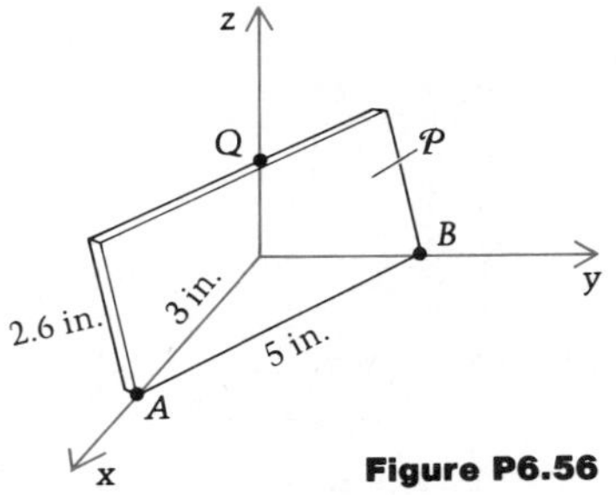

Figure P6.56

6.57 The angular velocity of a rigid body $\mathcal{B}$, in motion in frame $\mathcal{F}$, is $\boldsymbol{\omega} = 3\hat{\mathbf{i}} + 2\hat{\mathbf{j}}$ rad/sec. If possible, locate (from P) a point Q of $\mathcal{B}$ with zero velocity when:

a. $\mathbf{v}_{P/\mathcal{F}} = 5\hat{\mathbf{i}}$ in./sec
b. $\mathbf{v}_{P/\mathcal{F}} = 6\hat{\mathbf{i}} - 9\hat{\mathbf{j}} + 3\hat{\mathbf{k}}$ in./sec

In both cases explain why Q exists or not in light of the discussion about points of zero velocity in the preceding section.

6.58 Disk $\mathcal{D}$ in Figure P6.58 spins relative to the bent shaft $\mathcal{B}$ at constant angular speed Ω_1 rad/sec; $\mathcal{B}$ rotates in the reference frame $\mathcal{F}$ at the constant rate Ω_2 rad/sec. (The directions are indicated in the figures.) Using the rigid-body equations (6.54 and 6.55), find $\mathbf{v}_{Q/\mathcal{F}}$ and $\mathbf{a}_{Q/\mathcal{F}}$ for point Q on the periphery of the disk. Express the result in terms of components along the $(\mathbf{x}, y, z)$ axes, which are fixed in $\mathcal{B}$.

6.59 Rework the preceding problem, this time using the moving-frame concept of Section 6.5. Let bar $\mathcal{B}$ be the frame in motion with respect to $\mathcal{F}$, and let Q be the point moving relative to both $\mathcal{B}$ and $\mathcal{F}$.

6.60 The two shafts $\mathcal{S}_1$ and $\mathcal{S}_2$ are fixed to bevel gears $\mathcal{A}$ and $\mathcal{B}$ as shown in Figure P6.60. (a) Prove that if the velocities of each pair of contacting points are to match along the line of contact of the gears, then points A, B, and C must coincide. (b) Let A, B, and C coincide and find the ratio of ω to ω'.

6.61 (a) In the preceding problem show that if $0 < \beta < 90°$, the result is still true about A, B, and C coinciding. (b) Find ω/ω' for this case.

6.62 The bevel gear $\mathcal{A}$ in Figure P6.62 is fixed to a reference frame in which the mating gear $\mathcal{B}$ moves. The axis OC of $\mathcal{B}$ turns about the z axis at the constant rate $\Omega = 0.2$ rad/sec. Find the angular velocity of $\mathcal{B}$ in $\mathcal{A}$.

6.63 In the preceding problem, find the angular acceleration of $\mathcal{B}$ in $\mathcal{A}$ for the same defined motion.

6.64 A differential friction gear can be made with either bevel gears, as shown at the top of Figure P6.64, or friction disks, as shown at the bottom. In each case, body $\mathcal{D}$ rolls on $\mathcal{A}$ and $\mathcal{B}$ and may turn without resistance on the crank arm $\mathcal{C}$. Find $\boldsymbol{\omega}_{\mathcal{C}/\mathcal{F}}$, $\boldsymbol{\omega}_{\mathcal{D}/\mathcal{C}}$, and the velocities of points A and B of $\mathcal{D}$.

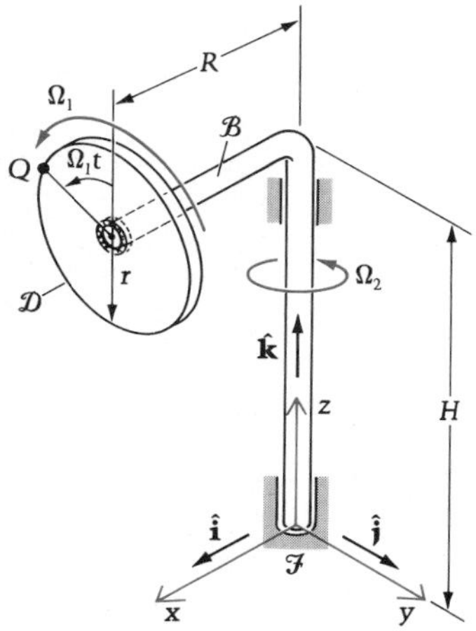

Figure P6.58

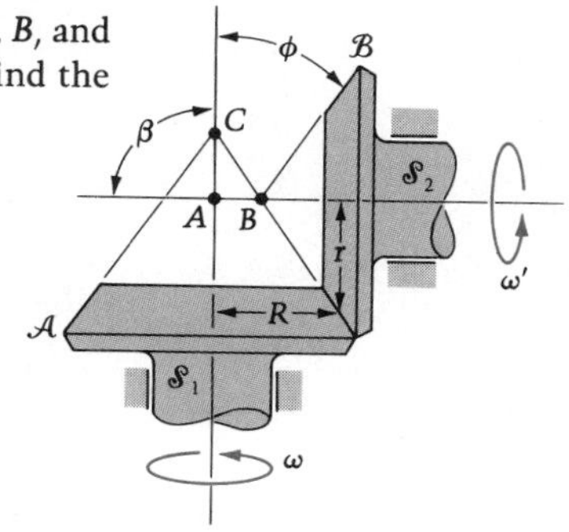

Figure P6.60

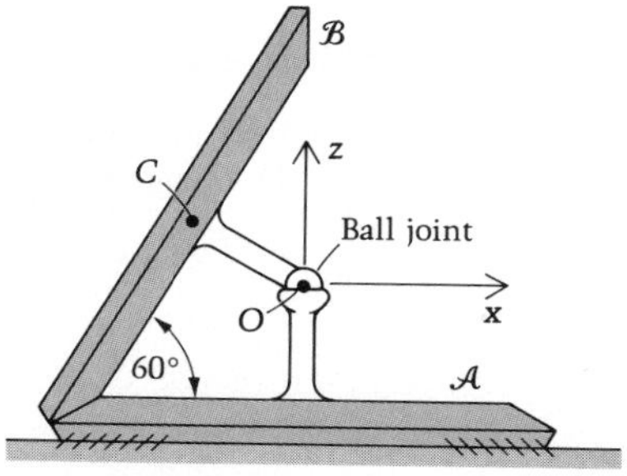

Figure P6.62

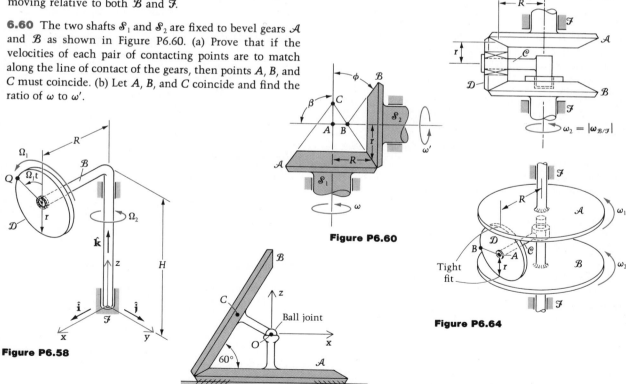

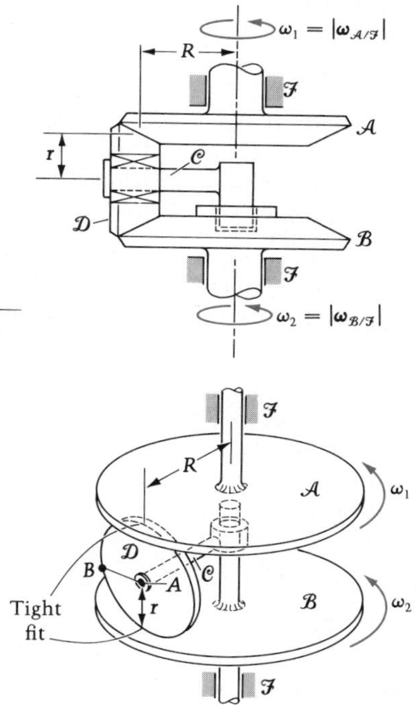

Figure P6.64

6.65 The uniform, solid, right circular cone $\mathcal{C}$ in Figure P6.65 rolls on the horizontal plane $\mathcal{F}$. Let $\mathcal{P}$ represent a frame in which the vertical axis z and the cone's axis ℓ are fixed. (Hence $\mathcal{P}$ has a simple angular velocity in $\mathcal{F}$ about the vertical.) Show that the cone can roll so that $|\boldsymbol{\omega}_{\mathcal{C}/\mathcal{F}}| = \omega$ and $|\boldsymbol{\omega}_{\mathcal{P}/\mathcal{F}}| = n$ are constants and

$$\omega \sin \alpha = n \cos \alpha$$

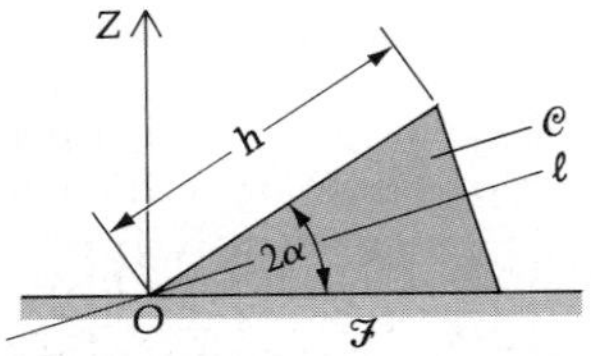

Figure P6.65

6.66 Depicted in Figure P6.66a are the main features of an automobile differential. The left and right axles, $\mathcal{L}$ and $\mathcal{R}$, are keyed to the bevel gears $\mathcal{B}_1$ and $\mathcal{B}_2$. Gear $\mathcal{G}$ is fixed to the case $\mathcal{C}$, and the combination is free to turn in bearings around line ℓ. Gear $\mathcal{G}$ meshes with gear $\mathcal{D}$ attached to the car's drive shaft. As the casing turns about the common axis ℓ of $\mathcal{L}$ and $\mathcal{R}$, its pins bear against the other two bevel gears within $\mathcal{C}$, which are $\mathcal{B}_3$ and $\mathcal{B}_4$. (Observe that these two gears do not turn about their axes at all on a straight road.)

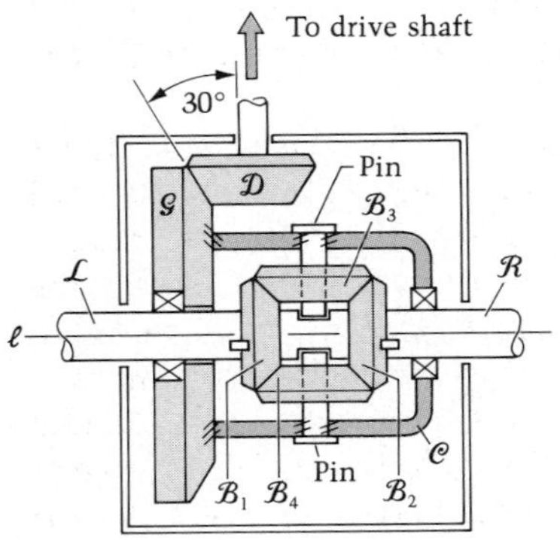

Figure P6.66a

Suppose the car makes a 30-ft turn at a speed of 20 mph (Figure P6.66b). If the tire radius is 14 in., find the angular velocities of $\mathcal{L}$ and $\mathcal{R}$ and use them to compute the angular velocity of $\mathcal{D}$. In the process note that the differential allows the driven wheels to turn at different angular speeds.

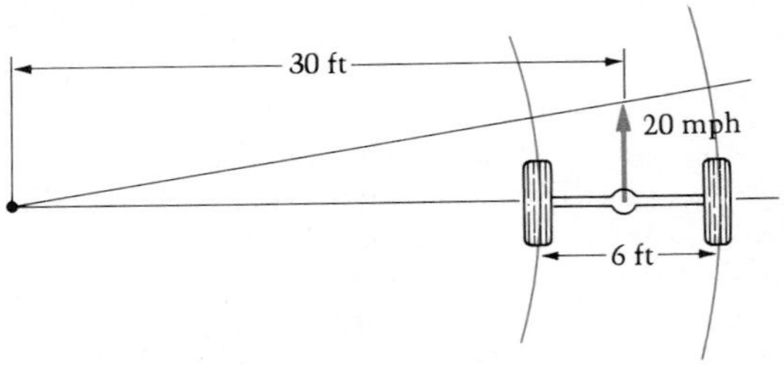

Figure P6.66b

6.67 The center C of the bevel gear $\mathcal{C}$ in Figure P6.67 rotates in a horizontal circle at a constant speed of 40 mm/s (clockwise when viewed from above). The mating gear $\mathcal{D}$ is fixed to the reference frame $\mathcal{F}$; shaft $\mathcal{S}$, rigidly attached to $\mathcal{C}$, is connected to $\mathcal{F}$ through a ball and socket joint at O. Find the angular velocity vector of $\mathcal{C}$ in $\mathcal{F}$.

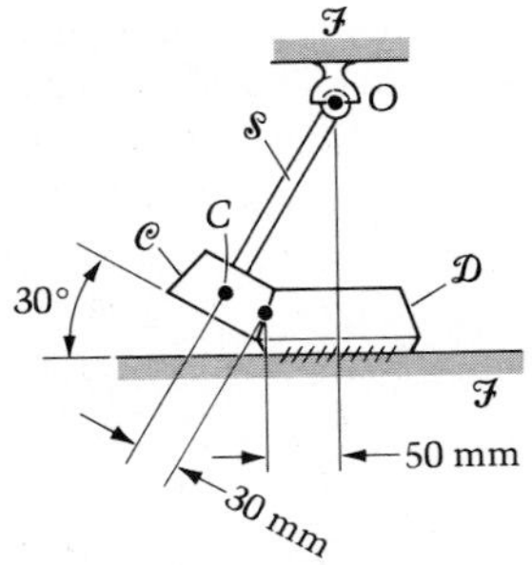

Figure P6.67

6.68 Find $\boldsymbol{\alpha}_{\mathcal{C}/\mathcal{F}}$ for the gear in Problem 6.67.

6.69 Find the angular acceleration of the plate of Problem 6.56 at the same time.

6.7 Describing the Orientation of a Rigid Body

In the case of plane motion of a rigid body $\mathcal{B}$, if we know the angular velocity

$$\boldsymbol{\omega} = \omega\hat{\mathbf{k}} = \dot{\theta}\hat{\mathbf{k}}$$

as a function of time, we may clearly integrate to find the orientation of $\mathcal{B}$ at any time t:

$$\int_0^t \dot{\theta}(\xi)\, d\xi = \theta(t) - \theta(0) \tag{6.56}$$

Thus in plane motion we may completely specify the position of $\mathcal{B}$ by giving the **xy** coordinates of a point (usually the mass center C is chosen to be the point) and the orientation angle θ.

A major difference between planar and general motion is that the angles which yield a body's orientation in space in three-dimensional motion are *not* the integrals by simple quadratures of the angular velocity components of the body. In fact, finding (in general) the orientation of a body $\mathcal{B}$ in closed form, given $\boldsymbol{\omega}(t)$ and the orientation of $\mathcal{B}$ at $t = 0$, is an unsolved fundamental problem in rigid-body kinematics. We now introduce the **Eulerian angles** in order to show the difficulty of determining a body's orientation in space when the motion is nonplanar.

We begin with the body $\mathcal{B}$ oriented so that the body-fixed axes $(\mathbf{x}, \mathbf{y}, \mathbf{z})$ initially coincide respectively with axes (X, Y, Z) embedded in the reference frame $\mathcal{I}$. Let $(\hat{\mathbf{i}}, \hat{\mathbf{j}}, \hat{\mathbf{k}})$ and $(\hat{\mathbf{I}}, \hat{\mathbf{J}}, \hat{\mathbf{K}})$ be sets of unit vectors respectively parallel to $(\mathbf{x}, \mathbf{y}, \mathbf{z})$ and (X, Y, Z). Three successive rotations about specific axes will now be described that will orient $\mathcal{B}$ in $\mathcal{I}$. (See Figure 6.12a.)

In Figure 6.12b the first rotation is through the angle ϕ about the Z axis. Let the new positions of $(\mathbf{x}, \mathbf{y}, \mathbf{z})$ after this first rotation be denoted by $(\mathbf{x}_1, \mathbf{y}_1, \mathbf{z}_1)$ as shown; these positions are embedded in an intermediate frame $\mathcal{I}_1$. Note that axes Z and $\mathbf{z}_1$ are identical and that $\mathcal{I}_1$ has the simple angular velocity $\dot{\phi}\hat{\mathbf{K}}$ in $\mathcal{I}$. Note also from Figure 6.12 that $\hat{\mathbf{n}}_{11}$, $\hat{\mathbf{n}}_{12}$, and $\hat{\mathbf{n}}_{13}$ are unit vectors that are respectively and always parallel to $\mathbf{x}_1, \mathbf{y}_1$, and $\mathbf{z}_1$. Next (Figure 6.12c) a rotation through the angle θ about axis $\mathbf{y}_1$ moves the body axes into the coordinate directions $(\mathbf{x}_2, \mathbf{y}_2, \mathbf{z}_2)$ of a second intermediate frame $\mathcal{I}_2$ having unit vectors $(\hat{\mathbf{n}}_{21}, \hat{\mathbf{n}}_{22}, \hat{\mathbf{n}}_{23})$. A final rotation, this time of amount ψ about $\mathbf{z}_2$ (Figure 6.12d), turns the body axes into their final positions in $\mathcal{B}$, indicated by $(\mathbf{x}, \mathbf{y}, \mathbf{z})$.

It is clear that we may use the addition theorem to express the angular velocity of $\mathcal{B}$ in $\mathcal{I}$ as follows:

$$\boldsymbol{\omega}_{\mathcal{B}/\mathcal{I}} = \boldsymbol{\omega}_{\mathcal{B}/\mathcal{I}_2} + \boldsymbol{\omega}_{\mathcal{I}_2/\mathcal{I}_1} + \boldsymbol{\omega}_{\mathcal{I}_1/\mathcal{I}} \tag{6.57}$$

$$\boldsymbol{\omega}_{\mathcal{B}/\mathcal{I}} = \dot{\psi}\hat{\mathbf{k}} + \dot{\theta}\hat{\mathbf{n}}_{12} + \dot{\phi}\hat{\mathbf{K}} \tag{6.58}$$

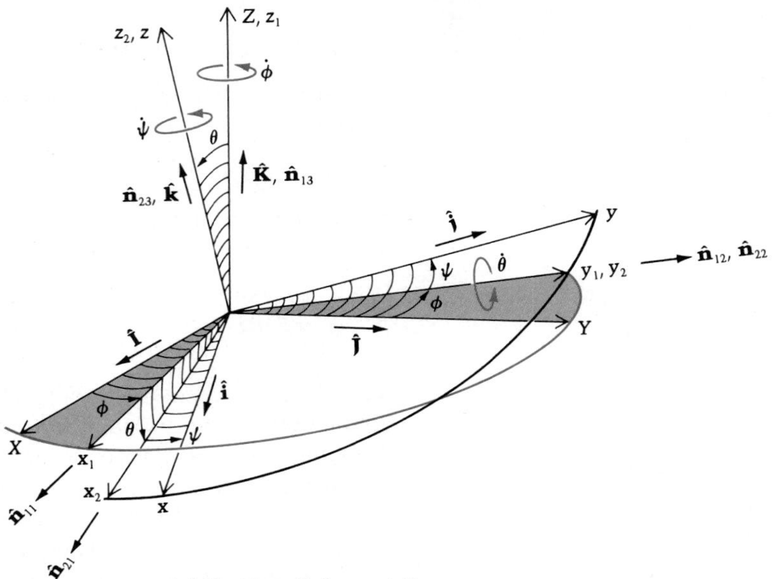

(a) Showing all three rotations

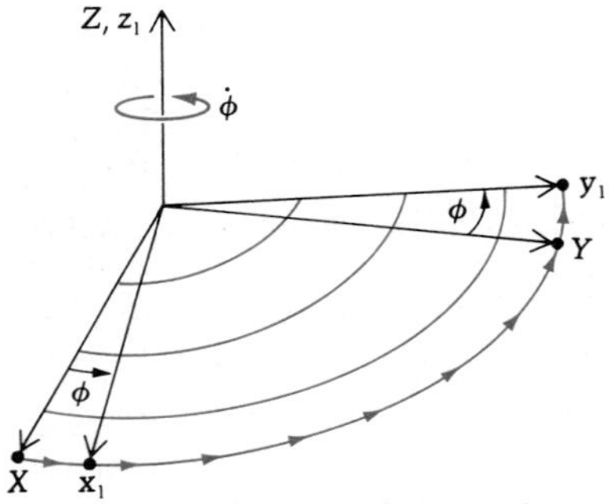

(b) First Eulerian angle rotation (ϕ), about Z

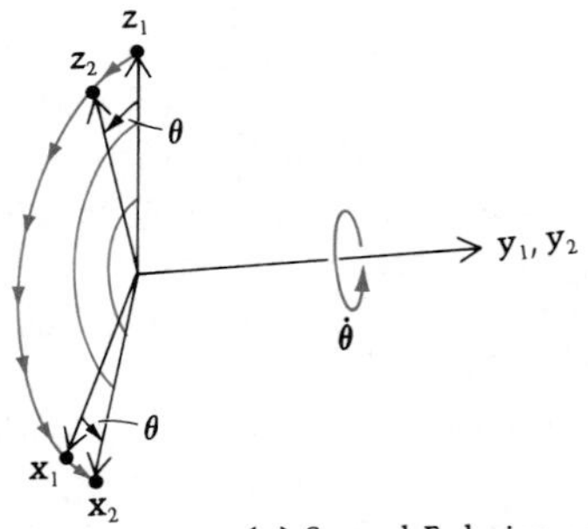

(c) Second Eulerian angle rotation (θ), about y_1

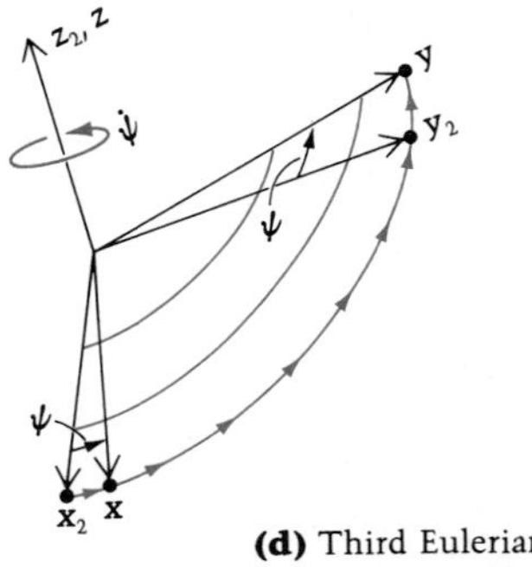

(d) Third Eulerian angle rotation (ψ), about z_2

Figure 6.12 Eulerian angles.

in which we again remark that $\hat{\mathbf{n}}_{ij}$ is a unit vector in the jth coordinate direction of frame $\mathcal{F}_i$ $(i = 1, 2)$.

If the preceding expression for $\boldsymbol{\omega}_{\mathcal{B}/\mathcal{F}}$ is to be functionally useful, its three terms should all be expressed in the *same frame*. Now in practice, as we shall see, they are sometimes expressed by components associated with body-fixed directions, at other times with space-fixed directions, and at still other times with directions fixed in intermediate frames such as $\mathcal{F}_1$ or $\mathcal{F}_2$. For example, let us write $\boldsymbol{\omega}_{\mathcal{B}/\mathcal{F}}$ in body components. This means, looking at Equation (6.58), that we must express $\hat{\mathbf{n}}_{12}$ and $\hat{\mathbf{K}}$ in terms of $\hat{\mathbf{i}}$, $\hat{\mathbf{j}}$, and $\hat{\mathbf{k}}$. We see from Figure 6.12 that

$$\hat{\mathbf{n}}_{12} = \sin\psi\hat{\mathbf{i}} + \cos\psi\hat{\mathbf{j}} \tag{6.59}$$

and that

$$\hat{\mathbf{K}} = -\sin\theta\hat{\mathbf{n}}_{21} + \cos\theta\hat{\mathbf{k}}$$

$$= -\sin\theta(\cos\psi\hat{\mathbf{i}} - \sin\psi\hat{\mathbf{j}}) + \cos\theta\hat{\mathbf{k}}$$

or

$$\hat{\mathbf{K}} = -s_\theta c_\psi\hat{\mathbf{i}} + s_\theta s_\psi\hat{\mathbf{j}} + c_\theta\hat{\mathbf{k}} \tag{6.60}$$

where $s_\theta = \sin\theta$, $c_\psi = \cos\psi$, and so forth. Substituting these expressions into Equation (6.58) then gives $\boldsymbol{\omega}_{\mathcal{B}/\mathcal{F}}$ in body components:

$$\boldsymbol{\omega}_{\mathcal{B}/\mathcal{F}} = (s_\psi\dot{\theta} - s_\theta c_\psi\dot{\phi})\hat{\mathbf{i}} + (c_\psi\dot{\theta} + s_\theta s_\psi\dot{\phi})\hat{\mathbf{j}} + (\dot{\psi} + c_\theta\dot{\phi})\hat{\mathbf{k}} \tag{6.61}$$

Alternatively, we may express $\boldsymbol{\omega}_{\mathcal{B}/\mathcal{F}}$ in terms of its reference (or space) components by writing $\hat{\mathbf{k}}$ and $\hat{\mathbf{n}}_{12}$ in terms of $\hat{\mathbf{I}}$, $\hat{\mathbf{J}}$, and $\hat{\mathbf{K}}$. Again referring to the figure, we see that

$$\hat{\mathbf{n}}_{12} = -s_\phi\hat{\mathbf{I}} + c_\phi\hat{\mathbf{J}} \tag{6.62}$$

and

$$\hat{\mathbf{k}} = \sin\theta\hat{\mathbf{n}}_{11} + \cos\theta\hat{\mathbf{K}}$$

$$= \sin\theta(\cos\phi\hat{\mathbf{I}} + \sin\phi\hat{\mathbf{J}}) + \cos\theta\hat{\mathbf{K}}$$

or

$$\hat{\mathbf{k}} = s_\theta c_\phi\hat{\mathbf{I}} + s_\theta s_\phi\hat{\mathbf{J}} + c_\theta\hat{\mathbf{K}} \tag{6.63}$$

so that, substituting into Equation (6.58), we have $\boldsymbol{\omega}_{\mathcal{B}/\mathcal{F}}$ written in $\mathcal{F}$:

$$\boldsymbol{\omega}_{\mathcal{B}/\mathcal{F}} = (-s_\phi\dot{\theta} + s_\theta c_\phi\dot{\psi})\hat{\mathbf{I}} + (c_\phi\dot{\theta} + s_\theta s_\phi\dot{\psi})\hat{\mathbf{J}} + (\dot{\phi} + c_\theta\dot{\psi})\hat{\mathbf{K}} \tag{6.64}$$

Finally, we notice that expressing $\boldsymbol{\omega}_{\mathcal{B}/\mathcal{F}}$ in the *intermediate* frames is easier still. To write it in terms of its components in $\mathcal{F}_1$, we note that

$$\hat{\mathbf{K}} \equiv \hat{\mathbf{n}}_{13} \tag{6.65}$$

and

$$\hat{\mathbf{k}} = c_\theta\hat{\mathbf{n}}_{13} + s_\theta\hat{\mathbf{n}}_{11} \tag{6.66}$$

so that from Equation (6.58) we get

$$\boldsymbol{\omega}_{\mathcal{B}/\mathcal{F}} = s_\theta\dot{\psi}\hat{\mathbf{n}}_{11} + \dot{\theta}\hat{\mathbf{n}}_{12} + (c_\theta\dot{\psi} + \dot{\phi})\hat{\mathbf{n}}_{13} \tag{6.67}$$

All these expressions for $\boldsymbol{\omega}_{\mathcal{B}/\mathcal{F}}$ appear different, but of course they all represent the ***same vector*** written in different frames. The components may vary from frame to frame, but the vector is the same.

The angles (ϕ, θ, ψ) are known as the Eulerian angles. They represent one way of orientating a rigid body in space. Unfortunately the Eulerian angles (ϕ, θ, ψ) do not carry the same symbol from one book to the next; worse still, the order and even the directions of the rotations vary from writer to writer. Obviously, then, it is important to choose a set to work with and then be consistent.

A physical feel for the Eulerian angles may be gained by considering a gyroscope $\mathcal{B}$ spinning in a Cardan suspension as shown in Figure 6.13. In this system the Eulerian angles (ϕ, θ, ψ) may be used as follows to pinpoint the orientation of the rotor $\mathcal{B}$ in space:

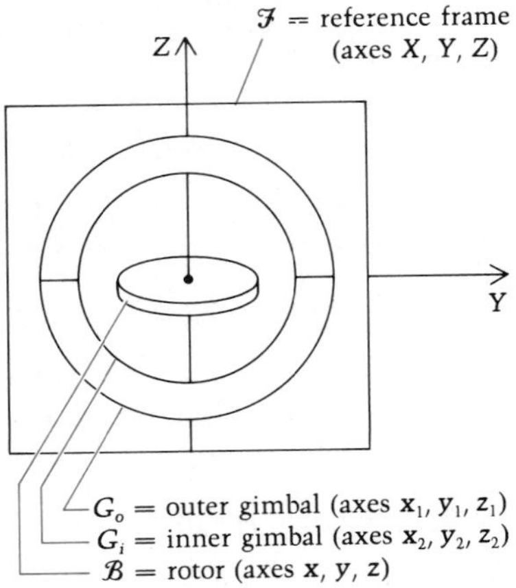

Figure 6.13

1. First Rotation: With respect to frame $\mathcal{F}$, we rotate the plane of the outer gimbal G_o (frame $\mathcal{F}_1$ in the earlier theory) about axis Z through angle ϕ. Axis X is turned into $\mathbf{x}_1$ and axis Y into $\mathbf{y}_1$; frames (bodies) G_i and $\mathcal{B}$ are not shown yet, but they move rigidly with G_o in this first rotation. (See Figure 6.14.)
2. Second Rotation: Next we turn the plane of the inner gimbal G_i $(\mathcal{F}_2)$ about axis $\mathbf{y}_1$, through angle θ, thereby tilting G_i with respect to G_o. Axis $\mathbf{z}_1$ is thereby rotated into $\mathbf{z}_2$ and axis $\mathbf{x}_1$ into $\mathbf{x}_2$. Body $\mathcal{B}$ (not shown in Figure 6.15) goes along for the ride.
3. Third Rotation: The third and last rotation turns the rotor $\mathcal{B}$ about axis $\mathbf{z}_2$ through angle ψ. (See Figure 6.16.) This allows $\mathcal{B}$ to spin relative to G_i. Axis $\mathbf{x}_2$ is turned into $\mathbf{x}$ and axis $\mathbf{y}_2$ into $\mathbf{y}$.

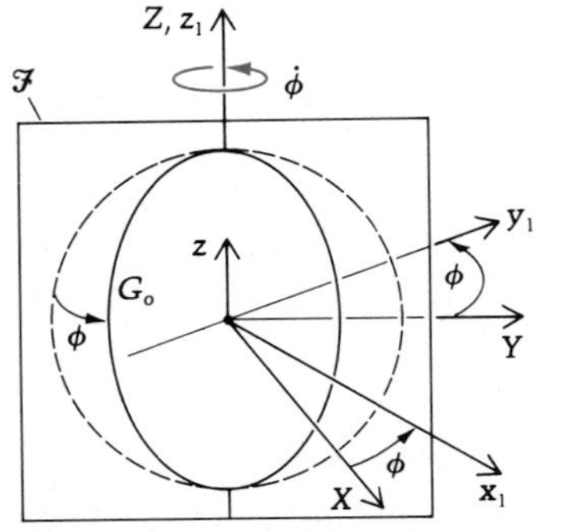

Figure 6.14 First rotation.

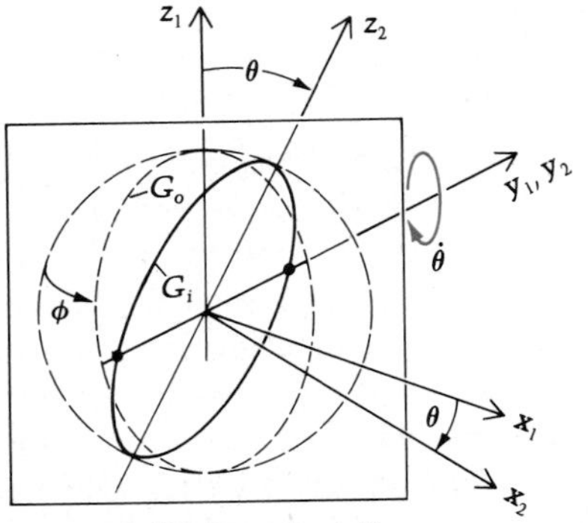

Figure 6.15 Second rotation.

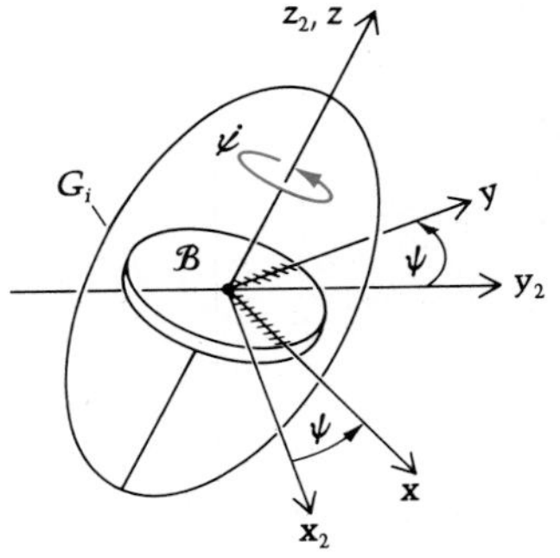

Figure 6.16 Third rotation.

Through these three Eulerian angle rotations, the body ($\mathcal{B}$) can be positioned in any desired orientation in space ($\mathcal{I}$). We are now able to see the difficulty of solving for the orientation of a rigid body in general motion. If we write $\boldsymbol{\omega}_{\mathcal{B}/\mathcal{I}}$ as $\omega_1\hat{\mathbf{i}} + \omega_2\hat{\mathbf{j}} + \omega_3\hat{\mathbf{k}}$, then Equation (6.61) is equivalent to

$$\begin{aligned} s_\psi\dot{\theta} - s_\theta c_\psi\dot{\phi} &= \omega_1 \\ c_\psi\dot{\theta} + s_\theta s_\psi\dot{\phi} &= \omega_2 \\ \dot{\psi} + c_\theta\dot{\phi} &= \omega_3 \end{aligned} \tag{6.68}$$

Solving for the rates of change of the Eulerian angles gives

$$\begin{aligned} \dot{\theta} &= \omega_1 s_\psi + \omega_2 c_\psi \\ \dot{\phi} &= \frac{\omega_2 s_\psi - \omega_1 c_\psi}{s_\theta} \\ \dot{\psi} &= \frac{\omega_3 s_\theta - \omega_2 c_\theta s_\psi + \omega_1 c_\theta c_\psi}{s_\theta} \end{aligned} \tag{6.69}$$

We see from these equations that even if we knew the $\boldsymbol{\omega}$ components as functions of time in closed form,* it would still be a formidable task to integrate Equations (6.69) analytically to obtain the Eulerian angles and thus to know the body's orientation in space. This is usually not even possible, and resort is made to computers that can numerically carry out integrations with a step-by-step scheme such as Runge-Kutta.

Incidentally, the sin θ denominators in Equations (6.69) present serious obstacles in the dynamics of space vehicles; whenever θ is zero or a multiple of π, the equations develop a singularity. Sophisticated programming or, in some cases, completely different mathematical schemes for orienting the body are required to overcome such difficulties.

We mention that use of the preceding set of Eulerian angles as defined requires that we maintain the *order* of rotation. To illustrate the importance of rotation order, we remark that if this book is rotated through two $\pi/2$ rotations about the space axes Y and Z, in opposite orders as suggested by Figure 6.17, it will end up in a different position. We should point out, however, that there are ways of setting up the axes and angles which make a body's final orientation independent of order. For instance, just as in our Eulerian angle development, let Z be fixed in $\mathcal{I}$, let z be fixed in $\mathcal{B}$, and let y be always perpendicular to both Z and z. But now restrict Z and z to be nonparallel. (Let z, the axis of $\mathcal{B}$, lie along X initially, for example.) In this case, the angles (ϕ, θ, ψ) as defined

*The equations governing these three components of angular velocity are the Euler equations of rigid-body kinetics. They themselves are also nonlinear and unsolvable in closed form in general three-dimensional motion except for a few special cases. We shall be studying these equations in Chapter 7.

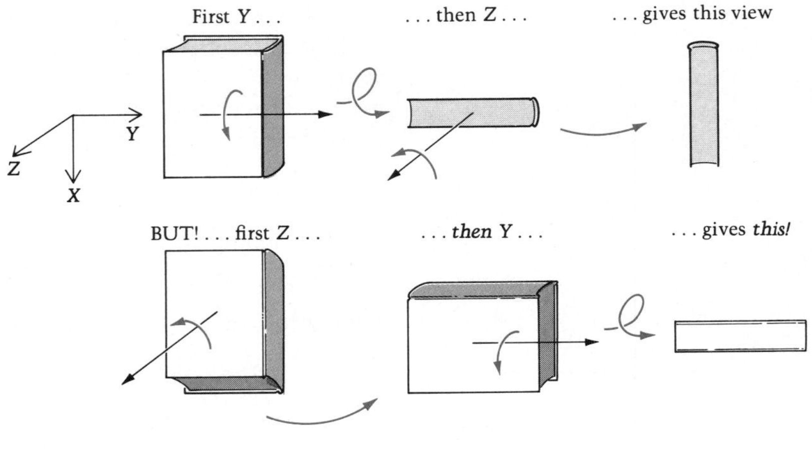

Figure 6.17 Finite rotations.

earlier may be performed in any of the six possible orders and the body's resulting orientation in $\mathcal{F}$ will be the same each time!*

An alternative set of three rotations has become popular in the literature in recent years:

1. Rotate through θ_1 about X.
2. Then rotate through θ_2 about y_1.
3. Then rotate through θ_3 about z_2.

This sequence results in angular velocity components in $\mathcal{B}$—that is, in the body-axis system $(\mathbf{x}, \mathbf{y}, \mathbf{z})$—of

$$\begin{aligned}\omega_1 &= \dot{\theta}_1 \cos\theta_2 \cos\theta_3 + \dot{\theta}_2 \sin\theta_3 \\ \omega_2 &= -\dot{\theta}_1 \cos\theta_2 \sin\theta_3 + \dot{\theta}_2 \cos\theta_3 \\ \omega_3 &= \dot{\theta}_1 \sin\theta_2 + \dot{\theta}_3\end{aligned} \tag{6.70}$$

Although (6.70) is not made up of classic Eulerian angles, the result is an equally valid set of relations between the $\boldsymbol{\omega}_{\mathcal{B}/\mathcal{F}}$ components and the angles θ_i of rotation. The order and axes of rotations of the θ_i in this system are quite simple to remember.

There are alternatives to using the Eulerian angles or the angles in Equation (6.70) to orient a body in space. One such alternative is to use the quaternions of Hamilton, which are free of the singular points caused by zero denominators at certain values of the Eulerian angles. (For this reason, quaternions were used in the Skylab Orbital Assembly's attitude control system.) Another approach to the orientation of a body in space is to determine the direction cosines of a unit vector, fixed in direction in space $\mathcal{F}$, with respect to a set of axes fixed in the body $\mathcal{B}$. Let this unit vector (call it $\hat{\mathbf{u}}$) have direction cosines (p, q, r) with respect to

*Also, subsequent reorientations will likewise be order-independent; see "Successive Finite Rotations," by T. R. Kane and D. A. Levinson, *Journal of Applied Mechanics*, Dec. 1978, Vol. **45**, pp. 945–946.

axes $(\mathbf{x}, \mathbf{y}, \mathbf{z})$ fixed in $\mathcal{B}$. Let $(\hat{\mathbf{i}}, \hat{\mathbf{j}}, \hat{\mathbf{k}})$ be unit vectors, always respectively parallel to $(\mathbf{x}, \mathbf{y}, \mathbf{z})$. Then

$$\hat{\mathbf{u}} = p\hat{\mathbf{i}} + q\hat{\mathbf{j}} + r\hat{\mathbf{k}}$$

and, differentiating in $\mathcal{I}$, we obtain

$$\begin{aligned} {}^{\mathcal{I}}\dot{\hat{\mathbf{u}}} = \mathbf{0} &= {}^{\mathcal{B}}\dot{\hat{\mathbf{u}}} + \boldsymbol{\omega}_{\mathcal{B}/\mathcal{I}} \times \hat{\mathbf{u}} \\ &= (\dot{p}\hat{\mathbf{i}} + \dot{q}\hat{\mathbf{j}} + \dot{r}\hat{\mathbf{k}}) + (\omega_x\hat{\mathbf{i}} + \omega_y\hat{\mathbf{j}} + \omega_z\hat{\mathbf{k}}) \times (p\hat{\mathbf{i}} + q\hat{\mathbf{j}} + r\hat{\mathbf{k}}) \\ &= (\dot{p} + \omega_y r - \omega_z q)\hat{\mathbf{i}} + (\dot{q} + \omega_z p - \omega_x r)\hat{\mathbf{j}} + (\dot{r} + \omega_x q - \omega_y p)\hat{\mathbf{k}} \end{aligned}$$

This vector equation has the following scalar component equations:

$$\begin{aligned} \dot{p} &= \omega_z q - \omega_y r \\ \dot{q} &= \omega_x r - \omega_z p \\ \dot{r} &= \omega_y p - \omega_x q \end{aligned} \tag{6.71}$$

If the $\boldsymbol{\omega}$ components are either prescribed or else found from kinetics equations (to be studied in Chapter 7), then Equations (6.71) may be solved for the direction cosines, thereby orienting $\mathcal{B}$ in $\mathcal{I}$. Equations (6.71) are known as the Poisson equations. They are alternatives to equations such as (6.68) and (6.70).

PROBLEMS / Section 6.7

6.70 Using the Eulerian angles (ϕ, θ, ψ) discussed in this section, express $\boldsymbol{\omega}_{\mathcal{B}/\mathcal{I}}$ in (terms of its components in) the frame $\mathcal{I}_2$.

6.71 Show that the magnitudes of $\boldsymbol{\omega}_{\mathcal{B}/\mathcal{I}}$ are all the same as expressed (a) in $\mathcal{B}$ in Equation (6.61); (b) in $\mathcal{I}$ in Equation (6.64); and (c) in $\mathcal{I}_1$ in Equation (6.67).

6.72 Derive Equations (6.70).

6.73 Write $\boldsymbol{\omega}_{\mathcal{B}/\mathcal{I}}$ in $\mathcal{I}$ by using the successive rotations θ_1, θ_2, and θ_3 that resulted in Equation (6.70) when expressed in $\mathcal{B}$.

6.74 Euler's Theorem for finite rotations is stated as follows: The most general rotation of a rigid body $\mathcal{B}$ with respect to a point A is equivalent to a rotation about some axis through A. Prove the theorem. *Hint:* Let A be considered fixed in the reference frame in which $\mathcal{B}$ moves. (See Figure P6.74.) Let point P be at P_1 prior to the rotation and at P_2 afterward; assume the same for point Q (Q_1 before, Q_2 after). Bisect angle P_1AP_2 with a plane normal to the plane of the angle. Do the same for Q_1AQ_2 and consider the intersection of the two planes.

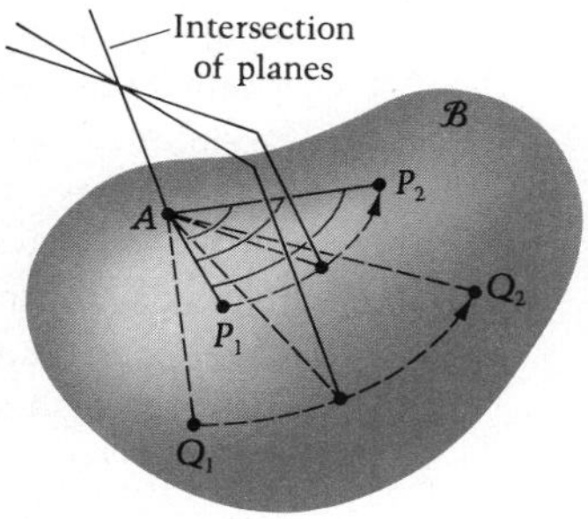

Figure P6.74

6.75 Chasle's Theorem states: The most general displacement of a rigid body is equivalent to the translation of some point A followed by a rotation about an axis through A. Show that this result follows immediately from the previous problem.

6.76 The circular drum of radius R in Figure P6.76 is pivoted to a support at O, where O is a distance $R/2$ from the center C of the drum. A weight $\mathcal{W}$ (particle) hangs from a cord wrapped around the drum. The drum is slowly rotated $\pi/2$ rad clockwise about O. Find the displacement of $\mathcal{W}$.

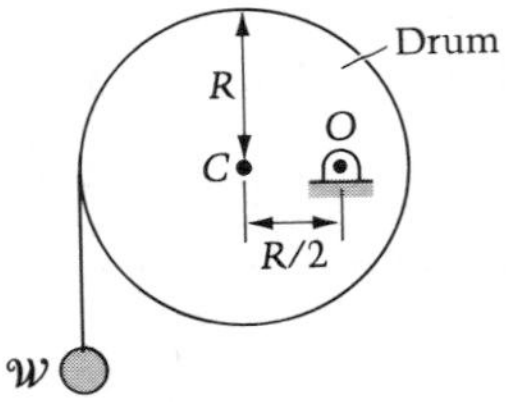

Figure P6.76

6.77 The sphere $\mathcal{S}$ rolls on the plane, and its angular velocity in the reference frame $\mathcal{I}$, in which (x, y, z) are fixed, is given by Equation (6.64). Noting that $\mathbf{v}_C = \dot{x}\hat{\mathbf{i}} + \dot{y}\hat{\mathbf{j}}$, write the constraint equations (the "no-slip" conditions) relating $\dot{x}$ and $\dot{y}$ to the Eulerian angles ϕ, θ, ψ and their derivatives.

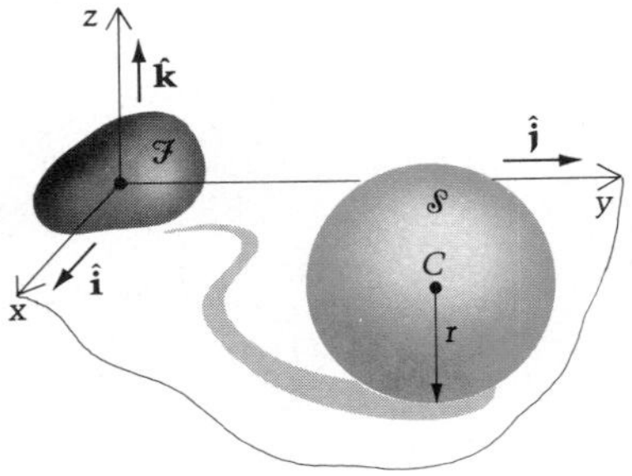

6.8 Rotation Matrices

Suppose that a vector **Q** is written in a frame $\mathcal{I}$ and that its components are the elements of a column matrix $\{Q\}_{\mathcal{I}}$. It is possible to develop a set of 3 × 3 matrices $[T_x]$, $[T_y]$, and $[T_z]$ each of which, when postmultiplied by $\{Q\}_{\mathcal{I}}$, gives the components of **Q** in a new frame rotated about an x, y, or z axis, respectively, of $\mathcal{I}$. These matrices are handy work-savers. For example, they reduce the work involved, in going from Equation (6.58) to (6.61) or (6.64), to a pair of simple matrix multiplications.

We shall develop $[T_z]$ and then state the results for $[T_x]$ and $[T_y]$, which are derived similarly. Let $Q = Q_x\hat{\mathbf{i}} + Q_y\hat{\mathbf{j}} + Q_z\hat{\mathbf{k}}$, in which ($\hat{\mathbf{i}}$, $\hat{\mathbf{j}}$, $\hat{\mathbf{k}}$) are a triad of unit vectors having fixed directions along axes (x, y, z) of frame $\mathcal{I}$. Suppose further that $\mathcal{B}$ is a frame whose orientation may be obtained from that of $\mathcal{I}$ by a rotation through the angle θ_z about z. The rotated axes, which were aligned with (x, y, z) prior to the rotation, will be denoted (x_1, y_1, z_1) with associated unit vectors ($\hat{\mathbf{i}}_1$, $\hat{\mathbf{j}}_1$, $\hat{\mathbf{k}}_1$).

We note that in the "new" frame $\mathcal{B}$, we may write $\mathbf{Q} = Q_{x_1}\hat{\mathbf{i}}_1 + Q_{y_1}\hat{\mathbf{j}}_1 + Q_{z_1}\hat{\mathbf{k}}_1$, where $Q_{z_1} = Q_z$ and $\hat{\mathbf{k}}_1 = \hat{\mathbf{k}}$ since the rotation is about this axis, common to both frames. Figure 6.18 shows the projection of Q into the xy (x_1y_1) plane. We have used trigonometry to indicate the sides of the two shaded triangles in Figure 6.18. It can be seen that

$$Q_{x_1} = Q_x \cos\theta_z + Q_y \sin\theta_z$$

$$Q_{y_1} = Q_y \cos\theta_z - Q_x \sin\theta_z$$

in which Q_{x_1} and Q_{y_1} are the components of **Q** in $\mathcal{B}$ and Q_x and Q_y were its components back in $\mathcal{I}$. Now we are ready to observe that if the matrix $[T_z]$ is defined as

$$[T_z] = \begin{bmatrix} \cos\theta_z & \sin\theta_z & 0 \\ -\sin\theta_z & \cos\theta_z & 0 \\ 0 & 0 & 1 \end{bmatrix}$$

then the same results for the components of **Q** in the rotated frame $\mathcal{B}$ are obtained from the matrix product $[T_z]\,\{Q\}_{\mathcal{F}}$:

$$\{Q\}_{\mathcal{B}} = [T_z]\,\{Q\}_{\mathcal{F}} = \begin{bmatrix} \cos\theta_z & \sin\theta_z & 0 \\ -\sin\theta_z & \cos\theta_z & 0 \\ 0 & 0 & 1 \end{bmatrix} \begin{Bmatrix} Q_x \\ Q_y \\ Q_z \end{Bmatrix}$$

$$= \begin{Bmatrix} Q_x\cos\theta_z + Q_y\sin\theta_z \\ -Q_x\sin\theta_z + Q_y\cos\theta_z \\ Q_z \end{Bmatrix}$$

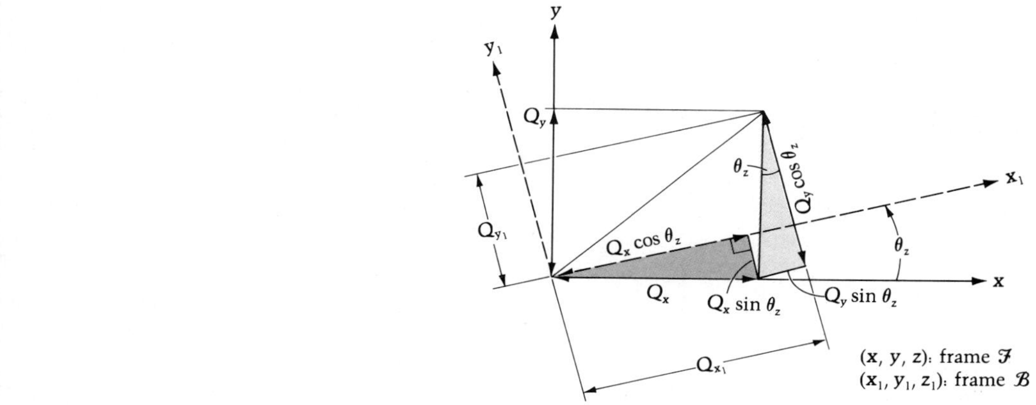

Figure 6.18 Development of rotation matrix about z axis.

The rotation matrices for rotations of θ_x about **x**, and θ_y about *y*, are respectively given by $[T_x]$ and $[T_y]$:

$$[T_x] = \begin{bmatrix} 1 & 0 & 0 \\ 0 & \cos\theta_x & \sin\theta_x \\ 0 & -\sin\theta_x & \cos\theta_x \end{bmatrix} \qquad [T_y] = \begin{bmatrix} \cos\theta_y & 0 & -\sin\theta_y \\ 0 & 1 & 0 \\ \sin\theta_y & 0 & \cos\theta_y \end{bmatrix}$$

The student may wish to verify one or both of these matrices, as we did for $[T_z]$. Note the change in the "sign of the sine" in the $[T_y]$ matrix. Moreover, if we must turn about an axis through a negative angle, we need only change the signs of both sine terms; this follows from the fact that $\cos(-\theta) = \cos\theta$, while $\sin(-\theta) = -\sin\theta$. We now consider examples of the use of the rotation matrices.

EXAMPLE 6.8

Use rotation matrices to obtain the components of $\boldsymbol{\omega}_{\mathcal{B}/\mathcal{F}}$ in body coordinates, given its representation (Equation 6.64) in the reference or space frame $\mathcal{F}$.

SOLUTION

We premultiply $\boldsymbol{\omega}_{\mathcal{B}/\mathcal{F}}$, expressed in $\mathcal{F}$ in matrix form, with rotation matrices of ϕ about the 3-axis, then θ about the new 2-axis, and then ψ about the new and final 3-axis:

$$\{\omega_{\mathcal{B}/\mathcal{F}}\}_{\mathcal{B}} = \underset{(\text{angle } \psi)}{[T_z]} \quad \underset{(\text{angle } \theta)}{[T_y]} \quad \underset{(\text{angle } \phi)}{[T_z]} \quad \{\omega_{\mathcal{B}/\mathcal{F}}\}_{\mathcal{F}}$$

$$= \begin{bmatrix} c_\psi & s_\psi & 0 \\ -s_\psi & c_\psi & 0 \\ 0 & 0 & 1 \end{bmatrix} \begin{bmatrix} c_\theta & 0 & -s_\theta \\ 0 & 1 & 0 \\ s_\theta & 0 & c_\theta \end{bmatrix} \underbrace{\begin{bmatrix} c_\phi & s_\phi & 0 \\ -s_\phi & c_\phi & 0 \\ 0 & 0 & 1 \end{bmatrix} \overbrace{\begin{Bmatrix} s_\theta c_\phi \dot{\psi} - s_\phi \dot{\theta} \\ c_\phi \dot{\theta} + s_\theta s_\phi \dot{\psi} \\ \dot{\phi} + c_\theta \dot{\psi} \end{Bmatrix}}^{\text{components of } \boldsymbol{\omega}_{\mathcal{B}/\mathcal{F}} \text{ in } \mathcal{F}}}_{\text{This gives the components of } \boldsymbol{\omega}_{\mathcal{B}/\mathcal{F}} \text{ in } \mathcal{F}_1}$$

$$= \begin{bmatrix} c_\psi & s_\psi & 0 \\ -s_\psi & c_\psi & 0 \\ 0 & 0 & 1 \end{bmatrix} \underbrace{\begin{bmatrix} c_\theta & 0 & -s_\theta \\ 0 & 1 & 0 \\ s_\theta & 0 & c_\theta \end{bmatrix} \begin{Bmatrix} s_\theta \dot{\psi} \\ \dot{\theta} \\ \dot{\phi} + c_\theta \dot{\psi} \end{Bmatrix}}_{\text{This yields the components of } \boldsymbol{\omega}_{\mathcal{B}/\mathcal{F}} \text{ in } \mathcal{F}_2}$$

$$= \begin{bmatrix} c_\psi & s_\psi & 0 \\ -s_\psi & c_\psi & 0 \\ 0 & 0 & 1 \end{bmatrix} \begin{Bmatrix} -s_\theta \dot{\phi} \\ \dot{\theta} \\ \dot{\psi} + c_\theta \dot{\phi} \end{Bmatrix}$$

Finally, we obtain the components of $\boldsymbol{\omega}_{\mathcal{B}/\mathcal{F}}$ in $\mathcal{B}$:

$$= \begin{Bmatrix} -s_\theta c_\psi \dot{\phi} + s_\psi \dot{\theta} \\ s_\psi s_\theta \dot{\phi} + c_\psi \dot{\theta} \\ \dot{\psi} + c_\theta \dot{\phi} \end{Bmatrix}$$

Comparing the elements of this matrix with the components in Equation (6.61), we see that rotation matrices indeed furnish us with a rapid means of "converting" a vector from one frame to another. Note also that the bracket in the second line above contains $\boldsymbol{\omega}_{\mathcal{B}/\mathcal{F}}$ expressed in $\mathcal{F}_1$, previously derived as Equation (6.67). The bracket in the third line gives the components of $\boldsymbol{\omega}_{\mathcal{B}/\mathcal{F}}$ in $\mathcal{F}_2$.

EXAMPLE 6.9

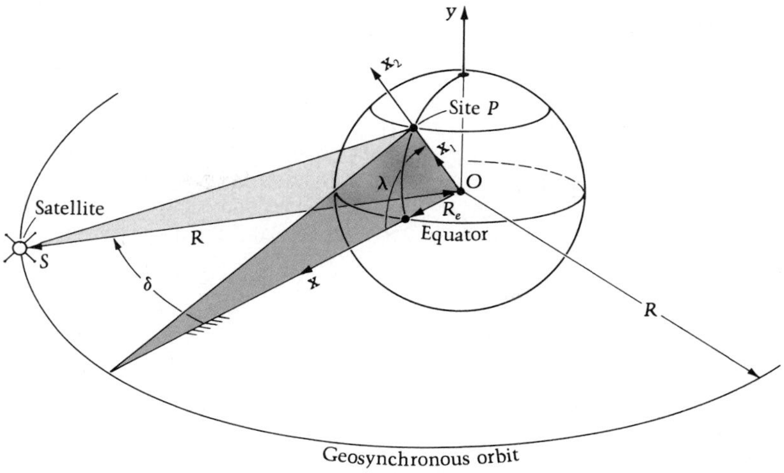

Using rotation matrices, compute the angles A (azimuth) and E (elevation) through which an antenna must respectively turn about (a) the negative of local vertical and then (b) the new, rotated position of the elevation axis in order to sight a satellite in geosynchronous orbit (see Problem 2.21). Angles A and E are called ***look angles***, and an antenna that performs azimuth followed by elevation in this manner is said to have "el over az" positioning. The azimuth angle A is a function of the local latitude λ and the relative west longitude δ of the satellite; the elevation angle E depends additionally on R_e/R, the ratio of the earth and orbit radii. (See the diagram.)

SOLUTION

The frame $\mathcal{F}$ ($\mathbf{x}, y, z$) has origin at the center of the earth as shown; the **xy** plane contains the site P and its meridian. The coordinates of the satellite in this frame are seen to be given by the position vector

$$\mathbf{r}_{OS} = R\cos\delta\,\hat{\mathbf{i}} + R\sin\delta\,\hat{\mathbf{k}}$$

First we rotate the frame $\mathcal{F}$ through the latitude angle λ about $\mathbf{z}$ in order to line up the new axis $\mathbf{x}_1$ with the local vertical at the site P. We call the resulting rotated frame $\mathcal{F}_1$ and obtain the following for the new components of $\mathbf{r}_{OS}$ ($c_\lambda = \cos\lambda$, $s_\delta = \sin\delta$, and so forth):

$$\{r_{OS}\}_{\mathcal{F}_1} = \underset{(\text{angle }\lambda)}{[T_z]}\ \{r_{OS}\}_{\mathcal{F}}$$

$$\{r_{OS}\}_{\mathcal{F}_1} = \begin{bmatrix} c_\lambda & s_\lambda & 0 \\ -s_\lambda & c_\lambda & 0 \\ 0 & 0 & 1 \end{bmatrix} \begin{Bmatrix} Rc_\delta \\ 0 \\ Rs_\delta \end{Bmatrix}$$

$$= \begin{Bmatrix} Rc_\lambda c_\delta \\ -Rs_\lambda c_\delta \\ Rs_\delta \end{Bmatrix}$$

(Continued)

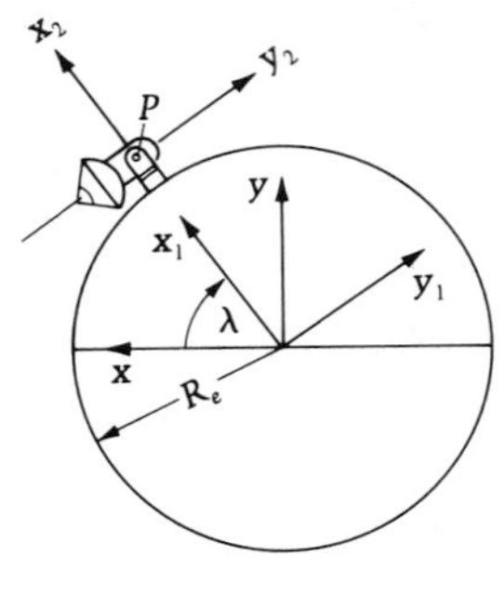

Next we translate the axes to the site, as shown in the diagram. The only component of $\{r_{OS}\}_{\mathcal{F}_1}$ that changes is $\mathbf{x}_1$, and we see by inspection that

$$\{r_{PS}\}_{\mathcal{F}_2} = \begin{Bmatrix} Rc_\lambda c_\delta - R_e \\ -Rs_\lambda c_\delta \\ Rs_\delta \end{Bmatrix}$$

Note that we must subtract the earth radius R_e from $\mathbf{x}_1$ to get the proper $\mathbf{x}_2$ coordinate of the satellite.

The second rotation is the azimuth rotation about $-\mathbf{x}_2$; if we call the rotated frame $\mathcal{F}_3$, the coordinates $(\mathbf{x}_3, \mathbf{y}_3, \mathbf{z}_3)$ of S in this frame are given by

$$\{r_{PS}\}_{\mathcal{F}_3} = \underset{(\text{angle } -A)}{[T_x]} \{r_{PS}\}_{\mathcal{F}_2}$$

$$= \begin{bmatrix} 1 & 0 & 0 \\ 0 & c_A & -s_A \\ 0 & s_A & c_A \end{bmatrix} \begin{Bmatrix} Rc_\lambda c_\delta - R_e \\ -Rs_\lambda c_\delta \\ Rs_\delta \end{Bmatrix}$$

$$= \begin{Bmatrix} Rc_\lambda c_\delta - R_e \\ -Rs_\lambda c_\delta c_A - Rs_\delta s_A \\ -Rs_\lambda c_\delta s_A + Rs_\delta c_A \end{Bmatrix}$$

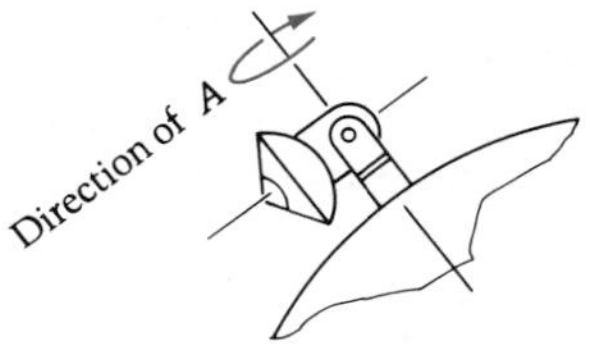

Here we take an important step. We want the z_3 component of $\mathbf{r}_{PS}$ to be zero because we wish to rotate next in elevation about z_3 and end up with the "boresight" (axis) of the antenna aiming at the satellite. Thus angle A is determined by setting the third element of the preceding matrix to zero:

$$-Rs_\lambda c_\delta s_A + Rs_\delta c_A = 0$$

$$\tan A = \tan \delta \csc \lambda$$

$$A = \tan^{-1}(\tan \delta \csc \lambda) \tag{1}$$

Finally we rotate through angle E about the z_3 axis:

$$\{r_{PS}\}_{\mathcal{F}_4} = \underset{(\text{angle } E)}{[T_z]} \{r_{PS}\}_{\mathcal{F}_3}$$

$$= \begin{bmatrix} c_E & s_E & 0 \\ -s_E & c_E & 0 \\ 0 & 0 & 1 \end{bmatrix} \begin{Bmatrix} Rc_\lambda c_\delta - R_e \\ -Rs_\lambda c_\delta c_A - Rs_\delta s_A \\ 0 \end{Bmatrix}$$

$$= \begin{Bmatrix} c_E(Rc_\lambda c_\delta - R_e) - s_E(Rs_\lambda c_\delta c_A + Rs_\delta s_A) \\ -s_E(Rc_\lambda c_\delta - R_e) - c_E(Rs_\lambda c_\delta c_A + Rs_\delta s_A) \\ 0 \end{Bmatrix}$$

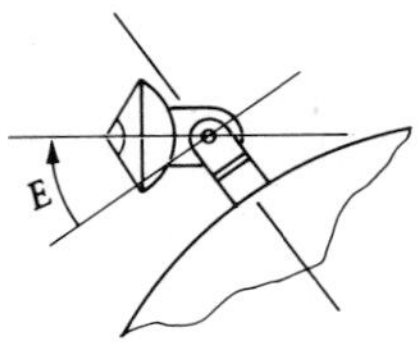

Now we come to the condition that will allow us to determine the value of angle E: We wish the antenna to aim directly at the satellite. Since the antenna boresight is now in the $-y_3$ direction, we wish the elevation rotation to stop when the $\mathbf{x}_3$ coordinate is zero:

(Continued)

$$c_E(Rc_\lambda c_\delta - R_e) - s_E(Rs_\lambda c_\delta c_A + Rs_\delta s_A) = 0$$

$$E = \tan^{-1}\left(\frac{Rc_\lambda c_\delta - R_e}{Rs_\lambda c_\delta c_A + Rs_\delta s_A}\right) \tag{2}$$

If $r = R_e/R$ ($\approx 1/6.61$), then (2) becomes

$$E = \tan^{-1}\left(\frac{c_\lambda c_\delta - r}{s_\lambda c_\delta c_A + s_\delta s_A}\right)$$

in which the azimuth angle A is given by Equation (1), so that

$$E = \tan^{-1}\left(\frac{c_\lambda c_\delta - r}{\sqrt{1 - c_\delta^2 c_\lambda^2}}\right) \tag{3}$$

There is a single circle in the sky in which geosynchronous satellites can exist. This circle, which was examined in Problem 2.21, has rapidly become very crowded, however. Figure 6.19, courtesy of Scientific-Atlanta (an antenna manufacturer), shows the situation in early 1982 regarding North American satellites. (The satellites authorized for launch but not yet operational in early 1982 are given in parentheses.)

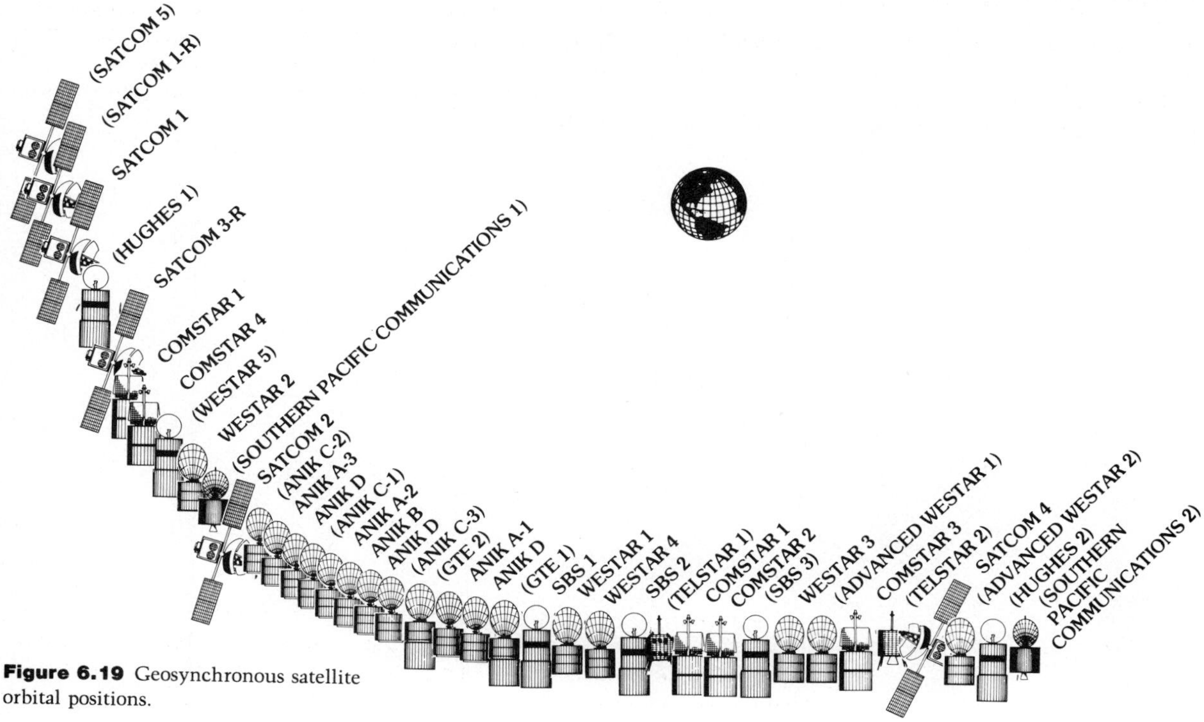

Figure 6.19 Geosynchronous satellite orbital positions.

PROBLEMS / Section 6.8

6.78 For the United States the eastern and western limits of usable satellite positions in the geosynchronous arc (see Problem 2.21) are about 70° and 143°W longitude, respectively. Find the ranges in azimuth and elevation that are required if the antenna in Example 6.9 is to sweep from the eastern to the western limit for a site at: (a) 34°N latitude and 84°W longitude; (b) your home town. (Select a city in the contiguous United States if you are from another country.)

6.79 Use Equations (6.70) together with rotation matrices to compute the angular velocity components in space-fixed axes.

6.80 Plot the elevation angle E versus the satellite angle δ (see Example 6.9) for the following values of λ (on the same graph): $\lambda = 0°, 20°, 40°, 60°$, and $80°$. What do the crossings of the δ axis of these curves physically represent?

6.81 Calculate the look angles (see Example 6.9) for the case in which the elevation rotation is performed *prior* to azimuth ("az over el" positioning).

6.82 It takes six *orbital parameters* to establish the location of a planet with respect to a frame fixed in space. (See Figure P6.82.) To find its orbital path, we first turn through the angle Ω in the ecliptic plane (the plane containing the path of the sun as we see it from earth) to the *ascending node* (the intersection of the ecliptic plane with the planet's path when going north). Next we turn in inclination through the angle i about $\mathbf{x}_1'$ to obtain the tilt of the planet's plane. (Thus earth's inclination is defined as zero.) Finally, the angle θ_o locates the perihelion of the planet's orbit. This is the closest point of the orbit to the sun's center S. Two other quantities give the orbit's shape, and a sixth one locates the planet in its orbit with respect to the perihelion. If $\mathbf{v} = (v_x, v_y, v_z)$ is a vector defined in the space frame $\mathcal{F}$, use rotation matrices to obtain the components of $\mathbf{v}$ in the frame $\mathcal{B}$ (x, y, z), located as shown in the orbital path at P, in terms of Ω, i, and θ_o.

6.83 An antenna $\mathcal{P}$ has three rotational degrees of freedom (see Figure P6.83):

1. Azimuth angle A about local vertical z
2. Elevation angle E about an axis originally parallel to $\mathbf{x}$
3. Polarization angle P about the axis of symmetry of the dish (originally parallel to y)

Use the addition theorem together with rotation matrices to calculate $\boldsymbol{\omega}_{\mathcal{P}/\mathcal{F}}$ in terms of its components in $\mathcal{F}$.

6.84 In the preceding problem calculate $\boldsymbol{\omega}_{\mathcal{P}/\mathcal{F}}$ in terms of its components in $\mathcal{P}$.

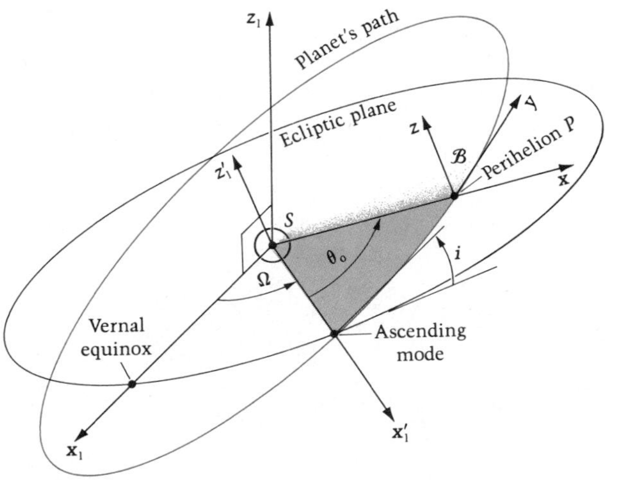

Figure P6.82

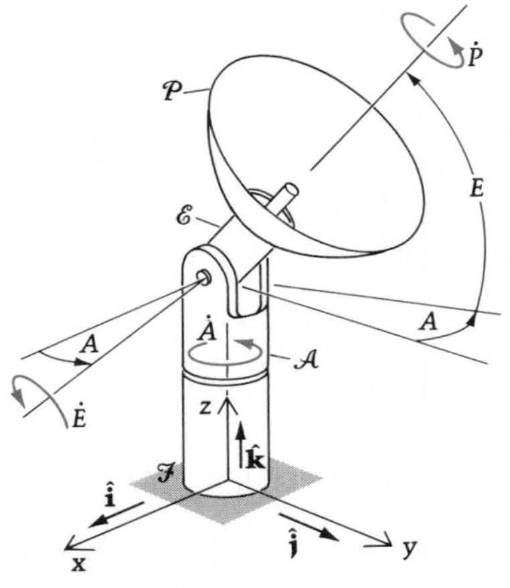

Figure P6.83

Answers to Questions / Chapter 6

Q6.1 Sure, as we have seen in Section 1.6.

Q6.2 Since $\alpha_1 \hat{\mathbf{b}}_1 \times \hat{\mathbf{b}}_1 = \mathbf{0}$, then α_1 can be anything and not affect the first of (6.6).

Q6.3 The (**xy**) plane of the paper can be chosen to be the plane containing $\hat{\mathbf{n}}_1$ and $\hat{\mathbf{n}}_2$ without loss of generality.

Q6.4 For θ_1 in the second quadrant, and with $0 < \cos\alpha \leq 1$, we see from (12) that $\tan\theta_2$ is a more negative number than $\tan\theta_1$ (unless $\alpha = 0$, in which case $\theta_2 = \theta_1$). Therefore, θ_1 and θ_2 are angles between 90° and 180°, and $\theta_2 < \theta_1$.

Q6.5 If we differentiate Equation (6.47) in $\mathcal{F}$, we encounter the term ${}^{\mathcal{F}}\dot{\mathbf{v}}_{P_{\mathcal{B}}/\mathcal{F}}\ (\neq \mathbf{a}_{P_{\mathcal{B}}/\mathcal{F}})$, which is not recognizable as a standard kinematic quantity; this is because $P_{\mathcal{B}}$ denotes a *succession* of points $\mathcal{B}$ which are at each instant coincident with P.

Q6.6 Because $\mathbf{r}_{PT} = \mathbf{r}_{PB}/2$.

Q6.7 Each point of $\mathcal{C}$ (extended) which lies on line $\overline{PO}$ has zero velocity. Thus the velocity of B is its distance from $\overline{PO}$ times $|\boldsymbol{\omega}_{\mathcal{C}}|$, and the same is true for point A. The former is seen to be coming out of the paper, and the latter going into it, both about $\overline{PO}$. Therefore $\boldsymbol{\omega}_{\mathcal{A}}$, as determined by the direction of $\mathbf{v}_A$, is in the negative $\mathbf{x}$ direction, opposite to $\boldsymbol{\omega}_{\mathcal{B}}$.

Review Questions / Chapter 6

True or False?

1. The angular velocity of body $\mathcal{B}$ in frame $\mathcal{F}$ depends only upon the changes in orientation of $\mathcal{B}$ with respect to $\mathcal{F}$.
2. The addition theorem for angular velocity applies equally well to angular acceleration.
3. The formula relating the velocities of two points of a rigid body in plane motion, $\mathbf{v}_B = \mathbf{v}_A + \boldsymbol{\omega} \times \mathbf{r}_{AB}$, applies to three-dimensional problems provided that $\boldsymbol{\omega}$, $\mathbf{r}_{AB}$, and the $\mathbf{v}$'s become three-dimensional vectors.
4. The formula relating the accelerations of two points of a rigid body in plane motion, $\mathbf{a}_B = \mathbf{a}_A + \boldsymbol{\alpha} \times \mathbf{r}_{AB} - \omega^2\mathbf{r}_{AB}$, applies to three-dimensional problems provided that $\boldsymbol{\alpha}$, $\mathbf{r}_{AB}$, $\boldsymbol{\omega}$, and the $\mathbf{a}$'s become three-dimensional vectors.
5. The equation $\omega_z = \dot{\theta}$ in plane motion extends to three similar linear equations in general motion for determining the orientation angles.
6. For a point P to have a nonvanishing Coriolis acceleration, there must be both a relative velocity of P with respect to the "moving frame" and an angular velocity of the moving frame relative to the reference frame.
7. If we premultiply a vector $\{v\}$ by a rotation matrix $[T]$, the 3×1 vector we get contains the "new" components of $\mathbf{v}$ in the rotated frame.
8. The Eulerian angles are used to orient a body in three-dimensional space.
9. The Eulerian angles are three rotations ϕ, θ, and ψ about axes which were originally distinct and orthogonal.
10. It is possible, for any moving point P, to choose a moving frame such that the Coriolis acceleration of P vanishes identically.

11. The angular velocity vector is used to relate the derivatives of a vector in two frames.

12. If one yoke of a universal (Hooke's) joint turns at constant angular speed, so does the other.

13. In general motion of a rigid body $\mathcal{B}$, as long as $\boldsymbol{\omega} \neq \mathbf{0}$ there is a point of zero velocity of $\mathcal{B}$ or $\mathcal{B}$ extended.

14. The order of rotations is important in orienting a body if the Eulerian angles are used as defined in this chapter (in conjunction with the Cardan suspension of the gyroscope).

Answers: T, F, T, F, F, T, T, T, F, T, T, F, F, T

CHAPTER

7

Kinetics of a Rigid Body in General Motion

chapter outline ►

7.1 Introduction/Motion of the Mass Center of a Body Near the Rotating Earth

In Chapters 2 and 4 we postulated the existence of (inertial, Newtonian, or Galilean) reference frames in which the motion of a body is governed by the equations

$$\mathbf{F}_r = \frac{d\mathbf{L}}{dt} \tag{7.1}$$

$$\mathbf{M}_{r_C} = \frac{d\mathbf{H}_C}{dt} \tag{7.2}$$

These general equations were specialized to ***plane*** motion of a ***rigid*** body $\mathcal{B}$ in Chapter 4 and will now be used to study the general motion of $\mathcal{B}$ in three dimensions.

As we indicated in Chapter 2, the first of the two vector equations given above describes the mass center motion of ***any*** system.* It is applicable, for example, to rigid or deformable solids, systems of small masses, liquids, and gases. For a body in general (three-dimensional) motion, Equation (7.1) now possesses three nontrivial scalar component equations; these take the following form when written in terms of coordinate axes (X, Y, Z) embedded in an inertial frame $\mathcal{I}$ in which $\mathcal{B}$ moves:

$$F_{r_x} = m\ddot{X}_C \tag{7.3a}$$

$$F_{r_y} = m\ddot{Y}_C \tag{7.3b}$$

$$F_{r_z} = m\ddot{Z}_C \tag{7.3c}$$

We note that, as was the case with plane motion, the mass center moves independently of the body's changing orientation (provided that the external forces do not themselves depend on the body's angular motion, which is frequently the case). We also note that if the components of the external force $(F_{r_x}, F_{r_y}, F_{r_z})$ acting on $\mathcal{B}$ are independent of the position vector (X_C, Y_C, Z_C) of C and its derivatives,† then Equations (7.3) are linear differential equations. Their solution yields the position vector of C, shown in Figure 7.1. We emphasize that such a simple and natural extension from two to three dimensions will not occur with the orientation (or angular) motion of $\mathcal{B}$, as we shall see in Section 7.5. The reason is that $\dot{\mathbf{H}}_C$ in Equation (7.2) *cannot* be written as the sum of three terms of the form $I_{z_C}\ddot{\theta}\hat{\mathbf{k}}$.

*Excluding throughout, of course, relativistic effects occurring when velocities are not small compared to the speed of light.

†This is not the case for a pair of orbiting bodies, in which the gravity force depends on the radius. In this case the equations for the mass center motion can still be solved, however.

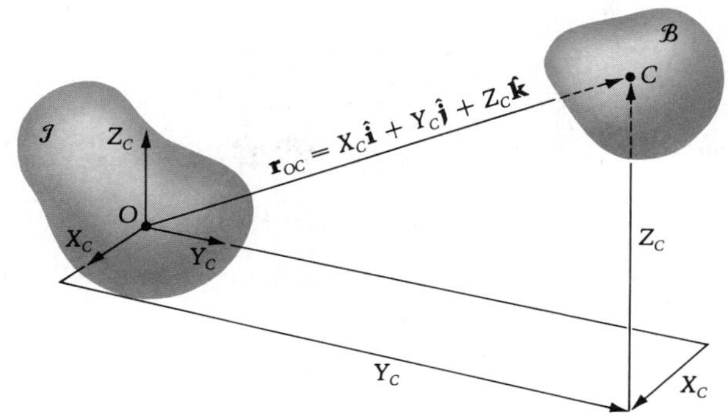

Figure 7.1 Mass center position in general motion.

We now make use of Equation (6.53) to set up the differential equation governing the position of the mass center C of a body $\mathscr{B}$ that is in motion near the earth. This equation will allow us to measure the position of C with respect to a desired site O' at latitude λ, which is itself in motion as the earth turns on its axis from west to east. We assume that for describing certain motions near the earth, a frame $\mathscr{I}$ with origin at the earth's mass center O is "sufficiently fixed" to be justifiably called inertial. The frame $\mathscr{I}$ does not turn with the earth, so the site O' has an acceleration $\ddot{\mathbf{R}}$ in $\mathscr{I}$.

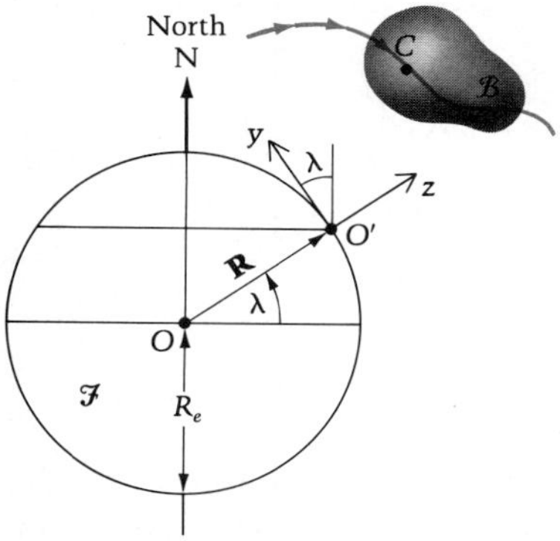

Figure 7.2

We set up the moving frame $\mathscr{F}$ as shown in Figure 7.2. The frame $\mathscr{F}$ *is* the turning earth, and the (x, y, z) axes are embedded in it at O' with x pointing east, y north, and z in the direction of local vertical. The acceleration of the mass center C of a body $\mathscr{B}$, moving near the earth and whose position is desired relative to O', is known from Equation (6.53) to be

$$\mathbf{a}_{C/\mathscr{I}} = \ddot{\mathbf{R}} + \boldsymbol{\alpha} \times \mathbf{r} + \boldsymbol{\omega} \times (\boldsymbol{\omega} \times \mathbf{r}) + 2\boldsymbol{\omega} \times \mathbf{v}_{\text{rel}} + \mathbf{a}_{\text{rel}} \tag{7.4}$$

where $\mathbf{r}$ is the position vector of C in $\mathscr{F}$, and $\mathbf{v}_{\text{rel}}$ and $\mathbf{a}_{\text{rel}}$ are the velocity and acceleration vectors of C in $\mathscr{F}$. Further, $\boldsymbol{\omega} = \boldsymbol{\omega}_{\mathscr{F}/\mathscr{I}}$ and $\boldsymbol{\alpha} = \boldsymbol{\alpha}_{\mathscr{F}/\mathscr{I}}$. We now use the mass center equation of motion

$$\mathbf{F}_r = m\mathbf{a}_{C/\mathscr{I}}$$

to obtain

$$\mathbf{F} - mg\hat{\mathbf{k}} = m[\ddot{\mathbf{R}} + \boldsymbol{\alpha} \times \mathbf{r} + \boldsymbol{\omega} \times (\boldsymbol{\omega} \times \mathbf{r}) + 2\boldsymbol{\omega} \times \mathbf{v}_{\text{rel}} + \mathbf{a}_{\text{rel}}] \tag{7.5}$$

where $\mathbf{F}$ represents all external forces on $\mathscr{B}$ besides gravity, which is written separately.

Question 7.1 This is a good place to ask: Why is Equation (7.5) restricted to bodies in motion near the earth?

We now proceed to compute the various terms in Equation (7.5). First we note that

$$\mathbf{r} = x\hat{\mathbf{i}} + y\hat{\mathbf{j}} + z\hat{\mathbf{k}}$$

$$\mathbf{v}_{\text{rel}} = \dot{x}\hat{\mathbf{i}} + \dot{y}\hat{\mathbf{j}} + \dot{z}\hat{\mathbf{k}}$$

$$\mathbf{a}_{\text{rel}} = \ddot{x}\hat{\mathbf{i}} + \ddot{y}\hat{\mathbf{j}} + \ddot{z}\hat{\mathbf{k}}$$

$$\boldsymbol{\omega} = \omega_e(\cos\lambda\hat{\mathbf{j}} + \sin\lambda\hat{\mathbf{k}}) \qquad \left(\text{where } \omega_e = 2\pi \text{ rad/day} \approx 0.0000727 \text{ rad/sec}\right)$$

$$\boldsymbol{\alpha} = \mathbf{0}$$

$$\mathbf{R} = R_e\hat{\mathbf{k}}$$

Next we compute the acceleration of the site (O'), using the rigid-body equation (6.55):

$$\ddot{\mathbf{R}} = \mathbf{a}_{O'/\mathcal{I}} = \cancelto{\mathbf{0}}{\mathbf{a}_{O/\mathcal{I}}} + \cancelto{\mathbf{0}}{\boldsymbol{\alpha}} \times \mathbf{R} + \boldsymbol{\omega} \times (\boldsymbol{\omega} \times \mathbf{R})$$

Therefore

$$\mathbf{F} - mg\hat{\mathbf{k}} = m\{\boldsymbol{\omega} \times [\boldsymbol{\omega} \times (\mathbf{R} + \mathbf{r})] + 2\boldsymbol{\omega} \times (\dot{x}\hat{\mathbf{i}} + \dot{y}\hat{\mathbf{j}} + \dot{z}\hat{\mathbf{k}}) + (\ddot{x}\hat{\mathbf{i}} + \ddot{y}\hat{\mathbf{j}} + \ddot{z}\hat{\mathbf{k}})\} \tag{7.6}$$

Neglecting $|\mathbf{r}|$ with respect to $|\mathbf{R}|$ and expressing $\mathbf{F}$ in terms of its components (F_x, F_y, F_z), we arrive at the differential equations governing the motion of C:

$$\begin{aligned} \ddot{x} &= 2\omega_e(\dot{y}\sin\lambda - \dot{z}\cos\lambda) + \frac{F_x}{m} \\ \ddot{y} &= -2\omega_e\dot{x}\sin\lambda - R_e\omega_e^2\cos\lambda\sin\lambda + \frac{F_y}{m} \\ \ddot{z} &= 2\omega_e\dot{x}\cos\lambda + R_e\omega_e^2\cos^2\lambda - g + \frac{F_z}{m} \end{aligned} \tag{7.7}$$

It will be left as a problem at the end of the section to show that the $R_e\omega_e^2$ terms are ordinarily small in comparison with the terms resulting from the Coriolis acceleration. Neglecting these small terms and setting $F_x = F_y = F_z = 0$ gives the mass center equations governing free fall of a body near the rotating earth:

$$\ddot{x} = 2\omega_e(\dot{y}\sin\lambda - \dot{z}\cos\lambda) \tag{7.8a}$$

$$\ddot{y} = -2\omega_e\dot{x}\sin\lambda \tag{7.8b}$$

$$\ddot{z} = 2\omega_e\dot{x}\cos\lambda - g \tag{7.8c}$$

We illustrate a solution of these equations in the following example.

EXAMPLE 7.1

Neglecting air resistance, find the location at which a falling rock will strike the earth if dropped from rest (say from a tower) from a height H.

SOLUTION

Equations (7.8b and c) may be integrated one time immediately, giving

$$\dot{y} = -2\omega_e \mathbf{x} \sin \lambda \tag{7.9a}$$

$$\dot{z} = 2\omega_e \mathbf{x} \cos \lambda - gt \tag{7.9b}$$

(The integration constants vanish since the rock is released from rest.) Substituting into Equation (7.8a) then gives

$$\ddot{\mathbf{x}} = -(2\omega_e)^2 [\mathbf{x}(\sin^2 \lambda + \cos^2 \lambda)] + 2\omega_e gt \cos \lambda$$

The first term is again small, so that

$$\ddot{\mathbf{x}} = 2\omega_e gt \cos \lambda \tag{7.10a}$$

Integrating, we get

$$\dot{\mathbf{x}} = \omega_e gt^2 \cos \lambda \tag{7.10b}$$

$$\mathbf{x} = \frac{\omega_e gt^3}{3} \cos \lambda \tag{7.10c}$$

(Again the integration constants are zero because of release from rest with $\mathbf{x} = 0$.)

Substituting $\mathbf{x}$ from Equation (7.10c) into (7.9a and b) and again neglecting the ω_e^2 terms (we must always be consistent with our initial assumptions) gives

$$\dot{y} = 0 \qquad \text{and} \qquad \dot{z} = -gt \tag{7.11}$$

from which integration yields

$$y = 0 \qquad \text{and} \qquad z = \frac{-gt^2}{2} + H \tag{7.12}$$

Setting $z = 0$ gives the time of fall to be $t_{\text{strike}} = \sqrt{2H/g}$.

Question 7.2 Unless the rock drops straight down, the value of z is actually slightly negative when the rock strikes the spherical earth—why?

Substituting this time of fall into Equation (7.10c) then gives

$$\mathbf{x}_{\text{strike}} = \frac{2\omega_e H}{3} \sqrt{\frac{2H}{g}} \cos \lambda \tag{7.13}$$

It is seen that the rock lands to the *east* of the origin by a small amount.

EXAMPLE 7.2

Due to the earth's rotation, the resultant force it exerts on a particle $\mathcal{P}$ at rest on its surface is not quite directed toward its center of mass. Use Equations (7.7) to find this deviation, assuming a spherical earth.

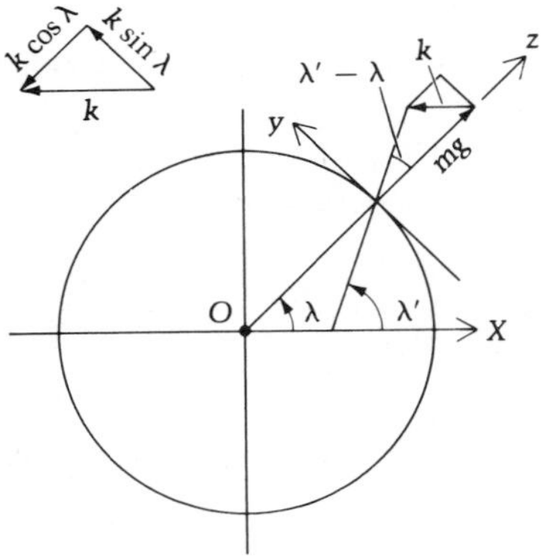

Figure 1

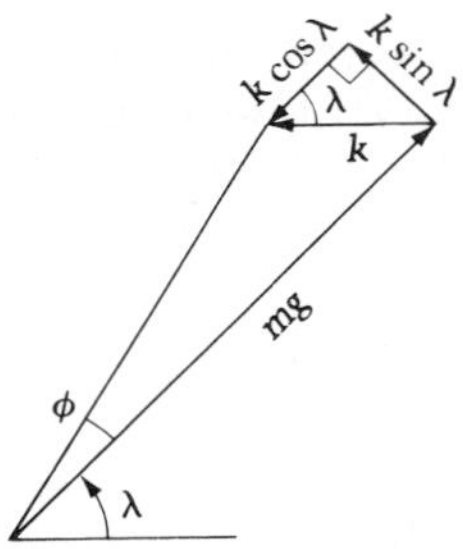

Figure 2

SOLUTION

We have $\dot{x}$, $\dot{y}$, $\dot{z}$, $\ddot{x}$, $\ddot{y}$, and $\ddot{z}$ all zero, so that the equations of motion of $\mathcal{P}$ (which moves!) in the inertial frame $\mathcal{I}$ (defined exactly as in the text preceding Equation (7.4)) are

$$F_x = 0$$

$$F_y = mR_e\omega_e^2 \cos\lambda \sin\lambda = k\sin\lambda$$

$$F_z = -mR_e\omega_e^2 \cos^2\lambda + mg = mg - k\cos\lambda$$

Calling $mR_e\omega_e^2 \cos\lambda = k$, we see (Figure 1) that the earth must push on $\mathcal{P}$ at a small angle $(\lambda' - \lambda)$ with the "*geometric vertical*" in order for $\mathcal{P}$ to remain at rest in the "moving frame" $\mathcal{F}$, rigidly attached to the surface of the planet. It is the angle λ', *not* the angle λ, that defines "*local vertical*"; this is because λ' is the angle that a plumb bob string makes with the axis X in the equatorial plane.

Figure 2 shows that

$$\tan\phi = \frac{k\sin\lambda}{mg - k\cos\lambda}$$

We note that k, in comparison with mg, is quite small:

$$\frac{k}{mg} = \frac{R_e\omega_e^2\cos\lambda}{g} \approx \frac{6370(0.0000727)^2\cos\lambda}{(9.81/1000)}$$

$$= \frac{6370(1000)(0.0000727)^2\cos\lambda}{9.81}$$

$$\approx 0.00343\cos\lambda$$

This means that we can use small angle approximations on the angle ϕ:

$$\phi \approx \tan\phi \approx \frac{k\sin\lambda}{mg} = \frac{R_e\omega_e^2\sin\lambda\cos\lambda}{g} = 0.00343\sin\lambda\cos\lambda$$

so that the deviation of the local "plumb-line" vertical from the "geometric vertical" is

$$\lambda' - \lambda = \phi \approx 0.00343\sin\lambda\cos\lambda$$

Of course there is *no* deviation in direction at the poles (where $\cos\lambda = 0$) or the equator (where $\sin\lambda = 0$). The maximum, at 45° latitude, is 0.0017 rad, or about 0.1°.

In the remainder of the chapter we shall first develop the expression for the moment of momentum of a rigid body $\mathcal{B}$ in general motion. This will then lead us into a study of the inertia properties of $\mathcal{B}$. Having partially examined the concept of inertia in Chapter 4, we shall extend this study to include transformations at a point as well as principal moments and axes of inertia. Then and only then shall we be fully prepared to derive the Euler equations that govern the rotational motion of a rigid body. We shall also examine, as we did for the plane case in Chapter 5, some special integrals of the equations of motion, which are known as the principles of impulse and momentum, angular impulse and angular momentum, and work and kinetic energy.

PROBLEMS / Section 7.1

7.1 If it were possible for a train to travel continuously around the world on a meridional track as shown in Figure P7.1a, one side of the track would wear out in time due to the Coriolis acceleration. Explain which side will wear out in each of the four numbered quadrants of the circular path. Figure P7.1b shows how the train's wheels rest on the track.

7.2 Explain in detail how the Coriolis acceleration produces a deflection of the air rushing toward a low-pressure area, thereby forming a hurricane. (See Figure P7.2.) Do the problem for each hemisphere!

7.3 A horizontal wheel is rotating about its fixed axis at a rate of 10 rad/sec, and this angular speed is increasing at the given time at $\dot{\omega} = 5$ rad/sec^2. (See Figure P7.3.) At this same instant, a bead is sliding inward relative to the spoke on which it moves at 5 ft/sec; this speed is slowing down at this time at 2 ft/sec^2. If the bead weighs 0.02 lb and is 1 ft from the center in the given configuration, find the external force exerted on the bead. Is it possible that this force can be exerted solely by the spoke and not in part by other external sources?

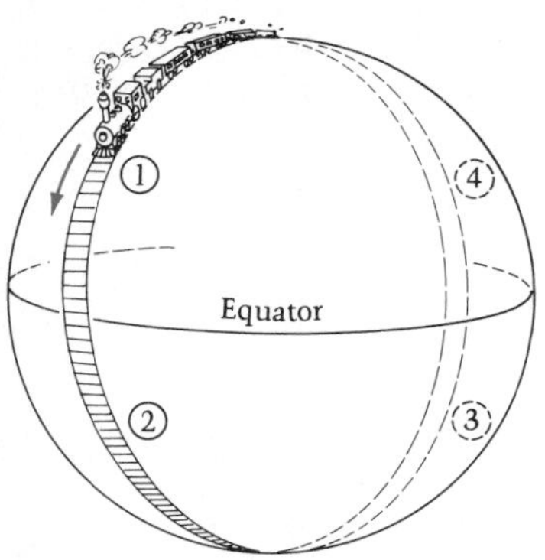

Figure P7.1a

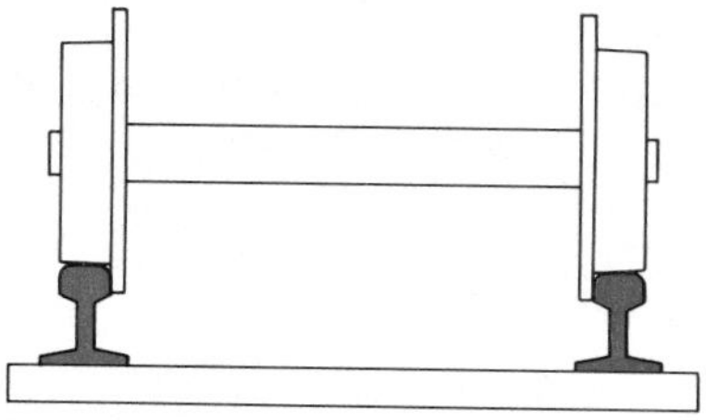

Figure P7.1b

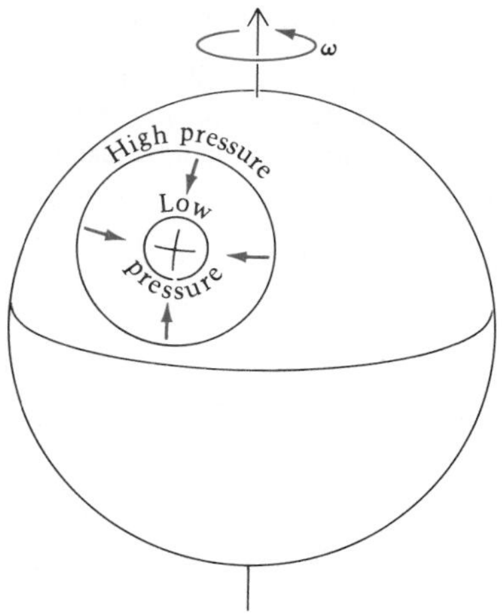

Figure P7.2

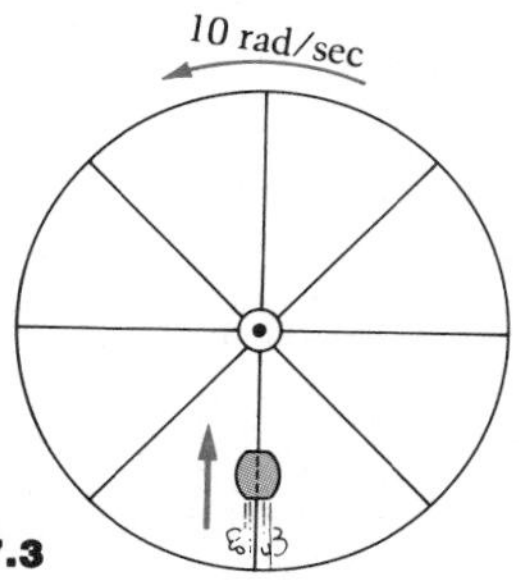

Figure P7.3

7.4 A bead slides down a smooth circular hoop that, at a certain instant, has $\omega = 2$ rad/sec and $\dot{\omega} = 3$ rad/sec^2 in the direction shown in Figure P7.4. The angular speed of line OP at this time is $\dot{\theta} = 0.5$ rad/sec and $\theta = 45°$. Find the value of $\ddot{\theta}$ and the force exerted on the bead by the hoop at the given instant.

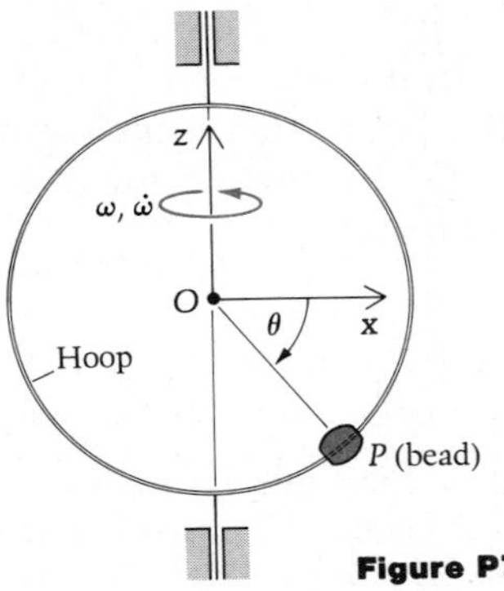

Figure P7.4

7.5 A rock is dropped from the top of a 200-ft tower at 45° latitude. Calculate the eastward deviation of its landing point from the point just beneath it at release.

7.6 Retain the $R_e\omega_e^2$ terms in the development of Equation 7.8, and show in the end that they produce a negligible change in the answer to Example 7.1.

7.7 A projectile is fired from a site at latitude λ, with the initial velocity components $\mathbf{v}(t = 0) = \dot{x}_i, \dot{y}_i, \dot{z}_i$). Determine the maximum height reached by the projectile, neglecting air resistance.

7.8 In the preceding problem let the projectile's initial velocity be $\dot{z}_i\hat{\mathbf{k}}$—that is, vertically upward. Find the distance between the firing and landing points.

7.9 A particle at latitude λ is projected with speed v_0 at an elevation β in the direction θ east of south, as shown in Figure P7.9. (a) Find the increase in the time of flight due to the earth's rotation with angular velocity $\boldsymbol{\omega}$. (b) Show that the particle falls a distance $4\omega \sin^2\beta(\cos\beta \sin\lambda + \frac{1}{3}\sin\beta \cos\lambda \cos\theta)v^3/g^2$ to the western side of the vertical plane of projection.

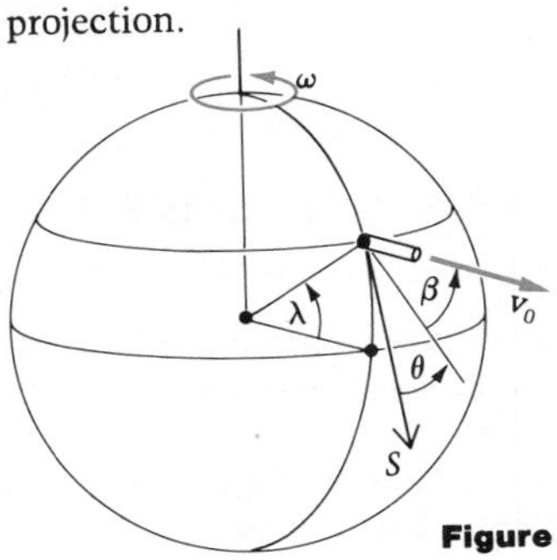

Figure P7.9

7.10 Several years ago, an article in a national sports magazine theorized that the Coriolis acceleration should result in more basketball free throws being missed on courts oriented on an east-west axis than on north-south courts. Assuming the dimensions in Figure P7.10, calculate the deviation in the trajectory of the ball on an east-west court caused by the Coriolis acceleration at 30°N latitude. Assume that the ball is released 6 ft directly above the free-throw line as shown at 12 ft/sec.

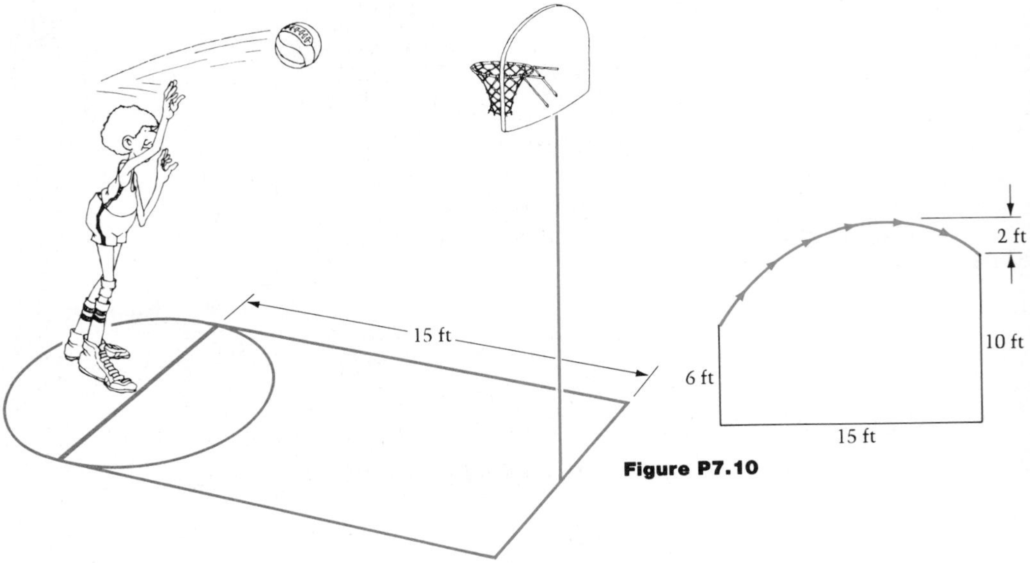

Figure P7.10

7.2 Moment of Momentum (Angular Momentum) in Three Dimensions

Whenever it is reasonable to treat a body $\mathcal{B}$ as being rigid, two important things happen that greatly simplify the equations of motion of $\mathcal{B}$. First, the mass center C becomes locked into the body; that is, it occupies the same point in $\mathcal{B}$ (or $\mathcal{B}$ extended) throughout its motion. Second, the **moment of momentum** with respect to any point P of $\mathcal{B}$ may be expressed in terms of the angular velocity of the body and a finite number of measures of mass distribution (the inertia properties of $\mathcal{B}$). Because of this intimate connection between the moment of momentum and angular velocity vectors for a rigid body, the moment of momentum is often called the **angular momentum.**

Let us now investigate this relationship by introducing a system of rectangular axes (x, y, z) which have their origin at P. The angular velocity of $\mathcal{B}$ in reference frame $\mathcal{I}$ may then be expressed in terms of its components along these axes by

$$\begin{aligned}\boldsymbol{\omega}_{\mathcal{B}/\mathcal{I}} = \boldsymbol{\omega} &= (\boldsymbol{\omega} \cdot \hat{\mathbf{i}})\hat{\mathbf{i}} + (\boldsymbol{\omega} \cdot \hat{\mathbf{j}})\hat{\mathbf{j}} + (\boldsymbol{\omega} \cdot \hat{\mathbf{k}})\hat{\mathbf{k}} \\ &= \omega_x\hat{\mathbf{i}} + \omega_y\hat{\mathbf{j}} + \omega_z\hat{\mathbf{k}}\end{aligned} \tag{7.14}$$

The location, relative to P, of a typical point in the body is given by

$$\mathbf{r} = x\hat{\mathbf{i}} + y\hat{\mathbf{j}} + z\hat{\mathbf{k}} \tag{7.15}$$

The moment of momentum of $\mathcal{B}$ relative to P is now defined to be

$$\mathbf{H}_P = \int \mathbf{r} \times \mathbf{v}\, dm \tag{7.16}$$

in which $\mathbf{v}$, the velocity of the mass element dm, is not the derivative of $\mathbf{r}$ but rather of the position vector *to* the element *from* a point fixed in the reference frame $\mathcal{I}$, as shown in Figure 7.3.

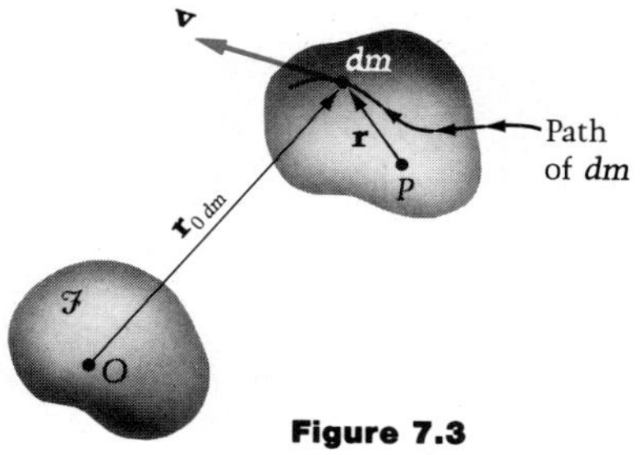

Figure 7.3

Since $\mathcal{B}$ is a rigid body, we know from Equation (6.54) that $\mathbf{v} = \mathbf{v}_P + \boldsymbol{\omega} \times \mathbf{r}$, and we may substitute this expression into Equation (7.16) to obtain

$$\begin{aligned}\mathbf{H}_P &= \int \mathbf{r} \times \mathbf{v}_P\, dm + \int \mathbf{r} \times (\boldsymbol{\omega} \times \mathbf{r})\, dm \\ &= \left(\int \mathbf{r}\, dm\right) \times \mathbf{v}_P + \int \mathbf{r} \times (\boldsymbol{\omega} \times \mathbf{r})\, dm\end{aligned} \tag{7.17}$$

Using the mass center definition, the integral in the first term on the right-hand side is $m\mathbf{r}_{PC}$. Therefore

$$\mathbf{H}_P = m\mathbf{r}_{PC} \times \mathbf{v}_P + \int \mathbf{r} \times (\boldsymbol{\omega} \times \mathbf{r})\, dm \tag{7.18}$$

In the cases where either (a) P is chosen to be the mass center C, or (b) $\mathbf{v}_P = \mathbf{0}$, or (c) $\mathbf{r}_{PC} \parallel \mathbf{v}_P$, the first term on the right side of Equation (7.18) vanishes. For these cases,

$$\mathbf{H}_P = \int \mathbf{r} \times (\boldsymbol{\omega} \times \mathbf{r})\, dm \tag{7.19}$$

Substituting $\boldsymbol{\omega}$ and $\mathbf{r}$ from Equations (7.14) and (7.15) and using the identity

$$\mathbf{A} \times (\mathbf{B} \times \mathbf{C}) \equiv (\mathbf{A} \cdot \mathbf{C})\mathbf{B} - (\mathbf{A} \cdot \mathbf{B})\mathbf{C} \tag{7.20}$$

we obtain the result

$$\begin{aligned} \mathbf{H}_P = {} & [\omega_x \textstyle\int (y^2 + z^2)\, dm - \omega_y \int xy\, dm - \omega_z \int xz\, dm]\hat{\mathbf{i}} \\ & + [-\omega_x \textstyle\int xy\, dm + \omega_y \int (x^2 + z^2)\, dm - \omega_z \int yz\, dm]\hat{\mathbf{j}} \\ & + [-\omega_x \textstyle\int xz\, dm - \omega_y \int yz\, dm + \omega_z \int (x^2 + y^2) dm]\hat{\mathbf{k}} \end{aligned} \tag{7.21}$$

Question 7.3 Why may the $\boldsymbol{\omega}$ components be brought outside the various integrals in Equation (7.21)?

Once we recognize the inertia properties (see Section 4.3),* this angular momentum expression becomes, for the case when $P = C$,

$$\begin{aligned} \mathbf{H}_C = {} & (I_{xx}^C \omega_x + I_{xy}^C \omega_y + I_{xz}^C \omega_z)\hat{\mathbf{i}} \\ & + (I_{xy}^C \omega_x + I_{yy}^C \omega_y + I_{yz}^C \omega_z)\hat{\mathbf{j}} \\ & + (I_{xz}^C \omega_x + I_{yz}^C \omega_y + I_{zz}^C \omega_z)\hat{\mathbf{k}} \end{aligned} \tag{7.22}$$

The form of the equation is identical if point P is not C, but rather either $\mathbf{v}_P = \mathbf{0}$ or $\mathbf{r}_{PC} \parallel \mathbf{v}_P$; the only difference is that the inertia properties are calculated with respect to axes at P instead of at the mass center:

$$\begin{aligned} \mathbf{H}_P = {} & (I_{xx}^P \omega_x + I_{xy}^P \omega_y + I_{xz}^P \omega_z)\hat{\mathbf{i}} \\ & + (I_{xy}^P \omega_x + I_{yy}^P \omega_y + I_{yz}^P \omega_z)\hat{\mathbf{j}} \\ & + (I_{xz}^P \omega_x + I_{yz}^P \omega_y + I_{zz}^P \omega_z)\hat{\mathbf{k}} \end{aligned} \tag{7.23}$$

Both of these forms for the angular momentum vector (Equations 7.22 and 7.23) will prove important to us in the sections to follow.

*In this chapter we shall denote moments of inertia as, for example, I_{zz}^C (instead of I_{z_C} as in Chapter 4); likewise, products of inertia will be expressed as I_{xy}^C (instead of I_{xy_C} as in Chapter 4).

PROBLEMS / Section 7.2

7.11 Find the angular momentum vector $\mathbf{H}_O$ of the wagon wheel of Problem 6.42.

7.12 Find the angular momentum vector of the disk $\mathcal{B}$ in Problem 6.28 about (a) C and (b) O.

7.13 Find the angular momentum vector for the bent bar of Example 4.27 about the mass center.

7.14 A thin homogeneous disk $\mathcal{D}$ of mass M and radius r rotates with constant angular speed ω_2 about the shaft $\mathcal{S}$ (Figure P7.14). This shaft is cantilevered from the vertical shaft $\mathcal{R}$ and rotates with constant angular speed ω_1 about the axis of $\mathcal{R}$. Find the angular momentum of the disk about point Q, and show the direction of the vector in a sketch.

7.15 Depicted in Figure P7.15 is a grinder in a grinding mill that is composed of three main parts:

1. The vertical shaft $\mathcal{S}$, which rotates at constant angular speed Ω
2. The slanted shaft $\mathcal{B}$ of length ℓ, which is pinned to $\mathcal{S}$ and turns with it
3. The grinder $\mathcal{D}$ of radius r, turning in bearings at C about $\mathcal{B}$, and rolling on the inner surface of $\mathcal{F}$.

As shaft $\mathcal{S}$ gets up to speed Ω, body $\mathcal{B}$ swings outward, and then the angle ϕ remains constant during operation. Treat the grinder $\mathcal{D}$ as a disk and find its angular momentum vector $\mathbf{H}_C$ in convenient coordinates. (A suggested set is shown.)

7.16 In the preceding problem note that point O is a fixed point of all three bodies $\mathcal{S}$, $\mathcal{B}$, and $\mathcal{D}$ extended. Compute the angular momentum $\mathbf{H}_O$ of $\mathcal{D}$, and verify that $\mathbf{H}_O = \mathbf{H}_C + \mathbf{r}_{OC} \times \mathbf{L}$.

7.17 In Problem 2.126 find: (a) the angular momentum of the system with respect to the origin; (b) the angular momentum of the system with respect to the mass center.

The *relative angular momentum* of a body $\mathcal{B}$ with respect to a point P was defined in Problem 4.3 to be

$$\mathbf{H}_{P_{\text{rel}}} = \int \mathbf{R} \times (\mathbf{v} - \mathbf{v}_P)\, dm$$

(See Figure P7.17, where the velocity $\mathbf{v}$ of dm is the derivative of $\mathbf{r}$ in frame $\mathcal{I}$.) Note that what makes it 'relative' is that the velocity in the integral is the difference between $\mathbf{v}$ (of dm) and $\mathbf{v}_P$. Now solve the following problems.

7.18 Show that $\mathbf{H}_P = \mathbf{H}_{P_{\text{rel}}} + m\mathbf{r}_{PC} \times \mathbf{v}_P$. (Thus $\mathbf{H}_C = \mathbf{H}_{C_{\text{rel}}}$ always!)

7.19 Show that $\mathbf{M}_{r_P} = \dot{\mathbf{H}}_P + \mathbf{v}_P \times m\mathbf{v}_C$. (Thus $\mathbf{M}_{r_P}$ is not generally equal to $\dot{\mathbf{H}}_P$!)

7.20 Show that $\mathbf{M}_{r_P} = \dot{\mathbf{H}}_{P_{\text{rel}}} + \mathbf{r}_{PC} \times m\mathbf{a}_P$.

7.21 Show that $\dot{\mathbf{H}}_P = \dot{\mathbf{H}}_{P_{\text{rel}}}$ if and only if $\mathbf{v}_P \times \mathbf{v}_C = \mathbf{r}_{PC} \times \mathbf{a}_P$.

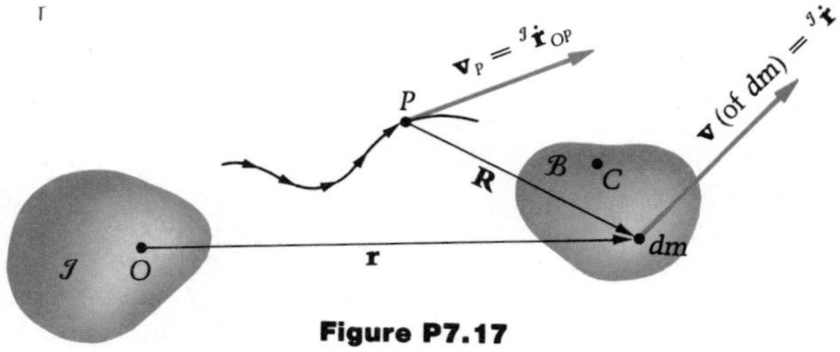

Figure P7.17

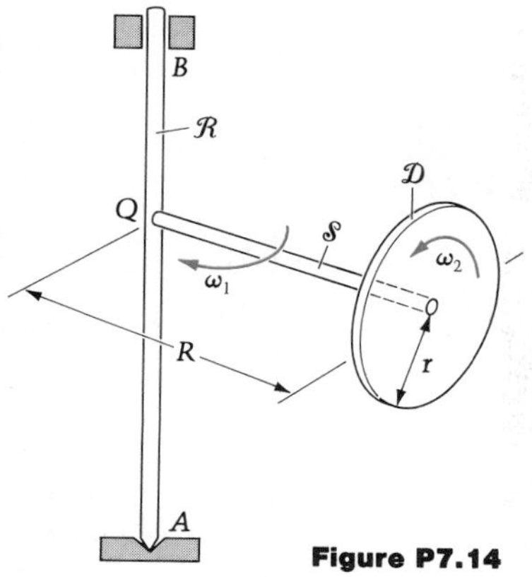

Figure P7.14

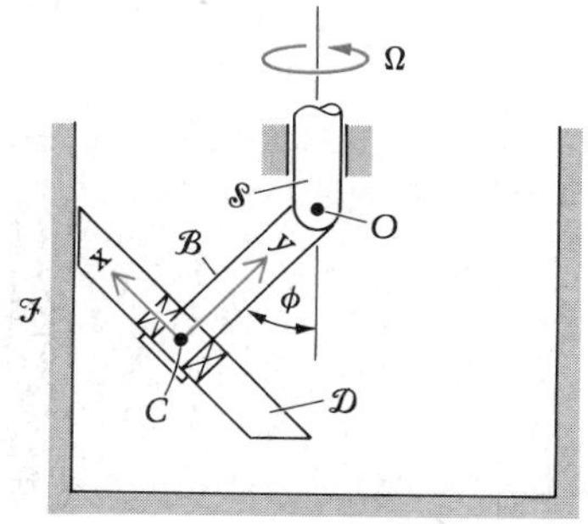

Figure P7.15

7.22 If velocities are varying continuously (no shock waves, for example), Euler's laws may be written in terms of accelerations as

$$\mathbf{F}_r = \dot{\mathbf{L}} = \frac{d}{dt}\int \mathbf{v}\, dm = \int \mathbf{a}\, dm$$

and

$$\mathbf{M}_{r_C} = \dot{\mathbf{H}}_C = \frac{d}{dt}\int \mathbf{r} \times \mathbf{v}\, dm = \int \mathbf{r} \times \mathbf{a}\, dm$$

Show that although $\mathbf{M}_{r_P} \neq \dot{\mathbf{H}}_P$, it *is* true that

$$\mathbf{M}_{r_P} = \int \mathbf{R} \times \mathbf{a}\, dm$$

in which P need not be fixed in the inertial frame $\mathcal{I}$! (See Figure P7.22.)

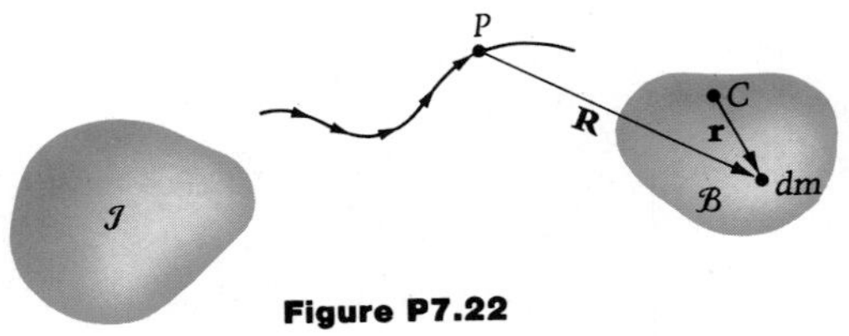

Figure P7.22

7.3 Transformations of Inertia Properties

Sometimes we need the moments and products of inertia at points other than the mass center C of a rigid body $\mathcal{B}$. These properties can be found without further integration by using the parallel-axis theorems, which were derived in Chapter 4. These are restated below, where $(\bar{x}, \bar{y}, \bar{z})$ are the coordinates of the mass center C relative to axes at P. For the moments of inertia at P:

$$I_{xx}^P = I_{xx}^C + m(\bar{y}^2 + \bar{z}^2) \quad (7.24a)$$

$$I_{yy}^P = I_{yy}^C + m(\bar{z}^2 + \bar{x}^2) \quad (7.24b)$$

$$I_{zz}^P = I_{zz}^C + m(\bar{x}^2 + \bar{y}^2) \quad (7.24c)$$

And for the products of inertia at P:

$$I_{xy}^P = I_{xy}^C - m\bar{x}\bar{y} \quad (7.25a)$$

$$I_{yz}^P = I_{yz}^C - m\bar{y}\bar{z} \quad (7.25b)$$

$$I_{zx}^P = I_{zx}^C - m\bar{z}\bar{x} \quad (7.25c)$$

EXAMPLE 7.3

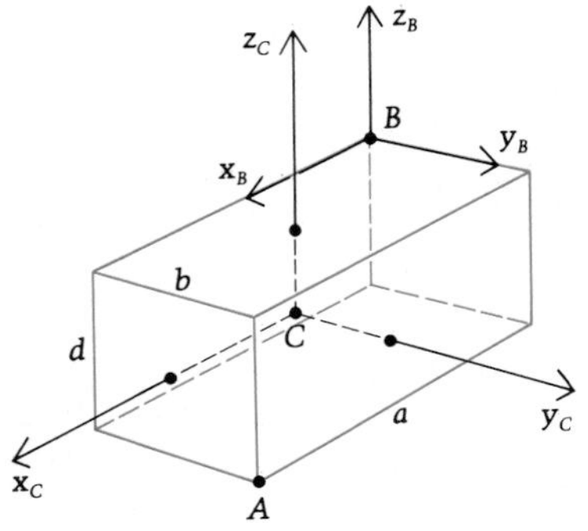

As a review example, compute the inertia properties at the corner B of the uniform rectangular solid of mass m shown in the diagram.

(Continued)

SOLUTION

For the moments of inertia we obtain

$$I_{xx}^{B} = I_{xx}^{C} + m(\bar{y}^2 + \bar{z}^2) = \frac{m}{12}(b^2 + d^2) + m\left[\left(\frac{b}{2}\right)^2 + \left(\frac{d}{2}\right)^2\right] = \frac{m}{3}(b^2 + d^2)$$

Note that the distance between $\mathbf{x}_B$ and $\mathbf{x}_C$ is $\sqrt{(b/2)^2 + (d/2)^2}$. In the same way,

$$I_{y_B} = \frac{m}{3}(d^2 + a^2) \qquad \text{and} \qquad I_{z_B} = \frac{m}{3}(a^2 + b^2)$$

For the products of inertia,

$$I_{xy}^{B} = I_{xy}^{C} - m\bar{x}\bar{y} = 0 - m\left(\frac{a}{2}\right)\left(\frac{b}{2}\right) = \frac{-mab}{4}$$

$$I_{yz}^{B} = I_{yz}^{C} - m\bar{y}\bar{z} = 0 - m\left(\frac{b}{2}\right)\left(-\frac{d}{2}\right) = \frac{mbd}{4}$$

$$I_{xz}^{B} = I_{xz}^{C} - m\bar{x}\bar{z} = 0 - m\left(\frac{a}{2}\right)\left(-\frac{d}{2}\right) = \frac{mad}{4}$$

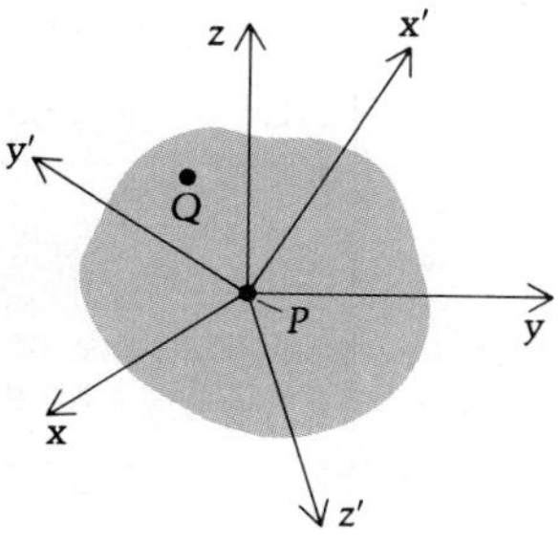

Figure 7.4

We now consider a second and equally important transformation, which will demonstrate that if we know the moments and products of inertia associated with a set of orthogonal axes through a point P, we can easily compute the moments and products of inertia associated with any *other* set of axes having the same origin. Consider two sets of axes with a common origin at P. (See Figure 7.4.) Let ℓ_x, ℓ_y, ℓ_z be the direction cosines of $\mathbf{x}'$ relative to $\mathbf{x}$, y, and z, respectively. Then the rectangular coordinate $\mathbf{x}'$ of a point Q in the body is related to the rectangular coordinates $\mathbf{x}$, y, z by

$$\mathbf{x}' = \mathbf{r}_{PQ} \cdot \underbrace{(\ell_x\hat{\mathbf{i}} + \ell_y\hat{\mathbf{j}} + \ell_z\hat{\mathbf{k}})}_{\text{unit vector along } \mathbf{x}' \text{ axis}}$$

$$= (\mathbf{x}\hat{\mathbf{i}} + y\hat{\mathbf{j}} + z\hat{\mathbf{k}}) \cdot (\ell_x\hat{\mathbf{i}} + \ell_y\hat{\mathbf{j}} + \ell_z\hat{\mathbf{k}})$$

$$\mathbf{x}' = \mathbf{x}\ell_x + y\ell_y + z\ell_z \tag{7.26}$$

We seek a formula for I_{xx}^{P} in terms of the inertia properties written with respect to the $(\mathbf{x}, y, z)$ axes. The definition of $I_{x'x'}^{P}$ is

$$I_{x'x'}^{P} = \int (y'^2 + z'^2)\, dm \tag{7.27}$$

Since $\mathbf{x}'^2 + y'^2 + z'^2 = \mathbf{x}^2 + y^2 + z^2$ (each is the square of the length of $\mathbf{r}_{PQ}$), we may add and subtract $\mathbf{x}'^2$ to produce this quantity in Equation (7.27):

$$I^P_{xx} = \int [(x'^2 + y'^2 + z'^2) - x'^2]\, dm$$
$$= \int [(x^2 + y^2 + z^2) - x'^2]\, dm \qquad (7.28)$$

Substituting $\mathbf{x}'$ from Equation (7.26) into (7.28) gives

$$I^P_{x'x'} = \int [x^2 + y^2 + z^2 - (x\ell_x + y\ell_y + z\ell_z)^2]\, dm$$

Expanding the trinomial and rearranging, we get

$$I^P_{x'x'} = \int [(1 - \ell_x^2)x^2 + (1 - \ell_y^2)y^2 + (1 - \ell_z^2)z^2$$
$$-2xy\ell_x\ell_y - 2xz\ell_x\ell_z - 2yz\ell_y\ell_z]\, dm \qquad (7.29)$$

Since the ℓ_i's are the direction cosines of the vector in the direction of the $\mathbf{x}'$ axis, we know that

$$\ell_x^2 + \ell_y^2 + \ell_z^2 = 1$$

Using this relation in the first three terms of the integrand in Equation (7.29) gives

$$I^P_{x'x'} = \int [(\ell_y^2 + \ell_z^2)x^2 + (\ell_x^2 + \ell_z^2)y^2 + (\ell_x^2 + \ell_y^2)z^2$$
$$-2xy\ell_x\ell_y - 2xz\ell_x\ell_z - 2yz\ell_y\ell_z]\, dm$$

Rearranging, we have

$$I^P_{x'x'} = \ell_x^2 \int (y^2 + z^2)\, dm + \ell_y^2 \int (x^2 + z^2)\, dm + \ell_z^2 \int (x^2 + y^2)\, dm$$
$$+ 2\ell_x\ell_y\left(-\int xy\, dm\right) + 2\ell_x\ell_z\left(-\int zx\, dm\right) + 2\ell_y\ell_z\left(-\int yz\, dm\right) \qquad (7.30)$$

Recognizing the six integrals in Equation (7.30) as the inertia properties associated with the $(\mathbf{x}, \mathbf{y}, \mathbf{z})$ directions at P, we arrive at our goal:

$$I^P_{x'x'} = \ell_x^2 I^P_{xx} + \ell_y^2 I^P_{yy} + \ell_z^2 I^P_{zz} + 2\ell_x\ell_y I^P_{xy} + 2\ell_x\ell_z I^P_{xz} + 2\ell_y\ell_z I^P_{yz} \qquad (7.31)$$

This formula allows us to compute the moment of inertia of the mass of $\mathcal{B}$ about any line through P if we know the properties at P for any set of orthogonal axes. We now illustrate its use with an example.

EXAMPLE 7.4

Compute the moment of inertia about the diagonal BA of the rectangular solid of Example 7.3.

SOLUTION

We define the axis $\mathbf{x}'$ to emanate from B, pointing toward A. The inertia properties at B were computed in the prior example. The direction cosines of $\mathbf{x}'$ are seen by inspection to be

(Continued)

$$\ell_x = \frac{a}{H} \qquad \ell_y = \frac{b}{H} \qquad \ell_z = \frac{-d}{H}$$

in which

$$H = \sqrt{a^2 + b^2 + d^2} = |\mathbf{r}_{BA}|$$

Substituting the ℓ_i's and the inertia properties at B into Equation (7.31) gives

$$\begin{aligned} I^B_{x'x'} &= \frac{m}{3H^2}[(b^2 + d^2)a^2 + (d^2 + a^2)b^2 + (a^2 + b^2)d^2] \\ &\quad + \frac{2m}{H^2}\left[(ab)\left(\frac{-ab}{4}\right) + (-ad)\left(\frac{ad}{4}\right) + (-bd)\left(\frac{bd}{4}\right)\right] \\ &= \frac{m(a^2b^2 + b^2d^2 + d^2a^2)}{6(a^2 + b^2 + d^2)} \end{aligned}$$

In the preceding example we observe that the line BA also passes through the mass center C. The moment of inertia about line BA is of course the same no matter which point on the line one uses to make the calculation. (It would actually be easier to compute at C in this case, because the ℓ_i's are the same while the ***products*** of inertia vanish!)

A result similar to Equation (7.31) for products of inertia will now be derived. Let n_x, n_y, and n_z be the direction cosines of axis y'. Note that the rectangular coordinate y' may then be written, in the same way as Equation (7.26), as follows:

$$y' = xn_x + yn_y + zn_z \tag{7.32}$$

By definition,

$$\begin{aligned} I^P_{x'y'} &= -\int x'y'\,dm \\ &= -\int (x\ell_x + y\ell_y + z\ell_z)(xn_x + yn_y + zn_z)\,dm \end{aligned}$$

Therefore, expanding and recognizing the product of inertia integrals,

$$\begin{aligned} I^P_{x'y'} = \int -(x^2\ell_x n_x + y^2\ell_y n_y + z^2\ell_z n_z)\,dm + (\ell_x n_y + \ell_y n_x)I^P_{xy} \\ + (\ell_x n_z + \ell_z n_x)I^P_{xz} + (\ell_y n_z + \ell_z n_y)I^P_{yz} \end{aligned} \tag{7.33}$$

Since (ℓ_x, ℓ_y, ℓ_z) and (n_x, n_y, n_z) are components of unit vectors along the mutually perpendicular axes x' and y', we may dot these vectors together and obtain

$$\ell_x n_x + \ell_y n_y + \ell_z n_z = 0$$

The integrand in Equation (7.33) may therefore be written as

$$\begin{aligned} &-(x^2\ell_x n_x + y^2\ell_y n_y + z^2\ell_z n_z) \\ &\quad = x^2(\ell_y n_y + \ell_z n_z) + y^2(\ell_x n_x + \ell_z n_z) + z^2(\ell_x n_x + \ell_y n_y) \\ &\quad = \ell_x n_x(y^2 + z^2) + \ell_y n_y(x^2 + z^2) + \ell_z n_z(x^2 + y^2) \end{aligned} \tag{7.34}$$

Substituting Equation (7.34) into (7.33) then yields the desired transformation equation for the products of inertia:

$$I^P_{x'y'} = \ell_x n_x I^P_{xx} + \ell_y n_y I^P_{yy} + \ell_z n_z I^P_{zz} + (\ell_x n_y + \ell_y n_x) I^P_{xy} + (\ell_x n_z + \ell_z n_x) I^P_{xz} + (\ell_y n_z + \ell_z n_y) I^P_{yz} \tag{7.35}$$

EXAMPLE 7.5

In Examples 7.3 and 7.4 let the solid be a cube ($a = b = d$) and let y' be defined as follows:

1. y' is perpendicular to $\mathbf{x}'$.
2. y' is in the same plane as z_B and $\mathbf{x}'$.

Find $I^B_{x'y'}$.

SOLUTION

From the equations in Example 7.4 we have $\hat{\boldsymbol{\ell}} = (\ell_x, \ell_y, \ell_z) = (1/\sqrt{3}, 1/\sqrt{3}, -1/\sqrt{3})$ for the unit vector along $\mathbf{x}'$; we now force the components of $\hat{\mathbf{n}}$—that is, (n_x, n_y, n_z)—to be such that conditions 1 and 2 are satisfied:

1. $\hat{\boldsymbol{\ell}} \perp \hat{\mathbf{n}} \Rightarrow \dfrac{1}{\sqrt{3}} n_x + \dfrac{1}{\sqrt{3}} n_y - \dfrac{1}{\sqrt{3}} n_z = 0 \Rightarrow n_x + n_y - n_z = 0$
2. $\underbrace{(\hat{\boldsymbol{\ell}} \times \hat{\mathbf{n}})}_{\uparrow} \cdot \hat{\mathbf{k}} = 0 \Rightarrow \dfrac{1}{\sqrt{3}} n_y - \dfrac{1}{\sqrt{3}} n_x = 0 \Rightarrow n_y - n_x = 0$

(a vector perpendicular to the plane of $\mathbf{x}'$ and $\mathbf{y}'$)

These two equations give $n_y = n_x$ and $n_z = 2n_x$. We must also ensure that $\hat{\mathbf{n}}$ is a unit vector:

$$1 = n_x^2 + n_y^2 + n_z^2 = n_x^2(1 + 1 + 4) = 6n_x^2$$

Thus $n_x = 1/\sqrt{6}$, so

$$n_x = \frac{1}{\sqrt{6}} \qquad n_y = \frac{1}{\sqrt{6}} \qquad n_z = \frac{2}{\sqrt{6}}$$

Substituting the components of $\hat{\boldsymbol{\ell}}$ and $\hat{\mathbf{n}}$ and the inertia properties at B (letting the general point P in (7.35) be B in this problem) into Equation (7.35) gives

$$I^B_{x'y'} = \frac{2ma^2}{3}\left[\frac{1}{\sqrt{3}}\frac{1}{\sqrt{6}} + \frac{1}{\sqrt{3}}\frac{1}{\sqrt{6}} - \frac{1}{\sqrt{3}}\left(\frac{2}{\sqrt{6}}\right)\right] + \frac{ma^2}{4}\left[-\left(\frac{1}{\sqrt{3}}\frac{1}{\sqrt{6}} + \frac{1}{\sqrt{3}}\frac{1}{\sqrt{6}}\right) + \left(\frac{1}{\sqrt{3}}\frac{2}{\sqrt{6}} - \frac{1}{\sqrt{3}}\frac{1}{\sqrt{6}}\right) + \left(\frac{1}{\sqrt{3}}\frac{2}{\sqrt{6}} - \frac{1}{\sqrt{3}}\frac{1}{\sqrt{6}}\right)\right] = 0$$

We note that in Example 7.5 the zero result is not obvious at this point in our study. While it is true that, for this case of $a = b = d$, the $\mathbf{x}'\mathbf{y}'$ plane is a plane of symmetry, this guarantees (see Section 4.3) that $I^B_{x'z'}$ and $I^B_{y'z'}$ are zero but not necessarily $I^B_{x'y'}$. Note also that there are *two* directions (180° apart) for $\mathbf{y}'$ that both satisfy conditions 1 and 2 in the preceding example.

Question 7.4 Where in the solution did we choose one of these directions? (And does it matter?)

We close this section by noting that Equations (7.31) and (7.35) are the transformation equations satisfied by a symmetric second-order tensor; thus the inertia properties do indeed form such a tensor. We also note from these two equations that only if the products of inertia are defined with the minus sign (see Equation 4.24) do we get the correct tensor transformation equations.

PROBLEMS / Section 7.3

7.23 Compute the moment of inertia with respect to line *AB* for the bent bar in Figure P7.23. The bar lies in a plane and has mass 4 m.

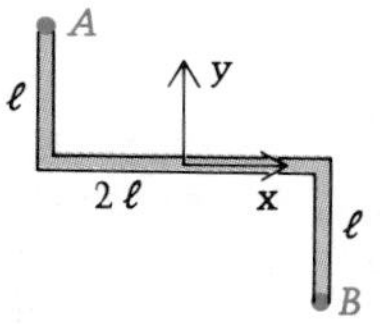

Figure P7.23

Find the product of inertia I^C_{yz} for the hoop $\mathcal{H}$ of mass m and radius R in Figure P7.24. The plane of $\mathcal{H}$ is misaligned with the $\mathbf{xy}$ plane by angle ϕ.

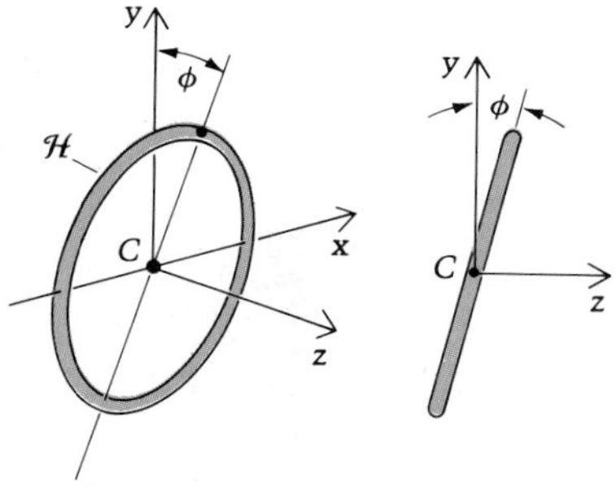

Figure P7.24

7.25 Find $I^B_{x'y'}$ for the problem of Example 7.5 if $b = d = a/2$. The axes have their origin at B just as in the example.

7.26 Compute the moment of inertia about line *BA* in Example 7.4 by using Equation (7.31) at C instead of B. Show that if $a = b = d$, the answer becomes equal to I^C_{xx} (which equals $I^C_{yy} = I^C_{zz}$ in this case).

7.27 The centroidal moments of inertia for the solid ellipsoid $\mathcal{E}$ in Figure P7.27 are

$$I^C_{xx} = \frac{m}{5}(b^2 + c^2)$$

$$I^C_{yy} = \frac{m}{5}(a^2 + c^2)$$

$$I^C_{zz} = \frac{m}{5}(a^2 + b^2)$$

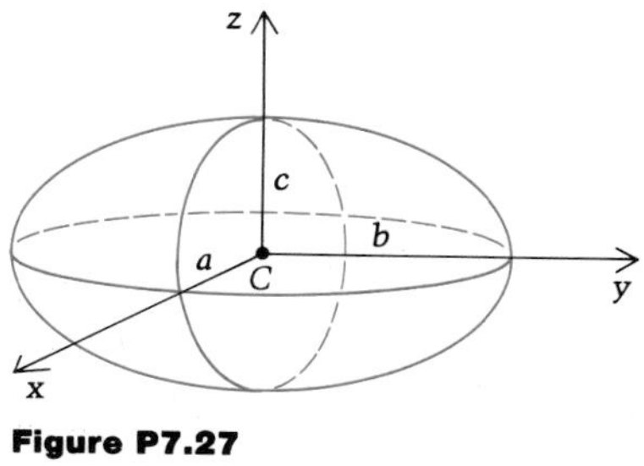

Figure P7.27

Further, the mass of $\mathcal{E}$ is $(4\pi/3)\rho abc$, where ρ is the mass density. Find the moment of inertia of the mass of $\mathcal{E}$ about the line making equal angles with $\mathbf{x}$, $\mathbf{y}$, and $\mathbf{z}$.

7.28 Calculate the inertia properties at O of the three circular fan blades connected by light rods in Figure P7.28. The blades are tilted 30° with respect to the axes OC_1, OC_2, and OC_3; the shaded halves of each are behind the plane of the drawing and the unshaded halves are in front of it.

7.29 Find the inertia properties at O for the body shown in Figure P7.29, which is composed of a rod and ring that have equal cross sections and densities. The rod is perpendicular to the plane of the ring.

7.30 Part of a special-purpose, dual-driven antenna system consists of an octagonal rotator as shown in Figure P7.30. Each of the eight equal sections is a square steel tube with the indicated dimensions and thickness $\frac{1}{8}$ in. Find the moment of inertia of the rotator about the axis of rotation. (Consider each section to have squared-off ends at the average 18-in. length and ignore the small overlaps. Use a density of 15.2 slug/ft^3.)

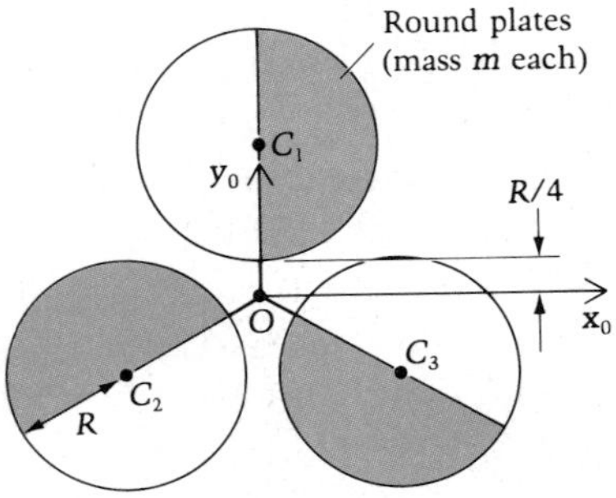

Figure P7.28

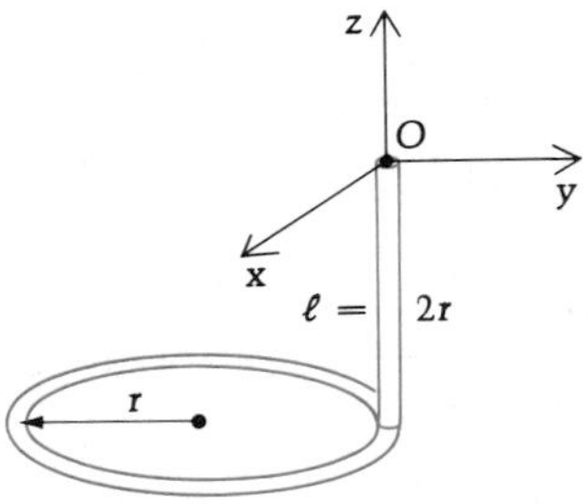

Figure P7.29

7.31 For the rigid body $\mathcal{B}$ in Figure P7.31 let the inertia properties at P be known $(I_{xx}^P, \ldots, I_{yz}^P)$. Further, let there be measured along the same axes the quantities

$$X = \frac{\ell_x}{\sqrt{I_{x'x'}^P}}$$

$$Y = \frac{\ell_y}{\sqrt{I_{x'x'}^P}}$$

$$Z = \frac{\ell_z}{\sqrt{I_{x'x'}^P}}$$

where $I_{x'x'}^P$ and (ℓ_x, ℓ_y, ℓ_z) are as defined in Section 7.3. Show that Equation (7.31) then implies

$$I_{xx}^P X^2 + I_{yy}^P Y^2 + I_{zz}^P Z^2 + 2XYI_{xy}^P + 2XZI_{xz}^P + 2YZI_{yz}^P = 1$$

This is the equation of an ellipsoid centered at P. Developed by Cauchy in 1827, it is called the ***ellipsoid of inertia***. Show that the moment of inertia about any line $\mathbf{x}'$ through P equals the reciprocal of the square of the distance from P to the point where X' intersects the ellipsoid.

7.32 Referring to the preceding problem, the equation of the ellipsoid of inertia written in terms of the principal moments of inertia (I_1^P, I_2^P, I_3^P) is

$$I_1^P X_P^2 + I_2^P Y_P^2 + I_3^P Z_P^2 = 1$$

Show that not all ellipsoids of the form $ax^2 + by^2 + cz^2 = 1$ can be ellipsoids of inertia. *Hint:* The sum of any two principal moments of inertia must always exceed the third, as shown in the next problem.

7.33 Show that the sum of any two of I_{xx}^P, I_{yy}^P, and I_{zz}^P always exceeds the third.

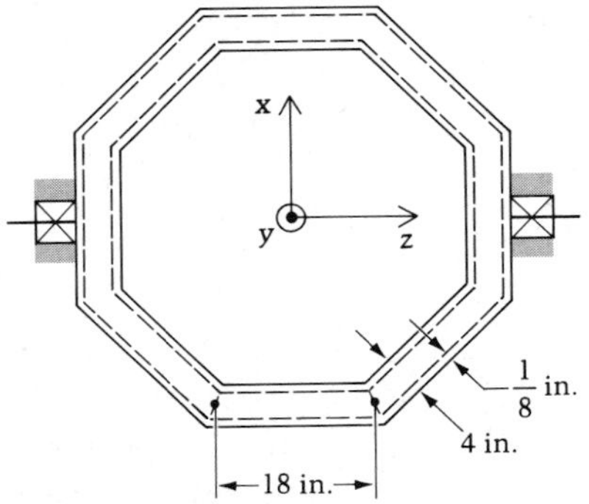

Figure P7.30

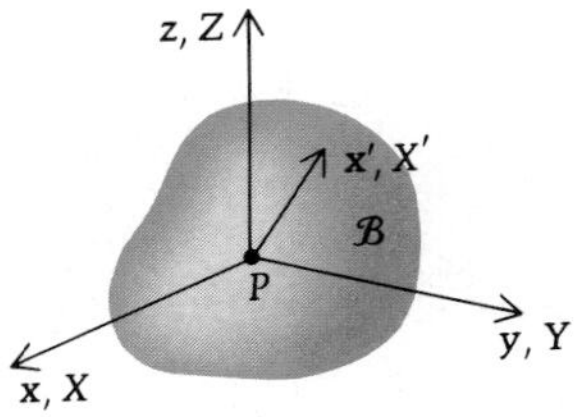

Figure P7.31

7.4 Principal Axes and Principal Moments of Inertia

In this section we demonstrate a particularly useful way of describing the inertia characteristics of a rigid body $\mathcal{B}$. It happens that at any point P of $\mathcal{B}$, it is always possible to find a set of rectangular axes so that the products of inertia at P with respect to these axes all vanish. These axes are called the **principal axes of inertia** at P, and the moments of inertia with respect to them are known as the **principal moments of inertia** for the point.

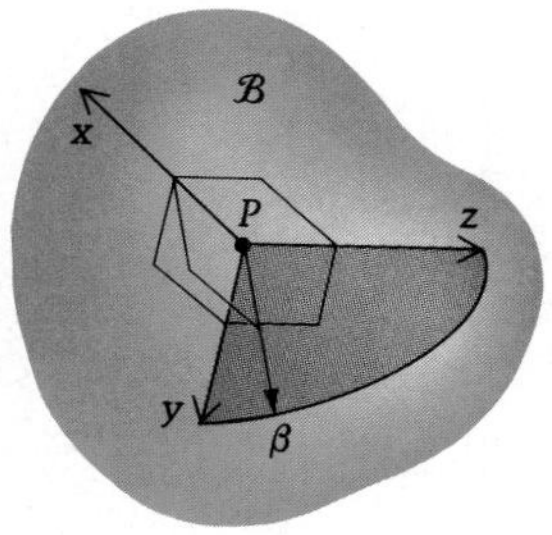

Figure 7.5

Specifically, an axis $\mathbf{x}$ is a principal axis at P if $I_{x\beta}^P = 0$, where β is any axis through P that is perpendicular to $\mathbf{x}$. (See Figure 7.5.) We can show that if $I_{xy}^P = I_{xz}^P = 0$, in which $(\mathbf{x}, y, z)$ form a triad of rectangular axes at P, then $\mathbf{x}$ is a principal axis at P. For the proof, we shall show that $I_{xy}^P = 0 = I_{xz}^P$ implies $I_{x\beta}^P = 0$, where β is the arbitrary axis through P normal to $\mathbf{x}$. Using Equation (7.35) we may write

$$I_{x\beta}^P = \ell_x n_x I_{xx}^P + \ell_y n_y I_{yy}^P + \ell_z n_z I_{zz}^P + (\ell_x n_y + \ell_y n_x) I_{xy}^P + (\ell_x n_z + \ell_z n_x) I_{xz}^P + (\ell_y n_z + \ell_z n_y) I_{yz}^P$$

In this equation $(\ell_x, \ell_y, \ell_z) = (1, 0, 0)$ are the direction cosines of $\mathbf{x}$ (the first subscript in $I_{x\beta}$) with respect to $\mathbf{x}$, y, and z, respectively, while (n_x, n_y, n_z) are the direction cosines of β (the second subscript in $I_{x\beta}$), also with respect to $\mathbf{x}$, y, and z. Since $\beta \perp \mathbf{x}$, we have $(n_x, n_y, n_z) = (0, n_y, n_z)$. Substituting these ℓ's and n's, we get

$$I_{x\beta}^P = 0 + 0 + 0 + (\ell_x n_y + 0) I_{xy}^P + (\ell_x n_z + 0) I_{xz}^P + 0$$

But $I_{xy}^P = I_{xz}^P = 0$, so that $I_{x\beta}^P = 0$ and we see that all it takes to make an axis, such as $\mathbf{x}$, a principal axis at a point is to show that $I_{xy}^P = 0$ and $I_{xz}^P = 0$, where $(\mathbf{x}, y, z)$ form an orthogonal triad at P. We shall need this result in what is to follow.

We now proceed to a computational procedure for finding the principal axes and moments of inertia. We shall do the derivation when $P = C$ and then later explain how it applies equally well to *all* points of $\mathcal{B}$.

For the moment let $(\mathbf{x}, y, z)$ be centered at C and at some instant let $\mathbf{x}$ be parallel to the angular velocity $\boldsymbol{\omega}_{\mathcal{B}/\mathcal{I}} = \boldsymbol{\omega}$ of $\mathcal{B}$ in a reference frame $\mathcal{I}$. Then $\boldsymbol{\omega} = \omega\hat{\mathbf{i}}$ and Equation (7.22) gives, for the angular momentum of $\mathcal{B}$ about C, the simplified expression

$$\mathbf{H}_C = I_{xx}^C \omega \hat{\mathbf{i}} + I_{xy}^C \omega \hat{\mathbf{j}} + I_{xz}^C \omega \hat{\mathbf{k}} \tag{7.36}$$

We note from Equation (7.36) that $\mathbf{H}_C$ is parallel to $\boldsymbol{\omega}$ if and only if $I_{xy}^C = I_{xz}^C = 0$—that is, if $\mathbf{x}$ is a principal axis at C. Thus we see, as did Leonhard Euler himself in the middle of the eighteenth century, that the principal axes have the property that when the angular velocity lies along one of them, so will the angular momentum. Euler was seeking an axis through C for which, when $\mathcal{B}$ was set spinning about it, the motion would continue about this axis without any need for external moments to maintain it. Note further that when $\mathbf{H}_C \parallel \boldsymbol{\omega}$ ("$\mathbf{H}_C$ is parallel to $\boldsymbol{\omega}$"), the proportionality constant is necessarily the moment of inertia I about their common axis.

Now we are ready for the big step. We let $\hat{\mathbf{n}} = n_x\hat{\mathbf{i}} + n_y\hat{\mathbf{j}} + n_z\hat{\mathbf{k}}$ be a unit vector and we seek the direction of $\hat{\mathbf{n}}$ that will ensure its being a principal axis. In other words, we want to find the values of the direction cosines of $\hat{\mathbf{n}}$ (n_x, n_y, n_z) such that if $\boldsymbol{\omega} = \omega\hat{\mathbf{n}}$, then $\mathbf{H} = I\boldsymbol{\omega}$. Writing $\boldsymbol{\omega}$ in component form gives

$$\boldsymbol{\omega} = \omega n_x\hat{\mathbf{i}} + \omega n_y\hat{\mathbf{j}} + \omega n_z\hat{\mathbf{k}} \tag{7.37}$$

Substituting from Equation (7.22) for $\mathbf{H}_C$, the vector relation $\mathbf{H}_C = I\boldsymbol{\omega}$ then gives the following three scalar component equations:

$$\begin{aligned} I_{xx}^C\omega_x + I_{xy}^C\omega_y + I_{xz}^C\omega_z &= I\omega n_x \\ I_{xy}^C\omega_x + I_{yy}^C\omega_y + I_{yz}^C\omega_z &= I\omega n_y \\ I_{xz}^C\omega_x + I_{yz}^C\omega_y + I_{zz}^C\omega_z &= I\omega n_z \end{aligned} \tag{7.38}$$

If we divide all three of Equations (7.38) by ω and note that since $\boldsymbol{\omega} = \omega\hat{\mathbf{n}}$ we have $n_x = \omega_x/\omega$, $n_y = \omega_y/\omega$, and $n_z = \omega_z/\omega$, then we may rearrange the equations as follows:

$$\begin{aligned} (I_{xx}^C - I)n_x + I_{xy}^C n_y + I_{xz}^C n_z &= 0 \\ I_{xy}^C n_x + (I_{yy}^C - I)n_y + I_{yz}^C n_z &= 0 \\ I_{xz}^C n_x + I_{yz}^C n_y + (I_{zz}^C - I)n_z &= 0 \end{aligned} \tag{7.39}$$

We now have a set of three equations that are algebraic, linear, and homogeneous in the three variables n_x, n_y, n_z. Such a system is known to have a nontrivial solution if and only if the determinant of the coefficients of the variables is zero.* In this case, we may drop the "nontrivial" adjective because the trivial solution ($n_x = n_y = n_z = 0$) fails to satisfy the side condition

$$n_x^2 + n_y^2 + n_z^2 = 1 \tag{7.40}$$

which must always be true for the direction cosines of a vector.

Setting the determinant of the coefficients equal to zero below will lead first to the special values of I for which the three equations have a solution. Each special value I is called an ***eigenvalue*** or ***characteristic value*** and will be a principal moment of inertia; the corresponding $\hat{\mathbf{n}}$ (with components n_x, n_y, n_z) is called the ***eigenvector*** associated with

*Cramer's rule clearly gives $n_x = n_y = n_z = 0$ as the only solution if the equations are independent, in which case the determinant D of the coefficients is not zero. If $D = 0$, then Cramer's rule yields the indeterminate form 0/0 for the n's and there is a chance for other solutions as the equations are then dependent.

this eigenvalue I. The unit vector $\hat{\mathbf{n}}$ points in the direction of a principal axis of inertia at C. The determinant is equated to zero below:

$$\begin{vmatrix} I_{xx}^C - I & I_{xy}^C & I_{xz}^C \\ I_{xy}^C & I_{yy}^C - I & I_{yz}^C \\ I_{xz}^C & I_{yz}^C & I_{zz}^C - I \end{vmatrix} = 0 \tag{7.41}$$

If we expand this characteristic determinant, we clearly get a cubic polynomial in I:

$$I^3 + a_1 I^2 + a_2 I + a_3 = 0 \tag{7.42}$$

The a_i are of course functions of the inertia properties. Now we know from algebra that if polynomials with real coefficients have any complex roots, they must occur in conjugate pairs. Thus the polynomial derived above has at this point at least one real root I_1. (It is positive by the definition of the quantity "moment of inertia" that it represents.) We now are guaranteed at least one principal moment of inertia and corresponding principal axis of inertia.

In order to show that there are two others, we next reorient our orthogonal triad of reference axes so that one of them ($\mathbf{x}$) coincides with the already identified principal axis; this then allows us to write $I_{xy}^C = 0$, $I_{xz}^C = 0$, and $I_{xx}^C = I_1$, where y and z are now a new pair of axes normal to our new (principal) $\mathbf{x}$ axis. Equations (7.39) now appear as

$$\begin{aligned} (I_1 - I)n_x &+ 0n_y &&+ 0n_z &&= 0 \\ 0n_x &+ (I_{yy}^C - I)n_y &&+ I_{yz}^C n_z &&= 0 \\ 0n_x &+ I_{yz}^C n_y &&+ (I_{zz}^C - I)n_z &&= 0 \end{aligned} \tag{7.43}$$

and the determinantal equation becomes:

$$\begin{vmatrix} I_1 - I & 0 & 0 \\ 0 & I_{yy}^C - I & I_{yz}^C \\ 0 & I_{yz}^C & I_{zz}^C - I \end{vmatrix} = 0 \tag{7.44}$$

This time the resulting cubic becomes factorable. Expanding the determinant, we have

$$(I_1 - I)[(I_{yy}^C - I)(I_{zz}^C - I) - I_{yz}^{C^2}] = 0 \tag{7.45}$$

The principal moments of inertia at C are the roots of Equation (7.45). The first root (the one we already knew) is reaffirmed by setting the first factor to zero:

$$I = I_1 \qquad (= I_{xx}^C)$$

The others will be seen to come from equating the second factor to zero:

$$I^2 - (I_{yy}^C + I_{zz}^C)I + (I_{yy}^C I_{zz}^C - I_{yz}^{C^2}) = 0 \tag{7.46}$$

This is, of course, a quadratic equation in I. Recalling that the two roots to

$$aI^2 + bI + c = 0$$

are

$$\frac{-b \pm \sqrt{b^2 - 4ac}}{2a}$$

we see that we shall have two (more) real roots I_2 and I_3 if the discriminant is positive or zero:

$$\begin{aligned} b^2 - 4ac &= (I_{yy}^C + I_{zz}^C)^2 - 4(1)(I_{yy}^C I_{zz}^C - I_{yz}^{C^2}) \\ &= I_{yy}^{C^2} - 2I_{yy}^C I_{zz}^C + I_{zz}^{C^2} + 4I_{yz}^{C^2} \\ &= (I_{yy}^C - I_{zz}^C)^2 + 4I_{yz}^{C^2} \geq 0 \end{aligned} \tag{7.47}$$

Therefore all three roots of the characteristic cubic equation are real (and positive), and so we always have three principal moments of inertia at C, each with its own corresponding principal axis.*

We mention at this point the procedure for obtaining the principal direction, given by n_x, n_y, and n_z, for ***each*** of the principal moments of inertia (I_1, I_2, or I_3). Equations (7.43), being dependent, may not be solved for the three components of each $\hat{\mathbf{n}}$ in themselves; however, together with the identity $n_x^2 + n_y^2 + n_z^2 = 1$, a solution may be found. The idea is to solve for, say, n_y and n_z in terms of n_x from two of Equations (7.43); then we substitute into Equation (7.40) and solve for n_x. Either sign may be used in taking the final square root, because there are obviously two legitimate sets of direction cosines. These two sets are negatives of each other, and each yields the correct principal axis. In Figure 7.6, either $\hat{\mathbf{n}}$ or $-\hat{\mathbf{n}}$ defines a principal axis through C. The principal axis is an undirected line.

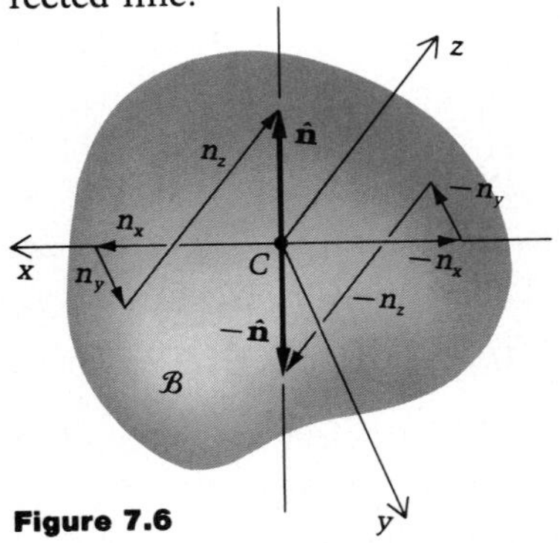

Figure 7.6

In the first of two examples to follow, we shall again see (as in the preceding discussion) that if at least two of the products of inertia vanish, then the cubic equation (7.42) becomes factorable. In this case, we do not have to solve it numerically.

*This was proved in 1755 for the first time by Segner, a contemporary of Euler. Segner also showed that the principal axes (for distinct principal moments of inertia) are orthogonal.

EXAMPLE 7.6

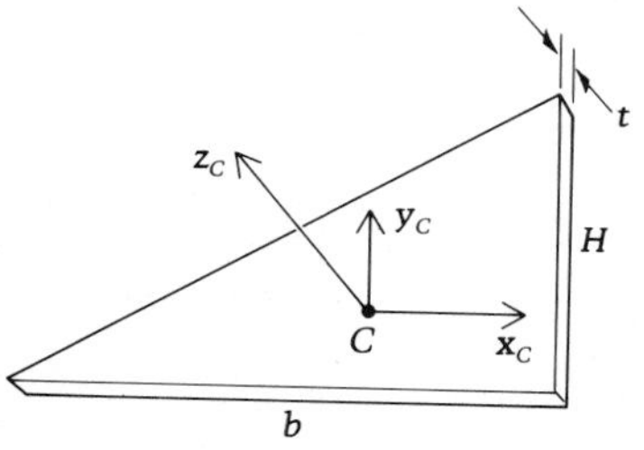

The inertia properties of a right triangular plate (see the diagram) are

$$I_{xx}^C = \frac{mH^2}{18} \qquad I_{xy}^C = \frac{-mbH}{36}$$

$$I_{yy}^C = \frac{mb^2}{18} \qquad I_{xz}^C = 0$$

$$I_{zz}^C = \frac{m(H^2 + b^2)}{18} \qquad I_{yz}^C = 0$$

Find the principal moments of inertia of the plate. Then find their associated principal axes when $b = H$.

SOLUTION

Equations (7.39), which lead to the principal moments and axes of inertia, become, for the plate,

$$\left(\frac{mH^2}{18} - I\right)n_x - \frac{mbH}{36}n_y + 0n_z = 0 \tag{1}$$

$$\frac{-mbH}{36}n_x + \left(\frac{mb^2}{18} - I\right)n_y + 0n_z = 0 \tag{2}$$

$$0n_x + 0n_y + \left(\frac{m(b^2 + H^2)}{18} - I\right)n_z = 0 \tag{3}$$

The algebra is simplified by dividing by $mH^2/36$ and defining

$$B = \frac{b}{H} \qquad \text{and} \qquad I^* = \frac{I}{(mH^2/36)} \tag{4}$$

This gives, in terms of the nondimensional parameters B and I^*,

$$(2 - I^*)n_x + (-B)n_y + 0n_z = 0 \tag{5}$$

$$(-B)n_x + (2B^2 - I^*)n_y + 0n_z = 0 \tag{6}$$

$$0n_x + 0n_y + [2(B^2 + 1) - I^*]n_z = 0 \tag{7}$$

Therefore the determinantal equation becomes

$$\begin{vmatrix} 2 - I^* & -B & 0 \\ -B & 2B^2 - I^* & 0 \\ 0 & 0 & 2(B^2 + 1) - I^* \end{vmatrix} = 0 \tag{8}$$

Expanding across the third row (or down the third column), we get

$$[2(B^2 + 1) - I^*][(2 - I^*)(2B^2 - I^*) - B^2] = 0 \tag{9}$$

Therefore one of the brackets must vanish and the roots come from

$$I^* = 2(B^2 + 1) \tag{10}$$

(Continued)

and

$$I^{*2} - 2I^*(B^2 + 1) + 3B^2 = 0 \tag{11}$$

Equation (10) gives, using Equations (4),

$$I_3 = \frac{m(b^2 + H^2)}{18} \tag{12}$$

Equation (11) gives, by the quadratic formula,

$$I^*_{1,2} = \frac{2(B^2 + 1) \pm \sqrt{4^2(B^2 + 1)^2 - 12B^2}}{2} \tag{13}$$

Thus

$$\begin{aligned} I_{1,2} &= \frac{mH^2}{36}\left(\frac{b^2}{H^2} + 1 \pm \sqrt{\frac{b^4}{H^4} - \frac{b^2}{H^2} + 1}\right) \\ &= \frac{m(b^2 + H^2)}{36} \pm \frac{m}{36}\sqrt{b^4 - b^2H^2 + H^4} \end{aligned} \tag{14}$$

The three principal moments of inertia of the plate are given by Equations (12) and (14). In the case when the right triangle is isosceles ($b = H$), we have $B = 1$ so that, from Equation (13),

$$I^*_{1,2} = 2 \pm \sqrt{4 - 3} = 2 \pm 1$$

or

$$I^*_1 = 3 \qquad \text{and} \qquad I^*_2 = 1 \tag{15}$$

Also, from Equation (10),

$$I^*_3 = 2(1^2 + 1) = 4$$

Changing back to dimensional inertias by Equation (4), we see that for a right isosceles triangular plate,

$$I_1 = \frac{mH^2}{12} \qquad I_2 = \frac{mH^2}{36} \qquad I_3 = \frac{mH^2}{9} \tag{16}$$

We shall now determine the principal axis associated with each of these principal moments of inertia. We first substitute $I^*_1 = 3$ (with $B = 1$) in each of Equations (5) to (7) and get

$$\begin{aligned} -n_x - n_y &= 0 \\ -n_x - n_y &= 0 \\ n_z &= 0 \end{aligned} \tag{17}$$

The third of these equations says that the principal axis for I_1 is in the plane of the plate (xy); the other two equations both give

$$n_x = -n_y \tag{18}$$

(Continued)

Substituting this result into

$$n_x^2 + n_y^2 + n_z^2 = 1 \tag{19}$$

gives

$$(-n_y)^2 + n_y^2 = 1 \tag{20}$$

$$n_y^2 = \frac{1}{2}$$

$$n_y = \pm \frac{1}{\sqrt{2}} \tag{21}$$

Thus, by Equation (18),

$$n_x = \mp \frac{1}{\sqrt{2}} \tag{22}$$

so that either

$$(n_x, n_y, n_z) = \left(-\frac{1}{\sqrt{2}}, \frac{1}{\sqrt{2}}, 0\right) \tag{23}$$

or

$$(n_x, n_y, n_z) = \left(\frac{1}{\sqrt{2}}, -\frac{1}{\sqrt{2}}, 0\right) \tag{24}$$

The lines ℓ_1 defined by these two sets of direction cosines are shown in the following diagrams. It is seen that the two preceding results represent the *same line;* its positive direction is opposite but unimportant. The inertia value, being the integral of $r^2\,dm$, is independent of the directivity of the line.

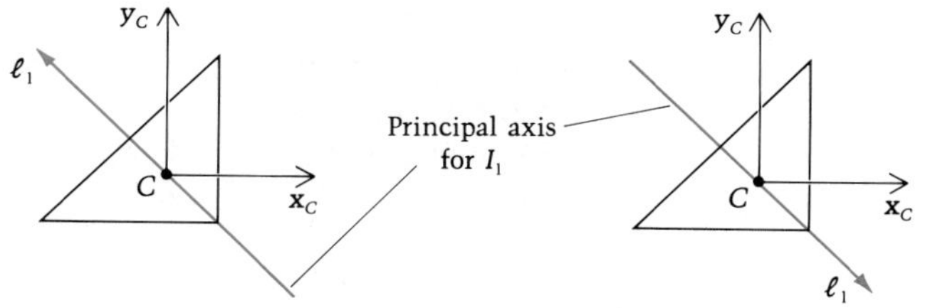

(a) Line of Eq. (7.23) **(b)** Line of Eq. (7.24)

For I_2^*, Equations (5) to (7) become

$$n_x - n_y = 0$$

$$-n_x + n_y = 0 \tag{25}$$

$$n_z = 0$$

This time $n_x = n_y$ with n_z again equaling zero, so this principal axis makes equal angles with $\mathbf{x}_C$ and y_C. (See the following diagrams.) Note from the preceding diagrams that there is much more inertia about the principal axis for I_1 than there is for I_2; the mass is more closely clustered about ℓ_2 than it is about ℓ_1.

(Continued)

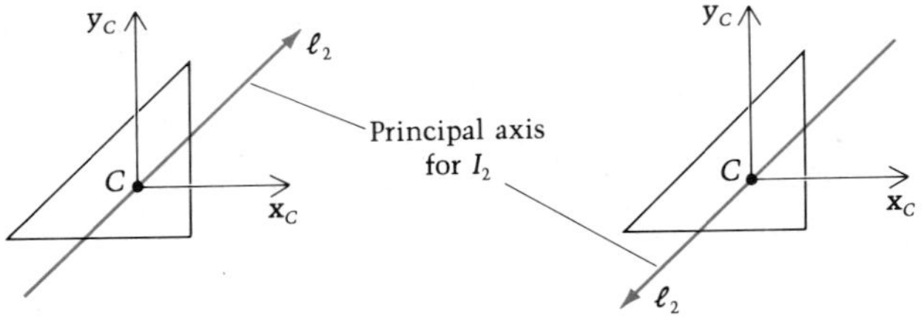

$n_x = n_y = 1/\sqrt{2}$ $\qquad$ $n_x = n_y = -1/\sqrt{2}$

The third principal axis is found from Equations (5) to (7) when, for $B = 1$,

$$I^* = I_3^* = 2(B^2 + 1) = 4 \tag{26}$$

These equations are

$$\begin{aligned} -2n_x - n_y &= 0 \\ -n_x - 2n_y &= 0 \\ [2(2) - 4]n_z &= 0 \end{aligned} \tag{27}$$

The first two of these equations have the solution $n_x = n_y = 0$. The third leaves n_z indeterminate. But from Equation (19) we have $n_z = 1$ or -1. Thus the principal axis for I_3 is the line normal to the plate at C. This will in fact always be true: When a body is a ***plate*** (that is, flat with negligible thickness compared to its other dimensions), the moment of inertia about the axis normal to the plate at any point is principal for that point. It is even true that it is the sum of the other two and therefore the largest.

We now wish to remark that a set of principal axes exists at ***every*** point of $\mathcal{B}$, not just at the mass center C. To show this we first recall Equations (7.18) and (7.23), which give the angular momentum $\mathbf{H}_P$ about any point P of body $\mathcal{B}$:

$$\begin{aligned} \mathbf{H}_P = m\mathbf{r}_{PC} \times \mathbf{v}_P &+ [I_{xx}^P\omega_x + I_{xy}^P\omega_y + I_{xz}^P\omega_z]\hat{\mathbf{i}} \\ &+ [I_{xy}^P\omega_x + I_{yy}^P\omega_y + I_{yz}^P\omega_z]\hat{\mathbf{j}} \\ &+ [I_{xz}^P\omega_x + I_{yz}^P\omega_y + I_{zz}^P\omega_z]\hat{\mathbf{k}} \end{aligned}$$

Whenever the cross-product term in $\mathbf{H}_P$ vanishes, the terms that remain are identical to those of Equation (7.22) if P replaces C. Therefore, for cases in which $\mathbf{r}_{PC} \times \mathbf{v}_P = \mathbf{0}$, we need only recall our arguments made for C and we shall be led, through an identical procedure, to the principal axes and moments of inertia for ***any*** point P of $\mathcal{B}$.

Therefore we only need to imagine that at some instant our point P of interest has either $\mathbf{v}_P = \mathbf{0}$ or $\mathbf{v}_P \parallel \mathbf{r}_{PC}$. In either case, $\mathbf{r}_{PC} \times \mathbf{v}_P = \mathbf{0}$ and $\mathbf{H}_P$ then has the form of Equations (7.23). Retracing our steps from that point, we arrive at the three axes (through P this time) for which all three products of inertia are zero; these are the same three axes through P for which $\mathbf{H}_P \parallel \boldsymbol{\omega}$ whenever $\boldsymbol{\omega}$ is aligned with one of them and $\mathbf{v}_P$

either vanishes or is parallel to $\mathbf{r}_{PC}$. Of course, all references to the motion conditions that led to the determinantal equation are again lost (as they were for C), so that the principal moments and axes of inertia depend only on the body's mass distribution.

EXAMPLE 7.7

Find the principal moments of inertia at O and the directions of their associated principal axes for the body shown in the diagrams. It is made up of three rigid, identical slender rods welded together at right angles to form a single rigid body $\mathcal{B}$.

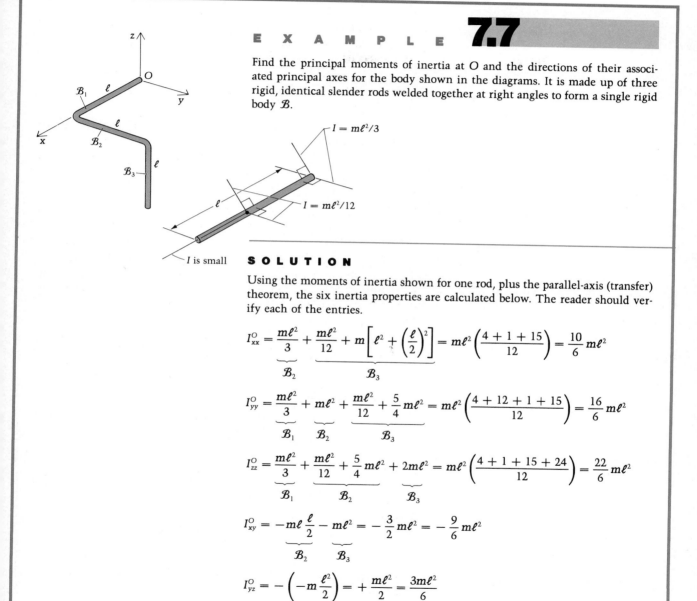

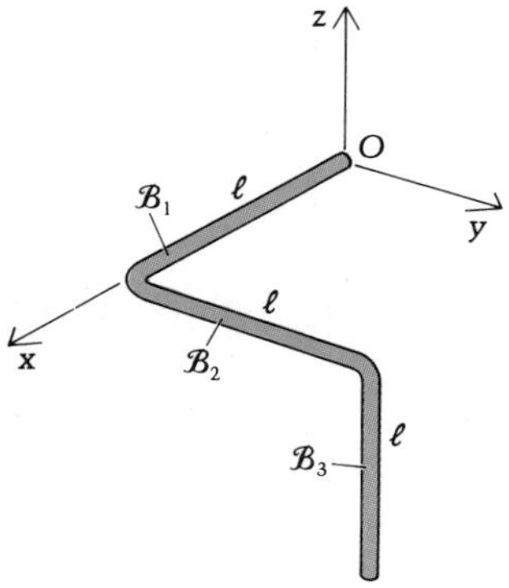

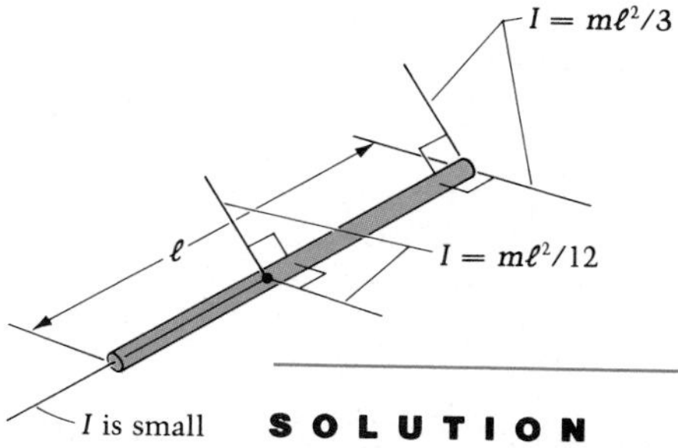

SOLUTION

Using the moments of inertia shown for one rod, plus the parallel-axis (transfer) theorem, the six inertia properties are calculated below. The reader should verify each of the entries.

$$I_{xx}^O = \underbrace{\frac{m\ell^2}{3}}_{\mathcal{B}_2} + \underbrace{\frac{m\ell^2}{12} + m\left[\ell^2 + \left(\frac{\ell}{2}\right)^2\right]}_{\mathcal{B}_3} = m\ell^2\left(\frac{4+1+15}{12}\right) = \frac{10}{6}m\ell^2$$

$$I_{yy}^O = \underbrace{\frac{m\ell^2}{3}}_{\mathcal{B}_1} + \underbrace{m\ell^2}_{\mathcal{B}_2} + \underbrace{\frac{m\ell^2}{12} + \frac{5}{4}m\ell^2}_{\mathcal{B}_3} = m\ell^2\left(\frac{4+12+1+15}{12}\right) = \frac{16}{6}m\ell^2$$

$$I_{zz}^O = \underbrace{\frac{m\ell^2}{3}}_{\mathcal{B}_1} + \underbrace{\frac{m\ell^2}{12} + \frac{5}{4}m\ell^2}_{\mathcal{B}_2} + \underbrace{2m\ell^2}_{\mathcal{B}_3} = m\ell^2\left(\frac{4+1+15+24}{12}\right) = \frac{22}{6}m\ell^2$$

$$I_{xy}^O = \underbrace{-m\ell\frac{\ell}{2}}_{\mathcal{B}_2} - \underbrace{m\ell^2}_{\mathcal{B}_3} = -\frac{3}{2}m\ell^2 = -\frac{9}{6}m\ell^2$$

$$I_{yz}^O = \underbrace{-\left(-m\frac{\ell^2}{2}\right)}_{\mathcal{B}_3} = +\frac{m\ell^2}{2} = \frac{3m\ell^2}{6}$$

(Continued)

$$I_{xz}^{O} = \underbrace{-\left(-m\frac{\ell^2}{2}\right)}_{\mathcal{B}_3} = +\frac{m\ell^2}{2} = \frac{3m\ell^2}{6}$$

For this problem, then, Equations (7.39) may be expressed as

$$\begin{aligned}
\left(10 - \frac{6I}{m\ell^2}\right)n_x - 9n_y + 3n_z &= 0 \\
-9n_x + \left(16 - \frac{6I}{m\ell^2}\right)n_y + 3n_z &= 0 \\
3n_x + 3n_y + \left(22 - \frac{6I}{m\ell^2}\right)n_z &= 0
\end{aligned} \tag{1}$$

in which we have multiplied the three equations by $(6/m\ell^2)$; we may replace $6I/m\ell^2$ by $\mathcal{I}$ and write the determinant as

$$\begin{vmatrix} 10-\mathcal{I} & -9 & 3 \\ -9 & 16-\mathcal{I} & 3 \\ 3 & 3 & 22-\mathcal{I} \end{vmatrix} = 0$$

Expanding gives the characteristic cubic equation:

$$f(\mathcal{I}) = -\mathcal{I}^3 + 48\mathcal{I}^2 - 633\mathcal{I} + 1342 = 0$$

If a computer or programmable calculator is not available,* we can always solve a cubic by trial and error rather quickly. Noting that $f(\mathcal{I})$ is 1342 at $\mathcal{I} = 0$ and is negative at $\mathcal{I} = 3$, for example, on a calculator we may proceed and within 15 minutes obtain the root between these values.† The procedure is as follows:

$\mathcal{I}$	$f(\mathcal{I})$	
3	−206	(thus a root is probably just past $\mathcal{I} = 2.5$)
2	260	
2.6	3.10	(still positive)
2.61	−0.93	(so it is >2.6 and <2.61, closer to the latter)
2.608	−0.123	(back up slightly!)
2.607	0.2802	(so it is about two-thirds of the way from 2.607 to 2.608)
2.6076	0.03835	(so just a little farther)
2.6077	−0.00195	(the root is close to this number!)
2.60769	0.00208	(should be halfway between the last one and this one. . .)
2.607695	0.00006	(now double-check)
2.607696	−0.00034	(so $\mathcal{I}_1 = 2.607695$ to seven figures)

(Continued)

*See Appendix C for a numerical solution to this problem using the Newton–Raphson method with a programmable calculator.

†We abandon our three-digit consistency in numerical analyses like this one in order to illustrate the speed of convergence.

Next we use synthetic division to obtain the reduced quadratic:

$$\begin{array}{r|rrrr} & -1 & 48 & -633 & 1342 \\ 2.607695 & & -2.607695 & 118.369287 & -1341.999938 \\ \hline & -1 & 45.392305 & -514.630713 & 0.000062 \approx 0 \end{array}$$

$$-\mathcal{J}^2 + 45.392305\mathcal{J} - 514.630713 = 0$$

Using the quadratic formula, we obtain

$$\mathcal{J}_2 = 22.000002$$

$$\mathcal{J}_3 = 23.392303$$

The value of $\mathcal{J}_2$ strongly hints that 22 might be a rational root. Synthetic division shows that it is, and refined values are then

$$\mathcal{J}_1 = 2.607695$$

$$\mathcal{J}_2 = 22$$

$$\mathcal{J}_3 = 23.392305$$

Since $\mathcal{J} = 6I/m\ell^2$, our dimensional principal moments of inertia are, to five digits,

$$I_1 = 0.43462m\ell^2$$

$$I_2 = 3.66667m\ell^2$$

$$I_3 = 3.89872m\ell^2$$

We next illustrate the computation of the direction cosines, which locate for us the principal axes of inertia. We find them from Equations (1), for which $\mathcal{J}_1$, $\mathcal{J}_2$, and $\mathcal{J}_3$ are the only special values (eigenvalues) of $\mathcal{J}$ for which these equations have a solution. First we seek the principal axis associated with $\mathcal{J}_1 = 2.607695$. The first of Equations (1) becomes

$$7.392305n_x - 9n_y + 3n_z = 0$$

Solving for n_z in terms of n_x and n_y and substituting the result into the second of Equations (1) gives

$$n_y = 0.732051n_x$$

Thus

$$n_z = -0.267950n_x$$

Substituting these expressions for n_y and n_z into $n_x^2 + n_y^2 + n_z^2 = 1$ yields

$$\text{Unit vector } \hat{\mathbf{n}}_1 \begin{cases} n_x = 0.788675 \\ n_y = 0.577350 \\ n_z = -0.211325 \end{cases}$$

(Continued)

Thus the angles that the principal axis of minimum moment of inertia makes with $\mathbf{x}$, $\mathbf{y}$, and $\mathbf{z}$ are, respectively, 37.94°, 54.74°, and 102.20°. This axis has to be the one to which, loosely, the mass finds itself closest. Examine the figure together with these angles and observe that they make sense.

Next we may follow the same procedure for the principal axis of maximum moment of inertia ($I_3 = \mathcal{I}_3 m\ell^2/6 = 3.898718 m\ell^2$). The results are, as the reader may verify,

$$\text{Unit vector } \hat{\mathbf{n}}_3 \begin{cases} n_x = 0.211325 & (\theta_x = \cos^{-1} n_x = 77.80°) \\ n_y = -0.577350 & (\theta_y = 125.26°) \\ n_z = -0.788675 & (\theta_z = 142.06°) \end{cases}$$

And for the intermediate moment of inertia ($I_2 = \mathcal{I}_2 m\ell^2/6 = 3.666667 m\ell^2$), the principal axis is defined by

$$\text{Unit vector } \hat{\mathbf{n}}_2 \begin{cases} n_x = 0.577350 & (\theta_x = 54.74°) \\ n_y = -0.577350 & (\theta_y = 125.26°) \\ n_z = 0.577350 & (\theta_z = 54.74°) \end{cases}$$

Note in the preceding example that $\hat{\mathbf{n}}_1 \cdot \hat{\mathbf{n}}_2 = \hat{\mathbf{n}}_2 \cdot \hat{\mathbf{n}}_3 = \hat{\mathbf{n}}_3 \cdot \hat{\mathbf{n}}_1 = 0$ and that $\hat{\mathbf{n}}_1 \times \hat{\mathbf{n}}_2 = \hat{\mathbf{n}}_3$.* These are very good checks on the solution since the principal axes, when the principal moments of inertia are distinct ($I_1 \neq I_2 \neq I_3 \neq I_1$), are orthogonal. To prove this in general, let $I_1 \neq I_2$ or I_3, and let $\mathbf{x}$ lie along the axis of I_1. Then from the first of Equations (7.43), which is

$$(I_1 - I)n_x = 0$$

we see that if I is either I_2 or I_3, then $n_x = 0$; that is, the cosine of the angle between $\mathbf{x}$ and the corresponding principal axis has to be zero. This means that $\mathbf{x}$ is perpendicular to the other two principal axes. In turn, these two axes are normal to each other; this follows from reorienting the axes once more so that $\mathbf{x}$ still lies along the axis of I_1 but now $\mathbf{y}$ lies along the axis of I_2. This time I_{yz}^C is also zero, so the new second equation becomes

$$(I_2 - I)n_y = 0$$

This shows that for $I = I_3$ ($I_2 \neq I_3$), the value of n_y for the third principal axis vanishes and it is then normal not only to $\mathbf{x}$ (which it still is, since we have only rotated it about $\mathbf{x}$) but also to $\mathbf{y}$. Thus the three principal axes are orthogonal if the principal moments of inertia are all different. We now turn our attention to what happens if they are not.

A commonly occurring case is for two of the principal moments of inertia at a point P to be equal to each other but different from the

*This could have been $\hat{\mathbf{n}}_1 \times \hat{\mathbf{n}}_2 = -\hat{\mathbf{n}}_3$—equally correct!

third. When this happens, we can show that *every* line through P in the plane of the two axes (call them $\mathbf{x}$ and $\mathbf{y}$) with equal moments of inertia (say $I_1 = I_2$) is a principal axis having this same value for its moment of inertia.

To do this, we first show that if $I_1 = I_2 \neq I_3$, then the axis associated with I_3 (call it $\mathbf{z}$) is perpendicular to those $(\mathbf{x}, \mathbf{y})$ of I_1 and I_2. The third equation of (7.43) gives

$$(I_3 - I)n_z = 0$$

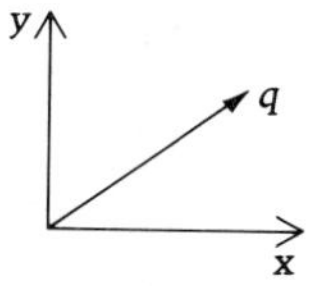

Figure 7.7

Thus when I is I_1 or I_2, then $n_z = 0$. Hence the angle between $\mathbf{z}$ and $\mathbf{x}$ (and between $\mathbf{z}$ and $\mathbf{y}$) is 90°. It does not follow in like manner from the equations, however, that axes $\mathbf{x}$ and $\mathbf{y}$ are perpendicular. To handle this case, we begin by showing that if the axes of the other two principal moments of inertia ($I_1 = I_2$) are *not* presumed to be orthogonal (say they are q and $\mathbf{x}$ in Figure 7.7 with $I_{qq}^P = I_1 = I_{xx}^P$), then all is well because I_{yy} is *also* equal to I_1.

To prove this, we use Equation (7.31):

$$I_{qq}^P = I_{xx}^P \ell_x^2 + I_{yy}^P \ell_y^2 + I_{zz}^P \underbrace{\ell_z^2}_{0} + \overbrace{0 + 0 + 0}^{\mathbf{x}\text{ and }\mathbf{z}\text{ are principal!}}$$

$$I_1 = I_1 \ell_x^2 + I_{yy}^P \ell_y^2$$

$$\underbrace{I_1(1 - \ell_x^2)}_{\ell_y^2} = I_{yy}^P \ell_y^2$$

Thus we see that

$$I_{yy} = I_1$$

Next we let $\hat{\boldsymbol{\ell}}$ be a unit vector in the direction of an arbitrary axis $\mathbf{x}'$ in the plane of the perpendicular axes $\mathbf{x}$ and $\mathbf{y}$. Then using Equation (7.31) again gives

$$I_{x'x'}^P = \ell_x^2 I_1 + \ell_y^2 \overset{= I_1}{I_2} + \overset{0}{\ell_z^2} I_3 + \underbrace{0 + 0 + 0}_{I_{xy}^P = I_{xz}^P = I_{yz}^P = 0}$$

$$= I_1(\ell_x^2 + \ell_y^2) = I_1$$

And Equation (7.35) yields (with $\mathbf{y}'$ perpendicular to $\mathbf{x}$ and lying in the plane of $\mathbf{x}'$, $\mathbf{x}$, and $\mathbf{y}$ as in Figure 7.8 on the next page):

$$I_{x'y'}^P = \ell_x n_x I_1 + \ell_y n_y \overset{= I_1}{I_2} + \overset{0}{\ell_z} \overset{0}{n_z} I_3 + 0 + 0 + 0$$

$$= I_1(\ell_x n_x + \ell_y n_y)$$

$$= I_1(\hat{\boldsymbol{\ell}} \cdot \hat{\mathbf{n}}) = 0$$

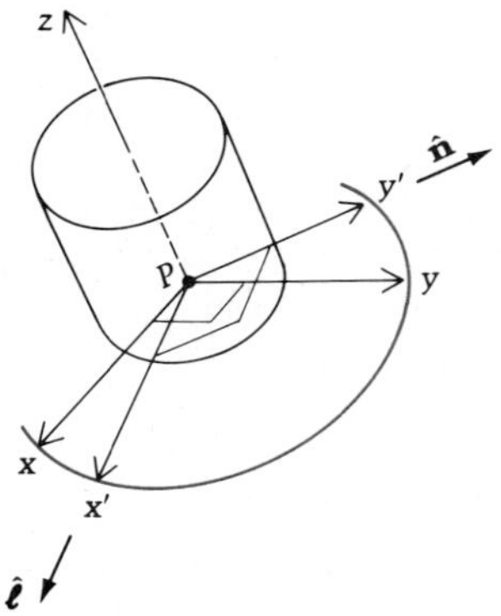

Figure 7.8

Noting that z is perpendicular to **x**′ and that $I^P_{x'z} = 0$ since z is principal, we then have the result that **x**′ is principal with the same moment of inertia as **x** and y, and this is what we wanted to prove. Note then, for a body with an axis of symmetry, for example, that any axis which passes through and is normal to the symmetry axis of $\mathcal{B}$ is always principal; this is true *even if the axis is not fixed in the body,* which will prove useful to us later.

Finally, if all three principal axes (**x**, y, z) through P have the same corresponding principal moment of inertia I_1, then *every* axis through P is principal with principal moment of inertia I_1. Let $\hat{\boldsymbol{\ell}}$ be the unit vector along an arbitrary axis **x**′ through P in this case of three equal principal moments of inertia. Also let y′ and z′ complete an orthogonal triad with **x**′, with $\hat{\mathbf{m}}$ and $\hat{\mathbf{n}}$ being unit vectors in the respective y′ and z′ directions. Thus (ℓ_x, ℓ_y, ℓ_z), (m_x, m_y, m_z), and (n_x, n_y, n_z) are the respective sets of direction cosines of **x**′, y′, and z′ with respect to (**x**, y, z). Equation (7.35) then gives

$$I^P_{x'y'} = (\ell_x m_x + \ell_y m_y + \ell_z m_z)I_1 + \underbrace{0 + 0 + 0}_{\substack{I^P_{xy} = I^P_{yz} = I^P_{xz} = 0 \text{ since} \\ (x,\, y,\, z) \text{ are principal axes}}}$$

$$= 0 \qquad (\text{since } \hat{\boldsymbol{\ell}} \perp \hat{\mathbf{m}})$$

In the same way, $I^P_{x'z'} = 0$ since $\hat{\boldsymbol{\ell}} \perp \hat{\mathbf{n}}$ as well. Thus the arbitrary axis **x**′ through P is principal, and Equation (7.31) shows that its moment of inertia is also I_1:

$$I^P_{x'x'} = \underbrace{(\ell_x^2 + \ell_y^2 + \ell_z^2)}_{= 1 \text{ (since } \hat{\boldsymbol{\ell}} \text{ is a unit vector)}} I_1 = I_1$$

Consider these examples of the preceding results regarding two and three equal principal moments of inertia:

Two Equal I's

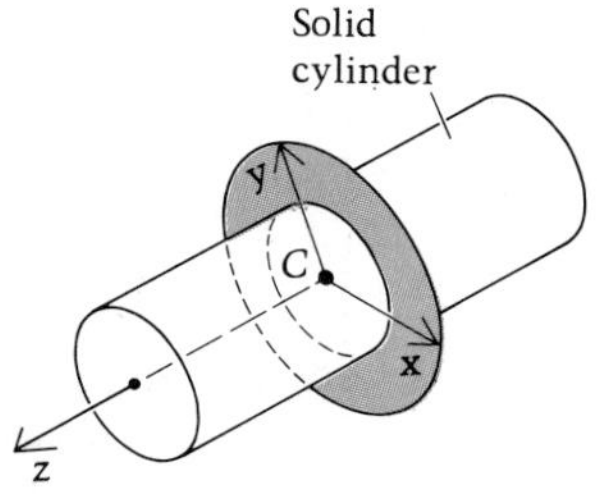

$$I_{xx}^C = I_{yy}^C = \frac{mR^2}{4} + \frac{m\ell^2}{12}$$

All axes through C in the shaded plane have this same inertia and are principal. Note that $I_{zz}^C = mR^2/2$ and is generally not equal to I_{xx} and I_{yy}; if, however, $\ell = \sqrt{3}R$, then ***every*** axis through C is principal with the same principal moment of inertia!

Three Equal I's

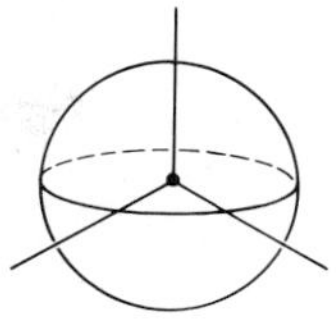

Solid sphere:

$$I = \frac{2}{5}mR^2$$

in any direction through C.

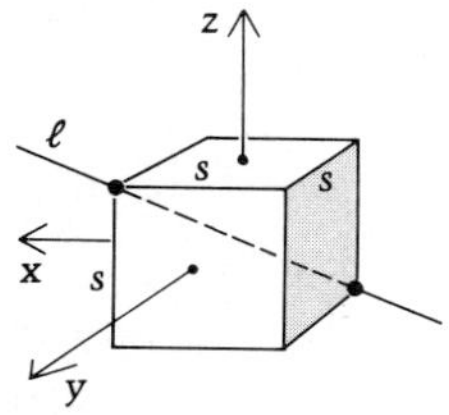

Solid cube:

$$I_x = I_y = I_z = \frac{m}{12}(s^2 + s^2) = \frac{ms^2}{6}$$

in any direction through C. Therefore if line ℓ is a diagonal of the cube, then $I_{\ell\ell} = ms^2/6$ even though it would be formidable to obtain this result by integration. Note that the direction cosines of ℓ are $(\ell_x, \ell_y, \ell_z) = (1/\sqrt{3}, 1/\sqrt{3}, 1/\sqrt{3})$ since it makes equal angles with **x**, **y**, and **z**. Thus Equation (7.31) gives

$$I_{\ell\ell}^C = \frac{1}{3}I_x + \frac{1}{3}I_y + \frac{1}{3}I_z + 0 + 0 + 0$$

$$= \left(\frac{1}{3} + \frac{1}{3} + \frac{1}{3}\right)\frac{ms^2}{6} = \frac{ms^2}{6}$$

An important property of principal moments of inertia is that the largest and smallest of these are the largest and smallest moments of inertia associated with *any* axis through the point in question. To show that this is true, let $I_1 \le I_2 \le I_3$ be the principal moments of inertia at P and let the corresponding principal axes be **x**, **y**, and **z**. The moment of inertia about some other axis, **x**′, is given by Equation (7.31):

$$I_{x'x'}^P = I_1\ell_x^2 + I_2\ell_y^2 + I_3\ell_z^2$$

or

$$\frac{I^P_{x'x'}}{I_1} = \ell_x^2 + \left(\frac{I_2}{I_1}\right)\ell_y^2 + \left(\frac{I_3}{I_1}\right)\ell_z^2 \geq 1$$

since

$$\frac{I_3}{I_1} \geq \frac{I_2}{I_1} \geq 1 \qquad \text{and} \qquad \ell_x^2 + \ell_y^2 + \ell_z^2 = 1$$

Thus $I^P_{x'x'} \geq I_1$, so that *no* line through P has a smaller moment of inertia than the smallest ***principal*** moment of inertia. By a similar argument $I^P_{x'x'}/I_3 \leq 1$, which demonstrates that $I^P_{x'x'} \leq I_3$ so that no line through P has a larger associated moment of inertia than the largest ***principal*** moment of inertia. Thus the largest and smallest moments of inertia at a point P are found among the principal moments of inertia for P. It may now be shown quite easily that the smallest moment of inertia at the mass center is the minimum I for ***any*** line through ***any point*** of $\mathcal{B}$ or $\mathcal{B}$ extended.

Question 7.5 Write a one-sentence proof of this statement by using the preceding results together with the parallel-axis theorem.

PROBLEMS / Section 7.4

7.34 Find the ratio of radius r to length L of a solid circular cylinder for which the moments of inertia about all axes through the mass center are equal.

7.35 Show that the moment of inertia of a spherical shell about any line through its mass center is $\frac{2}{3}mr^2$. (See Figure P7.35.)

7.36 Use the definitions of the moments of inertia to prove that if a body lies essentially in the **xy** plane (that is, it has very small dimensions normal to it), then $I^P_{zz} = I^P_{xx} + I^P_{yy}$, where P is ***any point*** in the plane, and that z is a principal axis at P.

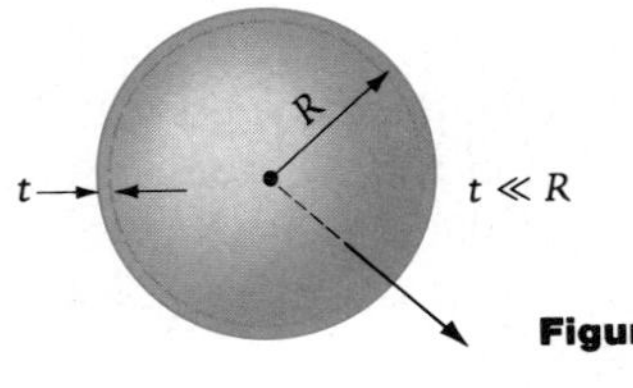

Figure P7.35

7.37 Find the moments of inertia I^P_{xx}, I^P_{yy}, and I^P_{zz} for the body depicted in Example 4.13. Then find the principal moments of inertia and corresponding principal axes at P.

7.38 Find the vector from O to the mass center of the bent bar in Example 7.7. Observe that it does not lie along any of the three principal axes at O.

7.39 Calculate the principal moments of inertia at O, and the direction cosines of their respective principal axes, for body $\mathcal{B}$ in Figure P7.39. It is made up of three bent bars welded together; all legs are either along, or parallel to, the coordinate axes.

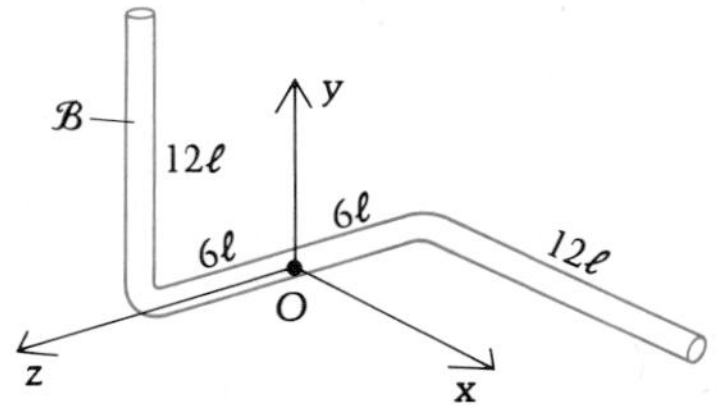

Figure P7.39

7.40 At the origin, find the principal moments of inertia and associated principal axes for a body consisting of three square plates welded along their edges as shown in Figure P7.40. (The axis of the smallest value of I should be the one that the mass lies closest to in an overall sense. Make this rough check on your solution.) Mass $= 3m$, side $= a$.

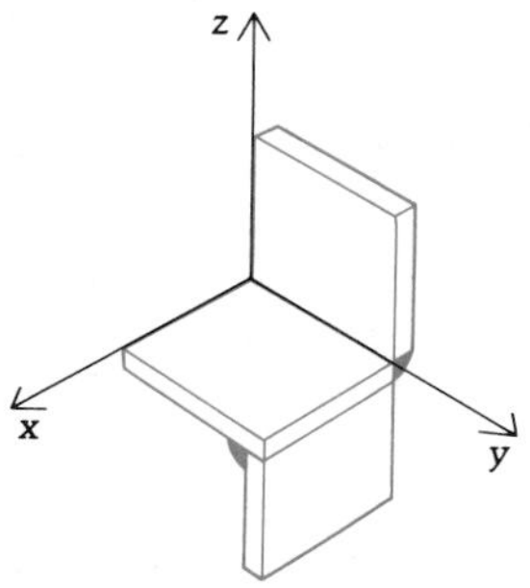

Figure P7.40

7.41 Find the principal moments of inertia and related principal axes at the origin for the body in Figure P7.41.

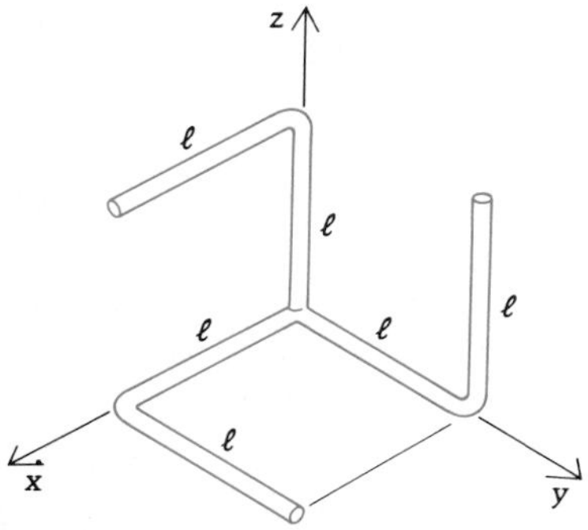

Figure P7.41

7.42 Show that if an axis through the mass center C of body $\mathcal{B}$ is principal at C, then it is a principal axis for *every point* on that axis. *Hint:* Use the transfer theorem for products of inertia, together with the orthogonality of principal axes. (See Figure P7.42.) Transfer I_{yz}^C and I_{xz}^C to P!

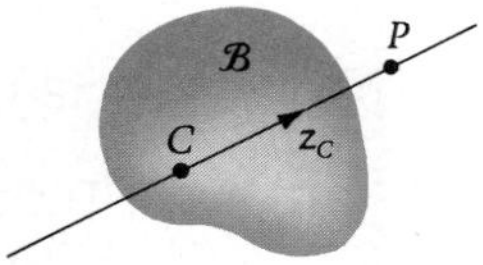

Figure P7.42

7.43 Show that if a principal axis for a point (such as P in the preceding problem) passes through C, then it is also principal for C. (Same hint!)

7.44 Show that if a line is a principal axis for two of its points, then it is a principal axis for the mass center.

7.45 Show that the three principal axes for any point lying on a principal axis for C are parallel to the principal axes for C.

7.46 In Problem 7.29 extend the problem and find the principal moments of inertia, and their principal axes, at point O.

7.47 For the homogeneous rectangular solid shown in Figure P7.47, find the smallest of the three angles between line AB and the principal axes of inertia at A.

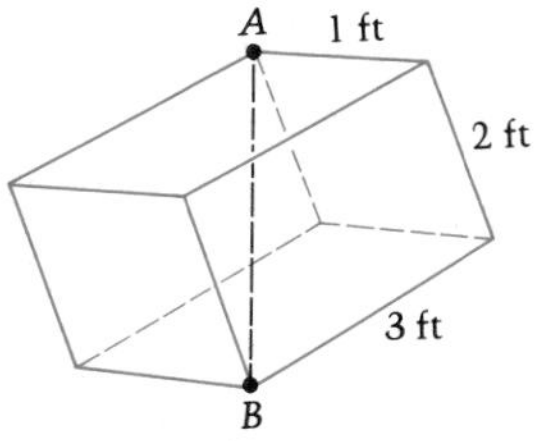

Figure P7.47

7.48 Four slender bars, each of mass m and length ℓ, are welded together to form the body shown in Figure P7.48. Find: (a) the inertia properties at the mass center C; (b) the principal axes and principal moments of inertia at C.

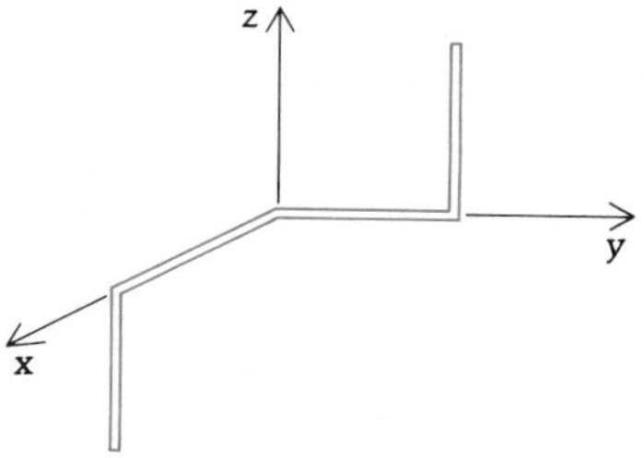

Figure P7.48

7.49 In Example 7.7 find the principal moments and axes of inertia at the mass center C.

7.50 Find the principal moments of inertia and their associated principal axes for the plate of Example 7.6 when $b = 2H$.

7.51 Find the principal moments and axes of inertia if the body of Example 7.6 has depth L instead of being a thin plate. (See Figure P7.51.) *Hint:* The xy plane is still one of symmetry, so $I_{xz}^C = I_{yz}^C = 0$ again!

7.52 Find the principal moments of inertia and their associated axes at O for the thin plate (Figure P7.52) in terms of its density ρ and thickness t.

7.53 Figure P7.53 shows part of a space station being constructed in orbit. Find the principal axes and moments of inertia at C_3. The modules have 33 ft diameters, but due to the material within they are not hollow. For the purposes of this problem, treat each as a uniform hollow shell with a radius of gyration about its axis of 12 ft.

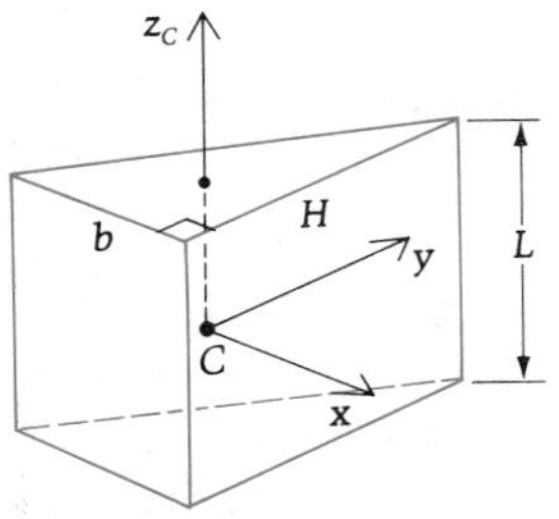

Figure P7.51

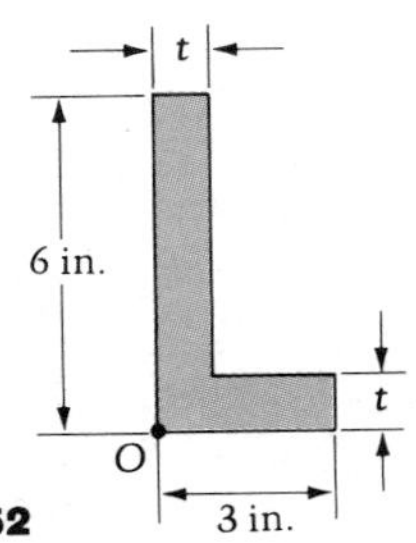

Figure P7.52

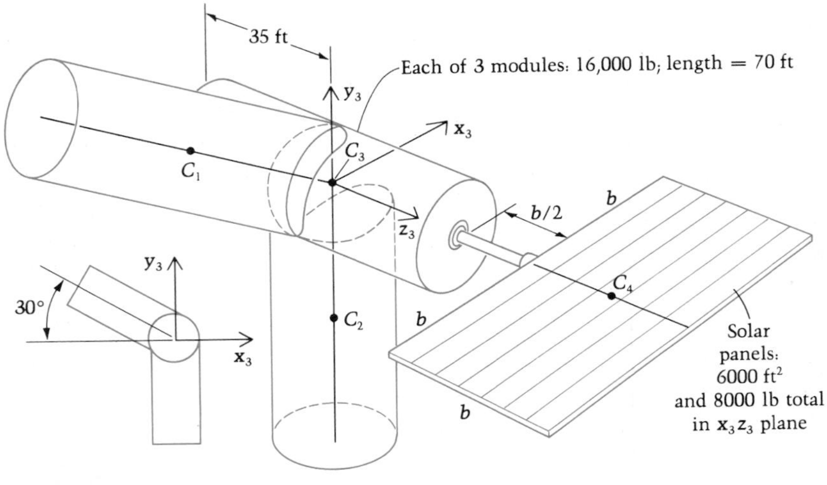

Figure P7.53

7.5 The Euler Equations Governing the Rotational Motion of a Rigid Body

In this section we derive the three differential equations governing the angular motion of a rigid body $\mathcal{B}$. Their solution, which is difficult to obtain in closed form in most cases, yields the three angular velocity components of $\mathcal{B}$ in an inertial frame $\mathcal{I}$. We begin with Equation (7.2), noting that the time derivative is taken in $\mathcal{I}$. However, the angular momentum $\mathbf{H}_C$ was most conveniently expressed in body $\mathcal{B}$ in Equation (7.22). Thus we shall use Equation (6.20) to move the derivative in (7.2) from $\mathcal{I}$ to $\mathcal{B}$:

$$\mathbf{M}_{\mathcal{I}_C} = {}^{\mathcal{I}}\dot{\mathbf{H}}_C = {}^{\mathcal{B}}\dot{\mathbf{H}}_C + \boldsymbol{\omega}_{\mathcal{B}/\mathcal{I}} \times \mathbf{H}_C \tag{7.48}$$

We now fix the axes (x, y, z) to body $\mathcal{B}$ so that, relative to $\mathcal{B}$, the inertia properties are constant. Using Equation (7.22), the first term on the right side of (7.48) is

$$\begin{aligned}{}^{\mathcal{B}}\dot{\mathbf{H}}_C = &(I_{xx}^C\dot{\omega}_x + I_{xy}^C\dot{\omega}_y + I_{xz}^C\dot{\omega}_z)\hat{\mathbf{i}} \\ &+ (I_{xy}^C\dot{\omega}_x + I_{yy}^C\dot{\omega}_y + I_{yz}^C\dot{\omega}_z)\hat{\mathbf{j}} \\ &+ (I_{xz}^C\dot{\omega}_x + I_{yz}^C\dot{\omega}_y + I_{zz}^C\dot{\omega}_z)\hat{\mathbf{k}}\end{aligned} \tag{7.49}$$

and the second term, after computing the cross product, is

$$\begin{aligned}\boldsymbol{\omega}_{\mathcal{B}/\mathcal{I}} \times \mathbf{H}_C = &[(I_{zz}^C - I_{yy}^C)\omega_y\omega_z + I_{yz}^C(\omega_y^2 - \omega_z^2) + \omega_x(\omega_y I_{xz}^C - \omega_z I_{xy}^C)]\hat{\mathbf{i}} \\ &+ [(I_{xx}^C - I_{zz}^C)\omega_z\omega_x + I_{xz}^C(\omega_z^2 - \omega_x^2) + \omega_y(\omega_z I_{xy}^C - \omega_x I_{yz}^C)]\hat{\mathbf{j}} \\ &+ [(I_{yy}^C - I_{xx}^C)\omega_x\omega_y + I_{xy}^C(\omega_x^2 - \omega_y^2) + \omega_z(\omega_x I_{yz}^C - \omega_y I_{xz}^C)]\hat{\mathbf{k}}\end{aligned} \tag{7.50}$$

The sum of Equations (7.49) and (7.50) yields the right side of Equation (7.48), which in turn equals the moment about the mass center C of all the external forces and couples acting on $\mathcal{B}$.

It is clear that this equation is extremely lengthy and complicated. If we select the body-fixed axes ($\mathbf{x}$, $\mathbf{y}$, $\mathbf{z}$) to be the ***principal*** axes through C, however, then all product of inertia terms vanish and we obtain

$${}^{\mathcal{B}}\dot{\mathbf{H}}_C = I_{xx}^C\dot{\omega}_x\hat{\mathbf{i}} + I_{yy}^C\dot{\omega}_y\hat{\mathbf{j}} + I_{zz}^C\dot{\omega}_z\hat{\mathbf{k}} \tag{7.51}$$

and

$$\begin{aligned}\boldsymbol{\omega}_{\mathcal{B}/\mathcal{I}} \times \mathbf{H}_C = &(I_{zz}^C - I_{yy}^C)\omega_y\omega_z\hat{\mathbf{i}} + (I_{xx}^C - I_{zz}^C)\omega_z\omega_x\hat{\mathbf{j}} \\ &+ (I_{yy}^C - I_{xx}^C)\omega_x\omega_y\hat{\mathbf{k}}\end{aligned} \tag{7.52}$$

so that, substituting into Equation (7.48) and equating the respective coefficients of $\hat{\mathbf{i}}$, $\hat{\mathbf{j}}$, and $\hat{\mathbf{k}}$, we obtain the **Euler equations:**

$$\begin{aligned}M_{\tau_{C_x}} &= I_{xx}^C\dot{\omega}_x - (I_{yy}^C - I_{zz}^C)\omega_y\omega_z \\ M_{\tau_{C_y}} &= I_{yy}^C\dot{\omega}_y - (I_{zz}^C - I_{xx}^C)\omega_z\omega_x \\ M_{\tau_{C_z}} &= I_{zz}^C\dot{\omega}_z - (I_{xx}^C - I_{yy}^C)\omega_x\omega_y\end{aligned} \tag{7.53}$$

These equations govern the rotational motion of $\mathcal{B}$ in the frame $\mathcal{I}$. We note that even when the principal axes through C are employed, the Euler equations are nonlinear in the $\boldsymbol{\omega}$ components—a major difference between plane and general motion. There is no natural extension of Equation (4.40), $M_{\tau_{C_z}} = I_{z_C}\ddot{\theta}$, to general rotational motion, as was the case for the mass center equations of motion (7.3). Note that in the case of plane motion we have $\omega_x \equiv 0 \equiv \omega_y$, so that Equations (7.53) reduce to the simple forms $M_{\tau_{C_x}} = 0$, $M_{\tau_{C_y}} = 0$, and $M_{\tau_{C_z}} = I_{zz}^C\alpha$.

Now if we have plane motion but ($\mathbf{x}$, $\mathbf{y}$, $\mathbf{z}$) are *not* principal, we obtain the following from Equations (7.48) to (7.50):

$$\begin{aligned}M_{\tau_{C_x}} &= I_{xz}^C\dot{\omega}_z - I_{yz}^C\omega_z^2 \\ M_{\tau_{C_y}} &= I_{yz}^C\dot{\omega}_z + I_{xz}^C\omega_z^2 \\ M_{\tau_{C_z}} &= I_{zz}^C\dot{\omega}_z\end{aligned} \tag{7.54}$$

In Equations (7.54), $\omega_z = \omega$ and $\dot{\omega}_z = \alpha$ in the usual plane motion notation; the z component of $\boldsymbol{\omega}$ (and $\boldsymbol{\alpha}$) is the sole nonvanishing component in this case.

Known solutions of Equations (7.53) in closed form for the angular velocity components are few; furthermore, to know the orientation of the body in space completely, one must ***further*** integrate a set of three nonlinear kinematics equations such as Equations (6.69) to get either the orientation angles of $\mathcal{B}$ or the direction cosines in space of a body-fixed axis.

EXAMPLE 7.8

Determine whether there are any axes through C about which an initial spin of body $\mathcal{B}$ at angular speed ω_0 will sustain itself (that is, remain constant) under the zero-moment (known as torque-free) condition.

SOLUTION

We are asking here whether the body, once started, will ***remain*** spinning about an axis fixed in the inertial reference frame $\mathcal{I}$. Without loss of generality, let the axis initially be z, so that Equations (7.54) apply. From the third of these equations, we see that if $\mathcal{B}$ is torque-free, $M_{r_{C_z}} = 0 = I_{zz}^C \dot{\omega}_z$ and that ω_z will then remain constant at its initial value ω_0. However, the other two equations give two further restrictions:

$$M_{r_{C_x}} = 0 \qquad \text{only if } I_{yz}^C = 0$$

$$M_{r_{C_y}} = 0 \qquad \text{only if } I_{xz}^C = 0$$

But $I_{yz}^C = 0 = I_{xz}^C$ means that z (the axis of initial spin) is principal! Therefore the spin will persist in the absence of external moment if and only if the axis of initial spin is a principal axis. This investigation is what led Euler to discover the principal-axis concept in 1750!

EXAMPLE 7.9

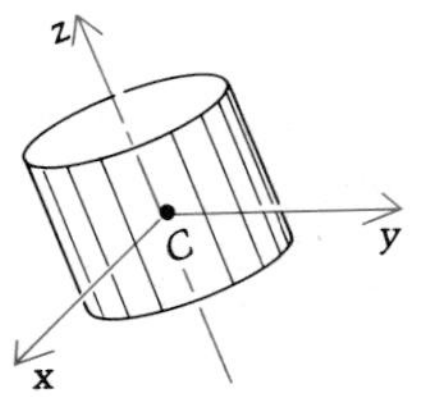

A satellite* is moving through deep space far from the influence of atmospheric drag and gravity. (See the diagram.) If the z axis is one of symmetry and if at some instant called $t = 0$ we have $\boldsymbol{\omega} = (\omega_{x_i}, \omega_{y_i}, \omega_{z_i})$ along the body-fixed centroidal axes, find $\boldsymbol{\omega}(t)$. Assume the satellite to be a rigid body.

(Continued)

*Such as the Voyager spacecraft, which left our solar system in 1983.

SOLUTION

The Euler equations, if $I_{x_C} = I_{y_C} = I$ and $I_{z_C} = J$, are

$$I\dot{\omega}_x - (I - J)\omega_y\omega_z = 0 \tag{1}$$

$$I\dot{\omega}_y - (J - I)\omega_z\omega_x = 0 \tag{2}$$

$$J\dot{\omega}_z - (I - I)\omega_x\omega_y = 0 \tag{3}$$

in which the moment components are zero in the absence of external forces and couples. Equation (3) gives

$$\omega_z = \text{constant} = \omega_{z_i} \tag{4}$$

so that Equations (1) and (2) become linear and are:

$$I\dot{\omega}_x - (I - J)\omega_y\omega_{z_i} = 0 \tag{5}$$

$$I\dot{\omega}_y - (J - I)\omega_{z_i}\omega_x = 0 \tag{6}$$

Differentiating Equation (6) and solving for $\dot{\omega}_x$, we get

$$\dot{\omega}_x = \frac{I\ddot{\omega}_y}{(J - I)\omega_{z_i}} \tag{7}$$

This expression may be substituted into (5) to yield an equation free of ω_x:

$$\ddot{\omega}_y + \left(\frac{(J - I)\omega_{z_i}}{I}\right)^2 \omega_y = 0$$

or

$$\ddot{\omega}_y + p^2\omega_y = 0$$

in which $p = (J - I)\omega_{z_i}/I$. The solution to this equation is harmonic:

$$\omega_y = A\cos pt + B\sin pt$$

Since $\omega_y = \omega_{y_i}$ at $t = 0$, we see that $A = \omega_{y_i}$. Finally, Equation (6) gives

$$\omega_x = \frac{\dot{\omega}_y}{p} = \frac{-\omega_{y_i}\, p\sin pt + Bp\cos pt}{p}$$

or

$$\omega_x = -\omega_{y_i}\sin pt + B\cos pt$$

The initial condition for ω_x gives us

$$\omega_{x_i} = 0 + B \Rightarrow B = \omega_{x_i}$$

so that the other two components (besides $\omega_z = \omega_{z_i}$) of $\boldsymbol{\omega}(t)$ are

$$\omega_x = \omega_{x_i}\cos pt - \omega_{y_i}\sin pt$$

$$\omega_y = \omega_{y_i}\cos pt + \omega_{x_i}\sin pt$$

This example illustrates a closed-form solution to Euler's equations. Actually, such known solutions are rare, and, in truth, problems are not always solved using Euler's equations in the form (7.53). Three reasons for this are:

1. We are often interested in expressing the external moments and the angular momentum of $\mathcal{B}$ in terms of components associated with directions fixed in a frame other than the body itself.
2. We often have one or two products of inertia to deal with, and it is in most cases more trouble to compute the principal moments and axes of inertia than to just go ahead and include the product terms in the angular momentum and differentiate it.
3. In certain stability studies of systems of connected rigid bodies, terms need to be added to the Euler equations to account for the relative motions.

These ideas and others suggest that we shall often find it more advantageous to begin our solutions with Euler's second law:

$$\mathbf{M}_{r_C} = {}^{\mathcal{I}}\dot{\mathbf{H}}_C$$

and to "move the derivative" to the frame (say, $\mathcal{F}$) in which we have $\mathbf{M}_{r_C}$ and $\mathbf{H}_C$ conveniently written (or expressed):

$$\mathbf{M}_{r_C} = {}^{\mathcal{F}}\dot{\mathbf{H}}_C + \boldsymbol{\omega}_{\mathcal{F}/\mathcal{I}} \times \mathbf{H}_C$$

In this way, we may use any frame we choose in order to express our vectors; if there is a nonzero product of inertia, it is simply included in $\mathbf{H}_C$. We now examine these ideas with a number of examples followed by two applications to the gyroscope.

EXAMPLE 7.10

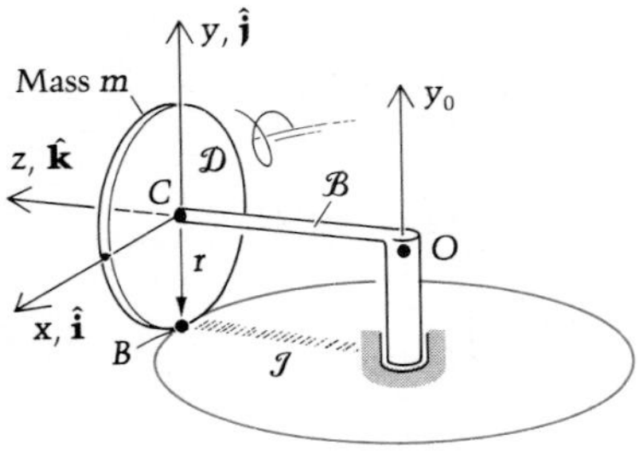

A disk $\mathcal{D}$ of mass m rolls in a circle as shown in the diagram; it is driven by a light but rigid arm $\mathcal{B}$ of length L that is made to rotate (by forces not shown) in a horizontal circle about O. The arm is connected to $\mathcal{D}$ at C by a ball joint. If the center of mass C of $\mathcal{D}$ travels at constant speed v_C, find the tensile force in the arm and determine the reaction of the floor onto $\mathcal{D}$. (The floor is to be the inertial frame $\mathcal{I}$ for the problem.) Assume line contact of $\mathcal{D}$ with the plane.

(Continued)

SOLUTION

We shall conveniently express our angular velocities and angular momentum in a frame that *is* the arm $\mathcal{B}$. Axes $(\mathbf{x}, \mathbf{y}, \mathbf{z})$ (see the figure) are locked into $\mathcal{B}$ at C. Moreover, $\mathbf{x}$ is always in the direction of $\mathbf{v}_C$, $\mathbf{y}$ is always vertically upward, and $\mathbf{z}$ is directed along the line from O to C. Note that $\mathcal{B}$ moves around the centerline y_0 with $\mathcal{D}$ but does not roll (about $\mathbf{z}$) with it.

The various angular velocities of interest to us here are related via the addition theorem:

$$\boldsymbol{\omega}_{\mathcal{D}/\mathcal{I}} = \boldsymbol{\omega}_{\mathcal{D}/\mathcal{B}} + \boldsymbol{\omega}_{\mathcal{B}/\mathcal{I}} \tag{1}$$

$$= \omega_r(-\hat{\mathbf{k}}) + \frac{v_C}{r_{OC}}(\hat{\mathbf{j}}) \tag{2}$$

where ω_r, the "roll" part of $\boldsymbol{\omega}_{\mathcal{D}/\mathcal{I}}$, may be found by relating the velocities of C and the contact point B of $\mathcal{D}$:

$$\mathbf{v}_C = \mathbf{v}_B + \boldsymbol{\omega}_{\mathcal{D}/\mathcal{I}} \times \mathbf{r}_{BC} \tag{3}$$

$$v_C\hat{\mathbf{i}} = \mathbf{0} + \left(-\omega_r\hat{\mathbf{k}} + \frac{v_C}{L}\hat{\mathbf{j}}\right) \times r\hat{\mathbf{j}}$$

so that

$$v_C = r\omega_r \Rightarrow \omega_r = \frac{v_C}{r} \tag{4}$$

Thus from Equation (2),

$$\boldsymbol{\omega}_{\mathcal{D}/\mathcal{I}} = \frac{v_C}{L}\left(\hat{\mathbf{j}} - \frac{L}{r}\hat{\mathbf{k}}\right) \tag{5}$$

Next we write the angular momentum $\mathbf{H}_C$ of $\mathcal{D}$. We shall use the fact that $(\mathbf{x}, \mathbf{y}, \mathbf{z})$ are all permanently principal axes for $\mathcal{D}$ at C—***even though they are not body-fixed in*** $\mathcal{D}$. This means that we may still write

$$\mathbf{H}_C = I_{xx}^C\omega_x\hat{\mathbf{i}} + I_{yy}^C\omega_y\hat{\mathbf{j}} + I_{zz}^C\omega_z\hat{\mathbf{k}} \tag{6}$$

in which $(\omega_x, \omega_y, \omega_z)$ are components in $\mathcal{B}$ of $\boldsymbol{\omega}_{\mathcal{D}/\mathcal{I}}$. $\mathbf{H}_C$ is thus legitimately expressed in terms of components associated with directions fixed in frame $\mathcal{B}$ in this case, even though it is the angular momentum of $\mathcal{D}$ moving in $\mathcal{I}$. At the end of the example, the reader will be asked to ponder the advantage gained by expressing $\mathbf{H}_C$ in an "intermediate frame" such as $\mathcal{B}$ in this problem.

Since $\omega_x = 0$, $I_{yy}^C = mr^2/4$ and $I_{zz}^C = mr^2/2$, we have the following from Equation (6):

$$\mathbf{H}_C = \frac{mr^2}{4}\frac{v_C}{L}\hat{\mathbf{j}} + \frac{mr^2}{2}\left(\frac{-v_C}{r}\hat{\mathbf{k}}\right)$$

$$= \frac{mr^2 v_C}{4L}\left(\hat{\mathbf{j}} - \frac{2L}{r}\hat{\mathbf{k}}\right) \tag{7}$$

(Continued)

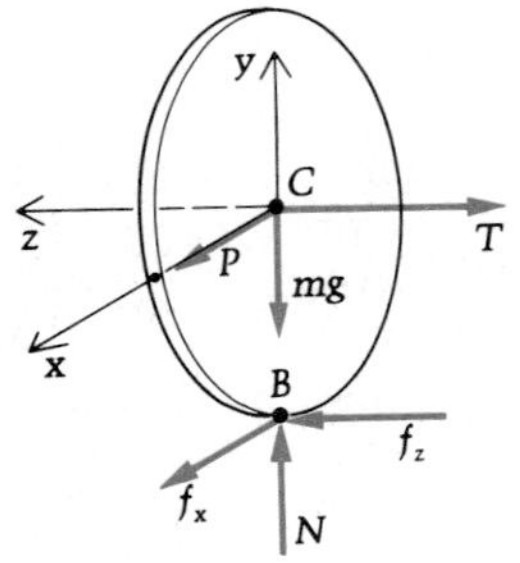

Next we apply the equations of motion. First, using the free-body diagram shown here, we apply the law of motion of the mass center (the first law of Euler):

$$\mathbf{F}_r = \dot{\mathbf{L}} = m\mathbf{a}_C$$

from which

$$F_{r_x} = f_x + P = ma_{C_x} = 0 \qquad \text{(since the speed of } C \text{ is constant)} \tag{8}$$

$$F_{r_y} = N - mg = ma_{C_y} = 0 \qquad \text{(since the mass center travels in a horizontal circle)} \tag{9}$$

$$F_{r_z} = f_z - T = ma_{C_z} = -\frac{mv_C^2}{L} \qquad \text{(which is the mass times the normal, or centripetal, acceleration of } C\text{)} \tag{10}$$

The second of Euler's laws is

$$\mathbf{M}_{r_C} = {}^{\mathcal{I}}\dot{\mathbf{H}}_C$$

but since we have $\mathbf{H}_C$ written in $\mathcal{B}$, we shall move the derivative there:

$$\mathbf{M}_{r_C} = {}^{\mathcal{I}}\dot{\mathbf{H}}_C = {}^{\mathcal{B}}\dot{\mathbf{H}}_C + \boldsymbol{\omega}_{\mathcal{B}/\mathcal{I}} \times \mathbf{H}_C \tag{11}$$

We note that ${}^{\mathcal{B}}\dot{\mathbf{H}}_C = \mathbf{0}$ since all entries in $\mathbf{H}_C$ (Equation 7) are constants there. Thus

$$\mathbf{M}_{r_C} = \left(\frac{v_C}{L}\hat{\mathbf{j}}\right) \times \frac{mr^2 v_C}{4L}\left(\hat{\mathbf{j}} - \frac{2L}{r}\hat{\mathbf{k}}\right)$$

$$= \frac{-mrv_C^2}{2L}\hat{\mathbf{i}} \tag{12}$$

The scalar component equations of (12) are

$$M_{r_{C_x}} = -f_z r = \frac{-mrv_C^2}{2L} \Rightarrow f_z = \frac{mv_C^2}{2L} \tag{13}$$

$$M_{r_{C_y}} = 0 = 0; \text{ hence identically satisfied.} \tag{14}$$

$$M_{r_{C_z}} = f_x r = 0 \Rightarrow f_x = 0 \tag{15}$$

Equations (15) and (8) show us that not only f_x, but also P, is zero.

Equation (13) gives us the other friction force acting on $\mathcal{D}$ at B, and Equation (9) shows that the normal force N is mg in this case. Substituting f_z from Equation (13) into (10) then gives the tension in the arm $\mathcal{B}$:

$$T = f_z + \frac{mv_C^2}{L} = \frac{3mv_C^2}{2L} \tag{16}$$

and the tension is three times the friction force. Notice that none of the forces depends on the radius r of the disk. If the disk has a mass of 20 kg, for example, and if $L = 0.9$ m and $v_C = 0.6$ m/s, then

$$f_z = \frac{20(0.6^2)}{2(0.9)} = 4 \text{ N} \qquad \text{and} \qquad T = 12 \text{ N}$$

Question 7.6 What advantage was gained in this example by writing $\mathbf{H}_C$ in terms of its components in $\mathcal{B}$?

Question 7.7 What happens to our solution if there is not enough friction to permit f_z—that is, if (assuming Coulomb friction) $\mu N < f_z$?

Question 7.8 Why is the force in the y direction at the ball joint able to be taken as zero at the beginning of the problem?

The results of the above example depend strongly on how $\mathcal{B}$ and $\mathcal{D}$ are joined. See Problem 7.72 for a different set of forces caused by a different type of connection between $\mathcal{B}$ and $\mathcal{D}$.

EXAMPLE 7.11

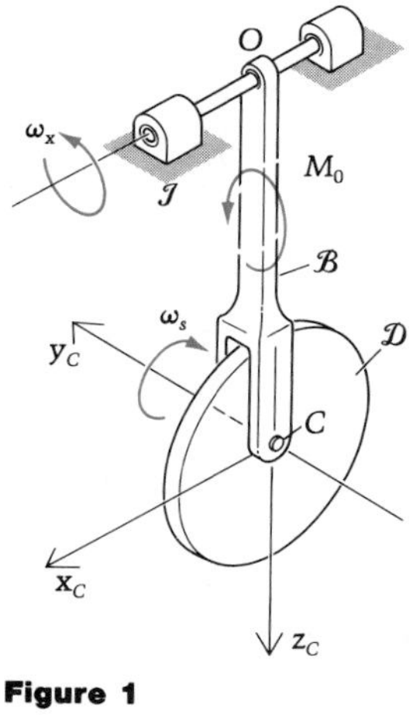

Figure 1

A symmetric wheel $\mathcal{D}$ spins at angular speed ω_s about its axis, which is a line fixed in body $\mathcal{B}$ as well as in $\mathcal{D}$. (See Figure 1.) The moments of inertia of $\mathcal{D}$ at C are $I_{yy}^C = J$ and $I_{xx}^C = I_{zz}^C = I$. Body $\mathcal{B}$ has negligible mass and rotates at angular speed ω_x about the $\mathbf{x}$ axis at O. If a moment M_0 is applied to $\mathcal{B}$ about the $\mathbf{x}_C$ axis, find, at the instant shown, the rates of change of ω_s and ω_x and the force and moment exerted on $\mathcal{D}$ at C by the pin.

SOLUTION

Let us denote the reference frame (in which $\mathcal{B}$ moves) by $\mathcal{I}$. The addition theorem for angular velocity then gives, for the wheel,

$$\boldsymbol{\omega}_{\mathcal{D}/\mathcal{I}} = \boldsymbol{\omega}_{\mathcal{D}/\mathcal{B}} + \boldsymbol{\omega}_{\mathcal{B}/\mathcal{I}}$$
$$= \omega_s\hat{\mathbf{j}} + \omega_x\hat{\mathbf{i}}$$

(Continued)

where the axes ($\mathbf{x}_C, y_C, z_C$) are fixed in $\mathcal{B}$ and their associated unit vectors ($\hat{\mathbf{i}}, \hat{\mathbf{j}}, \hat{\mathbf{k}}$) are fixed in direction in $\mathcal{B}$. Note that we may use the equation

$$\mathbf{H}_C = I^C_{xx}\omega_x\hat{\mathbf{i}} + I^C_{yy}\omega_y\hat{\mathbf{j}} + I^C_{zz}\omega_z\hat{\mathbf{k}}$$

for the angular momentum of $\mathcal{D}$ because even though ($\mathbf{x}_C, y_C, z_C$) are not fixed in body $\mathcal{D}$, they are nonetheless permanently principal. Therefore

$$\mathbf{H}_C = I\omega_x\hat{\mathbf{i}} + J\omega_s\hat{\mathbf{j}}$$

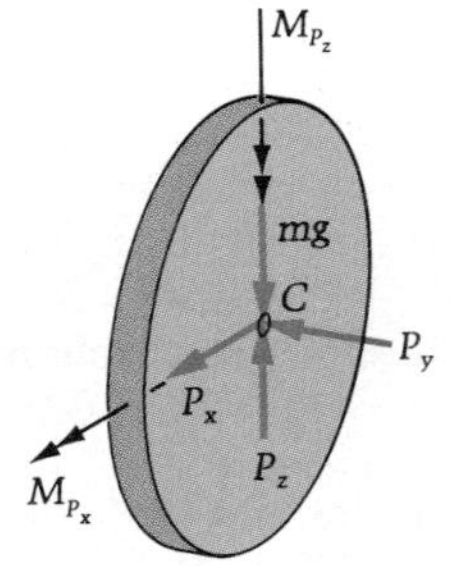

Figure 2

The second law of Euler then yields

$$\mathbf{M}_{r_C} = {}^{\mathcal{I}}\dot{\mathbf{H}}_C = {}^{\mathcal{B}}\dot{\mathbf{H}}_C + \boldsymbol{\omega}_{\mathcal{B}/\mathcal{I}} \times \mathbf{H}_C$$

where for convenience we differentiate $\mathbf{H}_C$ in frame $\mathcal{B}$ since the vector has been written in terms of its components there. Continuing,

$$\begin{aligned}\mathbf{M}_{r_C} &= I\dot{\omega}_x\hat{\mathbf{i}} + J\dot{\omega}_s\hat{\mathbf{j}} + (\omega_x\hat{\mathbf{i}}) \times (I\omega_x\hat{\mathbf{i}} + J\omega_s\hat{\mathbf{j}})\\ &= I\dot{\omega}_x\hat{\mathbf{i}} + J\dot{\omega}_s\hat{\mathbf{j}} + J\omega_x\omega_s\hat{\mathbf{k}}\end{aligned}$$

From Figure 2, a free-body diagram of $\mathcal{D}$, we get the components of $\mathbf{M}_{r_C}$ so that

$$M_{P_x}\hat{\mathbf{i}} + M_{P_z}\hat{\mathbf{k}} = I\dot{\omega}_x\hat{\mathbf{i}} + J\dot{\omega}_s\hat{\mathbf{j}} + J\omega_x\omega_s\hat{\mathbf{k}}$$

Thus we see that

$$M_{P_x} = I\dot{\omega}_x \quad (1)$$

$$0 = J\dot{\omega}_s \Rightarrow \omega_s = \text{constant} \quad (2)$$

$$M_{P_z} = J\omega_x\omega_s \quad (3)$$

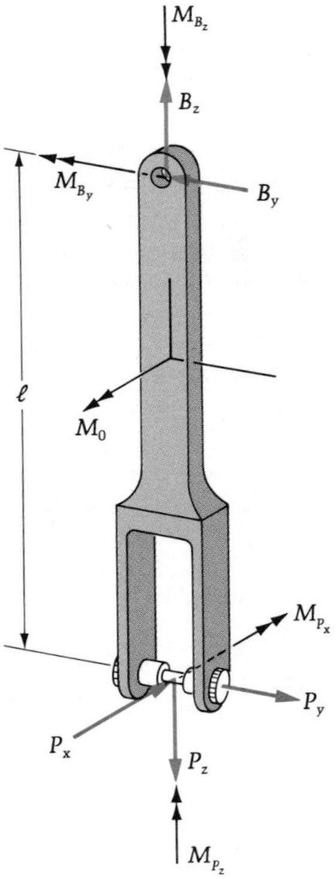

Figure 3

In addition, $\mathbf{F}_r = m\mathbf{a}_C$ for the disk yields

$$P_x\hat{\mathbf{i}} + P_y\hat{\mathbf{j}} - P_z\hat{\mathbf{k}} + mg\hat{\mathbf{k}} = m(-\ell\dot{\omega}_x\hat{\mathbf{j}} - \ell\omega_x^2\hat{\mathbf{k}})$$

so that

$$P_x = 0 \quad (4)$$

$$P_y = -m\ell\dot{\omega}_x \quad (5)$$

$$P_z = mg + m\ell\omega_x^2 \quad (6)$$

Turning now to Figure 3, a free-body diagram of the *light* body $\mathcal{B}$, we have $M_{r_{O_x}} \approx 0$ so that

$$\ell P_y + M_0 - M_{P_x} = 0 \quad (7)$$

Combining Equations (1), (5), and (7)

$$\dot{\omega}_x = \frac{M_0}{I + m\ell^2} \quad (8)$$

(Continued)

and

$$P_y = \frac{-m\ell M_0}{I + m\ell^2} \tag{9}$$

and

$$M_{P_x} = \frac{IM_0}{I + m\ell^2} \tag{10}$$

The results (4), (6), (8), (9), and (10) are recognizable as what would have been obtained had the disk been frozen in its bearings, in which case $\mathcal{B}$ and $\mathcal{D}$ would constitute a single rigid body in plane motion.

Equation (3), however, does not follow intuitively from the study of plane motion. The term $J\omega_x\omega_s$ is sometimes called a gyroscopic moment, and the equation says that a moment of this magnitude must act on $\mathcal{D}$ about z_C if the given motion is to occur. Note that in this case the body $\mathcal{D}$ is *not allowed* to turn about z_C as it spins (about y_C). If it were, say by means of a bearing between C and O, then the moment component M_{P_z} would become zero, and a third ω component (about z_C) would appear. We shall see this in the next example. Note also from Figure 3 that the gyroscopic moment *twists* the shaft of $\mathcal{B}$. This, for example, is a consideration in the retraction of the wheels of some airplanes.

EXAMPLE 7.12

Suppose that in the preceding example a frictionless bearing is inserted along the shaft so that $\mathcal{B}$ becomes two bodies $\mathcal{B}_1$ and $\mathcal{B}_2$ which are able to rotate with respect to each other about the z_C axis. If ϕ measures the relative rotation of $\mathcal{B}_2$ with respect to $\mathcal{B}_1$ (see the figures), use the rewritten equations of rotational motion to find $\ddot{\phi}$ and $\dot{\omega}_s$ at the instant given in the preceding problem. The relative rotation of $\mathcal{B}_2$ and $\mathcal{B}_1$ is initiated with the same starting conditions as in the preceding example, with, additionally, $\phi = \dot{\phi} = 0$.

SOLUTION

The angular velocity of $\mathcal{D}$ now carries an additional component, which is $\boldsymbol{\omega}_{\mathcal{B}_2/\mathcal{B}_1}$:

$$\begin{aligned}
\boldsymbol{\omega}_{\mathcal{D}/\mathcal{I}} &= \boldsymbol{\omega}_{\mathcal{D}/\mathcal{B}_2} + \boldsymbol{\omega}_{\mathcal{B}_2/\mathcal{B}_1} + \boldsymbol{\omega}_{\mathcal{B}_1/\mathcal{I}} \\
&= \omega_s\hat{\mathbf{j}} + \omega_z\hat{\mathbf{k}} + \omega_x\hat{\mathbf{I}} \\
&= \omega_s\hat{\mathbf{j}} + \omega_z\hat{\mathbf{k}} + \omega_x(\cos\phi\,\hat{\mathbf{i}} + \sin\phi\,\hat{\mathbf{j}})
\end{aligned}$$

(Continued)

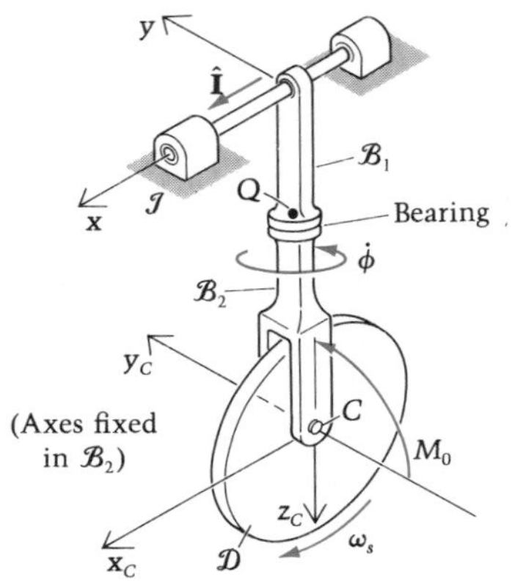

Next we obtain the angular momentum of $\mathcal{D}$ where, as before, $\hat{\mathbf{i}}$, $\hat{\mathbf{j}}$, and $\hat{\mathbf{k}}$ are fixed in $\mathcal{B}_2$ and are in permanently principal directions for $\mathcal{D}$ at C:

$$\mathbf{H}_C = I_{xx}^C \omega_x \hat{\mathbf{i}} + I_{yy}^C \omega_y \hat{\mathbf{j}} + I_{zz}^C \omega_z \hat{\mathbf{k}}$$

$$= I(\omega_x \cos\phi)\hat{\mathbf{i}} + J(\omega_s + \omega_x \sin\phi)\hat{\mathbf{j}} - I\dot{\phi}\hat{\mathbf{k}}$$

Again using Euler's second law,

$$\mathbf{M}_{r_C} = {}^{\mathcal{I}}\dot{\mathbf{H}}_C = {}^{\mathcal{B}_2}\dot{\mathbf{H}}_C + \boldsymbol{\omega}_{\mathcal{B}_2/\mathcal{I}} \times \mathbf{H}_C$$

we obtain the three equations of rotational motion of $\mathcal{D}$:

$$M_{r_{C_x}} = I\dot{\omega}_x \cos\phi + (J - 2I)\dot{\phi}\omega_x \sin\phi + J\omega_s\dot{\phi} \tag{1}$$

$$M_{r_{C_y}} = J\frac{d}{dt}(\omega_s + \omega_x \sin\phi) \tag{2}$$

$$M_{r_{C_z}} = -I\ddot{\phi} + \omega_x \cos\phi[(J - I)\omega_x \sin\phi + J\omega_s] \tag{3}$$

Noting that $M_{r_{C_z}} = 0$, Equation (3) yields, at $t = 0$

$$\ddot{\phi} = J\omega_s\omega_x/I \tag{4}$$

which indicates that $\mathcal{D}$ (and $\mathcal{B}_2$ with it) will immediately begin to turn about the negative z_C direction.

Continuing, once again we never have a moment component exerted on $\mathcal{D}$ about y_C, and thus Equation (2) may be expanded to yield

$$\dot{\omega}_s + \omega_x\dot{\phi}\cos\phi + \dot{\omega}_x \sin\phi = 0 \tag{5}$$

At $t = 0$ (since $\phi = \dot{\phi} = 0$ then), we see that

$$\dot{\omega}_s = 0$$

We note that since ϕ will begin to increase from zero (see Equation (4)), $\dot{\omega}_s$ will this time not remain zero, and, in contrast to Example 7.11, the spin rate ω_s ***will not stay constant.***

It can also be shown using free-body diagrams of $\mathcal{B}_1$ and $\mathcal{B}_2$, as well as $\mathcal{D}$, that $\dot{\omega}_x$ is initially equal to $M_0/(I + m\ell^2)$ as was the case in Example 7.11.

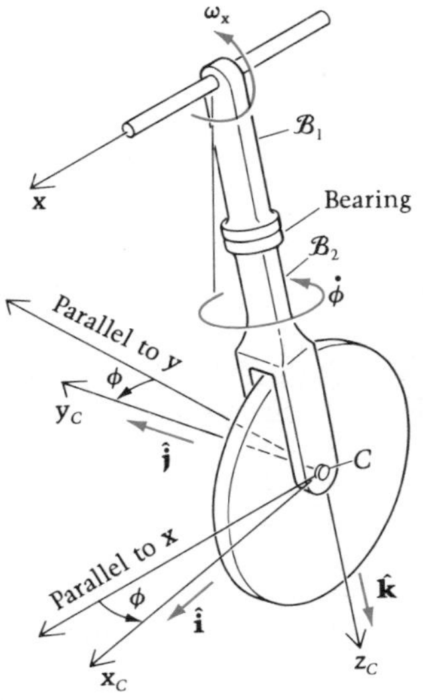

It is not always feasible to construct a rotating system so that the successive rotations of the body are each about a principal axis. This means that we sometimes have to deal directly with the products of inertia. In this case we must abandon the Euler equations in the form of (7.53) and return to the more general Euler's law: $\mathbf{M}_{r_O} = {}^{\mathcal{I}}\dot{\mathbf{H}}_O$. The following example illustrates these remarks.

EXAMPLE 7.13

For reasons of interference with other bodies, an antenna was recently designed and built with an offset axis as shown in the diagram. The antenna is composed of a 12-ft, 1200-lb parabolic reflector $\mathcal{R}$, a counterweight $\mathcal{W}$, a reflector support structure $\mathcal{S}$, and a positioner. The positioner consists of (1) a pedestal $\mathcal{P}$ that is fixed to the (inertial) reference frame; (2) an azimuth bearing at O and ring gear by means of which the housing $\mathcal{A}$ is made to rotate about the vertical; and (3) an elevation torque motor at E that rotates the support structure $\mathcal{S}$ with respect to $\mathcal{A}$.

a. Find the weight W of the counterweight $\mathcal{W}$.
b. Find the inertia properties at point O of the (assumed rigid) body $\mathcal{B}$ composed of $\mathcal{R} + \mathcal{W} + \mathcal{S}$.
c. Write the equations of rotational motion of $\mathcal{B}$.
d. Typical values for angular speed and acceleration for a tracking antenna like this one are 30°/sec and 30°/sec^2. In proper units, these are $\pi/6$ rad/sec and $\pi/6$ rad/sec^2. Determine the moments that the pedestal must exert about the $\mathbf{x}$, $\mathbf{y}$, and $\mathbf{z}$ axes to produce an angular acceleration of $\boldsymbol{\alpha} = (\pi/6)\hat{\mathbf{i}}$ rad/sec^2 when the angular velocity is $\boldsymbol{\omega} = (\pi/6)\hat{\mathbf{K}}$ rad/sec and the antenna is in the position shown in the figure.

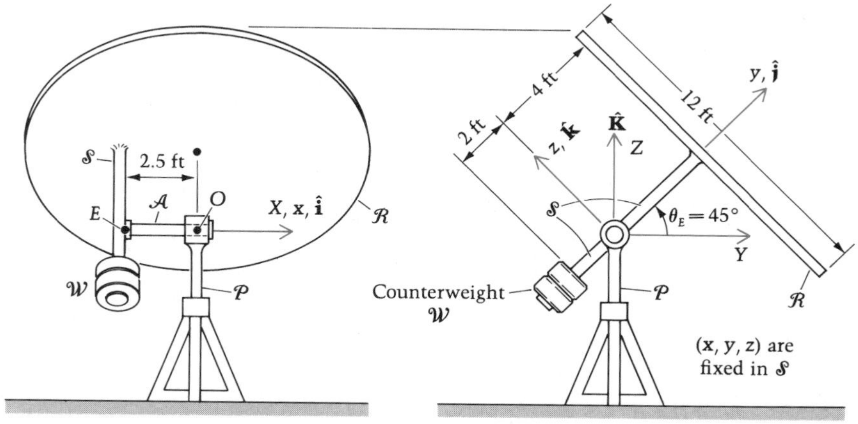

SOLUTION

Let us model the antenna as follows. The reflector is treated as a thin disk, the counterweight as a point mass, and the support structure as being rigid but light. Axes (X, Y, Z) are fixed in the inertial frame $\mathcal{J}$ while $(\mathbf{x}, \mathbf{y}, \mathbf{z})$ are attached to $\mathcal{B}$. Rotations of the reflector consist of an azimuth angle θ_A about Z, and an elevation angle θ_E about $\mathbf{x}$.

In part (a) the purpose of the counterweight is to place the mass center of $\mathcal{B}$ on its elevation axis. Thus

(Continued)

$$1200(4) = W(2)$$

$$W = 2400 \text{ lb}$$

For part (b) we generate the inertia properties of $\mathcal{B}$ about the point O. (See the following table.)

	Reflector	+	Counterweight	=	Total
I_{xx}^O	$= m\left(\frac{r^2}{4} + d^2\right)$	+	$\frac{2400}{32.2} d^2$		
	$= \frac{1200}{32.2}\left(\frac{6^2}{4} + 4^2\right)$	+	$\frac{2400}{32.2} 2^2$		
	= 932	+	298		= 1230 slug-ft^2
I_{yy}^O	$= m\frac{r^2}{2}$	+	md^2		
	$= \frac{1200}{32.2}\frac{6^2}{2}$	+	$\frac{2400}{32.2} 2.5^2$		
	= 671	+	466		= 1140 slug-ft^2
I_{zz}^O	$= m\left(\frac{r^2}{4} + d^2\right)$	+	md^2		
	$= \frac{1200}{32.2}\left(\frac{6^2}{4} + 4^2\right)$	+	$\frac{2400}{32.2}(2^2 + 2.5^2)$		
	= 932	+	764		= 1700 slug-ft^2
I_{xy}^O	= 0*	+	$(-m\overline{xy})$		
	= 0	+	$-\frac{2400}{32.2}(2.5)(2)$		
	= 0	+	−373		= −373 slug-ft^2
I_{yz}^O	= 0	+	$(-m\overline{yz})$		
	= 0	+	$-\frac{2400}{32.2}(2)(0)$		
	= 0	+	0	=	0
I_{xz}^O	= 0	+	$(-m\overline{xz})$		
	= 0	+	$-\frac{2400}{32.2}(2.5)(0)$		
	= 0	+	0	=	0

*The y axis through O is an axis of symmetry of $\mathcal{R}$; hence $I_{xy}^O = 0 = I_{yz}^O$. And the third product of inertia of $\mathcal{R}$, I_{xz}^O, vanishes because zy is a plane of symmetry for body $\mathcal{R}$.

(Continued)

Therefore the *inertia matrix* may be written as

$$\begin{bmatrix} 1230 & -373 & 0 \\ -373 & 1140 & 0 \\ 0 & 0 & 1700 \end{bmatrix} \quad \text{slug-ft}^2 \tag{1}$$

For part (c), one way to proceed is to compute the principal axes and moments of inertia from this matrix and then use the Euler equations (7.53). This would be a very unwise approach in this case, however. Not only is it tedious to locate the principal axes, but following this we would have to break up the angular velocity into its components along these directions; we would then obtain not-so-useful moment components about axes skewed with respect to the rotation axes. It is much simpler to use the second of Euler's laws:*

$$\mathbf{M}_{r_O} = {}^{\mathcal{I}}\dot{\mathbf{H}}_O$$

The only prices we will have to pay are (1) to retain and deal with the nonzero product of inertia I_{xy}^O and (2) to move the derivative from frame $\mathcal{I}$ to $\mathcal{B}$. The angular momentum of $\mathcal{B}$ about O is, for our problem,

$$\begin{aligned} \mathbf{H}_O = {} & (I_{xx}^O \omega_x + I_{xy}^O \omega_y + \cancelto{0}{I_{xz}^O} \omega_z)\hat{\mathbf{i}} \\ & + (I_{yx}^O \omega_x + I_{yy}^O \omega_y + \cancelto{0}{I_{yz}^O} \omega_z)\hat{\mathbf{j}} \\ & + (\cancelto{0}{I_{zx}^O} \omega_x + \cancelto{0}{I_{zy}^O} \omega_y + I_{zz}^O \omega_z)\hat{\mathbf{k}} \end{aligned} \tag{2}$$

The angular velocity of $\mathcal{B}$ in frame $\mathcal{I}$ is found by the addition theorem. The reflector and its supporting structure rotate in elevation with a simple angular velocity $\dot{\theta}_E \hat{\mathbf{i}}$ with respect to the housing $\mathcal{A}$; likewise, $\mathcal{A}$ rotates in azimuth with a simple angular velocity $\dot{\theta}_A \hat{\mathbf{K}}$ with respect to the pedestal (which is rigidly fixed to the reference frame $\mathcal{I}$). Therefore

$$\begin{aligned} \boldsymbol{\omega}_{\mathcal{B}/\mathcal{I}} &= \boldsymbol{\omega}_{\mathcal{B}/\mathcal{A}} + \boldsymbol{\omega}_{\mathcal{A}/\mathcal{I}} \\ &= \omega_{\mathcal{B}/\mathcal{A}} \hat{\mathbf{i}} + \omega_{\mathcal{A}/\mathcal{I}} \hat{\mathbf{K}} \\ &= \dot{\theta}_E \hat{\mathbf{i}} + \dot{\theta}_A (\sin\theta_E \hat{\mathbf{j}} + \cos\theta_E \hat{\mathbf{k}}) \end{aligned} \tag{3}$$

Substituting the $\boldsymbol{\omega}$ components into Equation (2) gives the angular momentum of $\mathcal{B}$ in $\mathcal{I}$:

$$\begin{aligned} \mathbf{H}_O = {} & (I_{xx}^O \dot{\theta}_E + I_{xy}^O \dot{\theta}_A \sin\theta_E)\hat{\mathbf{i}} \\ & + (I_{xy}^O \dot{\theta}_E + I_{yy}^O \dot{\theta}_A \sin\theta_E)\hat{\mathbf{j}} + I_{zz}^O \dot{\theta}_A \cos\theta_E \hat{\mathbf{k}} \end{aligned} \tag{4}$$

Next we use Equation (6.20) to differentiate $\mathbf{H}_O$ (note that point O is fixed in $\mathcal{B}$ and $\mathcal{I}$):

$$\mathbf{M}_{r_O} = {}^{\mathcal{I}}\dot{\mathbf{H}}_O = {}^{\mathcal{B}}\dot{\mathbf{H}}_O + \boldsymbol{\omega}_{\mathcal{B}/\mathcal{I}} \times \mathbf{H}_O \tag{5}$$

(Continued)

*All that is required of O for this equation to be valid is that it be a point of the inertial frame; in what follows, however, it needs to be and is a pivot of $\mathcal{B}$.

Taking the derivative and performing the cross product, we find that the three component equations are as follows. (Note that the inertia properties do not change in $\mathcal{B}$.)

$$\begin{aligned}
M_{r_{O_x}} &= I^O_{xx}\ddot{\theta}_E + I^O_{xy}(\ddot{\theta}_A \sin\theta_E + \dot{\theta}_A\dot{\theta}_E \cos\theta_E) + \dot{\theta}_A \sin\theta_E(I^O_{zz}\dot{\theta}_A\cos\theta_E) \\
&\quad - \dot{\theta}_A \cos\theta_E(I^O_{xy}\dot{\theta}_E + I^O_{yy}\dot{\theta}_A\sin\theta_E) \\
&= I^O_{xx}\ddot{\theta}_E + I^O_{yy}(-\dot{\theta}_A^2 \sin\theta_E\cos\theta_E) + I^O_{zz}(\dot{\theta}_A^2\sin\theta_E\cos\theta_E) \\
&\quad + I^O_{xy}(\ddot{\theta}_A\sin\theta_E) \qquad \text{(6a)}
\end{aligned}$$

$$\begin{aligned}
M_{r_{O_y}} &= I^O_{xy}\ddot{\theta}_E + I^O_{yy}(\ddot{\theta}_A\sin\theta_E + \dot{\theta}_A\dot{\theta}_E\cos\theta_E) + \dot{\theta}_A\cos\theta_E(I^O_{xx}\dot{\theta}_E + I^O_{xy}\dot{\theta}_A\sin\theta_E) \\
&\quad - \dot{\theta}_E(I^O_{zz}\dot{\theta}_A\cos\theta_E) \qquad \text{(6b)}
\end{aligned}$$

$$\begin{aligned}
M_{r_{O_z}} &= I^O_{zz}(\ddot{\theta}_A\cos\theta_E - \dot{\theta}_A\dot{\theta}_E\sin\theta_E) + \dot{\theta}_E(I^O_{xy}\dot{\theta}_E + I^O_{yy}\dot{\theta}_A\sin\theta_E) \\
&\quad - \dot{\theta}_A\sin\theta_E(I^O_{xx}\dot{\theta}_E + I^O_{xy}\dot{\theta}_A\sin\theta_E) \qquad \text{(6c)}
\end{aligned}$$

As an indication of the increased difficulty of three-dimensional dynamics problems, note that ***all four*** of the nonvanishing inertia properties contribute to ***each component*** of the external moment acting on $\mathcal{B}$ at O!

In the indicated position, $\theta_E = 45°$. Therefore

$$M_{r_{O_x}} = I^O_{xx}\ddot{\theta}_E + I^O_{yy}\left(\frac{-\dot{\theta}_A^2}{2}\right) + I^O_{zz}\left(\frac{\dot{\theta}_A^2}{2}\right) + I^O_{xy}\left(\frac{\ddot{\theta}_A}{\sqrt{2}}\right) \qquad \text{(7a)}$$

$$M_{r_{O_y}} = I^O_{xx}\left(\frac{\dot{\theta}_A\dot{\theta}_E}{\sqrt{2}}\right) + I^O_{yy}\left(\frac{\ddot{\theta}_A + \dot{\theta}_A\dot{\theta}_E}{\sqrt{2}}\right) + I^O_{zz}\left(\frac{-\dot{\theta}_A\dot{\theta}_E}{\sqrt{2}}\right) + I^O_{xy}\left(\ddot{\theta}_E + \frac{\dot{\theta}_A^2}{2}\right) \qquad \text{(7b)}$$

$$M_{r_{O_z}} = I^O_{xx}\left(\frac{-\dot{\theta}_A\dot{\theta}_E}{\sqrt{2}}\right) + I^O_{yy}\left(\frac{\dot{\theta}_A\dot{\theta}_E}{\sqrt{2}}\right) + I^O_{zz}\left(\frac{\ddot{\theta}_A - \dot{\theta}_A\dot{\theta}_E}{\sqrt{2}}\right) + I^O_{xy}\left(\dot{\theta}_E^2 - \frac{\dot{\theta}_A^2}{2}\right) \qquad \text{(7c)}$$

In part (d), for the case specified, $\boldsymbol{\alpha} = \ddot{\theta}_E\hat{\mathbf{i}}$ and $\ddot{\theta}_E = \pi/6$ while $\ddot{\theta}_A = 0$. Also, $\boldsymbol{\omega} = \dot{\theta}_A\hat{\mathbf{K}}$ and $\dot{\theta}_A = \pi/6$ while $\dot{\theta}_E = 0$. This case physically corresponds to the antenna, at 45° elevation, swinging around the vertical at 30°/sec and suddenly sensing an object traveling toward zenith; the controls activate a motor whose torque produces an angular acceleration that will send the antenna upward in elevation. The angular accelerations are large because the need is to get there quickly.

Substituting these values of $\dot{\theta}_E$, $\ddot{\theta}_E$, $\dot{\theta}_A$, and $\ddot{\theta}_A$ into Equations (7), along with the inertia values, gives our answer:

$$\begin{aligned}
M_{r_{O_x}} &= 1230\left(\frac{\pi}{6}\right) + (-1140 + 1700)\frac{(\pi/6)^2}{2} + (-373)(0) \\
&= 721 \text{ lb-ft} \\
M_{r_{O_y}} &= 1230(0) + 1140(0) + 1700(0) + (-373)\left[\frac{\pi}{6} + \frac{(\pi/6)^2}{2}\right] \qquad \text{(8)} \\
&= -246 \text{ lb-ft} \\
M_{r_{O_z}} &= 1230(0) + 1140(0) + 1700(0) + (-373)\left(\frac{-(\pi/6)^2}{2}\right) \\
&= 51 \text{ lb-ft}
\end{aligned}$$

(Continued)

These are the moments exerted by $\mathcal{P}$ onto $\mathcal{A}$, excluding that required to balance the dead weight of the antenna. In the inertial reference frame, the moments are

$$M_{r_{O_X}} = M_{r_{O_x}} = 721 \text{ lb-ft}$$
$$M_{r_{O_Y}} = M_{r_{O_y}} \cos\theta_E + M_{r_{O_z}}(-\sin\theta_E) = -210 \text{ lb-ft} \tag{9}$$
$$M_{r_{O_Z}} = M_{r_{O_y}} \sin\theta_E + M_{r_{O_z}}(\cos\theta_E) = -138 \text{ lb-ft}$$

In the opposite case when the antenna is tracking in elevation, say $\dot\theta_E = \pi/6$ rad/sec, and receives a sudden command resulting in $\ddot\theta_A = \pi/6$ rad/sec² at $\theta_E = 45°$, the moment components become (here $\ddot\theta_E = 0 = \dot\theta_A$):

$$M_{r_{O_x}} = 0 + 0 + 0 + (-373)\frac{\pi}{6}\frac{1}{\sqrt{2}} = -138 \text{ lb-ft}$$
$$M_{r_{O_y}} = 0 + 1140\left(\frac{\pi/6 + 0}{\sqrt{2}}\right) + 0 + 0 = 422 \text{ lb-ft} \tag{10}$$
$$M_{r_{O_z}} = 0 + 0 + 1700\left(\frac{\pi/6 - 0}{\sqrt{2}}\right) + (-373)\left[\left(\frac{\pi}{6}\right)^2 - 0\right]$$
$$= 629 - 102 = 527 \text{ lb-ft}$$

Once again we see the considerable effect of the product of inertia term.

The negatives of the X, Y, Z components respectively bend, bend, and twist the pedestal and are considerations in its design; far larger and more important moments, however, arise from the wind blowing against the "dish" and from gravity. There are also forces exerted on $\mathcal{A}$ at O due to gravity and the mass center acceleration.

EXAMPLE 7.14

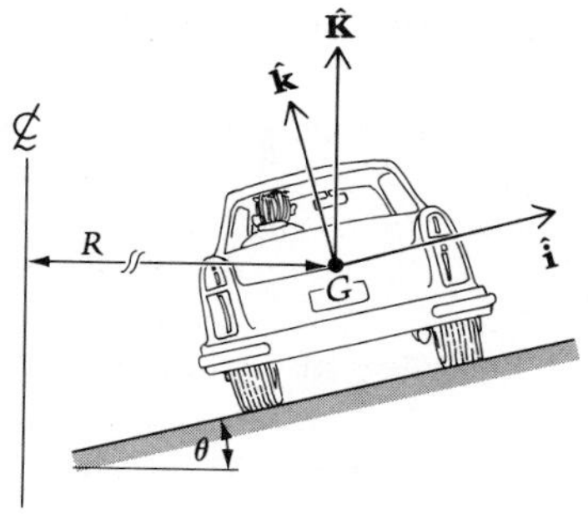

A car travels on a banked curve at constant speed v_G. (See the diagram.) Find the friction and normal forces acting on the tires as functions of the speed v_G, the bank angle θ, the radius R of the turn, and the moments of inertia I_1 and I_3 about centroidal axes parallel to $\hat{\mathbf{i}}$ and $\hat{\mathbf{k}}$ in the figure. Neglect the rolling inertia of the wheels. Note the difference in the forces acting on inner and outer tires compared with the flat-track, straight-path case ($\theta = 0$, $R \to \infty$) in which they are of course equal.

(Continued)

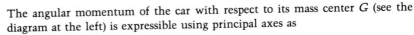

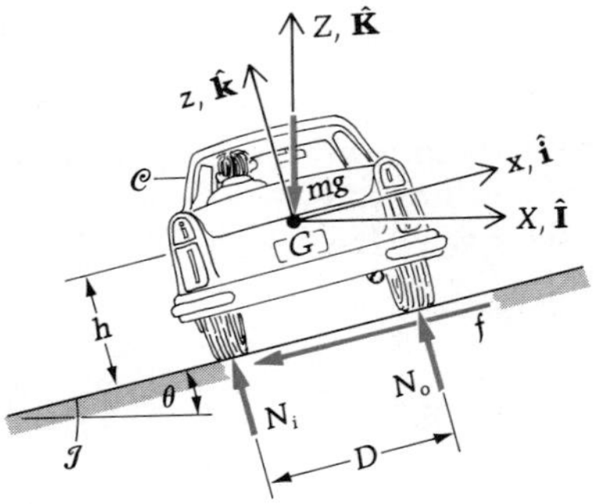

S O L U T I O N

The angular momentum of the car with respect to its mass center G (see the diagram at the left) is expressible using principal axes as

$$\mathbf{H}_G = I_1 \omega_x \hat{\mathbf{i}} + I_3 \omega_z \hat{\mathbf{k}} \tag{1}$$

in which the angular velocity vector of the car $\mathcal{C}$ is

$$\boldsymbol{\omega}_{\mathcal{C}/\mathcal{J}} = \frac{v_G}{R} \hat{\mathbf{K}} = \frac{v_G}{R} (\sin\theta\, \hat{\mathbf{i}} + \cos\theta\, \hat{\mathbf{k}}) \tag{2}$$

so that

$$\omega_x = \frac{v_G}{R} \sin\theta \qquad \omega_y = 0 \qquad \omega_z = \frac{v_G}{R} \cos\theta \tag{3}$$

Using Equation (7.2) and moving the derivative (see Equation 6.20) from $\mathcal{J}$ to $\mathcal{C}$ (where we have $\mathbf{H}_G$ expressed!) gives

$$\mathbf{M}_{r_G} = {}^{\mathcal{J}}\dot{\mathbf{H}}_G = {}^{\mathcal{C}}\dot{\mathbf{H}}_G^{\nearrow \mathbf{0}} + \boldsymbol{\omega}_{\mathcal{C}/\mathcal{J}} \times \mathbf{H}_G \tag{4}$$

Substituting the angular velocity components and using the free-body diagram to compute the moment vector,* we obtain

$$2(N_i - N_o) \frac{D}{2} \hat{\mathbf{j}} + 2fh\hat{\mathbf{j}} = \left(\frac{v_G}{R}\right)^2 (\sin\theta\, \hat{\mathbf{i}} + \cos\theta\, \hat{\mathbf{k}}) \times (I_1 \sin\theta\, \hat{\mathbf{i}} + I_3 \cos\theta\, \hat{\mathbf{k}}) \tag{5}$$

which reduces to the scalar equation

$$(N_i - N_o)D + 2fh = -\left(\frac{v_G}{R}\right)^2 (I_3 - I_1) \sin\theta \cos\theta \tag{6}$$

To complete the solution, we use the known, simple motion of G in the equation for the mass center motion:

$$\mathbf{F}_r = m\mathbf{a}_G = m \frac{v_G^2}{R} (-\hat{\mathbf{I}}) \tag{7}$$

$$= m \frac{v_G^2}{R} (-\cos\theta\, \hat{\mathbf{i}} + \sin\theta\, \hat{\mathbf{k}})$$

Writing the forces in vector form and substituting, we get

$$\hat{\mathbf{i}}(-2f - mg\sin\theta) + \hat{\mathbf{k}}(2N_i + 2N_o - mg\cos\theta) = m\frac{v_G^2}{R}(-\cos\theta\, \hat{\mathbf{i}} + \sin\theta\, \hat{\mathbf{k}}) \tag{8}$$

The scalar component equations of Equation (8) are easily solved:

(Continued)

*We are letting N_i (and N_o) be the pressing force beneath *each* inner (and outer) tire, while f is the resultant inward friction force beneath *each* of the front and back *pairs*.

$$f = \frac{mv_G^2}{2R}\cos\theta - \frac{mg\sin\theta}{2} \tag{9}$$

$$N_i + N_o = \frac{mv_G^2}{2R}\sin\theta + \frac{mg}{2}\cos\theta \tag{10}$$

Note that the friction resultant as well as the sum of the normal forces both contribute to the acceleration of the mass center; moreover, each helps to hold up the weight.

If we now substitute f from Equation (9) into (6) and then solve Equations (6) and (10) for the normal forces N_i and N_o, we get

$$N_i = \frac{-v_G^2}{2DR^2}(I_3 - I_1)\sin\theta\cos\theta + \frac{mg}{4}\left(\cos\theta + \frac{2h}{D}\sin\theta\right)$$
$$+ \frac{mv_G^2}{4R}\left(\sin\theta - \frac{2h}{D}\cos\theta\right) \tag{11}$$

$$N_o = \frac{v_G^2}{2DR^2}(I_3 - I_1)\sin\theta\cos\theta + \frac{mg}{4}\left(\cos\theta - \frac{2h}{D}\sin\theta\right)$$
$$+ \frac{mv_G^2}{4R}\left(\sin\theta + \frac{2h}{D}\cos\theta\right) \tag{12}$$

Note that if the path of the car is straight, the limits of f, N_o, and N_i as $R \to \infty$ give

$$f = \frac{-mg\sin\theta}{2}$$

$$N_o = \frac{mg}{4}\left(\cos\theta - \frac{2h}{D}\sin\theta\right) \tag{13}$$

$$N_i = \frac{mg}{4}\left(\cos\theta + \frac{2h}{D}\sin\theta\right)$$

In this case the car travels in a straight line at constant speed. The reader should verify that this is the case, showing that $\mathbf{F}_r = \mathbf{0} = \mathbf{M}_{r_G}$ with the preceding results. Note too that $N_i - N_o = (mgh/D)\sin\theta$, and the inside wheels take more force in this case.

Returning to the general case, we note next that the difference between an inner normal force and an outer one is

$$N_i - N_o = \frac{-v_G^2}{DR^2}(I_3 - I_1)\sin\theta\cos\theta + \frac{mgh}{D}\sin\theta - \frac{mv_G^2 h}{RD}\cos\theta \tag{14}$$

The middle term is the "static" term examined above. The other two are seen (from the presence of v_G and R) to be caused by the curved path. Note that both these terms, being negative, tend to make N_o larger than N_i as the path's curvature causes an increase in the forces acting on the outer tires (as the "weight shifts outward").

We now return our attention to the gyroscope whose orientation was examined in Section 6.7. First we shall derive the equations of rotational motion of such a gyroscope $\mathcal{G}$.* We begin by expressing its angular velocity $\boldsymbol{\omega}_{\mathcal{G}/\mathcal{I}}$ in the frame $\mathcal{F}_2$ (see the third brace in Example 6.8):

$$\boldsymbol{\omega}_{\mathcal{G}/\mathcal{I}} = -\dot{\phi}\sin\theta\,\hat{\mathbf{i}}_2 + \dot{\theta}\,\hat{\mathbf{j}}_2 + (\dot{\psi} + \dot{\phi}\cos\theta)\,\hat{\mathbf{k}}_2 \tag{7.55}$$

The axes ($\mathbf{x}_2$, $\mathbf{y}_2$, $\mathbf{z}_2$) of the inner gimbal (refer to the figures in Section 6.7) are not fixed in $\mathcal{G}$ because of its spin $\dot{\psi}$ — *but they are nonetheless permanently principal*. This important fact allows us to write the angular momentum of the gyroscope in the frame $\mathcal{F}_2$ as

$$\mathbf{H}_C = -I\dot{\phi}\sin\theta\,\hat{\mathbf{i}}_2 + I\dot{\theta}\,\hat{\mathbf{j}}_2 + J(\dot{\psi} + \dot{\phi}\cos\theta)\,\hat{\mathbf{k}}_2 \tag{7.56}$$

Euler's second law, together with Property (6.20) of the angular velocity vector, then gives the equations of motion of the gyroscope as follows:

$$\begin{aligned}\mathbf{M}_{r_C} &= {}^{\mathcal{I}}\dot{\mathbf{H}}_C = {}^{\mathcal{F}_2}\dot{\mathbf{H}}_C + \boldsymbol{\omega}_{\mathcal{F}_2/\mathcal{I}} \times \mathbf{H}_C \\ &= [-I(\ddot{\phi}\sin\theta + \dot{\phi}\dot{\theta}\cos\theta) + \dot{\theta}\dot{\phi}\cos\theta(J - I) + \dot{\theta}\dot{\psi}J]\hat{\mathbf{i}}_2 \\ &\quad + [I\ddot{\theta} + (J - I)\dot{\phi}^2\sin\theta\cos\theta + J\dot{\phi}\dot{\psi}\sin\theta]\hat{\mathbf{j}}_2 \\ &\quad + \left[J\frac{d}{dt}(\dot{\psi} + \dot{\phi}\cos\theta)\right]\hat{\mathbf{k}}_2\end{aligned} \tag{7.57}$$

In the preceding calculations we have used the addition theorem to observe the following:

$$\begin{aligned}\boldsymbol{\omega}_{\mathcal{G}/\mathcal{I}} &= \boldsymbol{\omega}_{\mathcal{G}/\mathcal{F}_2} + \boldsymbol{\omega}_{\mathcal{F}_2/\mathcal{I}} \\ &= \dot{\psi}\hat{\mathbf{k}}_2 + \boldsymbol{\omega}_{\mathcal{F}_2/\mathcal{I}}\end{aligned} \tag{7.58}$$

so that $\boldsymbol{\omega}_{\mathcal{F}_2/\mathcal{I}}$ is the same vector as in Equation (7.55) if $\dot{\psi}$ is omitted. The equations of motion of $\mathcal{G}$ are therefore

$$\begin{aligned}M_{r_{C_{x_2}}} &= -I(\ddot{\phi}\sin\theta + 2\dot{\phi}\dot{\theta}\cos\theta) + J\dot{\theta}(\dot{\psi} + \dot{\phi}\cos\theta) \\ M_{r_{C_{y_2}}} &= I(\ddot{\theta} - \dot{\phi}^2\sin\theta\cos\theta) + J\dot{\phi}\sin\theta(\dot{\psi} + \dot{\phi}\cos\theta) \\ M_{r_{C_z}} &= J\frac{d}{dt}(\dot{\psi} + \dot{\phi}\cos\theta) = J\frac{d\omega_z}{dt}\end{aligned} \tag{7.59}$$

where ω_z is the component of $\boldsymbol{\omega}_{\mathcal{G}/\mathcal{I}}$ about the spin axis of symmetry of the gyroscope. Note that it is made up of part of the precession speed as well as all of the spin.

The gyroscope equations are seen to be nonlinear, including not only products of the angles' derivatives but also trigonometric functions of them. Their general solution is an unsolved problem; however, there are two special solutions that are quite worthy of study. The first of

*We are taking $\mathcal{G}$ to be the rotor and are considering it as heavy with respect to the inner and outer gimbals, whose mass we then neglect. We also assume $\mathcal{G}$ to be symmetric about its axis. Of course, a gyroscope does not have to possess *any* gimbals; the earth is a massive gyro, as will be seen in an example to follow.

these is steady precession; the second is torque-free motion. We shall have a look at each in turn.

Steady precession is defined by the nutation angle θ, the precession speed $\dot{\phi}$, and the spin speed $\dot{\psi}$ each being constant throughout the motion. Let us call these constants θ_0, $\dot{\phi}_0$, and $\dot{\psi}_0$, and substitute them into Equations (7.59) to obtain:

$$\begin{aligned} M_{r_{C_{x_2}}} &= 0 \\ M_{r_{C_{y_2}}} &= -I\dot{\phi}_0^2 \sin\theta_0 \cos\theta_0 + J\dot{\phi}_0 \sin\theta_0(\dot{\psi}_0 + \dot{\phi}_0 \cos\theta_0) \\ M_{r_{C_z}} &= 0 \end{aligned} \tag{7.60}$$

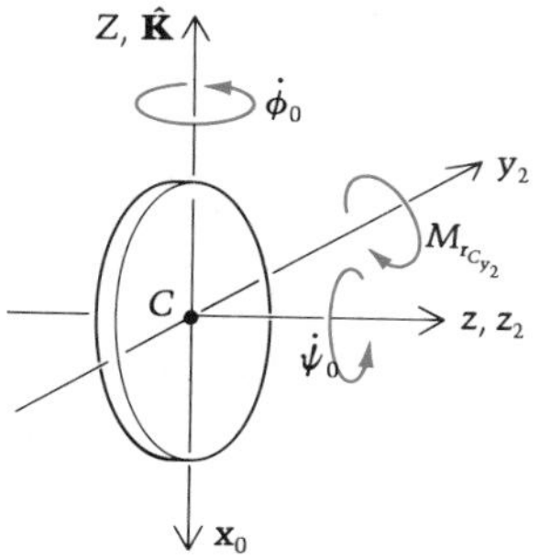

Figure 7.9

We see that only a moment about the y_2 axis is needed to sustain steady precession. Also, $M_{r_{C_z}} = 0 = J(d\omega_z/dt)$ means that ω_z is a constant.*

In the case in which $\theta = 90°$, we have the precession and spin axes orthogonal; the situation is shown in Figure 7.9. If the gyro is spinning and the torque is applied, there will simultaneously occur a precession that tends to turn the spin vector toward the torque vector. This is sometimes called the law of gyroscopic precession. The torque in this case, from Equation (7.60), is

$$M_{r_{C_{y_2}}} = J\dot{\psi}_0\dot{\phi}_0 \tag{7.61}$$

and it is seen to be the product of the spin momentum $J\dot{\psi}_0$ and the precessional angular speed $\dot{\phi}_0$.

We turn now to an illustration of the law of gyroscopic precession. We have just seen that when a freely spinning body is torqued about an axis normal to the spin axis, it precesses about a third axis that forms an orthogonal triad with the spin and torque vectors. The direction of the precession is such that it turns the spin vector toward the torque vector. This law of gyroscopic precession is responsible for the lunisolar precession of the equinoxes.

What is the lunisolar precession? Because of the billions of years of gravitational pull from the sun and moon, the earth is slightly bulged instead of round. It is in fact about 27 mi shorter across the poles than it is across the equator. This bulge, plus the fact that its axis is tilted $23\frac{1}{2}°$ to the ecliptic, causes the sun (and moon) to torque the earth in addition to the gravity pull, as shown in Figure 7.10.

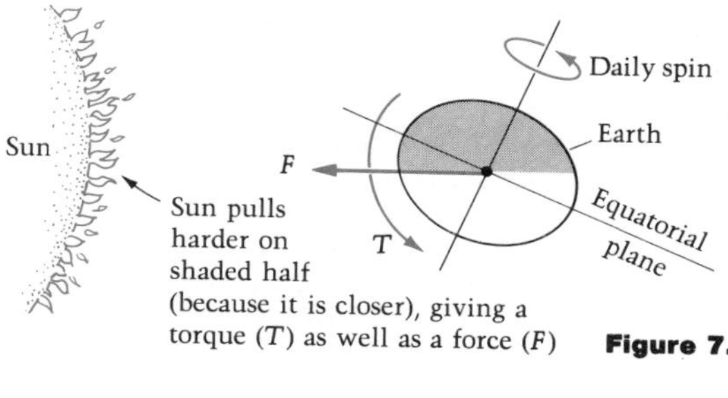

Figure 7.10

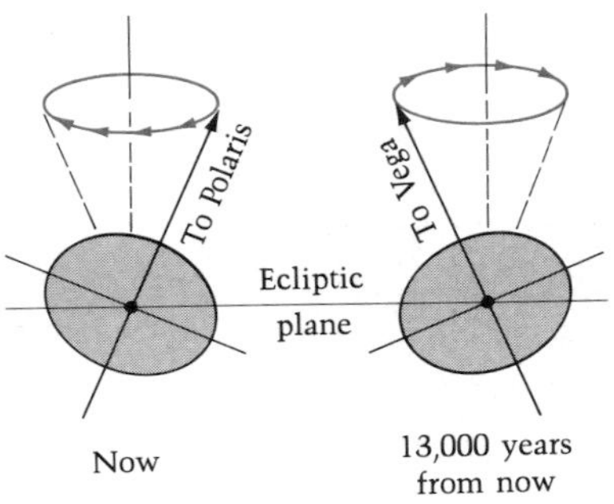

Figure 7.11

We see, then, that we live on the surface of a spinning gyroscope that is constantly being acted on by an external torque, and a precession is thus ongoing. This precession, shown in Figure 7.11, turns the spin axis of the earth out of the plane of the paper toward the torque vector.* This motion results in counterclockwise movement, on the celestial sphere, of the celestial pole to which the earth's rotation axis is directed. (This point is currently close to Polaris, the North Star.) The period of this rotation is about 26,000 years, and it is interesting that the moon's effect is 2.2 times that of the sun's because it is much closer.

We now take up the other example of a solution to the gyroscope equations: the case of **torque-free motion.** "Torque-free" means that $\mathbf{M}_{r_C}$ vanishes, so that $\mathbf{H}_C$ is a constant. This follows from Euler's second law:

$$\mathbf{M}_{r_C} = \mathbf{0} = {}^{\mathcal{I}}\dot{\mathbf{H}}_C \Rightarrow \mathbf{H}_C = \text{constant vector in the inertial frame } \mathcal{I}$$

We shall conveniently let the direction of the Z axis (see Figure 7.12), which is arbitrary, coincide with the constant direction of $\mathbf{H}_C$. Then Z becomes the precession axis of the motion, and the $(\mathbf{x}_2, \mathbf{y}_2, \mathbf{z}_2)$ axes appear as shown in Figure 7.12. It is seen that since

$$\begin{aligned}\mathbf{H}_C &= H_{C_{x_2}}\hat{\mathbf{i}}_2 + H_{C_{y_2}}\hat{\mathbf{j}}_2 + H_{C_{z_2}}\hat{\mathbf{k}}_2 = \text{constant}\\ &= -I\dot{\phi}\sin\theta\,\hat{\mathbf{i}}_2 + I\dot{\theta}\,\hat{\mathbf{j}}_2 + J(\dot{\psi} + \dot{\phi}\cos\theta)\,\hat{\mathbf{k}}_2 \qquad (7.62)\end{aligned}$$

and since $\mathbf{H}_C$ is seen always to lie in the $\mathbf{x}_2\mathbf{z}_2$ plane, then $H_{C_{y_2}}$ must vanish:

$$H_{C_{y_2}} = 0 = I\dot{\theta} \Rightarrow \theta = \text{constant} \qquad (7.63)$$

Note that $\boldsymbol{\omega}$ lies in the $\mathbf{x}_2\mathbf{z}_2$ plane along with $\mathbf{H}_C$, since its y_2 component, $\dot{\theta}$, vanishes. Furthermore, we see that

$$\begin{aligned} H_{C_{x_2}} &= -I\dot{\phi}\sin\theta \qquad \text{and} \qquad & H_{C_{z_2}} &= J(\dot{\psi} + \dot{\phi}\cos\theta)\\ &= I\omega_{x_2}{}^{\dagger} & &= J\omega_{z_2} \qquad (7.64)\end{aligned}$$

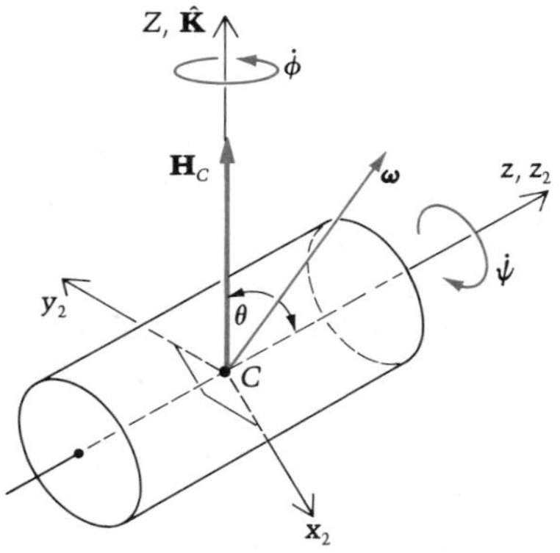

Figure 7.12

But also, as we can see from Figure 7.12,

$$H_{C_{x_2}} = -H_C\sin\theta \qquad \text{and} \qquad H_{C_{z_2}} = H_C\cos\theta \qquad (7.65)$$

Therefore, equating the first of Equations (7.64) and (7.65) for $H_{C_{x_2}}$, we obtain

$$\begin{aligned}\omega_{x_2} &= \frac{-H_C\sin\theta}{I} = \text{constant} \qquad (\text{since } H_C,\ \theta, \text{ and } I \text{ are constants}) \qquad (7.66)\\ &= -\dot{\phi}\sin\theta \qquad (7.67)\end{aligned}$$

*The spin axis of the earth is aligned with its angular velocity vector $\boldsymbol{\omega}$. The point where the $\boldsymbol{\omega}$ vector, placed at C, cuts the surface of the earth is the real meaning of the north pole. The North Pole wanders about the geometric pole (on the symmetry axis) as time passes; it has remained within a few feet of it in this century.

†Note again that the $(\mathbf{x}_2, \mathbf{y}_2, \mathbf{z}_2)$ axes are permanently principal, even though not body-fixed, and this lets us write $\mathbf{H}_C$ in terms of the $I\omega$'s along these axes.

and we see that

$$\dot{\phi} = \frac{H_C}{I} \qquad \text{(a constant)} \tag{7.68}$$

Similarly, equating the two preceding values of $H_{C_{z_2}}$ gives

$$\omega_{z_2} = \frac{H_C \cos\theta}{J} = \text{constant} \tag{7.69}$$

$$= \dot{\psi} + \dot{\phi}\cos\theta \tag{7.70}$$

so that

$$\dot{\psi} = \frac{H_C \cos\theta}{J} - \left(\frac{H_C}{I}\right)\cos\theta = H_C \cos\theta\left(\frac{I-J}{IJ}\right) \tag{7.71}$$

$$= \text{constant}$$

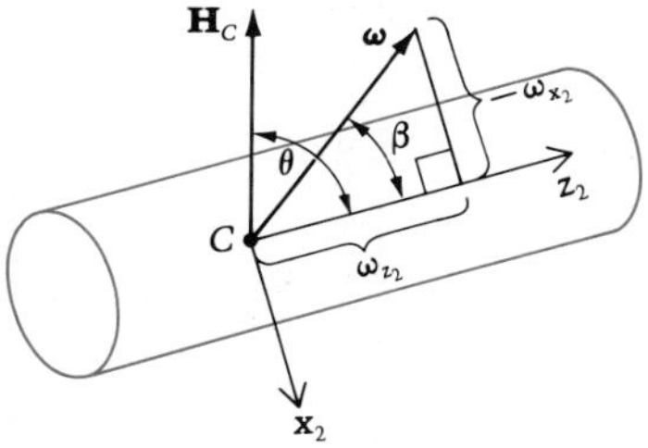

Figure 7.13

Therefore all conditions are satisfied for the body to be in a state of steady precession about the z axis fixed in $\mathcal{I}$!

Dividing Equation (7.66) by (7.69) leads to

$$\frac{\omega_{x_2}}{\omega_{z_2}} = -\frac{J}{I}\tan\theta \tag{7.72}$$

and Figure 7.13 shows that

$$\frac{-\omega_{x_2}}{\omega_{z_2}} = \tan\beta \tag{7.73}$$

where β is the angle between z_2 and $\boldsymbol{\omega}$. Therefore

$$\tan\beta = \frac{J}{I}\tan\theta$$

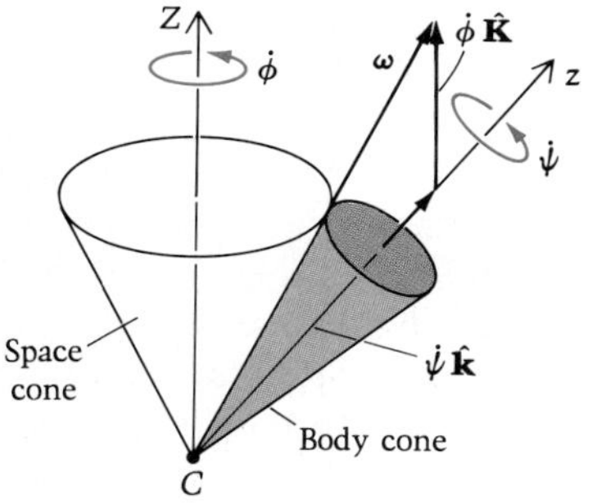

Figure 7.14

and we see that the answer to whether β is larger or smaller than θ depends on the ratio of J to I. If $J < I$, as in the elongated shape in Figure 7.13, then $\beta < \theta$ and the angular velocity vector lies inside of $\mathbf{H}_C$ and z_2, making a constant angle with each. Two cones may be imagined—one fixed to the body, the other in space ($\mathcal{I}$). The body cone is seen to roll on the fixed space cone as its spin and precession vectorially add to the vector $\boldsymbol{\omega}$, which changes only in direction.

This precession (Figure 7.14) is called ***direct*** because $\dot{\phi}$ and $\dot{\psi}$ have the same counterclockwise sense when observed from the $\boldsymbol{\omega}$ vector outside the cones. If, however, $J > I$, then $\beta > \theta$ and $\boldsymbol{\omega}$ lies *outside* the angle ZCz. This situation is harder to depict, but just as important. This time the body cone rolls around the *outside* of the fixed space cone (Figure 7.15), and the two rotations $\dot{\phi}$ and $\dot{\psi}$ have opposite senses. This precession is called ***retrograde.***

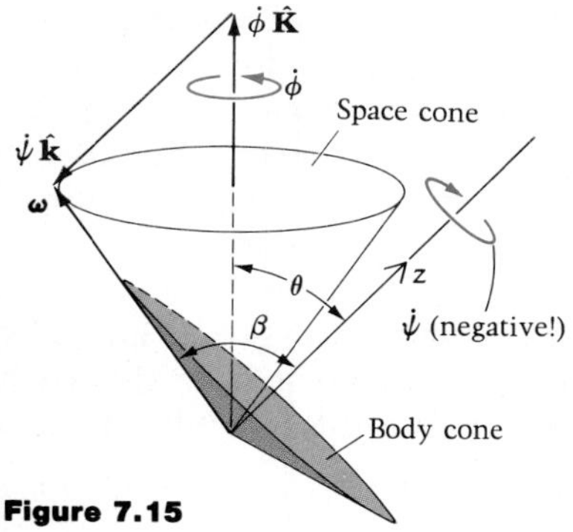

Figure 7.15

A final note on the theory of the torque-free body: If the constant value of θ is either 0 or 90°, there is *no* precession and the gyroscope is simply in a state of pure rotation, planar motion:

$\theta = 0$:

Here $z = Z$, so that $\omega_{x_2} = 0$ and $\omega_{z_2} = H_C/I$. If $\theta = 0°$, the body spins about its axis without precessing. In this case, the rates $\dot{\phi}$ and $\dot{\psi}$ cannot be distinguished.

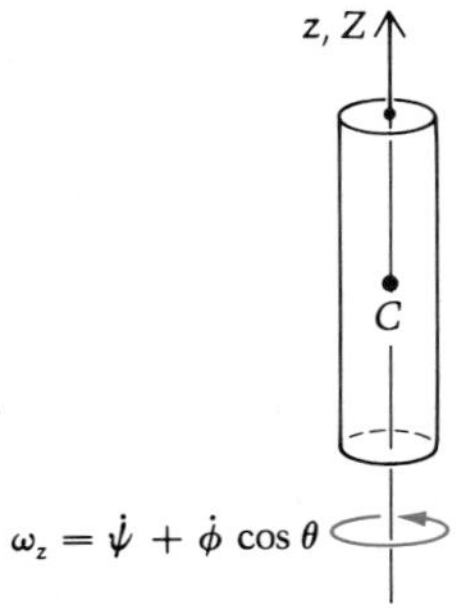

Question 7.9 Why can we not use Equations (7.68) and (7.71) to get $\dot{\phi}$ and $\dot{\psi}$ in this case?

$\theta = 90°$:

Here we have $\omega_{x_2} = -H_C/I$ and $\omega_{z_2} = 0$. Thus:

$$\dot{\phi} = \frac{H_C}{I}$$

$$\dot{\psi} = 0$$

If $\theta = 90°$, the body spins about a transverse axis without precessing.

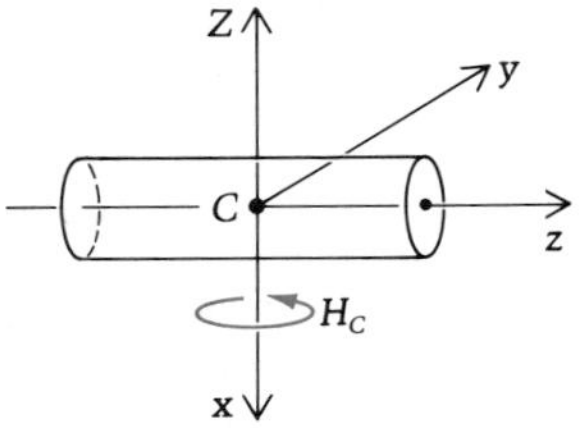

It requires a great many terms to describe the complex motion of the earth. We have seen one example of this in the lunisolar precession caused by the gravity torque exerted on the earth by the sun and moon. This motion is analogous to a differential equation's particular solution, which it has whenever the equation has a nonzero right-hand side. The complementary, or homogeneous, solution is analogous to the torque-free part of the solution to the earth's rotational motion. This part, called the free precession of the earth, is in fact retrograde. Both the space and body cones are very thin as the $\boldsymbol{\omega}$, $\mathbf{H}_C$, and z axes are all quite close together; each lies about $23\frac{1}{2}°$ off the normal to the ecliptic plane.

As we did in Chapter 5 for the case of plane motion, we could apply the principles of impulse and momentum and those of angular impulse and angular momentum to the three-dimensional motion of a rigid body $\mathcal{B}$. As we saw in Section 5.3, however, these applications are really nothing more than time integrations of the equations of motion.

There is one type of problem, however, in which these two principles furnish us with a means of solution—problems involving impact. Some three-dimensional aspects are sufficiently different from the planar case to warrant an example. But first we use the integrals of the Euler laws to derive the needed relations:

$$\int_{t_i}^{t_f} \mathbf{F}_r \, dt = \mathbf{L}_f - \mathbf{L}_i \tag{7.74}$$

$$\int_{t_i}^{t_f} \mathbf{M}_{r_C} \, dt = \mathbf{H}_{C_f} - \mathbf{H}_{C_i} \tag{7.75}$$

An alternative to the rotational equation (7.75) is to integrate the equally general equation

$$\mathbf{M}_{r_O} = \dot{\mathbf{H}}_O \tag{7.76}$$

where O is now a fixed point of the inertial frame $\mathcal{I}$:

$$\int_{t_i}^{t_f} \mathbf{M}_{r_O} \, dt = \mathbf{H}_{O_f} - \mathbf{H}_{O_i} = (\mathbf{H}_{C_f} - \mathbf{H}_{C_i}) + (\mathbf{r}_{PC} \times m\mathbf{v}_C)\Big|_i^f \tag{7.77}$$

To use either Equation (7.75) or (7.77) in an impact situation, we use Equation (7.22) for the body's angular momentum before the deformation starts (at t_i) and then again after it ends (at t_f). The following example illustrates the procedure.

EXAMPLE 7.15

The bent bar $\mathcal{B}$ of Examples 4.13 and 4.27 is dropped from a height H and strikes a rigid, smooth surface on one end of $\mathcal{B}$ as shown in the diagram. If the coefficient of restitution is e, find the angular velocity of $\mathcal{B}$, as well as the velocity of C, just after the collision.

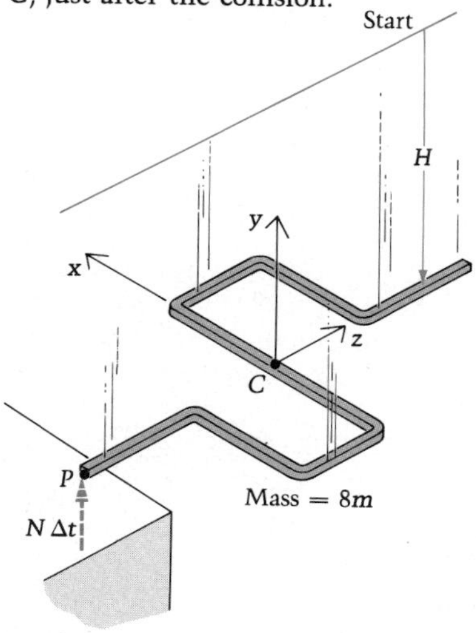

(Continued)

SOLUTION

Using the y component equation of (7.74) yields

$$N\Delta t = 8m\dot{y}_{C_f} - 8m(-\sqrt{2gH}) \tag{1}$$

in which the impulse of the gravity force is neglected as small in comparison with the impulsive upward force exerted by the surface over the short time interval Δt.

Next we write the component equations of (7.75); we first need the inertia properties of the body, which can be computed to be

$$\begin{aligned} I_{xx}^{C} &= \frac{22}{3}m\ell^2 \qquad & I_{xy}^{C} &= 0 \\ I_{yy}^{C} &= \frac{32}{3}m\ell^2 \qquad & I_{yz}^{C} &= 0 \\ I_{zz}^{C} &= \frac{10}{3}m\ell^2 \qquad & I_{zx}^{C} &= -2m\ell^2 \end{aligned} \tag{2}$$

We then obtain, from Equation (7.75),

$$\begin{aligned} 2\ell N\Delta t\hat{\mathbf{i}} &= m\ell^2\left[\frac{22}{3}\omega_x + 0\omega_y - 2\omega_z\right]\hat{\mathbf{i}} \\ &+ m\ell^2\left[0\omega_x + \frac{32}{3}\omega_y + 0\omega_z\right]\hat{\mathbf{j}} \\ &+ m\ell^2\left[-2\omega_x + 0\omega_y + \frac{10}{3}\omega_z\right]\hat{\mathbf{k}} \end{aligned} \tag{3}$$

in which the initial angular velocity components vanish and the desired final components are $(\omega_x, \omega_y, \omega_z)$. The component equations of (3) are

$$\frac{22}{3}\omega_x - 2\omega_z = \frac{+2N\Delta t}{m\ell} \tag{4}$$

$$\frac{32}{3}\omega_y = 0 \tag{5}$$

$$-2\omega_x + \frac{10}{3}\omega_z = 0 \tag{6}$$

At this point we have four equations in the five unknowns $\dot{y}_C$, ω_x, ω_y, ω_z, and the impulse $N\Delta t$. We get a fifth equation from the definition of the coefficient of restitution together with the y component of the rigid-body velocity relationship between P and C:

$$e = \frac{\dot{y}_{P_f} - 0}{0 - (-\sqrt{2gH})} \Rightarrow \dot{y}_{P_f} = \sqrt{2gH}\,e \tag{7}$$

and

(Continued)

$$\mathbf{v}_C = \mathbf{v}_P + \boldsymbol{\omega} \times \overset{2\ell\hat{\mathbf{k}}}{\mathbf{r}_{PC}} \tag{8}$$

which has the y-component equation

$$\dot{y}_{C_f} = \dot{y}_{P_f} - 2\ell\omega_x \tag{9}$$

Using Equation (7), we obtain

$$\dot{y}_{C_f} = \sqrt{2gH}\, e - 2\ell\omega_x \tag{10}$$

The solution to the five equations (1, 4, 5, 6, 10) is

$$\begin{aligned} \omega_x &= \frac{60(1+e)\sqrt{2gH}}{143\ell} & \dot{y}_{C_f} &= \frac{(23e - 120)\sqrt{2gH}}{143} \\ \omega_y &= 0 \\ \omega_z &= \frac{36(1+e)\sqrt{2gH}}{143\ell} & N\Delta t &= \frac{184m(1+e)\sqrt{2gH}}{143} \end{aligned} \tag{11}$$

Returning to Equation (8), we find that the $\mathbf{x}$ and z components of $\mathbf{v}_{C_f}$ vanish:

$$\begin{aligned} \mathbf{x} \text{ components} &\Rightarrow \dot{x}_{C_f} = 0 + 2\omega_y\ell = 0 \\ z \text{ components} &\Rightarrow \dot{z}_{C_f} = 0 + 0 = 0 \end{aligned} \tag{12}$$

The results in Equations (12) are obvious, since if there is no friction at the point of contact there can be no impulsive forces in the horizontal plane to change the momentum (from zero) in the $\mathbf{x}$ or z directions.

It is seen that the single nonzero product of inertia causes a coupling between ω_x and ω_z (see Equations 4 and 6), which prevents ω_z from vanishing—even though the only moment component with respect to C is about the $\mathbf{x}$ axis!

We shall now see with another example the advantages of Equation (7.77), which may be used to eliminate undesired forces from moment equations, just as was done in our study of statics.

EXAMPLE 7.16

Rework the preceding example by using Equation (7.77) instead of the combination of Equations (7.74) and (7.75). Find the value of $\boldsymbol{\omega}$ after impact.

(Continued)

SOLUTION

Equation (7.77) allows us to eliminate the impulse $N\Delta t$ by summing moments about the point (P) of impact:

$$\int_{t_i}^{t_f} \mathbf{M}_{r_P}\,dt = \mathbf{0} = (\mathbf{H}_{C_f} - \overset{\mathbf{0}}{\cancel{\mathbf{H}_{C_i}}}) + (\mathbf{r}_{PC} \times m\mathbf{v}_C)\Big|_i^f$$

$$= \left[\left(\frac{22}{3}\omega_x - 2\omega_z\right)m\ell^2\hat{\mathbf{i}} + \frac{32}{3}\omega_y m\ell^2\hat{\mathbf{j}} + \left(-2\omega_x + \frac{10}{3}\omega_z\right)m\ell^2\hat{\mathbf{k}}\right]$$

$$+ 2\ell\mathbf{k} \times [m\dot{y}_{C_f}\hat{\mathbf{j}} - m\sqrt{2gH}(-\hat{\mathbf{j}})]$$

We still have to use the coefficient of restitution and relate $\mathbf{v}_P$ and $\mathbf{v}_C$ exactly as before; making this substitution for $\dot{y}_{C_f}$ leads to the following three scalar component equations:

$$\frac{34}{3}\omega_x - 2\omega_z = \frac{2\sqrt{2gH}}{\ell}(1 + e)$$

$$\frac{32}{3}\omega_y = 0$$

$$-2\omega_x + \frac{10}{3}\omega_z = 0$$

These equations, of course, have the same solution as $(\omega_x, \omega_y, \omega_z)$ in the preceding example.

PROBLEMS / Section 7.5

The following six problems refer to Example 7.14.

7.54 Determine the speed v_G for which the normal forces N_i and N_o are equal for the following parameters:

$H = 2$ ft	$g = 32.2$ ft/sec^2	$\theta = 5°$
$D = 6$ ft	$I_1 = mk_1^2 = 3m$ slug-ft^2	
$R = 300$ ft	$I_3 = mk_3^2 = 5m$ slug-ft^2	

7.55 Determine the speed at which the car will overturn for the data in the preceding problem.

7.56 Determine the bank angle θ for which $N_i = N_o$ if

$H = 1.5$ ft	$g = 32.2$ ft/sec^2	$v_G = 55$ mph
$D = 5$ ft	$I_1 = 2m$ slug-ft^2	
$R = 200$ ft	$I_3 = 4m$ slug-ft^2	

7.57 Note that in Equation (1) (the formula for the angular momentum vector about G), $I_3 > I_1$ and $\omega_z > \omega_x$. Thus $\mathbf{H}_G$ lies between the z and Z axes. Use this fact to argue that the direction of $\mathbf{M}_{r_G}$ is ↺ without any calculations.

7.58 Determine the effect of the inertia of the car's wheels (mass m, radius r, and radius of gyration k_W about the axis of each wheel) on $N_i - N_o$ when $\theta = 0$. *Hint:* The moment of momentum about G is that of the car and wheels moving in the turn as if locked together *plus* that of the four wheels in their relative (turning) motion with respect to the chassis.

7.59 Writing Euler's laws as $\mathbf{F}_r + (-m\mathbf{a}_G) = \mathbf{0}$ and $\mathbf{M}_{r_G} + (-\dot{\mathbf{H}}_G) = \mathbf{0}$ results in what is known as the *reversed effective force* $(-m\mathbf{a}_G)$ and the *inertia torque* $(-\dot{\mathbf{H}}_G)$. If these quantities are added to the free-body diagram, the object may be treated as though it were in equilibrium. For the car at the instant of overturning, such a diagram would appear as shown in Figure P7.59.

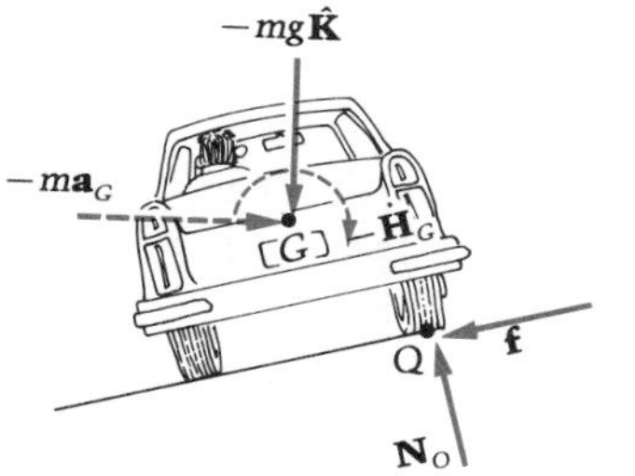

Figure P7.59

Calculate the moment about the point Q and, noting that $\mathbf{M}_{r_Q} = \mathbf{0}$ for this free-body diagram, compare the effects of the reversed effective force (or inertia force) $-m\mathbf{a}_G$ with that of the inertia torque $-\dot{\mathbf{H}}_G$ on the overturning tendency. Show that your solution predicts that the following make the car more likely to turn over: (a) higher values of v_G, H, $(I_3 - I_1)$, and m; (b) lower values of R, D, and θ.

7.60 There is a relationship among v_C, g, r, R, and θ such that the disk can roll around in a circle as shown in Figures P7.60a and b, with v_C and θ remaining constant. Find this relationship.

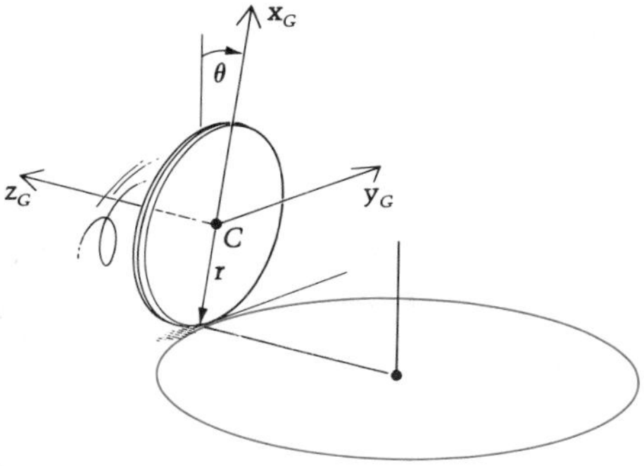

Figure P7.60a

7.61 A bicycle wheel weighing 5 lb and having a 14-in. radius is misaligned by 1° with the vertical. The top of the wheel tilts toward the right with the bike moving forward. If the bike is driven along a straight path at 15 mph, find the ***rocking couple*** (bearing moment) acting on the shaft. Use the result of Problem 7.24, neglecting the spokes and hub.

7.62 In the preceding problem, suppose instead that the bicycle is in a 50-ft-radius turn to the (a) right and (b) left. Determine the new values of the bearing moment acting on the misaligned wheel. Neglect the lean angle.

7.63 A disk $\mathcal{D}$ (mass m, radius r) rolls in a circle; it is driven by arm $\mathcal{A}$, which is connected to $\mathcal{D}$ by a ball joint at C. (See Figure P7.63.) Forces turn arm $\mathcal{A}$ at constant angular speed Ω about the centerline. Find the force in $\mathcal{A}$ and the reactions exerted onto $\mathcal{D}$ by the floor. The problem differs from Example 7.10 in that this one has a constant tilt angle $\theta_0 \neq 0$ as shown. Compare your results as $\theta_0 \rightarrow 0$ with those of that example.

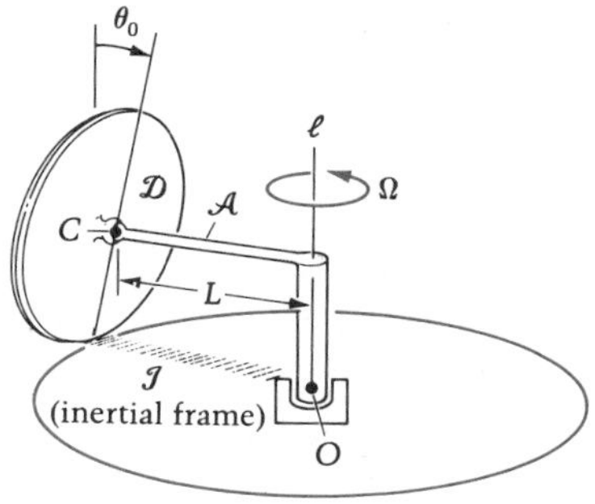

Figure P7.63

7.64 Suppose that the components of the mass center velocity $\mathbf{v}_C$ are written in body $\mathcal{B}$ instead of in an inertial frame $\mathcal{I}$. Use the property (6.20) of $\boldsymbol{\omega}_{\mathcal{B}/\mathcal{I}}$ to derive the scalar equations of motion of the mass center from Euler's first law: $\mathbf{F}_r = {}^{\mathcal{I}}\dot{\mathbf{L}}$.

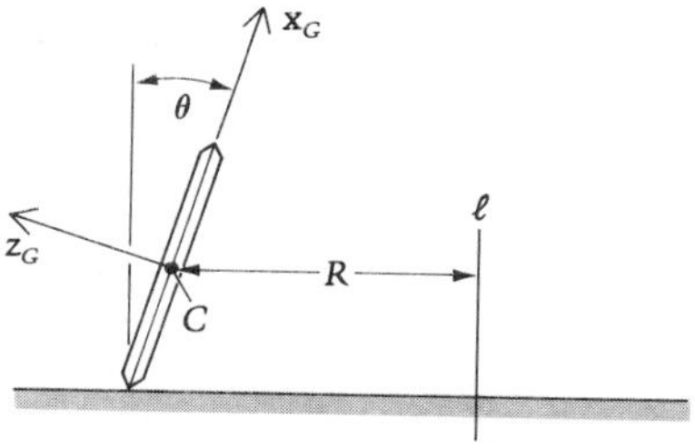

Figure P7.60b

7.65 A heavy disk $\mathcal{D}$ of mass m and radius r spins at the angular rate ω_3 $(= |\boldsymbol{\omega}_{\mathcal{D}/\mathcal{B}}|)$ with respect to the rigid, but light, bent bar $\mathcal{B}$. (See Figure P7.65.) Body $\mathcal{B}$ turns at rate ω_2 $(= |\boldsymbol{\omega}_{\mathcal{B}/\mathcal{I}}|)$ about a vertical axis through O, a point of both $\mathcal{B}$ and the inertial frame $\mathcal{I}$. Find the force and couple that must be acting on $\mathcal{B}$ at O to produce a motion of the system for which ω_2 and ω_3 are constants. Both sets of axes in the figure are fixed in $\mathcal{B}$, and note that $(\mathbf{x}_C, \mathbf{y}_C, \mathbf{z}_C)$ are always principal axes for $\mathcal{D}$ at G even though they are not fixed in $\mathcal{D}$.

7.66 Compute the moment M applied to the shaft $\mathcal{S}$ in Figures P7.66a and b as a function of the angle β if ω_1 and ω_2 $(= \dot{\beta})$ are constants.

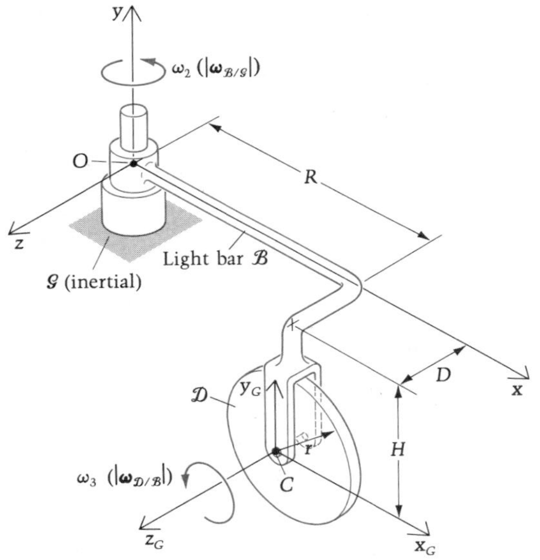

Figure P7.65

7.67 A bike rider enters a turn of radius R at a constant speed of v_C. (See Figure P7.67.) Other quantities are defined below:

r = radius of wheel

d = distance between axle and C

I_1, I_2 = principal moments of inertia of entire bike plus rider with respect to $\hat{\mathbf{n}}_1$ and $\hat{\mathbf{n}}_2$ directions through C

i = moment of inertia, with respect to $\hat{\mathbf{n}}_2$ direction, of one wheel about its axis of symmetry

m = total mass

ϕ = angle shown

Solve for the resultant force $\mathbf{F}_r$ and moment $\mathbf{M}_{r_G}$ in terms of these quantities. Compare the effects of the *D'Alembert force* $(-m\mathbf{a}_C)$ and the *inertia torque* $(-\dot{\mathbf{H}}_C)$ in righting the bike when ϕ is small. Neglect the products of inertia.

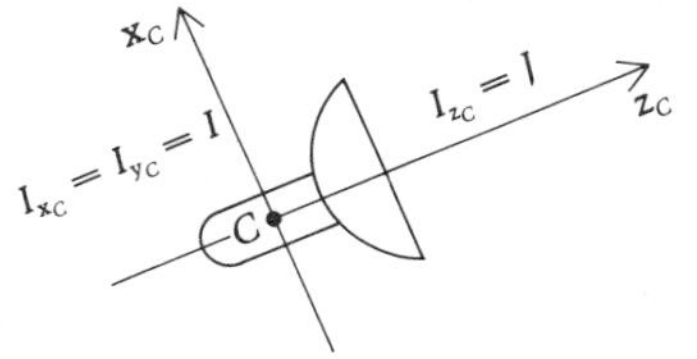

Figure P7.66a

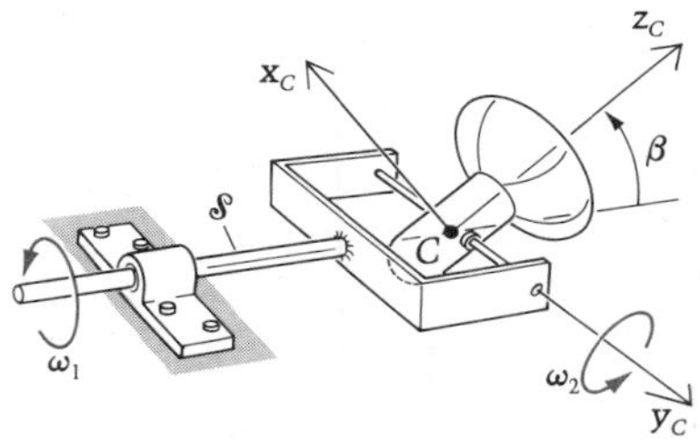

Figure P7.66b

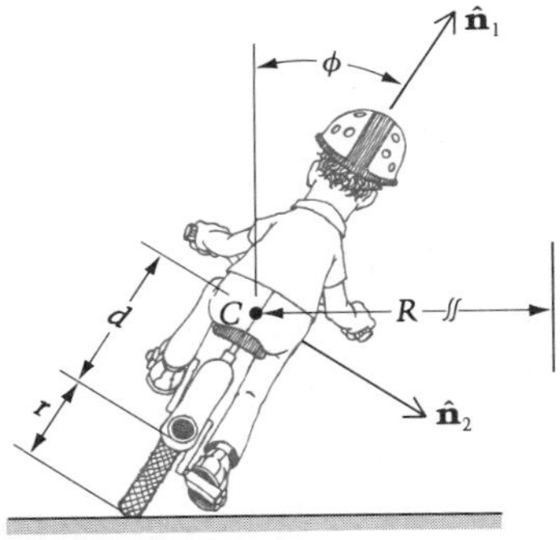

Figure P7.67

7.68 A ship's turbine has a mass of 2500 kg and a radius of gyration about its axis (y_C in Figure P7.68) of 0.45 m. It is mounted on bearings as indicated and turns at 5000 rpm clockwise when viewed from the stern (rear) of the boat.

a. If the ship is in a steady turn to the right of radius 500 m and is traveling at 15 knots, what are the reactions exerted on the shaft by the bearings? (1 knot = 1.15 mph = 1.85 km/hr)

b. If the ship on a straight course in rough seas pitches sinusoidally at ±12° amplitude with a 6-s period, what are the maximum bearing reactions then?

7.69 Disk $\mathcal{D}$ in Figure P7.69 turns in bearings at C at angular rate ω_s about the light rod $\mathcal{R}$, and both precess about axis z_O at angular rate ω_P as shown. Show with a free-body diagram how it is possible for the mass center C to remain in a horizontal plane. Then find the reactions exerted onto $\mathcal{R}$ by the socket at O. Is there any difference in the solution if $\mathcal{D}$ and $\mathcal{R}$ are rigidly connected?

7.70 Tell why a perfectly thrown, spiraling football will never start wobbling in midflight, just as a "wounded duck" wobbling one will never lose its wobble before coming down. (See Figure P7.70.)

7.71 Explain the spiraling football pass shown in Figure P7.71.

a. Does its axis remain aligned with its trajectory because the ball is torque-free and an initial angular momentum about z is imparted to it by the quarterback's arm motion?

b. Or is it because the air resistance imparts a small drag torque that causes the ball to precess?

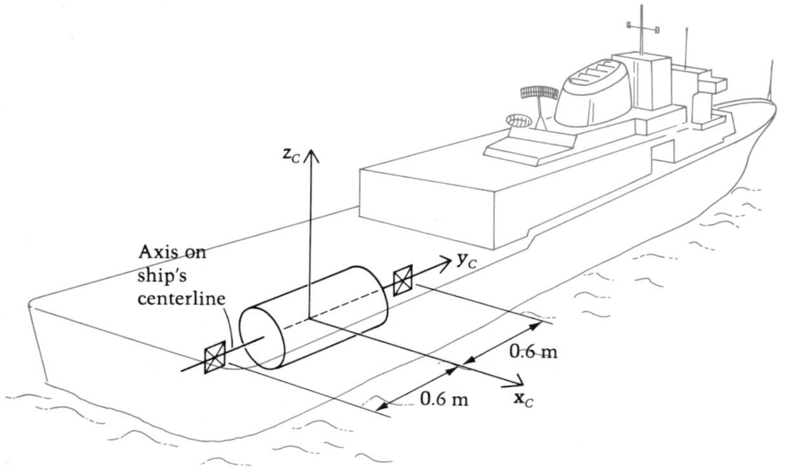

Figure P7.68

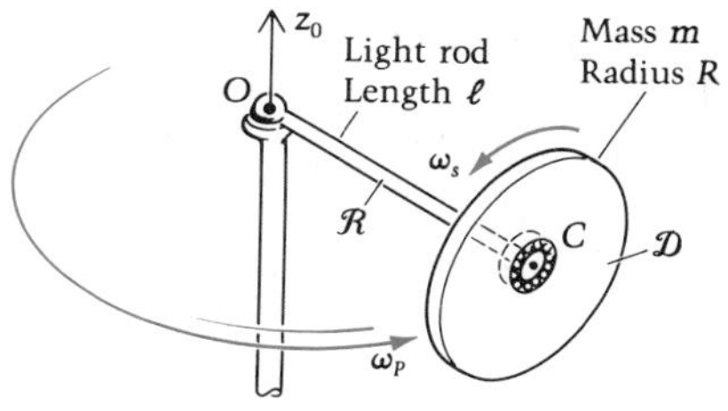

Figure P7.69

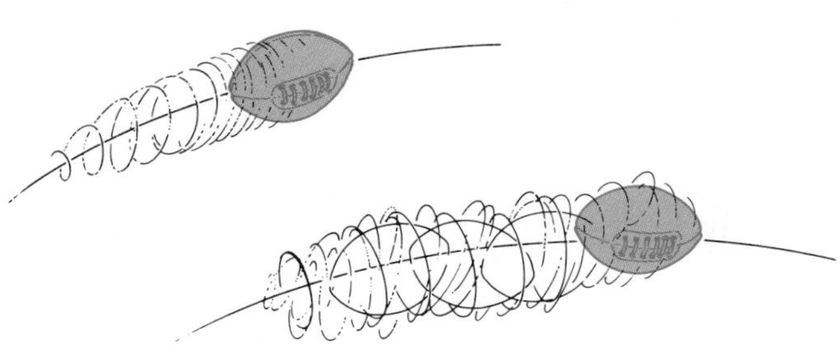

Figure P7.70

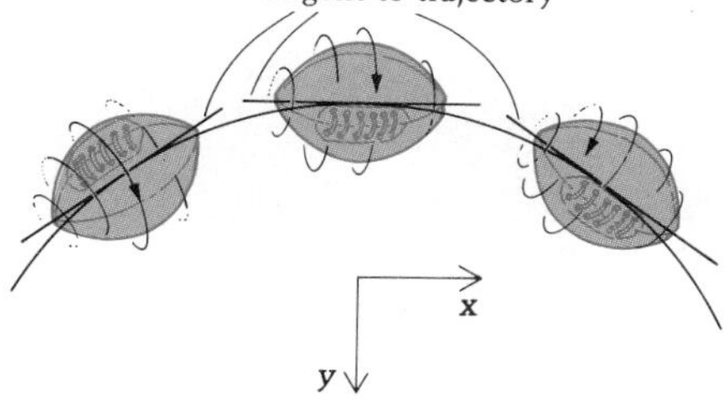

Figure P7.71

7.72 Disk $\mathcal{D}$ in Figure P7.72 turns in bearings around rod $\mathcal{R}$ as it rolls on the ground $\mathcal{G}$. Find all the forces acting on $\mathcal{D}$, and note that the normal force from the ground is increased above the gravity force $m\mathbf{g}$ due to gyroscopic action. Assume that forces applied to $\mathcal{R}$ force the mass center C of $\mathcal{D}$ to travel at constant speed v_C in a horizontal circular path.

7.73 A diver $\mathcal{D}$ leaves a diving board in a straight, symmetric position with angular velocity and angular momentum vectors each in the $\mathbf{x}$ direction as indicated in Figure P7.73a. Since $\mathbf{M}_{r_C}$ is zero, there will be no change in the angular momentum $\mathbf{H}_C$ in the inertial frame (the swimming pool) as long as the diver is in the air. Therefore, as long as he remains in the straight position, his constant angular momentum is expressed by

$$I^C_{xx}\omega_x + \overset{0}{\cancel{I^C_{xy}}}\,\overset{0}{\cancel{\omega_y}} + \overset{0}{\cancel{I^C_{xz}}}\,\overset{0}{\cancel{\omega_z}} = H_{C_1} = \text{constant} \qquad (1)$$

$$\overset{0}{\cancel{I^C_{yx}}}\,\omega_x + I^C_{yy}\,\overset{0}{\cancel{\omega_y}} + \overset{0}{\cancel{I^C_{yz}}}\,\overset{0}{\cancel{\omega_z}} = 0 \qquad (2)$$

$$\overset{0}{\cancel{I^C_{zx}}}\,\omega_x + \overset{0}{\cancel{I^C_{zy}}}\,\overset{0}{\cancel{\omega_y}} + I^C_{zz}\,\overset{0}{\cancel{\omega_z}} = 0 \qquad (3)$$

where we assume the body to be sufficiently internally symmetric so that the products of inertia all vanish. Now suppose the diver instantaneously moves his arms as shown in Figure P7.73b to initiate a twist. Following the maneuver, he may again be treated as a rigid body and we may use the same body-fixed axes as before. (Note that the mass center changes very little.)

a. From Figure P7.73c argue that the indicated changes in the products of inertia occur. (Only the shaded arms contribute to the products of inertia.) Argue also that I^C_{yz} is smaller than I^C_{xz} and also less than $|I^C_{xy}|$. Observe that all three products of inertia are small compared with the three moments of inertia and that $I^C_{yy} < I^C_{xx} < I^C_{zz}$, with I^C_{yy} being very much smaller than the other two moments of inertia. Note that $(\mathbf{x}, \mathbf{y}, \mathbf{z})$ are no longer principal, but this does not matter since we are not making use of principal axes here.

As the diver's body begins to twist and turn, the right sides of Equations (1) to (3) will change and *none* of the quantities on the left will remain zero. But the right sides will constitute the components *in the body frame* $\mathcal{D}$ of the vector $\mathbf{H}_C$, which will still vectorially add to $H_{C_1}\hat{\mathbf{I}}$, where $\hat{\mathbf{I}}$ is the original direction in $\mathcal{I}$ of $\mathbf{H}_C$ after the diver leaves the diving board (to the right in the first sketch).

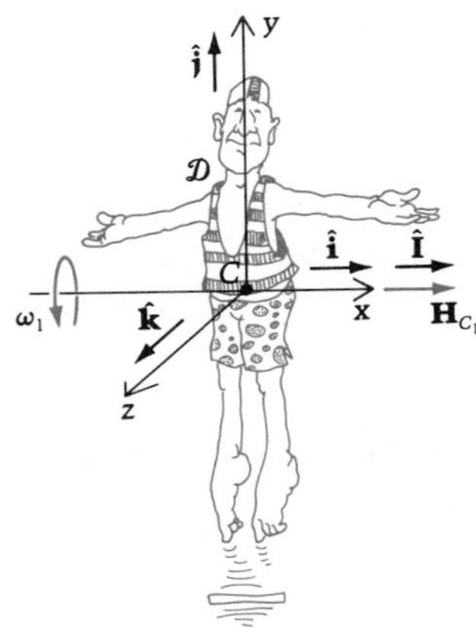

Figure P7.73a

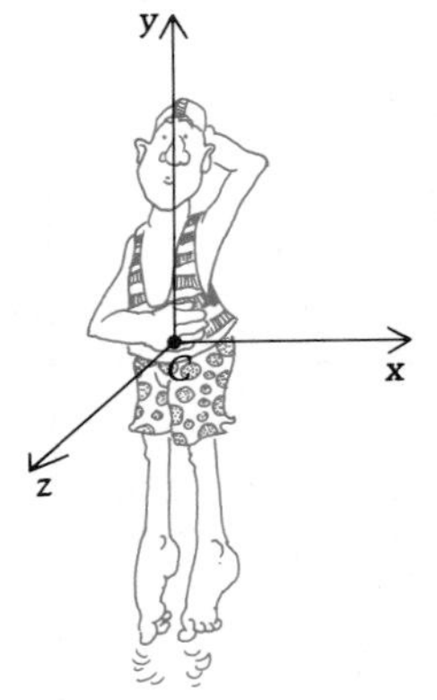

Figure P7.73b

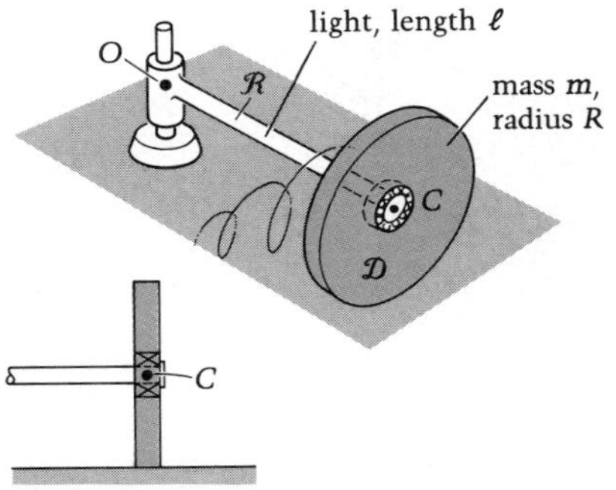

Figure P7.72

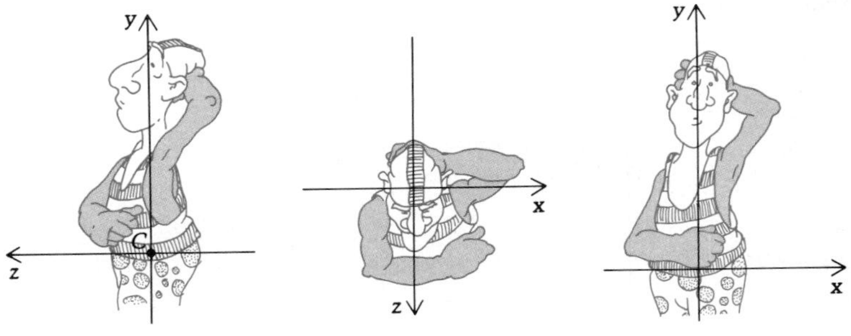

$I^C_{yz} > 0$ $\quad I^C_{xz} > 0$ $\quad I^C_{xy} < 0$

Figure P7.73c

b. After the rapid twist maneuver, but *before* the diver begins to twist, his axes are still instantaneously aligned with those of the frame $\mathcal{I}$. Use Equations (1) to (3), with the right sides $(H_{C_1}, 0, 0)$ and the now nonzero products of inertia, to show that:

i. There will be a small (compared to the original ω_x) angular velocity developed about the $-z_C$ direction (negative ω_z).

ii. There will be an angular velocity of twist developed about y_C (positive ω_y).

iii. There will be an increase in the somersaulting angular velocity component ω_x.

In arguing statements (i) to (iii), assume nothing about the ω's following the maneuver except that ω_x is still in the same direction as before.

7.74 Obtain the results of Example 7.10 by using the Euler equations (7.53). *Hint:* This time the axes are body-fixed, and the v_C/L part of $\boldsymbol{\omega}$ changes direction in $\mathcal{D}$; therefore the components ω_x and ω_y have derivatives that were formerly zero in $\mathcal{B}$. The v_C/L component of $\boldsymbol{\omega}_{\mathcal{D}/\mathcal{I}}$ is

$$\frac{v_C}{L}(\cos\theta_r\hat{\mathbf{j}} - \sin\theta_r\hat{\mathbf{i}})$$

where θ_r is the angle of roll as shown in Figure P7.74. Differentiate this expression, substitute $\dot{\theta}_r = v_C/r$, and *then* take $\theta_r = 0$. Finally, go to Equations (7.53) and substitute your results.

7.75 Find the magnitude and direction of the force and/or couple exerted on disk $\mathcal{D}$ by the shaft $\mathcal{S}$ in Problem 7.14.

7.76 A screwdriver-like motion between the planar case and general (three-dimensional) motion is defined as follows. All points of the body $\mathcal{B}$ have, at any time, identical z components of velocity in a reference frame $\mathcal{I}$. The unit vector $\hat{\mathbf{k}}$ of this $\dot{z}\hat{\mathbf{k}}$ component is constant in both $\mathcal{B}$ and $\mathcal{I}$, though $\dot{z}$ can vary with time. Thus the angular velocity vector is still expressible as $\boldsymbol{\omega} = \dot{\theta}\hat{\mathbf{k}}$. Derive a moment equation for $\mathcal{B}$ that is valid for this motion.

7.77 In the grinding mill of Problem 7.15, suppose that the wall is absent. (See Figure P7.77.) Find, for a given Ω (constant angular speed of $\mathcal{S}$), the angle ϕ that the axis of the grinder $\mathcal{D}$ will make with the vertical. Observe that with the wall present and ϕ fixed, larger speeds than this Ω will allow the grinder to work. In particular, show that the following set of parameters is satisfactory: $r = 2.5$ ft, $\ell = 6$ ft, $\Omega = 2\pi$ rad/sec, and $\phi = 60°$. Neglect the mass of body $\mathcal{B}$ in comparison with the heavy grinding disk $\mathcal{D}$.

7.78 Find the grinding force N produced at the wall of the grinding mill of Problems 7.15 and 7.77 for the given parameters.

7.79 A slender rod $\mathcal{A}$ and a small ball $\mathcal{B}$ each weigh 0.3 lb. (See Figure P7.79.) The bodies rotate about the vertical along with the slender shaft $\mathcal{S}$ and are supported by a smooth step or thrust bearing at D and by the cord $\mathcal{C}$. Find the tension in the cord if the angular speed of the system is 20 rad/sec.

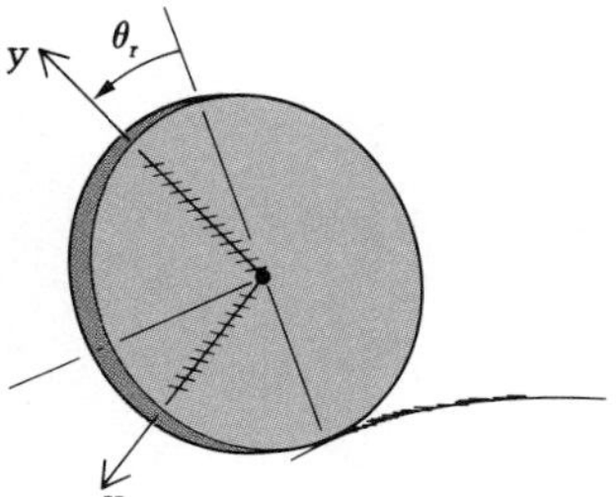

Figure P7.74

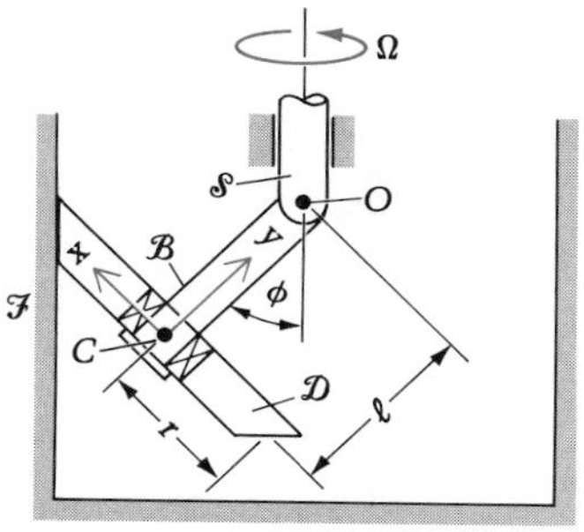

Figure P7.77

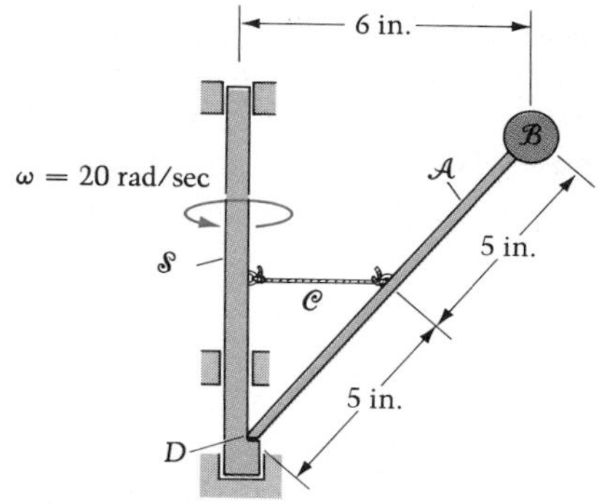

Figure P7.79

7.80 Find the reactions exerted by the bearings on the shaft to which a 30-kg thin plate $\mathcal{P}$ is welded. (See Figure P7.80.) The assembly is turning at the constant angular speed of 30 rad/s. Work the problem by using

$$\mathbf{M}_{r_C} = {}^{\mathcal{I}}\dot{\mathbf{H}}_C = {}^{\mathcal{P}}\dot{\mathbf{H}}_C + \boldsymbol{\omega}_{\mathcal{P}/\mathcal{I}} \times \mathbf{H}_C$$

that is, by expressing $\mathbf{H}_C$ using principal directions in $\mathcal{P}$ (which omits the need for computing the nonzero product of inertia I_{xz}).

7.81 Rework and check the results of the preceding problem by using Equation (7.35) to calculate I_{xz}; then use Equations (7.54) to obtain the bearing reactions.

7.82 Suppose that body $\mathcal{B}$ has a pivot O (that is, a point that never moves in either $\mathcal{B}$ or the inertial frame throughout the motion). Show that the Euler equations (7.53) are equally valid if C is replaced by O. (This means that the moments, the principal moments of inertia, and the $\boldsymbol{\omega}$ components are all taken along the principal axes of inertia of $\mathcal{B}$ at O instead of at the mass center.)

7.83 The blades of a fan turn at 1750 rpm, and the fan oscillates about the vertical axis z (Figure P7.83a) at the rate of one cycle every 10 sec. Assuming that the fan travels at the constant angular velocity of 0.2 rad/sec except when it is reversing direction (Figure P7.83b), find the moment exerted by the base on the arm section $\mathcal{A}$ at the $\frac{1}{4}$-cycle point due to gyroscopic action. For the calculation (*only*!) consider the blades (Figure P7.83c) to be 4-in.-diameter circular aluminum plates all in the same plane and $\frac{1}{32}$ in. thick. Use a density of 0.1 lb/in^3.

7.84 The disk (mass m, radius r) in Figure P7.84 is rigidly attached to the shaft, and the assembly is spun up to angular speed Ω about the z_C axis. Determine the bearing reactions at A and B in terms of m, r, g, β, Ω, and L.

7.85 Explain why, in the absence of air resistance, a baseball pitcher could not throw a curve ball.

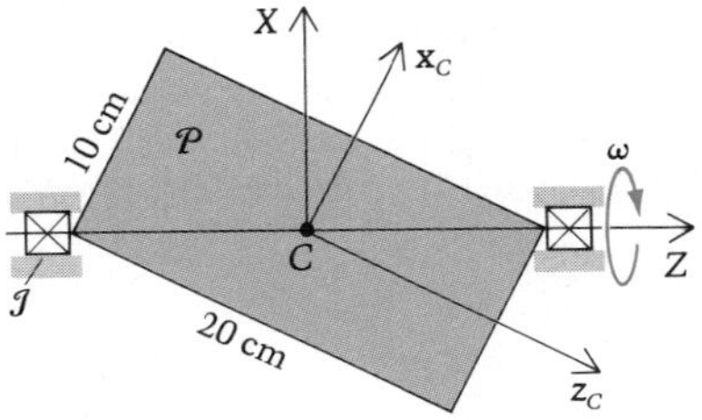

Figure P7.80

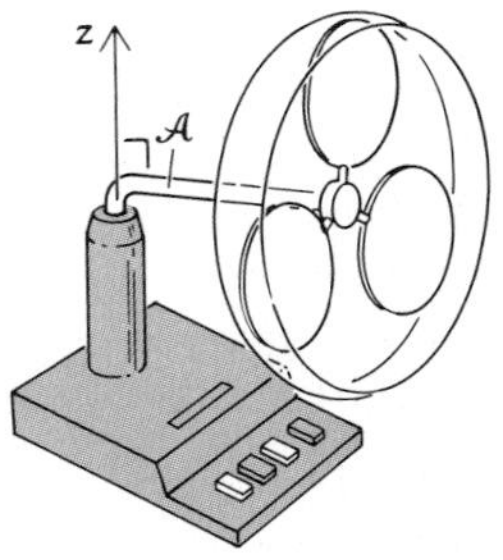

Figure P7.83a

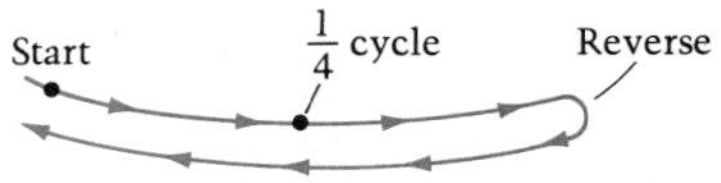

Figure P7.83b

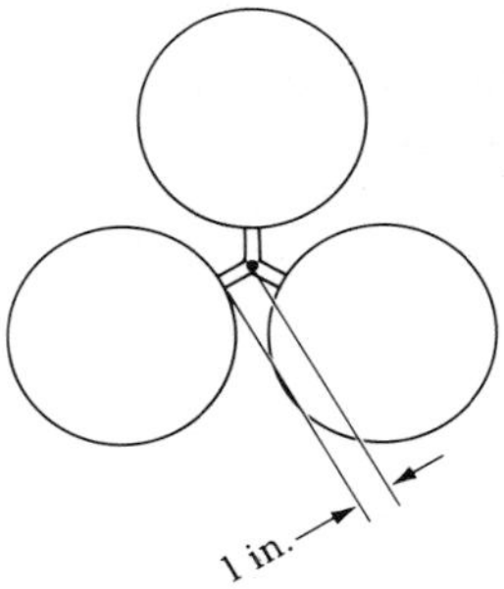

Figure P7.83c

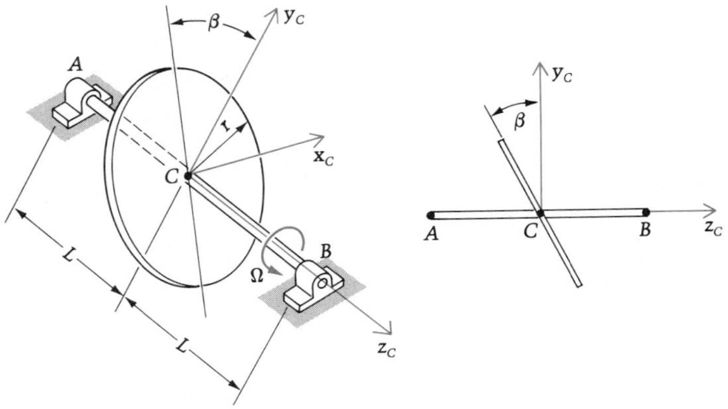

Figure P7.84

7.86 Show that if the solution $\boldsymbol{\omega}(t)$ of Example 7.9 is projected into the **xy** plane, the tip of the projection vector travels on a circle of radius $\sqrt{\omega_{x_i}^2 + \omega_{y_i}^2}$ at the frequency $(J - I)\omega_{z_i}/I$.

7.87 In Example 7.11 define the (x, y, z) axes at O as principal for $\mathcal{B}$, with associated respective moments of inertia $\bar{I}$, $\bar{J}$, and $\bar{K}$. Rework the problem without assuming that $\mathcal{B}$ has negligible mass. The two sets of axes are respectively parallel prior to the application of M_O.

7.88 A single-engine aircraft has a two-bladed propeller weighing 128 lb with a radius of gyration about its center of mass of 3 ft. It rotates counterclockwise at 2000 rpm when viewed from the rear. Find the gyroscopic moment on the propeller shaft when the plane is at the bottom of a vertical loop of 2000-ft radius with a speed of 500 mph. In which direction will the tail of the plane tend to move because of this moment?

7.89 Extend Problem 6.65 to show that the described motion is only possible if

$$3hn^2 \sin\alpha \cos\alpha(1 + 5\cos^2\alpha) < 5g(4 - 3\cos^2\alpha)$$

7.90 A top steadily precesses about the fixed direction Z at 60 rpm. (See Figure P7.90.) Treating the top as a cone of radius 1.2 in. and height 2.0 in., find the rate of spin $\dot{\psi}$ of the top about its axis of symmetry.

7.91 Using the fact that the sum of any two moments of inertia at a point is always larger than the third, show that for a torque-free axisymmetric body undergoing retrograde precession, $\dot{\phi} \geq 2|\dot{\psi}|$ and that the z axis of the body is always outside the space cone.

7.92 Find the angular acceleration in $\mathcal{I}$ of the gyroscope in the case of steady precession.

7.93 Show that if a rigid body $\mathcal{B}$ undergoing torque-free motion in an inertial frame $\mathcal{I}$ has three equal principal moments of inertia at its mass center, then its angular velocity is constant in $\mathcal{I}$.

7.94 Cone $\mathcal{C}$ in Figure P7.94 has radius 0.2 m and height 0.5 m. It is precessing about the vertical axis through the ball joint, in the direction shown, at the rate of $\dot{\phi} =$ 0.5 rad/s. If the angle θ is observed to be 20° and unchanging, what must be the rate of spin $\dot{\psi}$ of the cone?

7.95 In the preceding problem, suppose that $\dot{\psi}$ is given to be 400 rad/s in the same direction as given in the figure and that the cone's height H is not given. Find the value of H for which this steady precession will occur.

7.96 Show that the observer in Figure P7.96, fixed to the body cone $\mathcal{B}$ of a body undergoing torque-free motion, sees the tip of the angular velocity vector $\boldsymbol{\omega}_{\mathcal{B}/\mathcal{S}}$ moving to the right if the steady precession is direct (and to the left if retrograde). *Hint:* Express $\boldsymbol{\omega}_{\mathcal{B}/\mathcal{S}} = \dot{\varphi}\hat{\mathbf{K}} + \dot{\psi}\hat{\mathbf{k}}$, which in the intermediate frame $\mathcal{F}$ between $\mathcal{S}$ and $\mathcal{B}$ is

$$\boldsymbol{\omega}_{\mathcal{B}/\mathcal{S}} = -\dot{\varphi}\sin\theta\,\hat{\mathbf{i}} + (\dot{\varphi}\cos\theta + \dot{\psi})\hat{\mathbf{k}}$$

Then simply differentiate $\boldsymbol{\omega}_{\mathcal{B}/\mathcal{S}}$ in $\mathcal{B}$ and observe the direction of the result.

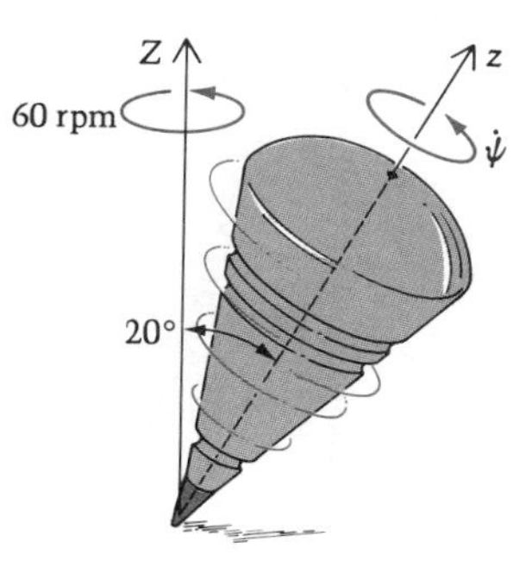

Figure P7.90

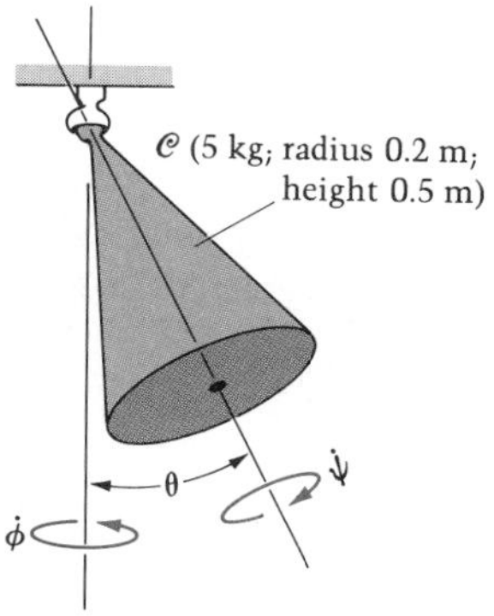

Figure P7.94

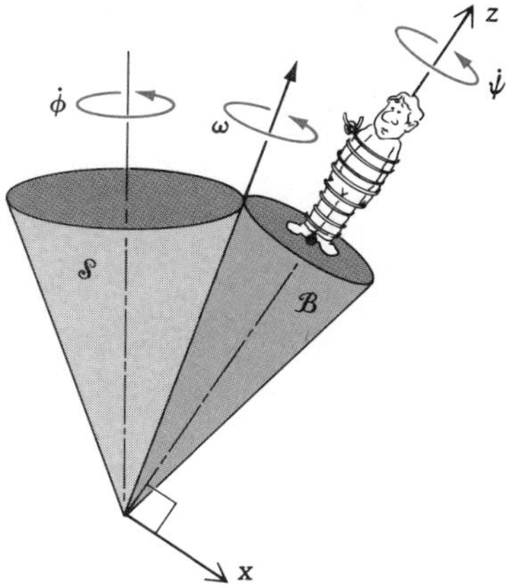

Figure P7.96

7.97 The rod $\mathcal{L}$ is rigidly attached to shaft $\mathcal{S}$ which is free to turn in the two bearings as indicated in Figure P7.97. The y and y' axes point into the page at C. Show that the moment with respect to C that must be supplied by the bearings to shaft $\mathcal{S}$ to sustain the motion must have the components:

$$M_{r_{C_{x'}}} = -\frac{m\ell^2}{12}(\sin\beta\cos\beta)\alpha$$

$$M_{r_{C_{y'}}} = -\frac{m\ell^2}{12}(\sin\beta\cos\beta)\omega^2$$

Do this in two ways: (1) Use the Euler equations (7.53) with the principal axes (x, y, z); (2) Use the Equations (7.54) with the axes (x', y', z') in the figure.

7.98 A result of the earth's bulge is that $I/J = 0.997$. Use this result to compute the period of one revolution of the earth's angular velocity vector (North Pole!) about its axis of symmetry. (The answer, obtained by Euler in 1752, is about 4 months less than the actual period first observed by S. Chandler in 1891. The difference is attributed to the nonrigidity of the earth. Although energy dissipation should damp out this "wobble," in fact it does not. The ongoing cause of the wobble is an unsolved problem in geodynamics at this time. See *Science News*, 24 October 1981.)

7.99 The spinning top (Figure P7.99) is another example of a gyroscope. Show that if the top's peg is not moving across the floor, the condition for steady precession is given by

$$mgd = J\dot{\psi}\dot{\phi} + (J - I)\dot{\phi}^2\cos\theta$$

7.100 The disk in Figure P7.100 is spinning about the light axle, and the axle is precessing at the constant rate of $32\hat{\mathbf{i}}$ rad/s. If the axle is observed to remain horizontal, find the magnitude and direction of the spin of the disk.

7.101 The equilateral triangular dinner bell in Figure P7.101 is struck with a horizontal force in the y direction that imparts an impulse $F\Delta t\hat{\mathbf{j}}$ to the bell. Find the angular velocity of the bell immediately after the blow is struck. Is the answer the same if the bell is an equilateral triangular *plate* of the same mass? Why or why not?

7.102 In the preceding problem, suppose the hammer is replaced by a bullet of mass m and speed v_b that rebounds straight back with a coefficient of restitution $e = 0.1$. Determine the resulting angular velocity of the bell.

7.103 Repeat Problem 7.101, but this time suppose the bell hangs from a string instead of from a ball and socket joint.

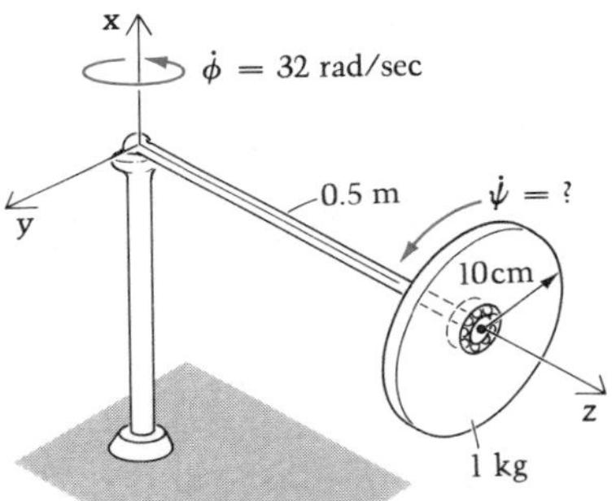

Figure P7.100

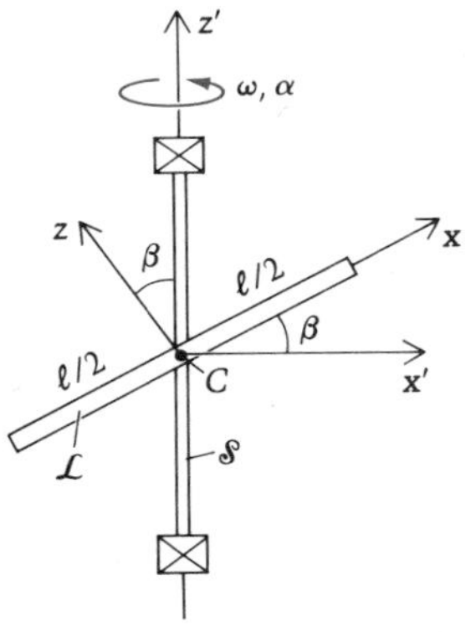

Figure P7.97

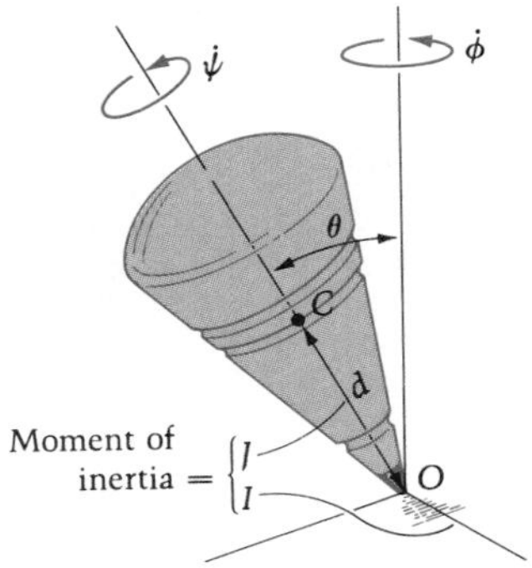

Figure P7.99

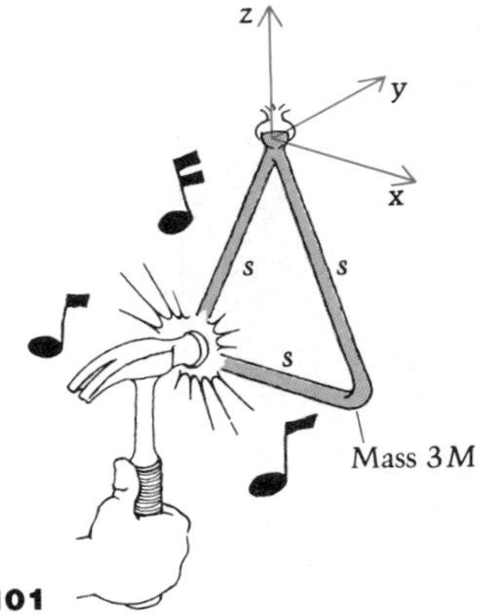

Figure P7.101

7.104 Bend a coat hanger or pipe cleaner into the shape of the bent bar of Example 7.15. Drop it onto the edge of a table as in the example and observe that the angular velocity direction following impact agrees with the results of the example.

7.105 The graph in Figure P7.105a depicts the stability of symmetrical satellites spinning about the axis z_C normal to the orbital plane. The abscissa is the ratio of I_{z_C} to the moment of inertia I_ℓ about any lateral axis (they are all the same for what is called a "symmetrical" satellite—it need not be *physically* symmetric about z_C). The ordinate is the ratio of the spin speed ω_s (about z_C) *in* the orbit to the orbital angular speed ω_o.

a. For a satellite equivalent to four solid cylinders each of mass m, radius R, and height $3R$, find I_{z_C} and I_ℓ. The distance from C to any cylinder's center is $2R$, and the connecting cross is light. The cylinders' axes are normal to the orbital plane. (See Figure 7.105b.)

b. Determine whether the station is stable for the following cases:

i. The station's orientation is fixed in inertial space.

ii. The station travels around the earth as the moon does.

iii. The station has twice the angular velocity of the orbiting frame.

iv. The same as (iii), but the spin is opposite in direction to the orbital angular speed.

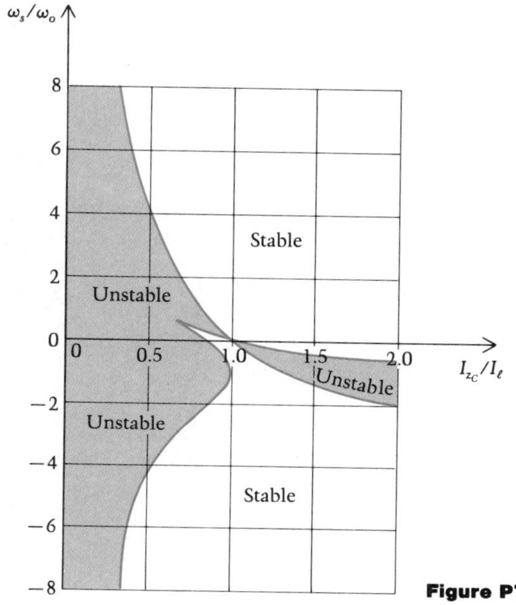

Figure P7.105a

The following five problems are advanced looks at statics of rigid bodies that depend on our study of dynamics.

7.106 It is possible for a spinning top to "sleep," meaning that its axis remains vertical and its peg stationary as it spins on a floor. (See Figure P7.106.) In the absence of a friction couple about the axis of the top, note that the spin speed ω_z is constant and that the equations of motion then reduce to $\mathbf{M}_{r_C} = \mathbf{0}$ and $\mathbf{F}_r = \mathbf{0}$. It thus follows that $\mathbf{M}_{r_O}$ is also zero. None of the particles of the top not on the z_C axis are in equilibrium, however, because they all have (inward) accelerations $r\omega^2$. A body is in equilibrium if and only if all its particles are in equilibrium, so the sleeping top cannot be in equilibrium. Explain this statement in light of $\mathbf{F}_r = \mathbf{0}$ and $\mathbf{M}_{r_O} = \mathbf{0}$, which were the equilibrium equations for a rigid body in statics. *Hint:* If O is a fixed point of rigid body $\mathcal{B}$ in an inertial frame $\mathcal{I}$, then

$$\mathbf{M}_{r_O} = {}^{\mathcal{I}}\dot{\mathbf{H}}_O = {}^{\mathcal{I}}\frac{d}{dt}(\mathbf{H}_C + \mathbf{r}_{OC} \times \mathbf{L})$$

$$= {}^{\mathcal{B}}\dot{\mathbf{H}}_C + \boldsymbol{\omega}_{\mathcal{B}/\mathcal{I}} \times \mathbf{H}_C + \mathbf{r}_{OC} \times \underbrace{m\mathbf{a}_C}_{\mathbf{F}_r}$$

Therefore show that just because $\mathbf{F}_r = \mathbf{0}$ and $\mathbf{M}_{r_O} = \mathbf{0}$, $\boldsymbol{\omega}_{\mathcal{B}/\mathcal{I}}$ need not be zero. Use the top as a counterexample and explain why the first two terms on the right side of the preceding equation vanish. Thus $\mathbf{F}_r = \mathbf{M}_{r_O} = \mathbf{0}$ are necessary but not *sufficient* conditions for equilibrium of a rigid body.

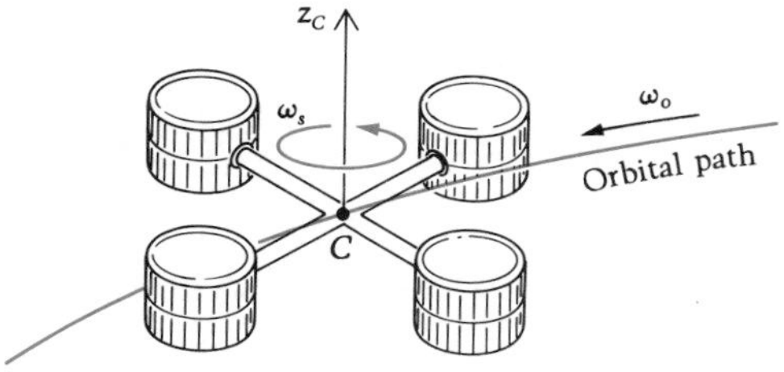

Figure P7.105b

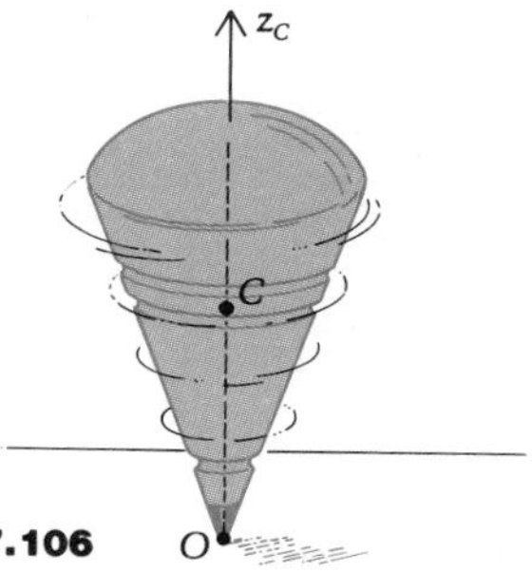

Figure P7.106

7.107 In the sleeping top counterexample of Problem 7.106, the terms ${}^{\mathcal{B}}\dot{\mathbf{H}}_C$ and $\boldsymbol{\omega}_{\mathcal{B}/\mathcal{I}} \times \mathbf{H}_C$ both vanish independently. Show that there are more complicated counterexamples in which $\boldsymbol{\omega}_{\mathcal{B}/\mathcal{I}}$ is not constant in direction in $\mathcal{B}$ or $\mathcal{I}$ and in which the two terms *add* to zero. *Hint:* $\mathbf{M}_{r_O} - \mathbf{r}_{OC} \times \mathbf{F}_r = \mathbf{M}_{r_C}$. What is $\mathbf{M}_{r_C}$ for the torque-free body?

7.108 Show that if $\boldsymbol{\omega}_{\mathcal{B}/\mathcal{I}} = \mathbf{0}$ at all times, then so is $\mathbf{M}_{r_C}$. Is the converse true?

7.109 If a frame $\mathcal{B}$ is moving relative to an inertial frame $\mathcal{I}$, it can be shown that $\mathcal{B}$ is also an inertial frame if and only if $\boldsymbol{\omega}_{\mathcal{B}/\mathcal{I}} = \mathbf{0}$ at all times *and* the acceleration in $\mathcal{I}$ of at least one point of $\mathcal{B}$ is zero at all times. Use this theorem to show that if a rigid body $\mathcal{B}$ is in equilibrium in an inertial frame $\mathcal{I}$, then $\mathcal{B}$ is *itself* an inertial frame. Is the converse true?

7.110 Show that a rigid body $\mathcal{B}$ is in equilibrium in an inertial frame $\mathcal{I}$ if and only if (a) at least one point of $\mathcal{B}$ is fixed in $\mathcal{I}$ and (b) $\boldsymbol{\omega}_{\mathcal{B}/\mathcal{I}} = \mathbf{0}$ at all times. What is the minimum number of constraints on $\mathcal{B}$ that will satisfy (a) and (b)? Describe one set of physical constraints that will assure equilibrium.

7.6 Work and Kinetic Energy in General Motion

A special integral of the equations of motion of a rigid body $\mathcal{B}$ yields a relationship between the work of the external forces (and/or couples) and the change in the kinetic energy of $\mathcal{B}$. To develop this relation, we must first explore expressions for the kinetic energy of the rigid body. **Kinetic energy** is usually denoted by the letter T and is defined by

$$T = \frac{1}{2}\int \mathbf{v} \cdot \mathbf{v}\, dm \tag{7.78}$$

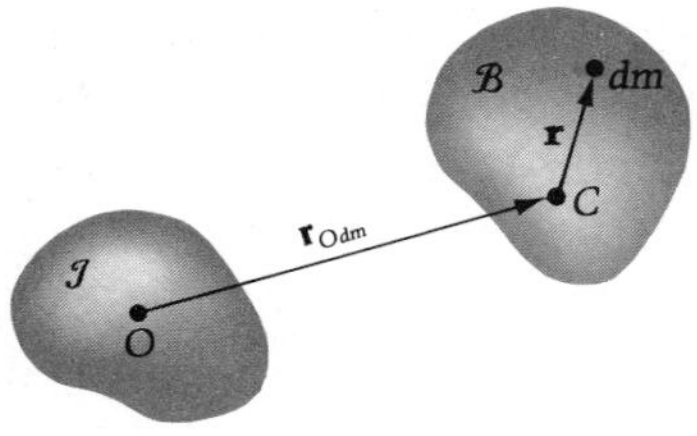

Figure 7.16

in which $\mathbf{v}$ is the derivative of the position vector from O (fixed point in the inertial frame $\mathcal{I}$ in Figure 7.16) to the differential mass element dm. In this section all time derivatives, velocities, and angular velocities are taken in $\mathcal{I}$ unless otherwise specified.

Since $\mathcal{B}$ is a rigid body, we may relate $\mathbf{v}$ to the velocity $\mathbf{v}_C$ of the mass center C of $\mathcal{B}$:

$$\mathbf{v} = \mathbf{v}_C + \boldsymbol{\omega} \times \mathbf{r} \tag{7.79}$$

in which $\boldsymbol{\omega}$ is $\boldsymbol{\omega}_{\mathcal{B}/\mathcal{I}}$ and $\mathbf{r}$ is the position vector from C to dm as shown in Figure 7.16. Substituting Equation (7.79) into (7.78), we get

$$T = \frac{1}{2}\mathbf{v}_C \cdot \mathbf{v}_C \int_{\mathcal{B}} dm + \frac{1}{2}\int_{\mathcal{B}} (\boldsymbol{\omega} \times \mathbf{r}) \cdot (\boldsymbol{\omega} \times \mathbf{r})\, dm + \mathbf{v}_C \cdot \left[\boldsymbol{\omega} \times \int_{\mathcal{B}} \mathbf{r}\, dm\right] \tag{7.80}$$

where $\mathbf{v}_C$ and $\boldsymbol{\omega}$ do not vary over the body's volume and can thus be taken outside the integrals. The integral in the last term is zero by virtue of the definition of the mass center:

$$\int_{\mathcal{B}} \mathbf{r}\, dm = m\mathbf{r}_{CC} = \mathbf{0} \tag{7.81}$$

The integral in the first term on the right side of Equation (7.80) is of course the mass m of $\mathcal{B}$. The integrand of the remaining term may be simplified by the vector identity:*

$$(\mathbf{A} \times \mathbf{B}) \cdot (\mathbf{C} \times \mathbf{D}) = \mathbf{A} \cdot [\mathbf{B} \times (\mathbf{C} \times \mathbf{D})] \tag{7.82}$$

Therefore Equation (7.80) becomes

$$T = \frac{m}{2}(\mathbf{v}_C \cdot \mathbf{v}_C) + \frac{1}{2}\boldsymbol{\omega} \cdot \int [\mathbf{r} \times (\boldsymbol{\omega} \times \mathbf{r})]\, dm \tag{7.83}$$

As we have already seen in Section 7.2, the integral in Equation (7.83) is the angular momentum (moment of momentum) of the body with respect to C, and thus we can write

$$T = \overbrace{\frac{m}{2}\mathbf{v}_C \cdot \mathbf{v}_C}^{T_v} + \overbrace{\frac{1}{2}\boldsymbol{\omega} \cdot \mathbf{H}_C}^{T_\omega} \tag{7.84}$$

It is seen that the kinetic energy can be represented as the sum of two terms:

1. A part $T_v = (m/2)\mathbf{v}_C \cdot \mathbf{v}_C$ that the body possesses if the mass center is in motion
2. A part $T_\omega = \frac{1}{2}\boldsymbol{\omega} \cdot \mathbf{H}_C$ that is due to the difference between the velocities of the points of $\mathcal{B}$ and the velocity of its mass center.

The term T_ω can be interpreted quite simply if at an instant we let $\boldsymbol{\omega} = \omega\hat{\mathbf{i}}$—that is, if we align the reference axis $\mathbf{x}$ with the angular velocity vector at that instant. In this case, using Equations (7.22), we obtain

$$\mathbf{H}_C = I_{xx}^C \omega \hat{\mathbf{i}} + I_{xy}^C \omega \hat{\mathbf{j}} + I_{xz}^C \omega \hat{\mathbf{k}} \tag{7.85}$$

so that

$$\frac{1}{2}\boldsymbol{\omega} \cdot \mathbf{H}_C = \frac{1}{2} I_{xx}^C \omega^2 \tag{7.86}$$

This means that the rotational part of T is *instantaneously* of the same form as it was for the plane case in Chapter 4. The difference, of course, is that the direction of the angular velocity vector $\boldsymbol{\omega}$ changes in the general (three-dimensional) case.

Suppose the body $\mathcal{B}$ has a point P with zero velocity. (This is not always the case in general motion as we have already seen in Chapter 6.) Then if $\mathbf{v}$ in Equation (7.78) is replaced by $\mathbf{v}_P + \boldsymbol{\omega} \times \mathbf{r}' = \boldsymbol{\omega} \times \mathbf{r}'$, where $\mathbf{r}'$ extends from P to the mass element dm, we obtain

$$T = \int (\boldsymbol{\omega} \times \mathbf{r}') \cdot (\boldsymbol{\omega} \times \mathbf{r}')\, dm \tag{7.87}$$

*Which is nothing more than interchanging the dot and cross of the scalar triple product $(\mathbf{A} \times \mathbf{B}) \cdot \mathbf{E}$, where $\mathbf{E}$ is the vector $\mathbf{C} \times \mathbf{D}$.

The identical steps that produced the second term of Equation (7.84) from the middle term of (7.80) then give

$$T = \frac{1}{2}\boldsymbol{\omega} \cdot \mathbf{H}_P \tag{7.88}$$

and the two terms of Equation (7.84) have collapsed into one if $\mathbf{H}$ is expressed relative to a point of zero velocity instead of C.

In Sections 2.4 and 5.2 we demonstrated one work and kinetic energy principle that remains true for the general case. This result came from integrating $\mathbf{F}_r = m\mathbf{a}_C$:

$$\int_{t_1}^{t_2} \mathbf{F}_r \cdot \mathbf{v}_C\, dt = \frac{1}{2} m|\mathbf{v}_C(t_2)|^2 - \frac{1}{2} m|\mathbf{v}_C(t_1)|^2$$

$$= \frac{1}{2} m(v_{C_2}^2 - v_{C_1}^2) \tag{7.89}$$

A second principle will now be deduced from the moment equation*

$$\mathbf{M}_{r_C} = \dot{\mathbf{H}}_C \tag{7.90}$$

but first we need to show that

$$\dot{\boldsymbol{\omega}} \cdot \mathbf{H}_C = \boldsymbol{\omega} \cdot \dot{\mathbf{H}}_C$$

To do this, we first recall that

$$\mathbf{H}_C = \int [\mathbf{r} \times (\boldsymbol{\omega} \times \mathbf{r})]\, dm \tag{7.91}$$

If ${}^{\mathcal{B}}\dot{\mathbf{H}}_C$ is the derivative of $\mathbf{H}_C$ taken in the body $\mathcal{B}$, then the derivative relative to the inertial frame can be written

$$\dot{\mathbf{H}}_C = {}^{\mathcal{B}}\dot{\mathbf{H}}_C + \boldsymbol{\omega} \times \mathbf{H}_C \tag{7.92}$$

Dotting $\boldsymbol{\omega}$ with both sides of Equation (7.92) shows that

$$\boldsymbol{\omega} \cdot \dot{\mathbf{H}}_C = \boldsymbol{\omega} \cdot {}^{\mathcal{B}}\dot{\mathbf{H}}_C \tag{7.93}$$

and since $\mathbf{r}$ is constant in time relative to body $\mathcal{B}$, we can differentiate Equation (7.91) there and obtain

$${}^{\mathcal{B}}\dot{\mathbf{H}}_C = \int \mathbf{r} \times (\dot{\boldsymbol{\omega}} \times \mathbf{r})\, dm \tag{7.94}$$

In Equation (7.94) we have used the property of $\boldsymbol{\omega}$ that its derivatives in $\mathcal{I}$ and $\mathcal{B}$ are the same; that is,

$${}^{\mathcal{I}}\dot{\boldsymbol{\omega}} = {}^{\mathcal{I}}\dot{\boldsymbol{\omega}}_{\mathcal{B}/\mathcal{I}} = {}^{\mathcal{B}}\dot{\boldsymbol{\omega}}_{\mathcal{B}/\mathcal{I}} + \boldsymbol{\omega}_{\mathcal{B}/\mathcal{I}} \times \boldsymbol{\omega}_{\mathcal{B}/\mathcal{I}} = {}^{\mathcal{B}}\dot{\boldsymbol{\omega}}_{\mathcal{B}/\mathcal{I}} = {}^{\mathcal{B}}\dot{\boldsymbol{\omega}}$$

Substituting Equation (7.94) into (7.93) then gives

*Derivatives such as $\dot{\boldsymbol{\omega}}$ are taken in the inertial frame $\mathcal{I}$ in this section unless the letter $\mathcal{B}$ appears beside the dot, in which case the derivative is taken in the body.

$$\begin{aligned}\boldsymbol{\omega} \cdot \dot{\mathbf{H}}_C &= \boldsymbol{\omega} \cdot \int \mathbf{r} \times (\dot{\boldsymbol{\omega}} \times \mathbf{r})\, dm \\ &= \int \boldsymbol{\omega} \cdot [\mathbf{r} \times (\dot{\boldsymbol{\omega}} \times \mathbf{r})]\, dm \\ &= \int (\boldsymbol{\omega} \times \mathbf{r}) \cdot (\dot{\boldsymbol{\omega}} \times \mathbf{r})\, dm \\ &= \int [(\dot{\boldsymbol{\omega}} \times \mathbf{r}) \cdot (\boldsymbol{\omega} \times \mathbf{r})]\, dm \\ &= \dot{\boldsymbol{\omega}} \cdot \int \mathbf{r} \times (\boldsymbol{\omega} \times \mathbf{r})\, dm\end{aligned}$$

Hence

$$\boldsymbol{\omega} \cdot \dot{\mathbf{H}}_C = \dot{\boldsymbol{\omega}} \cdot \mathbf{H}_C \tag{7.95}$$

We are now in a position to observe that

$$\boldsymbol{\omega} \cdot \mathbf{M}_{r_C} = \boldsymbol{\omega} \cdot \dot{\mathbf{H}}_C = \frac{d}{dt}\left(\frac{\boldsymbol{\omega} \cdot \mathbf{H}_C}{2}\right) \tag{7.96}$$

Integrating Equation (7.96), we have

$$\int_{t_1}^{t_2} \mathbf{M}_{r_C} \cdot \boldsymbol{\omega}\, dt = \frac{1}{2}\boldsymbol{\omega}(t_2) \cdot \mathbf{H}_C(t_2) - \frac{1}{2}\boldsymbol{\omega}(t_1) \cdot \mathbf{H}_C(t_1) \tag{7.97}$$

Note that the right sides of Equations (7.89) and (7.97) each represents the change, occurring in the time interval $t_1 \le t \le t_2$, of part of the kinetic energy of the body. The left sides of these equations are usually called a form of *work*.

While the relationships between work and kinetic energy that have been developed are important, another relationship that combines them is often more useful. We can differentiate Equation (7.84) and get

$$\frac{dT}{dt} = m\mathbf{v}_C \cdot \mathbf{a}_C + \frac{1}{2}\dot{\boldsymbol{\omega}} \cdot \mathbf{H}_C + \frac{1}{2}\boldsymbol{\omega} \cdot \dot{\mathbf{H}}_C$$

Using Euler's laws and (7.95), this may be put into the form

$$\frac{dT}{dt} = \mathbf{F}_r \cdot \mathbf{v}_C + \mathbf{M}_{r_C} \cdot \boldsymbol{\omega} \tag{7.98}$$

If we now let $\mathbf{F}_1, \mathbf{F}_2, \ldots$ represent the external forces acting on the body, and $\mathbf{C}_1, \mathbf{C}_2, \ldots$ represent the moments of the external couples, then

$$\mathbf{F}_r = \mathbf{F}_1 + \mathbf{F}_2 + \cdots \tag{7.99a}$$

$$\mathbf{M}_{r_C} = \mathbf{r}_1 \times \mathbf{F}_1 + \mathbf{r}_2 \times \mathbf{F}_2 + \cdots + \mathbf{C}_1 + \mathbf{C}_2 + \cdots \tag{7.99b}$$

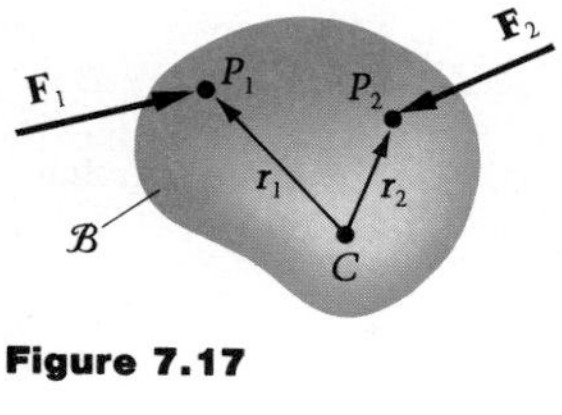

Figure 7.17

where $P_1, P_2, \ldots$ are the points of $\mathcal{B}$ where $\mathbf{F}_1, \mathbf{F}_2, \ldots$ are respectively applied and where $\mathbf{r}_1 = \mathbf{r}_{CP_1}$, $\mathbf{r}_2 = \mathbf{r}_{CP_2}$, and so forth, as shown in Figure 7.17. Recall from statics that a couple has the same moment about any point in space, so that the $\mathbf{C}_i$'s are simply added into the moment equation (7.99b).

Substituting Equations (7.99) into (7.98), we obtain

$$\frac{dT}{dt} = \mathbf{F}_1 \cdot \mathbf{v}_C + \mathbf{F}_2 \cdot \mathbf{v}_C + \cdots + \boldsymbol{\omega} \cdot (\mathbf{r}_1 \times \mathbf{F}_1) + \boldsymbol{\omega} \cdot (\mathbf{r}_2 \times \mathbf{F}_2) + \cdots + \boldsymbol{\omega} \cdot \mathbf{C}_1 + \boldsymbol{\omega} \cdot \mathbf{C}_2 + \cdots \tag{7.100}$$

However,

$$\boldsymbol{\omega} \cdot (\mathbf{r}_i \times \mathbf{F}_i) = \mathbf{F}_i \cdot (\boldsymbol{\omega} \times \mathbf{r}_i) \qquad (i = 1, 2, \ldots)$$

so that

$$\frac{dT}{dt} = \mathbf{F}_1 \cdot (\mathbf{v}_C + \boldsymbol{\omega} \times \mathbf{r}_1) + \mathbf{F}_2 \cdot (\mathbf{v}_C + \boldsymbol{\omega} \times \mathbf{r}_2) + \cdots + \boldsymbol{\omega} \cdot \mathbf{C}_1 + \boldsymbol{\omega} \cdot \mathbf{C}_2 + \cdots \tag{7.101}$$

We note that $\mathbf{v}_C + \boldsymbol{\omega} \times \mathbf{r}_1$ is the velocity of point P_1, the point of application of $\mathbf{F}_1$. Therefore

$$\frac{dT}{dt} = \dot{T} = \Sigma\, \mathbf{F}_i \cdot \mathbf{v}_i + \boldsymbol{\omega} \cdot \Sigma \mathbf{C}_i \tag{7.102}$$

Equation (7.102) leads us to define the **power,** or **rate of work,** as follows:

$$\begin{aligned} &\text{Power (rate of work) of a force } (\mathbf{F}_1) = \mathbf{F}_1 \cdot \mathbf{v}_1 \\ &\text{Power (rate of work) of a couple } (\mathbf{C}_1) = \boldsymbol{\omega} \cdot \mathbf{C}_1 \end{aligned} \tag{7.103}$$

Therefore

$$\dot{T} = \text{rate of work of external forces and couples}$$

Integrating Equation (7.102), we get

$$\int_{t_1}^{t_2} (\text{rate of work})\, dt = T(t_2) - T(t_1) = \Delta T \tag{7.104}$$

The integral on the left side of Equation (7.104) is called the **work** done on $\mathcal{B}$ between t_1 and t_2 by the external forces and couples. Hence

$$\text{Work} = \int (\Sigma \mathbf{F}_i \cdot \mathbf{v}_i + \boldsymbol{\omega} \cdot \Sigma \mathbf{C}_i)\, dt = \Delta T \tag{7.105}$$

That is, the work done on $\mathcal{B}$ equals its change in kinetic energy.

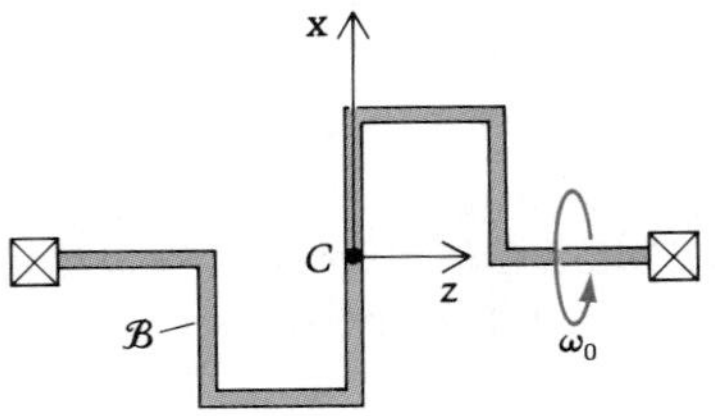

EXAMPLE 7.17

Find the work done on the bent bar of Examples 4.13, 4.27 by a motor that brings it up to speed ω_0 from rest. (See the diagram.)

(Continued)

SOLUTION

The mass center C does not move, so that Equation (7.84) gives, in this case,

$$T_f = \frac{1}{2}\boldsymbol{\omega} \cdot \mathbf{H}_C \tag{1}$$

Since $\boldsymbol{\omega}$ has only a $\hat{\mathbf{k}}$ component, $\omega_0\hat{\mathbf{k}}$, we may substitute Equation (7.22) into (1) and get

$$T_f = \frac{1}{2}\omega_0 (I^C_{xz}\overset{0}{\cancel{\omega_x}} + I^C_{yz}\overset{0}{\cancel{\omega_y}} + I^C_{zz}\overset{\omega_0}{\cancel{\omega_z}}) \tag{2}$$

We note that even though I^C_{xz} is not zero, it has no effect on the kinetic energy of $\mathcal{B}$ since it is multiplied by ω_x, which is forced to vanish by the bearings aligned with z.

Thus the work done by the motor on $\mathcal{B}$ is given simply by Equation (7.105):

$$\begin{aligned} W = \Delta T &= T_f - \overset{0}{\cancel{T_i}} \\ &= \frac{1}{2} I^C_{zz}\omega_0^2 \\ &= \frac{5}{3} m\ell^2\omega_0^2 \end{aligned} \tag{3}$$

where $I^C_{zz} = (10/3)m\ell^2$ from Example 7.15. The motor would, of course, have to do additional work besides that given by (3) to overcome its own armature inertia, bearing and belt friction, and air resistance.

We now consider an example in three dimensions in which the products of inertia do play a role in the kinetic energy calculation.

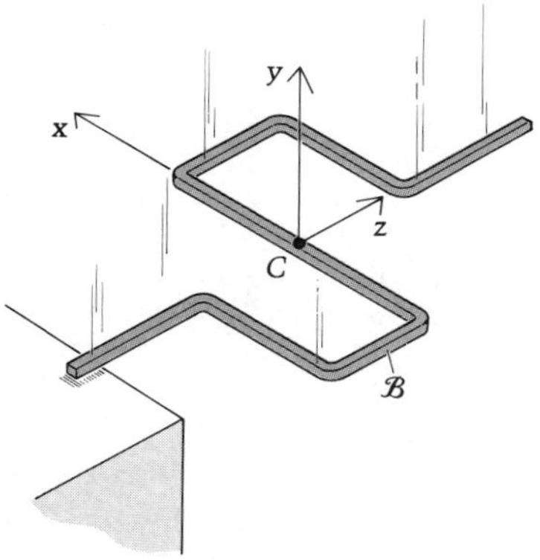

EXAMPLE 7.18

Find the kinetic energy lost by the bent bar of Example 7.15 when it strikes the table top as shown in the figure.

(Continued)

SOLUTION

During the impact with the table top, the bodies do not behave rigidly. The kinetic energy lost by bar $\mathcal{B}$ is transformed into noise, heat, vibration, and both elastic and permanent deformation. In Example 7.15 we found $\mathbf{v}_C$ and $\boldsymbol{\omega}_f$ just before and after impact; we now use these vectors to find the kinetic energy lost by $\mathcal{B}$. Just after impact we have

$$T_f = \frac{1}{2} m v_{C_f}^2 + \frac{1}{2} \boldsymbol{\omega}_f \cdot \mathbf{H}_{C_f}$$

The term $\boldsymbol{\omega}_f \cdot \mathbf{H}_{C_f}$ can be written just after impact, using Equation (7.22), as follows. (Note that ω_y and two of the products of inertia are zero here.)

$$\begin{aligned}
\boldsymbol{\omega}_f \cdot \mathbf{H}_{C_f} &= \omega_x(I_{xx}^C \omega_x + \cancelto{0}{I_{xy}^C} \cancelto{0}{\omega_y} + I_{xz}^C \omega_z) \\
&\quad + \omega_y(\cancelto{0}{I_{yx}^C} \omega_x + I_{yy}^C \cancelto{0}{\omega_y} + \cancelto{0}{I_{yz}^C} \omega_z) \\
&\quad + \omega_z(I_{zx}^C \omega_x + \cancelto{0}{I_{zy}^C} \cancelto{0}{\omega_y} + I_{zz}^C \omega_z) \\
&= I_{xx}^C \omega_x^2 + 2I_{xz}\omega_x\omega_z + I_{zz}^C \omega_z^2
\end{aligned}$$

Using this result and Equations (11) and (12) from Example 7.15, we obtain

$$\begin{aligned}
T_f = \frac{1}{2}(8m)\left[\frac{(23e - 120)\sqrt{2gh}}{143}\right]^2 &+ \frac{1}{2}\left\{\left[\frac{60(1+e)\sqrt{2gh}}{143\ell}\right]^2 \frac{22}{3} m\ell^2\right. \\
&+ 2\left[\frac{60(1+e)\sqrt{2gh}}{143\ell}\right]\left[\frac{36(1+e)\sqrt{2gh}}{143\ell}\right](-2m\ell^2) \\
&\left.+ \left[\frac{36(1+e)\sqrt{2gh}}{143\ell}\right]^2 \frac{10}{3} m\ell^2\right\}
\end{aligned}$$

which, after simplification, equals

$$T_f = mgh(1.29e^2 + 6.71)$$

The initial kinetic energy (just prior to the collision) was

$$T_i = \frac{1}{2}(8m)(\sqrt{2gh})^2 = 8mgh$$

Thus the change in kinetic energy of the bent bar is given by

$$\Delta T = T_f - T_i = mgh(1.29e^2 - 1.29)$$

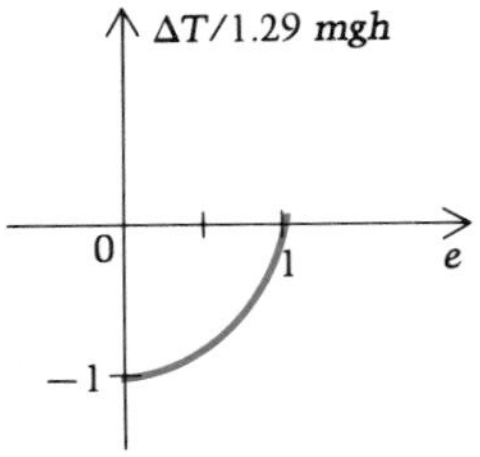

We see that if $e = 1$ (purely elastic collision), no loss in kinetic energy occurs and hence no work is done in changing T. The energy lost is seen in the figure to vary quadratically, with a maximum percentage loss (when $e = 0$) of

$$\frac{1.29mgh}{8mgh} \cdot 100 = 16.1\%$$

in this case. Because the point of striking is the end of the bar, 83.9 percent of the kinetic energy is retained. If the mass center of the bar were the point that struck the table, however, *all* the kinetic energy would have been lost if $e = 0$.

PROBLEMS / Section 7.6

7.111 A thin equilateral triangular plate $\mathcal{P}$ of side s is welded to the vertical shaft at A in Figure P7.111. The shaft is brought up to speed ω_0 from rest by a motor.

- **a.** How much work is done in bringing the system up to speed?
- **b.** Find the force and couple system acting on the plate at A after it is turning at the speed ω_0 and the motor is turned off.

7.112 Find the kinetic energy of the wagon wheel in Problem 6.42 and use it to deduce the work done by the boy in getting it up to its final speed from rest.

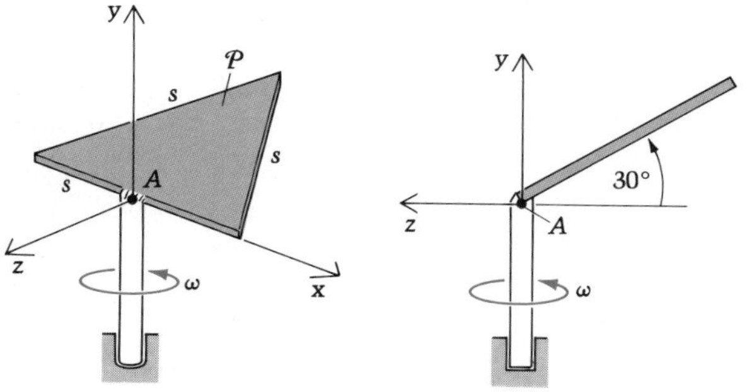

Figure P7.111

7.113 Find the kinetic energy of disk $\mathcal{B}$ in Problem 6.28.

7.114 A thin rectangular plate (Figure P7.114) is brought up from rest to speed ω_0 about a horizontal axis Y.

- **a.** Find the work that is done.
- **b.** If two concentrated masses of $m/2$ each are added on the $\mathbf{x}_C$ axis, one on each side of the mass center, find their distances d from the mass center that will eliminate the bearing reactions.

7.115 The center of mass C of a gyroscope $\mathcal{G}$ is fixed. Show that the kinetic energy of $\mathcal{G}$ is

$$\tfrac{1}{2}A(\dot{\theta}^2 + \dot{\phi}^2 \sin^2 \theta) + \tfrac{1}{2}C(\dot{\phi} \cos \theta + \dot{\psi})^2$$

where ϕ, θ, ψ are the Eulerian angles and A, A, C are the principal moments of inertia of $\mathcal{G}$ at C.

7.116 A ring is welded to a rod at a point A as shown in Figure P7.116. The cross sections and densities of the rod and ring are the same. The combined body is released with a gentle nudge with end B of the rod connected to the smooth plane by a ball joint and with point A at its highest point as shown. At the instant when A reaches its lowest point, find the relationship between the horizontal and vertical angular velocity components of the body.

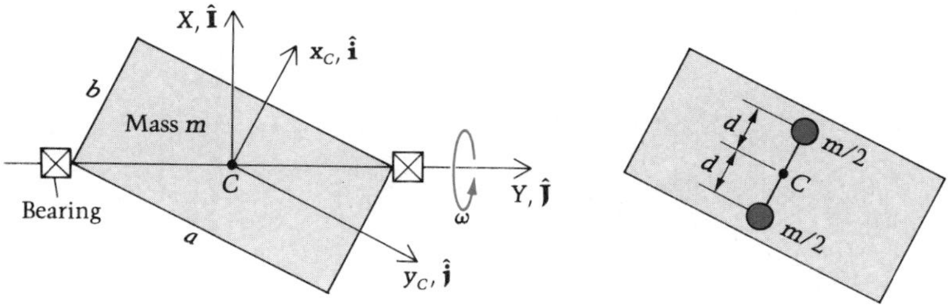

Figure P7.114

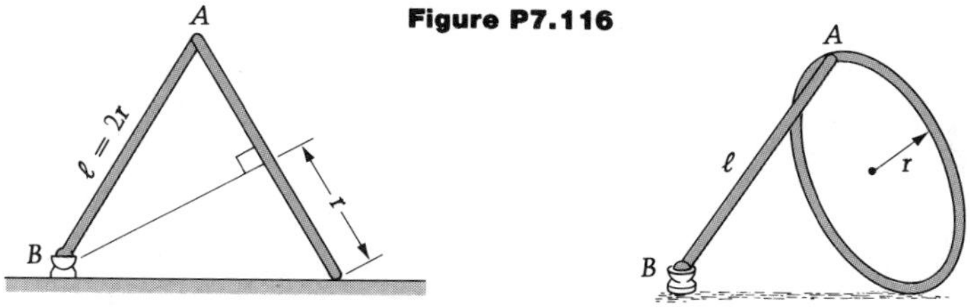

Figure P7.116

7.117 If in the preceding problem the plane is rough enough to prevent slipping, find the magnitude of the angular velocity when A reaches the floor.

7.118 Figure P7.118 shows a thin homogeneous triangular plate of mass m, base a, and height $2a$. It is welded to a light axle that can turn freely in bearings at A and B. Given:

$$I_{xx}^A = \frac{2ma^2}{3} \qquad I_{yy}^A = \frac{ma^2}{6} \qquad I_{xz}^A = I_{yz}^A = 0$$

$$I_{zz}^A = \frac{5ma^2}{6} \qquad I_{xy}^A = \frac{-ma^2}{6}$$

a. If the plate is turning at constant angular speed ω, find the torque that must be applied to the axle, and find the dynamic bearing reactions. .

b. Find the principal axes at A and the principal moments of inertia there. Draw the axes on a sketch.

c. Give the radius of a hole that, when drilled at C, will eliminate the bearing reactions. Give the answer in terms of m and ρt (density times thickness) of the plate.

d. Find the work done in bringing the plate up to speed ω from rest.

7.119 Find the kinetic energy of the grinder in Problem 7.78. Is this equal to the work done by a motor on $\mathcal{S}$ which brings the system up to speed? (Neglect the masses of $\mathcal{S}$ and $\mathcal{B}$.)

7.120 Two concentrated masses $m_1 = 10$ kg and $m_2 = 20$ kg are connected by a 15-kg slender rod m_3. As shown in Figure P7.120, $(\hat{\mathbf{I}}, \hat{\mathbf{J}}, \hat{\mathbf{K}})$ are unit vectors fixed in direction in the inertial frame $\mathcal{I}$ and $(\hat{\mathbf{i}}, \hat{\mathbf{j}}, \hat{\mathbf{k}})$ are parallel to principal axes fixed at C in the combined body. At two times t_1 and t_2, the velocities of C and the angular velocities of the combined body are

$$\mathbf{v}_C(t_1) = \hat{\mathbf{I}} + 2\hat{\mathbf{J}} \text{ m/s} \qquad \boldsymbol{\omega}(t_1) = \hat{\mathbf{i}} + 2\hat{\mathbf{j}} - 4\hat{\mathbf{k}} \text{ rad/s}$$

$$\mathbf{v}_C(t_2) = 3\hat{\mathbf{J}} - 4\hat{\mathbf{K}} \text{ m/s} \qquad \boldsymbol{\omega}(t_2) = 3\hat{\mathbf{j}} - \hat{\mathbf{k}} \text{ rad/s}$$

Find the total work done on the system between t_1 and t_2.

7.121 A disk $\mathcal{D}$ of mass 10 kg and radius 25 cm is welded at a 45° angle to a vertical shaft $\mathcal{S}$. (See Figure P7.121.) The shaft is then spun up from rest to a constant angular speed $\omega_f = 10$ rad/s.

a. How much work is done in bringing the assembly up to speed?

b. Find the force and couple system acting on the plate at C after it is turning at the constant speed ω_f.

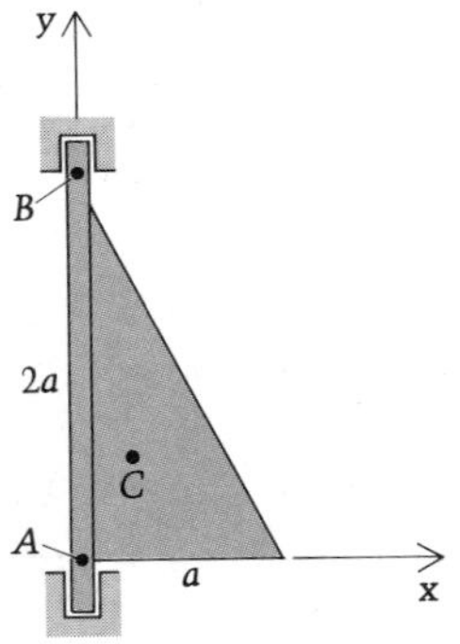

Figure P7.118

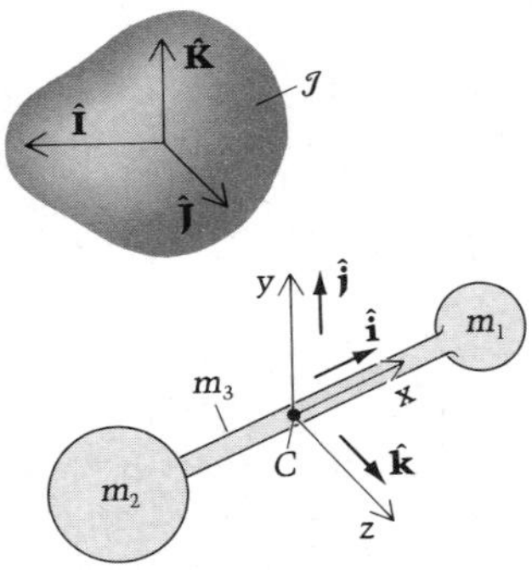

Figure P7.120

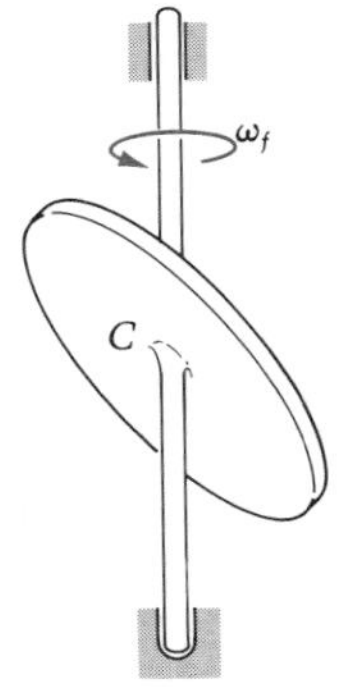

Figure P7.121

Answers to Questions / Chapter 7

Q7.1 It has been assumed that the strength of the gravitational field is constant, but of course gravity decreases as C moves farther and farther from the earth's surface.

Q7.2 Because of the earth's curvature; note of course that this assumes a perfectly spherical earth.

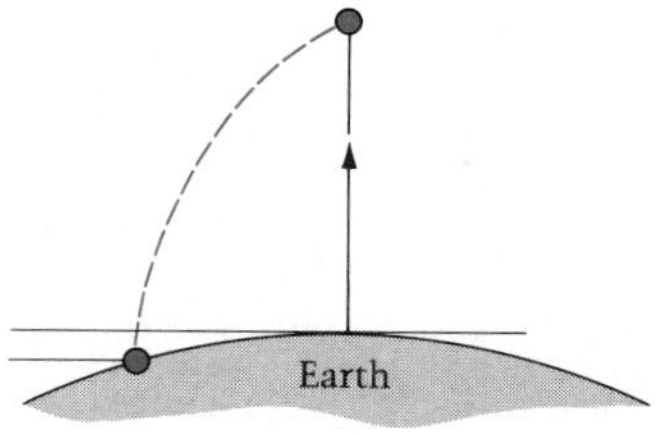

Q7.3 As we have seen in Chapter 6, $\boldsymbol{\omega}_{\mathcal{B}/\mathcal{I}}$ depends only on how a set of unit vectors, locked into $\mathcal{B}$, change their directions in $\mathcal{I}$. The angular velocity is a constant with regard to integration at a particular instant over the body's volume.

Q7.4 When we said that $n_x = +\sqrt{1/6}$, that is, took the positive square root, we chose y' to be in the direction making an acute angle with $\mathbf{x}_B$; had we chosen $-\sqrt{1/6}$, we would have gotten the opposite direction for y'. And had $I_{x'y'}$ been nonzero, the sign of the answer would have been opposite also.

Q7.5 The moment of inertia about any line ℓ through any point P other than C is larger than the moment of inertia about the line through C parallel to ℓ, by the transfer term md^2, and thus the smallest I at C is the smallest of all.

Q7.6 In $\mathcal{B}$, everything in $\mathbf{H}_C$ (Equation 7) was constant so that $\mathbf{M}_{r_C} = \boldsymbol{\omega}_{\mathcal{B}/\mathcal{I}} \times \mathbf{H}_C$. Had we used the Euler equations, there would have been derivatives of the $\boldsymbol{\omega}$ components to consider; specifically, the v_C/L component is not constant in direction if written in the frame (body) $\mathcal{D}$.

Q7.7 If f_z cannot exist in accordance with Equation (13), the motion cannot take place. The bottom of the disk will slip inward toward the axis y_O and will therefore leave the floor.

Q7.8 Because $\mathcal{B}$ can be settled onto bearings along the axis y_O and then the ball joint can be connected without vertical force at the ball joint. The small tolerance will then preclude development of such a force as time passes.

Q7.9 In deriving (7.68), if $\theta = 0$ we have divided both sides of an equation by zero. This result is then used in getting $\dot{\psi}$ in (7.71).

Review Questions / Chapter 7

True or False?

1. Products of inertia associated with principal axes always vanish, but only at the mass center.

2. If the principal moments of inertia at a point are distinct, then the principal axes of inertia associated with them are orthogonal.

3. The maximum moment of inertia about any line through point P of rigid body $\mathcal{B}$ is the largest principal moment of inertia at P.

4. General motion is a much more difficult subject than plane motion. A major reason for this is that neither the kinematics nor kinetics differential equations governing the orientation motion of the body are linear.

5. If we solve the Euler equations (7.53), we immediately know the orientation of the rigid body in space.

6. The sun and the moon exert gravity torques on the earth, and they cause the axis of our planet to precess.

7. If at a certain instant the moment of inertia of the mass of body $\mathcal{B}$ about an axis through C parallel to the angular velocity vector is I, then the kinetic energy of $\mathcal{B}$ at that instant is $\frac{1}{2}m\mathbf{v}_C^2 + \frac{1}{2}I\omega^2$.

8. The earth's lunisolar precession is the result of *both* the bulge at the equator *and* the tilt of the axis.

9. The kinetic energy lost during a collision of two bodies does not depend on the angular velocities of the bodies prior to impact.

10. The work-energy and impulse-momentum principles are general integrals of the equations of motion for a rigid body.

11. Sometimes it is better to use the products of inertia in $\mathbf{M}_{r_C} = {}^{\mathcal{I}}\dot{\mathbf{H}}_C$ than to take the time to compute principal moments and axes of inertia so as to be able to utilize Euler's equations (7.53).

12. In steady precession with the nutation angle θ equaling 90°, the spin vector always precesses away from the torque vector.

Answers: F, T, T, T, F, T, T, T, F, T, T, F

CHAPTER

8

Special Topics

chapter outline ►

8.1 Introduction to Vibrations

Vibration is a term used to describe oscillatory motions of a body or system of bodies. These motions may be caused by isolated disturbances as when the wheel of an automobile strikes a bump or by fluctuating forces as in the case of the fuselage panels in an airplane vibrating in response to engine noise. Similarly, the oscillatory ground motions resulting from an earthquake cause vibrations of buildings. In each of these cases the undesirable motion may cause discomfort to occupants; moreover, the oscillating stresses induced within the body may lead to a fatigue failure of the structure, vehicle, or machine.

Free Vibration

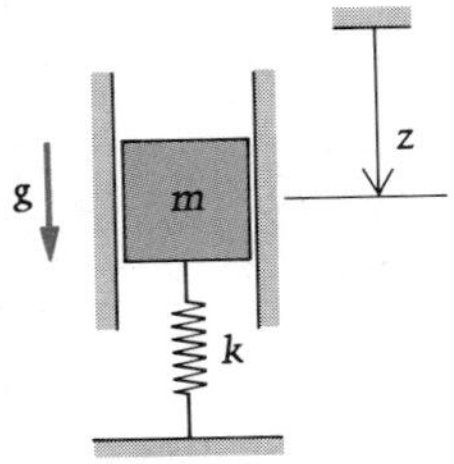

Figure 8.1

For perhaps the simplest example of a mechanical oscillator consider the rigid block and linear spring shown in Figure 8.1. The block is constrained to translate vertically; thus a single parameter (scalar) is sufficient to establish position and hence the system is called a **single-degree-of-freedom system.** We choose z to be the parameter and let $z = 0$ correspond to the configuration in which the spring is neither stretched nor compressed.

Using a free-body diagram of the block in an arbitrary position (Figure 8.2), Euler's first law yields

$$m\ddot{z} = mg - kz$$

or

$$m\ddot{z} + kz = mg \tag{8.1}$$

which is a second-order linear differential equation with constant coefficients describing the motion of the block. The fact that the differential equation is nonhomogeneous (the right-hand side is not zero) is a consequence of our choice of datum for the displacement parameter z. For if we make the substitution $y = z - mg/k$, the governing equation (8.1) becomes

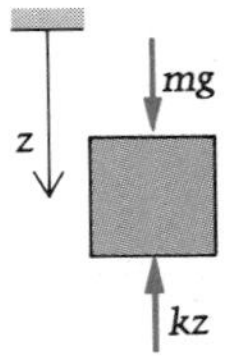

Figure 8.2

$$m\ddot{y} + ky = 0 \tag{8.2}$$

which is a *homogeneous* differential equation. It is not a coincidence that this occurs when the displacement variable is chosen so that it vanishes when the block is in the equilibrium configuration—that is, when the spring is compressed mg/k.

Motion described by an equation such as (8.2) is called a **free vibration** since there is no external force (external, that is, to the spring-mass system) stimulating it.

Rewriting Equation (8.2), we obtain

$$\ddot{y} + \frac{k}{m}y = 0$$

or, defining $\omega_n = \sqrt{k/m}$,

$$\ddot{y} + \omega_n^2 y = 0 \tag{8.3}$$

which has as its general solution

$$y = A \sin \omega_n t + B \cos \omega_n t \tag{8.4}$$

or

$$y = C \sin(\omega_n t + \varphi) \tag{8.5}$$

where

$$C = \sqrt{A^2 + B^2} \quad \text{and} \quad \tan \varphi = \frac{B}{A}$$

Whether expressed in the form of (8.4) or (8.5), y is called a **simple harmonic** function of time, ω_n is called the **natural circular frequency,** C is called the **amplitude** of the displacement y, and φ is said to be the **phase angle** by which y *leads* the reference function, $\sin \omega_n t$. The simple harmonic function is *periodic* and its **period** is $\tau_n = 2\pi/\omega_n$. Another quantity called **frequency** is $f_n = 1/\tau_n = \omega_n/2\pi$, which gives the number of cycles in a unit of time. When the unit of time is the second, the unit for f_n is the hertz (Hz); 1 Hz is 1 cycle per second.

The constants A and B in (8.4), or equivalently C and φ in (8.5), are determined from initial conditions of position and velocity. Thus if

$$y(0) = y_0$$

and

$$\dot{y}(0) = v_0$$

then

$$B = y_0$$

and

$$A = \frac{v_0}{\omega_n}$$

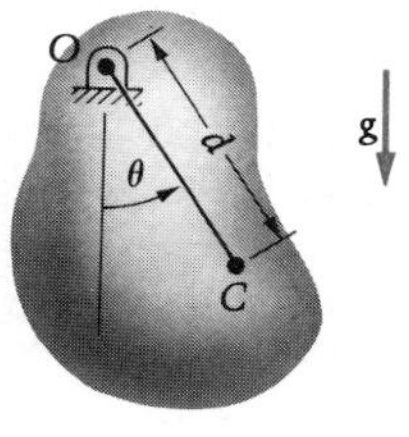

Figure 8.3

Now let us investigate what might seem an entirely different situation—that of a rigid body constrained to rotate about a fixed horizontal axis (through O as in Figure 8.3). Since the only kinematic freedom the body has is that of rotation, a single angle is sufficient to describe a configuration of the body. Let the angle be θ as shown, where we note that when $\theta = 0$ the mass center C is located directly below the pivot O.

Neglecting any friction at the axis of rotation, the free-body diagram appropriate to an arbitrary instant during the motion is shown in Figure 8.4. Summing moments about the axis of rotation, we get

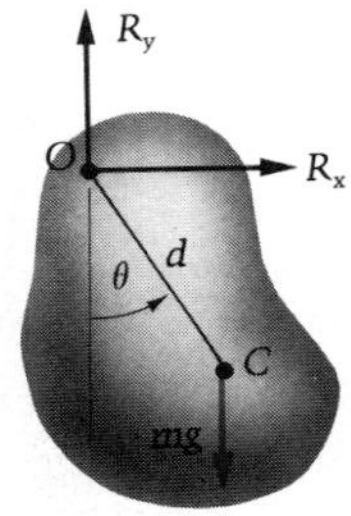

Figure 8.4

$$-mg\,d\sin\theta = I_0\ddot{\theta} \tag{8.6}$$

where I_0 is the mass moment of inertia about the axis of rotation. Equation (8.6) is a nonlinear differential equation because $\sin\theta$ is a nonlinear function of θ, but if we restrict our attention to sufficiently small angles so that $\sin\theta \approx \theta$, Equation (8.6) becomes

$$I_0\ddot{\theta} + (mgd)\theta = 0 \tag{8.7}$$

That is, θ is a simple harmonic function:

$$\theta = A\sin\omega_n t + B\cos\omega_n t$$

where now

$$\omega_n^2 = \frac{mgd}{I_0}$$

The two preceding examples have an important feature in common: Motion near the equilibrium configuration is governed by a homogeneous, second-order, linear differential equation with constant coefficients, and in each case the motion is simple harmonic. A point of difference is that in the block-spring case the gravitational field plays no role other than establishing the equilibrium configuration; in particular the natural frequency does not depend on the strength (g) of the field. In the second case where the body is basically behaving as a pendulum, the gravitational field provides the "restoring action" and the natural frequency is proportional to $\sqrt{g}$.

EXAMPLE 8.1

Find the natural frequency of oscillation about the equilibrium position of a uniform ball (sphere) rolling on a cylindrical surface.

SOLUTION

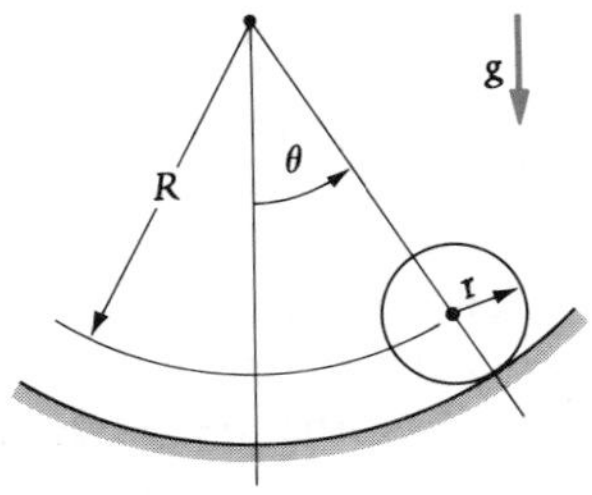

Figure 1

Let m be the mass of the ball, let R be the radius of the path of its center, and let θ be the polar coordinate angle locating the center as shown in Figure 1. Thus

$$\mathbf{a}_C = -R\dot{\theta}^2\hat{\mathbf{e}}_R + R\ddot{\theta}\hat{\mathbf{e}}_\theta$$

and the angular acceleration of the ball is $\boldsymbol{\alpha} = -(R\ddot{\theta}/r)\hat{\mathbf{k}}$ because of the no-slip condition. We shall now use $\mathbf{a}_C$ and $\boldsymbol{\alpha}$ in the equations of motion:

$$\mathbf{F}_r = m\mathbf{a}_C \tag{1}$$

(Continued)

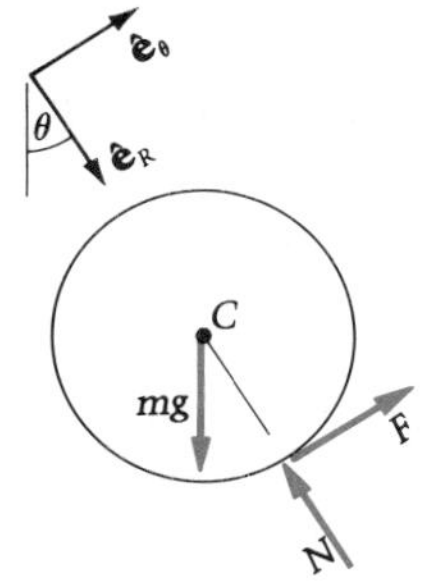

Figure 2

Hence, from the free-body diagram below (Figure 2), the $\hat{\mathbf{e}}_R$ and $\hat{\mathbf{e}}_\theta$ component equations of (1) are:

$$F - mg\sin\theta = mR\ddot{\theta} \tag{2}$$

and

$$N - mg\cos\theta = mR\dot{\theta}^2 \tag{3}$$

Also, from summing moments about C, we have

$$Fr = \left(\frac{2}{5}mr^2\right)\left(-\frac{R\ddot{\theta}}{r}\right)$$

or

$$F = -\frac{2}{5}mR\ddot{\theta} \tag{4}$$

Eliminating the friction force F between Equations (2) and (4), we obtain the differential equation

$$\frac{7}{5}mR\ddot{\theta} + mg\sin\theta = 0$$

For small θ so that $\sin\theta \approx \theta$,

$$\frac{7}{5}R\ddot{\theta} + g\theta = 0$$

from which we see that

$$\omega_n^2 = \frac{5g}{7R}$$

or

$$\omega_n = 0.845\sqrt{\frac{g}{R}}$$

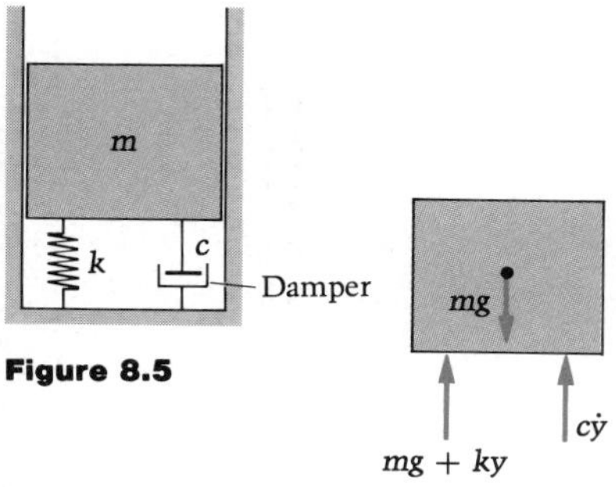

Figure 8.5

Damped Vibration

The simple harmonic motion in our examples of free vibration has a feature that conflicts with our experience in the real world; that is, the motion calculated persists forever unabated. Intuition would suggest decaying oscillations and finally the body coming to rest. Of course the problem here is that we have not incorporated any mechanism for energy dissipation in the analytical model. To do that, we shall return to the simple block-spring system and introduce a new element: a viscous damper (Figure 8.5). The *rate* of extension of this element is proportional to the force applied, through a damping constant c, so that the force is c times the rate of extension.

Referring to the free-body diagram in Figure 8.5 and letting $y = 0$ designate the equilibrium position as before, we have

$$m\ddot{y} = mg - (mg + ky) - c\dot{y}$$

or

$$m\ddot{y} + c\dot{y} + ky = 0 \tag{8.8}$$

The appearance of the $c\dot{y}$ term in (8.8) has a profound effect on the solution to the differential equation and hence on the motion being described. Solutions to (8.8) may be found from

$$y = Ae^{rt} \tag{8.9}$$

where A is an arbitrary constant and r is a characteristic parameter. Substituting (8.9) into (8.8), we obtain

$$(mr^2 + cr + k)Ae^{rt} = 0 \tag{8.10}$$

which is satisfied nontrivially (i.e., for $A \neq 0$) with

$$mr^2 + cr + k = 0 \tag{8.11}$$

This characteristic equation has two roots given by

$$r = -\frac{c}{2m} \pm \sqrt{\left(\frac{c}{2m}\right)^2 - \frac{k}{m}} \tag{8.12}$$

Except for the case in which $(c/2m)^2 = k/m$, the roots are distinct; if we call them r_1 and r_2, then the general solution to (8.8) is

$$y = A_1e^{r_1t} + A_2e^{r_2t}$$

In the exceptional case $(c/2m)^2 = k/m$, there is only the one repeated root $r = -c/2m$, but direct substitution will verify that there is a solution to (8.8) of the form $te^{-(c/2m)t}$ so that the general solution in that case is

$$y = A_1e^{-(c/2m)t} + A_2te^{-(c/2m)t} \tag{8.13}$$

With initial conditions

$$y(0) = y_0$$

and

$$\dot{y}(0) = v_0$$

we find that

$$A_1 = y_0$$

and

$$A_2 = v_0 + \left(\frac{c}{2m}\right)y_0$$

Since

$$\left(\frac{c}{2m}\right)^2 = \frac{k}{m} = \omega_n^2$$

the solution is

$$y = e^{-\omega_n t}[y_0 + (v_0 + \omega_n y_0)t] \tag{8.14}$$

Displacements given by (8.14) are plotted in Figure 8.6 for several representative sets of initial conditions (positive y_0 but positive and negative v_0). Two features of the motion are apparent:

1. $y \to 0$ (the equilibrium position) as $t \to \infty$.
2. The motion is not oscillatory; the equilibrium position is "overshot" at most once and only then when the initial speed is sufficiently large and in the direction opposite to that of the initial displacement.

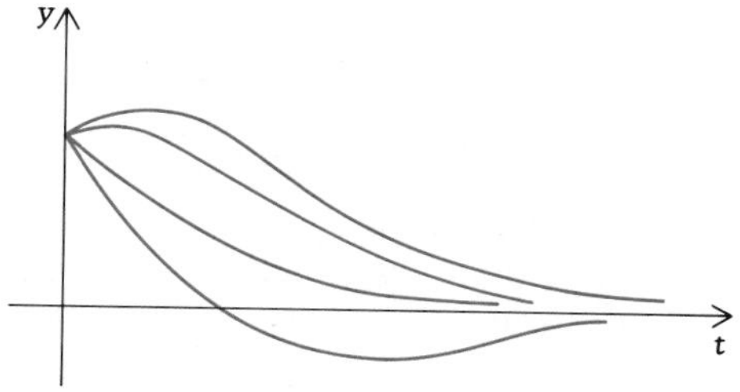

Figure 8.6 Motion of a critically damped system.

In the case we have just studied, the damping is called **critical damping,** because it separates two quite different mathematical solutions: for greater damping the roots of the characteristic equation (8.11) are both real and negative, and for smaller damping the roots are complex conjugates. If we let the critical damping be denoted by c_{crit} then we have seen that

$$c_{crit} = 2\sqrt{km} = 2m\omega_n \tag{8.15}$$

Now let us consider the case for which $c > c_{crit}$; the mechanical system is then said to be **overdamped** or the damping is said to be **supercritical.** In this case the roots given by (8.12) are both real and negative since $(c/2m)^2 > k/m$; if we call these roots $-a_1$ and $-a_2$, with $a_2 > a_1 > 0$, then the general solution to the differential equation of motion is

$$y = A_1 e^{-a_1 t} + A_2 e^{-a_2 t} \tag{8.16}$$

The motion described here is in no way qualitatively different from that for the case of critical damping, which we have just discussed. For a given set of initial conditions, Equation (8.16) yields a slower approach to $y = 0$ than does (8.13). That is, the overdamped motion is more "sluggish" than the critically damped motion as we would anticipate because of the greater damping.

Finally we consider the case in which the system is said to be **underdamped** or **subcritically** damped; that is, $c < c_{crit}$. The roots given by (8.12) are the complex conjugates

$$-\frac{c}{2m} \pm i\sqrt{\frac{k}{m} - \left(\frac{c}{2m}\right)^2}$$

where $i = \sqrt{-1}$. It is possible to express the general solution to the governing differential equation as

$$y = e^{-(c/2m)t}(A_1 \sin \omega_d t + A_2 \cos \omega_d t) \tag{8.17}$$

where $\omega_d = \sqrt{k/m - (c/2m)^2}$. A typical displacement history corresponding to (8.17) is shown in Figure 8.7. We note that, just as in the preceding cases, $y \rightarrow 0$ as $t \rightarrow \infty$; however, here the motion is oscillatory. We see that the simple harmonic motion obtained for the model without damping is given by (8.17) with $c = 0$. Moreover, we see that with light damping (small c) the analytical model that does not include damping adequately describes the motion during the first several oscillations. It is this case—subcritical damping—that is of greatest practical importance in studies of vibration.

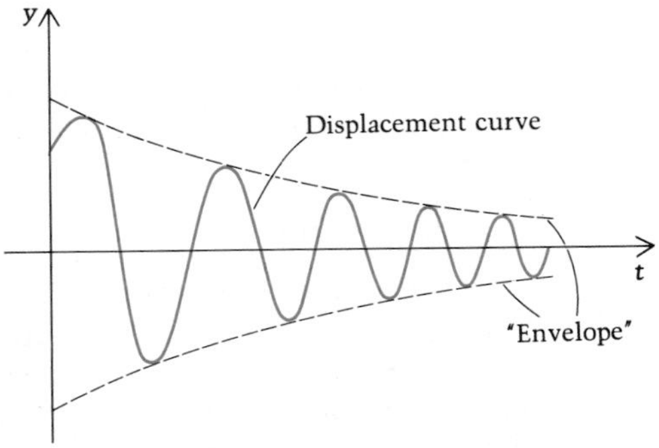

Figure 8.7 Motion of an underdamped system.

EXAMPLE 8.2

Find the damping constant c that gives critical damping of the rigid bar executing motions near the equilibrium position shown in the diagram.

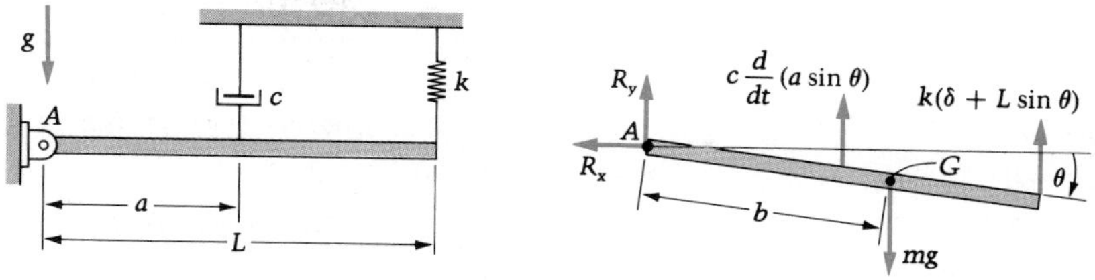

SOLUTION

We are going to restrict our attention to small angles θ, and thus we may ignore any tilting of the damper or the spring. However, it may help us develop the equation of motion in an orderly way if we assume that the upper ends of the spring and damper slide along so that each remains vertical as the bar rotates through the angle θ. Without further restriction, if we sum moments about A we obtain, with I_A the moment of inertia of the mass of the bar about the axis of rotation at A,

$$I_A\ddot{\theta} = mg(b\cos\theta) - \left[c\frac{d}{dt}(a\sin\theta)\right](a\cos\theta) - k(\delta + L\sin\theta)(L\cos\theta) \tag{1}$$

where δ is the spring stretch at equilibrium. Thus for small θ (that is, $\sin\theta \approx \theta$, $\cos\theta \approx 1$) we linearize Equation (1) and obtain

$$I_A\ddot{\theta} = mgb - ca^2\dot{\theta} - kL\delta - kL^2\theta$$

Of course, $\theta = 0$ is the equilibrium configuration so that

$$mgb = kL\delta$$

The linear governing differential equation is then

$$I_A\ddot{\theta} + ca^2\dot{\theta} + kL^2\theta = 0$$

and for critical damping we get, associating the coefficients of θ, $\dot{\theta}$, $\ddot{\theta}$ with those of y, $\dot{y}$, $\ddot{y}$ in Equation (8.8),

$$c_{\text{crit}}a^2 = 2\sqrt{(kL^2)I_A}$$

or

$$c_{\text{crit}} = 2\frac{L}{a^2}\sqrt{kI_A}$$

Any c less than this critical value will result in oscillations of decreasing amplitude.

Forced Vibration

Fluctuating external forces may have destructive effects on mechanical systems; this is perhaps the primary motivation for studying mechanical vibration. It is common for the external loading to be a periodic function of time, in which case the loading may be expressed as a series of simple harmonic functions (Fourier series). Consequently it is instructive to consider the case in which the loading is simple-harmonic. For the mass-spring-damper system shown in Figure 8.8, the differential equation of motion is

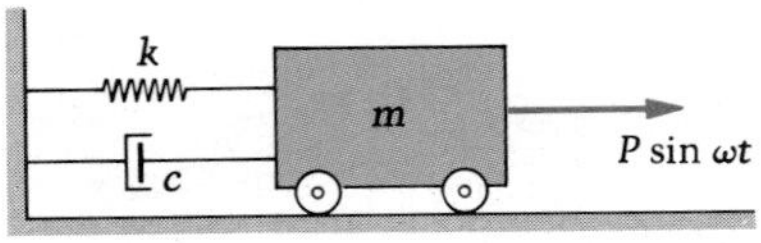

Figure 8.8

$$m\ddot{x} + c\dot{x} + kx = P \sin \omega t \tag{8.18}$$

The general solution is composed of two parts: a ***particular*** solution (anything that satisfies the differential equation) and what is called the ***complementary*** solution (the general solution to the homogeneous differential equation). A particular solution of the form $x = X \sin(\omega t - \varphi)$ may be found. If we substitute this expression in (8.18) we obtain

$$-m\omega^2 X \sin(\omega t - \varphi) + c\omega X \cos(\omega t - \varphi) + kX \sin(\omega t - \varphi) = P \sin \omega t$$

or

$$(k - m\omega^2)X(\sin \omega t \cos \varphi - \sin \varphi \cos \omega t) + c\omega X(\cos \omega t \cos \varphi + \sin \omega t \sin \varphi) = P \sin \omega t$$

or

$$[(k - m\omega^2) \cos \varphi + c\omega \sin \varphi]X \sin \omega t - [(k - m\omega^2) \sin \varphi - c\omega \cos \varphi]X \cos \omega t = P \sin \omega t$$

For this to be satisfied at every instant of time,

$$[(k - m\omega^2) \cos \varphi + c\omega \sin \varphi]X = P \tag{8.19}$$

and

$$-c\omega \cos \varphi + (k - m\omega^2) \sin \varphi = 0 \tag{8.20}$$

From (8.20) we get

$$\tan \varphi = \frac{c\omega}{k - m\omega^2} \tag{8.21}$$

so that

$$\sin \varphi = \frac{c\omega}{\sqrt{(k - m\omega^2)^2 + (c\omega)^2}}$$

and

$$\cos\varphi = \frac{k - m\omega^2}{\sqrt{(k - m\omega^2)^2 + (c\omega)^2}}$$

Substituting these expressions for $\sin\varphi$ and $\cos\varphi$ into (8.19), we obtain

$$\left[\frac{(k - m\omega^2)^2}{\sqrt{(k - m\omega^2)^2 + (c\omega)^2}} + \frac{(c\omega)^2}{\sqrt{(k - m\omega^2)^2 + (c\omega)^2}}\right]X = P$$

so that

$$X = \frac{P}{\sqrt{(k - m\omega^2)^2 + (c\omega)^2}} \tag{8.22}$$

We may now write the complete solution to the differential equation (8.18):

$$\mathbf{x} = \mathbf{x}_c(t) + X\sin(\omega t - \varphi) \tag{8.23}$$

where $\mathbf{x}_c$ is the complementary solution and is one of the three cases enumerated in the preceding section. That is, the form of $\mathbf{x}_c$ depends on whether the system is overdamped, critically damped, or underdamped. However, in each of these cases the negative exponential causes the function to approach zero as time becomes large. Thus for large time $\mathbf{x}_c$ tends to zero and $\mathbf{x}(t)$ tends to the particular solution. For this reason the simple-harmonic particular solution is called the **steady-state displacement,** since it represents the long-term behavior of the system.

We note that the steady-state motion is a simple harmonic function having amplitude X and lagging the excitation (force) function by the phase angle φ. We may put these in a convenient form by dividing numerator and denominator of (8.21) and (8.22) by k, so that

$$\tan\varphi = \frac{c\omega/k}{1 - \omega^2/\omega_n^2} \qquad \left(\text{where } \omega_n^2 = \frac{k}{m}\right) \tag{8.24}$$

and

$$X = \frac{P/k}{\sqrt{(1 - \omega^2/\omega_n^2)^2 + (c\omega/k)^2}} \tag{8.25}$$

Investigating the dimensionless quantity $c\omega/k$, we find

$$\frac{c\omega}{k} = \frac{c\omega}{m\omega_n^2}$$
$$= \frac{2c}{2m\omega_n}\left(\frac{\omega}{\omega_n}\right)$$

But we know that $2m\omega_n = c_{\text{crit}}$, the critical damping, so if we let ζ be the damping ratio (c/c_{crit}),

$$\frac{c\omega}{k} = 2\zeta\frac{\omega}{\omega_n} \tag{8.26}$$

and

$$\tan\varphi = \frac{2\zeta(\omega/\omega_n)}{1 - \omega^2/\omega_n^2} \tag{8.27}$$

and

$$X = \frac{P/k}{\sqrt{(1 - \omega^2/\omega_n^2)^2 + [2\zeta(\omega/\omega_n)]^2}} \tag{8.28}$$

The phase angle φ and the dimensionless displacement amplitude kX/P are plotted against the frequency ratio ω/ω_n in Figures 8.9 and 8.10, respectively, for various values of the damping ratio ζ. We see that, with small damping, large amplitudes of displacement occur when the excitation frequency ω is near the natural frequency ω_n. This phenomenon is called **resonance,** and the desire to avoid it has led to the de-

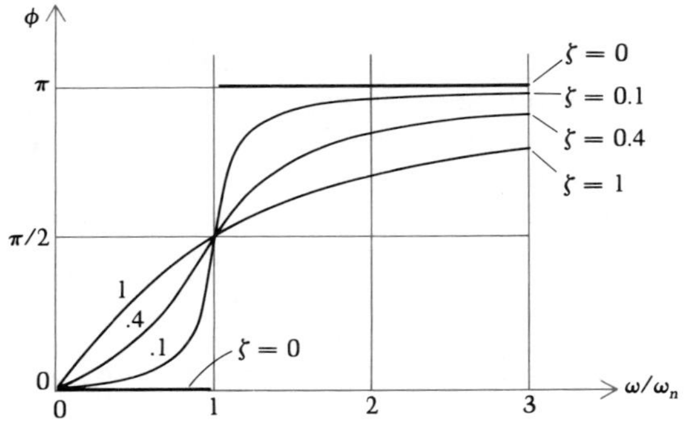

Figure 8.9

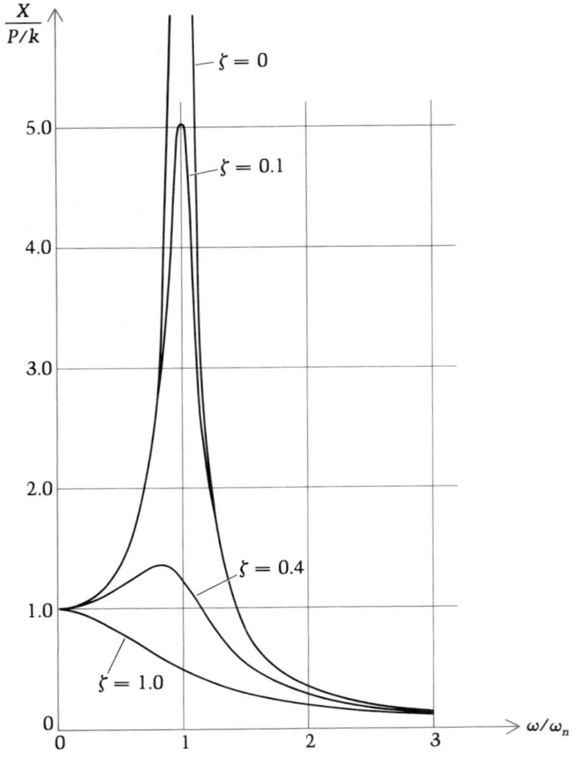

Figure 8.10

velopment of methods for estimating natural frequencies of mechanical systems. Note that the steady-state response curves are insensitive to the damping for sufficiently small damping (say $\zeta < 0.1$) provided that we are not in the near vicinity of $\omega/\omega_n = 1$. This is an important observation because often in engineering practice we have reason to believe that the damping is small but we do not have accurate quantitative information about it.

We close this section by discussing the usual source of a simple harmonic external loading—an imbalance in a piece of rotating machinery. Let the machine be made up of two parts. The first, of mass m_1, is a rigid body constrained to rotate about an axis fixed in the second body (mass m_2), which translates relative to the inertial frame of reference. Let the mass center of the rotating body lie off the axis of rotation a distance e and let the body rotate at constant angular speed ω. (See Figure 8.11.)

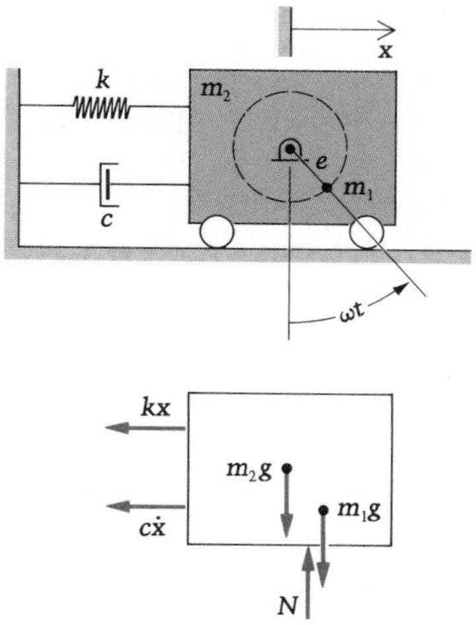

Figure 8.11

Referring to the free-body diagram in Figure 8.11, we have

$$-kx - c\dot{x} = m_2\ddot{x} + m_1\frac{d^2}{dt^2}(x + e\sin\omega t)$$

$$= (m_1 + m_2)\ddot{x} - m_1 e\omega^2 \sin\omega t$$

If we denote the total mass of the machine by m, then $m = m_1 + m_2$ and

$$m\ddot{x} + c\dot{x} + kx = m_1 e\omega^2 \sin\omega t$$

Thus the amplitude of the apparent "external" sinusoidal loading is $m_1 e\omega^2$ and its frequency is the angular speed of the rotating element.

EXAMPLE 8.3

A piece of machinery weighing 200 lb has a rotating element with imbalance ($m_1 e$ times the acceleration of gravity, which is $32.2 \times 12 = 386$ in./sec^2) 5 lb-in. and an operating speed of 1200 rpm. There are four springs, each of stiffness 1500 lb/in., supporting the machine whose frame is constrained to translate vertically. The damping ratio is $\zeta = 0.3$. Find the steady-state displacement of the frame.

SOLUTION

The effective spring stiffness is

$$k = 4(1500) = 6000 \text{ lb/in.}$$

so that

$$\omega_n = \sqrt{\frac{6000}{200/386}} = 108 \text{ rad/sec}$$

$$\omega = \frac{1200}{60}(2\pi) = 126 \text{ rad/sec}$$

The effective external force amplitude is

$$P = m_1 e\omega^2 = \left(\frac{5}{386}\right)(126)^2 = 206 \text{ lb}$$

From Equation (8.28) we get

$$X = \frac{P/k}{\sqrt{(1 - \omega^2/\omega_n^2)^2 + (2\zeta\omega/\omega_n)^2}}$$

$$= \frac{206/6000}{\sqrt{[1 - (126/108)^2]^2 + [2(0.3)(126/108)]^2}}$$

$$= 0.0436 \text{ in.}$$

The phase angle φ is given by

$$\tan\varphi = \frac{2\zeta\omega/\omega_n}{1 - \omega^2/\omega_n^2}$$

$$= \frac{2(0.3)(126/108)}{1 - (126/108)^2} = -1.94$$

so that $\varphi = 2.05$ rad (117°).

EXAMPLE 8.4

The machine of Example 8.3 (weight = 200 lb, imbalance = 5 lb-in., operating speed = 1200 rpm) is to be supported by springs with negligible damping. If the machine were bolted directly to the floor, the amplitude of force transmitted to the floor would be

$$(m_1 e)\omega^2 = 206 \text{ lb}$$

What should the stiffness of the support system be so that the amplitude of the force transmitted to the floor is less than 20 lb?

SOLUTION

The force exerted on the floor is transmitted through the supporting springs and is of amplitude kX, where X is the amplitude of displacement of the machine. From Equation (8.28) we have

$$kX = \frac{m_1 e\omega^2}{\sqrt{(1 - \omega^2/\omega_n^2)^2 + (2\zeta\omega/\omega_n)^2}}$$

or with negligible damping (i.e., $\zeta \approx 0$)

$$kX = \frac{m_1 e\omega^2}{\sqrt{(1 - \omega^2/\omega_n^2)^2}} = \frac{m_1 e\omega^2}{|1 - \omega^2/\omega_n^2|}$$

Thus for

$$\frac{1}{|1 - \omega^2/\omega_n^2|} = \frac{kX}{m_1 e\omega^2} < \frac{20}{206} = 0.0971$$

it is clear that $1 - \omega^2/\omega_n^2$ is negative. Note that only when $\omega^2/\omega_n^2 > 2$ is

$$\frac{1}{|1 - \omega^2/\omega_n^2|} < 1$$

Therefore we inquire into the condition for which

$$-\frac{1}{1 - \omega^2/\omega_n^2} < -0.0971$$

or

$$\frac{\omega^2}{\omega_n^2} > 1 + \frac{1}{0.0971} = 11.3$$

or

$$\omega_n^2 < \frac{(126)^2}{11.3} = 1400$$

since $\omega = 126$ rad/sec. But

(Continued)

$$k = m\omega_n^2 < \frac{200}{386}(1400) = 725 \text{ lb/in.}$$

Thus to satisfy the given conditions the support stiffness must be *less than* 725 lb/in.

If the only springs available give a greater stiffness, the problem may be solved by increasing the mass; particularly we might mount the machine on a block of material, say concrete, and then support the machine and block by springs. For example, if the only springs available were those of Example 8.3 for which $k = 6000$ lb/in., then we need m to be *at least* that given by

$$m = \frac{k}{\omega_n^2} = \frac{6000}{1400}$$

$$= 4.29 \text{ lb-sec}^2\text{/in.}$$

for which the weight is

$$(4.29)(386) = 1660 \text{ lb}$$

Therefore we need a slab or block weighing

$$1660 - 200 = 1460 \text{ lb}$$

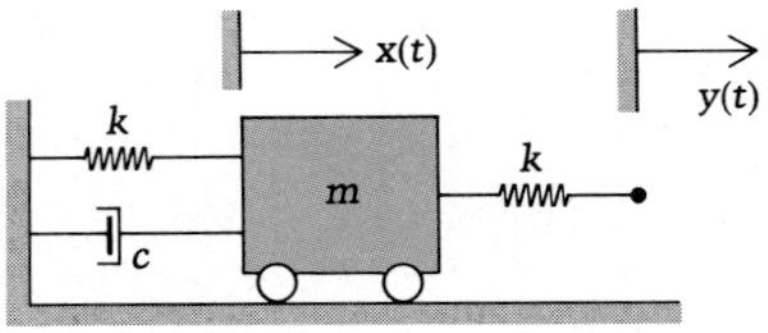

EXAMPLE 8.5

Find the steady-state displacement $x(t)$ if $y(t) = 0.1 \cos 120t$ inch, where t is in seconds, $m = 0.01$ lb-sec^2/in., $k = 100$ lb/in., and $c = 2$ lb-sec/in. In particular: (a) What is the amplitude of $x(t)$? (b) What is the angle by which $x(t)$ leads or lags $y(t)$?

SOLUTION

The differential equation of motion of the mass is seen to be

$$m\ddot{x} = -kx - c\dot{x} - k(x - y)$$

or

$$m\ddot{x} + c\dot{x} + 2kx = kY \cos \omega t$$

where $Y = 0.1$ in. and $\omega = 120$ rad/sec. Using Equation (8.18), we see that kY is playing the same role as the oscillating force P, so that the steady-state amplitude is

$$X = \frac{kY}{\sqrt{(2k - m\omega^2)^2 + (c\omega)^2}} = \frac{100(0.1)}{\sqrt{[200 - 0.01(120)^2]^2 + [2(120)]^2}}$$

(Continued)

or

$$X = 0.0406 \text{ in.}$$

The phase angle is

$$\varphi = \tan^{-1}\left(\frac{c\omega}{2k - m\omega^2}\right) = \tan^{-1}\left[\frac{2(120)}{200 - 0.01(120^2)}\right]$$

$$= 76.9° \text{ or } 1.34 \text{ rad} \qquad \text{(lagging)}$$

Thus the steady-state motion is

$$x_{ss} = 0.0406 \cos(120t - 1.34) \text{ in.}$$

PROBLEMS / Section 8.1

8.1 It is possible to determine experimentally the moments of inertia of large objects, such as the rocket shown in Figure P8.1. If the rocket is turned through a slight angle about z_C and released, for example, it oscillates with a period of 2.8 sec. Find the radius of gyration k_{z_C}.

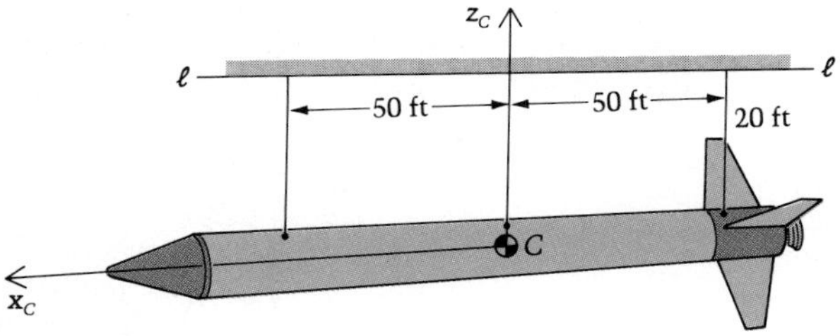

Figure P8.1

8.2 In the preceding problem, when the rocket is caused to swing with small angles about axis ℓℓ as shown, the period is observed to be 8 sec. Find from this information the value of k_{x_C}.

8.3–8.5 Find the equations of motion and periods of vibration of the systems shown in Figures P8.3 to P8.5. In each case, neglect the mass of the rigid bar to which the ball (particle) is attached.

8.6 Find the frequency of small vibrations of the round wheel $\mathscr{C}$ as it rolls back and forth on the cylindrical surface in Figure P8.6. The radius of gyration of $\mathscr{C}$ with respect to the axis through C normal to the plane of the figure is k_C. Verify the result of Example 8.1 with your answer.

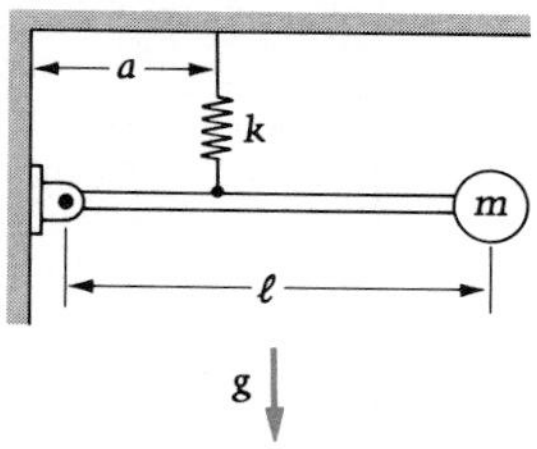

Figure P8.3

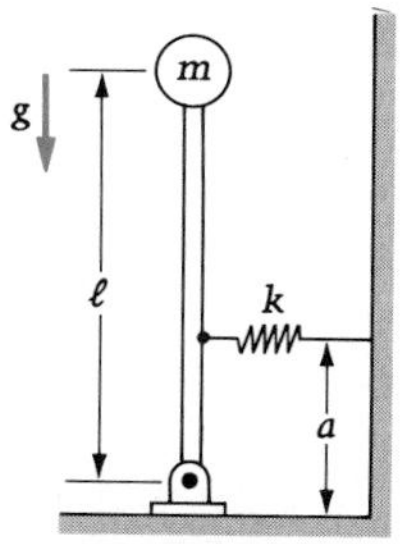

Figure P8.4

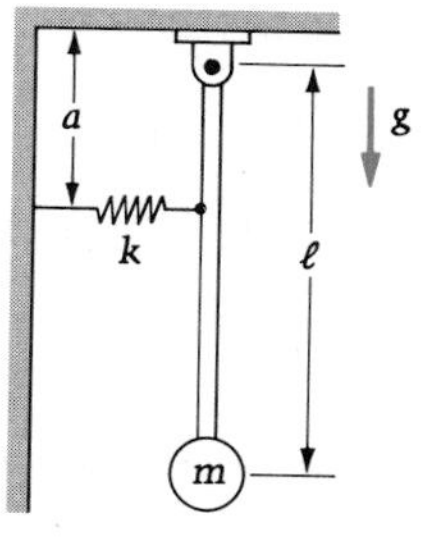

Figure P8.5

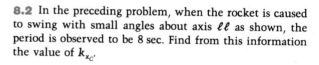

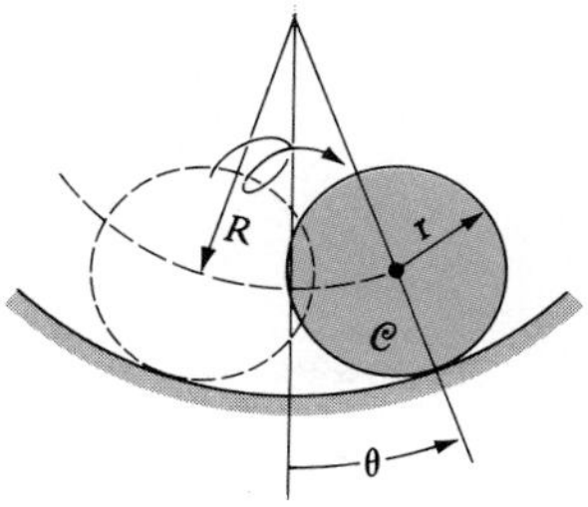

Figure P8.6

8.7 A uniform cylinder of mass m and radius R is floating in water. (See Figure P8.7.) The cylinder has a spring of modulus k attached to its top center point. If the specific weight of the water is γ, find the frequency of the vertical bobbing motion of the cylinder. *Hint:* The upward (buoyant) force on the bottom of the cylinder equals the weight of water displaced at any time (Archimedes' principle).

8.8 Find the frequency of small-amplitude oscillations of the uniform half-cylinder near the equilibrium position shown in Figure P8.8. Assume that the cylinder rolls on the horizontal plane.

8.9 Find the natural frequency of free vertical vibrations of the mass in Figure P8.9, which is suspended by two springs at angle ϕ with the vertical as shown.

8.10 The masses in Figure P8.10 are connected by an inextensible string. Find the frequency of small oscillations if mass m is lowered slightly and released.

8.11 The cylinder in Figure P8.11 is in equilibrium in the position shown. For no slipping, find the natural frequency of free vibration about this equilibrium position.

8.12 A block weighing 1 lb is dropped from height $H = 0.1$ in. (See Figure P8.12.) If $k = 2.5$ lb/in., find the time interval for which the ends of the springs are in contact with the ground.

8.13 The turntable in Figure P8.13 rotates in a horizontal plane at a *constant* angular speed ω. The particle P (mass $= m$) moves in the frictionless slot and is attached to the spring (modulus k, free length ℓ) as shown.

a. Derive the differential equation describing the motion $y(t)$ of the particle relative to the slot.
b. What is the extension of the spring such that P does not accelerate relative to the slot?
c. Suppose the motion is initiated with the spring unstretched and the particle at rest relative to the slot. Find the ensuing motion $y(t)$.

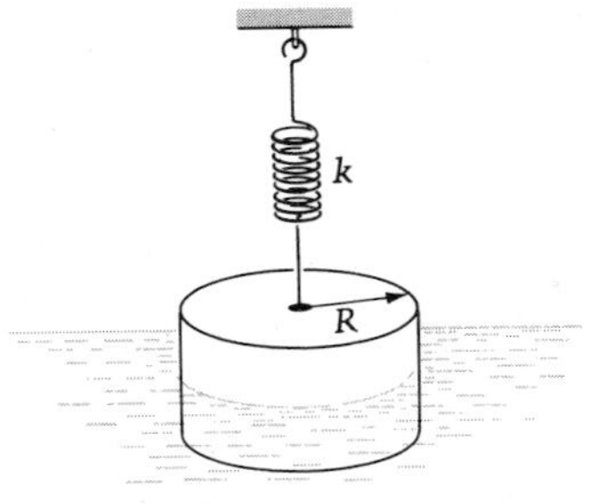

Figure P8.7

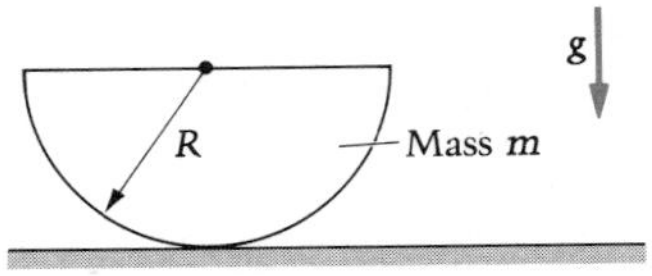

Figure P8.8

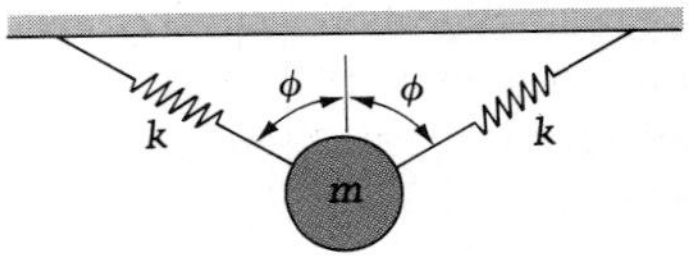

Figure P8.9

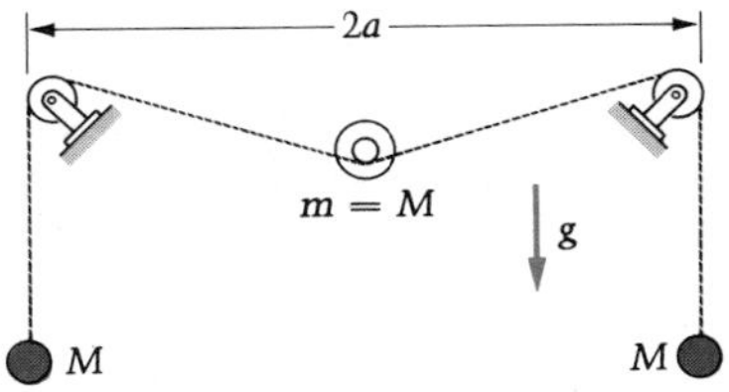

Figure P8.10

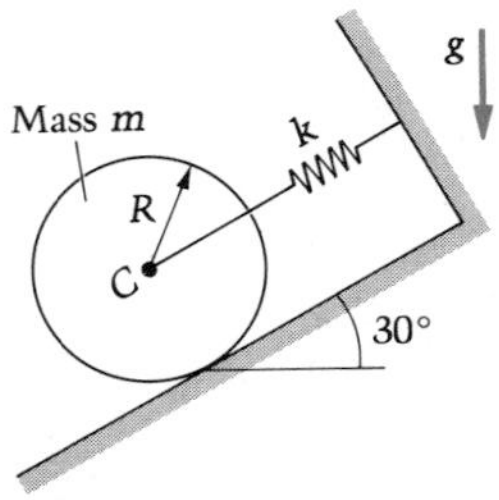

Figure P8.11

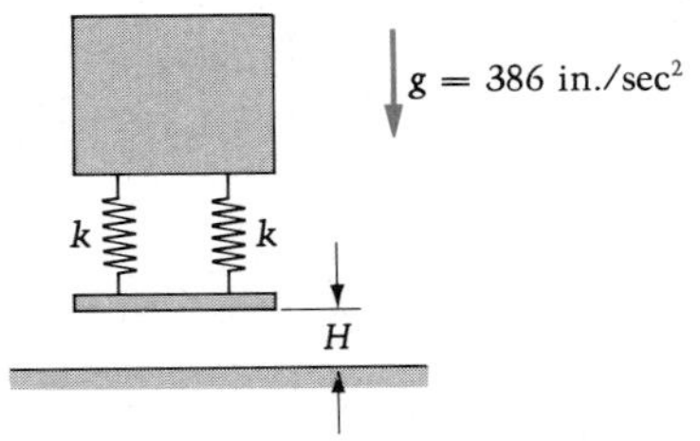

Figure P8.12

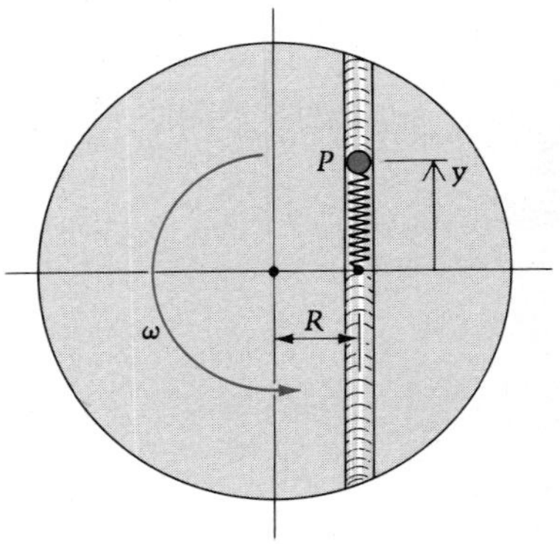

Figure P8.13

8.14 Assume that the slender rigid bar $\mathcal{B}$ in Figure P8.14 undergoes only small angles of rotation. Find the angle of rotation $\theta(t)$ if the bar is in equilibrium prior to $t = 0$, at which time the constant force P begins to traverse the bar at constant speed v.

8.15 Refer to the preceding problem: (a) Find the work done by P in traversing the bar $\mathcal{B}$; (b) show that this work equals the change in mechanical energy (which is the kinetic energy of $\mathcal{B}$ plus the potential energy stored in the spring).

8.16 A sack of cement of mass m is to be dropped on the center of a simply supported beam as shown in Figure P8.16. Assume that the mass of the beam may be neglected, so that it may be treated as a simple linear spring of stiffness k. Estimate the maximum deflection at the center of the beam.

8.17 The solid homogeneous cylinder in Figure P8.17 weighs 200 lb and rolls on the horizontal plane. When the cylinder is at rest, the springs are each stretched 2 ft. The modulus of each spring is 15 lb/ft. The mass center C is given an initial velocity of $\frac{1}{2}$ ft/sec to the right.

a. How far to the right will C go?
b. How long will it take to get there?
c. How long will it take to go halfway to the extreme position?

8.18 A spring with modulus 120 lb/in. supports a 200-lb block. (See Figure P8.18.) The block is fastened to the spring. A 400-lb downward force is applied to the top of the block at $t = 0$ when the block is at rest. Find the maximum deflection of the spring in the ensuing time.

8.19 A particle $\mathcal{P}$ of mass m moves on a rough, horizontal rail with friction coefficient μ. (See Figure P8.19.) It is attached to a fixed point on the rail by a linear spring of modulus k. The initial stretch of the spring is $7\,\mu gm/k$. Describe the subsequent motion if it is known that the particle starts from rest. Show that the mass stops for good when $t = 3\pi/\sqrt{k/m}$.

8.20 Find the value of c to give critical damping of the pendulum in Figure P8.20. Neglect the mass of the rigid bar to which the particle of mass m is attached.

8.21 Consider free oscillations of a subcritically damped oscillator. Do local maxima in the response occur periodically?

8.22 If $k = 100$ lb/in. and the mass of the uniform, slender, rigid bar in Figure P8.22 is 0.03 lb-sec²/in., what damping constant c results in critical damping?

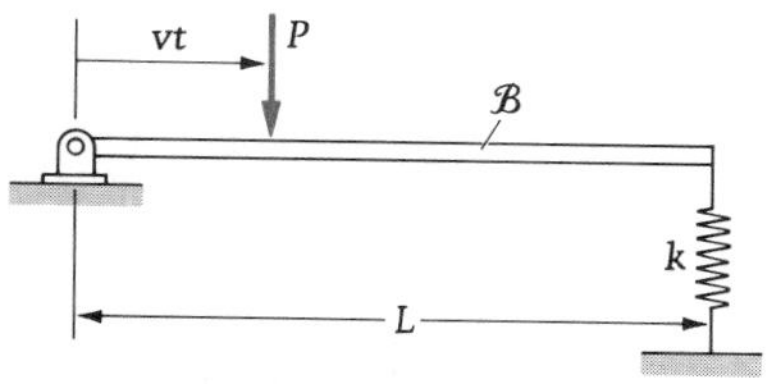

Figure P8.14

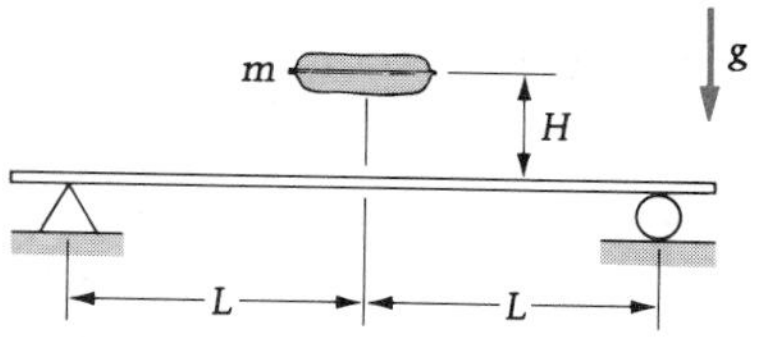

Figure P8.16

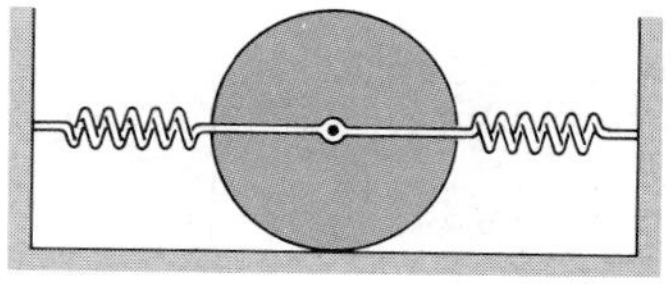
Figure P8.17

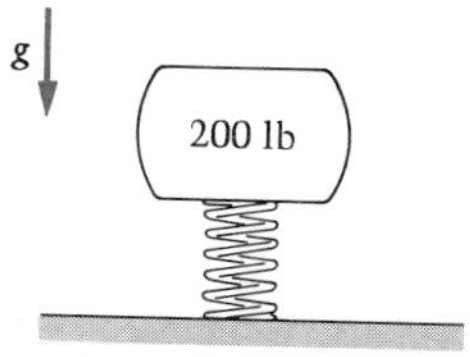

Figure P8.18

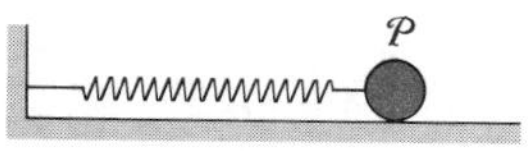

Figure P8.19

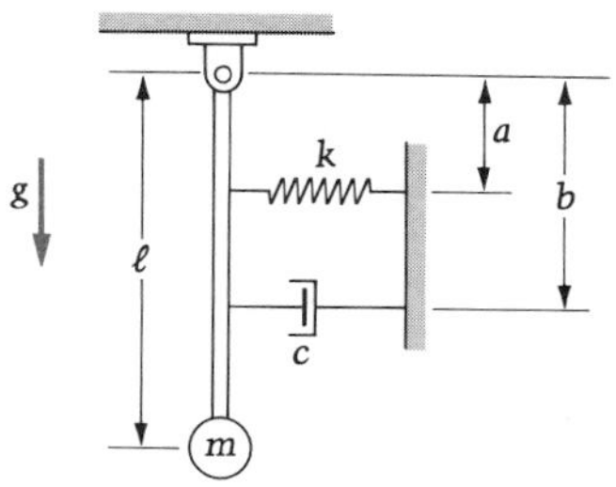

Figure P8.20

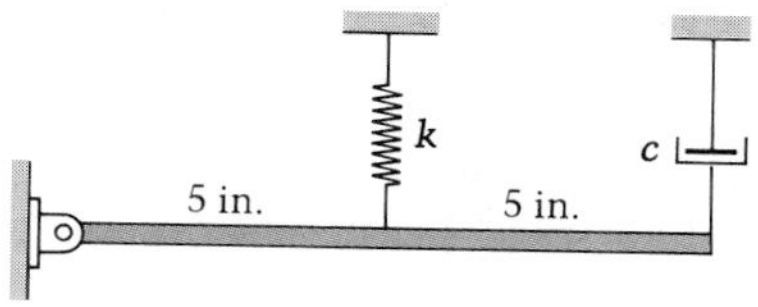

Figure P8.22

8.23 A cannon weighing 1200 lb shoots a 100-lb cannonball at a velocity of 600 ft/sec. (See Figure P8.23.) It then immediately comes into contact with a spring of stiffness 149 lb/ft and a dashpot that is set up to critically damp the system. Assuming that there is no friction between the wheels and the plane, find the displacement toward the wall after $\frac{1}{4}$ sec has elapsed.

Figure P8.23

8.24 If $k = 100$ lb/in. and the mass of the uniform, slender, rigid bar in Figure P8.24 is 0.03 lb-sec²/in., what damping modulus c results in critical damping? Compare with the c from Problem 8.22. For this damping, find $\theta(t)$ if the bar is turned through a small angle θ_0 and then released from rest. If the dashpot were removed, what would be the period of free vibration?

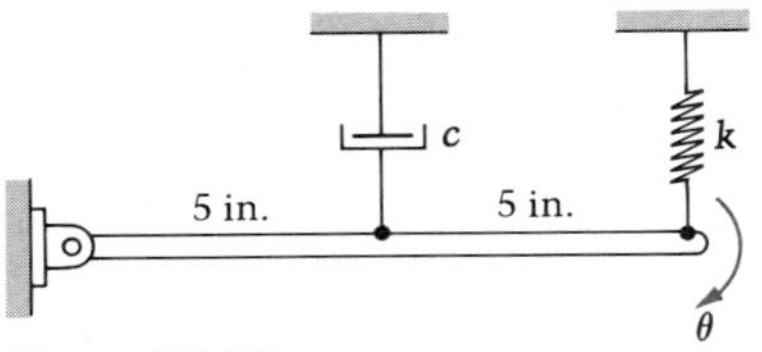

Figure P8.24

8.25 In Figure P8.25 find the response $\mathbf{x}_1(t)$ for the initial conditions $\mathbf{x}_1(0) = \dot{\mathbf{x}}_1(0) = 0$ if

$$k = 100 \text{ lb/in.}$$

$$m = 0.01 \text{ lb-sec}^2\text{/in.}$$

$$c = 1.0 \text{ lb-sec/in.}$$

$$X_2 = 0.05 \text{ in.}$$

$$\omega = 100 \text{ rad/sec}$$

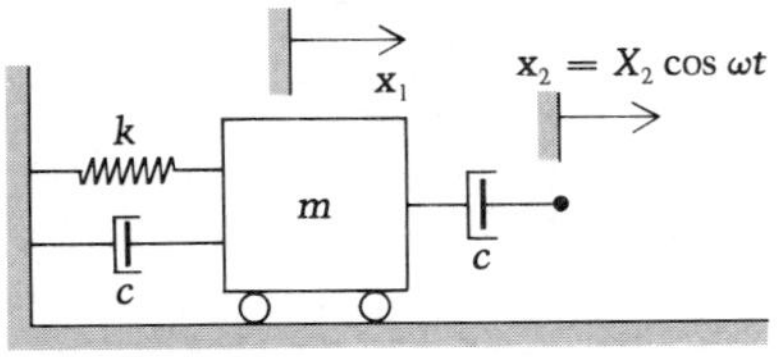

Figure P8.25

8.26 The cart in Figure P8.26 is given the motion $y(t) = 0.2 \cos 200t$ inch. Find the response $\mathbf{x}(t)$ if $\mathbf{x}(0) = \dot{\mathbf{x}}(0) = 0$ and

$$k_1 = k_2 = 100 \text{ lb/in.}$$

$$k_3 = 50 \text{ lb/in.}$$

$$m = 0.01 \text{ lb-sec}^2\text{/in.}$$

$$c = 2.0 \text{ lb-sec/in.}$$

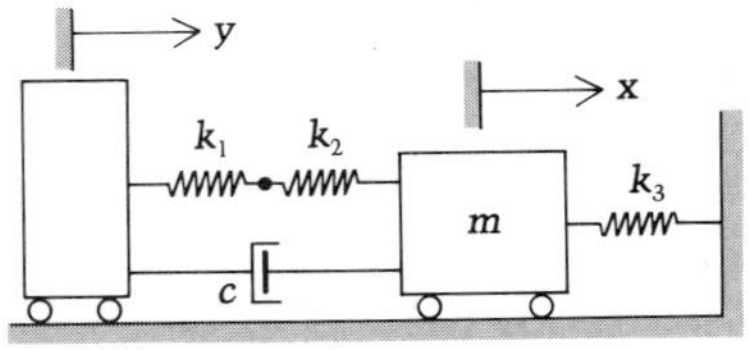

Figure P8.26

8.27 In Figure P8.27 find the steady-state displacement $\mathbf{x}(t)$ if $y(t) = 0.1 \sin 100t$ inch, where t is in seconds, $m = 0.01$ lb-sec²/in., $k = 100$ lb/in., and $c = 2$ lb-sec/in. In particular:

a. What is the amplitude of $\mathbf{x}(t)$?
b. What is the angle by which $\mathbf{x}(t)$ leads or lags $y(t)$?

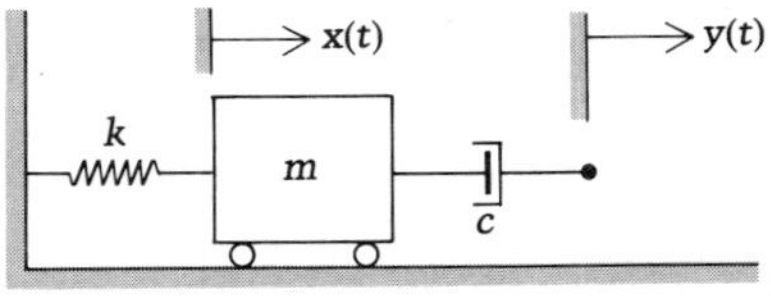

Figure P8.27

8.28 In Figure P8.28 find the steady-state displacement $\mathbf{x}(t)$ if $y(t) = 0.2 \sin 90t$ inch, where t is in seconds, $m = 0.01$ lb-sec²/in., $k = 50$ lb/in., and $c = 1$ lb-sec/in. In particular:

a. What is the amplitude of $\mathbf{x}(t)$?
b. What is the phase angle by which $\mathbf{x}(t)$ leads or lags $y(t)$?

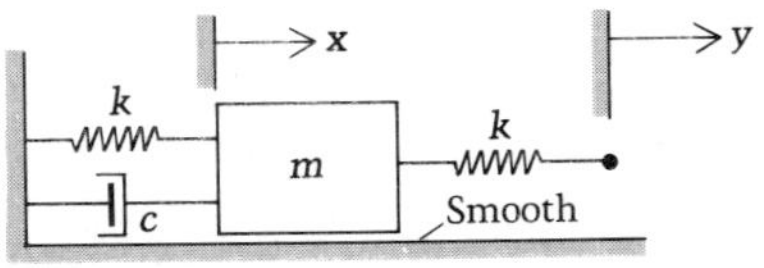

Figure P8.28

8.29 The block in Figure P8.29 is at rest in equilibrium prior to the application of the constant force $P = 50$ lb at $t = 0$. If $k = 100$ lb/in., $m = 0.01$ lb-sec^2/in., and the system is critically damped, find $x(t)$.

8.30 The cart in Figure P8.30 is at rest prior to $t = 0$, at which time the right end of the spring is given the motion $y = vt$, where v is a constant. Find $x(t)$.

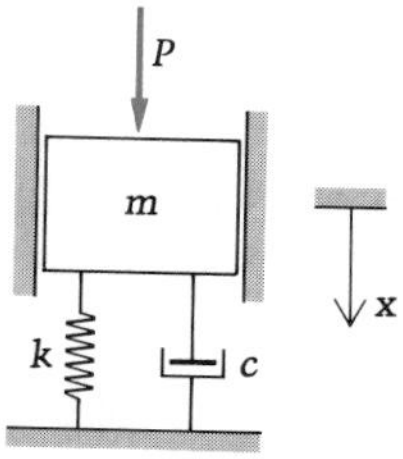

Figure P8.29

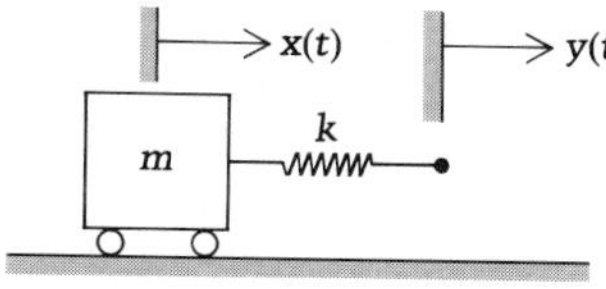

Figure P8.30

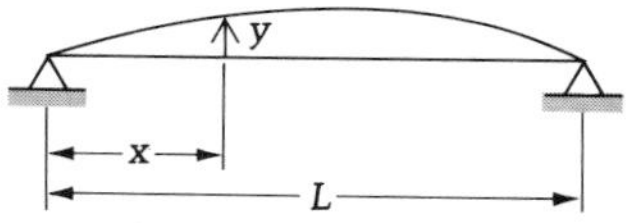

Figure P8.31

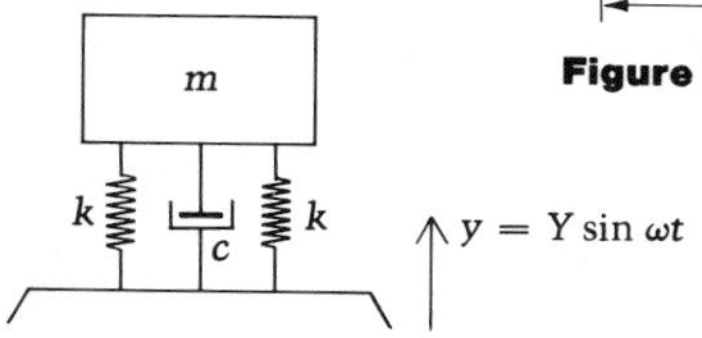

Figure P8.32

8.31 A uniform slender bar of density ρ, cross-sectional area A, and length L undergoes small-amplitude, free, transverse vibrations according to $y(x, t) = Y \sin(\pi x/L) \sin \omega t$, where y is the displacement perpendicular to the axis (x) of the bar. (See Figure P8.31.) Neglecting other components of displacement (and hence acceleration), calculate the maximum force generated at one of the supports during the motion. *Hint:* Assume that symmetry demands that the support reactions be the same; try using Equation (2.11).

8.32 The block of mass m in Figure P8.32 is mounted through springs k and damper c on a vibrating floor. Derive an expression for the steady-state acceleration of the block (whose motion is vertical translation). Show that the amplitude of the acceleration is less than that of the floor, regardless of the value of c, provided that $\omega > \sqrt{2}\omega_n$, where ω_n is the frequency of free undamped vibrations of the block. Show further that if $\omega > \sqrt{2}\omega_n$, then the smaller the damping the better the isolation.

8.33 Optical equipment is mounted on a table whose four legs are pneumatic springs. If the table and equipment together weigh 700 lb, what should be the stiffness of each spring so that the amplitude of steady, simple-harmonic, vertical displacement of the table will not be greater than 5 percent of a corresponding motion of the floor? Neglect damping in your calculations.

8.2 Euler's Laws for a Control Volume

Euler's laws describe the relationship between external forces and the motion of any body whether it be a solid, liquid, or gas. Sometimes, however, it is desirable to focus attention on some region of space (control volume) through which material may flow rather than on the fixed collection of particles that constitute a body. Examples of this sort are abundant in the field of fluid mechanics and include the important problem of describing and analyzing rocket-powered flight. Our purpose in this section is to discuss the forms taken by Euler's laws when the focus of attention is the control volume rather than the body.

We take as self-evident what might be called the "law of accumulation, production, and transport"—that is, the rate of accumulation of something within a region of space is equal to the rate of its production within the region plus the rate at which it is transported into the region.*

*The mathematical statement of this is known as the Reynolds Transport Theorem.

Thus, for example, the rate of accumulation of peaches in Georgia equals the rate of production of peaches in the state plus the net rate at which they are shipped in. This idea can be applied in mechanics whenever we are dealing with a quantity whose measure for a body is the sum of the measures for the particles making up the body. Thus we can apply this principle to things such as mass, momentum, moment of momentum, and kinetic energy.

Suppose that at an instant a closed region V (control volume) contains material (particles) making up body $\mathcal{B}$. Let $m_{\mathcal{B}}$ denote the mass of body $\mathcal{B}$ and m_V denote the mass associated with V (that is, the mass of whatever particles happen to be in V at some time). Instantaneously $m_V = m_{\mathcal{B}}$, but because some of the material of $\mathcal{B}$ is flowing out of V and some other material is flowing in, $\dot{m}_V \neq \dot{m}_{\mathcal{B}}$. In fact by the accumulation principle stated above

$$\dot{m}_V = \dot{m}_{\mathcal{B}} + (\text{net rate of mass flow into } V) \tag{8.29}$$

since clearly $\dot{m}_V$ represents the rate of buildup (accumulation) of mass in V and since $\dot{m}_{\mathcal{B}}$, the rate of change of mass of the material instantaneously within V, represents the production term. Of course a body, being a specific collection of particles, has constant mass; thus $\dot{m}_{\mathcal{B}} = 0$ and (8.29) becomes $\dot{m}_V =$ (rate of mass flow into V), which is often called the **continuity equation.**

For momentum $\mathbf{L}$, the statement corresponding to Equation (8.29) is

$$\dot{\mathbf{L}}_V = \dot{\mathbf{L}}_{\mathcal{B}} + (\text{net rate of flow of momentum into } V) \tag{8.30}$$

But Euler's first law applies to a body (such as $\mathcal{B}$) so that $\mathbf{F}_r = \dot{\mathbf{L}}_{\mathcal{B}}$, where $\mathbf{F}_r$ is the resultant of the external forces on $\mathcal{B}$—or, in other words, the resultant of the external forces acting on the material instantaneously in V. Thus Equation (8.30) becomes

$$\dot{\mathbf{L}}_V = \mathbf{F}_r + (\text{net rate of flow of momentum into } V) \tag{8.31}$$

which is the **control volume form of Euler's first law.** The momentum flow rate in the right side of (8.31) is calculated by summing up (or integrating) the momentum flow rates across infinitesimal elements of the boundary of V, where the momentum flow rate per unit of boundary area is the product of the mass flow rate per unit of area and the instantaneous velocity of the material as it crosses the boundary.

A similar derivation produces a **control-volume form of Euler's second law,** for which the result is

$$(\dot{\mathbf{H}}_O)_V = \mathbf{M}_{r_O} + \begin{pmatrix}\text{net rate of flow into } V \text{ of moment} \\ \text{of momentum with respect to } O\end{pmatrix} \tag{8.32}$$

where O is a point fixed in the inertial frame of reference.

It is important to realize that nothing in our derivations here has restricted the control volume except that it be a closed region in space. It may be moving relative to the frame of reference in almost any imaginable way, and it may be changing in shape or volume with time. We conclude this section with examples of two of the most common applications of Equation (8.31).

EXAMPLE 8.6

A fluid undergoes steady flow in a pipeline and encounters a bend at which the cross-sectional area of pipe changes from A_1 to A_2. At inlet 1 the density is ρ_1 and the velocity (approximately uniform over the cross section) is $v_1\hat{\mathbf{i}}$. At outlet 2 the density is ρ_2. Find the resultant force exerted on the pipe bend by the fluid. (See the diagram.)

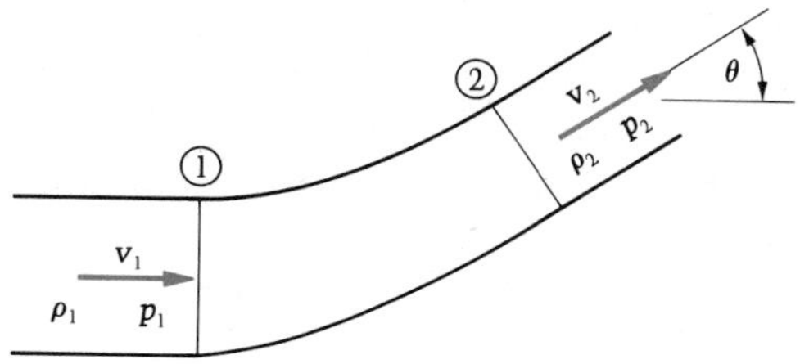

SOLUTION

Let the velocity of flow at the outlet be given by $v_2(\cos\theta\hat{\mathbf{i}} + \sin\theta\hat{\mathbf{j}})$. Then for steady flow the rate of mass flow at the inlet section is the same as that at the outlet section:

$$\rho_1 A_1 v_1 = \rho_2 A_2 v_2$$

so that

$$v_2 = \frac{\rho_1 A_1}{\rho_2 A_2} v_1$$

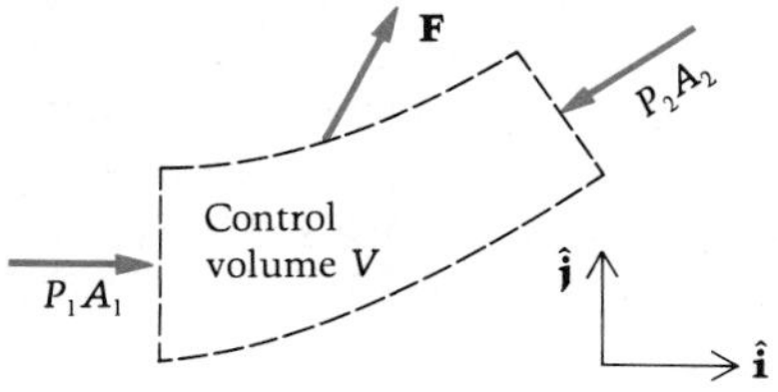

Let the control volume be the region bounded by the inner surface of the pipe bend and the inlet and outlet cross sections. (See the diagram.) A consequence of the condition of steady flow is that within the control volume the distributions of velocity and density are independent of time. Thus the total momentum associated with V is a constant and

$$\frac{d\mathbf{L}_V}{dt} = \mathbf{0}$$

But

$$\frac{d\mathbf{L}_V}{dt} = \mathbf{F}_r + (\text{net rate of momentum flow into } V)$$

(Continued)

Therefore

$$\mathbf{F}_r = (\text{net rate of momentum flow out of } V)$$

$$= \rho_2 A_2 v_2 (v_2 \cos\theta \hat{\mathbf{i}} + v_2 \sin\theta \hat{\mathbf{j}}) - \rho_1 A_1 v_1 (v_1 \hat{\mathbf{i}})$$

$$= \rho_1 A_1 v_1 (v_2 \cos\theta \hat{\mathbf{i}} + v_2 \sin\theta \hat{\mathbf{j}} - v_1 \hat{\mathbf{i}})$$

$$= \rho_1 A_1 v_1 \left[\left(\frac{\rho_1 A_1}{\rho_2 A_2} v_1 \cos\theta - v_1 \right) \hat{\mathbf{i}} + \left(\frac{\rho_1 A_1}{\rho_2 A_2} v_1 \sin\theta \right) \hat{\mathbf{j}} \right]$$

$$= \rho_1 A_1 v_1^2 \left[\left(\frac{\rho_1 A_1}{\rho_2 A_2} \cos\theta - 1 \right) \hat{\mathbf{i}} + \frac{\rho_1 A_1}{\rho_2 A_2} \sin\theta \hat{\mathbf{j}} \right]$$

And if $\mathbf{F}$ is the force exerted on the fluid by the bend, then

$$\mathbf{F}_r = \mathbf{F} + p_1 A_1 \hat{\mathbf{i}} + p_2 A_2 (-\cos\theta \hat{\mathbf{i}} - \sin\theta \hat{\mathbf{j}})$$

where p_1 and p_2 are the inlet and outlet fluid pressures respectively. Therefore

$$\mathbf{F} = \mathbf{F}_r - p_1 A_1 \hat{\mathbf{i}} + p_2 A_2 (\cos\theta \hat{\mathbf{i}} + \sin\theta \hat{\mathbf{j}})$$

and the force exerted on the bend by the fluid is $-\mathbf{F}$ with

$$-\mathbf{F} = \left[p_1 A_1 - p_2 A_2 \cos\theta + \rho_1 A_1 v_1^2 \left(1 - \frac{\rho_1 A_1}{\rho_2 A_2} \cos\theta \right) \right] \hat{\mathbf{i}}$$

$$- \left[p_2 A_2 \sin\theta + \rho_1 A_1 v_1^2 \left(\frac{\rho_1 A_1}{\rho_2 A_2} \right) \sin\theta \right] \hat{\mathbf{j}}$$

EXAMPLE 8.7

To illustrate how the control volume form of Euler's first law is used to describe the motion of a rocket vehicle, consider such a vehicle climbing in a vertical rectilinear flight. Let $v\hat{\mathbf{j}}$ be the velocity of the vehicle from which combustion products are being expelled at velocity $-v_e\hat{\mathbf{j}}$ relative to the rocket. Further let $M(t)$ be the mass at time t of the vehicle and its contents, let μ be the rate of mass flow of the ejected gases, and let p be the gas pressure at the nozzle exit of cross section A.

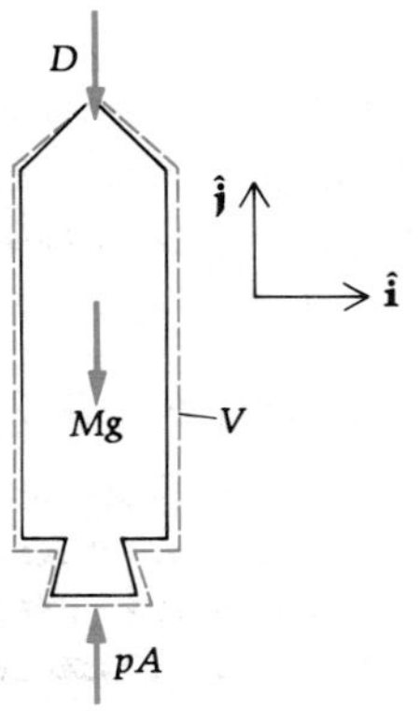

Figure 1 Free-body diagram of rocket

(Continued)

SOLUTION

Force D in the free-body diagram, representing the drag or resistance to motion, is the resultant of (1) all the shear stresses acting on the surface of the vehicle and (2) all the pressure on the surface. Thus $(pA - D)\hat{\mathbf{j}}$ represents the resultant of all the surface-distributed forces on the control volume. The two terms have been separated so that we may point out that the force pA remains even after the rocket has cleared the atmosphere. That is, p is pressure exerted by the gas particles *about to pass across* the nozzle exit plane* on the particles that have *just passed across*, and vice-versa.

If we let the control volume V surround the vehicle, Equation (8.31) becomes

$$\frac{d}{dt}\mathbf{L}_V = (pA - D)\hat{\mathbf{j}} - Mg\hat{\mathbf{j}} - \mu(v - v_e)\hat{\mathbf{j}}$$

At this point we must approximate $\mathbf{L}_V$ by $Mv\hat{\mathbf{j}}$; this is an approximation because some of the products of combustion inside the rocket are of course moving relative to the vehicle. Therefore

$$\frac{d}{dt}(Mv\hat{\mathbf{j}}) = (pA - D)\hat{\mathbf{j}} - Mg\hat{\mathbf{j}} - \mu(v - v_e)\hat{\mathbf{j}}$$

or

$$v\frac{dM}{dt} + M\frac{dv}{dt} = pA - D - Mg - \mu v + \mu v_e$$

But of course

$$\frac{dM}{dt} = -\mu$$

so that

$$M\frac{dv}{dt} = pA - D - Mg + \mu v_e$$

which is of the form of force = mass × acceleration, where one of the "forces" is the "thrust" μv_e.

*This term may be neglected if exhaust gases have expanded to atmospheric pressure or nearly so.

PROBLEMS / Section 8.2

8.34 Let dm_i/dt and dm_o/dt be the respective rates at which mass enters and leaves a system. Show that Equation (8.31) may be expressed in terms of these rates as

$$\mathbf{F}_r = m\frac{d\mathbf{v}}{dt} + \left(\frac{dm_i}{dt} - \frac{dm_o}{dt}\right)\mathbf{v} + \frac{dm_o}{dt}\mathbf{v}_o - \frac{dm_i}{dt}\mathbf{v}_i$$

where $m\mathbf{v} = \mathbf{L}_V$ and it is assumed that all the incoming particles have a common velocity $\mathbf{v}_i$ (in an inertial frame) and that all the exiting particles have a common velocity $\mathbf{v}_o$.

8.35 Liquid of specific weight w flows out of a hole in the side of a tank in a jet of cross section A. If the velocity of the jet is $\mathbf{v}$, determine the force exerted on the tank by the supporting structure that holds the tank at rest. Note that the pressure in the jet will be atmospheric pressure.

8.36 A child aims a garden hose at the back of a friend. (See Figure P8.36.) If the water (specific weight 62.4 lb/ft^3) stream has a diameter of $\frac{1}{4}$ in. and a speed of 50 ft/sec, estimate the force exerted on the "target" if: (a) he is stationary; (b) he is running away from the stream at a speed of 10 ft/sec. Assume the flow in contact with the boy's back to be vertical relative to him; that is, neglect any splashback.

8.37 A steady jet of liquid is directed against a smooth rigid surface and the jet splits as shown in Figure P8.37. Assume that each fluid particle moves in a plane parallel to that of the figure and ignore gravity. Ignoring gravity and friction, it can be shown that the particle speed after the split is still v as depicted. Estimate the fraction of the flow rate occurring in each of the upper and lower branches. *Hint:* Use the fact that no external force tangent to the surface acts on the liquid.

8.38 Air flows into the intake of a jet engine at mass flow rate q (slug/sec or kg/s). If v is the speed of the airplane flying through still air and u is the speed of engine exhaust relative to the plane, derive an expression for the force (thrust) of the flowing fluid on the engine. Neglect the fact that the rate of exhaust is slightly greater than q because of the addition of fuel in the engine.

8.39 Revise the analysis of the preceding problem to account for the mass of fuel injected into the engine. Let f be the mass flow rate of the fuel, and assume that the fuel is injected with no velocity relative to the engine housing.

8.40 Air at rest is sucked into a turbojet aircraft at the flow rate W. This indrawn air is accelerated in the turbine by means of fuel that is burned at a rate kW. The burned fuel and the air are ejected with constant velocity u relative to the aircraft, and air resistance is av^2, where a is a constant. If $W = 6.44$ lb/sec, $k = 0.05$, $u = 2000$ ft/sec, and the aircraft is climbing at an angle of 30° to the horizontal at a constant velocity of 500 ft/sec, determine the value of the air-resistance coefficient a.

8.41 In a quarry, rocks slide onto a conveyor belt at the constant mass flow rate k_i and at speed v_{rel} relative to the ground. (See Figure P8.41.) The belt is driven by a motor (with torque M applied to the drum on the right) at constant speed v_B. Find the power that the motor must deliver, neglecting friction in the shaft bearings and assuming the belt does not slip. *Hint:* Use the control volume indicated by the dashed lines to compute the difference in belt tensions, neglecting any sag of the belt due to the weight of the rocks.

Figure P8.36

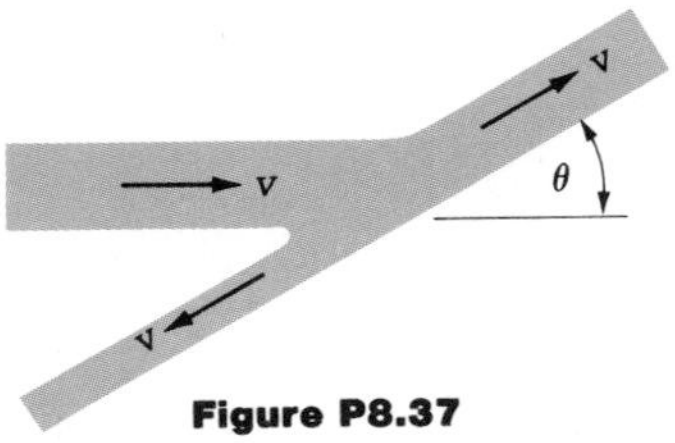

Figure P8.37

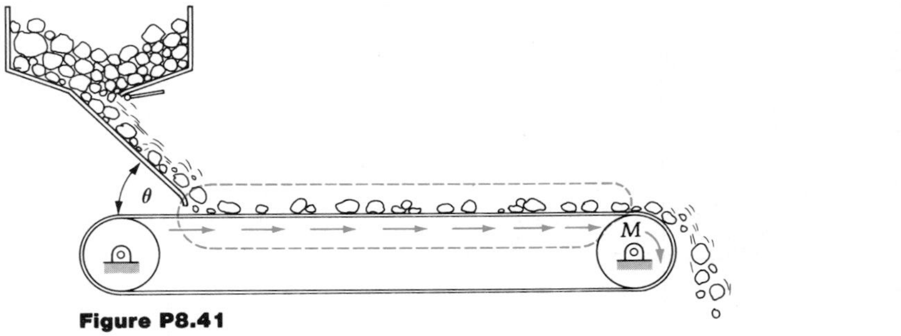

Figure P8.41

8.42 A black box with an initial mass of m_0 (of which 10 percent is box and 90 percent is fuel) is released from rest on the inclined plane in Figure P8.42. The coefficient of friction is μ between the box and the plane.

a. Show that with $\tan \alpha > \mu$, the box will begin to slide down the plane.

b. Assume now that $\tan \alpha > \mu$ and that a mechanism in the box is able to sense its velocity and eject particles rearward (up the plane) at a constant mass flow rate of k_0, and at a relative velocity always equal to the negative of the velocity of the box. Find the velocity of the box at the time t_f when the last of the fuel leaves.

c. Show that the box is going 5.5 times faster at $t = t_f$ than it would have gone if no fuel had been ejected.

8.43 Sand is being dumped on a flatcar of mass M at the constant mass flow rate of q. (See Figure P8.43.) The car is being pulled by a constant force P, and friction is negligible. The car was at rest at $t = 0$. Determine the car's acceleration as a function of P, M, q, and t.

8.44 A coal truck weighs 5 tons when empty. It is pushed under a loading chute by a constant force of 500 lb. The chute, inclined at 60° as shown in Figure P8.44, delivers 100 lb of coal per second to the truck at a velocity of 30 ft/sec. When the truck contains 10 tons of coal, its velocity is 10 ft/sec to the right. (a) What is its acceleration at this instant? The wheels are light and all horizontal frictional forces may be taken as included in the 500-lb force. (b) What is the horizontal component of force on the truck from the coal at this instant?

8.45 A raindrop falls through a region of fog, collecting mist from the region through which it drops. If it starts from rest with zero radius and stays round at all times (see Figure P8.45a) show that its acceleration is $g/7$. *Hint:* In spherical coordinates, the component of vertical velocity normal to dA is $\dot{x}\cos\phi$ (Figure P8.45b). Integrating, show then that the rate of increase of mass of the drop with time is

$$\dot{m} = \int_{\phi=0}^{\pi/2}\int_{\theta=0}^{2\pi} \rho\dot{x}\cos\phi \underbrace{r^2\sin\phi\, d\theta\, d\phi}_{dA} = \rho\pi r^2\dot{x}$$

where ρ is the mass density (assumed constant). But $\dot{m}$ also equals (ρ)(surface area)$(\dot{r}) = \rho 4\pi r^2\dot{r}$, so that $\dot{x} = 4\dot{r}$, or $x = 4r$ plus a constant that is zero. Noting that $m = (4/3)\pi r^3\rho$, substitute into the equation of Problem 8.34, with $\dot{m} = dm_i/dt - dm_o/dt$. You should obtain $x\ddot{x} + 3\dot{x}^2 = gx$, and trying the substitution $\dot{x} = c\sqrt{x}$ will lead to the solution.

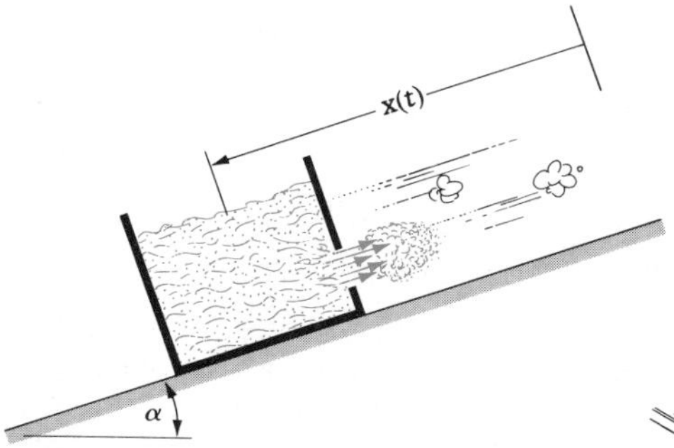

Figure P8.42

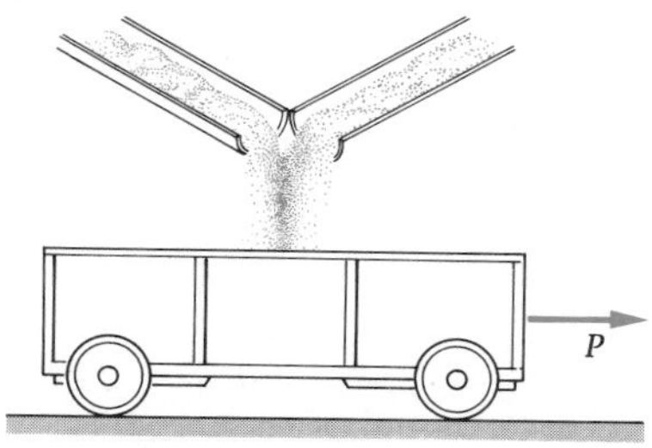

Figure P8.43

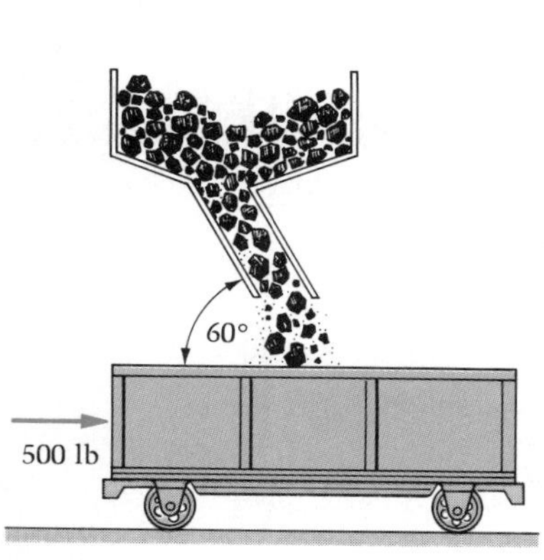

Figure P8.44

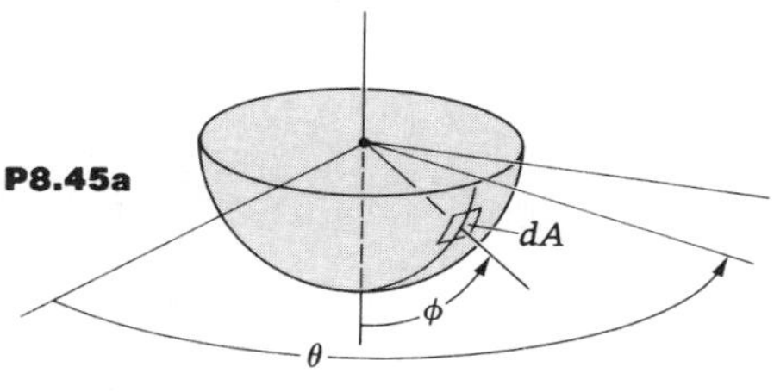

Figure P8.45a

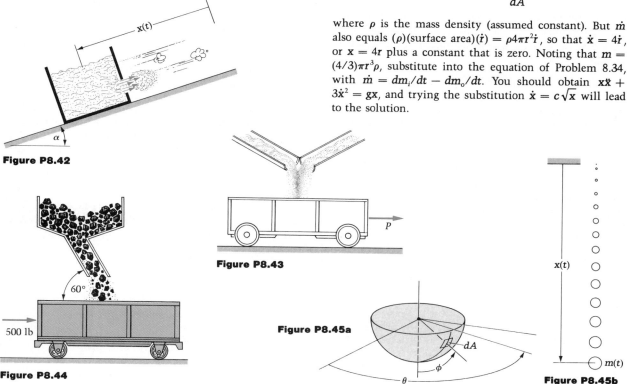

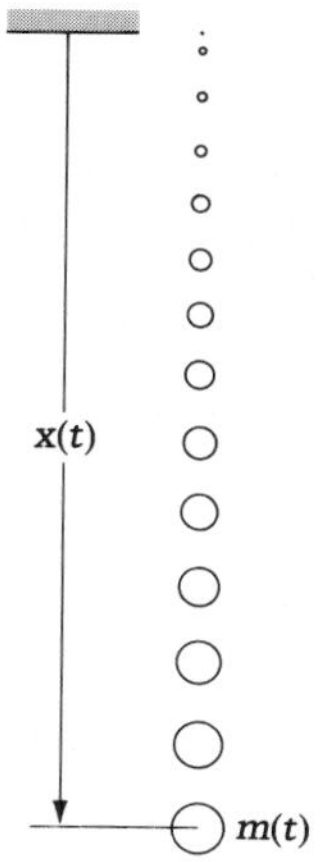

Figure P8.45b

8.46 Santa Claus weighs 450 lb and drops down a 20-ft chimney (Figure P8.46). He gains mass in the form of ashes and soot at the rate of 3 slugs/sec from a *very* dirty chimney.

a. Find Santa's velocity as a function of time. (Neglect friction.)

b. Calculate the velocity v_b and the time t_b at which he would hit bottom *without* adding mass and then compare v_b with his "ashes and soot velocity" at the same t_b.

8.47 Spherical raindrops produced by condensation are precipitated from a cloud when their radius is a. They fall freely from rest, and their radii increase by accretion of moisture at a uniform rate k. Find the velocity of a raindrop at time t, and show that the distance fallen in that time is

$$\frac{gt^2}{8}\left(\frac{2a + kt}{a + kt}\right)^2$$

8.48 The machine gun in Figure P8.48 has mass M exclusive of its bullets, which have mass M' in total. The bullets are fired at the mass rate of K_0 "slugs" per second, with velocity u_0 relative to the ground. If the coefficient of friction between the gun's frame and the ground is μ, find the velocity of the gun at the instant the last bullet is fired.

8.49 Bonnie and Clyde are making a getaway in a cart with negligible friction beneath its wheels. (See Figure P8.49.) Clyde is killing two birds with one stone by using his machine gun to propel the car as well as to ward off pursuers. He fires 500 rounds (shots) per minute with each bullet weighing 1 oz and exiting the muzzle with a speed relative to the car of 2500 ft/sec. The bullets originally comprised 2 percent of an initial total mass of $m_0 = 20$ slugs. If the system starts from rest at $t = 0$, find: (a) the maximum speed of Bonnie and Clyde; (b) how long it takes to attain this speed.

8.50 A 5-ton car is at rest, with brakes at the point of slipping on light wheels. (See Figure P8.50.) It suddenly begins raining hard, 1 in. accumulating every 10 min. The rain comes straight down at 8 ft/sec. Find the velocity of the car after 3 minutes. (Assume the water level to be horizontal in the calculations.)

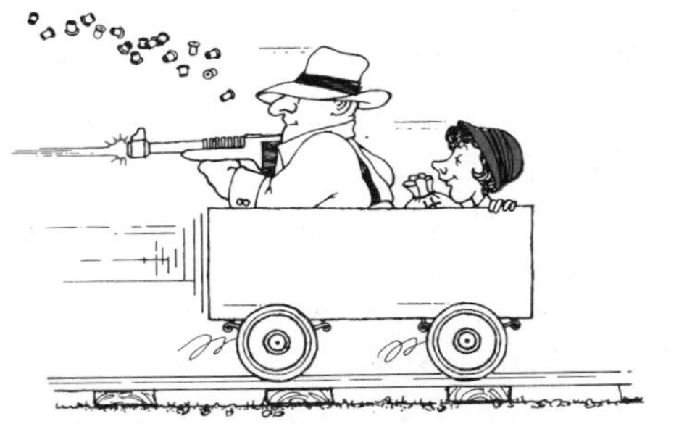

Figure P8.49

Figure P8.46

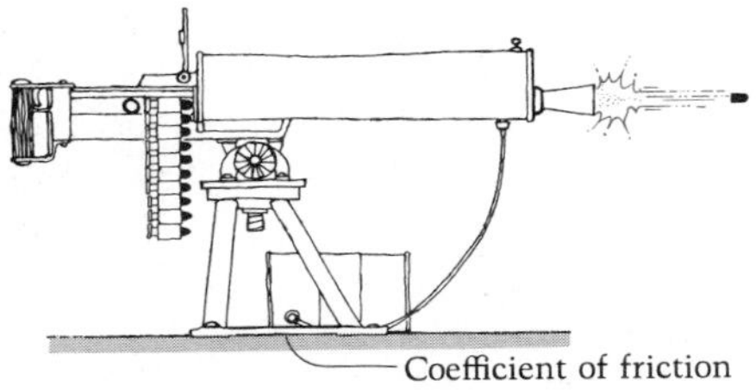

Figure P8.48

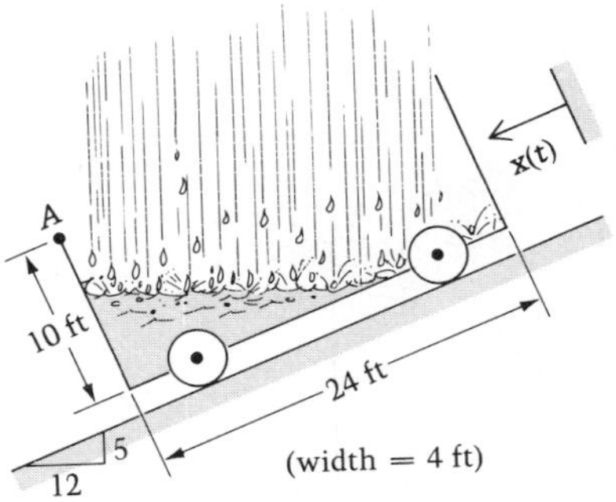

Figure P8.50

8.51 Solve the final equation of Example 8.7 for the velocity $v\hat{\mathbf{j}}$ of the rocket as a function of time in the case in which the pressure force pA and the drag D are negligible, the initial mass of the rocket is m_0, and the gravitational acceleration g, the rate μ, and the relative velocity v_e are all constants. Initially, the rocket is at rest.

8.52 Repeat the preceding problem, but this time include a drag force of $-kv$, where k is a positive constant.

8.53 (a) Extending Problem 8.51, find the height of the rocket as a function of time. (b) If the fraction of m_0 which is fuel is f, find the rocket's "burnout" velocity and position when all the fuel is spent.

8.54 From a rocket that is free to move vertically upward, matter is ejected downward with a constant relative velocity gT at a constant rate $2M/T$. Initially the rocket is at rest and has mass $2M$, half of which is available for ejection. Neglecting air resistance and variations in the gravitational attraction, (a) show that the greatest upward speed is attained when the mass of the rocket is reduced to M, and determine this speed. (b) Show also that the rocket rises to a height

$$\tfrac{1}{2}gT^2(1 - \ln 2)^2$$

8.55 A small rocket is fired vertically upward. Air resistance is neglected. Show that for the rocket to have constant acceleration upward, its mass m must vary with time t according to the equation

$$\frac{dm}{dt} = -\frac{a + g}{u}m$$

where a is the acceleration of the rocket and u is the velocity of the escaping gas relative to the rocket.

8.56 The end of a chain of length L and weight per unit length w, which is piled on a platform, is lifted vertically by a variable force P so that it has a constant velocity v. (See Figure P8.56.) Find P as a function of x. *Hint:* Choose a control-volume boundary so that material crosses the boundary (with negligible velocity) just *before* it is acted on by the moving material already in the control volume. That is, there is no force transmitted across the boundary of the control volume. The solution will be an approximation to reality because of assuming arbitrarily small individual links; but the more links having the common velocity of the fully engaged links within the volume, the better the approximation will be.

8.57 A chain of length L weighing γ per unit length begins to fall through a hole in a ceiling. (See Figure P8.57.) Referring to the hint in the previous problem:

- **a.** Find $v(x)$ if $v = 0$ when x and t are zero.
- **b.** Show that the falling chain's acceleration is the constant $g/3$.
- **c.** Show that when the last link has left the ceiling, the chain has lost more potential energy than it has gained in kinetic energy, the difference being $\gamma L^2/6$. Explain the reason for this loss.

8.58 A particle of mass m, initially at rest, is projected with velocity $\mathbf{v}_0$ at an angle α to the horizontal and moves under gravity. (See Figure P8.58.) During its flight, it gains mass at the uniform rate k. If air resistance is neglected, show that its equation of motion is

$$(m + kt)\ddot{\mathbf{r}} + k\dot{\mathbf{r}} = (m + kt)g\hat{\mathbf{k}}$$

and that the equation of its path is

$$\mathbf{r} = \frac{m^2}{4k^2}\left[\left(1 + \frac{kt}{m}\right)^2 - 1 - 2\log\left(1 + \frac{kt}{m}\right)\right]g\hat{\mathbf{k}} + \frac{m}{k}\log\left(1 + \frac{kt}{m}\right)\mathbf{v}_0$$

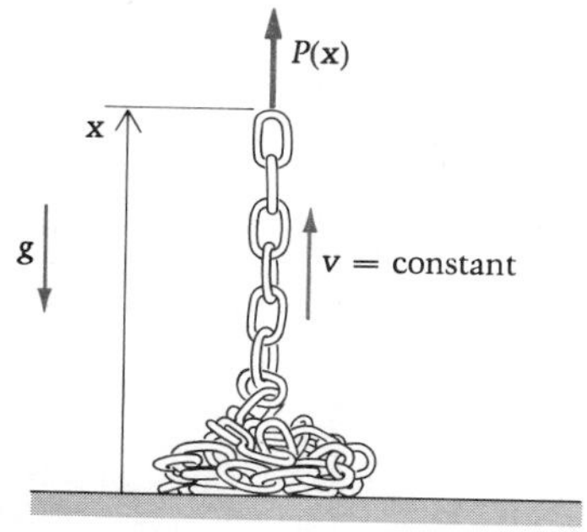

Figure P8.56

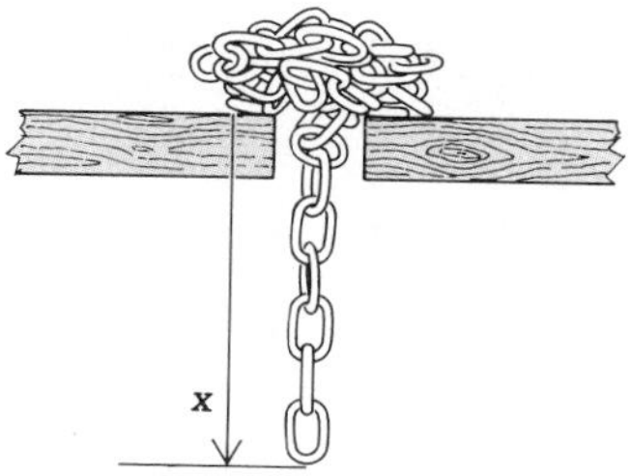

Figure P8.57

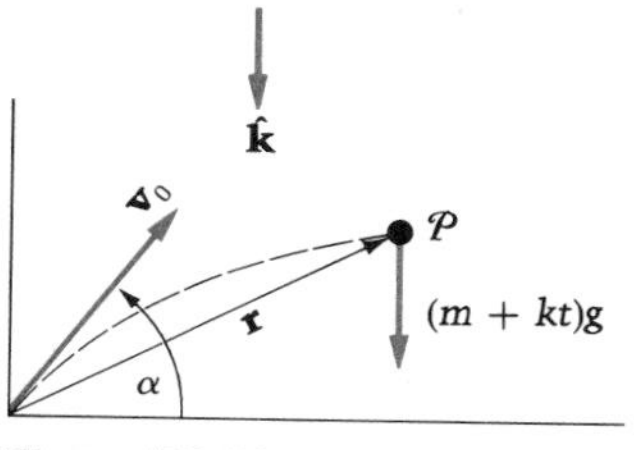

Figure P8.58

8.59 If in the preceding problem the air offers a resistance $-c\dot{\mathbf{r}}$, determine the equation of the path.

8.60 With the same notation and conditions as in Problem 8.34, show that Equation (8.32) may be written as

$$\mathbf{M}_{r_o} = (\dot{\mathbf{H}}_o)_v + \frac{dm_o}{dt}(\mathbf{r}_o \times \mathbf{v}_o) - \frac{dm_i}{dt}(\mathbf{r}_i \times \mathbf{v}_i)$$

where $\mathbf{r}_i$ and $\mathbf{r}_o$ are position vectors for the mass centers of the incoming and exiting particles.

8.61 A pinwheel of radius a, which can turn freely about a horizontal axis, is initially of mass M and moment of inertia I about its center. A charge is spread along the rim and ignited at time $t = 0$. While the charge is burning, the rim of the wheel loses mass at a constant rate m_1 mass units per second, and at the rim a mass m_2 of gas is taken up per second from the atmosphere, which is at rest. The total mass $m_1 + m_2$ is discharged per second tangentially from the rim, with velocity v relative to the rim. Prove that if θ is the angle through which the wheel has turned after t sec, then

$$\theta = \frac{v}{a(\mu - \lambda)}[\mu t - 1 + (1 - \lambda t)^{\mu/\lambda}]$$

where

$$\lambda = \frac{m_1 a^2}{I} \qquad \mu = \frac{(m_1 + m_2)a^2}{I}$$

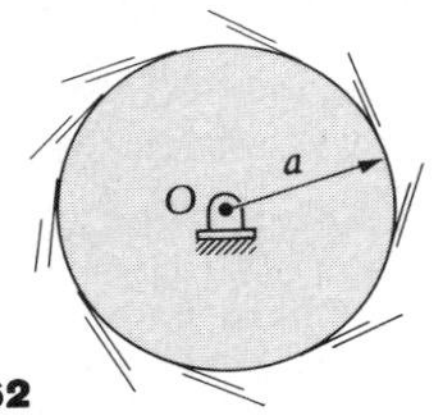

Figure P8.62

8.62 A wheel of radius a starts from rest and fires out matter at uniform rate from all points on the rim (Figure P8.62). The matter leaves tangentially with relative speed v and at such a rate that the mass decreases at the rim by m mass units per second. Show that the angle θ turned through by the wheel is given by

$$\theta = \frac{vI_O}{ma^3}\left[\left(1 - \frac{ma^2t}{I_O}\right)\ln\left(1 - \frac{ma^2t}{I_O}\right) + \frac{ma^2t}{I_O}\right]$$

in which I_O is the initial moment of inertia of the wheel about its axis.

8.3 Central Force Motion

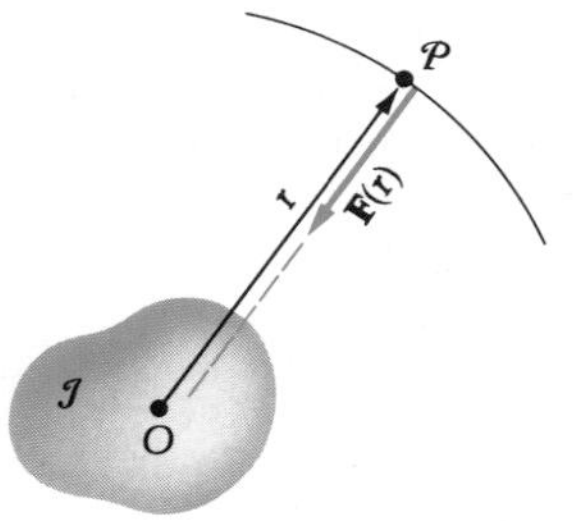

Figure 8.12

In Chapter 2 we defined a **central force** acting on a particle $\mathcal{P}$* as one which (1) always passes through a certain point O fixed in the inertial reference frame $\mathcal{I}$ and (2) depends only on the distance r between O and $\mathcal{P}$. (See Figure 8.12.) In this section we are going to treat the central force in more detail. We shall go as far as we can without specializing $\mathbf{F}(r)$—that is, without saying how $\mathbf{F}$ depends on r. In the second part of the section we shall study the most important of central forces: gravitational attraction.

If the central force $\mathbf{F}$ is the only force acting on the particle, then $\mathbf{F} = m\mathbf{a}$; and since the central force always passes through point O, $\mathbf{r} \times \mathbf{F}$ is identically zero. These two facts allow us to write:

$$\mathbf{r} \times \mathbf{F} = \mathbf{r} \times m\mathbf{a} = \mathbf{0}$$

or, since $\dot{\mathbf{r}} = \mathbf{v}$,

$$\frac{d}{dt}(\mathbf{r} \times m\mathbf{v}) = \mathbf{0}$$

Therefore for a particle acted on only by a central force,

$$\mathbf{r} \times \mathbf{v} = \text{constant vector in } \mathcal{I} = \mathbf{h}_O \tag{8.33}$$

*We use the term *particle* in this section for the sake of brevity. We are of course investigating the motion of the mass center of a body.

Dotting this equation with $\mathbf{r}$, we find, since $\mathbf{r} \times \mathbf{v}$ is perpendicular to $\mathbf{r}$,

$$\mathbf{r} \cdot (\mathbf{r} \times \mathbf{v}) = 0 = \mathbf{r} \cdot \mathbf{h}_O$$

and we see that $\mathbf{r}$ is always perpendicular to a vector that is constant in $\mathcal{I}$; therefore $\mathcal{P}$ moves in a plane in $\mathcal{I}$. Using polar coordinates to then describe the motion of $\mathcal{P}$ in this plane, the equations are:

$$F_r = -F(r) = m(\ddot{r} - r\dot{\theta}^2) \tag{8.34}$$

and

$$F_\theta = 0 = m(r\ddot{\theta} + 2\dot{r}\dot{\theta}) = \frac{m}{r}\frac{d}{dt}(r^2\dot{\theta}) \tag{8.35}$$

From Equation (8.35) we see immediately that

$$r^2\dot{\theta} = \text{constant} = h_O \tag{8.36}$$

where h_O is the magnitude of the constant vector $\mathbf{h}_O$ of Equation (8.33) because, expressing $\mathbf{r}$ and $\mathbf{v}$ in polar coordinates, we find

$$\mathbf{h}_O = \mathbf{r} \times \mathbf{v} = \text{constant} = r\hat{\mathbf{e}}_r \times (\dot{r}\hat{\mathbf{e}}_r + r\dot{\theta}\hat{\mathbf{e}}_\theta) = r^2\dot{\theta}\hat{\mathbf{k}}$$

so that

$$|\mathbf{h}_O| = h_O = r^2\dot{\theta} = \text{constant} \tag{8.37}$$

Equation (8.37) is a statement of the conservation of angular momentum of $\mathcal{P}$; the constant h_O is the magnitude of the angular momentum $\mathbf{H}_O$ of $\mathcal{P}$ divided by its mass m. Thus we shall call h_O the angular momentum (magnitude) per unit mass.

We can use the previous pair of results to show that the second of Kepler's three laws of planetary motion is in fact valid for any central force. This law states that the radius vector from the sun to a planet sweeps out equal areas in equal time intervals. From Figure 8.13 the incremental planar area ΔA swept out by $\mathcal{P}$ between θ (at t) and $\theta + \Delta\theta$ (at $t + \Delta t$) is approximately given by the area of the triangle OBB'* (see Figure 8.14):

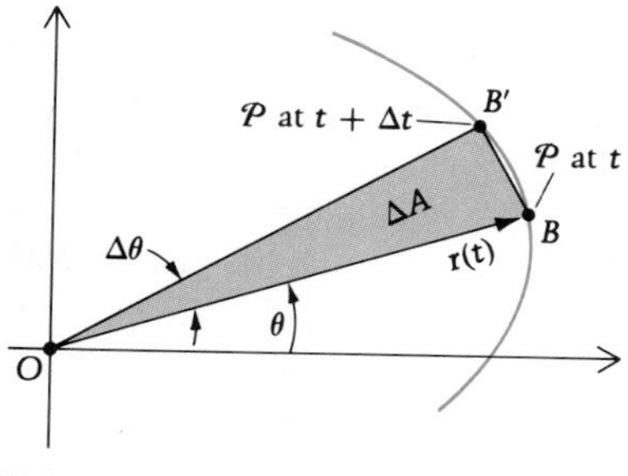

Figure 8.13

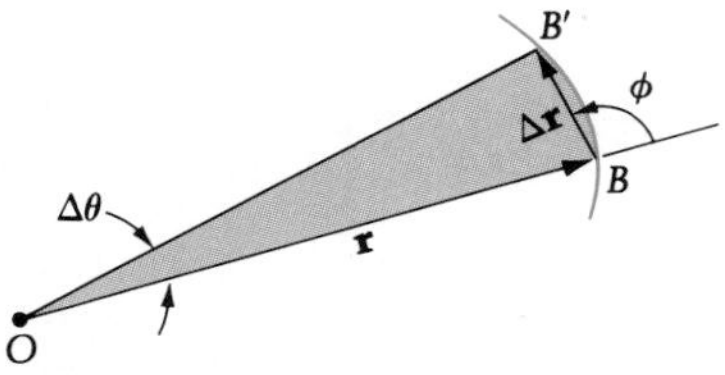

Figure 8.14

$$\begin{aligned}\Delta A &\approx \tfrac{1}{2} \times \text{base} \times \text{height} \\ &\approx \tfrac{1}{2} r\,(\Delta r \sin\phi) \\ &\approx \tfrac{1}{2}|\mathbf{r} \times \Delta\mathbf{r}|\end{aligned}$$

Dividing by the time increment Δt and taking the limit as $\Delta t \to 0$, we have

$$\lim \frac{\Delta A}{\Delta t} = \frac{dA}{dt} = \frac{1}{2}\lim_{\Delta t \to 0}\left|\mathbf{r} \times \frac{\Delta\mathbf{r}}{\Delta t}\right| = \frac{1}{2}\overbrace{|\mathbf{r} \times \mathbf{v}|}^{\mathbf{h}_O}$$

or

$$\frac{dA}{dt} = \frac{h_O}{2} \qquad \text{(a constant that is } r^2\dot{\theta}/2) \tag{8.38}$$

*"Approximately" because the area between arc and chord is outside the triangle.

Thus the rate of sweeping out area is a constant. This is why a satellite or a planet in elliptical orbit (Figure 8.15) has to travel faster when it is near the *perigee* than the *apogee*—the same area must be swept out in the same period of time.* We emphasize again that this result is valid for *all* central force trajectories, not just elliptical orbits and not just if the central force is gravity.

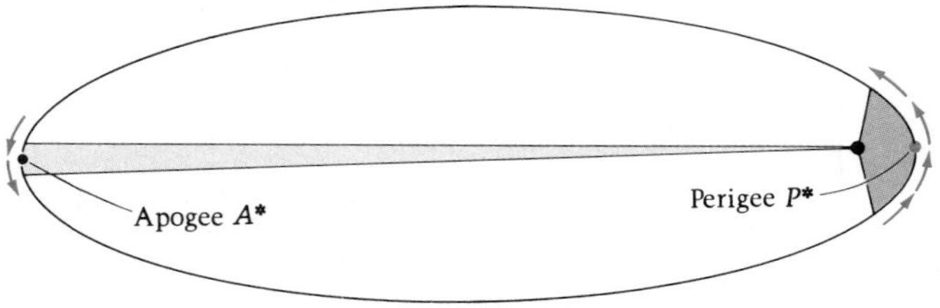

Figure 8.15

Next we focus our attention on the most important central force: gravitational attraction. If G is the universal gravitational constant and M and m are the masses of what we are considering to be the attracting and attracted bodies,† then the central force acting on m for this case is

$$F(r) = \frac{GMm}{r^2} \tag{8.39}$$

and Equation (8.34) becomes

$$m(\ddot{r} - r\dot{\theta}^2) = -\frac{GMm}{r^2} \tag{8.40}$$

Canceling m and inserting h_O for $r^2\dot{\theta}$ gives

$$\ddot{r} - \frac{h_O^2}{r^3} = -\frac{GM}{r^2} \tag{8.41}$$

Multiplying Equation (8.39) by $\dot{r}$ will allow us to integrate it:

$$\dot{r}\ddot{r} - h_O^2 r^{-3}\dot{r} = -GMr^{-2}\dot{r} \tag{8.42}$$

Integrating, we get

$$\frac{\dot{r}^2}{2} + h_O^2\frac{r^{-2}}{2} = \frac{-GMr^{-1}}{-1} + C_1 \tag{8.43}$$

If we multiply Equation (8.43) by m and replace h_O by $r^2\dot{\theta}$, we see that

*We use *perigee* and *apogee* in a general sense; technically these words refer to the nearest and farthest points, respectively, for the moon and artificial satellites. For the orbits of planets, the proper terms are *perihelion* and *aphelion*.

†Actually, of course, both are attracting and both are attracted—each to the other! The constant GM is, for the sun, 4.68×10^{21} ft^3/sec^2.

$$\frac{m}{2}[\dot{r}^2 + (r\dot{\theta})^2] - \frac{GMm}{r} = C_1 m \tag{8.44}$$

and the left side of Equation (8.44) is seen to be the total energy of $\mathcal{P}$, kinetic plus potential. Thus we shall replace C_1 by E, the energy of $\mathcal{P}$ per unit mass, and obtain

$$\dot{r}^2 + h_O^2 r^{-2} = +2GMr^{-1} + 2E \tag{8.45}$$

This equation will be helpful to us later. But now we are interested in studying the trajectory of particle $\mathcal{P}$—that is, in finding r as a function of θ. By the chain rule,

$$\frac{dr}{dt} = \frac{dr}{d\theta}\frac{d\theta}{dt} = \dot{\theta}\frac{dr}{d\theta}$$

and since $\dot{\theta} = h_O/r^2$ from Equation (8.36),

$$\frac{dr}{dt} = \frac{h_O}{r^2}\frac{dr}{d\theta} \tag{8.46}$$

We need the second derivative of r in Equation (8.41), so we apply the chain rule once more:

$$\begin{aligned}\frac{d^2r}{dt^2} &= \left[\frac{d}{d\theta}\left(\frac{h_O}{r^2}\frac{dr}{d\theta}\right)\right]\frac{d\theta}{dt} \\ &= \frac{h_O}{r^2}\left[-\frac{2h_O}{r^3}\left(\frac{dr}{d\theta}\right)^2 + \frac{h_O}{r^2}\frac{d^2r}{d\theta^2}\right] \\ &= \frac{h_O^2}{r^4}\frac{d^2r}{d\theta^2} - \frac{2h_O^2}{r^5}\left(\frac{dr}{d\theta}\right)^2 = \frac{h_O^2}{r^2}\frac{d}{d\theta}\left(\frac{1}{r^2}\frac{dr}{d\theta}\right)\end{aligned} \tag{8.47}$$

Substituting into Equation (8.41), we get

$$\frac{h_O^2}{r^2}\frac{d}{d\theta}\left(\frac{1}{r^2}\frac{dr}{d\theta}\right) - \frac{h_O^2}{r^3} = -\frac{GM}{r^2}$$

or

$$\frac{d}{d\theta}\left(\frac{1}{r^2}\frac{dr}{d\theta}\right) - \frac{1}{r} = -\frac{GM}{h_O^2} \tag{8.48}$$

The following simple change of variables will make the solution to this differential equation immediately recognizable:

$$u = \frac{1}{r} \tag{8.49}$$

Substituting Equation (8.49) into (8.48) along with

$$\frac{dr}{d\theta} = \frac{dr}{du}\frac{du}{d\theta} = \frac{-1}{u^2}\frac{du}{d\theta} \tag{8.50}$$

gives

$$\frac{d}{d\theta}\left[u^2\left(\frac{-1}{u^2}\frac{du}{d\theta}\right)\right] - u = \frac{-GM}{h_O^2} \tag{8.51}$$

or

$$\frac{d^2u}{d\theta^2} + u = \frac{GM}{h_O^2} \tag{8.52}$$

The solution to Equation (8.52), from elementary differential equations, consists of a homogeneous (or complementary) part plus a particular part:

$$u = \overbrace{A_1 \cos\theta + B_1 \sin\theta}^{u_H} + \overbrace{\frac{GM}{h_O^2}}^{u_P} \tag{8.53}$$

Switching variables back from u to r by Equation (8.49), we obtain

$$r = \frac{h_O^2/GM}{1 + (h_O^2/GM)(A_1 \cos\theta + B_1 \sin\theta)} \tag{8.54}$$

This solution for $r(\theta)$ is the equation of a **conic;** it can be put into a more recognizable form after a brief review of conic sections. For every point P on a conic, the ratio of the distances from P to a fixed point (O: the focus) and to a fixed line (ℓ: the directrix) is a constant called the **eccentricity** of the conic:

$$e = \frac{OP}{L} \tag{8.55}$$

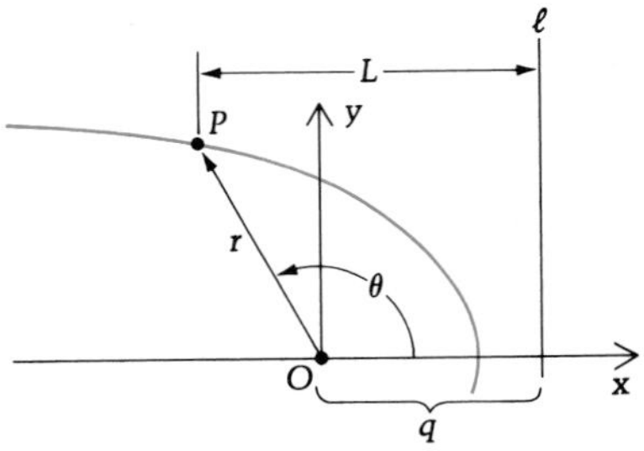

Figure 8.16

Therefore, in terms of the parameters in Figure 8.16,

$$e = \frac{r}{q - r\cos\theta} \tag{8.56}$$

or, solving for r,

$$r = \frac{eq}{1 + e\cos\theta} \tag{8.57}$$

The conic specified by Equation (8.57) is a:

$$\begin{aligned} &\text{Hyperbola if } |e| > 1 \\ &\text{Parabola if } e = 1 \\ &\text{Ellipse if } -1 < e < 1 \end{aligned} \tag{8.58}$$

The ellipse becomes a circle if $e = 0$. It is an ellipse with perigee (closest point to O) at $\theta = 0$ if $0 < e < 1$ and an ellipse with apogee (farthest point from O) at $\theta = 0$ if $-1 < e < 0$; this latter type is called a subcircular ellipse.

Returning to our solution (8.54) for $r(\theta)$, it is customary to select one of the constants A_1 and B_1 so that, as suggested by Figure 8.16, $dr/d\theta = 0$ when $\theta = 0$. This condition easily gives $B_1 = 0$, as the reader may wish to demonstrate using calculus. The result simply means that we are measuring θ from the perigee of the conic. At this point we should compare Equations (8.57) and (8.54) with $B_1 = 0$:

$$r = \frac{h_O^2/GM}{1 + (h_O^2/GM)A_1 \cos\theta} \tag{8.59}$$

and

$$r = \frac{eq}{1 + e\cos\theta} \tag{8.60}$$

By direct comparison of these two expressions for r, we see that

$$A_1 = \frac{eGM}{h_O^2} \qquad \text{and} \qquad eq = \frac{h_O^2}{GM}$$

It is more customary, however, to express the constant A_1 (as well as the eccentricity) in terms of the energy E of the orbit. To do this, Equations (8.45) and (8.46) give

$$\frac{h_O^2}{r^4}\left(\frac{dr}{d\theta}\right)^2 + \frac{h_O^2}{r^2} - \frac{2GM}{r} = 2E \tag{8.61}$$

At the point r_P where $\theta = 0$ and $dr/d\theta = 0$, we see that

$$\frac{h_O^2}{r_P^2} - \frac{2GM}{r_P} = 2E \tag{8.62}$$

Thus not all of h_O, r_P, and E are independent. We shall eliminate r_P. Multiplying Equation (8.62) by r_P^2, we get

$$2Er_P^2 + 2GMr_P - h_O^2 = 0 \tag{8.63}$$

Solving via the quadratic formula, we have

$$r_P = \frac{-2GM + \sqrt{4G^2M^2 + 8Eh_O^2}}{4E} \tag{8.64}$$

in which we use the plus sign since r_P is positive.

Returning to our solution (8.59), when $\theta = 0$ then

$$r_P = \frac{h_O^2/GM}{1 + (h_O^2/GM)A_1} \tag{8.65}$$

Equating the two expressions for r_P, Equations (8.64) and (8.65), we can solve for A_1.

We see by comparing Equations (8.59) and (8.60) that the eccentricity e of our conic will be $(h_O^2/GM)A_1$. Equating the right sides of Equations (8.64) and (8.65) and solving for this quantity, we get

$$\frac{h_O^2}{GM} A_1 = e = \sqrt{1 + \frac{2Eh_O^2}{G^2M^2}} \tag{8.66}$$

Therefore

$$r = \frac{h_O^2/GM}{1 + \sqrt{1 + (2Eh_O^2/G^2M^2)} \cos \theta} \tag{8.67}$$

which expresses r as a function of θ, the constant GM, the energy E, and the angular momentum per unit mass h_O. Note that by again comparing Equations (8.59) and (8.60) we can obtain the distance q between the focus O and the directrix ℓ:

$$\frac{h_O^2}{GM} = eq \Rightarrow q = \frac{h_O^2/GM}{\sqrt{1 + 2Eh_O^2/G^2M^2}} \tag{8.68}$$

The first of **Kepler's three laws of planetary motion** states that the planets travel in elliptical orbits with the sun at one focus.* These ellipses are very nearly circular for most of the planets; the eccentricity of earth is $e = 0.017$. To obtain the third of Kepler's laws, we return once more to our equations and obtain for elliptic orbits, from (8.67), the distance r when $\theta = 90°$:

$$r_{90} = \frac{h_O^2/GM}{1 + 0} = \ell \tag{8.69}$$

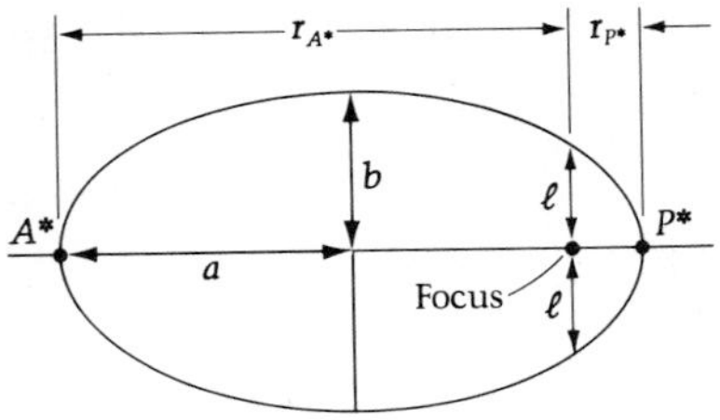

Figure 8.17

This distance, the ***semilatus rectum***, may be used to express the distance r_A between the focus O and apogee A^*, and the distance r_P between O and the perigee P^*. (See Figure 8.17.) At apogee, $\theta = \pi$ and Equations (8.59), (8.60), and (8.69) give

$$r_{A*} = \frac{h_O^2/GM}{1 - e} = \frac{\ell}{1 - e} \tag{8.70}$$

and at perigee ($\theta = 0$),

$$r_{P*} = \frac{\ell}{1 + e} \tag{8.71}$$

The semimajor axis length of the ellipse is

*Kepler's laws, based on his astronomical observations and set forth in 1609 and 1619, were studied by Newton before the Englishman published the *Principia*, which contained his own laws of motion.

$$a = \frac{r_{A*} + r_{P*}}{2} = \frac{\ell}{1 - e^2} \tag{8.72}$$

and the semiminor axis length is, from analytic geometry,

$$b = a\sqrt{1 - e^2} = \frac{\ell}{\sqrt{1 - e^2}} \tag{8.73}$$

An ellipse has area

$$A_T = \pi ab = \pi\left(\frac{\ell}{1 - e^2}\right)\left(\frac{\ell}{\sqrt{1 - e^2}}\right)$$

or

$$A_T = \frac{\pi\ell^2}{(1 - e^2)^{3/2}} = \pi a^2\sqrt{1 - e^2} \tag{8.74}$$

With these results in hand, we shall now prove Kepler's third law. Since dA/dt is constant,

$$\frac{dA}{dt} = \frac{h_O}{2} \Longrightarrow A = \frac{h_O t}{2} \tag{8.75}$$

where we take $A = 0$ when $t = 0$, say at the perigee. Over one orbit we have, with T being the orbit period,

$$\begin{aligned} A_T &= \text{area of ellipse} \\ &= \pi a^2\sqrt{1 - e^2} = \frac{h_O T}{2} \end{aligned} \tag{8.76}$$

Since (from Equations (8.69) and (8.72))

$$h_O = \sqrt{GM\ell} = \sqrt{GMa(1 - e^2)} \tag{8.77}$$

we obtain the following from Equation (8.76):

$$\pi a^2\sqrt{1 - e^2} = \frac{\sqrt{GMa(1 - e^2)}}{2}\,T \tag{8.78}$$

so that

$$T = \frac{2\pi a^{3/2}}{\sqrt{GM}}$$

or

$$T^2 = \frac{4\pi^2}{GM}a^3 \tag{8.79}$$

Equation (8.79) states the third of Kepler's laws: The squares of the planets' orbital periods are proportional to the cubes of the semimajor axes of their orbits.

EXAMPLE 8.8

Calculate the semimajor axis length of an earth satellite with a period of 90 min.

SOLUTION

We can solve this problem by using Kepler's third law. The weight of a particle (mass m) on the earth's (mass M) surface is both mg and GMm/r_e^2; thus

$$mg = \frac{GMm}{r_e^2} \Rightarrow GM = gr_e^2$$

and we see that the product of the unwieldy constants G and M is

$$GM = gr_e^2 = \frac{32.2}{5280}(3960)^2 = 95{,}600 \text{ mi}^3/\text{sec}^2$$

Therefore

$$T^2 = (90 \times 60)^2 = \frac{4\pi^2}{95{,}600} a^3$$

$$a = 4130 \text{ mi}$$

which is about 170 mi above the earth's surface.

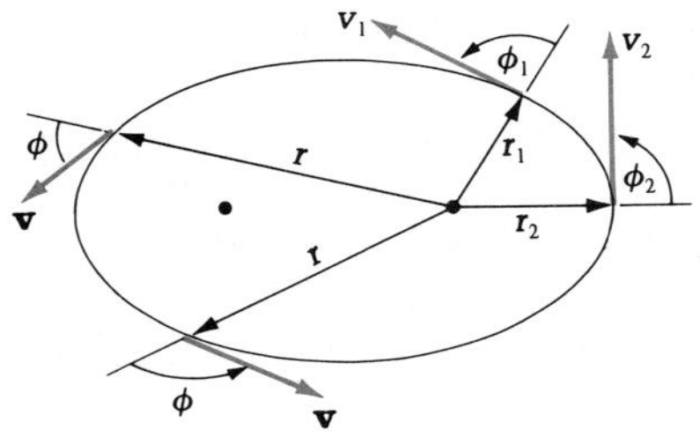

Figure 8.18

We shall present one more example on elliptical orbits under gravity, but first we need two equations relating the velocities v_1 and v_2 at any two points P_1 and P_2 with radii r_1 and r_2 on the orbit. The first of these comes from Equation (8.37), which states that $r^2\dot{\theta}$ = constant. From Figure 8.18, since the velocity $\mathbf{v}$ is always tangent to the path, we see that if ϕ is the angle between $\mathbf{r}$ and $\mathbf{v}$, then in cylindrical coordinates

$$v \sin \phi = (\text{transverse component of } \mathbf{v}) = r\dot{\theta}$$

so that

$$r(r\dot{\theta}) = rv \sin \phi = \text{constant}$$

or, for two points P_1 and P_2 on the orbital path,

$$r_1 v_1 \sin \phi_1 = r_2 v_2 \sin \phi_2 \tag{8.80}$$

Note that at apogee and perigee, $\phi = 90°$. Thus letting P_1 and P_2 be these two points, we get from Equation (8.80)

$$r_{A*} v_{A*} = r_{p*} v_{p*} \tag{8.81}$$

and the two velocities are inversely proportional to the radii, with v being faster at perigee as we have already seen from Kepler's second law.

The other equation relating v_1 and v_2 comes from the potential for gravity, which from Equation (2.27) and Example 8.8 is

$$\varphi = -\frac{gr_e^2 m}{r} = -\frac{GMm}{r}$$

Using conservation of energy between P_1 and P_2,

$$T_1 + \varphi_1 = T_2 + \varphi_2$$

$$\frac{mv_1^2}{2} - \frac{GMm}{r_1} = \frac{mv_2^2}{2} - \frac{GMm}{r_2}$$

$$v_2^2 - v_1^2 = 2GM\left(\frac{1}{r_2} - \frac{1}{r_1}\right) \tag{8.82}$$

If we let point P_2 represent the perigee P^*, as suggested in Figure 8.18, then Equation (8.80) becomes

$$v_2 = v_{P*} = \frac{r_1 v_1 \sin\phi_1}{r_P} \tag{8.83}$$

where $\sin\phi_2 = \sin 90° = 1$. Now if r_1, v_1, and ϕ_1 are initial (launch) values of r, v, and ϕ, then we may consider these as given quantities. Substituting Equation (8.83) into (8.82), we can obtain an equation for the perigee radius r_2 $(= r_{P*})$:

$$\frac{r_1^2 v_1^2 \sin^2\phi_1}{r_{P*}^2} - v_1^2 = 2GM\left(\frac{1}{r_{P*}} - \frac{1}{r_1}\right)$$

Multiplying through by $-r_{P*}^2/(r_1^2 v_1^2)$ and rearranging, we get

$$\left(\frac{r_{P*}}{r_1}\right)^2\left(1 - \frac{2GM}{r_1 v_1^2}\right) + \left(\frac{r_{P*}}{r_1}\right)\frac{2GM}{r_1 v_1^2} - \sin^2\phi_1 = 0 \tag{8.84}$$

We see that Equation (8.84) is simply a quadratic equation in the ratio (r_{P*}/r_1) and that $[2GM/(r_1 v_1^2)]$ is a nondimensional parameter of the orbit. We now illustrate the use of this important equation in an example.

EXAMPLE 8.9

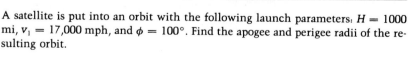

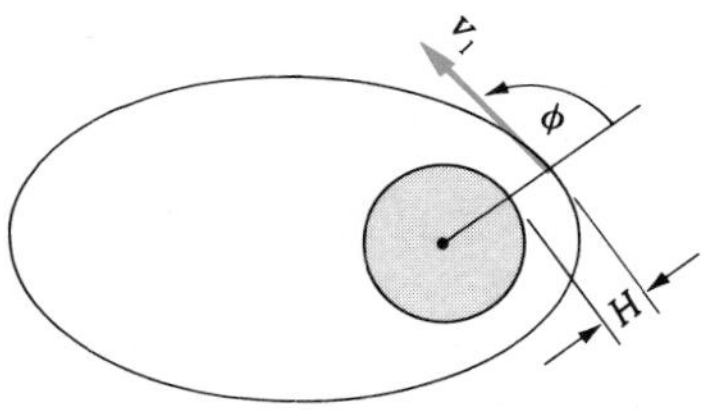

A satellite is put into an orbit with the following launch parameters: $H = 1000$ mi, $v_1 = 17{,}000$ mph, and $\phi = 100°$. Find the apogee and perigee radii of the resulting orbit.

(Continued)

SOLUTION

We need GM in mi^3/hr^2; therefore

$$GM = gr_e^2 = 32.2(3960)^2 \text{ ft-mi}^2/\text{sec}^2$$

$$= 32.2(3960)^2 \times \frac{1}{5280} \times 3600^2 \text{ mi}^3/\text{hr}^2$$

$$= 124 \times 10^{10} \text{ mi}^3/\text{hr}^2$$

The parameter $2GM/(r_1 v_1^2)$ in Equation (8.84) is therefore

$$\frac{2GM}{r_1 v_1^2} = \frac{2(124 \times 10^{10})}{(1000 + 3960)17{,}000^2}$$

$$= 1.73$$

Equation (8.84) becomes

$$\left(\frac{r_{P*}}{r_1}\right)^2(-0.73) + 1.73\left(\frac{r_{P*}}{r_1}\right) - 0.970 = 0$$

The quadratic formula gives

$$\left(\frac{r_{P*}}{r_1}\right)_{1,2} = \frac{-1.73 \pm \sqrt{1.73^2 - 4(-0.73)(-0.970)}}{2(-0.73)}$$

$$= 0.911 \text{ and } 1.46$$

Therefore

$$r_{P_1} = r_{P*} = 0.911(4960) = 4520 \text{ mi}$$

The other root corresponds to the apogee. (Since the starting condition of $\sin\phi = 1$ is the same for apogee and perigee, both answers are produced by the quadratic formula!)

$$r_{P_2} = r_{A*} = 1.46(4960) = 7240 \text{ mi}$$

The altitudes are

Perigee height = 4520 − 3960 = 560 mi

Apogee height = 7240 − 3960 = 3280 mi

To pin down the orbit in space, we need to know the angle to the perigee point from the launch point and also the orbit's eccentricity. A pair of problems to follow will be concerned with finding these two quantities given initial values of r, v, and ϕ.

PROBLEMS / Section 8.3

8.63 Show that if a satellite is in a circular orbit at radius r around a planet of mass M, the velocity to which it must increase to *escape* the planet's gravitational attraction is given by

$$v_{\text{escape}} = \sqrt{\frac{2GM}{r}}$$

Repeat the problem if "to which" is replaced by "by which."

8.64 A rocket is in a 200-mi-high circular parking orbit above a planet. What velocity boost at point P will result in the new, elliptical orbit shown in Figure P8.64?

8.65 A large meteorite approaches the earth. (See Figure P8.65.) Measurements indicate that at a given time it has a speed of 8000 mph at a radius of 100,000 mi. Will it orbit the earth? If so, what is the period? If not, what is the maximum velocity v that would have resulted in an orbit?

8.66 Show that if the launch velocity in Example 8.9 is 15,000 mi/hr, the satellite will fail to orbit the earth.

8.67 A satellite has $r_{A*} = 8000$ mi and $r_{P*} = 5000$ mi. If it was launched with a velocity of 15,000 mi/hr, what was its launch radius? What was the angle ϕ between $\mathbf{r}$ and $\mathbf{v}$ at launch?

8.68 What is the period of the satellite in the preceding problem?

8.69 Show that a satellite in orbit has a period T given by

$$T = \frac{2\pi ab}{r_{A*}v_{A*} \text{ (or } r_{P*}v_{P*})}$$

8.70 Classify the various orbits according to values of the dimensionless parameter $GM/(r_0 v_0^2)$ for a satellite launched with the conditions of Figure P8.70.

8.71 Show that, for a body in elliptical orbit (Figure P8.71), $b = \sqrt{r_{A*}r_{P*}}$.

8.72 Find the form of the central force $\mathbf{F}(r)$ for which all circular orbits of a particle about an attracting center O have the same angular momentum (and the same rate of sweeping out area).

8.73 Find the kinetic energy increase needed to move a satellite from radius R to $nR (n > 1)$.

8.74 A particle of mass m moves in the xy plane under the influence of an attractive central force that is proportional to its distance from the origin $(F(r) = kr)$. It has the same initial conditions as Problem 8.70. Find the largest and smallest values of r in the ensuing motion.

8.75 Halley's comet, due back in 1986 at the time of this writing (20 October 1982), was spotted today for the first time on the current pass. The comet orbits the sun in an elongated ellipse every 74 to 79 years; the period varies due to perturbations in its orbit caused by the four largest (Jovian) planets. (Its passages have been recorded since 240 B.C.!) What is the approximate semimajor axis length of Halley's comet? (Use 76 years as the period.)

8.76 Find the minimum period of a satellite in circular orbit about the earth. Upon what assumption is your answer based?

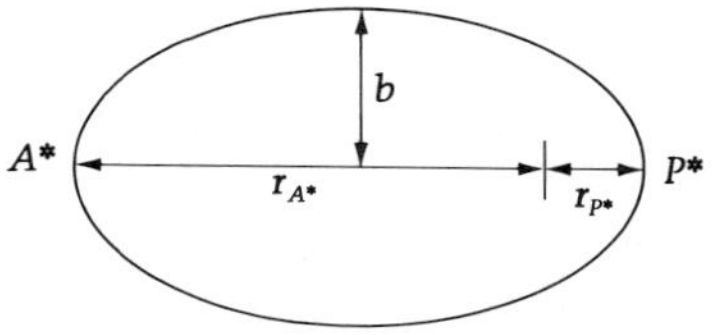

Figure P8.71

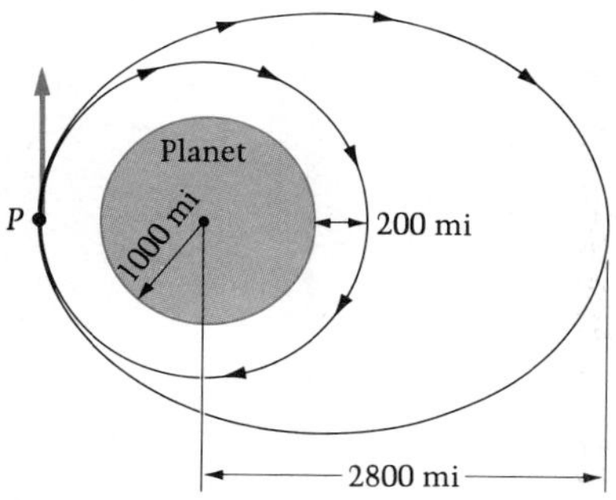

Figure P8.64

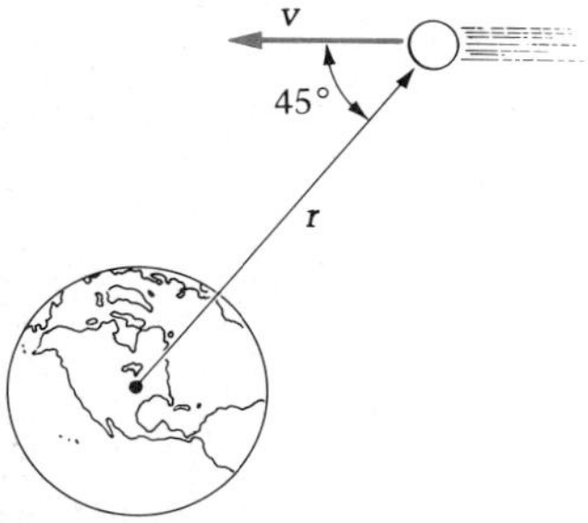

Figure P8.65

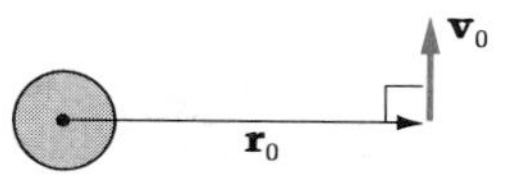

Figure P8.70

8.77 Repeat the preceding problem if the satellite orbits the moon. Assume

$$g_{\text{moon}} = \tfrac{1}{6}g_{\text{earth}}$$

$$r_{\text{moon}} = 0.27r_{\text{earth}}$$

8.78 A satellite has apogee and perigee points 1000 and 180 mi, respectively, above the earth's surface. Compute the satellite's period.

8.79 In the preceding problem find the speeds of the satellite at perigee and at apogee.

8.80 Show that for circular orbits around an attracting body of mass M, $rv^2 = GM$. Then use the 93×10^6 mi average orbital radius of earth, and the fact that its orbit is nearly circular, to find the constant GM_s for the sun as attracting center (heliocentric system).

8.81 A satellite is in a circular orbit of radius R_1. (See Figure P8.81.) Find the (negative) velocity increment that will send the satellite to position A, at radius R_2 ($< R_1$), 180° away. Then find the second negative velocity increment, this time applied at A, that will put the satellite in a circular orbit of radius R_2.

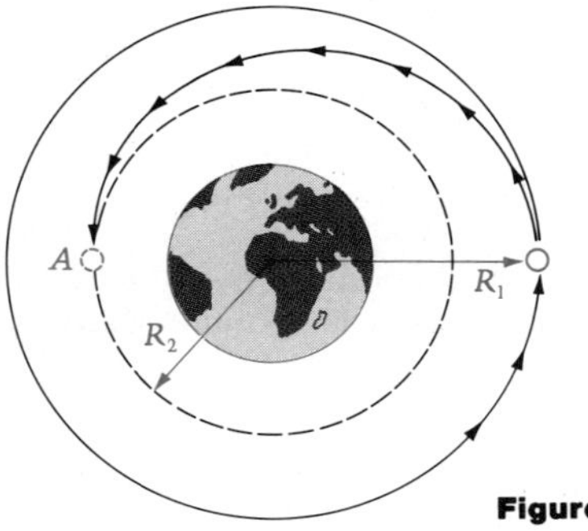

Figure P8.81

8.82 Show that, in terms of the radius r_{p*} and speed v_{p*} at perigee, the energy and eccentricity of the orbit may be expressed as

$$\frac{Er_{p*}}{GM} = \frac{r_{p*}v_{p*}^2}{2GM} - 1 \quad \text{and} \quad e = \frac{r_{p*}v_{p*}^2}{GM} - 1$$

8.83 The first artificial satellite to orbit the earth was the Russians' Sputnik I. Following insertion into orbit it had a period of 96.2 min. Find the semimajor axis length. If the initial eccentricity was 0.0517, find the maximum and minimum distances from earth following its injection into orbit.

8.84 Find the angle θ from perigee to the launch point of a satellite in orbit if the launch parameters r_1, v_1, and ϕ_1 are known.

8.85 Find the eccentricity e of the orbit if r_1, v_1, and ϕ_1 at launch are known.

8.86 Use Equations (8.81) and (8.82) to find the velocities at apogee and perigee in terms of the known radii r_{A*} and r_{p*}.

8.87 Prove that Equation (8.66) follows from (8.64) and (8.65).

8.88 Using Equation (8.54), show that $B_1 = 0$ follows from the condition $dr/d\theta = 0$ when $\theta = 0$.

8.89 In Figure P8.89, $\mathcal{P}$ is a particle moving in an inertial frame $\mathcal{I}$ and the position vector $\mathbf{r}$ locates it with respect to the point O of $\mathcal{I}$.

a. Show that

$$\mathbf{r} \times \mathbf{F}_r = \mathbf{r} \times m\dot{\mathbf{v}}$$

where $\mathbf{F}_r$ is the resultant force acting on $\mathcal{P}$ and $\mathbf{v} = \dot{\mathbf{r}}$ = velocity of $\mathcal{P}$ in $\mathcal{I}$.

b. Show further that

$$\mathbf{r} \times m\dot{\mathbf{v}} = \frac{d}{dt}(\mathbf{r} \times m\mathbf{v})$$

and thus, since $\mathbf{r} \times \mathbf{F}_r = \mathbf{M}_{r_O}$ = moment about O of the forces acting on $\mathcal{P}$, that

$$\mathbf{M}_{r_O} = \frac{d}{dt}(\mathbf{r} \times m\mathbf{v}) = \frac{d}{dt}(\mathbf{H}_O)$$

where $\mathbf{H}_O$ = angular momentum of $\mathcal{P}$ with respect to O. Thus we see that for the particle, $\mathbf{M}_{r_O} = {}^{\mathcal{I}}\dot{\mathbf{H}}_O$ is a *mathematical* result requiring no hypotheses past Newton's second law or Euler's first law for its existence.

Hint: For part (b) interpret the general equation $\mathbf{H}_O = \mathbf{H}_C + \mathbf{r}_{OC} \times \mathbf{L}$ for the case when the body is a particle.

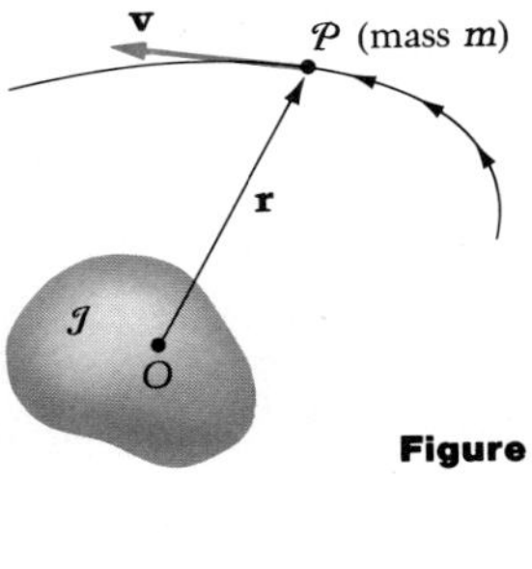

Figure P8.89

Appendices

APPENDIX A Units

The numerical value assigned to a physical entity expresses the relationship of that entity to certain standards of measurement called **units.** There is currently an international set of standards called the International System (SI) of Units, a descendant of the meter-kilogram-second (mks) metric system. In the SI system the unit of time is the **second** (s), the unit of length is the **meter** (m), and the unit of mass is the **kilogram** (kg). These independent (or *basic*) units are defined by physical entities or phenomena. The second is defined by the period of a radiation occurring in atomic physics; the meter is defined by the wavelength of a different radiation; the kilogram is defined to be the mass of a certain body of material stored in France. Any other SI units we shall need are *derived* from these three basic units. For instance, the unit of force, the **newton** (N), is a derived quantity in the SI system, as we shall see.

Until recently almost all engineers in the United States used the system (sometimes called British gravitational or U.S.) in which the basic units are the second (sec) for time, the **foot** (ft) for length,* and the **pound** (lb) for force. The pound is the weight, at a standard gravitational condition (location), of a certain body of material stored in the United States. In this system the unit of mass, the **slug,** is a derived quantity. It is a source of some confusion that sometimes there is used a unit of mass called the pound (the mass whose weight is one pound of force at standard gravitational conditions); also, particularly in Europe, the term *kilogram* has sometimes been used for a unit of force.† Grocery shoppers in the United States are exposed to this confusion by the fact that packages are marked by weight (or is it mass?) both in pounds and in kilograms. Throughout this text, without exception, *the pound is a unit of force* and *the kilogram is a unit of mass.*

*Sometimes, particularly in the field of mechanical vibrations, the inch is used as the unit of length; in that case the unit of mass is 1 lb-sec^2/in., which equals 12 slugs.

†A kilogram *was* a force unit in one of two mks systems, compounding the misunderstanding.

The United States is currently in the painful process of gradual changeover to the metric system of units after more than 200 years of attachment to the U.S. system. The new engineers who begin practicing their profession in the 1980s and 1990s will doubtless encounter both systems, and thus it is crucial to master both (including ***thinking*** in terms of the units of either) and to be able to convert from one to the other. The units mentioned here are summarized in Table A–1 for the SI and the U.S. systems.

Table A–1

Quantity	SI (Standard International or "Metric") Unit	U.S. Unit
force	newton (N)	pound (lb)
mass	kilogram (kg)	slug
length	meter (m)	foot (ft)
time	second (s)	second (sec)

We now examine how the newton of force is derived in SI units and the slug of mass is derived in U.S. units. Let the dimensions of the four basic dimensional quantities be labeled as F (force), M (mass), L (length), and T (time). From the first law of motion (discussed in detail in Chapter 2), $\mathbf{F} = m\mathbf{a}$, we observe that the four basic units are always related as follows:

$$F = \frac{ML}{T^2}$$

This means, of course, that we may select three of the units as basic and derive the fourth. Two ways in which this has been done are the ***gravitational*** and the ***absolute*** systems. The former describes the U.S. system; the latter describes SI. (See Table A–2.)

Table A–2

Gravitational System	**Absolute System**
The basic units are force, length, and time, and mass is derived: $M = \frac{FT^2}{L}$	The basic units are mass, length, and time, and force is derived: $F = \frac{ML}{T^2}$
This system has traditionally been more popular with engineers.	This system has traditionally been more popular with physicists.
As an example, in the U.S. system of units the pound, foot, and second are basic. Thus the mass unit, the slug, is derived: $1 \text{ slug} = 1 \frac{\text{lb-sec}^2}{\text{ft}}$	As an example, in the SI (metric) system of units the kilogram, meter, and second are basic. Thus the force unit, the newton, is derived: $1 \text{ newton} = 1 \frac{\text{kg} \cdot \text{m}}{\text{s}^2}$
This is summed up by: A slug is the quantity of mass that will be accelerated at 1 ft/sec^2 when acted upon by a force of 1 lb.	This is summed up by: A newton is the amount of force that will accelerate a mass of 1 kg at 1 m/s^2.

Therefore, in U.S. units the mass of an object weighing W lb is $W/32.2$ slugs. Similarly, in SI units the weight of an object having a mass of M kg is $9.81M$ newtons.

In the SI system the unit of moment of force is the newton · meter (N · m); in the U.S. system it is the pound-foot (lb-ft). Work and energy have this same dimension; the U.S. unit is the ft-lb whereas the SI unit is the joule (J), which equals 1 N · m. In the SI system the unit of power is called the watt (W) and equals one joule per second (J/s); in the U.S. system it is the ft-lb/sec. The unit of pressure or stress in the SI system is called the pascal (Pa) and equals 1 N/m^2; in the U.S. system it is the lb/ft^2, although often the inch is used as the unit of length so that the unit of pressure is the lb/in.^{2} (or psi). In both systems the unit of frequency is called the hertz (Hz), which is one cycle per second. Other units of interest in dynamics include those in Table A–3.

Table A–3

Quantity	SI Unit	U.S. Unit
velocity	m/s	ft/sec
angular velocity	rad/s	rad/sec
acceleration	m/s^2	ft/sec^2
angular acceleration	rad/s^2	rad/sec^2
mass moment of inertia	kg · m^2	slug-ft^2
momentum	kg · m/s	slug-ft/sec
moment of momentum	kg · m^2/s	slug-ft^2/sec
impulse	N · s (= kg · m/s)	lb-sec
angular impulse	N · m · s (= kg · m^2/s)	lb-ft-sec
mass density	kg/m^3	slug/ft^3
specific weight	N/m^3	lb/ft^3

Moreover, in the SI system there are standard prefixes to indicate multiplication by powers of 10. For example, kilo (k) is used to indicate multiplication by 1000, or 10^3; thus 5 kilonewtons, written 5 kN, stands for 5×10^3 N. Other prefixes that commonly appear in engineering are shown in Table A–4.

Table A–4

tera	T	10^{12}	centi	c	10^{-2}
giga	G	10^9	milli	m	10^{-3}
mega	M	10^6	micro	μ	10^{-6}
kilo	k	10^3	nano	n	10^{-9}
hecto	h	10^2	pico	p	10^{-12}
deka	da	10^1	femto	f	10^{-15}
deci	d	10^{-1}	atto	a	10^{-18}

We reemphasize that for the foreseeable future American engineers will find it desirable to know both the U.S. and SI systems well; for that reason we have used both sets of units in examples and problems throughout this book.

We turn now to the question of unit conversion. The conversion of units is quickly and efficiently accomplished by multiplying by equivalent fractions until the desired units are achieved. Suppose we wish to know how many newton-meters (N · m) of torque are equivalent to 1 lb-ft. Since we know there to be 3.281 ft per meter and 4.448 N per pound,

$$1 \text{ lb-ft} = 1 \not{\text{lb}}\text{-}\not{\text{ft}} \left(\frac{1 \text{ m}}{3.281 \not{\text{ft}}}\right)\left(\frac{4.448 \text{ N}}{1 \not{\text{lb}}}\right) = 1.356 \text{ N} \cdot \text{m}$$

Note that if the undesired unit (such as lb in this example) does not cancel, the conversion fraction is upside-down!

For a second example, let us find how many slugs of mass there are in a kilogram:

$$1 \text{ kg} = 1 \frac{\not{\text{N}} \cdot \text{s}^2}{\not{\text{m}}} \cdot \left(\frac{1 \text{ lb}}{4.448 \not{\text{N}}}\right) \cdot \left(\frac{1 \not{\text{m}}}{3.281 \text{ ft}}\right) = \frac{1}{14.59} \frac{\text{lb-sec}^2}{\text{ft}} = 0.06852 \text{ slug}$$

Inversely, 1 slug = 14.59 kg. A set of conversion factors to use in going back and forth between SI and U.S. units is given in Table A–5.*

Table A–5

To Convert From	To	Multiply By	Reciprocal (to Get from SI to U.S. Units)
Length, area, volume			
foot (ft)	meter (m)	0.30480	3.2808
inch (in.)	m	0.025400	39.370
statute mile (mi)	m	1609.3	6.2137×10^{-4}
foot^2 (ft^2)	meter^2 (m^2)	0.092903	10.764
inch^2 (in.^{2})	m^2	6.4516×10^{-4}	1550.0
foot^3 (ft^3)	meter^3 (m^3)	0.028317	35.315
inch^3 (in.^{3})	m^3	1.6387×10^{-5}	61024
Velocity			
feet/second (ft/sec)	meter/second (m/s)	0.30480	3.2808
feet/minute (ft/min)	m/s	0.0050800	196.85
knot (nautical mi/hr)	m/s	0.51444	1.9438
mile/hour (mi/hr)	m/s	0.44704	2.2369
mile/hour (mi/hr)	kilometer/hour (km/h)	1.6093	0.62137
Acceleration			
feet/second^2 (ft/sec^2)	meter/second^2 (m/s^2)	0.30480	3.2808
inch/second^2 (in./sec^2)	m/s^2	0.025400	39.370
Mass			
pound-mass (lbm)	kilogram (kg)	0.45359	2.20462
slug (lb-sec^2/ft)	kg	14.594	0.068522
Force			
pound (lb) or pound-force (lbf)	newton (N)	4.4482	0.22481

(Continued)

*Rounded to the five digits cited. Note, for example, that 1 ft = 0.30480 m, so that

$$(\text{Number of feet}) \times \left(\frac{0.30480 \text{ m}}{1 \text{ ft}}\right) = \text{number of meters}$$

Table A–5 (continued)

To Convert From	To	Multiply By	Reciprocal (to Get from SI to U.S. Units)
Density			
pound-mass/inch^3 (lbm/in.^{3})	kg/m^3	2.7680×10^4	3.6127×10^{-5}
pound-mass/foot^3 (lbm/ft^3)	kg/m^3	16.018	0.062428
slug/foot^3 (slug/ft^3)	kg/m^3	515.38	0.0019403
Energy, work, or moment of force			
foot-pound or pound-foot (ft-lb) (lb-ft)	joule (J) or newton · meter (N · m)	1.3558	0.73757
Power			
foot-pound/minute (ft-lb/min)	watt (W)	0.022597	44.254
horsepower (hp) (550 ft-lb/sec)	W	745.70	0.0013410
Stress, pressure			
pound/inch^2 (lb/in.^{2} or psi)	N/m^2 (or Pa)	6894.8	1.4504×10^{-4}
pound/foot^2 (lb/ft^2)	N/m^2 (or Pa)	47.880	0.020886
Mass moment of inertia			
slug-foot^2 (slug-ft^2 or lb-ft-sec^2)	kg · m^2	1.3558	0.73756
Momentum (or linear momentum)			
slug-foot/second (slug-ft/sec)	kg · m/s	4.4482	0.22481
Impulse (or linear impulse)			
pound-second (lb-sec)	N · s (or kg · m/s)	4.4482	0.22481
Moment of momentum (or angular momentum)			
slug-foot^2/second (slug-ft^2/sec)	kg · m^2/s	1.3558	0.73756
Angular impulse			
pound-foot-second (lb-ft-sec)	N · m · s (or kg · m^2/s)	1.3558	0.73756

Note that the units for time (s or sec), angular velocity (rad/s or 1/s), and angular acceleration (rad/s^2 or 1/s^2) are the same for the two systems. To five digits, the acceleration of gravity at sea level is 32.174 ft/s^2 in the U.S. system and 9.8067 m/s^2 in SI units.

We wish to remind the reader of the care that must be exercised in numerical calculations involving different units. For example, if two lengths are to be summed in which one length is 2 ft and the other is 6 in., the simple sum of these measures, 2 + 6 = 8, does not of course provide a measure of the desired length. It is also true that we may not add or equate the numerical measures of different types of entities; thus it makes no sense to attempt to add a mass to a length. These are said to have different dimensions. A dimension is the name assigned to the *kind* of measurement standard involved as contrasted with the choice of a particular measurement standard (unit). In science and engineering we attempt to develop equations expressing the relationships among various physical entities in a physical phenomenon. We express these equations in symbolic form so that they are valid regardless of the choice of a system of units, but nonetheless they must be *dimensionally consistent*. In the following equation, for example, we may check that the units on the left and right sides agree; r is a radial distance, P is a force, and dots denote time derivatives:

$$P - mg\cos\theta = m(\ddot{r} - r\dot{\theta}^2)$$

Dimensions of . . .	. . . are
P	F
$mg\cos\theta$	$M\left(\frac{L}{T^2}\right)(1) = F$
$m\ddot{r}$	$M\frac{L}{T^2} = F$
$-mr\dot{\theta}^2$	$ML\left(\frac{1}{T}\right)^2 = F$

Therefore the units of (every term in) the equation are those of force. If such a check is made prior to the substitution of numerical values, much time can be saved if an error has been made.

PROBLEMS / Appendix A

A.1. Find the units of the universal gravitational constant G, defined by

$$F = \frac{GMm}{r^2}$$

in (a) the SI system and (b) the U.S. system.

A.2. Find the weight in pounds of 1 kg of mass.

A.3. Find the weight in newtons of 1 slug of mass.

A.4. One pound-mass (lbm) is the mass of a substance that is acted on by 1 lb of gravitational force at sea level. Find the relationship between (a) 1 lbm and 1 slug; (b) 1 lbm and 1 kg.

A.5. The momentum of a body is the product of its mass m and the velocity v_C of its mass center. A child throws an 8-oz ball into the air with an initial speed of 20 mph. Find the magnitude of the momentum of the ball in (a) slug-ft/sec; (b) kg · m/s.

A.6. Is the following equation dimensionally correct?

$$\int_0^5 Fv\,dt = \frac{mv^2}{2} + ma \qquad (v = \text{velocity};\ a = \text{acceleration})$$

A.7. The equation for the distance r_s from the center of the earth to the geosynchronous satellite orbit is

$$gr_e^2 = r_s^3\omega^2 \qquad (\omega = \text{angular speed of earth};\ r_e = \text{earth radius})$$

a. Show that the equation is dimensionally correct.

b. Use the equation to find the ratio of the orbit radius to earth radius.

A.8. The universal gravitational constant is $G = 6.67 \times 10^{-11}$ N · m²/kg². Express G in units of lb-ft²/slug².

APPENDIX B

Demonstration That Newton's Third Law (Action and Reaction) Follows from Euler's Laws

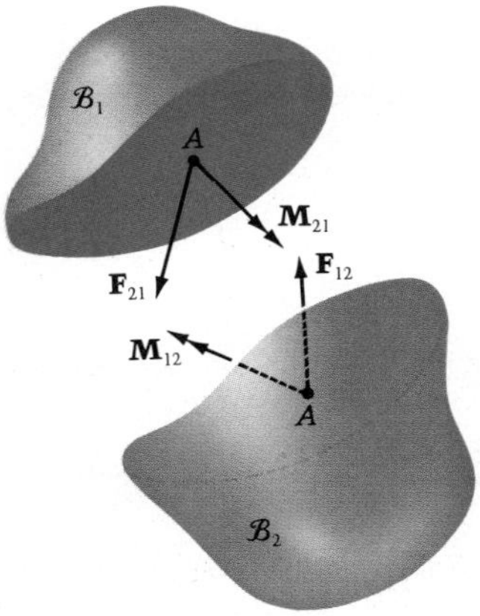

Figure B.1

From Euler's laws of motion we can deduce an important feature of the interaction of two bodies: The mutual mechanical actions are self-equilibrating. Consider a body $\mathcal{B}$ divided into two parts, $\mathcal{B}_1$ and $\mathcal{B}_2$, as suggested by Figure B.1. Among the external forces acting on $\mathcal{B}$, let $\mathbf{F}_{r_1}$ be the sum of those that act directly on $\mathcal{B}_1$ and let $\mathbf{F}_{r_2}$ be the sum of those that act on $\mathcal{B}_2$. Furthermore, let $\mathbf{F}_{12}$, $\mathbf{M}_{12}$ be the force-couple pair at a point A that represents the resultant of the mechanical action of $\mathcal{B}_1$ on $\mathcal{B}_2$. Similarly, let $\mathbf{F}_{21}$, $\mathbf{M}_{21}$ be a force-couple pair at the same point representing the resultant of the mechanical action of $\mathcal{B}_2$ on $\mathcal{B}_1$. This point A is drawn as a material point of both bodies in their interface, but in fact it could be any point in space.

If $\mathbf{L}_1$ and $\mathbf{L}_2$ are the momenta of $\mathcal{B}_1$ and $\mathcal{B}_2$, then by applying Euler's first law to each we get

$$\mathbf{F}_{r_1} + \mathbf{F}_{21} = \frac{d}{dt}\mathbf{L}_1 = \dot{\mathbf{L}}_1 \tag{B.1}$$

and

$$\mathbf{F}_{r_2} + \mathbf{F}_{12} = \frac{d}{dt}\mathbf{L}_2 = \dot{\mathbf{L}}_2 \tag{B.2}$$

Adding Equations (B.1) and (B.2), we obtain

$$\begin{aligned}\mathbf{F}_{r_1} + \mathbf{F}_{r_2} + \mathbf{F}_{12} + \mathbf{F}_{21} &= \dot{\mathbf{L}}_1 + \dot{\mathbf{L}}_2 \\ &= \dot{\mathbf{L}}\end{aligned} \tag{B.3}$$

But for the body $\mathcal{B}$,

$$\mathbf{F}_r = \mathbf{F}_{r_1} + \mathbf{F}_{r_2} = \dot{\mathbf{L}} \tag{B.4}$$

Subtracting Equation (B.4) from (B.3), we have

$$\mathbf{F}_{12} + \mathbf{F}_{21} = \mathbf{0} \tag{B.5}$$

or

$$\mathbf{F}_{12} = -\mathbf{F}_{21} \tag{B.6}$$

Thus we see that the forces exerted by $\mathcal{B}_1$ and $\mathcal{B}_2$ on each other are equal in magnitude and opposite in direction.

We wish next to show that the same is true for the couples $\mathbf{M}_{12}$ and $\mathbf{M}_{21}$. For this purpose let us now name a common line of action ℓ through A for $\mathbf{F}_{12}$ and $\mathbf{F}_{21}$ as shown in the separated sketch of $\mathcal{B}_1$ and $\mathcal{B}_2$ depicted in Figure B.2.

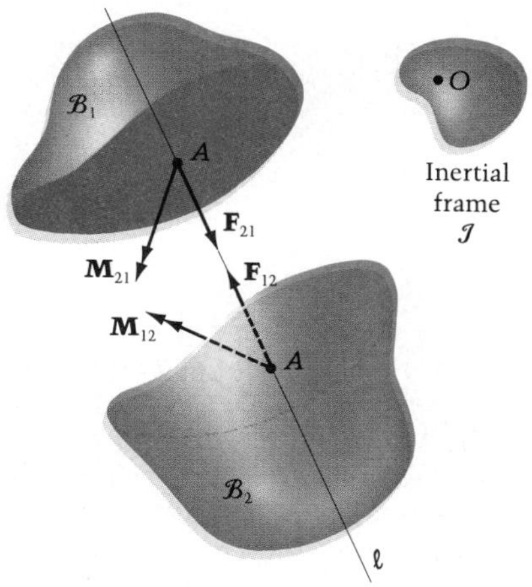

Figure B.2

Next we let $\mathbf{M}_{r_{O1}}$ be the sum of moments, with respect to O (a fixed point in the inertial frame $\mathcal{J}$), of those forces external to $\mathcal{B}$ that act directly on $\mathcal{B}_1$. Then, if $\mathbf{H}_{O1}$ is the moment of momentum with respect to O of $\mathcal{B}_1$, Euler's second law gives, for this body,

$$\mathbf{M}_{r_{O1}} + \mathbf{M}_{21}{}^{\dagger} + \mathbf{r}_{OA} \times \mathbf{F}_{21} = \frac{d}{dt}\mathbf{H}_{O1} = \dot{\mathbf{H}}_{O1} \tag{B.7}$$

In the same way, for $\mathcal{B}_2$,

$$\mathbf{M}_{r_{O2}} + \mathbf{M}_{12} + \mathbf{r}_{OA} \times \mathbf{F}_{12} = \frac{d}{dt}\mathbf{H}_{O2} = \dot{\mathbf{H}}_{O2} \tag{B.8}$$

But for body $\mathcal{B}$ $(= \mathcal{B}_1 + \mathcal{B}_2)$, we know that

$$\mathbf{M}_{r_O} = \mathbf{M}_{r_{O1}} + \mathbf{M}_{r_{O2}} = \frac{d}{dt}(\mathbf{H}_{O1} + \mathbf{H}_{O2}) = \dot{\mathbf{H}}_O \tag{B.9}$$

†Remember that a couple has the same moment about any point in space!

Adding Equations (B.7) and (B.8) and then subtracting (B.9), we get

$$\mathbf{M}_{21} + \mathbf{M}_{12} + \mathbf{r}_{OA} \times (\mathbf{F}_{21} + \mathbf{F}_{12}) = \mathbf{0} \tag{B.10}$$

The cross-product term vanishes by Equation (B.5), leaving

$$\mathbf{M}_{12} = -\mathbf{M}_{21} \tag{B.11}$$

Equations (B.6) and (B.11), taken together, state that the mechanical action of $\mathcal{B}_1$ on $\mathcal{B}_2$ is the negative of that of $\mathcal{B}_2$ on $\mathcal{B}_1$. This result, ***derived*** from Euler's laws, is an extension of what is commonly called "Newton's third law," when it is stated for particles.

The final picture, with corrected and coaligned $\mathbf{F}_{12}$ and $\mathbf{F}_{21}$, as well as $\mathbf{M}_{12}$ and $\mathbf{M}_{21}$, is shown in Figure B.3. The pairs of forces and moments at A have been seen to self-equilibrate, or cancel. Note that if the two bodies $\mathcal{B}_1$ and $\mathcal{B}_2$ are not normally thought of as constituting a combined body $\mathcal{B}$,† nonetheless they could be, so our development is valid for any pair of bodies.

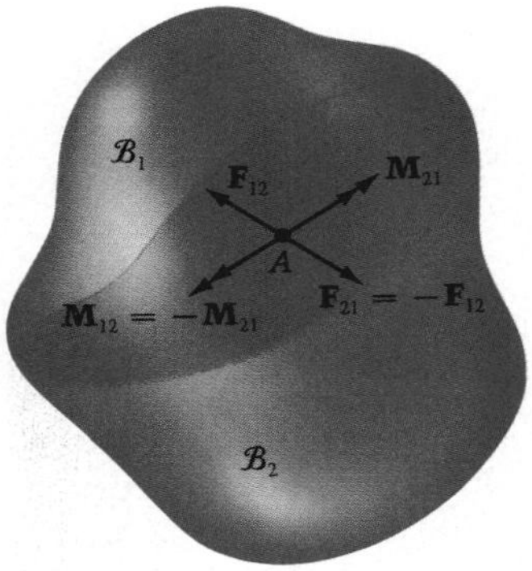

Figure B.3

†For example, $\mathcal{B}_1$ is a truck and $\mathcal{B}_2$ is the earth.

APPENDIX C

Using Newton–Raphson and Programmable Calculators to Solve Some Sticky Nondifferential Equations Occurring in the Book

There are a few places in this book where equations arise whose solutions are not easily found by elementary algebra; they are either polynomials of degree higher than 2 or else transcendental equations. In this appendix we explain in brief the fundamental idea behind the Newton–Raphson numerical method for solving such equations. We shall do this while applying the method to the solution for one of the roots of a cubic polynomial equation that occurs in Chapter 7. We then present calculator programs that greatly facilitate the solution to this problem and two others in Chapters 2 and 5.

To solve the cubic equation of Example 7.7,

$$f(\mathcal{I}) = -\mathcal{I}^3 + 48\mathcal{I}^2 - 633\mathcal{I} + 1342 = 0$$

we could, alternatively, use the Newton–Raphson algorithm. This procedure finds a root of the equation $f(\mathcal{I}) = 0$ (it need not be a polynomial equation, however) by using the slope of the curve. The algorithm, found in more detail in any book on numerical analysis, works as follows. If $\mathcal{I}_{1_0}$ is an initial estimate of a root $\mathcal{I}_1$, then a better approximation is

$$\mathcal{I}_{1_1} = \mathcal{I}_{1_0} - \frac{f(\mathcal{I}_{1_0})}{f'(\mathcal{I}_{1_0})}$$

Figures C.1 and C.2 indicate what is happening. The quantity $f(\mathcal{I}_{1_0})/f'(\mathcal{I}_{1_0})$ causes a backup in the $\mathcal{I}_1$ approximation—in our case from the *initial* value of 3 to the *improved* estimate $\mathcal{I}_{1_1}$:

$$\begin{aligned}\mathcal{I}_{1_1} = 3 - \frac{f(3)}{f'(3)} &= 3 - \frac{-152}{-372}\\ &= 3 - 0.408602150\\ &= 2.591397850\end{aligned}$$

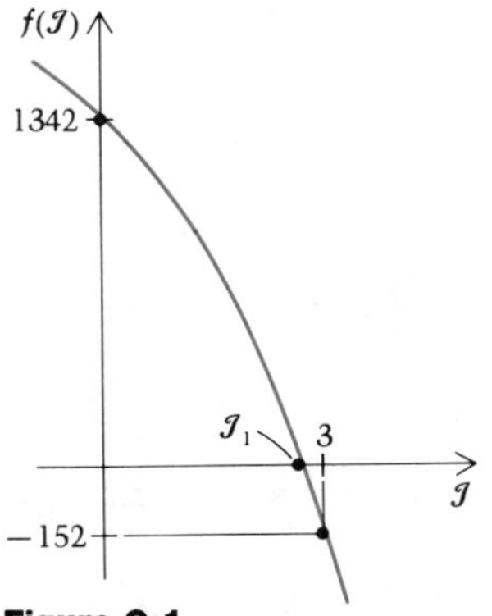

Figure C.1

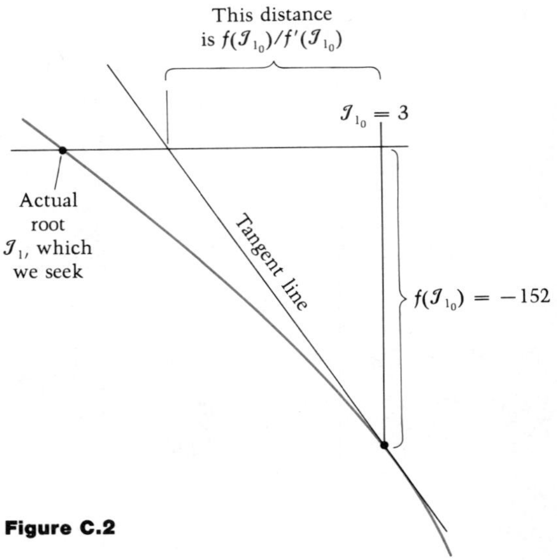

Figure C.2

where

$$f'(\mathcal{J}) = -3\mathcal{J}^2 + 96\mathcal{J} - 633$$

so that $f'(3) = -372$. Repeating the algorithm, we get

$$\begin{aligned}\mathcal{J}_{1_2} &= \mathcal{J}_{1_1} - \frac{f(\mathcal{J}_{1_1})}{f'(\mathcal{J}_{1_1})}\\ &= 2.591397850 - \frac{6.579491260}{-404.3718348}\\ &= 2.591397850 + 0.016270894\\ &= 2.607668744\end{aligned}$$

And one more time:

$$\begin{aligned}\mathcal{J}_{1_3} &= \mathcal{J}_{1_2} - \frac{f(\mathcal{J}_{1_2})}{f'(\mathcal{J}_{1_2})}\\ &= 2.607668744 - \frac{0.010645000}{-403.0636094}\\ &= 2.607668744 + 0.000026410\\ &= 2.607695154\end{aligned}$$

This algorithm is easily programmed on a calculator or home computer. For example, here is a set of statements for a Hewlett-Packard calculator with the program labeled N:

```
01 LBL ᵀN        Name of program
02 STO 01        Store the most recent estimate of 𝒥₁, say 𝒥₁ₙ, in register 1
03 96       ⎫
04 *        ⎬ 96𝒥₁ₙ
05 633      ⎫
06 -        ⎬ Subtract 633      ⎫ Compute f'(𝒥₁ₙ) and store it in register 2
07 RCL 01   ⎫
08 X↑2      ⎪
09 3        ⎬ Subtract 3𝒥₁ₙ²
10 *        ⎪
11 -        ⎭
12 STO 02
13 RCL 01   ⎫
14 X↑2      ⎪
15 48       ⎬ 48𝒥₁ₙ²
16 *        ⎭
17 RCL 01   ⎫
18 633      ⎪
19 *        ⎬ Subtract 633𝒥₁ₙ    ⎫ Compute f(𝒥₁ₙ) and store it in register 3
20 -        ⎭
21 1342     ⎫
22 +        ⎬ Add 1342
23 RCL 01   ⎫
24 3        ⎪
25 Y↑X      ⎬ Subtract 𝒥₁ₙ³
26 -        ⎭
27 STO 03
28 RCL 02   ⎫
29 /        ⎬ Divide to form f(𝒥₁ₙ)/f'(𝒥₁ₙ)
30 CHS      ⎫
31 RCL 01   ⎪ Subtract from 𝒥₁ₙ to form
32 +        ⎬ new estimate 𝒥₁ₙ₊₁
33 END
```

To run this program, we start with $\mathcal{J}_{1_0} = 3$, enter it, run the program and obtain $\mathcal{J}_{1_1} = 2.591397850$. We then simply enter this *new* estimate, run *it*, obtain $\mathcal{J}_{1_2} = 2.607668744$, and continue. Results are very rapidly obtained to be

$$\mathcal{J}_{1_0} = 3$$

$$\mathcal{J}_{1_1} = 2.591397850$$

$$\mathcal{J}_{1_2} = 2.607668744$$

$$\mathcal{J}_{1_3} = 2.607695154$$

$$\mathcal{J}_{1_4} = 2.607695156$$

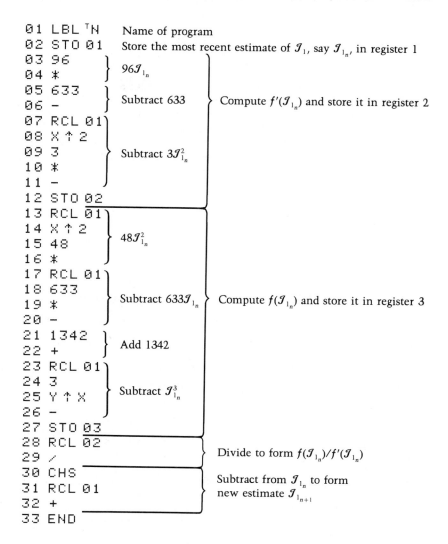

$$\left.\begin{aligned} \mathcal{I}_{1_5} &= 2.607695153 \\ \mathcal{I}_{1_6} &= 2.607695153 \\ \mathcal{I}_{1_7} &= 2.607695153 \end{aligned}\right\} \quad \text{convergence!}$$

which is in agreement with the results in Example 7.7. Note from Figures C.3 to C.5 that adding $(-f/f')$ to form the new estimate works equally well for the three other sign combinations of f and f'. Note also that if the estimate is *too far* from the root, such as P in Figure C.4, the procedure might not converge; the tangent at Q in this case would send us far from the desired root.

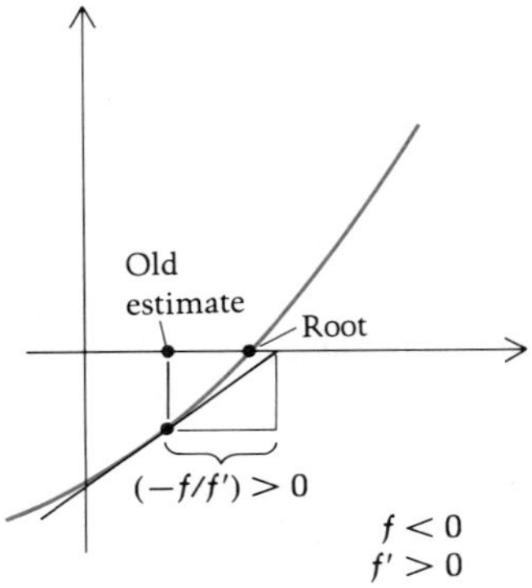

Figure C.3

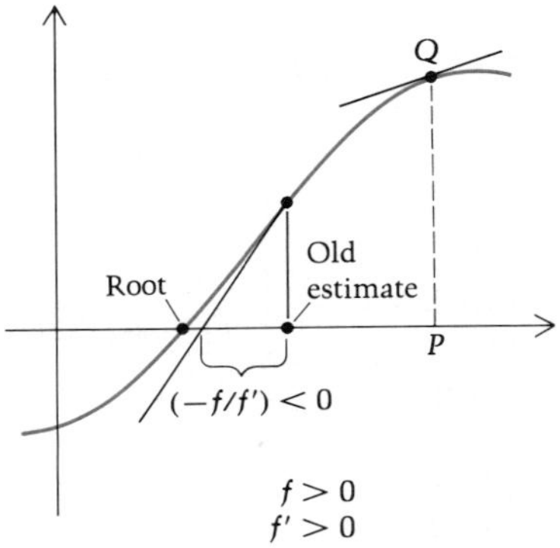

Figure C.4

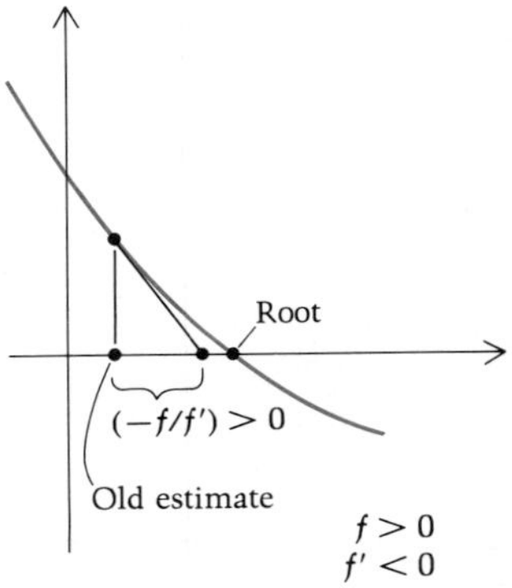

Figure C.5

We next consider the equation from Example 5.5:

$$f(\theta) = \sin\theta - \frac{\theta}{2} = 0 \tag{C.1}$$

with the derivative of f being

$$f'(\theta) = \cos\theta - \frac{1}{2}$$

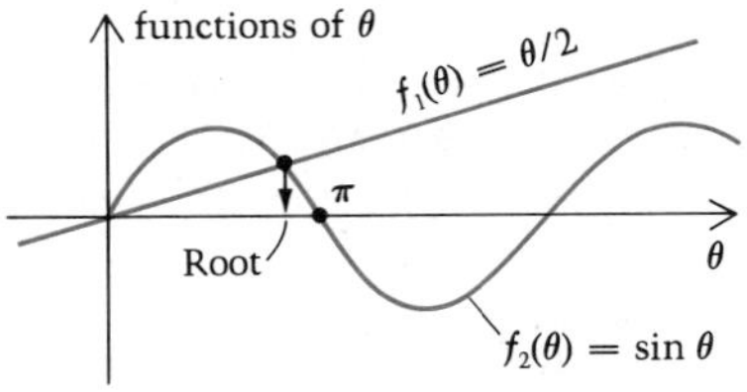

Figure C.6

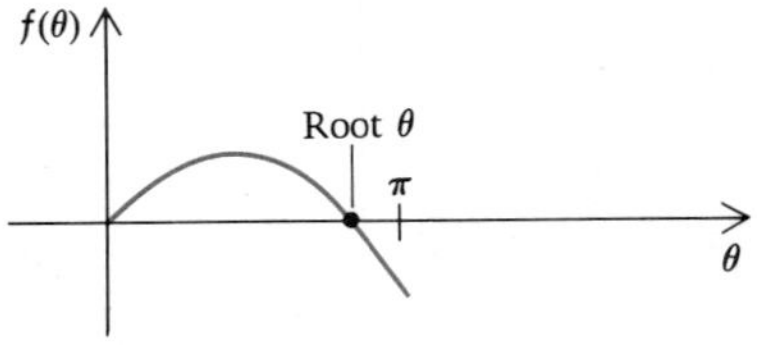

Figure C.7

There is but one root of Equation (C.1) for $\theta > 0$, as can be seen from Figure C.6, which shows the two functions making up $f(\theta)$. To find this root, we can use Newton–Raphson as previously described. Figure C.7 suggests that π might serve as a good first guess at the root. The following calculator program shows that it is and finds the root very quickly:

Program

```
01 LBL ᵀN
02 STO 01
03 SIN
04 RCL 01
05 2
06 /
07 -
08 STO 02
09 RCL 01
10 COS
11 0.5
12 -
13 STO 03
14 RCL 02
15 X≷Y
16 /
17 CHS
18 RCL 01
19 +
20 END
```

Results

$$\theta_0 = 3.141592654$$
$$\theta_1 = 2.094395103$$
$$\theta_2 = 1.913222955$$
$$\theta_3 = 1.895671752$$
$$\theta_4 = 1.895494285$$
$$\left.\begin{array}{l}\theta_5 = 1.895494267\\ \theta_6 = 1.895494267\\ \theta_7 = 1.895494267\end{array}\right\} \text{convergence!}$$

The last example in this appendix will be to solve the equation

$$\cos\left(\frac{\pi}{4} - q\right) = 0.373q$$

from Example 2.8. We write this equation as

$$f(q) = \cos\left(\frac{\pi}{4} - q\right) - 0.373q = 0$$

with

$$f'(q) = \sin\left(\frac{\pi}{4} - q\right) - 0.373$$

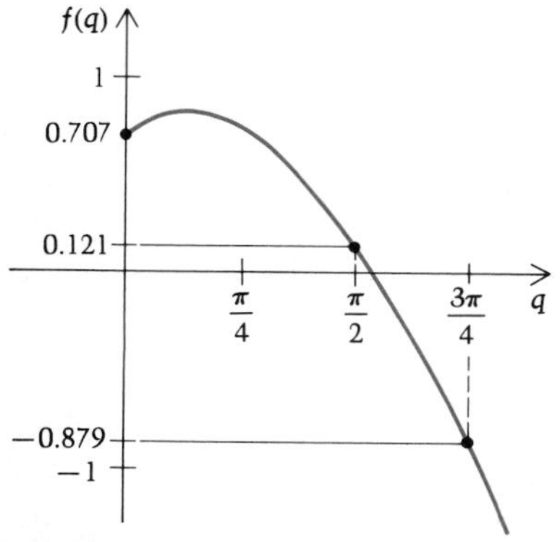

Figure C.8

The rough plot in Figure C.8 shows a few points which indicate that $\pi/2$ is fairly close to the root. Here are the results of a calculator program, which uses the Newton–Raphson method as in the first two examples, to narrow down on the root quickly and accurately:

Program

```
01 LBL ᵀN
02 STO 01
03 CHS
04 π
05 4
06 /
07 STO 04
08 +
09 COS
10 RCL 01
11 .373
12 *
13 -
14 STO 02
15 RCL 04
16 RCL 01
17 -
18 SIN
19 .373
20 -
21 RCL 02
22 X≷Y
23 /
24 CHS
25 RCL 01
26 +
27 END
```

Results

$q_0 = 1.570796327$

$q_1 = 1.683007224$

$q_2 = 1.679300543$

$$\left.\begin{array}{l} q_3 = 1.679296821 \\ q_4 = 1.679296821 \\ q_5 = 1.679296821 \end{array}\right\} \quad \text{convergence!}$$

APPENDIX E Answers to Odd-Numbered Problems

In the solutions to problems in Chapters 1–5, unless identified otherwise below, $\hat{\mathbf{i}}$, $\hat{\mathbf{j}}$, and $\hat{\mathbf{k}}$ are unit vectors in the respective directions →, ↑, and out of the page. In Chapters 6–8, the unit vectors are respectively parallel to axes defined in the problems.

CHAPTER 1

1.1 $18\hat{\mathbf{j}} - 8\hat{\mathbf{k}}$ kg · m/s² **1.3** $20\hat{\mathbf{j}}$ kg · m/s²
1.5 $82.1\hat{\mathbf{i}} + 83.6\hat{\mathbf{j}}$ kg · m/s²
1.7 $0.00420\hat{\mathbf{i}} + 29.7\hat{\mathbf{j}}$ kg · m/s²
1.9 $6\hat{\mathbf{i}} + 117\hat{\mathbf{j}} - 84\hat{\mathbf{k}}$ N · s **1.11** $150\hat{\mathbf{i}} + 220\hat{\mathbf{j}}$ N · s
1.13 $587\hat{\mathbf{i}} + 583\hat{\mathbf{k}}$ N · s
1.15 $-0.0183\hat{\mathbf{i}} + 213\hat{\mathbf{j}}$ N · s
1.17 Answer given in problem.
1.19 $-1.64\hat{\mathbf{i}} + 12.9\hat{\mathbf{j}}$ ft/sec² **1.21** $\hat{\mathbf{i}} - (\pi/2)\hat{\mathbf{j}}$ ft/sec²
1.23 $20\hat{\mathbf{i}} - 152\hat{\mathbf{k}}$ m; 153 m
1.25 $0.405\hat{\mathbf{i}} - 0.0417\hat{\mathbf{j}} + 0.00579\hat{\mathbf{k}}$ m; 0.407 m
1.27 $6.08\hat{\mathbf{i}} - 1.11\hat{\mathbf{k}}$ m; 6.18 m
1.29 29.3 sec and 1290 ft
1.31 30.8 sec from passing of police car by speeder.
1.33 20 ft/sec² **1.35** $2.5/\sin\theta$ ft/sec
1.37 1.09 → m/s²

1.39

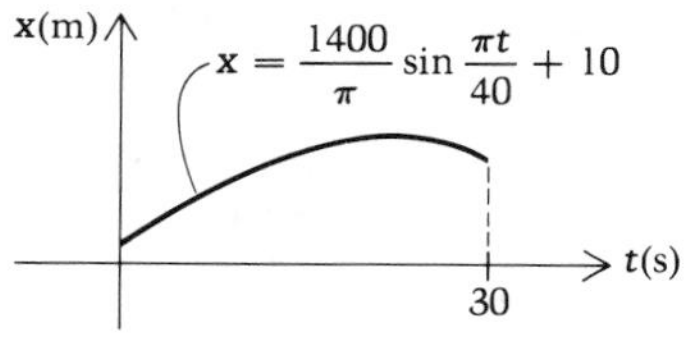

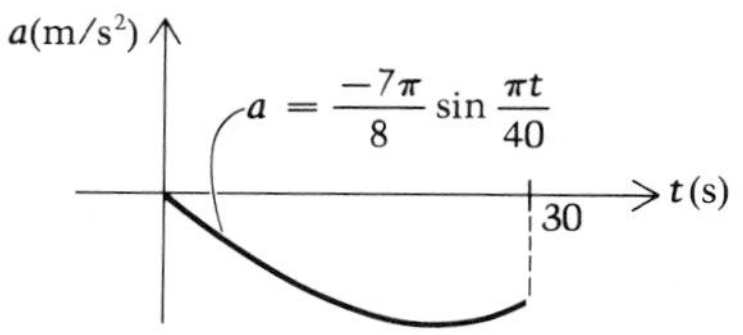

1.41 710 m
1.43 $\mathbf{v}_A = 10$ ↓ m/s and $\mathbf{v}_B = 20$ ↑ m/s (at $t = 2$ s)
1.45 11.3 m **1.47** 6.21 ↑ m/s
1.49 10.0 sec after the acceleration begins.
1.51 387 ft **1.53** 0.469 s **1.55** 530 ft
1.57 0.537 sec **1.59** $\sqrt{5x^2 - 0.0900}$ m/s
1.61 15 sec
1.63 $C_4 = C_2, C_3 = C_1/K; C_6 = C_1/K, C_5 = KC_2 - C_1^2/2$
1.65 $136\hat{\mathbf{i}} + 628\hat{\mathbf{j}} - 41\hat{\mathbf{k}}$ m; 641 m
1.67 $\dot{x} = 80.3$ ft/sec; $\dot{y} = 7.29$ ft/sec
1.69 zero, at $x = 750$ ft and 2250 ft
1.71 (a) $\mathbf{v}_P = -6\pi\sin\frac{\pi t}{2}\hat{\mathbf{i}} + 4\pi\cos\frac{\pi t}{2}\hat{\mathbf{j}}$ m/s;
$\mathbf{a}_P = -3\pi^2\cos\frac{\pi t}{2}\hat{\mathbf{i}} - 2\pi^2\sin\frac{\pi t}{2}\hat{\mathbf{j}}$ m/s²
(b) $\mathbf{r}_{OP} = 12\hat{\mathbf{i}}$ m; $\mathbf{v}_P = 4\pi\hat{\mathbf{j}}$ m/s; $\mathbf{a}_P = -3\pi^2\hat{\mathbf{i}}$ m/s²
(c) $(x/12)^2 + (y/8)^2 = 1$ (an ellipse)

1.73 $0.671\hat{\mathbf{i}} + 2.30\hat{\mathbf{j}} + 0.500\hat{\mathbf{k}}$ m/s
1.75 $3\hat{\mathbf{i}} + 8.71 \times 10^{13}\hat{\mathbf{j}}$ in./sec
1.77 $3\hat{\mathbf{i}} - 0.00312\hat{\mathbf{j}}$ in./sec
1.79 $0.0800\, t^3/\sqrt{L^2 - 0.0400\, t^4} \uparrow$ m/s
1.81 $\mathbf{v}_P = -10 \sin 5t\hat{\mathbf{i}} + 10 \cos 5t\hat{\mathbf{j}}$ m/s;
$\mathbf{a}_P = -50 \cos 5t\hat{\mathbf{i}} - 50 \sin 5t\hat{\mathbf{j}}$ m/s^2
1.83 The path is the straight line $6x - 2.5y = 19.5$.
1.85 $24.6\hat{\mathbf{j}}$ ft/sec^2
1.87 First,

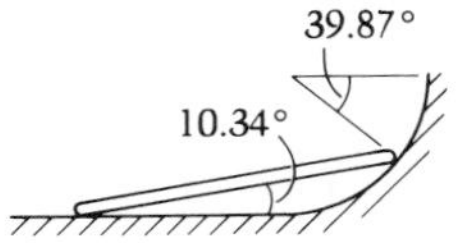

Then, $\mathbf{v}_B = 0.640 \angle 50.1°$ ft/sec.

1.89 $x^2 - y^2 = C^2$ **1.91** $(8a_0/15)\hat{\mathbf{j}}$
1.93 In order, $\mathbf{a}_P = -20\pi^2\hat{\mathbf{i}}$; $-20\pi^2\hat{\mathbf{j}}$; $20\pi^2\hat{\mathbf{j}}$; and $-20\pi^2\hat{\mathbf{i}}$ m/s^2 **1.95** $y = x^2$
1.97 Five significant digits here: at $t = 10$,
$\mathbf{r}_{OP} = 1021\hat{\mathbf{i}} + 10003\hat{\mathbf{j}} - 191\hat{\mathbf{k}}$ m;
10,054 m from starting point.
1.99 $2.61 \times 10^{14}\hat{\mathbf{j}}$ in./s^2 **1.101** $0.000604\hat{\mathbf{j}}$ in./s^2
1.103 14.9 m at 0.690 s **1.105** $624\hat{\mathbf{i}} - 556\hat{\mathbf{j}}$ ft/sec^2
1.107 (a) $(2k - c^2kt^4)\hat{\mathbf{e}}_r + (5ckt^2)\hat{\mathbf{e}}_\theta$ (b) $(2/c^2)^{1/4}$
1.109 (a) $-535\hat{\mathbf{e}}_r + 560\hat{\mathbf{e}}_\theta$ m/s^2 (at $\theta = 114°$, $t = 0.446$ s) (b) $-1380\hat{\mathbf{e}}_r + 160\hat{\mathbf{e}}_\theta$ m/s^2
1.111 (a) $\mathbf{r}_{OB} = R\hat{\mathbf{e}}_r + z\hat{\mathbf{k}}$ m;
$\mathbf{v}_B = R\dot{\theta}\hat{\mathbf{e}}_\theta + [p\dot{\theta}/(2\pi)]\hat{\mathbf{k}}$ m/s;
$\mathbf{a}_B = -R\dot{\theta}^2\hat{\mathbf{e}}_r + R\ddot{\theta}\hat{\mathbf{e}}_\theta + [p\ddot{\theta}/(2\pi)]\hat{\mathbf{k}}$ m/s^2
(Note $z = p\theta/(2\pi)$.)
(b) $\mathbf{v}_B = 0.180\hat{\mathbf{e}}_\theta + 0.0191\hat{\mathbf{k}}$ m/s; $\mathbf{a}_B = -0.108\hat{\mathbf{e}}_r$ m/s^2
1.113 $-0.000660\hat{\mathbf{i}} - 0.000452\hat{\mathbf{j}}$ m/s^2
1.115 (a) Never, but *magnitudes* equal at $r = 500\sqrt{5/\pi}$ ft; (b) Never, but *magnitudes* equal at $r = 2500$ ft; (c) $(r, \theta, z) = (159.15$ ft, 31.384 rad, 4994.9 ft), using five digits here.
1.117 Just one, $\ddot{z}$, which vanishes when $r = 631$ ft. Right *at* the top, the other two components are zero.
1.119 Answer given in problem.
1.121 $3110/(361 + x^2)^{3/2}$
1.123 Answer given in problem.
1.125 At $t = 1.7$ s, $\mathbf{r}_{OP} = 15.9\hat{\mathbf{i}} + 5.10\hat{\mathbf{j}} + 2.24\hat{\mathbf{k}}$ m, and $\dot{s}_P = 43.0$ m/s. At $t = 5.6$ s, $\rho = 4.89 \times 10^6$ m
1.127 $a_t = \ddot{s} = \dfrac{48t^3 + 18t}{\sqrt{16t^2 + 9}}$;

$$a_n = \frac{\dot{s}^2}{\rho} = \frac{12t^2}{\sqrt{16t^2 + 9}}; \text{ 91,200 m}$$

1.129 Using the hint, $\rho = 5.66$ m, while the distance = 2.83 m. **1.131** 0.931 s
1.133 $t = 0$ with $\theta = 0$;
$t = 3$ sec with $\theta = 0.750$ rad
1.135 $\mathbf{a}_P = 5.89\hat{\mathbf{e}}_t + 2.33\hat{\mathbf{e}}_n$ ft/sec^2; $\rho = 6.01$ ft
1.137 $-1.85\hat{\mathbf{i}} - 1.91\hat{\mathbf{j}} - 3.41\hat{\mathbf{k}}$ ft; 28.9 ft/sec^2
1.139 $(x, y) = (2.88, -0.584)$ m **1.141** $(-\dot{x}_0^2/a)\hat{\mathbf{j}}$
1.143 Answer given in problem.
1.145 $s = 20\pi t$ m; at $t = 2$ s, $s = 40\pi$ m;
circumference $= 2\pi\rho = 40\pi$ m
1.147 $s = 6.5t^2 + c$, where c is a constant defining the position of the particle on the straight line at some initial time.
1.149 $\mathbf{v}_B - \mathbf{v}_A = 46.2\hat{\mathbf{i}} + 64.5\hat{\mathbf{j}}$ mph;
$\mathbf{a}_B - \mathbf{a}_A = -7530\hat{\mathbf{i}} + 2400\hat{\mathbf{j}}$ mph^2
1.151 $y_M = 2D/3$ at $t = 2D/(3V_0)$

CHAPTER 2

2.1 $y_C = -0.856$ ft, $x_C = 0 = z_C$
2.3 On the axis of the shell, 2/3 of the way from vertex to base.
2.5 Answer given in problem. (Use a contradiction argument. Suppose it did lie on one side of a plane, and consider an axis normal to the plane through C.)
2.7 $a(n + 1)/(n + 2)$ **2.9** Answer given in problem.
2.11 $2R/\pi$ from the center of the circle along the line of symmetry in the plane of the wire.
2.13

$$\left(0, \frac{1370\rho_1 - 150\rho_2 - 2270\rho_3}{85.3\rho_1 + 250\rho_2 + 195\rho_3}, \frac{32\rho_3}{85.3\rho_1 + 250\rho_2 + 195\rho_3}\right) \text{ in.;}$$

$(0, -1.98, 0.0603)$ in.
2.15 Answer given in problem.
2.17 Neither the frame nor the equations is of any value without the other. Only by positive correlation of the results of a large number of experiments with $\mathbf{F} = m\mathbf{a}$ can we be confident of having an inertial frame. **2.19** 30°
2.21 (a) Answer to both questions is that, if not, the satellite can't remain stationary with respect to a point on the earth.
(b) 6.61, using the sun as the inertial frame so that

$$\omega_{\text{earth}} = 2\pi\left(1 + \frac{1}{365.25}\right)\frac{\text{rad}}{\text{day}}; \text{ also } g = 32.17 \text{ ft/sec}^2$$

2.23 $\sqrt{g(R - \delta)/\mu}$ **2.25** $\sqrt{2gR}$ **2.27** 1.5 m/s
2.29 $\phi = \tan^{-1}\left[\dfrac{2Hv_0^2 + g(D - d)^2}{2dv_0^2}\right]$;

$$v_i = \left[\frac{(d^2 + H^2)v_0^2}{(D - d)^2} + gH + \frac{g^2(D - d)^2}{4v_0^2}\right]^{1/2}; \; t_f = \frac{D - d}{v_0};$$

air resistance

2.31 88.1 ft/sec; 14.9°; 0.705 sec
2.33 mv_0/k; $v_0 e^{-Kt/m}$
2.35 28.2 N; 5.76 m/s² **2.37** 39.4 lb **2.39** 90.0 lb
2.41 164 N **2.43** 552 ft
2.45 $H = 3.15$ m; $D = 16.5$ m **2.47** 35.2 ft/sec
2.49 $v = \sqrt{25K/(24\,mS)}$, where m = mass of one dog.
2.51 $v < \sqrt{\rho g \cos\alpha}$
2.53 31.7°, compared to 26.6° for the static case.
2.55 592 ft; 621 ft
2.57 Answer given in problem.
2.59 Answer in problem for first part;
$x\big|_{t=\infty} = \dfrac{mv_i \cos\phi_i}{k}$; $\dot{y}\big|_{t=\infty} = \dfrac{-mg}{k}$
2.61 Answer given in problem.
2.63 Answer given in problem. **2.65** 8.63 m/s; 49.7°
2.67 540 m **2.69** Answers given in problem.
2.71 $m\ddot{x} + kx = mg$, with x measured downward from the point of release; $x(0) = 0$ and $\dot{x}(0) = 0$; $F_{\max} = 2mg$ at $t = \pi\sqrt{m/k}$ **2.73** $g\mu$
2.75 0.916 m/C; 156 ft
2.77 1.85 sec; 24.6 ft/sec **2.79** 1540 ft/sec
2.81 125 ft; 88.6 ft/sec **2.83** 0.0192 lb-sec²/ft²
2.85 11.2 ft (a bit small!) **2.87** 0.997 W
2.89 Answers given in problem. **2.91** 11.4 ft
2.93 Problem 2.27: 1.5 m/s; Problem 2.28: 3.04 m/s
2.95 Answer given in problem. (Show that the normal force never goes to zero!)
2.97 Answer given in problem. **2.99** 20.9 ft
2.101 272 lb **2.103** $-0.849\, mgr_e$
2.105 1.41 miles/sec
2.107 Answers given in problem. (It leaves at $\cos^{-1}(H/a)$ from the vertical.)
2.109 3.65 ft; 0.5% friction and 99.5% collision; 0.872 sec
2.111 7.45 sec required. With just $\frac{1}{6}$ his weight on earth available on the moon to produce friction, he has just $\frac{1}{6}$ the maximum friction force available on earth to stop the same forward momentum.
2.113 0.364 ft/sec
2.115 0.816 lb-sec; 8.16 lb
2.117 $v/3$ **2.119** 2.00 ft
2.121 7.86 sec; 491 ft/sec **2.123** 1.07 lb
2.125 17.5 ft/sec **2.127** 0.997 ft

CHAPTER 3

3.1 a,c,d,e **3.3** $0.1\hat{\mathbf{i}} + 0.1\hat{\mathbf{j}}$ m/s **3.5** $-4\hat{\mathbf{j}}$ ft/sec
3.7 $0.069\hat{\mathbf{i}} - 0.886\hat{\mathbf{j}}$ m/s
3.9 $\boldsymbol{\omega}_{\mathcal{B}} = 22.7$ ↺ rad/s; $\boldsymbol{\omega}_{\mathcal{C}} = 13.3$ ↺ rad/s
3.11 $\boldsymbol{\omega}_{\mathcal{B}} = 0.640$ ↺ rad/s; $\boldsymbol{\omega}_{\mathcal{A}} = 2$ ↺ rad/s
3.13 2.73 ↻ rad/s **3.15** $2\hat{\mathbf{j}}$ ft/sec; $(2/3)\hat{\mathbf{k}}$ rad/sec
3.17 1.08 → m/s
3.19 1.20 ↺ rad/sec; 25.8 → ft/sec
3.21 $0.3\hat{\mathbf{i}} - 1.3\hat{\mathbf{j}}$ m/s **3.23** 0.138 ↻ rad/s
3.25 (a) 0.0766 → m/s; (b) 0; (c) 0.0702 ← m/s
3.27 The plots can be constructed from the answer to 3.26, which is $[(r\cos\theta)/\sqrt{\ell^2 - r^2\sin^2\theta}\,]\,\dot{\theta}$
3.29 8.12 → cm/s; $\boldsymbol{\omega}_{\mathcal{A}} = 0.205$ ↺ rad/s; $\boldsymbol{\omega}_{\mathcal{L}} = 0.472$ ↻ rad/s
3.31 $v_B = \dot{y}_B = \dfrac{12\cos\theta}{\sqrt{169 - 120\sin\theta}}$;
swinging component of $v_A = \dfrac{12\sin\theta - 5}{\sqrt{169 - 120\sin\theta}}$
3.33 $\boldsymbol{\omega}_{\mathcal{B}_1} = 1.13$ ↻ rad/s; $\boldsymbol{\omega}_{\mathcal{B}_2} = 2.00$ ↺ rad/s
3.35 At $\theta = 90°$: $\boldsymbol{\omega}_{\mathcal{B}} = 0.2$ ↺ rad/sec $= \boldsymbol{\omega}_{\mathcal{C}}$; and at $\theta = 180°$: $\boldsymbol{\omega}_{\mathcal{B}} = 0.440$ ↺ rad/sec, $\boldsymbol{\omega}_{\mathcal{C}} = 0.0400$ ↺ rad/sec
3.37 $\boldsymbol{\omega}_{\mathcal{B}} = 0.0400$ ↺ rad/s; $\mathbf{v}_B = \mathbf{0}$
3.39 2.11 → ft/sec; 0.201 ft above P;

$\mathbf{v}_Q = 60.7$ 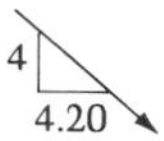

ft/sec;

$\mathbf{v}_S = 60.7$ 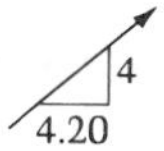

ft/sec;

$\mathbf{v}_R = 85.9$ → ft/sec
3.41 In each case, Ⓘ is at the intersection of the radial line OA and the normal to the slot at B; $\mathbf{v}_B = \mathbf{0}$ when $\overline{OAB}$ and $\overline{AOB}$ are straight lines.

3.43 9.87 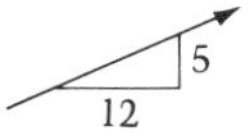

in./sec

3.45 $7.50\hat{\mathbf{i}} - 4.33\hat{\mathbf{j}}$ ft/sec **3.47** See 3.9.
3.49 See 3.13. **3.51** See Example 3.7.
3.53 0.275 ↺ rad/s **3.55** 3.91 ↺ rad/sec²
3.57 $-0.0938\hat{\mathbf{i}} - 0.225\hat{\mathbf{j}}$ m/s²
3.59 84.5 → in./sec; 38.1 ← in./sec²; 565 ← in./sec²
3.61 (a) 4 ↻ rad/sec²; (b) $\theta = 45°$; $\overline{PA} = \sqrt{2}/2$ ft

3.63 0.779 ↻ rad/sec²; 6.56 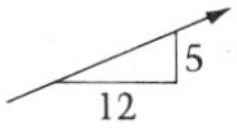

in./sec²

3.65 $\boldsymbol{\alpha}_{\mathcal{B}_1} = 5.70 \times 10^{-6}$ ↻ rad/s²; $\boldsymbol{\alpha}_{\mathcal{B}_2} = 16.1 \times 10^{-6}$ ↺ rad/s² **3.67** $v_0^2/2\hat{\mathbf{j}}$ ft/sec²
3.69 $\dfrac{[(r^2 - \ell^2)r\sin\theta]\dot{\theta}^2}{(\ell^2 - r^2\sin^2\theta)^{3/2}}$ ↻ **3.71** $0.186\hat{\mathbf{k}}$ rad/sec²
3.73 $24\hat{\mathbf{i}} - 27\hat{\mathbf{j}}$ ft/sec² **3.75** $-1.05\hat{\mathbf{i}} - 0.785\hat{\mathbf{j}}$ m/s²
3.77 $-5\sqrt{3}\hat{\mathbf{i}} + 5\hat{\mathbf{j}}$ ft/sec; $(100/L)\hat{\mathbf{i}} - (57.7/L)\hat{\mathbf{j}}$ ft/sec² (with L in feet)
3.79 0.0735 ↑ m/s²; 0.0172 ↻ rad/s²

3.81 $-51.7\hat{\mathbf{i}} - 30.5\hat{\mathbf{j}}$ ft/sec^2
(Note: y positive down here.) **3.83** 15.1 ↺ rad/sec^2
3.85 $2.73\hat{\mathbf{i}} + 1.13\hat{\mathbf{j}}$ ft/sec
3.87 $\mathbf{v}_A = 0.1\hat{\mathbf{i}} + 0.1\hat{\mathbf{j}}$ m/s; $\mathbf{v}_B = 0.2\hat{\mathbf{i}}$ m/s;
$\mathbf{v}_D = 0.1\hat{\mathbf{i}} - 0.1\hat{\mathbf{j}}$ m/s; $\mathbf{v}_E = \mathbf{0}$; 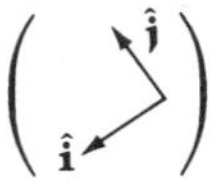

3.89 $\mathbf{v}_C = 1\hat{\mathbf{i}}$ m/s; $\mathbf{v}_B = 1\hat{\mathbf{i}} + 4\hat{\mathbf{j}}$ m/s
(Note: radius superfluous.)
3.91 $1.75\hat{\mathbf{i}}$ m/s; $1.75\hat{\mathbf{i}} - 0.25\hat{\mathbf{j}}$ m/s
3.93 In order, $\mathbf{v}_A = 2v_C \leftarrow$; $1.24\ v_C \leftarrow$; $\mathbf{0}$;
$0.758\ v_C \leftarrow$ cm/s
3.95 The short one gets run over at $t = 5R/v_0$
3.97 $\boldsymbol{\omega}_{\mathcal{B}} = \dfrac{2\pi}{3}$ ↺ rad/s; $\boldsymbol{\omega}_{\mathcal{S}} = \dfrac{2\pi}{9}$ ↺ rad/s
3.99 $\mathbf{v}_A = 117 \leftarrow$ in./sec; $\mathbf{v}_B = \mathbf{0}$; $\mathbf{v}_D = 2820 \rightarrow$ in./sec; $\mathbf{v}_E = 2930 \rightarrow$ in./sec; A is going backwards (to the left) since it's below Ⓘ, and the wheel turns ↻.
3.101 $\boldsymbol{\omega}_{\mathcal{B}} = 1.15$ ↺ rad/s; $\mathbf{v}_A = 4.89 \rightarrow$ m/s
3.103 $\frac{1}{2}$ **3.105** $\dot{x}_C =$ any positive constant k;
$\dot{\theta} = k/R$ (Directions are $\rightarrow$ and ↺.)
3.107 In order, $2v_0\hat{\mathbf{i}}$; $1.71v_0\hat{\mathbf{i}} - 0.707v_0\hat{\mathbf{j}}$;
$v_0\hat{\mathbf{i}} - v_0\hat{\mathbf{j}}$; $\mathbf{0}$; $v_0\hat{\mathbf{i}} + v_0\hat{\mathbf{j}}$
3.109 Let $\mathbf{x}$ and $\mathbf{y}$ respectively be directed down and toward the plane, with origin at the center of the disk. Then the point has $(\mathbf{x}, \mathbf{y}) = (4.80, 3.60)$ ft.
3.111 $\boldsymbol{\omega}_{\mathcal{C}} = 0.381$ ↺ rad/s; $\boldsymbol{\alpha}_{\mathcal{C}} = 2.39$ ↻ rad/s^2
3.113 $-\dfrac{v_0^2}{R-r}\hat{\mathbf{i}} - \dfrac{v_0^2}{r}\hat{\mathbf{j}}$
3.115 $-1.24\hat{\mathbf{j}}$ ft/sec^2
(Note wheel radius superfluous here.)
3.117 (a) $-8\hat{\mathbf{i}}$ cm/s; $6\hat{\mathbf{i}} + 8\hat{\mathbf{j}}$ cm/s^2
(b) $-8\hat{\mathbf{i}} + 8\hat{\mathbf{j}}$ cm/s; $-26\hat{\mathbf{i}} + 2\hat{\mathbf{j}}$ cm/s^2
(c) 19.8 cm/s^2
3.119 Let $\hat{\mathbf{i}}$ be from the center of $\mathcal{A}$ along $\mathcal{B}$, and $\hat{\mathbf{k}}$ be out of the page. Answers are then $-24\pi^2\hat{\mathbf{i}} + 3\pi\hat{\mathbf{j}}$ in./sec^2 ($\mathcal{A}$), and $138\pi^2\hat{\mathbf{i}} + 3\pi\hat{\mathbf{j}}$ in./sec^2 ($\mathcal{C}$).
3.121 $9 \leftarrow$ cm/s; 30.2 ↻ rad/s^2
3.123 (a) $-145\hat{\mathbf{i}}$ in./sec^2; $\mathbf{0}$
(b) $50.3\hat{\mathbf{i}}$ in./sec^2; 5.59 ↺ rad/sec^2 (c) $55\hat{\mathbf{i}}$ in./sec^2; $\mathbf{0}$
(d) $50.3\hat{\mathbf{i}}$ in./sec^2; 5.59 ↻ rad/sec^2
3.125 21.6 m/s^2 (It is the highest point of $\mathcal{P}$.)
3.127 Answer given in problem.
3.129 $\boldsymbol{\alpha}_{\mathcal{D}} = 0.0889\,\omega_0^2$ ↻; $\boldsymbol{\alpha}_{\mathcal{B}} = 0.178\,\omega_0^2$ ↻
3.131 $-0.889\hat{\mathbf{i}} - 5.41\hat{\mathbf{j}}$ m/s^2
3.133 $0.347\hat{\mathbf{i}} + 0.0198\hat{\mathbf{j}}$ m/s^2
3.135 $0.6\hat{\mathbf{i}} + 0.6\hat{\mathbf{j}}$ m/s; $0.75\hat{\mathbf{i}} - 0.45\hat{\mathbf{j}}$ m/s^2
3.137 41.7 ↺ rad/s^2 **3.139** 1.22 ↺ rad/sec^2
3.141 (a) $-4\hat{\mathbf{i}} + 3\hat{\mathbf{j}}$ in./sec^2 (b) -4 in./sec^2 ($\hat{\mathbf{j}}$ along y)
3.143 In order, $9890\hat{\mathbf{j}}$; $9130\hat{\mathbf{j}}$; $\mathbf{0}$; $-9130\hat{\mathbf{j}}$;
$-9890\hat{\mathbf{j}}$ ft/sec^2
3.145 $(11.3\alpha + 10.6)\hat{\mathbf{i}}$ in./sec^2
3.147 (a) Answer given in problem.
(b) Curve is concave downward.
(c) First is $a\alpha\hat{\mathbf{i}} + a\alpha\hat{\mathbf{i}} - \omega^2 a\hat{\mathbf{j}}$, and second is $(2a\alpha)\hat{\mathbf{i}} - [(2a\omega)^2/(4a)]\hat{\mathbf{j}}$, the same.
3.149 Answer given in problem.
3.151 $\boldsymbol{\omega}_{\mathcal{B}} = 5\omega_0$ ↻; $\boldsymbol{\omega}_{\mathcal{C}} = \omega_0$ ↺; $\boldsymbol{\alpha}_{\mathcal{B}} = 5\alpha_0$ ↻;
$\boldsymbol{\alpha}_{\mathcal{C}} = \alpha_0$ ↺
3.153 46.2 in., plus clearance and allowance for belt thickness **3.155** 0.120 ↺ rad/s
3.157 Answer given in problem.
3.159 $0.0600\hat{\mathbf{i}}$ m/s; 0.400 ↻ rad/s
3.161 0.2 ↻ rad/s; 0.293 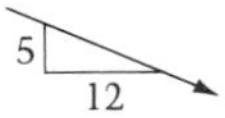

m/s
3.163 $(D\dot{\theta}\sin\theta/\cos^2\theta)(\cos\theta\,\hat{\mathbf{i}} + \sin\theta\,\hat{\mathbf{j}})$
3.165 0.781 ↻ rad/sec **3.167** 0.296 ↻ rad/sec
3.169 9.56 sec **3.171** 206 mph
3.173 $\mathbf{v}_{A/\mathcal{A}} = 0.240\hat{\mathbf{i}} + 0.180\hat{\mathbf{j}}$ ft/sec;
$\mathbf{v}_{A/\mathcal{B}} = -0.480\hat{\mathbf{i}} + 0.640\hat{\mathbf{j}}$ ft/sec;
$\mathbf{a}_{A/\mathcal{A}} = 0.512\hat{\mathbf{i}} + 0.384\hat{\mathbf{j}}$ ft/sec^2;
$\mathbf{a}_{A/\mathcal{B}} = -0.180\hat{\mathbf{i}} + 0.240\hat{\mathbf{j}}$ ft/sec^2
3.175 $\mathbf{a}_{P/\mathcal{B}} = -.0643\hat{\mathbf{i}} + .0343\hat{\mathbf{j}}$ m/s^2;
$\boldsymbol{\alpha}_{\mathcal{B}} = 0.104$ ↺ rad/s^2
3.177 Following the hint, $\mathbf{a}_{P/\mathcal{J}} = \mathbf{a}_{0'/\mathcal{J}} + \ddot{\theta}\hat{\mathbf{k}} \times \mathbf{r}_{0'P} - \dot{\theta}^2\mathbf{r}_{0'P}$, which is the same as Eq. (3.19) with P and $0'$ being the points of the body.
3.179 $\boldsymbol{\alpha}_{\mathcal{B}} = 0.0771$ ↺ rad/s^2
3.181 $\mathbf{v}_P = (\dot{\mathbf{x}} - R\dot{\phi})\hat{\mathbf{i}} + \mathbf{x}\dot{\phi}\hat{\mathbf{j}}$;
$\mathbf{a}_P = (\ddot{\mathbf{x}} - R\ddot{\phi} - \mathbf{x}\dot{\phi}^2)\hat{\mathbf{i}} + (\mathbf{x}\ddot{\phi} - R\dot{\phi}^2 + 2\dot{\mathbf{x}}\dot{\phi})\hat{\mathbf{j}}$
($\hat{\mathbf{i}}$ and $\hat{\mathbf{j}}$ are parallel to $\mathbf{x}$ and $\mathbf{y}$)
3.183 1.89 ft/sec^2 upward
3.185 $[D\dot{\theta}^2(1 + \sin^2\theta)/\cos^3\theta](\cos\theta\,\hat{\mathbf{i}} + \sin\theta\,\hat{\mathbf{j}})$
3.187 $-6\hat{\mathbf{i}} + 10.4\hat{\mathbf{j}}$ ft/sec^2

CHAPTER 4

4.1 Answer given in problem.
4.3 Answer given in problem.
4.5 16π slug-ft^2 **4.7** $\pi R^2L^3(17\rho_1 + 35\rho_0)/60$
4.9 187 kg · m^2
4.11 From the corner, 1.56 ft $\leftarrow$ and 0.562 $\uparrow$;
$I_{z_C} = 4.45$ slug-ft^2.
4.13 $m(b^2 + H^2)/18$ **4.15** Answer given in problem.
4.17 (a) Answer given in problem;
(b) $3m(H^2 + 4R^2)/80$
4.19 $0.0549\rho t r^4$ **4.21** $\rho abL(0.131L^2 + 0.393a^2)$

4.23 1.00 slug-ft^2 **4.25** 3.44 slug-ft^2
4.27 3220 kg · m^2; it's the same line!
4.29 Noting that $m = 4\rho t R\phi L$, the answer is $m[R^2(2\phi - \sin 2\phi)/(4\phi) + L^2/12]$.
4.31 $mR^2[1 - (\sin\alpha)/\alpha]/4$, where $m = \rho t R^2\alpha/2$
4.33 Answer given in problem. **4.35** 23.0 kg · m^2
4.37 Answer given in problem.
4.39 Answer given in problem.
4.41 Answer given in problem.
4.43 T_L(left) = 114 N; T_R(right) = 413 N; $\mathbf{a}_B = 1.55 \rightarrow$ m/s^2
4.45 $-12.9 \le a \le 8.05$ ft/sec^2
4.47 $(gb/\ell)/\sqrt{b^2 + H^2}$ **4.49** $0.427g$
4.51 $\mathbf{a}_C = 12.9 \rightarrow$ ft/sec^2; $\mathbf{N}_{\text{left}} = 93.3 \uparrow$ lb; $\mathbf{N}_{\text{right}} = 6.67 \uparrow$ lb
4.53 $8g/15 \rightarrow$ **4.55** 42.4 lb; $\mathbf{F}_A = 30\hat{\mathbf{i}} + 30\hat{\mathbf{j}}$ lb
4.57 $0.847 \le \ddot{x}_C \le 33.9$ ft/sec^2
4.59 (a) $\mu g(b - d)/(b - \mu h)$; (b) $d = \mu h$; $a_{\max} = \mu g$
4.61 Time $= v(a + b + \mu h)/(\mu g a)$; distance $= v^2(a + b + \mu h)/(2\mu g a)$
4.63 $6\pi^2$ ↺ lb-ft **4.65** 1.29 lb **4.67** 78.6 m
4.69 2.23 s, without having slipped
4.71 Center moves 12.3 ft.
4.73 Final equation is:

$$[I_{z_C} + m\delta^2\cos^2\theta + m(r - \delta\sin\theta)^2]\ddot{\theta} - (m\delta r\cos\theta)\dot{\theta}^2 - mg\delta\cos\theta = 0,$$ where $\delta = (4r/3\pi)$ and $I_{z_C} = (mr^2/2) - m\delta^2$

4.75 Each force $= \dfrac{WL\omega^2}{2g} + \dfrac{W}{2}\sin\theta$ (tension);

$\mathbf{a}_C = L\omega^2$ 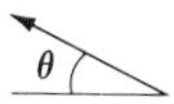 $+ g\cos\theta$ 

4.77 No, it rolls; $\mathbf{a}_C = 0.585\hat{\mathbf{i}}$ ft/sec^2
4.79 (a) 13.3 ↻ rad/sec^2 (b) 1.11 ↻ rad/sec^2 (c) 0.0197
4.81 0.450 ↺ rad/sec^2 **4.83** $7.16 \leftarrow$ ft/sec^2
4.85 17.4 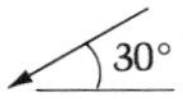 ft/sec^2
4.87 12.1 rad/sec^2, AB ↻ and BD ↺; 8.13 ↑ lb each
4.89 43.3 lb
4.91 $[mg\sin\theta(0.180\cos\theta - 0.424)/(0.849\cos\theta - 1.5)]\hat{\mathbf{i}} + mg[1 + 0.180\sin^2\theta/(0.849\cos\theta - 1.5)]\hat{\mathbf{j}}$
4.93 At A, $16.0\hat{\mathbf{i}} + 160\hat{\mathbf{j}}$ lb; at B, $-76.0\hat{\mathbf{i}}$ lb
4.95 41.6 lb
4.97 On the section to the left of the cut, $V = 3WL/64 \downarrow$ and $M = 9WL^2/256$ ↺.
4.99 $x = L/\sqrt{12}$
4.101 $t = \omega_i R(1 + \mu^2)/[\mu g(1 + \mu)]$; $\theta = R\omega_i^2(1 + \mu^2)/[2\mu g(1 + \mu)]$
4.103 $1.59\hat{\mathbf{j}}$ lb
4.105 Answer given in problem to first part; $9mg/64$
4.107 1.48 sec (I_C is 18.2 lb-in.-sec^2)
4.109 Answer given in problem.
4.111 In order, $\boldsymbol{\alpha} = 0.805$ ↻, 2.28 ↻ and 2.82 ↻ rad/sec^2; $\theta = 163°$
4.113 5.41 ↻ rad/sec^2
4.115 $mg(B^2 + H^2)/(4B^2 + H^2)$
4.117 $\mathbf{a}_C = 3g/4 \downarrow$; $\boldsymbol{\alpha} = 3g/(2s)$ ↻
4.119 $\mathbf{a}_C = g \downarrow$; $\boldsymbol{\alpha} = 6g/s$ ↻
4.121 $\mathbf{a}_C = 2g/3 \downarrow$; $\boldsymbol{\alpha} = 2g/s$ ↻
4.123 Answer is the magnitude of the horizontal force on the door, which is $-0.75ma_0\sin\beta\cos\beta\hat{\mathbf{i}} + ma_0(1 - 0.75\sin^2\beta)\hat{\mathbf{j}}$
4.125 (a) T is the only external force with a non-zero moment about the contact point. Since the conditions outlined in this section's text apply, the left wheel will start to roll to the left and the right wheel to the right. (b) $\frac{1}{2}$ **4.127** $\alpha_{\mathcal{C}} = 2.5$ ↻ rad/s^2; $\alpha_{\mathcal{R}} = 1.25$ ↻ rad/s^2
4.129 16.9 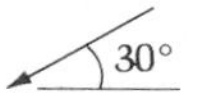 ft/sec^2; yes
4.131 In $\mathcal{A}$, $(x, y) = (-56/3, -21)$ in.; in $\mathcal{B}$, $(x, y) = (-70/3, -35)$ in.
4.133 24.0 rad/sec
4.135 Force is $M\ell\omega^2\sin\theta\hat{\mathbf{i}} - Mg\hat{\mathbf{j}}$; moment is $Mg\ell\sin\theta + M\ell^2\omega^2\sin\theta\cos\theta$ clockwise.
4.137 $(x_A, y_A) = (-\frac{5}{6}, -\frac{3}{4})$ in.; $(x_B, y_B) = (-\frac{7}{8}, -\frac{3}{16})$ in.
4.139 $(x_A, y_A) = (0.366, 0.691)$ ft; $(x_B, y_B) = (0.0488, 0.171)$ ft
4.141 $(m_{\mathcal{P}} + m_{\mathcal{C}} + m_{\mathcal{A}}/2)g \downarrow$ (each), plus $[m_{\mathcal{C}}\ell_{\mathcal{A}}\ell_{\mathcal{C}}/2 + m_{\mathcal{P}}(\ell_{\mathcal{A}} + \ell_{\mathcal{P}})\ell_{\mathcal{C}}]\omega^2/(2d)$, up on left and down on right.
4.143 $13mr^2\omega^2\sin\psi\cos\psi/(96\ell)$, down on left and up on right onto the shaft, and rotating with it.
4.145 $-0.698\hat{\mathbf{i}}$ m/s^2; 8.73 m to the left
4.147 (a) $a_0 \le 7.85$ m/s^2 (b) 20.0 m
4.149 It was the driven wheel and the larger the wheel, the farther the bike would travel for each revolution of the pedal crank. Distance = 9.42 ft.
4.151 $0.3g$; 1.79 ft/sec^2 $\rightarrow$
4.153 $3T_0/[(2R^2)(9M + 2m)]$ in whatever direction T_0 acts
4.155 (a) 20 kg; (b) The rope breaks first.
4.157 Let $\hat{\mathbf{i}}$ lie along OA and $\hat{\mathbf{j}}$ along AC. (C is mass center of bar.) Then the force is $-W(0.540 + 0.500\beta^2L/g)\hat{\mathbf{i}} - W(0.841 - \beta^2 r/g)\hat{\mathbf{j}}$ and the moment is $WL(0.270 + 0.333\beta^2L/g)\hat{\mathbf{k}}$
4.159 $(k_C^2 + r^2 + R^2 + 2Rr\cos\theta)\ddot{\theta} - (rR\sin\theta)\dot{\theta}^2 - g[R\sin\beta + r\sin(\beta + \theta)] = 0$
4.161 $(mg\sin\beta)/5$ (tensile)

4.163 (a) $H_1 =$

$$\left[\frac{v_0(5\mu\cos\alpha + \sin\alpha) - r\omega_0(\sin\alpha + \mu\cos\alpha)}{2g(3\mu\cos\alpha + \sin\alpha)^2}\right](r\omega_0 + v_0);$$

$$t_1 = \frac{r\omega_0 + v_0}{(3\mu\cos\alpha + \sin\alpha)g};$$

(b) $H_2 =$

$$3\left[\frac{2\mu v_0\cos\alpha - r\omega_0(\sin\alpha + \mu\cos\alpha)}{3\mu\cos\alpha + \sin\alpha}\right]^2/(4g\sin\alpha);$$

$$t_2 = 3\left[\frac{2\mu v_0\cos\alpha - r\omega_0(\sin\alpha + \mu\cos\alpha)}{3\mu\cos\alpha + \sin\alpha}\right]/(2g\sin\alpha)$$

Note: If $\mathbf{v}_C$ is zero or down the plane when slipping stops, the answers to (b) are both zero (reached highest point while slipping).

4.165 300 ft

4.167 Answer to first question given in problem. It is only the direction of $\ddot{\mathbf{x}}_C$ that matters here, and it is not dependent upon the size of ω_0. ($\mathbf{x}_C$ parallel to plane)

4.169 $b = \mu H$ **4.171** (a) 4.00 → m/s² (b) 0.23 m

4.173 $2g/3R$; $5g/7R$; $g/2R$

4.175 $md^2f'(1-f)^2/(1-f')$

4.177 (a) $F/(M + 3mn/8)$ (b) $Ft^2/(2M + 3mn/4)$

4.179 $\mathbf{a}_{\mathcal{A}} = 0.0204g \downarrow$, $\mathbf{a}_{\mathcal{B}} = 0.224g \uparrow$, $\mathbf{a}_{\mathcal{C}} = 0.184g \downarrow$; $\mathcal{C}$; 2.95 s ($g = 9.81$ m/s²)

4.181 (a) 5.50 kg · m² (b) $4.04\hat{\mathbf{i}}$ m/s²

4.183 Answer given in problem.

4.185 9.32 ← in./sec; 3.11 rad/sec

4.187 Answer given in problem (actually, 43 ft).

4.189 Answer given in problem.

4.191 3.39 ↑ ft/sec² (C) and 3.39 ↓ ft/sec² (B)

4.193 The mass center moves vertically downward; $mg/4 \downarrow$.

4.195 $q = \sqrt[3]{R^2r}$; $d = 2\sqrt[3]{R^2r}$.

4.197 160.3 lb, 0.43%; 160.5 lb, 0.31%

CHAPTER 5

5.1 16 ft-lb **5.3** $6.53\hat{\mathbf{i}}$ ft/sec

5.5 (a) 4/17 m; (b) 50 *N increase*, $\mu_{min} = 0.220$

5.7 100 J **5.9** $41.5\hat{\mathbf{j}}$ lb

5.11 It starts out to the right, the spring goes slack, and then it leaves on the right. (It would need one more foot of plane to stay on.)

5.13 At A, $3mg/2$ ←; at B, $mg/4$ ↑. **5.15** 5.69 mg

5.17 (a) 0.188 (b) 53.1° **5.19** See 4.177 (b)

5.21 (a) 2.32 ↺ rad/sec (b) 3.98 ↻ rad/sec²

5.23 13.2 rad/s **5.25** 3.42 ↻ rad/s **5.27** 140 N/m

5.29 15.2 rad/sec at the bottom.

5.31 $1.53\sqrt{g/R}$ (↻ for the left ring and ↺ for the right ring)

5.33 $\mathbf{v}_C = 2.57\sqrt{\mathbf{x}_C}$ ↙ 30°;

$\mathbf{v}_C = 3.30t$ ↙ 30°

5.35 $\dot{\mathbf{x}}_C(\mathbf{x}_C) = \sqrt{16F\mathbf{x}_C/(8M + 3mn)}$;
$\dot{\mathbf{x}}_C(t) = 8Ft/(8M + 3mn)$

5.37 $\tan^{-1}[\mu/(1 + 36k^2)]$; 5.27° **5.39** 30.6 N/m

5.41 (a) $2g/3 \downarrow$ (each); (b) $\sqrt{4gD/3}$

5.43 35 lb/ft; yes **5.45** 3.82 m

5.47 15.3 ↙ (slope 5:12) ft/sec

5.49 $\sqrt{24\pi gr\sin\beta/13}$ ↙ β

5.51 Answer given in problem. **5.53** 103,000 lb/ft

5.55 $\left[\dfrac{8ga(a^2 + b^2)}{a^4 + 10a^2b^2 + b^4}\left(1 - b\sqrt{\dfrac{a^2 + b^2}{a^4 + b^4}}\right)\right]^{1/2}$

5.57 47.4° **5.59** 4.91 ↺ rad/sec; $\frac{1}{3}$ ft

5.61 3.75 ↻ rad/sec

5.63 (a) 1.56 ↺ rad/sec; (b) 1.40 ↺ rad/sec

5.65 2.23 sec **5.67** 1.01 rad **5.69** $2gt_0/(R\pi)$

5.71 (a) 14.6°; (b) 1.38 s

5.73 $\dot{x}_C = gt/\sqrt{3}$ and $x_C = gt^2/(2\sqrt{3})$, both down the plane

5.75 0.753 slug-ft²; 38.0 → ft

5.77 $x = 5t^2$; $y = 4.91t^2$

5.79 (a) 1.88 ↓ m/s; (b) 4.51 ↑ m/s **5.81** 28.2 sec

5.83 See 4.101(c) **5.85** See 4.107

5.87 4.14 m, using P_E **5.89** 11.5 ft upward

5.91 $2d/9$ upward

5.93 $v_0m/(4m + M)$ parallel to bird's approach; $6v_0m/[\ell(4m + M)]$, counterclockwise looking down.

5.95 Answer given in problem.

5.97 Answer given in problem.

5.99 Answer given in problem. **5.101** $-0.540\sqrt{gh}\,\hat{\mathbf{j}}$

5.103 5.94 times, or 494% increase

5.105 0.167 lb-ft, opposite to the original rotation direction of $\mathcal{A}$

5.107 10.1 ft (spring compressed 9.6 ft when $\mathcal{B}$ is at its highest point)

5.109 0.446 **5.111** $-0.25v_0\hat{\mathbf{i}} + 0.75v_0\hat{\mathbf{j}}$

5.113 $6v_0\sin\beta\cos\beta/[L(1 + 3\cos^2\beta)]$

5.115 (a) 1040 ft-lb (b) 16.4 ft/sec

5.117 Answer given in problem.

5.119 (a) $\mathbf{v}_{G_\mathcal{A}} = \mathbf{0}$; $\mathbf{v}_{G_\mathcal{B}} = v_0 \rightarrow$; $\boldsymbol{\omega}_\mathcal{A} = v_0/r$ ↻; $\boldsymbol{\omega}_\mathcal{B} = \mathbf{0}$; (b) $\mathbf{v}_{G_\mathcal{A}} = 2v_0/7 \rightarrow$; $\mathbf{v}_{G_\mathcal{B}} = 5v_0/7 \rightarrow$; (c) If $\mu = 0$, final motion is given by (a).

5.121 $e = 0.8$; $\mu = 0.32$

5.123 (a) 1.66 rad/sec for both; (b) 17%
5.125 5.06 $m\hat{\mathbf{j}}$ lb-sec; 6.3 sec **5.127** 4.38 m
5.129 $(W_1 + W_2)/k$

$$+ \sqrt{[W_1^2(W_1 + W_2) + 2HW_1^2k]/[k^2(W_1 + W_2)]}$$

(total compression)
5.131 With $\mathbf{v}_{C_f} = 19.0 \downarrow$ ft/sec and $\boldsymbol{\omega}_f = 13.4 \circlearrowleft$ rad/sec, the percentage lost is 62.0%.
5.133 $R\omega_0 > 3v_0/(2\sqrt{2})$
5.135 0.96 ft to the left of the spring's initial position.
5.137 $\omega_f = \dfrac{3\ell\, v_{C_i}(1 + e)}{2\ell^2 + 9r^2}$ clockwise;
$\mathbf{v}_{C_f} = (-\ell\omega_f/2 + ev_{C_i})\hat{\mathbf{i}} + r\omega_f\hat{\mathbf{j}}$
(see Example 5.18 for $\hat{\mathbf{i}}$, $\hat{\mathbf{j}}$); $\omega_f(e = 0) = \frac{1}{2}\,\omega_f(e = 1)$;
$\dot{x}_{C_f}(e = 0) = 1.91\dot{x}_{C_f}(e = 1)$; $\dot{y}_{C_f}(e = 0) = \frac{1}{2}\,\dot{y}_{C_f}(e = 1)$
5.139 Answer given in problem.
5.141 Let $\hat{\mathbf{i}}$ be along AD and $\hat{\mathbf{k}}$ be upward.
Let $K = M_0\Delta t/(am)$. Then for AB: $\mathbf{v}_G = (-2K/5)\hat{\mathbf{j}}$ and $\boldsymbol{\omega} = (9K/5a)\hat{\mathbf{k}}$; for BC: $\mathbf{v}_G = (K/2)\hat{\mathbf{j}}$ and $\boldsymbol{\omega} = (-9K/10a)\hat{\mathbf{k}}$; and for CD: $\mathbf{v}_G = (-K/10)\hat{\mathbf{j}}$ and $\boldsymbol{\omega} = (3K/10a)\hat{\mathbf{k}}$.
5.143 Answer given in problem.
5.145 Let $\mathbf{x}$ be parallel to AC, and $\mathbf{y}$ to DB.
Then the mass center velocities are: for AB, $[7F\Delta t/(16m)]\hat{\mathbf{i}} + [3F\Delta t/(16m)]\hat{\mathbf{j}}$ and for BC, $[F\Delta t/(16m)]\hat{\mathbf{i}} + [3F\Delta t/(16m)]\hat{\mathbf{j}}$. The angular velocities are $3\sqrt{2}F\Delta t/(8m\ell)$, $\circlearrowleft$ for AB and $\circlearrowright$ for BC.
5.147 $3I^2/(2M)$
5.149 $5I\,[(a - c)m - 3cM]/[Ma^2(6M + 7m)]$
5.151 Answer given in problem.

CHAPTER 6

6.1 Answer given in problem.
6.3 Answer given in problem. **6.5** $\dot{\theta}\hat{\mathbf{i}}$
6.7 The solution is Equation (6.70) with

$$\theta_1 = \theta_{\text{pitch}} = \frac{\pi}{18}\sin\frac{2\pi t}{6};\ \theta_2 = \theta_{\text{roll}} = \frac{\pi}{6}\sin\frac{2\pi t}{8};$$

and $\theta_3 = \theta_{\text{yaw}} = \dfrac{\pi}{22.5}\sin\dfrac{2\pi t}{50}$.
6.9 To the components in 6.7, respectively, add $-\dot{P}\cos R$, $-\dot{R}$, and $-\dot{P}\sin R$.
6.11 $\boldsymbol{\omega}_{\mathcal{A}/\mathcal{G}} = 13.7\hat{\mathbf{j}}_1 + \hat{\mathbf{k}}_1$; $\boldsymbol{\alpha}_{\mathcal{A}/\mathcal{G}} = 1.12\hat{\mathbf{i}}_1 + 13.4\hat{\mathbf{j}}_1$
6.13 $\ddot{\theta}_1\cos\alpha/(1 - \sin^2\alpha\cos^2\theta_1)$
$\qquad - \dot{\theta}_1^2\cos\alpha\sin^2\alpha\sin 2\theta_1/(1 - \sin^2\alpha\cos^2\theta_1)^2$
6.15 Components are:
$(-0.00710, 0.00331, 0.0749)$ rad/sec² in frame $\mathcal{S}$.
6.17 (a) $(7\cos t + 8t - 14t^2)\hat{\mathbf{b}}_1 + (-7\sin t + 2 + 28t^3)\hat{\mathbf{b}}_2 + (2t\sin t - 4t^2\cos t + 7)\hat{\mathbf{b}}_3$;
(b) $7\hat{\mathbf{b}}_1 + 2\hat{\mathbf{b}}_2 + 7\hat{\mathbf{b}}_3$; (c) $6.64\hat{\mathbf{b}}_1 + 2.14\hat{\mathbf{b}}_2 + 6.60\hat{\mathbf{b}}_3$ rad/sec²

6.19 $0.458\hat{\mathbf{i}} + 0.200\hat{\mathbf{k}}$ rad/sec; $0.1\hat{\mathbf{i}} + 0.0915\hat{\mathbf{j}}$ rad/sec²;
$-0.8\hat{\mathbf{i}} - 19.8\hat{\mathbf{j}} + 11.4\hat{\mathbf{k}}$ ft/sec;
$7.93\hat{\mathbf{i}} - 10.7\hat{\mathbf{j}} - 6.57\hat{\mathbf{k}}$ ft/sec²
6.21 36.5 ft **6.23** $1.34\hat{\mathbf{i}} + 0.00905\hat{\mathbf{j}} - 1.41\hat{\mathbf{k}}$ ft/sec²
6.25 $17.3\hat{\mathbf{i}} + 15.7\hat{\mathbf{j}} - 10\hat{\mathbf{k}}$ in./sec;
$-147\hat{\mathbf{i}} + 218\hat{\mathbf{j}} - 64.3\hat{\mathbf{k}}$ in./sec²
6.27 $6\hat{\mathbf{j}}$ rad/sec²; $-32.0\hat{\mathbf{i}} + 1.20\hat{\mathbf{j}} + 8.90\hat{\mathbf{k}}$ in./sec²
6.29 $3.67\hat{\mathbf{i}}$ ft/sec; $-2.10\hat{\mathbf{i}} - 1.78\hat{\mathbf{j}}$ ft/sec²
6.31 $\omega_x\hat{\mathbf{i}} + \omega_y\cos\omega_x t\hat{\mathbf{j}} + (\omega_z + \omega_y\sin\omega_x t)\hat{\mathbf{k}}$
6.33 6.60 rad/sec
6.35 $84.8\hat{\mathbf{i}} + 294\hat{\mathbf{j}}$ ft/sec; $-24900\hat{\mathbf{i}} + 14000\hat{\mathbf{j}}$ ft/sec²
6.37 2 s
6.39 $\ddot{\theta} = \alpha_r s_\phi - \alpha_p c_\phi + (\omega_r c_\phi + \omega_p s_\phi)^2/\tan\theta$;
$\ddot{\phi} = (\alpha_r c_\phi + \alpha_p s_\phi)/\tan\theta + [(\omega_p^2 - \omega_r^2)\,s_\phi c_\phi + \omega_r\omega_p\cos 2\phi]$
$\qquad \cdot (1 + c_\theta^2)/s_\theta^2$
6.41 Answer given in problem.
6.43 Each $= -45\hat{\mathbf{i}} + 30\hat{\mathbf{k}}$ rad/sec
6.45 (a) Answer given in problem.
(b) Answer same if $\boldsymbol{\alpha}$ is replaced by $\boldsymbol{\omega}$.
6.47 $[-4\pi^2\sin\beta/(T^2\cos\beta)]\,\hat{\mathbf{i}}$
6.49 $\boldsymbol{\omega}_{\mathcal{C}_1/\mathcal{I}} = -\sqrt{3}\,(\Omega_1 + \Omega_2)\hat{\mathbf{i}} + \Omega_2\hat{\mathbf{k}}$;
$\boldsymbol{\alpha}_{\mathcal{C}_1/\mathcal{I}} = -\sqrt{3}\,\Omega_2\,(\Omega_1 + \Omega_2)\hat{\mathbf{j}}$;
$\mathbf{a}_{A/\mathcal{I}} = -b\Omega_2(2\Omega_1 + 3\Omega_2)\hat{\mathbf{i}} - \sqrt{3}\,b\,(\Omega_1 + \Omega_2)^2\hat{\mathbf{k}}$
6.51 (a) $\boldsymbol{\omega} = -(R\omega_1/r)\hat{\mathbf{i}} + \omega_1\hat{\mathbf{k}}$; $\boldsymbol{\alpha} = -(R\omega_1^2/r)\hat{\mathbf{j}}$
(b) $R\omega_1^2\hat{\mathbf{i}} + (R^2\omega_1^2/r)\hat{\mathbf{k}}$
6.53 $43\frac{1}{3}$ cm
6.55 $-0.000432t^5/\sqrt{363 - 0.000144t^6}\;\hat{\mathbf{k}}$ m/s
6.57 (a) There is no point. (b) All points in the plane $z = -3$ which lie on the line $y = (2/3)x - 1$.
6.59 $\mathbf{v}_{Q/\mathcal{I}} = r\Omega_2\sin\Omega_1 t\hat{\mathbf{i}} + (R\Omega_2 - r\Omega_1\cos\Omega_1 t)\hat{\mathbf{j}}$
$\qquad - r\Omega_1\sin\Omega_1 t\hat{\mathbf{k}}$;
$\mathbf{a}_{Q/\mathcal{I}} = (2r\Omega_1\Omega_2\cos\Omega_1 t - R\Omega_2^2)\hat{\mathbf{i}}$
$\qquad + [r(\Omega_1^2 + \Omega_2^2)\sin\Omega_1 t]\,\hat{\mathbf{j}} - r\Omega_1^2\cos\Omega_1 t\hat{\mathbf{k}}$
6.61 (a) Answer given in problem.
(b) $\sin\beta\cot\phi - \cos\beta$
6.63 $(\sqrt{3}\,\Omega^2/2)\hat{\mathbf{j}}$
6.65 Since each point of the contact line is at rest, $\boldsymbol{\omega}_{\mathcal{C}/\mathcal{I}}$ lies along this line. A triangle may be sketched consisting of a horizontal line, vertical line, and a line along ℓ—each leg of which represents one of the vectors in $\boldsymbol{\omega}_{\mathcal{C}/\mathcal{I}} = \boldsymbol{\omega}_{\mathcal{C}/\mathcal{P}} + \boldsymbol{\omega}_{\mathcal{P}/\mathcal{I}}$. The equation follows immediately.
6.67 1.36 rad/s, directed from O through the line of contact between $\mathcal{C}$ and $\mathcal{D}$.
6.69 $-11.8\hat{\mathbf{i}} + 25.0\hat{\mathbf{j}} - 5.33\hat{\mathbf{k}}$ rad/sec²
6.71 $|\boldsymbol{\omega}_{\mathcal{B}/\mathcal{I}}| = \sqrt{\dot{\phi}^2 + \dot{\theta}^2 + \dot{\psi}^2 + 2\dot{\phi}\dot{\psi}\cos\theta}$
6.73 Components are:
$(\dot{\theta}_1 + s_2\dot{\theta}_3,\ c_1\dot{\theta}_2 - s_1c_2\dot{\theta}_3,\ s_1\dot{\theta}_2 + c_1c_2\dot{\theta}_3)$

6.75 Let point A be displaced from its original to its final position. Then, using Euler's Theorem, all other points of the body may be placed in their final positions via a single rotation about an axis through A.
6.77 $\dot{x} = r(\dot{\psi}s_\phi s_\theta + \dot{\theta}c_\phi)$; $\dot{y} = -r(\dot{\psi}c_\phi s_\theta - \dot{\theta}s_\phi)$
6.79 Components are:
$(\dot{\theta}_1 + \dot{\theta}_3 s_2, \dot{\theta}_2 c_1 - \dot{\theta}_3 s_1 c_2, \dot{\theta}_2 s_1 + \dot{\theta}_3 c_1 c_2)$,
the same as Problem 6.73.
6.81 $E = \tan^{-1}[(c_\lambda c_\delta - r)/(s_\lambda c_\delta)]$;
$A = \tan^{-1}\left[s_\delta/\sqrt{c_\delta^2 - 2rc_\lambda c_\delta + r^2}\right]$, where $r = R_e/R$
6.83 Components are:
$(\dot{E}c_A - \dot{P}c_E s_A, \dot{E}s_A + \dot{P}c_E c_A, \dot{A} + \dot{P}s_E)$

CHAPTER 7

7.1 Sitting on the train facing forward,
(1) right side (2) left side (3) left side (4) right side.
7.3 $-0.0590\hat{\mathbf{i}} + 0.0609\hat{\mathbf{j}}$ lb. No, because the friction would have to be in the same direction as the relative velocity! ($\hat{\mathbf{i}} \rightarrow$ and $\hat{\mathbf{j}} \uparrow$)
7.5 0.290 in. **7.7** $\dot{z}_i^2(g/2 - \omega\dot{x}_i \cos\lambda)/(g - 2\omega\dot{x}_i \cos\lambda)^2$
7.9 (a) Increase is $(2v_0^2\omega \sin 2\beta \sin\theta \cos\lambda)/g^2$;
(b) Answer given in problem.
7.11 With $\hat{\mathbf{i}}$ along the axle from O through the wheel center, and $\hat{\mathbf{k}}$ out of the page,

$$\mathbf{H}_O = \frac{2\pi mb^2}{T}\left[-\cos\beta \sin^2\beta\,\hat{\mathbf{i}} + \left(1 - \frac{\cos^2\beta}{2}\right)\sin\beta\,\hat{\mathbf{j}}\right]$$

7.13 $m\ell^2\omega_0[2\hat{\mathbf{i}} + (10/3)\hat{\mathbf{k}}]$
7.15 With $\hat{\mathbf{i}}$ and $\hat{\mathbf{j}}$ parallel to x and y of the figure,

$$\mathbf{H}_C = \frac{mr^2}{4}\Omega \sin\phi\left(\hat{\mathbf{i}} - \frac{2\ell}{r}\hat{\mathbf{j}}\right)$$

7.17 (a) $\mathbf{H}_O = -18\hat{\mathbf{j}}$ sl-ft²/sec;
(b) $\mathbf{H}_C = -6\hat{\mathbf{j}} - 12\hat{\mathbf{k}}$ sl-ft²/sec
7.19 Answer given in problem.
7.21 Answer given in problem.
7.23 $2m\ell^2/3$ **7.25** $-0.0186ma^2$
7.27 $2m(a^2 + b^2 + c^2)/15$
7.29 $I_{xx}^O = \rho Ar^3(\frac{8}{3} + 11\pi) = 37.2\rho Ar^3$,
$I_{yy}^O = \rho Ar^3(\frac{8}{3} + 9\pi) = 30.9\rho Ar^3$;
$I_{zz}^O = 0 + 4\pi\rho Ar^3 = 12.6\rho Ar^3$,
$I_{yz}^O = 4\pi\rho Ar^3 = 12.6\rho Ar^3$; $I_{xy}^O = I_{xz}^O = 0$
7.31 Answer given in problem.
7.33 Write the three definitions, add two of them, and compare with the third.
7.35 Answer given in problem.
7.37 In order, $(22/3)m\ell^2$, $(32/3)m\ell^2$, and $(10/3)m\ell^2$.
$I_1 = 3.23m\ell^2$ with $(n_x, n_y, n_z) = (0.160, 0, -0.987)$;
$I_2 = 7.44m\ell^2$ with $(0.987, 0, 0.160)$;
$I_3 = 10.7m\ell^2$ with $(0, 1, 0)$.
7.39 $60m\ell^2$ with direction numbers $\pm(-1, 1, 2)$;
$132m\ell^2$ with $\pm(1, 1, 0)$; and
$168m\ell^2$ with $\pm(-1, 1, -1)$. (Total mass $= 3m$.)

7.41 $I_1 = \dfrac{7}{3}m\ell^2$ with $\hat{\mathbf{n}}_1 = \left(\dfrac{1}{\sqrt{3}}, \dfrac{1}{\sqrt{3}}, \dfrac{1}{\sqrt{3}}\right)$;

$I_2 = I_3 = \dfrac{23}{6}m\ell^2$ with $\hat{\mathbf{n}}$ in any direction normal to $\hat{\mathbf{n}}_1$.
7.43 Answer given in problem.
(Use the parallel axes theorem!)
7.45 Answer given in problem.
(Use the parallel axes theorem!)
7.47 3.58°, working with six digits and rounding at the end.
7.49 $I_1 = m\ell^2/4$ and $\hat{\mathbf{n}}_1 = (0.408, 0.816, -0.408)$;
$I_2 = 11m\ell^2/12$ and $\hat{\mathbf{n}}_2 = (0.707, 0, 0.707)$;
$I_3 = m\ell^2$ and $\hat{\mathbf{n}}_3 = (0.577, -0.577, -0.577)$

7.51 $I_{1,2} = \dfrac{m}{36}\left[(b^2 + H^2 + 3L^2) \pm \sqrt{b^4 - b^2H^2 + H^4}\right]$;

$I_3 = \dfrac{m(b^2 + H^2)}{18}$; principal directions are same as those of Example 7.6.
7.53 $I_1 = 34.3 \times 10^5$ slug-ft², $\hat{\mathbf{n}} = (0, 0, 1)$
$I_2 = 33.7 \times 10^5$ slug-ft², $\hat{\mathbf{n}} = (0.566, -0.824, 0)$
$I_3 = 47.5 \times 10^5$ slug-ft², $\hat{\mathbf{n}} = (0.824, 0.566, 0)$
Note: There is a precision problem here because
(1) I_{xx} and I_{yy} are so much larger than I_{xy}, and
(2) I_{xx} and I_{yy} are nearly equal.
7.55 90.6 mph
7.57 $\mathbf{M}_{r_C}$ is here equal to $\boldsymbol{\omega}_{\mathcal{C}/\mathcal{I}} \times \mathbf{H}_C$, and $\boldsymbol{\omega}_{\mathcal{C}/\mathcal{I}}$ is vertical. Since the z component of $\mathbf{H}_C$ is larger than the x component (with both positive), the direction of $\mathbf{M}_{r_C}$ is that of $\boldsymbol{\omega}_{\mathcal{C}/\mathcal{I}} \times \hat{\mathbf{k}}$, which is ↺.

7.59 $-\dot{\mathbf{H}}_G = \dfrac{v_G^2}{R^2}(I_3 - I_1)\sin\theta\cos\theta\,\hat{\mathbf{j}}$ and

$$\mathbf{r}_{QG} \times (-m\mathbf{a}_G) = \frac{mv_G^2}{R}\left(H\cos\theta - \frac{D}{2}\sin\theta\right)\hat{\mathbf{j}}$$

(Note that $\mathbf{M}_{r_Q} = \dot{\mathbf{H}}_G + \mathbf{r}_{QG} \times m\mathbf{a}_G$.) Higher values of v_G, $(I_3 - I_1)$, m, and H increase the danger of tipping, as do lower values of R, D, and θ (since θ is small).
7.61 0.328 lb-ft, clockwise if viewed from front of bike.
7.63 Tensile force in $\mathcal{A}$ is
$T = mL\Omega^2 - mg\tan\theta_0 + m\Omega^2(r\sin\theta_0 + 2L)/4$;
friction force is $T - mL\Omega^2$ outward;
normal force is $mg\uparrow$. As $\theta_0 \rightarrow 0$, answers agree.
7.65 $\mathbf{F} = (-mR\omega_2^2, -mD\omega_2^2, mg)$; $\mathbf{M}_O =$

$$\left(-mgD + mHD\omega_2^2 + \frac{mr^2}{2}\omega_2\omega_3, 0, mgR - mRH\omega_2^2\right)$$

7.67 Moment about bottom of wheel, of $-m\mathbf{a}_C$ and $-\dot{\mathbf{H}}_C$, are, respectively, $(\cos\phi)(d + r)\left(m\dfrac{v_C^2}{R}\right)$ ↺ and $\left[\dfrac{v_C^2}{R^2}\sin\phi\cos\phi\,(I_1 - I_2) + \dfrac{2iv_C^2}{rR}\cos\phi\right]$ ↺. For small ϕ, the moment of $-m\mathbf{a}_C$ is much larger than $-\dot{\mathbf{H}}_C$.

7.69

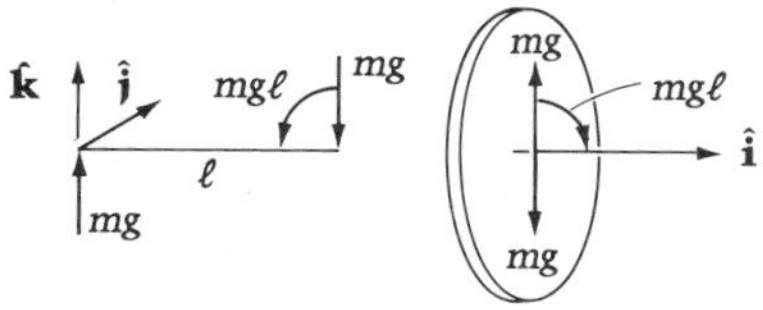

$\mathbf{F}_r = mg\hat{\mathbf{k}}$;
$\mathbf{M}_{r_O} = mg\ell\hat{\mathbf{j}} = \omega_P\hat{\mathbf{k}} \times (I_{x_O}\omega_s\hat{\mathbf{i}} + I_{z_O}\omega_P\hat{\mathbf{k}})$

$$mg\ell = \omega_P\omega_s mR^2/2$$

So if $2g\ell = \omega_P\omega_s R^2$, the motion will occur; this equation is the same if $\mathcal{D}$ and $\mathcal{R}$ are connected.

7.71 Answer is (a). If you throw the ball with a translating arm motion, its axis will tend to translate.

7.73 (a) See the sketches, and note that $I_{yz}^C > 0$ because the same arm mass is distributed with roughly the same "$|z|$" but with the left arm having much more "y" in an overall sense. $I_{xz}^C > 0$ since nearly all the mass of the two arms is in quadrants where $-xz$ is positive. $I_{xy}^C < 0$ due to roughly equal "$|x|$" with the left arm having much larger "y" overall. These products of inertia are much smaller than the three moments of inertia, since only the arms are involved in the products. I_{yz}^C is smallest since the z's are not large and the y's of the separate arms are of the same sign. (b) Then, Equation (3) gives (i), then (2) gives (ii), and finally, (1) yields (iii).

7.75 With x along $\mathcal{S}$ from Q and y upward,

$$\mathbf{F} = (-mR\omega_1^2,\ mg,\ 0) \text{ and } \mathbf{M}_C = \left(0, 0, \frac{mr^2}{2}\,\omega_1\omega_2\right)$$

7.77 $\phi = \cos^{-1}\{g\ell/[\Omega^2(\ell^2 - r^2/4)]\}$;
$\Omega_{\min} = 3.35$ rad/sec $< 2\pi$ rad/sec

7.79 5.64 lb

7.81 $I_{XZ}^C = 300$ kg · m^2; reactions are 12100 N, up on left and down on right onto the shaft.

7.83 0.123 lb-in.

7.85 Of course, the gravity trajectory is curved, but in a vertical plane. This problem addresses the ball curving out of this plane. Since the phenomena is caused by the varying air pressures resulting from the ball's velocity and spin, clearly it would not curve to the side without the air.

7.87 $(I + m\ell^2)$ becomes $(\bar{I} + I + m\ell^2)$.

7.89 Answer given in problem.

7.91 Answer given in problem. The second part is equivalent to showing that $\theta > \beta/2$.

7.93 $\mathbf{M}_{r_C} = \mathbf{0} = {}^{\mathcal{I}}\dot{\mathbf{H}}_C \Rightarrow \mathbf{H}_C =$ constant
$= I(\omega_x\hat{\mathbf{i}} + \omega_y\hat{\mathbf{j}} + \omega_z\hat{\mathbf{k}}) \Rightarrow \boldsymbol{\omega}_{\mathcal{B}/\mathcal{I}} =$ constant in $\mathcal{I}$.

7.95 0.328 m (Also, 51.9 m is a solution!)

7.97 Answer given in problem.

7.99 Answer given in problem. (Write the rotational equations at the fixed point O instead of C!)

7.101 $\left(\dfrac{6\sqrt{3}F\Delta t}{5ms}, 0, \dfrac{-6F\Delta t}{ms}\right)$, s being the length of a side and m the total mass. No, because both I_x and I_z at the origin differ for the two bodies.

7.103 $\left(\dfrac{2\sqrt{3}F\Delta t}{ms}, 0, \dfrac{-6F\Delta t}{ms}\right)$; No (same reason!)

7.105 $I_{z_C} = 18mR^2$; $I_\ell = 12mR^2$; ratio $= 1.5$; (i) is unstable and the other three are stable.

7.107 For a torque-free body $\mathcal{B}$ in general motion in an inertial frame $\mathcal{I}$, with $\boldsymbol{\omega}_{\mathcal{B}/\mathcal{I}}$ not parallel to $\mathbf{H}_C$, we have $\mathbf{M}_{r_C} = \mathbf{0} = {}^{\mathcal{B}}\dot{\mathbf{H}}_C + \boldsymbol{\omega}_{\mathcal{B}/\mathcal{I}} \times \mathbf{H}_C$, and the two terms *add* to zero.

7.109 If $\mathcal{B}$ is in equilibrium in $\mathcal{I}$, all its points are stationary there; thus $\mathbf{a} = \mathbf{0}$ for all these points, and also $\boldsymbol{\omega}_{\mathcal{B}/\mathcal{I}} = \mathbf{0}$. Hence $\mathcal{B}$ is an inertial frame. But *if* $\mathcal{B}$ is an inertial frame, it can at most translate at constant velocity with respect to another inertial frame $\mathcal{I}$. Thus it need not be stationary in $\mathcal{I}$, i.e., need not be in equilibrium in $\mathcal{I}$ even though none of its points accelerates in $\mathcal{I}$!

7.111 (a) $13ms^2\omega_0^2/192$; (b) $\mathbf{F} = (0, mg, ms\omega_0^2/4)$; $\mathbf{M}_A = \sqrt{3}ms^2\omega_0^2/32\,\hat{\mathbf{i}}$

7.113 $815m$ ft-lb, where m is the mass in slugs

7.115 Answer given in problem.

7.117 $\sqrt{8\sqrt{3}g/(13\pi r)}$

7.119 $823m$ ft-lb, where m is the mass of the disk in slugs; No, work must also be done in overcoming friction and gravity.

7.121 (a) 11.7 J; (b) 98.1 N and 7.81 N · m

CHAPTER 8

8.1 40.0 ft **8.3** $m\ell^2\ddot{\theta} + ka^2\theta = 0$; $2\pi\sqrt{m/k}\,\ell/a$

8.5 $m\ell^2\ddot{\theta} + (ka^2 + mg\ell)\theta = 0$; $2\pi\sqrt{m\ell^2/(ka^2 + mg\ell)}$

8.7 $\omega_n = \sqrt{(k + \pi\gamma R^2)/m}$

8.9 $\sqrt{k\cos^2\phi/(2\pi^2 m)}$ **8.11** $\sqrt{2k/(12m\pi^2)}$

8.13 (a) $m\ddot{y} + (k - m\omega^2)y = k\ell$
(b) $m\ell\omega^2/(k - m\omega^2)$
(c) $y(t) = (k\ell - m\ell\omega^2\cos\sqrt{(k - m\omega^2)/m}\,t)/(k - m\omega^2)$

8.15 (a) Work $= \int_0^{L/v} Pvt\,[\dot{\theta}(t)]\,dt =$

$$(P^2/k)\left[\frac{1}{2} - \frac{v}{\omega L}\sin\frac{\omega L}{v} - \frac{v^2}{\omega^2 L^2}\cos\frac{\omega L}{v} + \frac{v^2}{\omega^2 L^2}\right], \text{(b)}$$

which agrees with $T + \phi$.
8.17 (a) 0.278 ft (b) 0.876 sec (c) 0.291 sec
8.19 (1) moves to left with

$$\mathbf{x} = \frac{6\mu g m}{k}\left(1 - \cos\sqrt{\frac{k}{m}}\,t\right)$$

(2) moves to right with $\mathbf{x} = \dfrac{4\mu g m}{k}\left(1 - \cos\sqrt{\dfrac{k}{m}}\,t\right)$

(3) moves to left with $\mathbf{x} = \dfrac{2\mu g m}{k}\left(1 - \cos\sqrt{\dfrac{k}{m}}\,t\right)$ and

stops for good $\mu gm/k$ to left of unstretched position. Time in *each* interval is $\pi\sqrt{m/k}$, and total distance traveled $= 24\ \mu gm/k$.
8.21 Yes, they do. (Start with $\mathbf{x} = Ae^{-\zeta\omega_n t}\sin(\omega_d t - \varphi)$ and investigate when $\dot{\mathbf{x}} = 0$!) **8.23** 7.58 ft
8.25 $0.025\,[\cos 100t - e^{-100t}(1 + 100t)]$ in.
8.27 (a) 0.1 inch (b) zero (It is *in* phase, neither leading nor lagging.)
8.29 $\mathbf{x}(t) = 0.5\,[1 - (1 + 100t)e^{-100t}]$ in.
8.31 $\rho ALY\omega^2/\pi$
8.33 k(lb/in.) $< 0.0216\omega^2$ where ω is the excitation frequency in rad/sec.
8.35 $wA|\mathbf{v}|\mathbf{v}/g - \mathbf{W}$, where $\mathbf{W}$ is the weight of tank plus fluid.
8.37 $Q_u/Q = (1 + \cos\theta)/2$; $Q_\ell/Q = (1 - \cos\theta)/2$
8.39 $q(u - v) + fu$ **8.41** $k_i v_B(v_B - v_{rel}\cos\theta)$
8.43 $MP/(M + qt)^2$ **8.45** Answer given in problem.
8.47 $v = [(a + kt)^4 - a^4]g/[4k(a + kt)^3]$, which upon integration yields the distance given in the problem.
8.49 50.5 ft/sec; 24.7 sec
8.51 $v(t) = -gt + v_e\ln[m_0/(m_0 - \mu t)]$, valid until fuel gone.

8.53 $$\mathbf{x}(t) = -gt^2/2 + v_e\left[\left(\frac{m_0}{\mu} - t\right)\ln\left(1 - \frac{\mu t}{m_0}\right) + t\right];$$

$$v_{\text{burnout}} = \frac{-gfm_0}{\mu} - v_e\ln(1 - f);$$

$$\mathbf{x}_{\text{burnout}} = \frac{-gf^2m_0^2}{2\mu^2} + \frac{v_e m_0}{\mu}[(1 - f)\ln(1 - f) + f]$$

8.55 Answer given in problem.

8.57 (a) $v = \sqrt{2gx/3}$; (b) $a = g/3$;
(c) Mechanical energy is lost (to heat, deformation, vibration, etc.) as the links suddenly join the falling part of the chain.
8.59

$$\frac{(m + kt)^2 g\hat{\mathbf{k}}}{(2k + c)2k} + \frac{m\mathbf{v}_0}{c} - \frac{m^2 g\hat{\mathbf{k}}}{2kc} + \left[\frac{m^2 g\hat{\mathbf{k}}}{(2k + c)c} - \frac{m\mathbf{v}_0}{c}\right] \Big/ \left(1 + \frac{kt}{m}\right)^{c/k}$$

8.61 Answer given in problem.
8.63 Answer given in problem. Increment is $0.414\sqrt{GM/r}$. **8.65** No; 4980 mph
8.67 5960 mi; 77.5°
8.69 Answer given in problem. (Use Kepler's laws!)
8.71 Answer given in problem.
8.73 $GMm\,(n - 1)/(2Rn)$ **8.75** 1.67×10^9 mi
8.77 107 min ("No air resistance" needn't be assumed this time!)
8.79 $v_{P*} = 18100$ mph;
$v_{A*} = 15100$ mph

8.81 $$\sqrt{\frac{GM}{R_1}}\left(\sqrt{\frac{2R_2}{R_1 + R_2}} - 1\right);$$

$$\sqrt{\frac{GM}{R_2}}\left(1 - \sqrt{\frac{2R_1}{R_1 + R_2}}\right)$$

8.83 4320 mi; 583 mi and 137 mi

8.85 $$e = \sqrt{\left(\frac{r_1 v_1^2}{GM} - 1\right)^2\cos^2\phi_1 + \sin^2\phi_1}$$

8.87 Answer given in problem.
8.89 Answer given in problem.

APPENDIX A

A.1 (a) lb-ft²/slug²; (b) N · m²/kg² **A.3** 143 N
A.5 (a) 0.456 slug-ft/sec; (b) 2.03 kg · m/s

A.7 (a) $$\frac{L}{T^2}L^2 = L^3\left(\frac{1}{T}\right)^2;$$

(b) 6.61, using $g = 32.17$ ft/sec²,

$$\omega = 2\pi\left(1 + \frac{1}{365}\right) \text{ rad/day, and } r_e = 3960 \text{ mi.}$$